PHYSICS FOR DEGREE STUDENTS

B.Sc. Second Year

As per UGC Model Curriculum (For All Indian Universities)

C.L. ARORA
M.Sc.
Formerly Principal
D.A.V. College, Jalandhar, *and*
Dean of Colleges
Guru Nanak Dev University, Amritsar

Dr. P.S. HEMNE
M.Sc. (Physics), Ph.D.
Principal and Head
Postgraduate Department of Physics
Nevjabai Hitkarini College, Bramhapuri
(RTM Nagpur University)

S. CHAND
PUBLISHING

S Chand And Company Limited
(ISO 9001 Certified Company)

S Chand And Company Limited

(ISO 9001 Certified Company)

Head Office: D-92, Sector–2, Noida – 201301, U.P. (India), Ph. 91-120-4682700

Registered Office: A-27, 2nd Floor, Mohan Co-operative Industrial Estate, New Delhi – 110 044, Phone: 011-49731800

www.**schandpublishing.com**; e-mail: **info@schandpublishing.com**

Marketing Offices:

Chennai	:	Ph: 23632120; chennai@schandpublishing.com
Guwahati	:	Ph: 2738811, 2735640; guwahati@schandpublishing.com
Hyderabad	:	Ph: 40186018; hyderabad@schandpublishing.com
Jalandhar	:	Ph: 4645630; jalandhar@schandpublishing.com
Kolkata	:	Ph: 23357458, 23353914; kolkata@schandpublishing.com
Lucknow	:	Ph: 4003633; lucknow@schandpublishing.com
Mumbai	:	Ph: 25000297; mumbai@schandpublishing.com
Patna	:	Ph: 2260011; patna@schandpublishing.com

First Edition 2012
Second Revised Edition 2013
Reprints 2021, 2022

Reprint 2023

ISBN : 978-81-219-4059-7 **Product Code :** H6PDY68PHYS10ENAB13O

PRINTED IN INDIA

By Vikas Publishing House Private Limited, Plot 20/4, Site-IV, Industrial Area Sahibabad, Ghaziabad – 201 010 and Published by S Chand And Company Limited, A-27, 2nd Floor, Mohan Co-operative Industrial Estate, New Delhi – 110 044.

Preface To The Second Revised Edition

The general response to the first edition of the book was very encouraging. We, the authors feel that our work has been amply rewarded and wish to express our deep sense of gratitude, in common to the large number of readers who have used it, and in particular to those of them who have sent helpful suggestions from time to time for the improvement of the book.

We feel great pleasure in presenting the book ***"Physics for Degree Students"*** for the students of B.Sc. Second Year as per the revised syllabus in Physics framed by U.G.C. This Revised Edition is prepared keeping in view the syllabi framed by different Indian universities on the basis of U.G.C. Model Curriculum.

We are thankful to the Management Team and the Editorial Department of S. Chand & Company Pvt. Ltd., New Delhi for all help and support in the publication of this book.

Authors believe that every student and teacher will appreciate the method of presentation of the subject in this book. Any correction of errors and suggestions for further improvement of this book will be highly appreciated.

C. L. ARORA
Dr. P. S. HEMNE

Preface

We feel great pleasure in presenting the book *"Physics for Degree Students"* for the students of B.Sc. Second Year as per the revised syllabus in Physics framed by U.G.C. This textbook is prepared keeping in view the syllabi framed by different Indian universities on the basis of U.G.C. Model Curriculum. The present book has two sections. **Section I** covers Course 4 which includes chapters on Kinetic Theory, Thermodynamics and Statistical Physics. **Section II** covers Course 5 which includes chapters on Waves, Acoustics and Optics. The complete syllabus of B.Sc. Second Year is covered in 30 chapters. Some salient features of the book are:

- The material is presented in a comprehensive way using simple language.
- The sequence of articles in each chapter enables the students to understand the gradual development of the subject.
- A large number of illustrations, pictures and interesting examples have been given to illustrate the basic principles involved so as to make reading interesting and understandable.
- Solved examples as well as numerical questions frequently asked in different universities are incorporated, including summer 2011 examination at the end of each chapter.
- Exercises with large number of questions and Multiple Choice Questions are given at the end of each chapter to facilitate the students to prepare for their examination as per the latest pattern of setting question papers in most of the Indian universities.
- And many more is supplemented even for a last student who feels Physics as a killer subject.

Authors believe that every student and teacher will appreciate the method of presentation of the subject in this book. Any correction of errors and suggestions for further improvement of this book will be highly appreciated.

C. L. ARORA
Dr. P. S. HEMNE

UGC MODEL SYLLABUS
PHYSICS, B.Sc. SECOND YEAR

COURSE 4: KINETIC THEORY, THERMODYNAMICS AND STATISTICAL PHYSICS

4.1 KINETIC THEORY OF MATTER 25

Ideal Gas: Kinetic model, deduction of Boyle's law, interpretation of temperature, estimation of rms speeds of molecules. Brownian motion, estimate of the Avogadro number. Equipartition of energy, specific heat of monatomic gas, extension to di- and triatomic gases, Behaviour at low temperatures. Adiabatic expansion of an ideal gas, applications to atmospheric physics. (7)

Real Gas: Van der Waals gas, equation of state, nature of Van der Waals forces, comparison with experimental P-V curves. The critical constants, gas and vapour. Joule expansion of ideal gas, and of a Van der Waals gas, Joule coefficient, estimates of J-T cooling. (6)

Liquefaction of gases: Boyle temperature and inversion temperature. Principle of regenerative cooling and of cascade cooling, liquifaction of hydrogen and helium. Refrigeration cycles, meaning of efficiency. (5)

Transport phenomena in gases: Molecular collisions, mean free path and collision cross sections. Estimates of molecular diameter and mean free path. Transport of mass, momentum and energy and interrelationship, dependence on temperature and pressure. (7)

4.2 THERMODYNAMICS 25

The laws of Thermodynamics: The Zeroth law, various indicator diagrams, work done by and on the system, first law of thermodynamics, internal energy as a state function and other applications. Reversible and irreversible changes, Carnot cycle and its efficiency, Carnot theorem and the second law of thermodynamics. Different versions of the second law, practical cycles used in internal combustion engines. Entropy, principle of increase of entropy. The thermodynamic scale of temperature; its identity with the perfect gas scale. Impossibility of attaining the absolute zero; third law of thermodynamics. (8)

Thermodynamic Relationships: Thermodynamic variables; extensive and intensive, Maxwell's general relationships, application to Joule-Thomson cooling and adiabatic cooling in a general system, Van der Waals gas, Clausius-Clapeyron heat equation. Thermodynamic potentials and equilibrium of thermodynamical systems, relation with thermodynamical variables. Cooling due to adiabatic demagnetization, production and measurement of very low temperatures (10)

Blackbody Radiation: Pure temperature dependence, Stefan-Boltzmann law, pressure of radiation. Special distribution of BB radiation, Wien's displacement law, Rayleigh-Jean's law, the ultraviolet catastrophy, Planck's quantum postulates, Planck's law, complete fit with experiment. Interpretation of behaviour of specific heats of gases at low temperature. (7)

4.3 STATISTICAL PHYSICS **25**

The statistical basis of Thermodynamics: Probability and thermodynamic probability, principle of equal a priori probabilities, probability distribution and its narrowing with increase in number of particles. The expressions for average properties. Constraints, accessible and inaccessible states, distribution of particles with a given total energy into a discrete set of energy states. (6)

Some universal laws: The mu space representation, division of mu space into energy sheets and into phase cells of arbitrary size, applications to one-dimensional harmonic oscillator and free particles. Equilibrium before two systems in thermal contact, bridge with macroscopic physics. Probability and entropy, Boltzmann entropy relation. Statistical interpretation of second law of thermodynamics. Boltzmann canonical distribution law and its applications; rigorous from of equipartition of energy. (9)

Maxwellian distribution of speeds in an ideal gas: Distribution of speeds and of velocities, experimental verification, distinction between mean, rms and most probable speed values. Doppler broadening of spectral lines. (4)

Transition to Quantum Statistics: 'h' as a natural constant and its implications, cases of particle in a one-dimensional box and one-dimensional harmonic oscillator. Indistinguishability of particles and its consequences, Bose-Einstein, and Fermi-Dirac conditions; applications to liquid helium, free electrons in a metal, and photons in blackbody chamber. Fermi level and Fermi energy. (6)

COURSE 5: WAVES, ACOUSTICS AND OPTICS

5.1 WAVES **11**

Waves in Media: Speed of transverse waves on a uniform string, speed of longitudinal waves in a fluid, energy density and energy transmission in waves, typical measurements. Waves over liquid surface: gravity waves and ripples. Group velocity and phase velocity, their measurements. (5)

Superposition of Waves: Linear homogeneous equations and the superposition principle, nonlinear superposition and consequences. (3)

Standing Waves: Standing waves as normal modes of bounded systems, examples, Harmonics and the quality of sound; examples. Chladni's figures and vibrations of a drum. Production and detection of ultrasonic and infrasonic waves and applications. (4)

5.2 ACOUSTICS **13**

Noise and Music: The human ear and its responses; limits of human audibility, intensity and loudness, bel and decibel, the musical scale, temperament and musical instruments. (4)

Reflection, Refraction and Diffraction of sound: Acoustic impedance of a medium, percentage reflection and refraction at a boundary, impedance matching for transducers, diffraction of sound, principle of a sonar system, sound ranging. (4)

Applied Acoustics: Transducers and their characteristics. Recording and reproduction of sounds, various systems, Measurements of frequency, waveform, intensity and velocity. The acoustics of halls, reverberation period, Sabine's formula. (5)

5.3 GEOMETRICAL OPTICS 18

Fermat's Principle: Principle of extremum path, the aplanatic points of a sphere and other applications. (4)

General theory of image formation: Cardinal points of an optical system, general relationships, thick lens and lens combinations. Lagrange equation of magnification, telescopic combinations, telephoto lenses and eyepieces. (5)

Aberration in images: Chromatic aberrations, achromatic combination of lenses in contact and separated lenses. Monochromatic aberrations and their reductions; aspherical mirrors and Schmidt corrector plates, aplanatic points, oil immersion objectives, meniscus lens. (4)

Optical Instruments: Entrance and exit pupils, need for a multiple lens eyepiece, common types of eyepieces. (5)

5.4 PHYSICAL OPTICS 36

Interference of light: The principle of superpositions, two-slit interference, coherence requirement for the sources, optical path retardations, lateral shift of fringes, Rayleigh refractometer and other applications. Localised fringes; thin films, applications for precision measurements for displacements. (5)

Haidinger Fringes: Fringes of equal inclination. Michelson interferometer, its application for precision determination of wavelength, wavelength difference and the width of spectral lines. Twymann-Green interferometer and its uses. Inensity distribution in multiple beam interference, Tolansky fringes, Fabry-Perot interferometer and etalon. (8)

Fresnel Diffraction: Fresnel half-period zones, plates, straight edge, rectilinear propagation (2)

Fraunhofer Diffraction: Diffraction at a slit, half-period zones, phasor diagram and integral calculus methods, the intensity distribution, diffraction at a circular aperture and a circular disc, resolution of images, Rayleigh criterion, resolving power of telescope and microscopic systems, outline of phase contract microscopy. (9)

Diffraction Gratings: Diffraction at N parallel slits, intensity distribution, plane diffraction grating, reflection grating and blazed gratings. Concave grating and different mountings. Resolving power of a grating and comparison with resolving powers of prism and of a Fabry-Perot etalon (6)

Double refraction and Optical rotation: Refraction, in uniaxial crystals, its electromagnetic theory. Phase retardation plates, double image prism. Rotation of plane of polarisation, origin of optical rotation in liquids and in crystals. (3)

5.5 LASERS 12

Laser system: Purity of a special line, coherence length and coherence time, spatial coherence of a source, Einstein's A and B coefficients, Spontaneous and induced emissions, conditions for laser action, population inversion. (4)

Application of lasers: Pulsed lasers and tunable lasers, spatial coherence and directionality, estimates of beam intensity, temporal coherence and spectral energy density. (5)

Lasers and Non-linear Optics: Polarization P including higher order terms in E and generation of harmonics, momentum mismatch and choice of the right crystal and direction for compensation. (3)

Contents

SECTION – I

KINETIC THEORY OF MATTER

SECTION – II

THERMODYNAMICS

SECTION – III

STATISTICAL PHYSICS

SECTION – IV

WAVES

SECTION – V
ACOUSTICS

SECTION – VI

GEOMETRICAL OPTICS

SECTION – VIII

LASERS

SECTION I

Kinetic Theory of Matter

1. IDEAL GAS
2. BEHAVIOUR OF REAL GASES
3. TRANSPORT PHENOMENA IN GASES

KINETIC THEORY OF MATTER

The kenetic theory of matter states that matter is composed of a large number a small particles — individual atoms or molecules—that are in constant motion. The theory is also called the kinetic molecular theory of matter. By making some simple assumptions, which are made from observations and experiments, the theory helps to explain the behaviour of matter. Two important areas explained by the theory are the flow or tranfer of heat and the relationship between pressure, temperature and volume properties of gases. The temperature of an object or collection of matter is the average kinetic energy of the particles. Faster particles means a higher temperature. A thermometer is used to measure the temperature and put it into temperature degrees instead of kinetic energy units. The heat is the transfer of thermal energy from an object of higher temperature to one of lower temperature. For example, an object feels warm or ot if its temperature is higher than your skin tempeature. The kinetic theory of matter explains heat transfer by conduction, where thermal energy seems to move through a material, warming up cooler areas. This is called heat transfer or heat flow. Processes not covered in this theory are heat transfer by convection and by radiation.

The kinetic theory of matter states that the material's particles at higher temperatures move faster and have greater kinetic energy. When a fast moving particle collides with a slower moving particle, it transfers some of its energy to the slower moving particle, increasing the speed of that particle. If that particle then collides with another particle that is moving faster, its speed will be increased even more. But if it hits a slow moving particle, then it will speed up the third particle. With billions of moving particles colliding into each other, an area of high energy or high heat willl slowly diffuse across the material, making other areas warm too. By the conservation of energy, the total energy or total heat of the object will remain the same, but the heat will be evenly distributed throughout the object. The rate at which the kinetic or thermal energy is transferred from one particle to another depends on the separation of the particles and their freedom to move.

IDEAL GAS

1
Chapter

INTRODUCTION

An ideal gas is a *theoretical gas* composed of a set of randomly-moving, non-interacting point particles. The ideal gas *concept* is useful because it obeys the ideal gas law, a simplified equation of state, and is amenable to analysis under statistical mechanics.

At normal conditions such as standard temperature and pressure, most *real gases* behave qualitatively like an ideal gas. Many gases such as air, nitrogen, oxygen, hydrogen, noble gases and some heavier gases like carbon dioxide can be treated like ideal gases within reasonable *tolerances.* Generally, deviation from an ideal gas tends to decrease with higher temperature and lower density (*i.e,* lower pressure), as the work performed by intermolecular forces becomes less significant compared with the particles' kinetic energy, and the size of the molecules becomes less significant compared to the empty space between them.

The ideal gas model *tends to fail* at lower temperatures or higher pressures, when *intermolecular forces* and *molecular size,* both become important. It also fails for most heavy gases, such as water vapour and many refrigerants. At some point of lower temperature and higher pressure, real gases undergo a *phase transition,* such as to a liquid or a solid. The model of an ideal gas, however, does not describe or allow phase transitions. These must be modeled by more complex equations of state.

The ideal gas model has been explored in both, the Newtonian dynamics (as in kinetic theory, discussed in this chapter) and in quantum mechanics (as a 'gas in a box'). The ideal gas model has also been used to model the behaviour of electrons in a metal (in the 'Drude model' and the 'free electron model'), and it is one of the most important models in statistical mechanics. There are three basic classes of ideal gas:

(*i*) the classical or *Maxwell-Boltzmann ideal gas,*

(*ii*) the ideal quantum *Bose gas,* composed of *bosons**, and

(*iii*) the ideal quantum *Fermi gas,* composed of *fermions***.

1.1 CLASSICAL THERMODYNAMIC IDEAL GAS

The thermodynamic properties of an ideal gas can be described by two equations:

(*i*) the equation of state of a classical ideal gas is

$$PV = nRT \qquad \text{... } (i)$$

which is also termed as the ideal gas law***, and

(*ii*)The internal energy at constant volume of an ideal gas given by

$$U = C_v nRT \qquad \text{... } (ii)$$

where, P is the pressure, V is the volume, n is the amount of substance of the gas (in moles), R is the

* Governed by Bose-Einstein statistics.

** Governed by Fermi-Dirac statistics.

*** The ideal gas law is an extension of experimentally discovered gas laws. Real fluids at low density and high temperature approximate the behaviour of a classical gas. However, at lower temperatures or a higher density a real fluid deviates strongly from the behaviour of an ideal gas, particularly as it condenses from a gas into a liquid or solid. The deviation is expressed as a compressibility factor.

gas constant (8.314 $J.K^{-1}$ $mole^{-1}$), T is the absolute temperature, U is the internal energy and C_V is the dimensionless specific heat capacity at constant volume (which is ≅ 3/2 for monoatomic gas, 5/2 for diatomic gas and 3 for more complex (polyatomic) molecules.

Internal energy in differential form. For a real gas, the internal energy (U) is a function of pressure and temperature. The ideal gas should satisfy the equation

$$\left(\frac{dU}{dP}\right)_T = 0 \qquad \text{... (iii)}$$

where dU is a small change in internal energy (see first law of thermodynamics for details) of an ideal gas with pressure dP at constant temperature T.

$$\left(\frac{dU}{dV}\right)_T = \left(\frac{dU}{dP}\right)_T \left(\frac{dP}{dV}\right)_T$$

From equation (i), [For one gram-molecule (mole) of an ideal gas, *i.e.*, $n = 1$]

$$\frac{dP}{dV} = -\frac{RT}{V^2}$$

$$= -\left[\frac{RT}{V}\right]\left(\frac{l}{V}\right) = -\frac{P}{V}$$

$$\therefore \quad \left(\frac{dU}{dV}\right)_T = \left(\frac{dU}{dP}\right)_T \left(-\frac{P}{V}\right)_T$$

As $\left(-\frac{P}{V}\right)$ is not equal to zero and hence

$$\left(\frac{dU}{dP}\right)_T = 0$$

$$\therefore \quad \left(\frac{dU}{dV}\right)_T = 0$$

Since, for an ideal gas $\left(\frac{dU}{dP}\right)_T$ and $\left(\frac{dU}{dV}\right)_T$, both are equal to zero, the internal energy (U) of an ideal gas is a function of temperature only.

$$\therefore \quad U = f(T)$$

In a large number of calculations, a real gas can be taken approximately as an ideal gas. This involves an error that can be tolerated in the calculations. For all practical purposes, real gases below a pressure of two atmospheres can be considered as ideal gases. A saturated vapour in equilibrium with its own liquid can be considered to have the properties of an ideal gas. The error involved, however, is law.

1.2 KINETIC MODEL

The kinetic theory of matter explains two most important areas: (i) the flow or transfer of heat and (ii) the relationship between pressure, temperature and volume properties of gases.

The kinetic theory of matter, which takes into account the microscopic structure of matter, states that matter is composed of a large number of a small particles — individual atoms or molecules-that are in constant motion. By making some simple assumptions, such as the idea that matter is made of widely spaced particles in constant motion, the theory helps to explain the behaviour of matter. The best example in a gas enclosed in a vessel. It contains very large number of atoms and molecules moving in all directions with all possible velocities in a haphazard manner, thereby colliding with one another. The well-known laws of classical mechanics are applicable to the motion of these molecules.

The assumptions are made from observations and experiments, such as the fact that materials consists of (small) molecules or atoms. Here the size of the particles is assumed to be so small that it can be considered as a point.

1.3 KINETIC THEORY OF GASES

The kinetic theory of matter has been highly successful in explaining the behaviour of gases. The theory leads to a successful interpretation of *macroscopic* properties of ideal gas like pressure, temperature and volume in terms of *microscopic* properties of gas molecules like the molecular mass, diameter and velocity.

Postulates. The kinetic theory of gases is based upon the following important postulates:

(*i*) All gases are composed of molecules. The molecules of a gas are all alike and differ from the molecules of other gases.

(*ii*) The molecules are mere point masses, their dimensions being negligible as compared to the distance between them.

(*iii*) The molecules are in constant, *random* motion. They have velocities in all directions, ranging from zero to infinity. The velocity of molecules increases with temperature.

(*iv*) In a steady state the molecules are continuously colliding against each other and with the walls of the containing vessel. Between two collisions a molecule moves in a straight line. The distance is called the *free path* of the molecule.

(*v*) The time during which a collision takes place is negligible as compared to time required to traverse the free path, *i.e.*, collisions are *instantaneous*.

(*vi*) The molecules are perfect elastic spheres and exert negligible force of attraction or repulsion on one another or on the walls of the containing vessel. Newtonian mechanics applies to molecular collisions. *Hence in a direct impact the molecules begin to move in the opposite direction with the same velocity.*

(*vii*) The number of molecules per unit volume of the gas is very large and remains constant, on the average, in course of time.

1.4 EXPRESSION FOR THE PRESSURE EXERTED BY A GAS

Let us suppose we have a hollow cubical vessel of each side l with *perfectly elastic* walls, containing a very large number of molecules, say n. Consider one of the molecules having a velocity c_1.

The velocity can be resolved into three rectangular components u_1, v_1, w_1 parallel to X-axis, Y-axis and Z-axis respectively. The co-ordinate axes are supposed to be parallel to the sides of the cube, then

$$c_1^2 = u_1^2 + v_1^2 + w_1^2$$

Consider the motion in a direction perpendicular to the face A (fig. 1.1). The molecule strikes the face A with a velocity u_1 and rebounds with the same velocity in the backward direction, since the collision is perfectly elastic. If m is the mass of the molecule, the change in momentum during the collision is

$$mu_1 - (-mu_1) = 2mu_1$$

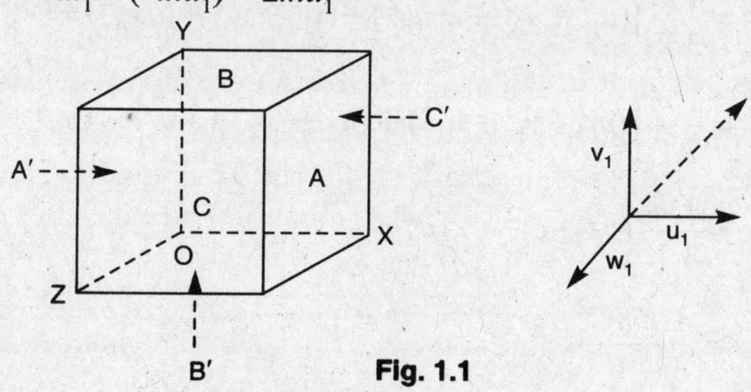

Fig. 1.1

The molecule now travels from *A* to *A'*. It strikes *A'* and travels back again to *A* after covering a distance $2l$. The time taken by the molecule to go to the other face and, then to come back is $\frac{2l}{u_1}$.

$$\therefore \text{Number of impacts per second} = \frac{u_1}{2l}$$

$$\text{and the rate of change of momentum} = 2mu_1 \cdot \frac{u_1}{2l} = \frac{mu_1^2}{l}$$

According to Newton's second law of motion the rate of change of momentum is equal to the impressed force. If F_1 is the force, then

$$F_1 = \frac{mu_1^2}{l}$$

Similarly the force F_2 due to another molecule, having a velocity c_2 and components u_2, v_2, w_2 is given by

$$F_2 = \frac{mu_2^2}{l}$$

and so on. Hence

$$\text{Total force } F \text{ on the face } A = \frac{mu_1^2}{l} + \frac{mu_2^2}{l} \ldots\ldots + \frac{mu_n^2}{l}$$

$$= \frac{m}{l}(u_1^2 + u_2^2 + \ldots. + u_n^2)$$

$\therefore$ Pressure on the face *A*,

$$P_a = \frac{F}{\text{area}} = \frac{m}{l^3}(u_1^2 + u_2^2 + \ldots. u_n^2)$$

$$= \frac{m}{V}(u_1^2 + u_2^2 + \ldots u_n^2) \qquad \ldots [\because l^3 = V]$$

Similarly for the face *B* perpendicular to the Y-axis, the pressure

$$P_b = \frac{m}{V}(v_1^2 + v_2^2 + \ldots. v_n^2)$$

and for the face *C*, perpendicular to the *Z*-axis, the pressure

$$P_c = \frac{m}{V}(w_1^2 + w_2^2 + \ldots w_n^2)$$

But the pressure is the same in all directions

$$\therefore \qquad P_a = P_b = P_c = P \text{ (say)}$$

$$\text{or} \qquad P = \frac{1}{3}(P_a + P_b + P_c)$$

$$= \frac{1}{3}\frac{m}{V}[(u_1^2 + u_2^2 + \ldots u_n^2) + (v_1^2 + v_2^2 + \ldots v_n^2) + (w_1^2 + w_2^2 + \ldots w_n^2)]$$

$$= \frac{1}{3}\frac{m}{V}[(u_1^2 + v_1^2 + w_1^2) + (u_2^2 + v_2^2 + w_2^2) \ldots (u_n^2 + v_n^2 + w_n^2)]$$

$$= \frac{1}{3}\frac{m}{V}(c_1^2 + c_2^2 + \ldots c_n^2)$$

$$= \frac{1}{3}\frac{mnC^2}{V} \qquad \ldots (i)$$

where $$C^2 = \frac{c_1^2 + c_2^2 + \dots c_n^2}{n}$$

Here C is called the **root mean square velocity** of the molecules and is equal to *the square root of the mean of the squares of the velocities of the various molecules.*

Now $mn = M$ the total mass of the gas, and $\frac{M}{V} = \rho$, the density of the gas

$$\therefore \qquad P = \frac{1}{3}\rho C^2 \quad \text{or} \quad C = \sqrt{\frac{3P}{\rho}} \qquad \dots (ii)$$

Relation between pressure and kinetic energy. According to relation (*ii*), the pressure P of the gas is given by

$$P = \frac{1}{3}\rho C^2 = \frac{2}{3}\cdot\frac{1}{2}\rho C^2$$

But $\frac{1}{2}\rho C^2$ is the mean kinetic energy per unit volume of the gas and let us denote it by E

$$\therefore \qquad P = \frac{2}{3}\cdot\frac{1}{2}\rho C^2 = \frac{2}{3}E$$

Hence the pressure of a gas is equal to $\frac{2}{3}$ *of the translational kinetic energy of the molecules in a unit volume.*

Table 1.1. Molecular r.m.s. velocities at 0°C

Gas	*Molecular Weight*	*r.m.s. velocity in cm/s*
Hydrogen	2.016	18.4×10^4
Helium	4	13.1×10^4
Nitrogen	28	4.95×10^4
Oxygen	32	4.61×10^4
Argon	40	4.14×10^4
Carbon dioxide	44	3.95×10^4
Chlorine	71	3.11×10^4

1.5 ESTIMATION OF R.M.S. SPEEDS OF MOLECULES

(*a*) For hydrogen.

The density of hydrogen at N.T.P. is 0.000089 gm/cc. Therefore, C for hydrogen can be calculated as follows :

$$C = \sqrt{\frac{3P}{\rho}}$$

$$= \sqrt{\frac{3 \times 76 \times 13.6 \times 981}{0.000089}}$$

$$= 18.4 \times 10^4 \text{ cm/s}$$

(*b*) For oxygen, the density at N.T.P. is 16 × 0.000089 gm/cc.

$$\therefore \qquad C \text{ for oxygen} = \sqrt{\frac{3 \times 76 \times 13.6 \times 981}{16 \times 0.000089}}$$

$$= 4.61 \times 10^4 \text{ cm/s}$$

(c) For air, the density at N.T.P. is 0.001293 g/cc.

$$\therefore \quad C \text{ for air} = \sqrt{\frac{3\times 76\times 13.6\times 981}{0.001293}}$$

$$= 4.85 \times 10^4 \text{ cm/s.}$$

1.6 R.M.S. VELOCITY AS A FUNCTION OF ABSOLUTE TEMPERATURE

The pressure of a gas, according to kinetic theory, is

$$P = \frac{1}{3}\rho C^2$$

or,

$$P = \frac{1}{3}\frac{MC^2}{V}$$

$$PV = \frac{1}{3} MC^2$$

For 1 mole of an ideal gas at temperature T K,

$$PV = RT$$

$$\therefore \quad \frac{1}{3}MC^2 = RT$$

$$\frac{1}{2}MC^2 = \frac{3}{2}RT$$

Let m be the mass of each gas molecule and N be Avogadro's number.

$$M = m \times N$$

$$\frac{1}{2}mNC^2 = \frac{3}{2}RT$$

$$\frac{1}{2}mC^2 = \frac{3}{2}\frac{R}{N}T = \frac{3}{2}kT \qquad \text{...(1.8)}$$

where k is called Boltzmann's constant.

Thus, from equation (1.8), the mean kinetic energy of a molecule is directly proportional to the absolute temperature (T) of a gas. When the temperature of the gas is increased, the mean kinetic energy of the molecules increases. When heat is withdrawn from a gas, the mean kinetic energy of the molecules decreases. Thus, at absolute zero temperature, the kinetic energy should be zero. It means, at $T = 0$ K, the molecules are in a perfect state of rest and have no kinetic energy. But, before the absolute zero temperature is reached, all gases change their states to liquids and solids.

Also from equation (1.8), $C^2 \propto T$ or $C \propto \sqrt{T}$. It means that the root mean square (r.m.s) velocity of the molecules is also directly proportional to the square root of the absolute temperature.

1.7 DERIVATION OF PERFECT GAS EQUATION

The perfect gas equation or equation of state for an ideal gas is given by $PV = RT$.

The heat energy of a body is the energy of motion of its molecules. At the absolute zero of temperature the heat energy of the molecules of a gas must be zero. Hence the kinetic energy and the root mean square velocity of the molecules must be zero.

When a certain amount of heat energy is added to the gas, it will appear as the kinetic energy of motion of its molecules and is given by

$$\frac{1}{2}mnC^2$$

where C is the root mean square velocity acquired.

The rise of temperature T is also proportional to the amount of heat added. Hence

$$\frac{1}{2}mnC^2 \propto T$$

The pressure of a gas

$$P = \frac{1}{V}\frac{1}{3}mnC^2$$

$$\therefore \quad PV = \frac{1}{3}mnC^2 \propto T$$

or $$\frac{PV}{T} = \text{constant} \quad \text{... } (i)$$

If V is the gram-molecular volume, then

$$\frac{PV}{T} = R \quad \text{or} \quad PV = RT$$

R is known as gram-molecular gas constant and has a value 8.314 J/gm mole K = 1.986 cal/gm mole K. In general, if V contains n gm molecules $PV = nRT$. Hence the S.I. unit of molar gas constant is Joule/mole K.

1.8 DERIVATION OF GAS LAWS

(*i*) **Boyle's law**

The Boyle's law can be proved using kinetic theory of gases.

According to the kinetic theory of gases the pressure P of a gas is given by

$$P = \frac{1}{V}\cdot\frac{1}{3}mnC^2$$

where n is the number of molecules contained in volume V.

Hence $$PV = \frac{1}{3}mnC^2 = \text{constant, at constant temperature}$$

Since mnC^2 is constant, PV is constant at constant temperature T.

This is Boyle's law.

(*ii*) **Charles' law.** It has been proved above that

$$PV = RT$$

If the pressure is kept constant, $\frac{V}{T}$ is constant. In other words, *the volume of a gas is directly proportional to the absolute temperature, which is Charles' law.*

(*iii*) **Dalton's law of partial pressures.** If a number of gases having densities ρ_1, ρ_2, ρ_3, etc. and mean square velocities C_1, C_2, C_3 etc. are mixed up in the same volume, the resultant pressure P, considering each set of molecules separately is given by

$$P = \frac{1}{3}\rho_1 C_1^2 + \frac{1}{3}\rho_2 C_2^2 + \frac{1}{3}\rho_3 C_3^2 \ldots$$

$$= p_1 + p_2 + p_3$$

where p_1, p_2, p_3 etc. are the pressures which each gas would exert if it alone occupied the whole volume, which is Dalton's law.

(*iv*) **Regnault's law**

According to kinetic theory,

$$P = \frac{1}{3}PC^2$$

$$= \frac{1}{3}\frac{M}{V}C^2$$

$$PV = \frac{1}{3}MC^2$$

Consider one gram molecule of a gas at absolute temperature T.

$$M = mN$$

$$\therefore \quad PV = \frac{1}{3}mNC^2 \quad \text{... }(i)$$

The mean kinetic energy of a molecule

$$\frac{1}{2}mC^2 = \frac{3}{2}kT$$

$$mC^2 = 3kT \quad \text{... }(ii)$$

Substituting this value in equation (i).

$$PV = NkT \quad \text{... }(iii)$$

where N is the Avogadro's number and k is the Boltzmann's constant.

When V is constant, from equation (iii),

$$P \propto T$$

It means for a given mass of a gas, the pressure is directly proportional to its absolute temperature provided the volume remains constant. This is Regnault's law.

(v) **Avogadro's hypothesis.** Suppose there are two gases and m_1, n_1 and C_1 represent the mass per molecule, number of molecules per unit volume and root mean square velocity respectively for one gas and m_2, n_2 and C_2 the corresponding values for the other gas. If the two gases exert the *same pressure P,* then

$$P = \frac{1}{3}m_1n_1C_1^{\,2} = \frac{1}{3}m_2n_2C_2^{\,2} \quad \text{... }(i)$$

If the two gases are also at the *same temperature* there will be no transfer of heat energy from one to the other when they are mixed up. *This is possible only if the mean kinetic energy per molecule in the two gases is equal.* In other words

$$\frac{1}{2}m_1C_1^{\,2} = \frac{1}{2}m_2C_2^{\,2} \quad \text{... }(ii)$$

Combining equations (i) and (ii), we have

$$n_1 = n_2$$

i.e., equal volumes of two gases at the same temperature and pressure have the same number of molecules, which is Avogadro's hypothesis.

(vi) **Graham's Law of diffusion of gases**

According to kinetic theory,

$$P = \frac{1}{3}\rho C^2$$

or $$C = \sqrt{\frac{3P}{\rho}}$$

or $$C \propto \frac{1}{\sqrt{\rho}} \text{ at constant pressure.}$$

At constant pressure, the root mean square (r.m.s) velocity of the gas molecules is inversely proportional to the square root of its density. Consider two gases whose root mean square velocities are C_1 and C_2. Then

$$\frac{C_1}{C_2} = \sqrt{\frac{\rho_2}{\rho_1}}$$

or

$$\frac{r_1}{r_2} = \sqrt{\frac{\rho_2}{\rho_1}}$$

Here r_1 and r_2 represent the rates of diffusion of the two gases.

1.9 TEMPERATURE DEPENDENCE OF PRESSURE

On the basis of kinetic theory of gases it has been proved that $\frac{PV}{T}$ = constant, since R is constant.

For a gas in a closed container volume ramains constant. For constant V, we have $\frac{P}{T}$ = constant. In other words, *volume remaining constant the pressure of a gas is directly proportional to the absolute temperature.*

1.10 ENERGY OF A GAS

On the kinetic theory of gases it is assumed that there is no force of attraction or repulsion between the molecules of a gas. Hence no work is done in changing the distance between the molecules and the gas possesses *no potential energy.* The internal energy of a gas is, therefore, wholly kinetic.

Kinetic energy per gram molecule. The pressure P of a gas is given by

$$P = \frac{1}{3}\rho C^2 = \frac{1}{3}\frac{mn}{V}C^2$$

$$\therefore \quad PV = \frac{1}{3} mnC^2 \qquad \text{... } (i)$$

Now, according to perfect gas equation

$$PV = RT$$

where V is the volume per gram molecule and R the universal gas constant. If in relation (i) V is also gram-molecular volume, then

$$mn = M$$

will be the mass per gram molecule. Hence this relation becomes

$$\frac{1}{3}MC^2 = RT \qquad \text{... } (ii)$$

or

$$\frac{1}{2}MC^2 = \frac{3}{2}RT$$

Hence the kinetic energy per gram molecule of any gas at the absolute temperature T is equal to $\frac{3}{2}RT$.

If N is the total number of molecules per gram molecule, then kinetic energy per molecule.

$$= \frac{1}{2}\frac{M}{N}C^2 = \frac{3}{2}\frac{R}{N}T = \frac{3}{2}kT$$

where k is Boltzmann's gas constant per molecule and has a value 1.38×10^{-23} J/K.

Hence average translational kinetic energy per molecule

$$\frac{1}{2}mC^2 = \frac{3}{2}kT \qquad \text{... } (iii)$$

and Root mean square velocity, $C = \sqrt{\frac{3kT}{m}}$ ...(*iv*)

Kinetic Energy Per Unit Volume of a gas.

According to Kinetic theory,

$$P = \frac{1}{3}\rho C^2$$

$$= \frac{2}{3} \times \frac{1}{2}\rho C^2$$

$$= \frac{2}{3}E \quad \text{... }(v)$$

where $E = \frac{1}{2}\rho C^2$ and is equal to the kinetic per unit volume of the gas. Here ρ is the mass per unit volume. Hence the pressure of a gas is numerically equal to two-thirds of the mean kinetic energy of translation of a unit volume of the molecules.

1.11 KINETIC INTERPRETATION OF TEMPERATURE

Consider a system of N molecules of monoatomic gas. If E is the total energy of the system, then some of the energy will be in the form of translational energy and the remaining may be rotational, vibrational potential energies of the molecules. Let E_t be the translational energy, then according to the law of equipartition of energy

$$\frac{E_t}{N} = \frac{3}{2}kT \quad \text{... }(i)$$

where k is Boltzmann's constant and T is temperature of the system in Kelvin. But $\frac{E_t}{N}$ is nothing but the average translational energy per molecule, denoted by $\bar{\varepsilon}_i$.

$$\therefore \quad \bar{\varepsilon}_i = \frac{E_t}{N} = \frac{3}{2}kT \quad \text{... }(ii)$$

or

$$\bar{\varepsilon}_i \propto T \quad \text{... }(iii)$$

Thus, the average kinetic energy of translation per molecule of an ideal monoatomic gas in a given mass is directly proportional to absolute temperature of the gas.

Hence, the average kinetic energy of translation of each molecule of the gas is zero, if $T = 0$.

Conclusion

The absolute zero temperature is that temperature at which the average kinetic energy of translation of each molecule of the gas is zero. *i.e.*, the temperature at which the molecular translational velocities of a gas are reduced to zero.

Thus, we may say that large, small and medium temperature of a system means large, small and medium average kinetic energy of translation of each of its molecules. Such an interpretation of temperature has been supported by Maxwell, who established theoretically that in a mixture of gases, there is an exchange of energy between different kinds of molecules untill the average kinetic energy of translation of each molecule is the same. Thus, if a number of gases having the same kinetic energy of translation are mixed together, there will be no net exchange of energy. Thus, the gases having same kinetic energies of translation of the molecules are at the same temperature. As a result, we may say that the average kinetic energy of translation of the molecules is proportional to the absolute temperature.

This is kinetic interpretation of temperature.

1.12 BROWNIAN MOTION

Brownian motion was first observed in 1827 by Brown, a botanist. The phenomenon of Brownian movement can be easily observed in a laboratory if a colloidal solution is examined under an ultramicroscope. As the direction of illumination is perpendicular to the axis of the microscope, the suspended particles in the solution look like bright illuminated spots. These illuminated particles continuously move to and fro in a random haphazard way. The particles spin, rise, sink and rise again. The movement of the particles is continuous and spontaneous. This non-stop random and haphazard motion is perpetual and spontaneous.

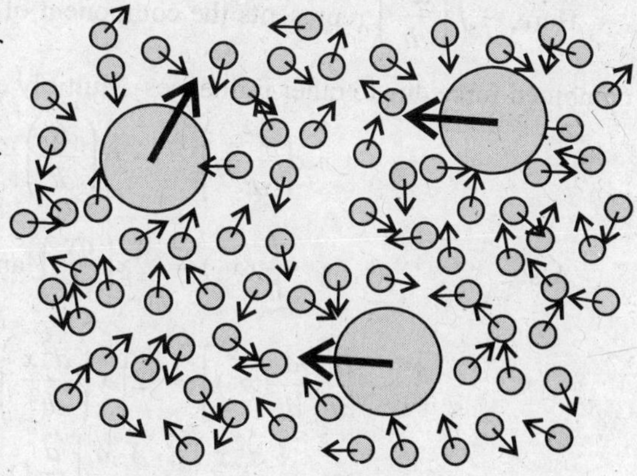

Fig. 1.2

This non-stop, random and haphazard motion is called Brownian motion.

Essential features.

(*i*) The motion of each particle is completely irregular and random. No two particles are found to execute the same motion.

(*ii*) The motion is continuous and takes place for ever.

(*iii*) The smaller particles appear to be more agitated than the larger ones.

(*iv*) The motion is independent of the nature of the suspended particles. Two particles of the same weight and size move equally fast at the same temperature.

(*v*) The motion becomes more vigorous when the temperature is increased.

(*vi*) The motion is more conspicuous in a liquid of lower viscosity.

(*vii*) The motion is not modified due to the shaking of the containing vessel.

Basis of Brownian motion. Brownian motion is due to the bombardment of the dispersed particles by molecules of the medium which are themselves in a state of continuous haphazard motion. When the particles are sufficiently large, the forces due to molecular impact almost completely balance. This is why Brownian motion cannot be observed with particles of large size. But when the size of the particles is very small, each particle will be acted upon by a resultant unbalanced force and will consequently move in a haphazard manner in response to the magnitude and direction of this force.

In other words, the particles move in response to the molecules of the liquid in a completely haphazard manner. The motion of the molecules of a gas is similar to Brownian movement of suspended particles in colloial solution. Thus the laws of kinetic theory of gases are applicable to Brownian motion.

1.13 LANGEVIN'S THEORY OF BROWNIAN MOTION

According to Langevin, the force experienced by a suspended particle is of two kinds :

(*i*) Frictional force proportional to the velocity

$$f\left(\frac{dx}{dt}\right) = 6\pi\eta av$$

(*ii*)Force due to all external influences of the surrounding fluid.

Consider the motion of a particle in a specified direction say x – direction. The force on the particle is given by the equation.

$$m\left(\frac{d^2x}{dt^2}\right) = -f\left(\frac{dx}{dt}\right) + F \qquad \text{... (i)}$$

Here, $-f\left(\frac{dx}{dt}\right)$ represents the component of the frictional force in the x – direction. F is the combined force due to other influences. Multiply equation (i) by x,

$$mx\left(\frac{d^2x}{dt^2}\right) = -fx\left(\frac{dx}{dt}\right) + Fx \quad \text{... (ii)}$$

Also
$$\frac{d}{dt}(x^2) = 2x\left(\frac{dx}{dt}\right) \text{ and } x\left(\frac{dx}{dt}\right) = \frac{1}{2}\frac{d}{dt}(x^2)$$

$$\frac{d}{dt}\left[\frac{d}{dt}(x^2)\right] = 2\left[x\left(\frac{d^2x}{dt^2}\right) + \left(\frac{dx}{dt}\right)^2\right]$$

$$\therefore \quad x\left(\frac{d^2x}{dt^2}\right) = \frac{1}{2}\frac{d}{dt}\left[\frac{d}{dt}(x^2)\right] - \left(\frac{dx}{dt}\right)^2 \quad \text{... (iii)}$$

Substituting these values in equation (ii)

$$\frac{m}{2}\frac{d}{dt}\left[\frac{d}{dt}(x^2)\right] - m\left(\frac{dx}{dt}\right)^2 = -\frac{f}{2}\frac{d}{dt}(x^2) + Fx \quad \text{... (iv)}$$

Equation (iv) represents the motion of a single particle. For all the particles, the mean value can be written as

$$\frac{m}{2}\frac{d}{dt}\left[\frac{d}{dt}\overline{(x^2)}\right] - m\overline{\left(\frac{dx}{dt}\right)^2} = -\frac{f}{2}\frac{d}{dt}\overline{(x^2)} + \overline{Fx} \quad \text{... (v)}$$

As the force F varies completely in an irregular manner, it can be assumed that

$$\overline{Fx} = 0$$

Further, from the law of equipartition of velocities,

$$m\overline{\left(\frac{dx}{dt}\right)^2} = kT$$

$$\text{Also take } \frac{d}{dt}\overline{(x^2)} = \frac{d}{dt}\left(\overline{x^2}\right) = U$$

Substituting these values in equation (v)

$$\frac{m}{2}\left(\frac{dU}{dt}\right) + \frac{fU}{2} = kT$$

or
$$\frac{dU}{dt} + \left(\frac{f}{m}\right)U = \frac{2kT}{m} \quad \text{... (vi)}$$

The general, solution of equation (vi) is

$$U = \frac{2kT}{f} + A\left[e^{-\left(\frac{f}{m}\right)t}\right] \quad \text{... (vii)}$$

As the value of m is very small, the value of (f/m) is very large and hence $e^{\left(\frac{-f}{m}\right)t}$ is negligibly small

$$\therefore \quad U = \frac{2kT}{f} = \frac{d\,\overline{x^2}}{dt} \quad \text{... (viii)}$$

For a time interval $t = 0$ to $t = \tau$, integrating equation (*viii*), we get

$$\overline{x^2} - \overline{x_0^2} = \left(\frac{2kT}{f}\right)\tau$$

At $t = 0$, $x_0 = 0$ and for small values, x^2 can be written as Δx^2

$$\therefore \quad \overline{\Delta x^2} = \left(\frac{2kT}{f}\right)\tau \qquad \text{... (ix)}$$

For each interval of time τ, the displacement Δx of the particle in the x – direction is determined. The mean square value $\overline{\Delta x^2}$ is calculated. In actual practice the particle makes millions of collisions and moves along zig-zag paths. The value $\overline{\Delta x^2}$ is only loosely related to the actual path.

According to Stokes's formula,

$$\because \quad f\left(\frac{dx}{dt}\right) = 6\pi\eta av = 6\pi\eta a\left(\frac{dx}{dt}\right)$$

$$\therefore \quad f = 6\pi\eta a$$

$$\therefore \quad \overline{\Delta x^2} = \frac{2kT\tau}{6\pi\eta a} = \frac{kT\tau}{3\pi\eta a} \qquad \text{... (x)}$$

or

$$\left[\overline{\Delta x^2}\right]^{\frac{1}{2}} = \frac{(kT)^{\frac{1}{2}}(\tau)^{\frac{1}{2}}}{(3\pi\eta a)^{\frac{1}{2}}} \qquad \text{... (xi)}$$

This theory indicates that $\overline{\Delta x^2}$ is not dependent on the mass of the particle. In the experiments of Perrin, the masses of the particles varied in the ratio 1 to 15000. However, within the limits of experimental error, the value of Boltzmann's constant k is the same.

Further,

$$\left(\overline{\Delta x^2}\right)^{\frac{1}{2}} \propto \tau^{\frac{1}{2}}$$

and

$$\left(\overline{\Delta x^2}\right)^{\frac{1}{2}} \propto \frac{1}{\eta^{\frac{1}{2}}} \qquad \text{... (xii)}$$

The effect of temperature is not very large because $\left(\overline{\Delta x^2}\right)^{\frac{1}{2}} \propto T^{\frac{1}{2}}$. However, viscosity decreases rapidly with increase in temperature. Thus, the pure temperature effect is negligibly small in comparison to the effect of viscosity.

Diffusion, fluctuations in concentration and Brownian motion represent a single phenomenon. Diffusion is a macroscopic phenomenon whereas Brownian motion is a microscopic phenomenon.

1.14 EINSTEIN'S THEORY OF BROWNIAN MOTION

According to Einstein's theory of transitional Brownian motion, the particles tend to diffuse into the medium in course of time. Consequently, the diffusion coefficient must be related to the Brownian movement.

The diffusion coefficient can be calculated in two different ways :

(1) From the irregular motion of the suspended particles.

(2) From the difference in osmotic pressure caused by the differences in concentration of the suspended particles.

Albert Einstein

Let D be the diffusion coefficient. Consider an imaginary cylinder with its axis along the x-axis. The end faces P and Q are separated by a distance Δ. Let n_1 and n_2 be molecular concentration at the end faces of the cylinder and A the area of cross-section.

The number of particles crossing the surface P to the right in time

$$T = \frac{1}{2} n_1 \, A\Delta$$

Similarly, the number of particles crossing the surface Q in the opposite direction $= \frac{1}{2} n_2 \, A\Delta$

It should be noted that half the particles contained in the imaginary cylinder move towards right whereas the other half move towards left.

The excess number of particles crossing a middle layer to the right $= \frac{1}{2}(n_1 - n_2)\, A\Delta$

From the definition of diffusion coefficient, the number,

$$\frac{1}{2}(n_1 - n_2)\, A\Delta = -D\frac{dn}{dx}\tau A$$

Here, $\left(\frac{dn}{dx}\right)$ is the concentration gradient.

But $$(n_1 - n_2) = -\Delta\frac{dn}{dx}$$

$$\therefore \quad -\frac{1}{2}\Delta^2\left(\frac{dn}{dx}\right) = -D\left(\frac{dn}{dx}\right)\tau$$

$$\therefore \quad \Delta^2 = 2\,D\tau$$

or $$D = \frac{\Delta^2}{2\tau} \qquad \text{...}(i)$$

Now, D is to be calculated from the concept of osmotic pressure. If p_1 and p_2 are the osmotic pressures at the end P and Q, then from the gas laws

$$p_1 = n_1\, kT$$
$$p_2 = n_2\, kT$$

Thus the cylinder experiences a resultant force, $(p_1 - p_2)\, A = (n_1 - n_2)\, kTA$ along the $+ve$ x – direction. This force is experienced by the particles contained in the cylinder.

The number of particles in the cylinder

$$= n\,A\,\Delta$$

where n is the mean concentration.

Therefore, the force acting on a single particle

$$f' = \frac{(n_1 - n_2)\, kTA}{nA\,\Delta} \qquad \text{...}(ii)$$

Substituting the value of $(n_1 - n_2) = -\Delta\,\frac{dn}{dx}$ in equation (ii)

$$f' = -\left(\frac{dn}{dx}\right)\left(\frac{kT}{n}\right)$$

$$= 6\pi\eta a v$$

$$\therefore \qquad = -\left(\frac{kT}{n}\right)\left(\frac{dn}{dx}\right)$$

or
$$nv = -\left(\frac{kT}{6\,\pi\eta a}\right)\left(\frac{dn}{dx}\right) \qquad ...(iii)$$

Here, nv is the number of particles moving to the right per unit area per second

$$\therefore \qquad nv = -D\left(\frac{dn}{dx}\right) \qquad ...(iv)$$

Equating (*iii*) and (*iv*)

$$-D\left(\frac{dn}{dx}\right) = -\left(\frac{kT}{6\pi\eta a}\right)\left(\frac{dn}{dx}\right)$$

or
$$D = \left(\frac{kT}{6\pi\eta a}\right) = \left(\frac{RT}{N}\right)\left(\frac{1}{6\pi\eta a}\right) \qquad ...(v)$$

From equations (*i*) and (*v*)

$$\frac{\Delta^2}{2\tau} = \left(\frac{RT}{N}\right)\left(\frac{1}{6\pi\eta a}\right)$$

$$\Delta^2 = \frac{RT}{N}\left(\frac{1}{3\pi\eta a}\right)\tau \qquad ...(vi)$$

Equations (*i*) and (*v*) have been verified experimentally. Also, with the help of equation (*v*), the value of Avogadro's number is calculated.

1.15 DISTRIBUTION OF BROWNIAN PARTICLES IN A VERTICAL COLUMN

The Brownian particles form a gas in equilibrium under the action of earth's gravitation. The concentration of the particles in a vertical column decreases with height due to gravity. Consider two layers of the particles at heights h and $(h + dh)$. Let P and $(P + dp)$ be the pressures respectively and ρ the density of the gas at a height h. Consider a unit area of the layer. The force due to gravity acting on the particles between two layers

$$= (1 \times dh)\,\rho g = \rho g dh$$

The net force on the layer

$$= (P + dP) - P + [\rho g dh]$$

In the equilibrium state, the net force must be equal to zero. Therefore,

$$(P + dP) - P + [\rho g dh] = 0$$

or
$$dP + \rho g dh = 0$$

or
$$dP = -\rho g dh \qquad ...(i)$$

For a perfect gas,

$$P = nkT$$

and
$$dP = kT\,dn$$

and
$$\rho = mn$$

Here, n is the number of molecules per unit volume and m is the mass of each molecule.

Substituting, we get

$$kTdn = -mngdh$$

$$\frac{dn}{n} = -\left(\frac{mg}{kT}\right)dh$$

Also $k = \frac{R}{N}$, here N is Avogadro's number.

$$\therefore \qquad \frac{dn}{n} = -\left(\frac{Nmg}{RT}\right)dh \qquad ...(ii)$$

Integrating

$$\log n = -\left(\frac{Nmg}{RT}\right)h + K \qquad ...(iii)$$

Here K is a constant.

At $h = h_0$, $n = n_0$

$$\therefore \qquad \log n_0 = -\left(\frac{Nmg}{RT}\right)h_0 + K$$

or

$$K = \log n_0 + \left(\frac{Nmg}{RT}\right)h_0$$

Substituting this value of K in equation (*ii*),

$$\log n = -\left(\frac{Nmg}{RT}\right)h + \left(\frac{Nmg}{RT}\right)h_0 + \log n_0$$

or,

$$\log\left(\frac{n}{n_0}\right) = -\left(\frac{Nmg}{RT}\right)(h - h_0) \qquad ...(iv)$$

$$\therefore \qquad n = n_0 \cdot e^{-\left(\frac{Nmg}{RT}\right)(h-h_0)} \qquad ...(v)$$

Determination of Avogadro's number

From equation (*iv*),

$$N = \frac{RT \log\left(\frac{n_0}{n}\right)}{mg\,(h - h_0)} \qquad ...(vi)$$

The effective mass of the suspended particle,

$$m = \frac{4}{3}\pi a^3 (d_1 - d_2)$$

where a is the radius of the particle, d_1 is the density of the particle and d_2 is the density of the intervening fluid.

$$\therefore \qquad N = \frac{3RT \log\left(\frac{n_0}{n}\right)}{4\pi a^3 g\,(d_1 - d_2)(h - h_0)} \qquad ...(vii)$$

Thus, from equation (*vii*), the value of Avogadro's number (N) can be calculated if the values of a and d_1 are known.

Determination of d_1: Perrin used the method of finding density (d_1) with a specific gravity bottle. Let m_1 and m_2 be the masses of water and emulsion filling the specific gravity bottle (volume V) and m_3 the mass of the particles left over after dessication in an oven. Taking d as the density of water,

Volume of granules,

$$= \left(\frac{m_1}{d}\right) - \left(\frac{m_2 - m_3}{d}\right)$$

$\therefore$ Density of granules,

$$d_1 = \frac{m_3}{\left(\frac{m_1}{d}\right) - \left(\frac{m_2 - m_3}{d}\right)}$$

or
$$d_1 = \frac{m_3\, d}{(m_1 + m_3 - m_2)} \qquad \text{...(viii)}$$

Determination of *a* : The radius of the particles is determined by Stoke's formula.

The terminal velocity v of the particles is determined by experiment.

$$\therefore \qquad 6\,\pi\eta\, a\, v = \frac{4}{3}\,\pi\, a^3\,(d_1 - d_2)\, g$$

$$\therefore \qquad a = \left[\frac{9\eta v}{2(d_1 - d_2)\, g}\right]^{\frac{1}{2}} \qquad \text{...(ix)}$$

Substituting these values of d_1 and a in equation (*vii*), the value of Avogadro's number can be calculated.

To measure the values of n and n_0 at heights h and h_0, photographs of the suspended particles are taken and the number of particles in these photographs is counted. The value of Avogadro's number determined by Perrin was found to be 6.82×10^{26} moles/kg–mole which is slightly higher than the present accepted value of 6.023×10^{26} molecules/kg–mole.

1.16 DEGREE OF FREEDOM

Definition. *The total number of independent variables required to describe completely the state of motion of a body are called its degrees of freedom.*

An ant moving along a straight line *i.e.*, a wire, it has only one degree of freedom (x). If it moves on a plane, then it has two degrees of freedom (x, y). A flying insect in air, say a mosquito has 3 degrees of freedom (x, y, z). A rigid body, however, not only moves but also rotates about any axis passing through itself. Hence, it has 3 degrees of freedom due to rotational motion. Thus, it has in all 6 degrees of freedom.

Monoatomic molecule

A Monoatomic molecule, *e.g.*, Neon (Ne), Helium (He) etc., has 3 degrees of freedom (all are translational) as shown in Fig. 1.3.

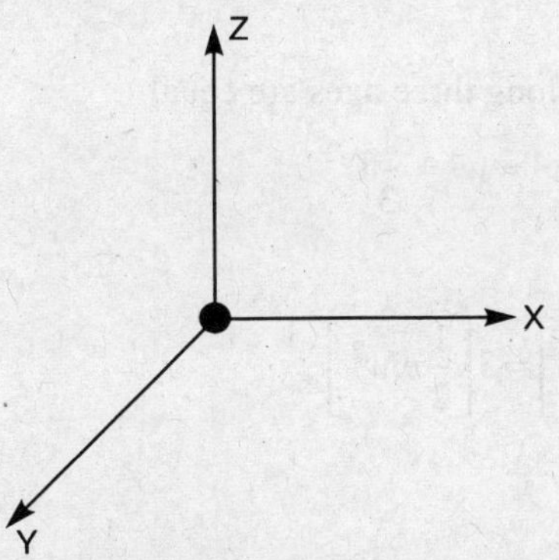

Fig. 1.3 Monoatomic molecule.

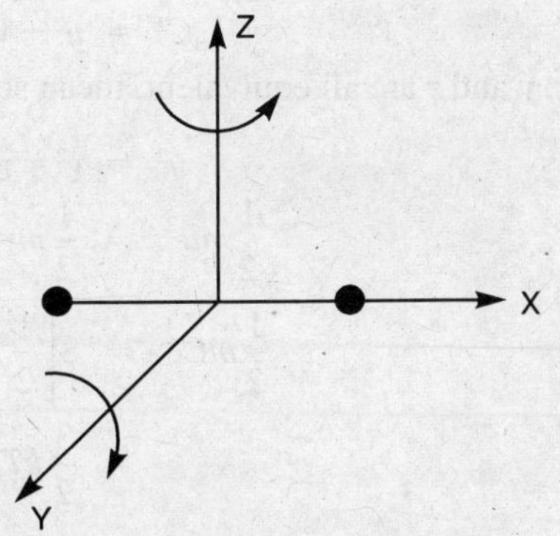

Fig. 1.4 Diatomic molecule.

Diatomic molecule

A diatomic molecule having a dumb-bell shape *e.g.*, HCl, Cl_2, O_2, CO etc. has 5 degrees of freedom (3 translational + 2 rotational), as shown in Fig. 1.4.

At high temperature

At high temperatures, in addition to the translational and rotational degrees of freedom, the vibrational degrees of freedom are also excited and in that case a diatomic molecule possesses 7 degrees of freedom in all.

Triatomic Molecule

A triatomic malecule, if it is non-linear *e.g.*, H_2S, SO_2 *etc.* has 6 degrees of freedom (3 translational + 3 rotational) Fig. 1.5 shows a triatomic non-linear molecule.

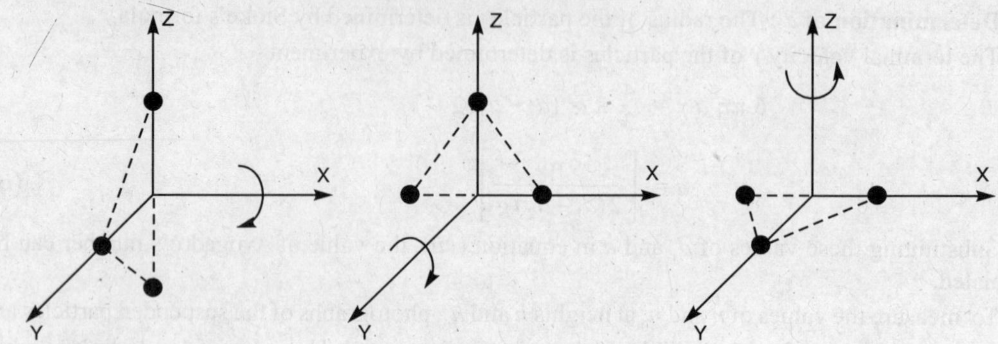

Fig. 1.5. Triatomic non-linear molecule.

A triatomic molecule, if it is linear *e.g.*, CO_2, CH_2 etc., has 5 degrees of freedom, similar to diatomic molecule (3 translational + 2 rotational). Fig. 1.6 shows a triatomic linear molecule.

In addition to translational and rotational motion as discussed above, the molecule also possesses vibratory motion, giving rise to vibrational degrees of freedom, particularly at high temperature.

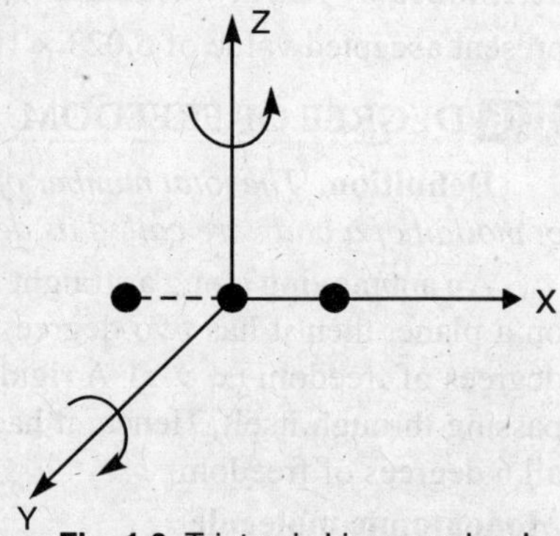

Fig. 1.6. Triatomic Linear molecule

1.17 MAXWELL'S LAW OF EQUIPARTITION OF ENERGY

According to kinetic theory of gases, the mean kinetic energy of a gas molecule at a temperature T is given by

$$\frac{1}{2}mC^2 = \frac{3}{2}kT.$$

But $$C^2 = u^2 + v^2 + w^2$$

As x, y and z are all equivalent, mean square velocities along three axes are equal.

$$\therefore \quad u^2 = v^2 = w^2 \quad \text{Hence, } u^2 = v^2 = w^2 = \frac{1}{3}C^2$$

or $$\frac{1}{2}mu^2 = \frac{1}{2}mv^2 = \frac{1}{2}mw^2$$

$$\therefore \quad \frac{1}{2}mC^2 = 3\left[\frac{1}{2}mu^2\right] = 3\left[\frac{1}{2}mv^2\right] = 3\left[\frac{1}{2}mw^2\right]$$

$$= \frac{3}{2}kT$$

$$\therefore \quad \frac{1}{2}mu^2 = \frac{1}{2}kT$$

$$\frac{1}{2}mv^2 = \frac{1}{2}kT$$

$$\frac{1}{2}mw^2 = \frac{1}{2}kT \qquad ...(i)$$

Therefore, the average kinetic energy associated with each degree of freedom $= \frac{1}{2}kT$.

Thus, the energy associated with each degree of freedom (whether translatory or rotatory) is $\frac{1}{2}kT$. This represents the theorem of equipartition of energy.

Statement. *It states that for any dynamical system in thermal equilibrium, the total energy is divided equally among all the degrees of freedom and the energy associated per molecule per degree of freedom is* $\frac{1}{2}kT$ *; where k is Boltzmann's constant and T is absolute temperature of the gas.*

1.18 RELATION BETWEEN MOLAR SPECIFIC HEATS AND DEGREES OF FREEDOM

For an ideal gas, there is no force of attraction between the molecules, hence the internal energy U of the gas is wholly kinetic. If f is the number of degrees of freedom, then according to the law of equipartition of energy,

$$\text{Average energy per molecule} = \frac{1}{2} fkT$$

$\therefore$ Total energy per gm. molecule $U = N \times \frac{1}{2} fkt$

where N is Avogadro's number.

$$\therefore \quad U = \frac{1}{2} fRT \qquad \left(\because \frac{R}{N} = k\right)$$

The molar specific heat of a gas at constant volume,

$$C_v = \frac{dU}{dT} = \frac{d}{dT}\left(\frac{1}{2} fRT\right) = \frac{1}{2} fR = R\left(\frac{f}{2}\right)$$

and molar specific heat of the gas at constant pressure,

$$C_p = C_v + R = \frac{1}{2} fR + R = R\left(\frac{f}{2} + 1\right)$$

$\therefore$ Ratio of two specific heats,

$$\gamma = \frac{C_p}{C_v} = \frac{R\left(\frac{f}{2} + 1\right)}{R\left(\frac{f}{2}\right)}$$

$$\therefore \quad \gamma = 1 + \frac{2}{f} \qquad ...(i)$$

1.19 SPECIFIC HEATS OF MONO-, DI- AND POLYATOMIC GAS

***(i)* Mono-atomic gas.** A mono-atomic gas molecule has one atom. Each molecule has three degrees of freedom due to translatory motion only.

Energy associated with each degree of freedom $= \frac{1}{2}kT$. Therefore energy associated with three degrees of freedom

$$= \frac{3}{2}kT$$

Consider one gram molecule of a gas.

Energy associated with one gram molecule of a gas

$$= N \times \frac{3}{2} kT = \frac{3}{2}(N \times k)T$$

$$\therefore \quad U = \frac{3}{2}RT \qquad (\because N \times k = R) \quad ...(i)$$

This energy of the gas is due to the energy of its molecules. It is called internal energy U. For an ideal gas, it depends upon temperature only.

$\therefore$ Specific heat at constant volume, C_V is given by

$$C_V = \frac{dU}{dT} = \frac{3}{2}R \qquad \text{... (ii)}$$

($\frac{dU}{dT}$ *is the increase in internal energy per unit degree rise of temperature.*)

But
$$C_P - C_V = R$$
$$C_P = C_V + R$$
$$= \frac{3}{2}R + R = \frac{5}{2}R$$

For a mono-atomic gas, ratio of specific heats

$$\gamma = \frac{C_P}{C_V}$$
$$= \frac{\frac{5}{2}R}{\frac{3}{2}R} = \frac{5}{3} = 1.67 \qquad \text{... (iii)}$$

The value of γ is found to be true experimentally for mono-atomic gases like argon and helium.

***(ii)* Diatomic gas.** A diatomic gas molecule has two atoms. Such a molecule at ordinary temperatures has three degrees of freedom of translation and two degrees of freedom of rotation.

Energy associated with each degree of freedom = $\frac{1}{2}kT$.

$\therefore$ Energy associated with 5 degrees of freedom = $\frac{5}{2}kT$.

Consider one gram molecule of gas.

Energy associated with 1 gram molecule of a diatomic gas

$$= N \times \frac{5}{2}kT = \frac{5}{2}N\,kT$$
$$U = \frac{5}{2}RT$$
$$C_V = \frac{dU}{dT}$$
$$= \frac{5}{2}R \qquad \text{... (iv)}$$

But
$$C_P - C_V = R$$
or
$$C_P = C_V + R$$
$$= \frac{5}{2}R + R = \frac{7}{2}R$$

Therefore, the ratio of specific heats

$$\gamma = \frac{C_P}{C_V}$$

$$= \frac{\frac{7}{2}R}{\frac{5}{2}R} = \frac{7}{5} = 1.40 \qquad \text{... }(v)$$

The value of $\gamma = 1.40$ has been found to be true experimentally for diatomic gases like hydrogen, oxygen, nitrogen etc.

Variation of γ with temperature

At high temperature, the vibrational degree of freedom also becomes prominent and hence taking this into account, the total degrees of freedom in a diatomic molecule possesses (3 translational + 2 rotational + 1 vibrational) 7 degrees of freedom.

$$\therefore \qquad \gamma = \frac{C_p}{C_v} = 1 + \frac{2}{f} = 1 + \frac{2}{7} = 1.29$$

***(a)* At low temperatures.** In case of **hydrogen molecule (H_2)** only *translatory* motion is possible at low temperatures of the order of 70 K and has only three degrees of freedom like that of a monoatomic gas. Hence,

$$C_v = \frac{3}{2}R \quad \text{and} \quad \gamma = 1.67$$

***(b)* At ordinary temperatures.** At ordinary temperatures of the order of 250 K, the hydrogen molecule which is like a dumb-bell shape has a rotational motion in addition to translational motion. It has, therefore, (3 translational + 2 rotational), five degrees of freedom and, hence

$$C_v = \frac{5}{2}R \quad \text{and} \quad \gamma = 1.40$$

***(c)* At high temperatures.** At high temperatures beyond 750 K, the collisions in the molecules cause the atoms to vibrate. Therefore, in addition to translational and rotational degrees of freedom, the hydrogen molecule also possesses vibrational degree of freedom. The hydrogen molecule in all possesses 7 degrees of freedom. Hence

$$C_v = \frac{7}{2}R \quad \text{and} \quad \gamma = 1.29$$

It should be noted that different gases show these effects at different temperatures.

***(iii)* Triatomic gas**

***(a)*Non-linear molecule.** A triatomic non-linear gas molecule has 6 degrees of freedom (3 translational + 3 rotational) neglecting the vibrational energy. Energy associated with 1 gram molecule

$$U = N \times \frac{6}{2} kT = 3RT$$

$$C_V = \frac{dU}{dT} = 3R \qquad \text{...}(vi)$$

But

$$C_P - C_V = R$$

$$C_P = C_V + R$$

$$= 3R + R = 4R$$

The ratio of two specific heats

$$\gamma = \frac{C_P}{C_V}$$

$$= \frac{4R}{3R} = 1.33 \qquad \text{...}(vii)$$

In addition to 3 translational and 3 rotational degrees of freedom, it also possesses energy due to vibrational motion at high temperature. The total number of effective degrees of freedom in such a molecule are *seven*. Hence,

$$C_v = \frac{7}{2}R \quad \text{and} \quad \gamma = 1 + \frac{2}{7} = 1.29$$

***(b)* Linear molecule.** A triatomic linear gas molecule has 5 degrees of freedom. This is similar to diatomic molecule and possess $\gamma = 1.40$.

Taking into consideration vibrational degrees of freedom, a triatomic linear molecule possesses in all 6 degrees of freedom. Hence

$$C_v = \frac{6}{2}R \quad \text{and} \quad \gamma = 1 + \frac{2}{6} = 1.33$$

Thus, the values of C_p, C_V and γ can be calculated depending upon the degrees of freedom of a gas molecule.

1.20 EXPERIMENTAL AND THEORETICAL AGREEMENT OF C_P, C_V AND γ

The experimental values of C_p, C_V and γ for monoatomic gases agree with the theoretical values. But the values of C_P and C_V for some diatomic gases and for most of the polyatomic gases are larger than the theoretical values. In addition, experimental values are found to increase with the temperature.

Explanation. In finding the theoretical values for diatomic and polyatomic gases, we have not considered the energy due to vibrations of atom. At high temperatures the atoms in each molecule can vibrate relative to each other. This increases the number of degrees of freedom.

1.21 BEHAVIOUR OF SPECIFIC HEATS AT LOW TEMPERATURE (FREEZING OF DEGREES OF FREEDOM)

Monoatomic gas. In a monoatomic gas, the molecule (having only one atom) behaves like a point mass. It has only kinetic energy of translation. Hence total energy of such a molecule is divided among only 3 degrees of freedom. Thus

$$C_V = \frac{3}{2}R = 1.5\,R$$

$$C_p = C_V + R = \frac{3}{2}R + R = \frac{5}{2}R = 2.5\,R$$

and
$$\gamma = \frac{C_P}{C_V} = \frac{5}{3} = 1.67$$

This is in good agreement with the experimental values of C_V and γ. Furthermore, the specific heats of mono-atomic gases are found to be practically independent of temperature, in agreement with the theory.

Diatomic gas. A diatomic gas molecule has a dumb-bell structure. For example, a hydrogen molecule.

***(i)* At high temperature.** Beyond 750 K, the collisions causes the atoms in the molecule to vibrate. In addition to translational and rotational motion, vibrational motion also takes place and the molecule has in all 7 degrees of freedom (3 translational + 2 rotational + 2 vibrational). The theory predicts

$$C_V = \frac{7}{2}R = 3.5\,R$$

and
$$\gamma = \frac{9}{7} = 1.29$$

These values are in good agreement with experimental values.

***(ii)* At room temperature.** The molecule has only rotational motion in addition to translatory motion. Thus, it has 5 degrees of freedom (3 translational + 2 rotational). The theory predicts

$$C_V = \frac{5}{2}R = 2.5\,R$$

and $$\gamma = \frac{7}{5} = 1.40$$

These values are also in good agreement with experimental values.

***(iii)* At low temperature.** At low temperature, say about 20 K, the specific heat capacities of hydrogen (the only diatomic gas that remains a gas at very low temperatures) decreases to $\frac{3}{2}R$; the value predicted by theory for a monoatomic gas.

Thus, at low temperature, neither the rotational nor the vibrational degrees of freedom contribute. In other words, the degrees of freedom go on freezing as the temperature lowers down and ultimately at 0K; the translational degree of freedom reduces to zero and the molecule completely comes to rest. This is known as **freezing of degrees of freedom.**

1.22 ADIABATIC EXPANSION OF AN IDEAL GAS

We consider 1 mole of an ideal gas undergoing an adiabatic change in its state. In an adiabatic process neither the heat is allowed to enter nor to leave the system. Hence $\delta Q = 0$. For an ideal gas, the internal energy U depends only on the temperature.

$$\therefore \quad dU = C_V\, dT \qquad \left(\because \quad C_V = \frac{dU}{dT}\right)$$

For a small change in the state of a system

$$\delta Q = dU + PdV$$

For adiabatic change, $0 = C_V dT + PdV$

$$\therefore \quad dT = -\frac{PdV}{C_V}$$

By definition, the ideal gas should satisfy the equation

$$PV = RT$$

or $$P = \frac{RT}{V}$$

Substituting for P, we get

$$C_V dT + \frac{RT}{V} dV = 0$$

or $$C_V \frac{dT}{T} + R\frac{dV}{V} = 0$$

or $$C_V \frac{dT}{T} + (C_P - C_V)\frac{dV}{V} = 0 \qquad (\because \quad C_P - C_V = R)$$

Dividing by C_V and substituting $\frac{C_P}{C_V} = \gamma$, we get

$$\frac{dT}{T} + (\gamma - 1)\frac{dV}{V} = 0$$

Integrating

$$\log_e T + (\gamma - 1)\log_e V = \text{constant}$$

or $$\log_e T V^{\gamma-1} = \text{constant}$$

or $$TV^{\gamma-1} = \text{constant} \qquad \text{... }(i)$$

Similarly, using relation $PV = RT$ for an ideal gas, we can obtain

$$PV^{\gamma} = \text{constant} \qquad \text{... }(ii)$$

$$\text{and} \qquad \frac{P^{\gamma-1}}{T^{\gamma}} = \text{constant} \qquad \text{... (iii)}$$

The above equations (*i*), (*ii*) and (*iii*) give the inter-relation between the macroscopic parameters *P*, *V* and *T* for an ideal gas expanded adiabatically.

1.23 CHANGE OF PRESSURE WITH HEIGHT

The pressure of the atmosphere decreases with increase in height from sea level. The density of air is not the same at all levels. If the density of air were the same at all levels, the atmospheric pressure at sea level would correspond to a vertical column of 8 km of air only. But is has been found that air exists even at a height of 200 km.

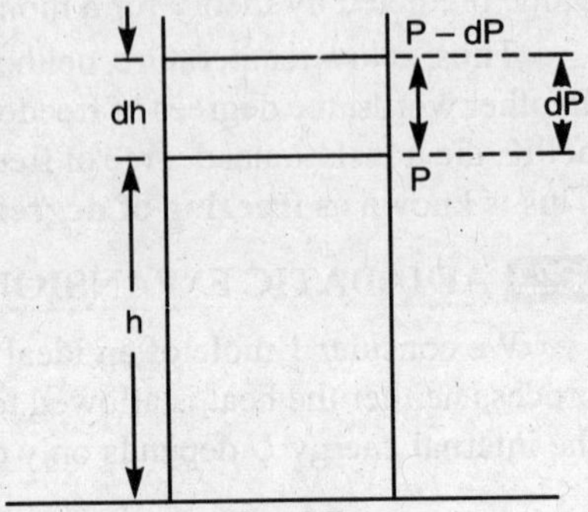

Fig. 1.7

Consider that the pressure of air at a height $h = P$. For an increase in height *dh*, the decrease in pressure is *dP* (Fig. 1.7).

$$dP = -(dh)\,\rho g \qquad \text{...(i)}$$

[–*ve* sign shows that the pressure decreases with increase in height]

Let *M* be the molecular weight of the gas and *V* its volume.

Then,
$$\rho = \frac{M}{V}$$

and
$$dP = -(dh)\,\frac{M}{V}\,g$$

But for a perfect dry air

$$PV = RT$$

$$V = \frac{RT}{P}$$

$$\therefore \qquad dP = -(dh)\,\frac{M.Pg}{RT}$$

$$\frac{dP}{P} = -\frac{Mg}{RT}\,(dh) \qquad \text{...(ii)}$$

Integrating

$$\log P = -\frac{Mg}{RT}\,h + C \quad \text{[where } C \text{ is a constant]}$$

When
$$h = 0,$$
$$P = P_0$$

$$\therefore \qquad C = \log P_0$$

Substituting, we get
$$\log P = -\frac{Mg}{RT}\,h + \log P_0$$

$$\log \frac{P}{P_0} = -\frac{Mg}{RT}\,h \qquad \text{... (iii)}$$

$$P = P_0\, e^{-\frac{Mg}{RT}h} \qquad \text{... (iv)}$$

Equation (*ii*) will be true only under isothermal conditions. But air temperature decreases uniformly with height. Assuming that the lapse rate is α in the lower portion of the atmosphere (troposphere),

$$T = T_0 - \alpha h$$

where T_0 is the temperature of the ground.

Substituting this value in equation (*ii*)

$$\therefore \qquad \frac{dP}{P} = -\frac{Mg}{R}\frac{dh}{(T_0 - \alpha h)}$$

Integrating, $$\log P = \frac{Mg}{R\alpha}\log(T_0 - \alpha h) + K \qquad ...(v)$$

When $$h = 0,$$

$$P = P_0$$

$$\log P_0 = \frac{Mg}{R\alpha}\log T_0 + K$$

or $$K = \log P_0 - \frac{Mg}{R\alpha}\log T_0$$

Substituting this value of K in equation (*v*)

$$\log P = -\frac{Mg}{R\alpha}\log(T_0 - \alpha h) + \log P_0 - \frac{Mg}{R\alpha}\log T_0$$

$$\log\frac{P}{P_0} = \frac{Mg}{R\alpha}\log\left(\frac{T_0 - \alpha h}{T_0}\right) \qquad ...(vi)$$

$$P = P_0\, e^{\frac{Mg}{R\alpha}\log\left(\frac{T_0 - \alpha h}{T_0}\right)} \qquad ...(vii)$$

1.24 APPLICATION TO ATMOSPHERIC PHYSICS (CONVECTIVE EQUILIBRIUM OF THE ATMOSPHERE)

The atmospheric temperature decreases with altitude upto a height of about 20 km. The rate of fall of temperature is about 5°C per Km height and is known as the lapse rate. The atmosphere is divided into two regions troposphere and stratosphere.

1. Troposphere or the convective zone. It is the region in which the temperature falls with increase in height from the ground.

2. Stratosphere or advective zone. It is the region in which the decrease in temperature does not take place with increase in height.

The layer that separates these two regions is known as *tropopause.* The height of tropopause is different at different places on the earth. It is about 14 km at the equator and about 10 km at the poles.

The fall in temperature with altitude in the troposphere is due to convection. When the heat radiations from the sun fall on the surface of the earth, it gets heated. The atmospheric air surrounding the earth gets heated by conduction and radiation. The heated air moves up and convection currents are set up. The heated air that moves up from the regions of higher to lower pressure is adiabatically cooled. Similarly when the colder air moves down from lower to higher pressure regions it is adiabatically heated. Thus a *convective equilibrium is set up* and there is a gradual fall of temperature with increase in height.

The earth gets heated up due to the heat radiations from the sun. The layers of the atmosphere close to the ground get heated up and convection currents start in the atmosphere. As the warm air rises up into region of low pressure, it **expands adiabatically.** For an atmosphere, in convective equilibrium, there is a vertical fall of temperature from the earth's surface. The **dry adiabatic lapse rate** is calculated as follows :

The relation between pressure and temperature for an adiabatic process is

$$\frac{P^{\gamma-1}}{T^{\gamma}} = \text{const.}$$

or $$P^{\gamma-1}\,T^{-\gamma} = \text{const.}$$

Differentiating

$$(\gamma-1)\,P^{\gamma-2}\,dP.T^{-\gamma} + P^{\gamma-1}\,(-\gamma)\,T^{-\gamma-1}\,dT = 0$$

Dividing by $P^{\gamma-1}\,T^{-\gamma}$

$$(\gamma-1)\,\frac{dP}{P} + (-\gamma)\,\frac{dT}{T} = 0$$

$$\frac{dP}{P} = \frac{\gamma}{\gamma-1}\,\frac{dT}{T} \qquad \text{... }(i)$$

For an increase in height *dh*, the decrease in pressure is *dP*

$$\therefore \qquad dP = \rho g\,dh = -\frac{M}{V}\,g\,dh$$

But $$PV = RT$$

or $$V = \frac{RT}{P}$$

$$\therefore \qquad dP = -\frac{MP}{RT}\,g\,dh$$

or $$\frac{dP}{P} = -\frac{Mg}{RT}\,dh \qquad \text{... }(ii)$$

Equating (*i*) and (*ii*)

$$\left(\frac{\gamma}{\gamma-1}\right)\frac{dT}{T} = -\frac{Mg}{RT}\,dh$$

$$\frac{dT}{dh} = -\frac{Mg}{R}\left(\frac{\gamma-1}{\gamma}\right) \qquad \text{... }(iii)$$

This equation holds good only for perfect dry air. The lapse rate $\frac{dT}{dh}$ can be calculated from equation (*iii*).

Substituting the values,

$$M = 29\ g$$

$$g = 981\ \text{cm/s}^2$$

$$R = 8.31\times10^7 \text{ ergs per gram mol per } K,$$

and $$\gamma = 1{\cdot}4$$

$$\frac{dT}{dh} = -\frac{29\times981}{8.31\times10^7}\left(\frac{1.40-1}{1.40}\right)$$

$$= 10^{-4}\,K \text{ per cm}$$

$$= 10^{-4}\,^{\circ}C \text{ per cm}$$

Thus, the lapse rate for a height of 1 km = $10^{-4}\times 10^5 = 10°C$. But it has been found that the average lapse rate is lower than this value. Under average conditions the lapse rate has been found to be between 5°*C* and 6·5°*C* per km.

However at night, when the earth is radiating heat to the atmosphere, the layers of air nearer the surface may cool to a temperature lower than that of the upper layers. Thus a *negative lapse rate is set up.*

SOLVED EXAMPLES

Example 1.1. *Calculate the number of molecules in one c.c. of oxygen at N.T.P. using the following data :*

Density of mercury = 13.6 g/cm³

r.m.s. velocity of oxygen molecules at 0°C

= 4.62 × 10⁴ cm/s

Mass of one molecule of oxygen = 52.8 × 10⁻²⁴ g.

Solution. The pressure is given by

$$P = \frac{1}{3}\rho C^2$$

Let the mass of each molecule be m and the number of molecules in one cc = n

$$\therefore \quad P = \frac{1}{3}mnC^2$$

or
$$n = \frac{3P}{mC^2}$$

Here,
$$P = 76 \times 13.6 \times 980 \text{ dynes/cm}^2$$
$$m = 52.8 \times 10^{-24} \text{ g}$$
$$C = 4.62 \times 10^4 \text{ cm/s}$$

$$\therefore \quad n = \frac{3 \times 76 \times 13.6 \times 980}{52.8 \times 10^{-24} \times (4.62 \times 10^4)^2}$$

$$\mathbf{n = 2.697 \times 10^{19}}$$

Example 1.2. *At what celsius temperature will oxygen molecules have the same root mean square velocity as that of hydrogen molecules at – 100°C ?*

Solution. The energy ot a gas molecule is,

$$\frac{1}{2}mC^2 = \frac{3}{2}kT$$

For hydrogen molecules

$$\frac{1}{2}m_1C_1^2 = \frac{3}{2}kT_1 \quad \text{... (i)}$$

For oxygen molecules

$$\frac{1}{2}m_2C_2^2 = \frac{3}{2}kT_2 \quad \text{... (ii)}$$

Dividing (*i*) by (*ii*),

$$\frac{m_1C_1^2}{m_2C_2^2} = \frac{T_1}{T_2} \quad \text{...(iii)}$$

Here,
$$C_1 = C_2 \text{ and } T_1 = -100°\text{C} = -100 + 273 = 173 \text{ K}$$
$$T_2 = ?$$
$$\frac{m_2}{m_1} = 16$$

From equation (*iii*),

$$T_2 = \frac{m_2C_2^2T_1}{m_1C_1^2}$$

$$= 16 \times 173$$
$$= 2768 \text{ K}$$

or $\quad T_2 = (2768 - 273) = \mathbf{2495°C.}$

Example 1.3. *Calculate the root mean square velocity of oxygen molecules at 27°C. Density of oxygen at N.T.P. = 1.43 kg m^{-3}.* **(Bang. U., 2004)**

Solution. Density of oxygen at N.T.P. *i.e.*, at $T_0 = 273$ K, $\rho_0 = 1.43$ kg m^{-3}

$\therefore$ Density of oxygen at 27°C *i.e.*, at $T = 273 + 27 = 300\ K = \rho$

Now
$$\frac{\rho}{\rho_0} = \frac{T_0}{T} \quad \therefore \rho = \frac{\rho_0 T_0}{T}$$

or
$$\rho = 1.43 \times \frac{273}{300} = 1.30 \text{ kg m}^{-3}$$

Normal pressure
$$P = 0.76 \times 13.6 \times 10^3 \times 9.8 = 1.013 \times 10^5 \text{ Nm}^{-2}$$

According to kinetic theory $P = \dfrac{1}{3}\rho C^2$

$\therefore$ Root mean square velocity $C = \sqrt{\dfrac{3P}{\rho}} = \sqrt{\dfrac{3 \times 1.013 \times 10^5}{1.30}}$

$$= 4.835 \times 10^2 \text{ ms}^{-1}$$

Example 1.4. *Calculate the volume occupied by 3.2 grams of oxygen at 76 cm of Hg and 27°C.*

Solution.
$$P = 76 \times 13.6 \times 980 \text{ dynes/sq.cm}$$
$$T = (27 + 273) = 300 \text{ K}$$
$$R = 8.31 \times 10^7 \text{ erg/g mol–K}$$

For an ideal gas,
$$PV = RT$$
$$V = \frac{RT}{P} = \frac{8.31 \times 10^7 \times 300}{76 \times 13.6 \times 980}$$
$$= 24610 \text{ cc per g mol}$$

Volume for 3.2 g of oxygen $= \dfrac{24610 \times 3.2}{32}$

$$= \mathbf{2461 \text{ cc.}}$$

Example 1.5. *Calculate the number of molecules in one cubic metre of an ideal gas at N.T.P.* **(Purvanchal 2000)**

Solution. For one gram molecule of a gas,
$$PV = RT$$

But
$$R = Nk$$
$$PV = NkT$$

Let n be the number of molecules per cc. Then
$$P = nkT \qquad [\because \text{when } V = 1,\ N = n]$$

or
$$n = \frac{P}{kT}$$

But
$$k = \frac{R}{N}$$

$$\therefore \quad n = \frac{PN}{RT}$$

Number of molecules in one cubic metre volume

$$x = n \times 10^6$$

$$x = \frac{PN \times 10^6}{RT}$$

Here,

$$P = 76 \times 13.6 \times 980 \text{ dynes/cm}^2$$

$$N = 6.023 \times 10^{23}$$

$$R = 8.31 \times 10^7 \text{ erg/g mol–K}$$

$$T = 273 \text{ K}$$

$$\therefore \quad x = \frac{76 \times 13.6 \times 980 \times 6.023 \times 10^{23} \times 10^6}{8.31 \times 10^7 \times 273}$$

$$\mathbf{x = 2.688 \times 10^{25}.}$$

Example 1.6. *The number of molecules per cc of a gas is 2.7 × 10^{19} at N.T.P. Calculate the number of molecules per cc of the gas.*

(i) at 0°C and 10^{-6} mm pressure of mercury and

(ii) at 39°C and 10^{-6} mm pressure of mercury.

Solution. For a unit volume of a gas

$$P = \frac{1}{3} mnC^2$$

(*i*) At 0°C, r.m.s. velocity is euqal to C.

At N.T.P.

$$P_1 = \frac{1}{3} mn_1 C^2 \quad \text{... (i)}$$

At 0°C and 10^{-6} pressure

$$P_2 = \frac{1}{3} mn_2 C^2 \quad \text{... (ii)}$$

From (*i*) and (*ii*)

$$\frac{P_1}{P_2} = \frac{n_1}{n_2}$$

or

$$n_2 = \frac{n_1 \times P_2}{P_1} \quad \text{... (iii)}$$

Here

$$n_1 = 2.7 \times 10^{19}$$

$$P_1 = 76 \times 13.6 \times 980 \text{ dynes/cm}^2$$

$$P_2 = 10^{-6} \text{ mm of Hg}$$

$$= 10^{-7} \text{ cm of Hg}$$

$$= 10^{-7} \times 13.6 \times 980 \text{ dynes/cm}^2$$

Using (*iii*),

$$n_2 = \frac{2.7 \times 10^{19} \times 10^{-7} \times 13.6 \times 980}{76 \times 13.6 \times 980}$$

$$\mathbf{n_2 = 3.553 \times 10^{10}}$$

(*ii*) Let the r.m.s. velocity at 0°C be C_2 and at 39°C be C_3. Number of molecules per cc at 0°C and 10^{-6} mm of Hg pressure

$$= n_2$$

and at 39°C and 10^{-6} mm of Hg pressure

$$= n_3$$

For unit volume, $$P = \frac{1}{3} mnC^2$$

Here pressure is the same in both the cases

$$\therefore \quad \frac{1}{3} mn_2 C_2^2 = \frac{1}{3} mn_3 C_3^2$$

$$\therefore \quad n_2 C_2^2 = n_3 C_3^2$$

$$n_3 = n_2 \times \frac{C_2^2}{C_3^2}$$

But $$C \propto \sqrt{T}$$

$$\therefore \quad \frac{C_2^2}{C_3^2} = \frac{T_2}{T_3}$$

$$\therefore \quad n_3 = \frac{n_2 \times T_2}{T_3}$$

Here $$n_2 = 3.553 \times 10^{10}$$

$$T_2 = 0°C = 273 \text{ K}$$

$$T_3 = 39°C = (273 + 39)$$

$$= 312 \text{ K}$$

$$n_3 = \frac{3.553 \times 10^{10} \times 273}{312}$$

$$\mathbf{n_3 = 3.109 \times 10^{10}.}$$

Example 1.7. *Calculate the mean translational kinetic energy per molecule of a gas at 727°C, given R = 8.32 joules/mole–K and Avogadro's number N = 6.06 × 10^{23}.* **(Ranchi 2000)**

Solution. Given : $$T = 727 + 273 = 1000 \text{ K}$$

$$R = 8.320 \text{ joules/mole–K}$$

$$N = 6.06 \times 10^{23}$$

$$k = \frac{R}{N} = \left(\frac{8.32}{6.06 \times 10^{23}} \right) \text{joules/mole–K}$$

Mean translational K.E. per molecule,

$$= \frac{3}{2} kT$$

$$= \frac{3 \times 8 \cdot 32 \times 1000}{2 \times 6.06 \times 10^{23}}$$

$$= \mathbf{2.059 \times 10^{-20} \text{ joule.}}$$

Example 1.8. *Calculate the r.m.s. speed of Carbon dioxide molecule at 27°C. Given molecular weight of CO_2 = 44 gm.* **(Guhati 2004)**

Solution. Molcular weight of CO_2, M = 44 gm = 0.044 kg

Gram molecular volume at N.T.P. = 22.4 litre = 22.4 × 10^{-3} m³ = 0.0224 m³

Density of CO_2 at N.T.P., $$\rho_0 = \frac{0.044}{0.0224} = 1.964 \text{ kg m}^{-3}$$

Density of CO_2 at 27°C, $\rho_{27} = \dfrac{\rho_0 \times 273}{(273+27)} = \dfrac{1.964 \times 273}{300}$

$= 1.7875$ kg. m^{-3}

Root mean square velocity at 27°C and normal pressure

$$= \sqrt{\frac{3P}{\rho_{27}}} = \sqrt{\frac{3 \times 1.013 \times 10^3}{1.7875}}$$

$$= 4.12 \times 10^2 \text{ ms}^{-1}$$

Example 1.9. *If the root mean square speed of molecules of hydrogen is 1.84 km s^{-1}, calculate the root mean square speed of oxygen at N.T.P. The molecular weight of hydrogen and oxygen are 2 and 32 respectively.* **(Bangalore 2006, IIT Entrance 2000)**

Solution. Let C_1 and C_2 be the root mean square spped of hydrogen and oxygen respectively, then

$$C_1 = 1.84 \text{ kms}^{-1} = 1.84 \times 10^3 \text{ ms}^{-1}$$

Also
$$\frac{C_2}{C_1} = \sqrt{\frac{\rho_1}{\rho_2}} = \sqrt{\frac{\text{Molecular wt. of hydrogen}}{\text{Molecular wt. of oxygen}}}$$

$$= \sqrt{\frac{2}{32}} = \frac{1}{4}$$

$\therefore$
$$C_2 = \frac{C_1}{4} = \frac{1.84 \times 10^3}{4} = 0.46 \times 10^3 \text{ ms}^{-1} = 460 \text{ ms}^{-1}$$

Example 1.10. *Calculate the root mean square velocity of a molecule of mercury vapour at 300 K. (Given: molecular weight of mercury = 221).* **(Rohilkhand 2000)**

Solution. Mean kinetic energy of one molecule of mercury

$$= \frac{1}{2} mC^2 = \frac{3}{2} kT$$

Let N be the Avogadro's number.

Then
$$\frac{1}{2} mNC^2 = \frac{3}{2} kNT$$

$$\frac{1}{2} MC^2 = \frac{3}{2} RT$$

$$C = \sqrt{\frac{3RT}{M}}$$

Here the molecular weight of mercury, $M = 221$.

$\therefore$
$$C = \sqrt{\frac{3 \times 8.3 \times 10^7 \times 300}{221}} = \mathbf{1.83 \times 10^4 \text{ cm/s.}}$$

Example 1.11. *Calculate the temperature at which the r.m.s. velocity of a hydrogen molecule will be equal to the speed of the earth's first satellite (i.e., v = 8 km/s)* **(Kanpur 2001)**

Solution. Energy for 1 gram molecule of hydrogen

$$= \frac{1}{2} Mv^2 = \frac{3}{2} RT$$

$$T = \frac{Mv^2}{3R}$$

$$= \frac{2 \times (8 \times 10^5)^2}{3 \times 8.3 \times 10^7}$$

$$\mathbf{T = 5.14 \times 10^3 \text{ K.}}$$

Example 1.12. *At what temperature, pressure remaining constant, will the r.m.s. velocity of a gas be half its value at 0°C ?*

Solution. $$\frac{C_2}{C_1} = \sqrt{\frac{T_2}{T_1}}$$

Here $$C_2 = \frac{C_1}{2}$$

$$T_2 = ?$$

$$T_1 = 273 \text{ K}$$

$$\therefore \quad \frac{1}{2} = \sqrt{\frac{T_2}{273}}$$

$$\frac{1}{4} = \frac{T_2}{273}$$

$$T_2 = \frac{273}{4} = \mathbf{68.25\ K}$$

or $$T_2 = \mathbf{-204.75°C}$$

Example 1.13. *At what temperature, pressure remaining unchanged, will the r.m.s. velocity of hydrogen be double of its value at N.T.P.?* **(*Nagpur 2008*)**

Solution. The r.m.s. velocity is given by

$$C = \sqrt{\frac{3RT}{M}}$$

$$\therefore \quad C \propto \sqrt{T}$$

Let the r.m.s. velocity at T_0 (*i.e.*, at 273 K) be C and at T be $2C$, then

$$\frac{2C}{C} = \sqrt{\frac{T}{T_0}}$$

or $$\frac{T}{T_0} = 4$$

$$T = 4\,T_0 = 4 \times 273 = 1092 \text{ K}$$

or $$t = (1092 - 273) = \mathbf{819°C.}$$

Example 1.14. *Calculate the total random kinetic energy of one gram of nitrogen at 300 K.* **(*Kerala 2002*)**

Solution. Total random energy for one gm molecule of nitrogen

$$= \frac{3}{2}RT$$

∴ total random energy for one gram of nitrogen

$$E = \frac{3}{2}\frac{RT}{M} \quad \text{where } M = 28 \text{ g}$$

$$= \frac{3 \times 8.3 \times 10^7 \times 300}{2 \times 28}$$

$$= 133.4 \times 10^7 \text{ erg}$$

$$\mathbf{E = 133.4\ joule.}$$

Example 1.15. *Calculate the total random kinetic energy of 2g of helium at 200 K. (Given R = 8.3 × 10⁷ erg/g mol – K)* **(Patna 2000)**

Solution. The total random energy for 1*g* of helium

$$= \frac{3RT}{2M}$$

∴ The total random energy of 2*g* of helium

$$= \frac{2\times 3RT}{2M}$$

$$= \frac{3RT}{M}$$

$$= \frac{3\times 8.3\times 10^7 \times 200}{4}$$

$= 1245 \times 10^7$ ergs

= 1245 joules.

Example 1.16. *Calculate the total random kinetic energy of one gram-mole of oxygen at 330 K. [R = 8.31 J/gmole-K]* **(Madurai 2002)**

Solution. Total random kinetic energy of one gram mole of oxygen.

$$U = \frac{3}{2}RT$$

$$U = \frac{3}{2}\times 8.31\times 300$$

$$\mathbf{U = 3.739 \times 10^3\ J.}$$

Example 1.17. *Calculate the number of molecules in one litre of an ideal gas at 136.5°C temperature and 3 atmospheric pressure.*

Solution. Let the number of molecules per cc = *n*

$$n = \frac{PN}{RT}$$

Number of molecules in one litre,

$$x = n \times 10^3$$

$$x = \frac{PN\times 10^3}{RT}$$

Here $P = 3 \times 76 \times 13.6 \times 980$ dynes/cm²

$N = 6.023 \times 10^{23}$

$R = 8.31 \times 10^7$ erg/g mol–K

$T = (273 + 136.5)$

$= 409.5$ K

$$x = \frac{3\times 76\times 13.6\times 980\times 6.023\times 10^{23}\times 10^3}{8.31\times 10^7\times 409.5}$$

$$\mathbf{x = 5.376 \times 10^{22}.}$$

Example 1.18. *Calculate the total random kinetic energy of one gm-molecule of oxygen at 300 K. [R = 8.3 × 10⁷ erg/gmol–K]* **(Madurai 2003)**

Solution. Total random kinetic energy per gram-molecule of oxygen

$$= \frac{1}{2}mC^2 \times N$$

$$= \frac{3}{2} k\, T.N$$

$$= \frac{3}{2} \cdot \frac{R}{N} T.N$$

$$= \frac{3}{2} RT$$

$$= \frac{3}{2} \times 8.3 \times 10^7 \times 300$$

$$= 3.735 \times 10^{10} \text{ ergs}$$

$$= \mathbf{3735\ joules.}$$

Note. The total random kinetic energy for one gram-molecule of any gas *i.e.*, hydrogen, oxygen, helium, nitrogen, air etc. is the same at the same temperature.

Example 1.19. *Calculate the average kinetic energy of molecule of a gas at a temperature of 300 K.* **(Meerut 2004)**

Solution. Average K.E. of a molecule

$$= \frac{1}{2} mC^2 = \frac{3}{2} kT$$

Here k is Boltzmann's constant.

$$k = 1.38 \times 10^{-16} \text{ erg/molecule-deg}$$

$$\text{Average K.E. of a molecule} = \frac{3}{2} \times 1.38 \times 10^{-16} \times 300$$

$$= \mathbf{6.21 \times 10^{-14}\ ergs.}$$

Note. The average kinetic energy of a molecule of any gas *i.e.,* hydrogen, oxygen, helium, nitrogen, air, etc., is the same at the same temperature.

Example 1.20. *The mean kinetic energy of molecules of hydrogen at 0°C is 5.64 × 10^{-21} J and molar gas constant is 8.32 J/mole–K. Calculate Avogadro's number.* **(Calcutta 1997)**

Solution. K.E. per molecule $E = \frac{3}{2} kT$

$$E = \frac{3}{2} \left(\frac{R}{N} \right) T$$

or
$$N = \frac{3RT}{2E}$$

Here
$$E = 5.64 \times 10^{-21} \text{ J}$$
$$R = 8.32 \text{ J/mole–K}$$
$$T = 273 \text{ K}$$

$$\therefore \quad N = \frac{3 \times 8.32 \times 273}{2 \times 5.64 \times 10^{-21}}$$

$$N = \mathbf{6.182 \times 10^{23}.}$$

Example 1.21. *Calculate the energy of 5 gram molecules of ozone (O_3) at 27°C. Given R = 8.3 J/g-mole/K.* **(Agra 2008, Lucknow 2000)**

Solution. Here
$$U = n \left[\frac{3}{2} RT \right]$$
$$n = 5$$
$$R = 8.3 \text{ J/g–mole/K}$$

$$T = 27 + 273 = 300 \text{ K}$$

$$U = 5 \times \left(\frac{3}{2}\right) \times 8.3 \times 300$$

$$\mathbf{U = 1.8675 \times 10^4 \text{ J.}}$$

Example 1.22. *Prove that*

$$\frac{K}{\eta\, C_v} = 1 \textit{ for a gas.}$$

Solution. The expression for coefficient of viscosity

$$\eta = \frac{m\, C}{3\lambda} \qquad ...(i)$$

The coefficient of thermal conductivity

$$K = \frac{mC\, C_v}{3\lambda} \qquad ...(ii)$$

Dividing (*ii*) by (*i*)

$$\frac{K}{\eta} = C_V$$

$$\therefore \qquad \frac{\boldsymbol{K}}{\boldsymbol{\eta C_v}} = \mathbf{1.}$$

Example 1.23. *Calculate RMS speed of molecule of a gas which has density of 1.24 × 10^{-5} gm/cm^3 at a temperature of 273 K and pressure of 10^{-2} atm. Identify the gas.* **(Nagpur 2008)**

Solution. The RMS speed at 273 K is

$$C = \sqrt{\frac{3P}{\rho}}$$

where $P = 10^{-2}$ atm $= 76 \times 10^{-2} \times 13.6 \times 980$ dynes/cm^2

and $\rho = 1.24 \times 10^{-5}$ gm/cm^3.

$$\therefore \qquad C = \sqrt{\frac{3 \times 76 \times 10^{-2} \times 13.6 \times 9.80}{1.24 \times 10^{-5}}} = 4.949 \times 10^4 \text{ cm/s}$$

The gas is nitrogen whose RMS value at NTP is 4.95 × 10^4 cm/s.

Example 1.24. *Show that for a gas possessing f degrees of freedom, the ratio of two specific heats is* $1 + \frac{2}{f}$ *and the atomicity of a molecule is given by* $f = \frac{2}{\gamma - 1}$.

(Garhwal 2000, Utkal 2003, Calcutta 2003, Guwahati 2000, Nagpur 2001, 2003)

Solution. Let us consider 1 mole of a gas. Let f be the number of degrees of freedom of each molecule. The gas contains the number of molecules equal to the Avogadro's number N. According to the law of equipartition of energy, the K.E. associated with each degree of freedom is $= \frac{1}{2} kT$.

∴ The total internal energy of 1 mole of the gas

$$U = N\left(f \times \frac{1}{2} kT\right)$$

$$= \frac{1}{2} Nf \frac{R}{N} T \qquad \left(\because \quad k = \frac{R}{N}\right)$$

$$= \frac{1}{2} f\, RT$$

Now the molar specific heat at constant volume C_V is

$$C_V = \left(\frac{\partial U}{\partial T}\right)_V = \frac{\partial}{\partial T}\left(\frac{1}{2} f RT\right)$$

$$= \frac{fR}{2}$$

For a perfect gas, $\quad C_P - C_V = R$

$$\therefore \quad C_P = C_V + R = \frac{fR}{2} + R$$

$$= \frac{(f+2)\,R}{2}$$

Hence, $\quad \gamma = \frac{C_P}{C_V} = \frac{(f+2)\,R}{fR} = 1 + \frac{2}{f}$

or $\quad$ **atomicity** $= \boldsymbol{f = \frac{2}{\gamma - 1}}$.

Example 1.25. *Calculate the values of the molar heat capacities C_V and C_P of a gas, if the ratio of the heat capacities is 1.33. What is the atomicity of the gas? Given R = 8.31 J/mol – K.*

(K.U. 2001)

Solution. We have $\quad C_P - C_V = R$ and $\frac{C_P}{C_V} = \gamma$

or $\quad C_P = \gamma\, C_V$

Substituting, $\quad \gamma\, C_V - C_V = R$

$$C_V = \frac{R}{\gamma - 1}$$

(*i*) $\quad C_V = \frac{R}{\gamma - 1} = \frac{8.31}{1.33 - 1} = \mathbf{25.18\ J/mol - K}$

(*ii*) $\quad C_P = \gamma\, C_V = 1.33 \times 25.18 = \mathbf{33.49\ J/mol - K}$

(*iii*) Atomicity : The number of degrees of freedom f of each molecule is

$$f = \frac{2}{\gamma - 1} = \frac{2}{1.33 - 1} = \frac{2}{0.33}$$

$$\boldsymbol{f = 6.}$$

The number of degrees of freedom is 6.

$\therefore$ the gas is polyatomic.

Example 1.26. *Given Avogadro's number = 6.02 × 10^{23} and Boltzmann's constant 1.37 × 10^{-23} JK^{-1}, calculate (a) kinetic energy of translation of an oxygen molecule at 27°C, (b) the total kinetic energy of an oxygen molecule at 27°C and (c) the total K.E. of a gram molecule of oxygen at 27°C.*

(Calcutta 2003)

Solution. Oxygen is a diatomic molecule. At room temperature *i.e.*, at 27°C, it has 5 degrees of freedom (3 translational + 2 rotational). Here $T = (273 + 27) = 300$ K.

(*a*) K.E. of translation per molecule $= \frac{3}{2} kT$

$$= \frac{3}{2} \times 1.37 \times 10^{-23} \times 300$$

$$= \mathbf{6.165 \times 10^{-21}\ joules.}$$

(*b*) Total K.E. per molecule

$$= \frac{5}{2} kT$$

$$= \frac{5}{2} \times 1.37 \times 10^{-23} \times 300$$

$$= \mathbf{10.275 \times 10^{-21}\ joules.}$$

(*c*) Total K.E. per gram mole

$$= N \times \frac{5}{2} kT$$

$$= 6.02 \times 10^{23} \times 10.275 \times 10^{-21}$$

$$= \mathbf{6185.55\ joules.}$$

Example 1.27. *Particles of mass 6.2 × 10^{-17} kg are suspended in a liquid at 27°C and the rms speed is found to be 1.4 × 10^{-2} m/s. Find Avogadro's number from the equipartition theorem. (R = 8.314 J/mole K).* **(*Nagpur 2005/W*)**

Solution. According to the law of equipartition of energy, the average K.E. of a particle per degree of freedom = $\frac{1}{2}$ *kT*. But a particle has 3 dregrees of freedom (translational only).

$\therefore$ Average K.E. of the particle = $\frac{3}{2} kT$

But from kinctic theory of gases, average K.E. of the particle = $\frac{1}{2} mC^2$

$$\therefore \quad \frac{3}{2} kT = \frac{1}{2} mC^2 \quad \text{or} \quad \frac{3RT}{2N} = \frac{1}{2} mC^2 \qquad \left(\because k = \frac{R}{N}\right)$$

or $$N = \frac{3RT}{mC^2} = \frac{3 \times 8.31 \times 300}{6.2 \times 10^{-17} \times (1.4 \times 10^{-2})^2} = \mathbf{6.155 \times 10^{23}/mol.}$$

Example 1.28. *The mean kinetic energy of a molecule of hydrogen gas at 0°C is 5.62 × 10^{-23} J and the molar gas constant is 8.31JK^{-1}. Calculate Avogadros's number.* **(*Bang. U. 2001, Cal. U. 2003*)**

Solution. Kinetic energy per molecule = $\frac{3}{2} kT = \frac{3}{2} \frac{R}{N} T$

where *k* is Boltzmann gas constant, *R* gas constant per gm molecule, N the Avogadro's number and T the absolute temperature.

$$\therefore \quad N = \frac{3RT}{2 \times \text{K.E. per molecule}}$$

$$= \frac{3 \times 8.31 \times 273}{2 \times 5.62 \times 10^{-21}} = 6.05 \times 10^{23}$$

Example 1.29. *Compute the average translational kinetic energy per molecule in a gas at room temperature and hence calculate the temperature needed to excite hydrogen atom. The excitation potential of hydrogen atom is 10.2 eV, room temperature = 27°C and Boltzmann's constant = 1.38 × 10^{-23} joule/molecule K. 1 eV = 1.6 × 10^{-19} Joule.*

Solution. Room temperature = 27°C = 273 + 27 = 300 K

Average translational kinetic energy per molecule at room temperature

$$= \frac{3}{2}kT = \frac{3}{2} \times 1.38 \times 10^{-23} \times 300 = 6.21 \times 10^{-31} \text{ joules.}$$

Excitation potential of hydrogen = 10.2 eV

Now $\quad 1 \text{ eV} = 1.6 \times 10^{-19}$ joule

$\therefore$ Energy required to excite a hydrogen atom = $10.2 \times 1.6 \times 10^{-19}$ joule

Let T be the absolute temperature for exciation, then

$$\frac{3}{2}kT = 10.2 \times 1.6 \times 10^{-19}$$

or
$$T = \frac{2 \times 10.2 \times 1.6 \times 10^{-19}}{3 \times 1.38 \times 10^{-23}} = 78840.6 \text{ K}$$

Example 1.30. *If $C_p = 29.08$ J mole^{-1} K^{-1} and $C_v = 20.77$ J mole^{-1} K^{-1}, find the number of degrees of freedom. What is the atomicity of gas molecule ?* **(*Nagpur U. 2003*)**

Solution. $C_p = 29.08$ J mole^{-1} K^{-1}; $C_v = 20.77$ J mole^{-1} K^{-1}

$$\therefore \quad \gamma = \frac{C_p}{C_v} = \frac{29.08}{20.77} = 1.4$$

Now
$$\gamma = 1 + \frac{2}{f} \quad \therefore \frac{1}{f} = \frac{\gamma - 1}{2}, \text{ or } \frac{1}{f} = \frac{1.4 - 1}{2} = 0.2 = \frac{1}{5}$$

$\therefore$ Number of degrees of freedom, $f = 5$

Hence it is a diatomic molecule because a diatomic molecule has 3 degrees of freedom for translatory motion and 2 degrees of freedom for rotatory motion *i.e.*, in all 3 + 2 = 5.

Example 1.31. *Explain on the basis of kinetic theory, why upper atmosphere is cooled while its kinetic temperature is of the order of 1000 K.* **(*Purvanchal 2004*)**

Solution. The quantity of heat (*i.e.*, heat content) of an individual molecule in the upper atmosphere is increased about 3.5 times than that at the earths's surface. Kinetic temperature on earth's surface is about 300 K., while in the upper atmosphere it is of the order of 1000 K. But the pressure and hence the density is very low in the upper atmosphere, *i.e.*, in the upper atmosphere the number of molecules per cc is very small. Thus obviously the total heat content in the upper atmosphere is very small as compared to the heat content on the earth in the same volume of the gas. This results **the temperature of the upper atmosphere very low.**

EXERCISE

1.1 (*a*) Describe kinetic model of a gas with basic postulates. Write down the expression for pressure exerted by a gas contained in a vessel. **(*M.S. Uni. Tamil Nadu 2007, Lucknow 2004, Guwahati 2000, Calcutta; 2003, Nagpur 2011, 2008, 2004, Patna 2004*)**

(*b*) State and prove Boyle's Law. **(*Nagpur, 2011*)**

1.2 Deduce laws of ideal gas on the basis of kinetic theory of gases. Derive the perfect ges equation. **(*Agra 2007, G.N.D.U. 2002, Patna 2004*)**

1.3 Give kinetic interpretation of temperature on the basis of kinetic theory of gases and define absolute zero temperature. **(*Calcutta 2000, Nagpur 2007*)**

1.4 On the basis of kinetic theory of gases, derive an expression for average translational kinetic energy per molecule. **(*Nagpur 2000*)**

1.5 Give any four postulates of kinetic theory of gases. **(*Madurai 2003*)**

1.6 Explain the basic principles of the kinetic theory of gases and show that the pressure of an ideal gas is proportional to its density. **(*Nagpur 2003, Patna 2001, Ranchi 2002*)**

1.7 In terms of kinetic theory of gases, obtain an expression for the temperature dependence of pressure of a gas in closed container. ***(M.D.U. 2000)***

1.8 What are essential features of Brownian motion? Explain its basis. Obtain the expression for the distribution of Brownian particles in a vertical isothermal column. Describe how its can be used for experimental determination of Avogadro's number.

(Garhwal 2000, Meerut 2002, Calcutta 2003)

1.9 Show that for an ideal gas, the internal energy is a function of temperature alone. Derive the relation for adiabatic expansion of an ideal gas. ***(Calcutta 2003, 2004)***

1.10 Explain Brownian motion. Discuss Langevin's theory for Brownian motion.

(M.S. Uni. Tamil Nadu 2007, Patna 2005)

1.11 What is Brownian motion? Derive Einstein's relation for Brownian motion.

(Agra 2007, Calcutta 2004, M.S. Uni. Tamil Nadu 2007)

1.12 Prove that Avogadro's number

$$N = \frac{3RT}{4\pi a^3 g(d_1 - d_2)(h - h_0)} \log\left(\frac{n_0}{n}\right)$$

and Einstein's equation

$$\Delta^2 = \frac{RT}{N}\left(\frac{1}{3\pi\eta a}\right)$$

(Garhwal 2002)

1.13 Explain the concept of degrees of freedom by taking suitable examples.

(Nagpur 2004/s, Meerut 2004)

1.14 A vessel contains mixture of H_2 and O_2 gas. Compare their (*i*) Average K.E. (*ii*) Root mean square speed. ***(Nagpur Uni. summer 2011)***

1.15 Give the theory of estimation of Avogadro's number. ***(Agra 2002, 2000)***

1.16 Explain the term : "Degrees of freedom". How many the degrees of freedom will monoatomic, diatomic and triatomic molecule have in the absence of vibratory motion?

(Alagappa 2004, Kuvempa 2005, Nagpur Uni. 2011, 2008, 2007, 2006)

1.17 Show that the average kinetic energy per molecule of a perfect gas per degree of freedom is kT. ***(Purvanchal 2002, Meerut 2005, Nagpur 2004, Alagappa 2004)***

1.18 Show that root mean square speed of the molecule is inversely proportional to square root of molecular weight. ***(Nagpur 2008, 2004)***

1.19 Explain how many degrees of freedom a non-linear tri-atomic molecule will have, including vibratory motion. ***(Nagpur 2005/W)***

1.20 State and prove Maxwell's law of equipartition of energy.

(Meerut 2003, 2004, Nagpur 2011/s, 2009, 2005/w, 2005/s, 2004/s, Alagappa 2004, Agra 2004, Bangalore 2004)

1.21 (*a*) Apply the kinetic theory of gases to account for the simple gas laws. Report on the equation of state for an ideal gas. Give the S.I. units of molar gas constant.

(G.N.D.U. 2002; Bang. U. 2004)

(*b*) In terms of kinetic theory of gases obtain an expression for the temperature dependence of pressure of a gas in closed container. ***(M.D.U. 2000)***

1.22 Using the law of equipartition of energy, show that for a gas possessing *f* degrees of freedom, the ratio of two specific heats (γ) is $\left(1+\frac{2}{f}\right)$. Estimate the value of γ for monoatomic, diatomic and polyatomic gases. Compare the values with the experimental results.

(Alagappa 2004/s, Meerut 2005, 2003, 2001, Nagpur 2004/5, Garhwal 2000, 1997, Rohilkhand 2000, 1995)

1.23 What do you mean by freezing of degrees of freedom? Explain behaviour of specific heats at low temperature.

1.24 Explain the term atomicity of gases. Obtain the atomicity for mono–, di– and triatomic molecules.

1.25 State the law of equipartition of energy. Apply this law to find an expression for the total internal energy of one gm-mole of a monoatomic ideal gas at an absolute temperature T. Find also the expression for the molar specific heat at constant volume. ***(Calcutta 2005)***

1.26 Explain low temperatures behaviour of specific heats of gases. ***(Nagpur 2005/s)***

1.27 Do the experimental values of C_p, C_v and γ agree with the theoretical values ? Explain. ***(Nagpur 2004)***

1.28 State the law of equipartition of energy. Prove the law for a perfect gas, whose molecules have *n* degrees of freedom. What are the expected values of γ for Neon and nitrogen. ***(Nagpur 2006, Purvanchal 2004, Ranchi 2005, Mysore 2004, Kuvempa 2005)***

1.29 Show from the kinetic theory of gases that the mean kinetic energy of translation of a molecule of a gas is $\frac{3}{2}kT$, where *k* is the Boltzmann's constant. ***(Meerut 2001, Ranchi 2001)***

1.30 Explain what do you mean by degrees of freedom. State the law of equipartition of energy. Prove that for a perfect gas whose molecules have *f* degrees of freedom

$$\frac{C_P}{C_V}=1+\frac{2}{f}$$

Hence show that for a mono-atomic gas $\gamma = 1.67$, for a diatomic gas $\gamma = 1.4$ and for triatomic gas $\gamma = 1.33$. ***(Purvanchal 2005, 2003, Nagpur, 2010, 2006, Garhwal 2002)***

1.31 Describe the 'stability of the atmosphere'. What is meant by lapse rate? ***(Alagappa 2004)***

1.32 Hot air rises up; Why is it cooler at the top of a mountain than near sea level ***(Nagpur 2009)***

1.33 Assuming that air behaves like a perfect gas and its temperature is constant throughout, derive a relation between pressure of air at ground level and pressure at height *h*. ***(Agra 2005)***

1.34 Write short notes on :

(*i*) Degrees of freedom. ***(Alagappa 2004, Nagpur 2005)***

(*ii*) Law of equipartition of energy. ***(Patna 2004)***

(*iii*) Avogadro's number. ***(Lucknow 2004, 2003)***

(*iv*) Brownian motion ***(M.S. Uni. Tamil Nadu 2007, Agra 2003, Lucknow 2003)***

(*v*) Kinetic interpretation of temperature. ***(Madras 2001 Calcutta 2000, Madurai 2001)***

(*vi*) Freezing of degrees of freedom. ***(Nagpur 2001)***

(*vii*) Adiabatic expansion of a gas. ***(Nagpur 2005)***

1.35 With what speed would 1 gram molecule of hydrogen at 27°C be moving in order that the translational kinetic energy of its centre of mass is equal to the total random kinetic energy of all its molecules ? (Molecular weight of hydrogen = 2) [**Ans.** 1.93×10^5 cm/s] ***(Nagpur 2000)***

1.36 Calculate the temperature at which r.m.s. velocity of a helium molecule will be equal to the speed of the earth's first satellite, *i.e.*, $v = 8$ km/s. [**Ans.** 10.28×10^3 K] ***(Madras 2001)***

1.37 Calculate the mean kinetic energy of a molecule of a gas at 1000°C. Given:

$R = 8.31 \times 10^7$ ergs/gram mol–K

$N = 6.02 \times 10^{23}$ [**Ans.** 2.07×10^{-13} ergs] ***(Kuvempa 2002)***

1.38 Calculate the r.m.s. velocity of the molecules of hydrogen at 0°C. Molecular weight of hydrogen = 2.016 and $R = 8.31 \times 10^7$ ergs/gram mole °C. [**Ans.** 18.4×10^4 cm/s] ***(Alagappa 2001)***

1.39 (*a*) Find an expression for the energy of a gas on kinetic theory of gases. What is the interpretation of temperature on this theory ?

(*b*) Calculate the kinetic energy of one gm. mole of oxygen at 300 K, $R = 8.36$ J/mole K.

[Ans. 3.76×10^3 J] ***(Madurai U, 2003)***

1.40 Calculate the values of molar heat capacity c_v and c_p of a diatomic gas. Given: R = 8.31 J/mol K. Calculate also the value of γ.

[Ans. $c_p = 2.9074$, $c_v = 2.077$**]** ***(Nagpur 2011/s, 2009, 2007, 2005/s)***

[Hint: $\frac{C_p}{C_v} = 1 + \frac{2}{f}$, Here $f = 5$ for diatomic gas $C_P - C_V = R$]

1.41. A particle under Brownian motion at 27°C has r.m.s. speed of 10^{-2} m/s. Find the mass of the partide. Boltzmann canstant k = 1.38×10^{-23} J/K. **[Ans.** 1.24×10^{-15} kg] ***(Kerala 2003)***

1.42. Calculate the mean kinetic energy of a molecule of a gas at 1,000°C. Given,

$R = 8.31 \times 10^7$ ergs/gram mol–K

$N = 6.02 \times 10^{23}$

[Ans. 2.07×10^{-15} ergs] ***(Meerut, 2001)***

1.43 If the density of nitrogen is 1.25 g/litre at N.T.P., calculate the R.M.S. velocity of its molecules. **[Ans.** 4.95×10^4 cm/s] ***(Punjab, 1999)***

1.44 At what temperature is the R.M.S. speed of oxygen molecules twice their R.M.S. Speed at 27°C ?

[Ans. 927°C] ***(Kerala, 2000)***

1.45 Calculate the R.M.S. velocity of the hydrogen molecules at room temperature, given that one litre of the gas at room temperature and normal pressure weigh's 0.086 g.

[Ans. 1.88×10^5 cm/s)]

1.46 At what temperature the molecular velocity (r.m.s) of oxygen will become half as that of hydrogen at NTP? Molecular wt. of oxygen and hydrogen are 32 and 2 respectively.

[Ans. 1092 K] ***(Nagpur 2008, 2004)***

1.47 State the law of equipartition of energy. Calculate the total molecular translational and rotational energy of 1 gm of hydrogen gas at 100 K and 300 K. Assuming the molecules to be dumb-bell shaped. Will the specific heat at constant volume be the same at these two temperatures?

(Calcutta 2003)

1.48 If $c_P = 29.08$ J mole^{-1} K^{-1} and $c_V = 20.77$ J mole^{-1} K^{-1}, find the number of degrees of freedom. What is the atomicity of gas molecule? **[Ans.** $f = 5$, Diatomic] ***(Nagpur 2003)***

1.49 Show that the pressure of an ideal gas is equal to 2/3 of the translational kinetic energy of the molecules in a unit volume. ***(Nagpur 2009)***

1.50 Calculate the r.m.s speed of oxygen molecules at 27°C. **[Ans.** 4.838×10^2 ms^{-1}]

1.51 Calcualte the molecular kinetic energy of 1 gm of hydrogen gas at 50°C, given that molecular weight of hydrogen is 2 and R = 8.3 Joule per gram. **[Ans.** 2×10^3 J]

1.52 The diameter of the molecules of a gas is 2×10^{-10} metre. Calculate the mean free path at N.T.P. Given $k = 1.38 \times 10^{-23}$ Joule/K. **[Ans.** 2.1×10^{-7} m]

MULTIPLE CHOICE QUESTIONS

1.1 The molecule of an ideal gas have:

(*a*) Only kinetic energy (*b*) Only potential energy
(*c*) Both K.E. and P.E. (*d*) None of them

1.2 The value of Boltzmann's constant is:

(*a*) 1.38×10^{-16} JK^{-1} (*b*) 1.38 JK^{-1}
(*c*) 1.38×10^{-23} JK^{-1} (*d*) 8.314 JK^{-1}

1.3 According to kinetic theory of gases the relation between pressure *P* density ρ and mean square velocity *C* is:

(*a*) $P = \frac{1}{2}\rho C^2$ (*b*) $P = \frac{1}{3}\rho C^2$ (*c*) $P = \frac{1}{2}\rho C$ (*d*) $P = \frac{1}{3}\rho C$

1.4 The temperature of a gas is doubled, then its root mean square speed becomes

(*a*) double (*b*) $\sqrt{2}$ times (*c*) half (*d*) $\frac{1}{\sqrt{2}}$ times

1.5 The average velocity of the molecules of a gas in equilibrium is:

(*a*) proportional $\sqrt{T}$ (*b*) proportional to T^2
(*c*) proportional to T (*d*) equal to zero

1.6 The average energy of the molecules of monoatomic gas at temperature *T* is :

(*a*) $\frac{1}{2}kT$ (*b*) $\frac{3}{2}kT$ (*c*) kT (*d*) $\frac{5}{2}kT$

1.7 A diatomic gas molecule has translational, rotational and vibrational degrees of freedom. Then the ratio $\frac{c_p}{c_v}$ at high temperature is :

(*a*) 1.29 (*b*) 1.33 (*c*) 1.4 (*d*) 1.67

1.8 The ratio of two specific heats $\left(\frac{c_p}{c_v}\right)$ of a diatomic gas at ordinary temperature is:

(*a*) 1.67 (*b*) 1.40 (*c*) 1.00 (*d*) 2 **(*Rohilkhand 2003*)**

1.9 The kinetic energy per unit volume of a perfect gas is equal to:

(*a*) $\frac{2}{3}p$ (*b*) $\frac{3}{2}p$ (*c*) $\frac{p}{3}$ where *p* is pressure

(*Garhwal 2005*)

1.10 For a gas the r.m.s speed at 800 K is:

(*a*) four times the value of 200 K (*b*) twice the value of 200 K
(*c*) half the value of 200 K (*d*) same as at 200 K

ANSWERS

1.1 (*a*)	**1.2** (*c*)	**1.3** (*b*)	**1.4** (*b*)	**1.5** (*d*)	**1.6** (*b*)
1.7 (*a*)	**1.8** (*b*)	**1.9** (*b*)	**1.10** (*b*)		

BEHAVIOUR OF REAL GASES

Chapter 2

INTRODUCTION

An ideal or perfect gas is that which strictly obeys gas laws like Boyle's law, Charles' law as well as gas equation $PV = RT$ under all conditions. But all real gases *viz.*, hydrogen, helium, nitrogen, oxygen etc., show some deviation from the gas laws particularly at low temperatures and high pressures. At very low pressure, the inter-molecular forces are negligible. Therefore, at low pressures, all real gases obey the gas equation $PV = RT$. An ideal gas can thus be defined as a real gas at low pressure. The internal energy of an ideal gas is a function of temperature only but for a real gas, internal energy (U) is a function of pressure and temperature. In fact, no gas is perfect or ideal in the true sense.

2.1 FOUR STATES OF MATTER

Gas is one of the four classical states of matter. Near absolute zero (–273.16° C), a substance exists as a **solid.** As heat is added to this substance, it melts into a **liquid** at its melting point, boils at its boilding point, and if heated high enough would enter a **plasma** state in which the electrons are so energized that they leave their parent atoms from within the gas. The interconversion of above four states *i.e.* solid, liquid, gas and plasma is shown in Fig. 2.1.

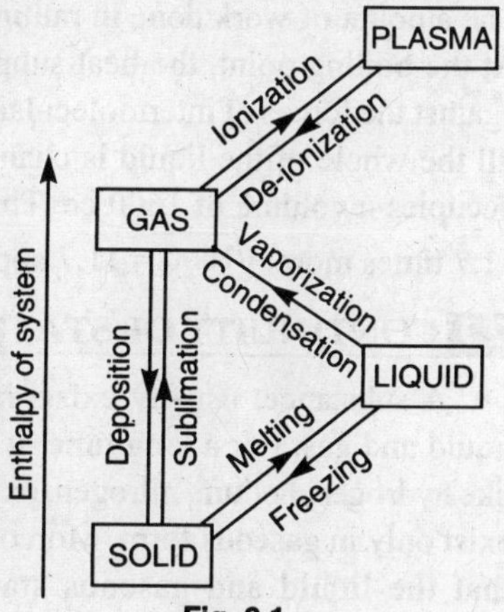

Fig. 2.1

A pure gas may be made up of individual atoms (*e.g.* a noble gas like neon, argon etc), **elemental** molecules made from one type of atom (*e.g.* oxygen, O_2), or **compound** molecules made from a variety of atoms (*e.g.* carbon dioxide, CO_2). A gas mixture would contain a variety of pure gases much like the air. What distinguishes a gas from liquids and solids is the vast separation of the individual gas particles. This separation usually makes a colourless gas invisible to the human observer. The interaction of gas particles in the presence of electric and gravitational fields are considered negligible.

The gaseous state of matter is found between the liquid and plasma states, the latter of which provides the upper temperature boundary of gases. Bounding the lower end of the temperature scale lie degenerative quantum gases which are gaining increased attention these days. High density atomic gases super cooled to incredibly low temperatures are classified by their statistical behaviour as either a Bose gas or a Fermi gas.

2.2 CHANGE OF STATE

The molecules in a liquid have random motion. The average kinetic energy of the molecules depends upon the temperature. The molecules are held together to occupy a particular volume due to intermolecular attraction.

Consider water kept in a tray and exposed to the atmosphere. Generally, the atmosphere is not saturated with water vapours. The molecules of the liquid which are in a state of constant random motion collide with each other resulting in exchange of kinetic energies. Thus, there exists the possibility of a particular molecule in the vicinity of water air interface, acquiring sufficiently high kinetic energy. This energy of the molecule may be high enough to enable the molecule to break away from the forces of attraction of the surrounding molecules. In fact, a molecule on the interface experiences minimum force of attraction. Such a molecule flies off into vapour phase. When this molecule leaves the liquid, the mean kinetic energy of the remaining liquid molecules decreases and there is fall in temperature. *This explains cooling caused by evaporation.* Some vapour molecules in the atmosphere, also coalesce and get absorbed by the water surface. This is called *condensation*. Evaporation and condensation continue simultaneously but rate of evaporation is higher than the rate of condensation.

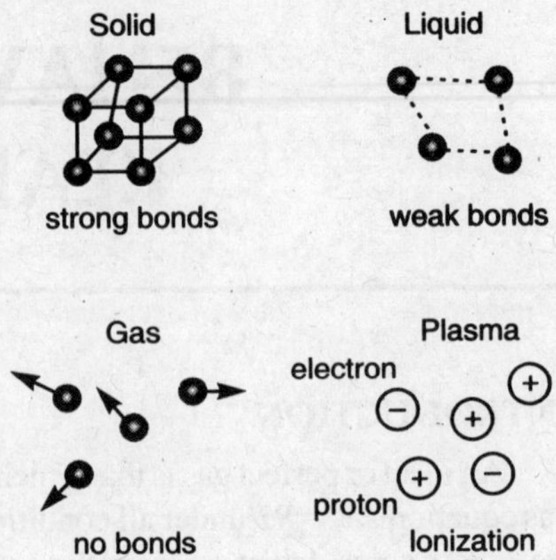

Fig. 2.2

An essential pre-condition for cooling due to evaporation is the capacity of air to absorb more moisture.

If the liquid is heated, and if the rise in temperature of the liquid is more than the cooling caused by evaporation, the temperature of the liquid increases in addition to the increase in the rate of evaporation. A stage will come when the rate of heat supply is exactly equal to the energy needed for evaporation. At this stage, evaporation takes place without change in temperature. This is *boiling*. At the boiling point, the molecules are pulled further apart and their positions are no longer restricted to the space occupied by the liquid. This explains the change of state from liquid to gas or vapour. The amount of work done in pulling the molecules apart measures the **latent heat of vaporisation.** At the boiling point, the heat supplied to the liquid is used up only in pulling the molecules apart against the forces of intermolecular attraction and hence no rise in temperature of the liquid is noticed till the whole of the liquid is changed to the gaseous state. 1 cc of water when converted to steam occupies a volume of 1600 cc. This means that the mean intermolecular distance is approximately 11.7 times more $\sqrt[3]{1600} = 11.7$ approximately) in a gas than the corresponding liquid.

2.3 CONTINUITY OF STATE

A substance, usually exists in three states *viz.*, solid, liquid and gas. For a long time, it was supposed that gases like hydrogen, helium, nitrogen, oxygen, carbon dioxide etc., exist only in gaseous form. Moreover, it was also supposed that the liquid and gaseous states are non-continuous and are distinctly separate from one another. But at high pressure and low temperature, there are marked deviations from Boyle's law. At high pressure, the molecules of a gas come close to each other and the forces of intermolecular attraction becomes appreciable. The distance between two gas molecules become so small that it is comparable or equal to those of the corresponding liquid, the gas gets *liquified.*

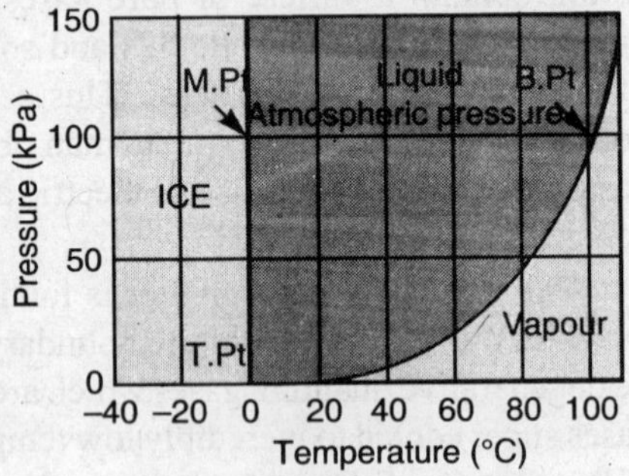

Fig. 2.3

If ice is heated, its molecules become more free to move and acquires a liquid state i.e., water. Further heating, the intermolecular forces become so small, that at a particular temperature the water gets converted in gaseous (i.e., vapour) state Fig. 2.3 The reverse process is also possible by withdrawing heat continuously. Thus all the three states of matter are continuous. In other words, solid, liquid and gaseous states are the three different stages of a continuous phenomenon Fig. 2.3.

2.4 ANDREW'S EXPERIMENTS ON CARBON DIOXIDE

Fig. 2.4 shows the apparatus used by Andrew in 1869, to study the isothermals of carbon dioxide at different temperatures.

The apparatus consists of two similar glass tubes *A* and *B*, having thick capillary tubes at the top and bulbs in the middle. Initially in tube *A* pure dry air is passed for a long time and then sealed at both the ends. Similarly, in tube B, carbon dioxide is passed for a long time (nearly 24 hours) and then both the ends are sealed. The lower ends of both the tubes are immersed in mercury. A small pellet of mercury is drawn in both the tubes by alternately heating and cooling the tubes. Both the tubes are fixed in H shaped copper vessel having steel stoppers S and S′. The vessel is filled with water. By screwing in the plunger, the water is compressed and the pressure is transmitted to the enclosed gases. Pressures of the order of 400 atmospheres can thus be applied.

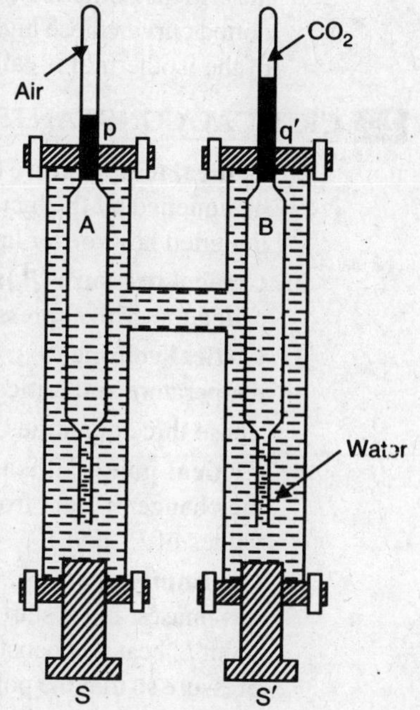

Fig. 2.4

Since the pressures of air in *A* and CO_2 in *B* are the same, from the volume of air in *A*, pressure of CO_2 can be calculated assuming air to obey Boyle's law. The volume of CO_2 is read from tube *B*, as it is calibrated to read the volume directly. Both the tubes are surrounded by water baths (not shown). Air is kept at constant temperature. The temperature of CO_2 could be maintained at any desired temperature between 0°C to 100°C. In Andrew's actual experiments, isothermals were drawn at temperatures 13.1°C, 21.5°C, 31.1°C, 32.5°C, 35.5°C and 48.1°C as shown in Fig. 2.5.

Discussion of Results

From the isothermals of CO_2 at various temperatures (Fig 2.5), following conclusions can be drawn.

1. At 13.1°C, the portion AB represents the gaseous states of CO_2 up to the point B, it obeys Boyle's law from B to C, volume decreases suddenly (enormous decrease) with slight increase in pressure. The portion BC represents the change of CO_2 from the gaseous to liquid state. At C, the gas has been liquefied completely. The portion CD represents the liquid state of CO_2.
2. At 21.5°C, the curve is similar. The horizontal portion BC has decreased further to FG.
3. At 31.1°C, the horizontal portion is just vanished. This is called critical isothermal for CO_2. Above this temperature i.e., 35.5°C and 48.1°C the horizontal portion does not appear at all and the gas obeys Boyle's law from one end to the other. These isothermals are similar to the isothermals for air as shown in Fig. 2.4.

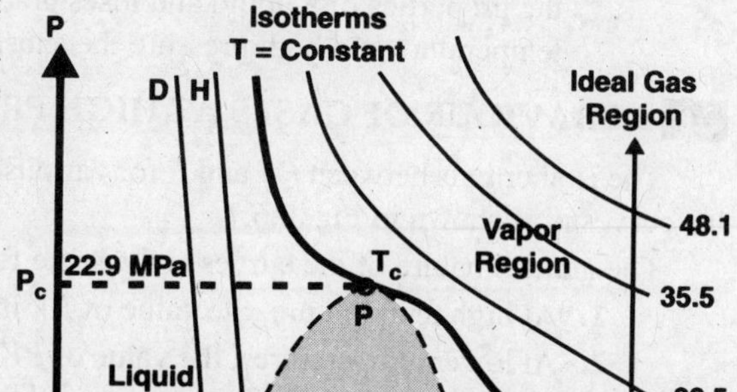

Fig. 2.5. Isothermals of CO_2 and air

4. Thus, for CO_2, the critical isothermal divides into two regions. Above the critical isothermal, the CO_2 gas cannot be liquefied howsoever large the applied pressure may be. Below the critical isothermal, CO_2 gas can be liquefied by the application of pressure alone. The dotted curve whose highest point P is called the critical point and the temperature 31.1°C of the isothermal is called the critical temperature.

2.5 CRITICAL CONSTANTS

1. **Critical temperature (T_c):** It is defined as the maximum temperature at which a gas can be liquefied by the increase of pressure alone. Above this temperature, the gas cannot be liquefied howsoever large the applied pressure may be.
2. **Critical pressure (P_c):** The pressure necessary to liquefy a gas at critical temperature is called the critical pressure.
3. **Critical volume (V_c):** It is the volume which unit mass of a gas occupies at the critical temperature and critical pressure.
   These three quantities, T_c, P_c and V_c are called critical constants of a gas.
4. **Critical point:** It is that point on the isothermal for the critical temperature at which the change of state from the gaseous state to the liquid state takes place under constant values of P_c and V_c.
5. **Continuity of states.** There exists a continuity of state between the liquid and the gaseous phases. If we start with a given mass of the substance in the state represented by the point *L*, heat it at constant volume until the point *M* is reached and then cool it at constant pressure so that the point *N* is reached, then the substance changes from gaseous to liquid form, without any discontinuity occurring [Fig (2.5)].
   From the above discussion it is clear that the **gaseous and liquid states may be regarded as distinct states of a series of continuous physical changes.** In other words, a gas and a liquid are widely separated forms of the same condition of matter, which can be made to pass from one to the other by a series of changes so graded that no breach of continuity occurs. Thus at high temperatures and low pressures, the substance approximates to the conditions of an ideal gas obeying Boyles law. As the temperature is lowered and the pressure is increased deviation form Boyle's law takes place, the substance slowly acquires the properties of a liquid and loses gradually the properties of a gas. At sufficiently low temperature and high pressure the substance entirely becomes a liquid.

2.6 BEHAVIOUR OF GASES AT HIGH PRESSURE

The isothermals between *PV* and *P* for various gases have been drawn and the general nature of the curves is as shown in Fig. 2.6.

The general nature of the curves is the same for all gases.

1. At high temperature, the value of *PV* increases with increase in *P*.
2. At lower temperatures, the value of *PV* decreases initially with increase in *P*. It becomes minimum at a particular pressure and then increases with increase in pressure. The locus of these minima is shown by the dotted curve *A*.
3. At temperatures below the critical temperature, there is a sudden decrease in the value of PV with increase in pressure. This corresponds to the change of state from gas to liquid. When the liquefaction is complete, *PV* gradually increases with increase in *P*. The shaded area represents the region of liquefaction.

Thus, the Boyle's law $PV = RT$ is obeyed only under ideal conditions, *i.e.*, at high temperature and low pressure

Attempts were made to modify Boyle's law. Based on the experimental results of Andrew, Amagat and Holborn, *K.* Onnes suggested for all real gases at high pressure and at constant temperature, an empirical relation

$$PV = A + BP + CP^2 + DP^3 + \ldots\ldots \quad \ldots (2.1)$$

where *A, B, C, D*, etc. are constants for a given temperature depending on the nature of the gas and are called *virial coefficients*. A is called the first virial coefficient, B the second virial coefficient, C the third virial coefficient, and so on. These coefficients are in decreasing order of magnitude, *i.e.*, $A > B > C > D$ and so on. The ratio B/A is of the order of 10^{-3} while C/A is of the order of 10^{-5}. A is simply equal to RT because as $P \rightarrow 0$, equation (*ii*) reduces to $PV = RT$. The second, third and fourth terms become negligible in comparison to first at very low pressure and then gases obey Boyle's law accurately–the lower the pressure, the higher is the degree of accuracy.

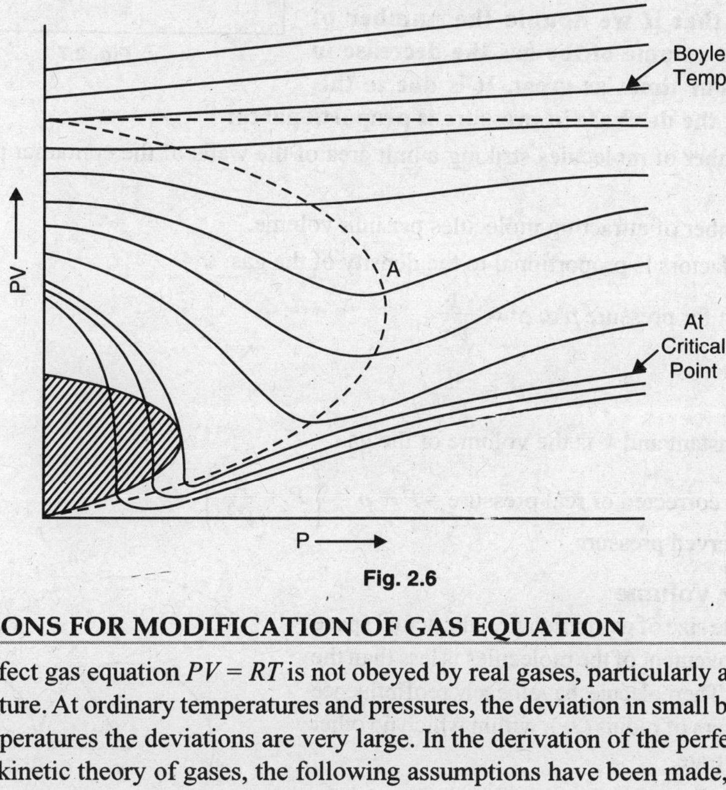

Fig. 2.6

2.7 REASONS FOR MODIFICATION OF GAS EQUATION

The perfect gas equation $PV = RT$ is not obeyed by real gases, particularly at high pressure and low temperature. At ordinary temperatures and pressures, the deviation in small but at high pressures and low temperatures the deviations are very large. In the derivation of the perfect gas equation, on the basis of kinetic theory of gases, the following assumptions have been made, which do not hold true in case of real gases. These are (*i*) the size of the molecule of the gas is negligible, and (*ii*) the forces of intermolecular attraction are absent. But in actual practice, at high pressure the size of the molecule of the gas becomes significant and cannot be neglected, in comparison with the volume of the gas. Also at high pressure, the molecules come closer and the forces of molecular attraction are appreciable. Hence, a perfect gas equation needs modification.

2.8 VAN DER WAALS EQUATION OF STATE

In 1879 Van der Waals modified the perfect gas equation $PV = RT$ by applying corrections for

1. Intermolecular force of attraction and

2. Finite size of molecules.

1. Correction for Pressure:

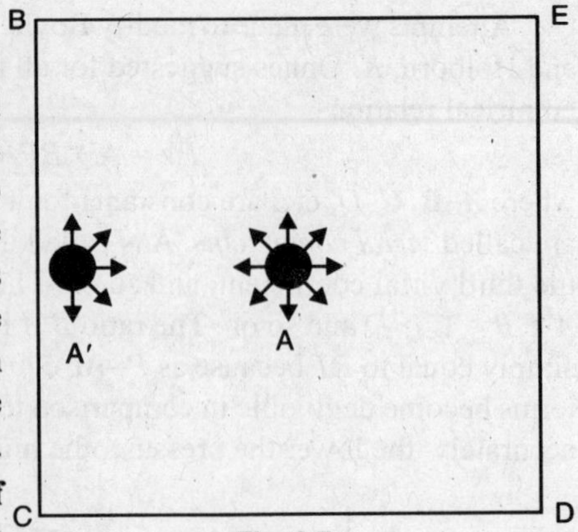

Fig. 2.7

Consider a molecule A of the gas, well inside the vessel as shown in Fig. 2.7. It is attracted by other molecules in all directions with the same force and the net force acting on it is zero. But when it strikes the wall of the vessel, (position A') it is pulled back by other molecules. Its velocity and hence momentum with which it will strike the wall would be less than the momentum with which it will strike in the absence of the force of attraction. This reduction in momentum results in decrease of pressure. In other words, the observed pressure of the gas is less than the actual pressure.

It is evident that if we double the number of molecules per unit volume of the gas the decrease in pressure will be four times as great. It is due to this simple reason that the decrease in pressure is proportional to:

(*i*) the number of molecules striking a unit area of the walls of the container per unit time and

(*ii*) the number of attracting molecules per unit volume.

Each of these factors is proportional to the density of the gas.

$\therefore$ Correction for pressure $p \propto \rho^2 \propto \dfrac{1}{V^2}$

$$\therefore \qquad p = \frac{a}{V^2}$$

Here a is a constant and V is the volume of the gas.

Therefore, the corrected or real pressure $= P + p = \left(P + \dfrac{a}{V^2}\right)$ ... (2.2)

where P is the observed pressure.

2. Correction for Volume

Due to the finite size of gas molecules, the actual space available for the movement of the molecules is less than the volume of the vessel. The molecules have the sphere of influence (Fig. 2.8) around them of radius ($2r$), within which no other molecule can penetrate.

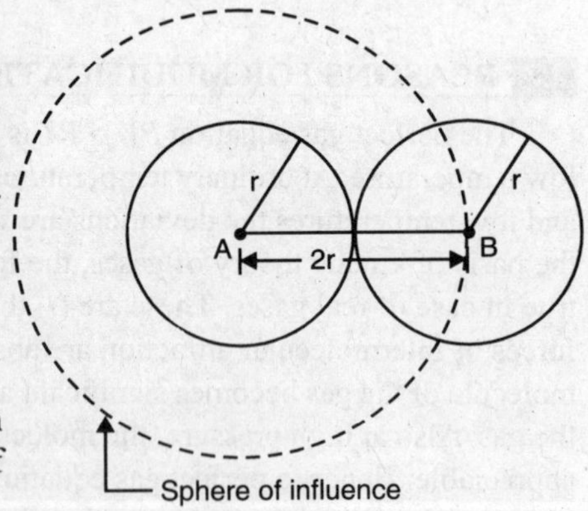

Fig. 2.8

Here r is the radius of each gas molecule.

Volume of the molecule $= x = \dfrac{4}{3}\pi r^3$

The centre of any two molecules can approach each other only by a minimum distance of $2r$. The volume of sphere of influence of each molecule

$$s = \frac{4}{3}\pi\,(2r)^3 = 8 \times \frac{4}{3}\pi r^3 = 8\,x$$

Let us fill the whole space of the volume V with n molecules one by one.

The volume available for the first molecule $= V$

The volume available for the second molecule $= V - 8x$

$= V - s$

The volume available for *n*th molecule $= [V-(n-1)s]$.

$\therefore$ Average space available for each molecule

$$= \frac{V+(V-s)+(V-2s)+\ldots\ldots+[V-(n-1)s]}{n}$$

$$= \frac{nV}{n} - \frac{s}{n}\{1+2+3+\ldots\ldots+(n-1)\}$$

$$= V - \frac{s}{n}\cdot\frac{(n-1)n}{2}$$

$$= V - \frac{ns}{2} + \frac{s}{2}$$

As the number of molecules is very large $\frac{s}{2}$ can be neglected.

$\therefore$ Average space available of each molecule

$$= V - \frac{ns}{2}$$

$$= V - \frac{n(8x)}{2} \qquad (\because \quad s = 8x)$$

$$= V - 4(nx)$$

$$= V - b \qquad \ldots (2.3)$$

Where $b = 4(nx)$ = four times the actual volume of the molecules.

Thus, the Van der Waals equation of state for a gas is

$$\left(P + \frac{a}{V^2}\right)(V-b) = RT \qquad \ldots (2.4)$$

where a and b are Van der Waals constants.

Dimensions. It should be noted that the constants a and b depend upon temperature and volume. Since a/V^2 has the dimensions of pressure, it is easy to determine the units of a. In the C.G.S system, if the pressure is expressed in atmospheres and volume in c.cs., the units of a are atmo. c.c^2 or atmo. cm^6. The unit of b is obviously c.c. The values of these constants are generally quoted for a gram molecule.

Hence in C.G.S. system a is given in atmo. c.c^2 mole^{-2} or atmo cm^6 mole^{-2} and b in c.c. mole^{-1} or cm^3 mole^{-1}. In S.I. units a is expressed in Jm^6 mole^{-2} and b in m^3 mole^{-1}.

2.9 COMPARISON WITH EXPERIMENTAL PV CURVES

Van der Waals equation is not perfect. It is only reaching to perfection. We will test the theoretical results obtained from the Van der Waal's *equation* by drawing the isothermals for carbon dioxide and comparing them with the experimental results of Andrews.

The Van der Waals equation of state is given by

$$\left(P + \frac{a}{V^2}\right)(V-b) = RT$$

or $$P = \frac{RT}{(V-b)} - \frac{a}{V^2} \qquad \ldots (2.5)$$

we see that if $V \to \infty$, $P \to 0$. Again if V is very small and $V \to b$, then $P \to \infty$.

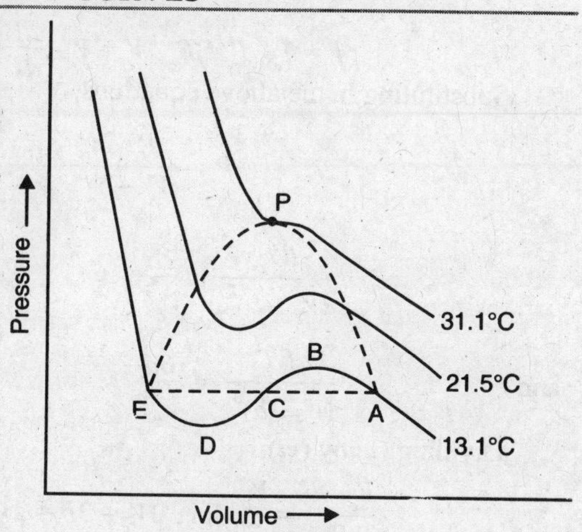

Fig. 2.9

A family of plots *i.e.*, curves between pressure and volume at various temperatures are drawn using equation 2.4. The graphs are as shown in Fig. 2.9. In the graph, the horizontal portion is absent. But in place of it, the curve *ABCDE* is obtained. This does not agree with the experimental isothermals for CO_2 as obtained by Andrew. However, the portion *AB* has been explained as due to super cooling of the vapours and the portion *ED* is due to super-heating of the liquid. But the portion *BCD* cannot be explained because it shows decrease in volume with decrease in pressure, which is not true in actual practice. At higher temperatures, however, both theoretical and practical *PV* curves are similar.

2.10 ESTIMATION OF CRITICAL CONSTANTS

Consider the critical isothermal *ACB* as shown in Fig. 2.10. At the critical point *C*, the curve is horizontal. Therefore, at point *C*, $\left(\frac{dP}{dV}\right) = 0$. At this point, the tangent also crosses the curve. Therefore, the tangent at such a point is said to be stationary and the point is called '**the point of inflexion**'. At the point of inflexion, we have

$$\frac{d^2P}{dV^2} = 0$$

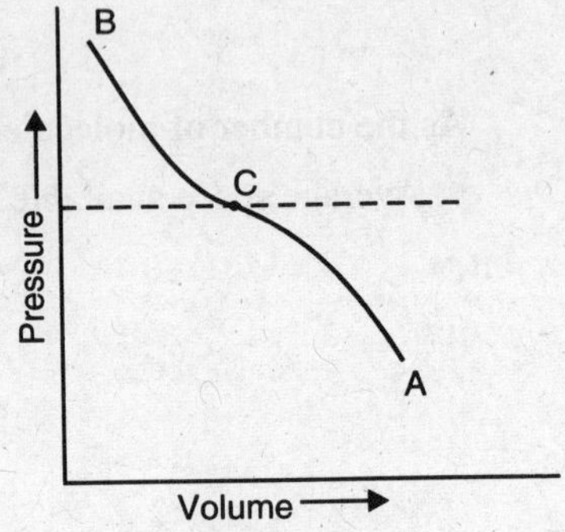

Fig. 2.10

Hence, at critical point *C*, we have $\left(\frac{dP}{dV}\right) = 0$

and
$$\frac{d^2P}{dV^2} = 0$$

The Van der Waals equation is

$$P = \frac{RT}{(V-b)} - \frac{a}{V^2} \qquad \text{... (i)}$$

$$\therefore \qquad \frac{dP}{dV} = \frac{RT}{(V-b)^2} + \frac{2a}{V^3} \qquad \text{... (ii)}$$

and
$$\frac{d^2P}{dV^2} = +\frac{2RT}{(V-b)^3} - \frac{6a}{V^4} \qquad \text{... (iii)}$$

At the critical point *C*, we have

$$T = T_c,\ P = P_c,\ V = V_c,\ \frac{dP}{dV} = 0 \text{ and } \frac{d^2P}{dV^2} = 0$$

Substituting in the above equations,

$$P_c = \frac{RT_c}{(V_c - b)} - \frac{a}{V_c^2} \qquad \text{... (iv)}$$

$$\frac{RT_c}{(V_c - b)^2} = \frac{2a}{V_c^3} \qquad \text{... (v)}$$

and
$$\frac{2RT_c}{(V_c - b)^3} = \frac{6a}{V_c^4} \qquad \text{... (vi)}$$

Dividing (*v*) by (*vi*),

$$\frac{V_c - b}{2} = \frac{V_c}{3} \quad \text{or} \quad 3V_c - 3b = 2V_c$$

$$\therefore \qquad V_c = 3b \qquad \text{... (2.6)}$$

Substituting the value of V_c in equation (*v*), we get

$$\frac{RT_c}{(3b-b)^2} = \frac{2a}{(3b)^3}$$

$$\frac{RT_c}{4b^2} = \frac{2a}{27b^3}$$

$$\therefore \qquad T_c = \frac{8a}{27bR} \qquad \text{... (2.7)}$$

Substituting the value of V_c and T_c in equation (*iv*), we get

$$P_c = \frac{8aR}{27bR(3b-b)} - \frac{a}{(3b)^2} = \frac{8a}{27\times 2b^2} - \frac{a}{9b^2}$$

$$= \frac{a}{9b^2}\left[\frac{8}{6}-1\right] = \frac{a}{9b^2}\times\frac{1}{3}$$

$$P_c = \frac{a}{27b^2} \qquad \text{... (2.8)}$$

Table 2.1 Critical Temperature and Pressure of Common Gases

Substance	*Critical Temperature °C*	*Critical Pressure (atmosphere)*
Helium	– 268°C	2.26
Hydrogen	– 240°C	12.80
Nitrogen	– 146°C	33.50
Air	– 140°C	39.00
Argon	– 122°C	48.00
Oxygen	– 118°C	50.00
Carbon dioxide	31°C	73.00
Ammonia	130°C	15.00
Chlorine	146°C	76.00
Water	374°C	218.00

2.11 CONSTANTS OF VAN DER WAALS EQUATION

From equation (2.7),

$$T_c = \frac{8a}{27bR}$$

$$\therefore \qquad \frac{a}{b} = \frac{27RT_c}{8} \qquad \text{... (vii)}$$

But

$$P_c = \frac{a}{27b^2}$$

$$\therefore \qquad \frac{a}{b^2} = 27P_c \qquad \text{... (viii)}$$

Dividing (*vii*) by (*viii*),

$$\frac{a}{b} \times \frac{b^2}{a} = \frac{27\,RT_c}{8} \times \frac{1}{27\,P_c}$$

$$\therefore \qquad b = \frac{RT_c}{8P_c} \qquad \text{... (2.9)}$$

Substituting for b in equation (*vii*), we get

$$\frac{a}{\frac{RT_c}{8P_c}} = \frac{27\,RT_c}{8}$$

$$\therefore \qquad a = \frac{RT_c}{8\,P_c} \times \frac{27\,RT_c}{8}$$

$$\text{or} \qquad a = \frac{27\,R^2\,T_c^{\,2}}{64\,P_c} \qquad \text{... (2.10)}$$

Equation (2.9) and (2.10) gives the value of constants a and b respectively.

***R* in Terms of P_c, V_c and T_c**

(*b*) Critical volume $V_c = 3b$

Critical pressure $P_c = \dfrac{a}{27b^2}$

Critical temperatue $T_c = \dfrac{a}{27Rb}$

$$\therefore \qquad \frac{P_c.V_c}{T_c} = 3b.\frac{a}{27b^2}.\frac{27Rb}{8a} = \frac{3}{8}R$$

$$\text{Or} \qquad R = \frac{8}{3}\frac{P_c V_c}{T_c} \qquad \text{... (2.11)}$$

Equation 2.11 is independent of Van der Waal's constants and does not involve a and b.

2.12 CRITICAL COEFFICIENT

The quantity $\dfrac{RT_c}{P_c V_c}$ is called the *critical coefficient*. Its value is obtained by substituting the values of P_c, V_c and T_c.

$$\therefore \text{ Critical coefficient } = \frac{RT_c}{P_c V_c} = \frac{R \times \frac{8a}{27\,bR}}{\frac{a}{27\,b^2} \times 3b}$$

$$= \frac{8\,a}{27\,b} \times \frac{27\,b^2}{3\,ab} = \frac{8}{3}$$

$$= 2.667 \qquad \text{... (2.12)}$$

The calculated value of critical coefficient $\dfrac{RT_c}{P_c V_c}$ is $\dfrac{8}{3}$ and it is the same for all gases.

The experimental values of critical coefficient for different gases are given in Table 2.2.

Table 2.2. Experimental values of Critical Coefficient $\frac{RT_c}{P_cV_c}$.

Substance	*T_c in K*	*P_c in atm. in cm³/g*	*Specific Volume*	$\frac{RT_c}{P_cV_c}$
Helium	5.1	2.25	15.4	3.13
Hydrogen	33.1	12.8	32.2	3.28
Nitrogen	125.9	33.5	3.21	3.42
Oxygen	154.2	49.7	2.32	3.42
Carbon dioxide	304	72.8	2.17	3.48
Water	647	218	3.181	4.30

Note: Here V_c = Molecular wt. × Specific volume.

The experimental value of the critical coefficient of all gases is greater than the theoretical value of 2.667.

2.13 LIMITATIONS OF VAN DER WAALS EQUATION

1. The values of constants '*a*' and '*b*' obtained by different methods differ considerably. Although Van der Waals theory assumes them to be constant. The value of *a* is found to depend on the temperature. At very high temperatures, it tends to zero.
2. According to theory, $V_c = 3b$, but it is found to depend on the nature of gas. Experimentally, it is found that $V_c \cong 2b$.
3. The theoretical value of critical coefficient $\frac{RT_c}{P_c V_c}$ is 2.667 for all gases. However, it varies from gas to gas with an average value of 3.7 for most of the gases. This constant appears to depend on the molecular structure of the gas.

Thus no gas obeys Van der Waals equation closely in the vicinity of the critical point.

2.14 REDUCED EQUATION OF STATE

When the actual pressure, volume and temperature of a fixed mass of a gas are divided by the critical values of these quantities, the resulting ratios are called the reduced pressure (P_r), reduced volume (V_r) and reduced temperature (T_r) respectively.

Thus, we have

$$P_r = \frac{P}{P_c}, \quad V_r = \frac{V}{V_c} \text{ and } T_r = \frac{T}{T_c}$$

$$\therefore \quad P = P_c P_r = \frac{a}{27b^2} P_r$$

$$V = V_c V_r = 3b\, V_r$$

$$T = T_c T_r = \frac{8a}{27\,Rb} T_r$$

Substituting these values of *P*, *V* and *T* in equation

$$\left(P + \frac{a}{V^2}\right)(V - b) = RT$$

we get

$$\left(\frac{a}{27b^2}P_r + \frac{a}{9b^2V_r^2}\right)(3bV_r - b) = R\frac{8a}{27Rb}T_r$$

or
$$\frac{a}{27b^2}\left(P_r + \frac{3}{V_r^2}\right)3b\left(V_r - \frac{1}{3}\right) = \frac{8a}{27b}T_r$$

or
$$\left(P_r + \frac{3}{V_r^2}\right)\left(V_r - \frac{1}{3}\right) = \frac{8}{3}T_r \quad ...(2.12)$$

This equation does not contain the constants a and b; which are characteristic of a particular gas. The equation is called Van der Waals reduced equation of State.

2.15 PROPERTIES OF MATTER NEAR CRITICAL POINT

Based on experiments of Andrews, Amagat and others, the state of matter near the critical point can be summarised as follows:

1. The densities of the vapour and the liquid generally approach each other and their densities are equal at the critical point.
2. At the critical point or just near the critical point, the line of demarcation between the liquid and the vapour disappears. Consequently, there must exist mutual diffusion between the two phases and the surface tension must be zero. This also means that the forces of intermolecular attraction in the liquid and vapour states must be equal at the critical point.
3. The whole mass of the substance exhibits a flickering experience which indicates that there may be variations of density inside mass. This has been experimentally verified.

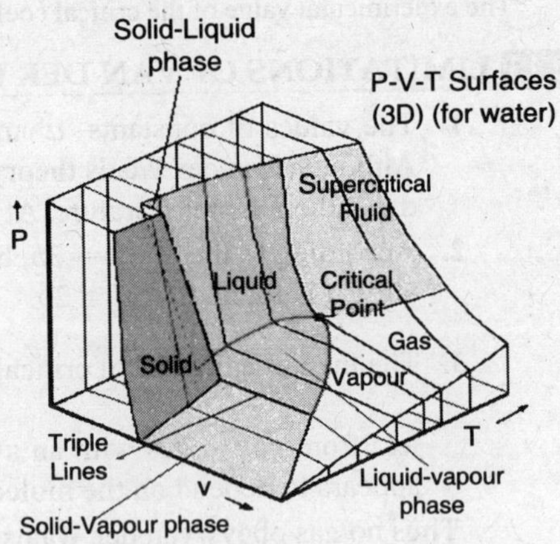

Fig. 2.11

4. The compressibility of the vapour becomes infinite at the critical temperature.

The experiments performed by Callendar and others have shown that the densities of the liquid and vapour are not equal at the critical temperature. The densities are equal at a temperature slightly higher than the critical temperature. The critical temperature of water is 374°C. It was found that the densities are equal at 380°C. At 374°C, the latent heat of water was found to be 72.4 calories per gram. At 380°C, the latent heat of water is zero.

To conclude, the critical point for any substance is not a particular temperature but it is a narrow region (critical region) (Fig.2.11). **The critical point is that point at which the properties of the two phases become identical.**

2.16 EXPERIMENTAL DETERMINATION OF CRITICAL CONSTANTS

1. The critical constants can be determined by plotting the isothermals between P and V at different temperatures of the gas. The critical point is found on the graph, where the horizontal portion of the curve (change from gaseous to the liquid state) just vanishes. The pressure, volume and temperature corresponding to the critical point give the *critical constants of the gas*.

2. With the help of Cagniard de la Tour's apparatus, the temperatures at which the liquid meniscus disappears and reappears are carefully noted. The mean of these two temperatures gives the critical temperature of the gas. The bulb is surrounded by a bath whose temperature is controlled by a sensitive thermostat and the temperature is varied by small amounts (say 0.1°C). The pressure is read with the help of a manometer. The pressure corresponding to the critical temperature gives the critical pressure. However, the measurement of critical volume offers a practical difficulty because even if the temperature is slightly below the critical temperature, the change in volume is enormous even for a small change in pressure.
3. Cailetet and Mathias have suggested a method for the determination of critical volume. Density of the liquid is determined accurately at different temperatures. The density of the saturated vapour is also determined accurately for the same temperature range. A graph is plotted between density and temperature (Fig. 2.12).

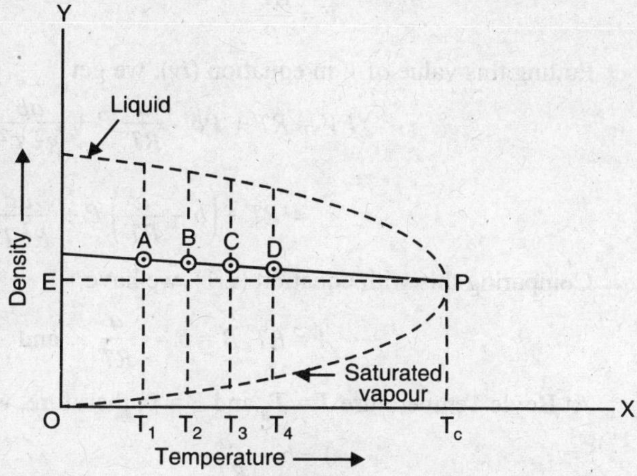

Fig. 2.12

The density of the liquid decreases with rise in temperature and that of the saturated vapour increases with rise in temperature. On the graph the points *A, B, C* and *D* represent the mean density of liquid and the vapour at temperatures T_1, T_2, T_3, T_4 etc. respectively.

The points lie on a straight line. The line is produced and the curves are extrapolated to meet the line at the point *P*. The density corresponding to the point *P* is obtained from the graph. Let this density be ρ.

The specific volume *i.e.* volume per unit mass

$$V_c = \frac{1}{\rho} \qquad \text{...(2.14)}$$

This represents the critical volume. The temperature corresponding to the point *P* gives the critical temperature.

2.17 BOYLE TEMPERATURE

The virial coefficient *B* given in equation 2.1 has a special importance. For all gases, it varies in a similar manner. Its value is negative at low temperatures, but gradually increases to zero and becomes positive as the temperature is increased. The temperature at which coefficient *B* is zero is called the **Boyle temperature**, represented by T_B. At this temperature Boyle's law holds good over a wide range of pressure provided the constants *C, B*, ... are all negligible. Thus, at Boyle temperature

$$B = \frac{d}{dP}(PV) = 0 \qquad \text{... (i)}$$

and

$$PV = A = \text{constant.} \qquad \text{... (ii)}$$

Form Van der Waals' equation of state for a real gas

$$\left(P + \frac{a}{V^2}\right)(V - b) = RT \qquad \text{... (iii)}$$

where *a* and *b* are Van der Waals' constants.

Solving eq. (*iii*), we get

$$PV - Pb + \frac{a}{V} - \frac{ab}{V^2} = RT$$

or $$PV = RT + Pb - \frac{a}{V} + \frac{ab}{V^2} \quad \text{... (iv)}$$

we can write approximately,

$$V = \frac{RT}{P} \quad \text{... (v)}$$

Putting this value of V in equation (*iv*), we get

$$PV = RT + Pb - \frac{a}{RT}P + \frac{ab}{R^2T^2}P^2 \quad \text{...(vi)}$$

$$= RT + \left(b - \frac{a}{RT}\right)P + \frac{ab}{R^2T^2}P^2$$

Comparing this with equation (2.1), we have

$$A = RT, B = b - \frac{a}{RT} \text{ and } C = \frac{ab}{R^2T^2} \quad \text{...(vii)}$$

At Boyle Temperature $T = T_B$ and $B = 0$, therefore, we have

$$0 = b - \frac{a}{RT_B}$$

or $$T_B = \frac{a}{Rb} \quad \text{... (2.2)}$$

This equation gives the value of Boyle's temperature for all gases obeying Van der Waals' equation of state. Clearly Boyle temperature (T_B) depends upon a and b which are different for different gases. Hence Boyle **temperature is different for different gases**. For nitrogen T_B is 50°C, for hydrogen it is –169°C while for helium it is – 254°C.

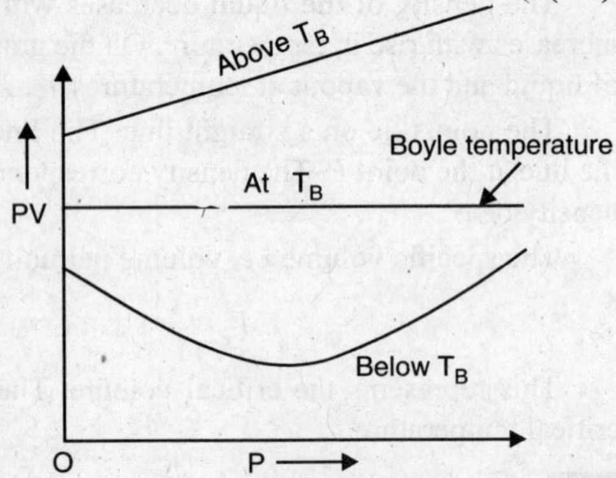

Fig. 2.13

The behaviour of gases above, below and at Boyle temperature is expressed by plotting the product PV against P for different temperatures (Fig. 2.13). The general nature of the curves is the same for all gases. Following conclusions can be drawn:

1. **Below Boyle Temperature, ($T < T_B$)** : The value of PV first decreases (as B is –*ve*) with increase in P, reaches a minimum at a particular pressure and then begins to increase. This is because at high pressures the terms CP^2, DP^3 etc. of equation (2.1) become more effective and are positive and thus PV increases with increase in P after reaching a minimum. This suggests the existence of inter-molecular attraction because below T_B, the gases are highly compressible.
2. **Above Boyle Temperature ($T > T_B$)** : Above Boyle temperature, PV continuously increases with increase in P. This is because, the coefficient B is +ve above Boyle temperature and thus the terms on R.H.S. of equation (2.1) are positive. It now indicates that the intermolecular attraction is now less significant.
3. **At Boyle temperature ($T = T_B$)** : The product PV is almost constant for a long range of pressure. Thus, Boyle's law is obeyed over a wide range of pressures.

Thus, below the Boyle temperature, the gases are highly compressible and this suggests the existence of intermolecular attraction. Beyond Boyle temperature, Boyle's law is obeyed and intermolecular attractions are less significant.

Definition: *Boyle temperature is that temperature above which a real gas behaves like an ideal gas and obeys Boyle's law.*

Relation: Relation between Boyle temperature and critical temperature

$$\frac{T_B}{T_C} = \frac{a}{R_b} \times \frac{27R_b}{8a} = \frac{27}{8} = 3.38$$

or $$T_B = 3.38\ T_C \qquad \text{... (2.15)}$$

2.18 TRIPLE POINT

Consider a cylinder fictted with a gas tight piston and filled with a substance partly in the liquid state and partly in the state of saturated vapour at a certain temperature. If the temperature is increased, the saturation vapour pressure increases.

If we plot a graph between maximum vapour pressure and temperature, we get a curve *PS* as shown in Fig. 2.14 (*i*). It is called the *curve of vaporisation* or **steam line** in case of water and steam. For any point above the line the substance is entirely in the liquid state and for any point below the line the substance is entirely in the vapour state.

A liquid and a solid may similary exist together in equilibrium at the temperature at which the solid melts under a given pressure. The relation between pressure and temperature is represented by the curve *PI* in Fig. 2.14 (*ii*). It is called the *curve of fusion* or **ice line** in case of ice and water. There are two kinds of substances.

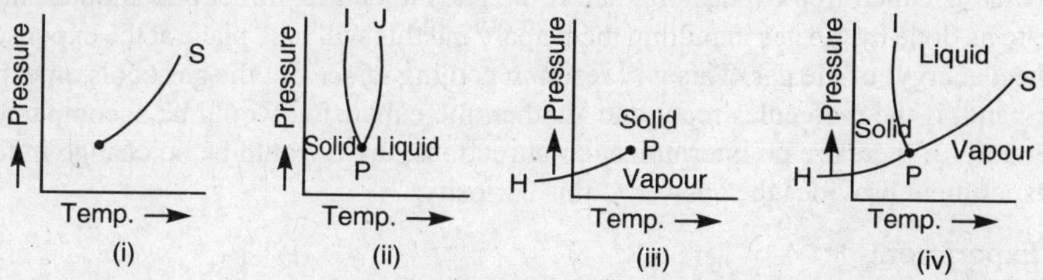

Fig . 2.14

(*i*) Those which *expand* on **solidification** *e.g.*, ice. In such a case the melting point is lowered by increase of pressure. This case is represented by the curve *PI* which slopes towards the left.

(*ii*) *Those which contract on* **solidification** *e.g.* wax. In such a case the melting points is raised by increase of pressure. This case is represented by the curve *PJ*, which slopes towards the right.

A solid and its vapour may also exist together in equilibrium *e.g.*, ice and water vapour. The graph representing the relation between pressure and temperature of a substance partly in the solid and partly in the vapour state is called the *curve of sublimation or* **hoarfrost line** in case of ice. It is shown in Fig. 2.14 (*iii*).

When all these three curves are drawn on the same diagram they meet in a **single point *P*** called the **triple point** as shown in Fig. 2.14 (*iv*).

It may be defined as the point where, for a particular temperature and pressure solid, liquid and vapour exist simultaneously in the same space in equilibrium without any change in their relative proportion. Thus tirple point is independent of pressure.

The substance is entirely in the solid state in the region *HPI*, entirely in the liquid state in the region *IPS* and entirely in the vapour state in the region below the lines *HP* and *PS*.

Only one triple point: It should be carefully noted that triple point is a single point. This can be proved in a very simple way.

If the three curves do not meet at a single point, let them intersect each other forming a small triangle *ABC* as shown in Fig. 2.15.

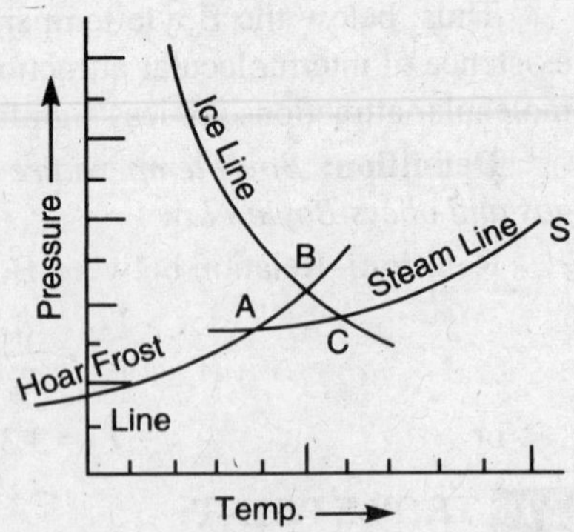

Fig. 2.15

Then, since the space *ABC* is above the steam line points in it will correspond to the liquid state alone in equilibrium. But as the space *ABC* is below the Hoarfrost line, points in its must correspond to the vapour state alone in equilibrium. Also because *ABC* is below the ice line, points in it will correspond to solid state alone. Thus the assumption that the three curves do not meet at a point leads to contradictory results. Hence these curves must meet at a single point.

Effect of pressure: The triple point of a substance like ice, which expands on solidification is above its melting point at normal pressure and the substance cannot exist in the solid state above the triple point. The triple point of a substance like wax which contracts on solidification is below its melting point under normal pressure and the substance cannot exist in the liquid state above the tripple point.

2.19 JOULE'S LAW FOR A PERFECT GAS (INTERMOLECULAR ATTRACTION)

If a fixed mass of a gas is allowed to expand without doing any external work under adiabatic condition *i.e.*, no heat is allowed to enter or leave the gas. In other words, the gas is allowed to undergo what is called free expansion, then if the gas molecules attract one another, internal work will have to be done by the gas in pulling them apart and this will take place at the expense of internal energy (heat energy) of the gas. This will result in cooling *effect i.e.*, the gas cools on expansion. On the other hand, if the molecules repel one another, the expansion would be accompanied by a rise of temperature. If there are no intermolecular attractions, there would be no change in temperature. Joule's experiment provides the answer to this curiosity.

Joule's Experiment

Joule performed an experiment to study the intermolecular attraction between the molecules of a gas. He designed an apparatus to measure the change in energy of a gas when its pressure and volume are changed.

A and *B* are two flasks connected by a tube and fitted with a stop-cock *S*. *A* was filled with the gas at high pressure and *B* was perfectly evacuated (Fig. 2.16). The bulbs were surrounded by a water bath and its temperature was noted with a sensitive thermometer. The stop-cock was opened and the gas rushed from *A* to *B*. If work was done during expansion, cooling was expected.

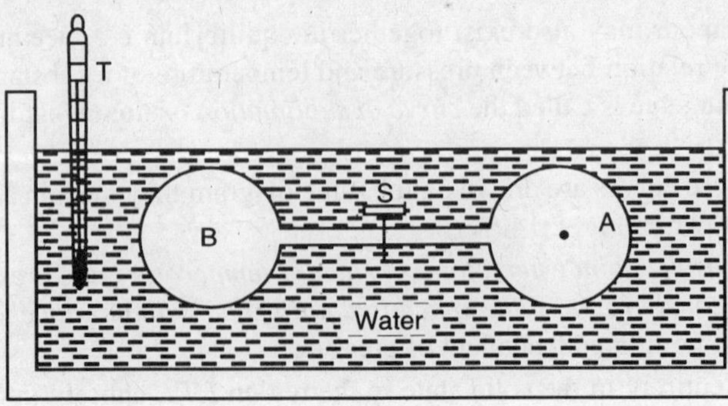

Fig. 2.16

But in this experiment no cooling was observed. This was puzzling as the work had been done by the gas in overcoming the forces of inter-molecular attractions. Joule was able to explain the drawback in his experiment by assuming that the gas has expanded into vacuum and so no cooling was observed.

In this experiment only the pressure and volume of the gas changed and as the bath did not show any change in temperature, it is evident that the temperature of the gas remained constant.

According to Joule, this experiment also shows that the internal energy of a gas is a function of temperature only, irrespective of the changes in pressure and volume.

Thus, so long as the temperature of the gas remains constant, its internal energy remains unchanged. This experimental result is often referred as **Joule's law** and may be stated that the internal energy (U) of a gas depends only on its temperature (T) and is quite independent of its volume (V) and pressure (P). Mathematically, internal energy (U) is a function of temperature (T) alone, *i.e.*

$$U = f(T)$$

$$\therefore \quad \left(\frac{\partial U}{\partial V}\right)_T = 0 \text{ and } \left(\frac{\partial U}{\partial P}\right)_T = 0 \qquad \text{... } (i)$$

$$\text{and} \quad \left(\frac{\partial U}{\partial T}\right)_P = \left(\frac{\partial U}{\partial T}\right)_V \qquad \text{... } (ii)$$

The equation (*i*) implies that the internal energy of a system does not change with volume or pressure so long as the temperature remains constant. The equation (*ii*) indicates that the increase of internal energy per degree of temperature is the same, no matter whether the pressure or the volume is kept constant during the process.

It should be noted that the Joule's law is obeyed strictly by a perfect gas only.

The porous plug experiment designed by Joule and Kelvin, however, showed cooling due to the work done by a gas in over-coming the inter-molecular attraction when it was passed from the high pressure to the low pressure side.

2.20 EXPRESSION FOR JOULE'S COEFFICIENT (η)

Let us consider the free expansion of a real gas in Joule's experiment. In a real gas, there exist intermolecular forces and hence the gas is allowed to expand freely. In Joule's experiment, the intermolecular separation changes and the gas has to do some internal work against the intermolecular forces in separating the molecules further away from each other. This increases the potential energy of the gas and since no heat is allowed to enter or leave the gas (due to rapid expansion), it will happen at the cost of its kinetic energy. Thus the K.E. of the gas decreases resulting in a fall of its temperature. Thus, if the intermolecular forces are attractive, a real gas will show a cooling effect when allowed to suffer free expansion. On the similar reasoning, we may say that if intermolecular forces are repulsive, the expansion will result on the heating of the gas. But experimentally it is seen that Joule's expansion of a real gas always produces cooling and hence the intermolecular forces in a gas are attractive in nature.

The result of a free expansion is measured in terms of Joule's coefficient, defined by

$$\text{Joule's coefficient, } \eta = \left(\frac{\partial T}{\partial V}\right)_U$$

Using thermodynamical treatment and Van der Waals equation of state for real gas (or Van der Waals gas) we get the value of Joule's coefficient as

$$\eta = \left(\frac{\partial T}{\partial V}\right)_U = -\frac{1}{C_V}\frac{a}{V^2}$$

Proof:

Consider 1 kg mole of a real gas which undergoes free expansion in Joule's experiment. The change in temperature in free expansion is expressed in terms of **Joule's coefficient** defined by

$$\eta = \left(\frac{\partial T}{\partial V}\right)_U \qquad \text{... } (i)$$

Here ∂T is the change in temperature due to change in volume ∂V in free expansion of the gas.

In Van der Waals gas, due to intermolecular forces of attraction, an internal pressure developed is given by $p = \frac{a}{V^2}$. Therefore, the internal work done against this internal pressure in Joule's free expansion is

$$dW = p.dV = \frac{a}{V^2}dV \qquad \text{... } (ii)$$

where a is Van der Waals' constant and V is the initial volume of the gas. This internal work is done at the expense of its internal kinetic energy and consequently its internal potential energy increases. Thus,

Decrease in internal K.E. = Increase in internal P.E.

$$= -dW$$

$$= -\frac{a}{V^2}dV \qquad \text{... } (iii)$$

The decrease in internal *K.E.* results in decrease in temperature (cooling effect). Let dT be the decrease in temperature due to free expansion of the real gas and C_v be its molar specific heat at constant volume, then

Decrease in internal K.E. $= C_v\, dT$ ... (*iv*)

Equating (*iii*) and (*iv*), we get

$$C_v\, dT = -\frac{a}{V^2}dV$$

or

$$\frac{dT}{dV} = -\frac{1}{C_v}.\frac{a}{V^2}$$

Using equation (*i*), the Joule's coefficient (η) will be

$$\eta = \left(\frac{\partial T}{\partial V}\right)_U$$

or

$$\eta = -\frac{1}{C_v}.\frac{a}{V^2} \qquad \text{... } (v)$$

Alternative Method

Consider 1 kg mole of a real gas which underdoges free expansion in Joule's experiment. Let ∂T be the change in temperature, corresponding to the change in volume ∂V, then the Joule's coefficient is defined as

$$\eta = \left(\frac{\partial T}{\partial V}\right)_U \qquad \text{... } (i)$$

The variation of internal energy U with its volume V at any temperature T is obtained by using laws of thermodynamics. According to first law of thermodynamics

$$dQ = dU + dW = dU + PdV \qquad \text{... } (ii)$$

From 2nd law of thermodynamics, in terms of entropy

$$dQ = TdS \qquad \text{... } (iii)$$

Equating (*ii*) and (*iii*), we have

$$dU + PdV = TdS$$

$$dU = TdS - PdV$$

For isothermal change, we can write

$$\left(\frac{\partial U}{\partial V}\right)_T = T\left(\frac{\partial S}{\partial V}\right)_T - P \qquad \text{... } (iv)$$

Using Maxwell's second thermodynamic relation

$$\left(\frac{\partial S}{\partial V}\right)_T = \left(\frac{\partial P}{\partial T}\right)_V \qquad \text{... } (v)$$

Substituting in equation (*iv*), we get

$$\left(\frac{\partial U}{\partial V}\right)_T = T\left(\frac{\partial P}{\partial T}\right)_V - P \qquad \text{...} (vi)$$

According to Van der Waals' equation of state for a real gas (also called as Van der Waals' gas),

$$\left(P+\frac{a}{V^2}\right)(V-b) = RT$$

or

$$P = \frac{RT}{V-b} - \frac{a}{V^2}$$

Differentiating w.r.to *T*, we get

$$\left(\frac{\partial P}{\partial T}\right)_V = \frac{R}{V-b} \qquad (\because \text{ Assuming } V \text{ constant})$$

Substituting, in equation (*vi*), we get

$$\left(\frac{\partial U}{\partial V}\right)_T = T.\frac{R}{V-b} - \frac{RT}{V-b} + \frac{a}{V^2} = \frac{a}{V^2} \qquad \text{... } (vii)$$

The internal energy *U* is also a function of volume in addition to temperature.

Hence, for a real gas, $U = f(T, V)$ ... (*viii*)

Therefore, $dU = \left(\frac{\partial U}{\partial V}\right)_T dV + \left(\frac{\partial U}{\partial T}\right)_V dT$

But during free expansion of a real gas in Joule's experiment, the internal energy of gas remains constant i.e. $dU = 0$

$$\therefore \qquad 0 = \left(\frac{\partial U}{\partial V}\right)_T dV + \left(\frac{\partial U}{\partial T}\right)_V dT$$

or

$$\left(\frac{\partial T}{\partial V}\right)_U = -\frac{\left(\frac{\partial U}{\partial V}\right)_T}{\left(\frac{\partial U}{\partial T}\right)_V} \qquad \text{... } (ix)$$

But the quantity $\left(\frac{\partial U}{\partial T}\right)_V$ is defined as C_v, the specific heat at constant volume.

$$\therefore \qquad \left(\frac{\partial T}{\partial V}\right)_U = -\frac{1}{C_V}.\left(\frac{\partial U}{\partial V}\right)_T$$

$$= -\frac{1}{C_V}\frac{a}{V^2} \qquad [\because \text{Eq. } (vii)]$$

Therefore, the Joule's coefficient defined by eq. (*i*),

$$\eta = \left(\frac{\partial T}{\partial V}\right)_U = -\frac{1}{C_V}\frac{a}{V^2} \qquad \text{... } (2.16)$$

Discussion. Since the Van der Waals' constant *a* is a positive quantity, the Joule's coefficient (η) from equation (2.16) for a real gas is a negative quantity. Thus, when a real gas is allowed to suffer Joule's free expansion *i.e. a throttle, a cooling effect is produced which is in quite agreement with the experimental results.*

2.21 DIFFERENTIATION BETWEEN GAS AND VAPOUR

The dotted curve shown in Fig. 2.5, joining the horizontal part of the isothermals is called the **border curve**. The border curve clearly indicates the different states of substance. To its right, the substance is in gaseous state, within it a mixture of saturated vapour and liquid and to its left, it is entirely in the liquid state.

At critical temperature, the volume and density of the saturated vapour becomes equal to the volume and density of the liquid. It is not possible to distinguish or differentiate the vapour (or gas) from the liquid.

Below the critical temperature, the gas is termed as vapour and above the critical temperature it is called a gas.

2.22 JOULE-THOMSON EFFECT

If a gas initially at a constant high pressure is allowed to suffer throttle expansion through the porous plug of silk wool or cotton wool having a number of fine pores, to a region of constant lower pressure adiabatically, a change of temperature of the gas (either cooling or heating) is observed. This effect is called Joule-Thomson or Joule-Kelvin effect.

Joule in collaboration with William Thomson [Lord Kelvin] devised a very sensitive technique known as **porous plug experiment** and performed number of experiments from 1852 to 1862 and established beyond doubt the existence of intermolecular attraction. In this effect, the total heat function $H = U + PV$ remains constant.

At ordinary temperatures, all gases except hydrogen show cooling effect on passing through the porous plug, but hydrogen shows a heating effect. At sufficiently low temperatures all gases show a cooling effect.

2.23 JOULE-THOMSON POROUS PLUG EXPERIMENT

The experimental set up for the porous plug experiment to study the Joule-Thomson effect is shown in Fig. 2.17. It consists of the following main parts:

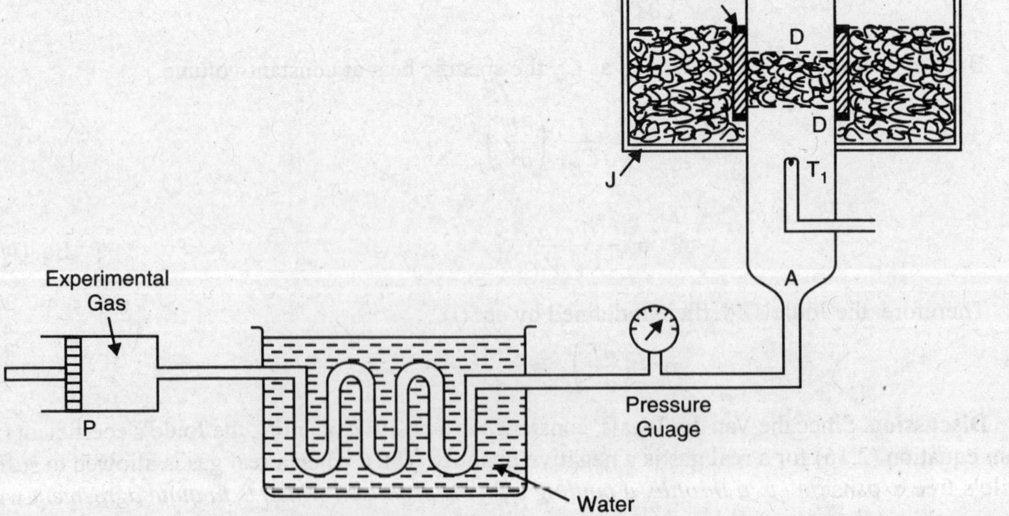

Fig. 2.17. J-T Porous Plug Experiment

1. The apparatus consists of a porous plug having two perforated brass discs *D, D*.
2. The space between *D, D* is packed with cotton wool or silk fibers.
3. The porous plug is fitted in a cylindrical box-wood *W* which is surrounded by a vessel containing cotton wool (Fig. 2.17). This is done to avoid loss or gain of heat from the surroundings.
4. T_1 and T_2 are two sensitive platinum resistance thermometers and they measure the temperatures of the incoming and outgoing gas.
5. The gas is compressed to a high pressure with the help of piston P and it is passed through a spiral tube immersed in water bath maintained at a constant temperature. If there is any heating of the gas due compression, this heat is absorbed by the circulating water in the water bath.

Experimental procedure

By means of the compression pump *P*, the experimental gas is passed slowly and uniformly through the porous plug keeping the high pressure constant, read by pressure gauge. During its passage through the porous plug, the gas is throttled, *i.e.* the separation between the molecules increases. On passing through the porous plug, the volume of the gas increases against the atmospheric pressure. As there is no loss or gain of heat during the whole process, the expansion of the gas takes place adiabatically.

The initial and final temperatures are noted by means of the platinum resistance thermometers T_1 and T_2.

Experimental Results

Inversion Temperature: The behaviour of a large number of gases was studied at various inlet temperatures of the gas and the results obtained are as follows:

(*i*) At sufficiently low temperatures, all gases show a cooling effect.

(*ii*) At ordinary temperatures, all gases except hydrogen and helium show cooling effect. Hydrogen and helium show heating instead of cooling at room temperature.

(*iii*) The fall in temperature is directly proportional to the difference in pressure on the two sides of the porous plug.

(*iv*) The fall in temperature for a given difference of pressure decreases with rise in the initial temperature of the gas. It was found that the cooling effect decreased with the increase of initial temperature and became zero at a certain temperature and at a temperature higher than this temperature, instead of cooling, heating was observed. This particular temperature at which the Joule-Thomson effect changes sign is called the *temperature of inversion*.

Definition: *The temperature at which Joule-Thomson effect is zero and changes sign is known as the temperature of inversion.*

It is denoted by T_i and at this temperature

$$\frac{2a}{RT_i} = b$$

or

$$T_i = \frac{2a}{Rb} \qquad \text{... (2.17)}$$

when $T < T_i$, cooling takes place, and

when $T > T_i$, heating takes place.

for helium $T_i = 35$ K, and for hydrogen $T_i = 193$ K.

2.24 REGENERATIVE COOLING

In case of hydrogen and helium, heating was observed at room temperature because it was at a temperature far higher than its temperature of inversion. The temperature of inversion for hydrogen is – 80°C and for helium is –258°C. If helium is passed through the porous plug at a temperature lower than –258°C, it will also show cooling effect. It means any gas below its temperature of invession shows a cooling effect when it is passed through the porous plug or a throttle valve. This is called regenerative cooling or Joule-Kelvin cooling. This principle is used in the liquefaction of the so called permanent gases like nitrogen, oxygen, hydrogen and helium.

2.25 ESTIMATES OF *J-T* COOLING (THEORY OF POROUS PLUG EXPERIMENT)

Theory: The simple arrangement of the Joule-Thomson porous plug experiment is shown in Fig. 2.18. The gas is allowed to pass through the porous plug from the high pressure side to the low pressure side. Due to this large difference of pressure, the gas flows through the fine pores and becomes throttled or wire drawn, *i.e.* the molecules of the gas are further drawn apart from one another.

Consider one gram molecule of a gas to the left and to right of the porous plug. Let P_1, V_1 and P_2, V_2 represent the pressure and volume on the two sides of the porous plug. When the piston A is moved through a certain distance dx, the piston B also moves through the same distance dx. The work done on the gas by the piston $A = P_1A_1dx = P_1V_1$. The work done by the gas on the piston $B = P_2A_2dx = P_2V_2$. Thus, the net external work done by the gas is $P_2V_2 – P_1V_1$.

If w is the work done by the gas in separating the molecules against their intermolecular attractions, the total amount of work done by the gas is

$$(P_2V_2 - P_1V_1) + w$$

No heat is gained or lost to the surroundings. There are three possible cases :

(*i*) Below the Boyle temperature,

$$P_1V_1 < P_2V_2$$

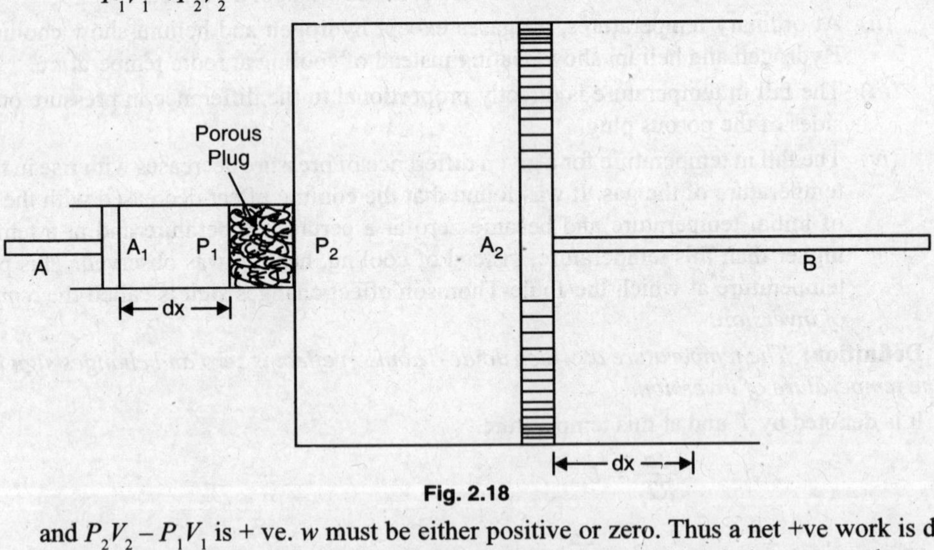

Fig. 2.18

and $P_2V_2 - P_1V_1$ is + ve. w must be either positive or zero. Thus a net +ve work is done by the gas and there must be cooling when the gas passes though the porous plug.

(*ii*) At the Boyle temperature if P_1 is not very high

$$P_1V_1 = P_2V_2$$

and $$P_2V_2 - P_1V_1 = 0$$

The total work done by the gas in this case is w. Therefore colling effect at this temperature is only due to the work done by the gas in overcoming intermolecular attractions.

(*iii*) Above the Boyle temperature

$$P_1V_1 > P_2V_2$$

or $$P_2V_2 - P_1V_1 \text{ is } -\text{ve}$$

Thus, the observed effect will depend upon whether $(P_2V_2 - P_1V_1)$ is greater or less than w.

If $w > (P_1V_1 - P_2V_2)$, cooling will be observed.

If $w < (P_1V_1 - P_2V_2)$, heating will be observed.

Thus, the cooling or heating of a gas due to free expansion through a porous plug from a high pressure to a low pressure side will depend on (*i*) the deviation from Boyle's law and (*ii*) work done in overcoming intermolecular attractions.

2.26 JOULE-KELVIN EFFECT–TEMPERATURE OF INVERSION

Assuming that the Van der Waals equation is obeyed, the attractive forces between the molecules are equivalent to an internal pressure $\frac{a}{V^2}$.

When the gas expands from V_1 to V_2, the work done in overcoming intermolecular attractions

$$w = \int_{V_1}^{V_2} P.dV$$

But $$P = \frac{a}{V^2}$$

$\therefore$ $$w = \int_{V_1}^{V_2} \frac{a}{V^2} dV$$

$$= -\frac{a}{V_2} + \frac{a}{V_1}$$

If V_1 and V_2 represent the gram molecular volumes on the high and the low pressure sides respectively, the external work done by the gas is

$$(P_2V_2 - P_1V_1)$$

Hence the total work done by the gas

$$W = (P_2V_2 - P_1V_1) + w$$

$$= (P_2V_2 - P_1V_1) - \frac{a}{V_2} + \frac{a}{V_1}$$

Van der Waals equation of state for a gas is

$$\left(P + \frac{a}{V^2}\right)(V - b) = RT$$

or $$PV + \frac{a}{V} - bP - \frac{ab}{V^2} = RT$$

or $$PV = RT + bP - \frac{a}{V} \qquad \left[\because \frac{ab}{V^2} \text{ is negligible}\right]$$

$$W = \left[RT + bP_2 - \frac{a}{V_2}\right] - \left[RT + bP_1 - \frac{a}{V_1}\right] - \frac{a}{V_2} + \frac{a}{V_1}$$

$$= b[P_2 - P_1] + 2a\left[\frac{1}{V_1} - \frac{1}{V_2}\right]$$

But $V_1 = \dfrac{RT}{P_1}$ and $V_2 = \dfrac{RT}{P_2}$

$$\therefore \quad W = b[P_2 - P_1] + 2a\left[\frac{P_1}{RT} - \frac{P_2}{RT}\right]$$

or
$$W = -b[P_1 - P_2] + \frac{2a}{RT}[P_1 - P_2]$$

or
$$W = (P_1 - P_2)\left(\frac{2a}{RT} - b\right) \quad \text{... (i)}$$

Suppose the fall in temperature is δT

$$W = JH$$
$$= J[MC_P\, \delta T]$$

where M is the gram-molecular weight of the gas

$$\therefore \quad JMC_P\, \delta T = (P_1 - P_2)\left(\frac{2a}{RT} - b\right)$$

or
$$\delta T = \left[\frac{P_1 - P_2}{JMC_P}\right] \times \left(\frac{2a}{RT} - b\right) \quad \text{... (ii)}$$

(*i*) Since $P_1 - P_2$ is + *ve*

δT will be + *ve* if $\left(\dfrac{2a}{RT} - b\right)$ is + *ve*

i.e.,
$$\frac{2a}{RT} > b \text{ or } T < \frac{2a}{Rb} \quad \text{... (iii)}$$

Therefore, cooling will take place if the temperature of the gas is less than $\dfrac{2a}{Rb}$.

(*ii*) For δT to be zero, from equation (*ii*),

$$\frac{2a}{RT} - b = 0$$

or
$$T = \frac{2a}{Rb}$$

This temperature is called the temperature of inversion and is represented by T_i

$$T_i = \frac{2a}{Rb} \quad \text{... (iv)}$$

(*iii*) δT will be negative, if

$$\left(\frac{2a}{RT} - b\right) \text{ is – ve}$$

i.e.,
$$b > \frac{2a}{RT}$$

or
$$T > \frac{2a}{Rb}$$

$$T > T_i$$

Therefore, heating will take place if the temperature of the gas is more than the temperature of inversion.

Results

(*i*) If the gas is at the temperature of inversion, then no cooling or heating is observed when it is passed through the porous plug.

(*ii*) If the gas is at a temperature lower than the temperature of inversion, cooling will take place when it is passed through the porous plug. This is called *regenerative cooling* or *Joule-Kelvin cooling*. This principle has been used in the liquefaction of the so-called permanent gases like nitrogen, oxygen, hydrogen and helium.

(*iii*) If the gas is at a temperature higher than the temperature of inversion, instead of cooling, heating is observed when the gas is passed through the porous plug.

2.27 RELATION BETWEEN BOYLE TEMPERATURE, TEMPERATURE OF INVERSION AND CRITICAL TEMPERATURE

Temperature of inversion,

$$T_i = \frac{2a}{Rb} \quad \text{... (2.18)}$$

Boyle temperature,

$$T_B = \frac{a}{Rb} \quad \text{... (2.19)}$$

Critical temperature,

$$T_c = \frac{8a}{27Rb} \quad \text{... (2.20)}$$

From (2.18) and (2.19), we have

$$T_i = 2\,T_B$$

From (2.18) and (2.20), we have

$$\frac{T_i}{T_c} = \frac{2a}{Rb} \cdot \frac{27Rb}{8a}$$

$$= \frac{27}{4} = 6.75 \quad \text{... (2.21)}$$

The experimental value of $\frac{T_i}{T_c}$ for actual gases is just less than 6.

It means that the temperature of inversion is very much higher than the critical temperature. For hydrogen $T_i = 190$ K and $T_c = 33$ K. As $T_i \geq T_c$, the methods employing regenerative cooling (Joule-Kelvin cooling) are preferred to those employing the initial cooling of the gas below the critical temperature.

Since ratio $\frac{T_i}{T_c}$ is a number (*i.e.*, 6.75) from equation 2.21, the ratio of temperature of inversion and critical temperature does not depend upon the nature of the gas.

2.28 DISTINCTION BETWEEN JOULE EXPANSION, JOULE-THOMSON EXPANSION AND ADIABATIC EXPANSION

In all the three processes *i.e.*, Joule expansion, Joule-Thomson expansion and adiabatic expansion, the system under consideration is thermally isolated from external systems, *i.e.*, $dQ = 0$. In other words, the heat is neither allowed to enter nor to leave the system. Hence, according to definition, all the three processes are adiabatic. But in **Joule expansion**, the system is also isolated *mechanically* from the external system and hence the net external work done by or on the gas is zero, *i.e.*, $W = 0$.

In **Joule-Thomson expansion**, on the contrary, the system is not isolated mechanically from the external systems and both the pistons can move keeping the pressures P_1 and P_2 constant. The net external work done by the gas in passing through the porous-plug

$$W = P_2V_2 - P_1V_1$$

In **adiabatic expansion,** the system is not mechanically isolated from the external systems and also there is no attempt to keep the pressure of the gas constant. The expanding gas performs external work against the pressure of the atmosphere, which is given by

$$W = \int P.dV$$

Here, the energy required for doing this work against the constant external pressure is drawn from the internal energy of the gas. Hence there is a decrease in internal energy, resulting in fall of temperature of the gas.

Applying the first law of thermodynamics

$dQ = dU + dW$ to all these three processes, keeping in mind that $dQ = 0$, we have

(*i*) in Joule expansion, $dU = 0$

(*ii*) in Joule-Thomson expansion, $dU = -dW = P_1V_1 - P_2V_2$

(*iii*) in adiabatic expansion, $dU = -\int P.dv$

In the first case *i.e.*, Joule expansion, the cooling effect is produced. Here the gas has to perform internal work against intermolecular forces of attraction during *free expansion*, at the expense of its internal kinetic energy. In Joule expansion, where $dU = 0$, a real gas will always show *some cooling effect*.

In the second case *i.e.*, Joule-Thomson expansion, the net external work done by the gas passing through porous, plug may produce a cooling effect or a heating effect depending upon whether P_2V_2 is greater or smaller than P_1V_1.

However, in the third case *i.e.*, adiabatic expansion, external work is done at the expense of internal energy of the gas itself. Hence, there will always be large fall of temperature. In other words, in adiabatic expansion cooling is mainly due to external work at the cost of its internal energy (U).

2.29 NATURE OF VAN DER WAALS FORCES

The matter is made up of atoms and molecules and has the tendency to occupy minimum space at low temperatures through transforming into liquid and solid state. From this we conclude that the atoms or molecules must exert a force of attraction on each other. As the atoms or molecules do not lose their individuality inspite of this attractive force, we may further conclude that there must be repulsive forces at short intermolecular distances. The simplest model giving rise to such forces was assumed by Van der Waal, according to which, "*The molecules are hard and impenetrable surrounded by attractive field*".

This model does not explain the observed compressibility of solids and does not agree with the modern concept of electronic structure of atom suggested by Bohr and Rutherford according to which, "The atom consists of a positively charged nucleus surrounded by negatively charged electrons in shells". Hence, it may be assumed that the forces between atoms and molecules are due to interaction between the electrons and the nuclei of the two systems. Hence, we are led to the conclusion that intermolecular forces are essentially *electrostatic*. However, there are also magnetic forces which are ordinarily small and may be neglected. These forces of attraction are referred to Van der Waals forces.

Except for atoms having closed shells *viz*. Helium (He), Argon (Ar), Neon (Ne) etc., almost all other atoms are electrically unsymmetrical giving rise to electric dipole. The molecules formed of such atoms will also behave as dipoles. The dipole moment of such atoms continuously varies in magnitude and direction. Therefore, the average dipole moment is always zero. However, in the case of molecules, this average is not necessarily zero. Hence, it posseses permanent dipole moment. Such molecules are called polar molecules. On the other hand, if the average dipole moment of a molecule vanishes, the molecule is said to be non-polar molecule.

If such non-polar molecules are placed in an electric field, we get induced dipole. Thus, the Van der Waals forces between the molecules of a matter (gas), more conveniently the potential energy of the system, may be due to:

(*i*) Dipole - dipole interaction.

(*ii*) Dipole - induced dipole interaction.

(*iii*) Dispersion interaction.

Thus, there are three types of Van der Waal's forces. These are

1. Orientation forces between polar molecules because of their permanent electric dipole moment.
2. Polarization forces between polar and non-polar molecules; these forces arise because of polarization of a non-polar molecule in the electric field of a polar molecule, and
3. Dispersion forces between non-polar molecules; these forces arise because of the occurrence of momentary electric dipole moments due to the rotational motion of the electrons.

The mutual potential energy of two molecules, at a distance of seperation r, due to Van der Waals' forces can be shown to be $-\frac{A}{r^6}$, where A is a constant.

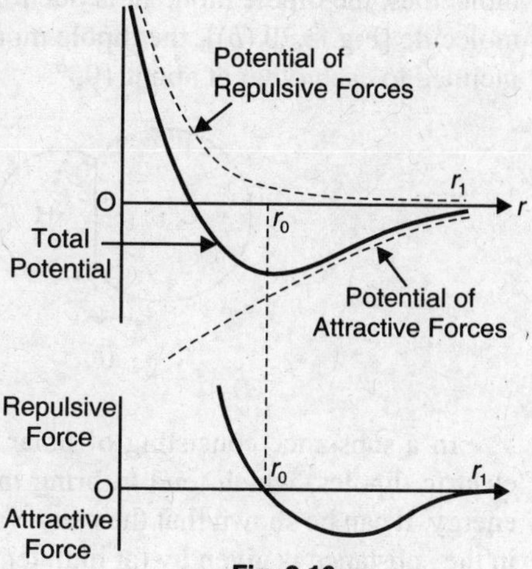

Fig. 2.19

If two molecules, are very close to each other so that their charge distributions overlap, the intermolecular force becomes repulsive. The mutual potential energy of two molecules due to the repulsive interaction is represented by the empirical expression $\frac{B}{r^{12}}$. Hence, the total potential energy of two molecules is expressed by

$$U = \frac{B}{r^{12}} - \frac{A}{r^6}$$

The exponential form $ce^{-r/\rho}$ where ρ is a measure of the range of interaction is also used for the repulsive interaction.

If we plot the potential energies as function of the separation r between two molecules, we get curves as shown in Fig. 2.19. The curves for the force

$$F = -\frac{dU}{dr}$$

is shown in Fig. 2.19. It is seen that the attractive force between two molecules is effective within the distance of separation from r_0 at which the total potential energies is minimum to r_1 at which the force becomes negligible. In the case of simple molecules of diameter of about 2.5×10^{-10} m, the value of the equilibrium distance r_0 is about 3.5×10^{-10} m and that of r_1 is about 10^{-9} m which is about 4 times the molecular diameter.

Van der Waals' forces are much weaker than those which exist in ionic and covalent bonds. Consequently molecular crystals and crystals of inert gases have low melting and boiling points and little mechanical strength. These forces account for many familiar phenomena such as friction, surface tension, viscosity, cohesion, adhesion, liquefaction of gases, solidification liquids etc.

2.30 ORIGIN OF VAN DER WAALS' FORCES

1. In Polar Molecules

In a molecule, the protons can be considered as equivalent to a positive charge at some point in the molecule. This point is called the "centre of gravity" of the protons. Similarly the electrons are quivalent to an equal negative charge at the "centre of gravity" of the electrons. In a polar molecule, these points are separated by a short distance and therefore these molecules have electric dipole moments. All diatomic molecules, such as HCl, CO, etc. consisting of different atoms are polar.

In HCl molecule [Fig. 2.20 (*a*)], the dipole moment is due to some excess of positive charge at the hydrogen end and a corresponding excess of negative charge at the chlorine end. In some molecules, the dipole moment is due to unsymmetrical geometrical structure. For example, in H_2O molecule [Fig. 2.20 (*b*)], the dipole moment is due to the unsymmetric structure; the O-H axes are inclined to each other at about 105°.

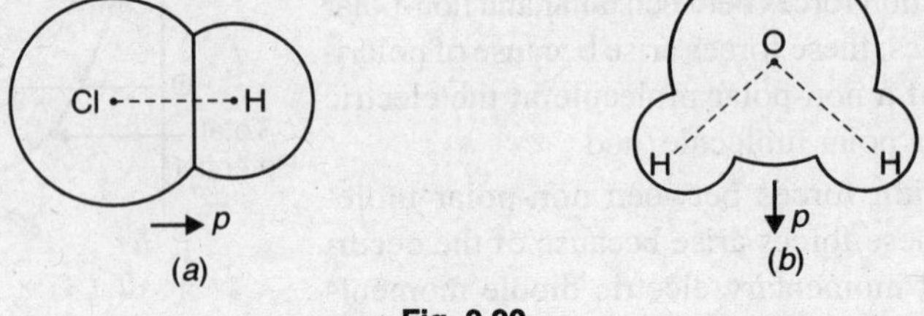

Fig. 2.20

In a substance consisting of polar molecules, there are orientation forces between molecular electric dipoles, which tend to bring the interacting system into a position of minimum potential energy. It can be shown that the mean value of the mutual potential energy of two molecular dipoles in the substance is given by (at high temperatures).

$$\bar{U} = -\frac{1}{4\pi\varepsilon_0}\cdot\frac{2}{3}\cdot\frac{p^4}{kT\,r^6} \quad \text{(In S.I. system)} \quad \ldots (2.22)$$

where p is the electric dipole moment, ε_0 is the permittivity of free space, k is Boltzmann's constant, T is the absolute temperature and r is the distance of separation.

The –*ve* sign indicates that the force is attractive. The force is given by

$$F = -\frac{d\bar{U}}{\partial r} = \frac{1}{4\pi\varepsilon_0}\cdot\frac{12}{3}\cdot\frac{p^4}{kT\,r^7} \quad \ldots (2.23)$$

Thus, the force is proportional to $\frac{1}{r^7}$.

2. Interaction between a polar and a nonpolar molecules :

The electric field E at a distance r from a molecular dipole of moment p is given by

$$E = \frac{1}{4\pi\varepsilon_0}\cdot\frac{p}{r^3}\sqrt{1+3\cos^2\theta} \quad \ldots (2.24)$$

where θ is the angle between the direction of r and the axis of the dipole.

The field E induces an electric dipole moment P_1 in the non-polar molecule. The induced dipole moment is given by

$$p_1 = \alpha E = \frac{1}{4\pi\varepsilon_0}\cdot\frac{\alpha p}{r^3}\sqrt{1+3\cos^2\theta} \quad \ldots (2.25)$$

where α is the polarizability of the molecule.

The potential energy of the induced dipole in the field E is given by

$$U = -p_1 E = -(\alpha E)E = -\alpha E^2$$

or

$$U = -\left(\frac{1}{4\pi\varepsilon_0}\right)^2\cdot\frac{\alpha p^2}{r^6}(1+3\cos^2\theta) \quad \ldots (2.26)$$

This is the expression for the mutual potential energy of the two molecules. The negative sign shows the force between them is attractive. The force is given by

$$F = -\frac{\partial \bar{U}}{\partial r} = \left(\frac{1}{4\pi\varepsilon_0}\right)^2 . \frac{6\alpha\, p^2}{r^7}(1 + 3\cos^2\theta) \qquad ...(2.27)$$

Thus, the force is proportional to $\frac{1}{r^7}$.

3. Interaction between non-polar molecules

The time average electric moments of non-polar molecules are zero. Since the electrons are in the rotational motion around the nuclei even in the lowest electronic state, this motion gives rise to momentary electric dipole moment in a non-polar molecule. A momentary dipole moment of one molecule produces an electric field at another molecule. Consequently the electric dipoles give rise to the so called *dispersion forces*. By the procedure similar to that used in the previous case, it can be shown that this force is also proportional to the seventh power of the distance of separation between two molecules.

SOLVED EXAMPLES

Example 2.1. *Calculate the Van der Waals constants for dry air, given that $T_c = 132$ K, $P_c = 37.2$ atmospheres and R per mole = 82.07 cm³ atmos K^{-1}.*

(Nagpur 2005, S/2009, S/2011, W/2010)

Solution. Given: $P_c = 37.2$ atmospheres

$T_c = 132$ K

$R = 82.07$ cm³ atm K^{-1}

(*i*)
$$a = \frac{27}{64} \times \frac{R^2 T_c^{\,2}}{P_c}$$

$$= \left(\frac{27}{24}\right)\frac{(82.67)^2 \times (132)^2}{37.2}$$

$$\boldsymbol{a = 13.31 \times 10^6 \text{ atm cm}^6}$$

(*ii*)
$$b = \frac{RT_c}{8P_c} = \frac{82.07 \times 132}{8 \times 37.2}$$

$$\boldsymbol{b = 36.41 \text{ cm}^3}$$

Example 2.2. *Calculate critical temperature for CO_2, given that a = 0.00874 atmos, cm⁶ and b = 0.0023 cm³.* **(Nagpur 2007 Similar)**

Solution. Critical temperature is given by

$$T_c = \frac{8a}{27\,Rb}$$

To evaluate the value of gram molecular gas constant *R* at a pressure of 1 atmosphere for a gram molecular volume at 0°C or 273 K from Van der Waals equation

$$\left(P + \frac{a}{V_2}\right)(V - b) = RT$$

Substituting the values,

$$\left(1 + \frac{0.0874}{1}\right)(1 - 0.0023) = R \times 273$$

$$\therefore \qquad R = \frac{1.00874 \times 0.9977}{273}$$

Thus $T_c = \dfrac{8a}{27\,Rb} = \dfrac{8\times 0.0874\times 273}{27\times 1.00874\times 0.9977\times 0.023}$

$= 305.7$ K

$\therefore$ $\mathbf{T_c = 32.7°C}$

Example 2.3. *The Van der Waals constants a and b for 1 gram molecule of hydrogen are a = 0.245 atms-litre²-mole⁻² and b = 2.67 × 10⁻² litre-mole⁻¹. Calculate the critical temperature.* **(*Kerala 2001*)**

Solution. Critical temperature

$$T_c = \frac{8a}{27\,Rb}$$

Here, $a = 0.245 \times 76 \times 13.6 \times 980 \times 10^6$ dynes-cm⁴-mole⁻²

$b = 2.67 \times 10^{-2} \times 10^3$ cm³ mole⁻¹

$R = 8.31 \times 10^7$ ergs/mole–K

$$\therefore \quad T_c = \frac{8\times 0.245\times 76\times 13.6\times 980\times 10^6}{27\times 8.31\times 10^7\times 2.67\times 10^{-2}\times 10^3}$$

$\mathbf{T_c = 33.14\ K = -239.86\ °C}$

Example 2.4. *Van der Waals' constants of a gas are a = 0.245 litre² atms mole⁻², b = 2.67 × 10⁻¹⁰ litre mole⁻¹. The initial volume and pressure of the gas are 2 litre and 4 atmosphere respectively. The gas is made to suffer a free expansion to tripple its volume. Find the change in temperature of the gas. Given C_v = 5 cals. deg⁻¹ mole⁻¹.* **(*Rohilkhand 1996*)**

Solution. The Joule's coefficient for a real gas is given by

$$\eta = \left(\frac{\partial T}{\partial V}\right)_U = -\frac{1}{C_V}\cdot\frac{a}{V^2}$$

or $$dT = -\frac{1}{C_V}\cdot\frac{a}{V^2}dV$$

Here, $a = 0.245$ litre² . atm . mole⁻²

$= 0.245 \times (10^3)^2 \times 10^{-12} \times 10^5$ Nm⁴ mole⁻²

$= 0.245 \times 10^{-1}$ Nm⁴ mole⁻²

$= 0.0245$ Nm⁴ mole⁻²

$C_v = 5$ cals . deg⁻¹ . mole⁻¹

$V = 2$ litre $= 2 \times 10^3$ cm³ $= 2 \times 10^{-3}$ m³

$dV = 2 \times 3 - 2 = 4$ litre $= 4 \times 10^{-3}$ cm³ $= 4 \times 10^{-3}$ m³

Substituting, the change in temperature will be

$$dT = -\frac{1}{5\times 4.2}\times\frac{0.0245}{(2\times 10^{-3})^2}\times 4\times 10^{-3}$$

$$= -\frac{0.0245\times 4\times 10^3}{5\times 4.2\times 4}$$

$= \mathbf{-1.17\ K = -1.17°C}$

Thus, the temperature will fall by 1.17°C.

Example 2.5. *Van der Waal's constants for a gas are a = 6.9 × 10⁻² J m³ mole⁻² and b = 2.9 × 10⁻⁵ m³ mole⁻¹. The universal gas constant R = 8.31 J mole⁻¹ K⁻¹. Calculate the critical temperature of the gas.* **(*Nagpur 2004, Kerala 2001*)**

Solution. Critical temperature

$$T_c = \frac{8a}{27Rb}$$

$$= \frac{8\times 6.9\times 10^{-2}\ J\ m^3\ mole^{-2}}{27\times 8.31\ J\ mole^{-1}\ K^{-1}\times 2.9\times 10^{-5}\ m^3\ mole^{-1}}$$

$$= \mathbf{84.83\ K}$$

Example 2.6. *Calculate the values of Van der Waal constants a and b in Van der Waals' equation for He, when critical pressure is* $0.23\times 10^6 N/m^2$ *and critical volume is* $58\times 10^{-8}\ m^3/mol$, $P_c = 37.2$ *atm, R per mole =82.07* cm^3 *atoms* K^{-1}. **(Nagpur 2006)**

Solution. Given $P_c = 0.23\times 10^6\ N/m^2$

$V_c = 58\times 10^{-8}\ m^3/mol$

Critical volume $V_c = 3b$

$$b = \frac{V_c}{3} = \frac{58\times 10^{-8}}{3} = 19.33\times 10^{-8}\ m^3\ /\ mole \quad \text{... (i)}$$

and $\frac{a}{b^2} = 27P_c = 27\times 0.23\times 10^6$

$\therefore$ $a = b^2\times 27\times 0.23\times 10^6$

$= (19.33\times 10^{-8})^2\times 27\times 0.23\times 10^6$

or $a = 2.32\times 10^{-7}\ Jm^3\ mole^{-2}$ ... (ii)

Example 2.7. *Van der Waal's constants for a gas are;* $a = 1.32$ *litre*^{2} *atm. mole*$^{-2}$; $b = 3.13\times 10^{-2}$ *litre mole*$^{-1}$. *The gas is stored at temperature 300 K and has volume 1.2 litre mole*$^{-1}$. *Find the pressure of the gas.* **(Kanpur 2002)**

Solution. Van der Waal's constant $a = 1.32$ litre^2 atm. mole^{-2}

and $b = 3.13\times 10^{-2}$ litre mole^{-1}

If we take the unit of volume as litre and unit of pressure as one atmosphere, the gas constant R per gram molecule is given by

$$R = \frac{PV}{T} = \frac{1\times 22.4}{273} = 0.082\ \text{atm litre } K^{-1}$$

Temperature $T = 300$ K, volume $V = 1.2$ lite mole^{-1}

If P is the corresponding pressure, then according to Van der Waal's equation

$$\left(P + \frac{a}{V^2}\right)(V - b) = RT$$

or
$$\left(P + \frac{1.32}{(1.2)^2}\right)(1.2 - 3.13\times 10^{-2}) = 0.082\times 300$$

or
$$P + \frac{1.32}{1.44} = \frac{0.082\times 300}{1.1687} = 21.05$$

$\therefore$ $P = 21.05 - 0.92 = 20.13$ Atm

Example 2.8. *Van der Waal's constant b for hydrogen is* $118\times 10^{-5}\ cm^3$ *when referred to 1 c.c.of the gas at N.T.P. Calculate the molecular diameter. The gram molecular volume at N.T.P.=22400 cc and Avogadro's number =* 6.02×10^{23} *mole*$^{-1}$ **(Luck U. 2000)**

Solution. Van der Waal's constant $b = 118 \times 10^{-5}$ cm³

Now $$b = 4\,nv$$

Where n is number molecules per cc. and v is the volume per molecule.

Avogadro's number = 6.02×10^{23} mole⁻¹; Gram molecular volume = 22400 c.c.

$$\therefore \quad n = \frac{6.02\times10^{23}}{22400} = 2.6875\times10^{19}$$

$$\therefore \quad 4 \times 2.6875 \times 10^{19} \times v = 118 \times 10^{-5}$$

or $$v = \frac{118\times10^{-5}}{4\times2.6875\times10^{19}} = 10.977\times10^{-24}\ c.c$$

Now $$v = \frac{4}{3}\pi r^3$$

$$\therefore \quad r = \left(\frac{3v}{4\pi}\right)^{1/3} = \left[\frac{3\times10.977\times10^{-24}}{4\times3.142}\right]^{1/3}$$

$$= [2.62 \times 10^{-24}]^{1/3} = 1.38 \times 10^{-8}\ \text{cm}$$

∴ Diameter of the molecules $d = 2r = 2.76 \times 10^{-10}$ m

Example 2.9. *Calculate the critical constants of a gas, given $b=10^{-3}$, the unit of volume being the gram molecular volume at N.T.P., and $a = 10^{-5}$, the unit of pressure being one standard atmosphere*

Solution. According to Van der Waals' equation

$$\left(P+\frac{a}{V^2}\right)(V-b) = RT$$

Substituting $P = 1$ atmosphere and $a = 10^{-5}$

$V = 1$ gram molecular volume ; $b = 10^{-3}$ and $T = 273$ K, we have

$$\left(1+\frac{10^{-5}}{1}\right)(1-10^{-3}) = R\times273$$

$$\therefore \quad R = \frac{1.00001\times0.999}{273} = 0.00366$$

Now critical temperature $$T_c = \frac{8a}{27\,Rb} = \frac{8\times10^{-5}}{27\times0.00366\times10^{-3}}$$

$$= 0.81\text{ K} = 0.81 - 273 = -272.19°\text{C}$$

Critical pressure $$P_c = \frac{a}{27\,b^2} = \frac{10^{-5}}{27\times10^{-6}} = 0.37 \text{ atmosphere}$$

Critical volume $$V_c = 3b = 3 \times 10^{-3} = 0.003 \text{ gm. mol. vol.}$$

Example 2.10. *The critical temperature and pressure of CO_2 is 31 °C and 73 atmospheres respectively. Assuming that CO_2 obeys van der Waal's equation compute the critical volume of CO_2 and estimate the diameter of CO_2 molecule.* **(Banaras University 2001)**

Solution. Critical temperature $T_c = 31°\text{C} = 31 + 273 = 304$ K

Critical pressure $P_c = 73$ atm

$R = 82.07$ cm³ atm K⁻¹

For a gas which obeys van der Waal's equation

$$R = \frac{8}{3}\frac{P_c V_c}{T_c}$$

∴ Critical volume $$V_c = \frac{3}{8}\frac{RT_c}{P_c} = \frac{3 \times 82.07 \times 304}{8 \times 73} = 128.16\,\text{cm}^3$$

Now $$V_c = 3b \text{ or } b = \frac{V_c}{3} = 42.72\,\text{cm}^3$$

But $b = 4\,nv$ where n is number of molecules per cm³ and v is the volume per molecule. If Avogadro's number = 6.02×10^{23} mole^{-1} and gram molecular volume = 22400 cm³, Then

$$n = \frac{6.02 \times 10^{23}}{22400} = 2.6875 \times 10^{19}$$

∴ $$4 \times 2.6875 \times 10^{19}\, v = 42.72$$

or $$v = \frac{42.72}{4 \times 2.6875 \times 10^{19}} = 3.97 \times 10^{-19}$$

Now $v = \frac{4}{3}\pi r^3$ where r is the radius of the molecule

$$r = \left[\frac{3v}{4\pi}\right]^{1/3} = \left[\frac{3 \times 3.97 \times 10^{-19}}{4 \times 3.142}\right]^{1/3} = \left[0.948 \times 10^{-19}\right]^{1/3}$$

$$= 4.56 \times 10^{-7}\text{ cm}$$

or diameter $d = 2r = 9.12 \times 10^{-7}$ cm $= 9.12 \times 10^{-9}$ m.

Example 2.11. *Calculate the temperature for triple point of water in absence of air. Given that the vapour pressure at 0 °C and 1 °C are 4.60 and 4.94 mm, respectively, and the lowering of melting point per atmosphere is 0.0072°C.*

Solution. Water exapnds on solidification, hence its triple point is above 0 °C, which is the freezing point under normal pressure of 760 mm.

Let the triple point correspond to a temperature t °C and a pressure P.

Vapour pressure at 0 °C= 4.60 mm

Vapour pressure 1 °C = 4. 94 mm.

Increase in vapour pressure per °C = 0.34 mm.

∴ Increase in vapour pressure for t °C = 0.34 t mm.

Hence $V.P$ at triple point t°C= 4.60 + 0.34 t mm.

Since there is no air, the pressure above water is only due to its own vapur pressure at the triple point.

Melting point of ice under 760 mm pressure = 0°C

Lowering of melting point per atmosphere increase of pressure = 0.0072°C

But vapour pressure = 4.60 + 0.34t mm.

∴ Decrease in pressure = 760 – 4.60 – 0.34t = 755.4 – 0.34t mm.

Hence increase in melting point = $\frac{0.0072}{760} \times (755.4 - 0.34t)$

or Melting point = $0 + \frac{0.0072}{760} \times (755.4 - 0.34t)$ = The triple point = t °C

$$\text{or} \qquad \frac{0.0072 \times 755.4}{760} = t + \frac{0.34 \times 0.0072}{760} t = \left(\frac{760.00245}{760} \right) t$$

$$\therefore \qquad t = \frac{0.0072 \times 755.4}{760.00245} = 0.007156\ °C$$

EXERCISE

2.1 In what way a real gas differs from an ideal gas. ***(Meerut 2000, G.N.D.U. 2000)***

2.2 (*a*) Describe Andrew's experiments on carbon dioxide. Discuss the results obtained. Show that the liquid and the gaseous states are two distant stages of a continuous phenomenon. ***(Kerala 2001, Meerut 2000, Bangalore 2000)***

(*b*) How internal energy of a real gas differs from an ideal gas ***(P.U.2007, 2007, 2005)***

2.3 Differentiate between gas and vapour using Andrew's experimental isothermals for CO_2.

2.4 What are critical constants of a gas? ***(Alagappa 2004)***

Define: (*i*) Critical temperature T_c

(*ii*) Critical pressure P_c

(*iii*) Critical volume V_c

2.5 Define Boyle temperature and inversion temperature. Explain the behaviour of real gases above and below the Boyle temperature. ***(Nagpur 2009, 2007, Calcutta 2004, Pune 2010)***

2.6 Is the Boyle temperature different for different gases? Explain why.

2.7 Explain the reasons for modification of perfect gas equation. ***(Nagpur 2005, 2004)***

2.8 Derive and discuss the Van der Waals equation of state of a gas. Compare the results with Andrew's experimental PV curves. ***(Purvanchal 2006, Patna 2005, Nagpur 2010, 2005, 2004)***

2.9 What do you understand by Virial coefficients? What is the value of first Virial coefficient? How does the second Virial coefficient vary with temperature? ***(Calcutta 2001)***

2.10 (*a*) What consideration led van der Waals to modify gas equation of state? Deduce Van der Waals' gas equation $\left(P + \frac{a}{V^2}\right)(V - b) = RT$. What are the dimensions of a and b? Discuss how far this equation is in keeping with experimental facts. What are its limitations? Give three defects of van der Waals' equation. ***(Kerala U. 2001; Nag. U.2001, M.D.U.2003,2002, 2001; Luck U 2001, Garhawal U.2000, Meerut U.2007, 2006, 2005, 2003, 2000; K.U. 2000, Gauhati U.2000; Mithila U.2001, Kan U. 2002, Cal. U. 2003)***

(*b*) Compare van der Waal's isothermals with Andrews' practical isothermals for CO_2 and account for the difference. ***(Meerut. U.2005, Mithila U. 2001, Utkal U. 2003)***

2.11 Distinguish between a perfect gas and a real gas. Derive Van der Waals' equation of state and use it to obtain the expressions for the critical constants in terms of the constants of the Van der Waals' equation.

(Calcutta 2005, Ranchi 2004, M.D.U. 2003, 2001, 2000, 1999, Kerala 2001, K.U. 2001, Utkal 2003, Nagpur 2003, 2001)

2.12 What are the critical constants of a gas? Calculate the values of these constants in terms of the constants of the Van der Waals equation.

(Nagpur 2010/W, 2005, Alagappa 2004, Calcutta 2003)

2.13 Define critical coefficient of a gas. Is it same for all gases? Does experimental values agree with the theoretical value? ***(Nagpur 2005)***

2.14 Mention limitations of Van der Waals equation. ***(Nagpur 2004, Garhwal 2001)***

2.15 Derive Van der Waals reduced equation of state of a gas. ***(Kerala 2001, Garhwal 2000)***

2.16 What is Joule-Thomson effect? How is it experimentally established? How will you interpret the effect? What is throttling process? ***(Nagpur 2009, 2008, 2007, 2006, 2004, Bangalore 2005, Mysore 2005, Patna 2005, Madras 2004)***

2.17 Describe the porous plug experiment. What conclusions have been drawn from it? What is inversion temperature?
(Bangalore 2004, 2005, Alagappa 2004, Ranchi 2004, Madras 2005)

2.18 What is temperature of inversion? Derive the expression for inversion temperature for Van der Waals gas, $T_i = \dfrac{2a}{Rb}$
(G.N.D.U. 2007, 2006, 2004, Garhwal 2000, Meerut U., 2007, 2003, 2002, P.U. 2007, 2006, 2002, Nagpur 2001, Purvanchal 2006, 2003, Pbi 2005)

2.19 Define Joule-Thomson coefficient for the real gas. Give its expression
(Garhwal 2005, Nagpur 2011/5, 2007, 2006, 2004, Purvanchal 2006)

2.20 Give the theory of Joule-Thomson effect. Hence explain why hydrogen and helium show heating effect at ordinary temperatures. ***(Nagpur 2004, Garhwal 2000, 2002, 2005, Bangalore 2002)***

2.21 Give the theory of the porous plug experiment. Obtain conditions for heating and cooling of a gas. ***(Gujrat 2005, Alagappa 2005, Bangalore 2002, 2004, Nagpur 2009, 2007)***

2.22 Show that in Porous-plug experiment the total heat function (ethalpy) remains constant.
(Pbi U. 2008, P.U. 2002, GNDU 2006, 2003, Kerale U.2001, NDU 2007, Nagpur 2009, 2007, 2003, 2001)

2.23 What are the critical constants of a gas? State and explain Van der Waals equation. Calculate the Van der Waals constants *a* and *b* in terms of V_c, P_c and T_c ***(Nagpur U.S/2011, 2007)***

2.24 What is Joule-Thomson effect? Distinguish between adiabatic expansion and Joule-Thomson effect. ***(Nagpur, 2008, 2009, Kuvempa Uni.2005)***

2.25 How is the ideal gas equation modified when mutual attraction between molecules and the finite size of the molecules are taken into consideration? Obtain relations between critical constants and the constants appearing in the Van der Waals equation.

2.26 Derive the reduced equation of state for a gas starting from Van der Waals equation of state. Show that if the two gases have the same reduced pressure and volume, they also have the same reduced temperature? ***(Kerala 2001, Garhwal 2000)***

2.27 Starting from elementary ideas, set up Van der Waals equation of state and show how the constants of the equation can be expressed in terms of the critical constants. What is the experimental bearing of this equation? ***(Meerut 2000)***

2.28 Explain the corrections introduced by Van der Waals in the gas equation. Show that for a gas obeying Van der Waals equation

$$\frac{RT_c}{P_c V_c} = \frac{8}{3}$$

(Utkal 2003, K.U. 2001, Nagpur 2003, 2004, Meerut 2002, 2000, Kerala 2001, Calcutta 2003)

2.29 Draw Andrew's curves for CO_2 at different temperatures and obtain expression for critical constants from them. Show that

$$\frac{T_i}{T_c} = \frac{27}{4}$$

2.30 Derive and discuss the Van der Waals' equation of state of a gas. Mention its defects. ***(Calcutta 2003)***

2.31 Explain the principle of regenerative cooling ***(Nagpur 2008, Garhwal 2005, Bangalore 2004, Kuvempa 2005)***

2.32 Obtain equation of Joule-Kelvin coefficient η by discussing Joule-Thomson porous plug experiment in detail. ***(Calcutta 2004, Nagpur 2004, Gujarat 2005, 2004, Mumbai 2005)***

2.33 Deduce Van dar Waals equation of state. How far does it conform to Andrews' experimental results on carbon dioxide? What is the importance of Andrews' experiments in the problem of liquefaction of gases? ***(Punjab 2003)***

2.34 Define Boyle point, temperature of inversion and critical temperature of a gas. How Boyle temperature is related to critical temperature? ***(Pune 2010, Nagpur 2009, 2005, winter)***

2.35 Prove that the ratio of temperature of inversion and critical temperature does not depend upon the nature of the gas. ***(Ranchi 2000)***

2.36 Distinguish between ideal and real gases. ***(Bangalore 2004, Mysore 2004)***

2.37 Describe briefly the porous plug experiment. Deduce a theoretical expression for the Joule-Thomson cooling for 1 gram molecule of a gas obeying Van der Waals equation of state. Why does hydrogen show a negative Joule-Thomson effect? ***(Mysore 2004, Madias 2004)***

2.38 Describe experiment to find the value of the critical constants of a gas

2.39 State limitations of the ideal gas equation ***(Nagpur U. Summer 2011, 2005)***

2.40 Explain Joule's law of expansion. Discuss the results of Joule's experiment after describing the experimental details of the experiment. How this expansion differs from adiabatic and Joule-Thomson expansion? ***(Purvanchal 2007, Rohilkhand 2000)***

2.41 Derive an expression for Joule coefficient for real gas. ***(Nagpur 2009, 2004)***

2.42 Write short notes on:

(*i*) Van der Waals' gas equation ***(Kanpur 1995, Rohilkhand 2000)***

(*ii*) Behavior of gases at high pressure.

(*iii*) J-T cooling ***(Nagpur 2011/S, 2005)***

(*iv*) Critical point.

(*v*) Temperature of inversion. ***(Nagpur 2004, 2005, Mumbai 2005, Mysore 2005)***

(*vi*) Reduced equation of state for a gas.

(*vii*) Porous plug experiment. ***(Rohilkhand 2001)***

(*viii*) Boyle's temperature ***(Nagpur 2004, 2005)***

(*ix*) Adiabatic cooling and Joule-Kelvin cooling. ***(Bangalore 2002)***

2.43 The Van der Waals constants for carbon dioxide are $a = 1.32 \times 10^4$ N m^4 mole^{-2} and $b = 3.64 \times 10^{-5}$ m^3 mole^{-1}. Calculate critical pressure and temperature.

[Ans. $P_c = 3.68 \times 10^{11}$ atm; $T_c = 29.42 \times 10^9$ K**]** ***(Kuvempa Uni. 2005)***

2.44 Calculate the fall in temperature produced by adiabatic throttling process of a gm. mole of oxygen when pressure is reduced by 50 atmospheres. The initial temperature of gas is 27°C. Given for oxygen $C_p = 7.03$ cal/mole K, $a = 1.32$ litre^2 × atms mole^{-2}, $b = 3.12 \times 10^{-2}$ litre mole^{-1} and $R = 8.3$ Joule/K mole. **[Ans.** 12.7°C**]** ***(Nagpur 2008, Punjab 2005)***

2.45 Calculate the change in temperature when helium gas suffers Joule-Thomson expansion at – 173°C, the pressure difference on the two sides of the porous plug being 20 atmospheres. Does the gas show a heating effect or a cooling effect in this expansion at – 173°C. Given R = 8.3 Joule/K mole and for helium Van der Waal's constant $a = 0.0341$ litre^2 atms mole^{-2} and b = 0.0237 litre mole^{-1}. **[Ans.** $dT = -1.5$ K, Since $T_i = -240$°C, heating effect**]**

2.46 (*a*) Define critical constants from Van der Waals' equation of state and obtain expression for the critical temperature, pressure and volume in terms of the constants of Van der Waals' equation.

(*b*) Show that for a van der Waals' gas $R = \frac{8}{3}\frac{P_c V_c}{T_c}$ where P_c, V_c and T_c are critical pressure, volume and temperature respectively. Hence calculate the critical coefficients.

(M.D.U. 2005, 2003, 2001, 2000, Kerala U. 2001, K.U. 2001; Utkal U. 2003; Nag. U. 2003, 2001: Meerut U. 2007, 2006, 2002, 2000)

2.47 Derive the reduced equation of a gas starting from van der Waal's equation of state. Show that if two gases have the same reduced pressure and volume, they also have the same reduced temperature. What is law of corresponding states?

(Kerala U. 2001, Gharwal. U. 2000, Luck U. 2001)

2.48 Starting from van der Waals' equation derive an equation which does not involve van der Waals' constants *a* and *b*. ***(Luck U. 2001)***

2.49 What is Boyle temperature? How is it related to critical temperature? Calculate the ratio of Boyle temperature to critical temperature. ***(Luck U. 2001; Cal U. 2002)***

2.50 Calculate critical volume of 1 mole of hydrogen

Given : $T_C = -240°C$, $\rho_C = 12.8 \times 10^5$ N/m² and $R = 8.3$ J/mole K.

(Nagpur U. Summer 2011)

Hint: $R = \frac{8}{3}\frac{P_c V_c}{T_c}$ Hence $V_c = \frac{3}{8}\frac{RT_c}{P_c} = \frac{3}{8} \times \frac{8.3 \times 33}{12.8 \times 10^5} = 8.02 \times 10^{-5}\, m^3$

MULTIPLE CHOICE QUESTIONS

2.1 The force between the molecules of a gas is that of

(*a*) attraction (*b*) repulsion

(*c*) neither attraction nor repulsion

2.2 The dimensions of the constant *b* in van der Waals' gas equation are that of

(*a*) Pressure (*b*) Volume

(*c*) Pressure/Volume

2.3 The Van der Waals' equation of states for a real gas is

(*a*) $\left(P + \frac{a}{V^2}\right)(V - b) = RT$ (*b*) $PV = RT$

(*c*) $\left(P - \frac{a}{V^2}\right)(V - b) = RT$ (*d*) $\left(P - \frac{a}{V^2}\right)(V + b) = RT$ ***(Kanpur 2001)***

2.4 Choose the wrong statement for Joule expansion

(*a*) Joule expansion is a free expansion.

(*b*) In Joule-expansion, gas expands into vacuum.

(*c*) No external work is done in Joule expansion.

(*d*) No internal work is done in Joule expansion.

2.5 Choose the wrong statement from the following

(*a*) Joule expansion always produces a cooling effect

(*b*) Joule-Thomson effect is always a cooling effect

(*c*) In Joule expansion, only internal work is done

(*d*) In Joule-Thomson expansion, external work is done on the gas as well as by the gas.

2.6 In general gas equation $PV = RT$, Van der Waal introduced a correction factor $\frac{a}{V^2}$ in pressure. This term represents

(*a*) Effective area of molecules (*b*) Mean velocity of gas molecules
(*c*) Volume occupied by molecules (*d*) Attractive force between molecules

2.7 The dimensions of the constant *b* in Van der Waals' gas equation are that of

(*a*) Pressure (*b*) Volume
(*c*) Pressure/Volume (*d*) Volume/Pressure

2.8 The value of critical volume V_c according to Van der Waals' gas equation is

(*a*) $V_c = b$ (*b*) $V_c = 3b$
(*c*) $V_c = 2b$ (*d*) none of these

2.9 According to Van der Waals' gas equation, critical coefficient $\frac{RT_c}{P_cV_c}$ is equal to

(*a*) 8 (*b*) $\frac{8}{3}$
(*c*) 8.3 (*d*) 1

2.10 Two gases have equal reduced pressure and reduced temperature. The gas having higher molecular weight

(*a*) shall have larger reduced volume
(*b*) shall have smaller reduced volume
(*c*) their reduced volume is the same
(*d*) no relationship between them, sometimes decreases

2.11 In Joule-Thomson effect: ***(Garhwal 2005)***

(*a*) Temperature decreases (*b*) Temperature increases
(*c*) some times temperature increases (*d*) None of these

2.12 For an ideal gas Joule's law is:

(*a*) internal energy of an ideal gas does not depend on its pressure and volume and is a function of temperature only
(*b*) internal energy of an ideal gas does not depend on its temperature and volume and is a function of pressure only.
(*c*) internal energy of an ideal gas does not depend on its pressure and temperature and is a function of volume only.
(*d*) internal energy of an ideal gas depends upon pressure, volume and temperature.

ANSWERS

2.1 (*a*)	**2.2** (*b*)	**2.3** (*a*)	**2.4** (*d*)	**2.5** (*b*)	**2.6** (*d*)
2.7 (*b*)	**2.8** (*b*)	**2.9** (*b*)	**2.10** (*c*)	**2.11**(*a*)	**2.12** (*a*)

TRANSPORT PHENOMENA IN GASES

3

Chapter

INTRODUCTION

In the kinetic theory of gases, it is assumed that the molecules of a gas are all indentical and perfectly elastic spheres of negligible but finite size. Due to haphazard motion, the molecules constantly collide on each other. At room temperature, an air molecule suffers about 4×10^9 collisions per second and travels a mean free path about 10^{-7}m. When the gas is not in equilibrium, the random *i.e.*, haphazard motion of gas molecules give rise to viscosity, conductivity and diffusion, when it tries to attain an equilibrium state by transporting momentum, heat and mass respectively.

3.1 MOLECULAR COLLISIONS

According to kinetic theory of gases, the molecules are of finite size. During the random motion of gas molecules in all possible directions and with all possible velocities, they collide with each other. Between two successive collisions, a molecule moves with a constant speed, along a straight line as shown in Fig. 3.1. These collisions are known as molecular collisions.

3.2 MEAN FREE PATH (λ)

At a given temperature, the molecules suffer continuous collisions with one another. Between any two collisions, the molecule travels freely a certain distance in a straight line. This distance is known as **free path.** Thus, the path covered by a gas molecule between any two consecutive collisions in a straight line is called the free path. The direction of the molecule is changed after every collision. After a number of collisions, the total path appears to be *zig-zag* and free path is not constant as shown in Fig. 3.1. Therefore, a term mean free path is used and is defined as the average distance travelled by a gas molecule between two successive collisions. It is denoted by λ. If the total distance travelled after N collisions is S, then mean free path (λ) is given by

$$\lambda = \frac{S}{N}$$

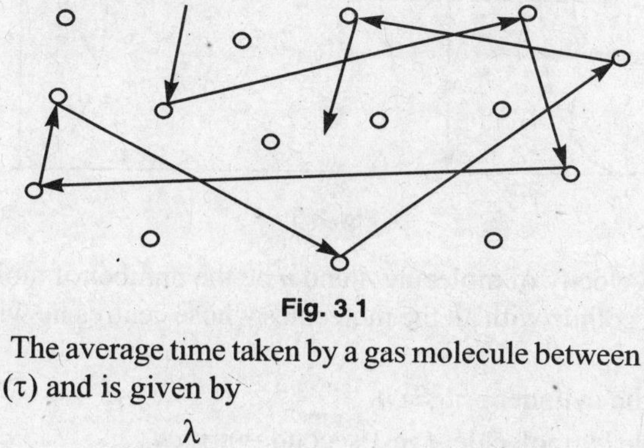

Fig. 3.1

Mean free time (τ): The average time taken by a gas molecule between two successive collisions is called mean free time (τ) and is given by

$$\tau = \frac{\lambda}{C}$$

where C is the average velocity of a gas molecule.

3.3 SPHERE OF INFLUENCE

In the kinetic theory for an ideal gas, it is assumed that all the gas molecules are identical and perfectly elastic spheres moving randomly in all possible directions. In doing so, they are continuously colliding against each other. For the sake of simplicity of calculations, it is assumed that the molecule under consideration say A, is in motion, all other molecules are at rest Fig. (3.2). Let σ be the diameter of each gas molecule. Taking the centre of molecule A as a centre, if we draw a sphere of radius σ, then molecule A will collide with all those molecules whose centres lie within this sphere. Such a sphere is called 'sphere of influence'.

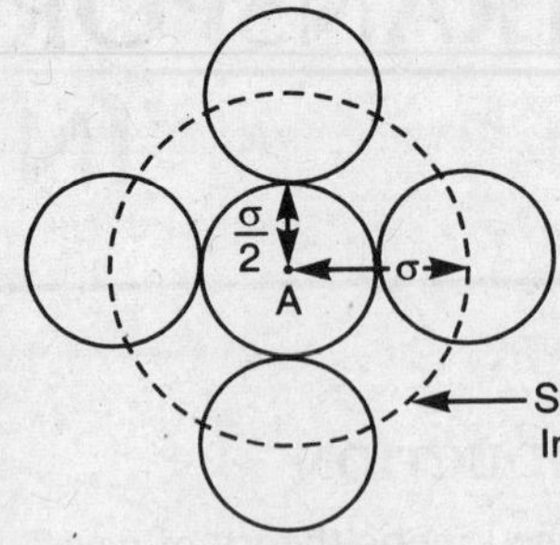

Fig. 3.2

3.4 COLLISION CROSS-SECTION

If we take a cross-section along the diameter of a sphere of influence, we get a collision cross-section. The area = $\pi\sigma^2$ is called the collision cross-section as shown by dotted lines in Fig. 3.2. It is the effective area which determines the probability that a molecule of diameter σ will collide on another molecule of the same diameter.

3.5 EXPRESSION FOR MEAN FREE PATH (CLAUSIUS' EXPRESSION)

Assumptions. To simplify the calculations we make the following assumptions.

(*i*) Only the molecule under consideration is in motion and all other molecules are at rest.

(*ii*) The sphere of influence of the molecule has a diameter 26. In other words, it means that the molecule can collide only with those molecules the centres of which lie at this distance or less.

Suppose the molecules of a gas are all identical, perfectly elastic spheres and each of diameter σ. The sphere of influence of the molecule has a radius σ *i.e.* equal to the diameter of the molecule as shown in Fig. 3.3. It is assumed that the molecules do not exert any force on each other.

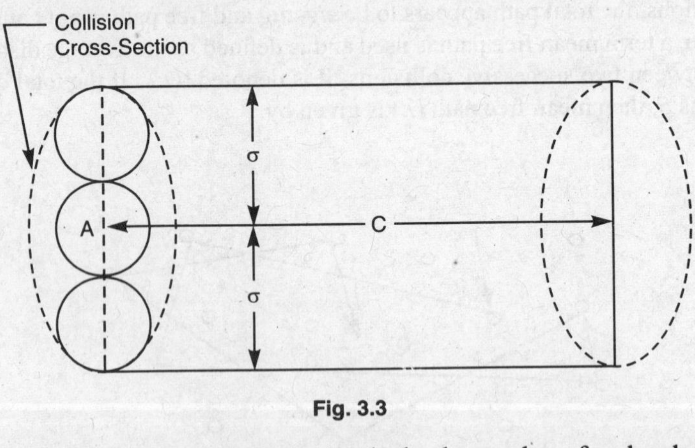

Fig. 3.3

Let c be the average velocity of molecule A and n be the number of molecules per unit volume, then in one second, A will collide with all the molecules whose centres lie within a cylinder of radius σ and length c.

No. of molecules in the cylinder = $\pi\sigma^2\, cn$

No. of collisions made by molecule A in 1 second = $\pi\sigma^2\, cn$

$\therefore$ 1 Collision takes place in $\dfrac{1}{\pi r^2 cn}$ seconds.

The time interval between two successive collisions

$$= \frac{1}{\pi\sigma^2 cn} \text{ seconds}$$

$\therefore$ Distance travelled between two successive collisions

$$= \text{speed} \times \text{time}$$

$$= c \times \frac{1}{\pi\sigma^2 cn} = \frac{1}{\pi\sigma^2 n}$$

$\therefore$ Mean free path,

$$\lambda = \frac{1}{\pi\sigma^2 n} \qquad \text{... (3.1)}$$

This is Clausius' expression for the mean free path.

Maxwell's Formula for Mean Free Path

The above expression (3.1) is not exact because it is based on the assumption that only one molecule under consideration is in motion while all other are at rest. If the motion of all the molecules is taken into account and the Maxwell's law of distribution of molecular velocities is applied, which must hold in steady state, the corrected formula for mean free path is

$$\lambda = \frac{1}{\sqrt{2}\,\pi\sigma^2 n} \qquad \text{... (3.2)}$$

This is Maxwell's formula for the mean free path.

If λ and n are known, the molecular diameter σ can be calculated. Collision frequency is $f = \dfrac{c}{\lambda}$ where c is average velocity of the gas molecule.

3.6 VARIATION OF λ WITH TEMPERATURE AND PRESSURE

Consider 1 mole of an ideal gas. Then

$$PV = RT$$

Dividing both sides by N, the Avogadro's number, we have,

$$\frac{PV}{N} = \frac{RT}{N}$$

or,

$$P = \frac{N}{V}.\frac{R}{N}.T$$

$\therefore$ The pressure P of a gas is given by

$$P = n\,k\,T \qquad \text{... (3.3)}$$

Where, $n = \dfrac{N}{V}$, the no. of molecules per unit volume and $k = \dfrac{R}{N}$, the Boltzmann's constant.

Equation (3.3) gives,

$$n = \frac{P}{kT}$$

Substituting for n in equation (3.2), the mean free path

$$\lambda = \frac{1}{\sqrt{2}\,\pi\sigma^2 n}$$

or

$$\lambda = \frac{kT}{\sqrt{2}\,\pi\sigma^2 P} \qquad \text{... (3.4)}$$

Thus, the mean free path varies directly as the absolute temperature (T) and inversely as the pressure.

3.7 TRANSPORT PHENOMENA

The transport phenomena are characterised by the transport of momentum, energy or mass of gas molecules or atoms under non-equilibrium state of the gas. These pheuomena are explained on the basis of **movement** of gas atoms or molecules and during the process momentum, energy or mass is transported from one region* to the other. According to the kinetic theory of gases, the molecules of a gas are in a constant state of thermal agitation. The gas, therefore, has a tendency to come to a steady or equilibrium state**. When a gas is not in a steady state, the following three cases arise:

(*i*) **Transport of momentum.** The different layers of the gas may have *different velocities.* In such a case there will be a relative motion of the layers of the gas with respect to each other and to bring about a steady state the layers moving faster will *transfer momentum* to the layers moving slower. This is the phenomenon of **viscosity** and arises due to transport of momentum.

(*ii*) **Transport of energy.** The different layers of the gas may be at *different temperatures.* The molecules at a higher temperature will have greater energy. To bring about a steady state the molecules will *transfer energy* from regions of higher temperature to regions of lower temperature. This energy is in the form of heat and gives rise to phenomenon of **thermal conductivity.** Thus the phenomenon of thermal conductivity is due to transport of energy.

(*iii*) **Transport of mass.** The different layers of the gas may have *different concentrations.* In such a case the molecules of the gas will move from regions of higher concentration (number of molecules per unit volume) to regions of lower concentration to come to a steady state, thereby *transferring mass* from regions of higher concentration to regions of lower concentration. This is the **phenomenon of diffusion** and arises due to transport of mass.

Thus *viscosity, conduction* and *diffusion* respresent the transport of *momentum, energy* and *mass* respectively and are called *transport phenomena.*

3.8 VISCOSITY: TRANSPORT OF MOMENTUM

The different layers of a non-equilibrium gas may have different velocities. This will result in the relative motion of the different layers with respect to one another. In such a case, the layer moving faster will impart momentum to the layer moving slower to bring about an equilibrium state. Thus, the transport of momentum gives rise to the phenomenon of viscosity.

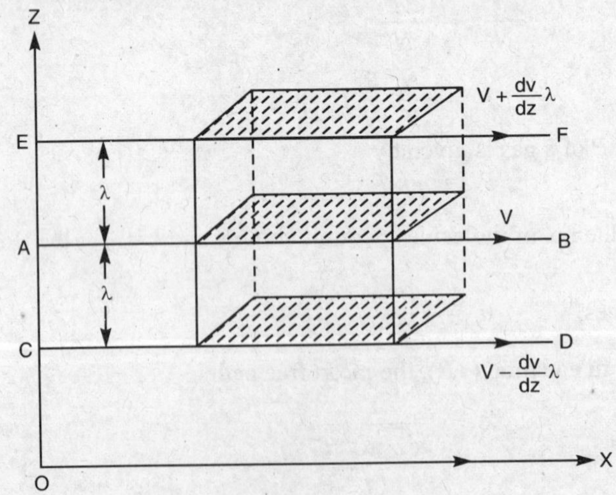

Fig. 3.4

* Means layers

** When mass, momentum energy transfer ceases.

Consider a gas flowing over a horizontal surface *OX* from left to right. The velocity of the layer in contact with the surface *OX* is zero and goes on increasing along *OZ* at a uniform rate $\frac{dv}{dz}$, known as velocity gradient (Fig 3.4).

Let us consider a layer *AB* moving with a drift velocity *V*, situated at a distance *Z* from *O*. All the molecules in this layer have the same drift velocity. Consider two more layers *EF* and *CD* just above and below *AB* respectively at a distance λ equal to the mean free path of the molecules, so that the molecules moving vertically up or down do not suffer any collision while moving between the two layers.

The velocity of the gas in the layer $EF = V + \frac{dv}{dz}\lambda$

and the velocity of the gas in the layer $CD = V - \frac{dv}{dz}\lambda$

Due to thermal agitation, the gas molecules are moving in all directions, thus, the average number of molecules along any one (+ ve or – ve) axis will be $\frac{1}{6}$th of the total number of molecules in the gas.

Let *n* be number of molecules per unit volume,

m be the mass of each gas molecule,

c be the average velocity of a molecule at a given temperature of the gas, then

Number of molecules passing downwards from *EF* to *CD* per unit area of the *AB* in one second

$$= \frac{nc}{6}$$

$\therefore$ Forward momentum lost per unit area per second by the layer *EF*

$$= m \times \frac{nc}{6} \times \left(V + \frac{dv}{dz}\lambda \right)$$

Similarly, the number of molecules passing upwards from *CD* to *EF* per unit area of the layer *AB* in one second is also

$$= \frac{nc}{6}$$

$\therefore$ Forward momentum gained per unit area per second by the layer *EF*

$$= m \times \frac{nc}{6} \times \left(V - \frac{dv}{dz}\lambda \right)$$

$\therefore$ Net momentum lost by the layer *EF* per unit area per second

$$= \frac{mnc}{6}\left\{\left(V + \frac{dv}{dz}\lambda\right) - \left(V - \frac{dv}{dz}\lambda\right)\right\}$$

$$= \frac{1}{3}\, mnc\lambda \frac{dv}{dz}$$

The layer *CD* below *AB* gains the same amount of momentum. Hence, the layer *EF* above *AB* tends to accelerate its motion and the layer *CD* below *AB* tends to retard its motion.

The backward dragging force per unit area = gain or loss of momentum per unit area per second.

$$\therefore \quad F = \frac{1}{3}\, mnc\lambda \frac{dv}{dz} \qquad \text{... (3.5)}$$

This must be equal to the viscous or tangential force $\left(\eta \frac{dv}{dz}\right)$ acting per unit area of the layer *AB*, η being the coefficient of viscosity of the gas.

$$\therefore \qquad \eta \frac{dv}{dz} = \frac{1}{3} mnc\lambda \frac{dv}{dz}$$

or
$$\eta = \frac{1}{3} mnc\lambda$$

or
$$\eta = \frac{1}{3} \rho c\lambda \qquad \text{... (3.6)}$$

where $\rho = mn$, the density of the gas.

3.9 EFFECT OF TEMPERATURE ON η

The density (ρ) of the gas decreases with increase in temperature but λ, the mean free path increases in the same proportion so that ρλ remains constant. Since the average molecular speed c is directly proportional to the square root of its absolute temperature $\left(c \propto \sqrt{T}\right)$, the coefficient of viscosity (η) will also be proportional to $\sqrt{T}$

Therefore,

$$\eta \propto \sqrt{T} \qquad \text{... (3.7)}$$

In other words, the coefficient of viscosity of a gas increases with increase in temperature.

3.10 EFFECT OF PRESSURE ON η

The density (ρ) of a gas increases with increase in pressure but λ, the mean free path decreases in the same proportion, so that ρλ remains constant. Moreover, the average molecular speed c is independent of pressure. Thus, using relation

$$\eta = \frac{1}{3} \rho c\lambda$$

η, the coefficient of viscosity is independent of pressure.

3.11 THERMAL CONDUCTIVITY: TRANSPORT OF THERMAL ENERGY

The different layers of a gas may have different temperatures. The molecules will carry kinetic energy (heat) from regions of higher temperature to the region of lower temperature to bring about an equilibrium state. Thus, the transport of thermal energy gives rise to the phenomenon of thermal conductivity.

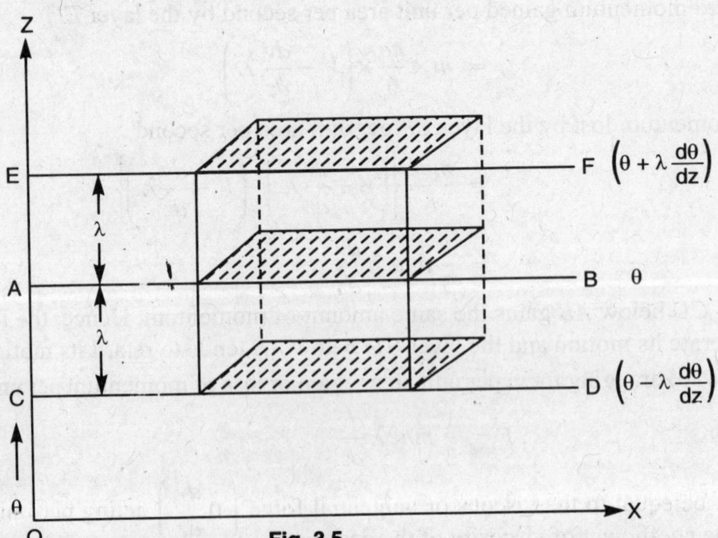

Fig. 3.5

Consider a fixed mass of a gas at rest as shown in Fig. 3.5. Suppose there is a uniform temperature gradient $\frac{d\theta}{dz}$ along OZ. In the steady state, the temperature of each horizontal layer parallel to OX is constant. Let θ be the temperature of layer AB. The temperature of the layer EF which is at a distance λ (the mean free path) above AB is $\left(\theta+\lambda\frac{d\theta}{dz}\right)$ and the layer CD at the distance λ below AB will be $\left(\theta-\lambda\frac{d\theta}{dz}\right)$. This means the gas above the layer AB is at higher temperature than the gas below AB. Hence, molecules of gas coming from EF and passing downward across AB possesses more K.E. than the molecules coming from CD and passing upward across AB. Thus, the average energy of molecules below AB is continuously increasing and that of molecules above AB is continuously decreasing, till the equilibrium is reached.

Number of molecules crossing unit area of the layer AB upward or downward per second $=\frac{nc}{6}$, where n is the number of gas molecules per unit volume and c is the average molecular velocity.

If m is the mass of each gas molecule, then the mass of gas crossing unit area of layer AB upward or downward per second $=\frac{mnc}{6}$.

If C_v be the specific heat of the gas at constant volume, then

Heat energy carried by the molecules in crossing unit area of the layer AB in the downward direction per second

$$= \text{Mass} \times \text{Specific heat} \times \text{Temperature}$$

$$= \frac{mnc}{6}\times C_v \times\left(\theta+\lambda\frac{d\theta}{dz}\right)$$

Similarly, heat energy carried by the molecules in crossing unit area of the layer AB in the upward direction per second

$$= \frac{mnc}{6}\times C_v \times\left(\theta-\lambda\frac{d\theta}{dz}\right)$$

$\therefore$ Net transfer of heat energy per unit area of layer AB in the downward direction per second

$$Q = \frac{mnc}{6}\times C_v \times\left(\theta+\lambda\frac{d\theta}{dz}\right)-\frac{mnc}{6}\times C_v \times\left(\theta-\lambda\frac{d\theta}{dz}\right)$$

$$= \frac{mnc}{6}\times C_v\times 2\lambda\frac{d\theta}{dz}=\frac{1}{3}mnc\times C_v\lambda\frac{d\theta}{dz}$$

Since $mn = \rho$, the density of the gas,

$$Q = \frac{1}{3}\rho c\lambda\, C_v\frac{d\theta}{dz} \quad \text{... (3.8)}$$

The coefficient of thermal conductivity (K) of the gas is defined as the quantity of heat that flows per unit area per second per unit temperature gradient.

or
$$Q = K\frac{d\theta}{dz} \quad \text{... (3.9)}$$

Equating equation (3.8) and (3.9),

$$K\frac{d\theta}{dz} = \frac{1}{3}\rho c\lambda C_v\frac{d\theta}{dz}$$

or, $$K = \frac{1}{3}\rho c\lambda C_v \qquad \text{... (3.10)}$$

3.12 RELATION BETWEEN η AND K

The coefficient of viscosity of a gas is given by

$$\eta = \frac{1}{3}\rho c\lambda$$

and, the coefficient of thermal conductivity is

$$K = \frac{1}{3}\rho c\lambda\, C_v = \eta\, C_v \qquad \text{... (3.11)}$$

The relation however, does not hold good with the experimental values since it is true only if all the heat supplied is converted into kinetic energy of translation.

According to Chapman and Enskog, the corrected value is

$$K = \varepsilon\eta\, C_v$$

where $\varepsilon = \frac{1}{4}(9\gamma - 5)$, where γ is ratio of two specific heats of the gas.

$$K = \frac{1}{3}\varepsilon\rho c\lambda\, C_v \qquad \text{... (3.12)}$$

3.13 EFFECT OF TEMPERATURE ON K

The coefficient of thermal conductivity (K) is given by equation (3.10) as

$$K = \frac{1}{3}\rho c\lambda\, C_v$$

$$= \frac{1}{3} mnc.\, C_v\lambda \qquad (\because\ \rho = mn)$$

$$K = \frac{1}{3} mnc.\, C_v \times \frac{1}{\sqrt{2}\,\pi\,\sigma^2 n}$$

$$K = \frac{mcC_v}{3\sqrt{2}\,\pi\sigma^2} \qquad \text{... (3.13)}$$

But $$C_v \propto \sqrt{T}$$

$\therefore$ $$K \propto \sqrt{T}$$

Thus, the coefficient of thermal conductivity of a gas is directly proportional to the square root of its absolute temperature. In other words, coefficient of thermal conductivity increases with increase in the temperature.

3.14 EFFECT OF PRESSURE ON K

In the relation, $K = \varepsilon\eta\, C_v$, all the quantities ε, η and C_v are independent of pressure. Hence the coefficient of conductivity K is also independent of pressure.

3.15 LARGEST THERMAL CONDUCTIVITY OF HYDROGEN

The coefficient of thermal conductivity of a gas is given by

$$K = \frac{1}{3}\rho c\lambda\, C_v$$

Substituting for $$\lambda = \frac{1}{\sqrt{2}\,\pi\sigma^2 n}$$

we have, $$K = \frac{1}{3\sqrt{2}} \frac{mcC_v}{\pi\sigma^2} \qquad (\because \rho = mn)$$

If M is the molecular weight and N, the Avogadro's number, then $m = \frac{M}{N}$ and C_v its molar specific heat, then

$$K = \frac{1}{3\sqrt{2}} \cdot \frac{c}{\pi\sigma^2}\left(\frac{M}{N}\right)\frac{C_v}{M}$$

$$= \frac{1}{3\sqrt{2}} \cdot \frac{C_v}{\pi N\sigma^2} c \qquad \text{... (3.14)}$$

In this relation, N is constant, being Avogadro's number. For a diatomic gas, the value of gram molecular specific heat C_v is the same for all gases being equal to $\frac{5}{2}R$. Assuming equal molecular diameters (σ) for all gases

$$K \propto c$$

But c = average speed of the molecules

$$= \sqrt{\frac{8kT}{\pi m}}$$

Where K is Boltzmann's constant.

$$K \propto \frac{1}{\sqrt{m}}$$

Thus, coefficient of thermal conductivity (K) is inversely proportional to the square root of the mass of the molecule of the gas.

As the mass of molecule of hydrogen is the least among diatomic gases, the thermal conductivity (K) of hydrogen gas will be greatest at a given temperature.

3.16 SELF DIFFUSION: TRANSPORT OF MASS

The phenomenon of self diffusion is due to the transport of mass from a region of higher concentration to a region of lower concentration to bring about an equilibrium.

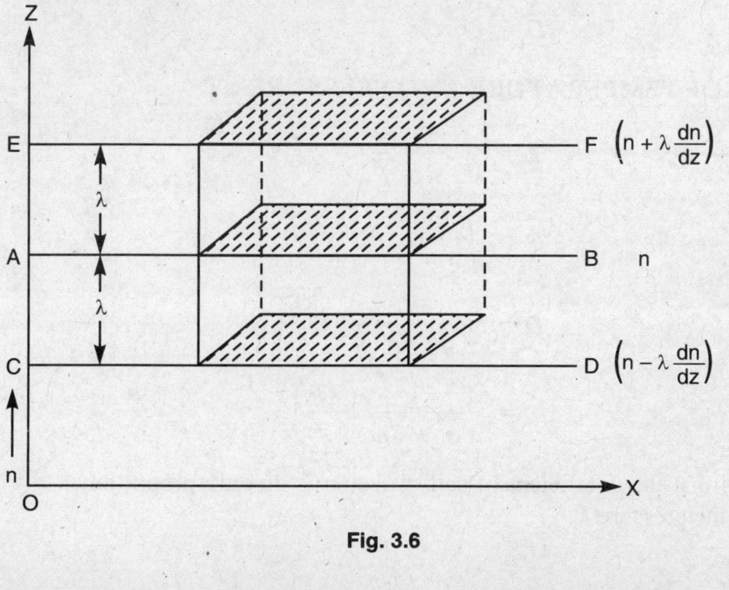

Fig. 3.6

Consider a gas having a concentration gradient $\frac{dn}{dz}$ along *OZ*. Let *n* be the concentration of layer *AB*.

Consider two more layers *EF* and *CD* above and below that of *AB* at a distance λ equal to the mean free path of the molecules of the gas, so that the molecules moving vertically upward or downward do not collide bewteen these two layers (Fig. 3.6).

Concentration at the layer $EF = n + \lambda \frac{dn}{dz}$

and concentration at the layer $CD = n - \lambda \frac{dn}{dz}$

Therefore, the number of molecules coming from layer *EF* and crossing *AB* downward per unit area per second

$$= \frac{1}{6} c \left(n + \lambda \frac{dn}{dz} \right)$$

and number of molecules coming from layer *CD* and crossing *AB* upward per unit area per second

$$= \frac{1}{6} c \left(n - \lambda \frac{dn}{dz} \right)$$

∴ Net number of molecules crossing per unit area per second of layer *AB* in downward direction

$$= \frac{1}{6} c \left(n + \lambda \frac{dn}{dz} \right) - \frac{1}{6} c \left(n - \lambda \frac{dn}{dz} \right)$$

$$= \frac{1}{3} c \lambda \frac{dn}{dz}$$

Definition. *The coefficient of diffusion is defined as the ratio of the number of molecules crossing per unit area in one second to the rate of change of concentration with distance.*

$$\therefore \text{ Coefficient of diffusion } D = \frac{\frac{1}{3} c \lambda \frac{dn}{dz}}{\frac{dn}{dz}}$$

$$D = \frac{1}{3} c \lambda \qquad \text{... (3.15)}$$

3.17 EFFECT OF TEMPERATURE AND PRESSURE

We have $\lambda = \frac{kT}{\sqrt{2}\, \pi \sigma^2 P}$ (from Eq. 3.4)

and $c = \sqrt{\frac{8kT}{\pi m}}$

$$D = \frac{1}{3} \cdot \frac{kT}{\sqrt{2}\, \pi \sigma^2 P} \cdot \sqrt{\frac{8kT}{\pi m}}$$

$$= \frac{2}{3} \frac{1}{\sigma^2 P \sqrt{m}} \cdot \left(\frac{kT}{\pi} \right)^{\frac{3}{2}} \qquad \text{... (3.16)}$$

This shows that the coefficient of self diffusion is directly proportional to $T^{3/2}$ and inversely proportional to the pressure *P*.

3.18 RELATION BETWEEN η AND D

From the relations for

$$\eta = \frac{1}{3}\rho c \lambda$$

and $$D = \frac{1}{3} c\lambda$$

we have $$D = \frac{\eta}{\rho} \quad \text{... (3.17)}$$

This relation between D and η does not agree well with the experimental values for a gas. A more detailed analysis yields the relation

$$D = f'\frac{\eta}{\rho}$$

where f' is a numerical factor whose value is nearly 1.4.

The values of the density (ρ), average speed (C), molecular diameter (σ), mean free path (λ), coefficient of viscosity (η), coefficient of conductivity (K) at N.T.P. and specific heat capacity at constant volume (C_v) for some gases are given in the following table. All the quantities are in *SI* units.

Table 3.1

Gas	*σ m*	*ρ kg/m³*	*C m/s*	*λ m*	*η Ns/m²*	*K W/mk*	*C_v J/kgK*
Helium	2.18×10^{-10}	0.179	1202.1	28.5×10^{-4}	18.6×10^{-6}	1415×10^{-4}	3157
Hydrogen	2.47×10^{-10}	0.090	1699.9	18.3×10^{-8}	8.35×10^{-6}	1684×10^{-4}	10142
Nitrogen	3.5×10^{-10}	1.25	454.4	9.44×10^{-8}	16.7×10^{-6}	243×10^{-4}	741
Oxygen	3.5×10^{-10}	1.429	424.9	9.95×10^{-8}	19.2×10^{-6}	244×10^{-4}	652
Carbon dioxide	4.18×10^{-10}	1.977	362.4	6.3×10^{-8}	14.3×10^{-6}	145×10^{-4}	640
Air	–	1.293	–	–	18.325×10^{-6}	241×10^{-4}	708

NUMERICAL EXAMPLES

Example 3.1. *The mean free path of nitrogen molecule at 0°C and 1 Atmosphere pressure is 8×10^{-8} m. The molecular diameter is 3.23×10^{-10} m. Find the number of molecules per unit volume at this temperature and pressure.* **(Nagpur U. 2003)**

Solution. $$\sigma^2 = \frac{1}{\pi\lambda n} \qquad \therefore n = \frac{1}{\pi\lambda\sigma^2}$$

$$= \frac{1}{3.142 \times 8 \times 10^{-8} \times 3.23 \times 3.23 \times 10^{-10} \times 10^{-10}}$$

$$= 3.8 \times 10^{25} \text{ per m}^3$$

Example 3.2. *The diameter of nitrogen molecule is 3.2×10^{-10} m. The number of molecules at 0°C and 1 atm. pressure is 2.69×10^{25} per m³. Calculate mean free path for nitrogen molecules.* **(Nagpur U. 2004)**

Solution. The mean free path λ is given by

$$\lambda = \frac{1}{\sqrt{2}\,\pi\sigma^2 n}$$

$$= \frac{1}{\sqrt{2}\times 3.14\times(3.2\times10^{-10})^2\times 2.69\times10^{25}}$$

$$= \frac{1}{122.32}\times10^{-5}$$

$$= \mathbf{0.008175 \times 10^{-5}\ m}$$

Example 3.3. *Calculate the molecular diameter of Nitrogen molecules if n = 2.7 × 10^{25} molecules per m^3 and mean free path λ for nitrogen is 8 × 10^{-8} m.* **(Nagpur 2007, 2004)**

Solution. The mean free path λ is given by

$$\lambda = \frac{1}{\sqrt{2}\,\pi\sigma^2 n}$$

$$\therefore \quad \sigma^2 = \frac{1}{\sqrt{2}\,\pi\lambda n}$$

$$= \frac{1}{\sqrt{2}\times 3.14\times 8\times10^{-8}\times 2.7\times10^{25}}$$

$$= \frac{1}{95.918\times10^{17}} = \frac{1}{959.18\times10^{16}}$$

$$\therefore \quad \sigma = \mathbf{0.03229 \times 10^{-8}\ m}$$

Example 3.4. *Calculate the mean free path of a gas molecule, given that the molecular diameter is 2 × 10^{-8} cm and the number of molecule per cc is 3 × 10^{19}.*

(Purvanchal 2005, Kuvempa Uni. 2005)

Solution.
$$\lambda = \frac{1}{\sqrt{2}\,\pi\sigma^2 n}$$

$$\lambda = \frac{1}{1.44\times 3.14\times(2\times10^{-8})^2\times 3\times10^{19}}$$

$$\lambda = \mathbf{2.12 \times 10^{-5}\ cm}$$

[**Note:** The mean free path is less than the wavelength of light in the visible spectrum.]

Example 3.5. *Calculate the mean free path of gas molecules in a chamber of 10^{-6} mm of mercury pressure, assuming the molecular diameter to be 2A°. One gram molecule of the gas occupies 22.4 liters at N.T.P. Take the temperature of the chamber to be 273 K.* **(Calcutta 2002)**

Solution. At 760 mm Hg pressure and 273 K temperature, the number of molecules in 22.4 litres of a gas is

$$= 6.023 \times 10^{22}$$

Therefore, the number of molecules per cm^3 in the chamber at 10^{-6} mm of pressure and 273 K temperature

$$n = \frac{6.023\times10^{22}\times10^{-6}\times10^{-1}}{22400\times760\times10^{-1}}$$

$$n = 3.538 \times 10^{10}\ \text{molecules/cm}^3$$

$$\sigma = 2\ \text{Å} = 2 \times 10^{-8}\ \text{cm}$$

$$\text{Mean free path} = \lambda = \frac{1}{\sqrt{2}\,\pi\sigma^2 n}$$

$$= \frac{1}{1.44 \times 3.14 \times (2 \times 10^{-8})^2 \times 3 \times 538 \times 10^{10}}$$

$$\lambda = \mathbf{1.97 \times 10^4\ cm}$$

Example 3.6. *Calculate the mean free path of nitrogen molecule at 27°C temperature and one atmospheric pressure. The molecular diameter of nitrogen is 3.5 × 10⁻⁸ cm.*

(Agra 1995, Sagar 1998, Garhwal 1999)

Solution. One atmospheric pressure = 76 cm height of mercury

$$P = 76 \times 10^{-3} \times 13.6 \times 10^3 \times 9.8$$

$$P = 1.013 \times 10^5\ Nm^{-2}$$

Temperature, $T = 273 + 27 = 300$ K

Boltzmann's constant, $k = 1.38 \times 10^{-23}$ J/k

Molecular diameter, $\sigma = 3.5 \times 10^{-8}$ cm $= 3.5 \times 10^{-10}$ m

Mean free path, $\lambda = \dfrac{kT}{\sqrt{2}\,\pi P \sigma^2}$

$$= \frac{kT}{\sqrt{2}\,\pi P \sigma^2}$$

$$= \frac{1.38 \times 10^{-23} \times 300}{1.414 \times 3.14 \times 1.013 \times 10^5 \times (3.5 \times 10^{-10})^2}$$

$$\lambda = \mathbf{7.5 \times 10^{-8}\ m}$$

Example 3.7. *The diameter of the molecule of a gas is 2 × 10⁻⁸ cm and Boltzmann's constant is 1.38 × 10⁻²³ J/K. Calculate the mean free path at NTP.* **(*Garhwal 2000*)**

Solution. Here $d = 2 \times 10^{-8}$ cm $= 2 \times 10^{-10}$ m

$$k = 1.38 \times 10^{-23}\ J/K$$

Let n be the number of the molecules per cubic meter.

$$PV = RT$$

$$PV = NkT$$

$$n = \left(\frac{N}{V}\right) = \frac{P}{kT}$$

At NTP $n = \dfrac{0.76 \times 13.6 \times 10^3 \times 9.8}{1.38 \times 10^{-23} \times 273}$

$$n = 2.688 \times 10^{25}$$

Mean free path $\lambda = \dfrac{1}{\sqrt{2}\,\pi d^2 n} = \dfrac{1}{3.14 \times (2 \times 10^{-10})^2 \times 2.688 \times 10^{25} \times 1.414}$

$$\lambda = \mathbf{2.094 \times 10^{-7}\ m}$$

Example 3.8. *For oxygen at standard conditions, calculate (a) Collision frequency of the molecules, and (b) the mean free path. Given at standard conditions for oxygen,*

(i) the number of molecules per cubic metre = 3 × 10²⁵

(ii) Diameter of oxygen molecule = 3.6 × 10⁻¹⁰ m **(*Pune 2001*)**

Solution.

Here $k = 1.38 \times 10^{-23}$

and
$$m = \frac{32}{6.023 \times 10^{26}} = 5.31 \times 10^{26}\ kg$$

Average speed ot oxygen molecule at 0°C (273 K)

$$v = \sqrt{\frac{8\,k\,T}{\pi\,m}}$$

$$v = \sqrt{\frac{8 \times 1.38 \times 10^{-23} \times 273}{3.14 \times 5.31 \times 10^{-26}}}$$

v = **426 m/s**

Also $d = 3.6 \times 10^{-10}$ m

$n = 3 \times 10^{25}$ molecules/m^3

(*a*) Collision frequency (number of collisions per second)

$= \pi d^2 vn$

$= 3.14\,(3.6 \times 10^{-10})^2 \times 426 \times 3 \times 10^{25}$

$= \mathbf{5.2 \times 10^9}$

(*b*) Mean free path

$$\lambda = \frac{1}{\sqrt{2}\,\pi d^2\ n} = \frac{1}{3.14 \times (3.6 \times 10^{-10})^2 \times 3 \times 10^{25} \times 1.414}$$

$\lambda = \mathbf{5.792 \times 10^{-8}}$ **m.**

Example 3.9. *The molecular diameter of an ideal gas is* 2×10^{-10} *m at a temperature of 20°C and pressure 1 atmosphere. Calculate the*

(a) mean free path

(b) collision frequency

Velocity of molecules at

20°C = 511 m/s,

Take I atmospheric pressure

= 1.01 × 10⁵ N/m² **(Garhwal 1998)**

Solution.

(*a*) Here T = 293 K

As the gas is ideal, the equation

$PV = N\,k\,T$ can be appiied

$$\therefore \quad n = \frac{N}{V} = \frac{P}{kT} = \frac{1.01 \times 10^5}{(1.38 \times 10^{-23}) \times 293}$$

$n = 2.5 \times 10^{25}$ molecules/m^3

mean free path

$$\lambda = \frac{1}{1.414 \times \pi d^2 n}$$

$$\lambda = \frac{1}{1.414 \times \pi\,(2 \times 10^{-10})^2 \times (2.5 \times 10^{25})}$$

$\lambda = \mathbf{2.25 \times 10^{-7}}$ **m**

(*b*) collision frequency

$$f = \frac{\bar{v}}{\lambda} = \frac{511}{2.25 \times 10^{-7}}$$

$$\mathbf{f = 2.27 \times 10^9 \text{ per second}}$$

Example 3.10. *Calculate the number of collisions per second of a molecule of a gas having mean free path 1.876 × 10⁻⁷ m. Take average speed of the molecule as 511 m/s.* **(*Madras 2000*)**

Solution. Mean free path $\lambda = 1.876 \times 10^{-7}$ m

$v = 511$ m/s

Collision frequency,

$$f = \frac{\bar{v}}{\lambda} = \frac{511}{1.876 \times 10^{-7}} = 2.72 \times 10^9 \text{ per second}$$

Example 3.11. *Calculate the frequency of sound at which the wavelength of sound wave becomes equal to the mean free path in oxygen at 0°C and 1.0 atmospheric pressure. Take diameter of oxygen molecule as 3 × 10⁻⁸ cm.* **(*Calcutta 2000*)**

Solution. The mean free path is given by

$$\lambda = \frac{1}{\sqrt{2}\,(\pi \sigma^2 n)}$$

At 0°C and 1 atmospheric pressure means at N.T.P.,

$$\eta = \frac{6 \times 10^{23}}{22400} \quad \text{and} \quad \sigma = 3 \times 10^{-8} \text{ cm},$$

$$\therefore \quad \lambda = \frac{1}{1.414 \times 3.14 \times (3 \times 10^{-16})^2 \times (6 \times 10^{23} / 22400)}$$

$$= \frac{22400}{1.414 \times 3.14 \times 9 \times 10^{-16} \times 6 \times 10^{23}}$$

$$= \mathbf{9.37 \times 10^{-6} \text{ cm.}}$$

This wavelength, as per given problem, represents the wavelength of sound waves. Therefore, the frequency, corresponding to this wavelength will be

$$v = \frac{V}{\lambda} = \frac{332 \times 100 \text{ cm/sec}}{9.37 \times 10^{-6} \text{ cm}}$$

$$= \mathbf{3.54 \times 10^9 \text{ Hertz}}$$

Example 3.12. *Calculate the diameter of a molecule of benzene, if n = 2.79 × 10¹⁹ molecules per c.c. and mean free path λ for benzene = 2.2 × 10⁻⁶ cm.* **(*Nagpur 2008*)**

Solution. The mean free path λ is given by

$$\lambda = \frac{1}{\sqrt{2}\,(\pi \sigma^2 n)}$$

$$\therefore \quad \sigma^2 = \frac{1}{\sqrt{2}\,(\pi \lambda n)}$$

$$= \frac{1}{1.414 \times 3.14 \times 2.2 \times 10^{-6} \times 2.79 \times 10^{19}}$$

$$= \frac{1}{2.725 \times 10^{14}} = 36.2404 \times 10^{-16}$$

$\therefore$ Molecular diameter, $\sigma = \mathbf{6.02 \times 10^{-8}}$ **cm.**

Exmaple 3.13. *The mean free path of molecules of a certain gas at pressure p and temperature T is 2 × 10^{-5} cm. Deduce the mean free path under the following conditions:*

(i) pressure p × 10^{-6}, temperature T.

(ii) pressure p/2, temperature 2T.

Also calculate the probability that under condition, pressure p and temperature T, molecules vould traverse 6 × 10^{-5} cm. without collision.

Solution. The mean free path at pressure p and temperature T is given by

$$\lambda = \frac{kT}{(\sqrt{2})\,\pi\sigma^2\,p} \qquad ...(i)$$

(*i*)The mean free path at pressure $p \times 10^{-6}$ and temperature T will be

$$\lambda_1 = \frac{kT}{(\sqrt{2})\,\pi\sigma^2\,p\times 10^{-6}} \qquad ...(ii)$$

Dividing equation (*ii*) by equation (*i*), we get

$$\frac{\lambda_1}{\lambda} = 10^6 \quad \text{or} \quad \lambda_1 = 10^6\,\lambda = 10^6 \times 2 \times 10^{-5}\text{ cm}$$

$$\lambda_1 = \mathbf{20\ cm.} \qquad ...(I)$$

(*ii*) The mean free path at pressure $p/2$ and temperature 2 T will be

$$\lambda_2 = \frac{k.2T}{\sqrt{2}\,\pi\sigma^2\,(p/2)} = \frac{4\,kT}{\sqrt{2}\,\pi\sigma^2\,p} \qquad ...(iii)$$

Dividing eqn. (*iii*) by eqn. (*i*), we get

$$\frac{\lambda_2}{\lambda} = 4 \quad \text{or} \quad \lambda_2 = 4\,\lambda = 4 \times 2 \times 10^{-5}$$

$$\lambda_2 = \mathbf{8 \times 10^{-5}\ cm.} \qquad ...(II)$$

Let the probability that a molecule will collide in traversing a distance dx be μdx, where μ is some constant related to λ. Then the fraction of molecules colliding within a distance dx after traversing a distance x will be

$$\frac{dN}{N} = -\,\mu\,dx \qquad ...(iv)$$

where N is the number of molecules which traverse freely a distance x. Negative sign indicates the decrease of probability with increase in distance.

Integrating (*iv*), we get

$$\log_e N = -\,\mu\,x + C \qquad \text{where } C \text{ is a constant of integration.}$$

and is evaluated by boundary conditions.

At $\quad x = 0, N = N_0$

Substituting, $\quad \log_e N_0 = C$

$\therefore \quad \log_e N = -\mu\,x + \log_e N_0$

or
$$\log_e\left(\frac{N}{N_0}\right) = -\,\mu x$$

$$\frac{N}{N_0} = e^{-\mu x} \qquad ...(v)$$

The required probability is given by

$$P = e^{-\mu x_1} \qquad ...(vi)$$

To deduce the ratio μ and λ, from equation (*v*), which gives

$$dN = N_0\, e^{-\mu x}\,(-\mu\, dx)$$

$$\therefore \qquad P\,(x_1 + dx > x > x_1) = -\frac{dN}{N_0} = \mu e^{-\mu x} dx$$

$$\therefore \qquad \lambda = \int_0^{\infty} x_1\, \mu\, e^{-\mu x} dx = \frac{1}{\mu}$$

$$\text{or} \qquad \mu = \frac{1}{\lambda}$$

Substituting in eqn. (*vi*), we get, the required probability as

$$P = e^{-x_1/\lambda}$$

Here $\lambda = 2 \times 10^{-5}$ cm, and $x_1 = 6 \times 10^{-5}$ cm.

Probability, $P = e^{(-6 \times 10^{-5}/2 \times 10^{-5})}$

$$\boldsymbol{P = e^{-3}}$$

Example 3.14. *Determine the temperature at which mean free path of oxygen at 1 atmospheric pressure is 9.95 × 10⁻⁸ m. Given the radius of oxygen molecule is 1.695 × 10⁻¹⁰ m.*

Solution. Mean free path $\lambda = \dfrac{kT}{\sqrt{2}\,\pi\, p\sigma^2} \qquad \therefore\; T = \dfrac{\sqrt{2}\pi p\sigma^2\lambda}{k}$

Here,

$$P = 1 \text{ atm} = 0.76 \times 13.6 \times 10^3 \times 9.8 = 1.013 \times 10^5 \text{ Nm}^{-2}$$

$$\sigma = 2 \times 1.695 \times 10^{-10} \text{ m} = 3.390 \times 10^{-10} \text{ m}$$

$$\lambda = 9.95 \times 10^{-8} \text{ m}$$

$$\therefore \qquad T = \frac{\sqrt{2} \times \pi \times 1.013 \times 10^5 \times 3.39 \times 3.39 \times 10^{-20} \times 9.95 \times 10^{-8}}{1.38 \times 10^{-23}}$$

$$= 373 \text{ K} = 373 - 273 = 100° \text{ C}.$$

Example 3.15. *Find the mean free path, frequency of collision and molecular diameter of nitrogen gas, given the viscosity of gas* $\eta = 166 \times 10^{-7}$ *Nm⁻² per unit velocity gradient, average velocity* $c = 4.5 \times 10^2$ *ms⁻¹, density* $\rho = 15.25$ *kg. m⁻³ and number of molecules per m³ = 2.705 × 10²⁵.*

(Nag. U. 2010/w, 2001; Madurai. U. 2003; Kan. U., 1997; K.U. 1992)

Solution. (*i*) The viscosity of a gas is given by the relation

$$\eta = \frac{1}{3}\rho c \lambda$$

∴ Mean free path

$$\lambda = \frac{3\eta}{\rho c} = \frac{3 \times 166 \times 10^{-7}}{1.25 \times 4.5 \times 10^2} = 8.85 \times 10^{-8} \text{ m}$$

(*ii*) **Frequency of collision.** Frequency of collision is the average number of collisions made by a molecule per second and is

$$= \frac{\text{Average velocity } c}{\text{Mean free path } \lambda} = \frac{4.5 \times 10^2}{8.85 \times 10^{-8}}$$

$$= 0.5085 \times 10^{10} \text{ collision per second.}$$

(*iii*) **Molecular diameter.** Mean free path

$$\lambda = \frac{1}{\pi \sigma^2 n}$$

where n is the number of molecules per m^3

$$\sigma^2 = \frac{1}{\pi\lambda n}$$

or Molecular diameter $$\sigma = \left(\frac{1}{\pi\lambda n}\right)^{\frac{1}{2}} = \left(\frac{1}{\pi\times 8.85\times 10^{-8}\times 2.705\times 10^{25}}\right)^{\frac{1}{2}}$$

$$= 3.646 \times 10^{-10} \text{ m}$$

Example 3.16. *Calculate the diameter of a molecule if number of molecules per cm³ in a gas is 3 × 10¹⁹ and mean free path is 2 × 10⁻⁸ cm.* **(Utkal. U. 2003)**

Solution. $$\sigma = \left[\frac{1}{\pi\lambda n}\right]^{\frac{1}{2}} = \left[\frac{1}{3.142\times 2\times 10^{-8}\times 3\times 10^{19}}\right]^{\frac{1}{2}}$$

$$= 7.28 \times 10^{-7} \text{ cm.}$$

Example 3.17. *At what temperature will the mean free path of a gas molecule at constant pressure be three times that at 27°C* **(Nag. U. 2010/W, 2001)**

Solution. $$\lambda_1 = \frac{k(273+27)}{\sqrt{2}\pi\, pd^2};\ \lambda_2 = \frac{kT}{\sqrt{2}\,\pi\, pd^2}$$ As p is constant

$$\frac{\lambda_2}{\lambda_1} = \frac{T}{300} = 3$$

$\therefore$ $$T = 900 \text{ K} = 627°\text{C}$$

Example 3.18. *The viscosity of a gas at NTP was found to be 1.66 × 10⁻⁴ dynes/cm² per unit velocity gradient. The r.m.s. velocity of the molecules is 4.5 × 10⁴ cm/s. The density of gas is 1.25 grams per litre.* **(Nagpur 2001, 2009; Madurai 2003)**

Calculate (i) mean free path of the molecules of the gas,

(ii) frequency of collision.

(iii) molecular diameter of the gas molecules.

Solution. Here $$\eta = 1.66 \times 10^{-4} \text{ units}$$

$$c = 4.5 \times 10^{-4}$$

$$\rho = 1.25 \times 10^{-3} \text{ g/cc}$$

(*i*) $$\lambda = \frac{3\eta}{\rho c}$$

$$= \frac{3\times 1.66\times 10^{-4}}{1.25\times 10^{-3}\times 4.5\times 10^{4}}$$

$$\lambda = \mathbf{9 \times 10^{-6}\ cm}$$

(*ii*) Frequency of collision = number of collisions per second

$$= \frac{\text{r.m.s. Velocity}}{\text{Mean free path}}$$

$$N = \frac{c}{\lambda}$$

$$= \frac{4.5\times 10^{4}}{9\times 10^{-6}}$$

$$= \mathbf{5 \times 10^{9}}$$

(*iii*) Avogadro's number = 6.023×10^{23}

Number of molecules per $cc = n = \dfrac{6.023 \times 10^{23}}{22400}$

According to Maxwell's relation

$$\lambda = \frac{1}{\sqrt{2}.\pi\sigma^2 n} = \frac{1}{1.44 \times \pi\sigma^2 n}$$

$$\sigma = \frac{1}{\sqrt{1.44 \times \pi \times n \times \lambda}}$$

$$\sigma = \frac{1}{\sqrt{1.44 \times 3.142 \times \left(\dfrac{6.023 \times 10^{23}}{22400}\right) \times 9 \times 10^{-6}}}$$

$$\sigma = \mathbf{3 \times 10^{-8}\ cm}$$

Example 3.19. *Find the coefficient of viscosity of nitrogen at N.T.P. from the following data:* $\rho = 1.25$ kg/m^3, $\bar{c} = 454.4$ m/s, $\lambda = 9.44 \times 10^{-8}$ m. **(*Nagpur 2005*)**

Solution. The mean free path is given by

$$\lambda = \frac{3\eta}{\rho\bar{c}}$$

$\therefore$ Coefficient of viscosity,

$$\eta = \frac{\lambda\rho\bar{c}}{3}$$

$$= \frac{9.44 \times 10^{-8} \times 1.25 \times 454.4}{3}$$

$$= \mathbf{1787 \times 10^{-8}\ Nm^{-2}}$$ **per unit velocity gradient.**

Example 3.20. *In an experiment, the viscosity of the gas was found to be 2.25×10^{-4} CGS units. The r.m.s. velocity of the molecules is 4.5×10^4 cm/s. The density of the gas is 1 gram per litre. Calculate the mean free path of the molecules.* **(*Bangalore 2001*)**

Solution. Here

$$\eta = 2.25 \times 10^{-4} \text{ CGS units}$$

$$c = 4.5 \times 10^4 \text{ cm/s}$$

$$\rho = 1 \text{ g/litre}$$

$$\rho = 10^{-3} \text{ g/cc}$$

and

$$\lambda = \frac{3\eta}{\rho c}$$

$$= \frac{3 \times 2.25 \times 10^{-4}}{10^{-3} \times 4.5 \times 10^4}$$

$$\lambda = \mathbf{15 \times 10^{-6}\ cm}$$

Example 3.21. *The average speed of molecules of a gas is 4.5×10^2 m/s. The density of gas is 1.25 kg/m^3 and the coefficient of viscosity of gas is 1.66×10^{-5} Ns/m^2. Find mean free path and frequency of collision.* **(*Nagpur 2005/w*)**

Solution. Given:

$$\eta = 1.66 \times 10^{-15} \text{ Ns/m}^2$$

$$\rho = 1.25 \text{ kg/m}^3$$

$$\bar{c} = 4.5 \times 10^2$$

Mean free path,

$$\lambda = \frac{3\eta}{\rho\bar{c}} = \frac{3 \times 1.66 \times 10^{-5}}{1.25 \times 4.5 \times 10^2}$$

$$= \mathbf{8.85 \times 10^{-8}\ m}$$

Frequency of collision,

$$f = \frac{\bar{c}}{\lambda} = \frac{4.5\times10^2}{8.85\times10^{-8}}$$

$$= \mathbf{5.085 \times 10^9\ collisions/s}$$

Example 3.22. *Calculate the radius of an oxygen molecule if its coefficient of thermal conductivity*

$K = 24 \times 10^{-3}$ *J/m-s-K at 0°C*

and $C_v = 20.9 \times 10^3$ *J/kilo-mole-K,*

Boltzmann's constant $k = 1.38 \times 10^{23}$ *J/K*

and mass of an oxygen molecule

$m = 5.31 \times 10^{-26}$ *kg* **(Rajasthan 2001)**

Solution. $$K = \frac{mc\,C_v}{3(\sqrt{2})\,\pi\sigma^2}$$

But $\frac{1}{2}mc^2 = \frac{3}{2}kT$ or $c = \sqrt{\frac{3\,kT}{m}}$

$$\sigma^2 = \frac{mc\,C_v}{3\times1.414\times3.142\times K}$$

$$\sigma^2 = \frac{(\sqrt{3\,kTm})\,C_v}{3\times1.414\times3.142\times K}$$

$$\sigma^2 = \frac{[3\times1.38\times10^{-23}\times273\times5.31\times10^{-26}]^{1/2}\times20.9\times10^3}{3\times1.414\times3.142\times24\times10^{-3}}$$

or $\sigma = \mathbf{3.036 \times 10^{11}\ m}$

Example 3.23. *The viscosity of a gas (oxygen) at a temperature of 16°C is 169 micro poise. Calculate the diameter of the molecules of the gas. Avogadro's number = 6.02 × 10²³. Molecular weight of oxygen = 32 and Boltzmann's constant, k = 1.38 × 10⁻²³ J/k.*

Solution. Viscosity $\eta = 169 \times 10^{-6}$ poise $= 169 \times 10^{-7}$ deca-poise

Boltzmann's constant $k = 1.38 \times 10^{-23}$ Joule/K

Temperature $T = 273 + 16 = 289$ K

Avogadro's number $N = 6.02 \times 10^{23}$

Now viscosity $\eta = \frac{1}{3}\rho c\lambda$ or $\lambda = \frac{3\eta}{\rho c}$

Also $\lambda = \frac{1}{\pi\sigma^2 n}$ where σ is the molecular diameter

$\therefore$ $\sigma^2 = \frac{1}{\pi\lambda n} = \frac{\rho c}{3\pi\eta n}$

Again $\frac{\rho}{n} = \frac{\text{Mass per c.c.}}{\text{Number of molecules per c.c}}$

$$= \frac{\text{Mass per gm. molecule}}{\text{Number of molecules per gm. molecule}}$$

$$= \frac{\text{Molecular weight}}{\text{Avogadro's number}} = \frac{M}{N} = m = \frac{32}{6.02\times10^{23}} = 5.3156\times10^{-23}\ \text{gm}$$

$$= 5.3156 \times 10^{-26} \text{ kgm}$$

Also root mean square velocity $c = \sqrt{\dfrac{3kT}{m}}$

$$\therefore \quad \sigma^2 = \frac{1}{3\pi\eta} m \sqrt{\frac{3kT}{m}} = \frac{1}{\pi\eta}\sqrt{\frac{mkT}{3}}$$

$$= \frac{1}{\pi \times 169 \times 10^{-7}} \sqrt{\frac{5.3156 \times 10^{-26} \times 1.38 \times 10^{-23} \times 289}{3}}$$

$$= 0.0158 \times 10^{-17}$$

$$\sigma = 0.397 \times 10^{-9} \text{ m} = 3.97 \times 10^{-10} \text{ m}$$

EXERCISE

3.1 What do you mean by mean free path, frequency of collision and mean free time of a gas molecule? Write their S.I. units. Explaining the assumptions made show that the mean free path is equal to $\dfrac{kT}{\sqrt{2}\pi\, pd^2}$ where k is Boltzmann's constant p the pressure and d the diameter of each molecule. Explain the variation of mean free path with temperature and pressure.
(Nagpur U. 2003, 2001; K.U. 2002, 2000; M.D.U. 2003, 2000; Garhwal U. 2002)

3.2 Show that the mean free path of the molecules of gas is inversely proportional to the density of the gas. ***(K.U. 2001)***

3.3 Define the terms :

(*i*) mean free path,

(Nagpur 2005, Bangalore 2002, 2004, Mysore 2004, Kamraj 2003, Calcutta 2005, M.S. Uni Tamil Nadu 2007, Agra 2003, Kuvempa 2005, Agra 2004)

(*ii*) mean free time of a gas molecule. ***(Nagpur 2007, 2006/s)***

3.4 Explain the term sphere of influence and collision cross-section.
(Nagpur 2011/s, 2007, 2004, 05)

3.5 Derive the Clausius expression for mean free path (λ) on the basis of kinetic theory of gases.
(M.S. Uni. Tamil Nadu 2007, Patna 2004, Kanpur 2002, Nagpur Uni. 2010, 2007)

3.6 Obtain Maxwell's expression for mean free path $\lambda = \dfrac{1}{\sqrt{2}\,\pi\sigma^2 n}$, where σ is the molecular diameter and n is the number of molecules per unit volume, on the basis of kinetic theory of gases.
(Nagpur 2011/s, 2007, 2004, 2003, 2001, Ranchi 2005, Bangalore 2002, 2005, Pune 2010, Mysore 2005, Purvanchal 2005, 2003)

3.7 Explain the variation of mean free path (λ) with temperature and pressure. ***(Bangalore 2002)***

3.8 What is meant by the mean free path of the molecules of a gas. Obtain an expression for it and show that it is inversely proportional, to the pressure and directly proportional to the absolute temperature of the gas. ***(Agra 2002, Nagpur 2009, Rohilkhand 2001)***

3.9 Calculate the diameter of a molecule of a gas if the number of molecules per cm^3 in a gas is 3 $\times$ 19^{19} and mean free path is 2×10^{-8} cm. **[Ans.** 7.28×10^{-7} cm] ***(Utkal 2003)***

3.10 The diameter of the molecules of a gas is 3×10^{-10} m. Find the mean free path λ at S.T.P.
[Ans. 1.03×10^{-7} m] ***(Garhwal 2000, Purvanchal 2004)***

3.11 Calculate the mean free path of argon molecules at 25°C and 1 atm. pressure. Given $\sigma = 2.56$ A° and $k = 1.38 \times 10^{-23}$ Joule/K. **[Ans.** 1.4×10^{-7} metre]

3.12 The mean free path of molecules of a gas at pressure p and temperature T is λ. What will be the mean free path if

(*i*) the pressure in kept constant and the temperature is doubled,

(*ii*) the temperature is kept constant and the pressure is doubled.

[Hint : the mean free path will be doubled in the first case and will be halved in the second case]

3.13 The mean free path of the molecules of a gas is 2×10^{-7} metre at pressure p and temperature 200 K. Calculate its value of (*i*) p, 400 *K*, (*ii*) $2p$, 200 *K*, (*iii*) $\frac{1}{2}$ p, 400 K.

[Ans. (*i*) 4×10^{-7} m (*ii*) 1×10^{-7} metre (*iii*) 8×10^{-7} metre]

3.14 The mean free path of a gas molecule at pressure p and temperature T is 5.2×10^{-6} metre. What will be the mean free path at (*i*) a pressure $2p$ and temperature $\frac{T}{4}$, (*ii*) pressure $p/2$ and temperature $2T$.

[Ans. (*i*) 0.15×10^{-6} m (*ii*) 4.8×10^{-6} m] ***(Nagpur 2001, Bangalore 2001, Lucknow 1998)***

3.15 Determine the temperature at which mean free path of oxygen at 1 atmospheric pressure is 9.95×10^{-8} m. Given: the radius of oxygen molecule is 1.695×10^{-10} m. **[Ans.** 100°C] ***(K.U. 1995)***

3.16 A cross-section for collision of slow neutrons in hydrogen is about 20×10^{-24} cm². What is the mean free path of such neutrons in hydrogen at N.T.P.? Normal pressure is 10^5 N/m² and Boltzmann constant $k = 1.38 \times 10^{-23}$ J/k.

[Hint : $n = \dfrac{p}{kT}$, collision cross-section = $\pi\sigma^2 = 20 \times 10^{-24}$ cm²] **[Ans.** 13.4 m]

3.17 Calculate the diameter of molecule of benzene if $n = 2.79 \times 10^{19}$ molecules per c.c. and mean free path λ for benzene is 2.2×10^{-6} cm. **[Ans.** 6.06×10^{-8} cm]

3.18 Calculate the mean free path of a gas molecule of diameter 3.2 Å. The number of molecules per unit volume is 2.5×10^{25} m^{-3}. **[Ans.** 8.79×10^{-8} m]

3.19 Calculate the fraction of molecules which will be travelling undeflected after traversing 0.693 times the mean free path. Calculate the distance after which only one percent of the incident molecules remain undeflected. ***(Calcutta 2003)***

3.20 Discuss the effect of temperature and pressure on mean free path. ***(Garhwal 2002)***

3.21 What are transport phenomena? Explain in brief the viscosity, conductivity and self diffusion on the basis of kinetic theory of gases.

(Agra 2006, Nagpur 2009, 2004, 05, Ranchi 2005, Patna 2004)

3.22 Derive an expression for the viscosity (η) of a gas in terms of mean free path of its molecules. Show that it is independent of pressure but depends upon the temperature of the gas. Discuss the effect of pressure and temperature on coefficient of viscosity.

(Nagpur 2010, 2007, 2005/w, 2005/s, Ranchi 2005, Osmania 2005, Alagappa 2004, Patna 2004, 2000, 2002, Agra 2002, Delhi 2000, Kanpur 2001, Bangalore 2000, Calcutta 2005, Purvanchal 2006, 2002, M.S. Uni. Tamil Nadu 2007)

3.23 Obtain the relation between coefficient of viscocity (η) and coefficient thermal conductvity (K). Why the experimental values do not follow this relationship? ***(Nagpur 2006)***

3.24 Write notes on

(*i*) Transport phenomenon ***(Patna 2005, Nagpur 2008, Purvanchal 2002)***

(*ii*) Relation between thermal and electrical conductivities ***(Patna 2004)***

(*iii*) Mean free path (λ). ***(Agra 2006, Kanpur 2000, Rohilkhand 2000, 2002)***

(*iv*) Frequency of collision. (*Nagpur 2010, Agra 2001*)
(*v*) Degrees of freedom (*Kumaon 1997*)

3.25 Obtain the relation between coefficient of viscosity and coefficient of diffusion.

3.26 Explain the effect of temperature and pressure on coefficient of viscosity. (*Nagpur 2011/s, 2008, 2004, Purvanchal 2004*)

3.27 Calculate the radius of an oxygen molecule if its coefficient of viscosity at 0°C is 19.2×10^{-6} N-s/m^2. Given: mass of an oxygen molecule = 5.31×10^{-26} kg. Boltzmann constant $k = 1.38 \times 10^{-23}$ joule/degree. [**Ans.** 1.5 R] (*Purvanchal 2006*)

3.28 The density of nitrogen at N.T.P. is 1.25 kg/m^3 and its coefficient of viscosity is 1.66×10^{-5} M.K.S. units. Determine mean free path of a molecule of a gas. [**Ans.** 8.084×10^{-5} m]

3.29 The viscosity of a gas (oxygen) at a temperature of 16°C is 169 micropoise. Calculate the diameter of the molecules of the gas. Avogadro's number = 6.02×10^{23}. Molecular weight of oxygen = 32 and Boltzmann's constant K = 1.38×10^{-23} J/K. [**Ans.** 3.97×10^{-10} m]

3.30 (*a*) Derive an expression for thermal conductivity (K) of a gas on the basis of kinetic theory of gases. Show that coefficient of thermal conductivity of hydrogen should be largest among all diatomic molecules.

(*Madras 2004, 05, Ranchi 2005, Alagappa 2004, Agra 2006, 2000, Meerut 2002, Nagpur 2009, 2006, M.S. Uni. Tamil Nadu 2007*)

(*b*) Show that it is independent of pressure and density of the molecules (*Nagpur 2010*)

3.31 What is the effect of temperature and pressure on thermal conductivity. (*Nagpur 2011/s, 2008, 2006, 2004*)

3.32 Derive the reletion for coefficient of self-diffusion *D* and show that it is directly proportional to $T^{3/2}$ and inversely proportional to the pressure. (*Calcutta 2004, Garhwal 2001, Meerut 2001*)

MULTIPLE CHOICE QUESTIONS

3.1 Viscosity of a gas is due to transport of
(*a*) momentum (*b*) energy (*c*) mass (*d*) none of these

3.2 Out of the following which has the highest thermal conductivity at the same temperature
(*a*) oxygen (*b*) helium (*c*) hydrogen (*d*) carbon dioxide

3.3 In diffusion, the transport of the following occurs:
(*a*) momentum (*b*) energy (*c*) mass (*d*) none of these

3.4 At very low temperatures, the coefficient of viscosity of a gas
(*a*) decreases with decrease of pressure. (*b*) increases with crease of pressure.
(*c*) is independent of pressure. (*d*) is equal to precssure.

3.5 The mean free path of a gas molecule is inersely proportional to (*Kanpur 2001*)
(*a*) square of the diameter of the molecule.
(*b*) square root of the diameter of the molecule.
(*c*) molecular diameter.
(*d*) fourth power of the molecular diameter.

3.6 The mean free path of molecules of a certain gas at pressure *p* and temperature *T* is 2×10^{-6} m. The mean free path at pressure $p \times 10^{-6}$ and temperature *T* is
(*a*) 2×10^{-1} m (*b*) 2×10^{-3} m (*c*) 2×10^{-4} m (*d*) 2×10^{-10} m

(*Rohilkhand 2003*)

3.7 The molecular density in a gas is n and the diameter of its molecule is d, then the mean free path of molecule is **(*Kanpur 2002*)**

(*a*) $\dfrac{\pi}{nd^2}$ (*b*) $\dfrac{1}{\pi nd}$ (*c*) $\dfrac{1}{\sqrt{2}\pi nd^2}$ (*d*) $\dfrac{1}{3\sqrt{2}\,\pi nd^3}$

3.8 Viscosity of a gas is directly proportional to: **(*Garhwal 2002*)**

(*a*) Temperature (*b*) Density of gas (*c*) $\sqrt{T}$ (*d*) T^2

3.9 The mean free path of gas at pressure p and temperature T is **(*Grahwal 2002*)**

(*a*) $\dfrac{kT}{(\sqrt{2})\sigma^2\pi p}$ (*b*) $\dfrac{p}{(\sqrt{2})\,\pi\sigma^2\,kT}$ (*c*) $\dfrac{\sqrt{kT}}{\pi\sigma^2 p}$ (*d*) $\dfrac{1}{\sigma}\sqrt{\dfrac{kT}{2\pi p}}$

ANSWERS

3.1 (*a*)	**3.2** (*c*)	**3.3** (*c*)	**3.4** (*c*)	**3.5** (*a*)	**3.6** (*a*)
3.7 (*c*)	**3.8** (*b*)	**3.9** (*a*)			

SECTION II

Thermodynamics

4. LAWS OF THERMODYNAMICS
5. ENTROPY
6. THERMODYNAMICAL RELATIONSHIPS
7. LIQUEFACTION OF GASES
8. RADIATION

THERMODYNAMICS

Thermodynamics is a branch of Physics that which mainly deals with the transformation of heat into mechanical work. In other words, it is the science of energy conversion involving heat and other forms of energy, most notably mechanical work. The principles of thermodynamics are very general and give the relation of heat to other forms of energy *e.g.* electrical, light, etc. It studies and interrelates the macroscopic variables, such as temperature, volume and pressure, which describe physical, thermodynamic systems. Today, the scope of thermodynamics has very much increased covering all the branches of physics, chemistry and engineering and has innumerable applications. Thermodynamics does not take into account the atomic constitution of matter, but it only deals with the macroscopic properties (large scale response) of the system that we can observe and measure in experiments.

The zeroth law of thermodynamics introduces the concept of temperature and the concept of thermodynamic equilibrium when two objects attain the same temperature when brought in thermal contact. During the process of reaching thermal equilibrium, heat is transferred between the objects. The first law of thermodynamics represents the relationship between heat and mechanical work while the second law depicts the manner in which these energy changes take place. In 1906, Nernst proposed a general principle, known as third law of thermodynamics, which explains the nature of bodies in the neighbourhood of absolute zero of temperature. At absolute zero temperature, the entropy tends to zero and the constituents of any matter are in a perfect order.

Historically, thermodynamics developed out of a desire to increase the efficiency of early steam engines, particularly through the work of French Physicist Nicolas Léonard Sadi Carnot (1824) who believed that engine efficiency was the key that could help France win the Napoleonic Wars. Scottish physicist Lord Kelvin was the first to formulate a concise definition of thermodynamics when he stated it in 1854.

LAWS OF THERMODYNAMICS

4

Chapter

INTRODUCTION

In thermodynamics, interactions between large ensembles of objects are studied and categorized. Central to this are the concepts of *system* and *surroundings.* A system is composed of particles, whose average motions define its properties, which in turn are related to one another through equations of state. Properties can be combined to express internal energy and thermodynamic potentials, which are useful for determining conditions for equilibrium and spontaneous processes. The starting point for most thermodynamic considerations are the laws of thermodynamics, which postulate that energy can be exchanged between physical systems as heat or work. They also postulate the existence of a quantity named entropy, which can be defined for any isolated system that is in thermodynamic equilibrium.

With these tools, thermodynamics can be used to describe how systems respond to changes in their environment. This can be applied to a wide variety of topics in science and engineering, such as engines, phase transitions, chemical reactions, transport phenomena, and even black holes. The results of thermodynamics are essential for other fields of physics and for chemistry, chemical engineering, aerospace engineering, mechanical engineering, cell biology, biomedical engineering materials science, and economics, to name a few.

4.1 THERMODYNAMIC SYSTEM

A system may be defined as a definite quantity of matter (solid, liquid or gases) bounded by some closed surface. The simplest example of a system is a gas contained in a cylinder with a movable piston. When a system is completely uniform throughout, such as gas, or mixture of gases, or a pure solid, or a liquid or a solution is called a homogeneous system. When a system consists of two or more phases, which are separated from one another by definite boundary surfaces, it is said to be a heterogeneous system. Examples of heterogeneous system are: a liquid and its vapour, two immiscible or partially miscible liquids. Anything outside this system which can exchange energy with it and has a direct bearing on the behaviour of the system is called as its **surroundings**.

A system may be separated from its surroundings by a real or imaginary boundary through which heat or mechanical energy may pass. The existence of the boundary is, however, essential to visualize the system distinctly from rest of the universe. A thermodynamic system may contain no substance at all, but may consist of radiant energy or electric and magnetic field. The combination of a system and its surroundings is called an *universe.*

Thermodynamic Variables and equation of state

The thermodynamic state or macroscopic state of a system is determined by four observable properties. These properties are the composition, pressure, volume and temperature, and are called the *variables of state*. For a homogeneous system, consisting of a single substance, the composition is fixed. Hence the state of the system is determined completely by the three variables: pressure (P), volume (V) and temperature (T).

For a system, in general, there is an equation of state

$$f(P, V, T) = 0$$

Therefore, the pressure, volume and temperature are not independent variables. Consequently, the thermodynamic state of a homogeneous system is determined completely by knowing two out of three variables.

For example: the equation of state for an ideal gas in a static condition is represented as

$$PV = RT$$

where R is universal gas constant.

For a Van der Waals' gas, the equation of state is

$$\left(P + \frac{a}{V^2}\right)(V - b) = RT$$

where a and b are Van der Waal's constants.

Limitations:

(*i*) The equation of state is not applicable to systems which are not in thermodynamic equilibrium.

(*ii*) Every thermodynamic system has its own equation of state independent of the others.

(*iii*) An equation of state expresses the peculiar behaviour of one individual system which distinguishes it from the others.

(*iv*) An equation of state is not a theoretical deduction from thermodynamics but is an experimental backing to it.

The thermodynamic systems in engineering are gas, vapour, steam, mixture of gasoline vapour and air, ammonia vapours and its liquid. In physics, thermodynamics includes besides the above, systems like stretched wires, thermocouples, magnetic materials, electrical condenser, electrical reversible cells, solid and surface films etc.

There are three classes of system.

(*i*) **Open system:** A system which can exchange matter and energy with the surroundings is called an *open system. e.g.* Air compressor : Air at low pressure enters and air at high pressure leaves the system i.e. there is an exchange of matter and energy with the surroundings.

(*ii*) **Closed system:** A system which can exchange only energy (and not matter) with the surroundings is called a *closed system. e.g.*, Gas enclosed in a cylinder expands when heated and drives the piston outwards. The boundary of the system moves but the matter (here gas) in the system remains constant.

(*iii*) **Isolated system:** A system which is thermally insulated and has no communication of heat or work with the surroundings is called *isolated system.*

4.2 FLOW OF HEAT

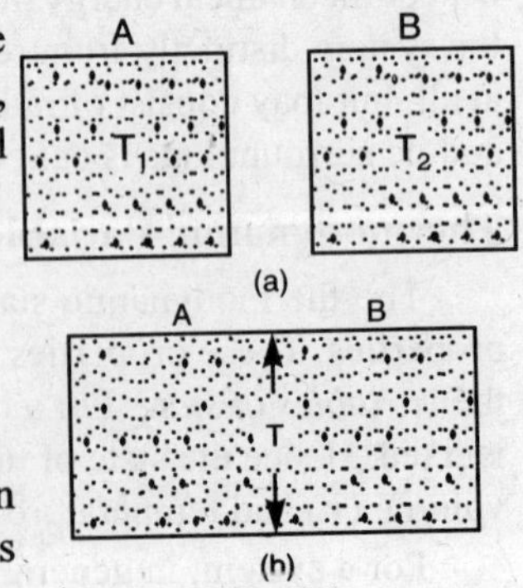

Fig. 4.1

Consider two systems A and B isolated from each other and from the surroundings as shown in Fig.4.1 (*a*). Let T_1 and T_2 be the temperatures, u_a and u_b the average kinetic energy per molecule and n_a and n_b the total number of molecules of the system A and B respectively, then

Heat content of the system A, $Q_a = n_n u_a$

Heat content of the system B, $Q_b = n_b u_b$

Total heat content of the two systems

$$Q = Q_a + Q_b = n_a u_a + n_b u_b$$

The two systems are now placed in thermal contact as shown in Fig. 4.1(*b*). The average kinetic energy u_a per molecule of system A is greater than the average kinetic energy u_b per molecule of the system

B because T_1, the temperature of system *A* is higher than T_2 the temperature of system *B* and average kinetic energy per molecule is proportional to temperature. As a result, the higher energy molecules in system *A* will collide with lower energy molecules in system *B* and in the process molecules of the system *A* will lose some kinetic energy to the molecules of system *B*. The process of transfer of kinetic energy of the molecules from system *A* to system *B* will go on till all molecules have the same average kinetic energy. As the temperature of the system is proportional to average kinetic energy of the molecules, the whole system will have the same common temperature *T*. As no further transfer of energy takes place the system is in equilibrium.

Thus we find that energy in the form of heat flows from system *A* to system *B* because of a difference of temperature. In other words, *heat must flow from a higher temperature to a lower temperature.*

The value of common temperature *T* is calculated as under:

Let *u* be the average kinetic energy per molecule at the equilibrium temperature *T*, then

$$u = \frac{\text{Total heat content of the two systems}}{\text{Total number of molecules}} = \frac{Q}{n_a + n_b} = \frac{Q_a + Q_b}{n_a + n_b} = \frac{u_a n_a + u_b n_b}{n_a + n_b}$$

$$= \frac{(n_a + n_b)u_a - n_b(u_a - u_b)}{n_a + n_b} = u_a - \frac{n_b}{n_a + n_b}(u_a - u_b)$$

Since, $u_a > u_b$, $(u_a - u_b)$ is a positive quantity and hence *u* the final average kinetic energy per molecule is less than u_a. Similarly it can be proved that *u*, the final average kinetic energy per molecule is greater than u_b.

Thus, $u_a > u > u_b$ and in terms of temperature $T_1 > T > T_2$.

Thus whenever two systems at different temperatures are brought into thermal contact, heat must flow from the system at higher temperature to the system at lower temperature till the whole system attains a common uniform temperature which lies between the temperatures of the two systems.

4.3. THERMAL AND THERMODYNAMIC EQUILIBRIUM

Thermal Equilibrium: When all the parts of a thermodynamic system are at the same temperature, which is the same as that of the surroundings, the system is said to be in thermal equilibrium. Thermal equilibrium is a macroscopic term. Microscopically, the molecules of a system are constantly in motion, colliding with each other and the walls of the containing vessel. Thus the energy and momentum of individual molecules goes on constantly changing but when the system is in equilibrium the average behaviour of the system taken as a whole does not change. Thus a system is in thermal equilibrium with itself, if the average kinetic energy of each molecule is the same. The average kinetic energy of a molecule at temperature *T* is given by $\frac{3}{2}kT$ where *k* is Boltzmann's constant.

Thermodynamic Equilibrium: *A system is said to be in thermodynamic equilibrium if it is simultaneously in mechanical, thermal and chemical equilibrium.*

A system is in mechanical equilibrium if there is no unbalanced force between different parts of the system or between the system and the surroundings.

This means that pressure of the system is constant everywhere.

A system is in thermal equilibrium with itself if the kinetic energy of each molecule is the same.

This means that temperature of the system is the same everywhere.

A system is in chemical equilibrium *if the composition of the system is the same everywhere.*

Thus for a system in thermodynamic equilibrium, pressure *P*, temperature *T* and volume *V* will be the same throughout the system irrespective of the fact how this equilibrium state has been reached. In other words, the state variables *P, V* and *T* are well (or uniquely) defined only if the system is in thermodynamic equilibrium.

4.4 MECHANICAL EQUIVALENT OF HEAT

As mentioned carrier, thermodynamics is a science that establishes a relation between the mechanical work done and the heat produced. Whenever work is transformed into heat or heat into work, the quantity of work is mechanically equivalent to the quantity of heat.

$$W \propto H$$

$$W = JH$$

where J is a constant known as **Joule's mechanical equivalent of heat**. The value of J is 4.186 Joules/calorie or 4.186×10^7 ergs per calorie. In S.I. system both W and H are measured in Joules. Therefore, in this system $J = 1$ and therefore $W = H$.

4.5 ZEROTH LAW OF THERMODYNAMICS

The zeroth law of thermodynamics deals with the thermal equilibrium of more than two macro systems. When a hot body A is brought in thermal contact with a cold body B, heat flow from A to B and after some time the flow stops. The bodies are then said to be in thermal equilibrium with each other.

The zeroth law of thermodynamics states that:

"If two systems are separately in thermal equilibrium with a third system, then they are also in thermal equilibrium with one another"

Consider three systems A, B and C at temperatures T_A, T_B and T_C respectively. The system A and B are separated from each other by an *adiabatic wall i.e.* a wall which conducts no thermal energy. Both the systems A and B are in contact with a third system C through a *diathermic wall i.e.,* a wall which allows heat to pass through.

Thus, the systems A and B are individually in equilibrium with the system C.

Explanation: We consider two systems A and B insulated from each other but in good thermal contact with a common system C, as shown in Fig. 4.2 (*a*) systems A and B will attain thermal equilibrium with third system C.

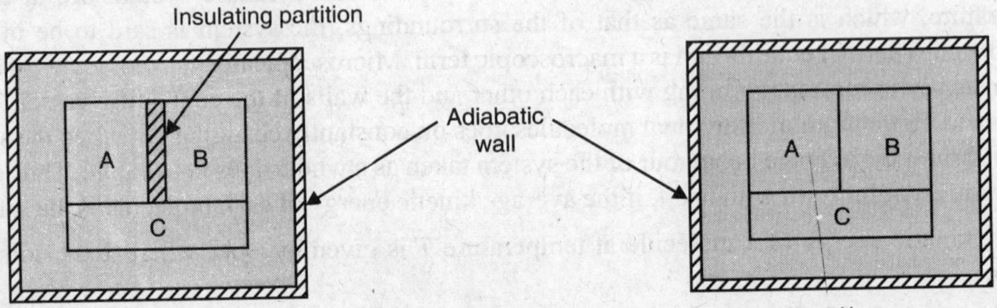

Fig. 4.2

Now if the insulating partition is removed and systems A and B are brought into thermal contact, we find that there is no further change. This means that the systems A and B are also in thermal equilibrium with another. This is known as zeroth law of thermodynamics.

Conversely the law can be stated as follows:

If three or more systems are in thermal contact, each to each, by means of diathermal walls and are all in thermal equilibrium together, then any two systems taken separately are in thermal equilibrium with one another.

Consider three fluids A, B and C. Let P_A, V_A represent the pressure and volume of A, P_B, V_B, the pressure and volume of B, and P_C, V_C are the pressure and volume of C.

If A and B are in thermal equilibrium, then

$$\phi_1 (P_A, V_A) = \phi_2 (P_B, V_B)$$

or $F_1 (P_A, V_A, P_B, V_B) = 0$ ... (*i*)

Expression (*i*) can be solved, and

$$P_B = f_1 (P_A, V_A, V_B) \qquad \text{... } (ii)$$

If B and C are in thermal equilibrium

$$\phi_2 (P_B, V_B) = \phi_3 (P_C, V_C)$$

or $F_2 (P_B, V_B, P_C, V_C) = 0$

Also $P_B = f_2 [V_B, P_C, V_C]$ ... (*iii*)

From equations (*ii*) and (*iii*) for A and C to be in thermal equilibrium separately,

$$f_1 (P_A, V_A, V_B) = f_2 (V_B, P_C, V_C) \qquad \text{... } (iv)$$

If A and C are in thermal equilibrium with B separately, then according to the zeroth law, A and C are also in thermal equilibrium with one another.

$\therefore$ $F_3 (P_A, V_A, P_C, V_C) = 0$ ... (*v*)

Equation (*iv*) contains a variable V_B, whereas equation (*v*) does not contain the variable V_B. It means

$$\phi_1 (P_A, V_A) = \phi_3 (P_C, V_C) \qquad \text{... } (vi)$$

In general,

$$\phi_1 (P_A, V_A) = \phi_2 (P_B, V_B) = \phi_3 (P_C, V_C) \qquad \text{... } (vii)$$

These three functions have the same numerical value though the parameters (P, V) of each are different. This numerical value is termed as *temperature* (T) of the body.

$\therefore$ $\phi (P, V) = T$ ... (*viii*)

Suppose T_A, T_B and T_C be the temperatures of three systems A, B and B. Since the systems A and C are in thermal equilibrium, the temperature of system A must be equal to the temperature of system C,

i.e. $T_A = T_C$ ... (*i*)

Since, the system B and C are in thermal equilibrium the temperature of system B must be equal to the temperature of system C,

i.e. $T_B = T_C$ ... (*ii*)

Comparing Eq. (*i*) with Eq. (*ii*), we get

$$T_A = T_B \qquad \text{... } (iii)$$

Eq. (*iii*) shows that the systems A and B are also in thermal equilibrium with each other.

Example. In a mercury glass thermometer, the pressure above the mercury column is zero and volume of mercury measures the temperature. If a thermometer shows a constant reading in two systems A and B separately, it will show the same reading even when A and B are brought in contact.

Significance and concept of temperature. The significance of Zeroth law of thermodynamics lies in the fact that the two systems in thermal equilibrium must have a common characteristic called *temperature.*

Hence ***temperature of a system is defined as the property which determines whether or not a system is in thermal equilibrium with another system.***

As a consequence, if a thermal contact between two systems A and B results in a net transfer of thermal energy from A to B, then the system A is said to be at a higher temperature than B.

Thus ***order of temperature is defined as the thermal state of a system which determines the direction of flow of thermal energy (heat) from one system to another when they are in thermal contact.***

4.6 WORK: A PATH DEPENDENT FUNCTION

A system can be taken from state A to state B by many ways as shown in fig. 4.3. Let the coordinates of state A and B be A (P_1 V_1) and B (P_2, V_2) respectively, where P_1 and V_1 represents pressure and volume of state A and P_2 and V_2 that of state B.

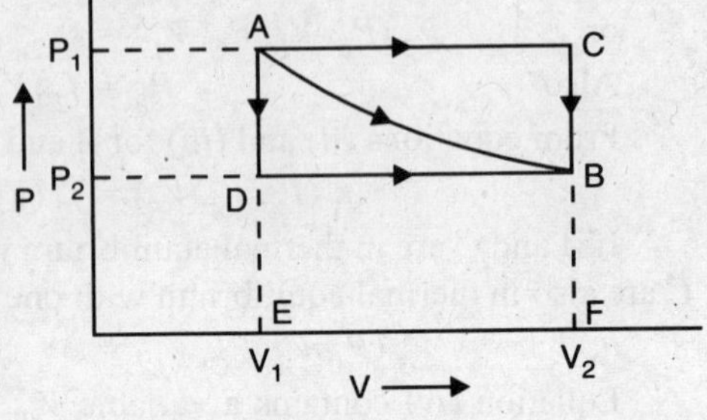

Fig. 4.3

1. Starting from point A, the pressure is continuously decreased from P_1 to P_2 along the curve AB so that volume increases from V_1 to V_2. The work done by the system is given by

$$W_1 = \int_{V_1}^{V_2} P dV = \text{area } ABFE$$

2. Starting from point A the volume V_1 is kept constant in going from A to D, the pressure decreases from P_1 to P_2 and then P_2 is kept constant from D to B. The work done in this process *i.e.* along ADB path given by

$$W_2 = P_2 (V_2 - V_1) = \text{area } DBFE$$

3. Again starting from point A, the pressure P_1 is kept constant in going from A to C and then the volume V_2 is kept constant from C to B. The work done in this process *i.e.* along path ACB is given by

$$W_3 = P_1 (V_2 - V_1) = \text{area } ACFE$$

$$\therefore \quad W_1 \neq W_2 \neq W_3$$

Thus, the work done depends on the path of the process when the system goes from one thermodynamic state to another.

4.7 CONCEPT OF INTERNAL ENERGY (U)

The energy content of a system is called its internal energy. It is the sum of following forms of energy of the system:

(*i*) kinetic energy due to translational, rotational and vibrational motion of the molecules, all of which depend only on the temperature,

(*ii*) potential energy due to intermolecular forces, which depends on the separation between the molecules, and

(*iii*) the energy of electrons and nuclei.

In practice, it is not possible to measure the total internal energy of a system in any given state. Only change in its value can be measured.

If the state of the system is changed from an initial state 1 to a final state 2, by supplying heat Q to the system and if W is the work done by the system during the change, then increase in the internal energy of the system is given by

$$(U_2 - U_1) = Q - W \qquad \text{... (4.1)}$$

where U_1 is the internal energy in state 1, and U_2 the internal energy in state 2.

Internal Energy of a Real gas and Ideal gas

The internal energy of a perfect (or ideal) gas is only dependent upon temperature where as that of a real gas, it is a function of both temperature and volume.

For a perfect gas, the internal energy is only kinetic energy of its molecules. It is proportional to kT and hence only depends upon temperature (T), k being Boltzmann's constant. The internal energy of a perfect gas is independent of its volume. Internal energy of a perfect gas at temperature T is $U = \frac{3}{2}NkT$, where N is the total number of molecules of the gas and k the Boltzmann constant.

For a real gas, say a gas obeying Van der Waals' equation, the internal energy is the sum of the kinetic and potential energies of molecules due to their mutual forces of attraction. The force of attraction between the molecules depends upon the intermolecular distance and is thus a function of volume. Hence, for a real gas, internal energy is a function of both, the temperature and volume.

Equation (4.1) is the mathematical statement of the first law of thermodynamics. In a convenient form, the law is stated below.

4.8 FIRST LAW OF THERMODYNAMICS

Statement: When a certain amount of heat Q is supplied to a system which does external work W in passing from state 1 to state 2, the amount of heat is equal to sum of the increase in the internal energy of the system and the external work done by the system. In symbols, the law is expressed as

$$Q = (U_2 - U_1) + W \qquad \text{... (4.2)}$$

For a very small change in the state of the system, the law is expressed in the differential form as

$$\delta Q = dU + \delta W \qquad \text{... (4.3)}$$

where δQ and δW are not perfect differentials but dU is a perfect differential because U is a function of the state of the system.

It should be remembered that all the quantities in equation (4.2) or (4.3) must be expressed in unit of energy. The amount of heat Q or δQ is taken positive if heat is supplied to the system, and negative if heat is removed from it. Similarly, the work W or δW is positive when the external work is done by the system in expansion and negative if the work is done on it in compression.

Any thermodynamic system in an equilibrium state possesses state variable called the internal energy (U). Between any two equilibrium states, the change in internal energy is equal to the difference of the heat transfered into the system and work done by the system, shown in Fig. 4.4.

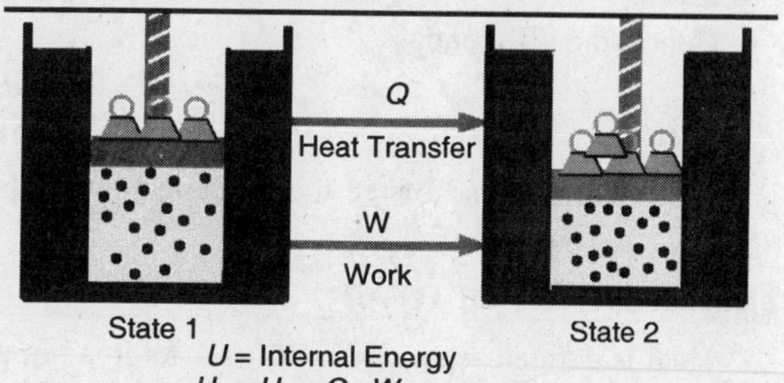

Fig. 4.4 First law of Thermodynamics

Physical Significance of first law: The first law of thermodynamics is based on the principle of conservation of energy of a system, in its most general form. It speaks about transfer of energy through either work or heat. It includes in the internal energy of the system *all* forms of energy due to *ordered* as well as *disordered* motion of the particles of the system. Work is the *macroscopic displacement* of a body and represents *ordered* kinetic and potential energy. Heat (or thermal energy) on the other hand measures energy of *disordered* motion of the particles of the system arising from a temperature difference. The law can, therefore, be stated as:

In all transformations the energy due to heat supplied to the system must be balanced by the external work done plus the increase in interal energy.

The law establishes the relation between heat and work. According to this law heat can be produced only by the expenditure of energy in some form or the other. Hence it follows directly from this law that it is impossible to make a perpetual motion machine or to derive work without any expenditure of energy.

Limitations of the First Law:

The first law of thermodynamics is based on the principle of conservation of energy of a system. Though it is applicable to every process in nature between the equilibrium states, it does not specify the condition under which a system can use its heat energy to produce a supply of mechanical work. It also does not say how much of the heat energy can be converted into work.

4.9 INTERNAL ENERGY AS A STATE FUNCTION

Suppose Q is the heat supplied to a working substance forming a system and W is the work done by it in taking the system from state 1 to state 2. then the increase in the internal energy of the system is given by first law of thermodynamics as

$$(U_2 - U_1) = Q - W$$

where U_1 is the initial internal energy of the system in state 1 and U_2 the final internal energy in state 2 along path A as shown in fig. 4.5.

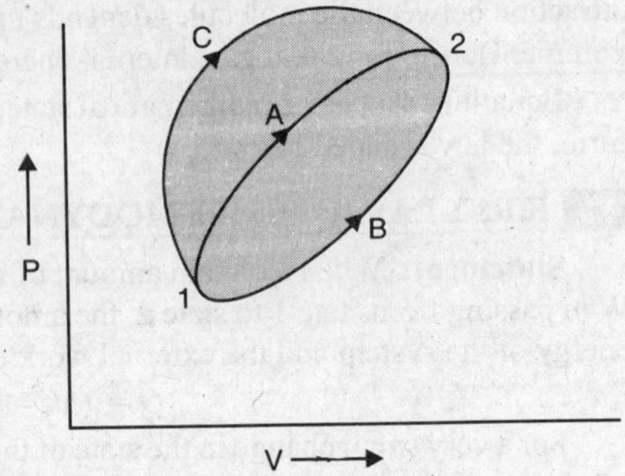

Fig. 4.5

If the state of the system is changed from the same initial state 1 to the same final state 2 by different paths, say B or C, it is found experimentally that the difference $(Q - W)$ is the same for all paths between the states 1 and 2. Hence we conclude:

When a system changes from one state to another, the change of internal energy of the system has a definite value, independent of the paths between the two states. Hence, internal energy is called as state function or point function and not a path function.

In other words, the internal energy is a function of any two of the thermodynamic variables P, V and T.

Thus, $\quad U = f(P, V)$ or $U = f(V, T)$ or $U = f(T, P)$

Hence, the differential

$$dU = \left(\frac{\partial U}{\partial P}\right)_V dP + \left(\frac{\partial U}{\partial V}\right)_P dV \qquad \text{... (4.4)}$$

is a perfect differential of the thermodynamic variables P and V. Internal energy is an extensive variable. (refer article 6.2)

4.10 CONCEPT OF HEAT

Heat is defined as *energy in transit*. As it is not possible to speak of work in a body, it is also not possible to speak of heat in a body. Work is either done on a body or by a body. Similarly, heat can flow from a body or to a body. If a body is at a constant temperature, it has both mechanical and thermal energies due to the molecular agitations and it is not possible to separate them. So, in this case, we cannot talk of heat energy. It means, if the flow of heat stops, the word heat cannot be used. It is only used when there is transfer of energy between two or more systems.

Consider two systems A and B in thermal contact with one another and surrounded by adiabatic walls.

For the system A,

$$Q = U_2 - U_1 + W \qquad \text{... } (i)$$

where Q is the heat energy transferred, U_1 is the initial internal energy, U_2 is the final internal energy and W is the work done.

Similarly for the system B,

$$Q' = U_2' - U_1' + W' \qquad \text{... } (ii)$$

Adding (i) and (ii)

$$\therefore \quad Q + Q' = (U_2 - U_1) + W + (U_2' - U_1') + W'$$

$$Q + Q' = [(U_2 + U_2') - (U_1 + U_1')] + (W + W') \qquad \text{... } (iii)$$

The total change in the internal energy of the composite system

$$= [(U_2 + U_2') - (U_1 + U_1')]$$

The work done by the composite system = $W + W'$

It means that the heat transferred by the composite system = $Q + Q'$. But the composite system is surrounded by adiabatic walls and the net heat transferred is zero.

$$\therefore \quad Q + Q' = 0$$

or $$Q = -Q' \quad \text{... } (iv)$$

Thus, for two systems A and B in thermal contact with each other, and the composite system surrounded by adiabatic walls, the heat gained by one system is equal to the heat lost by the other system.

4.11 TWO SPECIFIC HEATS OF A GAS

Heat capacity : Different materials have different capacity to absorb heat to produce a given change of temperature in a given mass. If a material of mass m absorbs heat Q, so that its temperature rises through ΔT, then

$$\text{Heat capacity} = \frac{Q}{\Delta T}$$

Specific heat (C) : Heat capacity per unit mass is known as specific heat and is denoted by C.

$$\therefore \quad \text{Specific heat} = \frac{\text{Heat capacity}}{\text{mass}}$$

$$C = \frac{Q}{m.\Delta T}$$

A liquid has only one specific heat. When a liquid is heated there is only an increase in volume, the change in pressure being almost zero and hence negligible. A liquid, therefore has only one specific heat. However, a *Gas* has two specific heats.

Two specific heats of a gas. When a gas is heated, ordinarily there is an increase in volume as well as pressure in addition to the rise of temperature. For the sake of simplicity, either the volume or the pressure may be kept constant. Therefore a gas has two specific heats.

(*i*) *Specific heat at constant volume, and*

(*ii*) *Specific heat at constant pressure.*

The specific heat of a gas at constant volume is the amount of heat required to raise the temperature of a unit mass of the gas through 1°C keeping the volume constant.

If dQ is the amount of heat required to raise the temperature of a unit mass of the gas through dT°C, keeping the volume constant, then

$$\textit{Specific heat at constant volume } C_v = \left(\frac{dQ}{dT}\right)_v$$

The specific heat of a gas at constant pressure is the amount of heat required to raise the temperatures of a unit mass of the gas through 1°C keeping the pressure constant.

If dQ is the amount of heat required to raise the temperature of a unit mass of the gas through dT°C keeping the pressure constant, then

$$\textit{Specific heat at constant pressure } c_p = \left(\frac{dQ}{dT}\right)_p$$

The C.G.S unit of mass is a gm and of heat is a calorie. Therefore, C.G.S. unit of specific heat is calories per gm per °C (cal. gm^{-1} C^{-1}). The S.I. unit of mass is a kg. and of heat (energy) is a Joule. Therefore the S.I. unit of specific heat is Joules per kg per degree kelvin (J kg^{-1} K^{-1}).

Specific heat of water = 1 cal/gm °C

$= 1$ kilo calories /kg°C

$= 4.18 \times 103$ Joule/kg °C

Gram molecular specific heat: *The amount of heat required to raise the temperature of 1 gram molecule (mole) of a gas through 1°C is called the gram-molecular specific heat.*

If ΔQ is the amount of heat required to raise the temperature of 1 gm. molecule of the gas through ΔT°C at constant volume, then gram-molecular specific heat at constant volume

$$C_v = \left(\frac{\Delta Q}{\Delta T}\right)_v$$

If ΔQ is the amount of heat required to raise the temperature of 1 gm. molecule of the gas through ΔT°C at constant pressure, then gram-molecular specific heat at constant pressure.

$$C_p = \left(\frac{\Delta Q}{\Delta T}\right)_p$$

For an infinitesimally small amount of heat dQ and rise of temperature dT, gram molecular specific heat at constant volume.

$$C_v = \left(\frac{dQ}{dT}\right)_v$$

and gram molecular specific heat at constant pressure,

$$C_p = \left(\frac{dQ}{dT}\right)_p$$

Why C_p is > C_v. In the first case, the heat supplied in used to increase the speed and the kinetic energy of the molecules. This results in a rise of temperature and an increase in the pressure of the gas. But as the volume is kept constant the gas does no external work.

In the second case, since the pressure is kept constant, therefore, the volume increases. The heat supplied now does two things:

(*i*) it raises the temperature of the gas, and

(*ii*) it does work in expanding the gas against the external pressure.

The amount of heat required to raise the temperature of the gas in this case is the same as i the first case when the volume is kept constant but in addition heat energy has to be supplied to do external work. *Hence the specific heat at constant pressure is greater than the specific heat at constant volume, by an mount which is thermal equivalent of the work done in expanding the gas against external pressure.*

Difference between two specific heats. Consider a gram molecule of a gas having a volume V at a pressure P and temperature TK contained in a cylinder closed by a piston as shown in Fig. 4.6.

The heat required to raise the temperature by a small amount dT at *constant volume* is given by

$$dQ_v = C_v\, dT \qquad \text{... } (i)$$

Where C_v is the gram molecular specific heat at constant volume (in units of work).

If, instead of the volume, the pressure is kept constant then heat required to raise the temperature by a small amount dT at *constant pressure* is given by

$$dQ_p = C_p\, dT \qquad \text{... } (ii)$$

Where C_p is the gram molecular specific heat at constant pressure (in units of work)

As the pressure remains constant, the volume increases and work is done by the gas. If A is the area of the piston which moves outward due to the expansion of the gas by a distance dx, then

Work done $\qquad dW = P\,Adx = P\,dV$

According to the first law of thermodynamics for an isochoric process *i.e.*, when the process takes place at constant volume, we have

$$dQ_v = dU_v + dW_v$$

As the volume remains constant

$$dW_v = P\,dV = 0$$

$$\therefore \qquad dQ_v = dU_v \qquad \dots (iii)$$

From (*i*) and (*iii*), we have $dU_v = C_v\,dT$ ... (*iv*)

For an *isobaric process i.e.*, when the process takes place at constant pressure

$$dW_p = P\,dV_p$$

and according to the first law of thermodynamics

$$dQ_p = dU_p + P\,dV_p \qquad \dots (v)$$

From (*ii*) and (*v*), we have

$$C_p\,dT = dU_p + PdV_p \qquad \dots (vi)$$

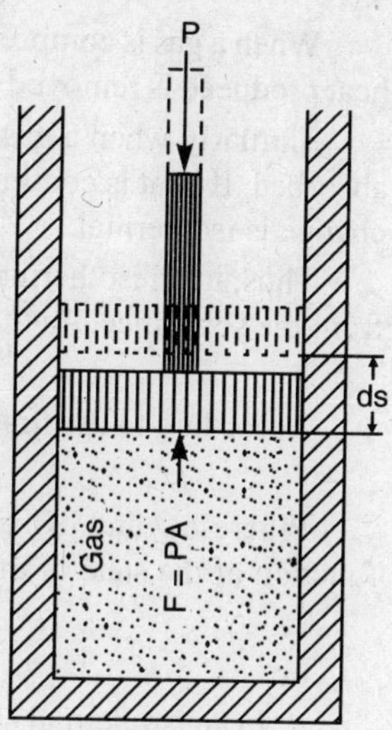

Fig. 4.6

As the internal energy of an ideal gas only depends upon temperature and the temperature difference in the two processes, one *isochoric* (at constant volume) and the other *isobaric* (at constant pressure) is the same

$$\therefore \qquad dU_p = dU_v$$

Hence from (*iv*) and (*vi*), we have

$$C_p\,dT = C_v\,dT + P\,dV_p \qquad \dots (vii)$$

or

$$(C_p - C_v)\,dT = P\,dV_p$$

For an ideal gas $PV = RT$, if the pressure is kept constant, then

$$P\,dV_p = R\,dT$$

Substituting in (*vii*), we have

$$(C_p - C_v)\,dT = R\,dT$$

or

$$C_p - C_v = R \qquad \dots (viii)$$

Equation (*viii*) holds good for any number of moles of a perfect gas. According to this equation, it is clear that specific heat at constant pressure is always greater than the specific heat at constant volume. In this equation C_p, C_v and R are all measured in J mol^{-1} K^{-1}.

If C_p and C_v are measured in calories, then

$$\therefore \qquad C_p \quad C_v = \frac{R}{J} \qquad \dots (4.5)$$

If c_p and c_v denote the two specific heat for 1 gm. of the gas and r is the gas constant for one gm., then

$$c_p - c_v = r$$

All the quantities are measured in ergs gm^{-1} K^{-1}

If c_p and c_v are measured in calories, then

$$(c_p - c_v) = \frac{r}{J} \qquad \dots (4.6)$$

This formula is known as Mayer's relation.

4.12 APPLICATIONS OF FIRST LAW OF THERMODYNAMICS

First law of thermodynamics is applicable to the following processes:

4.12.1. Isothermal Process

When a change in pressure and volume of a substance takes place but the temperature remains constant, the change (process) is said to be isothermal.

When a gas is compressed suddenly, some heat is produced but if the compression is slow and the heat produced is removed at once, so that the temperature remains constant, the change is isothermal.

Similarly when a gas is allowed to expand suddenly, work is done by the gas and some heat is absorbed. If heat is continuously supplied from outside so that the temperature remains constant, the change is isothermal.

Thus, in an isothermal change the temperature is kept constant by adding heat or taking it away from the substance. As there is no change in temperature, there is no change in internal energy.

$$\therefore \quad dU = 0$$

According to the first law of thermodynamics

$$\delta Q = dU + \delta W$$

Where δQ and δW are not perfect differentials but dU is a perfect differential because U is a function of the state of the system,

$$\therefore \quad \delta Q = 0 + \delta W$$

$$\text{or} \quad \delta Q = \delta W$$

∴ In an isothermal change (or process or transformation). Heat added (or subtracted) = External work done by (or on) the gas.

For a perfect gas, an isothermal change is represented by Boyle's law given by equation. $PV =$ constant

Suppose the working substance undergoes isothermal expansion, starting from point A having initial pressure P_1 and volume V_1.

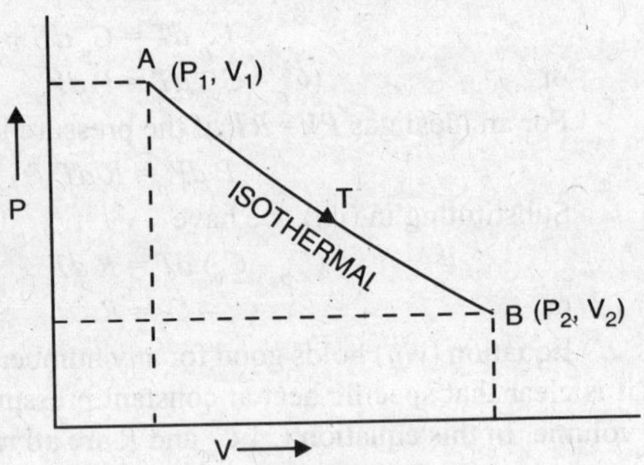

Fig. 4.7

Pressure is decreased in going from A to B and work is done by the working substance at the cost of its internal energy and suffers a fall in temperature. But the system is perfectly conducting to the surroundings. It absorbs heat from the surroundings and maintains a constant temperature.

Thus, from A to B, the temperature remains constant. The curve AB is called the isothermal curve or isothermal. In going from B to A back, the system gives out extra heat to the surroundings and maintains the temperature constant.

Thus, during isothermal process, the temperature of the working substance remains constant. It can absorb heat or give heat to the surroundings.

The equation for isothermal process is

$$PV = RT = \text{constant (for one gram molecule of a gas)}$$

For n gram molecules of a gas,

$$PV = nRT$$

For an ideal gas undergoing isothermal process,

$$(U_2 - U_1) = 0$$

∴ From the first law of thermodynamics, we get

$$Q = W$$

i.e., in an isothermal process the heat supplied to an ideal gas is equal to the work done by the gas.

4.12.2 Adiabatic process

When a change in pressure and volume of a substance takes place but no heat is allowed to enter or leave it, the change (process) is said to be adiabatic.

In the example given above, if no heat is taken away in the first case the temperature will rise and if no heat is supplied in the second case, the temperature will fall.

Hence in an adiabatic change, the temperature does not remain constant and no heat from outside is supplied to the system or taken away from it.

It does not mean that the heat content of the system remains constant. When a gas is allowed to expand, it does external work and the energy for this purpose is drawn from the internal energy of the gas. The internal energy of the gas therefore, decreases. When the gas is compressed, work is done on it and the energy gained appears as internal energy. The internal energy of the gas, therefore, increases.

During an adiabatic process no heat enters or leaves the gas

$\therefore \qquad \delta Q = 0$

According to the first law of thermodynamics

$$\delta Q = dU + \delta W$$

$\therefore \qquad 0 = dU + \delta W$

or $\qquad dU = -\delta W$

$\therefore$ In an adiabatic process

Increase (or decrease) in internal energy = External work done on (or by) the gas.

(*ii*) **Adiabatic expansion.** When the volume of a substance increases (or pressure decreases) in an adiabatic change (process) the process is known as adiabatic expansion.

(*iii*) **Adiabatic compression.** When the pressure of a substance increases (or volume decreases) in an adiabatic change (process) the process is known as adiabatic compression.

Adiabatic equation of state. Consider a gm molecule of a perfect gas contained in a perfectly non-conducting cylinder fitted with a non-conducting piston as shown in Fig. 4.8. Let the pressure, volume and temperature be P, V and T respectively. Suppose the gas is compressed adiabatically, so that the piston moves inwards through a distance dx. If A is the area of cross-section of the piston, then total force applied is $P \times A$ and

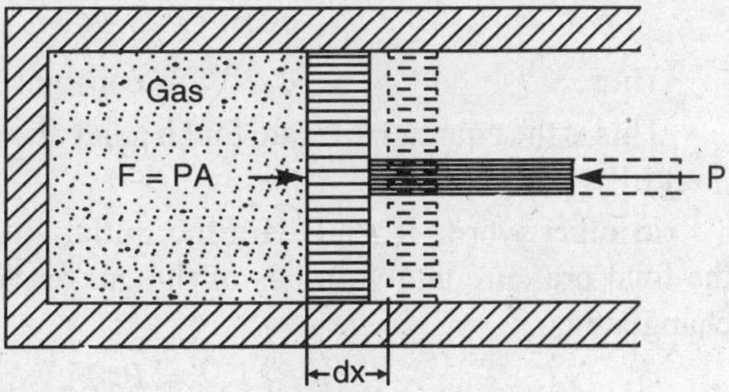

Fig. 4.8

Work done by piston = Force × Distance

$$= P \times A \times dx$$

or $\qquad dW = P\,dV$

as $A\,dx = dV$ gives the change in volume. The heat generated due to compression causes a rise of temperature dT and is given by $C_v\,dT$, where C_v is the *gram molecular specific heat at constant volume.*

According to the first law of thermodynamics

$$dQ = dU + dW = dU + P\,dV$$

For an adiabatic change $\qquad dQ = 0$

$\therefore \qquad dU + P\,dV = 0$

Now $\qquad C_v = \dfrac{dU}{dT}$

$\therefore \qquad dU = C_v\,dT$

Hence $\qquad C_v\,dT + P\,dV = 0 \qquad \ldots (i)$

The equation of state for one mole of a perfect gas is $PV = RT \qquad \ldots (ii)$

Differentiating (*ii*), We have

$$PdV + V\,dP = R\,dT$$

or $$dT = \frac{P\,dV + V\,dP}{R}$$

Substituting the value of dT in (i), we have

$$C_v \frac{P\,dV + V\,dP}{R} + P\,dV = 0$$

or $$C_v P\,dV + C_v V\,dP + RP\,dV = 0$$

or $$(C_v + R)\,P\,dV + C_v\,V\,dP = 0$$

But $$C_p - C_v = R \text{ or } C_v + R = C_p$$

(where C_p is the gram molecular specific heat at constant pressure)

Hence $$C_p P\,dV + C_v\,V\,dP = 0$$

Dividing by $C_v PV$ and putting $\frac{C_p}{C_v} = \gamma$, where γ is the ratio of specific heat at constant pressure to specific heat at constant volume, we have

$$\gamma \frac{dV}{V} + \frac{dP}{P} = 0$$

Integrating we get

$$\gamma \log_e V + \log_e P = \text{constant}$$

or $$\log_e PV^\gamma = \text{constant}$$

$$PV^\gamma = e^{\text{constant}} = \text{constant} = K$$

Thus, $$PV^\gamma = \text{constant} \qquad \text{... (4.7)}$$

This is the equation of state for an adiabatic change (process in terms of thermodynamic variables P and V.

In other words if $P_1\ V_1$ are the initial and P_2, V_2 the final pressure and volumes of the gas an adiabatic change, then

$$P_1 V_1^\gamma = P_2 V_2^\gamma$$

Fig 4.9 shows the adiabatic expansion and is represented by curve AB. Starting from initial state A (P_1, V_1, T_1), an expansion is carried out in such a way that no heat enters or leaves the system, being an adiabatic process and finally reaches to point B (P_2, V_2, T_2) in which, there is a decrease in temperature $(T_2 < T_1)$. Here work in done by the system at the cost of decrease in its internal energy.

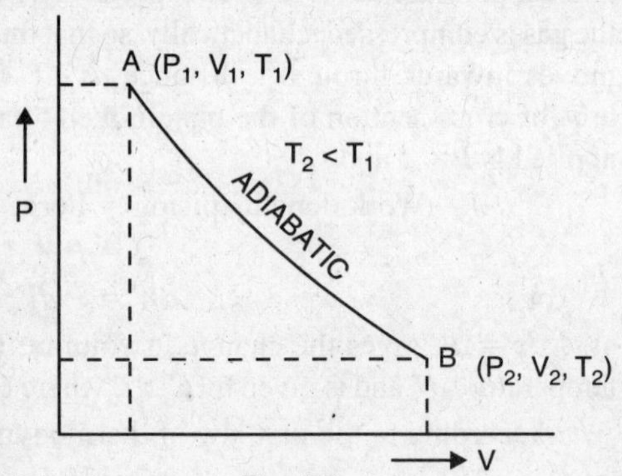

Fig. 4.9

Adiabatic relation between V and T

From eq (i), we have

$$C_v dT + PdV = 0$$

Subsisting $P = \frac{RT}{V}$ and dividing by C_v T, we have

$$\frac{dT}{T} + \frac{R}{C_v}\frac{dV}{V} = 0$$

or $$\int \frac{dT}{T} + \frac{R}{C_v}\int \frac{dV}{V} = 0 \text{ or } \log_e T + \frac{R}{C_v}\log_e V = \text{Constant}$$

or $$\log_e TV^{R/C_v} = \log_e TV^{(C_p - C_v)/C_v} = \log_e TV^{\gamma-1} = \text{Constant}$$

or $$TV^{\gamma-1} = \text{a constant} \qquad \text{... (4.8)}$$

This is the equation of state for an adiabatic change in terms of thermodynamic variables T and V.

Adiabatic relation between P and T

Now, for a perfect gas $PV = RT$

or $$V = \frac{RT}{P}$$

Substituting the value of V in the relation $PV^{\gamma} = K$ we have,

$$P\left(\frac{RT}{P}\right)^{\gamma} = K$$

or $$\frac{T^{\gamma}}{P^{\gamma-1}} = \frac{K}{R^{\gamma}} = K_1 \qquad \text{(another constant)}$$

or $$T^{\gamma} P^{1-\gamma} = \text{Constant}$$

$$\boxed{T P^{(1-\gamma)/\gamma} = \textbf{Constant}} \qquad \text{... (4.9)}$$

This is the equation of state for an adiabatic change in terms of thermodynamic variables P and T. Thus, during adiabatic process,

(*i*) PV^{γ} = constant (*ii*) $TV^{\gamma-1}$ = constant, and (*iii*) $TP^{(1-\gamma)/\gamma}$ = constant

4.12.3 Isochoric Process (*i.e.*, Volume constant)

If a system undergoes a change in which the volume remains constant, the process is called isochoric. At constant volume, no external work is done *i.e.* $\delta W = 0$

$\therefore$ Heat absorbed is given by $\delta Q = dU$

This expression may be used to define the internal energy of a system.

Thus, the increase in the internal energy of a system is equal to the heat absorbed by the system, at constant volume. The work done in isochoric process is zero.

4.12.4 Isobaric Process (*i.e.*, Pressure constant)

If a system undergoes a change in which the pressure is kept constant, the process is called isobaric. Suppose Q is the heat absorbed by a system at constant pressure P and suppose its volume increases from V_1 to V_2. Then from the first law, we have

$$Q = (U_2 - U_1) + W$$

or $$Q = (U_2 - U_1) + P(V_2 - V_1)$$

or $$Q = (U_2 + PV_2) - (U_1 + PV_1)$$

$$= H_2 - H_1 \qquad \text{... (4.10)}$$

where $$H = U + PV.$$

From equation (4.10), we conclude that the heat absorbed at constant pressure is equal to increase in quantity H, called as *enthalpy*. Like the internal energy U, the quantity H is a function of thermodynamic variables. The quantity H is also called the heat function at constant pressure.

4.12.5 Non-Isolated system (Internal energy of the Universe)

System and surroundings together constitutes the universe. In a non-isolated system, heat transfer between the system and the surroundings is allowed. When a heat δQ is supplied to the system its internal energy increases by dU and it performs an external work δW.

$$\delta Q = dU + \delta W$$

or $$dU = \delta Q - \delta W$$

This gives the change in internal energy of the system.

Here δQ is +*ve*, since heat enters the system.

δW is +ve, since work is done by the system.

On applying the first law of thermodynamic to the surroundings, which loses a heat δQ and receives a work δW *i.e.* a work δW is done on the surroundings; hence both δQ and δW are negative for the surroundings.

∴ Change in internal energy of the surroundings

$$dU' = -\delta Q - (-\delta W)$$

$$dU' = \delta W - \delta Q \quad \text{... (4.11)}$$

Net change in internal energy $= dU + dU'$

$$= (\delta Q - \delta W) + (\delta W - \delta Q)$$

$$= 0$$

Thus, the net change in internal energy is always zero.

4.13 THE INDICATOR DIAGRAM

It is very convenient to represent the behaviour of an engine by an indicator diagram. This helps to understand the performance of heat engines. Suppose that a system is taken from an initial equilibrium state A to the final equilibrium state B along path AB and the process is quasistatic (Fig 4.10).

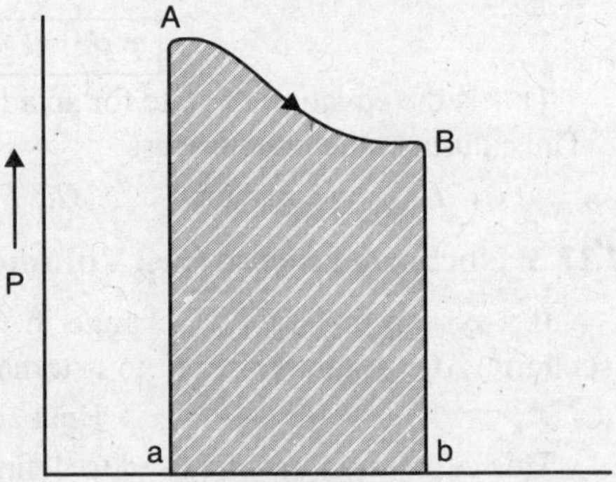

Fig. 4.10

Each point on the path traced represents the state of the system in terms of coordinates (P, V). Such a trace obtained in an actual engine is called indicator diagram or $P - V$ diagram.

The work done by the gas in expanding against pressure P is PdV, where dV is the infinitesimal change in volume. For finite process AB, the work done

$$= \int PdV = \text{ area } ABba$$

Thus, the indicator diagram directly indicates the work done by the engine during each cycle of operation, the work being equal to the area enclosed by the indicator diagram. The work done is +*ve* if the indicator diagram is traced in clock-wise direction and –*ve* if the direction is anticlock wise.

4.14 WORK DONE DURING AN ISOTHERMAL PROCESS

When a gas is allowed to expand isothermally, work is done by it.

Let the initial and final volumes be V_1 and V_2 respectively. Fig 4.11 represents the indicator diagram. The area of the shaded strip represents the work done for a small change in volume dV. When the volume changes from V_1 to V_2,

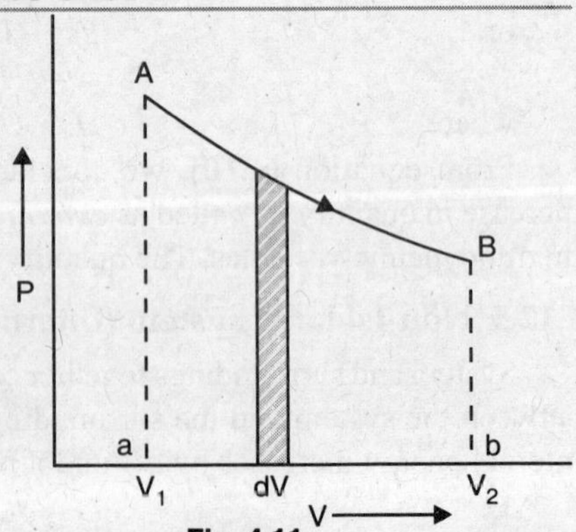

Fig. 4.11

$$\text{Work done} = \int_{V_1}^{V_2} P\,dV = \text{Area } ABba$$

Considering one gram molecule of the gas

$$PV = RT$$

or $$P = \frac{RT}{V}$$

$$\therefore \quad W = RT\int_{V_1}^{V_2} \frac{dV}{V}$$

$$W = RT \times 2.303 \log_{10}\frac{V_2}{V_1} \quad \text{... (4.12)}$$

Also $$P_1V_1 = P_2V_2$$

$$\frac{V_2}{V_1} = \frac{P_1}{P_2}$$

$$\therefore \quad W = RT \times 2.303 \log_{10}\frac{P_1}{P_2} \quad \text{... (4.13)}$$

Here, the change in the internal energy of the system is zero (because the temperature remains constant). So the heat absorbed by the system is equal to the work done by it.

4.15 WORK DONE DURING AN ADIABATIC PROCESS

During an adiabatic process, the system is thermally insulated from the surroundings. The gas expands from volume V_1 to V_2 as shown by indicator diagram [Fig. 4.12]. The work done by the gas for an increase in volume dV is $P\,dV$.

Work done when the gas expands from V_1 to V_2 is given by

$$W = \int_{V_1}^{V_2} P\,dV = \text{Area } ABba$$

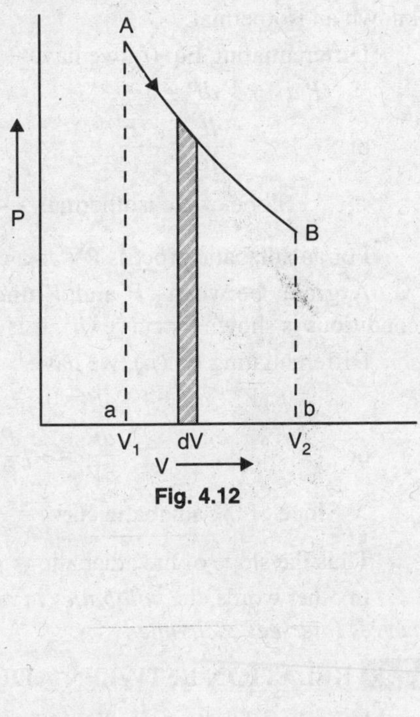

Fig. 4.12

During an adiabatic process

$$PV^{\gamma} = \text{constant} = K$$

or $$P = \frac{K}{V^{\gamma}}$$

$$\therefore \quad W = K\int_{V_1}^{V_2}\frac{dV}{V^{\gamma}}$$

$$= \frac{K}{1-\gamma}\left[\frac{1}{V_2^{\gamma-1}} - \frac{1}{V_1^{\gamma-1}}\right]$$

Since A and B lie on the same adiabatic,

$$P_1V_1^{\gamma} = P_2V_2^{\gamma} = K$$

$$W = \frac{1}{1-\gamma}\left[\frac{K}{V_2^{\gamma-1}} - \frac{K}{V_1^{\gamma-1}}\right]$$

$$= \frac{1}{1-\gamma}\left[\frac{P_2V_2^{\gamma}}{V_2^{\gamma-1}} - \frac{P_1V_1^{\gamma}}{V_1^{\gamma-1}}\right]$$

$$W = \frac{1}{1-\gamma}[P_2V_2 - P_1V_1] \quad \text{...(4.14)}$$

Taking T_1 and T_2 as the temperatures at points A and B respectively and considering one gram molecule of the gas

$$P_1V_1 = RT_1$$
$$P_2V_2 = RT_2$$

Substituting these values in equation (4.14)

$$W = \frac{1}{1-\gamma}[RT_2 - RT_1]$$

$$W = \frac{R}{1-\gamma}\left[T_2 - T_1\right] = \frac{R}{\gamma-1}[T_1 - T_2] \quad \text{... (4.15)}$$

Hence the work done in adiabatic process depends only upon the initial and final temperatures T_1 and T_2. Thus, *the work done along any adiabatic between two isothermals is independent of the particular adiabatic.*

4.16 SLOPES OF ADIABATIC AND ISOTHERMALS

For an isothermal process

$$PV = \text{a constant} \quad \text{... } (i)$$

A graph between P and V under isothermal conditions is shown by curve (i) in Fig. 4.13. It is known an isothermal.

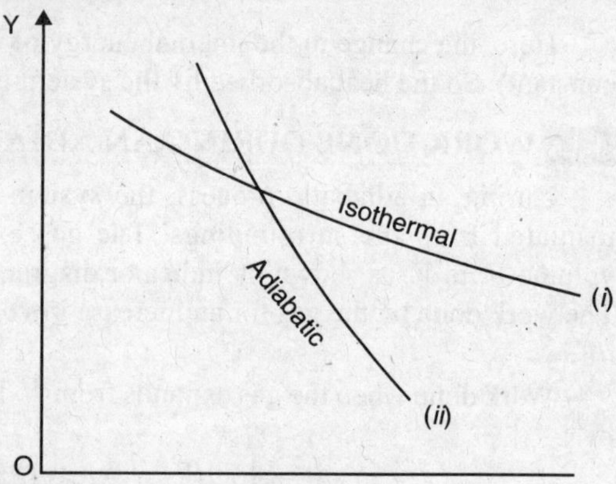

Fig. 4.13

Differentiating Eq. (i), we have

$$P\,dV + V\,dP = 0$$

or
$$\frac{dP}{dV} = -\frac{P}{V}$$

$\therefore$ Slope of the isothermal $= -\dfrac{P}{V}$

For an adiabatic process PV^{γ} = a constant ... (ii)

A graph between P and V under adiabatic conditions is shown by curve (ii). It is known as an adiabatic.

Differentiating Eq (ii), we have

$$P\gamma V^{\gamma-1}\,dV + V^{\gamma}\,dP = 0$$

or
$$\frac{dP}{dV} = -\gamma\frac{P}{V}$$

$\therefore$ Slope of the adiabatic curve $= -\gamma\dfrac{P}{V}$ ... (4.16)

Thus the slope of the adiabatic is γ times (C_p/C_v times) the slope of the isothermal.

In other words, the *adiabatic curve is steeper than the isothermal curve at a point where the two curves intersect each other.*

4.17 RELATION BETWEEN ADIABATIC AND ISOTHERMAL ELASTICITY

Volume elasticity (or bulk modules) E is defined as

$$E = -\frac{dP}{dV/V}$$

where dp is the applied pressure (stress) and $-dV/V$ is the

$$\frac{\text{change in volume}}{\text{original volume}} = \text{volumetric strain}$$

For an isothermal change, $PV = a$ constant

Differentiating, we get $P\,dV + V\,dP = 0$

or
$$P = -V\frac{dP}{dV} = -\frac{dP}{dV/V} = E_{iso}$$

For an adiabatic change PV^{γ} = Constant

Differentiating, we get

$$P\gamma V^{\gamma-1}\, dV + V^{\gamma}\, dP = 0$$

or
$$\gamma P = \frac{-dP\, V}{dV} = -\frac{dP}{dV/V} = E_{ad}$$

$$\therefore \quad \frac{E_{ad}}{E_{iso}} = \gamma \text{ or } E_{ad} = \gamma E_{iso} \quad \text{... (4.17)}$$

Thus The adiabatic elasticity of a gas is γ-times its isothermal elasticity.

Now, $\gamma = \dfrac{C_p}{C_v}$ where C_p is the specific heat of a gas at constant pressure and C_v at constant volume.

$$\therefore \quad \frac{E_{ad}}{E_{iso}} = \frac{C_p}{C_v}$$

Thus the ratio of adiabatic to isothermal elasticity is equal to the ratio of two specific heats.

4.18 CONVERSION OF WORK INTO HEAT

Work can be completely converted into heat

Consider an *ideal* gas enclosed in a cylinder fitted the a frictionless piston and completely isolated from its surroundings. Let T be the temperature of the gas. Suppose the gas is compressed quasi-statically from a state A to a state B under ***adiabatic conditions*** as shown in Fig.4.14 so that its volume decreases from V_A to V_B and pressure increases from P_A + P_B. In this process work is done **on** the gas but as the compression is adiabatic *i.e.* no heat is allowed to enter or leave the cylinder.

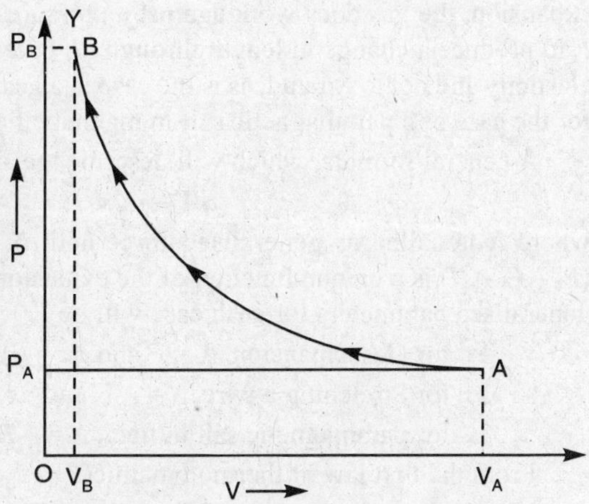

Fig. 4.14

$$\Delta Q = 0$$

If ΔW is the work done on the gas in this process and ΔU is the increase in thermal energy, then according to first law of thermodynamics.

$$\Delta Q = \Delta U + \Delta W$$

or
$$\Delta U + \Delta W = 0 \text{ or } \Delta U = -\Delta W$$

But as the work is done on the gas ΔW is itself a negative quantity

$$\therefore \quad \Delta Q = \Delta W$$

This equation shows that the *internal energy* of the gas increases by an amount *equal* to the work done on it. But internal energy is only a form of heat energy. Thus work done on the gas is wholly converted into heat.

The above discussion shows that work can be wholly converted into heat.

4.19 COOLING DUE TO ADIABATIC REVERSIBLE PROCESSES

On the basis of first law of thermodynamics, we can show that when a thermodynamic system such as a gas, stretched wire or a paramagnetic salt is made to undergo an adiabatic process reversibly, the temperature of the sytem changes. The thermodynamic coordinates taking part in the above three processes are discussed below:

1. **Gas expansion:** A system of gas is a function of three parameters, the pressure (P), the volume (V) and the temperature (T), *i.e.*, $f(P, V, T)$.
2. **Stretched wire:** System of stretched wire is a function of stretching force (F), the length of the stretching wire (l) and the temperature of the wire (T), *i.e.*, $f(F, l, T)$. Here pressure and volume are considered to be constant.
3. **Paramagnetic salt:** A system of paramagnetic salt is a function of the magnetic field (H), the intensity of magnetisation (I) and the temperature of the salt (T), *i.e.*, $f(H, I, T)$.

In all these cases, we can calculate the work done (dW) in a reversible infinitesimal change *i.e.*, the corresponding changes in above parameters. Let us write down the work done in each case.

1. In gas expansion, the work done by the system is

$$dW = P.dV$$

2. In stretching a wire, the work done is

$$dW = -F.dl$$

3. The paramagnetic salt in a magnetic field,

$$dW = -H\,dl$$

We presume here that the work done by the system is positive. For example: in case of gas expansion, the gas does work against a pressure, in stretching a wire, the work is done by tension F to produce a change in length through dl, because the wire by iteself wants to contract due to its elasticity and not to expand, as is the case in a gas. Hence $-ve$ sign in used with Fdl. The same is true for the case of a paramagnetic salt in magnetic field, *i.e.*, $dW = -Hdl$.

A general formula, which will describe the above three systems can be of the form

$$dW = X\,dx \qquad \text{... (i)}$$

where X is called as generalised force and dx as generalised extension. The generalised force ($P, -F, -H$) is a unique function of the extensions (V, l, I) respectively, at a fixed temperature. The generalised parameters for each case will be

1. for Gas expansion, $X = P$ and $dx = dV$
2. for Stretching a wire, $X = -F$ and $dx = dl$
3. for paramagnetic salt in field, $X = -H$ and $dx = dI$

From the first law of thermodynamics

$$\delta Q = dU + \delta W$$

In adiabatic process, $\delta Q = 0$

$$\therefore \qquad dU = -\delta W$$

In terms of generalised parameters

$$dU = -X\,dx$$

But in adiabatic process,

$$dU = C_v\,dT$$

$$\therefore \qquad C_v\,dT = -X\,dx$$

$$\text{or} \qquad dT = -\frac{1}{C_v}X\,dx \qquad \text{... (ii)}$$

In adiabatic process, P and V are related by

$$PV^{\gamma} = \text{constant}$$

Differentiating, we get

$$PV^{\gamma-1}\,dV + V^{\gamma}\,dP = 0$$

$$\text{or} \qquad PV = -\frac{V}{\gamma}dP \qquad \text{... (iii)}$$

for n gm. moles of a gas, the equation of state is

$$PV = nRT$$

At constant pressure, differentiating *w.r.* to T, we get

$$P\left(\frac{\partial V}{\partial T}\right)_P = nR$$

Multiplying both sides by T, we get

$$PT\left(\frac{\partial V}{\partial T}\right)_P = nRT = PV$$

or

$$T\left(\frac{\partial V}{\partial T}\right)_P = V$$

Substituting for V in eqn. (*iii*), we have

$$PdV = -\frac{T}{\gamma}\left(\frac{\partial V}{\partial T}\right)_P dP$$

Substituting for PdV in eqn. (*ii*), we get

$$dT = \frac{T}{\gamma C_V}\left(\frac{\partial V}{\partial T}\right)_P dP$$

But $\gamma = \frac{C_P}{C_V}$ or $\gamma C_V = C_P$

$$\therefore \quad dT = \frac{T}{C_P}\left(\frac{\partial V}{\partial T}\right)_P dP \quad \text{... (4.18)}$$

In terms of generalised coordinates, the change in temperature during adiabatic process may be written as

$$dT = \frac{T}{C_X}\left(\frac{\partial x}{\partial T}\right)_X dx \quad \text{...(4.19)}$$

This is the equation of state of the system in terms of generalised parameters. Substituting the corresponding values of parameters, the change in temperature of the system undergoing adiabatic process can be evaluated. Let us examine the same for the three cases:

1. Adiabatic expansion of gas

In case of a gas

$$X = P \quad \text{and} \quad x = V$$

substituting in eqn. (4.19), the change

$$dT = \frac{T}{C_P}\left(\frac{\partial V}{\partial T}\right)_P dP$$

During expansion, dP decreases *i.e.* falls. It means the temperature of the gas also decreases. Thus, if a highly compressed gas is suddenly allowed to expand adiabatically, the gas cools considerably.

If α is the coefficient of volume expansion, given by $\alpha = \frac{1}{V}\left(\frac{\partial V}{\partial T}\right)_P$, then

$$dT = \frac{TV\alpha}{C_P} dP \quad \text{...(4.20)}$$

2. Stretching of a wire

In case of adiabatic stretching of a wire,

$$X = -F \quad \text{and} \quad x = l$$

Substituting in Eqn. 4.19, the change in temperature

$$dT = \frac{-T}{C_F}\left(\frac{\partial l}{\partial T}\right)_F dF \quad \text{... (4.204)}$$

Since length increases with increase in temperature, $\left(\frac{\partial l}{\partial T}\right)$ is positive. From Eqn 4.20A, therefore, we conclude that wire of those substances which expand on heating, show a cooling effect when stretched adiabatically.

If β is coefficient of linear expansion, m the mass per unit length, then $\beta = \frac{1}{l}\left(\frac{\partial l}{\partial T}\right)_F$ or $\left(\frac{\partial l}{\partial T}\right)_F = \beta l$ and $C_F = mlC_P$

$$\therefore \qquad dT = -\frac{T}{mlC_P}\beta l d_F = \frac{-T\beta}{mC_P}dF \qquad \text{... (4.21)}$$

This gives fall in temperature due to adiabatic stretching of a wire.

3. Adiabatic demagnetisation of paramagnetic salt

In this case, $X = -H$ and $x = I$

Substituting in eqn. 4.19, the change in temperature

$$dT = -\frac{T}{C_H}\left(\frac{\partial I}{\partial T}\right)_H dH \qquad \text{... (4.22)}$$

where C_H is the specific heat of the substance at constant field.

Here $\left(\frac{\partial I}{\partial T}\right)_H$ is always negative. Hence R.H.S. is always positive. If H decreases, dH is – *ve* and hence dT is –*ve*. Hence, a cooling is produced during adiabatic demagnetisation of a para magnetic substance.

4.20 CLEMENT AND DESORMES METHOD-DETERMINATION OF γ

Clement and Desormes in 1819 designed an experiment to find γ, the ratio between the two specific heats of a gas.

The vessel A has a capacity of 20 to 30 litres and is fitted in a box containing cotton and wool. At the top end, three tubes are fitted as shown in Fig. 4.15. Through S_1, dry air is forced into the vessel A. The stop cock S_1 is closed when the pressure inside A is slightly greater than the atmospheric pressure. Let the difference in level on the two sides of the manometer be H and the atmospheric pressure be P_0. The pressure of air inside the vessel is P_1.

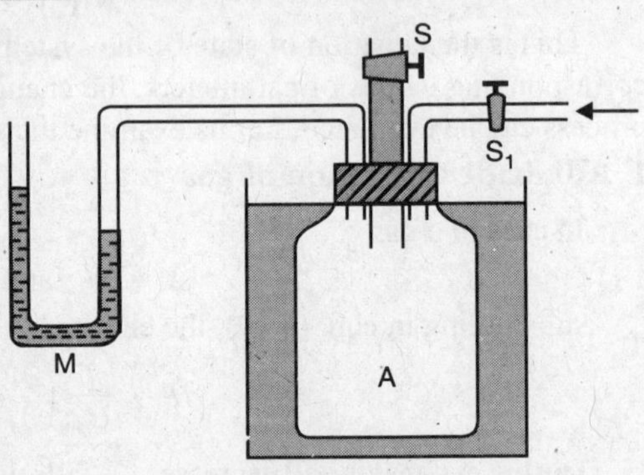

Fig. 4.15

The stop-cock S is suddenly opened and closed just at the moment when the levels of the liquid on the two sides of the manometer are the same. Some quantity of air escapes to the atmosphere. The air inside the vessel expands adiabatically. The temperature of air inside the vessel falls due to adiabatic expansion. The air inside the vessel is allowed to gain heat from the surroundings and it finally attains the temperature of the surroundings. Let the pressure at the end be P_2 and the difference in levels on the two sides of the manometer be h.

Theory. Consider a fixed mass of air left in the vessel in the end. This mass of air has expanded from volume V_1 (less than the volume of the vessel) at pressure P_1 to volume V_2 at pressure P_0. The process is adiabatic as shown by the curve AB (Fig. 4.16).

$$P_1V_1^{\gamma} = P_0V_2^{\gamma}$$

$$\frac{P_1}{P_0} = \left(\frac{V_2}{V_1}\right)^{\gamma} \quad \ldots (i)$$

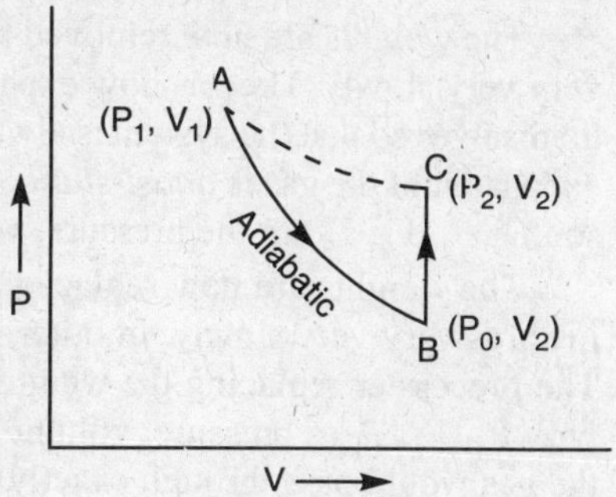

Fig. 4.16

Finally the point C is reached. The points A and C are at the room temperature. Therefore AC can be considered as an isothermal.

$$P_1V_1 = P_2V_2$$

$$\frac{V_2}{V_1} = \frac{P_1}{P_2} \quad \ldots (ii)$$

Substituting the value of $\frac{V_2}{V_1}$ in equation (i),

$$\frac{P_1}{P_0} = \left(\frac{P_1}{P_2}\right)^{\gamma}$$

Taking logarithms,

$$\log P_1 - \log P_0 = \gamma\,[\log P_1 - \log P_2]$$

$$\gamma = \frac{\log P_1 - \log P_0}{\log P_1 - \log P_2} \quad \ldots (iii)$$

But $\quad P_1 = P_0 + H$ and $P_2 = P_0 + h$

$$\therefore \quad \gamma = \frac{\log(P_0 + H) - \log P_0}{\log(P_0 + H) - \log(P_0 + h)}$$

$$= \frac{\log\left(\frac{P_0 + H}{P_0}\right)}{\log\left(\frac{P_0 + H}{P_0 + h}\right)} = \frac{\log\left(1 + \frac{H}{P_0}\right)}{\log\left(1 + \frac{H - h}{P_0 + h}\right)}$$

Approximately, $\quad \gamma = \dfrac{\frac{H}{P_0}}{\frac{H-h}{P_0}} = \dfrac{H}{H-h}$

Hence $\quad \gamma = \dfrac{H}{H-h} \quad \ldots (4.23)$

Similarly, γ for any gas can be determined by this method.

Drawbacks. When the stop-cock is opened, a series of oscillations are set up. This is shown by the up and down movement of the liquid in the manometer. Therefore, the exact moment when the stopcock should be closed is not known. The pressure may not be equal to the atmospheric pressure when the stop-cock is closed. It may be higher or less than the atmospheric pressure. Thus the result obtained will not be accurate.

4.21 REVERSIBLE AND IRREVERSIBLE PROCESSES

Reversible process

Consider a gas enclosed in a cylinder fitted with a frictionless piston, perfectly insulated from its surroundings so that no heat can leave or enter into it. Some removable weights are placed on the top of the piston and P_1, V_1, T_1 are the pressure,volume and temperature of the gas in the cylinder in the intitial state.

The weights are now removed by very small amount at a time so that the piston moves outward very very slowly. The gas now expands adiabatically and the pressure and volume of the gas adjust themselves so that the system is always in thermodynamic equilibrium. In other words, the adiabatic expansion of the gas is *quasi-static*. The process of removing the weights in small steps is continued so that P_2, V_2, T_2 are the pressure, volume and temperature of the gas in the final state.

The weights are now replaced again in very small amounts at a time so that the piston moves inwards very very slowly. In other words the adiabatic compression of the gas is also *quasi-static*. The process of replacing the weights in small steps is continued till the gas reaches the initial state with P_1, V_1, T_1 as pressure, volume and temperature. During this process of adiabatic compression the gas would pass through exactly the same intermediate steps through which it passed during the adiabatic expansion *i.e. all the intermediate states are retraced in the opposite direction.* Such a process is known as a reversible process.

Hence *A reversible process is that which can be retraced in the opposite direction, so that the working substances passes through exactly the same condition as it does in the direct process.*

In other words it means that in a step where heat is absorbed in the direct process, it is given out in the reverse process and *vice versa*. Similarly in a step where work is done by the working substance in the direct process an equal amount of work is done on the working substance in the reverse process..

Conditions for reversible process (*i*) For the process to be reversible, it should be carried out very very slowly so that the system remains in thermodynamic equilibrium at each stage of the process *i.e.* the process is *quasi-static*.

(*ii*) There should be no frictional forces present.

It is not possible to realise these condition in actual practice. Thus no process is reversible in the true sense of the word. However, some examples of reversible processes in practice are given below.

Examples of reversible process

1. All isothermal and adiabatic changes are reversible when performed *slowly*. In such a case it is assumed that there is no friction to be overcome as work done in overcoming friction is wasted. The process carried out is very slow, hence no energy is wasted in producing oscillations and eddy currents and no heat is lost by conduction, convection and radiation.

Under these conditions if heat is supplied to a given mass of a gas at constant pressure, it expands and does some external work. If the same amount of work is done on the gas in compressing it, the same quantity of heat will be given out.

2. Ice melts when a certain amount of heat is absorbed by it. The water so formed can be converted into ice if the same amount of heat is removed from it.

3. If the resistance of a thermocouple is neglected there will be no heat produced due to Joule heating effect. In such a case Peltier heating or cooling is reversible. At a junction where a cooling effect results due to Peltier effect when current flows in one direction an equal heating effect is produced when the current is reversed.

4. Electrolysis

Irreversible process: If in the experiment described above the weights on the piston are removed quickly the gas will undergo sudden expansion so that the intermediate states will not be equilibrium state *i.e.*, it will not be possible to define the value of pressure, volume and temperature for these states. Similarly when the weights are replaced quickly the gas will undergo sudden compression and the values of pressure, volume and temperature for the intermediate states will again be not defined. Thus during the process of sudden compression it will not be possible to reproduce the intermediate states through which the system passed during sudden expansion. Such a process is known as irreversible process.

Hence an irreversible process is that which cannot be retraced in the opposite direction so that the working substance does not pass through the same intermediate states through which it passes during the direct process.

Examples of irreversible process

1. Work done against friction is irreversible. It is because in reversing the direction of motion again work has to be done against friction.

2. Joule-Kelvin effect is irreversible because on reversing the flow of gas, a similar cooling or heating effect is not observed. Similarly diffusion, conductions and radiation are also irreversible.

3.Heat produced by the passage of a current through a resistance is irreversible. It is because even when the current is reversed the same effect is observed.

4. Rusting of iron.

4.22 QUASI-STATE PROCESS

A quasi-static process is a process carried out very very slowly so that the system under consideration remains at all times arbitrarily close to the equilibrium state.

The state of a system is uniquely determined by a point on a *P–V* diagram known as indicator diagram, each point on it representing an equilibrium state. A line joining a series of such points represents a succession of infinitely large number of equilibrium states infinitesimally close to each other and hence depicts a quasi-static process. This can be explained statistically as under:

Consider a gas enclosed in a cylinder fitted with a frictionless piston. Suppose the gas expands isothermally by an *infinitesimal* amount from the initial state (P, V, T) to the final state $(P - dP)$, $(V + dv, T)$, then the molecules of the gas would shift from the most probable distribution (P, V, T) to the new most probable distribution $(P - dP, V + dV, T)$. This readjustment takes a certain amount of time say $d\tau$. If this time $d\tau$ *is much less than the time dt that the gas takes to change its state by an infinitestimal amount, then the process is said to be quasi static.*

Thus, for a quasi static process

$$dt >> d\tau$$

When a system changes its state quasi-statically, the process is very very slow and the system is virtually in its most probable state, all the time.

Importance in thermodynamic process:

As explained above, the state of a system is uniquely determined by a point on a *P–V* diagram, each point on it representing equilibrium state. When a system changes its state quasi-statically the process is very, very slow and the system is in its most probable or equilibrium state all the time. This is why the thermodynamic process should be quasi-static for *P–V* diagrame to be meaningful.

4.23 CYCLIC PROCESS

A *cyclic process is a thermodynamic process in which the system returns to its initial state after undergoing a series of changes.*

A cyclic process may consist of (*i*) a closed cycle or (*ii*) an open cycle.

(*i*) **Closed cycle.** Consider a gas having a volume V_1 at a pressure P_1 as at A. Suppose it expands *quasi-statically* from its initial state A to the state B (pressure P_2, volume V_2) via the path AEB. The work done by the gas during expansion.

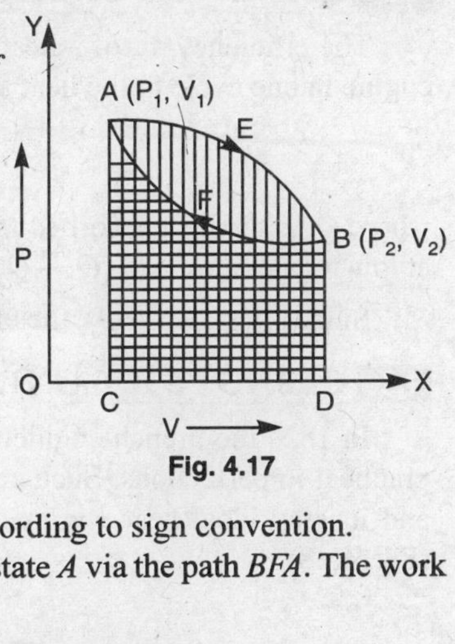

Fig. 4.17

$$W_1 = \int_{V_1}^{V_2} P\,dV = \text{Area } AEBDCA$$

As W_1 is the work done by the gas W_1 is taken as *positive* according to sign convention.

The gas is *compressed quasi-statically* from the state B to the state A via the path BFA. The work done on the gas during compression

$$W_2 = \int_{V_1}^{V_2} PdV = \text{Area } BDCAFB$$

As W_2 is the work done on the gas, W_2 is taken as *negative* according to sign convention.

∴ Net work done in traversing the cycle *AEBFA*, $W = W_1 - W_2$ = Area of the closed path *AFBFA* (shown by vertical lines).

Hence, network done during a closed cyclic process is equal to the area of the loop representing the cycle.

In the above case, the loop has been traversed in the *clockwise* direction and W is *positive i.e.* net work is done by the system.

If the gas is taken from the state A to B *via* the path *AFB* and brought back to A along the path *BEA i.e.* the loop is traversed in the *anti-clockwise* direction the work done will be = Area *AFBEA* = – Area *AEBFA* = $-W$ *i.e.* net work done is negative and work is done on the system. To sum up (*i*) *The work done per cycle is numerically equal to the area of the loop representing the closed cycle.* (*ii*) *When the loop is traversed clock-wise work is done by the system and is taken as positive where as when the loop is traversed anti-clockwise work is done on the system and is taken as negative.*

(*ii*) ***Open cycle.*** If the gas expands from the state A to the state B along the path *AEB* and is compressed back to the initial state A along the same path it is an *open cycle*. In such a cyclic process the work done on the system during compression is equal and opposite to the work done by the system during expansion. Hence the total work done in an open cycle is zero.

Non cyclic process. *It should be noted that a thermodynamic process in which the final state is not the same as the initial state is known as a non-cyclic process.*

4.24 HEAT ENGINES

Any practical machine which converts heat into mechanical work is called a heat engine. Heat engines in their operation absorb heat at a higher temperature, convert part of it into mechanical work, and reject the remaining heat at a low temperature. In this process, a working substance is used. In steam engines, the working substance is water vapour, and in all gas engines the working substance is a combustible mixture of gases.

In any heat engine, the working substance goes through certain changes of pressure, volume and temperature, and then returns to the initial state. The complete changes through which the working substance undergoes from its initial state and back to its starting state constitute one cycle of operation.

4.25 DEFINITION OF EFFICIENCY

The efficiency, η, of a heat engine is defined as the ratio of the mechanical work done by the engine in one cycle to the heat absorbed from the high temperature source. Thus

$$\eta = \frac{Q_1 - Q_2}{Q_1} \quad \text{... (4.24)}$$

where Q_1 is the heat absorbed from the source at high temperature, Q_2, is the heat rejected to a sink at low temperature and $(Q_1 - Q_2)$ is the mechanical work done by the engine in one cycle.

Since $(Q_1 - Q_2) < Q_1$, the efficiency can never be 100 %.

4.26 CARNOT'S IDEAL HEAT ENGINE

In 1824 the French engineer Sadi Carnot conceived a theoretical engine which is free from all practical imperfections. Such an engine can not be realised in practice. It has maximum efficiency and it is an ideal heat engine. Sadi Cornot's heat engine requires the following important parts (Fig. 4.18):

1. A **cylinder** having perfectly non-conducting walls, a perfectly conducting base and is provided with a perfectly non-conducting piston which moves without friction in the cylinder. The cylinder contains one mole of perfect gas as the working substance.
2. **Source.** A reservoir maintained at a constant temperature T_1 from which the engine can draw heat by perfect conduction. It has infinite thermal capacity and any amount of heat can be drawn from it at constant temperature T_1.
3. **Heat insulating stand.** A perfectly non-conducting platform acts as a stand for adiabatic processes.
4. **Sink.** A reservoir maintained at a constant lower temperature T_2 ($T_2 < T_1$) to which the heat engine can reject any amount of heat. The thermal capacity of sink is infinite so that its temperature remains constant at T_2, no matter how much heat is given to it.

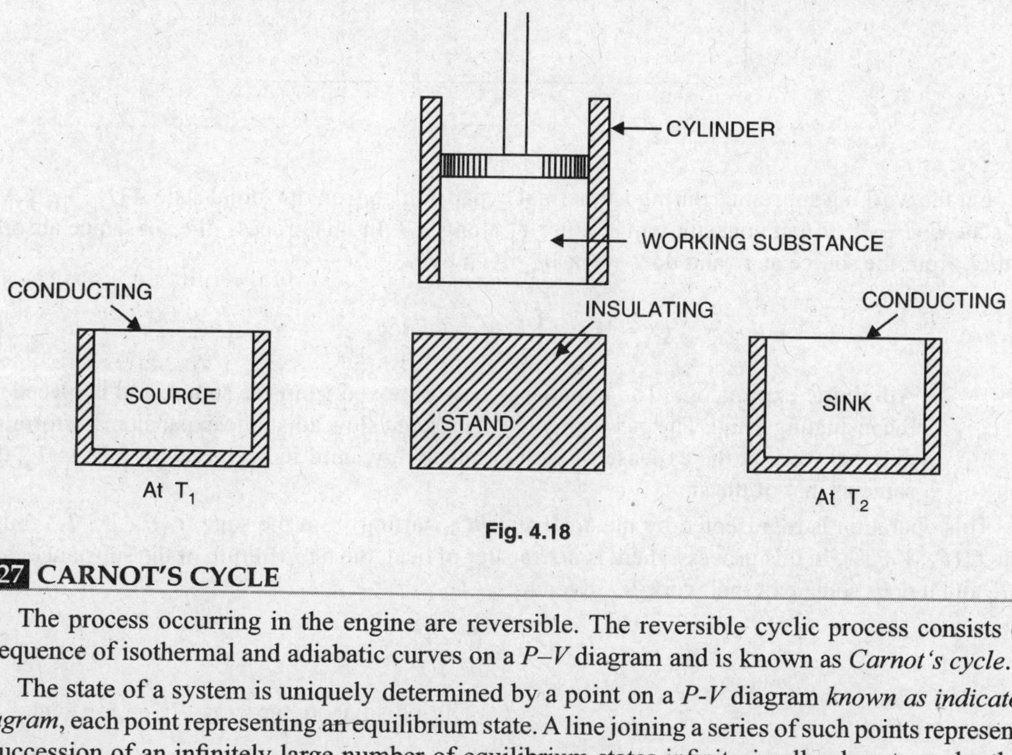

Fig. 4.18

4.27 CARNOT'S CYCLE

The process occurring in the engine are reversible. The reversible cyclic process consists of a sequence of isothermal and adiabatic curves on a P–V diagram and is known as *Carnot's cycle.*

The state of a system is uniquely determined by a point on a P-V diagram *known as indicator diagram*, each point representing an equilibrium state. A line joining a series of such points represents a succession of an infinitely large number of equilibrium states infinitesimally close to each other and hence depicts a *quasi-static process*. A quasi static process is a process carried out sufficiently slowly so that the system under consideration remains at all times arbitrarily close to equilibrium.

Work done in Carnot's cycle. In carnot's cycle the working substance is supposed to undergo the following four quasi-static operations, as shown in Fig. 4.19.

1. **Isothermal expansion.** Suppose the cylinder contains *one mole of perfect gas*. Let the initial temperature of the gas in the cylinder be T_1 *K*. Let the temperature of heat source is also T_1 *K*.

The cylinder is first placed on the source, so that the gas acquires the temperature T_1 of the source. It is then allowed to undergo quasi-static expansion. As the gas expands, its temperature tends to fall. Heat passes into the cylinder through the perfectly conducting base which is in contact with the source. The gas therefore, undergoes slow isothermal expansion at the constant temperature T_1.

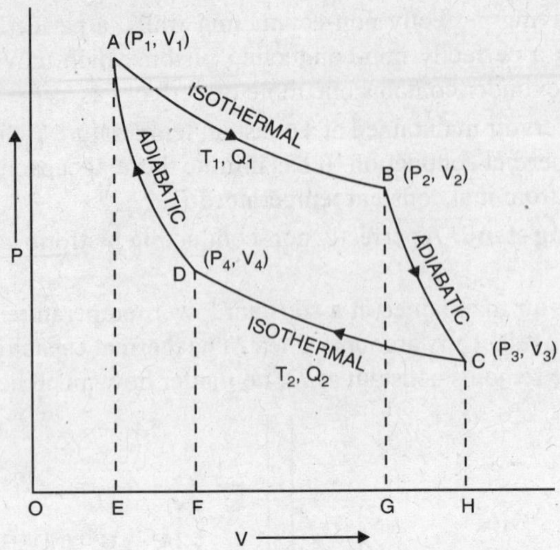

Fig. 4.19

Let the working substance during isothermal expansion go from its initial state $A\ (P_1, V_1, T_1)$ to the state $B\ (P_2, V_2, T_1)$ at constant temperature T_1 along AB. In this process, the substance absorbs heat Q_1 from the source at T_1 and does work W_1 given by

$$Q_1 = W_1 = \int_{V_1}^{V_2} PdV = RT_1 \log_e \frac{V_2}{V_1} = \text{area } ABGEA \qquad \ldots (i)$$

2. **Adiabatic expansion.** The cylinder is now removed from the source and is placed on the insulating stand. The gas is allowed to undergo slow adiabatic expansion, performing external work at the expense of its internal energy, until its temperature falls to T_2, the same as that of the sink.

This operation is represented by the adiabatic BC, starting from the state $B\ (P_2, V_2, T_1)$ to the state $C\ (P_3, V_3, T_2)$. In this process, there is no transfer of heat, the temperature of the substance falls to T_2 and it does some external work W_2 given by

$$W_2 = \int_{V_2}^{V_3} P.dV = K\int_{V_2}^{V_3} \frac{dV}{V^\gamma}$$

$$(\because \text{During adiabatic process, } PV^\gamma = \text{constant} = K)$$

$$= \frac{KV_3^{1-\gamma} - KV_2^{1-\gamma}}{1-\gamma}$$

$$= \frac{P_3V_3 - P_2V_2}{1-\gamma} \qquad (\because P_2\, V_2^{\gamma} = P_3\, V_3^{\gamma} = K)$$

$$= \frac{RT_2 - RT_1}{1-\gamma}$$

$$= \frac{R(T_1 - T_2)}{\gamma - 1} = \text{Area } BCHGB \qquad \ldots (ii)$$

3. **Isothermal Compression.** The cylinder is now removed from the insulating stand and is placed on the sink which is at a temperature T_2. The piston is now very slowly moved inwards so that the work is done on the gas. The temperature tends to increase due to heat produced by compression since the conducting base of the cylinder is in contact with the sink, the heat developed passes to the sink and the temperature of the gas remains

constant at T_2. Thus the gas undergoes isothermal compression at a constant temperature T_2 and gives up some heat to the sink.

This operation is represented by the isothermal *CD*, starting from the state *C* (P_3, V_3, T_2) to the state *D* (P_4, V_4, T_2). In this process, the substance rejects heat Q_2 to the sink at T_2 and work W_3 is done on the substance given by

$$Q_2 = W_3 = \int_{V_3}^{V_4} PdV = RT_2 \log_e \frac{V_4}{V_3} \quad \text{...(iii)}$$

$$= -RT_2 \log_e \frac{V_3}{V_4} = \text{area } CHFDC$$

(– *ve* signe indicates that work is done on the working substance)

4. **Adiabatic Compression.** The cylinder is now removed from sink and again placed on the insulating stand. The piston is slowly moved inwards so that the gas is adiabatically compressed and the temperature rises. The adiabatic compression is continued till the gas comes back to its original condition *i.e.* state *A* (P_1, V_1, T_1), thus completing one full cycle.

   This operation is represented by adiabatic *DA*, starting from *D* (P_4, V_4, T_2) to the final state *A* (P_1, V_1, T_1). In this process, work W_4 is done on the substance and is given by

$$W_4 = \int_{V_4}^{V_1} P.dV$$

$$= -\frac{R(T_1 - T_2)}{\gamma - 1} = \text{Area } DFEAD \quad \text{...(iv)}$$

(– *ve* sign indicates that work is done on the working substance. Since W_2 and W_4 are equal and opposite, they cancel each other.)

Work done by the engine per cycle

During the above cycle, the working substance absorbs an amount of heat Q_1 from the source and rejects Q_2 to the sink.

Hence, the net amount of heat absorbed by the gas per cycle

$$= Q_1 - Q_2$$

The net work done by the engine per cycle

$$= W_1 + W_2 + W_3 + W_4$$

$$= W_1 + W_3 \quad (\because W_2 = -W_4)$$

From the graph, the net work done per cycle

$$= \text{area } ABGEA + \text{area } BCHGB - \text{area } CHFDC - \text{area } DFEAD$$

$$= \text{area } ABCDA \quad \text{... (v)}$$

Thus, the area enclosed by the Carnot's cycle consisting of two isothermals and two adiabatics gives the net amount of work done per cycle.

In the cyclic process,

Net heat absorbed = Net work done per cycle.

$$Q_1 - Q_2 = W_1 + W_3$$

$$= RT_1 \log_e \frac{V_2}{V_1} - RT_2 \log_e \frac{V_3}{V_4} \quad \text{...(vi)}$$

Since the points *A* and *D* lie on the same adiabatic *DA*

$$T_1 V_1^{\gamma-1} = T_2 V_4^{\gamma-1}$$

$$\therefore \quad \frac{T_2}{T_1} = \left(\frac{V_1}{V_4}\right)^{\gamma-1} \quad \text{... (vii)}$$

Similarly, points B and C lie on the same adiabatic BC

$$T_1 V_2^{\gamma-1} = T_2 V_3^{\gamma-1}$$

$$\therefore \quad \frac{T_2}{T_1} = \left(\frac{V_2}{V_3}\right)^{\gamma-1} \quad \text{... (viii)}$$

From Equations (*vii*) and (*viii*),

$$\left(\frac{V_1}{V_4}\right)^{\gamma-1} = \left(\frac{V_2}{V_3}\right)^{\gamma-1}$$

or
$$\frac{V_1}{V_4} = \frac{V_2}{V_3}$$

or
$$\frac{V_2}{V_1} = \frac{V_3}{V_4}$$

Substituting in equation (*vi*), we get

$$\text{Net work done} = Q_1 - Q_2 = RT_1 \log_e \frac{V_2}{V_1} - RT_2 \log_e \frac{V_2}{V_1}$$

$$W = (Q_1 - Q_2) = R(T_1 - T_2) \log_e \frac{V_2}{V_1} \quad \text{... (ix)}$$

Efficiency

The efficiency of the heat engine is the ratio of quantity of heat converted into work (Useful output) per cycle to the total amount of heat absorbed per cycle.

$$\text{Efficiency, } \eta = \frac{\text{Useful Output}}{\text{Input}} = \frac{W}{Q_1}$$

$$= \frac{(Q_1 - Q_2)}{Q_1}$$

$$= \frac{R(T_1 - T_2)\log_e \frac{V_2}{V_1}}{RT_1 \log_e \frac{V_2}{V_1}}$$

$$\eta = \frac{T_1 - T_2}{T_1} \quad \text{... (4.25)}$$

or,
$$\eta = 1 - \frac{T_2}{T_1} \quad \text{... (4.26)}$$

Also we can write,
$$\frac{Q_1 - Q_2}{Q_1} = \frac{T_1 - T_2}{T_1}$$

or,
$$1 - \frac{Q_2}{Q_1} = 1 - \frac{T_2}{T_1}$$

or
$$\frac{Q_2}{Q_1} = \frac{T_2}{T_1} \quad \text{... (4.27)}$$

From equation (4.26), we conclude that the efficiency depends only upon the temperature of the source and sink and is always less than unity. The efficiency is independent of the nature of working substance. From equation, $\eta = 1 - \frac{T_2}{T_1}$, we get $\eta = 1$, if $T_2 = 0\text{K}$ *i.e.* the temperature of the sink is at absolute zero degrees. In practice, it is never possible to reach absolute zero and hence 100% conversion of heat energy into mechanical work is not possible, even though there is no wastage of energy in the Carnot's reversible engine.

Again, the efficiency is minimum or zero when $T_1 = T_2$ *i.e.* the temperature of the source is equal to the temperature of sink, then $\eta = 0$ *i.e.* the engine does not work.

Equation 4.27 shows that *the ratio of the heat absorbed from the source to the heat rejected to the sink is equal to the ratio of the absolute temperature of the source and the sink.*

Characteristics of Carnot's reversible engine

(i) The efficiency depends upon the temperature of the *heat source* and the *heat sink.*

(ii) The efficiency is independent of the nature of the working substance.

(iii) The reversible engines working between the same two temperature have the same efficiency.

(iv) As $T_1 - T_2 < T_1$, the efficiency of a Carnot't engine is always less than one.

Condition for 100% efficiency. For efficiency to be equal to unity *i.e.,* for the engine to be 100% efficient T_2 the sink temperature must be 0 K, which is not possible in practice. Hence the efficiency even of an ideal Carnot's engine is less than one. In other words it is not possible to extract heat from a body (heat source) and convert the whole of it into work.

Source, sink and working substance in a steam engine. In a steam engine the source is the boiler producing super heated steam, super heated steam is the *working substance and the sink is the atmosphere.*

Carnot's engine not a practical possibility. A Carnot's engine is not a practical possibility because

(i) A Carnot's engine consists of a cylinder having *perfectly non-conducting walls* and *perfectly conducting bottom.* In it moves a perfectly *non-conducting piston.*

(ii) The cylinder contains a *perfect gas* as a working substance

(iii) The source is a *hot body* at *constant temperature* T_1 *K* and sink a *cold body* at *constant temperature* T_2 *K. i.e.* the source and sink are large reservoirs of *infinite heat capacity* so that their temperature does not change during any transfer of heat to or from the cylinder.

(iv) The reversible cyclic process occurring in the engine consists of a sequence of *isothermal* and *adiabatic curves* on a *P–V* diagram.

All the above conditions are *ideal* and cannot be realised in practice. So a Carnot's engine is only an ideal engine and is not a practical possibility.

Reversibility of Carnot's heat engine

The Carnot's heat engine is perfectly reversible. It can be operated in the reverse direction also. Then it works as a refrigerator. The heat Q_2 is taken from the sink and external work is done on the working substance and heat Q_1 is rejected to the source at a higher temperature (principle of a refrigerator).

Moreover in the Carnot's heat engine, the process of isothermal and adiabatic expansions and compressions are carried out very-very slowly *i.e.* quasi-static. This is an ideal case. Any practical engine can not satisfy these conditions. Therefore, all practical engines have an efficiency less than the Carnot's engine.

Effective Ways to Increase Efficiency

The expression for the efficiency of a Carnot's engine is

$$\eta = 1 - \frac{T_2}{T_1}$$

The efficiency η can be increased by one of the following ways:

(i) Using a heat-source at constant temperature T_1 and heat-sink at temperature as low as possible, or

(ii) Using a heat-sink at constant temperature T_2 and a heat-source at temperature as high as possible.

In practice, it is not convenient to use heat sink at a temperature below that of the atmosphere. Therefore, the more effective way to increase the efficiency is to use a heat-source at temperature as high as possible.

4.28 LIMITATIONS OF FIRST LAW OF THERMODYNAMICS

The first law of thermodynamics is only a special case of the general law of conservation of energy. It states the exact relation between mechanical work done and heat energy produced. Whereas the whole of the work done can be converted into heat energy, the law does not state anything about the extent of heat that can be converted into work and the conditions under which this reverse process can take place.

4.29 SECOND LAW OF THERMODYNAMICS

The first law of thermodynamics states the equivalence of heat and energy. It simply tells that whatever work is obtained, an equivalent amount of heat is used up, or vice versa. It does not say anything about the direction in which the change might occur or about the range or limit to which it can be possible. The first law shows that perpetual motion of the first kind is impossible *i.e.* energy can not be created out of nothing or production of energy without disappearance of an equivalent energy of another form is not possible.

The first law has no answer why heat always flows from a body at a higher temperature to a body at a lower temperature and does not flow in the reverse direction. It can not explain why vast amount of a vailable heat cannot be converted into mechanical work. It we could control and make use of the limitless store of heat viz-solar energy, we could have an inexhaustible supply of useful energy. Thus, we could have a perpetual motion machine not forbidden by the first law. This is called perpetual motion of the second kind. In practice, however there is no engine which can convert the heat from the single source in to useful work without rejecting some heat to a heat sink at a lower temperature *i.e.* perpetual motion of the second kind *i.e.* production of useful energy from the internal energy of one body is impossible.

It was the quest of several such questions which led to the formulation of second law of thermodynamics. The second law is a generalization of certain experiences and observations and is concerned with the direction in which energy transfer takes place. The law has been stated in a number of ways, which means the same thing.

Lord Kelvin's Statement:

In a heat engine the working substance extracts heat from the source, converts a part of it into work and rejects the rest to a sink at a lower temperature. The temperature of the source must be higher than the surroundings and engine will not work when the temperatures of source and sink are the same. No engine has ever been constructed which converts all the heat absorbed from the source into work without rejecting a part of it to the cold body. As the engine absorbs more and more heat from the hot body, the latter suffers a continuous fall in temperature and if a continuous supply of work is desired, the hot body will in the long run become as cold as its surroundings. Then no heat flow will be possible, the engine will stop working and hence no mechanical work will be obtained. It means that we can not obtain a continuous supply of work from a single supply of heat *i.e.* the presence of colder body is a must for the continuous conversion of heat into work. Such considerations led Lord Kelvin to state that

"*It is impossible to get a continuous supply of work from a body by cooling it to a temperature lower than that of its surroundings.*"

Planck's Statement:

"It is impossible to construct an engine which, working in a complete cycle, will produce no effect other than the raising of a weight and the cooling of a heat reservoir."

Thus, it is impossible to construct an engine which working in a complete cycle, will produce no effect other than the absorption of heat from a reservoir and its conversion into an equivalent amount of work *i.e.* perpetual motion of the second kind is impossible. The engine must reject a part of the heat absorbed to a sink at lower temperature.

These two statements can be combined into one equivalent statement, known as the **Kelvin-Planck's statement** of the second law of thermodynamics.

Kelvin-Planck Statement:

"It is impossible to construct an engine which, operating in a cycle, has the sole effect of extracting heat from a reservoir and performing an* equivalent *amount of work."

Clausius's Statement:

According to Clausius "It is impossible for a self-acting machine working in a cyclic process, unaided by external agency, to transfer heat from a body at a lower temperature to a body at a higher temperature."

In other words it may be stated as "*Heat cannot flow of itself from a colder body to a hotter body.*"

This statement is based upon the performance of refrigerator – a heat engine working in the backward direction. This statement means that natural flow of heat is always from a hot body to a cold body. If heat is to be transferred from cold body to hot body, work will have to be done by external agency. A refrigerator is a device which transfers heat from a colder body to hotter body by doing external work on the working substance. The compression is brought about by an external agency *i.e.,* 'electricity' by performing work on the working substance.

4.30 CARNOT'S THEOREM

Statement: *From the second law of thermodynamics two important results are derived; these conclusions are taken together to constitute Carnot's theorem which may be stated in the following forms.*

(*a*) 'No engine can be more efficient than a perfectly reversible engine working between the same two temperatures.'

(*b*) 'The efficiency of all reversible engines, working between the same two temperatures is the same, whatever the working substance.'

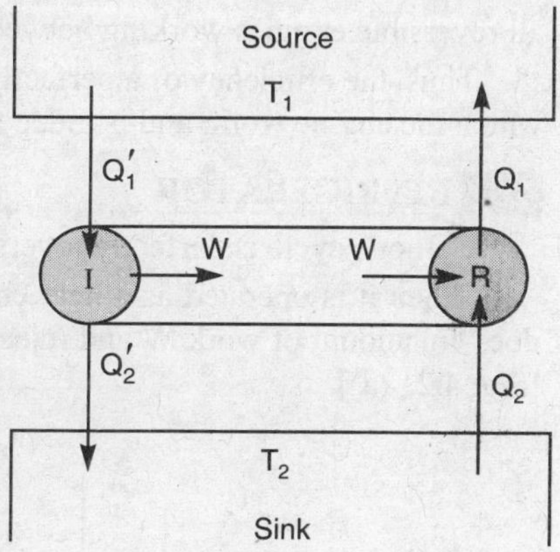

Fig. 4.20

Proof:

First Part: To prove the first part of the theorem, we consider two engines R and I working between the temperatures T_1 and T_2 where $T_1 > T_2$ [Fig. 4.20]. Of these two engines R is reversible and I is irreversible.

Suppose I is more efficient than R. Suppose in each cycle, R absorbs the quantity of heat Q_1 from the source at T_1 and rejects the quantity of heat Q_2 to the sink at T_2. Suppose in each cycle I absorbs the quantity of heat Q_1' from the source at T_1 and gives up the quantity of heat Q_2' to the sink at T_2. Let the two engines do the same amount of work W in each cycle. According to the assumption I is more efficient than R.

$$\therefore \qquad \frac{Q_1' - Q_2'}{Q_1'} > \frac{Q_1 - Q_2}{Q_1}$$

or

$$\frac{W}{Q_1'} > \frac{W}{Q_1}$$

or $$Q_1 > Q_1'$$

and $$Q_1' - Q_2' = Q_1 - Q_2$$

or $$Q_2 - Q_2' = Q_1 - Q_1'$$

Now because $$Q_1 > Q_1'$$

$$\therefore \quad Q_2 > Q_2'$$

Now suppose the two engines are coupled together so that I drives R backwards and suppose they use the same source and sink. The combination forms a self-acting machine in which I supplies external work W and R absorbs this amount of work in its reverse cycle. I in its cycle absorbs heat Q_1' from the source and gives up heat Q_2' to the sink. R in its reverse cycle, absorbs heat Q_2 from the sink and gives up heat Q_1 to the source.

The net result of the complete cycle of the coupled engines is given by

Gain of heat by the source at $T_1 = Q_1 - Q_1'$

Loss of heat by the sink at $T_2 = Q_2 - Q_2'$

External work done on the system = 0

Thus, the coupled engines forming a self-acting machine unaided by any external agency transfer heat continuously from a body at low temperature to a body at a higher temperature.

This conclusion is contrary to the second law of thermodynamics, according to which in a cyclic process heat cannot be transferred from one body to another at a higher temperature by a self-acting machine. Hence our assumption is incorrect and we conclude that no engine can be more efficient than a perfectly reversible engine working between the same temperatures.

Second Part: The second part of the theorem may be proved by the same arguments as before. For this purpose, we consider two reversible engines R_1 and R_2 and assume that R_2 is more efficient than R_1. Proceeding in the same way we can show that R_2 cannot be more efficient than R_1. Therefore, all reversible engines working between the same two temperatures have the same efficiency.

Thus, the efficiency of a perfectly reversible engine depends only on the temperatures between which the engine work, and is independent of the nature of the working substance.

4.31 REFRIGERATOR

Carnot's cycle is perfectly reversible. It can work as a heat engine and also as a refrigerator.

When it is operated as a heat engine, it absorbs heat Q_1 from the source at a temperature T_1, does an amount of work W and rejects heat Q_2 to the sink at temperature T_2, ($T_2 < T_1$) as shown in [Fig. 4.21 (a)].

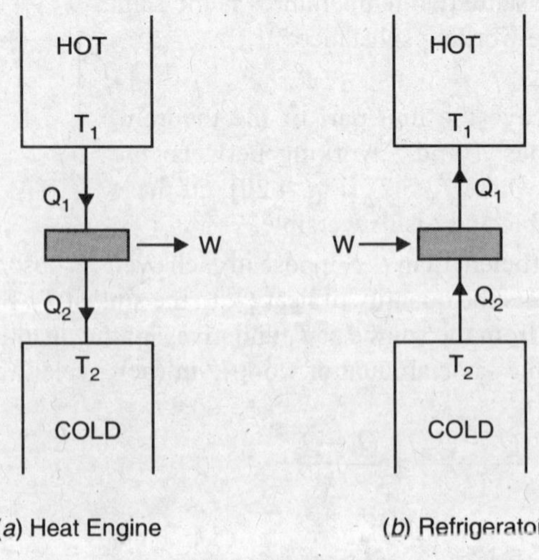

(a) Heat Engine (b) Refrigerator

Fig. 4.21

When it is operated as a refrigerator, it absorbs heat Q_2 from the sink at temperature T_2. W amount of work is done on it by some external means and rejects heat Q_1 to the source at temperature T_1, $(T_1 > T_2)$ as shown in [Fig. 4.21 (*b*)].

In the second case [Fig. 4.21 (*b*)], heat flows from a body at a lower temperature T_2 to a body at a higher temperature T_1 with the help of external work done on the working substance. This action is that of a refrigerator. In every cycle, heat Q_2 is extracted from the cold body. This will not be possible if the cycle is not completely reversible.

Difference between refrigerator and heat pump. A Carnot's cycle being reversible can be performed in the reverse direction. When all the operations of a Carnot's cycle are performed in the opposite direction, it will *absorb* an amount of heat Q_2 from the sink or the cold body at a temperature T_2 and *transfer* to the hot body or source an amount of heat Q_1 at a higher temperature T_1. According to second law of thermodynamics, heat cannot be transferred from a cold to a hot body, without the expenditure of energy. Therefore, mechanical energy has to be supplied to the engine from outside. If W is the amount of mechanical energy supplied to the engine, then

$$Q_1 = Q_2 + W$$

The Carnot's cycle working in the reverse direction is shown in Fig. 4.22. It can perform two functions.

(*i*) It can further cool the cold body and

(*ii*) It can further heat the hot body.

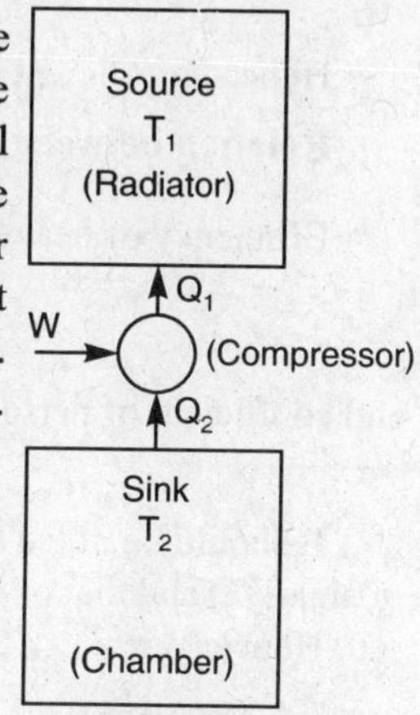

Fig. 4.22

When a reversed heat engine is used to further cool the colder system, the arrangement is called a refrigerator.

When a reversed heat engine is used to further heat the hotter system is it is called a **heat pump**

Essential parts of refrigerator. *A refrigerator consists of following main parts:*

(*i*) **A chamber to be cooled.** It acts as a sink at a lower temperature.

(*ii*) *An electric compressor.* It works on electric supply and acts as the external agency to supply mechanical energy from outside.

(*iii*) *A radiator* which acts as the source at a higher temperature.

(*iv*) *A thermostat* to control the temperature of the chamber.

The working substance, generally 'Freon' vapour absorbs a quantity of heat Q_2 from the chamber (sink) of the refrigerator. The compressor does an amount of work W on it and their sum $Q_2 + W$ is transferred (or rejected) to the atmosphere (source) with the help of the radiator. During each cycle the chamber loses an amount of heat Q_2 and its temperature goes on falling. As soon as the desired temperature is reached according to the setting of the thermostat, the thermostat switches off the compressor. The inside chamber of the refrigerator is thermally insulated but the insulation cannot be absolutely perfect. Therefore some heat slowly enters the chamber from outside and the temperature rises above that for which the thermostat is set. The thermostat then restarts the compressor and cooling begins.

4.32 COEFFICIENT OF PERFORMANCE

The coefficient of performance of a refrigerator is defined as the ratio of the heat drawn from the cold body to the work done on the machine.

Assuming that the machine performs the Carnot's cycle in the reverse direction if Q_2 is the quantity of heat absorbed from the cold body at a temperature T_2 and Q_1 is the quantity of heat given to the hot body at a temperature T_1, then the work done by the machine

$$= Q_1 - Q_2$$

For an ideal refrigerating machine, if W is the work done on the machine, then

$$W = Q_1 - Q_2$$

$\therefore$ Coefficient of performance $\beta = \dfrac{Q_2}{W} = \dfrac{Q_2}{Q_1 - Q_2}$

For a reversible Carnot's engine $\dfrac{Q_1}{T_1} = \dfrac{Q_2}{T_2}$

$\therefore$ $\dfrac{Q_2}{Q_1 - Q_2} = \dfrac{T_2}{T_1 - T_2}$

Hence co-efficient of performance $\beta = \dfrac{T_2}{T_1 - T_2}$

Relation between β and η

Efficiency of heat engine $\eta = 1 - \dfrac{T_2}{T_1}$ $\therefore$ $\dfrac{T_2}{T_1} = 1 - \eta$

and coefficient of performance $\beta = \dfrac{T_2}{T_1 - T_2} = \dfrac{\dfrac{T_2}{T_1}}{1 - \dfrac{T_2}{T_1}} = \dfrac{1-\eta}{1-1+\eta} = \dfrac{1-\eta}{\eta}$

It should be noted that the coefficient of performance β becomes infinite when the temperature of sink (T_2) and that of source (T_1) become equal. Using relation ($\beta = T_2/(T_1 - T_2)$, at $T_1 = T_2$, $T_1 - T_2 = 0$. Hence ($\beta = \infty$) *i.e.*, the coefficient of performance becomes infinite.

Now $\beta = \dfrac{T_2}{T_1 - T_2} = \dfrac{Q_2}{Q_1 - Q_2}$ $\therefore Q_1 - Q_2 = 0$ or $Q_1 = Q_2$

But $Q_1 = Q_2 + W$ where W is the external work done on the refrigerator or external energy drawn from it. When $Q_1 = Q_2$, $W = 0$ *i.e.*, no mechanical energy is drawn from outside and no external work is done on the machine.

As no energy is drawn by the refrigerator, the compressor may go on infinitely but no heat will be transferred from the sink to the source in accordance with second law of thermodynamics.

Relation between mechanical energy required and sink temperature. The coefficient of performance of a refrigerator $\beta = \dfrac{T_2}{T_1 - T_2}$

As the temperature T_2 of the cold body (sink) decreases $T_1 - T_2$ increases. Therefore, β decreases due to decrease in T_2 as well as increase in $T_1 - T_2$. As $\beta = \dfrac{Q_2}{W}$, for a given value of Q_2, W increases as β decreases. Hence the mechanical or electrical energy W required to extract a given amount of heat Q_2 from the cold body increases as the temperature T_2 of the cold body decreases for same inside cooling.

Effect of keeping refrigerator door open. If the door of a refrigerator is kept open in the hall, there will be no change in the temperature of the hall *i.e.*, it will make it neither warm nor cool.

As the door of the refrigerator is open, the chamber which acts as the sink is in contact with the air in the hall and attains its temperature. The radiator which acts as the source is also in contact with the air in the hall and so attains the temperature of the source T_1. Finally $T_1 = T_2$ and there is no change in the temperature of the hall.

Refrigerator works better in winter than in summer. The co-efficient of performance β of a refrigerator is given by $\beta = \dfrac{T_2}{T_1 - T_2}$ where T_2 is the temperature of the sink or the chamber where the body to be cooled is placed and T_1 the temperature of the source or radiator which is at atmospheric temperature. In winter T_1 the atmospheric temperature decreases. Therefore, $T_1 - T_2$ decreases and hence β the co-efficient of performance *increases* and, therefore, the refrigerator works better.

Defrosting of refrigerator is a must. When a refrigerator has been working for a sufficient time good amount of ice is formed in the freezer and on the cooling coils. It decreases temperature T_2 (*i.e.*, the temperature of the sink) and thus increases $T_1 - T_2$ (the difference of temperature between the source and the sink). The efficiency or co-efficient of performance of a refrigerator is given by

$$\beta = \frac{T_2}{T_1 - T_2}$$

The value of β, therefore, decreases with decrease of T_2 as well as increase of $T_1 - T_2$. Hence the efficiency of the refrigerator goes down and it produces very slow cooling.

Defrosting increases T_2 and decreases $T_1 - T_2$ thereby increasing the co-efficient of performance and hence better cooling. The refrigerator must be defrosted after 3-4 days to get proper cooling. This is the reason, why now a days, mostly all refrigerators undergo defrosting at regular intervals, by using automatic defrosting arrangement.

SOLVED EXAMPLES

Example 4.1 *A motor car tyre has a pressure of 2 atmospheres at the room temperature of 27°C. If the tyre suddenly bursts, find the resulting temperature.* ***(Purvanchal 2006, Garhwal 2001)***

Solution. Here

$$P_1 = 2 \text{ atmospheres}$$
$$T_1 = 273 + 27 = 300 \text{ K}$$
$$P_2 = 1 \text{ atmosphere}$$
$$T_2 = ?$$
$$\gamma = 1.4$$

$$\frac{P_1^{\gamma-1}}{T_1^{\gamma}} = \frac{P_2^{\gamma-1}}{T_2^{\gamma}} \quad \text{or} \quad \left(\frac{P_2}{P_1}\right)^{\gamma-1} = \left(\frac{T_2}{T_1}\right)^{\gamma}$$

$$\left(\frac{1}{2}\right)^{0.4} = \left(\frac{T_2}{300}\right)^{1.4}$$

$$2.304 \times 0.4 \log_{10}(0.5) = 1.4 [\log_{10} T_2 - \log_{10} 300] \times 2.304$$

$$-0.1204 = 1.4 \log_{10} T_2 - 3.4680$$

$$1.4 \log_{10} T_2 = 3.4680 - 0.1204 = 3.3476$$

$$\log_{10} T_2 = \frac{3.3476}{1.4} = 2.3911$$

$$T_2 = \text{Antilog}(2.3911) = 246.1 \text{ K}$$

$$\mathbf{T_2 = -26.9°C}$$

Example 4.2 *A quantity of air at 27°C and atmospheric pressure is suddenly compressed to half its original volume. Find the final (i) pressure and (i) temperature.* (Given $\gamma = 1.4$, $2^{1.4} = 2.64$)

(Rohilkhand 2000)

Solution.

(*i*) $P_1 = 1$ atmosphere; $P_2 = ?$, $\gamma = 1.4$

$$V_1 = V, \quad V_2 = \frac{V}{2}$$

During sudden compression, the process is adiabatic.

$$P_1 V_1^{\gamma} = P_2 V_2^{\gamma}$$

$$P_2 = P_1 \left[\frac{V_1}{V_2}\right]^{\gamma}$$

$$= 1\,[2]^{1.4} = \text{Antilog}\,[1.4 \log 2] = \text{Antilog}\,[1.4 \times 0.30]$$
$$= \text{Antilog}\,(0.42)$$
$$\mathbf{P_2 = 2.636\ atom}$$

(*ii*) $V_1 = V\,;\ V_2 = \dfrac{V}{2},\ T_1 = 300\ \text{K}\,;\ T_2 = ?,\ \gamma = 1.4$

$$T_1\,(V_1)^{\gamma-1} = T_2\,(V_2)^{\gamma-1}$$
$$T_2 = T_1\,[\,2\,]^{1.4-1}$$
$$= 300\,[\,2\,]^{0.4}$$
$$= 395.9\ \text{K}$$
$$\mathbf{T_2 = 122.9°C}$$

Example 4.3 *Air at N.T.P. is compressed adiabatically to half of its volume. Calculate the change in its temperature.* **(Bangalore 1998)**

Solution. Let the initial temperature be T_1 *K* and the final temperature T_2 *K*.

$$\text{Initial volume} = V_1$$
$$\text{Final volume} = V_2 = \frac{V_1}{2},\ T_1 = 273\ K$$

During an adiabatic process

$$T_1\,V_1^{\gamma-1} = T_2\,V_2^{\gamma-1}$$
$$T_2 = T_1\left[\frac{V_1}{V_2}\right]^{\gamma-1}$$
$$T_2 = T_1\,[\,2\,]^{\gamma-1}$$

But $\quad \gamma$ for air $= 1.4$

$$T_2 = T_1\,[\,2\,]^{0.4} = 273 \times [2]^{0.4}$$
$$\log_e T_2 = \log_e 273 + 0.4 \log_e{}^2 = 2.4362 + 0.1204$$
$$\log_{10} T_2 = \log_{10} 273 + 0.4 \log_{10}{}^2 = 2.5566$$
$$T_2 = 360.2\ \text{K} = 87°\text{C}$$

$\therefore \quad$ Change in temperature $= (360.2 - 273)$

$$= \mathbf{87.2°C}$$

Example 4.4 *1 gram molecule of a monoatomic* $\left(\gamma = \dfrac{5}{3}\right)$ *prefect gas at 27°C is adiabatically compressed in a reversible process from an initial pressure of 1 atmosphere to a final pressure of 50 atmospheres. Calculate the resulting difference in temperature.* **(Mysore 2000)**

Solution. In a reversible adiabatic process

$$\frac{P_1^{\gamma-1}}{T_1^{\gamma}} = \frac{P_2^{\gamma-1}}{T_2^{\gamma}}$$

or

$$\left(\frac{P_2}{P_1}\right)^{\gamma-1} = \left(\frac{T_2}{T_1}\right)^{\gamma}$$

Here

$$P_2 = 50$$
$$P_1 = 1$$
$$T_1 = 273 + 27 = 300\ \text{K}$$
$$T_2 = ?$$
$$\gamma = \frac{5}{3}$$

$$\therefore \quad \left(\frac{50}{1}\right)^{\frac{2}{3}} = \left(\frac{T_2}{300}\right)^{\frac{5}{3}}$$

$$\frac{2}{3}\log 50 = \frac{5}{3}[\log T_2 - \log 300]$$

$$\mathbf{T_2 = 1434\ K}$$

$$\mathbf{= 1161°C}$$

Example 4.5 *A quantity of dry air at 27°C is compressed (i) slowly and (ii) suddenly to $\frac{1}{3}$ of its volume. Find the change in temperature in each case, assuming γ to be 1.4 for dry air.*

Solution. ***(Rohilkhand 2002)***

1. When the process is slow, the temperature of the system remains constant. Therefore, there is no change in temperature.
2. When the compression is sudden, the process is adiabatic.

Here $\quad V_1 = V,\ V_2 = \frac{V}{3},\ T_1 = 300\text{ K},\ T_2 = ?,\ \gamma = 1.4$

$$T_2 (V_2)^{\gamma-1} = T_1 (V_1)^{\gamma-1}$$

$$T_2 = T_1\left(\frac{V_1}{V_2}\right)^{\gamma-1}$$

$$= 300\left[\frac{3V}{V}\right]^{\gamma-1} = 300\,[\,3\,]^{1.4-1} = 300\,[\,3\,]^{0.4}$$

$$T_2 = 465.5\text{ K}$$

$$\mathbf{= 192.5°C}$$

The temperature of air increases by

$$= 192.5 - 27$$

$$\mathbf{= 165.5°C = 165.5\ K}$$

Example 4.6 *A certain mass of gas at NTP is expanded to three times its volume under adiabatic conditions. Calculate the resulting temperature and pressure. γ for the gas is 1.40.* **(*Osmania 2001*)**

Solution.

(*a*) Here, $\quad V_1 = V; \quad V_2 = 3V$

$$T_1 = 273\text{ K}\ \ T_2 = ?$$

$$T_1 V_1^{\gamma-1} = T_2 V_2^{\gamma-1}$$

or

$$T_2 = T_1\left[\frac{V_1}{V_2}\right]^{\gamma-1}$$

$$T_2 = 273\left[\frac{1}{3}\right]^{1.4-1}$$

$$\mathbf{T_2 = 176\ K = -97°C}$$

(*b*) Here, $\quad V_1 = V, \quad V_2 = 3V$

$$P_1 = 1\text{ atm},\ P_2 = ?$$

$$P_1 V_1^{\gamma} = P_2 V_2^{\gamma}$$

$$P_2 = P_1\left[\frac{V_1}{V_2}\right]^{\gamma}$$

or $$P_2 = 1\left(\frac{1}{3}\right)^{1.4}$$

$$P_2 = \mathbf{0.2148\ atm}$$

Example 4.7 *A gas occupying 1 litre at 80 cm of Hg pressure is expanded adiabatically to 1190 c.c. If the pressure falls to 60 cm of Hg in the process, deduce the value of γ.* **(Garhwal 1999)**

Solution. Here $$P_1V_1^{\gamma} = P_2V_2^{\gamma}$$

$$\left(\frac{P_1}{P_2}\right) = \left(\frac{V_2}{V_1}\right)^{\gamma}$$

Taking logs,

$$\log\left(\frac{P_1}{P_2}\right) = \gamma\left(\log\frac{V_2}{V_1}\right)$$

or $$\gamma = \frac{\log\left(\frac{P_1}{P_2}\right)}{\log\left(\frac{V_2}{V_1}\right)}$$

$$\gamma = \frac{\log\left(\frac{80}{60}\right)}{\log\left(\frac{1190}{1000}\right)}$$

$$\gamma = \frac{0.1249}{0.0755}$$

$$\gamma = \mathbf{1.65}$$

Example 4.8 *The initial temperature of a gas is 27°C. Calculate the rise in temperature when the gas is compressed suddenly to 8 times its original pressure (γ = 1.5).*

(Osamania 2004, Nagpur 2004/s, Mysore 2004)

Solution. For an adiabatic process,

$$\frac{P_1^{\gamma-1}}{T_1^{\gamma}} = \frac{P_2^{\gamma-1}}{T_2^{\gamma}}$$

$$\left(\frac{P_2}{P_1}\right)^{\gamma-1} = \left(\frac{T_2}{T_1}\right)^{2}$$

Here, $$\frac{P_2}{P_1} = 8 \text{ and } T_1 = 273 + 27 = 300\text{ K}$$

$$\gamma = 1.5$$

∴ $$(8)^{\gamma-1} = \left(\frac{T_2}{300}\right)^{\gamma}$$

$$(8)^{0.5} = \left(\frac{T_2}{300}\right)^{1.5}$$

$$0.5 \log 8 = 1.5\,[\log T_2 - \log 300]$$

$$0.5\,(0.9031) = 1.5 \log T_2 - 1.5 \times 2.4771$$

$$1.5 \log T_2 = 0.5 \times 0.9031 + 1.5 \times 2.4771$$

$$1.5 \log T_2 = 0.4516 + 3.7157$$

$$1.5 \log T_2 = 4.1673$$

$$\log T_2 = \frac{4.1673}{1.5}$$

$$\log T_2 = 2.7782$$

$$T_2 = \mathbf{600\ K}$$

$$T_2 = \mathbf{327°C}$$

Rise in temperature = 327 – 27 = 300°C

Example 4.9 *Air in the cylinder of a combustion engine at 20°C is compressed from an initial pressure of 1 atmosphere and volume 8×10^{-4} m³ to a volume 6×10^{-5} m³. Assume that air behaves as an ideal gas (γ = 1.40) and the process is adiabatic, calculate* **(*Agra 2000*)**

(*a*) *final pressure* (*b*) *final temperature of the gas*

Solution. Process is adiabatic

(*a*)

$$P_1 V_1^{\gamma} = P_2 V_2^{\gamma}$$

$$P_2 = P_1 \left[\frac{V_1}{V_2}\right]^{\gamma}$$

Here

$$P_1 = 1 \text{ atmosphere}$$

$$V_1 = 8 \times 10^{-4} \text{ m}^3$$

$$V_2 = 6 \times 10^{-5} \text{ m}^3$$

$$T_1 = 273 + 20 = 293 \text{ K}$$

$$P_2 = 1\left[\frac{8\times10^{-4}}{6\times10^{-5}}\right]^{1.4}$$

$$P_2 = \mathbf{37.6\ atm}$$

(*b*) Since $\frac{PV}{T}$ is always constant during the process as no gas has escaped from the cylinder

$$\frac{PV}{T} = nR = \text{constant}$$

$$\frac{P_1V_1}{T_1} = \frac{P_2V_2}{T_2}$$

$$T_2 = \left(\frac{P_2}{P_1}\right)\left(\frac{V_2}{V_1}\right)T_1$$

$$T_2 = \left(\frac{37.6}{1}\right)\left(\frac{6\times10^{-5}}{8\times10^{-4}}\right)293$$

$$T_2 = \mathbf{826.2\ K}$$

$$T_2 = \mathbf{553.2°C}$$

Example 4.10 *Calculate the rise in temperature of a gas initially at 27°C, if its pressure is suddenly doubled, γ = 1.4.* **(*Kerala 2001*)**

Solution.

Here,

$$\frac{P_1^{\gamma-1}}{T_1^{\gamma}} = \frac{P_2^{\gamma-1}}{T_2^{\gamma}}$$

$$\left(\frac{P_2}{P_1}\right)^{\gamma-1} = \left(\frac{T_2}{T_1}\right)^{\gamma}$$

Here,

$$P_2 = 2\,P_1$$

$$T_1 = 27°C = 27 + 273 \text{ K}$$

$$= 300 \text{ K}$$
$$T_2 = ?$$
$$\gamma = 1.4$$
$$(2)^{1.4-1} = \left(\frac{T_2}{300}\right)^{1.4}$$
$$0.4 \log 2 = 1.4 [\log T_2 - \log 300]$$
$$0.4 \times 0.3010 = 1.4 [\log T_2 - 2.4771]$$
$$\log T_2 = 0.086 + 2.4771$$
$$= 2.5631$$
$$T_2 = 365.7 \text{ K}$$
$$= 365.7 - 273 = 92.7°\text{C}$$
$$\text{Rise in temperature} = (92.7 - 27)\ °\text{C}$$
$$= \mathbf{65.7°C}$$

Example 4.11 *Calculate the work done when one litre of a mono atomic perfect gas at N.T.P. is compressed adiabatically to half its volume γ = 1.67.* **(Calicut 2000)**

Solution. Here $V_1 = 1 \text{ litre} = 10^3 \text{ cm}^3 = 10^{-3} \text{ m}^3$

$$V_2 = 0.5 \text{ litre} = 500 \text{ cm}^3 = 5 \times 10^{-4} \text{ m}^3$$
$$\gamma = 1.67$$

Work done $$W = \left(\frac{1}{1-\gamma}\right)\left[\frac{1}{V_2^{\gamma-1}} - \frac{1}{V_1^{\gamma-1}}\right]$$

$$W = \left(\frac{1}{1-1.67}\right)\left[\left(\frac{1}{5\times10^{-4}}\right)^{0.67} - \left(\frac{1}{10^{-3}}\right)^{0.67}\right]$$
$$= -\left(\frac{1}{0.67}\right)[(2 \times 10^3)^{0.67} - (10^3)^{0.67}$$
$$= -\left(\frac{1}{0.067}\right)[162.8 - 102.3]$$
$$= -90.3 \text{ J}$$

Negative sign shows, work is done on the gas.

Hence work done on the gas = **90.3 J.**

Example 4.12 *At normal temperature (0°C) and at normal pressure (1.013 × 10^5 Newton/m^2) when one gm of water freezes its volume increases by 0.091 cm^3. Calculate the change in its internal energy. (Latent heat of melting ice = 80 cals/gm. 1 cal = 4.2 J).* **(Agra 2000)**

Solution. Mass of water = 1 gm

Volume of 1 gm of water = 1 cm^3

Volume of 1 gm of ice = 1.091 cm^3

Because when 1 gm of water freezes its volume increases by 0.091 cm^3

∴ Change in volume $dV = 1 - 1.091 = -0.091 \text{ cm}^3$

Normal pressure $P = 1.013 \times 10^5 \text{ Newton/m}^2$

$$= \frac{1.013\times10^5\times10^5}{10^4} = 1.013 \times 10^6 \text{ dynes cm}^{-2}$$

Work done when 1 gm of water freezes = $dW = PdV = 1.013 \times 10^6 \times (-.091)$ erg.

$$= -(9.2 \times 10^4) \text{ erg.}$$

Heat lost by 1 gm of water = 80 cal = $80 \times 4.2 \times 10^7$ ergs. = 336×10^7 ergs.

According to the First law of thermodynamics $dQ = dU + dW$

$\therefore$ Change in internal energy $dU = dQ - dW = 336 \times 10^7 + 9.2 \times 10^4$ erg.

$= (336 + 0.0092)\ 10^7$ erg.

$= 336.0092 \times 10^7$ erg. = 336.0092 Joule.

Example 4.13 *One gm of water becomes 1671 cc of steam when boiled at a constant pressure of 1 atmosphere. The heat of vaporisation at this temperature is 540 cal per gm. Compute the external work and increase in itnernal energy.*

Solution. Volume of 1 gm of water = 1 cc

Volume of 1 gm of steam = 1671 cc

$\therefore$ Increase in volume dV = $1671 - 1 = 1670$ cc = 1670×10^{-6} m^3

Pressure P = 1 atmosphere = $76 \times 13.6 \times 981$

$= 1.013 \times 10^6$ dynes cm^{-2} = 1.013×10^5 Nm^{-2}

External work done when one gm of water becomes steam

$$dW = PdV = 1.013 \times 10^5 \times 1670 \times 10^{-6} \text{ J} = 169.3 \text{ J}$$

Heat supplied $dQ = 540$ cal $= 540 \times 4.2 = 2268$ J

According to first law of thermodynamics,

$$dQ = dU + dW$$

or $dU = dQ - dW = 2268 - 169.3 = 2098.7$ Joule.

Example 4.14 *Calculate the change in internal energy when 0.004 kg of air is heated from 2°C to 10°C at constant volume. Specific heat of air at constant volume c_v = 0.172 k-calories/Kg°C.*

Solution. Mass of air m = 0.004 kg. Increase in temperature $\theta = 10 - 2 = 8$°C

Specific heat at constant volume C_v = 0.172 K cal/kg°C

According to the first law of thermodynamics $dQ = dU + dW$ or $dQ = dU + PdV$

At constant volume $dV = 0$

$\therefore$ $dQ = dU$

Now $dQ = mc_v\,\theta = 0.004 \times 0.172 \times 8$ kilo-cals

$= 0.004 \times 0.172 \times 8 \times 4.2 \times 10^3 = 23.1$ Joules

$\therefore$ Change in internal energy dU = 23.1 Joules.

Example 4.15 *When a system is taken from the state A to the state B, along the path ACB, 80 joules of heat flows into the system, and the system does 30* joules *of work (Fig. 4.23).*

(a) How much heat flows into the system along the path ADB, if the work done is 10 joules.

(b) The system is returned from the state B to the state A along the curved path. The work done on the system is 20 joules. *Does the system absorb or liberate heat and how much?*

(c) If $U_A = 0$, $U_D = 40$ joules, *find the heat absorbed in the process AD and DB.* ***(Nagpur 2000)***

Solution.

Along the path ACB,

$$Q_{ACB} = U_B - U_A + W$$

Here $Q = +80$ joules

$W = +30$ joules

$\therefore$ $+80 = U_B - U_A + 30$

$U_B - U_A = 80 - 30 =$ **50 joules**

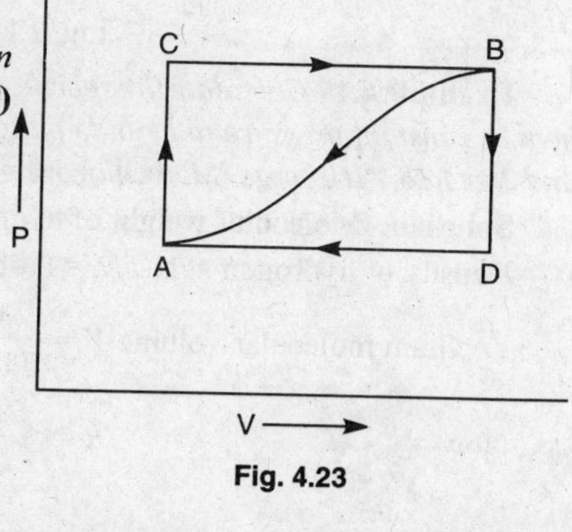

Fig. 4.23

(*a*) Along the path *ADB*,

$$W = +10 \text{ joules}$$
$$Q_{ADB} = U_B - U_A + W$$
$$Q = 50 + 10 = \mathbf{60\ joules}$$

(*b*) For the curved path from *B* to *A*,

$$W = -20 \text{ joules}$$
$$Q = (U_A - U_B) + W$$
$$= -50 - 20 = \mathbf{-70\ joules}$$

(– *ve* sign shows that heat is liberated by the system)

(*c*) $$U_A = 0, \qquad U_D = 40 \text{ joules}$$
$$U_B - U_A = 50$$
$$\therefore \qquad U_B = 50 \text{ joules}$$

In the process *ADB*, 10 joules of work is done. Work done from *A* to *D* is + 10 joules and from *D* to *B* is zero.

For *AD*,

$$Q_{AD} = (U_D - U_A) + W$$
$$= 40 + 10 = 50 \text{ joules}$$

For *DB*

$$Q_{DB} = U_B - U_D + W$$
$$= 50 - 40 + 0 = 10 \text{ joules}$$

Example 4.16 *Prove that if internal energy* $U = a + \frac{PV}{b}$, *then in reversible adiabatic process* $PV^{(1+b)}$ *is constant where a and b are constants* **(Gharwal 2001)**

Solution. According to the first law of thermodynamics

$$dQ = dU + dW = dU + P\,dV$$

For an adiabatic process $dQ = 0 \quad \therefore \quad dU + P\,dV = 0$ .. (*i*)

Here $U = a + \frac{PV}{b} \quad \therefore \quad dU = \frac{1}{b}(P\,dV + V\,dP)$

Substituting in (*i*), we get

$$\frac{1}{b}(P\,dV + V\,dP) + P\,dV = 0 \quad \text{or} \quad P\,dV + V\,dP = b\,P\,dV = 0$$

or $$(1 + b)\,P\,dV + V\,dP = 0$$

Dividing both dises by *PV*, we get

$$(1+b)\frac{dV}{V} + \frac{dP}{P} = 0$$

Integrating we get; $(1 + b)\log_e V + \log_e P = \text{constant}$

or $$\log_e PV^{(1+b)} = \text{constant or } PV^{(1+b)} = e^{\text{constant}} = \text{constant.}$$

Example 4.17 *Calculate the specific heat of hydrogen gas at constant volume, given that specific heat at constant pressure to be 6.85 cal/gm molecule, density of hydrogen at N.T.P. 0.0899 gm/litre and* $J = 4.18 \times 10^7$ *ergs, Molecular weight of hydrogen = 2.016 gm.* **(Cal. U. 2002)**

Solution. Molecular weight of hydrogen = 2.016 gm

Density of hydrogen at N.T.P. = 0.0899 gm/litre

$\therefore$ Gram molecular volume $V = \frac{2.016}{0.0899} = 22.42 \text{ litre} = 22.42 \times 10^3 \text{ c.c}$

Now $$R = \frac{PV}{T} = \frac{76 \times 13.6 \times 981 \times 22.42 \times 10^3}{273} = 8.33 \times 10^7 \text{ ergs/ mole}$$

$$= \frac{8.33\times10^7}{4.18\times10^7} = 1.99 \text{ cals/ mole}$$

Specific heat at constant pressure $C_p = 6.85$ cal/gm molecule

Again $C_p - C_v = R$

$\therefore$ $C_v = C_p - R = 6.85 - 1.99 = 4.86$ cal/gm molecule

Example 4.18 *A gas occupying one litre at 80 cm pressure is expanded adiabatically to 1190 c.c. If pressure falls to 60 cm. in the process, deduce the value of γ.* **(Kanpur U., 2002)**

Solution. Initial volume V_1 = one litre = 1000 c.c.; Initial pressure P_1 = 80 cm

Final volume V_2 = 1190 cc; Final pressure P_2 = 60 cm

$$\therefore \quad \left(\frac{P_1}{P_2}\right) = \left(\frac{V_2}{V_1}\right)^\gamma \quad \text{or} \quad \log\left(\frac{P_1}{P_2}\right) = \gamma\log\left(\frac{V_2}{V_1}\right)$$

or
$$\gamma = \frac{\log\left(\frac{P_1}{P_2}\right)}{\left(\frac{V_2}{V_1}\right)} = \frac{\log\left(\frac{80}{60}\right)}{\log\left(\frac{1190}{1000}\right)} = \frac{\log 1.333}{\log 1.190} = \frac{0.1248}{0.0755} = 1.65$$

Example 4.19 *Find the efficiency of the Carnot's engine working between the steam point and the ice point.* **(Bangalore 2001)**

Solution. Here
$$T_1 = 273 + 100 = 373 \text{ K}$$
$$T_2 = 273 + 0 = 273 \text{ K}$$
$$\eta = 1 - \frac{T_2}{T_1}$$
$$= 1 - \frac{273}{373} = \frac{100}{373}$$
$$\% \text{ efficiency} = \frac{100}{373}\times100$$
$$= \mathbf{26.81\%}$$

Example 4.20 *Find the efficiency of a Carnot's engine working between 127°C and 27°C. It absorbs 80 cals of heat. How much heat is rejected.* **(Nagpur 2008, 2001, Meerut 2005, 2004, 2003)**

Solution.
$$T_1 = 273 + 127 = 400 \text{ K}$$
$$T_2 = 273 + 27 = 300 \text{ K}$$
$$\eta = 1 - \frac{T_2}{T_1}$$
$$= 1 - \frac{300}{400} = 0.25$$
$$\% \eta = \mathbf{25\%}$$
$$W = n Q_1 = 0.25 \times 80 = 20 \text{ cals.}$$
$$\therefore \quad \text{Heat rejected } Q_2 = Q_1 - W = 80 - 20$$
$$= \mathbf{60 \text{ cals.}}$$

Example 4.21 *A Carnot's engine whose temperature of the source is 400 K takes 200 calories of heat at this temperature and rejects 150 calories of heat to the sink. What is the temperature of the sink? Also calculate the efficiency of the engine.* **(Nagpur 2005/w, Panjab 2005)**

Solution.
$$Q_1 = 200 \text{ Cal} ; \quad Q_2 = 150 \text{ Cal}$$
$$T_1 = 400 \text{ K} ; \quad T_2 = ?$$

$$\frac{Q_1}{T_1} = \frac{Q_2}{T_2} \qquad \text{Efficiency, } \eta = 1 - \frac{T_2}{T_1}$$

$$T_2 = \frac{Q_2}{Q_1} \times T_1 \qquad = 1 - \frac{300}{400} = 0.25$$

$$= \frac{150}{200} \times 400$$

$$\mathbf{T_2 = 300\ K\ ; \qquad \%\ Efficiency = 25\%}$$

Example 4.22 *A Carnot's engine working as a refrigerator between 260 K and 300 K receives 500 calories of heat from the reservoir at the lower temperature. Calculate the amount of heat rejected to the reservoir at the higher temperature. Calculate also the amount of work done in each cycle to operate the refrigerator. (1 calorie = 4.2 Joules).* ***(Kerala 2001)***

Solution. $Q_1 = ?$ $\qquad Q_2 = 500$ Cal

$T_1 = 300$ K $\qquad T_2 = 260$ K

$$\frac{Q_1}{Q_2} = \frac{T_1}{T_2}$$

$$Q_1 = Q_2 \frac{T_1}{T_2} = \frac{500 \times 300}{260} = \mathbf{576.92\ Cal}$$

Work done, $\quad W = Q_1 - Q_2 = 76.92$ Cal

$= 76.92 \times 4.2$ joules

$\mathbf{W = 323.08}$ **joules.**

Example 4.23 *A Carnot's engine whose lower temperature heat-sink is at 27°C has its efficiency 40%. What is the temperature of the heat sources. By how much should the temperature of the source be raised if the efficiency is to be raised to 70%.* ***(Kanpur 2001, Nagpur 2004/w)***

Solution. $T_2 = 273 + 27 = 300$ K, $\eta_1 = \frac{40}{100} = 0.4$, $\eta_2 = \frac{70}{100} = 0.7$

(*i*) **Temperature of the source:**

$$\eta_1 = 1 - \frac{T_2}{T_1} \quad \text{or} \quad \frac{T_2}{T_1} = 1 - \eta_1$$

$$\therefore \qquad T_1 = \frac{T_2}{1-\eta_1} = \frac{300}{1-0.4} = \frac{300}{0.6} = \mathbf{500\ K = 227°C}$$

(*ii*) **Required rise in source temperature:**

$$T_1' = \frac{T_2}{1-\eta_2} = \frac{300}{1-0.7} = \frac{300}{0.3} = 1000\ \text{K} = 727°\text{C}$$

$\therefore$ Required rise in the source temperature

$= T_1' - T_1 = 1000 - 500 =$ **500 K = 500°C**

Example 4.24 *A Carnot's refrigerator takes heat from water at 0°C and discards it to a room temperature at 27°C. 1 kg of water at 0°C is to be changed into ice at 0°C. How many calories of heat are discarded to the room? What is the work done by the refrigerator in this process? What is the coefficient of performance of the machine? (1 cal = 4.2 Joule).* ***(Patna 2000)***

Solution. Here $\quad Q_1 = ?\ Q_2 = 1000 \times 80 = 80{,}00$ cal

$T_1 = 300$ K,

$T_2 = 273$ K

(*i*) **Heat rejected to the room:**

$$\frac{Q_1}{Q_2} = \frac{T_1}{T_2}$$

$$\therefore \quad Q_1 = \frac{T_1}{T_2} Q_2 = \frac{300}{273} \times 80,000$$

$$Q_1 = 87,900 \text{ cal}$$

(*ii*) **Work done by the Refrigerator:**

$$W = J(Q_1 - Q_2)$$

$$= 4.2\,(87,900 - 80,000)$$

$$= 4.2 \times 7900$$

$$\mathbf{W = 3.183 \times 10^4 \text{ Joules}}$$

(*iii*) **Coefficient of performance:**

$$\beta = \frac{Q_2}{Q_1 - Q_2}$$

$$= \frac{80,000}{87,900 - 80,000} = \frac{80,000}{7900}$$

$$\beta = \mathbf{10.13}$$

Example 4.25 *A Carnot engine whose low temperature reservoir is at 7°C has an efficiency of 50%. It is desired to increase the efficiency to 70%. By how many degrees should the temperature of the high temperature reservoir be increased?*

Solution. In the first case

$$\eta = 50\% = 0.5,\ T_2 = 273 + 7 = 280 \text{ K}.$$

$$T_1 = ?$$

$$\eta = 1 - \frac{T_2}{T_1}$$

or $$0.5 = 1 - \frac{280}{T_1}$$

or $$T_1 = 560 \text{ K}$$

In the second case

$$\eta' = 70\% = 0.7,$$

$$T_2 = 280 \text{ K},$$

$$T_1' = ?$$

$$\eta' = 1 - \frac{T_2}{T_1'}$$

$$0.7 = 1 - \frac{280}{T_1'}$$

or $$T_1' = 840 \text{ K}$$

Increase in temperature = 840 – 560 = **280 K**

Example 4.26 *An inventor claims to have developed an engine working between 600 K and 300 K capable of having an efficiency of 52%. Comment on his claim.*

(*Nagpur 2009, Lucknow 2003, Purvanchal 2002*)

Solution. The efficiency of a Carnot engine working between 600 K and 300 K

$$\eta = 1 - \frac{T_1}{T_2}$$

$$= 1 - \frac{300}{600} = 0.5 = 50\%$$

The efficiency claimed is 52%. It means that the efficiency of the engine developed is more than the efficiency of Carnot's engine working between the same two temperature limits. But no engine can have an efficiency more than a Carnot's engine. Hence his claim is invalid.

Example 4.27 *Calculate the work done when a gram molecule of an ideal gas expands isothermally at 27°C to double its original volume. [R = 8.3 J/deg mole and $\log_{40}{}^2 = 0.3010$]*

(Rohilkhand 2001, Meerut 2003)

Solution.

Work done $= \int_{V_1}^{V_2} P\,dV$

But $PV = RT$

$$P = \frac{RT}{V}$$

$$W = RT \int_{V_1}^{V_2} \left(\frac{dV}{V}\right)$$

or $W = RT \log_e \left(\frac{V_2}{V_1}\right)$

Here $\frac{V_2}{V_1} = 2$

$$R = 8.3 \text{ J/deg mole}$$

$$T = 273 + 27 = \mathbf{300\ K}$$

$$W = 8.3 \times 300 \times \log_e 2 = 8.3 \times 300 \times 3.14 \log_{10} 2$$

$$= 8.3 \times 300 \times 3.14 \times 0.3010$$

$$W = \mathbf{2353.4\ J}$$

Example 4.28 *A Carnot engine has an efficiency of 30% when the temperature of the sink is 27°C. What must be the change in temperature of the source to make its efficiency 50%.*

(Purvanchal 2005, Calcutta 2001)

Solution. In the first case,

$$\eta = 30\% = 0.3,\ T_2 = 300\text{ K},\ T_1 = ?$$

$$\eta = 1 - \frac{T_2}{T_1}$$

$$0.3 = 1 - \frac{300}{T_1}$$

$$\mathbf{T_1 = 428.57\ K}$$

In the second case,

$$\eta' = 50\% = 0.5,\ T_2 = 300\text{ K},\ T_1' = ?$$

$$\eta' = 1 - \frac{T_2}{T_1'}$$

$$0.5 = 1 - \frac{300}{T_1'}$$

$$T_1' = 600\text{ K}$$

Increase in temperature $= \mathbf{171.43\ K}$

Example 4.29 *A 100 k-W engine is operating between 117°C and 17°C. Calculate*

(i) the amount of heat absorbed

(ii) the amount of heat rejected and

(iii) the efficiency of the engine.

(Osmania 2005)

Solution. Given: $W = 100\text{ kW} = 10^5\text{ watts} = 10^5\text{ J/s}$

$$T_1 = 117°\text{C} = 400\text{ K}$$

$$T_2 = 17°\text{C} = 300\text{ K}$$

Efficiency,
$$\eta = 1-\left(\frac{T_2}{T_1}\right)$$

$$\eta = 1-\frac{300}{400}$$

$$\eta = 0.25 = 25\%$$

Also
$$\frac{Q_1}{Q_2} = \frac{T_1}{T_2} = \frac{400}{300} = \frac{4}{3}$$

or
$$\left(\frac{Q_1}{Q_2}\right)-1 = \left(\frac{4}{3}\right)-1$$

$$\frac{Q_1-Q_2}{Q_2} = \frac{1}{3}$$

But
$$W = Q_1 - Q_2 = 10^5\text{ J/s}$$

$\therefore$
$$\frac{10^5}{Q_2} = \frac{1}{3}$$

$$Q_2 = 3\times 10^5\text{ J/s}$$

But
$$\frac{Q_1}{Q_2} = \frac{4}{3}$$

$$Q_1 = \left(\frac{4}{3}\right)Q_2 = \left(\frac{4}{3}\right)\times 3\times 10^5$$

or
$$Q_1 = 4\times 10^5\text{ J/s}$$

Hence (*i*) Amount of heat absorbed per second

$$= 4\times 10^5\text{ J/s}$$

(*ii*) Amount of heat rejected per second

$$= 3\times 10^5\text{ J/s}$$

(*iii*) Efficiency of the engine = 0.25 = 25%

Example 4.30 *A carnot engine has its source at 100°C and its sink is maintained at a constant temperature by means of ice at 0°C. If it is working at the rate of 100 watts, how much ice will melt in one minute?* **(*Rohilkhand 2002*)**

Solution. $P = 100$ watts = 100 J/s

Work done in one minute = 100 × 60 = 6000 J

$$T_1 = 100°\text{C} = 373\text{ K}$$

$$T_2 = 0°\text{C} = 273\text{ K}$$

$$\frac{Q_1-Q_2}{Q_1} = \frac{T_1-T_2}{T_1}$$

$$\frac{W}{Q_1} = \frac{T_1-T_2}{T_1} \quad ...(i)$$

$$\frac{Q_1}{Q_2} = \frac{T_1}{T_2} \quad ...(ii)$$

From (*i*) and (*ii*)

$$\frac{W}{Q_2} = \frac{T_1 - T_2}{T_2}$$

$$Q_2 = \left[\frac{WT_2}{T_1 - T_2}\right] = \frac{6000 \times 273}{100}$$

$$Q_2 = 16380 \text{ J}$$

Suppose, *m* kg of ice melts

$$Q_2 = mL$$

$$\boldsymbol{m = \frac{Q_2}{L} = \frac{16380}{80 \times 10^3 \times 4.2} = 0.04875 \text{ kg}}$$

Example 4.31 *Two Carnot engines X and Y are operating in series. The first one X receives heat at 1200* K *and rejects to a reservoir at temperature T* K. *The second engine Y receives the heat rejected by X and inturn rejects to a heat reservoir at 300* K. *Calculate the temperature T for the situation when.*

(*i*) *The work outputs of two engines are equal.*

(*ii*) *The efficiency of two engines are equal.* **(*Nagpur 2001*)**

Solution.

(*i*) When work outputs are equal.

For the first engine,

$$W_1 = Q_1 - Q_2$$

For the second engine,

$$W_2 = Q_2 - Q_3$$

Here

$$W_1 = W_2$$

$$Q_1 - Q_2 = Q_2 - Q_3$$

$$\left(\frac{Q_1}{Q_2}\right) - 1 = 1 - \left(\frac{Q_3}{Q_2}\right)$$

Also

$$\frac{Q_1}{Q_2} = \frac{1200}{T} \text{ and } \frac{Q_2}{Q_3} = \frac{T}{300} \text{ or } \frac{Q_3}{Q_2} = \frac{300}{T}$$

$\therefore$

$$\left(\frac{1200}{T}\right) - 1 = 1 - \frac{300}{T}$$

or

$$\boldsymbol{T = 750 \text{ K}}$$

(*ii*) When efficiencies are equal.

$$\eta_1 = 1 - \left(\frac{Q_2}{Q_1}\right)$$

$$\eta_2 = 1 - \left(\frac{Q_3}{Q_2}\right)$$

$$\eta_1 = \eta_2$$

$$1 - \left(\frac{Q_2}{Q_1}\right) = 1 - \left(\frac{Q_3}{Q_2}\right)$$

$$1 - \left(\frac{T}{1200}\right) = 1 - \left(\frac{300}{T}\right)$$

$$\frac{T}{1200} = \frac{300}{T}$$

$$\boldsymbol{T = 600 \text{ K}}$$

Example 4.32 *A Carnot engine working between a source at temperature 27°C and a sink at – 73°C delivers 300 calories of heat to the latter in one cycle. Calculate the work performed in joules by the engine per cycle.* **(*Mumbai 2001*)**

Solution. Here,

$$T_1 = 27 + 273 = 300 \text{ K}$$
$$T_2 = -73 + 273 = 200 \text{ K}$$
$$Q_2 = 300 \text{ calories.}$$
$$Q_1 = ?$$
$$\frac{Q_1}{Q_2} = \frac{T_1}{T_2}$$
$$\frac{Q_1}{300} = \frac{300}{200}$$
$$Q_1 = 450 \text{ calories}$$

Work performed per cycle,

$$W = Q_1 - Q_2$$
$$= 450 - 300$$
$$= 150 \text{ calories}$$
$$= 150 \times 4.2 \text{ J}$$
$$= \mathbf{630 \ J}$$

Example 4.33 *A Carnot's engine is operated between two reservoirs at temperatures of 450* K *and 350* K*. If the engine receives 1000 calories of heat from the source in each cycle, calculate the amount of heat rejected to the sink in each cycle. Calculate the efficiency of the engine and the work done by the engine in each cycle. (1* calorie = *4.2* joules) **(*Madurai Kamraj 2001, Nagpur 2007*)**

Solution.

$$T_1 = 450 \text{ K} ; \qquad T_2 = 350 \text{ K}$$
$$Q_1 = 1000 \text{ cal} ; \qquad Q_2 = ?$$
$$\frac{Q_2}{Q_1} = \frac{T_2}{T_1}$$
$$Q_2 = Q_1 \times \frac{T_2}{T_1}$$
$$= \frac{1000 \times 350}{450} = 777.77 \text{ cals}$$
$$\eta = 1 - \frac{T_2}{T_1} = 1 - \frac{350}{450} = \frac{100}{450}$$
$$= 0.2222$$
$$\% \text{ efficiency} = \mathbf{22.22\%}$$

Work done in each cycle

$$= Q_1 - Q_2$$
$$= 1000 - 777.77$$
$$= 222.23 \text{ cal}$$
$$= 222.23 \times 4.2 \text{ joules}$$
$$= \mathbf{933.33 \ joules}$$

Example 4.34 *Efficiency of a Carnot's cycle changes from $\frac{1}{6}$ to $\frac{1}{3}$ when source temperature is raised by 100 K. Calculate the temperature of the sink.* **(*Purvanchal 2004*)**

Solution.

$$\eta = 1 - \frac{T_2}{T_1} = \frac{1}{6}$$

or $$\frac{T_2}{T_1} = 1 - \frac{1}{6} = \frac{5}{6}$$

or $$T_1 = \frac{6}{5} T_2 \qquad ...(i)$$

or $$\eta' = 1 - \frac{T_2}{T_1 + 100} = \frac{1}{3}$$

or $$\frac{T_2}{T_1 + 100} = \frac{2}{3}$$

or $$T_2 = \frac{2}{3}(T_1 + 100) = \frac{2}{3} T_1 + \frac{2}{3} \times 100$$

or $$T_2 = \frac{2}{3} T_1 + \frac{200}{3}$$

$$= \left(\frac{2}{3} \times \frac{6}{5} T_2\right) + \frac{200}{3}$$

$$= \frac{4}{5} T_2 + \frac{200}{3}$$

or $$\frac{T_2}{5} = \frac{200}{3}$$

or $$T_2 = \frac{1000}{3}$$

$$= \mathbf{333.3\ K}$$

Example 4.35 *A heat engine working between two temperatures could theoretically convert one-eighth of the heat supplied into work. If the lower temperature is reduced by 95 °C, the theoretical efficiency would be doubled. Find the initial temperatures.* **(*Madras 2005, 2004/w*)**

Solution. (*i*) As the engine converts $\frac{1}{8}$ th of the heat input in to work, its efficiency $\eta = \frac{1}{8}$.

Let T_1 and T_2 be the temperatures of the source and sink respectively, then efficiency

$$\eta = 1 - \frac{T_2}{T_1}$$

or $$\frac{1}{8} = 1 - \frac{T_2}{T_1}$$

or $$\frac{T_2}{T_1} = 1 - \frac{1}{8} = \frac{7}{8}$$

(*ii*) When the temperature of the sink is reduced by 95 °C *i.e.*, 95 K, then

Efficiency $$\eta' = 1 - \frac{T_2 - 95}{T_1}$$

As the efficiency is doubled, $n' = 2 \times \frac{1}{8} = \frac{1}{4}$

$$\therefore \quad \frac{1}{4} = 1 - \frac{T_2}{T_1} + \frac{95}{T_1}$$

$$= 1 - \frac{7}{8} + \frac{95}{T_1}$$

or $$\frac{95}{T_1} = \frac{1}{4} - 1 + \frac{7}{8} = \frac{1}{8}$$

or $$T_1 = 95 \times 8$$
$$= \mathbf{760\ K}$$

and $$T_2 = \frac{7}{8} \times T_1 = \frac{7}{8} \times 760$$
$$= \mathbf{665\ K}$$

Example 4.36 *A refrigerator works between 260 K and 300 K when it takes 2100 Joules of heat from the freezer. What is the heat rejected to the atmosphere and what is the work done for this ?* **(Kerala 2001)**

Solution. Quantity of heat absorbed from the cold body = Q_2

Temperature of cold body (sin K) = T_2 = 260 K

Quantity of heat given to the hot body = Q_1

Temperature of the hot body (source) T_1 = 300 K

For a refrigerator, co-efficient of performance

$$\beta = \frac{Q_2}{Q_1 - Q_2} = \frac{T_2}{T_1 - T_2} = \frac{260}{300 - 260} = \frac{260}{40} = 6.5$$

Now $$Q_2 = 2100 \text{ J}$$

$$\therefore \quad \frac{2100}{Q_1 - 2100} = 6.5 \quad \text{or} \quad Q_1 = 2100 + \frac{2100}{6.5} = 2423 \text{ J}$$

$\therefore$ Work done for this $W = Q_1 - Q_2 = 2423 - 2100 = 323$ J.

Example 4.37 *An ideal refrigerator has to transfer an average of 200 Joule of heat per second from – 10°C to 27°C. Calculate the average power consumed.* **(P.U., 2003)**

Solution. $$\beta = \frac{Q_2}{W} = \frac{T_2}{T_1 - T_2}$$

$$\therefore \quad W = Q_2 \frac{T_1 - T_2}{T_2} = \frac{200\,(300 - 263)}{263} = 28.13 \text{ J/s}$$
$$= 28.13 \text{ Watt}$$

Example 4.38 *An ideal refrigerator takes heat from a body at –13°C and rejects it to a body at 27°C. Find its co-efficient of performance.* **(P.U. 2007)**

Solution. $$T_2 = 273 - 13 = 260 \text{ K}$$
$$T_1 = 273 + 27 = 300 \text{ K}$$
$$\therefore \quad \beta = \frac{T_2}{T_1 - T_2} = \frac{260}{40} = 6.5$$

Example 4.39 *A motor in a refrigerator has powe output of 200 watt. If the freezing compartment is at 270 K and outside air at 300 K, assuming ideal efficiency, what is the maximum amount of heat that can be extracted from the freezing compartment in 10 minutes ?* **(Banglore 2001)**

Solution. Temperature of the freezing compartment (sink) T_2 = 270 K

Temperature of outside air (source) T_1 = 300 K

Work done by the refrigerator motor per second W = 200 Joules

Co-efficient of performance $\beta = \frac{Q_2}{W} = \frac{T_2}{T_1 - T_2}$

where Q_2 is the heat extracted from the sink (freezing compartment) per second.

$$\therefore \quad Q_2 = W\beta = W\frac{T_2}{T_1 - T_2} = 200 \times \frac{270}{300 - 270} = 1800 \text{ Joules}$$

Maximum amount of heat that can be extracted from the freezing compartment in 10 minutes

$$= 1800 \times 10 \times 60 = 1.08 \times 10^6 \text{ Joules.}$$

EXERCISE

4.1 Explain the term 'Heat' and 'Temperature' ***(Nagpur 2009, Madras 2005)***

4.2 Explain thermodynamic system and thermodynamic variables. ***(Nagpur 2009)***

4.3 Explain the terms:

(*i*) open system

(*ii*) closed system

(*iii*) isolated system

4.4 State and explain zeroth law of thermodynamics. What is its importance. On the basis of this law introduce the concept of temperature.

(Nagpur 2008, 2007, Madras 2004/w, Rohilkhand 2001, Ranchi 2004, Calcutta 2003, 2005, Lucknow 2003, Purvanchal 2005, 2004, 2003, Osmania 2004, M.S. Uni. Tamil Nadu 2007)

4.5 What is meant by thermodynamic equilibrium. What are the basic requirements to be strictly followed? ***(Purvanchal 2007, Rohilkhand 2002, G.N.D.U.2000)***

4.6 What do you mean by quasi-static process? Can it be achieved in practice? State its importance? ***(Punjab 2000, Meerut 2003, Calcutta 2003, Purvanchal 2004)***

4.7 Show that heat and work both are path dependent functions with the help of indicator diagram. ***(Purvanchal 2007, 2003)***

4.8 What is internal energy of a system? "Internal energy is a state frunction and not a path function", Explain. ***(Agra 2005, Nagpur 2009, Kanpur 2002, Lucknow 2002, G.N.D.U. 2004)***

4.9 Explain the terms:

(*i*) Isochoric process ***(Rohilkhand 2003, Locknow 2000)***

(*ii*) Isobaric process ***(Rohilkhand 2003, Lucknow 2004)***

(*iii*) Cyclic process ***(Kerala 2001)***

(*iv*) Isothermal process

(Madras 2004/w, Rohilkhand 2003, Mysore 2004, Purvanchal 2004, Lucknow 2004)

(*v*) Adiabatic process

(Madras 2004/w, Rohilkhand 2003, Mysore 2004, Purvanchal 2004, Lucknow 2004)

4.10 Show that change in the internal energy of a non-isolated system is always zero. ***(Nagpur 2000)***

4.11 What is indicator diagram? State its importance. ***(Kuvempa 2005, Garhwal 2005)***

4.12 Explain the concepts of reversible and irreversible processes.

(Nagpur 2008, Agra 2006, 2005, 2003,2000, Madras 204/w, Kanpur 2001, Garhwal 2000)

4.13 Define a thermodynamic system. Distinguish between open and closed system.

4.14 Write notes on "Thermal equilibrium and zeroth law of thermodynamics." ***(Rohilkhand2003)***

4.15 What is equation of state? Give its limitations. ***(Garhwal 2001)***

4.16 What do you understand by variables of state of a system and equation of state?

4.17 The conduction of heat from a hot body to a cold body is reversible or irreversible process? Explain. ***(Kanpur 2000)***

4.18 Explain different forms of internal energy and show that the internal energy of a system is a function of state of the system. ***(Nagpur 2004 – Winter)***

4.19 State first law of thermodynamics. Give its physical significance. What are the limitations of first law? ***(Rohilkhand,2003, Nagpur 2009, 2008, 2007, 2004/s, 2005/w, 2001, Kerala 2001, Meerut 2001, 2000, 2002, Madras 2003, Bangalore 2005, Purvanchal 2003)***

4.20 Give applications of the first law to (*a*) isobaric (*b*) adiabatic processes and (*c*) isothermal process ***(Nagpur 2008, 2004/s, Purvanchal 2003)***

4.21 Explain how first law of thermodynamics leads to the concept of internal energy.

4.22 What is the first law of thermodynamics? Show that this law is a particular form of general law of conservation of energy. ***(Purvanchal 2004, 2007, Agra 2006, Rohilkhand 2000)***

4.23 What do you mean by isobaric process and isochoric process? Calculate the work done by an ideal gas in these processes. ***(Kanpur 2002, Utkal 2003, G.N.D.U. 2004)***

4.24 Obtain an expression for the external work done, when one mole of an ideal gas is suddenly expanded. ***(Agra 2005)***

4.25 From the first law of thermodynamics, prove that $C_P - C_V = R$. ***(Agra 2004, Garhwal 2004)***

4.26 Calculate the work done by an ideal gas during an isothermal expansion. ***(Rohilkhand 2002, Nagpour 2008, 2005/w, 2005/s, Kuvempa 2005, Mysore 2004, Bangalore 2005)***

4.27 Derive an expression for work done during an adiabatic process. ***(Kerala 2001, Nagpur 2001, G.N.D.U. 2004, Garhwal 2005, Purvanchal 2004, Alageppa 2005, M.S. uni, Tamil Nadu 2007, Bangalore 2005, Kuvempa 2005)***

4.28 Show that the slope of an adiabatic is γ times the slope of the isothermal. ***(Nagpur 2001, 2003, Kerala 2001)***

4.29 Prove that $PV^{\gamma} =$ constant in an adiabatic transformation. ***(Madras 2005/Summer, Rohilkhand 2001, Nagpur 2003)***

4.30 Derive the relation between volume and pressure of a gas undergoing adiabatic changes. ***(Bangalore 2000, G.N.D.U. 2000, 2001, Calcutta 2003)***

4.31 Why does the temperature of a gas drop when it is subjeted to adiabatic expansion? Explain.

4.32 Show that the work done during an adiabatic process depends only upon the initial and final temperatures. ***(Rohilkhand 2001 Imp.)***

4.33 Explain why the gases have two specific heats? Explain why the specific heat at constant pressure C_p is greater than that C_V; the specific heat at constant volume. Prove $C_P - C_V = R$. ***(Madurai 2003, Garhwal 2005, Rohilkhand 2000, 2001, 2002, 2003, Lucknow 2001, Kerala 2001)***

4.34 State the first law of thermodynamics. Express it mathematically and explain its physical significance. ***(Meerut 2001, Nagpur 2009, 2008, 2003, 2001 Kerala 2001, G.N.D.U. 2001, Alagappa 2005)***

4.35 Distinguish between isothermal and adiabatic changes. Show that for an adiabatic change in a perfect gas

$$PV^{\gamma} = \text{constant.}$$ ***(Delhi 2000)***

4.36 For a quasi-static adiabatic process of an ideal gas, prove $\dfrac{T}{(P)^{(\gamma-1)/\gamma}} =$ constant.

(Bangalore 2000, Utkal 2003, Purvanchal 2003, Madras 2004/w)

4.37 Show that the equation of an adiabatic change is $V^{\gamma-1}\, T =$ constant. ***(Madras 2004/Summer, Bangalore 2000)***

4.38 Show that the work done in adiabatic expansion of an ideal gas from a state (P_1, V_1) a state (P_2, V_2) is given by

$$W = \frac{1}{\gamma - 1}[P_1 V_1 - P_2 V_2]$$ ***(Meerut 2004)***

4.39 What is the principle of adiabatic demagnetisation? ***(Nagpur 2010, Pune 2010, Madras 2004/w)***

4.40 Define specific heat at constant volume and at constant pressure. Why a gas has two specific heats? Explain. ***(Alagappa 2003, 2004, Madras 2005, Madurai Kamraj 2003)***

4.41 Distinguish between isothermal and adiabatic process. ***(Mysore 2004/s, Bangalore 2004)***

4.42 Draw P – V diagrams representing isothermal and adiabatic processes. ***(Mysore 2004/w, Bangalore 2004)***

4.43 Obtain an expression for the ratio of heat capacities C_P/C_V. ***(Gujrat 2004, 2005, Alagappa 2004)***

4.44 Derive an expression for cooling produced by adiabatic demagnetisation of a paramagnetic salt. ***(Nagpur 2010, Punjab 2005, Meerut 2003)***

4.45 Why does a rubber string heat up on stretching? ***(Punjab 2005)***

4.46 A metallic wire is stretched reversibly and adiabatically. Calculate the temperature change in the process.
Show that when a rubber band is quickly stretched, it warms up. ***(Agra 2000)***

4.47 Prove that adiabatic elasticity of a gas is γ times the isothermal elasticity. or $\frac{E_{ad}}{E_{iso}} = \gamma$.
(Calcutta 2003, G.N.D.U. 2001, Madurai Kamraj 2003)

4.48 Derive thermodynamic relation for stretching of a wire treating it as a thermodynamic system. Interpret the relation. ***(Mumbai 2005)***

4.49 Define efficiency of a heat engine. Obtain an expression for the efficiency of a reversible Carnot's engine with a perfect gas as the working substance. Mention the effective ways to increase its efficiency. ***(Garhwal 2001, 1999, Rohilkhand 2001 Imp., Meerut 2005, 2003, Bangalore 2005, Nagpur 2004–Winter 2005, 2009)***

4.50 What is meant by a reversible process? Describe Carnot's cycle and show that all reversible engines working between the same two temperatures have the same efficiency. ***(Kanpur 1997, Nagpur 2009, 2007, 2001)***

4.51 Calculate the work done in a Carnot's cycle of operations. Deduce the efficiency of a Carnot's engine in terms of the temperatures between which it works. ***(Agra 2004, Nagpur 2007, 2006, H.P.U. 2001)***

4.52 Derive an expression for the efficiency of an ideal heat engine, working between the temperatures T_1 and T_2 K. Prove that the efficiency of a reversible engine is maximum for the given source and sink.
(Meerut 2004, 2002, 2001, Gawahati 2000, Pbi 2000, Kerala 2001, Punjab 2001, 1999)

4.53 State the Second Law of thermodynamics. Describe Carnot's cycle and deduce the efficiency of an ideal heat engine.
(Kanpur 2001, Allahabad 1999, Meerut 1996, Purvanchal 2005, Bangalore 2004, Alagappa 2004, Garhwal 2005, 2004)

4.54 Describe Carnot's cycle and obtain an expression for the efficiency of an ideal heat engine working between two temperature T_1 and T_2
(Rohilkhand 2002, 2003, Meerut 2000, Nagpur 2007)

4.55 Show that no engine working between two given temperatures can be more efficient than a reversible Carnot's engine working between the same two temperatures.
(Purvanchal 2003, Calcutta 2004)

4.56 What is the principle used in the working of a refrigerator. Define coefficient of performance. Is it greater than 1? Explain. ***(Garhwal 2001)***

4.57 Give two versions of second law of thermodynamics.
(Nagpur 2005/s, Meerut 2001, Mysore 2004/s, 2004/w)

4.58 State and prove Carnot's theorem. Show that the temperature below zero Kelving is not possible. ***(Kanpur 2001, Rohilkhand 2001, 2001 Imp., 2003. Nagpur 2010, 2007, 2005/w. Alagappa 2004, Meerut 2001, Bangalore 2005, Purvanchal 2007, 2005, 2002, Agra 2006)***

4.59 Explain the concepts of reversible and irreversible processes. Obtain an expression for the efficiency of a reversible Carnot engine with a perfect gas as the working substance. Show that for the same pair of working temperatures, the efficiency of all reversible engines is the same and is the maximum value attainable. ***(Garhwal 2001)***

4.60 State Carnot's Theorem and show that it is a necessary consequence of Second Law of Thermodynamics. Prove that the efficiency of a Carnot engine using an ideal gas as a working substance is $\eta = \frac{T_1 - T_2}{T_1}$. ***(Mysore 2004, Bangalore 2004, Kuvempa 2005, Agra 2003)***

4.61 What are Kelvin–Planck and Clausius statement of second law of thermodynamics? Prove that they are correct in terms of principle of increase of entropy
(Utkal 2003, Calcutta 2003, G.N.D.U. 2001, Calicut 2004)

4.62 State the second law of thermodynamics in atleast two distinctly different ways. Are these statements equivatent to each other? Justify your answer.
(Agra 2008, Kanpur 2002, Alagappa U. 2004)

4.63 Prove the equivalence of the Clausius and Kelvin–Planck's statement of the second law of thermodynamics. ***(Meerut 2005, Calcutta 2005, 2003)***

4.64 Describe with necessary theory, the construction and working of a Diesel engine. Explain its merits over otto engine. ***(Madras 2004/w, 2005, Calcutta 2004)***

4.65 Explain Clement and Desorme's method of determining the ratio of two specific heat capacities of a gas. ***(Alagappa 2004)***

4.66 Discuss with necessary theory the construction and working of a Otto engine. ***(Madras 2004)***

4.67 Write short notes on:

(*i*) Carnot cycle ***(Meerut 2001, 2000, Rohilkhand 2000)***

(*ii*) Carnot's theorem ***(Patna 2005)***

(*iii*) Efficiency of Carnot's Heat engine. ***(Alagappa 2005, Patna 2005)***

(*iv*) Second law of thermodynamics.
(Agra 2000, Rohilkhand 1999, 2001 Imp., Kanpur 1995)

(*v*) Zeroth law of thermodynamics. ***(Kanpur 2002, Rohilkhand 2000)***

(*vi*) First law of thermodynamics. ***(Meerut 2001, Patna 2004)***

(*vii*) Internal energy.

(*viii*) Quasi-static process. ***(Rohilkhand 2000)***

(*ix*) External and Internal work.

(*x*) Thermodynamic equilibrium.

(*xi*) Reversible and Irreversible process. ***(Punjab 2005, Kuvempa 2005, Ranchi 2005, Calcutta 2003, 2004, Madras 2005, Autonomus)***

4.68 (*a*) What is thermal equilibrium and thermodynamic equilibrium? ***(G.N. D.U.2000)***

(*b*) State zeroth law of thermodynamics and explain its significance. On the basis of this law introduce the concept of temperature.

(Meerut U. 2006, 2005; Cal. U. 2003, Luck U. 2001,Vid. S.U. 2002)

or Explain the concept of temperature on thermodynamic considerations. ***(Osm. U. 2004)***

4.69 (*a*) (*i*) Explain the terms adiabatic and isothermal changes OR Distinguish between adiabatic and isothermal process. ***(Bang. U. 2002; G.N.D.U. 2008, 2004, Kan. U. 2003 Nagpur U. 2003; Cal. U.2000, Burd U. 2002)***

(*ii*) What is adiabatic expansion? ***(Meerut U. 2007)***

(*iii*) What is adiabatic compression? ***(G.N. D.U. 2008)***

(*b*) Show that for an adiabatic change in a perfect gas PV^{γ} = constant, where γ is the ratio of specific heat at constant pressure and constant volume respectively.

(G.N.D.U. 2002, 2001, Bang. U. 2000, Cal. U., 2003, Gharwal U. 2002; A.U. 2002, Nagpur U. 2003, Kan. U. 2004, Burd U. 2005)

4.70 (*a*) Derive a relation between volume and temperature for a perfect gas during an adiabatic transformation. ***(Bang. U. 2000)***

(*b*) Air at N.T.P. is compressed adiabatically to half its volume. Calculate change in its temperature. **[Ans.** 122.9 °C**]**

(*c*) Can work be converted completely into heat. Discuss.

4.71 Discuss Carnot's reversible heat engine. What is Carnot's cycle. Describe. Show how the work done in each operation is represented on a pressure volume diagram. Calculate the work done per cycle. Explain the term efficiency of a heat engine and deduce an expression for the efficiency of Carnot's reversible heat engine (using one mole of ideal gas as working substance)

4.72 What are the characteristics of Carnot's engine? Under what conditions is it 100% efficient? What is the source, sink and working substance in a steam engine?

(P.U. 2007, 2006, 2001, G.N.D.U. 2008, 2001, Meerut U. 2005, 2002, 2001, 2001, Kerala U. 2001; M.D.U. 2000; Pbi U. 2000. Gauhati U. 2000)

4.73 (*a*) The co-efficient of performance of a refrigerator becomes infinite when the temperatures of the two bodies become equal. Why? ***(P.U. 2005)***

(*b*) A refrigerator works better in winter than in summer. Explain why ? ***(Pbi U. 2005)***

4.74 If the door of a refrigerator is kept open in the hall, will it make the hall warm or cool? ***(G.N.D.U. 2000)***

4.75 Why is defrosting done in a refrigerator? ***(Nagpur U. 2002)***

4.76 A certain volume of gas at 10^7 Newton per metre^2 pressure expands isothermally till its volume is three fold and then adiabatically till its volume is again three fold. Calculate the final pressure of the gas (γ =1.5) **[Ans.** 1.925×10^6 N/m^2**]** ***(Kan. U. 2002)***

4.77 A monoatomic ideal gas of volume 1 litre and pressure 8 atmosphere undergoes adiabatic expansion untill the pressure drops to 1 atmosphere. What is the final volume and how much work is done? Given : 1 atms. = 10^5 N/m^2. **[Ans.** 3.47×10^{-3} m^3, 686 Joules**]**

4.78 A motor car tyre has a pressure of 3 atmospheres at the room temperature of 27°C. If the tyre suddenly bursts what is the resulting temperature? **[Ans.** 218.6 K = – 54.4°C**]** ***(Ranchi 2000)***

4.79 A quantity of air (γ = 1.4) at 27°C is compressed suddenly to $\frac{1}{4}$ of its original volume. Find the final temperature.

[Ans. 522.3 K = 249.3°C**]** ***(Rohilkhand 2001, Alagappa 2004, Nagpur 2009)***

4.80 A quantity of air at 27°C and atmospheric pressure is suddenly compressed to $\frac{1}{5}$ of its original volume. Find (*i*) the final pressure and (*ii*) the final temperature.

[Ans. (*i*) 8.29 atmospheres **(*ii*)** 571.1 K = 298.1°C**]** **(*Agra 2001*)**

4.81 A certain mass of an ideal gas at 27°C temperature and 8 atmospheric pressure, is expanded suddenly to 4 times of its volume. Find final pressure and final temperature. (Given: $\gamma = 1.5$)

(*Nagpur 2009, 2004/w, 2003*)

4.82 A carnot engine has an efficiency of 30% when the temperature of the sink is 27°C. Find the temperature of the source. **[Ans.** 428.6 K**]** **(*Madras 2005 – Summer*)**

4.83 The temperature of 5 kg of air is raised by 1°C at constant volume. Calculate the increase in its internal energy. Given $C_p = 999 \text{ J kg}^{-1} \text{ K}^{-1}$, $\gamma = \frac{5}{3}$· **[Ans.** 2997 J**]** **(*Madras 2004 – Winter*)**

4.84 If the temperature of the sink is 20°C and that of the source 250°C, calculate the efficiency of the Carnot's engine. **[Ans.** 43.98%**]** **(*Madras 2004 – Summer*)**

4.85 Initial pressure of an ideal gas at 30 °C is 'P'. Calculate the rise in temperature at which the pressure suddenly increases to 10 times its original pressure. (Given: $\gamma = 1.4$).

(*Nagpur 2005/w*)

[Hint: $T_1 = 303$ K, $P_2 = 10$, P_1, $\gamma = 1.4$,

$$T_2 = T_1\left(\frac{P_2}{P_1}\right)^{\gamma-1/\gamma} = 303\times(10)^{\frac{0.4}{1.4}} = 303\times10^{\frac{2}{7}} = 303\times10^{0.2875} = 585\ K$$

$\therefore$ Rise in temperature = 585 – 303 = 282 K.]

4.86 Dry air at N.T.P. (normal temperature and pressure) is compressed adiabatically to one-third of its original volume. Calculate the resulting pressure. Given that the ratio C_p/C_v of air is 1.40.

[Ans. 4.655 atm**]** **(*Bangalore 2001*)**

4.87 A Carnot's engine whose temperature of the source is 400 K takes 500 calories of heat at this temperature and rejects 400 calories of heat to the sink. What is the temperature of the sink? Calculate the efficiency of the engine. **[Ans. (*i*)** 320 K, **(*ii*)** 20%**]** **(*Mysore* 2001)**

4.88 A Carnot's engine is operated between two reservoirs of 500 K and 400 K. If the engine receives 2000 calories of heat from the source in each cycle, calculate (*a*) the amount of heat rejected to the sink in each cycle, (*b*) the efficiency of the engine and (*c*) the work done by the engine in each cycle in (*i*) Joule (*ii*) Kilo-Watt hours

[Ans. (*a*) 1600 calories, **(*b*)** 20%, **(*c*) (*i*)** 1780 Joules **(*ii*)** 4.94×10^4 kWh**]** **(*Nagpur 2005/w*)**

4.89 A Carnot's engine takes in 1000 kilocalories of heat a reservoir at 627°C and exhausts it to a sink at 27°C. (*a*) What is its efficiency, (*b*) How much work does it perform? Express it in (*i*) ergs (*ii*) kilo-watt hours (*iii*) electron volts.

[Ans. (*a*) 66.67%, **(*b*) (*i*)** 14×10^{12} ergs **(*ii*)** 0.39 kWh **(*iii*)** 8.75×10^{24} eV**]** **(*Rajasthan 2001*)**

4.90 A Carnot's engine working as a refrigerator between 250 K and 300 K receives 1000 calories of heat from the reservoir at the lower temperature. (*i*) Calculate the amount of heat rejected to the reservoir at the higher temperature. (*ii*) Calculate also the amount of work done in each cycle to operate the refrigerator. **[Ans. (*i*)** 1200 cal, **(*ii*)** 840 joules**]** **(*Calcutta 2000*)**

4.91 A reversible engine works between two temperatures whose difference is 100 °C. If it absorbs 746 Joule of heat from the source and gives 546 Joule to sink, calculate the temperatures of the source and the sink. **[Ans.** 100°C, 0°C**]** **(*Meerut 1999*)**

4.92 A Caront engine has the same efficiency between 1000 K and 500 K and between x K and 1000 K (This being the temperature of the sink in this case). Calculate x.

[Ans. 2000 K**]** **(*Bangalore 1995*)**

4.93 The efficiency of a Carnot's cycle is 1/6. If on reducing the temperature of the sink by 65 °C, the efficiency becomes 1/3. Find the initial and final tempratures between which the cycle is working. **[Ans.** 117°C, 52°C] ***(Pune 2010)***

4.94 The efficiency of an ideal engine is 0.2. If the temperaure of the sink is lowered by 20°C, the efficiency becomes 0.25. Find the temperature of the source and sink.
[Ans. T_1 = 5860 K, T_2 = 4688 K] ***(Madras 2004)***

4.95 A Carnot engine working between a source at 400 K and a sink at T_2 K has an efficiency of 50%. If the temperature of both the source and the sink are increased by 100 K, what is the efficiency of the engine?

[Hint: $\eta = 1 - \frac{T_2}{400} = \frac{1}{2}$ and $\eta' = 1 - \frac{T_2 + 100}{500}$] **[Ans.** 40%] ***(Mysore 2004/w, Bangalore 2005)***

4.96 A Carnot's engine has an efficiency 0.5 when the temperature of the sink is 300 K. What is the temperature of the source? By how much should it be raised so that the efficieney becomes 0.75 ? **[Ans.** T_1 = 600 K, T_1' = 1200 K ΔT = 600 K] ***(Bangalore 2002)***

4.97 Two samples of air having same composition and initially at the same temperature and pressure are compressed from a volume V to a volume $V/2$, one isothermally and the other adiabatically. In which sample, is the final pressure greater?
[Ans. Sample which compressed adiabatically] ***(Lucknow 2003)***

4.98 A thermos bottle containing coffee is vigorously shaken and thereby the temperature of the coffee rises. Regard the coffee as system: (*i*) Has heat been transferred to coffee? (*ii*) Has work been done in coffee? ***(Lucknow 2003)***

4.99 When a refrigerator is switched off, the ice stored in a cold storage melts at the rate of 36 kg/hr when the external temperature is 30°C. Find the minimum output power of the motor of the refrigerator required to prevent the ice from melting. (Latent heat of fusion of ice = 80 cal/gm, J = 4.2 Joules/cal.) **[Ans.** 1197.6 kW] ***(Calcutta 2003)***

4.100 Find the work done to stretch a copper wire 1 m long and 2 mm in diameter through 1 cm. (Given $P = 12.5 \times 10^{10}$ N/m^2) **[Ans.** 3.925×10^3 J] ***(Mysore 2004/s, Bangalore 2004)***

MULTIPLE CHOICE QUESTIONS

4.1 Out of the following, the physical quantity that relates with first law of thermodynamics is:
(*a*) Temperature (*b*) Pressure
(*c*) Energy (*d*) Number of moles ***(Agra 2003)***

4.2 In producing cooling by adiabatic demagnetisation, we use:
(*a*) magnetic substance (*b*) paramagnetic substance
(*c*) diamagnetic substance (*d*) ferromagnetic substance ***(Agra 2003)***

4.3 Zeroth law of thermodynamics is:
(*a*) kinetic energy of molecules of a gas is zero.
(*b*) ideal gas does not contain molecules
(*c*) if two systems are separated in thermal equilibrium with a third system then they themselves are in thermal equilibrium with each other.
(*d*) absolute zero temperature cannot be attained. ***(Garhwal 2005)***

4.4 Transfer of heat from a body at low temperature to a body at high temperature:
(*a*) is impossible.
(*b*) is possible by keeping both the bodies in contact.
(*c*) is possible by doing some external work.
(*d*) none of these. ***(Garhwal 2005)***

4.5 In Carnot cycle, the first step is:

(*a*) Isothermal expansion. (*b*) Isothermal compression.
(*c*) Adiabatic expansion. (*d*) Adiabatic compression. **(*Garhwal 2005*)**

4.6 An engine works between the temperatures 30 K and 300 K. What is its efficiency?

(*a*) 50% (*b*) 47%
(*c*) 90% (*d*) 10% **(*Garhwal 2002*)**

4.7 The efficienay of Carnot engine working between steam point and ice point is:

(*a*) 1 (*b*) 0
(*c*) 26.81% (*d*) 16.81% **(*Garhwal 2000, Rohilkhand 2003*)**

4.8 An ideal mono atomic gas initially at 17°C is suddenty compressed to 1/8th of its initial volume. The temperature of gas after compression is:

(*a*) 17 °C (*b*) 136 °C
(*c*) 887 °C (*d*) 10^{10} K **(*Garhwal 2002*)**

4.9 The ratio of adiabatic and isothermal elasticities of a gas is:

(*a*) γ (*b*) γ^2
(*c*) γ^{-1} (*d*) $\left(d.\sqrt{\pi}\,\sqrt{\gamma}\right)$ **(*Garhwal 2002*)**

4.10 The efficiency of Carnot's engine working between 127 °C and 27 °C is:

(*a*) 25% (*b*) 50%
(*c*) 75% (*d*) 100% **(*Agra 2004, Kanpur 2001*)**

4.11 The efficiency of a reversible Carnot's engine working between temperature T_1 and T_2 $(T_1 > T_2)$ is:

(*a*) $\dfrac{T_2}{T_1}$ (*b*) $\dfrac{T_1}{T_2}$
(*c*) $\left(1-\dfrac{T_2}{T_1}\right)$ (*d*) $\left(\dfrac{T_1}{T_2}-1\right)$ **(*Ranchi 2005*)**

4.12 Internal energy of a real gas depends upon:

(*a*) only on the temperature of the gas.
(*b*) only on the volume of the gas.
(*c*) only on the pressure of the gas.
(*d*) size of the molecule **(*Rohilkhand 2002*)**

4.13 An adiabatic process occurs at constant

(*a*) temperature (*b*) pressure
(*c*) heat (*d*) none of these **(*Rohilkhand 2002*)**

4.14 A diatomic gas at 18 °C is compressed adiabatically to $\dfrac{1}{32}$ times of its original volume. The temperature after compression will be:

(*a*) 18 °C (*b*) 887 °C
(*c*) 891 °C (*c*) 144 °C **(*C.B.S.E. 1996*)**

4.15 The gas law $\dfrac{PV}{T}$ = constant is true for

(*a*) isothermal change only (*b*) adiabatic change only
(*c*) both isothermal and adiabatic changes (*d*) none of these **(*Rohilkhand 2002*)**

4.16 The work done during an adiabatic expansion of an ideal gas is:

(*a*) $W = \frac{1}{1-\gamma}(P_1V_1 - P_2V_2)$ (*b*) $W = \frac{1}{\gamma-1}(P_1V_1 - P_2V_2)$

(*c*) $W = R(T_2 - T_1)$ (*d*) $W = RT_1 \log_e \frac{V_2}{V_1}$ **(*Rohilkhand 2002*)**

4.17 A Carnot's engine is operating between 100° and 50 °C. It's efficiency will be:

(*a*) 13.4% (*b*) 15.2%

(*c*) 50% (*d*) 100% **(*Rohilkhand 2002*)**

4.18 A reversible heat engine can be 100% efficient, if the temperature of the sink is:

(*a*) less than that of source (*b*) equal to that of source

(*c*) 0 °C (*d*) 0 K

4.19 The efficiency of a Carnot engine is 0.4. If the temperaute of the sink is 27 °C, the temperature of source is:

(*a*) 127 °C (*b*) 500 °C

(*c*) 500 K (*d*) 400 K **(*Kanpur 2002*)**

4.20 The physics underlying the working of a refrigerator closely resembles the physics underlying:

(*a*) Ice formation (*b*) Heat engine

(*c*) Vapour compression machine (*d*) Vaporization of water

4.21 The door of running refrigerator inside a room is left open. Mark the correct statement:

(*a*) the room will be cooled slightly

(*b*) the room will be warmed up gradually

(*c*) the room will be cooled to the temperature inside the refrigerator

(*d*) the temperature of the room will remain unaffected

ANSWERS

4.1 (*c*)	**4.2** (*b*)	**4.3** (*c*)	**4.4** (*c*)	**4.5** (*a*)	**4.6** (*c*)
4.7 (*c*)	**4.8** (*c*)	**4.9** (*a*)	**4.10** (*a*)	**4.11** (*c*)	**4.12** (*a*)
4.13 (*c*)	**4.14** (*c*)	**4.15** (*c*)	**4.16** (*b*)	**4.17** (*a*)	**4.18** (*d*)
4.19 (*c*)	**4.20** (*b*)	**4.21** (*b*)			

ENTROPY

5

Chapter

INTRODUCTION

Entropy is a thermodynamic property that is a measure of the energy not available for useful work in a thermodynamic process, such as in energy conversion devices, engines, or machines. Such devices can only be driven by convertible energy, and have a theoretical maximum efficiency when converting energy to work. During this work entropy accumulates in the system, but has to be removed by dissipation in the form of waste heat (refer article 5.6) The concept of entropy is defined by second law of thermodynamics, which states that the entropy of a closed system always increases or remains constant. Thus, entropy is also a measure of the tendency of a process, *(viz.* chemical) reaction to proceed in a particular direction.

5.1 CONCEPT OF ENTROPY

The concept of entropy was first introduced by Rudolf Clausius in 1865 during his work on the formulation and application of the second law of thermodynamics. It is a very important thermodynamic quantity and has proved to be very useful in the study of behaviour of heat engines. In order to describe the condition of a working substance completely. Clausius felt the need of another variable of state, in addition to volume, pressure, temperature, internal energy etc. This new variable is entropy. The quantity entropy found to remain constant in a adiabatic process just as temperature remains constant in an isothermal process. Thus, entropy can be defined as the "***thermodynamic peoperty of a working substance which remains constant during an adiabatic process, when no energy is given to or removed from it***". Entropy plays a very vital role in the study of thermodynamics.

5.2 CHANGE IN ENTROPY

Consider a number of isothermals T_1, T_2, T_3 etc. drawn at small but equal temperature differences bounded by the same two adiabatics *L* and *M* as shown in an indicator diagram (Fig.5.1) for an ideal gas. Then all along the adiabatics *L* and *M*, there is a change in volume and temperature with change in pressure. Let *ABCD* and *DCEF* represent the Carnot's reversible cycles. During carnot's cycle *ABCD*, Q_1 amount of heat is absorbed in going from *A* to *B* at constant temperature T_1 and Q_2 amount of heat is rejected at constant temperature T_2. Then according to Carnot's theorem, the efficiency of Carnot's engine is given by

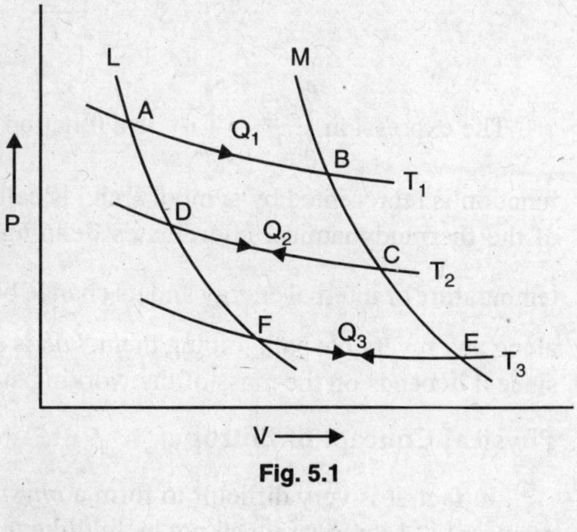

Fig. 5.1

$$\eta = \frac{Q_1 - Q_2}{Q_1} = \frac{T_1 - T_2}{T_1}$$

$$1 - \frac{Q_2}{Q_1} = 1 - \frac{T_2}{T_1}$$

$$\frac{Q_2}{Q_1} = \frac{T_2}{T_1}$$

$$\therefore \quad \frac{Q_1}{T_1} = \frac{Q_2}{T_2} \qquad \text{... (5.1)}$$

Similarly considering the Carnot's cycle *DCEF* in which an amount of heat Q_2 is absorbed at constant temperature T_2 and heat Q_3 is rejected at constant temperature T_3.

$$\therefore \quad \frac{Q_2}{T_2} = \frac{Q_3}{T_3} \qquad \text{... (5.2)}$$

From equations (5.1) and (5.2) we have

$$\frac{Q_1}{T_1} = \frac{Q_2}{T_2} = \frac{Q_3}{T_3} = \text{........ constant.}$$

In general, if Q is the amount of heat absorbed or rejected at a temperature T in going from one adiabatic to the other, then

$$\frac{Q}{T} = \text{constant}$$

If the two adiabatics are very close to each other and if δQ is the small quantity of heat absorbed at constant temperature T in going from one adiabatic to another, then

$$\frac{\delta Q}{T} = \text{constant}$$

This constant ratio is called the 'change in entropy' between the states represented by the two adiabatics. It is denoted by δS

$$\therefore \quad \delta S = \frac{\delta Q}{T}$$

The change in entropy for a finite reversible change in the state of working substance from A to B is given by

$$\int_{S_A}^{S_B} ds = \int_A^B \frac{dQ}{T}$$

$$\therefore \quad S_B - S_A = \int_A^B \frac{dQ}{T} \qquad \text{... (5.3)}$$

The expression $\int_A^B \frac{dQ}{T} = \int_{S_A}^{S_B} ds$ is a function of the thermodynamic coordinates of a system. This function is represented by symbol S and is called entropy. Hence "entropy of a system is a function of the thermodynamical coordinates defining the state of the system *viz*., the pressure, volume, temperature or internal energy and its change between two states is equal to $\int \frac{dQ}{T}$ between the states along any reversible path joining them." dS is an exact differential. Entropy is an extensive property since it depends on the mass of the working substance.

Physical Concept of Entropy

In fact, it is very difficult to form a *physical concept of entropy* as there is nothing physical to represent it. Moreover it can not be felt like temperature, pressure etc. But according to definition

$$\text{Change of entropy} = \frac{\text{Heat added or subtracted}}{\text{Absolute temperature}} = \frac{dQ}{T} \qquad \text{... (5.3a)}$$

One can say that heat has the same dimensions as the product of entropy and absolute temperature. Since the gravitational potential energy of a body is proportional to the product of its mass and height above some zero level, likewise we may take temperature analogous to height and entropy as analogous to mass or inertia. Thus, we may take *entropy as* ***thermal inertia*** *which bears to heat motion, a relation similar to that which mass bears to linear motion or moment of inertia bears to rotational motion.*

Unit of Entropy

Unit of entropy depends on the unit of heat used and the absolute temperature. It is measured in calories (or ergs or Joules) per degree Kelvin *i.e.*, cal./K. S.I. unit is Joule/K or Kcal/K.

5.3 CHANGE IN ENTROPY IN ADIABATIC PROCESS

In an adiabatic process, no heat is allowed to enter or leave the system. Hence $\delta Q = 0$

$$\therefore \quad \delta S = \frac{\delta Q}{T} = 0 \quad \text{... (5.4)}$$

Thus, there is no change of entropy during an adiabatic process, or the entropy remains constant during an adiabatic reversible process. Hence the adiabatic curves on the $P - V$ diagram are called as *isoentropics* or curves of constant entropy.

5.4 CHANGE OF ENTROPY IN REVERSIBLE CYCLE (CARNOT'S CYCLE)

Consider a complete reversible Carnot's cycle *ABCD* as shown in Fig. 5.2 for an ideal gas formed by two isothermals *i.e. AB* at a temperature T_1 and *CD* at temperature T_2 and the two adiabatics *BC* and *DA*.

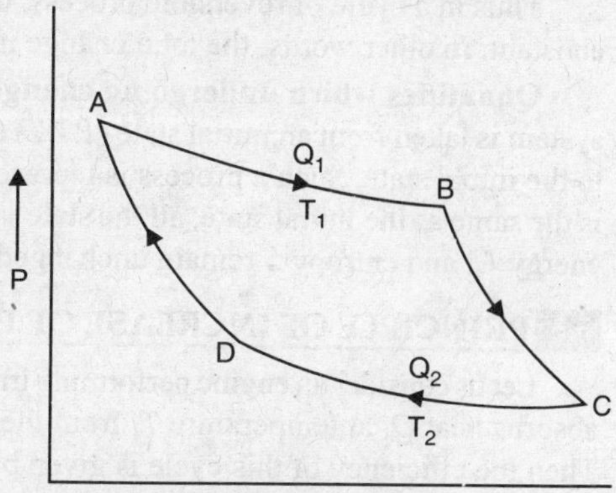

Fig. 5.2

(*i*) **Isothermal expansion AB:** Let Q_1 be the quantity of heat absorbed by the working substance in going from state *A* to state *B* during isothermal expansion *AB* at a constant temperature T_1. The increase in entropy of the working substance is given by

$$\int_A^B dS = +\frac{Q_1}{T_1}$$

(*ii*) **Adiabatic expansion *BC*:** In going from state *B* to state *C* along the adiabatic *BC*, there is no change in entropy of the working substance, but the temperature falls from T_1 to T_2 due to expansion.

$$\int_B^C dS = 0$$

(*iii*) **Isothermal compression *CD*:** In going from state *C* to state *D* along the isothermal *CD*, the working substance rejects heat Q_2 to the sink at temperature T_2. The entropy of the working substance decreases and the change in entropy is given by

$$\int_C^D dS = -\frac{Q_2}{T_2}$$

(*iv*) **Adiabatic compression *DA*:** In going from *D* to *A* along the adiabatic *DA*, there is no change in entropy but temperature rises from T_2 to T_1.

$$\therefore \quad \int_D^A dS = 0$$

Thus the net gain in entropy of the working substance in the whole cycle ABCDA

$$= \int_A^B dS + \int_B^C dS + \int_C^D dS + \int_D^A dS$$

$$\oint dS = \frac{Q_1}{T_1} - \frac{Q_2}{T_2}$$

But for a reversible Carnot's cycle

$$\frac{Q_1}{T_1} = \frac{Q_2}{T_2} \quad \text{or} \quad \frac{Q_1}{T_1} - \frac{Q_2}{T_2} = 0$$

Substituting, we get

$$\oint dS = \frac{Q_1}{T_1} - \frac{Q_2}{T_2} = 0 \qquad \text{... (5.5)}$$

where the integral sign with a circle refers to a complete cycle.

Thus in a cycle of reversible process, the entropy of the system remains unchanged or remains constant. In other words, the total change in entropy is always zero.

Quantities which undergo no change in cyclic process: A cyclic process is one in which a system is taken from an initial state (P, V, T) through a succession of states but is always brought back to the initial state. Such a process is represented by a closed path in a PV diagram. As the final state is the same as the initial state, all the state variables as pressure P, volume V, temperature T, internal energy U, and entropy S remain unchanged at the end of the cycle.

5.5 PRINCIPLE OF INCREASE OF ENTROPY

Let us consider an engine performing irreversible cycle of changes in which the working substance absorbs heat Q_1 at temperature T_1 from the source and rejects heat Q_2 to the sink at temperature T_2. Then the efficiency of this cycle is given by

$$\eta' = \frac{Q_1 - Q_2}{Q_1} = 1 - \frac{Q_2}{Q_1}$$

But according to Carnot's theorem, this efficiency is less than that of a reversible engine working between the same two temperatures T_1 and T_2 for which efficiency is given by

$$\eta = 1 - \frac{T_2}{T_1}$$

Thus,

$$\eta' < \eta$$

$$1 - \frac{Q_2}{Q_1} < 1 - \frac{T_2}{T_1}$$

$$\frac{Q_2}{Q_1} > \frac{T_2}{T_1} \quad \text{or} \quad \frac{Q_2}{T_2} > \frac{Q_1}{T_1}$$

or

$$\frac{Q_2}{T_2} - \frac{Q_1}{T_1} > 0 \qquad \text{... (5.6)}$$

Considering the whole system, the source loses (– *ve* sign) the entropy by an amount $\frac{Q_1}{T_1}$ and the sink gains (+ *ve* sign) an entropy $\frac{Q_2}{T_2}$. Therefore, the net change in entropy for the whole system is $\left(\frac{Q_2}{T_2}-\frac{Q_1}{T_1}\right)$, which is clearly greater than zero *i.e.*, positive (∵ equation 5.6). Thus, there is an increase in entropy of the system during an irreversible process.

To make it more clear, let us consider the natural process of conduction of heat from a body A at temperature T_1 to another body B at a lower temperature T_2. This process is irreversible. Since heat always flows from higher to a lower temperature the quantity of heat thus transmitted be Q, then

$$\text{Decrease in entropy of hotter body} = \frac{Q}{T_1}$$

$$\text{Increase in entropy of colder body} = \frac{Q}{T_2}$$

∴ Net incrrease in entropy of the system

$$dS = \frac{Q}{T_2} - \frac{Q}{T_1}$$

$$= +\text{ve quantity as } T_1 > T_2.$$

Thus, we conclude that the entropy of a system increases in all irreversible processes. This is known as the law or principle of increases of entropy. All natural processes taking place in the universe are irreversible. It means the entropy of the universe increases.

$$dS\ [\text{universe}] \geq 0$$

5.6 CHANGE OF ENTROPY IN IRREVERSIBLE PROCESS

The thermodynamical state of a system can be defined with the help of the thermodynamical coordinates of the system. The state of a system can be changed by altering the thermodynamical coordinates. Changing from one state to the other by changing the thermodynamical coordinates is called a *process*.

Consider two states of a system *i.e.*, state A and state B. Change of state from A to B or *vice versa* is a process and the direction of the process will depend upon a new thermodynamical coordinate called **entropy**. All processes are not possible in the universe.

Consider the following processes:

1. Let two blocks A and B at different temperatures T_1 and T_2 ($T_1 > T_2$) be kept in contact but the system as a whole is insulated from the surroundings. Conduction of heat takes place between the blocks, the temperature of A falls and the temperature of B rises and thermodynamical equilibrium will be reached.
2. Consider a flywheel rotating with an angular velocity ω. Its initial kinetic energy is ½ $I\omega^2$. After some time the wheel comes to rest and kinetic energy is utilised in overcoming friction at the bearings. The temperature of the wheel and the bearings rises and the increase in their internal energy is equal to the original kinetic energy of the fly wheel.
3. Consider two flasks A and B connected by a glass tube provided with a stop cock. Let A contain air at high pressure and B is evacuated. The system is isolated from the surroundings. If the stop cock is opened, air rushes from A to B, the pressure in A decreases and the volume of air increases.

All the above three examples though different, are thermodynamical processes involving change in thermodynamical coordinates. Also, in accordance with the first law of thermodynamics, the principle of conservation of energy is not violated because the total energy of the system is conserved. It is also clear that, with the initial conditions described above, the three processes will take place.

Let us consider the possibility of the above three processes taking place in the reverse direction. In the first case, if the reverse process is possible, the block B should transfer heat to A and initial conditions should be restored. In the second case, if the reverse process is possible, the heat energy must again change to kinetic energy and the fly wheel should start rotating with the initial angular velocity ω. In the third case, if the reverse process is possible the air in B must flow back to A the initial condition should be obtained.

But, it is a matter of common experience, that none of the above conditions for the reverse processes are reached. It means that the direction of the process cannot be determined by knowing the thermodynamical coordinates in the two end states. To determine the direction of the process a new thermodynamical coordinate has been devised by Clausius and this is called the entropy of the system. Similar to internal energy, entropy is also a function of the state of a system. For any possible process, the entropy of an isolated system should increase or remain constant. The process in which there is a possibility of decrease in entropy cannot take place.

If the entropy of an isolated system is maximum, any change of state will mean decrease in entropy and hence that change of state will not take place.

To conclude, *processes in which the entropy of an isolated system decreases do not take place or for all processes taking place in an isolated system the entropy of the system should increase or remain constant.* It means a process is irreversible if the entropy decreases when the direction of the process is reversed. A process is said to be irreversible if it cannot be retraced back exactly in the opposite direction. During an irreversible process, heat energy is always used to overcome friction. Energy is also dissipated in the form of conduction and radiation. This loss of energy always takes place whether the engine works in one direction or the reverse direction. Such energy cannot be regained. In actual practice all the engines are irreversible. If electric current is passed through a wire, heat is produced. If the direction of the current is reversed, heat is again produced. This is also an example of an irreversible process. All chemical reactions are irreversible. In general, all natural processes are irreversible.

Entropy in an irreversible cycle. Suppose in a simple irreversible cycle Q_1 is the quantity of heat absorbed at a temperature T_1 and Q_2 is the quantity of heat rejected at a temperature T_2.

Then the efficiency of this cycle is given by $\dfrac{Q_1 - Q_2}{Q_1}$

According to Carnot's theorem this efficiency is less than that of the reversible engine for which

$$\eta = \frac{T_1 - T_2}{T_1}$$

when the latter also works between the same two temperatures T_1 and T_2. Hence

$$\frac{Q_1 - Q_2}{Q_1} < \frac{T_1 - T_2}{T_1}$$

or

$$1 - \frac{Q_2}{Q_1} < 1 - \frac{T_2}{T_1}$$

$$\therefore \quad \frac{Q_2}{Q_1} > \frac{T_2}{T_1} \quad \text{or} \quad \frac{Q_2}{T_2} > \frac{Q_1}{T_1}$$

or

$$\frac{Q_2}{T_2} - \frac{Q_1}{T_1} > 0.$$

Considering the whole system the source loses an entropy $\frac{Q_1}{T_1}$ whereas the sink gains the entropy by an amount $\frac{Q_2}{T_2}$. The net change in entropy of the whole process is given by

$$\frac{Q_2}{T_2} - \frac{Q_1}{T_1} > 0 \qquad \text{... (5.7)}$$

which is greater than zer and hence is positive

Thus we see that in an irreversible cycle there is always a net gain of entropy of $\delta S > 0$

5.7 CARNOT'S THEOREM: ON THE BASIS OF ENTROPY

Carnot's theorem. Carnot's theorem consists of two parts and is stated as follows:

(*i*) *All reversible heat engines, working between the same limits of temperature, have the same efficiency.*

(*ii*) *No heat engine can be more efficient than a reversible heat engine, working between the same limits of temperature.*

The efficiency of a Carnot's reversible heat engine is $\left(1 - \frac{T_2}{T_1}\right)$ when working between the temperature T_1 and T_2. This is independent of the nature of the working substance. *Hence we see that the efficiency of all reversible heat engines, working between the same limits of temperature is the same, which is Carnot's theorem.*

On the basis of entropy. Both the parts can also be proved on the basis of entropy.

A reversible heat engine operates in a cyclic manner and absorbs an amount of heat Q_1 from the source at a temperature T_1, converts a part of it into mechanical work and delivers the rest *i.e.* Q_2 to the sink at a temperature T_2.

Decrease in entropy of the source = $\frac{Q_1}{T_1}$

Increase in entropy of the sink = $\frac{Q_2}{T_2}$

As the whole cycle is reversible, net change in entropy = 0

Hence increase in entropy of the sink = Decrease in entropy of the source.

$$\frac{Q_1}{T_1} = \frac{Q_2}{T_2} = \delta S \text{ (say)}$$

or $$Q_1 = T_1\, \delta S \text{ and } Q_2 = T_2\, \delta S$$

Work done $$W = Q_1 - Q_2 = (T_1 - T_2)\delta S = \frac{(T_1 - T_2)\, Q_1}{T_1}$$

$\therefore$ Efficiency $$\frac{W}{Q_1} = \frac{T_1 - T_2}{T_1} = 1 - \frac{T_2}{T_1} \qquad \text{... (5.8)}$$

which is the same as for Carnot's cycle.

Hence the efficiency of all reversible heat engines, working between the same limits of temperature, is the same and equal to that of Carnot's engine.

For an irreversible cycle $\delta S > 0$ *i.e.*, there is a net increase in entropy when an irreversible engine operates. In other words, for such an engine.

Increase in entropy of the sink $\delta S'$ > Decrease in entropy of the source δS.

Let the irreversible heat engine and the reversible heat engine absorb thc same amount of heat Q_1 from the source at temperature T_1, then

For both $$\delta S = \frac{Q_1}{T_1} \text{ or } Q_1 = T_1\, \delta S$$

Let the irreversible heat engine reject a quantity of heat Q_2' to the sink and the reversible heat engine a quantity Q_2 at a temperature T_2, then

$$Q_2' = T_2\, \delta S' \quad \text{and} \quad Q_2 = T_2\, \delta S$$

$$\therefore \qquad Q_2' > Q_2$$

If η and η' are the efficiencies of the reversible and irreversible engines respectively, then

$$\eta = \frac{Q_1 - Q_2}{Q_1} \quad \text{and} \quad \eta' = \frac{Q_1 - Q_2'}{Q_1}$$

As $$Q_2' > Q_2\,;\; Q_1 - Q_2' < Q_1 - Q_2$$

Hence the efficiency of an reversible heat engine is greater than that of any irreversible heat engine, working between the same two temperatures.

Importance: According the Carnot's theorem the efficiency of a reversible heat engine depends only upon the temperature of the source and the sink and is *independent* of the nature of the working substance. All practical heat engines will, therefore, have an efficiency less than that of a reversible heat engine. Since

$$\eta = 1 - \frac{T_2}{T_1}$$

where T_2 is the temperature of the sink and T_1 that of the source, it is not possible to convert all the heat available in the source into useful work, unless the sink is at the temperature of absolute zero. In other words, whereas, it is possible to convert all the work done into an equivalent amount of heat, given by the first law of thermodynamics, the reverse is not true, *i.e.* it is not possible to convert all the heat into work. Further, work is done when heat is a absorbed by a heat engine from the source at a higher temperature, and is given to the sink at a lower temperature. If, therefore, heat is to be taken from the sink at a lower temperature, work will have to be done on the engine. Thus it is not possible for a self-acting machine to convey heat from a body at a lower temperature to a body at a higher temperature, without the help of an external agency. This is precisely the statement of second law of thermodynamics. Carnot's theorem is, therefore, the basis for the second law and shows that it is impossible to derive continuous supply of work by cooling a body to a temperature lower than that of the coldest of its surrounding and thus debars perpetual motion.

5.8 SECOND LAW OF THERMODYNAMICS (IN TERMS OF ENTROPY)

In terms of entropy, the second law of thermodynamics can be stated as under:

'*Every physical or chemical process in nature takes place in such a manner that the total entropy increases*'

If two bodies A and B, at different temperatures, are placed in contact, the combined system is not in equilibrium. Heat energy flows from the body at the higher temperature to the body at the lower temperature, till the temperature is equalised *i.e.*, a state of equilibrium is reached. During the process of exchange of heat the thermodynamic probability W of the combined system goes on increasing, till it reaches a maximum value at equilibrium state. As W increases, entropy also increases and the combined system moves towards the state of maximum entropy. The probability of the system going back to initital state of the two bodies is practically negligible.

In order to put this law in mathematical form , let S_1 and S_2 be the entropies of a substance, in the initial and the final state, as measured from any arbitrary zero, then

The change in entropy $= S_2 - S_1 = \int_A^B \frac{dQ}{T}$

If two states are very close to each other, we can write the relation as

$$dS = \frac{dQ}{T}$$

or $$dQ = T.\, dS \qquad \text{... (5.10)}$$

This is the mathematical form of second law of thermodynamics.

5.9 THE T- S DIAGRAM

The temperature-entropy diagram is used in engineering and meteorology. Consider the Carnot's cycle *ABCDA* (Fig. 5.3).

Thermodynamic changes in the state of a substance can be represented by plotting entropy (*S*) along horizontal axis and temperature (*T*) along perpendicular axis. Such a diagram is called temperature entropy diagram (*i.e. T–S* diagram) for the substance.

A Carnot's reversible cycle is shown on *P–V* diagram (Fig. 5.4). It is composed of two isothermals *AB* and *CD* at constant temperatures T_1 and T_2 respectively.

On the *T–S* diagram (Fig 5.3), the isothermal curves are shown by two straight lines *AB* and *CD* parallel to S-axis and the adiabatic curves are shown by the straight lines *BC* and *DA* parallel to *T*-axis. The resulting *T–S* diagram is a rectangle *ABCD*.

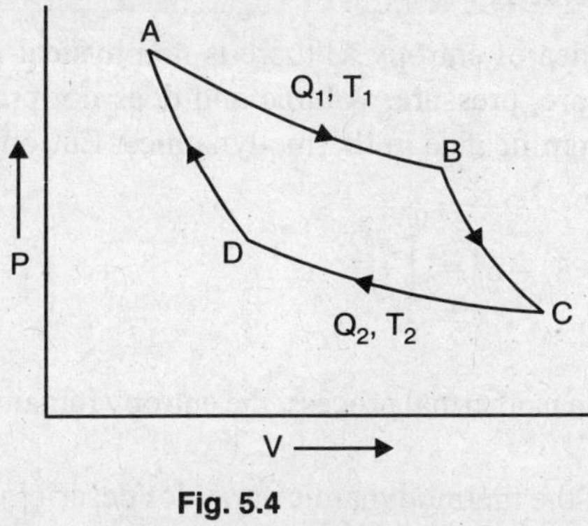

Fig. 5.4

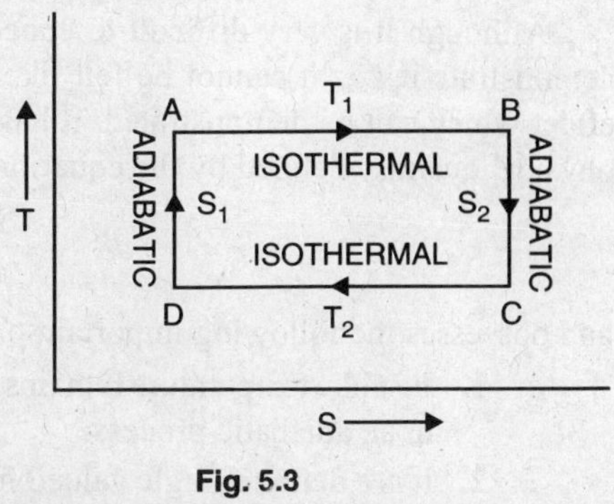

Fig. 5.3

Consider a Carnot's reversible cycle on *T–S* diagram (Fig 5.3). Suppose S_1 be the entropy of working substance in state *A*, S_2 the entropy in state *B*, Q_1 be the heat absorbed along *AB i.e.* during isothermal expansion at constant temperature T_1 and Q_2 is the heat rejected in the isothermal compression at constant temperature T_2 along *CD*.

In going from *A* to *B*, along isothermal expansion *AB*, the gain in the entropy of the working substance is given by

$$S_2 - S_1 = \frac{Q_1}{T_1} \qquad \text{... (5.11)}$$

In going from *B* to *C*, along adiabatic expansion *BC*, there is no change in the entropy.

In going from *C* to *D*, along isothermal compression *CD*, the loss in the entropy of the working substance is given by

$$S_2 - S_1 = \frac{Q_2}{T_2} \qquad \text{... (5.12)}$$

Along adiabatic compression *DA*, there is no change in the entropy. Therefore, from equations (5.11) and (5.12), we get

$$Q_1 = T_1(S_2 - S_1)$$

and
$$Q_2 = T_2(S_2 - S_1)$$

$$\therefore \quad Q_1 - Q_2 = (T_1 - T_2)(S_2 - S_1) \quad ...(5.13)$$

The quantity $(Q_1 - Q_2)$ represents the external work done in the cycle and $(T_1 - T_2)(S_2 - S_1)$ is the area of the rectangle on the *T–S* diagram.

Thus, the area of the rectangle on the *T–S* diagram represents the external work done in a reversible Carnot's cycle.

Efficiency of Carnot's engine

Using *T–S* diagram, we can write

$$\text{Efficiency of the engine} = \frac{\text{external work done in one cycle}}{\text{heat absorbed from the source}}$$

$$\eta = \frac{Q_1 - Q_2}{Q_1}$$

$$= \frac{(T_1 - T_2)(S_2 - S_1)}{T_1(S_2 - S_1)} = \frac{T_1 - T_2}{T_1}$$

$$\eta = 1 - \frac{T_2}{T_1} \quad ...(5.14)$$

5.10 PHYSICAL SIGNIFICANCE OF ENTROPY

Although it is very difficult to conceive the idea of entropy as there is no physical method to demonstrate it *i.e.*, it cannot be felt like temperature, pressure, volume and does not produce any effect which can be demonstrated; it has a great significance in thermodynamics. Entropy is a real physical quantity defined by the equation

$$dS = \frac{\delta Q}{T} \quad \text{or} \quad S_B - S_A = \int_A^B \frac{dQ}{T} \quad ...(5.15)$$

and possesses the following important properties:

1. Just like temperature remains constant in isothermal process, the entropy remains constant in an adiabatic process.
2. It is a definite single valued function of the thermodynamic variables describing the state of a working substance.
3. In every natural process (*i.e.* irreversible change) there is always an increase in entropy.
4. The second law of thermodynamics can be stated in terms of entropy of a system.
5. Due to increase in entropy, unavailable energy increases.
6. According to Freeman Dyson, the entropy is a measure of disorderness. This disorderness can be evaluated by using the relation.

$$S = k \log W \quad ...(5.16)$$

where k is the Boltzmann's constant and W is the disorder parameter. This is known as Boltzman entropy relation.

5.11 IMPORTANCE OF S–T DIAGRAM

(*i*) The *S–T* diagram also known as *tephigram* is extensively used in meterology.

(*ii*) The *S–T* diagram helps us to very easily calculate the work done per cycle because the *S–T* diagram is a rectangle. This is why it is used by auto engineers to study the working efficiency of the engines.

(*iii*) **To check the efficient working of actual engines.** The *S–T* diagram for a cycle can be used to check the efficient working of actual engines. The engineer has before him the standard diagram for the engine. Any deviation in *S–T* diagram from the standard showing a change in entropy during a cycle indicates immediately that the amount of energy available for useful work is on the decrease. This is due to some irreversible process that may have started in the engine and must be checked. Thus it is a timely warning against possible breakdown.

5.12 ENTROPY IS PATH INDEPENDENT

Consider two reversible paths on a *P-V* diagram, path 1 and path 2, as shown in Fig 5.5 which form a *closed reversible cycle.* The net change in entropy in going from *A* to *B* via path 1 and back from *B* to *A* via path 2 is zero.

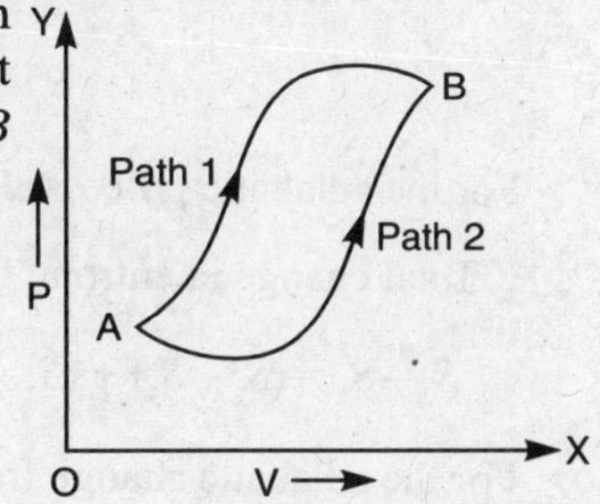

Fig. 5.5

or $$\int_A^B dS\ (\text{Path 1}) + \int_B^A dS\ (\text{Path 2}) = 0$$

As both the paths are reversible, we have

$$\int_A^B dS\ (\text{Path 1}) = -\int_B^A dS\ (\text{Path 2})$$

$$= +\int_A^B dS\ (\text{Path 2}) = S_b - S_a$$

wher S_b is the entropy of the system in the state *B* and S_a that in the state *A*. In other words *the change in the entropy $S_b - S_a$ in going from the equilibrium state A to the equilibrium state B is independent of the path chosen and depends only upon the positions of the initial and final states.*

Hence $$S_b - S_a = \int_A^B dS = \int_A^B \frac{\delta Q}{T} \quad \text{... (5.17)}$$

The integration is to be carried over any *reversible* path going from state *A* to *B* in a *quasi-static manner i.e.* by taking δQ to be an infinitesimally small quantity.

5.13 ISOENTROPIC PROCESS

(*i*) **Isoentropic process** is that process during which the entropy of a system remains constant. Thus a gas undergoing an adiabatic change is an examples of an isoentropic process. In an adiabatic change no heat is gained or lost *i.e.,*

$$dQ = 0$$

$$\therefore \quad \text{Change in entropy} = \int_1^2 dS = S_2 - S_1 = \int_1^2 \frac{dQ}{T} = 0 \quad \text{... (5.18)}$$

Thus the entropy is the same at all points along a reversible adiabatic path. In other words, entropy stays constant in an adiabatic process. Hence an adiabatic process is alos known as an *isoentropic process.*

(*ii*) **For Isothermal process.** In an isothermal process the temperature remains constant. In other words, there is no change in internal energy.

$$\therefore \quad dU = 0$$

and $$dQ = dU + dW = P\,dV$$

$$\therefore \quad \text{Change is entropy} = \int_1^2 dS = S_2 - S_1 = \int_1^2 \frac{dQ}{T}$$

$$= \int_1^2 \frac{P\,dV}{T} = \int_1^2 \frac{RT}{V}\frac{dV}{T}$$

$$S_2 - S_1 = R\int_1^2 \frac{dV}{V} = R\log_e \frac{V_2}{V_1} \qquad \text{... (5.19)}$$

(*iii*) **For isobaric process:** An isobaric process can be considered to be made up of two parts (*a*) an isothermal change from volume V_1 to V_3 at temperature T_1 and (*b*) an adiabatic change from V_3 to V_2 as shown in Fig. 5.6 because the change in entropy is independent of the path.

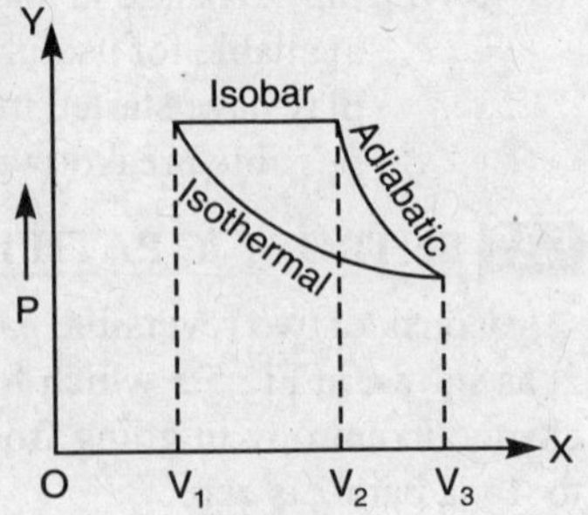

Fig. 5.6

For the isothermal process, change in entropy

$$S_3 - S_1 = R\log_e \frac{V_3}{V_1}$$

For the adiabatic process change in entropy $S_2 - S_3 = 0$

$\therefore$ Total change in entropy for the isobaric process

$$S_2 - S_1 = (S_2 - S_3) + (S_3 - S_1) = R\log_e \frac{V_3}{V_1}$$

For the adiabatic change from V_3 to V_2, we have

$$T_1 V_3^{\gamma-1} = T_2 V_2^{\gamma-1}$$

where T_2 is the temperature corresponding to volume V_2

or
$$\frac{T_2}{T_1} = \left(\frac{V_3}{V_2}\right)^{\gamma-1} \qquad \text{... }(i)$$

For a perfect gas

$$\frac{P_1V_1}{T_1} = \frac{P_2V_2}{T_2}$$

For an isobaric process $P_1 = P_2$

$$\therefore \quad \frac{V_1}{T_1} = \frac{V_2}{T_2} \text{ or } \frac{T_2}{T_1} = \frac{V_2}{V_1} \qquad \text{... }(ii)$$

From (*i*) and (*ii*), we have

$$\left(\frac{V_3}{V_2}\right)^{\gamma-1} = \frac{T_2}{T_1} = \frac{V_2}{V_1}$$

or
$$V_3 = \left(\frac{V_2^{\gamma}}{V_1}\right)^{\frac{1}{\gamma-1}} = \frac{V_2^{\gamma/\gamma-1}}{V_1^{1/\gamma-1}}$$

Hence
$$\frac{V_3}{V_1} = \frac{V_2^{\gamma/\gamma-1}}{V_1^{\gamma/\gamma-1}} = \left(\frac{V_2}{V_1}\right)^{\frac{\gamma}{\gamma-1}}$$

or change in entropy
$$S_2 - S_1 = R\log_e \left(\frac{V_2}{V_1}\right)^{\frac{\gamma}{\gamma-1}} = R\left(\frac{\gamma}{\gamma-1}\right)\log_e \frac{V_2}{V_1} \qquad \text{... (5.20)}$$

5.14 ENTROPY OF A PERFECT GAS

Consider one mole of a perfect gas at a pressure P and temperature T having a volume V. If dQ is the amount of heat given to the gas, then it is used partly in raising the internal energy by an amount dU and thus the temperature of the gas rises through an amount dT at constant volume and partly in doing work in expansion of the gas by an amount dV against the external pressure P. If C_v is the molar specific heat of the gas at constant volume, then

$$dQ = dU + dW = C_v\,dT + P\,dV$$

$\therefore$ Change of entropy $dS = \dfrac{dQ}{dT} = \dfrac{C_v dT + P\,dV}{T}$

or Entropy $S = \int dS = \int C_v \dfrac{dT}{T} + \int \dfrac{P}{T} dV$ ... (*i*)

(*i*) **Entropy in terms of volume and temperature.** For a pertect gas $PV = RT$

$\therefore$ $$\frac{P}{T} = \frac{R}{V}$$ ... (*ii*)

Substituting in Eq. (*i*) we hav, entropy for one mole of perfect gas at temperature T and volume V

$$S = \int dS = \int C_v \frac{dT}{T} + \int R \frac{dV}{V}$$ ... (*ii*) (*a*)

or $$= C_v \log_e T + R \log_e V + \text{constant}$$ ... (*iii*)

or $$S = C_v \log T + (C_p - C_v) \log_e V + \text{constant}$$ ... (*iii*) (*a*)

as $$R = (C_p - C_v)$$

(*ii*) *Entropy in terms of pressure and temperature*. Again $PV = RT$

$\therefore$ $$V = R\frac{T}{P}$$ ... (*iv*)

and $$PdV + VdP = RdT$$ ... (*v*)

Substituting $PdV = RdT - VdP$ in (*i*), we have

$$S = C_v \int \frac{dT}{T} + \int \left[\frac{RdT - VdP}{T} \right]$$

$$= C_v \int \frac{dT}{T} + \int \left[\frac{RdT - R\frac{T}{P}dP}{T} \right]$$

$$= C_v \int \frac{dT}{T} + R \int \left[\frac{dT}{T} - \frac{dP}{P} \right]$$

$$= (C_v + R) \int \frac{dT}{T} - R \int \frac{dP}{P}$$

$$= C_p \int \frac{dT}{T} - R \int \frac{dP}{P}$$

$\therefore$ $$S = C_p \log_e T - R \log_e P + \text{constant}$$ ... (*vi*)

$$= C_p \log_e T - (C_p - C_v) \log_e P + \text{constant}$$ ... (*vi*) (*a*)

(*iii*) **Entropy in terms of pressure and volume.** Since $PV = RT$

$$T = \frac{PV}{R}$$

and $$dT = \frac{PdV + VdP}{R}$$ ... (*viii*)

Substituting $dT = \dfrac{PdV + VdP}{R}$ in Eq. (*i*), we get

$$S = \int C_v \frac{PdV + VdP}{RT} + \int \frac{PdV}{T}$$

$$= C_v \int \frac{PdV + VdP}{PV} + \int \frac{PdV}{PV} R$$

$$= C_v \int \left(\frac{dV}{V} + \frac{dP}{P} \right) + (C_p - C_v) \int \frac{dV}{V}$$

$$= C_p \int \frac{dV}{V} + C_v \int \frac{dP}{P}$$

or $$S = C_p \log_e V + C_v \log_e P + \text{constant} \qquad \text{... }(ix)$$

Entropy for *n*-moles (*in terms of V and T*)

Entropy for n moles $= n\, C_v \ln T + n\, R \ln V + \text{constant}$

$= nC_v \ln T + n\, (C_p - C_v) \ln V + \text{constant}$

(*In terms of P and T*)

Entropy for n moles $= n\, C_p\, ln\, T = nR\, ln\, P + \text{constant}$

$= n\, C_p\, ln\, T = n\, (C_p - C_v) \ln P + \text{constant}$

(*Interms of P and V*)

Entropy for n moles $= nC_p \ln V + nC_v \ln P + \text{constant}$

Entropy per gram. If M is gram molecular mass, then entropy per gram (in terms of V and T)

$$S = \left[\frac{C_v}{M} \log_e T + \frac{R}{M} \log_e V \right] + \text{another constant}$$

$$= C_v \log_e T + r \log_e V + \text{constant}$$

which c_v is the specific heat per gram at constant volume and r is gas constant per gram

$$= C_v \log_e T + (C_p - C_v) \log_e V + \text{constant}$$

where c_p is specific heat per gram at constant pressure.

(*In terms of P and T*)

Entropy per gram $\quad S = c_p \log_e T - r \log_e P + \text{constant}$

(*In terms of P and V*)

Emtropy per gram $\quad S = c_p\, log_e V + c_v \log_e P + \text{constant}$

Entropy of *m* gms. For m grams of a gas, entropy

(*In terms of V and T*)

$$mS = m\, c_v \log_e T + mr \log_e V + \text{constant}$$

$$= mc_v \log_e T + m\, (C_p - C_v) \log_e V + \text{constant}$$

or $$mS = \frac{m}{M} C_V \log_e T + \frac{m}{M} (R) \log_e P + \text{constant}$$

$$= \frac{m}{M} C_p \log_e T - \frac{m}{M} (C_p - C_v) \log_e P + \text{constant}$$

(*In terms of P and T*)

$$mS = m\, c_p \log_e T - mr \log_e P + \text{constant}$$

$$= m\, c_p \log_e T - m\, (c_p - c_v) \log_e P + \text{constant}$$

or $$mS = \frac{m}{M} C_p \log_e T - \frac{m}{M} R \log_e P + \text{constant}$$

$$= \frac{m}{M} C_p \log_e T - \frac{m}{M} (C_P - C_V) \log_e P + \text{constant}$$

(In terms of P and V)

$$mS = m\log_e V + mc_v \log_e P + \text{constant}$$

$$= \frac{m}{M} C_p \log_e V + \frac{m}{M} C_v \log_e P + \text{constant}$$

Change of entropy. To find the change in entropy from a temperature T_a and voluem V_a to a temperature T_b and volume V_b integrate the expression (*ii*) (*a*) between the corresponding limits.

$$\therefore \text{ Change in entropy } S_b - S_a = C_v \int_{T_a}^{T_b} \frac{dT}{T} + R \int_{V_a}^{V_b} \frac{dV}{V}$$

$$= C_v \log_e \left(\frac{T_b}{T_a} \right) + R \log_e \left(\frac{V_b}{V_a} \right) = C_v \log_e \left[\frac{T_b}{T_a} \right] + (C_p - C_v) \log_e \left[\frac{V_b}{V_a} \right]$$

where S_b is the entropy of the gas at temperature T_b and volume V_b and S_a the entropy at temperature T_a and volume V_a.

For *n* moles, change in entropy

$$= n.C_v \log_e \left(\frac{T_b}{T_a} \right) + nR \log_e \left(\frac{V_b}{V_a} \right) = nC_v \log_e \left[\frac{T_b}{T_a} \right] + n(C_p - C_v) \log_e \left[\frac{V_b}{V_a} \right]$$

If *M* is the gram molecular mass, then entropy of *m* gram of the perfect gas is given by

$$\frac{m}{M}(S_b - S_a) = \frac{m}{M} \left\{ C_v \log_e \left(\frac{T_b}{T_a} \right) + R \log_e \left(\frac{V_b}{V_a} \right) \right\} = \frac{m}{M} \left\{ C_v \log_e \left[\frac{T_b}{T_a} \right] + (C_p - C_v) \log_e \left[\frac{V_b}{V_a} \right] \right\}$$

$$= mc_v \log_e \left(\frac{T_b}{T_a} \right) + mr \log_e \left(\frac{V_b}{V_a} \right) = mc_v \log_e \left[\frac{T_b}{T_a} \right] + m(c_p - c_v) \log_e \left[\frac{V_b}{V_a} \right]$$

where c_v is the specific heat per gram at constant volume, c_p the specific heat per gm at constant pressure and *r* the gas constant per gram.

Note. The value of $S_b - S_a$ in terms of *T* and *P* and in terms of *P* and *V* for one mole will be as under

(*i*) *In terms of T and P*

$$S_b - S_a = C_p \log_e \left[\frac{T_b}{T_a} \right] - R \log_e \left[\frac{P_b}{P_a} \right]$$

$$= C_p \log_e \left[\frac{T_b}{T_a} \right] - (C_p - C_v) \log_e \left[\frac{P_b}{P_a} \right]$$

(*ii*) *In terms of P and V*

$$S_b - S_a = C_p \log_e \left[\frac{V_b}{V_a} \right] + C_v \log_e \left[\frac{P_b}{P_a} \right]$$

Similarly corresponding expression for *n* moles and *m* gms can be written.

5.15 ENTROPY OF A STEAM

Consider 1 gm. of ice at temperature T_1. Let us calculate the total change in entropy when this ice changes into steam at absolute temperature T_2.

If *dQ* is the amount of heat supplied to a substance at temperature *T*, the change in entropy is given by

$$dS = \frac{dQ}{T}$$

In order to convert 1 gm. of ice at T_1 K into water at T_1 K, an amount of heat absorbed is equal to latent heat of ice, L_i, say.

$\therefore$ Change in entropy during this process $= \frac{L_i}{T_1}$ ... (*i*)

Now 1 gm. of water at temperature T_1 is heated to T_2 K. The change in entropy during the process

$$= \int_{T_1}^{T_2} \frac{dQ}{T} = \int_{T_1}^{T_2} C\frac{dT}{T} = C\log_e \frac{T_2}{T_1} \quad \text{... (ii)}$$

where C is the specific heat of water, a constant quantity.

Further, 1 gm of water at T_2 K is converted into steam at the same temperature T_2 K. It will absorb heat L_s, equal to the latent heat of steam. Thus,

Change of entropy during this process $= \frac{L_S}{T_2}$ ... (*iii*)

$\therefore$ Adding (*i*), (*ii*) and (*iii*), we get total change in entropy during the complete process,

$$\Delta S = \frac{L_i}{T_1} + C\log_e \frac{T_2}{T_1} + \frac{L_S}{T_2} \quad \text{... (iv)}$$

This equation gives the total entropy of 1 gm. of steam at temperature T_2 K. For *m* gm. of substance, we will multiply eqn. (*iv*) by *m*, to give the total change in entropy during the process. *i.e.*, for converting the ice into steàm.

5.16 KELVIN'S THERMODYNAMIC SCALE OF TEMPERATURE

In the investigation and measurement of temperature all scales depend upon the properties of a particular substance, *e.g.*, expansion of mercury, the change in resistance of platinum, etc., with rise of temperature. The nature of the substance and the property used for finding the temperature influence the value of temperature measured. So the temperatures measured are *arbitrary* and show a remarkable degree of disagreement.

Lord Kelvin, from a study of the efficiency of a reversible heat engine and application of second law of thermodynamics, obtained a new temperature scale which is independent of the properties of any particular substance and is, therefore, non-arbitrary. *Such a scale is called the absolute, work or Kelvin scale or thermodynamic scale of temperature.*

The efficiency of all reversible Carnot engines, working between the two temperatures is a function of theose two temperatures alone. If Q_1 is the quantity of heat absorbed at a higher temperature θ_1 and Q_2 is the quanity of heat given out at a lower temperature θ_2 where θ_1 and θ_2 are the temperatures measured on any arbitrary scale then efficiency η can be written as

$$\eta = \frac{Q_1 - Q_2}{Q_1} = f(\theta_1, \theta_2)$$

or

$$= 1 - \frac{Q_2}{Q_1} = f(\theta_1, \theta_2)$$

$$\therefore \quad \frac{Q_1}{Q_2} = F(\theta_1, \theta_2) \quad \text{... (i)}$$

where

$$F(\theta_1, \theta_2) = \frac{1}{1 - f(\theta_1, \theta_2)}$$

Similarly if we have another reversible heat engine working between the temperatures θ_2 and θ_3 ($\theta_2 > \theta_3$), then

$$\frac{Q_2}{Q_3} = F(\theta_2, \theta_3)$$

Multiplying (*i*) and (*ii*) we have

$$\frac{Q_1}{Q_3} = F(\theta_1, \theta_2) \times F(\theta_2, \theta_3) \qquad \text{... (iii)}$$

Now if we have a reversible heat engine, working between the temperature θ_1 and θ_3, such that it absorbs an amount of heat Q_1 at a temperature θ_1 and rejctcts an amount of heat Q_3 at a temperature θ_3, then

$$\frac{Q_1}{Q_3} = F(\theta_1, \theta_3) \qquad \text{... (iv)}$$

Comparing (*iii*) and (*iv*), we have

$$F(\theta_1, \theta_3) = F(\theta_1, \theta_2) \times F(\theta_1, \theta_3) \qquad \text{... (v)}$$

In equation (*v*) we find that θ_2 is not present on the left-hand side, but is present in both the factors on the right-hand side. Therefore the form of $F(\theta_1, \theta_2)$ must be given by

$$F(\theta_1, \theta_2) = \frac{\phi(\theta_1)}{\phi(\theta_2)} \qquad \text{... (vi)}$$

and

$$F(\theta_2, \theta_3) = \frac{\phi(\theta_2)}{\phi(\theta_3)} \qquad \text{(vii)}$$

where ϕ is another unknown function of θ. Multiplying (*vi*) and (*vii*), we get

$$F(\theta_1, \theta_2) \times F(\theta_2, \theta_3) = \frac{\phi(\theta_1)}{\phi(\theta_2)} \times \frac{\phi(\theta_2)}{\phi(\theta_3)} = \frac{\phi(\theta_1)}{\phi(\theta_3)} = F(\theta_1, \theta_3)$$

Representing $\phi(\theta)$ by τ, the temperature on the thermodynamic scale, we get

$$\frac{Q_1}{Q_2} = \frac{\tau_1}{\tau_2}$$

Thus thermodynamic or kelvin scale of temperature is that scale on which any two temperatures are in the same ratio as the quantities of heat absorbed and rejected by Carnot's reversible heat engine operating between these two temperatures.

Alternative Method (using Carnot's cycle)

Consider a Carnot's reversible cycle *ABCD* (Fig. 5.7). The two isothermals are *AB* and *CD*. The temperature on the work scale is θ_1 and the ideal scale is T_1 for the isothermal *AB* and the temperature on the work scale is θ_2 and the ideal gas scale is T_2 for the isothermal *CD*. Let the pressure and volume for the points *A, B, C* and *D* be $P_1, V_1; P_2, V_2; P_3, V_3$ and P_4, V_4 respectively. Consider that 1 gram of an ideal gas is used as the working substance.

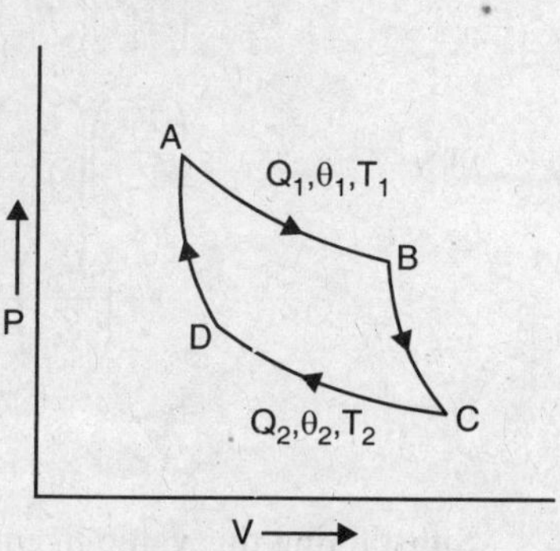

Fig. 5.7

From *A* to *B*, the process is isothermal.

$$\text{Work done} = \int_{V_1}^{V_2} P\,dV$$

$$= rT_1 \int_{V_1}^{V_2} \frac{dV}{V}$$

$$\therefore \qquad Q_1 = rT_1 \log \frac{V_2}{V_1} \qquad \text{... (i)}$$

Here r is expressed in heat units.

From C to D, the process is isothermal and a quantity of heat Q_2 is rejected by the gas.

$$Q_2 = -\int_{V_3}^{V_4} PdV$$

$$= -rT_2 \int_{V_3}^{V_4} \frac{dV}{V}$$

$$Q_2 = rT_2 \log \frac{V_2}{V_4} \qquad \text{... (ii)}$$

Dividing (*i*) by (*ii*)

$$\frac{Q_1}{Q_2} = \frac{rT_1 \log \frac{V_2}{V_1}}{rT_2 \log \frac{V_3}{V_4}} \qquad \text{... (iii)}$$

Also, A and B lie on the same isothermal

$$P_1 V_1 = P_2 V_2$$

or

$$\frac{P_2}{P_1} = \frac{V_1}{V_2} \qquad \text{... (iv)}$$

C and D lie on the same isothermal

$$P_3 V_3 = P_4 V_4$$

or

$$\frac{P_3}{P_4} = \frac{V_4}{V_3} \qquad \text{... (v)}$$

B and C lie on the same adiabatic

$$P_2 V_2^{\gamma} = P_3 V_3^{\gamma} \qquad \text{... (vi)}$$

A and D lie on the same adiabatic

$$P_1 V_1^{\gamma} = P_4 V_4^{\gamma} \qquad \text{(vii)}$$

Dividing (*vi*) by (*vii*)

$$\left(\frac{P_2}{P_1}\right)\left(\frac{V_2}{V_1}\right)^{\gamma} = \left(\frac{P_3}{P_4}\right)\left(\frac{V_3}{V_4}\right)^{\gamma}$$

or

$$\left(\frac{V_1}{V_2}\right)\left(\frac{V_2}{V_1}\right)^{\gamma} = \left(\frac{V_4}{V_3}\right)\left(\frac{V_3}{V_4}\right)^{\gamma}$$

$$\left(\frac{V_2}{V_1}\right)^{\gamma-1} = \left(\frac{V_3}{V_4}\right)^{\gamma-1}$$

or

$$\frac{V_2}{V_1} = \frac{V_3}{V_4} \qquad \text{... (viii)}$$

Substituting this value in equation (*iii*)

$$\frac{Q_1}{Q_2} = \frac{T_1}{T_2} \qquad \text{... (ix)}$$

From the work scale of temperature,

$$\frac{Q_1}{Q_2} = \frac{\theta_1}{\theta_2} \qquad \text{... (x)}$$

$$\therefore \quad \frac{\theta_1}{\theta_2} = \frac{T_1}{T_2} \qquad \text{... (xi)}$$

Special considerations

(*i*) If θ_1 is zero, T_1 is also zero. It means the two scales are identical at absolute zero temperature.

(*ii*) Consider that the Carnot's engine works between the steam point and the ice point and there are 100 degrees between the ice point and the steam point on the two scales.

$$\frac{\theta_{steam}}{\theta_{ice}} = \frac{T_{steam}}{T_{ice}}$$

$$\frac{\theta_{ice} + 100}{\theta_{ice}} = \frac{T_{ice} + 100}{T_{ice}}$$

or $$\theta_{ice} = T_{ice}$$

Therefore, the ice point is the same on the two scales.

Similarly $\theta_{steam} = T_{steam}$

Hence the two scales are completely identical.

Efficiency. The efficiency of a Carnot's engine,

$$\eta = 1 - \frac{Q_2}{Q_1}$$

$$= 1 - \frac{\theta_2}{\theta_1} = 1 - \frac{T_2}{T_1}$$

Let the adiabatic expansion ratio be ρ

$$\rho = \frac{V_3}{V_2}$$

The points B and C lie on the same adiabatic

$$\therefore \quad T_1 V_2^{\gamma-1} = T_2 V_3^{\gamma-1}$$

$$\frac{T_2}{T_1} = \left(\frac{V_2}{V_3}\right)^{\gamma-1}$$

$$= \left(\frac{1}{\rho}\right)^{\gamma-1}$$

$$\eta = 1 - \frac{T_2}{T_1} = 1 - \left(\frac{1}{\rho}\right)^{\gamma-1} \qquad \text{... (5.12)}$$

5.17 ZERO OF ABSOLUTE OR THERMODYNAMIC SCALE

The efficiency of Carnot's reversible heat engine is

$$\eta = \frac{Q_1 - Q_2}{Q_1} = 1 - \frac{Q_2}{Q_1} \qquad \text{... (i)}$$

On the thermodynamic scale,

$$\frac{Q_2}{Q_1} = \frac{T_2}{T_1}$$

$\therefore \qquad \eta = 1 - \frac{\tau_2}{\tau_1}$

Suppose $\tau_2 = 0$ then $\eta = 1$

Substiuting $\eta = 1$ in (*i*) we get $Q_2 = 0$

Thus zero of absolute or thermodynamic sclae is defined as that temperature of the sink at which the reversible heat engine undergoing Carnot's cycle rejects no heat to it.

Thermodynamic scale independent of working substance. The thermodynamic scale has been derived on the basis of the efficiency of a reversible Carnot's heat engine which depends only on the temperature of the source and sink but is independent of the working substance. Hence the thermodynamic scale of temperature is also independent of the properties of the working substance.

Relation between thermodynamic or perfect gas scale. As proved above the efficiency of a reversible heat engine is one when the temperature of the sink $\tau_2 = 0$. A temperature less than $\tau_2 = 0$ is not possible, because in that case the efficiency of the heat engine will be more than one, which is not possible.

On the gas scale also, the efficiency of a heat engine

$$\eta = \frac{T_1 - T_2}{T_1} = 1 - \frac{T_2}{T_1}$$

which will again be one, when T_2 = 0K, as a temperpature lower than 0K is not possible, because the molecules have no motion at absolue zero and consequently the molecules do not possess any thermal energy. At a temperature lower than absolue zero, the thermal energy will be negative, which is not possible. *Hence the zero of the perfect gas scale and the thermodynamic scale is the same.*

Since the efficiencies in both the cases have the same values, we have

$$\frac{T_2}{T_1} = \frac{\tau_1}{\tau_2}$$

This shows that the ratio of any two temperatures on the perfect gas scale and thermodynamic scale are equal.

If now T_1 is the absolute temperature of boiling water (*Steam point*) and T_2 that of melting ice (*ice point*), then

$$T_1 - T_2 = 100$$

Size of the degree. To determine the size of a degree on the thermodynamic scale, consider a Carnot's reversible engine, working between the temperatures of boiling water and melting ice at normal pressure. The Carnot cycle is represented by *ABCD* and the work done by the engine is numerically equal to area *ABCD* (Fig. 5.8). Le this area be divided into 100 equal parts by isothermals drawn parallel to *AB* and *CD*; then the area of each part represents one degree on the thermodynamic scale. *Hence a thermodynamic degree is defined as the difference in temperature between which a Carnot's engine should be worked to get energy equal to 1/100 th of the energy obtained from Carnot's engine working between the temperatures of boiling water and melting ice.*

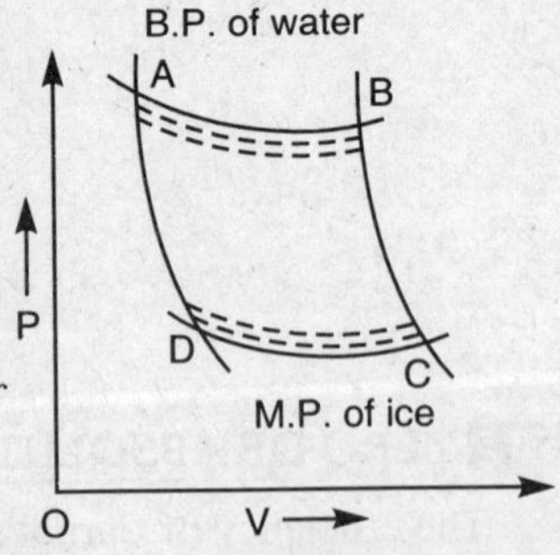

Fig. 5.8

Therefore on the thermodynamic scale, for the same difference of temperature, we have $\tau_2 - \tau_2 = 100$

Hence using the thermodynamic scale, we get

$$\eta = \frac{Q_1 - Q_2}{Q_1} = \frac{\tau_1 - \tau_2}{\tau_1} = \frac{100}{\tau_1}$$

and on the perfect gas scale, we have

$$\eta = \frac{Q_1 - Q_2}{Q_1} = \frac{T_1 - T_2}{T_1} = \frac{100}{T_1}$$

Since the efficiencies are the same

$$\frac{100}{\tau_1} = \frac{100}{T_1}$$

i.e. the absolute temperature of the boiling point of water is the same, whether measured on the perfect gas scale of temperature or on the thermodynamic scale of temperature.

In a similar manner it can be shown that any other temperature will have the same value on the two scales. Hence the two scales are identical.

5.18 THIRD LAW OF THERMODYNAMICS : NERNST'S HEAT THEOREM

In 1906, Nernst proposed a general principle supported by a series of experimental evidences on the problem of atomic heat at low temperature. Actually, it was proposed as '*the new heat theorem*' but its importance is so great that today it is recognized as 'the *third law of thermodynamics*'.

Statement: According to Nernst, "*The heat capacities of all solids tend to zero as the absolute zero of temperature is approached and that the internal energies and entropies of all substances become equal there, approching their common value asymptotically tending to zero*". This simple statement of Nernst follows neither from the first law (law of conservation of energy) nor from the second law (law of transmutability of energy) and is thus of the nature of a new law, usually called as *third law of thermodynamics*.

This theorem is useful in explaining the nature of bodies in the neighbour hood of absolute zero temperature. Its importance lies in the fact that it permits the calculations of absolute values of entropy, Helmholtz and Gibb's free energies, etc. In terms of entropy, the theorem may also be stated as:

It states that "at absolute zero temperature, the entropy tends to zero and the molecules of a substance or a system are in perfect order (well arranged)".

In all heat engines, there is always loss of heat in the form of conduction, radiation and friction. Therefore, in actual heat engines $\frac{Q_1}{T_1}$ is not equal to $\frac{Q_2}{T_2}$.

$\therefore \left(\frac{Q_1}{T_1} - \frac{Q_2}{T_2}\right)$ is not zero but it is a positive quantity. When cycle after cycle is repeated, the entropy of the system increases and attains a maximum value. When the system has attained the maximum value, a stage of stagnancy is reached and no work can be done by the engine at this stage. In the universe; the entropy is increasing and ultimately the universe will also reach a maximum value of entropy when no work will be possible. With the increase in entropy, the disorder of the molecules of a substance increases. The entropy is also a measure of the disorder of the system. With decrease in entropy, the disorder decreases. At absolute zero temperature, the entropy tends to zero and the molecules of a substance or a system are in perfect order (well arranged). This is known as the third law of thermodynamics.

Example: The molecules are more free to move in the gaseous state than in the liquid state. The entropy in the gaseous state is more than in the liquid state. The entropy in liquid state is more than in the solid state. Thus, when a substance is converted from solid to a liquid and then from the liquid to the gaseous state, the entropy increases and vice-versa. When ice is converted into water and then into steam, the entropy and disorder of the molecules increase. When steam is converted into water and then into ice, the entropy and disorder of the molecules decrease. Hence, entropy is a measure of the disorder of the molecules of the system.

By any ideal procedure, it is impossible to bring any system to absolute zero temperature performing a finite number of operations. This is called the principle of unattainability of absolute zero. Thus, according to Fowler and Guggenheim, the unattainability principle is called the third law of thermodynamics.

5.19 ZERO POINT ENERGY

According to Kinetic theory, the energy of a system at absolute zero should be zero. It means the molecules of the system do not possess any motion. But according to the modern concept, even at absolute zero, the molecules are not completely deprived of their motion and hence possess energy. The energy of the molecules at absolute zero temperature is called **zero point energy**.

5.20 NEGATIVE TEMPERATURES (NOT POSSIBLE)

From the validity point of view of the laws of thermodynamics, there cannot be any temperature less than zero Kelvin *i.e.*, a negative temperature is not possible on Kelvin's scale. This can be logically explained as under:

If at all negative temperature exists and is attainable, then let us assume the sink of a heat engine be at some negative temperature $T_2 = -T_2$ say. Then the efficiency would be

$$\eta = 1 - \left(\frac{-T_2}{T_1}\right) = 1 + \frac{T_2}{T_1}$$

$$\therefore \quad \eta > 1 \qquad \text{...(i)}$$

i.e., the efficiency of engine is greater than unity which is against the second law thermodynamics and hence impossible. In other words, the above conclusion obtained is in contradiction with the laws of thermodynamics, as it violates them. Hence the assumption made that the negative temperature exists is not true. Thus, there cannot be a negative temperature.

The laws of thermodynamics prohibits a heat engine to have 100% efficiency. The necessary condition for 100% efficiency is that the sink should be at the absolute zero and since nature does not provide us with a sink at absolute zero (which is unattainable) an engine with 100% efficiency is not possible.

The specific heat of a substance decreases with increase in temperature. However, the specific heat does not tend to be zero as the temperature tends to infinity. This shows that the temperature† has a + ve sign only.

5.21 HEAT DEATH OF UNIVERSE

In the universe, the natural processes that are taking place such as flow of heat, conduction, radiation, mixing, diffusion and so on are all irreversible and in these processes, the system tends to have maximum value of entropy by attaining a uniform state of temperature, pressure, composition etc. Therefore we conclude that the entire universe would attain a state of absolute uniformity at some distant future. In such a state all physical, chemical and biological processes would stop, although there

† However, experiments by Ramsey (1956) have shown that a part of a system *i.e.,* the nucleus of a solid, can have a negative temperature. This sub-system is considered isolated from the main system (*i.e.,* solid lattice). The specific heat of the sub-system tends to zero at high temperature. A small amount of heat energy tends to raise the temperature of the system to infinity. It is possible to add still more energy to the sub-system at infinity and it forces the sub-system into the negative temperature region. It has been shown by microscopic statistical analysis that there is no distinction between the temperature of $+\infty$ and $-\infty$. In thermodynamics, the parameter 1T is more significant than T. The negati ve temperatures are hotter than the positive temperatures.

The negative temperature is not possible with the system as a whole and is only an exception to the rule that only positive temperatures exist. The negative temperatures are possible only for isolated sub-systems. Fr all normal purposes the temperatures are always positive.

would be no change in the total energy of the universe. As there will be no temperature difference, it will not be possible to convert any amount of heat into even a small amount of useful work. This means that we are heading towards the **Heat death** of the universe. This state being inevitably approached; although it may take millions of billions of years. Today, we are facing with evidence the problems of global warming. The consequence of this will be heat death of universe.

SOLVED EXAMPLES

Example 5.1. *Calculate the change in entropy when 5 kg of water at 100°C is converted into steam at the same temperature. (Given; Latent heat of steam = 540 cal/gram)* ***(Alagappa 2004)***

Solution. Heat absorbed by 5 kg of water at 100°C when it is converted into steam at 100°C

$$\delta Q = \text{mass} \times \text{latent heat of vaporisation of steam}$$
$$= 5000 \times 540$$
$$\therefore \quad \delta Q = 2700000 \text{ cal}$$

The gain in entropy,
$$dS = \frac{\delta Q}{T}$$
$$= \frac{2700000}{373}$$
$$\mathbf{dS = 7240 \text{ cal/K}}$$

Example 5.2 *Calculate the change in entropy when 10 grams of ice at 0°C is converted into water at the same temperature. (Given: Latent heat of ice. = 80 cal/gram).*

(Rohilkhand 2001 Imp. Nagpur 2010, 2008, 2005/s, Meerut 2004, 2001, Purvanchal 2002)

Solution. Heat absorbed by 10 gram of ice at 0°C to convert it into water at 0°C is 10 × 80 = 800 cal.

$$\therefore \quad \delta Q = 800 \text{ cal}$$
$$T = 0°\text{C} = 273 \text{ K}$$

The gain in entropy,
$$dS = \frac{\delta Q}{T}$$
$$= \frac{800}{273}$$
$$\mathbf{dS = 2.93 \text{ cal/K}}$$

Example 5.3 *Calculate the increase in entropy when 1 gram of ice at – 10°C is converted into steam at 100°C. (Given: specific heat of ice = 0.5, Latent heat of ice = 80 cal/gram, Latent heat of steam = 540 cal/gram).* ***(Rohilkhand 2001, Lucknow 2004)***

Solution.

1. Increase in entropy when the temperature of 1 gram of ice increases from – 10°C to 0°C

$$dS = \int_{T_1}^{T_2} \frac{\delta Q}{T}$$
$$= ms \int_{T_1}^{T_2} \frac{dT}{T}$$
$$= ms \log_e \frac{T_2}{T_1}$$
$$= ms \times 2.303 \log_{10} \frac{T_2}{T_1}$$

$$= 1 \times 0.5 \times 2.303 \log_{10} \frac{273}{263}$$

$$= \mathbf{0.01865\ cal/K}$$

2. Increase in entropy when 1 gram of ice at 0°C is converted into water at 0°C

$$dS = \frac{\delta Q}{T}$$

$$= \frac{1 \times 80}{273} = \mathbf{0.293\ cal/K}$$

3. Increase in entropy when 1 gram of water is raised from 0°C to 100°C

$$dS = \int_{T_1}^{T_2} \frac{\delta Q}{T}$$

$$= ms \times 2.303 \log_{10} \frac{T_2}{T_1}$$

$$= 1 \times 1 \times 2.303 \times \log_{10} \frac{373}{273}$$

$$= \mathbf{0.312\ cal/K}$$

4. Increase in entropy when 1 gram of water at 100°C is converted into steam at 100°C

$$dS = \frac{\delta Q}{T}$$

$$= \frac{1 \times 540}{373} = 1.447 \text{ cal/K}$$

$\therefore$ Total increase in entropy = 0.01865 + 0.293 + 0.312 + 1.447

= 2.07065 cal/K

Example 5.4 *One gram molecule of a gas expands isothermally to four times of its volume. Calculate the change in its entropy in terms of the gas constant.* ***(Agra 2000)***

Solution. Work done $= \int_{V_1}^{V_2} PdV$

But $PV = RT$

or $P = \frac{RT}{V}$

$\therefore$ $W = RT \int_{V_1}^{V_2} \frac{dV}{V} = RT \log_e \frac{V_2}{V_1}$

Here, $\frac{V_2}{V_1} = 4$

$\therefore$ $W = RT \times 2.303 \times \log_{10} (4)$

Here W and R are in units of work

$\therefore$ Gain in entropy $= \frac{\delta Q}{T}$

$$= \frac{W}{JT} = \frac{2.303\,RT}{JT} \log_{10}^{(4)}$$

$$\mathbf{dS = 1.387\ \frac{R}{J}\ cal/K}$$

Example 5.5 *50 grams of water at 0°C is mixed with an equal mass of water at 80°C. Calculate the resultant increase in entropy.* **(*Calcutta 2001*)**

Solution.

(*i*) $m_1 = 50$ gram ; $T_1 = 273$ K

$m_2 = 50$ gram, $T_2 = 273 + 80 = 353$ K

Let the final temperature of the mixture be TK

$$m_1 \, s \times (T - T_1) = m_2 \, s \, (T_2 - T)$$

$$50 \times 1 \times (T - 273) = 50 \times 1 \times (353 - T)$$

$$\therefore \quad T = 313 \text{ K}$$

(*ii*) Change in entropy of 50 gm of water when its temperature rises from 273 K to 313 K will be

$$dS = \frac{\delta Q}{T}$$

$$= ms \int_{T_1}^{T} \frac{dT}{T}$$

$$= 50 \times 1 \log_e \frac{313}{273}$$

$$= 50 \times 2.303 \log_{10} \frac{313}{273}$$

$$= \mathbf{+\,6.829 \text{ cal/K}}$$

Here, the + ve sign indicates gain in entropy.

(*iii*) Change in entropy when temperature of water falls from 353 K to 313 K

$$dS = \frac{\delta Q}{T} = ms \int_{T_2}^{T} \frac{dT}{T}$$

$$= 50 \times 1 \times \log_e \frac{313}{353}$$

$$= 50 \times 2.3026 \times \log_{10} \frac{313}{353}$$

$$= \mathbf{-\,6.023 \text{ cal/K}}$$

Here, – ve sign indicates loss in entropy.

Therefore, the total gain in entropy of the system

$$= 6.829 - 6.023$$

$$= \mathbf{0.806 \text{ cal/K}}$$

Example 5.6 *1 gram of water at 20°C is converted into ice at – 10°C at constant pressure. Heat capacity at 1 gm of water is 4.2 J/g.K and that of ice is 2.1 J/g. K. Heat of fusion of ice at 0°C = 335 J/g. Calculate the total change in the entropy of the system.* **(*Bangalore 2000*)**

Solution.

(*i*) Change in entropy when the temperature of 1 gram of water at (273 + 20) = 293 K falls to 273 K.

$$dS = \frac{\delta Q}{T} = ms \int_{T_1}^{T_2} \frac{dT}{T}$$

$$= 1 \times 4.2 \int_{293}^{273} \frac{dT}{T}$$

$$= 4.2 \times 2.3026 \log_{10}\left(\frac{273}{293}\right)$$
$$= -0.2969 \text{ J/K}$$

(*ii*) Change in entropy when 1 gm of water at 273 K is converted into ice at 0°C *i.e.*, 273 K

$$dS = \frac{\delta Q}{T} = \frac{-1 \times 335}{273} = -1.227 \text{ J/K}$$

(*iii*) Change of entropy when the temperature of 1 gram of ice at 273 K falls to 263 K

$$dS = \frac{\delta Q}{T} = ms \int_{T_1}^{T_2} \frac{dT}{T}$$

$$= 1 \times 2.1 \times 2.3026 \times \log_{10}\left(\frac{263}{273}\right)$$
$$= -0.07834 \text{ J/K}$$

∴ Total change in entropy of the system

$$= -0.2969 - 0.07834 - 1.227$$
$$= \mathbf{-1.60224 \text{ J/K}}$$

Negative sign indicates that there is a decrease in entropy of the system.

Example 5.7 *1 kg of water at 273* K *is brought in contact with a heat reservoir at 373* K*. Then (1) what is the change in entropy of water when its temperature reaches 373* K*? (2) what is the change in entropy of (i) the reservoir and (ii) the universe.* **(*Purvanchal 2000*)**

Solution. 1. Increase in entropy when the temperature of 1000 gram of water is raised from 273 K to 373 K

$$dS = \int_{T_1}^{T_2} \frac{\delta Q}{T}$$

$$= ms \times 2.3026 \log_{10} \frac{T_2}{T_1}$$

$$= 1000 \times 1 \times 2.3026 \times \log_{10} \frac{373}{273}$$
$$= \mathbf{+312 \text{ cal/K}}$$

2. (*i*) Change in entropy of the reservoir

$$dS = -\frac{\delta Q}{T}$$

$$= -\frac{1000 \times 1 \times 100}{373} = \mathbf{-268.1 \text{ cal/K}}$$

Negative sign shows decrease in entropy.

(*ii*) Change in entropy of the universe

$$= 312 - 268.1 = \mathbf{43.9 \text{ cal/K}}$$

Therefore, the net increase in entropy of the universe

$$= \mathbf{43.9 \text{ cal/K}}$$

Example 5.8 *Calculate the change in entropy when 10 g of water at 60°C is mixed with 30 g of water at 20°C.* **(*Purvanchal 2005, Calcutta 2004*)**

Solution.

(*i*) $m_1 = 10\text{g}$ $\quad T_1 = 333 \text{ K}$

$m_2 = 30\text{g}$ $\quad T_2 = 293 \text{ K}$

Let the final temperature be T K

$$m_1 s (T_1 - T) = m_2 s (T - T_2)$$

$$10 \times 1 \times (333 - T) = 30 \times 1 (T - 293)$$

$$T = 303 \text{ K}$$

(*ii*) Change in entropy when the temperature of 10 g of water changes from 333 K to 303 K

$$\frac{\delta Q}{T} = ms \int_{T_1}^{T} \frac{dT}{T}$$

$$= 10 \times 1 \times \log_e \frac{303}{333}$$

$$= 10 \times 2.3026 \times \log_{10} \left(\frac{303}{333}\right)$$

$$= \mathbf{-0.9443 \text{ cal/K°}}$$

Negative sign indicates decrease in entropy.

(*iii*) Change in entropy when the temperature of 30 g of water change from 293 K° to 303 K°,

$$= \frac{\delta Q}{T} = ms \int_{T_2}^{T} \frac{dT}{T}$$

$$= 30 \times 1 \log_e \frac{303}{293}$$

$$= 30 \times 2.3026 \log_{10} \left(\frac{303}{293}\right)$$

$$= \mathbf{+1.002 \text{ cal/K°}}$$

The total gain in entropy of the system

$$= +1.0020 - 0.9443$$

$$= \mathbf{0.0577 \text{ cal/K°}}$$

Example 5.9 *Calculate the increase in entropy of 10 kg of water at 100°C when it changes to vapour.* **(*Lucknow 2001*)**

Solution. Increase in entropy when 10 kg of water at 100 °C is changed into vapour

$$dS = \frac{\delta Q}{T} = \frac{10 \times 10^3 \times 1 \times 540}{373}$$

$$dS = \mathbf{14{,}477 \text{ cal/K}}$$

Example 5.10 *When 50 g of water is heated from 10°C to 90°C, by how much does its entropy change?* **(*Rajasthan 2002*)**

Solution. Here $\quad T_1 = 273 + 90 = 363 \text{ K}$

$T_2 = 273 + 10 = 283 \text{ K}$

Increase in entropy when 50 g of water is heated from 10°C to 90°C

$$dS = \int_{T_1}^{T_2} \frac{\delta Q}{T}$$

$$dS = ms \times 2.3026 \log_{10}\left(\frac{T_2}{T_1}\right)$$

$$dS = 50 \times 1 \times 2.3026 \log_{10}\left(\frac{363}{283}\right)$$

$$dS = \mathbf{12.448\ cal/K}$$

Example 5.11 *Calculate the change in entropy when 10 grams of ice at 0°C is converted into steam at 100°C.* ***(Rohilkhand 1998, Kumaun 1998)***

Solution.

1. Increase in entropy when 10 grams of ice at 0°C is converted into water at 0°C.

$$dS = \frac{\delta Q}{T}$$

$$= \frac{10 \times 80}{273} = 2.93 \text{ cal/K}$$

2. Increase in entropy when the temperature of 10 g of water is raised from 0°C to 100°C

$$dS = \int_{T_1}^{T_2} \frac{\delta Q}{T}$$

$$= ms \times 1 \times 2.3026 \log_{10}\left(\frac{T_2}{T_1}\right)$$

$$= 10 \times 1 \times 2.3026 \log_{10}\left(\frac{373}{273}\right)$$

$$= \mathbf{3.12\ cal/K}$$

3. Increase in entropy when 10 g of water at 100°C is converted into steam at 100°C

$$dS = \frac{\delta Q}{T}$$

$$= \frac{10 \times 540}{373} = 14.47 \text{ cal/K}$$

Total increase in entropy = 2.93 + 3.12 + 14.47

= 20.52 cal/K

Example 5.12 *Calculate the change in entropy of 1 gram of nitrogen when its temperature rises from 50°C to 100°C while its volume is kept constant. Molar specific heat, C_v = 0.18 and molecular weight of nitrogen is 28.* ***(Madras 2001)***

Solution. Specific heat of nitrogen $= \frac{C_v}{28} = \left(\frac{0.18}{28}\right)$

Change in entropy, $$dS = \int_{T_1}^{T_2} \frac{\delta Q}{T}$$

$$dS = \frac{1 \times 0.18}{28} \int_{T_1}^{T_2} \frac{dT}{T}$$

$$dS = \left(\frac{0.18}{28}\right) \int_{323}^{373} \frac{dT}{T}$$

$$dS = \left(\frac{0.18}{28}\right) \times 2.3026 \ \log_{10}\left(\frac{373}{323}\right)$$

$$dS = \frac{0.18 \times 2.3026 \times 0.0625}{28}$$

$$\mathbf{dS = 9.253 \times 10^{-4}\ cal/K}$$

Example 5.13 *1 kg of water at 273* K *is brought in contact with a heat reservoir at 373* K*. What is the change in entropy of water as its temperature reaches 373* K*?* **(*Kanpur 2001*)**

Solution. Increase in entropy when the temperature of 1 kg of water is raised from 273 K to 373 K

$$dS = \int_{T_1}^{T_2} \frac{\delta Q}{T}$$

$$m = 1 \text{ kg}, \ s = 10^3 \text{ cal/kg-K}$$

$$dS = ms \times 2.3026 \log_{10}\left(\frac{T_2}{T_1}\right)$$

$$dS = 1 \times 10^3 \times 2.3026 \log_{10}\left(\frac{373}{273}\right)$$

$$dS = \mathbf{312\ cal/K.}$$

Example 5.14 *Calculate the change in entropy when 0.0273 kg of ice of 0°C is converted into water at the same temperature* **(*Garhwal 2003*)**

Solution. Here

$$m = 0.0273 \text{ kg}$$

$$L = 80 \text{ Cal/g} = 80 \times 10^3 \text{ Cal/kg}$$

$$T = 273 \text{ K}$$

$$dS = \frac{\delta Q}{T} \quad \text{or} \quad dS = \frac{mL}{T}$$

$$dS = \frac{0.0273 \times 80 \times 10^3}{273}$$

$$dS = 8 \text{ cal/K} = 8 \times 4.2 \text{ J/K}$$

$$= \mathbf{33.6\ J/K}$$

Example 5.15 *Calculate the increase in entropy of 1 kg. of ice when it is converted into steam. Given the specific heat of water is 1 k cal kg °C, latent heat of ice is 80 kcal/kg and the latent heat of steam is 540 k cal/kg.* **(*Rohilkhand 1999*)**

Solution.

1. Increase in entropy when 1 kg of ice at 0°C is converted into water at 0°C

$$dS = \frac{\delta Q}{T} = \frac{80}{273} = 0.293\,k \text{ cal/K}$$

2. Increase in entropy when the temperature of 1 kg of water is raised from 0°C to 100°C.

$$dS = \int_{T_1}^{T_2} \frac{\delta Q}{T} = ms \times 2.3026 \log_{10}\left(\frac{T_2}{T_1}\right)$$

$$= 1 \times 1 \times 2.3026 \log_{10}\left(\frac{373}{273}\right) = 0.312 \text{ k cal/K}$$

3. Increase in entropy when 1 kg of water at 100°C is converted into steam at 100°C.

$$dS = \frac{\delta Q}{T} = \frac{540}{373} = 1.447 \text{ k cal/K}$$

Total increase in entropy,

$$dS = 0.293 + 0.312 + 1.447$$

$$= \mathbf{2.052\ k\ cal/K.}$$

Example 5.16 *Calculate the change of entropy on converting a mole of a gas occupying 20 litres at a pressure of 2 × 10⁵ N/m² to occupying 50 litres at a pressure of 5 × 10⁵ N/m². Given*

$$R = 8.4 \text{ J mol}^{-1}\text{ K}^{-1}$$

$$C_v = 21 \text{ J mol}^{-1}\text{ K}^{-1}$$

(Kerala, 2001)

Solution. Here,

$$P_1 = 2 \times 10^5 \text{ N/m}^2$$

$$V_1 = 20 \text{ litres}$$

$$P_2 = 5 \times 10^5 \text{ N/m}^2$$

$$V_2 = 50 \text{ litres}$$

For a perfect gas,

$$\frac{P_1 V_1}{T_1} = \frac{P_2 V_2}{T_2}$$

$$\therefore \quad \frac{T_2}{T_1} = \frac{P_2 V_2}{P_1 V_1}$$

$$= \frac{5\times10^5 \times 50}{2\times10^5 \times 20}$$

$$= 6.25$$

Also,

$$\frac{V_2}{V_1} = \frac{50}{25} = 2.5$$

$$C_v = 21 \text{ J/mol} - \text{K}$$

$$R = 8.4 \text{ J/mol} - \text{K}$$

The change in entropy,

$$dS = C_v \times 2.3026 \log_{10}\left(\frac{T_2}{T_1}\right) + R \times 2.3026 \log_{10}\left(\frac{V_2}{V_1}\right)$$

$$= 21 \times 2.3026 \log_{10}(6.25) + 8.4 \times 2.3026 \log_{10}(2.5)$$

$$= 21 \times 2.3026 \times 0.7959 + 8.4 \times 2.3026 \times 0.3979$$

$$= 38.4852 + 7.6961$$

$$= 46.1813$$

$$= \mathbf{46.18 \text{ J/K}}$$

Therefore, increase in entropy = **46.18 J/K**

Example 5.17 *12.5 gm of ice at – 24°C is converted into steam at 100°C. Calculate the change in entropy. It is given that, latent heat of steam = 5.36 cal/gm, latent heat of ice = 80 cal/gm. specific heat of ice = 0.5 cal/gm–K.* **(Kumaun 2002)**

Solution. [In *SI* system],

$$m = 12.5 \text{ grams} = 12.5 \times 10^{-3} \text{ kg}$$

$$\text{Latent heat of ice} = 80 \text{ cal/gram} = 80 \times 10^3 \times 4.2 \text{ J/kg}$$

$$= 3.36 \times 10^5 \text{ J/kg}$$

$$\text{Latent heat of steam} = 536 \text{ cal/gram} = 536 \times 10^3 \times 4.2 \text{ J/kg}$$

$$= 2.25 \times 10^6 \text{ J/kg}$$

$$\text{Specific heat of ice} = 0.5 \text{ cal/gram} - \text{K}$$

$$= 0.5 \times 10^3 \times 4.2 \text{ J/kg} - \text{K}$$

$$= 2.1 \times 10^3 \text{ J/kg} - \text{K}$$

$$\text{Specific heat water} = 1 \text{ cal/gram} - \text{K}$$

$= 1 \times 10^3 \times 4.2$ J/kg – K

$= 4.2 \times 10^3$ J/kg – K

1. Increase in entropy when temperature of 12.5×10^{-3} kg of ice is increased from – 24°C to 0°C (249 K to 273 K)

$$dS = \int_{T_2}^{T_1} \frac{\delta Q}{T}$$

$$= ms \int_{T_1}^{T_2} \frac{dT}{T}$$

$$= ms \log_e \left(\frac{T_2}{T_1}\right)$$

$$= 12.5 \times 10^{-3} \times 2.1 \times 10^3 \log_e \left(\frac{273}{249}\right)$$

$$= 12.5 \times 2.1 \times 2.3026 \log_{10} \left(\frac{273}{249}\right)$$

$= \mathbf{2.42\ J/K}$

2. Increase in entropy when 12.5×10^{-3} kg of ice at 0°C is converted into water at 0°C

$$dS = \frac{\delta Q}{T} = \frac{mL}{T}$$

$$= \frac{12.5 \times 10^{-3} \times 3.36 \times 10^5}{273}$$

$= \mathbf{15.38\ J/K}$

3. Increase in entropy when the temperature of 12.5×10^{-3} kg of water is raised from 0°C to 100°C (273 K to 373 K)

$$dS = \int_{T_1}^{T_2} \frac{\delta Q}{T}$$

$$= ms \times 2.3026 \log_{10} \left(\frac{T_2}{T_1}\right)$$

$$= 12.5 \times 10^{-3} \times 4.2 \times 10^3 \times 2.3026 \log_{10} \left(\frac{373}{273}\right)$$

$= \mathbf{16.38\ J/K}$

4. Increase in entropy when 12.5×10^{-3} kg of water at 100°C is converted into steam at 100°C (373 K)

$$dS = \frac{\delta Q}{T} = \frac{mL}{T}$$

$$= \frac{12.5 \times 10^{-3} \times 2.25 \times 10^6}{373}$$

$= \mathbf{75.4\ J/K}$

Total increase in entropy

$= 2.42 + 15.38 + 16.38 + 75.4$

$= \mathbf{109.58\ J/K}$

Example 5.18 *How much time will it take for a layer of ice of thickness 20 cm to increase by 10 cm, on the surface of a pond, when the temperature of surroundings is – 15°C?*

K = 0.005 C.G.S. units; L = 80 Cal/g, ρ = 0.90 g/cm³ **(*Purvanchal 2003*)**

Solution. [In *CGS* system]

$$\int dt = \frac{\rho L}{K\theta}\int x\,dx$$

$$t = \frac{\rho L}{2K\theta}\left[x_2^2 - x_1^2\right]$$

Here,

$$\rho = 0.9 \text{ g/cm}^3$$
$$L = 80 \text{ cal/cm}^3$$
$$K = 0.005 \text{ C G S units}$$
$$\theta = 15°\text{C}$$
$$x_1 = 20 \text{ cm}$$
$$x_2 = 30 \text{ cm}$$

$$t = \left[\frac{0.9\times 80}{2\times 0.005\times 15}\right]\left[(30)^2-(20)^2\right]$$
$$= 2.4\times 10^5\ s$$
$$= \mathbf{66\ hrs - 40\ min}$$

Example 5.19 *Assuming the specific heat capacity of water, find the change in entropy when 0.1 kg. of water at 15°C is mixed with 0.16 kg. of water at 40°C.* **(Bhagalpur 2001)**

Solution. [In *SI* system],

Specific heat of water,

$$s = 1 \text{ k cal/kg} - °\text{C}$$
$$= 4.2\times 10^3 \text{ J/kg} - °\text{C}$$

Here,

$$m_1 = 0.1 \text{ kg}$$
$$T_1 = 15 + 273 = 288 \text{ K}$$
$$m_2 = 0.16 \text{ kg}$$
$$T_2 = 40 + 273 = 313 \text{ K}$$

Let the final temperature be T

$$m_1 \times s \times [T - T_1] = m_2 \times s \times [T_2 - T]$$
$$m_1\,[T - T_1] = m_2\,[T_2 - T]$$
$$0.1\,[T - 288] = 0.16\,[313 - T]$$
$$T = \mathbf{303.4\ K}$$

(*i*) Change in entropy when the temperature of 0.1 kg of water rises from 288 K to 303.4 K

$$\frac{\delta Q}{T} = ms\int_{T_1}^{T}\left(\frac{dT}{T}\right)$$
$$= 0.1\times 4.2\times 10^3\times 2.3026\log_{10}\left(\frac{303.4}{288}\right)$$
$$= \mathbf{21.88\ J/K}$$

(*ii*) Change in entropy when the temperature of 0.16 kg of water decreases from 313 K to 303.4 K

$$\frac{\delta Q}{T} = ms\int_{T_2}^{T}\left(\frac{dT}{T}\right)$$
$$= 0.16\times 4.2\times 10^3\times 2.3026\log_{10}\left(\frac{303.4}{313}\right)$$
$$= -20.93 \text{ J/K}$$

Therefore, the net change in entropy of the system = + 21.88 – 20.93

$$= \mathbf{0.95\ J/K}$$

Hence, net increase in entropy,

$$= \mathbf{0.95\ J/K}$$

Example 5.20 *Prove that in a T-S diagram the slop of isochoric curve is* $\frac{T}{C_V}$ *and that of isobaric curve is* $\frac{T}{C_P}$. **(Calcutta 2002)**

Solution. Suppose the heat supplied is dQ at temperature T. The increase in entropy is

$$dS = \frac{dQ}{T}$$

$$\therefore \quad dQ = TdS$$

1. In case of isochoric process, volume is constant,

$$\therefore \quad dQ = C_V\, dT$$

or

$$TdS = C_V\, dT$$

$$\therefore \quad \text{change in entropy, } dS = \frac{C_V dT}{T}$$

and, slope of isochoric curve, $\frac{dT}{dS} = \frac{\boldsymbol{T}}{\boldsymbol{C_V}}$ ... (*i*)

2. In case of isobaric process, pressure is constant,

$$\therefore \quad dQ = C_P\, dT$$

or

$$T.dS = C_P\, dT$$

$$\therefore \quad \text{change in entropy, } dS = \frac{C_P\, dT}{T}$$

and, slope of isobaric curve, $\frac{dT}{dS} = \frac{\boldsymbol{T}}{\boldsymbol{C_P}}$ ... (*ii*)

Example 5.21 *A mass m of a liquid at temperature T_1 is mixed with an equal mass of the same liquid at a lower temperature T_2. The system is thermally insulated. Show that when the system comes in equilibrium the entropy change of the universe will be*

$$ms \log_e \frac{(T_1+T_2)^2}{4T_1 T_2} = 2\,ms \log_e \left[\frac{(T_1+T_2)/2}{(T_1T_2)^{1/2}}\right]^2$$

where s is the mean specific heat of the liquid

(G.N.D.U. 2006, 2003, Kan U. 1994; Cal. U. 2002, Purvanchal 2004)

Solution: Let T be the common temperature after the two volumes of the liquid at temperatures T_1 and T_2 are mixed together, m the mass of each liquid and s the specific heat, then

$$ms\,(T_1 - T) = ms\,(T - T_2)$$

$$\therefore \quad 2T = T_1 + T_2$$

or

$$T = \frac{T_1 + T_2}{2} \qquad \text{... (i)}$$

Change in entropy of the liquid at initial temperature T_1

$$(\Delta S)_1 = \int_{T_1}^{T_2} \frac{dQ}{T}$$

$$= ms \int_{T_1}^{T_2} \frac{dQ}{T}$$

$$= ms \log_e \left(\frac{T}{T_2}\right)$$

Change in entropy of the liquid at initial temperature T_2

$$(\Delta S)_2 = \int_{T_1}^{T_2} \frac{dQ}{T}$$

$$= ms \int_{T_2}^{T} \frac{dT}{T}$$

$$= ms \log_e \left(\frac{T}{T_2}\right)$$

Total change in entropy = entropy change of the universe

$$= (\Delta S)_1 + (\Delta S)_2$$

$$= ms \log_e \left(\frac{T}{T_1}\right) + ms \log_e \left(\frac{T}{T_2}\right)$$

$$= ms \left[\log_e \left(\frac{T}{T_1}\right) + \log_e \left(\frac{T}{T_2}\right)\right]$$

$$= ms \log_e \left(\frac{T^2}{T_1 T_2}\right)$$

Substituting $T = \frac{T_1 + T_2}{2}$ from (*i*), we have

$$\text{Entropy change of the universe} = ms \log_e \frac{\left(\frac{T_1 + T_2}{2}\right)^2}{T_1 T_2} = ms \log_e \frac{(T_1 + T_2)^2}{4 T_1 T_2}$$

$$= ms \log_e \left[\frac{\frac{T_1 + T_2}{2}}{(T_1 T_2)^{1/2}}\right]^2$$

$$= 2\ ms\ \log_e \frac{(T_1 + T_2)/2}{(T_1 T_2)^{1/2}}$$

EXERCISE

5.1 Define entropy. Discuss the physical meaning of entropy.

(Garhwal 2005, 2001, Madras 2005, Rohilkhand 2003, Agra 2006, Nagpur 2007, 205/s, Lucknow 2004, Purvanchal 2007, 2005, 2003, M.S. Uni. Tamil Nadu 2007)

5.2 Show that during a reversible adiabatic process the entropy of the system remains constant.

(Nagpur 2004/s, 2005/s, Meerut 2005, Madras 2005, Madurai Kamraj 2003)

5.3 Prove that for a complete reversible cycle of change in the state of substance

$$\oint ds = 0$$ ***(Calcutta 2001)***

5.4 What happens to change in entropy of a system which undergoes

(*a*) a reversible process (*b*) an irreversible process

(*c*) an adiabatic process ***(Purvanchal 2005, M.S. University, Tamil Nadu 2007)***

5.5 State and prove the principle of increase of entropy. ***(Kanpur 1997)***

5.6 Show that entropy of an irreversible process, entropy of the universe increases.
(Nagpur 2009, 2007, 2004/w, Meerut 2001, Lucknow 2004, Patna 2005, Bangalore 2004, Mysore 2004)

5.7 Explain the term entropy and isoentropics. Show that the entropy of a system tends to be a maximum. Represent the Carnot's cycle on a temperature -entropy diagram and prove that its area represents available energy. ***(Nagpur 2005/s)***

5.8 Establish the relation for efficiency of a Carnot's engine using T–S diagram in

$$\eta = \frac{T_1 - T_2}{T_1}$$

(Garhwal 1997, 1999)

5.9 Explain the concept of entropy and disorder.
(Kanpur 2000, Alagappa 2004, Bangalore 2005)

5.10 What do you mean by entropy? Show that entropy remains constant in reversible process but increases in irreversible process.
(Meerut 2005, 1996, Kanpur 1998, 2001, Rohilkhand 1995, 1999, 2000; Patna 2004, Bangalore 2004, Mysore 2004, Nagpur 2005/w)

5.11 What is T-S diagram? What is the importance of T-S diagram? Find the expression for efficiency of a reversible Carnot's engine with the help of T-S diagram.
(Garhwal 1997, 1999, Rohilkhand 1998, Meerut 2003, Ranchi 2005, Punjab 2005, Nagpur 2010, 2005/w, Osmania 2004)

5.12 Define entropy. Derive an expression for entropy of an ideal gas in terms of its pressure, volume and specific heat. ***(Meerut 2005, 1997, Garhwal 2000, Agra 2003)***

5.13 Derive an expression for the change of entropy of a perfect gas. ***(Kanpur 2000)***

5.14 Obtain an expression for the change in entropy when ice changes in to steam.
(M.S. Uni. Tamil Nadu 2007, Meerut 2003)

5.15 Mention whether the change in entropy of a mole of an ideal gas is positive, zero or negative, when the gas undergoes free expansion. ***(Meerut 2003)***

5.16 Describe Carnot's reversible heat engine. Deduce an expression for its efficiency. Is Carnot's cycle a practical possibility? Explain ***(Agra 2005, Garhwal 2005)***

5.17 Explain how concept of entropy leads to 2nd law of thermodynamics. ***(Punjab 2005)***

5.18 What is entropy? Give the expression for entropy? What is the unit of entropy?
(Madras 2005)

5.19 (*a*) Explain Kelvin's thermodynamic scale of temperature.
(Nagpur 2009, 2005/s, Madras 2005, 2004, Purvanchal 2005, 2002)

(*b*) Show that the ratio of two temperatures of Kelvin's scale is the ratio of heats absorbed and rejected by a Carnot's engine operating between these temperatures. ***(Punjab 2001)***

5.20 Discuss absolute zero on the basis of Kelvin's thermodynamic scale of temperature. How size of 1 degree represents on a P-V indicator diagram. Hence define a thermodynamic degree.
(Agra 2001, Madras 2000)

5.21 Explain thermodynamic scale of temperature and show that the thermodynamic and the ideal gas scale are identical. Explain why?
(Rohilkhand 2001, Garhwal 2000, 1995, Agra 1998, Meerut 2000, Kanpur 1996, Nagpur 2004/s, 2005/w)

5.22 Why the zero on absolute scale is considered to be the lowest possible temperature? Is negative temperature possible on absolute scale of temperature?
(Purvanchal 2002, Rohilkhand 2001, 1999, Meerut 2003, 2004, Agra 2004, 2003)

5.23 How do you define zero on absolute scale of temperature? ***(Rohilkhand 1999)***

5.24 What do you mean by thermodynamical scale of temperature? How did Kelvin introduce this scale? How is this scale realised in practice? ***(Patna 2004)***

5.25 Assuming temperature to be a thermodynamic coordinate of a system, show how Kelvin derived a scale of temperature which is independent of the properties of the measuring system.
(Calcutta 2003)

5.26 State and explain with example the third law of thermodynamics. Prove that it is impossible to obtain absolute zero temperature.
(Nagpur 2005/s, Alagappa 2004, Madras 2005, Bangalore 2005, Nagpur 2005/w)

5.27 What is Nernst's heat theorem? State and explain. What do you infer from that?
(Nagpur 2010, Pune 2010, Garhwal 2005, Madras 2004)

5.28 Explain the concept of 'Heat Death of Universe'. ***(Mumbai 2000)***

5.29 Write notes on the following:

(*i*) Kelvin's absolute scale of temperature. ***(Meerut 2001, Bangalore 2005)***

(*ii*) Entropy and disorder. ***(Osmania 2005)***

(*iii*) T-S diagram. ***(Rohilkhand 2001 Imp.)***

(*iv*) Absolute scale of temperature. ***(Kanpur 1997, Rohilkhand 2000)***

(*v*) Principle of increase of entropy. ***(Osmania 2000)***

(*vi*) Entropy of a ges. ***(Agra 2000)***

(*vii*) Third law of thermodynamics. ***(Meerut 2000)***

(*viii*) Heat death of universe. ***(Kerala 1999)***

5.30 Calculate the change in entropy when 10 kg of water at 100°C is converted into steam at the same temperature. **[Ans. 14480 Cal/K]** ***(Calcutta 2001)***

5.31 Two gram molecules of a gas expands isothermally to four times its initial volume. Calculate the change in its entropy in terms of gas constant.

[Ans. 2.774 $\frac{R}{J}$ Cal/K] ***(Madurai 2002, Agra 2006)***

5.32 Calculate the change in entropy when 100 g of water at 15°C is mixed with 160 g of water at 40°C. Specific heat of water may be assumed to be equal to 1.

[Ans. – 0.130 Cal/K] ***(Kanpur 2000, Agra 2006)***

5.33 Calculate the change in entropy when one gram of ice is mixed with 2 millilitres of water at 55°C. Assume that specific heat and specific gravity of water remains constant (*i.e.* equal to 1 in both cases) between 0°C and 55°C. latent heat of ice = 80 Cal/g.

[Ans. + 0.03372 Cal/K] ***(Madras 2001)***

5.34 3 gm. of nitrogen doubles in volume at constant temperature. Calculate the change in entropy. Given for nitrogen, molecular weight = 28. **[Ans. 0.621 Joule/K]** ***(Kanpur 1996)***

5.35 10 gm. of water is heated from 40 °C to 80 °C. Calculate the change in entropy.

[Ans. 12.1 Cal/K] ***(Punjab 2005, Kanpur 1999)***

5.36 Find the change in entropy when 1 gm. of ice at 0°C changes to water at 10°C.
(Garhwal 2005)

5.37 The T-S diagram of a reversible engine is given below. Find its efficiency. ***(Calcutta 2004)***

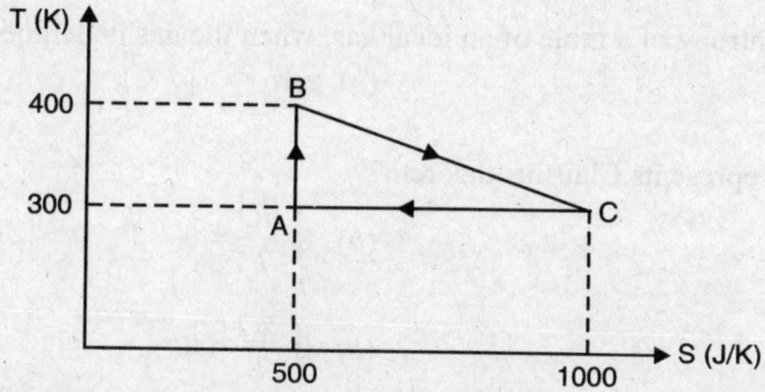

5.38 10 gm of steam at 100 °C is blown into 90 gms. of water at 0 °C contained in a calorimeter of water equivalent 10 gms. The whole of the steam is condensed. Calculate the increase in the entropy of the system. ***(Purvanchal 2003)***

5.39 Calculate the change in entropy when temperature of 1 kg of water is raised from 0 °C to 100°C. Specific heat capacity of water is 1 cal/gm. K. **[Ans.** 1305.4 J/K**]** ***(Nagpur 2005/w)***

5.40 State and prove Carnot's theorem on the basis of entropy for a reversible heat engine. Show that the efficiency of any reversible heat engine is greater than that of any irreversible heat engine, working between the same two temperatures. What is the importance of this theorem in Physics and Engineering?

(G.N.D.U. 2007, 2006, 2004; Calicut U. 2004, Meerut U. 2001; Nagpur U 2001; Kerala U. 2001, K.U. 2002, 2001; P.U. 2006, 2000, H.P.U. 2002, M.D.U. 2002, Meerut. U. 2007, 2002, Mithila U. 2001; Gharwal U. 2002)

5.41 (*a*) Define second law of thermodynamics, How does it follow from the law of increase of entropy? State it in the mathematical form.

(P.U. 2005, 2003; Pbi U., 2005, Cal. U. 2006 (Pbi. U.2002))

(*b*) What is thermodynamic definition of entropy? ***(Pbi. U. 2002)***

5.42 (*a*) Identify the thermodynamic quantities which undergo no change in cyclic process. ***(G.N. D.U. 2001)***

(*b*) n moles of an ideal gas are contained in a volume V within an insulated container. This volume is separated from an evacuated volume V with a partition. After the partition is removed the gas expands into the adjascent evacuated volume so that the final volume is $2V$. Show that the change in entropy is nR ln 2. ***(Cal. U. 2003)***

5.43 Explain what you understand by absolute or thermodynamic scale of temperature. Show that this scale is identical to the ideal gas temperature scale? Explain why negative temperature on this scale is not possible.

(P.U., 2002, 2001, 2000, Cal U. 2003, M.D.U.2008, 2001, 2000, K.U. 2002, 2001, 2000; Luck U. 2001, Nagpur U. 2001; Meerut U. 2000, Gharwal U. 2000)

MULTIPLE CHOICE QUESTIONS

5.1 In a reversible adiabatic process, entropy:

(*a*) increases (*b*) remains unchanged
(*c*) decreases (*d*) none of these ***(Garhwal 2005)***

5.2 Which of these are intensive variables?

(*a*) area (*b*) entropy
(*c*) temperature (*d*) volume ***(Punjab 2000)***

5.3 The change in entropy of a mole of an ideal gas, when the gas undergoes free expansion is

(*a*) positive (*b*) zero

(*c*) negative **(*Meerut 2003*)**

5.4 Which relation represents Clausius theorem?

(*a*) $\oint_R \frac{dQ}{T} = 0$ (*b*) $\oint_R \frac{dQ}{T} \neq 0$

(*c*) $\oint_R \frac{dQ}{T} > 0$ (*d*) $\oint_R \frac{dQ}{T} < 0$ **(*Rohilkhand 2003*)**

5.5 Entropy remains constant in:

(*a*) adiabatic process (*b*) isothermal process

(*c*) isochoric process (*d*) isolated process **(*Rohilkhand 2003*)**

5.6 In a reversible process, the entropy of the system:

(*a*) increases (*b*) decreases

(*c*) remains zero (*d*) remains constant **(*Kanpur 2002*)**

5.7 The entropy of a system in an irreversible process:

(*a*) increases (*b*) decreases

(*c*) remains constant (*d*) none of the above **(*Rohilkhand 2002, 2003*)**

5.8 The unit of entropy is

(*a*) Joule/Kelvin (*b*) Cal/Kelvin

(*c*) Both (*a*) and (*b*) (*d*) None of the above **(*Rohilkhand 2002*)**

5.9 Choose the correct statement:

(*a*) All quasi-static processes are reversible.

(*b*) All reversible processes are quasistatic.

(*c*) Adiabatic process is quasi-static.

5.10 A piece of ice is added to water in a cup. The entropy

(*a*) is increased (*b*) decreased

(*c*) undergoes no change (*d*) sometimes increases, sometimes not

5.11 When water vapour condenses into water, its entropy:

(*a*) increases (*b*) decreases

(*c*) remains unchnged (*d*) first decreases and then increases

5.12 Entropy is a measure of

(*a*) prefect order (*b*) available energy

(*c*) disorder (*d*) none of the above

5.13 Entropy is maximum in which state

(*a*) Solid (*b*) liquid

(*c*) gas (*d*) can be any

5.14 In a natural process entropy

(*a*) increases (*b*) decreases

(*c*) remains the same

5.15 Net entropy change of a system in Carnot's cycle

(*a*) zero (*b*) positive

(*c*) negative (*d*) more than 1

5.16 Which of the following represents a reversible process?

(*a*) $dS < 0$ (*b*) $dS = 0$

(*c*) $dS > 0$ (*d*) none of these

5.17 In the two gases at the same temperature:

(*a*) the average kinetic energy per molecule is equal.

(*b*) the internal energy is equal

(*c*) the entropy is equal.

ANSWERS

5.1 (*b*)	**5.2** (*c*)	**5.3** (*a*)	**5.4** (*a*)	**5.5** (*b*)
5.6 (*d*)	**5.7** (*a*)	**5.8** (*c*)	**5.9** (*b*)	**5.10** (*a*)
5.11 (*b*)	**5.12** (*c*)	**5.13** (*c*)	**5.14** (*a*)	**5.15** (*a*)
5.16 (*b*)	**5.17** (*b*)			

THERMODYNAMIC RELATIONSHIPS

Chapter 6

INTRODUCTION

There exists a useful thermodynamic variable, called entropy (S). The natural processes that starts in one equilibrium state and end in another will go in the direction that causes the entropy of the system plus the environment to increase for an irreversible process ($S_2 > S_1$) and to remain constant for a reversible process ($S_2 = S_1$). The thermodynamic state of any system is described by the variables P, V, T and S. All these variables are not independent. If any two of them are known, the others can be found out *e.g.*, the internel energy U is completely known, if the volume V and the temperature T are given. All the above thermodynamic variables are related by four relations, known as Maxwell's thermodynamic relations.

6.1 THERMODYNAMIC VARIABLES

The thermodynamic state of a substance is specified by some of its properties like pressure, volume, temperature, internal energy and entropy. These properties undergo a change when the system passes from one state to another. These variables are known as thermodynamic variables or co-ordinates. These are called macroscopic co-ordinates. They require the specification of few measurable properties of the system and do not require the knowledge of microscopic structure of matter composing the system.

6.2 EXTENSIVE AND INTENSIVE VARIABLES

Extensive variable

An *extensive variable* of a system is a macroscopic parameter which describes a system in equilibrium and which has a value equal to the sum of its values in each part of the system. The extensive variable depends upon the mass or the size of the substance present in the system.

Example: Familiar examples of extensive variables are mass, volume, internal energy, entropy, length, area, heat-capacity, electric charge, magnetisation etc. If we have two areas A_1 and A_2 and join them, the total area A of the system so obtained is given by $A = A_1 + A_2$. Entropy is also having additive property by its nature, *i.e.*, for a system consisting of a number of subsystems in equilibrium, the entropy of the composite (or combined) system is equal to the sum of the entropies of all subsystems. Hence,

$$S = S_1 + S_2 + S_3 + \ldots$$

Thus entropy is an extensive variable.

Intensive Variable

An *intensive variable* of a substance is a macroscopic parameter which describes the system in equilibrium and has the same value in any part of the system. It is independent of mass or size of the system. It is a characteristic of the substance present in the system.

Example: The examples of intensive variables are Pressure, temperature, viscosity, refractive index, density, specific volume, magnetic induction, surface tension, electromotive force etc. If we have two vessels containing water at the same temperature, and mix them, the temperature will remain the same.

Distinction between Extensive and Intensive variables

Consider a homogeneous system in equilibrium. Suppose the system is divided into many parts. If a macroscopic variable x of the system has the values x_1, x_2, x_3, ... in each of these parts of the system respectively, then

(*i*) x is said to be extensive variable, if $x = x_1 + x_2 + x_3 +$ ...., and

(*ii*) x is said to be intensive variable, if $x = x_1 = x_2 = x_3 =$ ....

An extensive variable may become an intense variable by specifying unit amount of substance. Thus, mass and volume are extensive variables but density $\left(\frac{M}{V}\right)_V$ and specific volume $\left(\frac{V}{M}\right)$ are intensive variables. Similarly heat capacity is extensive variable but specific heat is an intensive variable. The extensive and intensive variables of some systems are given in Table 6.1.

Table 6.1

System	*Extensive Variable*		*Intensive Variable*	
Chemical system	Volume	V	Pressure	P
Stretched wire	Length	L	Tension	F
Surface film	Area	A	Surface tension	T
Electric cell	Charge	Q	E.M.F.	E
Paramagnetic solid	Magnetisation	I	Magnetic intensity	H

Additive nature (extensive property) of Entropy. We can explain additive nature of entropy using thermodynamic probability (W) (described in ch. 9) as under:

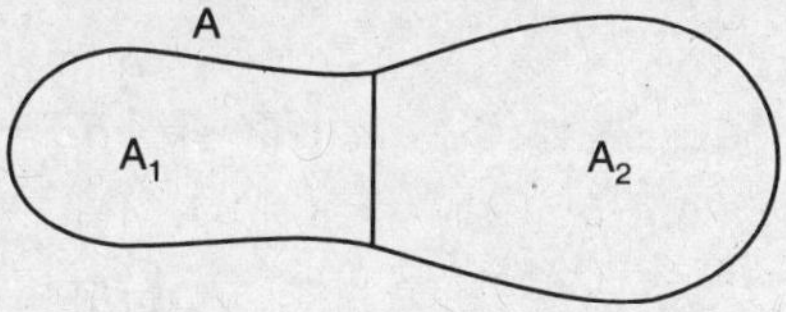

Fig. 6.1

Consider a system A in equilibrium, represented by Fig. 6.1 having entropy S corresponding to a thermodynamic probability W.

$$\therefore \qquad S = k \ln W$$

Suppose the system consists of two parts A_1, and A_2. Let the sub-system A_1 have an entropy S_1, corresponding to thermodynamic probability W_1 and the subsystem A_2 an entropy S_2 corresponding to a thermodynamic probability W_2, then

$$S_1 = k \ln W_1 \text{ and } S_2 = k \ln W_2$$

Now W_1 and W_2 represent the number of microstates (or arrangements) for the sub-system A_1 and A_2 respectively. According to the law of probability, their combined probability

$$W = W_1 \times W_2 \qquad \text{... (i)}$$

because any one of the W_1 arrangements of the system A_1 can be taken together with any one of the arrangements of the system A_2. Taking natural logarithms on both sides of equation (*i*) we have

$$\ln W = \ln W_1 + \ln W_2$$

$$\therefore \qquad k \ln W = k \ln W_1 + k \ln W_2$$

$$\text{or} \qquad S = S_1 + S_2 \qquad \text{... (6.1)}$$

Hence the entropy of the system is equal to the sum of the entropies of its two sub-systems.

The above argument can be extended to a system consisting of more than two sub-systems in equilibrium. In such a case the entropy of the compcsite or combined system is equai to the sum of the entropies of all the sub-systems.

$$S = S_1 + S_2 + S_3 \ldots \qquad \ldots (6.2)$$

In other words, entropy is an *extensive* parameter.

Before studying the Maxwell's thermodynamic relations let us get acquainted first with perfect and imperfect differentials, as they are essentially needed to be understood.

6.3 PERFECT DIFFERENTIAL

Cossider a quantity Z which is a function of two independent variables x and y *i.e.*,

$$Z = f(x, y)$$

Let the two variables be represented along two perpendicular axes x and y as shown in Fig. 6.2. Any point in the figure has fixed values of x and y and, therefore, has a definite or unique value of Z. *The function Z is, therefore, single valued and its differential dZ is called a perfect differential.* Let a point A on this figure have the value $Z = Z_A$. If we go from the point A to another point D having the value $Z = Z_D$, then change in the value of Z in going from A to $D = (Z_D - Z_A)$. *This change is independent of the path travelled.*

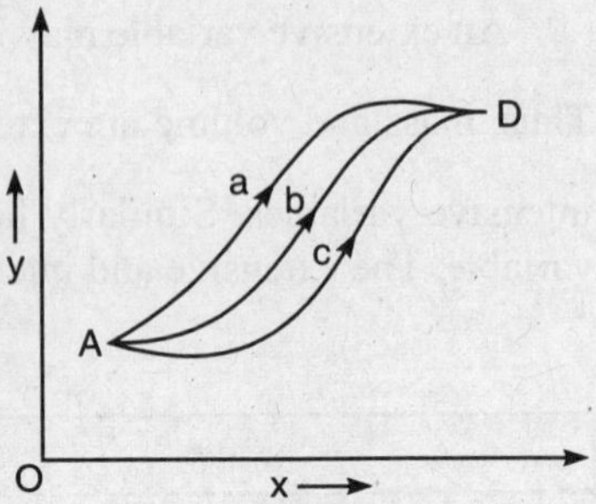

Fig. 6.2

For example, for the three paths, a, b and c shown in the figure, the quantity $(Z_D - Z_A)$ is the same and given by

$$Z_D - Z_A = a\int_A^D dZ = b\int_A^D dZ = c\int_A^D dZ \text{ and so on}$$

$$\therefore \qquad a\int_A^D dZ - b\int_A^D dZ = 0 \quad \text{or} \quad a\int_A^D dZ = b\int_D^A dZ \qquad \left[\because \int_A^D dZ = -\int_D^A dZ\right]$$

or

$$\oint dZ = 0$$

where the symbol $\oint$ denotes cyclic integral from A to D and back to A.

Therefore, *the integral of a perfect differential over a complete cycle is always zero.*

Condition for perfect differential. Let the co-ordinates of the point A be (x, y) as shown in Fig. 6.3.

Suppose the independent variables x and y change by infinitesimal amounts dx and dy respectively as shown by the rectangle $ABCD$ so that $AB = dx$ and $AC = dy$.

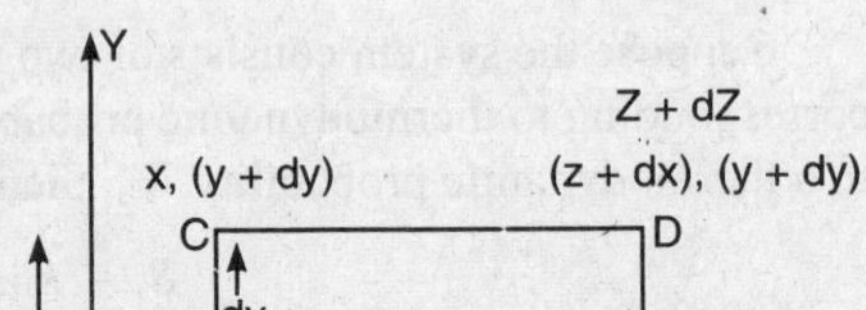

Fig. 6.3

Then, co-ordinates of various points are $A = (x, y)$; $B = [(x + dx), y]$; $C = [x, (y + dy)]$; $D = [(x + dx), (y + dy)]$.

If we wish to go to the point D from the point A, then two paths ABD or ACD are available.

For the path ABD, the value of Z at B is $Z + \frac{\partial Z}{\partial x}dx$ and at D it is $Z + \frac{\partial Z}{\partial x}dx + \frac{\partial}{\partial y}\left(Z + \frac{\partial Z}{\partial x}dx\right)dy.$

For the path ACD the value of Z at C is $Z + \frac{\partial Z}{\partial y}dy$ and at D it is $Z + \frac{\partial Z}{\partial y}dy + \frac{\partial}{\partial x}\left(Z + \frac{\partial Z}{\partial y}dy\right)dx.$

For dZ to be a perfect differential the value of Z at D must be the same whatever the path chosen from A to D *i.e.*,

$$Z+\frac{\partial Z}{\partial x}dx+\frac{\partial}{\partial y}\left(Z+\frac{\partial Z}{\partial x}dx\right)dy = Z+\frac{\partial Z}{\partial y}dy+\frac{\partial}{\partial x}\left(Z+\frac{\partial Z}{\partial y}dy\right)dx$$

or $$Z+\frac{\partial Z}{\partial x}dx+\frac{\partial Z}{\partial y}dy+\frac{\partial^2 Z}{\partial y\,\partial x}dx\,dy = Z+\frac{\partial Z}{\partial y}dy+\frac{\partial Z}{\partial x}dx+\frac{\partial^2 Z}{\partial x\,\partial y}dy\,dx$$

or $$\frac{\partial_z^2 Z}{\partial y\,\partial x}dx\,dy = \frac{\partial_z^2 Z}{\partial x\,\partial y}dy\,dx$$

or $$\frac{\partial^2 Z}{\partial y\,\partial x} = \frac{\partial^2 Z}{\partial x\,\partial y} \text{ or } \frac{\partial}{\partial y}\left(\frac{\partial Z}{\partial x}\right) = \frac{\partial}{\partial x}\left(\frac{\partial Z}{\partial y}\right) \quad ...(6.3)$$

This is the required condition for dZ to be a perfect differential.

Examples. (*i*) **Perfect differentials.** If we consider a $P-V$ diagram and go from a given point A to another point B and come back to A the net change in the value of P, V, T, S and U are all zero as these quantities are path independent.

$$\therefore \quad \oint dP = 0;\ \oint dV = 0;\ \oint dT = 0;\ \oint dS = 0;\ \oint dU = 0$$

$\therefore$ dP, dV, dT, dS and dU *are all perfect differentials.*

(*ii*) **Imperfect differentials.** Consider the relation

$$\delta W = P\,dV$$

The integral $\int_A^B P\,dV$ represents the work done by the system when its state changes from A to B.

As the work done depends upon the path for going from A to B, δW is not a *perfect differential.*

Further, according to the first law of thermodynamics

$$\delta Q = dU + \delta W \quad ...(6.4)$$

where dU is a perfect differential. As δW is not a perfect differential δQ *is also not a perfect differential.*

6.4 MAXWELL'S THERMODYNAMICAL RELATIONS

Maxwell's thermodynamical relations can be deduced by making use of the *first* and *second* laws of thermodynamics.

First law. ***If a substance absorbs a very small amount of heat δQ at a constant pressure, then part of this heat is used up to raise the temperature which results in the increase of internal energy dU. The rest of the heat is used in doing work in allowing the substance to increase in volume by an amount dv against the external pressure P.*** Thus according to the first law, we have

$$\delta Q = dU + \delta W = dU + PdV$$

$$dU = \delta Q - PdV \quad ...(i)$$

Second law. ***If a substance absorbs a very small amount of heat δQ at a temperature T and if all the changes that it undergoes are perfectly reversible, then the change in entropy dS is equal to $\frac{\delta Q}{T}$.***

or, $$\delta Q = T.dS$$

Substituting the value of δQ in equation (*i*), we have

$$dU = TdS - PdV \quad ...(ii)$$

From the first and second law of thermodynamics, Maxwell was able to derive four fundamental thermodynamical relations. The state of a system can be specified by any pair of quantities, *viz.* pressure (P), volume (V), temperature (T), and entropy (S). In solving any thermodynamical problem, the most suitable pair is chosen and the quantities constituting the pair are taken as independent variables.

Considering U, S and V to be function of two independent variables x and y [here, in general, x and y can be any two variables out of P, V, T and S],

$$dU = \left(\frac{\partial U}{\partial x}\right)_y dx + \left(\frac{\partial U}{\partial y}\right)_x dy$$

$$dS = \left(\frac{\partial S}{\partial x}\right)_y dx + \left(\frac{\partial S}{\partial y}\right)_x dy$$

and
$$dV = \left(\frac{\partial V}{\partial x}\right)_y dx + \left(\frac{\partial V}{\partial y}\right)_x dy$$

Substituting these values in equation (*ii*), we get

$$\left(\frac{\partial U}{\partial x}\right)_y dx + \left(\frac{\partial U}{\partial y}\right)_x dy = T\left[\left(\frac{\partial S}{\partial x}\right)_y dx + \left(\frac{\partial S}{\partial y}\right)_x dy\right] - P\left[\left(\frac{\partial V}{\partial x}\right)_y dx + \left(\frac{\partial V}{\partial y}\right)_x dy\right]$$

$$= \left[T\left(\frac{\partial S}{\partial x}\right)_y - P\left(\frac{\partial V}{\partial x}\right)_y\right]dx + \left[T\left(\frac{\partial S}{\partial y}\right)_x - P\left(\frac{\partial V}{\partial y}\right)_x\right]dy$$

Comparing the coefficients of dx and dy, we get

$$\left(\frac{\partial U}{\partial x}\right)_y = T\left(\frac{\partial S}{\partial x}\right)_y - P\left(\frac{\partial V}{\partial x}\right)_y \quad \text{... (iii)}$$

$$\left(\frac{\partial U}{\partial y}\right)_x = T\left(\frac{\partial S}{\partial y}\right)_x - P\left(\frac{\partial V}{\partial y}\right)_x \quad \text{... (iv)}$$

Differentiating equation (*iii*) with respect to y and equation (*iv*) with respect to x,

$$\frac{\partial^2 U}{\partial y.\partial x} = \left(\frac{\partial T}{\partial y}\right)_x \left(\frac{\partial S}{\partial x}\right)_y + T\frac{\partial^2 S}{\partial y.\partial x} - \left(\frac{\partial P}{\partial y}\right)_x \left(\frac{\partial V}{\partial x}\right)_y - P\frac{\partial^2 V}{\partial y.\partial x}$$

and
$$\frac{\partial^2 U}{\partial x.\partial y} = \left(\frac{\partial T}{\partial x}\right)_y \left(\frac{\partial S}{\partial y}\right)_x + T\frac{\partial^2 S}{\partial x.\partial y} - \left(\frac{\partial P}{\partial x}\right)_y \left(\frac{\partial V}{\partial y}\right)_x - P\frac{\partial^2 V}{\partial x.\partial y}$$

The change in internal energy brought about by changing V and T, whether V is changed by dV first and T by dT later or vice versa is the same.

It means dU is a perfect differential.

$$\therefore \quad \frac{\partial^2 U}{\partial y.\partial x} = \frac{\partial^2 U}{\partial x.\partial y} \text{ and}$$

$$\left(\frac{\partial T}{\partial y}\right)_x \left(\frac{\partial S}{\partial x}\right)_y + T\frac{\partial^2 x}{\partial y.\partial x} - \left(\frac{\partial P}{\partial y}\right)_x \left(\frac{\partial V}{\partial x}\right)_y - P\frac{\partial^2 V}{\partial y.\partial x}$$

$$= \left(\frac{\partial T}{\partial x}\right)_y \left(\frac{\partial S}{\partial y}\right)_x + T\frac{\partial^2 S}{\partial x.\partial y} - \left(\frac{\partial P}{\partial x}\right)_y \left(\frac{\partial V}{\partial y}\right)_x - P\frac{\partial^2 V}{\partial y.\partial x} \quad \text{... (v)}$$

Since dS and dV are also perfect differentials, we have

$$\frac{\partial^2 S}{\partial x.\partial y} = \frac{\partial^2 S}{\partial y.\partial x} \quad \text{and} \quad \frac{\partial^2 V}{\partial x.\partial y} = \frac{\partial^2 V}{\partial y.\partial x}$$

Equation (*v*), therefore reduces to

$$\left(\frac{\partial T}{\partial y}\right)_x \left(\frac{\partial S}{\partial x}\right)_y - \left(\frac{\partial P}{\partial y}\right)_x \left(\frac{\partial V}{\partial x}\right)_y = \left(\frac{\partial T}{\partial x}\right)_y \left(\frac{\partial S}{\partial y}\right)_x - \left(\frac{\partial P}{\partial x}\right)_y \left(\frac{\partial V}{\partial y}\right)_x \quad \text{... (6.5)}$$

This is the general expression for Maxwell's thermodynamical relations. In place of the independent variables x and y, any two of the four variables S, T, P and V can be substituted so that there may be one mechanical variable (P or V) and one thermal variable (S or T). Thus, there may be four sets of possible substitutions (S, V), (T, V), (S, P) and (T, P), providing the four Maxwell's thermodynamical relations.

First Relation:

Put $x = S$ and $y = V$ in equation (6.5), so that

$$\frac{\partial S}{\partial x} = 1, \quad \frac{\partial V}{\partial y} = 1$$

and
$$\frac{\partial S}{\partial y} = 0, \quad \frac{\partial V}{\partial x} = 0$$

Substituting in equation (6.5), we get

$$\left(\frac{\partial T}{\partial y}\right)_x = -\left(\frac{\partial P}{\partial x}\right)_y$$

But $\partial y = \partial V$ (as $y = V$) and $\partial x = \partial S$ (as $x = S$).

Hence
$$\left(\frac{\partial T}{\partial V}\right)_s = -\left(\frac{\partial P}{\partial S}\right)_V \quad \text{... (6.6)}$$

This is Maxwell' first thermodynamical relation.

Second Relation :

Put $x = T$ and $y = V$ in equation (6.5),

then
$$\frac{\partial T}{\partial x} = 1, \quad \frac{\partial V}{\partial y} = 1$$

and
$$\frac{\partial T}{\partial y} = 0, \quad \frac{\partial V}{\partial x} = 0$$

Substituting in equation (6.5), we get

$$\left(\frac{\partial S}{\partial V}\right)_T = \left(\frac{\partial P}{\partial T}\right)_V \quad \text{... (6.7)}$$

This shows that increase in entropy per unit increase in volume at constant temperature is equal to the increase in pressure per unit increase of temperature at constant volume.

This is known as Maxwell's **second** thermodynamical relation.

Multiplying both sides of (6.7) by T, we have

$$T\left(\frac{\partial S}{\partial V}\right)_T = T\left(\frac{\partial P}{\partial T}\right)_V$$

But
$$T\partial S = \partial Q$$

$\therefore$
$$\left(\frac{\partial Q}{\partial V}\right)_T = T\left(\frac{\partial P}{\partial T}\right)_V$$

Third Relation:

Put $x = S$ and $y = P$, in equation (6.5) then

$$\frac{\partial S}{\partial x} = 1, \frac{\partial P}{\partial y} = 1, \frac{\partial S}{\partial y} = 0, \frac{\partial P}{\partial x} = 0$$

Substituting these in equation (6.5), we get

$$\left(\frac{\partial T}{\partial P}\right)_S = \left(\frac{\partial V}{\partial S}\right)_P \qquad \text{... (6.8)}$$

This is Maxwell's third thermodynamical relation

Fourth Relation:

Put $x = T$ and $y = P$, then equation (6.5) gives

$$\frac{\partial T}{\partial x} = 1, \quad \frac{\partial P}{\partial y} = 1, \quad \frac{\partial T}{\partial y} = 0 \text{ and } \frac{\partial P}{\partial x} = 0$$

Substituting these values in equation (6.5), we get

$$\left(\frac{\partial S}{\partial P}\right)_T = -\left(\frac{\partial V}{\partial T}\right)_P \qquad \text{... (6.9)}$$

This shows that the decrease in entropy per unit increase of pressure during an isothermal transformation is equal to the increase in volume per unit increase of temperature during an isobaric process.

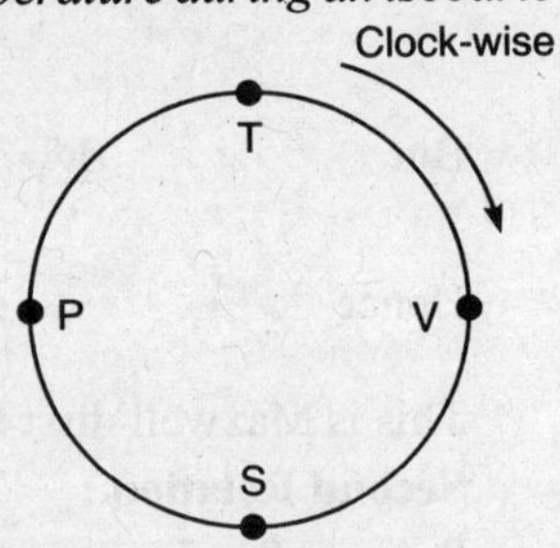

Fig. 6.4

This is known as Maxwell's **fourth** thermodynamical relation.

Multiplying both sides by T in 6.9, we have

$$T\left(\frac{\partial V}{\partial T}\right)_P = -T\left(\frac{\partial S}{\partial P}\right)_T$$

But $$T\partial S = \partial Q$$

$$\therefore \quad \left(\frac{\partial Q}{\partial P}\right)_T = -T\left(\frac{\partial V}{\partial T}\right)_P$$

Note. How to remember Maxwell's relations. Maxwell's relations are very often used in thermodynamics. A convenient method to remember these relations is as under.

(*i*) Draw a circle and write down the four variables (or parameters) *T*, *V*, *S* and *P* in clock-wise order as shown. Remember the order with the help of words.

***T V* Special Programme**

(*ii*) Start from *T* move clockwise for *First* relation and anticlock wise for *Third* relation. Start from *S* and move anticlockwise for *Second* relation and clockwise for *Fourth* relation.

(*iii*) In each case equate the partial derivative of variable with respect to next keeping still next constant with partial derivative of last variable with respect to last but one keeping last but two constant.

(*iv*) Give negative sign to partial derivative on the right hand side for clock-wise movement.

First relation. *Start from T and move clock-wise*

Starting variable is *T*, next *V* and still next S

$\therefore$ Partial derivation on L.H.S. = $\left(\frac{\partial T}{\partial V}\right)_S$

Last variable *P*, last but one S and last but two *V*

$\therefore$ Partial derivative on R.H.S = $\left(\frac{\partial P}{\partial S}\right)_V$

$\therefore$ Partial derivative on L.H.S = $\left(\frac{\partial S}{\partial V}\right)_T$

Last variable is *P*, last but one *T* and last but two *V*

$\therefore$ Partial derivative on R.H.S. = $\left(\frac{\partial P}{\partial T}\right)_V$

This gives the relation

$$\left(\frac{\partial S}{\partial V}\right)_T = \left(\frac{\partial P}{\partial T}\right)_V$$

Third relation. *Start from T and move anticlock-wise*

Starting variable is *T*, next *P* and still next *S*

$\therefore$ Partial derivative on L.H.S. = $\left(\frac{\partial T}{\partial P}\right)_S$

Last variable is *V*, last but one *S* and last but two *P*

$\therefore$ Partial derivative on R.H.S. = $\left(\frac{\partial V}{\partial S}\right)_P$

This gives the relation $\left(\frac{\partial T}{\partial P}\right)_S = \left(\frac{\partial V}{\partial S}\right)_P$

Fourth relation. *Start from S and move clock-wise*

Starting variable is *S* next *P* and still next *T*

$\therefore$ Partial derivative on L.H.S. = $\left(\frac{\partial S}{\partial P}\right)_T$

Last variable is *V*, last but one *T* and last but two *P*

$\therefore$ Partial derivative on R.H.S. = $\left(\frac{\partial V}{\partial T}\right)_P$

For clock-wise move give negative sign to R.H.S.

This gives the relation $\left(\frac{\partial S}{\partial P}\right)_T = -\left(\frac{\partial V}{\partial T}\right)_P$

Or

Another method

The four Maxwell's thermodynamical relations need not be memorised. These can be easily written by remembering the word SP or $T_i\ V_e$ (sportive). In this, *SPTV* represent entropy, pressure, temperature and volume. In deriving the first two relations, ∂S is written in the numerator on the left hand side of the equation and the rest of the quantities ∂P, ∂T and ∂V are written (*i*) in the clockwise order and (*ii*) in the anticlockwise order. In the clockwise direction, the right hand side expression is with a + sign and for the anticlockwise direction, it is – *ve*. These equations are

$$\left(\frac{\partial S}{\partial V}\right)_T = \left(\frac{\partial P}{\partial T}\right)_V \quad ...\ (i)$$

$$\left(\frac{\partial S}{\partial P}\right)_T = -\left(\frac{\partial V}{\partial T}\right)_P \quad ...\ (ii)$$

Similarly to write the other equations, ∂S is written in the denominator of the right hand side of the equation and the other quantities are written (*i*) in the anticlockwise direction and (*ii*) in the clockwise direction. These equations are

$$\left(\frac{\partial T}{\partial V}\right)_S = -\left(\frac{\partial P}{\partial S}\right)_V \quad ...\ (iii)$$

$$\left(\frac{\partial T}{\partial P}\right)_S = \left(\frac{\partial V}{\partial S}\right)_P \quad ...\ (iv)$$

Two more Relation

Further there are two more relations within the mechanical variables (*P, V*) and thermal variables (*T, S*).

Fifth Relation:

Put $x = P$ and $y = V$

$$\frac{\partial P}{\partial x} = 1, \frac{\partial V}{\partial y} = 1, \frac{\partial P}{\partial y} = 0 \text{ and } \frac{\partial V}{\partial x} = 0$$

Substituting these values in equation (6.5), we get

$$\left(\frac{\partial T}{\partial P}\right)_V \left(\frac{\partial S}{\partial V}\right)_P - \left(\frac{\partial T}{\partial V}\right)_P \left(\frac{\partial S}{\partial P}\right)_V = 1 \quad \text{... (6.10)}$$

Sixth Relation:

Put $x = T$ and $y = S$

$$\frac{\partial T}{\partial x} = 1, \frac{\partial S}{\partial y} = 1, \frac{\partial T}{\partial y} = 0 \text{ and } \frac{\partial S}{\partial x} = 0$$

Substituting in equation (6.5), we get

$$\left(\frac{\partial P}{\partial T}\right)_S \left(\frac{\partial V}{\partial S}\right)_T - \left(\frac{\partial P}{\partial S}\right)_T \left(\frac{\partial V}{\partial T}\right)_S = 1 \quad \text{... (6.11)}$$

Out of these six thermodynamic relations, the one suited for a particular problem is used and the problem is solved. Let us see, some of the important applications of these Maxwell's thermodynamic relations.

6.5 APPLICATIONS OF MAXWELL'S THERMODYNAMIC RELATIONS

6.5.1 Specific Heat Equation

The specific heat at constant pressure is given

as $$C_P = \left(\frac{\partial Q}{\partial T}\right)_P = T\left(\frac{\partial S}{\partial T}\right)_P \quad (\because \partial Q = T.\partial S)$$

and the specific heat at constant volume is

$$C_V = \left(\frac{\partial Q}{\partial T}\right)_V = T\left(\frac{\partial S}{\partial T}\right)_V$$

Now, if the entropy *S* is regarded as a function of *T* and *V* and since *dS* is a perfect differential,

$$dS = \left(\frac{\partial S}{\partial T}\right)_V dT + \left(\frac{\partial S}{\partial V}\right)_T dV$$

$$\therefore \quad \left(\frac{\partial S}{\partial T}\right)_P = \left(\frac{\partial S}{\partial T}\right)_V \left(\frac{\partial T}{\partial T}\right)_P + \left(\frac{\partial S}{\partial V}\right)_T \left(\frac{\partial V}{\partial T}\right)_P$$

or $$T\left(\frac{\partial S}{\partial T}\right)_P = T\left(\frac{\partial S}{\partial T}\right)_V + T\left(\frac{\partial S}{\partial V}\right)_T \left(\frac{\partial V}{\partial T}\right)_P$$

But $$\left(\frac{\partial S}{\partial V}\right)_T = \left(\frac{\partial P}{\partial T}\right)_V$$, from Maxwell's second relation (Eqn. 6.7)

$$\therefore \quad T\left(\frac{\partial S}{\partial T}\right)_P - T\left(\frac{\partial S}{\partial T}\right)_V = T\left(\frac{\partial P}{\partial T}\right)_V \left(\frac{\partial V}{\partial T}\right)_P$$

$$C_P - C_V = T\left(\frac{\partial P}{\partial T}\right)_V \left(\frac{\partial V}{\partial T}\right)_P \quad \text{... (6.12)}$$

Now $$T\left(\frac{\partial P}{\partial T}\right)_V = P + \left(\frac{\partial U}{\partial V}\right)_T \quad (\because \text{Eq. 6.4})$$

$$\therefore \quad C_P - C_V = \left[P + \left(\frac{\partial U}{\partial V}\right)_T\right]\left(\frac{\partial V}{\partial T}\right)_P \quad \text{... (6.12a)}$$

***(a)* For a perfect gas:** The equation of state is

$$PV = RT$$

$$\therefore \quad \left(\frac{\partial P}{\partial T}\right)_V = \frac{R}{V} \text{ and } \left(\frac{\partial V}{\partial T}\right)_P = \frac{R}{P}$$

Hence, equation (6.12) becomes

$$C_P - C_V = T\left(\frac{R}{V}\right)\left(\frac{R}{P}\right) = \frac{R^2T}{PV} = \frac{R^2T}{RT} \quad (\because PV = RT)$$

$$\therefore \quad C_P - C_V = R \quad \text{... (6.13)}$$

***(b)* Van der Waals :** The equation of state is

$$\left(P + \frac{a}{V^2}\right)(V - b) = RT$$

where a and b are constants.

$$\left(P + \frac{a}{V^2}\right) = \frac{RT}{(V - b)}$$

Differentiating with respect to T at constant volume, we get

$$\left(\frac{\partial P}{\partial T}\right)_V = \frac{R}{(V - b)}$$

and differentiating with respect to T at constant pressure we get

$$0 - \frac{2a}{V^3}\left(\frac{\partial V}{\partial T}\right)_P = -\frac{RT}{(V-b)^2}\left(\frac{\partial V}{\partial T}\right)_P + \frac{R}{(V-b)}$$

$$= \frac{R}{(V-b)} \quad \text{or} \quad \left(\frac{\partial V}{\partial T}\right)_P \left[\frac{RT}{(V-b)^2} - \frac{2a}{V^3}\right]$$

or $$\left(\frac{\partial V}{\partial T}\right)_P = \frac{\left[\frac{R}{(V-b)}\right]}{\left[\frac{RT}{(V-b)^2} - \frac{2a}{V^3}\right]}$$

Substituting these values of $\left(\frac{\partial P}{\partial T}\right)_V$ and $\left(\frac{\partial V}{\partial T}\right)_P$ in equation (6.12), we get

$$C_P - C_V = -\frac{T\left(\frac{R}{V-b}\right)\left(\frac{R}{V-b}\right)}{\left[\frac{RT}{(V-b)^2} - \frac{2a}{V^3}\right]} = \frac{R}{1 - \frac{2a}{V^3}\left[\frac{(V-b)^2}{RT}\right]}$$

Neglecting b in comparison to V,

$$C_P - C_V = \frac{R}{1-\dfrac{2a}{V^3}\cdot\dfrac{V^2}{RT}} = \frac{R}{1-\dfrac{2a}{VRT}} = R\left(1-\frac{2a}{VRT}\right)^{-1}$$

Expanding binomially and neglecting a in higher powers, as a is also small as compared to V, we have

$$C_P - C_V = R\left(1+\frac{2a}{VRT}\right) \qquad \text{... (6.14)}$$

[Here R is in heat units]

6.5.2 Joule-Thomson Cooling (Throttling Process)

As we know, in Joule-Thomson porous plug experiment, if a gas at constant high pressure is forced through a porous plug to a region of constant low pressure, the temperature of escaping gas changes. This is called Joule-Thomson effect.

Due to throttling, the gas suffers expansion. Although there is a pressure difference on two sides of the porous plug, the enthalpy H of the gas remains constant. This is the necessary condition i.e.

$$H = U + PV = \text{a constant}$$

$\therefore$ $$dH = dU + PdV + VdP = 0$$

But $$dQ = dU + PdV \qquad (\because \text{ First law of thermodynamics})$$

Also $$dQ = TdS \qquad (\because \text{ Second law of thermodynamics})$$

$\therefore$ $$dH = TdS + VdP = 0 \qquad \text{... (6.15)}$$

Now dS being a perfect differential and S is a function of P and T i.e. $S = f(P\ T)$, we have

$$dS = \left(\frac{\partial S}{\partial T}\right)_P dT + \left(\frac{\partial S}{\partial P}\right)_T dP$$

Substituting in equation (6.15),

$$T\left(\frac{\partial S}{\partial T}\right)_P dT + \left[T\left(\frac{\partial S}{\partial P}\right)_T + V\right]dP = 0$$

Now $$C_P = \left(\frac{\partial Q}{\partial T}\right)_P = T\left(\frac{\partial S}{\partial T}\right)_P$$

where C_P is specific heat at constant pressure.

$\therefore$ $$C_P\, dT = -\left[T\left(\frac{\partial S}{\partial P}\right)_T + V\right]dP \qquad \text{... (6.16)}$$

According to Maxwell's fourth thermodynamical relation

$$\left(\frac{\partial S}{\partial P}\right)_T = -\left(\frac{\partial V}{\partial T}\right)_P$$

$\therefore$ $$C_P\, dT = \left[T\left(\frac{\partial V}{\partial T}\right)_P - V\right]dP$$

or $$dT = \frac{1}{C_P}\left[T\left(\frac{\partial V}{\partial T}\right)_P - V\right]dP \qquad \text{... (6.17)}$$

6.5.3 Joule-Thomson Coefficient (μ)

The Joule-Thomson coefficient is given by

$$\mu = \left(\frac{\partial T}{\partial P}\right)_H = \frac{1}{C_P}\left[T\left(\frac{\partial V}{\partial T}\right)_P - V\right]$$

Now $\left(\frac{\partial V}{\partial T}\right)_P = \alpha V$, where α is the coefficient of increase of volume at constant pressure, because $\alpha = \frac{1}{V}\left(\frac{\partial V}{\partial T}\right)_P$.

∴ From equation (6.17), we have

$$dT = \frac{V}{C_P}[\alpha T - 1]\, dP \qquad \text{... (6.18)}$$

Equation (6.18) gives the change in temperature due to Joule-Thomson effect.

Here dP represents a fall in pressure in porous plug experiment and it is always a negative quantity. All other quantities in equation (6.18) are + ve. Hence

(*i*) There is cooling effect *i.e.* dT is negative if $(\alpha T - 1)$ is + ve or $\alpha T > 1$.

(*ii*) There is neither a cooling nor a heating effect *i.e.* $dT = 0$ when $(\alpha T - 1) = 0$ or $\alpha T = 1$

(*iii*) There is a heating effect *i.e.* dT is positive if $(\alpha T - 1)$ is negative or $\alpha T < 1$.

Let us apply the thermodynamical treatment of heating and cooling effect to a perfect gas and Van der Waals gas.

(*a*) For a perfect gas:

According to the gas equation.

$$PV = RT \qquad \text{(For perfect gas)}$$

At constant pressure,

$$P\partial V = R\partial T$$

$$\therefore \qquad \left(\frac{\partial V}{\partial T}\right)_P = \frac{R}{P}$$

$$\text{or} \qquad T\left(\frac{\partial V}{\partial T}\right)_P = \frac{RT}{P} = V$$

$$\text{Hence} \qquad \left[T\left(\frac{\partial V}{\partial T}\right)_P - V\right] = 0$$

Using equation (6.17), we have

For a perfect gas, $dT = 0$

In other words, there is no Joule-Thomson effect for a perfect gas. Hence, the porous plug experiment provides us a method to decide whether the given gas is perfect or not. For a perfect gas, the Joule-Thomson coefficient (μ) is zero.

$$\therefore \qquad \mu = \left(\frac{\partial T}{\partial P}\right)_H = \frac{1}{C_P}\left[T\left(\frac{\partial V}{\partial T}\right)_P - V\right] = 0$$

(*b*) For Van der Waals gas

Van der Waals equation for one mole of a real gas is given by

$$\left(P + \frac{a}{V^2}\right)(V - b) = RT \qquad \text{... (}i\text{)}$$

$$\text{or} \qquad PV - Pb + \frac{a}{V} - \frac{ab}{V^2} = RT$$

Since both the Van der Waals' constants a and b are very small quantities, their product ab is very small as compared to V^2 and hence $\frac{ab}{V^2}$ can be neglected.

$$\therefore \quad PV - Pb + \frac{a}{V} = RT$$

Differentiating both sides with respect to T, keeping P constant, we get.

$$P\left(\frac{dV}{dT}\right)_P - \frac{a}{V^2}\left(\frac{dV}{dT}\right)_P = R$$

or

$$\left(\frac{dV}{dT}\right)_P\left(P - \frac{a}{V^2}\right) = R$$

or

$$\left(\frac{dV}{dT}\right)_P = \frac{R}{\left(P - \frac{a}{V^2}\right)} \qquad \text{... (ii)}$$

From equation (i)

$$P = \frac{RT}{V-b} - \frac{a}{V^2}$$

Substituting for P in equation (ii), we have

$$\left(\frac{dV}{dT}\right)_P = \frac{R}{\left(\frac{RT}{V-b} - \frac{2a}{V^2}\right)} = \frac{R(V-b)}{\left[RT - \frac{2a(V-b)}{V^2}\right]}$$

$$= \frac{R(V-b)}{RT - \frac{2a}{V}} \qquad [\because \text{As } b \text{ is very small, } V - b = V]$$

$$\therefore \quad T\left(\frac{dV}{dT}\right)_P = \frac{RT(V-b)}{RT - \frac{2a}{V}} = \frac{V-b}{1 - \frac{2a}{VRT}}$$

$$= (V-b)\left(1 - \frac{2a}{VRT}\right)^{-1}$$

$$= (V-b)\left(1 + \frac{2a}{VRT}\right)$$

Neglecting squares and higher powers of $\frac{2a}{VRT}$ as $\frac{2a}{VRT} << 1$.

$$\therefore \quad T\left(\frac{dV}{dT}\right)_P = (V-b)\left(1 + \frac{2a}{VRT}\right)$$

$$= V - b + \frac{2a(V-b)}{VRT}$$

$$= V - b + \frac{2a}{RT} \qquad [\text{As } b \text{ is very small } V - b = V]$$

$$\therefore \quad T\left(\frac{dV}{dT}\right)_P - V = \frac{2a}{RT} - b \qquad \text{... (6.19)}$$

According to Joule-Thomson effect, the change in temperature

$$dT = \frac{1}{C_P}\left[T\left(\frac{\partial V}{\partial T}\right)_P - V\right] dP$$

where C_P is molar specific heat at constant pressure.

Substituting the value of $\left[T\left(\frac{\partial V}{\partial T}\right)_P - V\right]$ from equation (6.19), we get

$$dT = \frac{1}{C_P}\left[\frac{2a}{RT} - b\right]dP$$

$$dT = \frac{dP}{C_P}\left[\frac{2a}{RT} - b\right] \quad \text{... (6.20)}$$

Equation (6.20) gives the change in temperature for Van der Waals gas *i.e.* real gas due to Joule-Thomson effect.

Thus, we conclude that

(*i*) If $\frac{2a}{RT} > b$ *i.e.*, T is less than $\frac{2a}{Rb}$, then δT is positive. Since δP is $-ve$, dT will be $-ve$. Hence there will be a cooling effect.

(*ii*) If $\frac{2a}{RT} < b$ *i.e.*, T is greater than $\frac{2a}{Rb}$, then δT is negative, since δP is $-ve$, dT will be $+ve$. Hence there will be a heating effect.

(*iii*) If $\frac{2a}{RT} = b$ or $T = \frac{2a}{Rb}$, then $dT = 0$, there will be no change of temperature.

6.5.4 Temperature of Inversion

The temperature at which Joule-Thomson effect is zero and changes sign is known as *temperature of inversion* and is denoted by T_i

$$\therefore \quad T_i = \frac{2a}{Rb} \quad \text{... (6.21)}$$

When the room temperature is above the temperature of inversion ($T > T_i$) or $\frac{2a}{RT} < b$, there will be a heating effect.

Hence, during the expansion of the gas through a porous plug (1) cooling will take place if temperature of the gas is less than its temperature of inversion $\left[T < \frac{2a}{Rb}\right]$ and (2) heating will take place if temperature of the gas is more than its temperature of inversion $\left[T > \frac{2a}{Rb}\right]$.

6.5.5 Heating Effect for Hydrogen and Helium at Room Temperature

For most of the gases, the ordinary working temperatures are below the temperature of inversion and hence they show cooling effect.

The temperature of inversion for hydrogen and helium is much below the ordinary room temperature. Hence, at ordinary room temperatures, these gases show a heating effect. For hydrogen, at about 100 atmospheres pressure, the temperature of inversion (T_i) is – 80°C. In hydrogen, a cooling effect is, therefore, observed if it is cooled below – 80°C and then passed through the porous plug.

For helium, the temperature of inversion (T_i) is 35 K (*i.e.* – 238°C). Hence at ordinary temperature, it shows a heating effect.

6.5.6 Heating Produced by Adiabatic Compression

From first thermodynamical relation

$$\left(\frac{\partial T}{\partial V}\right)_S = -\left(\frac{\partial P}{\partial S}\right)_V$$

Multiplying and dividing the right hand side by T, we get

$$\left(\frac{\partial T}{\partial V}\right)_S = -T\left(\frac{\partial P}{T\partial S}\right)_V = -T\left(\frac{\partial P}{\partial Q}\right)_V \quad \text{... (6.22)}$$

An increase in the amount of heat (δQ) at constant volume always results in an increase in pressure. Therefore, $\left(\frac{\partial P}{\partial Q}\right)_V$ is positive. Hence the term $\left(\frac{\partial T}{\partial V}\right)_S$ should be negative. In other words, the temperature should decrease with an increase in volume at constant entropy (*i.e.* adiabatic process). Thus, *an adiabatic expansion must result in a fall of temperature.*

Recalling third thermodynamical relation

$$\left(\frac{\partial T}{\partial P}\right)_S = \left(\frac{\partial V}{\partial S}\right)_P$$

or

$$\left(\frac{\partial T}{\partial P}\right)_S = T\left(\frac{\partial V}{T\partial S}\right)_P = T\left(\frac{\partial V}{\partial Q}\right)_P \quad \text{... (6.23)}$$

Since, an increase in the quantity heat (∂Q) at constant pressure, always results in an increase in its volume the term $\left(\frac{\partial V}{\partial Q}\right)_P$ is positive, hence the term $\left(\frac{\partial T}{\partial P}\right)_S$ must also be positive i.e. the temperature must increase with increase in pressure at constant entropy (*i.e.* adiabatic process). Thus, *an adiabatic compression must result in an increase in temperature.*

6.5.7 Cooling Produced by Adiabatic Expansion

Using Maxwell's first thermodynamical relation

$$\left(\frac{\partial T}{\partial V}\right)_S = -\left(\frac{\partial P}{\partial S}\right)_V$$

During an adiabatic change, the entropy remains constant. The equation can now be put as

$$\left(\frac{\partial T}{\partial V}\right)_S = -T\left(\frac{\partial P}{\partial Q}\right)_V \qquad [\because T\,\partial S = \partial Q]$$

Now

$$\left(\frac{\partial P}{\partial Q}\right)_V = \left(\frac{\partial P}{\partial T}\right)_V\left(\frac{\partial T}{\partial Q}\right)_V$$

But $\left(\frac{\partial P}{\partial T}\right)_V = P\beta$ where β is the coefficient of increase of pressure at constant volume and $\left(\frac{\partial T}{\partial Q}\right)_V = \frac{1}{C_V}$ because $\left(\frac{\partial Q}{\partial T}\right)_V = C_V$ whre C_V denotes the gram molecular specific heat at constant volume

$$\therefore \qquad \left(\frac{\partial T}{\partial V}\right)_S = -\frac{T\,P\beta}{C_v}$$

If we have m gms of a substance and c_v denotes specific heat per gram, then

$$\left(\frac{\partial T}{\partial V}\right)_S = -\frac{TP\beta}{mc_v}$$

If dT is the change in temperature due to a change in volume dV then for a gram mole

$$dT = -\frac{TP\beta}{C_v} \cdot dV$$

For m gms of the substance

$$dT = -\frac{T\,P\beta}{mc_v} dV$$

For most of the substances β is positive quantity as these expand on heating, C_v and all other quantities are also positive. Therefore for all such substances dT is negative *i.e.*, they show a cooling effect when allowed to expand adiabatically.

If, however, the expansion is to be carried out isothermally a quantity of heat $\Delta Q = C_v$, $\Delta T = -TP\beta\,\Delta V$ must be supplied to the system from outside for gm. mole and for m gms

$$\Delta Q = mc_v \Delta T$$

$$= -TP\,\beta\,\Delta V$$

where ΔV is the change in volume of the m gms of the substance.

Note: Do all substances get cooled on adiabatic expansion ? The answer is No. The change in temperature dT due to change in volume dV on adiabatic expansion is given by $dT = -\frac{TP\beta}{C_v} dV$. In this expression T, P, C_v, and dV are all positive quantities. If β, the coefficient of pressure at constant volume is positive, then dT is negative. Hence, only substances having β positive show cooling effect on adiabatic expansion.

6.5.8 Clausius-Clapeyron's Equation (First Latent Heat Equation)

The equation $\frac{dP}{dT} = \frac{L}{T\,(V_2 - V_1)}$ was first derived by Clapeyron using Carnot's reversible cycle. Therefore, it is some times called **Clapeyron's equation.**

Whenever there is a change of state, either from a solid state to liquid state or from liquid state to vapour state, the temperature remains constant; as far as the change takes place. This temperature (either melting point or boiling point) depends upon the pressure and is a characteristic of each substance. The above relation shows how melting or boiling points vary with pressure and was derived by Clapeyron by applying the second law of thermodynamics. Later Clausius obtained the same relation by using the Maxwell's thermodynamic relations, as under :

Derivation :

From second thermodynamical relation

$$\left(\frac{\partial S}{\partial V}\right)_T = \left(\frac{\partial P}{\partial T}\right)_V$$

Multiplying both sides by T, we have

$$T\left(\frac{\partial S}{\partial V}\right)_T = T\left(\frac{\partial P}{\partial T}\right)_V$$

But $\qquad T\partial S = \partial Q$

$$\therefore \qquad \left(\frac{\partial Q}{\partial V}\right)_T = T\left(\frac{\partial P}{\partial T}\right)_V \qquad \text{... (6.24)}$$

The quantity $\left(\frac{\partial Q}{\partial V}\right)_T$ represents the quantity of heat absorbed or liberated per unit change in volume at constant temperature. As there is a change in volume due to the heat absorbed at constant temperature, the heat represents the latent heat used when a substance changes from solid to liquid (melting) or liquid to vapour (boiling) state when the temperature remains constant, during the change of state.

If L is the quantity of heat required to change the state of a unit mass of the substance, V_2 and V_1 the corresponding specific volumes (Volume per unit mass) then

$$\partial Q = L \text{ and } \partial V = V_2 - V_1.$$

Hence
$$\left(\frac{\partial Q}{\partial V}\right)_T = \frac{L}{V_2 - V_1}$$

Therefore,
$$\frac{L}{V_2 - V_1} = T\left(\frac{\partial P}{\partial T}\right)_V$$

or
$$\frac{dP}{dT} = \frac{L}{T(V_2 - V_1)} \qquad \text{... (6.25)}$$

Where dT is the change in melting point or boiling point due to a change in pressure dP. This is Clausius-Clapeyron latent heat equation.

Effect of Pressure on Boiling Point of Liquid

When a liquid converts into vapour state, there is an increase in volume. Hence $V_2 > V_1$; T and L being positive quantities, $\left(\frac{\partial P}{\partial T}\right)_V$ is + ve. This shows that the boiling point of the liquid is raised by increasing the pressure. This is known as elevation of boiling point.

Effect of Pressure on Melting Point of Solid

When a solid melts, there may be increase in volume, as in case of wax or there may be decrease in volume, as in case of ice. In melting of wax

$$(V_2 - V_1) = +\text{ve}$$

$$\therefore \quad \frac{dP}{dT} = +\text{ve}$$

Hence the melting point is raised by increase of pressure.

In case of ice, $(V_2 - V_1) = -$ ve, hence melting point is lowered by increase of pressure. This is known as depression in melting point. In general, substances which expands on melting, have their melting points raised by increase of pressure and the substance which contract on melting have their melting points lowered by increase of pressure.

6.6 THERMODYNAMIC POTENTIALS

The state of a system can be completely described by any two of the five state variables P, V, T, S and U. Out of these, U the internal energy state variable is determined by using the remaining four, as proved below:

According to first law of thermodynamics,

$$\delta Q = dU + PdV$$

and from the second law of thermodynamics,

$$\delta Q = TdS$$

$$\therefore \quad dU + PdV = TdS$$

or
$$dU = TdS - PdV$$

Thus, U can be eliminated and we are left with four state variables P, V, T and S only. This is the reason, why only four state variables are defined. Taking two of the four state variables P, V, T and S at a time, there are six possible pairs, *i.e.* (P, V), (P, T), (P, S), (V, T), (V, S) and (T, S). The pair (P, V) is connected with composite and in exact differential quantity δW as $\delta W = PdV$ and pair (T, S) with δQ as $\delta Q = TdS$. Hence these two pairs can be eliminated. Thus, we are left with four pairs of thermodynamical variables (P, T), (P, S), (V, T) and (V, S). Corresponding to each pair, we can write a thermodynamical relation. These four thermodynamical relations are known as Maxwell's thermodynamic relations.

There are four thermodynamic potentials:

(*i*) Internal energy, U

(*ii*) Helmholtz free energy, $F = U - TS$

(*iii*) Enthalpy $H = U + PV$

(*iv*) Gibbs function $G = U + PV - TS$

Each of Maxwell's four thermodynamical relations can be derived from one of these thermodynamic potentials U, F, H and G. Hence, only four thermodynamic potentials are defined.

1. Internal Energy *U*. The internal energy or the intrinsic energy is the total energy of a system. According to the first law of thermodynamics,

$$dQ = dU + dW = dU + PdV$$

$$\therefore \quad dU = dQ - PdV$$

and from second law of thermodynamics

$$dQ = TdS$$

Substituting for dQ, we get

$$dU = TdS - PdV \qquad \text{... (6.26)}$$

This equation gives the change in internal energy of the system in terms of four thermodynamical variables P, V, T and S. Internal energy (U) is called *first thermodynamical potential.*

(*a*) For an adiabatic process :

$$dQ = 0$$

$$\therefore \quad dU = -PdV$$

i.e. the work done by the system in an adiabatic process is at the expense of its internal energy.

(*b*) For an isochoric adiabatic process :

$$dV = 0 \text{ and } dQ = 0$$

$$\therefore \quad dU = 0 \text{ or } U = \text{a constant.}$$

i.e., the internal energy of system remains constant in an isochoric adiabatic process.

2. Helmholtz Free Energy *F*. Helmholtz free energy is defined as

$$F = U - TS$$

As U, T and S are state variables, F is also a state variable. F has dimensions of energy. According to first and second law of thermodynamics,

$$dU = TdS - dW$$

If the system is maintained at a constant temperature by exchanging heat continuously with the surroundings,

then, $$TdS = d(TS)$$

$$\therefore \quad dU = d(TS) - dW$$

or $$d(U - TS) = -dW$$

or $$dF = -dW \qquad \text{... (6.27)}$$

where $F = (U - TS)$ is known as *Helmholtz free energy or Helmholtz work function.*

Now $$dF = d(U - TS) = dU - TdS - SdT$$

But $$dU = TdS - PdV$$

$$\therefore \quad dF = -PdV - SdT \qquad \text{... (6.28)}$$

This equation gives the change in Helmholtz free energy during an infinitesimal reversible process.

***(a)* For reversible isothermal process:**

$$dT = 0$$

$$\therefore \quad dF = -PdV \quad \text{or} \quad PdV = -dF$$

Thus, the work done in a reversible isothermal process is equal to the decrease in Helmholtz free energy.

***(b)* For isothermal isochoric process:**

$$dT = 0 \text{ and } dV = 0$$

$$\therefore \quad dF = 0 \text{ or } F = \text{a constant.}$$

i.e., the Helmholtz free energy remains constant during isothermal isochoric process.

3. Enthalpy *H*. This is known as the total heat and is given by

$$H = U + PV$$

As *U, P* and *V* are state variables, *H* is also a state variable. *H* has dimensions of energy. If the system undergoes an infinitesimal reversible process, then

Change in enthalpy, $$dH = dU + PdV + VdP$$

But $$dU = TdS - PdV$$

$$\therefore \quad dH = TdS + VdF \qquad \text{... (6.29)}$$

***(a)* For reversible isobaric process :**

$$dP = 0$$

$$\therefore \quad dH = TdS = dQ$$

i.e., for an isobaric process, the change in enthalpy is equal to the heat absorbed.

***(b)* For an isobaric adiabatic process :**

$$dP = 0; \qquad dQ = 0$$

$$\therefore \quad dH = 0 \quad \text{or} \quad H = \text{a constant}$$

i.e., enthalpy remains constant in a reversible isobaric adiabatic process.

4. Gibbs Function G or Gibbs Free Energy. Gibbs function (*G*) was called by Duhem the thermodynamical potential at constant pressure. This is defined by the equation

$$G = U - TS + PV \qquad \text{... (6.30)}$$

As $F = U - TS$, therefore, we can write

$$G = F + PV \qquad \text{... (6.31)}$$

This is the relation between Gibbs function and Helmholtz function. An enthalpy $H = U + PV$, therefore, equation (6.30) becomes

$$G = H - TS$$

or $$H = G + TS \qquad \text{...(6.32)}$$

$\therefore$ Enthalpy = Gibbs free energy + latent heat

***(a)* For an isothermal process:** $TdS = d(TS)$

***(b)* For an isobaric process:** $dP = 0$

Hence, if the process is isothermal and isobaric then

$$dH = d(TS)$$

$$d(H - TS) = 0$$

$$dG = 0 \quad \text{or} \quad G = \text{a constant}$$

where $G = H - TS$ is known as thermodynamic potential at constant pressure; or Gibbs function or Gibbs free energy.

Thus, Gibbs function (G) or Gibbs free energy remains constant in an isothermal-isobaric process.

6.7 SIGNIFICANCE OF THERMODYNAMIC POTENTIALS

A mechanical system is said to be in stable equilibrium when the potential energy of the system is minimum. It means that the system must proceed in such a direction so as to acquire minimum potential energy. This is what we observe in nature, viz. water flows from a higher level to lower level, electric current flows from higher to lower potential, heat flows from higher to lower temperature, a body falls from higher to lower potential due to gravitational field and so on.

In thermodynamics, the behavior of internal energy (U), Helmholtz free energy (F), enthalpy (H), and Gibbs free energy (G) is similar to potential energy in *mechanics*. As we have seen, the direction of isothermal-isochoric process is to make Helmholtz free energy (F) minimum. In isothermal-isobaric process, Gibbs free energy (G) tends to be minimum. In an isobaric-adiabatic process, the enthalpy (H) tends to be minimum.

Since the four functions U, F, H and G play in thermodynamics the same role as played by potential energy in mechanics, hence, they are called thermodynamic potentials.

6.8 RELATION OF THERMODYNAMICAL POTENTIALS WITH THEIR VARIABLES

The four quantities $U(S, V)$, $F(T, V)$, $H(S, P)$ and $G(P, T)$ are called thermodynamic potentials because the thermodynamic variables S, T, P and V can be derived from them by their differentiations with respect to the independent variables associated with them. Let us derive them:

1. Thermodynamic Potential Internal Energy U (S, V)

According to first and second law of thermodynamics,

Change in internal energy, $dU = \delta Q - PdV$

$$= TdS - PdV \qquad (\because \delta Q = TdS)$$

Taking partial differentials of internal energy U with respect to variable S (entropy), we have

$$\left(\frac{\partial U}{\partial S}\right)_V = T \text{ and } \left(\frac{\partial U}{\partial V}\right)_S = -P \qquad \text{... (6.33)}$$

These are the relations connecting the internal energy U with thermodynamical variables S, V, T and P.

Since dU is a perfect differential (*i.e.* U is a single valued), we have

$$\left[\frac{\partial}{\partial V}\left(\frac{\partial U}{\partial S}\right)_V\right]_S = \left[\frac{\partial}{\partial S}\left(\frac{\partial U}{\partial V}\right)_S\right]_V$$

or
$$\left(\frac{\partial T}{\partial V}\right)_S = -\left(\frac{\partial P}{\partial S}\right)_V \qquad \text{... (6.34)}$$

This is Maxwell's *first* thermodynamical relation.

2. Thermodynamic Potential Helmholtz Free Energy F (T, V)

The Helmholtz function or Helmholtz free energy is given by

$$F = U - TS$$

$\therefore$
$$dF = d(U - TS) = dU - TdS - SdT$$

But
$$dU = TdS - PdV$$

$\therefore$
$$dF = -PdV - SdT$$

Taking partial differential of Helmholtz function F with respect to T and V, we get

$$\left(\frac{\partial F}{\partial T}\right)_V = -S \text{ and } \left(\frac{\partial F}{\partial V}\right)_T = -P \quad \text{... (6.35)}$$

Since dF is a perfect differential,

$$\left[\frac{\partial}{\partial V}\left(\frac{\partial F}{\partial T}\right)_V\right]_T = \left[\frac{\partial}{\partial T}\left(\frac{\partial F}{\partial V}\right)_T\right]_V$$

or
$$\left(\frac{\partial S}{\partial V}\right)_T = \left(\frac{\partial P}{\partial T}\right)_V \quad \text{... (6.36)}$$

This is Maxwell's *second* thermodynamical relation.

3. Thermodynamic Potential Enthalpy H (S, P)

The enthalpy (H) is given by

$$H = U + PV$$

Differentiating, $dH = dU + PdV + VdP$

But $dU = TdS - PdV$

$\therefore$ $dH = TdS + VdP$

Taking partial differential of enthalpy H with respect to S and P, we get

$$\left(\frac{\partial H}{\partial S}\right)_P = T \text{ and } \left(\frac{\partial H}{\partial P}\right)_S = V \quad \text{... (6.37)}$$

Since dH is a perfect differential,

$$\left[\frac{\partial}{\partial P}\left(\frac{\partial H}{\partial S}\right)_P\right]_S = \left[\frac{\partial}{\partial S}\left(\frac{\partial H}{\partial P}\right)_S\right]_P$$

or
$$\left(\frac{\partial T}{\partial P}\right)_S = \left(\frac{\partial V}{\partial S}\right)_P \quad \text{... (6.38)}$$

This is Maxwell's *third* thermodynamical relation.

4. Thermodynamic Potential Gibb's Function G (P, T)

The Gibbs function (G) is given by

$$G = H - TS$$

Differentiating, we get

$$dG = dH - TdS - SdT$$

But $dH = TdS + VdP$

$\therefore$ $dG = VdP - SdT$

Taking partial differential of Gibbs function G with respect to P and T, we get

$$\left(\frac{\partial G}{\partial P}\right)_T = V \text{ and } \left(\frac{\partial G}{\partial T}\right)_P = -S \quad \text{... (6.39)}$$

As dG is perfect differential,

$$\left[\frac{\partial}{\partial T}\left(\frac{\partial G}{\partial P}\right)_T\right]_P = \left[\frac{\partial}{\partial P}\left(\frac{\partial G}{\partial T}\right)_P\right]_T$$

or
$$\left(\frac{\partial V}{\partial T}\right)_P = -\left(\frac{\partial S}{\partial P}\right)_T \quad \text{... (6.40)}$$

This is Maxwell's *fourth* thermodynamical relation.

Thus, the thermodynamical variables S, T, P and V can be written by using equation (6.33), (6.35), (6.37) and (6.39) as

$$S = -\left(\frac{\partial G}{\partial T}\right)_P = -\left(\frac{\partial F}{\partial T}\right)_V$$

$$T = \left(\frac{\partial U}{\partial S}\right)_V = \left(\frac{\partial H}{\partial S}\right)_P$$

$$P = -\left(\frac{\partial U}{\partial V}\right)_S = -\left(\frac{\partial F}{\partial V}\right)_T$$

$$V = \left(\frac{\partial H}{\partial P}\right)_S = \left(\frac{\partial G}{\partial P}\right)_T \quad \text{... (6.41)}$$

These equations give the values of thermodynamical variables in terms of thermodynamical potentials.

The *Gbb's* potential G is given by $G = H - TS$

But
$$S = -\left(\frac{\partial G}{\partial T}\right)_P$$

Hence,
$$G = H + T\left(\frac{\partial G}{\partial T}\right)_P$$

6.9 RELATIONS FOR C_P, C_V AND μ

The specific heat at constant pressure C_p, the specific heat at constant volume C_V and the Joule-Kelvin coefficient μ are defined as follows:

$$C_P = \left(\frac{\partial H}{\partial T}\right)_P \quad \text{... }(i)$$

$$C_V = \left(\frac{\partial U}{\partial T}\right)_V \quad \text{... }(ii)$$

$$\mu = \left(\frac{\partial T}{\partial P}\right)_h \quad \text{... }(iii)$$

These three quantities are defined in terms of the thermodynamic properties *viz.* pressure, volume, temperature, internal energy and enthalpy. Hence C_P, C_V and μ are also thermodynamic properties of a substance.

6.10 VARIATION C_V WITH VOLUME

The specific heat at constant volume is given by the relation

$$C_V = \left(\frac{\partial Q}{\partial T}\right)_V$$

But
$$dQ = TdS$$

$\therefore$
$$C_V = T\left(\frac{\partial S}{\partial T}\right)_V$$

Differentiating with respect to volume V, keeping temperature T constant, we get

$$\left(\frac{\partial C_v}{\partial V}\right)_T = \frac{\partial}{\partial V}\left[T\left(\frac{\partial S}{\partial T}\right)_V\right]_T = T\frac{\partial^2 S}{\partial V\,\partial T} \quad \text{... }(i)$$

According to Maxwell's thermodynamical relation

$$\left(\frac{\partial S}{\partial V}\right)_T = \left(\frac{\partial P}{\partial T}\right)_V$$

Differentiating with respect to T keeping V constant, we get

$$\frac{\partial}{\partial T}\left[\left(\frac{\partial S}{\partial V}\right)_T\right]_V = \frac{\partial}{\partial T}\left[\left(\frac{\partial P}{\partial T}\right)_S\right]_V$$

or $$\frac{\partial^2 S}{\partial T\,\partial V} = \left(\frac{\partial^2 P}{\partial T^2}\right)_V \quad \text{or} \quad T\frac{\partial^2 S}{\partial T\,\partial V} = T\left(\frac{\partial^2 P}{\partial T^2}\right)_V \qquad \text{... }(ii)$$

Comparing (*i*) and (*ii*), we have

$$\left(\frac{\partial C_v}{\partial V}\right)_T = T\left(\frac{\partial^2 P}{\partial T^2}\right)_v \qquad \text{... (6.42)}$$

Equation 6.42 gives the variation of C_v with volume V during an isothermal process (constant T).

***(i)* For a perfect gas.** For one mole of a perfect gas

$$PV = RT \quad \text{or} \quad P = \frac{RT}{V}$$

Differentiating twice with respect to T, keeping V constant, we get

$$\frac{\partial P}{\partial T} = \frac{R}{V}$$

and $$\frac{\partial^2 P}{\partial T^2} = 0 \qquad \text{... }(iii)$$

Comparing 6.42 and (*iii*), we have

$$\left(\frac{\partial C_v}{\partial V}\right)_T = 0 \qquad \text{... }(iv)$$

This equation shows that C_v does not change with volume during isothermal expansion or compression for a perfect gas.

***(ii)* For a van der Waal gas.** For one mole of a Van der Waal gas

$$\left(P + \frac{a}{V^2}\right)(V - b) = RT$$

or $$P = \frac{RT}{V-b} - \frac{a}{V^2}$$

Differentiating twice with respect to T, keeping V constant, we get

$$\left(\frac{\partial P}{\partial T}\right)_v = \frac{R}{V-b}$$

and $$\left(\frac{\partial^2 P}{\partial T^2}\right) = 0 \qquad \text{... }(v)$$

Comparing (6.42) and (*iv*), we get

$$\left(\frac{\partial C_v}{\partial V}\right)_T = 0 \qquad \text{... }(vi)$$

This equation shows that C_v does not change with volume (i.e., independent of volume) during isothermal expansion or compression for a Van der Waal gas.

6.11 VARIATION OF C_P WITH PRESSURE

Thespecific heat at constant pressure

$$C_P = \left(\frac{\partial Q}{\partial T}\right)_P$$

But $\qquad dQ = T\,dS$

$$\therefore \qquad C_P = T\left(\frac{\partial S}{\partial T}\right)_p$$

Differentiating with respect to p keeping T constant, we get

$$\left(\frac{\partial C_P}{\partial P}\right)_T = \frac{\partial}{\partial_P}\left[T\left(\frac{\partial S}{\partial T}\right)_P\right]_T$$

or
$$\left(\frac{\partial C_p}{\partial P}\right)_T = T\frac{\partial^2 S}{\partial p\,\partial T} \qquad \text{... } (i)$$

According to Maxwell's thermodynamic relation

$$\left(\frac{\partial V}{\partial T}\right)_P = -\left(\frac{\partial S}{\partial P}\right)_T$$

Differentiating with respect to T keeping P constant, we have

$$\frac{\partial}{\partial T}\left[\left(\frac{\partial V}{\partial T}\right)_p\right]_p = \frac{\partial}{\partial T}\left[-\left(\frac{\partial S}{\partial P}\right)_T\right]_p$$

or
$$\left(\frac{\partial^2 V}{\partial T^2}\right)_p = -\frac{\partial^2 S}{\partial P\,\partial T} \quad \text{or} \quad T\left(\frac{\partial^2 V}{\partial T^2}\right)_p = -T\frac{\partial^2 S}{\partial P\,\partial T} \qquad \text{... } (ii)$$

Comparing (*i*) and (*ii*), we get

$$\left(\frac{\partial C_p}{\partial P}\right)_T = -T\left(\frac{\partial^2 V}{\partial T^2}\right)_p \qquad \text{... (6.43)}$$

For a perfect gas. For one mole of a perfect gas

$$PV = RT \qquad \text{or} \qquad V = \frac{RT}{P}$$

Differentiating twice with respect to T, keeping P constant

$$\left(\frac{\partial V}{\partial T}\right)_p = \frac{R}{P} \quad \text{and} \quad \left(\frac{\partial^2 V}{\partial T^2}\right)_p = 0 \qquad \text{... } (iii)$$

Comparing Eq. (6.43) and (*iii*), we have

$$\left(\frac{\partial C_p}{\partial P}\right)_T = 0$$

This shows that C_p does not vary with pressure for a perfect gas when the temperature is kept constant.

6.12 INTERNAL ENERGY OF IDEAL AND REAL GASES

The internal energy of a perfect gas is only dependent upon temperature whereas the internal energy of a real gas (or Van der Waals gas) is a function of both temperature and volume.

For a perfect gas, the internal energy is only kinetic energy of its molecules. It is proportional to kT and hence only depends upon temperature, k being Boltzmann's constant. The internal energy of a perfect gas is independent of its volume.

For a real gas, say a gas obeying Van der Waals equation, the internal energy is the sum of the kinetic and potential energies of molecules due to their mutual forces of attraction. The force of attraction between the molecules depends upon the intermolecular distance and is thus a function of volume. This is why the internal energy of a real gas or a gas obeying Van der Waals equation is a function of both volume and temperature. As the volume of an actual gas increases (at constant temperature) the internal energy increases. This is due to the fact that when the volume increases, the molecules move farther apart against intermolecular attraction. Hence work is done thereby causing an increase in potential energy. (for details refer 6.5.1)

6.13 CHANGE OF INTERNAL ENERGY WITH VOLUME

This is an application of Maxwell's thermodynamical relation

$$\left(\frac{\partial S}{\partial V}\right)_T = \left(\frac{\partial P}{\partial T}\right)_V$$

Multiplying both sides by T and substituting $T\,\partial S = \partial Q$

We have $$\left(\frac{\partial Q}{\partial V}\right)_T = T\left(\frac{\partial P}{\partial T}\right)_V$$

Now $$\partial Q = \partial U + PdV$$

$\therefore$ $$\left(\frac{\partial U + PdV}{\partial V}\right)_T = T\left(\frac{\partial P}{\partial T}\right)_V$$

or $$\left(\frac{\partial U}{\partial V}\right)_T + P = T\left(\frac{\partial P}{\partial T}\right)_V$$

$$\left(\frac{\partial U}{\partial V}\right)_T = T\left(\frac{\partial P}{\partial T}\right)_V - P \qquad \text{... (6.44)}$$

This relation gives the change of internal energy with volume under **isothermal conditions.**

For a perfect gas

$$PV = RT \qquad \therefore \left(\frac{\partial P}{\partial T}\right)_V = \frac{R}{V}$$

Hence, $$\left(\frac{\partial U}{\partial V}\right)_T = \frac{RT}{V} - P$$

$$= \frac{RT}{V} - \frac{RT}{V} = 0 \qquad \text{... (6.45)}$$

This means that the internal energy of an ideal gas is independent of its volume. In other words, the change in internal energy with volume of an ideal gas is zero. This is known as *Joule's law.*

This further shows that during an isothermal expansion or compression there is no change in internal energy. Since there is no rise in temperature there is no variation in kinetic energy also. In other words *there is no inter-molecular attraction in perfect gases.*

Integrating Eq. 6.45, we get

$$U = \text{a constant} = U(T) \qquad \text{... (6.46)}$$

This equation shows that the *internal energy of a perfect gas is only a function of temperature.*

For a gas obeying van der Waals equation. For such a gas

$$\left(P+\frac{a}{V^2}\right)(V-b) = RT$$

or
$$P = \frac{RT}{V-b} - \frac{a}{V^2}$$

$$\therefore \quad \left(\frac{\partial P}{\partial T}\right)_V = \frac{R}{V-b}$$

and from equation (6.44),

$$\left(\frac{\partial U}{\partial V}\right)_T = T\left(\frac{\partial P}{\partial T}\right)_V - P$$

$$= \frac{RT}{(V-b)} - \frac{RT}{V-b} + \frac{a}{V^2} = \frac{a}{V^2}$$

or
$$\Delta U = \frac{a}{V^2}\Delta V \quad \text{... (6.47)}$$

This relation indicates that as the volume of an actual gas increases (at constant temperature) the internal energy increases. This is due to the fact that when the volume increases the molecules move farther-apart against intermolecular attraction. Hence work is done thereby causing an increase of potential energy *i.e.*, of the internal energy. This, trherefore, proves the existence of intermolecular attraction in case of real gases.

Furrther, integrating the equation

$$\left(\frac{\partial U}{\partial V}\right)_T = \frac{a}{V^2}, \text{ we get}$$

$$U = -\frac{a}{V} + f(T) \quad \text{... (6.48)}$$

where $f(T)$ is some function of temperature. This equation shows that the internal energy of a real gas depends upon both, volume as well as temperature.

6.14 SPECIFIC HEATS IN TERMS OF *U* AND *H*

$$dQ = dU + PdV$$

When volume remains constant, $dV = 0$

Hence, $PdV = 0$

$\therefore$ $dQ = dU$

Hence,
$$C_V = \left(\frac{dQ}{dT}\right)_V = \left(\frac{dU}{dT}\right)_V \quad \text{... (i)}$$

Enthalpy is given by, $H = U + PV$

$\therefore$ $dH = dU + PdV + VdP$

Putting $dU = TdS - PdV$

$\therefore$ $dH = TdS + VdP$

When pressure remains constant, $dP = 0$ or $VdP = 0$

$\therefore$ $dH = TdS = dQ$

or
$$\left(\frac{dH}{dT}\right)_P = \left(\frac{dQ}{dT}\right)_P = C_P$$

6.15 THE *T.dS* EQUATIONS

1. *First T.dS Equation.* The entropy S of a pure substance can be taken as a function of temperature and volume.

$$\therefore \qquad dS = \left(\frac{\partial S}{\partial T}\right)_V dT + \left(\frac{\partial S}{\partial V}\right)_T dV$$

Multiplying both sides by T

$$T.dS = T\left(\frac{\partial S}{\partial T}\right)_V dT + T\left(\frac{\partial S}{\partial T}\right)_V dV \qquad \text{... } (ii)$$

$\therefore$ But

$$C_V = T\left(\frac{\partial S}{\partial T}\right)_V$$

and from Maxwell's relations

$$\left(\frac{\partial S}{\partial V}\right)_T = \left(\frac{\partial P}{\partial T}\right)_V$$

Substituting these values in equation (*ii*)

$$Tds = C_V dT + T\left(\frac{\partial P}{\partial T}\right)_V dV \qquad \text{... (6.49)}$$

Equation (6.49) is called the first *T.dS* equation.

2. *Second T.dS Equation.* The entropy S of a pure substance can also be regarded as a function of temperature and pressure

$$\therefore \qquad dS = \left(\frac{\partial S}{\partial T}\right)_P dT + \left(\frac{\partial S}{\partial P}\right)_T dP$$

Multiplying both sides by T

$$T.dS = T\left(\frac{\partial S}{\partial T}\right)_P dT + T\left(\frac{\partial S}{\partial P}\right)_T dP$$

But

$$C_P = T\left(\frac{\partial S}{\partial T}\right)_P$$

and from Maxwell's relations

$$\left(\frac{\partial S}{\partial P}\right)_T = -\left(\frac{\partial V}{\partial T}\right)_P$$

Substituting these values in equation (*v*)

$$T.dS = C_P dT - T,\left(\frac{\partial V}{\partial T}\right)_P dP \qquad \text{... (6.50)}$$

Equation (6.50) is called the second $T\,dS$ equation.

6.16 CLAPEYRON'S LATENT HEAT EQUATION USING MAXWELL'S *THERMODYNAMICAL RELATIONS*

From Maxwell's second thermodynamical relation

$$\left(\frac{\partial S}{\partial V}\right)_T = \left(\frac{\partial P}{\partial T}\right)_V \qquad (\because \text{eqn. 6.36})$$

Multiplying by T, we get

$$T\left(\frac{\partial S}{\partial V}\right)_T = T\left(\frac{\partial P}{\partial T}\right)_V$$

$$\left(\frac{\partial Q}{\partial V}\right)_T = T\left(\frac{\partial P}{\partial T}\right)_V \qquad (\because T\,\partial S = \partial Q)$$

Here $\left(\frac{\partial Q}{\partial V}\right)_T$ represents the quantity of heat absorbed per unit increase in volume at constant temperature. This quantity of heat absorbed at constant temperature is the latent heat (L).

Thus, $\partial Q = L$ and $\partial V = V_2 - V_1$, for unit mass of a substance.

Substituting,
$$\left(\frac{L}{V_2 - V_1}\right)_T = T\left(\frac{\partial P}{\partial T}\right)_V$$

or
$$\frac{L}{V_2 - V_1} = T\frac{dP}{dT}$$

or
$$\frac{dP}{dT} = \frac{L}{T(V_2 - V_1)} \quad \text{... (6.51)}$$

This is **Clapeyron's latent heat equation.**

6.17 CLAPEYRON LATENT HEAT EQUATION USING CARNOT'S CYCLE

Consider the isothermals FBAE at temperature $T + dT$ and *GCDH* at temperature T. Here *EA* and *HD* show the liquid state of the substance. At *A* and *D*, the substance is purely in the liquid state (Fig. 6.5). From *A* to *B* or *D* to *C*, the substance is in transition from the liquid to the gaseous state and *vice versa*. At *B* and *C*, the substance is purely in the gaseous state. From *B* to *F* or *C* to *G*, the substance is in the gaseous state. Join *A* to *D* and *B* to *C* by dotted lines.

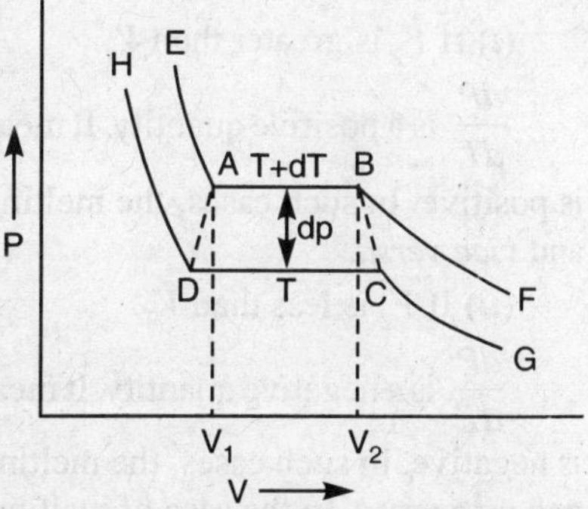

Fig. 6.5

The cycle *ABCD* represents a complete cycle and Carnot's theorem can be applied. Suppose the volume at the point *A* is V_1 and temperature is $T + dT$. The pressure is just below its saturation pressure and the liquid begins to evaporate and at the point *B* the volume is V_2. The substance is in the vapour state. Suppose the mass of the liquid at *B* is one gram the amount of heat absorbed is Q_1, here $Q_1 = L + dL$, where $L + dL$ is the latent heat of the liquid at temperature ($T + dT$).

At the point *B*, the pressure is decreased by *dP*. The vapour will expand and its temperature falls. The temperature at *C* is *T*. At this pressure and temperature *T*, the gas begins to condense and is converted into the liquid state. At the point *D*, the substance is in the liquid state From *C* to *D*, the amount of heat rejected (given out) is Q_2. Here $Q_2 = L$ where *L* is latent heat at temperature *T*. By increasing the pressure a little, the original point *A* is restored. The cycle *ABCDA* is completely reversible. Applying the principle of the Carnot's reversible cycle

$$\frac{Q_1}{T_1} = \frac{Q_2}{T_2}$$

or
$$\frac{Q_1}{Q_2} = \frac{T_1}{T_2}$$

$$\frac{Q_1 - Q_2}{Q_2} = \frac{T_1 - T_2}{T_2}$$

Here,
$$Q_1 = L + dL, Q_2 = L,$$
$$T_1 = T + dT, T_2 = T$$
$$Q_1 - Q_2 = L + dL - L = dL$$

$$T_1 - T_2 = T + dT - T = dT$$

$$\therefore \quad \frac{dL}{L} = \frac{dT}{T}$$

The area of the figure

$$ABCD = Q_1 - Q_2 = dL$$
$$= dP\,(V_2 - V_1)$$

$$\therefore \quad \frac{dP(V_2 - V_1)}{L} = \frac{dT}{T}$$

$$\frac{dP}{dT} = \frac{L}{T(V_2 - V_1)} \quad ...(6.52)$$

This is called the **Clapeyron's latent heat equation.**

Applications

1. *Effect of change of pressure on the melting point.*

When a solid is converted into a liquid, there is change in volume.

(*i*) If V_2 is greater than V_1.

$\frac{dP}{dT}$ is a positive quantity. It means that the rate of change of pressure with respect to temperature is positive. In such cases, the melting point of the substance will increase with increase in pressure and *vice versa.*

(*ii*) If V_2 is less than V_1.

$\frac{dP}{dT}$ is a negative quantity. It means that the rate of change of pressure with respect to temperature is negative. In such cases, the melting point of the substance will decrease with increase in pressure and *vice versa.* In the case of melting ice, the volume of water formed is less than the volume of ice taken. Hence $V_2 < V_1$.

Therefore, the melting point of ice decreases with increase in pressure. Hence ice will melt at a temperature lower than zero degree centigrade at a pressure higher than the normal pressure.

Ice melts at 0°C only at a pressure of 76 cm of Hg.

2. *Effect of change of pressure on the boiling point.*

When a liquid is converted into a gaseous state, the volume V_2 of the gas is always greater than the corresponding volume V_1 of the liquid *i.e.,* $V_2 > V_1$.

Therefore, $\frac{dP}{dT}$ is a + *ve* quantity.

With increase in pressure, the boiling point of a substance increases and *vice versa.* The liquid will boil at a lower temperature under reduced pressure. In the case of water, the boiling point increases with increase in pressure and *vice versa.* Water boils at 100°C only at 76 cm of Hg pressure. In the laboratories, while preparing steam, the boiling point is less than 100°C because the atmospheric pressure is less than 76 cm of Hg. In pressure cookers, the liquid boils at a higher temperature because the pressure inside is more than the atmospheric pressure.

6.18 SECOND LATENT HEAT EQUATION

Consider two isothermals *aABb* and *dDCc* at temperatures T and $T - dT$ respectively for a unit mass of a liquid. From A and B draw adiabatics meeting the second isothermal at D and C respectively as shown in Fig. 6.6. Now consider the cycle *ABCD*. If c_1 denotes the specific heat of the liquid in contact with its own vapour and c_2 the specific heat of the saturated vapour in contact with the liquid, then the amount of heat absorbed by the liquid during its passage from D to A when the temperature rises by dT

$$= c_1\, dT$$

The amount of heat absorbed from A to B when the liquid changes into vapour form at $B = L$.

The amount of heat given back, by the vapour along the path BC when its temperatures falls by

$$dT = c_2\, dT$$

The amount of heat given out by the vapour from C to D when it condenses into liquid at $D = L - dL$

$\therefore$ Total amount of heat taken during the cycle $DABC$

$$= c_1\, dT + L - c_2\, dT - (L - dL)$$
$$= (c_1 - c_2)\, dT + dL$$

If the two isothermals lie very close to each other, then in the limiting position the net heat absorbed by the system is equal to the work done by the working substance and is equal to the area $ABCD$. But according to Clausius Clapeyron's equation

$$\text{Area } ABCD = dp\,(v_2 - v_1) = L\frac{dT}{T}$$

where dp is the change in pressure and v_1 and v_2 are the volumes of the unit mass of the working substance at B and A respectively.

Thus $$(c_1 - c_2)\, dT + dL = L\frac{dT}{T}$$

or $$(c_1 - c_2)\, dT = L\frac{dT}{T} - dL$$

or $$(c_1 - c_2) = \frac{L}{T} - \frac{dL}{dT}$$

or $$\frac{dL}{dT} - \frac{L}{T} = (c_2 - c_1) \qquad \text{... (6.53)}$$

This is known as *second latent heat equation* and this equation gives us a method of calculating the specific heat of saturated vapour.

In Fig. 6.6 at the point B the substance is in the vaporous state at a temperature T and at point C it is still in the vaporous state but at a temperature $T - dT$. The dotted curve joining B and C is the **saturation curve.**

Specific heat of saturated vapour. *The specific heat of saturated vapour is the amount of heat required to raise the temperature of a unit mass of the vapour through 1°C keeping the vapour always saturated.*

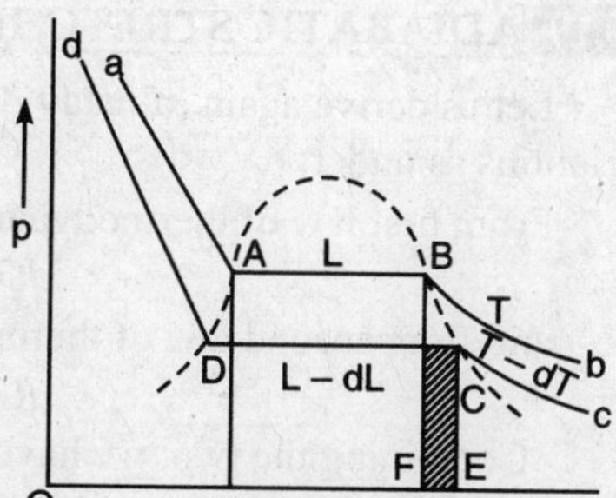

Fig. 6.6

The vapour pressure of a liquid increases with rise of temperature. When the temperature of a saturated vapour is raised the vapour becomes unsaturated. In order to keep it saturated its volume must be decreased by applying an external pressure, thus we see that neither the volume nor the pressure remains constant. The specific heat is, therefore, neither at constant pressure nor at constant volume.

When an external pressure is applied to the vapour the work is done on it and the energy appears as heat energy.

Consider two isothermals *aABb* and *dDCc* at temperatures T and $T - dT$ respectively for a unit mass of the substance as shown in Fig. 6.6 At the point C the substance is in the form of saturated vapour at a temperature $T - dT$ and at the point B it is still in the state of saturated vapour at a temperature T. In order to keep the vapour saturated throughout the passage from C to B the pressure and volume have to be adjusted in such a manner that it travels along the border curve CB shown by the dotted line. Such a curve is known as the *saturation curve.* In passing from C to B the volume decreases and the pressure increases hence work is done on the vapour.

The work done = average pressure × change in volume = area $BCEF$

This work done on the gas appears as heat. Let this quantity of heat be equal to Q_1 when the temperature rises by 1°C *i.e.*, when $dT = 1°C$.

Let Q_2 be the quantity of heat required to raise the temperature of the vapour through 1°C. When the quantity of heat Q_2 is supplied from outside, the temperature rises by 1°C but the vapour expands and thus becomes unsaturated. To keep the vapour saturated it has to be compressed and hence the quantity of heat Q_1 which is produced due to compression will have to be removed from the system if further rise in temperature is to be checked. Now three cases arise.

(*i*) **When $Q_2 = Q_1$.** In this case the net quantity of heat $(Q_2 - Q_1)$ supplied from outside is zero. *The specific heat in such a case is said to be zero.*

(*ii*) **When $Q_2 > Q_1$.** The net quantity of heat supplied $(Q_2 - Q_1)$ is positive. *In such a case the specific heat is said to be positive.* In other words, to raise the temperature of the vapour keeping the vapour just saturated some heat has to be supplied from outside.

(*iii*) **Nagative specific heat.** When $Q_2 < Q_1$. The net quantity of heat supplied $(Q_2 - Q_1)$ is negative. Hence some heat will have to be removed from such a system in order to raise the temperature through 1°C keeping the vapour just saturated. In such a case *the specific heat is said to be negative.*

Physical significance of negative specific heat. As explained above the heat generated by compression in the case of steam is consiiderably large as compared to the heat required to raise the temperature of the vapour through 1°C keeping the vapour just saturated. Hence in order to maintain the saturation at a higher temperature heat has to be withdrawn. This shows *why the specific heat of saturated steam is negative* and explains its physical significance.

6.19 ADIABATIC STRETCHING OF A WIRE

Let us derive again (already discussed in 4.19.2), the relation using Maxwell's thermodynamic relations as under:

From first law of thermodynamics,

$$dQ = dU + PdV$$

and from second law of thermodynamics, we have

$$dQ = TdS$$

Combining the two, we have

$$TdS = dU + PdV$$

In case of stretching a wire, the work done, $dW = -F.dl$ instead of $P.dV$. Here F is the tension applied for the change in length dl.

Therefore,

$$T.dS = dU - F.dl$$

or

$$dU = T.dS + F.dl \qquad \text{... (i)}$$

Now writting $-F$ and l in place of P and V respectively, in Maxwell's first thermodynamic relation

$$\left(\frac{\partial T}{\partial V}\right)_S = -\left(\frac{\partial P}{\partial S}\right)_V$$

we get $$\left(\frac{\partial T}{\partial l}\right)_S = \left(\frac{\partial F}{\partial S}\right)_l$$

$$= T\left(\frac{\partial F}{T\partial S}\right)_l = T\left(\frac{\partial F}{\partial Q}\right)_l$$

or $$\left(\frac{\partial T}{\partial l}\right)_S = T\left(\frac{\partial F}{\partial T}\right)_l\left(\frac{\partial T}{\partial Q}\right)_l \quad \text{... (ii)}$$

Now taking length l as a function of F and T, we put dl as sum of partial differentials *w.r.* to F and T.

Thus, $$dl = \left(\frac{\partial l}{\partial F}\right)_T dF + \left(\frac{\partial l}{\partial T}\right)_F dT$$

under condition of no elongation *i.e.*, $dl = 0$, we get

$$0 = \left(\frac{\partial l}{\partial F}\right)_T dF + \left(\frac{\partial l}{\partial T}\right)_F dT$$

or $$0 = \left(\frac{\partial l}{\partial F}\right)_T\left(\frac{\partial F}{\partial T}\right)_l + \left(\frac{\partial l}{\partial T}\right)_F$$

or $$\left(\frac{\partial F}{\partial T}\right)_l = -\frac{\left(\frac{\partial l}{\partial T}\right)_F}{\left(\frac{\partial l}{\partial F}\right)_T} \quad \text{... (iii)}$$

Substituting in equ. (*ii*), we get

$$\left(\frac{\partial T}{\partial l}\right)_S = -T\frac{\left(\frac{\partial l}{\partial T}\right)_F}{\left(\frac{\partial l}{\partial F}\right)_T} \times \left(\frac{\partial T}{\partial Q}\right)_l \quad \text{... (iv)}$$

But, the coefficient of linear expansion $\beta = \frac{1}{l}\left(\frac{\partial l}{\partial T}\right)_F$ and isothermal Young's modulus

$Y_T = \frac{l}{A}\left(\frac{\partial F}{\partial l}\right)_T$, so that $\left(\frac{\partial l}{\partial F}\right)_T = \frac{l}{Y_T . A}$, where A is the area of cross-section of the wire. Moreover,

$\left(\frac{\partial Q}{\partial T}\right)_l = C_l$, the specific heat at constant length in *erg*. $\text{deg}^{-1}\ \text{gm}^{-1}$. Putting these values in equation (*iv*), we get

$$\left(\frac{\partial T}{\partial l}\right)_S = -T\frac{\beta l Y_T A}{l} . \frac{1}{C_l} = -\frac{T\beta Y_T A}{C_l}$$

$$= -\frac{T\beta Y_T A}{mC} \quad \text{... (v)}$$

where m is the mass per unit length of the wire and C is the specific heat of the material of wire in ergs $\text{deg}^{-1}\ \text{gm}^{-1}$.

Equation (*v*) may be in the form $\left(\frac{\partial T}{\partial F}\right)_S$.

Therefore,
$$\left(\frac{\partial T}{\partial F}\right)_S = \left(\frac{\partial T}{\partial l}\right)_S \cdot \left(\frac{\partial l}{\partial F}\right)_S$$
$$= -\frac{T\beta l\, Y_T\, A}{mC}\left(\frac{1}{Y_S\, A}\right)$$

This is because
$$Y_S = \frac{l}{A}\left(\frac{\partial F}{\partial l}\right)_S$$

$\therefore$
$$\left(\frac{\partial T}{\partial F}\right)_S = -\frac{T\beta l}{mC} \qquad \text{... (vi)}$$

where we have put $Y_T = Y_S$, which is closely true for all solids.

Therefore,
$$dT = -\frac{T\beta l}{mC}dF \qquad \text{... (6.54)}$$

If β is +*ve*, then an increase in tension would cool the wire. Thus, the wires of substances which expand on heating should show a cooling effect when stretched adiabatically.

In case of *a rubber string*, β is negative *i.e.*, rubber contracts on heating. Therefore, for a positive value of *dF i.e.*, for an increase in stretching force, the temperature rises and there is a heating effect. This is why, a rubber string gets heated up on stretching.

6.20 ISOTHERMAL ELASTICITY AND BULK MODULUS

From equation (6.44), we have
$$\left(\frac{\partial U}{\partial V}\right)_T = T\left(\frac{\partial P}{\partial T}\right)_V - P$$

Now
$$\left(\frac{\partial P}{\partial T}\right) = \frac{\partial P}{\partial V/V} \cdot \frac{\partial V/V}{\partial T}$$

Again
$$\frac{\partial P}{\partial V/V} = \frac{\text{Applied pressure}}{\text{Change in volume/ original volume}}$$
$$= E_\theta, \text{ the isothermal elasticity}$$

and
$$\frac{\partial V/V}{\partial T} = \frac{\text{Change in volume per unit volume}}{\text{Rise in temperature}}$$
$$= \propto, \text{ the Bulk modulus.}$$

Substituting, we get
$$\left(\frac{\partial P}{\partial T}\right) = E_\theta . \propto$$

Hence,
$$\left(\frac{\partial U}{\partial V}\right)_T = TE_\theta \propto - P \qquad \text{... (6.55)}$$

6.21 BEHAVIOUR OF WATER BELOW AND ABOVE 4°C

This is an application of Maxwell's **fourth thermodynamical relation**
$$\left(\frac{\partial S}{\partial P}\right)_T = -\left(\frac{\partial V}{\partial T}\right)_P$$

Multiplying both sides by T and replacing $T\,\partial S = \partial Q$ we have

$$\left(\frac{\partial Q}{\partial P}\right)_T = -T\left(\frac{\partial V}{\partial T}\right)_P = -TV\alpha \qquad \text{... (6.56)}$$

where a is the coefficient of increase of volume at constant pressure given by $\alpha = \frac{1}{V}\left(\frac{\partial V}{\partial T}\right)$

If α is positive the substance expands onheating. In such a case $\left(\frac{\partial Q}{\partial P}\right)_T$ will be negative. Thus when ∂P is positive *i.e.*, the pressure is increased the temperature will remain constant only if ∂Q is negative *i.e.*, heat is taken away from the substance. In other words, heat is generated when a substance which expands on heating is compressed.

On the other hand a cooling effect will be produced when a substance which contracts on heating is compressed. In such a case α is negative and hence is $\left(\frac{\partial Q}{\partial P}\right)_T$ positive.

Thus when dP is positive *i.e.*, pressure is increased the temperature will ramain constant only if ∂Q is positive *i.e.*, heat is added to the substance. In other words there is a cooling effect.

These results have been verified by Joule in the case of water below 4°C and water above 4°C.

Thus Maxwell's thermodynamic relation $\left(\frac{\partial s}{\partial P}\right)_T = -\left(\frac{\partial V}{\partial T}\right)_P$ explains why water below 4°C cools by adiabatic compression although its internal energy increases.

6.22 ENTROPY AND THE SECOND LAW OF THERMODYNAMICS

Consider a closed system undergoing a reversible process from state 1 to state 2 along the path A and from state 2 to state 1 along the path B (Fig. 6.7). As this is a reversible cyclic process

$$\oint \frac{\delta Q}{T} = 0$$

$$\therefore \quad \int_{1A}^{2A} \frac{\delta Q}{T} + \int_{2B}^{1B} \frac{\delta Q}{T} = 0 \qquad \text{... (i)}$$

Now consider the reversible cycle from state 1 to state 2 along the path A and from state 2 to state 1 along the path C.

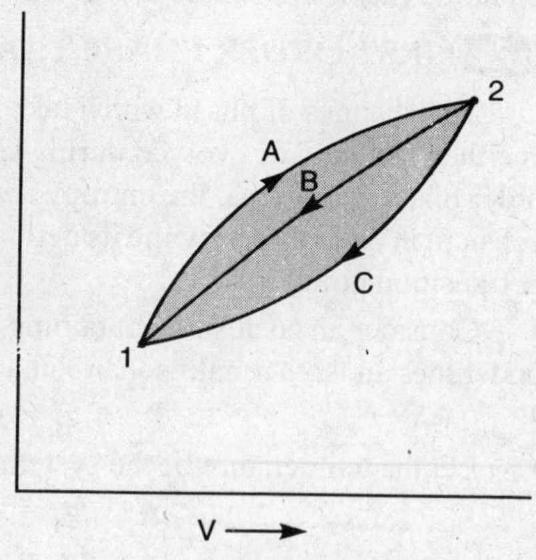

Fig. 6.7

For this reversible cyclic process

$$\int_{1A}^{2A} \frac{\delta Q}{T} + \int_{2C}^{1C} \frac{\delta Q}{T} = 0 \qquad \text{... (ii)}$$

From equations (*i*) and (*ii*)

$$\int_{2B}^{1B} \frac{\delta Q}{T} = \int_{2C}^{1C} \frac{\delta Q}{T} \qquad \text{... (iii)}$$

This shows that $\int \frac{\delta Q}{T}$ has the same value for all the reversible paths from state 2 to state 1. The quantity $\int \frac{\delta Q}{T}$ is independent of the path and is a function of the end states only, therefore it is a property.

This property is called entropy. Entropy is a thermodynamical property and is defined by the relation

$$dS = \frac{\delta Q}{T}$$

or
$$S_2 - S_1 = \int_1^2 \frac{\delta Q}{T} \quad \text{... (6.57)}$$

The quantity $S_2 - S_1$ represents the change in entropy of the system when it is changed from state 1 to state 2.

6.23 EQUILIBRIUM BETWEEN LIQUID AND ITS VAPOUR

Applying Gibbs potential (G), it is possible to investigate the equilibrium between a liquid and its vapour or between any two phases of a substance. Consider a closed system containing a liquid in equilibrium with its saturated vapour. The temperature and pressure are equal in both the phases. As each of the phases is in equilibrium, the temperature and pressure must remain constant throughout the phase and hence the thermodynamical coordinates V, S, U and G will be equal to the product of the specific value and the mass of the substance in that phase.

Suppose m_1 and m_2 are the masses in the liquid and vapour phases and g_1 and g_2 are the specific values of the Gibbs potential in the two phases. Then, for the whole system

$$G = m_1 g_1 + m_2 g_2 \quad \text{... (i)}$$

If a small quantity of liquid changes into vapour, differentiating equation (*i*)

$$\delta G = \delta m_1 g_1 + \delta m_2 g_2 \quad \text{... (ii)}$$

As the change takes place at constant temperature and pressure, the process is isothermal and isobaric

$$\therefore \quad \delta G = 0$$

From equation (*ii*)

$$\delta m_1 g_1 + \delta m_2 g_2 = 0$$

But
$$\delta m_1 = -\delta m_2$$

$$\therefore \quad g_1 = g_2 \quad \text{... (iii)}$$

This shows that the thermodynamical potential per unit mass will be equal in the two phases. Equation (*iii*) is applicable to the processes of evaporation, fusion and sublimation.

6.24 FIRST ORDER PHASE TRANSITIONS

The changes of phase which take place at constant temperature and pressure and in which heat is either absorbed or evolved during change of phase are called first order phase transitions. In first order phase transitions, the entropy and density (or volume) change. The Gibbs function G remains constant in both phases; while its derivative with respect to temperature and pressure is discontinuous at transition point.

Consider an enclosure containing a liquid and its saturated vapour in equilibrium. If this system undergoes an isothermal, isobaric change, then

$$g_1 = g_2 \quad \text{... (i)}$$

Let the temperature of the system be increased from T to $T + dT$. For equilibrium

$$g_1 + dg_1 = g_2 + dg_2 \quad \text{... (ii)}$$

or
$$dg_1 = dg_2 \quad \text{... (iii)}$$

If the condition of saturation is satisfied,

$$\left(\frac{dg_1}{dT}\right)_{sat.} = \left(\frac{dg_2}{dT}\right)_{sat.} \quad \text{... (iv)}$$

The pressure also changes from P to $P + dP$,

$$\therefore \quad dg_1 = \left(\frac{\partial g_1}{\partial T}\right)_P dT + \left(\frac{\partial g_1}{\partial P}\right)_T dP \quad \text{... (v)}$$

or
$$\left(\frac{\partial g_1}{\partial T}\right) = \left(\frac{\partial g_1}{\partial T}\right)_P + \left(\frac{\partial g_1}{\partial P}\right)_T \left(\frac{\partial P}{\partial T}\right) \quad \text{... (vi)}$$

But, for a unit mass,

$$dg = VdP - SdT \qquad \text{... (vii)}$$

$$\therefore \quad \left(\frac{\partial g}{\partial P}\right)_T = V$$

and

$$\left(\frac{\partial g}{\partial T}\right)_P = -S$$

Substituting these values in equation (*vi*)

$$\left(\frac{\partial g_1}{\partial T}\right)_{sat.} = -S_1 + V_1\left(\frac{\partial P}{\partial T}\right)_{sat.}$$

Similarly,

$$\left(\frac{\partial g_2}{\partial T}\right)_{sat.} = -S_2 + V_2\left(\frac{\partial P}{\partial T}\right)_{sat.}$$

Substituting these values in equation (*iv*)

$$-S_1 + V_1\left(\frac{dP}{dT}\right) = -S_2 + V_2\left(\frac{dP}{dT}\right)$$

or

$$\left(\frac{dP}{dT}\right)_{sat.} = \left(\frac{S_2 - S_1}{V_2 - V_1}\right) \qquad \text{... (viii)}$$

But

$$S_2 - S_1 = \frac{\delta Q}{T} = \frac{L}{T}$$

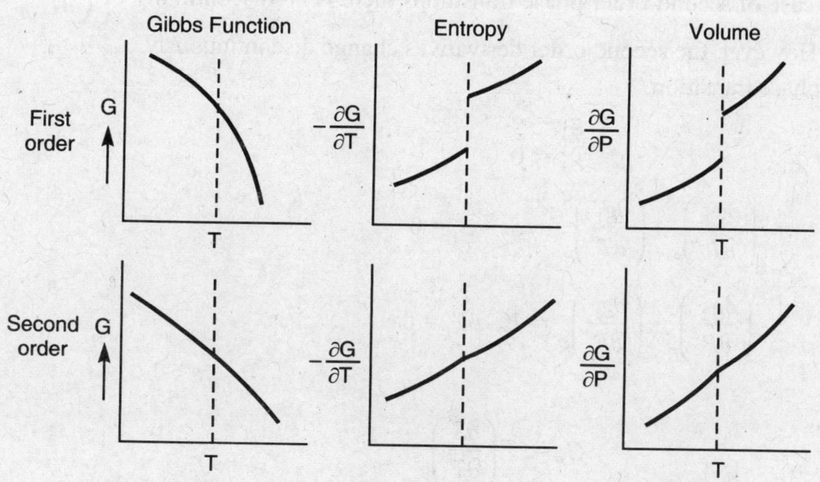

Fig. 6.8

Here L is latent heat of vaporization

$$\therefore \quad \left(\frac{dP}{dT}\right) = \frac{L}{T(V_2 - V_1)} \qquad \text{... (ix)}$$

This is **Clausius-Clapeyron latent heat equation.**

This equation holds good in the first order phase transitions. In these processes, there is transference of heat and hence there is change in entropy and volume. Therefore a first order phase transition can be defined as that one in which the Gibbs function with respect to pressure and temperature change discontinuously at the transition point. However, the value of the Gibbs function is the same in both the phases at equilibrium. These changes are represented graphically in (Fig. 6.8).

Examples: When heat is given to water at 100°C and 1 atmospheric pressure, it changes from liquid to vapour state. The density of water is 1000 kg/m^3, while that of vapour is 0.6 kg/m^3. Therefore, *the transformation of water into vapour at constant temperature and volume is first order phase transitions. Similarly the transformation of ice into water at 0°C and 1 atmospheric pressure is an example of first order phase transition.*

6.25 SECOND ORDER PHASE TRANSITIONS: EHRENFEST'S EQUATIONS

Some recent investigations have revealed that during phase change there is no transference of heat and there is no change of volume. It has been found in the case of transition from liquid helium I to liquid helium II, that there is no transfer of heat and no change in volume. Such transitions are called second order phase transitions. Second order phase transitions can be defined as the phenomenon that takes place with no change in entropy and volume at constant temperature and pressure.

Example: The examples of second order phase transitions are:

1. Transition of liquid helium I to liquid helium II, at λ point (2.19 K).
2. Transition of a ferromagnetic material to a paramagnetic material at the Curie point.
3. Transition of a superconducting metal into an ordinary conductor in the absence of a magnetic field.
4. Order-disorder transitions in chemical compounds and alloys.

In the case of second-order phase transitions there is no discontinuity of $\left(\frac{\partial G}{\partial T}\right)_P$ and $\left(\frac{\partial G}{\partial P}\right)_T$ [Fig. 6.8]. However, the second order derivatives change discontinuously.

For a phase transition,

$$g_1 = g_2$$

or

$$g_2 - g_1 = 0 \quad \text{... (i)}$$

$$-\left(\frac{\partial g_2}{\partial T}\right)_P + \left(\frac{\partial g_1}{\partial T}\right)_P = S_2 - S_1 = 0 \quad \text{... (ii)}$$

$$\left(\frac{\partial g_2}{\partial P}\right)_T - \left(\frac{\partial g_1}{\partial P}\right)_T = V_2 - V_1 = 0 \quad \text{... (iii)}$$

Also

$$C_P = T\left(\frac{\partial S}{\partial T}\right)_P$$

$$\left(\frac{C_P}{T}\right) = \left(\frac{\partial S}{\partial T}\right)_P = \frac{\partial}{\partial T}\left[-\left(\frac{\partial g}{\partial T}\right)_P\right]_P = -\left(\frac{\partial^2 g}{\partial T^2}\right)_P \quad \text{... (iv)}$$

$$K = -\frac{1}{V}\left(\frac{\partial V}{\partial P}\right)_T$$

$$KV = -\left(\frac{\partial V}{\partial P}\right)_T = -\frac{\partial}{\partial P}\left[\left(\frac{\partial g}{\partial P}\right)_T\right]_T$$

or
$$KV = -\left(\frac{\partial^2 g}{\partial P^2}\right)_T \quad \text{... }(v)$$

$$\alpha = \frac{1}{V}\left(\frac{\partial V}{\partial T}\right)_P$$

$$\alpha V = \left(\frac{\partial V}{\partial T}\right)_P = \frac{\partial}{\partial T}\left[\left(\frac{\partial g}{\partial P}\right)_T\right]_P$$

or
$$\alpha V = \left(\frac{\partial^2 g}{\partial T \partial P}\right) \quad \text{... }(vi)$$

Here K is the isothermal compressibility and α is the volume coefficient of expansion.

From equation (*iv*)

$$\frac{C_{P1}}{T} = -\left(\frac{\partial^2 g_1}{\partial T^2}\right)_P \quad \text{... }(vii)$$

and
$$\frac{C_{P2}}{T} = -\left(\frac{\partial^2 g_2}{\partial T^2}\right)_P \quad \text{... }(viii)$$

Subtracting (*viii*) from (*vii*)

$$\left(\frac{\partial^2 g_2}{\partial T^2}\right)_P - \left(\frac{\partial^2 g_1}{\partial T^2}\right)_P = \frac{C_{P1} - C_{P2}}{T} \quad \text{... }(ix)$$

Similarly from equations (*v*) and (*vi*)

$$\left(\frac{\partial^2 g_2}{\partial P^2}\right)_T - \left(\frac{\partial^2 g_1}{\partial P^2}\right)_T = V(K_1 - K_2) \quad \text{... }(x)$$

and
$$\left(\frac{\partial^2 g_2}{\partial T.\partial P}\right) - \left(\frac{\partial^2 g_2}{\partial T.\partial P}\right) = V(\alpha_2 - \alpha_1) \quad \text{... }(xi)$$

For second order phase transitions, there is no change in entropy and volume.

$\therefore \quad S_1 = S_2$ (at constant temperatures T and pressure P)

$\therefore \quad S_1 + dS_1 = S_2 + dS_2$ (at temp. $T + dT$ and pressure $P + dP$)

$\therefore \quad dS_1 = dS_2 \quad \text{... }(xii)$

But
$$dS = \left(\frac{\partial S}{\partial T}\right)_P dT + \left(\frac{\partial S}{\partial P}\right)_T dP$$

or
$$dS = \left(\frac{C_P}{T}\right) dT - \left(\frac{dV}{dT}\right)_P dP$$

$$dS = \left(\frac{C_P}{T}\right) dT - V\alpha_1\, dP$$

$\therefore$
$$dS_1 = \left(\frac{C_{P1}}{T}\right) dT - V\alpha_1\, dP$$

and
$$dS_2 = \left(\frac{C_{P2}}{T}\right) dT - V\alpha_2\, dP$$

$$\therefore \quad \left(\frac{C_{P1}}{T}\right) dT - V\alpha_1\, dP = \left(\frac{C_{P2}}{T}\right) dT - V\alpha_2\, dP$$

$$\therefore \quad \frac{dP}{dT} = \frac{C_{P2} - C_{P1}}{TV(\alpha_2 - \alpha_1)} \qquad \text{... (6.58)}$$

Similarly by assuming $V_1 = V_2$ for second order phase transition,

$$V_1 = V_2 \qquad \text{(at temp. } T \text{ and pressure } P\text{)}$$

$$V_1 + dV_1 = V_2 + dV_2 \qquad \text{(at temp. } T + dT \text{ and pressure } P + dP\text{)}$$

$$\therefore \quad dV_1 = dV_2$$

But
$$dV = \left(\frac{\partial V}{\partial T}\right)_P dT + \left(\frac{\partial V}{\partial P}\right)_T dP$$

$$dV = V\alpha dT - VKdP$$

$$\therefore \quad dV_1 = V\alpha_1 dT - VK_1 dP$$

and
$$dV_2 = V\alpha_2 dT - VK_2 dP$$

$$\therefore \quad V\alpha_1 dT - VK_1 dP = V\alpha_2 dT - VK_2 dP$$

$$\therefore \quad \frac{dP}{dT} = \left(\frac{\alpha_2 - \alpha_1}{K_2 - K_1}\right) \qquad \text{... (6.59)}$$

Equations (6.58) and (6.59) are called **Ehrenfest's Equations.**

These equations represent the condition of equilibrium between the two phases.

6.26 EMF OF A REVERSIBLE CELL

The first TdS equation is given by

$$TdS = C_V dT + T\left(\frac{\partial P}{\partial T}\right)_V dV \qquad \text{... (i)}$$

In the case of a reversible cell composed of solids and liquids in the above equation replace charge q for V and EMF, $-E$ for P.

$$\therefore \quad TdS = C_q\, dT - T\left(\frac{\partial E}{\partial T}\right)_q dq \qquad \text{... (ii)}$$

For a reversible cell, in which the E.M.F. depend only on temperature,

$$TdS = C_q\, dT - T\left(\frac{\partial E}{\partial T}\right)_q dq \qquad \text{... (iii)}$$

For a reversible cell, when a charge $(q_f - q_i)$ is transferred isothermally, taking $TdS = \delta Q$, we get

$$\delta Q = -T\left(\frac{\partial E}{\partial T}\right)_q (q_f - q_i) \qquad \text{... (iv)}$$

Here $(q_f - q_i)$ is negative when positive charge flows from positive to the negative electrode externally. The work done by the cell,

$$\delta W = -E\,(q_f - q_i) = -E\,\Delta q \qquad \text{... (v)}$$

Here
$$\Delta q = (q_f - q_i)$$

$$\therefore \quad \delta Q = -T\left(\frac{\partial E}{\partial T}\right)\Delta q \qquad \text{... (vi)}$$

From the first law of thermodynamics

$$\delta Q = dU + \delta W$$

$$\therefore \quad dU = \delta Q - \delta W$$

$$dU = \Delta q\left[E - T\left(\frac{\partial E}{\partial T}\right)\right] \quad \text{... (vii)}$$

For a thermodynamic process taking place at constant pressure and negligible change in volume, the change in enthalpy is obtained as follows

$$H = U + PV$$

$$dH = dU + PdV + VdP$$

But $\quad dP = 0$ and dV is negligible

$$\therefore \quad dH = dU \quad \text{... (viii)}$$

From equation (*vii*) and (*viii*)

$$dH = \Delta q\left[E - T\left(\frac{\partial E}{\partial T}\right)\right] \quad \text{... (ix)}$$

dH represents the net transfer of heat between the final and the initial states of the reaction.

From equation (*ix*)

$$\frac{dH}{\Delta q} = E - T\left(\frac{\partial E}{\partial T}\right)$$

or

$$E = \left(\frac{dH}{\Delta q}\right) + T\left(\frac{\partial E}{\partial T}\right)$$

Let

$$\left(\frac{dH}{\Delta q}\right) = Q$$

Here Q is the heat transferred per unit charge

$$\therefore \quad E = Q + T\left(\frac{\partial E}{\partial T}\right) \quad \text{... (6.60)}$$

Equation 6.60 is called **Gibbs Helmholtz equation.**

SOLVED EXAMPLES

Example 6.1 *Show that $C_P - C_V = TE\,\alpha^2 V$ where T is the absolute temperature, E the modulus of isothermal elasticity, α the coefficient of volume expansion and V the specific volume.*

(Meerut 2004; Agra 2006, 2000, 2002; Garhwal 2001; Rohilkhand 2000; Kerala 2002; Utkal 2003; Lucknow 2002; Madras 2004, 2005; Purvanchal 2003; Mumbai 2005; Calcutta 2005)

Solution. From Maxwell's second law of thermodynamical relation

$$\left(\frac{\partial S}{\partial V}\right)_T = \left(\frac{\partial P}{\partial T}\right)_V$$

and from the definitions of C_P and C_V as

$$C_P = \left(\frac{\partial Q}{\partial T}\right)_P = T\left(\frac{\partial S}{\partial T}\right)_P$$

or

$$C_V = \left(\frac{\partial Q}{\partial T}\right)_V = T\left(\frac{\partial S}{\partial T}\right)_V$$

Considering $S = f(T, V)$, we write

$$dS = \left(\frac{\partial S}{\partial T}\right)_V dT + \left(\frac{\partial S}{\partial V}\right)_T dV$$

$$\therefore \quad \left(\frac{\partial S}{\partial T}\right)_P = \left(\frac{\partial S}{\partial T}\right)_V\left(\frac{\partial T}{\partial T}\right) + \left(\frac{\partial S}{\partial V}\right)_T \cdot \left(\frac{\partial V}{\partial T}\right)_P$$

But from the equation

$$\left(\frac{\partial S}{\partial V}\right)_T = \left(\frac{\partial P}{\partial T}\right)_V$$

$$T\left(\frac{\partial S}{\partial T}\right)_P = T\left(\frac{\partial S}{\partial T}\right)_V + T\left(\frac{\partial P}{\partial T}\right)_V \cdot \left(\frac{\partial V}{\partial T}\right)_P$$

$$\therefore \quad C_P = C_V + T\left(\frac{\partial P}{\partial T}\right)_V\left(\frac{\partial V}{\partial T}\right)_P \quad \text{... }(i)$$

Again taking the general equation of state for a gas as $P = f(V, T)$

$$\therefore \quad dP = \left(\frac{\partial P}{\partial T}\right)_V dT + \left(\frac{\partial P}{\partial V}\right)_T dV$$

$$\frac{dP}{dT} = \left(\frac{\partial P}{\partial T}\right)_V + \left(\frac{\partial P}{\partial V}\right)_T\left(\frac{\partial V}{\partial T}\right)_P$$

At constant pressure

$$dP = 0$$

$$\therefore \quad \left(\frac{\partial P}{\partial T}\right)_V = -\left(\frac{\partial P}{\partial V}\right)_T\left(\frac{\partial V}{\partial T}\right)_P$$

Substituting this value in equation (*i*)

$$C_P - C_V = -T\left(\frac{\partial P}{\partial V}\right)_T\left(\frac{\partial V}{\partial T}\right)_P^2 \quad \text{... }(ii)$$

$$C_P - C_V = -T\left(\frac{\partial V}{\partial T}\right)_P^2\left(\frac{\partial P}{\partial V}\right)_T \quad \text{... }(iii)$$

But, the bulk modulus of elasticity is given by

$$E = -V\left(\frac{\partial P}{\partial V}\right)_T$$

and the coefficient of volume expansion is

$$\alpha = \frac{1}{V}\left(\frac{\partial V}{\partial T}\right)_P$$

Substituting these values in equation (*iii*), we get

$$C_P - C_V = -T E \alpha^2 V \quad \text{... }(iv)$$

Example 6.2. *Deduce the second latent heat equation of Clausius*

$$C_2 - C_1 = \frac{dL}{dT} - \frac{L}{T}$$

were C_1 and C_2 represent the specific heat of a liquid and its saturated vapour and L is the latent heat of the vapour.

(Meerut 2000, M.D.U. 2001, Kerala 2000, Nagpur 2003, Purvanchal 2003, 2004)

Solution. For a change of state from liquid to vapour

$$S_2 - S_1 = \frac{L}{T} \quad \text{... }(i)$$

Here S_1 and S_2 are the entropies in the liquid and vapour states respectively.

Differentiating equation (*i*) with respect to T

$$\frac{dS_2}{dT}-\frac{dS_1}{dT} = -L\frac{1}{T^2}+\frac{1}{T}\cdot\frac{dL}{dT}$$

$$T\left(\frac{dS_2}{dT}\right)-\left(\frac{dS_1}{dT}\right) = -\frac{L}{T}+\frac{dL}{dT}$$

$$\therefore \qquad C_2-C_1 = \frac{dL}{dT}-\frac{L}{T} \qquad \text{... (ii)}$$

This equation is known as Clausius latent heat equation.

Example 6.3. *Calculate under what pressure water would boil at 150°C if the change in specific volume when 1 gram of water is converted into steam is 1676 CC. Given latent heat of vaporization of steam = 540 cal per gram; J = 4.2 × 10^7 erg/cal and one atmosphere pressure = 10^6 dynes/cm^2.*

Solution. Here

$$\partial Q = 540 \text{ cal} = 540 \times 4.2 \times 10^7 \text{ erg}$$

$$\partial V = 1676 \text{ CC}$$

$$T = 373 \text{ K and } 150°\text{C} = 423 \text{ K}$$

$$\partial T = 423 - 373 = 50 \text{ K}$$

using these values in Maxwell's second thermodynamical relation

$$\left(\frac{\partial S}{\partial V}\right)_T = \left(\frac{\partial P}{\partial T}\right)_V$$

$$T\left(\frac{\partial S}{\partial V}\right)_T = T\left(\frac{\partial P}{\partial T}\right)_V$$

$$\left(\frac{\partial Q}{\partial V}\right)_T = T\left(\frac{\partial P}{\partial T}\right)_V$$

$$\frac{540\times4.2\times10^7}{1676} = 373\times\left(\frac{\partial P}{50}\right)$$

$$\partial P = \frac{540\times4.2\times10^7\times50}{373\times1676}$$

$$= \mathbf{1.814 \times 10^6 \text{ dynes/cm}^2}$$

$$= \mathbf{1.814 \text{ atmospheres}}$$

Therefore, the pressure at which water would boil at 150°C

$$= 1.814 + 1.000$$

$$= \mathbf{2.814 \text{ atmospheres}}$$

Example 6.4. *Calculate the specific heat of saturated steam given that the specific heat of water at 100°C = 1.01 and latent heat of vaporization decreases with rise in temperature at the rate of 0.64 cal/K. Latent heat of vaporization of steam is 540 cal.* ***(Kanpur 2000)***

Solution. Here

$$C_1 = 1.01, \quad C_2 = ? \quad T = 373 \text{ K}$$

$$\frac{dL}{dT} = -0.64 \text{ cal/K}$$

$$C_2 - C_1 = \frac{dL}{dT}-\frac{L}{T}$$

$$C_2 = C_1 + \frac{dL}{dT} - \frac{L}{T}$$

$$= 1.01 + (-0.64) - \frac{540}{373}$$

$$= \mathbf{-1.077\ cal/g}$$

Example 6.5. *Calculate the specific heat of saturated steam at 100°C from the following data:*

L at 90°C = 545.25 cal

L at 100°C = 539.30 cal

L at 110°C = 533.17 cal

specific heat of water at 100°C = 1.013 cal/g **(*Kerala 2001, 2002*)**

Solution. Here

$$\frac{dL}{dT} = \frac{533.17 - 545.25}{20}$$

$$= \frac{-12.08}{20} = -0.604 \text{ cal/K}$$

$$L = 539.3 \text{ cal}$$

$$T = 373 \text{ K}$$

$$C_1 = 1.013,\ C_2 = ?$$

$$C_2 - C_1 = \frac{dL}{dT} - \frac{L}{T}$$

$$\therefore \quad C_2 = C_1 + \frac{dL}{dT} - \frac{L}{T}$$

$$= 1.013 - 0.604 - \frac{539.3}{373}$$

$$\mathbf{C_2 = -1.036\ cal/g}$$

Example 6.6. *Calculate under what pressure ice freezes at 272* K *if the change in specific volume when 1 kg of water freezes is 91 × 10^{-6} m^3. Given latent heat of ice = 3.36 × 10^5 Jkg^{-1}.* **(*Calcutta 2000*)**

Solution. Latent heat of ice $L = 3.36 \times 10^5$ J kg^{-1}

Freezing point of ice under normal pressure T = 273 K, change in freezing point, dT = (273 – 272) = 1 K

Change in specific volume = $(V_2 - V_1) = 91 \times 10^{-6}$ m^3

According to Clapeyron's latent heat equation

$$\frac{dP}{dT} = \frac{L}{T(V_2 - V_1)}$$

$$\therefore \quad dP = \frac{dT \times L}{T(V_2 - V_1)} = \frac{1 \times 3.36 \times 10^5}{273 \times 91 \times 10^{-6}} = 1.352 \times 10^7 \text{ N/m}^2$$

Taking 1 Atm = 10^5 N, we have

$$dP = \frac{1.352 \times 10^7}{10^5} = 135.2 \text{ Atm.}$$

∴ Pressure under which ice would freeze at 272 K.

$$= 1 + 135.2$$

$$= \mathbf{136.2\ Atm.}$$

Example 6.7. *Calculate the change in boiling point of water when the pressure is increased by 1 atmosphere. Boiling point of water is 373 K. Specific volume of steam = 1.671 m^3 kg^{-1} and latent heat of steam 2.268 × 10^6 J kg^{-1}.* **(*Agra 2002, Kanpur 2001*)**

Solution. Specific volume of steam $V_2 = 1.671$ m^3 kg^{-1}, specific volume of water $V_1 = 1 \times 10^{-3}$ m^3 kg^{-1} = 0.001 m^3 kg^{-1}

$$\therefore \qquad V_2 - V_1 = (1.671 - 0.001) = 1.670 \text{ m}^3 \text{ kg}^{-1}$$

Boiling point of water $T = 373$ K

Latent heat of steam = 2.268×10^6 J kg^{-1}

$$\text{Now} \qquad \frac{dP}{dT} = \frac{L}{T(V_2 - V_1)} = \frac{2.268\times10^6}{373\times1.670} = 3.64 \times 10^3 \text{ N/m}^2\text{K}$$

Increase in pressure dP = 1 Atm = 10^5 N/m^2

$$\therefore \qquad dT = \frac{10^5}{3.64\times10^{-3}}$$

$$\mathbf{dT = 27.47 \text{ K}}$$

Example 6.8. *Calculate the depression of melting point of ice produced by one atmosphere increase of pressure. Given that latent heat of ice = 80 cal/gm and specific volume of ice and water at 0°C are 1.091 cm^3 and 1.0 cm^3 respectively.*

(Nagpur 2009, 2007, 2001, Lucknow 2001, Kanpur 1995, Purvanchal 2004, 2002, Kuvempa 2005)

Solution. Increase of pressure = $dP = 1 \times 76 \times 13.6 \times 981$ dynes/cm^2

Latent heat L = 80 cal = $80 \times 4.2 \times 10^7$ ergs

Change in volume when 1 gm of ice melts = $V_2 - V_1$

= (1.091 – 1.0) = 0.091 cm^3

Melting point of ice T = 0°C = 273 K

Let dT be the depression of melting point, then

$$\frac{dP}{dT} = \frac{L}{T\,(V_2 - V_1)}$$

$$\text{or} \qquad \frac{76\times13.6\times981}{dT} = \frac{80\times4.2\times10^7}{273\times0.091}$$

$$\therefore \qquad dT = \frac{76\times13.6\times981\times273\times0.091}{80\times4.2\times10^7}$$

$$= \mathbf{0.007497 \text{ K}}$$

$$= \mathbf{0.007497\ °C}$$

Example 6.9. *Find the increase in the boiling point of water at 100°C when the pressure is increased by one atmosphere. Latent heat of vaporisation of steam is 540 cal/gram and 1 gram of steam occupies a volume of 1677 cm^3.* **(Madras 2001)**

Solution.

$$dP = 76 \times 13.6 \times 980 \text{ dynes/cm}^2$$

$$T = 100 + 273$$

$$= 373 \text{ K}$$

$$L = 540 \times 4.2 \times 10^7 \text{ ergs}$$

$$V_1 = 1.00 \text{ cm}^3$$

$$V_2 = 1677 \text{ cm}^3$$

$$\frac{dP}{dT} = \frac{L}{T\,(V_2 - V_1)}$$

$$dT = \frac{dP \times T\,(V_2 - V_1)}{L}$$

$$= \frac{76 \times 13.6 \times 980 \times 373 \times 1676}{540 \times 4.2 \times 10^7}$$

$$= 27.92°C$$

Therefore, the increase in the boiling point of water with an increase in pressure of one atmosphere

$$= \mathbf{27.92°C}$$

$$= \mathbf{27.92\ K}$$

Example 6.10. *Prove the Clausius inequality* $\Sigma \frac{Q}{T} \leq 0$ *with examples.* **(*Rajasthan 2000*)**

Solution. 1. Consider two blocks at temperature 300 K and 200 K that are in contact. The block at higher temperature conducts heat to the block at the lower temperature. Suppose, 200 joules of heat is transferred.

$$Q_1 = 200 \text{ joules}; \quad T_1 = 300 \text{ K}$$

$$Q_2 = -200 \text{ joules}, \quad T_2 = 200 \text{ K}$$

$$\therefore \quad \Sigma \frac{Q}{T} = \frac{Q_1}{T_1} + \frac{Q_2}{T_2}$$

$$= \frac{200}{300} + \frac{(-200)}{200} = -\frac{1}{3}$$

$$= \mathbf{-0.33} \text{ joule/degree}$$

2. Consider an actual heat engine working between the temperatures 500 K and 300 K. Suppose its efficiency is 20% and it takes 1000 joules of heat from the high temperature reservoir. As the efficiency is 20%.

$$\eta = 1 - \frac{Q_2}{Q_1}$$

$$\frac{20}{100} = 1 - \frac{Q_2}{100}$$

$$\text{or} \quad Q_2 = 800 \text{ joules}$$

$$\therefore \quad \Sigma \frac{Q}{T} = \frac{Q_1}{T_1} + \frac{Q_2}{T_2}$$

Here

$$Q_1 = +1000 \text{ joules}$$

$$Q_2 = -800 \text{ joules (since heat is rejected)}$$

$$T_1 = 500 \text{ K}$$

$$T_2 = 300 \text{ K}$$

$$\Sigma \frac{Q}{T} = \frac{1000}{500} + \frac{-800}{300}$$

$$= -\frac{2}{3} \text{ joule/degree}$$

3. Consider a Carnot's reversible engine working between the temperatures 500 K and 300 K. Suppose 1000 joules of heat energy is drawn from the high temperature reservoir.

Here

$$\frac{Q_1}{T_1} = \frac{Q_2}{T_2}$$

$$\frac{1000}{500} = \frac{Q_2}{300}$$

or $$Q_2 = 600 \text{ joules}$$

$$\Sigma\frac{Q}{T} = \frac{Q_1}{T_1} + \frac{Q_2}{T_2}$$

Here
$$Q_1 = +1000 \text{ joules}$$
$$Q_2 = -600 \text{ joules}$$
$$T_1 = 500 \text{ K}$$
$$T_2 = 300 \text{ K}$$

$\therefore$ $$\Sigma\frac{Q}{T} = \frac{1000}{500} + \frac{(-600)}{300}$$

or $$\Sigma\frac{Q}{T} = 0$$

This example shows $\Sigma\frac{Q}{T} = 0$, only in the limiting case and in no case $\Sigma\frac{Q}{T}$ is greater than zero.

Example 6.11. *Calculate the change in temperature of boiling water when the pressure is increased by 27.12 mm of Hg. The normal boiling point of water at atmospheric pressure is 100°C.*

Latent heat of steam = *537* cal/g

and specific volume of steam = *1674* cm^3 **(*Bangalore 2002*)**

Solution.
$$dP = 2.712 \times 13.6 \times 980 \text{ dynes/cm}^2$$
$$T = 100 + 273 = 373 \text{ K}$$
$$L = 537 \times 4.2 \times 10^7 \text{ ergs}$$
$$V_1 = 1.000 \text{ cm}^3$$
$$V_2 = 1674 \text{ cm}^3$$
$$\frac{dP}{dT} = \frac{L}{T(V_2 - V_1)}$$
$$dT = \frac{dP \times T\,(V_2 - V_1)}{L}$$
$$= \frac{2.712 \times 13.6 \times 980 \times 373 \times 1673}{537 \times 4.2 \times 10^7}$$
$$= \mathbf{1°C \text{ or } 1 \text{ K}}$$

Example 6.12. *Calculate the change in the melting point of naphthalene for one atmosphere rise in pressure given that its melting point is 80°C. Latent heat of fusion is 4563 cal/mol and increase in volume on fusion is 18.7 cm^3 per* mol *and 1* calorie = *4.2 × 10^7 ergs.*

(*Meerut 2001, Kanpur 2000, Punjab 2002*)

Solution.
$$\frac{dP}{dT} = \frac{L}{T\,(V_2 - V_1)}$$
$$dP = 1 \text{ atmosphere}$$
$$= 76 \times 13.6 \times 980 \text{ dynes/cm}^2$$
$$dT = ?$$
$$L = 4563 \text{ cal/mol}$$
$$= 4563 \times 4.2 \times 10^7 \text{ ergs/mol}$$
$$V_2 - V_1 = 18.7 \text{ cm}^3/\text{mol}$$
$$T = 80 + 273$$

$$= 353 \text{ K}$$

$$dT = \frac{dP.T.(V_2 - V_1)}{L}$$

$$= \frac{76 \times 13.6 \times 980 \times 353 \times 18.7}{4563 \times 4.2 \times 10^7}$$

$$= \mathbf{0.03488\ K}$$

Therefore, the increase in the melting point of naphthalene with an increase in pressure of one atmosphere

$$= \mathbf{0.03488\ K}$$

$$= \mathbf{0.03488^\circ C}$$

Example 6.13. *Calculate the boiling point of benzene under a pressure of 80 cm of mercury. The normal boiling point is 80°C. Latent heat of vaporization is 380 joules/g, density of vapour at the boiling point is 4 g/litre and that of the liquid 0.9 g/cm³.* **(*Punjab 2000*)**

Solution. Here $dP = 80 - 76$

$$= 4 \text{ cm of Hg}$$

$$= 4 \times 13.6 \times 980 \text{ dynes/cm}^2$$

$$T = 80^\circ C = 80 + 273$$

$$= 353 \text{ K}$$

$$L = 980 \text{ joules/g}$$

$$= 380 \times 10^7 \text{ ergs/g}$$

$$V_1 = \frac{1}{0.9} = 1.11 \text{ cm}^3$$

$$V_2 = \frac{1{,}000}{4} = 250 \text{ cm}^3$$

$$\frac{dP}{dT} = \frac{L}{T[V_2 - V_2]}$$

$$dT = \frac{dP \times T[V_2 - V_1]}{L}$$

$$dT = \frac{4 \times 13.6 \times 980 \times 353[250 - 1.11]}{380 \times 10^7}$$

$$= \mathbf{1.233\ K}$$

$$= \mathbf{1.233^\circ C}$$

∴ The boiling point of benzene at a pressure of 80 cm of Hg

$$= 80 + 1.233$$

$$= \mathbf{81.233^\circ C}$$

Example 6.14. *Calculate the change in the boiling point of water when the pressure of steam on its surface is increased from 1 atmosphere to 1.10 atmospheres.* **(*Madurai 2001*)**

Latent heat of water at 100°C

$$= 537 \text{ cal/g}$$

Volume of one gram of steam at 100°C

$$= 1{,}676 \text{ cm}^3$$

Solution. $dP = 1.10 - 1$

$$= 0.10 \text{ atmosphere}$$

$$dP = 0.10 \times 76 \times 13.6 \times 980 \text{ dynes/cm}^2$$

$$T = 100 + 273$$
$$= 373 \text{ K}$$
$$L = 537 \text{ cals/g}$$
$$= 537 \times 4.2 \times 10^7 \text{ ergs/g}$$
$$V_1 = 1.00 \text{ cm}^3$$
$$V_2 = 1{,}676 \text{ cm}^3$$
$$\frac{dP}{dT} = \frac{L}{T[V_2 - V_1]}$$
$$dT = \frac{dP \times T[V_2 - V_1]}{L}$$
$$dT = \frac{0.1 \times 76 \times 13.6 \times 980 \times 373 \times (1{,}675)}{537 \times 4.2 \times 10^7}$$
$$= \mathbf{2.792\ K}$$
$$= \mathbf{2.792°C}$$

Therefore, the increase in the boiling point of water with an increase of 0.1 atmosphere pressure

$$= \mathbf{2.792\ K}$$
$$= \mathbf{2.792°C}$$

Example 6.15. *Calculate the change in the melting point of ice when it is subjected to a pressure of 100 atmospheres.* ***(Bundelkhand 2001)***

Density of ice = 0.917 g/cm^3 and
Latent heat of ice = 336 J/g

Solution.
$$\frac{dP}{dT} = \frac{L}{T\,(V_2 - V_1)}$$
$$dP = 100 - 1$$
$$= 99 \text{ atmospheres}$$
$$dP = 99 \times 76 \times 13.6 \times 980 \text{ dynes/cm}^2$$
$$L = 336 \text{ J/g}$$
$$= 336 \times 10^7 \text{ ergs/g}$$
$$T = 273 \text{ K}$$
$$(V_2 - V_1) = 1 - \frac{1}{0.917}$$
$$= -\frac{0.083}{0.917}$$
$$= -0.091 \text{ cm}^3$$
$$\therefore \quad dT = \frac{T\,dP\,(V_2 - V_1)}{L}$$
$$dT = \frac{273 \times 99 \times 76 \times 13.6 \times 980 \times (-0.091)}{336 \times 10^7}$$
$$dT = -0.7326 \text{ K}$$
$$= \mathbf{-0.7326°C}$$

The decrease in the melting point of ice with a pressure of 100 atmospheres

$$= \mathbf{0.7326°C}$$

Example 6.16. *Water boils at a temperature of 101°C at a pressure of 787 mm of* Hg. *1 gram of water occupies 1,601 cm³ on evaporation. Calculate the latent heat of steam,* $J = 4.2 \times 10^7$ ergs/cal. **(*Nagpur 2001*)**

Solution.

$$\frac{dP}{dT} = \frac{L}{T(V_2 - V_1)}$$

$$dP = 787 - 760$$
$$= 27 \text{ mm of Hg}$$
$$= 2.7 \text{ cm of Hg}$$
$$= 2.7 \times 13.6 \times 980 \text{ dynes/cm}^2$$
$$dT = 1°\text{C} = 1 \text{ K}$$
$$T = 373 \text{ K}$$
$$V_2 - V_1 = 1{,}601 - 1 = 1{,}600 \text{ cm}^3$$
$$L = ?$$
$$L = \frac{T\,dP(V_2 - V_1)}{dT}$$
$$L = \frac{373 \times 2.7 \times 13.6 \times 980 \times 1{,}600}{1} \text{ ergs/g}$$
$$L = \frac{373 \times 2.7 \times 13.6 \times 980 \times 1{,}600}{4.2 \times 10^7} \text{ cal/g}$$
$$\mathbf{L = 511.3 \text{ cal/g}}$$

Example 6.17. *When lead is melted at atmospheric pressure, (the melting point is 600 K) the density decreases from 11.01 to 10.65 g/cm³ and the latent heat of fusion is 24.5 J/g. What is the melting point at a pressure of 100 atmospheres?* **(*Bangalore 2001*)**

Solution.

$$\frac{dP}{dT} = \frac{L}{T(V_2 - V_1)}$$
$$dP = 99 \text{ atmospheres}$$
$$dP = 99 \times 76 \times 13.6 \times 980 \text{ dynes/cm}^2$$
$$L = 24.5 \text{ J/g}$$
$$= 24.5 \times 10^7 \text{ ergs/g}$$
$$V_1 = \frac{1}{11.01} \text{ cm}^3$$
$$V_2 = \frac{1}{10.65} \text{ cm}^3$$
$$V_2 - V_1 = \frac{1}{10.65} - \frac{1}{11.01}$$
$$= \left(\frac{0.36}{10.65 \times 11.01}\right) \text{cm}^3$$
$$dT = \frac{T dP\,(V_2 - V_1)}{L}$$
$$dT = \frac{600 \times 99 \times 76 \times 13.6 \times 980 \times 0.36}{10.65 \times 1.01 \times 24.5 \times 10^7}$$
$$dT = \mathbf{0.7539 \text{ K}}$$

$= \mathbf{0.7539°C}$

$\therefore$ Melting point of lead at 100 atmospheres pressure

$= 600 + 0.7539$

$= \mathbf{600.7539\ K}$

Example 6.18. *Calculate under what pressure water will boil at 120°C if the change in specific volume when 1 gram of water is converted into steam is 1,676 cm³.* **(Nagpur 2009, Delhi 2001)**

Latent heat of steam $= 540$ cal/g

$J = 4.2 \times 10^7$ *ergs/cal*

1 atmosphere pressure

$= 10^6$ *dynes/cm²*

Solution. Here

$$\frac{dP}{dT} = \frac{L}{T(V_2 - V_1)}$$

$$dT = 120 - 100$$
$$= 20\ K$$
$$T = 373\ K$$
$$(V_2 - V_1) = 1{,}676\ cm^3$$
$$L = 540\ cal/g$$
$$= 540 \times 4.2 \times 10^7\ ergs/g$$
$$dP = ?$$
$$dP = \frac{L.dT}{T(V_2 - V_1)}$$
$$dP = \frac{540 \times 4.2 \times 10^7 \times 20}{373 \times 1{,}676}\ dynes/cm^2$$
$$dP = \frac{540 \times 4.2 \times 10^7 \times 20}{373 \times 1{,}676 \times 10^6}\ atmospheres$$
$$dP = 0.7254\ atmosphere$$

Pressure required $= 1 + 0.7254$

$= \mathbf{1.7254\ atmospheres}$

Example 6.19. *Calculate the pressure required to lower the melting point of ice by 1°C. Given L = 79.6 cal/g. Specific volumes of water at 0°C = 1 cm³.*
Specific volume of ice, at 0°C = 1.091 cm². 1 atmospheric pressure = 1.013 × 10⁶ dyne/cm².

(Mumbai 2001)

Solution.

$$\frac{dP}{dT} = \frac{L}{T(V_2 - V_1)}$$
$$dT = -1\ K$$
$$T = 273\ K$$
$$V_2 - V_1 = -0.091\ cm^3$$
$$L = 79.6\ cal/g$$
$$= 79.6 \times 4.18 \times 10^7\ ergs/g$$
$$dP = \frac{L\,dT}{T(V_2 - V_1)} = \frac{79.6 \times 4.18 \times 10^7 \times 1}{273 \times 0.091}\ dynes/cm^2$$

$$dP = 1.33932 \times 10^8 \text{ dynes/cm}^2$$

$$dP = \frac{1.33932 \times 10^8}{1.013 \times 10^6} \text{ atmospheres}$$

$$dP = \mathbf{132.21} \text{ atmospheres}$$

$\therefore$ Pressure required $= 132.21 + 1$

$= \mathbf{133.21}$ **atmospheres**

Example 6.20. *Calculate the temperature at which ice will freeze if the pressure is increased by 135.2 atmospheres. Given that the increase in the specific volume when 1 gram of water freezes into ice is 0.091 cm³. Latent heat of fusion of ice is 80 cal/gram.* ***(Agra 2001)***

Solution.
$$\frac{dP}{dT} = \frac{L}{T(V_2 - V_1)}$$

Here
$$dP = 135.2 \text{ atmospheres}$$

$$dP = 135.2 \times 1.013 \times 10^6 \text{ dynes/cm}^2$$

$$V_2 - V_1 = -0.091 \text{ cm}^3$$

$$L = 80 \text{ cal/g}$$

$$L = 80 \times 4.18 \times 10^7 \text{ ergs/g}$$

$$T = 273 \text{ K}$$

$\therefore$
$$dT = \frac{dP \, . \, T(V_2 - V_1)}{L}$$

$$dT = \frac{135.2 \times 1.013 \times 10^6 \times 273 \times (-0.091)}{80 \times 4.18 \times 10^7}$$

$$dT = -1 \text{ K}$$

or
$$dT = -1°\text{C}$$

Therefore, the temperature at which ice will freeze is

$= \mathbf{-1°C}$ **or 272 K.**

Example 6.21. *If the initial temperature is 27°C, calculate the drop in temperature produced by adiabatic throttling in the case of oxygen when the pressure is reduced by 50 atmospheres. Given*

$$a = 13.2 \times 10^{-2} \text{ Nm}^4 \text{ (mole)}^{-2}$$

$$b = 31.2 \times 10^{-6} \text{ m}^3 \text{ (mole)}^{-1}$$

$$R = 8.3 \text{ JK}^{-1} \text{ (mole)}^{-1}$$

$$C_P = 3.5 R$$

(Kanpur 2002)

Solution.
$$\delta T = \left(\frac{\delta P}{JMC_P}\right)\left(\frac{2a}{RT} - b\right)$$

Here
$$\delta P = 50 \text{ atmospheres}$$

$$= 50 \times 0.76 \times 13.6 \times 10^3 \times 9.8 \text{ N/m}^2$$

$$J = 4.2 \text{ J/cal}$$

$$M = 32$$

$$C_P = 3.5 R = 3.5 \times 8.3 \text{ JK}^{-1} \text{ (mole)}^{-1}$$

$$T = 300 \text{ K°}$$

$$\delta T = \left(\frac{50 \times 0.76 \times 13.6 \times 10^3 \times 9.8}{4.2 \times 32 \times 3.5 \times 8.3}\right) \times \left(\frac{2 \times 13.2 \times 10^{-2}}{8.3 \times 300} - 31.2 \times 10^{-6}\right)$$

$$\boldsymbol{\delta T = 0.0969 \text{ K°}}$$

Example 6.22. *Water boils at a temperature of 101°C at a pressure of 787 mm of Hg. 1 gram of water occupies 1601 cm³ on evaporation. Calculate the lalent heat of steam. (J = 4.42 × 10⁷ ergs/cal)* **(*Kerala 2000*)**

Solution.
$$\frac{dP}{dT} = \frac{L}{T(V_2 - V_1)}$$
$$dP = 787 - 760 = 27 \text{ mm of Hg}$$
$$= 2.7 \text{ cm of Hg}$$
$$= 2.7 \times 13.6 \times 980 \text{ dynes/cm}^2$$
$$dT = 1°\text{C} = 1 \text{ K}$$
$$T = 373 \text{ K}$$
$$V_2 - V_1 = 1601 - 1 = 1600 \text{ cm}^3$$
$$L = ?$$
$$L = \frac{TdP(V_2 - V_1)}{dT}$$
$$= \frac{373 \times 2.7 \times 13.6 \times 980 \times 1600}{1} \text{ ergs/g}$$
$$= \mathbf{2147.6 \times 10^7} \text{ ergs/g}$$
$$L = \frac{2147.6 \times 10^7}{4.2 \times 10^7}$$
$$= \mathbf{511.33 \text{ cal/g}}$$

Example 6.23. *For a metallic copper disc at 300 K, the following values are known*

$$C_P = 24.5 \text{ J/mol K}$$ **(*Madras 2001*)**
$$\alpha = 50.4 \times 10^{-6} \text{ K}^{-1}$$
Isothermal compressibility, $K = 7.78 \times 10^{-12} \text{ N/m}^2$
$$V = 7.06 \text{ cm}^3/\text{mol}$$

Determine C_V.

Solution.
$$C_P - C_V = TE\alpha^2 V$$
and
$$E = \frac{1}{K}$$
$$C_P - C_V = \frac{TV\alpha^2}{K}$$
$$V = 7.06 \text{ cm}^3/\text{mol} = 7.06 \times 10^{-6} \text{ m}^3/\text{mol}$$
$$\therefore \quad C_P - C_V = \frac{300 \times 7.06 \times 10^{-6} \times (50.4 \times 10^{-6})^2}{7.78 \times 10^{-12}}$$
$$C_P - C_V = 0.6915 \text{ J/mol-K}$$
$$C_V = C_P - 0.6915$$
$$C_V = 24.5 - 0\ 0.6915$$
$$C_V = \mathbf{23.8085} \text{ J/mol-K}$$

Example 6.24. *Calculate the drop in temperature produced by adiabatic throttling process in the case of oxygen when the pressure is reduced by 50 atmospheres. Initial temperature of the gas is 27°C. Given that the gas obeys Van der Waals' equation and*

$$a = 1.32 \times 10^{12}\ cm^4\ dynes/mole^2$$

$$b = 31.2\ \text{cm}^3/\text{mole}$$

and $$C_P = 7\ cal/mole\text{-}K$$ **(*Punjab 2002*)**

Solution. Here $$\frac{\delta T}{\delta P} = \frac{1}{C_P}\left[\frac{2a}{RT} - b\right]$$

$$\delta T = \frac{\delta P}{C_P}\left[\frac{2a}{RT} - b\right]$$

$$\delta P = 50 \text{ atmospheres}$$

$$= 50 \times 76 \times 13.6 \times 980 \text{ dynes/cm}^2$$

$$C_P = 7 \text{ cal/mole-K}$$

$$= 7 \times 4.2 \times 10^7 \text{ ergs/mole-K}$$

$$T = 27°\text{C} = 300 \text{ K}$$

$$R = 8.31 \times 10^7 \text{ ergs/mole-K}$$

$$\therefore \quad \delta T = \left(\frac{50\times 76\times 13.6\times 980}{7\times 4.2\times 10^7}\right)\left(\frac{2\times 1.32\times 10^{12}}{8.31\times 10^7 \times 300} - 31.2\right)$$

$$\delta T = \left(\frac{0.5064}{7\times 4.2}\right)(317.7 - 31.2)$$

$$\delta T = \mathbf{4.935\ K}$$

$$= \mathbf{4.935°C}$$

Example 6.25. *In a cold storage ice melts at the rate of 3 kg per hour when the external temperature is 28°C. Find the minimum power output of the motor used to run the refrigerator just to prevent melting of ice.*

Specific latent heat of ice fusion = 3.3 × 10^5 J kg^{-1} **(*Calcutta 2002*)**

Solution. $$m = 3 \text{ kg}$$

$$L = 3.3 \times 10^5 \text{ J kg}^{-1}$$

$$s = 4.2 \times 10^3 \text{ J kg}^{-1} - °\text{C}$$

Heat taken by ice to melt into water and reach a temperature of 28°C in one hour.

$$Q = mL + ms\,\theta$$

$$= 3 \times 3.33 \times 10^5 + 3 \times 4.2 \times 10^3 \times 28$$

$$= 1.343 \times 10^6 \text{ J}$$

To prevent melting of ice, the refrigerator should have the power out put,

$$P = 1.343 \times 10^6 \text{ J/hour}$$

$$= \frac{1.343\times 10^6}{3600} \text{ J/s}$$

$$= \mathbf{373\ Watts.}$$

Example 6.26. *Calculate the depression of melting point of ice by one atmosphere increase of pressure, given, latent heat of ice = 3.35 × 10^5 J/kg and the specific volumes of 1 kg of ice and water at 0°C are 1.090 × 10^{-3} m^3 and 10^{-3} m^3 respectively.* **(*Nagpur 2007, Kerala 2000*)**

Solution. Here, $$L = 3.35 \times 10^5 \text{ J/kg}$$

$$T = 273 \text{ K}$$

$$V_1 = 1.090 \times 10^{-3} \text{ m}^3$$

$$V_2 = 10^{-3} \text{ m}^3$$

$$dP = 1 \text{ atmosphere}$$

$$= 0.76 \times 13.6 \times 10^3 \times 9.8$$
$$= 1.01 \times 10^5 \text{ N/m}^2$$
$$\frac{dP}{dT} = \frac{L}{T(V_2 - V_2)}$$
$$dT = \frac{dP \times T(V_2 - V_1)}{L}$$
$$= \frac{1.01 \times 10^5 \times 273\,(10^{-3} - 1.09 \times 10^{-3})}{3.35 \times 10^5}$$
$$= \mathbf{-7.4 \times 10^{-3}\ K}$$
$$= \mathbf{-7.4 \times 10^{-3}\ °C}$$

Therefore, decrease in melting point of ice with an increase in pressure of one atmosphere.

Example 6.27. *Considering mercury at 0°C and 1 atmospheric pressure, determine γ (C_P/C_V). Given C_P = 28.0 J/mol K, specific volume V = 1.47 × 10^{-5} m^3/mol, volume expansivity β = 1.81 × 10^6 K^{-1} and compressibility is 3.94 × 10^{-6} atm^{-1} (3.89 × 10^{-11} Pa^{-1}).* **(*Alagappa 2001*)**

Solution. $$C_P - C_V = \frac{TV\beta^2}{k}$$

Here, $$T = 273 \text{ K}, V = 1.47 \times 10^{-5} \text{ m}^3\text{/mol}$$
$$\beta = 1.81 \times 10^6 \text{ K}^{-1}$$
$$k = 3.89 \times 10^{-11} \text{ Pa}^{-1}$$

$\therefore$ $$\frac{TV\beta^2}{k} = \frac{273 \times 1.47 \times 10^{-5} \times (1.81 \times 10^{-6})}{3.89 \times 10^{-11}}$$
$$= 338 \times 10^{-6}$$

$\therefore$ $$C_V = C_P - 338 \times 10^{-6} = 28 - 338 \times 10^{-6}$$
$$= 27.999662 \text{ J mol K}$$
$$\gamma = \frac{C_P}{C_V} = \frac{\mathbf{28}}{\mathbf{27.999662}} = \mathbf{1} \text{ (approx.)}$$

Example 6.28. *Find the change in boiling point of water for 100 cm change in mercury pressure if the specific volume of steam is 1601 cm^3/gram at 100°C and 76 cm mercury pressure. Take latent heat as 5366 cal/gram.* **(*Rajasthan 2001*)**

Solution. [In SI system]

Here, $$dP = 100 \text{ cm Hg} = 1 \text{ m Hg}$$
$$= 1 \times 13.6 \times 10^3 \times 9.8 \text{ N/m}^2$$
$$= 1.33 \times 10^5 \text{ N/m}^2$$
$$V_1 = 1 \text{ cm}^3\text{/gram} = 1 \times 10^{-3} \text{ m}^3\text{/kg}$$
$$V_2 = 1601 \text{ cm}^3\text{/gram} = 1601 \times 10^{-3} \text{ m}^3\text{/kg}$$
$$L = 536 \text{ cal/gram}$$
$$= 536 \times 4.2 \times 10^3 \text{ J/kg}$$
$$= 2.25 \times 10^6 \text{ J/kg}$$
$$T = 100°\text{C} = 373 \text{ K}$$
$$dT = ?$$
$$\frac{dP}{dT} = \frac{L}{T(V_2 - V_1)}$$

$$dT = \frac{dP \times T\,(V_2 - V_1)}{L}$$

$$= \frac{1.33 \times 10^5 \times 373\,(1601 \times 10^{-3} - 1 \times 10^{-3})}{2.25 \times 10^6}$$

$$= \mathbf{35.28\ K}$$

$$= \mathbf{35.28°C}$$

Therefore, increase in boiling point of water with an increase of 100 cm in mercury pressure

$$= \mathbf{35.28\ K = 35.28°C}$$

Example 6.29. *The ratio of the densities of ice and water at 0°C is $\frac{10}{11}$ Calculate the decrease in the melting point of ice, given that latent heat of fusion of ice = 80 cal/gm and change in specific volume – 0.1 c.c./gm. The increase in pressure is one atmosphere.* ***(Osmania, 2000)***

Solution. [In SI system],

$$dP = 0.76 \times 13.6 \times 10^3 \times 9.8 = 1.01 \times 10^5\ \text{N/m}^2$$

$$L = 80\ \text{cal/gram}$$

$$= 80 \times 4.2 \times 10^3\ \text{J/kg}$$

$$= 3.36 \times 10^5\ \text{J/kg}$$

$$T = 273\ \text{K}$$

$$V_2 - V_1 = -\,0.1\ \text{cm}^3/\text{gram}$$

$$= -\,0.1 \times 10^{-3}\ \text{m}^3/\text{kg}$$

$$dT = ?$$

$$\frac{dP}{dT} = \frac{L}{T\,(V_2 - V_1)}$$

$$dT = \frac{dP \times T\,(V_2 - V_1)}{L}$$

$$= \frac{1.01 \times 10^5 \times 273\,(-0.1 \times 10^{-3})}{3.36 \times 10^5}$$

$$= \mathbf{-\,8.2 \times 10^{-3}\ K}$$

$$= \mathbf{-\,8.2 \times 10^{-3}\ °C}$$

Therefore, the decrease in the melting point of ice with an increase in pressure of one atmosphere

$$= \mathbf{8.2 \times 10^{-3}\ K = 8.2 \times 10^{-3}\ °C}$$

Example 6.30. *Using Maxwell's thermodynamical relations, prove that for any substance, the ratio of the adiabatic and isothermal elasticities is equal to the ratio of the two specific heats.* ***(Garhwal 2005, 2001, Lucknow 2002)***

Solution. From definition,

Isothermal elasticity $\quad E_T = -V\left(\frac{\partial P}{\partial V}\right)_T$

Adiabatic elasticity $\quad E_S = -V\left(\frac{\partial P}{\partial V}\right)_S$

$$\therefore \quad \frac{E_S}{E_T} = \frac{-V\left(\frac{\partial P}{\partial V}\right)_S}{-V\left(\frac{\partial P}{\partial V}\right)_T}$$

$$= \frac{\left(\frac{\partial P}{\partial V}\right)_S}{\left(\frac{\partial P}{\partial V}\right)_T}$$

$$= \frac{\left(\frac{\partial P}{\partial T}\right)_S \left(\frac{\partial T}{\partial V}\right)_S}{\left(\frac{\partial P}{\partial S}\right)_T \left(\frac{\partial S}{\partial V}\right)_T}$$

$$\frac{E_S}{E_T} = \frac{\left(\frac{\partial S}{\partial P}\right)_T \left(\frac{\partial T}{\partial V}\right)_S}{\left(\frac{\partial T}{\partial P}\right)_S \left(\frac{\partial S}{\partial V}\right)_T} \qquad \text{... } (i)$$

But from the first four Maxwell's equations

$$\left(\frac{\partial S}{\partial P}\right)_T = -\left(\frac{\partial V}{\partial T}\right)_P$$

$$\left(\frac{\partial T}{\partial V}\right)_S = -\left(\frac{\partial P}{\partial S}\right)_V$$

$$\left(\frac{\partial T}{\partial P}\right)_S = \left(\frac{\partial V}{\partial S}\right)_P$$

$$\left(\frac{\partial S}{\partial V}\right)_T = \left(\frac{\partial P}{\partial T}\right)_V$$

∴ Substituting these values in equation (*i*)

$$\frac{E_S}{E_T} = \frac{\left(\frac{\partial V}{\partial T}\right)_P \left(\frac{\partial P}{\partial S}\right)_V}{\left(\frac{\partial V}{\partial S}\right)_P \left(\frac{\partial P}{\partial T}\right)_V}$$

$$\frac{E_S}{E_T} = \frac{\left(\frac{\partial S}{\partial V}\right)_P \left(\frac{\partial V}{\partial T}\right)_P}{\left(\frac{\partial S}{\partial P}\right)_V \left(\frac{\partial P}{\partial T}\right)_V}$$

$$= \frac{\left(\frac{\partial S}{\partial T}\right)_P}{\left(\frac{\partial S}{\partial T}\right)_V}$$

$$= \frac{\left(\frac{\partial Q}{\partial T}\right)_P}{\left(\frac{\partial Q}{\partial T}\right)_V}$$

But $$\left(\frac{\partial Q}{\partial T}\right)_P = C_P$$

and $$\left(\frac{\partial Q}{\partial T}\right)_V = C_V$$

$$\therefore \qquad \frac{E_S}{E_T} = \frac{C_P}{C_V} = \gamma \qquad \text{... (ii)}$$

Example 6.31. *Using Maxwell's thermodynamical relations, prove that the ratio of the adiabatic to the isobaric coefficient of expansion is* **(*Nagpur 2001*)**

$$\frac{1}{(1-\gamma)}$$

Solution. Adiabatic volume coefficient of expansion,

$$\alpha_S = \frac{1}{V}\left(\frac{\partial V}{\partial T}\right)_S \qquad \text{... (i)}$$

Isobaric volume coefficient of expansion,

$$\alpha_P = \frac{1}{V}\left(\frac{\partial V}{\partial T}\right)_P \qquad \text{... (ii)}$$

$$\therefore \qquad \frac{\alpha_S}{\alpha_P} = \frac{\frac{1}{V}\left(\frac{\partial V}{\partial T}\right)_S}{\frac{1}{V}\left(\frac{\partial V}{\partial T}\right)_P} \qquad \text{... (iii)}$$

$$= \frac{\left(\frac{\partial V}{\partial T}\right)_S}{\left(\frac{\partial V}{\partial T}\right)_P} = \frac{1}{\left(\frac{\partial T}{\partial V}\right)_S\left(\frac{\partial V}{\partial T}\right)_P}$$

But $$\left(\frac{\partial T}{\partial V}\right)_S = -\left(\frac{\partial P}{\partial S}\right)_V$$

$$\therefore \qquad \frac{\alpha_S}{\alpha_P} = -\frac{1}{\left(\frac{\partial P}{\partial S}\right)_V\left(\frac{\partial V}{\partial T}\right)_P}$$

$$= \frac{1}{-\left[\left(\frac{\partial P}{\partial T}\right)_V\left(\frac{\partial T}{\partial S}\right)_V\left(\frac{\partial V}{\partial T}\right)_P\right]}$$

$$= \frac{\left(\frac{\partial S}{\partial T}\right)_V}{-\left(\frac{\partial P}{\partial T}\right)_V\left(\frac{\partial V}{\partial T}\right)_P}$$

$$= \frac{T\left(\frac{\partial S}{\partial T}\right)_V}{-T\left(\frac{\partial P}{\partial T}\right)_V\left(\frac{\partial V}{\partial T}\right)_P}$$

But $$C_P - C_V = T\left(\frac{\partial P}{\partial T}\right)_V\left(\frac{\partial V}{\partial T}\right)_P$$

and $$C_V = T\left(\frac{\partial S}{\partial T}\right)_V$$

$\therefore$ $$\frac{\alpha_S}{\alpha_P} = \frac{C_V}{-(C_P - C_V)}$$

or $$\frac{\alpha_S}{\alpha_P} = \frac{1}{1-\left(\frac{C_P}{C_V}\right)}$$

or $$\frac{\alpha_S}{\alpha_P} = \frac{1}{1-\gamma} \qquad \text{... (iv)}$$

Example 6.32. *Using Maxwell's thermodynamical relations, prove that the ratio of the adiabatic to the isochoric pressure coefficient of expansion is equal to*

$$\frac{\gamma}{(\gamma-1)}$$ **(*Calicut 2003, Calcutta 2003*)**

Solution. Adiabatic pressure coefficient of expansion,

$$\beta_S = \frac{1}{P}\left(\frac{\partial P}{\partial T}\right)_S \qquad \text{... (i)}$$

Isochoric pressure coefficient of expansion,

$$\beta_V = \frac{1}{P}\left(\frac{\partial P}{\partial T}\right)_V \qquad \text{... (ii)}$$

$\therefore$ $$\frac{\beta_S}{\beta_V} = \frac{\frac{1}{P}\left(\frac{\partial P}{\partial T}\right)_S}{\frac{1}{P}\left(\frac{\partial P}{\partial T}\right)_V} \qquad \text{... (iii)}$$

$$= \frac{\left(\frac{\partial P}{\partial T}\right)_S}{\left(\frac{\partial P}{\partial T}\right)_V}$$

$$= \frac{1}{\left(\frac{\partial T}{\partial P}\right)_S\left(\frac{\partial P}{\partial T}\right)_V}$$

But $$\left(\frac{\partial T}{\partial P}\right)_S = \left(\frac{\partial V}{\partial S}\right)_P \qquad \text{... (iv)}$$

$$\frac{\beta_S}{\beta_V} = \frac{1}{\left(\frac{\partial V}{\partial S}\right)_P \left(\frac{\partial P}{\partial T}\right)_V} \qquad \text{... (v)}$$

$$= \frac{1}{\left(\frac{\partial V}{\partial T}\right)_P \left(\frac{\partial T}{\partial S}\right)_P \left(\frac{\partial P}{\partial T}\right)_V}$$

$$= \frac{\left(\frac{\partial S}{\partial T}\right)_P}{\left(\frac{\partial V}{\partial T}\right)_P \left(\frac{\partial P}{\partial T}\right)_V}$$

$$= \frac{T\left(\frac{\partial S}{\partial T}\right)_P}{T\left(\frac{\partial V}{\partial T}\right)_P \left(\frac{\partial P}{\partial T}\right)_V}$$

$$= \frac{C_P}{C_P - C_V}$$

or
$$\frac{\beta_S}{\beta_V} = \frac{\left(\frac{C_P}{C_V}\right)}{\left(\frac{C_P}{C_V} - 1\right)} = \frac{\gamma}{\gamma - 1} \qquad \text{... (vi)}$$

Example 6.33. *Show that for a perfect gas*

$$\left(\frac{\partial U}{\partial V}\right)_T = 0$$

Solution. From the Maxwell's first equation,

$$\left(\frac{\partial S}{\partial V}\right)_T = \left(\frac{\partial P}{\partial T}\right)_V$$

$$\frac{1}{T}\left(\frac{\partial Q}{\partial V}\right)_T = \left(\frac{\partial P}{\partial T}\right)_V$$

or
$$\left(\frac{\partial Q}{\partial V}\right)_T = T\left(\frac{\partial P}{\partial T}\right)_V$$

But
$$dQ = dU + PdV$$

$\therefore$
$$\left(\frac{dU + PdV}{dV}\right) = T\left(\frac{\partial P}{\partial T}\right)_V$$

$$\frac{dU}{dV} + P = T\left(\frac{\partial P}{\partial T}\right)_V$$

or
$$\left(\frac{\partial U}{\partial V}\right)_T = T\left(\frac{\partial P}{\partial T}\right)_V - P$$

But from perfect gas equation,

$$PV = RT$$

$$\therefore \qquad \left(\frac{\partial P}{\partial T}\right)_V = \frac{R}{V}$$

$$\left(\frac{\partial U}{\partial V}\right)_T = T\left(\frac{R}{V}\right) - P$$

$$\text{or} \qquad \left(\frac{\partial U}{\partial V}\right)_T = P - P = 0$$

$$\left(\frac{\partial U}{\partial V}\right)_T = 0$$

Example 6.34. *If L = 800 – 0.705 T, show that the specific heat of steam is negative.*

(*Lucknow 2002; Kanpur 2002; Meerut 2000*)

Solution. Given :

$$L = 800 - 0.705\ T$$

$$\therefore \qquad \frac{dL}{dT} = -0.705$$

At normal boiling point of water *i.e.*, at 100 °C = 373 K, the value of L would be

$$L = 800 - 0.705 \times 373 = 537$$

The specific heat of water, in liquid state, $C_1 = 1$.
Using second latent heat equation

$$C_2 - C_1 = \frac{dL}{dT} - \frac{L}{T}$$

$$\therefore \qquad C_2 = C_1 + \frac{dL}{dT} - \frac{L}{T}$$

$$= 1 + (-0.705) - \frac{537}{373}$$

$$= \mathbf{-1.14467}$$

Thus, the specific heat of saturated vapour pressure at 100 °C is **negative**.

Example 6.35. *Calculate the specific heat of saturated steam from the equation* $\frac{dL}{dT} - \frac{L}{T} = (c_2 - c_1)$ *and explain the meaning of its negative value. Given L = 540 k.cal/kgm, T = 100°C,* $\frac{dL}{dT}$ = *– 0.640 k cal/Kg K.; Sp. heat of water at 100°C =* c_1 *= 1.01 k cal/kgK.* **(*Kanpur U., 2002*)**

Solution. L = 540 k.cal/kg = 540 cal/gm

$$T = 100\ °C = 373\ K$$

$$\frac{dL}{dT} = -0.64 \text{ k.cal/kg.K}$$

$$= -0.64 \text{ cal/gm.K}$$

Specific heat of water at 100°C = 1.01 k.cal/kg = 1.01 cal/gmK

$$\text{or} \qquad c_1 = 1.01 \text{ cal/gm K}$$

$$\text{Hence} \qquad c_2 = c_1 - \frac{L}{T} + \frac{dL}{dT}$$

$$= 1.01 - \frac{540}{373} - 0.64$$

$$= 1.01 - 2.088$$
$$= -1.078 \text{ k cal/kg.}$$

The meaning of negative sign is explained in article.

Example 6.36. *A one metre length of copper wire at 15°C is stretched adiabatically by 1 cm. Calculate the change in its temperature. For copper the following data is given.*

Density 8.9 gm/cm³; Specific heat 0.09 cal/gm/K; Co-efficient of linear expansion = 1.67 × 10⁻⁵/K; coefficient of adiabatic elasticity 1.2 × 10¹² dynes/cm². Prove the relation used.

(H.P.U 2000; P.U. 2001)

Solution. Temperature of the wire $T = 273 + 15 = 288$ K

Length of the wire $l = 1\text{m} = 100$ cm

Coefficient of linear expansion $\alpha = 1.67 \times 10^{-5}$ K

Adiabatic elasticity $y_a = 1.2 \times 10^{12}$ dynes/cm²

Increase in length $dl = 1$ cm.

Specific heat = 0.9 cal/gm/K or $c_p = .09 \times 4.2 \times 10^7$ erg/gm/K

Density of the wire $\rho = 8.9$ gm/cm³

Now if m is the mass of the wire and a its area of cross section $m = l\,a\,\rho$.

If dF is the applied force, then

$$y_a = \frac{dF / a}{dl / l} \quad \text{or} \quad dF = \frac{y_a \cdot dl.a}{l}$$

$$\therefore \qquad \frac{ldF}{m} = \frac{l.y_a.dl.a}{l.l.a.\rho} = \frac{y_a \cdot dl}{l\rho}$$

Hence
$$dT = -\frac{T\alpha}{c_p} \cdot \frac{y_a dl}{l.\rho}$$

$$= \frac{2.88 \times 1.67 \times 10^{-5} \times 1.2 \times 10^{12} \times 1}{.09 \times 4.2 \times 10^7 \times 100 \times 8.9}$$

$$= 1.716 \text{ K} = 1.716 \text{ °C}$$

Example 6.37. *Prove the thermodynamical relation*

$$\left(\frac{\partial Q}{\partial P}\right)_T = -T\left(\frac{\partial V}{\partial T}\right)_P = -TV\alpha$$

and show that heat is generated when a substance which expands on heating is compressed and for substance which contract a cooling takes place. **(Rohilkhand 2000, 2001, 2002, G.N.D.U. 2001, Punjab 2002, Meerut 2004, Garhwal 2003, Agra 2003)**

Solution. From Maxwell's fourth thermodynamical relation,

$$\left(\frac{\partial S}{\partial P}\right)_T = -\left(\frac{\partial V}{\partial T}\right)_P$$

Multiplying both sides by T, we get

$$\left(\frac{T.\partial S}{\partial P}\right)_T = -T\left(\frac{\partial V}{\partial T}\right)_P$$

But $\qquad T.\partial S = \partial Q$, substituting, we get

$$\left(\frac{\partial Q}{\partial P}\right)_T = -T\left(\frac{\partial V}{\partial T}\right)_P \qquad \text{... (i)}$$

which is the required relation.

Now, $\frac{1}{V}\left(\frac{\partial V}{\partial T}\right)_P$ represents the increase in volume per unit rise of temperature at constant pressure. This is nothing but the coefficient of volume expansion α. Hence

$$\frac{1}{V}\left(\frac{\partial V}{\partial T}\right)_P = \alpha$$

or

$$\left(\frac{\partial V}{\partial T}\right)_P = V\alpha \quad \text{... (ii)}$$

Substituting in eqn (*i*), we get

$$\therefore \quad \left(\frac{\partial Q}{\partial P}\right)_T = -TV\alpha \quad \text{... (iii)}$$

This is another form of required relation.

From this equation, it is very clear that if α is positive *i.e.*, if the substance expands on heating, then $\left(\frac{\partial Q}{\partial P}\right)_T$ is negative. It means heat must be withdrawn from the substance in order to keep its temperature constant when the pressure is increased *i.e.*, *the increase in pressure heats a body that expands on rise of temperature.*

On the other hand, if α is negative i.e., the substance contracts on heating, the $\left(\frac{\partial Q}{\partial P}\right)_T$ is positive, which means heat must be added to keep its temperature constant, when the pressure is increased *i.e., a cooling is produced when a substance, which contracts on heating, is suddenly compressed.*

EXERCISE

6.1 Explain thermodynamic system. What are thermodynamic variables? Define intensive and extensive variables. ***(Kerala 2002, Bangalore 2000, Nagpur 2004, 2005, 2008, Agra 2003)***

6.2 Define extensive and intensive variables. 'An extensive variable may become intensive variable, explain with example. ***(Nagpur 2007, 2005, Calcutta 2004)***

6.3 Derive Maxwell's thermodynamical general equations connecting the thermodynamic quantities.
(Meerut 2003; Garhwal 2001, 2005; Bangalore 2000; Kerala 2000, 2002; Utkal 2003; Nagpur 2009, 2007, Lucknow 2001; M.D.U. 2002; Madras 2005)

6.4 Show that for a homogeneous fluid.

$$C_P - C_V = T\left(\frac{\partial P}{\partial T}\right)_V \left(\frac{\partial V}{\partial T}\right)_P$$

where the symbols have their usual meaning. ***(Garhwal 2000, Madras 2004, Calcutta 2004, M.S. Uni. Tamil Nadu 2007, Purvanchal 2003, 2004)***

6.5 Use Maxwell's relations to obtain $C_P - C_V = R$ for an ideal gas where C_P and C_V are molar specific heats at constant pressure and constant volume respectively.
(Garhwal 2005; Lucknow 2004; Garhwal 2000, 2004)

6.6 Show that for a Van der Waals' gas

$$C_P - C_V = R\left(1 + \frac{2a}{VRT}\right)$$

(Kerala 2000, 2001, 2002; Purvanchal 2004, 2002)

6.7 Write down the Maxwell's four thermodynamic relations and prove the relation

$$\frac{E_S}{E_T} = \frac{C_P}{C_V}$$

(Rohilkhand 2000, Agra 2008, 2007)

6.8 Discuss the effect of change of pressure on melting point of a solid.

(Nagpur 2007, Rohilkhand 2001, Ranchi 2002)

6.9 Using Carnot's cycle derive Clausius-Clopeyron's latent heat equation

$$\frac{dP}{dT} = \frac{L}{T(V_2 - V_1)}$$

where T is the absolute temperature at which a body changes its state, L the latent heat and V_1 and V_2 the specific volumes in the first and second state respectively.

(Kerala 2000, 2001; Guwahati 2000, Meerut 2003, 2004)

6.10 Deduce and state first latent heat equation,

$$\frac{dP}{dT} = \frac{L}{T(V_2 - V_1)}$$

(Luck. U. 2001; Purvanchal 2005, Garhwal 2005; Meerut 2007; KU. 2002, 2001, 2000; P.U. 2008, 2007)

6.11 Explain Joule-Thomson effect using the Maxwell's thermodynamical relations.

(Nagpur 2005/w, Purvanchal 2003)

6.12 'There is no Joule-Thomson effect for a perfect gas' Explain. ***(Agra 2008, Madras 2002)***

6.13 Explain Joule-Thomson effect for Van der Waal's gas. Obtain the relation for temperature of inversion. ***(Kanpur 2001, Purvanchal 2003)***

6.14 Establish the Clausius-Clapeyron's equation

$$\frac{dP}{dT} = \frac{L}{T(V_2 - V_1)}$$

from Maxwell's thermodynamical relations and explain the effect of pressure on

(*a*) boiling point of a liquid, and

(*b*) melting point of a solid. ***(Agra 2006, 2004, Purvanchal 2007, 2003; Nagpur 2010, 2009, 2008, 2007, Patna 2005; Kanpur 1998; 2001; Ranchi 2004; Osmania 2004; Kerala 2000, 2001, 2002; Punjab 2001, 2002, 2005; Utkal 2003; Lucknow 2001, 2002; Meerut 2001, 2002, 2005; Bangalore 2002, 2004; Garhwal 2004)***

6.15 Prove that

$$\left(\frac{\partial T}{\partial P}\right)_S = \frac{TV\alpha}{C_P}$$

(Purvanchal 2000)

6.16 Prove the thermodynamic relation

$$\left(\frac{\partial S}{\partial V}\right)_T = \left(\frac{\partial P}{\partial T}\right)_V$$

(Kanpur 2001, Purvanchal 2007, Meerut 2005, Osmania 2002)

and $$\left(\frac{\partial S}{\partial P}\right)_T = \left(\frac{\partial P}{\partial T}\right)_P$$

(Purvanchal 2007)

6.17 Prove the thermodynamic relations

(*a*) $T.dS = C_V dT + T\left(\frac{\partial P}{\partial T}\right)_V dV$

(Calicut 2004, Garhwal 2004, Agra 2004, Purvanchal 2002, Gujarat 2004, 2005)

(b) $T.dS = C_P dT - T\left(\frac{\partial V}{\partial T}\right)_P dP$ ***(Garhwal 2000, 2001, Ranchi 2005, Gujarat 2004, 2005)***

6.18 What is the most important application of Joule-Kelvin effect? ***(Kanpur 2000; Utkal 2002)***

6.19 The internal energy of a perfect gas is only dependent on temperature whereas internal energy of Van der Waals' gas is a function of both temperature and volume. Why? ***(G.N.D.U. 2003)***

6.20 Prove Maxwell's second thermodynamic relation

$$\left(\frac{\partial S}{\partial V}\right)_T = \left(\frac{\partial P}{\partial T}\right)_V$$

from it, establish Clausius-Clapeyron relation

$$\left(\frac{\partial P}{\partial T}\right) = \frac{L}{T(V_2 - V_1)}$$

and find out the effect of pressure on boiling point of a liquid and melting point of a solid. ***(Rohilkhand 2001, Nagpur 2010, 2007, Meerut 2005)***

6.21 Derive Maxwell's four thermodynamical relations. Use one of these to obtain Clausius-Clapeyron's latent heat equation. ***(Purvanchal 2002, Meerut 2001)***

6.22 How internal energy of ideal gas differ from a real gas? ***(Punjab 2005, Calicut 2002)***

6.23 Give two statements of second law of thermodynamics. Prove the Clausius inequality $\sum_i \frac{Q_i}{T_i} \leq 0$.

Hence develop the concept of entropy. ***(Patna 2004, Mumbai 2001)***

6.24 Prove the thermodynamic relation

$$\left(\frac{\partial T}{\partial V}\right)_S = -\left(\frac{\partial P}{\partial S}\right)_V$$ ***(Calcutta 2003, Agra 2005, 2001, Meerut 2002)***

6.25 Derive an expression for the work done on a paramagnetic material filled in a current carrying solenoid and placed in an external field. ***(Mumbai 2005)***

6.26 Prove that $H = \left[\frac{\partial (G/T)}{\partial (1/T)}\right]_P$, where H is enthalpy and G is free energy. ***(Mumbai 2005)***

6.27 Derive Maxwell's four thermodynamical equations and hence find two $T.dS$ equations. ***(Purvanchal 2004)***

6.28 What do you mean by first order phase transition? Prove that $\left(\frac{dP}{dT}\right) = \frac{S_2 - S_1}{V_2 - V_1}$, where the terms have their usual meaning. ***(Magadh 2002, Purvanchal 2004, Nagpur 2001, 2003)***

6.29 Deduce Maxwell's four thermodynamical relations. Discuss the usefulness and importance of these relations. ***(Nagpur 2010, Lucknow 2003, Ranchi 2002, Patna 2002)***

6.30 Discuss about the first order phase transition and hence Clausius-Clapeyron's latent heat equation. Can latent heat become zero? If yes, under what condition ? ***(M.S. Uni. Tamil Nadu 2007, Lucknow 2003)***

6.31 Discuss about second order phase transition and hence derive Ehrenfest's theorem. ***(M.S. Uni. Tamil Nadu 2007)***

6.32 Prove the following thermodynamical relation

$$\left(\frac{\partial U}{\partial V}\right)_T = T\left(\frac{\partial P}{\partial T}\right)_V - P$$

Find also the value of $\left(\frac{\partial U}{\partial V}\right)_T$ for an ideal gas.

Also prove

$$\left(\frac{\partial C_V}{\partial V}\right)_T = T\left(\frac{\partial^2 P}{\partial T^2}\right)$$

(Agra 2002, M.S. Uni., Tamil Nadu 2007, Calcutta 2005)

6.33 Define thermodynamic potentials *U, F, H* and *G*. Derive their relationship with state variables. **(Pune 2010, Rohilkhand 2000, 2001; Kanpur 2000, Lucknow 2004)**

6.34 Obtain the values of thermodynamical variables *P, V* in terms of thermodynamic potentials. **(Kerala 2003)**

6.35 Define the Helmholtz function and for an isochoric process establish the relation

$$U = F - T\left(\frac{\partial F}{\partial T}\right)_V$$

(Agra 2008, Garhwal 2000, Sagar 2002)

6.36 Define the Gibbs potential function and show that for simultaneous isobaric and isothermal process, Gibbs free energy remains constant. **(Agra 2000)**

6.37 Deduce an expression for the change in temperature of wire thermodynamically when it is stretched adiabatically. **(Agra 2002, Punjab 2002, 2003, 2005; G.N.D.U. 2001; 2003)**

6.38 What do you mean by thermodynamic potentials? Give the physical significance. **(Garhwal 2001, Agra 2000)**

6.39 What do you mean by enthalpy *H* and Gibbs function G. Show that

$$\left(\frac{\partial H}{\partial T}\right)_P = C_P$$

[**Hint :** $dH = dU + P.dV + V.dP = T.dS - P.dV + P.dV + VdP = T.dS + VdP$

$$\therefore \left(\frac{\partial H}{\partial T}\right)_P = T\left(\frac{\partial S}{\partial T}\right)_P = \left(\frac{\partial Q}{\partial T}\right)_P = C_P]$$

(Rajasthan 2001, Kerala 2000)

6.40 Define the thermodynamic potential enthalpy and obtain Maxwell thermodynamic relation

$$\left(\frac{\partial T}{\partial P}\right)_S = \left(\frac{\partial V}{\partial S}\right)_P$$

(Kanpur 2002, Agra 2001)

6.41 What do you mean by order of phase. Distinguish between first and second order phase transitions. **(Purvanchal 2003, 2006)**

6.42 Starting from four thermodynamical potentials, derive Maxwell's thermodynamic relations. **(M.S. Uni. Tamil Nadu 2007, Pb. 2001; Kerala 2001, Garhwal 2000, Kanpur 2001)**

6.43 What do you understand by the thermodynamic potentials. What is the importance of these potentials? Deduce from thermodynamic potentials the corresponding Maxwell's thermodynamic relations. **(Purvanchal 2006, Rohilkhand 2002, Garhwal 2004, 2005; Ranchi 2005, Osmania 2005)**

6.44 Define Helmholtz function and Gibbs function and derive related Maxwell's thermodynamic relations. **(Garhwal 2000)**

6.45 Establish the following relation

$$C_P - C_V = -T\left(\frac{\partial V}{\partial T}\right)_P^2 \left(\frac{\partial P}{\partial V}\right)_T$$

(Calcutta 2004)

6.46 A measure of the result of a Joule-Kelvin expansion (throttling process) is provided by the Joule-Kelvin coefficient $\mu = \left(\frac{\partial T}{\partial P}\right)_H$. Derive the expression.

$$\mu = \frac{1}{C_P}\left[T\left(\frac{\partial V}{\partial T}\right)_P - V\right]$$

and show that for an ideal gas $\mu = 0$. Explain the existence of inversion temperature.

(*Kanpur 2000, Agra 2001*)

6.47 Deduce the expression for the Joule-Thomson coefficient

$$\mu = \left(\frac{\partial T}{\partial P}\right)_H = \frac{1}{C_P}\left[T\left(\frac{\partial V}{\partial T}\right)_P - V\right]$$

by using Maxwell's relations and show that for an ideal gas $\mu = 0$ and for a Van der Waals' gas

$$\mu = \frac{1}{C_P}\left(\frac{2a}{RT} - b\right)$$ **(*Nagpur 2007, Purvancnai 2003, G.N.D.U. 2001*)**

6.48 Explain enthalpy, Helmholtz and Gibbs functions in thermodynamics. Derive Gibbs Helmholtz equation. **(*Purvanchal 2002, Garhwal 2005, Patna 2005*)**

6.49 Heat is produced by adiabatic compression of a substance. Explain it by using Maxwell relations. **(*Punjab 2005, Kerala 2002*)**

6.50 Why does a rubber string heat up on stretching. **(*Punjab 2005, Garhwal 2000, 2001*)**

6.51 Show that for a system undergoing processes at constant pressure and constant temperature, the Gibbs free energy will be minimum at equilibrium. **(*Calcutta 2004, Agra 2001, 2003, 2008*)**

6.52 Establish the following relation between enthalpy (*H*), Helmholtz free energy (*F*) and Gibbs free energy (*G*),

$$G = H + T\left(\frac{dG}{dT}\right)_P$$ **(*Purvanchal 2007, 2006, 2003, Madras 2001*)**

6.53 Prove that enthalpy remains constant (isoenthalpic) in Joule-Kelvin expansion, *i.e.*, in a throttling process. **(*Nagur 2010, 2009, 2007, 2006, Purvanchal 2003, Lucknow 2003*)**

6.54 Give the difference between enthalpy and internal energy.

(*Gujarat 2005, 2004, G.N.D.U. 2001*)

6.55 Establish that for a system of constant volume and in contact with a heat reservoir, the equilibrium corresponds to the mimimum of Helmholtz free energy. **(*Calcutta 2005*)**

6.56 State Gibbs phase rule. Explain the rule with example. What is meant by a higher order phase transition?

(*Calcutta 2005*)

6.57 Derive the Gibbs-Helmholtz equation for e.m.f. of a reversible cell

$$E = H + T\left(\frac{dE}{dT}\right)$$

using *T–dS* equations. Examine the cases when,

$$(i)\ \frac{\partial E}{\partial T} = 0 \quad (ii)\ \frac{\partial E}{\partial T} > 0 \quad \text{and} \quad (iii)\ \frac{\partial E}{\partial T} < 0$$

6.58 Prove that for any substance $\left(\frac{\partial C_p}{\partial P}\right)_p = -T\left(\frac{\partial^2 V}{\partial T^2}\right)_p$. Hence show that for a perfect gas C_p does not vary with pressure when the temperature is kept constant. **(*M.D.U. 2001; P.U., 2001*)**

6.59 What is perfect differential? Give two examples each of perfect and imperfect differential.

(*G.N.D.U. 2006*)

6.60 Discuss the additive nature of entropy. Prove that the entropy of a combind system is equal to the sum of entropies of its sub-systems.

(*G.N.D.U. 2001, P.U. 2003, 2001; K.U. 2001; Pbi. U. 2007*)

6.61 (*a*) Show by mathematical analysis using Maxwell's equations that adiabatic expansion produces cooling and adiabatic compression produces heating. Obtain the expression for the change in temperature. ***(G.N.D.U. 2001, Pbi.U., 2001; P.U. 2005, 2002)***

(*b*) Using Maxwell's relation, prove that ratio of adiabatic to isothermal bulk modulus is equal to the ratio of specific heats, or $\dfrac{E_S}{E_T} = \dfrac{C_p}{C_v} = \gamma$

(Pbi. U. 2005; P.U. 2007, 2006, Meerut U. 2002; Luck. U. 2002, 2004)

6.62 (*a*) Define and explain four fundamental thermodynamic potentials; internal energy u, Helmholtz free energy F; Gibb's potential G; and enthalpy H. Explain what happens to Helmholtz free energy during isothermal isochoric process? Prove that in an isobaric process change in enthalphy is equal to the heat transferred during the process.

What is the significance of these functions and why they are called thermodynamic potentials? ***(Luck. U. 2001; Pbi U. 2008, 2000, M.D.U. 2007, 2000, K.U. 2001, 2000, Gharwal U. 2000; P.U. 2006, 2000, G.N.D.U. 2007, 2006)***

(*b*) Derive the relation (*i*) $C_p = -T\left(\dfrac{\partial^2 G}{\partial T^2}\right)_p$ ***(Luck. U., 2001)***

(*ii*) $F = u + \left(T\dfrac{dF}{dT}\right)_v$ ***(M.D.U. 2001, Gharwal U. 2000, Kan. U., 2002)***

(*iii*) $u = -T^2\left(\dfrac{\partial}{\partial T}\dfrac{F}{T}\right)_v$ ***(Cal. U., 2001)***

6.63 Show that for a homogeneous fluid

$C_p - C_v = T\left(\dfrac{\partial p}{\partial T}\right)_v\left(\dfrac{\partial v}{\partial T}\right)_p = \left[P + \left(\dfrac{\partial u}{\partial v}\right)_T\right]\left(\dfrac{\partial v}{\partial T}\right)_p$ and show that $C_p > C_v$ ***(Gharwal U. 2000)***

6.64 Derive the relation $\left(\dfrac{dQ}{dp}\right)_T = -Tv\alpha$. Show that heat is generated when a substance which expands on heating is compressed and for substances which contract on heating a cooling should take place. ***(M.D.U. 2001; Kan. U. 2000)***

6.65 (*a*) Prove the latent heat equation

$$\frac{dL}{dT} - \frac{L}{T} = (c_2 - c_1)$$

where $(c_2 - c_1)$ is the difference between the specific heat of a liquid and that of its saturated vapour. ***(M.D.U. 2008, 2001; Meerut U. 2009; K.U. 2000; Nag. U. 2003)***

(*b*) Define specific heat of saturated vapour and explain how it can be negative. What is the physical significance of negative specific heat. ***(Meerut U. 2007, 2003, 2001; M.D.U. 2001; K.U. 2000; Luck. U., 2002; Banaras U., 2003)***

6.66 Write short notes on

(*a*) Thermodynamic potentials. ***(Agra 2002)***

(*b*) Joule-Kelvin effect. ***(Rohilkhand 2001, Meerut 2003, Madras 2005, Purvanchal 2004)***

(*c*) Maxwell's thermodynamic relations. ***(Agra 2000)***

(*d*) Clausius-Clapeyron equation. ***(Rohilkhand 1995)***

(*e*) Specific heat of saturated vapour. ***(Meerut 2001)***

(*f*) First order phase transformation. ***(Magadh 2001)***

6.67 Calculate the pressure required to make ice freeze at –2°C. Change of specific volume when 1 gram of water freezes into ice = 0.091 cc. [**Ans.** *271.4 atmospheres*]

6.68 Calculate the change in temperature of boiling water when pressure is increased by 27.12 mm of Hg. The normal boiling point of water at atmospheric pressure is 100°C.
[**Ans.** = 1°C, 1K] ***(Kanpur 2000)***

6.69 Calculate the change in melting point of ice when it subjected to a pressure of 100 atmosphere. Density of ice = 0.917 g/cm^3, Latent heat of ice = 336 J/g.
[**Ans.** $dT = -0.7326$°C = – 0.7326 K] ***(Osmania 2001)***

6.70 One mole of an ideal gas at temperature T undergoes free expansion which doubles its volume. Show that change of entropy $\Delta S = Nk \log_e (2)$.

6.71 Calculate the temperature of inversion of *He*. Given: a = 0.0341 atm. lit^2/mole^3, b = 0.0237 litre/mole, R = 8.3 Joule/mole–K. [**Ans.** – 238 °C] ***(Mysore 2000)***

6.72 If sulphur has a specific gravity of 2.05 just before and 1.86 just after melting, the melting point being 115°C and the latent heat of fusion 9.3 calories/gm, find the alteration in melting point per atmospheric change of pressure. [**Ans.** 0.02485 °C]

6.73 Calculate the specific heat of saturated steam at 100 °C using the following data.
Specific heat of water at 100°C = 4.255 Kj/Kg and latent heat of vaporisation in J/kg at 90 °C, 100 °C and 110 °C respectively are 2.288 × 10^6, 2.268 × 10^6 and 2.238 × 10^6.
[**Ans.** –4.325 × 10^3 J/kg°C] ***(K.U. 2001, 2002)***

6.74 Calculate the specific heat of saturated vapour (steam) at 100 °C if the latent heat L varies with temperature T according to equation $L = 800 - 0.705\,T$, specific heat of water at 100°C = 1 cal/gm°C. [**Ans.** –1.145 cal/gm°C] ***(Meerut U. 2000; Luck. U. 2002)***

MULTIPLE CHOICE QUESTIONS

6.1 Four thermodynamic potentials are: ***(Garhwal 2005)***
(*a*) Pressure, volume, temperature and internal energy function.
(*b*) Pressure, volume, internal energy and Helmholtz function.
(*c*) Internal energy function, Helmholtz function, enthalpy and Gibbs function.
(*d*) None of these.

6.2 Specific heat of saturated vapur pressure: ***(Garhwal 2005)***
(*a*) zero (*b*) positive
(*c*) negative (*d*) sometimes positive sometimes negative.

6.3 From Maxwell's thermodynamic relations $\dfrac{E_S}{E_T} = ?$ ***(Agra 2004, Rohilkhand 2002)***
(*a*) 2 (*b*) $\dfrac{1}{\gamma}$
(*c*) γ (*d*) 2

6.4 The Clausius-Clapeyron equation is:
(*a*) $\dfrac{dP}{dT} = \dfrac{L}{T(V_2 - V_1)}$ (*b*) $\dfrac{dP}{dT} = \dfrac{L}{L(V_2 - V_1)}$
(*c*) $\dfrac{dP}{dT} = TL\,(V_2 - V_1)$ (*d*) $\dfrac{dT}{dP} = \dfrac{L}{T(V_2 - V_1)}$ ***(Rohilkhand 2002)***

6.5 Gibbs potential is defined as:

(*a*) $G = U - PV + TS$ (*b*) $G = U + PV + TS$

(*c*) $G = U - PV - TS$ (*d*) $G = U + PV - TS$ ***(Kanpur 2001)***

6.6 The value of $C_P - C_V$ for an ideal gas is:

(*a*) $TE\alpha V^2$ (*b*) $T^2 E\alpha V$

(*c*) $TE\alpha^2 V$ (*d*) $TE^2 \alpha^2 V^2$ ***(Rohilkhand 2002)***

6.7 Helmholtz free energy function is defined by:

(*a*) $F = U + TS$ (*b*) $F = U - TS$

(*c*) $F = U + PV$ (*d*) $F = U + PV - TS$ ***(Kanpur 2002)***

6.8 Hydrogen and helium at normal temperature show:

(*a*) Heating effect

(*b*) Cooling effect

(*c*) Sometimes cooling and some times heating

(*d*) None of these

6.9 Maxwell's thermodynamic relation is:

(*a*) $\left(\frac{\partial S}{\partial T}\right)_T = \left(\frac{\partial P}{\partial V}\right)_T$ (*b*) $\left(\frac{\partial T}{\partial V}\right)_S = -\left(\frac{\partial P}{\partial S}\right)_V$

(*c*) $\left(\frac{\partial T}{\partial P}\right)_P = \left(\frac{\partial V}{\partial P}\right)_P$ (*d*) None of the above ***(Rohilkhand 2003)***

6.10 Rice takes longest time to cook

(*a*) in submarine 100 m below the surface of the sea.

(*b*) at the sea level.

(*c*) at Simla.

(*d*) at Mount–Everest.

6.11. Paraffin wax contracts on solidification. The melting point of wax will

(*a*) increase with pressure (*b*) decrease with pressure

(*c*) no change with pressure (*d*) decrease linearly with pressure.

6.12. Total heat of a substance is known as

(*a*) internal energy (*b*) enthalpy

(*c*) entropy (*d*) thermal capacity ***(P.U. 2000)***

6.13. Maxwell's thermodynamic relations are based on

(*a*) two (*b*) three

(*c*) six (*d*) four thermodynamic potential

(H.P.U. 2000)

6.14. Which of the following is not a symbol for thermodynamical potential ?

(*a*) E (*b*) F

(*c*) G (*d*) H ***(P.U., 2002)***

6.15. When a substance which expands on heating is compressed

(*a*) heat is generated (*b*) heat is absorbed

(*c*) none of these

6.16. The co-efficient of volume expansion is

(*a*) $\frac{1}{v}\left(\frac{\partial v}{\partial T}\right)_p$ (*b*) $v\left(\frac{\partial v}{\partial T}\right)_p$

(*c*) $\frac{1}{v}\left(\frac{\partial T}{\partial v}\right)_p$ (*d*) $v\left(\frac{\partial T}{\partial v}\right)_p$ **(*H.P.U. 2001*)**

6.17. The variation of C_v with volume is zero for

(*a*) real gas (*b*) ideal gas

(*b*) both real and ideal

ANSWERS

6.1 (*c*)	**6.2** (*c*)	**6.3** (*c*)	**6.4** (*a*)	**6.5** (*d*)	**6.6** (*c*)
6.7 (*b*)	**6.8** (*a*)	**6.9** (*b*)	**6.10** (*d*)	**6.11** (*a*)	**6.12** (*b*)
6.13 (*d*)	**6.14** (*a*)	**6.15** (*a*)	**6.16** (*b*)	**6.17** (*c*)	

LIQUEFACTION OF GASES

7
Chapter

INTRODUCTION

It was believed since ancient time that nothing could be colder than ice and hence temperature of ice was considered as the zero (or starting point) of the scale used in temperature measurements in all the earlier systems. More over, it was also thought that air remains in the gaseous state permanently at all temperatures. Later on, Andrew's experiments on CO_2 gas in 1862 showed that below the critical temperature, a gas can be liquefied by mere application of pressure. The study of the laws of perfect gases categorically stated that the lowest conceivable temperature on centrigrades scale is – 273 °C, much below the temperature of ice (0°C) and now – 273 °C is universally taken as the zero of the absolute temperature scale, known as *absolute zero temperature*. Different methods are employed to reach the absolute zero temperature.

In 1877, Pictet liquefied oxygen by cascade process. Lande, was able to liquefy air in 1895 by Joule-Thomson effect. Using the principle of Joule-Thomson effect hydrogen and helium were also liquefied. The critical temperatures and boiling points of some important gases are given in Table 7.1. Moreover, a thermometer showing the significant temperatures of few gases are plotted by using the logarithmic thermodynamic scale (Fig. 7.1).

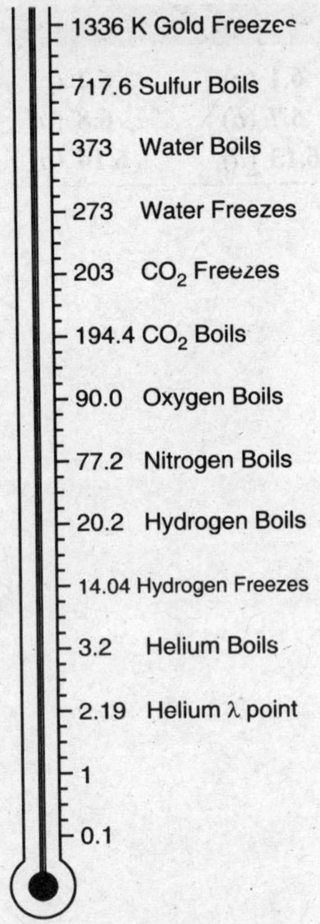

Fig.7.1 Thermometer showing temperatures on Kelvin scale.

Table 7.1

Gas	*Critical Temperature*	*Boiling Point*
CO_2	– 31.1 °C	– 78.6 °C
Oxygen	–118.8 °C	– 183 °C
Nitrogen	– 146 °C	– 195.8 °C
Hydrogen	– 240 °C	– 252.8 °C
Helium	– 267.8 °C	– 269.8 °C

7.1 DIFFERENT METHODS OF LIQUEFACTION OF GASES

The production of very low temperatures is essentially linked with the liquefaction of gases. The usual method is to first liquefy the suitable gas by purely mechanical process and then evaporating it. The latent heat of evaporation is obtained at the expense of the heat content of the liquid itself and an intense cooling results. The different methods employed to produce very low temperatures are:

(i) Method of freezing mixture.
(ii) Evaporation of liquid under reduced pressure.
(iii) Adiabatic expansion of a gas.
(iv) Joule-Thomson expansion.
(v) Regenerative cooling
(vi) Adiabatic demagnetisation of a paramagnetic salt.

The last method is useful in the region of liquid helium temperatures and is used in producing the lowest temperatures. Some of them are discussed below.

7.2 METHOD OF FREEZING MIXTURE

The easiest method for production of temperatures much lower than that of ice is adding a salt to ice. When a salt (NaCl say) is added to ice, the freezing mixture is formed. The salt is dissolved by the water adhering to ice and more ice melts. The heat necessary for this process *viz.* heat of solution and the latent heat necessary for melting ice is extracted from the mixture itself. As a result, the temperature of the mixture falls down. This is the principle involved in *freezing mixtures*. The process cannot go indefinitely. The lowest temperature that can be attained by the method of freezing mixture is definite and depends upon the salt used and is called the **eutectic temperature**. For different salts, the eutectic temperatures are as given in Table 7.2.

Table 7.2

Salt	*Amount of salt in 100 gm. of mixture*	*Eutectic temperature in °C*
$Mg\ SO_4$	19.0 gm.	– 3.9
KCl	19.7 gm.	– 11.1
$NH_4\ Cl$	18.6 gm.	– 15.8
$Na\ NO_3$	37.0 gm.	– 18.5
Na Cl	22.0 gm.	– 21.2
$Ca\ Cl_2$	29.8 gm.	– 55.0

The temperatures thus reached by the method of freezing mixtures are not low enough.

It was observed that substances which were gaseous at ordinary temperature could be converted into a liquid state if sufficiently cooled and simultaneously subjected to a high pressure. Faraday in 1825 attempted to liquefy certain gases and was successful in his attempt in liquefying ammonia, chlorine, carbon dioxide and sulphur dioxide. But attempts to liquefy O_2, N_2 and H_2 failed and thus these gases were called permanent gases.

Liquefaction of Chlorine. The apparatus employed is shown in Fig. 7.2. It consists of a strong *V* shaped glass tube closed at both ends. One end of the bent glass tube contains crystals of chlorine hydrate while the other end is immersed in a freezing mixture. On heating the end containing chlorine hydrate, chlorine is evolved in large quantities and a high pressure is developed. When cooling is caused by the freezing mixture it is liquefied into an oily liquid under its own pressure.

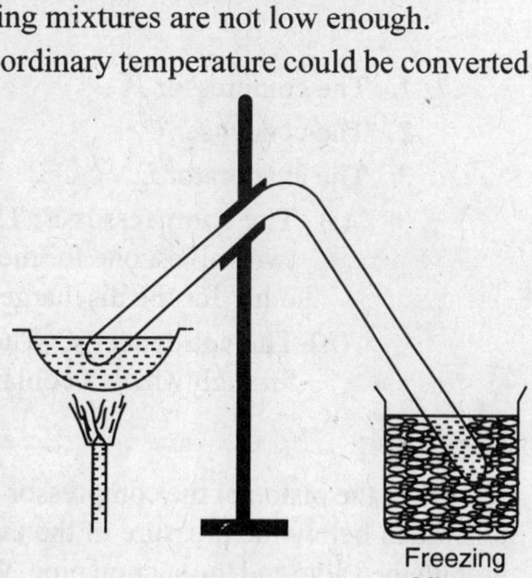

Fig. 7.2

7.3 COOLING BY EVAPORATION UNDER REDUCED PRESSURE

Whenever any liquid evaporates, some heat is required to convert it from liquid to gaseous state, which is equal to the latent heat of vaporisation. If the liquid is isolated, the latent heat of vaporisation is supplied at the expense of the total heat content of the liquid itself and hence an intensive cooling results. The amount of cooling depends upon the nature of liquid and the rate of evaporation. To produce very low temperatures generally liquefied gases, *e.g.*, ammonia, sulphur-dioxide etc. are used. Now-a-days, the contrivance utilising the process is employed in refrigerating machines used for industrial and domestic purpose. The machines used are:

(*i*) Vapour compression machine,

(*ii*) Vapour absorption machine.

Out of these two, the machines of first type are much popular as they have more efficiency and require less initial cost.

Vapour Compression Machine. Fig. 7.3 shows the schematic arrangement of vapour compression machine. It consists of the following main parts:

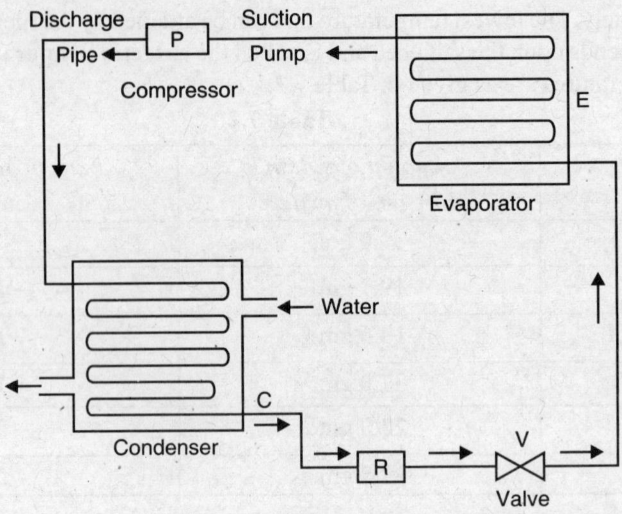

Fig. 7.3

1. The compressor *P*.
2. The condenser *C*.
3. The evaporator *E*.

(*i*) **The compressor *P*:** The electrically driven compressor *P* consists of a piston and two valves, one for the suction of low pressure vapour from the evaporator and the other for the discharge of the compressed vapour to the condenser.

(*ii*) **The condenser *C*:** The condenser is generally a copper coil embedded in a vessel through which a coolant like cold water or brine water is passed.

Working

When the piston of the compressor P moves upwards, the pressure in the cylinder of compressor falls much below the pressure in the evaporator. As a result, the low pressure vapour is sucked up through the valve and the suction pipe. When the piston moves downwards, the vapour is compressed and the compressed vapour is then delivered to the condenser *C* through the discharge valve and discharge pipe. The vapour at high pressure enters the condenser *C*, maintained at low temperature liquefies the vapour in condenser *C*. The liquid passes through the expansion or regulator valve *V*, which reduces the pressure of the liquid and delivers it to the evaporator E. Due to the low pressure, the liquid evaporates, thereby extracting its latent heat from the cold storage space. As a result, the temperature of cold storage space is cooled further. The low pressure vapour is again sucked in the compressor P and the cycle of operation continues.

The liquid used in the cycle of operation as refrigerant should have large latent heat of evaporation, smaller specific volume and sufficient vapour pressure at the temperature of the evaporator.

7.4 COOLING BY ADIABATIC EXPANSION

In adiabatic process, the working gas is perfectly insulated from the surroundings so that no heat is allowed to enter or leave the system. Under this condition, when the gas expands against some external pressure, the gas has to do external work at the expense of its internal energy. As the internal energy (U) of the gas is a function of its temperature, during adiabatic expansion of the gas, its internal energy and hence its temperature falls. If a highly compressed gas is suddenly allowed to expand adiabatically, it has to perform large amount of external work and an intense cooling results.

Theory

If P is the pressure and T the temperature of a perfect gas, then during adiabatic process

$$PV^{\gamma} = \text{constant} \quad \text{... (i)}$$

Here γ is the ratio of two specific heats ($\gamma = C_P/C_V$).

Using perfect gas equation, $PV = RT$, Equation (*i*) becomes

$$TP^{(1-\gamma)/\gamma} = \text{constant} \quad \text{... (ii)}$$

If P_1 and T_1 are the initial pressure and temperature of the given mass of gas, while P_2 and T_2 are its final pressure and temperature, we have

$$T_1 P_1^{(1-\gamma)/\gamma} = T_2 P_2^{(1-\gamma)/\gamma}$$

$$\therefore \quad T_2 = T_1\left(\frac{P_1}{P_2}\right)^{(1-\gamma)/\gamma} \quad \text{... (iii)}$$

From equation (*iii*), it is clear that a gas at initial ordinary temperature T_1 and compressed at a very high pressure P_1 is suddenly allowed to expand, the gas may be liquefied or even solidified. For example: if a CO_2 gas at room temperature compressed to a very light pressure (say 150 atmosphere) is suddenly released to atmospheric pressure, the final temperature T_2 reached is

$$T_2 = T_1\left(\frac{P_2}{P_1}\right)^{(\gamma-1)/\gamma}$$

$$= 300\left[\frac{1}{150}\right]^{(1.3-1)/1.3}$$

$$= 96 \text{ K}$$

The final temperature reached is so low that the expanding CO_2 gas gets even solidified, usually known as **dry ice**.

Gas Expansion Machine

A machine which uses above principle for cooling purpose is known as gas expansion machine and is schematically shown by Fig. 7.4. The machine consists of the following major parts:

1. Compressor P driven by an electric motor.
2. Water cooler C.
3. An expansion engine E.
4. A reservoir or cold storage space.

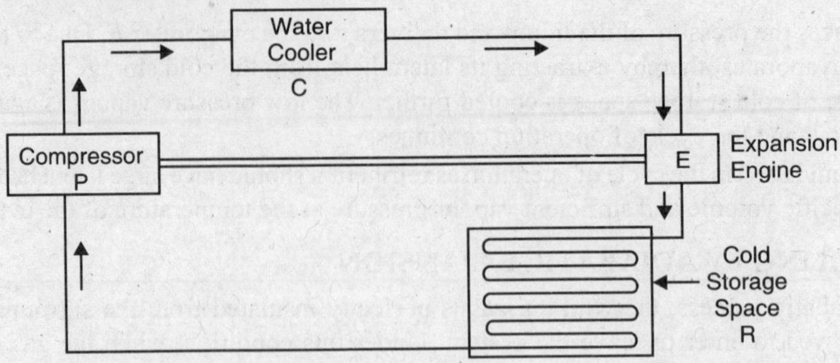

Fig. 7.4

Working

The working substance (*e.g.* gas or air) is first compressed in a compressor *P* adiabatically to a desired pressure. This compressed gas (or air) is passed through a water cooler *C* maintained at suitable temperature at constant pressure. The cooled gas is then delivered to expansion engine where it suffers adiabatic expansion to the pressure equal to that of the intake of the compressor. The gas gets cooled. This cold gas is further passed to the reservoir or the cold storage space where it absorbs heat at constant pressure and finally returns to the compressor *P* and a fresh cycle of operation begins.

7.5 JOULE-THOMSON EXPANSION

In 1852, Joule in collaboration with Lord Kelvin (Thomson) devised a very sensitive technique known as porous plug experiment. The principle involved is as under:

Principle

When a gas in allowed to escape adiabatically, through a porous plug, from the region of constant high pressure to the region at constant low pressure, it undergoes a *change* of temperature. This phenomenon is known as *Joule-Thomson effect* and the process is called the *Joule-Thomson* or *adiabatic throttling*.* The change of temperature may be positive or negative, depending upon the nature of gas and the initial temperature. This phenomenon is explained with theory in detail in Chapter 2. Let us summarize the results of the experiment:

Discussion of the results: After a series of experiments the following results were obtained by Joule and Thomson.

(*i*) In all gases except hydrogen a slight cooling effect was observed. In the case of hydrogen a slight heating effect was observed.

The magnitude of the effect can be judged from the following results. At 0°C for air a *cooling* of 0.275°C per atmosphere, for carbon dioxide a cooling of 1.39°C per atmosphere and for **hydrogen** a heating of 0.03°C per atmosphere was observed.

(*ii*) The fall in temperature was found to be proportional to the difference of pressure on the two sides of the porous plug. The greater the difference of pressure the higher was the temperature difference.

(*iii*) The fall in temperature for the given difference of pressure decreases with rise in the initial temperature of the gas. It was found that the cooling effect decreased with the increase of initial temperature and became zero at a certain temperature. The temperature at which the cooling effect becomes zero is called the **temperature of inversion.** If the gas is above this temperaute it shows a heating effect instead of cooling.

* **Throttling process.** In Joule Thomson experiment the porous plug provides a large number of very narrow paths for the gas molecules. While crossing these narrow orifices under high pressure the gas molecules are pulled apart or are *wire drawn.* This phenomenon is known as *throttling.*

For example, air shows a cooling effect of 0.152 °C per atmosphere at 93° C but 0.263°C per atmosphere at 1°C. Carbon dioxide shows a fall of 0.645 °C per atmosphere at 96°C and 1.23°C per atmosphere at 8°C. Hydrogen which shows a heating effect of 0.039°C per atmosphere at 30°C gives a heating effect of 0.004°C per atmosphere at –71°C. The heating effect is zero at –80°C which, therefore, is the temperature of inversion for hydrogen. Below this temperature hydrogen shows a cooling effect.

Conclusion: Any gas below its temperature of inversion will cool on suffering Joule-Thomson expansion.

The fall in temperature (Refer chapter 2) dT for a gas obeying Van der Waals' eqation of state is given by

$$dT = \frac{1}{C_P}\left(\frac{2a}{RT} - b\right)dP \quad \text{... (i)}$$

where dP is the difference of pressure on the two sides of the porous plug. From equation (*i*), we infer that

(*a*) dT will be positive, if $\frac{2a}{RT} > b$

or $$T < \frac{2a}{Rb} \quad \text{... (ii)}$$

Hence, cooling will result if the initial temperature of the gas is less than $\frac{2a}{Rb}$.

(*b*) dT will be zero, if $\frac{2a}{Rb} - b = 0$ *i.e.*, $T = \frac{2a}{Rb}$

This is called temperature of inversion and is represented by T_i. Thus,

$$\therefore \quad T_i = \frac{2a}{Rb} \quad \text{... (iii)}$$

(*c*) dT will be negative if $\left(\frac{2a}{Rb} - b\right)$ is negative.

or $$T > \frac{2a}{Rb}$$

or $$T > T_i \quad \text{... (iv)}$$

Hence, heating will take place if the temperature of the gas is greater than its temperature of inversion.

Inversion Curve: In order to plot first isenthalpic curve and then inversion curve, let us consider a series of Joule-Thomson experiments are performed on the same gas.

Let P_1 and T_1 be the initial pressure and temperature (arbitrarily chosen) on the high pressure side of the porous plug and P_2 be the pressure ($P_2 < P_1$) on the other side of the plug which is taken to be constant. The temperature of escaping gas after throttling is measured, let it be T_2. Now pressure P_2 is varied (keeping P_1, T_1 constant) on the other side and every time temperature T_2 in noted. If now a graph is plotted between various values of P_2 and the corresponding T_2, a curve is obtained as shown in Fig. 7.5. Since in Joule-Thomson expansion, the initial and final enthalpies are equal, the enthalpies at all the points on the curve, *i.e.*, *a, b, c, d, e, f* etc. is the same. Such a curve with constant enthalpy is called **isenthalpic curve**.

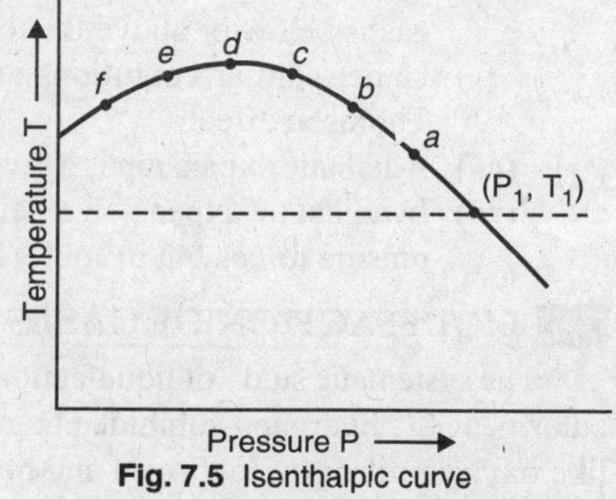

Fig. 7.5 Isenthalpic curve

Keeping Pressure P_1 constant, the temperature T_1 on higher pressure side is changed. Again pressure P_2 is varied and every time the temperature T_2 is noted. The experiment is repeated for various constant values of T_1. A family of curves corresponding to different values of enthalpy H are obtained as shown in Fig. 7.6. Every curve possesses a maximum temperature (inversion temperature). The locus of such inversion points is called the **inversion curve.** The dotted curve in Fig. 7.6 represents the inversion curve. The region inside the inversion curve represents the *cooling region* (in which gas shows *cooling effect*) and region out side the inversion curve represents the *region of heating* (in which gas shows *heating effect*).

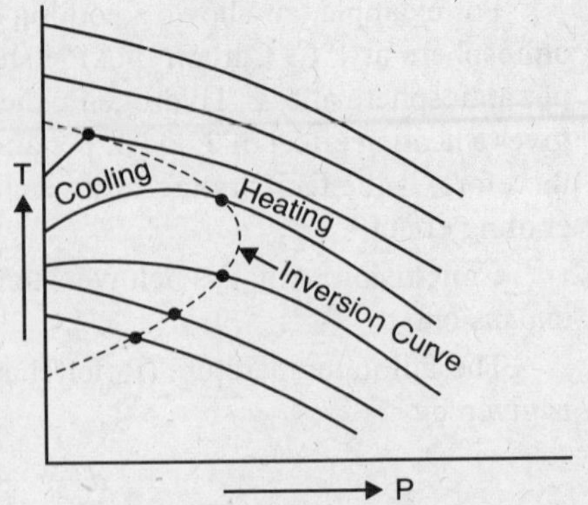

Fig. 7.6 Inversion curve

It can be noted that at any given pressure less than the extreme value there are two inversion temperatures. One is called the 'ordinary' inversion temperature (higher T_i) at which the heating effect gives place to cooling and the other (lower T_i) inversion temperature at which cooling gives place again to heating. The ratio of these two inversion temperatures for a Van der Waals' gas is approximately 9. Thus, in order to produce cooling effect the gas must be precooled first below the inversion temperature. For most of the gases, ordinary temperature is already below the maximum inversion temperatures, so precooling is not necessary. However, for hydrogen and helium, even the maximum inversion temperature are much below the ordinary temperatures, hence at ordinary temperatures, they show heating effect. These gases must, therefore, be precooled up to – 80 °C and – 243 °C respectively, then they will also show a cooling effect.

Note that the **inversion temperature is not constant** for a given gas but varies with the initial pressure of the gas.

Difference between Adiabatic Expansion and Joule-Thomson effect :

Adiabtic exapnsion differs from Joule -Thomson effect in the following respects:

(i) In adiabatic expansion *entropy* remains constant. This is why adiabtic expansion is called *isoentropic* expansion.
In Joule -Thomson effect *enthalpy* of the gas remains constant. This is why Joule-Thomson effect is known as *isoenthalpic expansion.*

(ii) In adiabatic epxpansion work is done by the gas but in Joule -Thomson expansion work is done *by* the gas or *on* the gas.

(iii) In adiabatic expansion cooling is produced even in case of a perfect gas. For a perfect gas Joule Thomson effect is zero *i.e.* no cooling is produced due to Joule-Thomson effect in a perfect gas.

(iv) In adiabatic expansion a *real* gas always shows cooling effect. In Joule-Thomson effect cooling or heating is produced depending upon whether the initial temperature of the real gas is below or above the temperature of inversion.

(v) Under similar conditions adiabatic expansion shows much larger cooling than Joule-Thomson effect.

(vii) Adiabatic expansion is a reversible process whereas Joule-Thomson effect is irreversible.

(vii) In adiabatic expansion external work is done by the gas which may be used to overcome outside forces but in Joule-Thomson effect work done by the gas is only used internally.

7.6 LIQUEFACTION OF GASES

The systematic study of liquefaction of gases started in 1823 when Faraday successfully liquefied chlorine, CO_2, hydrogen sulphide etc. only by cooling them and applying pressure. However-gases like oxygen, nitrogen, hydrogen and helium etc. could not be liquefied even by applying very high pressure. Hence, they were termed as **permanent gases.** Andrew's experiment on CO_2 in 1862 showed that a gas must be cooled below its critical temperature before being liquefied under pressure. Real success was obtained when Joule-Thomson effect is coupled with regenerative cooling.

Essential conditions for liquefication of gases. The most essential conditions for liquefaction of gases are as under.

(*i*) For liquefaction of a gas by pressure alone it must be cooled to a temperature below the *critical temperature.*

(*ii*) For liquefaction of a gas by Joule-Thomson expansion it must be cooled to a temperature below the *temperature of inversion.*

7.7 PRINCIPLE OF REGENERATIVE COOLING

The cooling produced in Joule-Thomson expansion of a gas depends on the difference of pressure on the two sides of the porous plug and the initial temperature. For most of the gases, the Joule-Thomson cooling is very small. For example: at a temperature of 20°C when the pressure on the two sides are 50 atmospheres and 1 atmosphere respectively, the air cools only by 11.7°C. In order to increase the cooling, the pressure difference was increased; but the cooling did not increase very much. For example, at a temperature of 20°C when the pressure on two sides were kept 210 atmospheres and 1 atmosphere, the air was cooled only by 42°C. However, the cooling effect can be intensified by employing the process called the regenerative cooling.

Principle: In this process a portion of the gas which suffers Joule-Thomson expansion and becomes cooled is employed to cool other portions of the incoming gas before the later reaches the nozzle to suffer Joule-Thomson expansion. After suffering the Joule-Thomson expansion at the nozzle, the incoming gas becomes further cooled. By continuous repetition of this process the temperature of the gas coming out of the nozzle falls continuously till the gas starts liquefying on the low pressure side. The commutative process used to cool the gas continuously is called the regenerative cooling.

Experimental Arrangement: The schematic arrangement of regenerative principle is shown in Fig. 7.7. The gas is compressed to a high pressure in the compressor. The heat of compression evolved is removed by passing the gas through spiral contained in the water cooled jacket A. The gas next enters the inner tube of regenerator R and suffers Joule-Thomson expansion at the nozzle V so that it cools by a small amount. This cooled gas returns by the outer tube and absorbs heat from the incoming high pressure gas before it reaches the compressor at the same temperature as the incoming gas at R. The gas is again compressed, cooled by A and re-enters at R. As time passes, the gas approaching V becomes more and more cooled till the Joule-Thomson cooling at V is sufficient to liquefy the gas. A portion of the escaping gas then condenses inside the Dewar flask F. At this stage, the temperature of the whole apparatus becomes steady and the temperature of the low pressure gas in R is less than the temperature of the adjacent high pressure gas at every point of the inner tube. Thus, the low pressure gas extracts heat from the incoming gas.

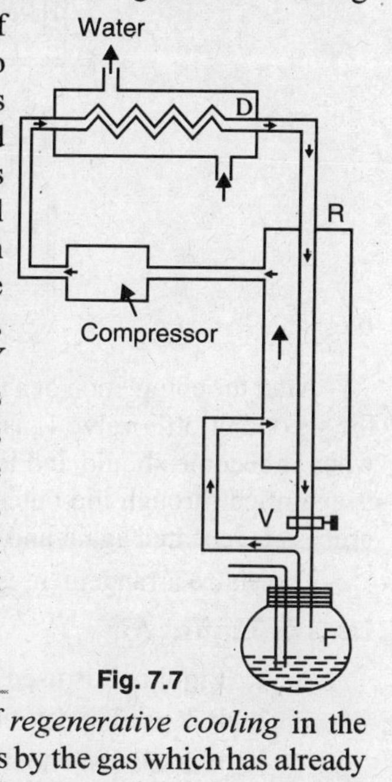

Fig. 7.7

7.8 LIQUEFACTION OF AIR-LINDE'S PROCESS

Linde in 1895 used *Joule-Thomson effect* and the *principle of regenerative cooling* in the liquefaction of air. The principle involved in is cooling the incoming gas by the gas which has already undergone cooling due to Joule-Thomson effect. By maintaining a continuous cycle of operation, the initial temperature can be progressively lowered and hence the cooling effect is enormously enhanced.

The apparatus employed by Linde for liquefaction of air is schematically shown in Fig. 7.8.

It consists of two-stage compressors P_1 and P_2. The compressor P_1 compresses the incoming air to a pressure of 10 atmospheres.

The compressed air is then cooled by passing through the water bath. It is then led through a caustic soda bath to free from carbon dioxide. If this is not done CO_2 will solidify and choke the valves.

The gas then enters the compressor P_2 where it is compressed to a pressure of about 200 atmospheres. The compressed air is then passed through the cooling coil kept immersed in a freezing mixture (cooling bath) so that the gas attains a temperature of about –20°C. The cooled air under a pressure of 200 atmospheres is passed through the inner pipe and suffers Joule-Thomson expansion at a throttle valve V_1 which can be operated by means of a handle. In the course of this expansion the pressure is suddenly lowered to 20 atmospheres and this causes the temperature to fall to about –78°C. The cooled air is then made to flow through the outer tube cooling the air flowing through the inner pipe and finally returning back to the compressor P_2 where it is again compressed to a pressure of 200 atmosphere. It is again circulated through the inner pipe.

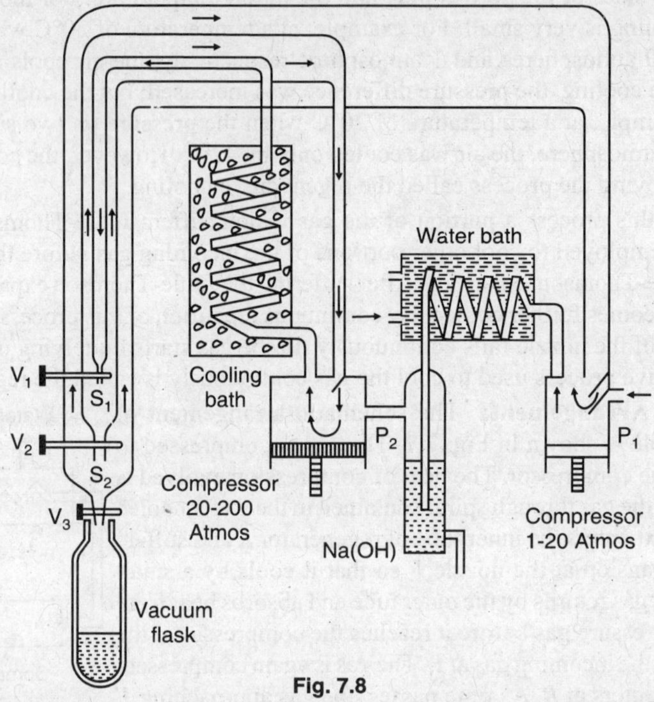

Fig. 7.8

After the completion of a few cycles, the temperature of the incoming gas falls sufficiently low and the second throttle valve V_2 is opened. The air is allowed to expand to a pressure of one atmosphere when it becomes liquid and is collected in a Dewar flask through the valve V_3. The unliquefied gas is again led through the outermost tube back to compressor P_1 as shown by the arrow heads. The process is repeated again and again.

The entire arrangement is enclosed in cotton wool to avoid any conduction or radiation.

Uses of Liquid Air

1. Liquid air is used to cool the substances to very low temperatures.
2. It is used in developing high vacuum by solidifying the gas less volatile than air.
3. It is used in chemical industries for separation and purification of gases.

Constituents of liquid air (oxygen, nitrogen, argon, neon etc.) can be easily seperated and are used commercially as:

(*i*) Oxygen is used in metal cutting such as breaking up of old ships and other machines. It is also used in breathing apparatus in aeroplanes, mines and diving suits.

(*ii*) Nitrogen is used in the manufacture of ammonia and fertilizers.

(*iii*) Argon is used in gas filled electric lamps.

(*iv*) Neon is used in discharge tubes required for illumination and advertising sign boards.

7.9 PRINCIPLE OF CASCADED COOLING–LIQUEFACTION OF OXYGEN

The principle of cascaded cooling can be well understood by considering process employed in liquefaction of oxygen. It was found that gases like oxygen, nitrogen, hydrogen, carbon monoxide, etc. even when subjected to a pressure of 3,000 atmospheres, could not be liquefied. The cause of thse failures was explained by Andrews' discovery of critical temperature for various gases. It was shown that a gas when above a certain temperature (called critical temperature) cannot be liquefied, however high a pressure it may be subjected to. It was found that the critical temperature of oxygen is –118°C. Oxygen, therefore, nust be cooled to a temperaure below this particular temperature if it was to be liquefied. The apparatus used by Kamerlingh-Onnes is shown in Fig. 7.9

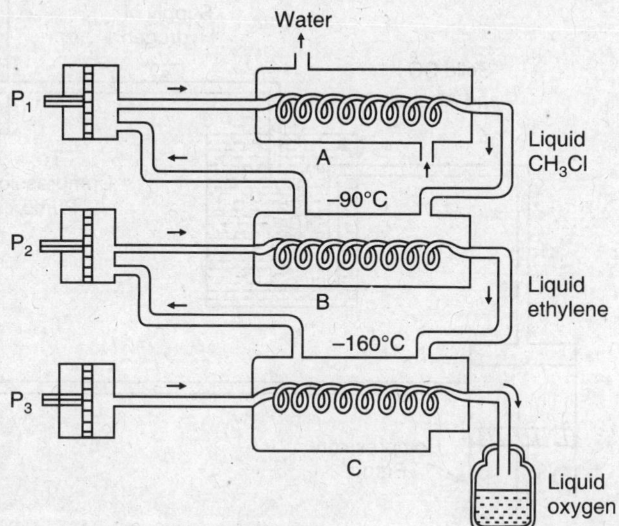

Fig. 7.9

It consists of three compressor machines *A,B* and *C* worked by three pumps P_1, P_2 and P_3 respectively. Machine *A* is used to liquefy methyl chloride (CH_3Cl) whose critical temperature is 143°C. Hence it can be liquefied by a pressure of a few atmospheres at the room temperature. The gas is brought to room temperature by making it flow through a spiral surrounded by water jacket.

Liquid methyl chloride is made to flow round a spiral in *B* and finally is connected to the suction side of the compressor P_1. Liquid methyl chloride evaporates in *B* under reduced pressure and this results in a fall of temperature to about –90°C in this chamber.

Inside the jacket *B* is placed a condenser coil through which ethylene passes from the compressor P_2 and is cooled to –90°C. It is thus liquefied as its critical temperature is 9.5°C. Liquid ethylene is then made to flow round the spiral in *C* and is allowed to evaporate under reduced pressure by connecting it to the suction side of the compressor P_2. The temperature falls to about –160°C which is much below the critical temperature of oxygen.

Through the spiral in *C* oxygen under pressure passes, which gets liquefied and is collected in the vacuum flask. The lowest temperature attained by boiling liquid oxygen under reduced pressure is –218°C.

7.10 LIQUEFACTION OF HYDROGEN

Hydrogen cannot be liquefied by cascade process because its critical temperature is – 240°C. Linde's ordinary apparatus used for the liquefaction of air cannot be used for hydrogen because the temperature of inversion for hydrogen is – 83°C. The gas must initially be cooled to a temperature lower than the temperature of inversion for the cooling to take place due to Joule - Kelvin effect. The original apparatus designed by Dewar (1898) was improved later by Travers, Olszewski, Nernst and others (Fig. 7.10). To have complete insulation, the whole apparatus in enclosed in an outer Dewar flask *L*.

Hydrogen under a pressure of 200 atmospheres is passed through a coil immersed in solid CO_2 and alcohol. It enters the coil in the chamber *A* where it is further cooled by the outgoing hydrogen.

The chamber B contains liquid air and cools hydrogen in the coil E. In the chamber C, liquid air is allowed to boil under reduced pressure (10 cm of Hg) and hydrogen in the coil F is cooled to a temperature of – 200°C. This cooled hydrogen passes through the regenerative coil G and nozzle N. Hydrogen is cooled further due to Joule-Kelivin effect. The cooled hydrogen coming from the nozzle N is allowed to circulate back to the pump as shown in Fig 7.10. The process of regenerative cooling continues and after some time hydrogen gets *liquefied* and is collected in the Dewar flask D.

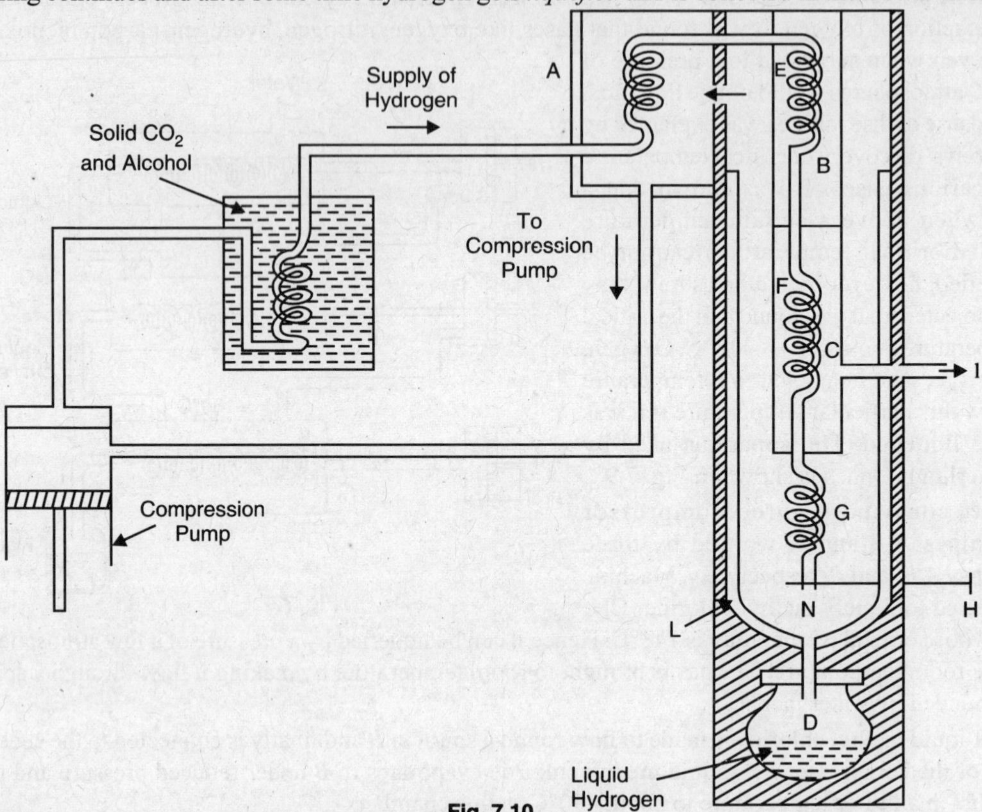

Fig. 7.10

7.11 LIQUEFACTION OF HELIUM (K. ONNES METHOD)

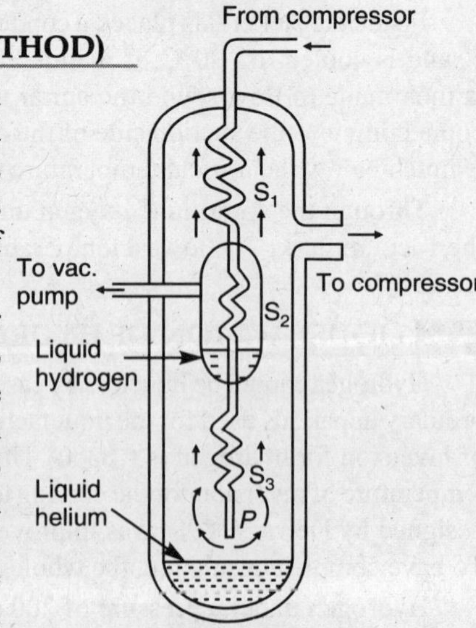

Fig. 7.11

Earlier attempts to liquefy Helium gas were unsuccessful because critical temperature for Helium is 5.3 K and temperature of inversion –240°C. Both these temperatures wer too low for earlier workers in the field. It was K. Onnes who was able to liquefy Helium in 1908 by the process of regenerative cooling.

The inversion temperature of helium is 35K and for practical reasons the gas must be cooled to 17K in order to obtain cooling by the throttle expansion. This temperature was obtained by Onnes by boiling hydrogen under reduced pressure as shown in Fig.7.11. Helium at a pressure of 36 atmospheres is passed through a spiral S_1, S_2, S_3, out of which S_2 is immersed in hydrogen boiling under reduced pressure. This helium cooled to below 17 K is allowed to undergo Joule Thomson expansion at the porous plug P. Helium, therefore, further cools down. This helium circulates round the spiral, thereby further cooling the gas inside it. By this process of *regenerative cooling,* the temperature of helium goes on falling till it liquefies and collects at the bottom of the Dewar flask.

In 1926 Keesum succeeded in obtaining solid helium. It was found that helium solidifies at 4.2 K under a pressure of 140 atmospheres and at a temperature of 1.1. K under a pressure of 2.3 atmosphers.

7.12 HELIUM I AND HELIUM II

Liquid Helium above a temperature of 2.19 K is called Helium I. Liquid He I behaves in a normal way. Below 2.19 K liquid Helium is known as He II. He II has abnormal properties

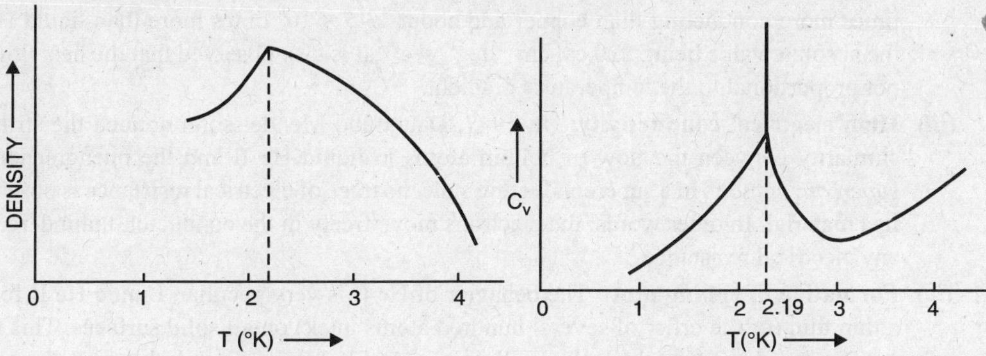

Fig. 7.12(a) Fig. 7.12(b)

The specific heat of liquid helium increases up to 2.19 K and *at this temperature there is a sudden* and abnormal increase in the specific heat. Beyond 2.19 K, the specific heat first decreases and then increases Fig. 7.12. (b)

The specific heat-temperature graph resembles λ and hence this temperature at which the specific heat (C_v) changes abruptly (2.19 K) is called the λ **point**.

Liquid helium above 2.19 K is called helium I because it behaves in a normal way and below 2.19 K, it is called helium II because of its abnormal properties.

No heat is evolved or absorbed during the transition from Helium I to Helium II and vice-versa. In other words, no latent heat is involved during this transition. Hence the following conclusions can be drawn.

Conclusions:

(*i*) The entropy is constant across the curve *i.e.*, entropy of He II is practically the same as that of He I.

(*ii*) The density of both the types of liquid is about the same *i.e.*, there is no change of density during transition from one form of Helium to another.

While viscosity of liquid He I decreases with decrease in temperature and this property is contrary to the property of a liquid but it resembles that of a gas. The viscosity of He II is almost zero and it can flow rapidly through narrow capillary tubes. Liquid He I is a normal liquid while He II presents a very anomalous behaviour. The thermal conductivity of He II is abnormally high, its internal friction is practically zero and that when it is forced through a capillary tube, the emerging liquid He II cools while that which remains behind warm up.

7.13 SOME PECULIAR PROPERTIES OF HELIUM II

(*i*) **Superfluidity:** One of the most remarkable properties of He II is its abnormally low viscosity, even less than that of hydrogen gas and its rate of flow through capillaries is independent of the pressure difference across their two ends. According to Poiseuille's law of liquid flow, the rate of flow of a liquid through a capillary tube is inversely proportional to its viscosity. In case of He II, it is found that at λ point ($T = 2.19$ K) the rate of flow increases abruptly and below it the flow is extraordinarily large. Thus, viscosity of He II is practically zero and can flow rapidly, almost without resistance, through narrow tubes. Hence, the liquid He II is called **superfluid**, a new frictionless state of matter. This property associated with He II is known as **superfluidity**.

(ii) **High heat conductivity:** Liquid He I has a quite normal heat conductivity, of the order of 10^{-4}, which is of the same order for gases at ordinary temperatures. But He II is found to have an extraordinarily high coefficient of thermal conductivity. He II is nearly 800 times more conducting than copper and about 13.5×10^6 times more than liquid He I, the absolute value being 820 cal cm^{-1} deg^{-1} sec^{-1}. It is also observed that the heat flow in not proportional to the temperature gradient.

(iii) **High electrical conductivity:** In 1947, Daunt and Mendelssohn noticed the striking similarity between the flow of helium atoms in liquid He II and the phenomenon of *superconductivity*. In a superconducting state, no trace of electrical resistance is observed in a material. In other words, the electrons move freely in the conductor, unhindered by any electrical resistance.

(iv) **Formation of rolling film:** The behavior of He II is very peculiar. Liquid He II forms a thin film (of the order of several hundred atoms thick) on all solid surfaces. This film is called **rolling film** (also called rolling-sim on film). It is through this film that He II flows from one vessel to another. If a tube containing He II be placed in a He II bath, it is observed that:

(*a*) If the liquid level inside the tube is lower than that out side, the liquid helium II from outside starts creeping in to the tube along its surface in the direction of arrows shown in Fig. 7.13 (*a*). The process of creeping continues till both the inside and outside levels of the tube become equal.

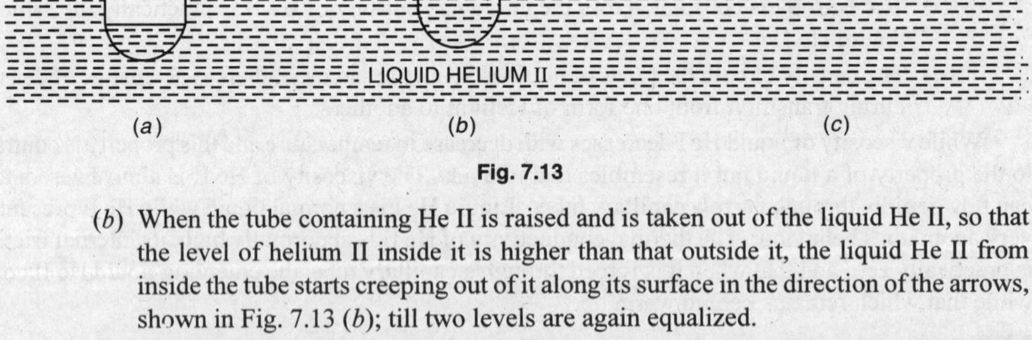

Fig. 7.13

(*b*) When the tube containing He II is raised and is taken out of the liquid He II, so that the level of helium II inside it is higher than that outside it, the liquid He II from inside the tube starts creeping out of it along its surface in the direction of the arrows, shown in Fig. 7.13 (*b*); till two levels are again equalized.

(*c*) If now the tube containing liquid He II is lifted entirely out of the liquid He II bath, the He II inside the tube now creeps out and flows down along the surface of the tube, collects at its bottom in the form of drops and falls into the liquid bath. It continues to flow out until the tube is empty Fig. 7.13(*c*).

(*v*) **The 'Fountain, Effect:** Liquid He II exhibits a peculiar property in the form of fountain effect as shown in Fig. 7.14. Heat radiation falling locally on a system containing liquid He II produces a pressure difference. The apparatus was first suggested, in 1938, by Allen and Misener. The apparatus consists of liquid He II taken in a Dewar flask. Liquid He II is also taken in a tube *AB* open at both the ends such that part A is in the form of capillary and is kept projecting out of the liquid He II bath. The lower part B has a small orifice O at the bottom of the tube. A fine *emery* powder is packed in part *B* of the tube. The interspacing between emery fine particles provide fine capillaries for the liquid He II to flow in the tube *AB*. When heat radiation from a light torch fall on the powder, liquid He II is found to spurt out of the capillary tube in the form of fountain which is even of the height 30 cm. This phenomenon is due to the absorption of heat energy by the fine emery powder and the capillary action of the fine emery particles.

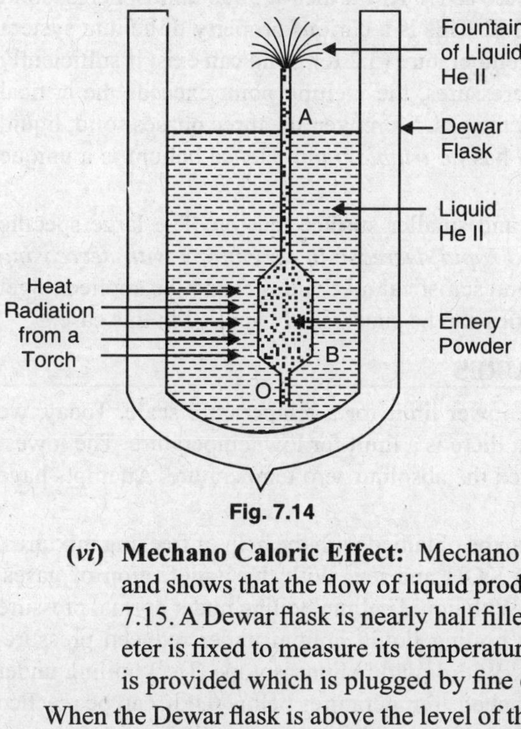

Fig. 7.14

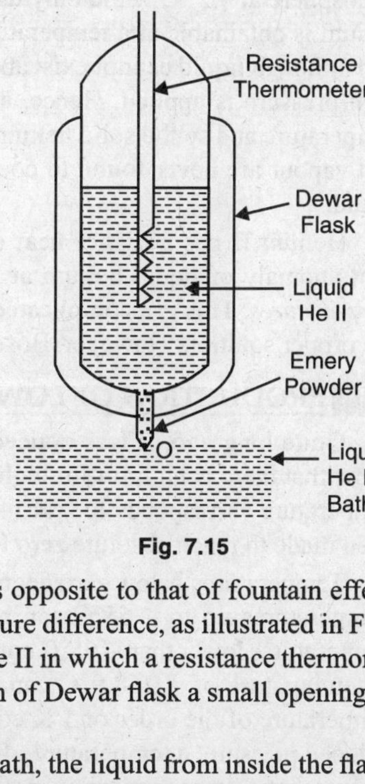

Fig. 7.15

(*vi*) **Mechano Caloric Effect:** Mechano caloric effect is opposite to that of fountain effect and shows that the flow of liquid produces a temperature difference, as illustrated in Fig. 7.15. A Dewar flask is nearly half filled with liquid He II in which a resistance thermometer is fixed to measure its temperature. At the bottom of Dewar flask a small opening O is provided which is plugged by fine emery powder.

When the Dewar flask is above the level of the liquid He II bath, the liquid from inside the flask drains out through the powder and its temperature is found to increase by about 0.1 °C. On the other hand, if the Dewar flasked is well dipped in the bath and the liquid He II is allowed to flow into the Dewar flask through the opening O via. the fine powder, the temperature inside the Dewar flask is observed to decrease.

Reasoning for Peculiar Properties of Liquid He II

The explanation for the peculiar properties mentioned above of liquid helium II is provided by stating that the liquid He II is generally a mixture of two different kinds of liquids. One liquid is composed of normal atoms with normal viscosity while the other one is composed of those atoms which are in the state of lowest energy (known as zero point energy). Hence its entropy is zero. Therefore, this part is capable of moving through the normal atoms without friction or viscosity. According to recent concepts, this part of liquid consists of 'condensed' atoms and appears at its λ point for the first time. This liquid is called a 'superfluid' which can flow through capillary tubes without showing any resistance. Such a fluid property is known as superfluidity. On lowering the temperature, the superfluid grows in quantity at the expense of normal component, untill at absolute zero, the whole liquid becomes superfluid.

7.14 HELIUM 'A UNIQUE LIQUID'

Helium was the last gas to be liquefied on account of having the lowest critical temperature (– 268.8° C) of all known gases. It is a colourless, transparent, very volatile liquid and has the lowest boiling point of 4.2 K at a pressure of 1 atmosphere. It has no normal freezing point and shows some of the peculiar properties (see Fig. 7.13) which single it out as almost an exceptional and unique substance.

An amazing property of helium system is that solid helium cannot be obtained merely by lowering the temperature of helium liquid. Kamerlingh Onnes failed to solidify helium despite the fact that he reached a temperature of 0.84 K. Solid helium was first obtained by Keesom at a pressure of 250 atmosphere at 4.2 K, while only under 23 atmosphere at 1.1 K. Further at 5800 atmospheres, solid helium is obtainable at a temperature as high as 42 K. This is a curious property of helium system that although liquid cannot exist above the critical temperature (5.2 K), solid can exist if sufficiently high pressure is applied. Hence, at high enough pressures, the melting point exceeds the critical temperature and so the solid helium 'melts' to form the gas. Moreover, the three phases solid, liquid and vapour are never found to coexist *i.e.*, helium has *no tripple point*. Hence helium is a unique liquid.

Helium II has a higher heat of vaporization and smaller surface tension. The large specific heat anomaly of liquid helium at 2.19 K is due to *rapid decrease in its entropy with decreasing temperatures*. This is a complicated phenomenon and statistical mechanics cannot be applied to get the proper solution. However, Bose-Einstein statistics can be successfully applied in this case.

7.15 PRODUCTION OF LOW TEMPERATURES

Quite long ago, it was noticed that there is a lower limit for a temperature scale. Today, we know that there is no limit for high temperature but there is a limit for low temperature. The lowest temperature corresponds to 0°K (– 273.16°C) called the absolute zero temperature. Attempts have been made to reach absolute zero temperature.

Temperatures below zero degree centrigrade can be obtained with the help of freezing mixtures. Temperatures up to – 65°C can be obtained with KOH and ice. With the liquefaction of gases, temperatures lower than – 65°C could be achieved. With liquid helium boiling under normal pressure at temperature of – 268.9°C can be reached. By boiling liquid helium under reduced pressure, temperature of the order of 1 K could be obtained. With liquid helium isotope (He^3) boiling under reduced pressure, a temperature of 0.4 K can be reached. Temperatures below 0.4 K can be reached by 'adiabatic demagnetisation methods' as per Debye (1926) and Giauque (1927).

7.16 ADIABATIC DEMAGNETISATION

Debye, Giauque and Macdougall were able to produce temperatures below 1 K with the help of gadolinium sulphate. Haas and Kramers used the magnetic balance for a number of paramagnetic substances. They found that potassium and chromium alum give a much lower temperature.

Method. The apparatus used is shown in (Fig 7.16). The paramagnetic specimen (salt) is suspended in a vessel A, which is surrounded by liquid helium. Liquid helium taken in Dewar flask *D*, is boiled under reduced pressure. It is surrounded by another Dewar flask D_2 containing liquid hydrogen. The salt is in contact with the helium gas. A magnetic field of the order of 30,000 Gauss is applied.

When the magnetic field is switched on the specimen (salt) is magnetised. The heat due to magnetisation is removed by first introducing hydrogen gas into A and then pumping it off with high vacuum pump, so that the specimen is thermally isolated. In the mean time, the specimen (salt) picks up the temperature of the liquid helium *i.e.*, the specimen and the liquid helium are at the same temperature. Now the magnetic field is switched off. Adiabatic demagnetisation of the specimen takes place and its temperature falls. The temperature of the specimen is determined by fitting a co-axial solenoid coil round the tube A and measuring the self-inductance and hence susceptibility of the substance at the beginning and at the end of experiment. Then temperature T is calculated by Curie law $\chi = \frac{C}{T}$.

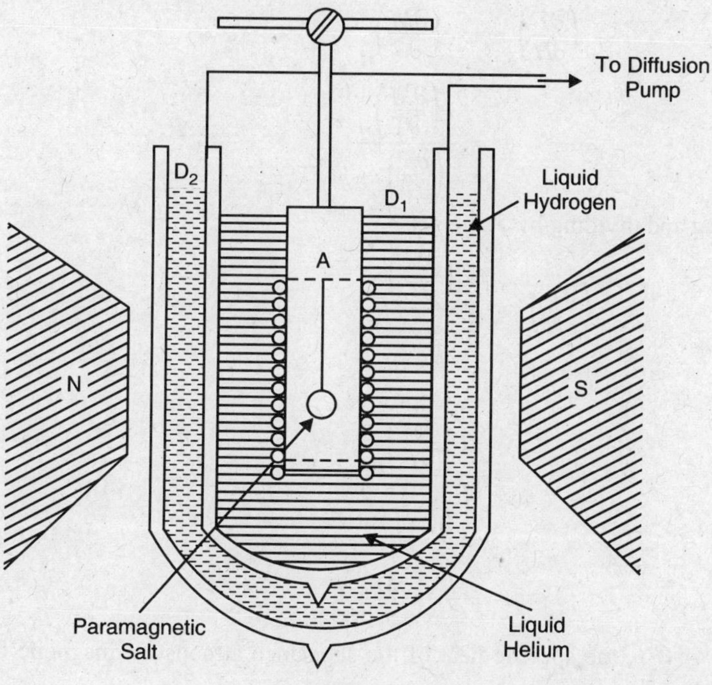

Fig. 7.16

Haas, in 1944, was able to produce temperatures upto 0.002 K using a double sulphate of potassium and aluminium. Klerk, Stenland and Gorter used powdered mixed crystals of chromium alum and aluminium alum and went down to a temperature of **0.0014 K**.

Theory. When a paramagnetic material is placed in a magnetising field H, its elementary magnetic dipoles get aligned parallel to the direction of the field. *The magnetic moment per unit volume* thus produced is called the **intensity of magnetisation (*I*)**. According to Curie's law, this intensity of magnetisation is directly proportional to the magnetising field H and inversely proportional to temperature T of the paramagnetic substance. Thus

$$I \propto \frac{H}{T}$$

$$I = C\left(\frac{H}{T}\right) \quad \text{... (i)}$$

where C is a constant, known as Curie constant.

If V is the volume of 1 mole (*i.e.*, molar volume), then intensity of magnetisation of 1 mole of paramagnetic substance

$$M = IV$$

Substituting for I from eqn (*i*), we get

$$M = CV\left(\frac{H}{T}\right) \quad \text{... (ii)}$$

In the experiment, let 1 mole of paramagnetic substance is placed in magnetising field H. Then its thermodynamic behaviour can be expressed in terms of thermodynamic variables P, V, T and S. In thermodynamic system, an increase in pressure P results in decrease in volume V, analogously in our case any increase in H results in an increase in M. Hence replacing P by – H and V by M in Maxwell's third thermodynamcal relation

$$\left(\frac{\partial T}{\partial P}\right)_S = \left(\frac{\partial V}{\partial S}\right)_P$$

We have
$$\left(\frac{\partial T}{\partial H}\right)_S = -\left(\frac{\partial M}{\partial S}\right)_H$$

$$= \frac{-\left(\frac{\partial M}{\partial T}\right)_H}{\left(\frac{\partial S}{\partial T}\right)_H}$$

Multiplying and dividing by T, we get

$$\left(\frac{\partial T}{\partial H}\right)_S = \frac{-T\left(\frac{\partial M}{\partial T}\right)_H}{T\left(\frac{\partial S}{\partial T}\right)_H}$$

$$= \frac{-T\left(\frac{\partial M}{\partial T}\right)_H}{\left(\frac{\partial Q}{\partial T}\right)_H} \qquad [\therefore T.\partial S = \partial Q]$$

$$= -\frac{T}{C_H}\left(\frac{\partial M}{\partial T}\right)_H \qquad \text{... }(iii)$$

where $\left(\frac{\partial Q}{\partial T}\right)_H = C_H$, the specific heat of the substance at constant magnetic field H. The unit is Joule/mole K.

Since the process is carried out adiabatically (in which entropy S remains unchanged, *i.e.*, constant), we may write,

$$dT = -\frac{T}{C_H}\left(\frac{\partial M}{\partial T}\right)_H .dH$$

Therefore, when a field changes from H_1 to H_2, the change in temperature

$$\Delta T = -\frac{T}{C_H}\int_{H_1}^{H_2}\left(\frac{\partial M}{\partial T}\right)_H .dH \qquad \text{... }(iv)$$

Differentiating eqn. (*ii*) *w.r.* to T, we get

$$\left(\frac{\partial M}{\partial T}\right)_H = -\frac{CVH}{T^2} \qquad \text{... }(v)$$

Substituting in equation (*iv*), we have

$$\Delta T = -\frac{T}{C_H}\int_{H_1}^{H_2}\left(-\frac{CVH}{T^2}\right)dH$$

$$= \frac{CV}{C_H.T}\int_{H_1}^{H_2} H.dH$$

or,
$$\Delta T = \frac{CV}{2C_H.T}\left(H_2^2 - H_1^2\right) \quad ...(vi)$$

If the magnetic field is reduced from $H_1 = H$ to $H_2 = 0$, then the change in temperature will be
$$\Delta T = -\frac{CV}{2C_H.T}.H^2 \quad ...(vii)$$

This is the desired relation. The following conclusions can be drawn from this equation (*vii*).

Conclusions

(*a*) The temperature of the paramagnetic substance decreases on decreasing the magnetising field *H* (indicated by negative sign of ΔT), and

(*b*) Greater is the initial field H and lower is the initial temperature *T*, greater is the temperature fall ΔT.

It should be noted that the quantity *CV* is the Curie constant per mole. If 1 gm of paramagnetic substance (salt) is taken, then *CV* would stand for Curie constant per gm.

***T–S* Diagram**

The cooling produced by adiabatic demagnetisation is explained by using *T–S i.e.*, temperature entropy diagram, as shown in Fig. 7.17.

In the diagram, the entropy of the substance is represented schematically as a function of temperature for two different magnetising fields constants, *i.e.*, $H_2 = 0$ and $H_1 = H$. The arrow *AB* represents the change in entropy during isothermal magnetisation (at constant temperature). This arrow shows the decrease in entropy of the substance during magnetisation (from $H = 0$ to *H*). Now if the field is switched off, the substance demagnetises adiabatically, so that the temperature falls while during this process entropy remains constant, as represented by an arrow *BC*. During this process, *i.e.*, adiabatic demagnetisation the temperature of the order of 10^{-4} can be reached. The main hurdle in this method is the measurement of such a low temperature. This problem is resolved by using the Curie law. Assuming the validity of Curie law, and if the susceptibility of the sample is measured, the **Curie temperature** $T_C = \frac{C}{\chi}$ (also known as **magnetic temperature**) can be calculated. Knowing the Curie temperature, the thermodynamical temperature T can be evaluated. The temperature measured by this method is of the order of 10^{-4} K. We can still go down up to the order of 10^{-6} by making use of *nuclear magnetism*. But it should be rembered that absolute zero can never be attained in actual practice.

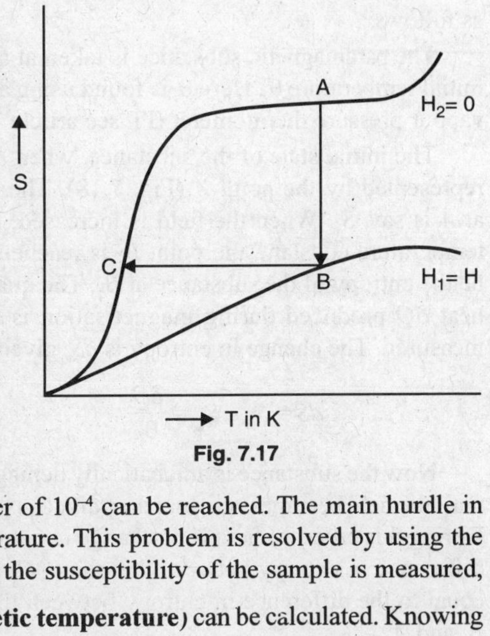

Fig. 7.17

7.17 CONVERSION OF MAGNETIC TEMPERATURE TO KELVIN TEMPERATURE

The thermodynamical behaviour of a paramagnetic crystal, placed in a magnetic field H, depends also on the strength of the magnetising field in addition to pressure and volume. Therefore, the usual relation
$$\delta H = dU + dW = dU + PdV$$
is modified in the form
$$\delta H = dU + PdV - BdI \quad ...(i)$$

Here *B* is the magnetic flux density and *dI* is the change in the intensity (or strength) of magnetisation per gram molecule. Therefore,

$$I = \chi\, VB$$

where χ is the susceptibility per unit volume.

Assuming the change in volume to be negligible *i.e.*, $dV = 0$

$$\delta H = dU - BdI$$

when the magnetic field is kept zero and constant,

we have $$\theta = \left(\frac{\partial U}{\partial S}\right)_{B=0}$$

or $$\theta = \frac{\left(\frac{\partial U}{\partial T}\right)_{B=0}}{\left(\frac{\partial S}{\partial T}\right)_{B=0}} \quad \text{... (ii)}$$

Here θ is the temperature expressed on the Kelvin scale and T is the magnetic temperature.

To determine the value of θ, the values of $\left(\frac{\partial U}{\partial T}\right)_{B=0}$ and $\left(\frac{\partial S}{\partial T}\right)_{B=0}$ are to be evaluated. It is done as follows :

The paramagnetic substance is taken at a known initial temperature θ_i. Here θ_i is found using a helium vapour pressure thermometer (Pl. see article 7.18).

The initial state of the substance, when $B = 0$, is represented by the point A (Fig. 7.18). The entropy at A is say, S_1. When the field is increased, keeping temperature constant, the point B_1 is reached. Let S_2 be the entropy of the substance at B_1. The quantity of heat δQ produced during magnetisation is directly measured. The change in entropy is dS, given by

$$dS = S_2 - S_1 = \frac{\delta Q}{\theta_i}$$

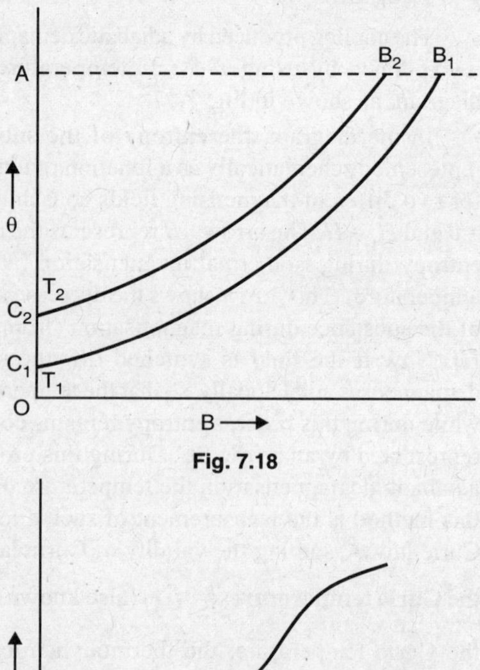

Fig. 7.18

Now the substance is adiabatically demagnetised until the field becomes zero and the point C_1 is reached. During this process, there is no change in entropy. The difference in entropy between the points C_1 and A is equal to the difference in entropy between the points B_1 and A.

Suppose, the temperature at the points C_1 is θ degrees Kelvin or T degrees magnetic. The values T and $(S_2 - S_1)$ will depend upon the value of the magnetising field B. The experiment is repeated with different initial fields and graph is plotted between dS and T (Fig 7.19)

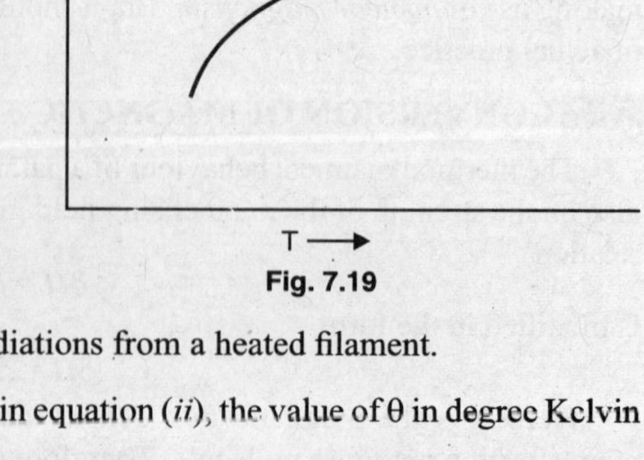

Fig. 7.19

From this graph, the value of $\left(\frac{\partial S}{\partial T}\right)_{B=0}$ is found. To determine the value of $\left(\frac{\partial U}{\partial T}\right)_{B=0}$, the salt is heated from a temperature T to T' and the heat absorbed is estimated directly. This can be done by introducing γ-rays or radiations from a heated filament.

Substituting the values of $\left(\frac{\partial S}{\partial T}\right)_{B=0}$ and $\left(\frac{\partial U}{\partial T}\right)_{B=0}$ in equation (*ii*), the value of θ in degree Kelvin can be calculated.

A lower temperature can be reached by using a mixed salt because magnetic interactions between the paramagnetic ions are weakened when the paramagnetic salt is mixed with a non-paramagnetic salt. A mixed crystal can also be used.

Haas and Wiersma (1935) were able to produce temperatures up to 0.003 K using ferric ammonium alum and potassium chrome alum.

7.18 MEASUREMENT OF VERY LOW TEMPERATURES: (HELIUM VAPOUR PRESSURE THERMOMETER)

Helium vapour pressure thermometer is used to measure temperatures up to **0.7 K**. The apparatus consists of a bulb A containing liquid helium. This bulb is connected to the manometer limbs M_1 and M_2 through a connecting tube C. The tube C is surrounded by a copper tube B to ensure uniform temperature of the vapour. R is reservoir containing mercury (Fig. 7.20).

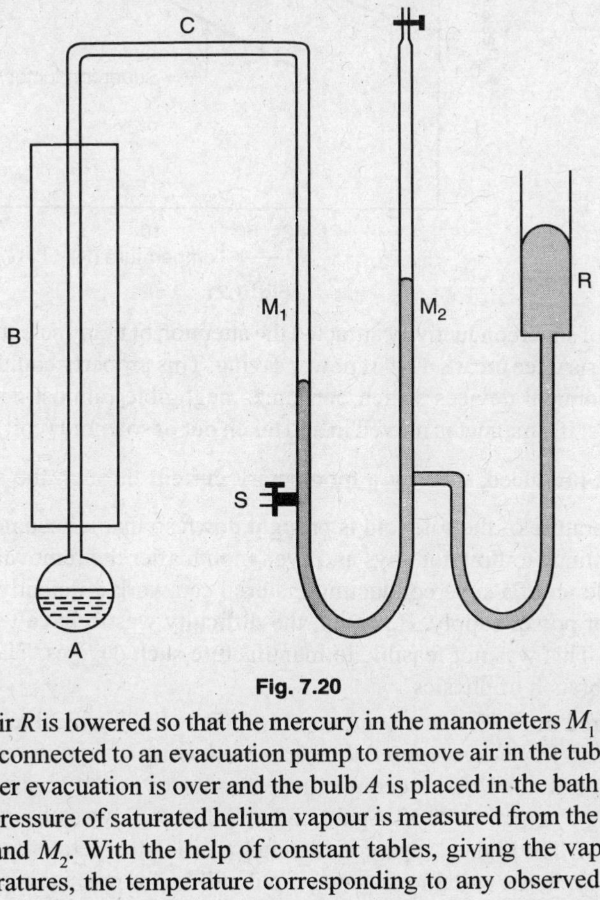

Fig. 7.20

Initially the reservoir R is lowered so that the mercury in the manometers M_1 and M_2 is below the stop-cock S. The tube is connected to an evacuation pump to remove air in the tube C and bulb A. The stop-cock S is closed after evacuation is over and the bulb A is placed in the bath whose temperature is to be measured. The pressure of saturated helium vapour is measured from the difference in levels of Hg in the limbs M_1 and M_2. With the help of constant tables, giving the vapour pressure of the liquid at various temperatures, the temperature corresponding to any observed vapour pressure is determined. For lower temperatures, the graph between saturated vapour pressure and temperature is extrapolated.

7.19 SUPERCONDUCTIVITY

The electrical resistivity of many metals and alloys drops suddenly to zero when cooled to a sufficiently low temperature, particularly in the *liquid helium range*. This phenomenon was first observed by Kamerling Onnes in 1911, while studying the electrical resistance of Mercury at very low temperatures. He observed that the resistance of mercury first decreases regularly like that of a metal but at 4.2 K, the resistivity *suddenly decreases to zero* (actually becomes one millionth of its original value at the melting point) as shown in Fig. 7.21. The temperature at which this phase transition occurs and the resistance suddenly drops to zero, is called the *critical temperature* (T_c). Similar results were obtained for other metals like Al, Pb, Sn, Cd, Zn, In etc. *This phenomenon of disappearance of electrical resistance of material below a certain temperature is called superconductivity and the material in this state is called a superconductor.*

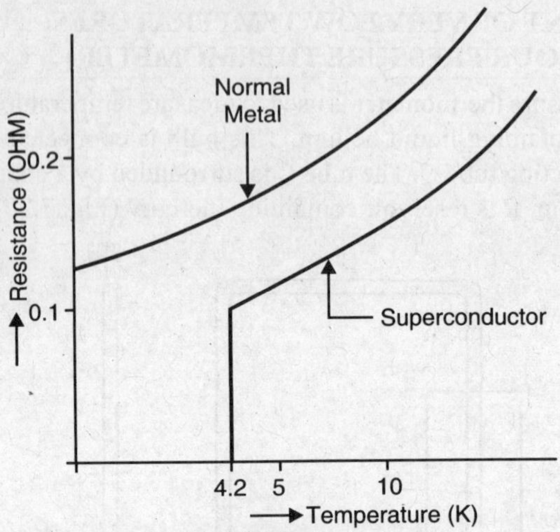

Fig. 7.21

The discovery of superconductivity attracted the attention of many scientists because the materials with no electrical resistance means a lot of power saving. This property could be exploited to fabricate powerful and economical devices which consumes negligible (almost zero) amount of electrical energy. For example: if a magnet is moved in and taken out of solenoid (coil) rapidly, then an induced *e.m.f.* $e = -\dfrac{d\phi}{dt}$ is produced, thereby a momentary current through the solenoid is changing. If, however, the temperature of the solenoid is brought down so that it becomes superconducting, then the current will continue to flow for days and even month after the removal of the magnet. Thus, an electromagnet made-up of a superconducting material can work efficiently for years together, even after the removal of power supply. However, the difficulty was to have very low temperatures for practical purposes. This was not feasible to manufacture such devices. Therefore, a lot of research is going on in this branch of physics.

It was discovered in 1977, that a compound Nb_3 Ge to have critical temperature (T_C) of 23.2 K. In 1986, Bednorz and Miller discovered the superconducting transition temperature for $La_{1.85}$ $Sr_{0.15}$ $CuSO_4$ was about 40 K. In 1987, a ceramic superconductor of the composition YBa_2 Cu_3 O_7 was discovered showing T_C equal to 90 K. Again higher value of T_C was observed for thallium cuprates giving T_C to be 125 K. The highest superconducting transition temperature reported till today is about 150 K. Higher the T_C, better suitable the compound will be for practical utility.

7.20 MEISSNER EFFECT

When a metal becomes superconductor, its *thermal conductivity decreases* considerably and tends to zero approximately as the cube of the temperature. Superconducting state being a distinct phase of matter having characteristic electrical, magnetic, thermodynamic and other physical properties. Superconductors have *zero electrical resistance*. In a superconducting state the *magnetic permeability* (χ) *is also zero* and shows *perfect diamagnetism*. This phenomenon is known as *Meissner effect.*

In 1933, Meissner and Ochsenfild found that if a long superconductor is cooled in a longitudinal magnetic field to below its critical temperature corresponding to that field, then the lines of magnetic induction B are pushed out of the body of the superconductor at the transition Fig. 7.22. This is the basis of Meissner effect.

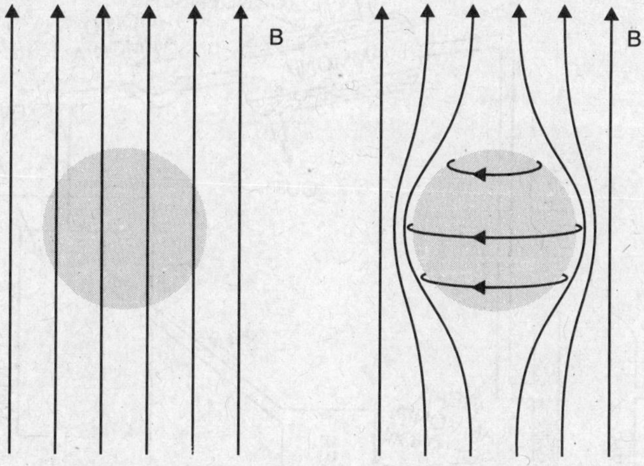

Fig. 7.22. The Meissner effect. (*a*) An applied magnetic field can exist inside a superconductor at temperatures above its critical temperature T_C. (*b*) When the superconductor is then cooled below $T_{C,\ face}$ currents appear whose effect is to expel the magnetic field from the interior of the superconductor.

If the field is first switched on after the material is in the perfectly conducting state then it cannot penetrate more than a thickness of order λ_L of the specimen; on the other hand if the material is transformed to the state of perfect conductivity after the field is applied then the field penetrates the specimen and remains frozen there, even if the source of the external field is removed.

Thus, the Meissner effect shows that a superconducting pure metal, in the chosen geometry, *always expels the magnetic flux irrespective of the sequence on which the field is applied or the material becomes superconducting. Thus, a superconductor is not a perfect conductor but a perfect diamagnetic material with zero electrical resistance.*

The perfect diamagnetism and zero resistivity are the two independent essential properties of the superconducting states.

7.21 REFRIGERATION CYCLE: ELECTROLUX REFRIGERATOR

An electrolux refrigerator produces sufficiently low temperature without using a compressor pump. The principle of refrigeration cycle can be explained by the schematic diagram of an electrolux refrigerator shown in Fig. 7.23.

The weak ammonia solution in the boiler is forced into the absorber. The strong ammonia solution goes into the boiler and the gaseous ammonia enters the condenser. It is cooled and is condensed. Liquid ammonia enters the evaporator and is mixed with hydrogen. Hydrogen reduces the partial pressure of ammonia below its saturation point and causes evaporation. This evaporator is surrounded by the chilled compartment. Hydrogen and gaseous ammonia leave the evaporator and enter the absorber. Here they meet with the weak ammonia solution. The ammonia gas is dissolved and hydrogen gas rises through the absorber and enters the evaporator. The strong ammonia solution is forced up into the boiler again. This cycle continues and a sufficiently low temperature is produced in the chilled compartment. The advantage of this apparatus is that no compressor is required and the circulation of the liquid and the gas is automatic.

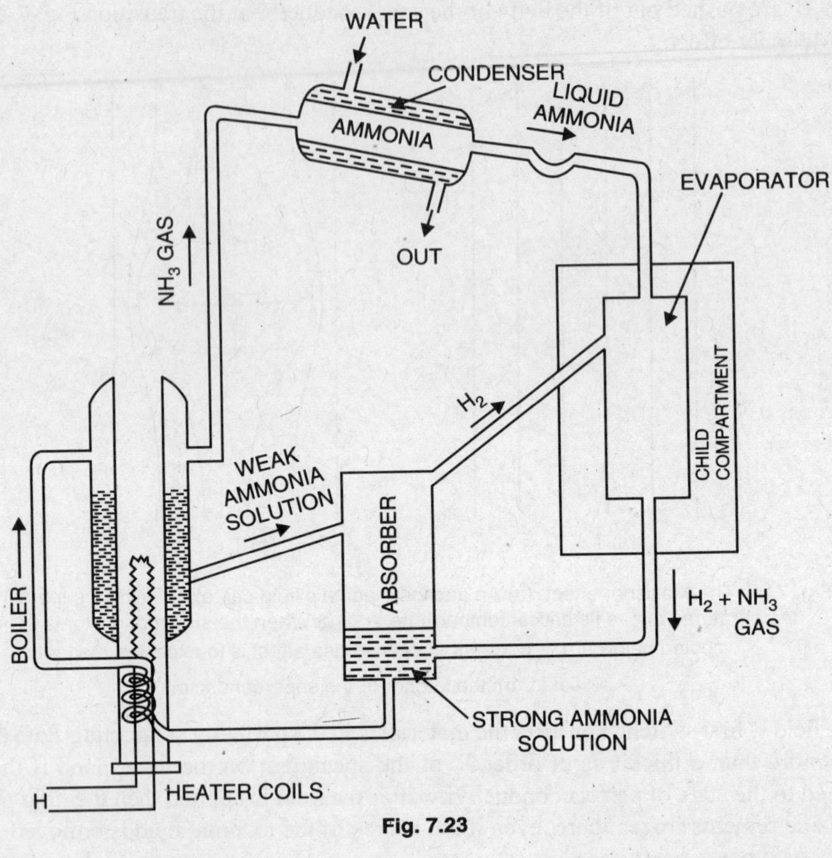

Fig. 7.23

SOLVED EXAMPLE

Example 7.1 *Calculate the cooling produced by adiabatic demagnetisation of a paramagnetic salt as the field is reduced from 10,000 oersted to zero at initial temperature of 2 K. (Given : Curie constant per gm. mol per c.c. = 0.042 erg degree/gm oersted^{-2} and C_H = 0.42 Joule gm^{-1} deg^{-1}.*

Solution. The cooling produced is given by

$$\Delta T = \frac{CV}{2C_H T}.H^2$$

Here, $T = 2$ K, C_H = 0.42 Joule gm^{-1} deg^{-1}

= 0.42 × 10^7 erg gm^{-1} deg^{-1}

Curie constant, CV = 0.042 erg deg gm^{-1} Oersted^{-2}.

Substituting, we get

$$\Delta T = -\frac{0.042\times(10{,}000)^2}{2\times(0.42\times10^7)\times 2}$$

$$= -\mathbf{0.25\ K}$$

EXERCISE

7.1 Explain the principle of regenerative cooling with neat diagram
(Nagpur 2010, 2009, 2008, 2007, 2005, 04, Garhawal 2004, Bangalore 2004)

7.2 Discuss the industrial process for liquefaction of hydrogen. Explain the principle on which it is based.
(Patna 2004, 05, Nagpur 2010, 2005/w, Garhwal 2000)

7.3 Explain K. Onnes method for the liquefaction of helium. What is the role of activated charcoal in liquefaction of helium used by Simon? ***(Madras 2004)***

7.4 Discuss the properties of Helium I and Helium II. What do you understand by λ-point?
(Agra 2002)

7.5 Explain the peculiar properties of Helium II.

7.6 What is rolling film. What is its role in Helium II. Which statistics is applicable to explain the behaviour of Helium II.

7.7 Can we produce refrigeration without using a compressor? Explain how.

7.8 Explain refrigeration cycle. How it is used to obtain sufficiently low temperature in electrolux refrigerator. ***(Osmania 2005)***

7.9 Describe a process for production of low temperature. ***(Patna 2004)***

7.10 Describe Linde's process for the liquefaction of air.
(Madras 2004, Madras Autonomous 2005, Mysore 2004, Bangalore 2004)

7.11 Discuss K. Onnes method for the liquefaction of helium and explain how the principle of regenerative cooling has been utilised in this method. ***(Purvanchal 2002)***

7.12 Discuss the properties of Helium I and Helium II. What do you understand by λ-point? Give the properties of Helium II. ***(Kanpur 2001, Meerut 2000, Agra 2003)***

7.13 Write a short account on the production and measurement of low temperatures.
(Nagpur 2010)

7.14 What are the practical uses and applications of low temperatures? ***(Madras 2005)***

7.15 Explain cooling due to adiabatic demagnetisation.
(Nagpur 2004, Mysore 2004, Alagappa 2004)

7.16 Describe fully one method of liquefying a gas. State clearly the principle underlying the method and explain how the temperature of liquefaction may be measured.

7.17 Give an account of liquefaction of gases. Explain regenerative cooling method.
(Kuvempa 2005)

7.18 What is Joule-Thomson effect? What is meant by adiabatic demagnetisation? ***(Madras 2004)***

7.19 What do you mean by liquefaction of gases? Give briefly an account of various methods used for liquefaction of gases. Discuss the difficulties encountered in liquefying hydrogen and Helium. ***(Nagpur 2004)***

7.20 When pressure is increased what happens on melting point of ice? Explain adiabatic demagnetisation with necessary theory. ***(Madras 2005)***

7.21 What are the practical uses and applications of low temperatures? ***(Madras 2004)***

7.22 Explain adiabatic demagnetisation with necessary theory. Explain how very low temperature is achieved using it.
(Madras 2004, Madras autonomous 2005, Punjab 2005, Kanpur 1999, Meerut 1999, M.S. Uni. Tamil Nadu 2007, Pbi U.2008, 2005)

7.23 Explain the principle of cascaded cooling
(Nagpur 2010, Mysore 2004, Purvanchal 2002, Kanpur 2001)

7.24 Explain the principle of adiabatic demagnetisation for reaching temperature near absolute zero. Explain superfluidity and superconductivity with example.

(Bangalore 2004, Garhwal 2000, 2001, 2002, Agra 2002, Purvanchal 2007, Meerut 2000, Agra 2003)

7.25 What is the principle of cooling produced by adiabatic expansion of gas? Give a schematic diagram of gas expansion machine. ***(Grahwal 2002)***

7.26 Explain why solidification of Helium is difficult. Explain why the temperature of a gas drops in an adiabatic expansion.

7.27 Compare Joule-Thomson expansion with adiabatic expansion. ***(Madras 2004)***

7.28 Explain briefly why early attempts to liquefy helium were not useful. Give a brief account of method of liquefying this gas. ***(Purvanchal 2003)***

7.29 Explain the theory of Joule-Thomson regenerative cooling. How can you obtain liquid helium by its application? ***(Nagpur 2010, Meerut 2001)***

7.30 What is adiabatic demagnetisation? Deduce thermodynamical expression for cooling produced by adiabatic demagnetisation of a paramagnetic salt. How will you employ the phenomenon to produce and measure very low temperatures?

(Purvanchal 2007, Alagappa 2004, Bangalore 2001, M.S. Uni. Tamil Nadu 2007)

7.31 How magnetic temperatures are measured on absolute scale?

(Calcutta 2000, Lucknow 2001)

7.32 What is the value of Joule-Thomson coefficient for ideal gas? Show that a real gas will always suffer a cooling effect when it undergoes a free Joule expansion.

(Lucknow 2004, Ranchi 2001)

7.33 Explain the principle of adiabatic demagnetisation and method of reaching temperature near absolute zero. How are such temperatures measured?

(Garhwal 2000, 2001, Lucknow 2003, Agra 2002)

7.34 Explain the principle of producing low temperatures by evaporation of a liquid under reduced pressure. ***(Garhwal 1996)***

7.35 Discuss the process of liquefaction of gases. Explain regenerative cooling method. Give uses of liquid air. ***(Kanpur 2000, Garhwal 2001)***

7.36 What is a refrigerator? Describe vapour compression machine. ***(Garhwal 2000, 2001)***

7.37 What is adiabatic demagnetisation? Deduce expression for change of temperature of paramagnetic salt due to adiabatic demagnetisation. Give its experimental details and then derive the expression for final temperature of the salt in terms of initial temperature and the strength of applied magnetic field. Show your cooling results by drawing curves for change of entropy against temperatures at constant magnetic fields. ***(Rohilkhand 2001)***

7.38 Find the expression of work done when the magnetisation of a magnetic solid in increased by an amount dM quasi-statically. ***(Calcutta 2005)***

7.39 What are liquid Helium I and II. Discuss the important properties of liquid Helium II.

(Purvanchal 2002, Agra 2002)

7.40 Describe the necessary theory of the method of production of low temperature by adiabatic demagnetisation of a paramagnetic salt. ***(Purvanchal 2007, Meerut 2002)***

Define plasma frequency w_s. When a material becomes a superconductor? Explain. Distinguish between Type-I, and Type II superconductors. ***(M.S.Uni. Tamil Nadu 2007)***

7.41 Write short notes on:

(*a*) Adiabatic demagnetisation.

(Agra 2000, Kanpur 2001, 1996, Meerut 2001, Alagappa 2004, Bangalore 2005)

(*b*) Cooling by evaporation.
(*c*) Joule-Thomson expansion. **(*Rohilkhand 2001, Meerut 2003*)**
(*d*) Temperature of inversion.
(*e*) Regenerative cooling. **(*Agra 2002*)**
(*f*) Helium I and Helium II. **(*Patna 2004*)**
(*g*) Superfluidity.
(*h*) Superconductivity. **(*M.S. Uni. Tamil Nadu 2007, Alagappa 2004, Kanpur 2000, 2001*)**
(*i*) Meissner effect.
(*j*) Electrolux refrigerator. **(*Madras 2005*)**
(*k*) Magnetic temperature.

7.42 Calculate the cooling produced by adiabatic demagnetisation of a paramagnetic salt as field is reduced from 0.5 weber/m^2 to zero at initial temperature of 20 K. Given: $C_H = 0.42$ J.gm^{-1} deg^{-1} and Curie constant = 0.06 C.G.S. units. [**Ans.** 0.0009 °C] **(*Mysore 2000, Kanpur 1999*)**

7.43 Calculate the fall in temperature produced by adiabatic demagnetisation of a paramagnetic salt at an initial temperature 3 K when magnetic field is reduced from 8400 oersted to zero. Curie constant per cm^3, K = 0.6 c.g.s. units, and $C_B = 0.2$ cal gm^{-1} K^{-1}.
(**Ans.** 1.01 K] **(*Purvanchal 2001*)**

7.44 Calculate the change in temperature of Helium at 100K when throttled at a pressure differences of 50 atmospheres. Given that for Helium $a = 0.0341$ litre^2 atmosphere per mol^2 and $b = 0.0234$ litre mole^{-1} and C_p for Helium = 20.75 J mole^{-1} K^{-1}. (**Ans.** – 3.71 K)

7.45 For one mole of hydrogen Van der Waal's constants are ; $a = 0.245$ litre^2 atm mol^{-2} $b = 2.67 \times 10^{-2}$ litre mol^{-1} $R = 2$ cal/mole K. Calculate its temperature of inversion.
(**Ans.** $T_i = 220$ K)

7.46 Calculate the fall in temperature produced by adiabatic demagnetisation of a paramagnetic salt at an initial temperature 4 K when the magnetic field is reduced from 10,000 oersted to zero. Given curie constant per c.c. $k = 0.005$ K/gm and $C_H = 0.1$ cal/gm K. (**Ans.** 0.15 K)

MULTIPLE CHOICE QUESTIONS

7.1 Using adiabatic demagnetisation, the minimum temperature produced is **(*Garhwal 2001*)**
(*a*) 1 K (*b*) 10^{-3} K
(*c*) 10^{-4} K (*d*) 10^{-5} K

7.2 The temperature of inversion (T_i) of a gas is
(*a*) $\frac{a}{Rb}$ (*b*) $\frac{2a}{Rb}$
(*c*) $\frac{8a}{27Rb}$ (*d*) $\frac{27}{8}\frac{Rb}{a}$

7.3 If T_i is temperature of inversion, T_B is Boyle temperature and T_C is critical temperature, then
(*a*) $T_i < T_B < T_C$ (*b*) $T_c > T_B < T_i$
(*c*) $T_B > T_i > T_c$ (*d*) $T_i = 2T_B = 6.75\, T_C$

7.4 The critical temperature of a gas is 31 °C. The room temperature is 27 °C. Then the gas in the room
(*a*) can be liquefied (*b*) cannot be liquefied
(*c*) will be solidified (*d*) will break into electrons.

7.5 The coefficient of performance of a refrigerator working between ice-point and room temperature 30 °C is

(*a*) 2 (*b*) 9
(*c*) 10 (*d*) 6

7.6 The rate of coefficient of viscosities of He II and He I is

(*a*) 10^{-2} (*b*) 10^{-1}
(*c*) 10^{-3} (*d*) 10^{3}

7.7 If *A*, *B*, *C*, *D* are the virial coefficients, then

(*a*) $A > B > C > D$ (*b*) $\frac{B}{A} = 10^{-3}$
(*c*) $\frac{C}{A} = 10^{-5}$ (*d*) All the above

7.8 The Boyle temperature T_B is given by the relation **(*Garhwal 2001*)**

(*a*) $T_B = \left(\frac{27}{8}\right) T_C$ (*b*) $T_B = \frac{a}{bR}$
(*c*) Both the above (*d*) None of the above

7.9 The last gas to be liquefied was

(*a*) Oxygen (*b*) Hydrogen
(*c*) Nitrogen (*d*) Helium

7.10 The decrease in temperature during adiabatic demagnetisation is

(*a*) more when initial magnetic field is high.
(*b*) more when diamagnetic solid is used.
(*c*) more when initial temperature is high.
(*d*) more when initial magnetic field is low

7.11 Solid helium

(*a*) can be obtained merely by lowering the temperature of the liquid.
(*b*) melts to form solid.
(*c*) melts to form gas.
(*d*) cannot exist above the critical temperature.

7.12 Liquid Helium is superconductor at **(*Garhwal 2002*)**

(*a*) above 2.19 K (*b*) below 4.2 K
(*c*) below 2.19 K (*d*) below ice point

7.13 The relation between critical temperature T_C and inversion temperature T_i is $T_C = X\, T_i$, then $X = ?$

(*a*) $\frac{27}{16}$ (*b*) $\frac{4}{27}$
(*c*) $\frac{8}{27}$ (*d*) $\frac{16}{27}$

7.14 Lowest temperature can be achieved by

(*a*) Joule -Thomson process (*b*) cascade process
(*c*) adiabatic expansion (*d*) adiabatic demagetisation

7.15 Joule and Thomson perfomred the experiment to study the behaviour of

(*a*) ideal gas (*b*) real gas
(*c*) both ideal and real gas

7.16. The temperaure at which gases show neither cooling nor heating effect is known as

(*a*) Boyle temperature (*b*) Critical temperature

(*c*) Temperature of inversion

7.17 Internal energy of an ideal gas depends upon

(*a*) volume (*b*) temperature

(*c*) pressure

7.18 The change in temperature of an ideal gas due to Joule-Thomson effect is

(*a*) $\frac{PV}{R}$ (*b*) zero

(*c*) $\frac{\left(\frac{2a}{RT}-b\right)}{C_p}dp$

7.19 Which is constant is Joule-Thomson expansion

(*a*) internal energy (*b*) enthalpy

(*c*) Gibb's function (*d*) pressure

7.20 In Joule-Thomson effect cooling occurs if

(*a*) $T_i < \frac{2a}{bR}$ (*b*) $T_i > \frac{2a}{bR}$

(*c*) $T_i = \frac{2a}{bR}$ (*d*) none

7.21 Cooling produced due to adiabatic demagnetisation depends upon

(*a*) initial temperature only (*b*) intensity of magnetic field only

(*c*) intial temperature and intensity of magnetic field

7.22 Viscosity of Liquid Helium II is

(*a*) greater than viscosity of Liquid Helium I

(*b*) negligible

(*c*) very large

ANSWERS

7.1 (*b*)	7.2 (*b*)	7.3 (*d*)	7.4 (*a*)	7.5 (*b*)	7.6 (*c*)
7.7 (*d*)	7.8 (*c*)	7.9 (*d*)	7.10 (*a*)	7.11 (*c*)	7.12 (*b*)
7.13 (*b*)	7.14 (*d*)	7.15 (*b*)	7.16 (*c*)	7.17 (*b*)	7.18 (*b*)
7.19 (*b*)	7.20 (*a*)	7.21 (*c*)	7.22 (*b*)		

RADIATION

8
Chapter

INTRODUCTION

Heat transfers from one point to another by three modes: conduction, convection and radiation. The process of conduction and convection necessarily requires the presence of material medium filling the space between the source and the receiver. However, in radiation, material medium is not necessary. It can travel through vacuum too. The energy radiated by a hot body is termed as radiant energy. It is well established and documented that heat radiations are electromagnetic in nature, like that of light. The position of heat radiation in electromagnetic spectrum is beyond the red end of the spectrum, *i.e.*, infrared region having wavelengths ranging from 8×10^{-5} cm to 0.04 cm. Heat is developed when these radiation are absorbed by matter or animal skin. Some substances do absorb radiant heat to a greater or less degree and thereby get heated to a more or less extent, e.g. glass, wood, metals etc. absorb radiant heat and become heated, while quartz, rock salt, fluorspar etc. allow radiant heat to pass through themselves, without getting much warmed. The heat radiations obey nearly all the laws of light, like rectilinear propagation, reflection, refraction, interference, diffraction, polarization etc. A concept of 'perfectly black body' and the results obtained from such studies have proved to be of far reaching importance in physics.

8.1 THERMAL RADIATION

According to Maxwell, thermal radiation is defined as the transfer of heat from a hot body to a cooler body without appreciable heating of the intervening medium (or space). Heat or thermal energy is propagated by radiation, just like the light energy propagates, it being an electromagnetic wave. The thermal radiation has the following properties:

1. Thermal radiation has electromagnetic wave nature and hence travels through empty space (no material medium is necessary) with the velocity of light.
2. Like light, thermal radiation travels in straight line.
3. Thermal radiation obeys the law of inverse square.
4. It exhibits reflection, refraction, interference, diffraction and polarisation phenomena.

In short, thermal radiation has the same nature as that of light. On wavelength spectrum, the position of thermal radiation is just beyond the red end of visible-light. The wavelength of thermal radiation is greater than red colour and is, therefore called as **infrared region of E-M spectrum**. Hence thermal radiation is frequently called as **infrared radiation (I.R.)**. Its wavelength is ranging from 8000 Å to 4000 Å. The thermal radiations cannot be detected by the eye or the photographic plate but they can be detected either by **bolometer**, which measures the change in resistance with temperature, or by using **thermopile**. These are discussed in detail at the end of this chapter.

8.2 SOME USEFUL DEFINITIONS

(*i*) **Total energy density (*u*):** Total energy density of radiations at any point is the total radiant energy per unit volume around that point for all wavelengths taken together. It is generally represented by u and has a unit in S.I. joule/m^3 and in C.G.S. erg/cm^3.

(ii) **Spectral energy density (u_λ):** Spectral energy density for a particular wavelength is the energy per unit volume per unit range of wavelength. This is denoted by u_λ.

(iii) **Total emissive power or emissivity (E):** The total emissive power of a body is defined as the radiant energy emitted per unit time per unit surface area of the body for all wavelengths taken together. It is denoted by E.

(iv) **Spectral emissive power (e_λ):** The spectral emissive power of a body at a particular wavelength is the radiant energy emitted per unit time per unit surface area of the body within a unit wavelength range. It is denoted by e_λ. For a perfectly black body, e_λ is maximum and is denoted by E_λ.

(v) **Absorptivity or absorptive power (a_λ):** The absorptive power of a body at a particular temperature and for a particular wavelength is defined as the ratio of the radiant energy absorbed per unit surface area per unit time to the total energy incident on the same area of the body in unit time within a unit wavelength range. It is denoted by a_λ.

We can easily write the two results from above definitions as :

$$u = \int_0^\infty u_\lambda \, d\lambda$$

and

$$E = \int_0^\infty E_\lambda \, d\lambda$$

(vi) **Relative emittance (*e*):** The ratio of emittance of a surface to the emittance of a blacl body is called relative emittance. It is denoted by *e*.

If E' is the emittance of a given surface and E the emittance of a black body, then relative emittance of the surface is given by

$$e = \frac{E'}{E}$$

8.3 PREVOST'S THEORY OF HEAT EXCHANGES

The ideas regarding *radiant energy*, prior to 1792, were very much confused. People used to talk of *hot radiations* emitted by hot bodies and *cold radiations* emitted by cold bodies like an ice block. Prevost of Geneva was the first to recognise that this was a loose way of talking. Prevost pointed out the conception, that the ice emits cold radiations is incorrect and proved that thermal radiation is essentially an exchange process. The rise or fall of temperature which is observed in a body is due to an exchange of radiant energy with the surroundings.

Statement: According to Prevost, every body emits radiations continuously to its surroundings at the rate which depends only on the nature of its surface and temperature and at the same time it receives radiations from its surroundings at the rate depending on the surface and the temperature of surroundings.

Thus, when we stand near a fire, we have the sensation of warmth because our body, *which is also a radiator,* is receiving more energy from the fire than it is losing by its own radiation. When we stand near a block of ice, we feel a sensation of cold because our body being at about 37 °C loses more energy by radiation than it receives from the ice, which is at a much lower temperature, *i.e.* 0°C. The exchange process continues till thermal equilibrium is reached.

Thus, all bodies at all temperatures are in a state of dynamical thermal equilibrium when they are at the same temperature. The quantity of heat radiated by a body decreases with the approach towards absolute zero of temperature. When the body is at absolute zero, it will not radiate energy.

8.4 APPLICATIONS OF HEAT RADIATION

1. White clothes are preferred in summer and dark coloured clothes in winter.

When heat radiations fall on white clothes, they are reflected back. No heat is absorbed by the clothes and a person does not get heat from outside in summer. Dark clothes in winter will absorb the heat radiations falling on them and keep the body warm.

2. Cooking utensils are blackened at the bottom and polished on the upper surface. Black surface will absorb the whole of the heat from the furnace and the upper surface will not allow the heat inside the utensil to flow out.
3. Hot water pipes and radiators used in rooms are painted black so that they can radiate maximum amount of heat to the room. The same pipes outside the rooms are painted white so that they do not lose heat to the surroundings.
4. The thermocouple junction exposed to heat is blackened to absorb maximum quantity of heat.
5. Polished reflectors are used in electric heaters to reflect maximum heat in the room.

8.5 UNIFORM TEMPERATURE ENCLOSER

In general, the nature of radiation and the amount of radiation per unit area per second *i.e.*, the quality and quantity of radiation depends on the temperature as well as on the nature of the surface. However, if we consider a non-conducting encloser maintained at a constant temperature, known as *non-conducting isothermal encloser,* the quality and quantity of radiation inside it depends on its temperature and is entirely independent of both the nature of the walls and the nature of the bodies which may be present inside it. Such an encloser is known as uniform temperature (isothermal) encloser.

8.6 BLACK BODY

A perfectly black body is one which absorbs totally all the radiation of any wavelength which fall on it. As it neither reflects nor transmits any radiation, it appears black; whatever be the colour of incident radiation. The main characteristic of such a body is that when heated to a suitable high temperature, it emits full or total radiation. As it is a perfect absorber, it is also a perfect radiator, its emission being the greatest possible for every wavelength at any given temperature.

8.7 BLACK BODY IN PRACTICE

In practice, a perfectly black body is not available. Lamp-black and Platinum black are the nearest approach to a black body. However, a body showing close approximation to a perfectly black body can be constructed.

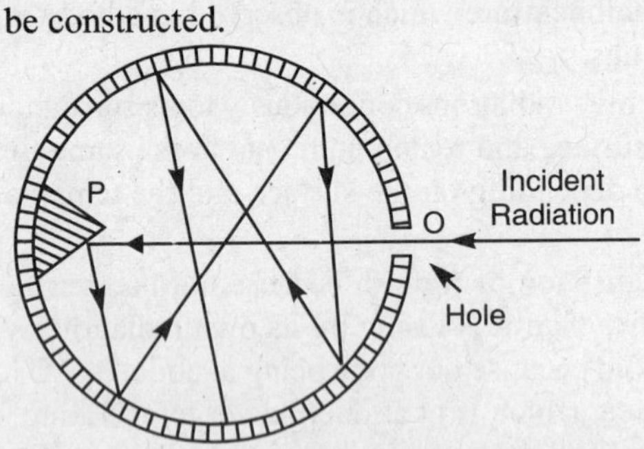

Fig. 8.1. Black body absorber

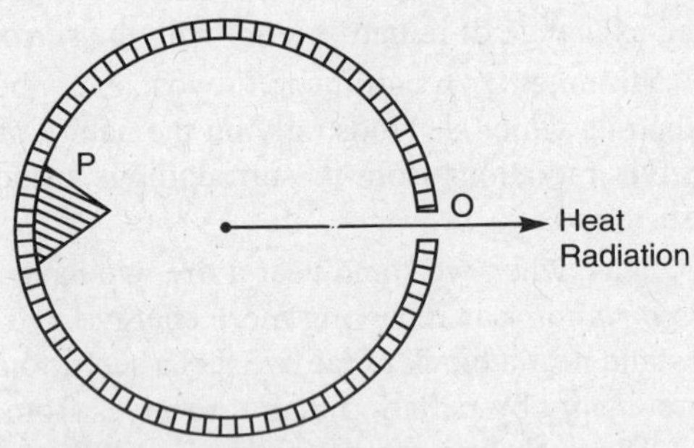

Fig. 8.2. Black body emitter

8.7.1 Fery's black body

A double walled hollow copper sphere is taken and coated with lamp black on its inner surface (Fig. 8.1). A fine hole *O* acts as a very narrow opening. When the radiation enter through the narrow opening they suffer multiple reflections after falling on a pointed projection *P* which is made just in front of the hole. After suffering multiple reflection, the radiation are completely absorbed. This type of black body shown in Fig. 8.1 was designed by Fery. This body acts as a black body absorber. When this body is placed in a bath at a fixed temperature, the heat radiation come out of the hole (Fig. 8.2). The hole acts a black body radiator. It should be remembered that only the hole and not the walls of the body, acts as the black body radiator. Lamp black absorbs about 96% of visible light, while platinum black absorbs about 98%.

8.7.2 Wien's black body

Wien also constructed a black body in the form of cylinder. This black body is commonly used nowadays, as it is more convenient to use.

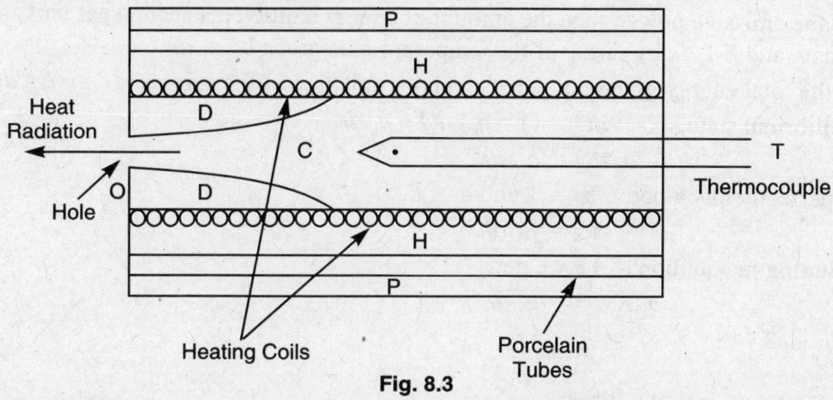

Fig. 8.3

It consists of a hollow cylindrical chamber C of brass or platinum having heating coil of thin platinum wire HH wound over it fig. 8.3. The cylinder is protected by co-axial porcelain tubes PP. When a suitable current is passed through the coil, the cylinder is heated to a desired temperature. The inner surface of the chamber C is painted black, so that it reaches the equilibrium position in a small time. The radiation from the cylinder is limited by blackened concave diaphragms D and emerges out of the hole O. The temperature of the chamber is measured by the thermocouple T. The hole O will act as a black body radiator.

8.8 BLACK BODY RADIATION AND ITS TEMPERATURE DEPENDENCE

When a black body is placed inside a uniform temperature (isothermal) encloser, it will emit full radiation of the encloser after it is in equilibrium with the encloser. These radiations are independent of the nature of substance, nature of walls of the encloser, and presence of any other body in the encloser but depends only on temperature. Since a perfectly black body is one which absorbs all the radiations that fall on it, of whatever wavelength they may be, the radiation emitted by it will possess a character independent of the property of any particular substance, purely dependent on the temperature to which it is raised. Hence it is solely temperature dependent.

Such radiation in a uniform temperature encloser are known as *black body radiation.*

8.9 EMISSIVE POWER AND ABSORPTIVE POWER

Emissive power: The emissive power e_λ of a body at a particular temperature (T) for a wavelength λ is defined as the energy emitted per second per unit surface area of the body within a unit wavelength range.

The energy radiated per second per unit surface area lying between λ and $\lambda + d\lambda$ is then given by $e_\lambda \, d\lambda$. The emissive power of a perfectly black body is maximum and is denoted by E_λ.

Absorptive power: The absorptive power a_λ of a body at a particular temperature T and for a particular wavelength λ is defined as the ratio of the amount of energy absorbed in a given time by the surface to the amount of energy incident on the surface in the same time.

8.10 KIRCHHOFF'S LAW

It states that the ratio of the emissive power to the absorptive power for a given wavelength at a given temperature is the same for all bodies and is equal to the emissive power of a perfectly black body at that temperature.

Proof

Suppose a_λ is the absorptive power of a body. Let dQ be the quantity of heat energy lying within the wavelengths λ and $\lambda + d\lambda$, incident on unit area of the surface in one second, then the quantity of heat energy absorbed by the surface is given by $a_\lambda\, dQ$. The remaining energy $(dQ - a_\lambda\, dQ) = dQ\,(1 - a_\lambda)$ will be either reflected or transmitted or both.

If e_λ is the emissive power, then the amount of energy emitted per second per unit area between wavelengths λ and $\lambda + d\lambda$ by virtue of the temperature, is given by $e_\lambda\, d\lambda$.

Thus, the total energy given out per second per unit area of the surface $= (1 - a_\lambda)\, dQ + e_\lambda\, d\lambda$

In equilibrium state, $dQ = (1 - a_\lambda)\, dQ + e_\lambda\, d\lambda$

or $$a_\lambda\, dQ = e_\lambda \cdot d\lambda \qquad \text{... (8.1)}$$

For a perfectly black body, $a_\lambda = 1$ and $e_\lambda = E_\lambda$

$\therefore$ $$dQ = E_\lambda\, d\lambda$$

Substituting in equation (8.1), we get

$$a_\lambda\, E_\lambda\, d\lambda = e_\lambda\, d\lambda$$

or $$E_\lambda = \frac{e_\lambda}{a_\lambda} \qquad \text{... (8.2)}$$

This is known as Kirchhoff's law.

Applications

Kirchhoff's law tells us that good absorbers are good emitters. If a body absorbs radiation of a particular wavelength strongly, it also emits the same radiation strongly. A sodium vapour which emits two yellow lines (D_1 and D_2 of wavelengths 5890 Å and 5896 Å) strongly is also a good absorber of light of these two wavelengths. The origin of Fraunhofer dark lines in sun's spectrum is successfully explained. When the radiation from the central glowing mass which gives continuous spectrum passes through the surrounding atmosphere containing various elements like hydrogen, nitrogen, sodium, caesium, copper etc., in the gaseous state, absorb those wavelengths which they can emit at a higher temperature. As a result, those wavelengths are missing from the sun's spectrum and dark lines are seen in their places. The nature and origin of characteristic spectra of elements is also successfully explained by the Kirchhoff's law.

8.11 PRESSURE OF RADIATION

1. Normal Incidence: The radiation possesses the properties of light. As we know, light exerts a small but definite pressure on the surface on which it is incident. Maxwell proved on the basis of his electromagnetic theory that the pressure is equal to the energy density (*i.e.*, the amount of radiation per unit volume) for normal incidence on a surface. The expression for the pressure of radiation is derived on the basis of quantum theory.

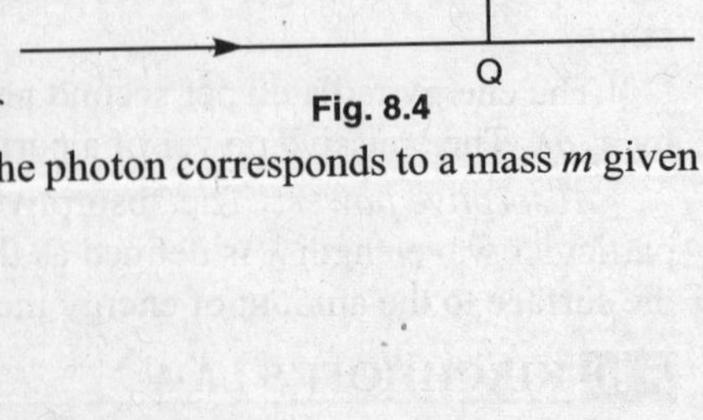

Fig. 8.4

Consider a photon of energy $h\nu$ moves with velocity of light c. According to the theory of relativity, this energy of the photon corresponds to a mass m given

$$mc^2 = h\nu$$

$$m = \frac{h\nu}{c^2}$$

Then, the momentum of the photon = mass × velocity

$$= \frac{h\nu}{c^2} \times c = \frac{h\nu}{c}$$

$$= \frac{e}{c}$$

e being energy of a photon.

Now momentum incident on the surface PQ per unit area per second is given by

$$P = \Sigma\frac{e}{c} = \frac{\Sigma e}{c} \qquad \text{... }(i)$$

but Σe = total energy incident on the surface per unit area per second = E (say).

Substituting in (*i*), we get

$$P = \frac{E}{c} \qquad \text{... }(ii)$$

If u is the energy density *i.e.*, energy per unit volume, then the total energy passing through area s of the surface normal to the incident radiation per second = ucs

∴ The energy flux, *i.e.*, energy radiation per unit area per second.

$$E = \frac{ucs}{s} = uc$$

Substituting in (*ii*), we get

$$P = \frac{uc}{c} = u \qquad \text{... }(iii)$$

If all of this momentum is absorbed, it will give rate of change of momentum per unit area *i.e.*, **the pressure of radiation** $\quad P = u \qquad$ ... (8.3)

Thus for normal incidence on the surface, the pressure of radiation is equal to the energy density.

2. Pressure of Diffuse Radiation: Suppose the radiation is diffuse, *i.e.*, the photons move in all possible directions. Let the incident beam of radiation make an angle θ with the normal to the surface LM.

The energy incident on the surface KL per unit area per second = uc.

If s is surface area of plane KL, the energy incident on the surface KL in 1 sec

$$= uc \,.\, s$$

But the surface area of the plane KL = surface area of the plane $LM \times \cos\theta$

$$= s' \cos\theta$$

where s' is surface area of plane LM.

∴ Energy incident on the surface KL per second

$$= uc \,.\, s' \cos\theta$$

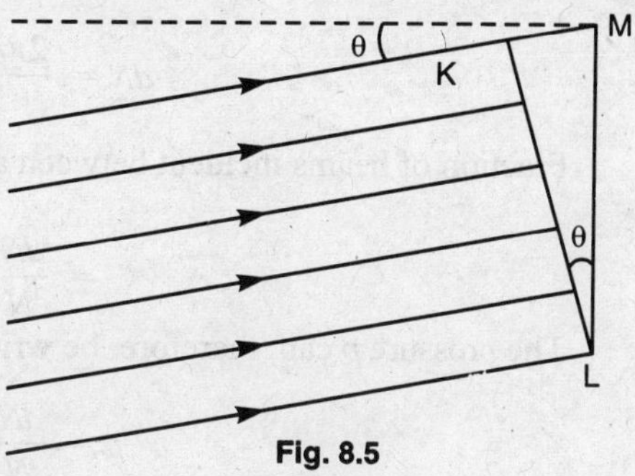

Fig. 8.5

From Fig. 8.5, it is obvious that the total energy crossing the plane LM per second is equal to the energy incident on the plane LM per seconc

∴ Energy incident on the plane LM per second

$$= uc \,.\, s' \cos\theta$$

∴ Energy incident on per unit area of the plane LM per second

$$= \frac{uc.\, s' \cos\theta}{\text{surface area of plane } LM}$$

$$= \frac{uc \,.\, s' \cos\theta}{s'} = uc\cos\theta \qquad \text{... }(iv)$$

Therefore, the momentum incident the surface LM per unit area per second

$$= \frac{uc\cos\theta}{c}$$

$= u \cos\theta$

The component of this momentum in a direction normal to the surface *LM*

$= (u \cos\theta) . \cos\theta$

$= u \cos^2\theta$ ... (*v*)

If all this momentum is absorbed, then equation (*v*) gives the total rate of change of momentum per unit area normal to *LM*, *i.e.*, the pressure of radiation on the surface *LM* is given by

$= u \cos^2\theta$ ... (*vi*)

In the case of diffuse radiations, the radiation is incident from all possible directions with equal probability. Therefore, if there are *N* beams of radiation meeting the surface, the total radiation pressure on the surface is given by

$= \Sigma u \cos^2\theta$ ... (*vii*)

In order to calculate $\Sigma u \cos^2\theta$, imagine a hemisphere of radius *r* (Fig. 8.6) around the centre *O* of the element *LM*.

Now from the surface of the hemisphere cut out a ring shaped element (*AB, CD*) by means of two cones of semi-angles θ and (θ + *d*θ) drawn from *O* as apex and *OY* (which is perpendicular to *LM*) as axis.

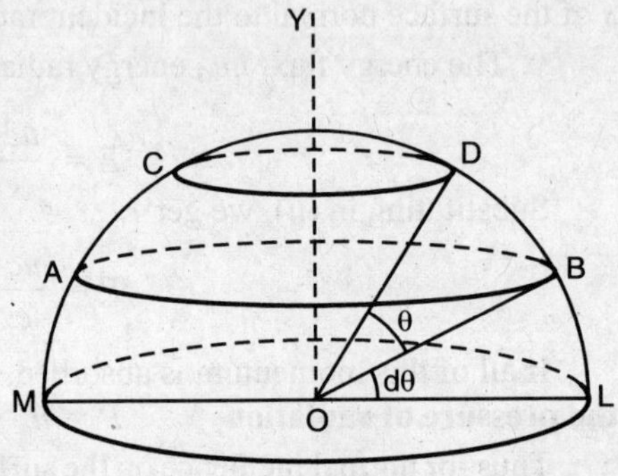

Fig. 8.6

The area of the ring

$= 2\pi r \sin\theta . r d\theta$

$= 2\pi r^2 \sin\theta \, d\theta$

As *N* beams are considered to be uniformly distributed over the surface of the hemisphere which is equal to $\left(\frac{1}{2} \times 4\pi r^2\right) = 2\pi r^2$, the number of beams crossing through the area of the ring

$$dN = \frac{2\pi r^2 \sin\theta \, d\theta}{2\pi r^2} N = N \sin\theta \, d\theta \quad \text{... (viii)}$$

Fraction of beams incident between angle θ and (θ + *d*θ)

$$= \frac{dN}{N} = \sin\theta . \, d\theta$$

The pressure *p* can, therefore, be written as

$$p = \frac{dN}{N} . \, u \cos^2\theta = \sin\theta \, d\theta . \, u \cos^2\theta$$

$$= u \cos^2\theta \sin\theta \, d\theta \quad \text{...(ix)}$$

For all possible rings within the hemisphere θ may vary from 0 to $\frac{\pi}{2}$, therefore equation (*ix*) in the integral form may be written as

$$P = \int_0^{\pi/2} u \cos^2\theta . \, \sin\theta \, d\theta$$

$$= u \int_0^{\pi/2} \cos^2\theta . \sin\theta \, d\theta$$

Now put $\cos\theta = x$

∴ $-\sin\theta \, d\theta = dx$

$$\therefore \quad P = u\int_1^0 x^2\,(-dx) = -u\int_0^1 x^2\,dx$$

$$= u\left[\frac{x^3}{3}\right]_0^1$$

$$P = \frac{u}{3} \qquad \text{... (8.4)}$$

$\therefore$ **Pressure of diffuse radiation = $\frac{1}{3}$ total energy density.**

8.12 STEFAN-BOLTZMANN LAW

It states that the rate of emission of radiant energy given out by unit area of a perfectly black body is directly proportional to the fourth power of its absolute temperature.

Thus,
$$E = \sigma T^4 \qquad \text{... (8.5)}$$
where σ is a constant called **Stefan's constant**. The value of σ is $5.67 \times 10^{-8}\ W m^{-2} k^{-4}$.

The Stefan's law refers to the emission of heat radiation only by the body itself, and not to the net loss of heat by the body after exchange with the surroundings. It simply deals with the amount of heat emitted by the body by virtue of its temperature irrespective of what it receives from the surroundings. The law can be extended to represent the net loss of heat and may be enunciated as follows:

If a black body at absolute temperature T is surrounded by another black body at absolute temperature T_o, the net rate of loss of heat energy per unit area of the surface is given by

$$E \propto (T^4 - T_o^4)$$

or
$$E = \sigma\,(T^4 - T_o^4) \qquad \text{... (8.6)}$$

This is in accordance with the Prevost's theory of heat exchanges, *i.e.*, the net loss of heat is really the difference in the heat radiated by hot body and that absorbed by it from its surroundings.

This law (given by eqn. 8.6) is known **Stefan Boltzmann's law** as Boltzmann, in 1884, gave a theoretical proof of Stefan's law on the basis of thermodynamics.

If a body has emissivity or relative emittance e, then total energy radiated by the body per second

$$= e\sigma\,(T^4 - T_o^4) \qquad \text{... (8.7)}$$

Thermodynamical proof

The black body radiation exerts pressure similar to a gas. Hence thermodynamical relations are also applicable to heat radiation.

Let u be the energy density of radiation inside a uniform temperature encloser at temperature T. Suppose P is the pressure and V is the volume of the encloser.

Applying the first law of thermodynamics $\delta Q = dU + PdV$

Applying the second thermodynamical relation

$$\left(\frac{\partial Q}{\partial V}\right)_T = T\left(\frac{\partial P}{\partial T}\right)_V$$

$$\left(\frac{\partial U + P\partial V}{\partial V}\right)_T = T\left(\frac{\partial P}{\partial T}\right)_V$$

$$\therefore \qquad \left(\frac{\partial U}{\partial V}\right)_T = T\left(\frac{\partial P}{\partial T}\right)_V - P \qquad \text{... } (i)$$

Since V is the volume of the encloser, then the total internal energy of radiation inside the encloser is given by

$$U = uV \qquad \text{... } (ii)$$

and, pressure exterted
$$P = \frac{u}{3} \qquad (\because \text{ eqn. 8.4}) \qquad \text{... } (iii)$$

or differentiating equation (*ii*) with respect to V, keeping T constant, we get

$$\left(\frac{\partial U}{\partial V}\right)_T = u$$

Here u is a function of temperature alone.

Substituting these values in equation (*i*), we get

$$u = \frac{T}{3}\frac{du}{dT} - \frac{u}{3}$$

$$\frac{4u}{3} = \frac{T}{3}\frac{du}{dT}$$

$$\frac{du}{u} = 4\frac{dT}{T}$$

Integrating, we get

$$\log u = 4 \log T + \text{constant} \qquad \text{... } (iv)$$

$$= 4 \log T + \log A$$

where $\log A$ being a constant of integration

$$\therefore \qquad \log u = \log T^4 + \log A$$

$$= \log AT^4$$

or
$$u = AT^4 \qquad \text{... } (v)$$

Also the total rate of emission per unit area of a black body is proportional to the energy density.

$$\therefore \qquad E \propto u \propto T^4$$

or
$$E = \sigma T^4 \qquad \text{... } (vi)$$

where σ is Stefan's constant. The value of Stefan's constant is 5.672×10^{-5} C.G.S. units and in M.K.S. system it is 5.672×10^{-8} M.K.S. units.

8.13 DISTRIBUTION OF ENERGY IN BLACK BODY SPECTRUM

The intensity of radiation emitted by a black body is not uniformly distributed over the whole range of wavelengths in the continuous spectrum emitted by it. It can be easily observed that when a blackened platinum wire is electrically heated, it becomes **dull red** at about 500°C, **cherry red** at about 900°C, **orange red** at 1100°C, **yellow** at 1250°C, and finally **white** at 1600°C. This shows that as the temperature rises the radiations emitted contain more and more of shorter wavelengths. In other words as *the temperature is raised the maximum intensity of emission shifts towards the shorter wavelength side.* In other words, the hotter an object is, the wavelength at which it will emit most of its radiation and the maximum (or peak) wavelength is found by dividing Wien's constant by the temperature in Kelvins.

When the principle of thermodynamics is applied to the black body radiation we obtain the following relations.

(*i*) $\qquad \lambda_m T = \text{a constant} \qquad$ ... (8.8) where λ_m is the wavelength corresponding to which the emissive power E_λ is a maximum at the absolute temperature T, the constant is also known as Wein's constant and

(*ii*) $$E_m T^{-5} = \text{constant} \qquad \text{... (8.8A)}$$

The first relation indicates in the mathematical form that as the temperature is raised the maximum intensity of radiation emitted is *displaced* towards the shorter wavelength side. This is, therefore, known as **Wein's displacement law.** The second relation is known as **Fifth power law.** The constant is called **Wien's displacement constant** and has a value 0.2896 cm K.

Wilhelm Wien in 1893 combined the two relations (8.8) and (8.8A) and proved that the energy distribution (for proof see article 8.14) in the wavelength range λ and $(\lambda + d\lambda)$ is given by

$$E_\lambda d\lambda = K\lambda^{-5} e^{-a/\lambda T} d\lambda$$

where $E_\lambda d\lambda$ represents the amount of energy dE associated with the spectral region lying in the wavelength range λ and $\lambda + d\lambda$ emitted by a black body at a temperature T

$$\therefore \qquad dE = K\lambda^{-5} e^{-a/\lambda T} d\lambda$$

This expression is known as **Wien's law of energy distribution.**

Experimental arrangement: Lummer and Pringsheim investigated the distribution of energy among the radiation emitted by a black body at different temperatures.

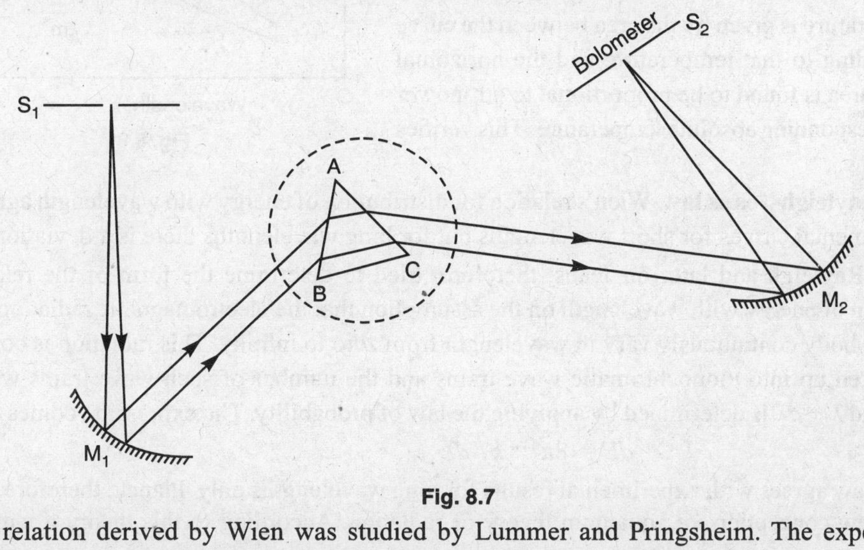

Fig. 8.7

The relation derived by Wien was studied by Lummer and Pringsheim. The experimental arrangement is shown in Fig. 8.7. Their experimental black body was a small aperture of an electrical heated chamber whose temperature was measured by a thermocouple.

The radiation from a black body pass through the slit S_1 and fall on the reflector M_1. After being reflected, the parallel beam of radiation fall on a rock-salt or fluorspar prism *ABC* placed on the rotating (or turn) table of the spectrometer. The emergent light is focused by the reflector (concave mirror) M_2 on a line bolometer placed behind the slit S_2. The bolometer is connected to a sensitive galvanometer. The turn table is rotated slowly so that the different parts of the radiation spectrum successively fall on the bolometer and the corresponding deflection in galvanometer connected in the bolometer circuit are read. The intensity of each line is proportional to the deflection in the galvanometer.

Practically they used a black body consisting of carbon plug at the middle of an electrically heated and thermally insulated carbon tube through which a slow stream of nitrogen was kept flowing to avoid quick oxidation. The thermal spectrum is obtained by dispersion through a Fluorspar prism, concave mirrors being used for collimating as well as converging the beam instead of lenses to avoid absorption. A linear bolometer consisting of a single strip of platinum foil 0.6 mm. in width, enclosed in an air tight case to minimise the absorption due to water vapour and carbon dioxide was used to measure the radiation energy corresponding to small ranges of wavelength along the spectrum. A number of observations were taken by maintaining the black body at various temperatures ranging from 620 K to 1650 K and E_λ the energy contained in a small range of wavelength was plotted against temperatures at which the black body was maintained.

Graph between E_λ and λ. The graph between E_λ and λ for various temperature (904 K, 1095 K, 1250 K, 1440 K and 1640 K) is shown in Fig. 8.8. The important results are discussed below.

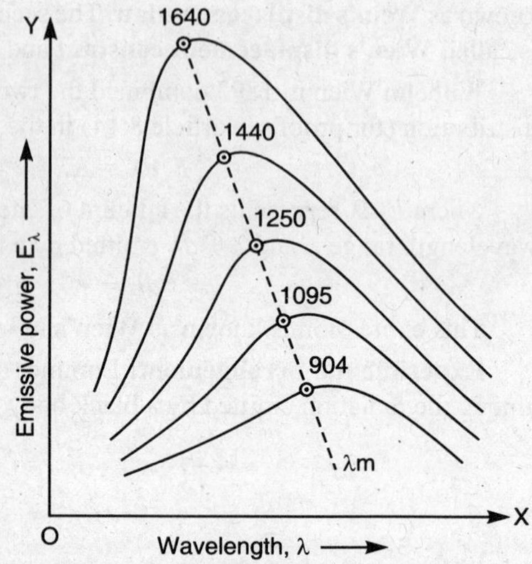

Fig. 8.8

Discussion. (*i*) Wien's displacement law. As shown in Fig. 8.8 the wavelength corresponding to maximum energy represented by the peak of the curve shifts to shorter wavelength side is as the temperature increases showing thereby that $\lambda_m . T$ = constant.

This confirms Wien's displacement law.

***(ii)* Stefan's law.** The total energy of radiation at any temperature is given by the area between the curve corresponding to that temperature and the horizontal axis. The area is found to be proportional to 4th-power of the corresponding absolute temperature. This verifies Stefan's law.

***(iii)* Rayleigh-Jeans law.** Wien's relation for distribution of energy with wavelength agrees with the experimental curves for short wavelengths but for long wavelengths there is a deviation.

Lord Rayleigh and later on Jeans, therefore, tried to determine the form of the relation for distribution of energy with wavelength on the assumption that the electromagnetic radiation emitted by a black body continuously vary in wavelength from zero to infinity. This radiation is considered to be broken up into monochromatic wave trains and the number of such wave trains within the range λ and $\lambda + d\lambda$ is determined by applying the law of probability. The expression comes out to be

$$dE = 8\pi\lambda^{-4}\, kT\, d\lambda$$

This law agrees with experimental results for long wavelengths only. Planck, therefore, gave an entirely new conception *i.e.*, quantum theory of radiation. According to this theory a radiation of frequency v has energy in discrete bundles or quanta each of value hv or $h\frac{c}{\lambda}$ where h is Planck's constant of value 6.65×10^{-27} ergs sec. and c is the velocity of light. Planck derived the relation

$$dE = \frac{8\pi hc\lambda^{-5}}{e^{hc/k\lambda T} - 1}\, d\lambda$$

This relation agrees with the experimental curve. The explanation of the distribution of energy with wavelength on the basis of quantum theory placed this theory on a very sound footing.

8.14 DERIVATION OF WIEN'S LAW OF ENERGY DISTRIBUTION

$\lambda_m T$ = constant is known as Wien's displacement law.

Wien has shown that the energy E_{max} is directly proportional to the fifth power of the absolute temperature

$$E_m \propto T^5$$

$$E_m = \text{Constant} \times T^5$$

or $$E_m . T^{-5} = \text{constant} \qquad ... (8.9)$$

The two relations (8.8) and (8.8A) can be combined into one and expressed as

$$E_\lambda = C\lambda^{-5} f(\lambda T) \qquad ... (8.10)$$

where E_λ is the emissiver power of the black body which has a maximum value at the absolute temperature T. Here C is a constant and $f(\lambda T)$ is some function of λT.

To determine the form of $f(\lambda T)$, Wien made some arbitrary but plausible assumptions concerning the actual mechanism of emission and absorption of radiation and concluded that the function should have the form $Ae^{-a/\lambda T}$.

$$\therefore \quad E_\lambda = C\lambda^{-5} Ae^{-a/\lambda T}$$
$$= K\lambda^{-5} e^{-a/\lambda T}$$

where K is another constant having value 4.94×10^{15} C.G.S. units and another constant and has a value 1.43 cm-degree. Multiplying both sides by $d\lambda$, we have

$$E_\lambda \, d\lambda = K\lambda^{-5} e^{-a/\lambda T} d\lambda$$

Now $E_\lambda \, d\lambda$ represents the amount of energy dE associated with the spectral region lying in the wavelength range λ and $(\lambda + d\lambda)$ emitted by a black body at a temperature T.

$$\therefore \quad dE = K\lambda^{-5} e^{-a/\lambda T} d\lambda \qquad ... (8.11)$$

This expression is known as **Wien's law of energy distribution.**

This law, however, was found not in accordance with the experimental curves of Lummer and Pringsheim. The success of Wien's law is that it holds good only in the region of shorter wavelengths at lower temperatures. It does not hold good at longer wavelengths at higher temperatures. This is shown in fig. 8.9.

8.15 RAYLEIGH-JEANS LAW

Lord Rayleigh and Sir J. Jeans tried to establish a relation for distribution of energy with wavelengths that could be made to fit with the experimental curves, on the assumption that the electromagnetic radiation spectrum emitted by a black body continuously vary in wavelengths from zero to infinity.

According to Rayleigh-Jeans law the energy distribution is given by the formula

$$E_\lambda \, d\lambda = \frac{8\pi kT}{\lambda^4} d\lambda \qquad ... (8.12)$$

where k is Boltzmann's constant.

This law, like Wien's law was found not in accordance with the experimental curves of Lummer and Pringsheim. The Rayleigh-Jeans law holds good in the region of longer wavelengths at higher temperatures but fails for shorter wavelengths. This was shown by Rubens and Kurlbaum. Thus, Wien's law and Rayleigh-Jeans law do not precisely agree with the experimental results. These are the short comings of both these laws.

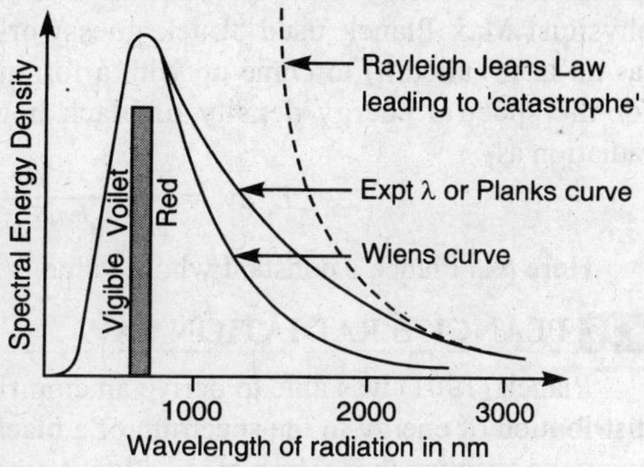

Fig. 8.9

8.16 THE ULTRAVIOLET CATASTROPHE

Rayleigh-Jeans law agrees well with the experimental results at low frequencies but near the maximum in the spectrum and at higher frequencies it is in violent disagreement (Fig. 8.10).

According to this law the energy density will be

$$E_\lambda\, d\lambda = \frac{8\pi\, kT}{\lambda^4}\, d\lambda$$

or $$dE = \frac{8\pi\, kT}{c^4} \times \nu^4 \times \frac{c}{\nu^2}\, d\nu \qquad \left(\because \lambda = \frac{c}{\nu},\ d\lambda = \frac{c}{\nu^2}\, d\nu\right)$$

$$dE = \frac{8\pi\, \nu^2\, kT\, d\nu}{c^3}$$

The energy density will continuously increase with increase in frequency ν and approaches ∞ when $\nu \to \infty$.

This is in contradiction to observed results. This fatal objection to the law has been known as ultraviolet catastrophe.

Further, at any temperature T, the total energy of radiation,

$$E = \int_0^\infty \frac{8\pi\, \nu^2\, kT\, d\nu}{c^3}$$

is predicted to be infinite which is against the Stefan's law.

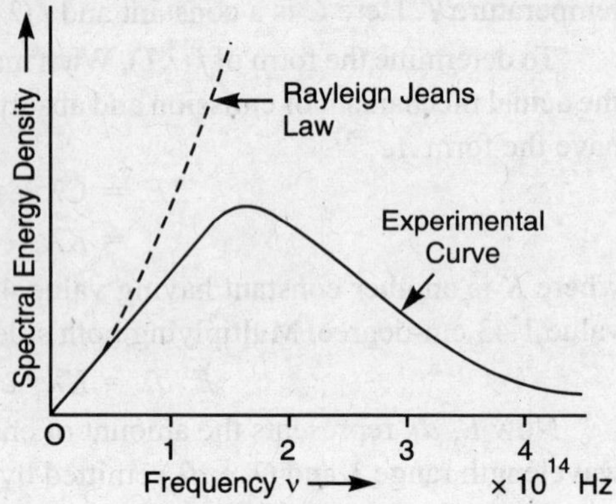

Fig. 8.10

Figure 8.10 shows the comparison of the Rayleigh-Jeans formula for the spectrum of the radiation from a black body at 1500 K with the observed spectrum. This discrepancy known as **ultraviolet catastrophe** (energy density increasing continuously with frequency) Fig. 8.11 was the fatal failure of classical physics. This failure of classical physics led Planck (in 1900) to the discovery that radiation is emitted in quanta whose energy is $h\nu$. In 1900, the German physicist Max Planck used "Luck guesswork" (as he later called it) to come up with a formula for the spectral energy density of black body radiation as

$$E_\nu\, d\nu = \frac{8\,\pi h}{c^3} \frac{\nu^3 d\nu}{e^{h\nu/kT} - 1}$$

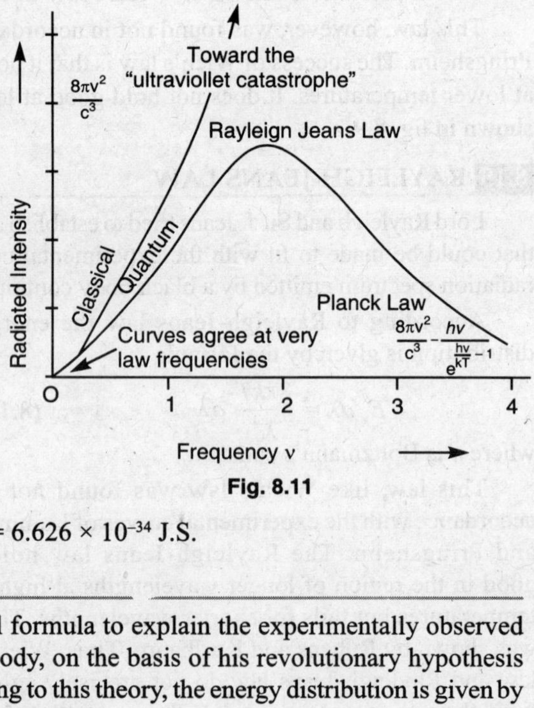

Fig. 8.11

Here h is Planck's constant whose value is $h = 6.626 \times 10^{-34}$ J.S.

8.17 PLANCK'S RADIATION LAW

Planck (1901) was able to derive an empirical formula to explain the experimentally observed distribution of energy in the spectrum of a black body, on the basis of his revolutionary hypothesis known as quantum theory heat of radiation. According to this theory, the energy distribution is given by

$$E_\lambda\, d\lambda = \frac{8\pi\, hc}{\lambda^5 \left[e^{hc/k\lambda T} - 1\right]}\, d\lambda \qquad \text{... (8.13)}$$

This relation agrees (and hence completely fit) with the experimental curves.

This formula of distribution of energy with wavelengths, on the basis of quantum theory, was deduced using following assumptions, which may be called as Planck's quantum postulates or Planck's hypothesis.

The planck's radiation formula in terms of frequency is given by

$$E_\nu \, d\nu = \frac{8\pi h\nu^3}{c^3} \cdot \frac{1}{e^{h\nu/KT} - 1} d\nu \qquad \text{... (8.13A)}$$

(This is proved is article 8.19)

8.18 PLANCK'S QUANTUM POSTULATES

1. A black body radiation chamber is filled up not only with radiation, but also with simple harmonic oscillators or resonators (energy emitters) of the molecular dimensions, known as Planck's oscillators or Planck's resonators, which can vibrate, with all possible frequencies. The vibration of the resonator entails one degree of freedom only.

2. The oscillators (or resonators) cannot radiate or absorb energy continuously, but energy is emitted or absorbed in the form of packets or quanta called photons. Planck assumed that each photon has an energy $h\nu$ where h is the Planck's constant, its value being equal to 6.625×10^{-34} Joule-sec, and ν is the frequency of radiation. This assumption is the most revolutionary in character. In other words, the theory states that the exchange of energy between radiation and matter cannot take place continuously but only in certain multiples of the fundamental frequency of the **resonator** (energy emitter). As the energy of a photon is $h\nu$, the energy emitted (or absorbed) is equal to 0, $h\nu$, $2h\nu$, $3h\nu$, ... $nh\nu$, *i.e.*, in multiplets of some small unit, called as **quantum**.

8.19 DERIVATION OF PLANCK'S RADIATION LAW

In order to derive Planck's radiation law, let us first derive the number of resonators per unit volume lying in the frequency range ν and $(\nu + d\nu)$ and then the average energy of Planck's resonator.

(*a*) Number of resonators per unit volume lying in the frequency range ν and $(\nu + d\nu)$: For this, consider the radiation to be enclosed in a hollow cubic encloser, the walls of which are perfectly reflecting. According to electromagnetic theory, the radiation is supposed to consists of a number of waves, travelling in all possible directions in the encloser and suffer multiple reflections from the various walls of the encloser. Inside the encloser, stationary waves are formed with the walls as nodal points, as a result of interference of the reflected waves with the corresponding incident waves. This is just like the vibrations of a stretched string with both end point fixed. With this analogy, it is well known that the nodes are formed at the end points of the string due to stationary vibrations and only certain discrete frequencies of vibrations are allowed. If l is the length of the string, the allowed wavelengths are given by

$$\lambda = \frac{2l}{n} \; ; \; n = 1, 2, 3, 4, ..., \infty. \qquad \text{... }(i)$$

The corresponding allowed frequencies are

$$\nu = \frac{c}{\lambda} = \frac{nc}{2l}; \quad n = 1, 2, 3, 4, ..., \infty.$$

where c is the speed of the waves.

Every allowed frequency is called a *mode of vibration*. The allowed modes of vibration inside a cube can be calculated just as in the case of string. But in this case, the waves are confined to a three-dimensional space. Let each side of the cube be a. If we take the three intersecting edges of the cube as X, Y and Z-axes of a Cartesian coordinate system and $\cos\alpha$, $\cos\beta$ and $\cos\gamma$ as direction cosines of the direction of propagation of a particular wave, then the projections of the edges of the cube on the direction of propagation of the wave are $a\cos\alpha$, $a\cos\beta$ and $a\cos\gamma$ respectively. In this case, only the waves will be allowed for which all the faces of the cube form nodal planes. For this, the allowed wavelengths have to satisfy three conditions of the kind of equation (*i*), in which l is replaced by $a\cos\alpha$, $a\cos\beta$ and $a\cos\gamma$ respectively. Thus, an allowed wavelength λ must satisfy

$$\lambda = \frac{2a\cos\alpha}{n_1}, \; \lambda = \frac{2a\cos\beta}{n_2} \text{ and } \lambda = \frac{2a\cos\gamma}{n_3} \qquad \text{... }(ii)$$

where n_1, n_2 and n_3 are positive integers.

Using trigonometric condition of direction cosines,

$$\cos^2\alpha + \cos^2\beta + \cos^2\gamma = 1 \qquad \text{... (iii)}$$

Substituting values of cos α, cos β and cos γ from equation (*ii*), we have

$$\frac{n_1^2\lambda^2}{4a^2} + \frac{n_2^2\lambda^2}{4a^2} + \frac{n_3^2\lambda^2}{4a^2} = 1$$

or

$$n_1^2 + n_2^2 + n_3^2 = \frac{4a^2}{\lambda^2} = \left(\frac{2a}{\lambda}\right)^2$$

As $\lambda = \frac{c}{\nu}$, we have

$$n_1^2 + n_2^2 + n_3^2 = \left(\frac{2a\nu}{c}\right)^2 \qquad \text{... (iv)}$$

This equation gives the allowed frequencies (or modes of vibration) inside the cube. The total number of modes of vibrations are the total number of possible sets (n_1, n_2, n_3). The number of modes of vibrations within the frequency interval ν and ν + *d*ν can be found with the help of equation (*iv*).

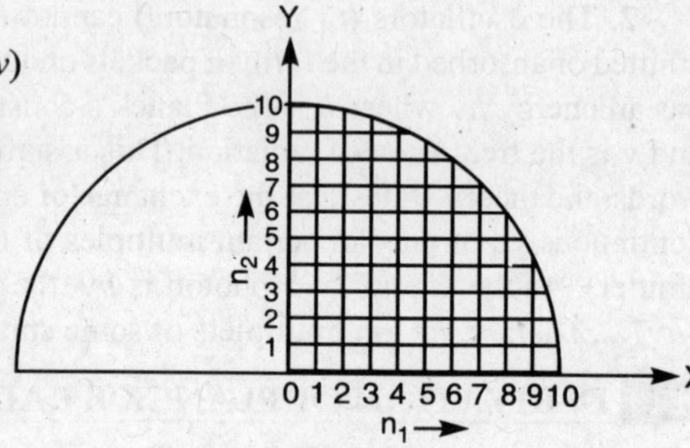

Fig. 8.12

For simplicity, let us first count the modes of vibration in an analogous two-dimensional problem. The two-dimensional equation will be

$$n_1^2 + n_2^2 = \left(\frac{2a\nu}{c}\right)^2 \qquad \text{.. (v)}$$

This represents a circle of radius $\left(\frac{2a\nu}{c}\right)$. Therefore, if we plot a graph with n_1 along X = axis and n_2 along Y = axis, for a given value of ν $\left(\textit{i.e.}, \text{ of } \frac{2a\nu}{c}\right)$, we get the points lying on a circle Fig. 8.12 shows a graph having n_1 and n_2 with all possible integral values upto 10.

The points on the circle correspond to the frequency *v*, while those lying inside the circle correspond to frequencies less than *v*. The lines drawn divide the quadrant in a number of unit squares. Every point of intersection can be associated with one square. Area of each square is unity.

∴ The number of squares = area inside the quadrant of the circle.

Thus, the number of modes of vibration within frequency interval *v* and *v* + *dv* is equal to the area in the positive quadrant lying between two circles of radii $\frac{2a\nu}{c}$ and $\frac{2a(\nu+d\nu)}{c}$.

The area is

$$= \frac{1}{4}\pi\left[\left\{\frac{2a(\nu+d\nu)}{c}\right\}^2 - \left(\frac{2a\nu}{c}\right)^2\right]$$

$$= \frac{\pi}{4} \times \frac{8a^2\nu d\nu}{c^2} = \frac{2\pi a^2\nu}{c^2}d\nu$$

Extending the idea to 3-dimensions, the number of modes of vibration within frequency range ν and ν + *d*ν is now equal to the volume of the constant $\left(\frac{1}{8}th\right)$ of the spherical shells with radii $\frac{2a\nu}{c}$ and $\frac{2a(\nu+d\nu)}{c}$.

$$\text{This volume is} = \frac{1}{8}\left[\frac{4}{3}\pi\left(\frac{2a\nu}{c}\right)^3 - \frac{4}{3}\pi\left\{\frac{2a(\nu+d\nu)}{c}\right\}^3\right]$$

$$= \frac{1}{8}\times\frac{4}{3}\pi\times\frac{8a^3}{c^3}[\nu^3 - \nu^3 + 3\nu^2 d\nu]$$

$$= \frac{4\pi\, a^3\, \nu^2}{c^3} d\nu$$

But $a^3 = V$ = volume of the cube.

Therefore, the modes of vibration inside cubical encloser with frequency range ν and $\nu + d\nu$ will be

$$= \frac{4\pi\, V\nu^2 d\nu}{c^3}$$

$\therefore$ Number of modes of vibration per unit volume with frequency range ν and $(\nu + d\nu)$ will be

$$= \frac{4\pi\, \nu^2 d\nu}{c^3}$$

Since, the black body radiation travel with velocity of light c and are transverse in character and hence modes of vibrations are double as that for longitudinal waves.

Therefore, for black body radiation or electromagnetic waves, the number of modes of vibration per unit volume within frequency range ν and $\nu + d\nu$ will be

$$= 2\times\frac{4\pi\, \nu^2 d\nu}{c^3} = \frac{8\pi\, \nu^2 d\nu}{c^3} \qquad \text{... (vi)}$$

***(b)* Average energy of Planck's oscillator:** If N is the total number of Planck's resonators and E their total energy, then average energy per Planck's oscillator is given by

$$\bar{\varepsilon} = \frac{E}{N} \qquad \text{... (vii)}$$

According to Maxwell's law of molecular motion, the energies 0, ε, 2ε, 3ε, ..., $r\varepsilon$, ... are in the ratio $1 : e^{-\varepsilon/kT} : e^{-2\varepsilon/kT} : e^{-3\varepsilon/kT}$: .... : $e^{-r\varepsilon/kT}$ .... etc.

If N_0 is the number of resonators having energy zero, then the number of resonators N_1 having energy ε will be $N_0\, e^{-\varepsilon/kT}$, the number of resonators N_2 having energy 2ε will be $N_0\, e^{-2\varepsilon/kT}$ and in general, the number of resonators N_r, having energy $r\varepsilon$ will $N_0\, e^{-r\varepsilon/kT}$ and so on.

$$\therefore \qquad N = N_0 + N_1 + N_2 + \dots\dots + N_r + \dots\dots$$

$$= N_0 + N_0 e^{-\varepsilon/kT} + N_0\, e^{-2\varepsilon/kT} + \dots\dots + N_0\, e^{-r\varepsilon/kT} + \dots\dots$$

$$= N_0\, [1 + e^{-\varepsilon/kT} + e^{-2\varepsilon/kT} + \dots\dots + e^{-r\varepsilon/kT} + \dots\dots]$$

Putting $e^{-\varepsilon/kT} = y$, we have

$$N = N_0\, [1 + y + y^2 + \dots\dots + y^r + \dots\dots]$$

or

$$N = \frac{N_0}{1-y}$$

The total energy of Planck's resonator will be

$$E = o \times N_0 + \varepsilon \times N_1 + 2\varepsilon \times N_2 + \dots\dots + r\varepsilon \times N_r + \dots\dots$$

$$= \varepsilon\, N_0\, e^{-\varepsilon/kT} + 2\varepsilon\, N_0\, e^{-2\varepsilon/kT} + \dots\dots + r\varepsilon\, N_0\, e^{-r\varepsilon/kT} + \dots\dots$$

$$= N_0\, \varepsilon\, [e^{-\varepsilon/kT} + \dots\dots + re^{-r\varepsilon/kT} \dots\dots]$$

$$= N_0\, \varepsilon\, \frac{y}{(1-y)^2}$$

Therefore, the average energy of a resonator will be

$$\bar{\varepsilon} = \frac{E}{N} = \frac{N_0\,\varepsilon\,.\,\dfrac{y}{(1-y)^2}}{\dfrac{N_0}{1-y}}$$

$$= \frac{\varepsilon y}{1-y} = \frac{\varepsilon e^{-\varepsilon/kT}}{1-e^{-\varepsilon/kT}} = \frac{\varepsilon}{e^{\varepsilon/kT}-1} \qquad \text{... }(viii)$$

According to Planck's hypothesis of quantum theory, $\varepsilon = h\nu$, therefore, the average energy of Planck's oscillator is given by

$$\bar{\varepsilon} = \frac{h\nu}{e^{h\nu/kT}-1} \qquad \text{... }(ix)$$

Therefore, the *energy density* belonging to the range $d\nu$ can be obtained by multiplying the average energy of Planck's oscillator by the number of oscillators per unit volume, in this frequency range ν and $(\nu + d\nu)$.

i.e.,

$$E_\nu\, d\nu = \left(\frac{8\pi\nu^2}{c^3}\, d\nu\right)\left(\frac{h\nu}{e^{h\nu/kT}-1}\right)$$

or

$$= \frac{8\pi\, h\nu^3}{c^3} \cdot \frac{1}{e^{h\nu/kT}-1}\, d\nu \qquad \text{... (8.14)}$$

where $E_\nu\, d\nu$ is energy density (*i.e.,* total energy per unit volume) belonging to the range $d\nu$. Equation (8.14) is called *Planck's radiation law.*

The energy density $E_\lambda\, d\lambda$ belonging to range $d\lambda$ can be obtained by using the relation $\nu = \dfrac{c}{\lambda}$ and hence $|d\nu| = \left|-\dfrac{c}{\lambda^2}\, d\lambda\right|$ we get

$$E_\lambda\, d\lambda = \frac{8\pi h}{c^3}\left(\frac{c^3}{\lambda^3}\right) \cdot \frac{1}{e^{hc/\lambda kT}-1} \cdot \left|-\frac{c}{\lambda^2}\, d\lambda\right|$$

$$= \frac{8\pi hc}{\lambda^5} \cdot \frac{d\lambda}{e^{hc/\lambda kT}-1} \qquad \text{... (8.15)}$$

This is *Planck's radiation law* in terms of wavelength.

8.20 DERIVATION OF STEFAN'S LAW

The fact that black body radiations exert pressure similar to a gas, helps in applying thermodynamics to heat radiation.

Let ψ be the energy density of radiation inside a uniform temperature encloser at temperature T. P is the pressure and V is the volume. Applying the first law of thermodynamics

$$\delta Q = dU + P\, dV \qquad \text{... }(i)$$

Applying thermodynamical relation

$$\left(\frac{\partial Q}{\partial V}\right)_T = T\left(\frac{\partial P}{\partial T}\right)_V \qquad \text{... }(ii)$$

$$\left(\frac{\partial U + P\partial V}{\partial V}\right)_T = T\left(\frac{\partial P}{\partial T}\right)_V$$

$$\left(\frac{\partial U}{\partial V}\right)_T = T\left(\frac{\partial P}{\partial T}\right)_V - P \qquad \text{... }(iii)$$

Now $$U = V\psi$$

and $$P = \frac{\psi}{3}$$

or $$\left[\frac{\partial U}{\partial V}\right]_T = \psi$$

Here ψ is a function of temperature alone.

Substituting these values in equation (*iii*),

$$\psi = \frac{T}{3}\frac{d\psi}{dT} - \frac{\psi}{3}$$

$$\frac{4\psi}{3} = \frac{T}{3}\frac{d\psi}{dT}$$

or $$\frac{d\psi}{\psi} = 4\frac{dT}{T}$$

Integrating, $$\log \psi = 4 \log T + \text{constant}$$

or $$\psi = KT^4 \qquad \text{... (iv)}$$

Here K is a constant

Also the total rate of emission per unit area of a black body is proportional to the energy density.

$\therefore$ $$R \propto \psi \propto T^4$$

$\therefore$ $$R = \sigma T^4 \qquad \text{... (8.16)}$$

where σ is Stefan's constant.

The value of Stefan's constant in C.G.S. system is 5.672×10^{-5} C.G.S. units and in M.K.S. system it is 5.672×10^{-8} M.K.S. units.

8.21 DERIVATION OF NEWTON'S LAW OF COOLING FROM STEFAN'S LAW

Stefan's law is applicable for all temperatures of a hot body. But Newton's law is applicable when the difference of temperature between the hot body and the surrounding is small. Consider a hot body at a temperature T_1 placed in a uniform temperature enclosure at T_2. According to Stefan's law,

$$R = e\sigma\,(T_1^4 - T_2^4)$$

Here e is the emissivity of the surface of the hot body

$$R = e\sigma\,(T_1 - T_2)\,(T_1^3 + T_1^2\,T_2 + T_1\,T_2^2 + T_2^3)$$

As $(T_1 - T_2)$ is small, T_1 can be taken approximately equal to T_2.

Then, $$R = e\sigma\,(T_1 - T_2)\,(T_2^3 + T_2^3 + T_2^3 + T_2^3)$$

$$R = 4\,e\sigma T_2^3\,(T_1 - T_2)$$

Taking $$4\,e\sigma T_2^3 = k$$

$$R = k\,(T_1 - T_2)$$

or $$R \propto (T_1 - T_2) \qquad \text{... (8.17)}$$

This equation represents Newton's law of cooling and is true when the difference of temperature is small.

8.22 EXPERIMENTAL VERIFICATION OF STEFAN'S LAW

In 1897, Lummer and Pringsheim experimentally verified Stefan's law over a wide range of temperature (100°C to 1,300°C).

The apparatus consisted of a black body C. For temperatures between 200°C and 600°C, a hollow copper sphere coated inside with platinum black was used. The fused nitrates of sodium and potassium having a melting point of 219°C were used as the bath surrounding the black body. For temperature between 900°C and 1,300°C, an iron cylinder coated inside with platinum black was used as a black body and it was enclosed in a double walled gas furnace. A thermocouple T was used as a thermometer. A bolometer B was used to measure the intensity of the emitted heat radiations. S_1, S_2 and S_3 were the water-cooling shutters (Fig. 8.13).

Another black body A at 100°C was used to standardize the bolometer. The double walled vessel of the black body A containing boiling water at 100°C. The bolometer B was allowed to face the opening of the black body A and the shutter S_3, raised. The deflections in the galvanometer of the bolometer at various distances were noted and it was found that the deflection was inversely proportional to the square of the distance between the bolometer and the opening of the black body A. Thus the deflection in the galvanometer was proportional to the intensity of heat radiations.

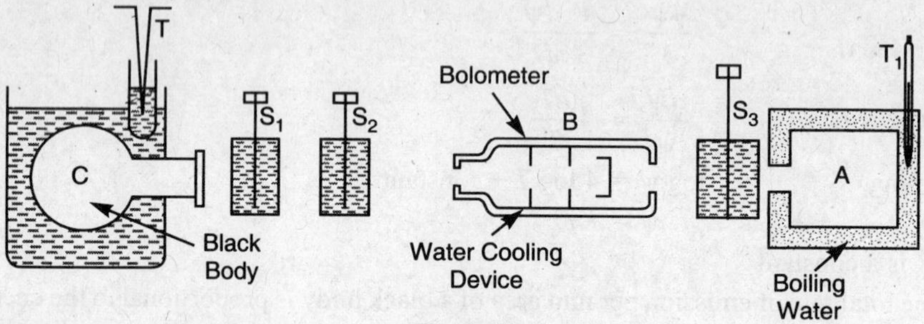

Fig. 8.13

The shutter S_3 was closed and the bolometer B was allowed to face the opening of the black body C. The shutters of S_1 and S_2 were raised. The bath surrounding the black body was maintained a constant temperature and the maximum deflection produced in galvanometer of the bolometer was noted. Thus at various constant temperatures of the black body, corresponding to constant deflections (in the galvanometer of the bolometer) were observed. Then the data was reduced to a common arbitrary unit in terms of the total radiations from the black body A at 100°C.

Let θ be the deflection in the galvanometer, T_1 the temperature of the black body and T_2, the temperature at the entrance of the bolometer. It was found that

$$R \propto (T_1^4 - T_2^4)$$

But

$$\theta \propto R$$

$\therefore$

$$\theta \propto (T_1^4 - T_2^4) \qquad \text{... (8.18)}$$

This verifies Stefan's law.

Recently Coblentz has verified Stefan's law more accurately. He took an electrically heated black body whose temperature was measured by an accurate thermocouple. An absolute bolometer was used to measure the amount of heat radiations emitted by the black body. He was able to show the correctness of Stefan's law experimentally up to 1,600°C.

8.23 DETERMINATION OF STEFAN'S CONSTANT (LABORATORY METHOD)

The laboratory apparatus used to determine the Stefan's constant is shown in Fig. 8.14. A hollow hemispherical metallic vessel A is enclosed in a wooden box W. The inner surface of A is coated with lamp black and the wooden box W is lined with tin plates. The whole apparatus is placed on a wooden base having a small hole at its centre. The vessel A is heated by passing steam inside the box and A acts as a black body radiator. The thermometers T T record the temperature of A.

A small silver disc B whose upper surface is coated with lamp black is placed at the central hole. The ebonite covering C is used to cover and uncover the disc B from the radiations of the enclosure. It can be arranged from outside with the help of the handle H. The disc B is connected to a thermocouple arrangement. One junction of the thermocouple is immersed in a tube containing oil. The tube is surrounded by a beaker containing water. A sensitive galvanometer G is used in the circuit. The leads connected to the terminals of the galvanometer are immersed in cotton wool in the box *F* to avoid any distribution effect due to the difference of temperature in the leads. A rheostat Rh can be used in the circuit to obtain the deflection within the range. The actual experiment consists of two parts.

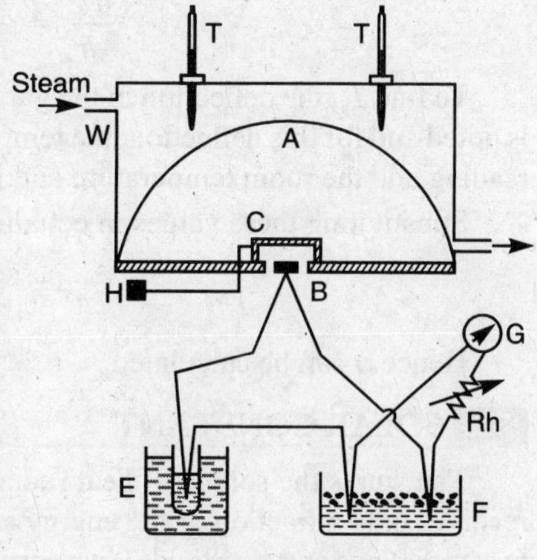

Fig. 8.14

1. The thermocouple is first standardized. Before passing steam into the chamber, the disc *B* is at the room temperature. The water bath *E* acts as a hot junction. It is heated and at various temperatures of the hot junction, the corresponding deflections in the galvanometer are noted. A graph between the difference of temperature of the hot junction and the room temperature along the *Y*-axis and galvanometer deflection along *X*-axis is plotted (Fig. 8.15). From the graph,

$$\frac{dT}{d\theta} = \tan\alpha = \frac{AB}{BC} \qquad \dots (i)$$

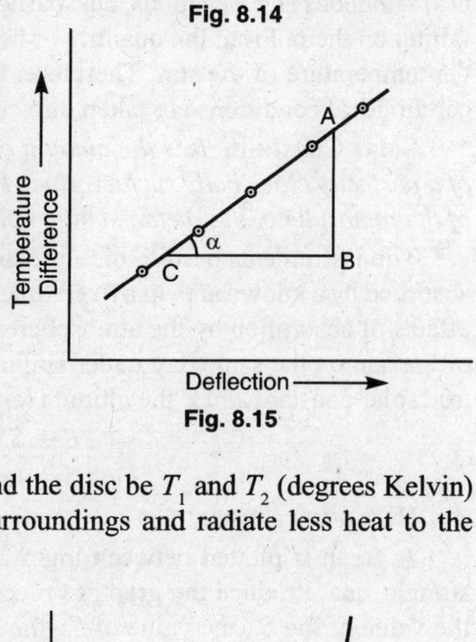

Fig. 8.15

2. The disc is completely covered with *C* and steam is passed into the chamber. After sometime, the thermometers *T T* show constant temperature. The bath *E* is kept at room temperature. With the help of the handle *H*, the cover *C* is tilted so that the upper surface of the disc *B* receives the radiations from the enclosure. The deflections in the galvanometer are observed after equal intervals of time (say 10 seconds). A graph is plotted between time and deflection (Fig. 8.16). A tangent is drawn on the curve at a point *D*.

$$\frac{dt}{d\theta} = \tan\beta = \frac{EF}{GF} \qquad \dots (ii)$$

Theory

Let, at any instant, the temperature of the enclosure and the disc be T_1 and T_2 (degrees Kelvin) respectively. The disc will absorb more heat from the surroundings and radiate less heat to the surroundings. Its temperature will rise. From Stefan's law,

$$R_1 = \sigma T_1^4 \text{ and } R_2 = \sigma T_2^4$$

$$\therefore \quad (R_1 - R_2) = \sigma(T_1^4 - T_3^4) \qquad \dots (iii)$$

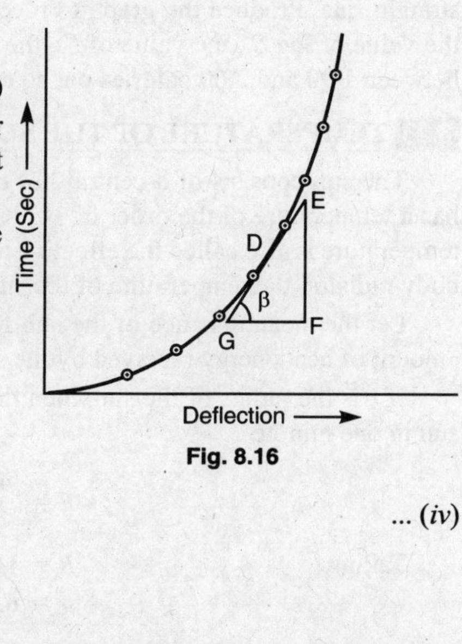

Fig. 8.16

Here, R_1 is the amount of heat radiation absorbed per unit area per second by the disc and R_2 is the amount of heat radiation emitted per unit area per second by the disc.

Let the mass of the disc be *m*, specific heat *S*, rate of rise of temperature *dT*/*dt*, and area of the upper surface *A*.

Then $$\frac{(R_1 - R_2)A}{J} = mS\frac{dT}{dt}$$

$$\frac{\sigma(T_1^4 - T_2^4)A}{J} = mS\frac{dT}{dt}$$

$$\sigma = \frac{JmS}{A(T_1^4 - T_2^4)} \times \frac{dT}{dt} \qquad \dots (iv)$$

To find $\frac{dT}{dt}$, equations (*i*) and (*ii*) are used.

$$\frac{dT}{dt} = \frac{dT}{d\theta} \times \frac{d\theta}{dt} = \frac{\tan\alpha}{\tan\beta}$$

To find T_2, the deflection in the galvanometer corresponding to the point D on graph in Fig. 8.16 is noted and for this deflection, the temperature difference from the graph (Fig. 8.15) is noted. To this reading add the room temperature and find T_2 in degrees Kelvin.

Substituting these values in equation (*iv*)

$$\sigma = \frac{JmS \tan\alpha}{A\,(T_1^4 - T_2^4)\tan\beta} \quad \text{... (8.19)}$$

Hence σ can be calculated.

8.24 SOLAR CONSTANT

The sun is the source of heat radiations and it emits heat radiations in all directions. The earth receives only a fraction of the energy emitted by the sun. The atmosphere also absorbs a part of the heat radiations and air, clouds, dust particles etc. in the atmosphere scatter the heat and light radiations falling on them. From the quantity of heat radiations received by the earth, it is possible to estimate the temperature of the sun. Therefore, to determine the value of a constant, called *solar constant,* certain ideal conditions are taken into consideration.

Solar Constant. *It is the amount of heat energy (radiation) absorbed per minute by the sq cm of a perfectly black body surface placed at a mean distance of the earth from the sun, in the absence of the atmosphere, the surface being held perpendicular to the sun's rays.*

The instruments used to measure the solar constant are called pyrheliometers. The heat energy absorbed by a known area in a fixed time is found with the help of the pyrheliometer. To eliminate the effects of absorption by the atmosphere, the value of the solar constant is found at various altitudes of the sun on the same day under similar sky conditions. If S is the observed solar constant, S_0 the true solar constant and Z the altitude (angular elevation) of the sun, then

$$S = S_0\, a^{\sec Z}$$

or

$$\log S = \log S_0 + \sec Z \log a \quad \text{... (8.20)}$$

Here a is a constant.

A graph is plotted between log S along the y-axis and sec Z along the x-axis. The graph is a straight line. Produce the graph to meet the y-axis. The intercept on the y-axis gives log S_0. From the value of log S_0, the value of S_0, the solar constant can be calculated. The value obtained varies between 1.90 and 2.60 calories per sq cm per minute.

8.25 TEMPERATURE OF THE SUN

The sun consists of a central hot portion surrounded by the photosphere. The central portion has a temperature of the order of 10^7 K. The photosphere has a temperature of about 6000 K. This temperature is also called the effective temperature of the sun. Considering the sun as a perfect black body radiator, the temperature of the sun can be calculated.

Let the mean distance of the sun from the earth be R and S the solar constant. Then, the total amount of heat energy received by the sphere of radius R in one minute = $4\pi R^2 S$.

If r is the radius of the sun, then the amount of heat energy radiated by 1 sq cm surface of the sun in one minute

$$E = \frac{4\pi R^2 S}{4\pi r^2} = \left(\frac{R}{r}\right)^2 \times S$$

Taking

$$R = 148.48 \times 10^7 \text{ km}$$

$$r = 6.928 \times 10^5 \text{ km}$$

The mean value of

$$S = 1.94 \text{ cals per cm}^2 \text{ per minute.}$$

$$E = \left(\frac{148.48\times10^7}{6.928\times10^5}\right)^2 \times \frac{1.94}{60} \text{ cals per second} \quad ...(i)$$

Also $\quad E = \sigma T^4$

But $\quad \sigma = 5.75 \times 10^{-5}$ ergs per cm² per second

$$= \frac{5.75\times10^{-5}}{4.2\times10^7} \text{ cals per cm}^2 \text{ per second}$$

$$E = \frac{5.75\times10^{-5}}{4.2\times10^7} . T^4 \quad ...(ii)$$

Equating (*i*) and (*ii*)

$$\left(\frac{5.75\times10^{-5}}{4.2\times10^7}\right)T^4 = \left(\frac{148.48\times10^7}{6.928\times10^5}\right)^2 \times \frac{1.94}{60} \quad ...(iii)$$

$$\therefore \quad T = 5730 \text{ K}$$

This temperature gives the effective temperature of the sun acting as a black body radiator. The actual temperature of the sun is higher than this value. The temperature of the sun is usually taken as 6000 K.

Temperature of the sun can also be calculated from Wien's displacement law,

$$\lambda_{max} T = 0.2892 \quad ...(8.21)$$

The wavelength of the radiations for which the energy is maximum in the spectrum is 4900×10^{-8} cm.

Substituting the value of λ_m, the value of T comes out to be 5902 K. This value is in agreement with the accepted value. Hence, the effective temperature of the sun (photosphere) is about 6000 K.

Let us calculate the black body temperature of the sun from the following data:

(Purvanchal 2003, Bangalore 2004, Mysore 2004)

Stefan's constant = 1.37×10^{-12} cal/cm²/s

Solar constant = 2.3 cal/cm²/minute

Radius of the sun = 7×10^{10} cm

Distance between the sun and the earth

$= 1.5 \times 10^{13}$ cm

$$E = \frac{4\pi R^2 S}{4\pi r^2} = \left(\frac{R}{r}\right)^2 . S$$

Here, $\quad R = 1.5 \times 10^{13}$ cm

$r = 7 \times 10^{10}$ cm

$S = 2.3$ cal/cm²/minute

$$\therefore \quad E = \left[\frac{1.5\times10^{13}}{7\times10^{10}}\right]^2 \times \frac{2.3}{60} \text{ cal/s} \quad ...(iv)$$

$$E = \sigma T^4 \quad ...(v)$$

Equating (*iv*) and (*v*), we get

$$\sigma T^4 = \left[\frac{1.5\times10^{13}}{7\times10^{10}}\right]^2 \times \frac{2.3}{60}$$

$$1.37 \times 10^{-12} \times T^4 = \left[\frac{1.5\times10^{13}}{7\times10^{10}}\right]^2 \times \frac{2.3}{60}$$

$$T = \mathbf{5987\ K}$$

8.26 TOTAL RADIATION PYROMETER

This pyrometer was first devised by Fery. It is used to measure the total energy of radiation of all wavelengths.

It consists of a concave mirror M made of copper and plated with nickel on the front surface. E is an eye-piece fitted at the pole of the mirror through a hole. D is a diaphragm placed in front of the mirror and at its focus. The position of M can be adjusted with the help of rack and pinion arrangement P. Here S is a metal strip painted black on the side facing the mirror and one of the junctions of a thermocouple T is connected to S. A sensitive calibrated galvanometer is connected in the thermocouple arrangement. A shield Q is used to surround S so that the radiation is not incident directly on S (Fig. 8.17).

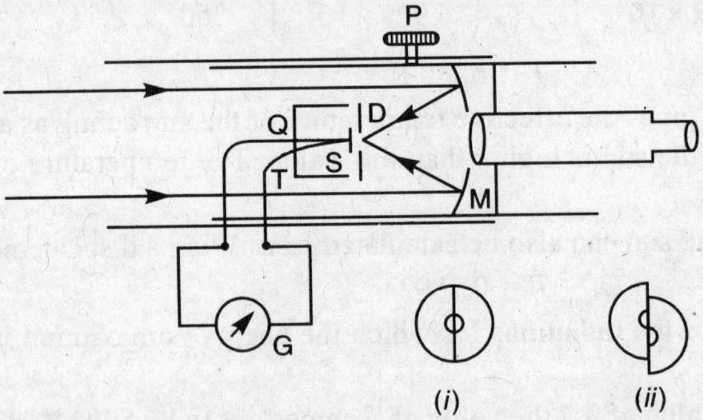

Fig. 8.17

When the radiation is incident on the mirror, the focussed image falls on the opening of D. Due to the incident radiations falling on S, the galvanometer shows a deflection. In order to focus the image accurately, the diaphragm D is made of two semicircular mirrors inclined at a small angle to each other. When the diaphragm D is viewed through the eyepiece it appears to be circular as shown in Fig. 8.17 (*i*), if the focussing is accurate. If focussing is not proper, the two halves appear displaced as shown in Fig. 8.17 (*ii*).

The image formed on D must completely cover the aperture. The reading of the galvanometer of the thermocouple is independent of the distance of the source provided the aperture of D is completely filled by the image of the source (furnace etc.). The reading of the galvanometer depends only on the intensity of the image. If the distance of the source is doubled, the amount of radiation received by the mirror becomes one-fourth. But the area of the image is also reduced to one-fourth. Thus the intensity of the image remains the same.

Calibration

If the temperature of the source is T and that of the receiving instruments is T_0 the deflections of the galvanometer are not proportional to $(T^4 - T_0^4)$. This is because T is large as compared to T_0. The instrument is directly calibrated with source at various temperatures. (The temperature of the source is measured with a standard thermocouple or an optical pyrometer.) A graph is obtained between the logarithm of the deflection of the galvanometer and the logarithm of the absolute temperature T.

$$\log \theta = \text{K} \log T$$

The graph will be a straight line and this method gives better results. Accurate experiments show that due to the conduction along the wires of the thermocouples and stray e.m.f. generated, Stefan's fourth power law is not strictly applicable. The power of T varies from 3.8 to 4.2.

Rotating Sector Device

To measure the temperature over a wide range, a rotating sector device is used (Fig. 8.18). It instrument is directly calibrated with source at various temperature.

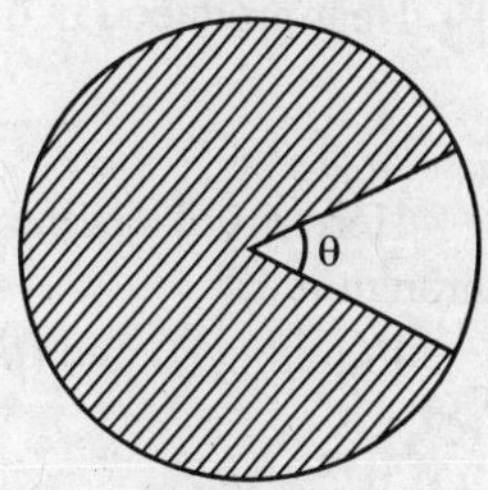

Fig. 8.18

To measure the temperature over a wide range, a rotating sector device is used (Fig. 8.18.) It consists of a circular disc with a sector of suitable angle θ cut from the disc. The disc is rotated about the axis of the pyrometer and it limits the amount of radiation incident on the mirror.

Let T be the temperature of the source and T_1 the temperature measured with the instrument. Then

$$\frac{T^4}{T_1^4} = \frac{360}{\theta}$$

$$T^4 = \frac{T_1^4 (360)}{\theta}$$

$$T = T_1 \left[\frac{360}{\theta}\right]^{\frac{1}{4}} \quad \text{... (8.17)}$$

8.27 ANGSTROM'S PYRHELIOMETER

Pyrheliometer is an instrument which is used to find the amount of incident heat radiations and the solar constant. Angstrom's pyrheliometer consists of two identical strips A and B of blackened platinum foil (Fig. 8.19).

The strip A is exposed to the sun and B is shielded by a cover C. A thermocouple having a sensitive galvanometer with A as one junction and B as the other junction is used. The strip B can be heated by an electrical arrangement and suitable current passing through B can be adjusted with the help of a rheostat.

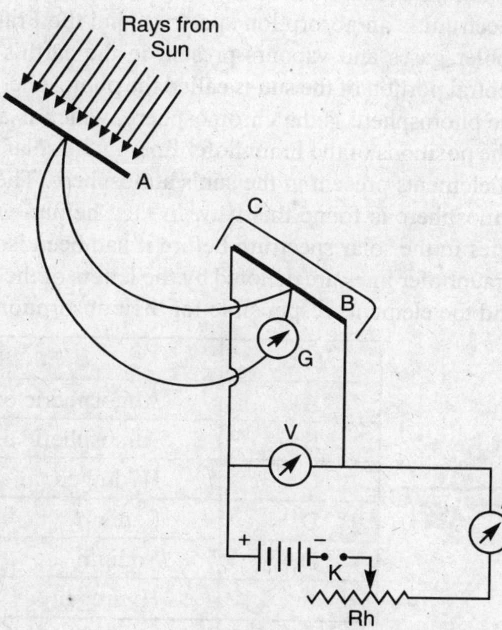

Fig. 8.19

When both the strips A and B are shielded from the sun, their junctions are at the same temperature and the galvanometer shows no deflection. The strip A is exposed to the sun and B is kept shielded. The strip A receives heat radiations from the sun and its temperature rises and the galvanometer shows deflection. The strip B is gradually heated by passing current through it and the current is adjusted so that there is no deflection in the galvanometer.

It means the strips A and B are again at the same temperature and they are receiving heat energy at the same rate.

Let H calories of heat be incident on one sq cm surface of the strip in one minute. The area of the plate $A = A$ sq cm and absorption coefficient = a.

The amount of heat radiations absorbed in one minute by the plate $A = H\,Aa$ calories. ... (*i*)

Heat produced in one minute in the strip B

$$= \frac{EI \times 60}{4.2} \text{ cal} \quad \text{... (ii)}$$

Here E volts is the potential difference across the strip B and I amperes is the current flowing through it.

Equating (*i*) and (*ii*),

$$H\,Aa = \frac{EI \times 60}{4.2} \quad \text{... (8.23)}$$

Hence H can be calculated.

8.28 SOLAR SPECTRUM

The radiation received from the sun is similar to that of a perfectly black body. In a solar spectrum, the wavelength corresponding to maximum energy is about 5000 Å for a surface temperature of 5750 K. The solar spectrum consists of a large number of dark lines called **Fraunhofer lines**. These lines were first observed by Wollaston in 1802 and later studied by Fraunhofer in 1814. These lines are observed in the complete range of the spectrum viz. ultraviolet, visible and infrared. Fraunhofer measured the wavelengths of many of these lines accurately and found that they occupied exactly the same positions as the bright lines emitted by different gases and vapours.

According to Kirchhoff's law of radiation, any substance at a lower temperature will absorb the radiations of those wavelengths, which it will emit when excited by electric discharge. Thus the solar spectrum is an absorption spectrum and the Fraunhofer lines are the absorption lines of relatively cooler gases and vapours present in the earth's atmosphere and the sun's outer atmosphere. The central portion of the sun is called the photosphere, which is at a very high temperature. Surrounding the photosphere is the chromosphere, which is at a much lower temperature than the photosphere. The positions of the Fraunhofer lines in the solar spectrum give the information regarding the nature of elements present in the sun's atmosphere. The presence of more than sixty elements in the sun's atmosphere is found this way. In fact, helium was first discovered by the study of the Fraunhofer lines in the solar spectrum before it had been isolated in the laboratory. The most prominent of the Fraunhofer lines are denoted by the letters of the alphabet. Some of the lines, with their wavelengths and the elements responsible for their absorption are given below :

Line	*Element*	*Wavelength in Å*
A	Atmospheric oxygen	7594
B	Atmospheric oxygen	6867
C	Hydrogen	6563
D_1	Sodium	5896
D_2	Sodium	5890
F	Hydrogen	4861
G_1	Hydrogen	4341
H	Calcium	3969
K	Calcium	3934

8.29 INFRARED SPECTRUM

Extending on either side of the visible spectrum, there are invisible radiations which do not cause the sensation of sight. The radiations beyond the red end of the visible spectrum are called infrared radiations and their wavelengths extend up to 400,000 Å. Sun is a powerful source of infrared radiation. Beyond the violet end of the visible spectrum, the radiations extending up to a wavelength of 100 Å are called ultraviolet radiations.

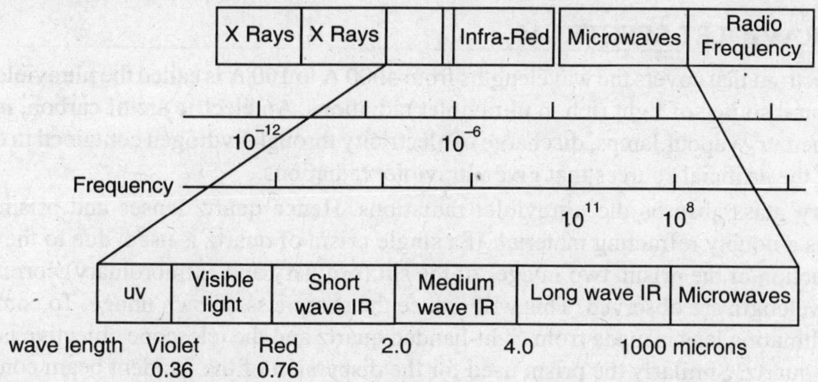

Fig. 8.20. Part of the Electro-Magnetic Spectrum

The heating effect of the infrared radiations is used in measuring the wavelength of the radiation. An infrared ray spectrometer is shown in Fig. 8.21. Light from a strong source of light such as an electric arc is rendered parallel by reflection from a concave stainless steel mirror M_1. This parallel beam is refracted through the rock salt prism P and the emergent dispersed beam after reflection from the mirrors M_2 and M_3 is incident on a thermopile or a bolometer. M_2 is a plane mirror and M_3 is a concave mirror. Wien's law of radiation is given by

$$\lambda_m T = 0.2892$$

where T is the absolute temperature and λ_m is the wavelength of the radiation. The thermopile readings help in the calculation of temperature and from the temperature, λ can be calculated.

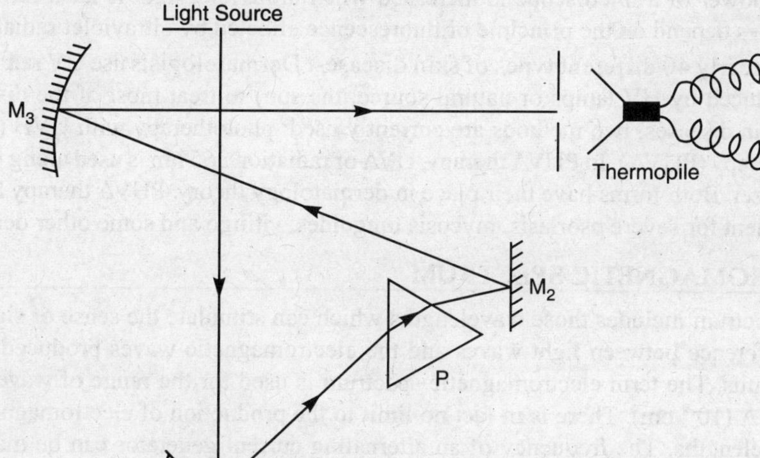

Fig. 8.21

As infrared radiations are not absorbed by air or thick fog, infrared ray photograph can be taken over long distances of fog and mist where visible light cannot penetrate. For this, specially designed photographic plates are used with suitable filters. A solution of iodine in alcohol is a suitable filter, because this transmits the infrared radiations and absorbs the visible light. In the World War II, infrared photography played a very useful part in detecting objects in the dark through mist, fog and clouds.

Infrared radiations have a wide application in the field of medicine, industry etc. Infrared radiations can penetrate deep into the human body and by their property of heating can dilate the blood vessel at the portion exposed to the radiations. This enables increased flow of blood.

8.30 ULTRAVIOLET SPECTRUM

The spectrum that covers the wavelengths from 4000 Å to 100 Å is called the ultraviolet spectrum. Sun is a natural source of light rich in ultraviolet radiations. An electric arc of carbon, iron or other materials, mercury vapour lamps, discharge of electricity through hydrogen contained in quartz tubes are some of the artificial sources that give ultraviolet radiations.

Ordinary glass absorbs the ultraviolet radiations. Hence quartz lenses and prisms are used. But quartz is a doubly refracting material. If a single prism of quartz is used, due to the property of double refraction of the prism, two images of the slit (ordinary and extraordinary) corresponding to a single wavelength are observed. This will reduce the sharpness of each image. To compensate for this, the collimating lens is made from right-handed quartz and the telescope objective is made from left-handed quartz. Similarly the prism used for the dispersion of the incident beam consists of two halves. The two halves are held together by glycerine. One half is made from right-handed quartz and the other half is made from left-handed quartz.

In Fig. 8.22, right angled *ABD* is a prism of right-handed quartz and *ADC* is of left-handed quartz. For recording an ultraviolet spectrum for wavelengths shorter than 1,200Å, a concave reflection grating is used. The diffracted beam is photographed. The grating and the photographic plate are enclosed in a metal chamber which is evacuated.

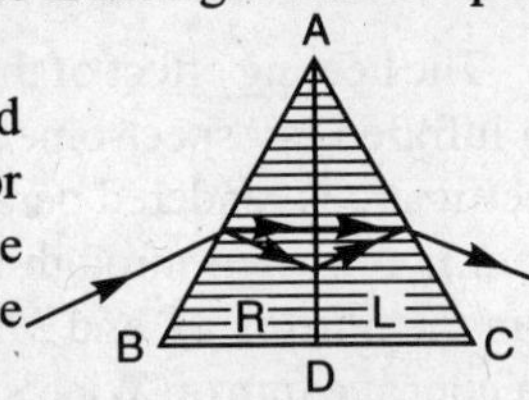

Fig. 8.22

Ultraviolet radiations have a variety of applications. Sterilisation of rooms in which blood plasma, drugs, vaccines etc. are prepared and sealed in the containers is done by ultraviolet radiations. Drugs, poisons, dyes etc. fluoresce under the action of ultraviolet rays. The resolving power of a microscope is increased when ultraviolet light is used for illumination. Fluorescent tubes depend on the principle of fluorescence effected by ultraviolet radiations.

There are nearly 40 different types of skin diseases. Dermatologists use UV radiations (either artificially produced by UV lamps or natural source, the sun) to treat most of the diseases. In the treatment of skin diseases, two methods are currently used: phototherapy with UVB (311 nm) and photochemotherapy (PUVA). In PUVA therapy, UVA of radiation 365 nm is used along with psoralen as photosensitizer. Both forms have their place in dermatology theory. PUVA therapy has become a standard treatment for severe psoriasis, mycosis tungoides, vitiligo and some other dermatoses.

8.31 ELECTROMAGNETIC SPECTRUM

Visible spectrum includes those wavelengths which can stimulate the sense of sight. But there is no basic difference between light waves and the electromagnetic waves produced by electrical oscillating circuits. The term electromagnetic spectrum is used for the range of wavelengths from 10^4 meters to 1Å (10^{-8} cm). There is in fact no limit to the production of electromagnetic waves of very long wavelengths. The frequency of an alternating current generator can be made as low as possible by decreasing the speed of the generator. The wavelength of waves transmitted by a 50 cycle transmission line is 5×10^8 cm. Waves of shorter wavelength can be produced by electrical oscillators. X-rays and gamma rays represent the waves of very short lengths. The complete electromagnetic spectrum is shown in the following table.

It is interesting to note that the visible range of the spectrum comprises only a small range of the electromagnetic spectrum extending approximately from 4000Å in the extreme violet region to 8000Å in the extreme red. Beyond the violet region of the visible spectrum is the ultraviolet, the X-rays and γ-rays. Beyond the red end of the visible spectrum are the short radio waves and radio broadcast waves.

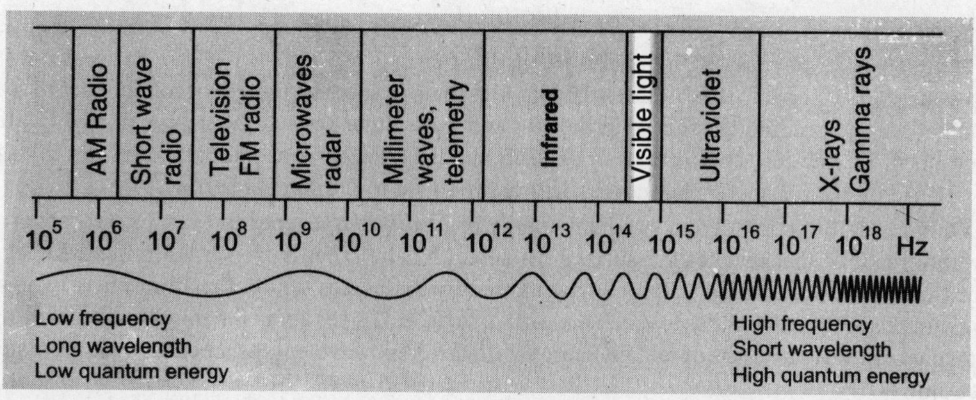

8.32 SOURCES OF SOLAR ENERGY: (SOME EVERYDAY APPLICATIONS)

Solar energy is worldwide easily available, major natural source of energy. Today, the exploration of this unlimited solar energy as useful electrical energy has become an essential national need to overcome the frequent power cut or load-shedding and to operate the electrical appliances in remote areas where electricity is not available.

The sun is powerful fusion reactor, where **hydrogen atoms** fuse to form helium and give off a tremendous amount of energy. The surface of the sun also known as photosphere has a temperature of 6000 K (or 10,000 °F). The temperature of the core region of nuclear fusion is 36,000,000 °F. A ball of coal, the size of sun would burn up completely in 3000 years, yet the sun has already been burning for 3 billion years and is expected to burn for another 4 billion of years. The power emitted by sun is 3.9×10^{26} Watts.

Solar energy reaches the earth at the rate of about 1.4 kW per square metre of surface perpendicular to the direction of the sun as shown in Figure 8.23. By how much does the mass of the sun decrease per second owing to this energy loss? The mean radius of the earth's orbit is 1.5×10^{11} m.

The solar energy reaching the earth's surface is estimated as under. The surface area of a sphere of radius r is $A = 4\pi r^2$. The total power radiated by the sun, which is equal to the power received by a sphere whose radius is that of the earth's orbit, is therefore

$$P = \frac{P}{A}A = \frac{P}{A}(4\pi r^2) = (1.4 \times 10^3 \text{ W/m}^2)(4\pi)(1.5 \times 10^{11} \text{ m})^2 = 4.0 \times 10^{26} \text{ W}$$

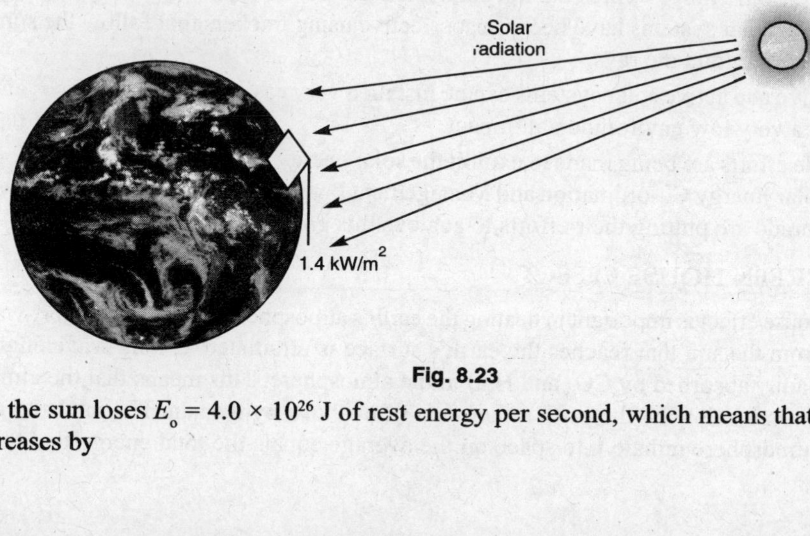

Fig. 8.23

Thus the sun loses $E_o = 4.0 \times 10^{26}$ J of rest energy per second, which means that the sun's rest mass decreases by

$$m = \frac{E_o}{c^2} = \frac{4.0\times10^{26}\ J}{(3.0\times10^8\ \text{m/s})^2} = 4.4 \times 10^9\ \text{kg}$$

per second. Since the sun's mass is 2.0×10^{30} kg, it is in no immediate danger of running out of matter. The chief energy-producing process in the sun and most other stars is the conversion of hydrogen to helium in its interior. The formation of each helium nucleus is accompanied by the release of 4.0×10^{-11} J of energy, so 10^{37} helium nuclei are produced in the sun per second.

Only a small fraction of sun's radiant energy or *insolation,* reaches the earth's atmosphere and only about half of that reaches the surface of the earth. The other half is either reflected back into space by clouds and ice or is absorbed or scattered by molecules within the atmosphere. The sun's energy travels 93,000,000 miles to reach the earth's surface. It takes about 8.5 minutes after leaving the photosphere in various forms of radiant energy with different wavelengths, known as electromagnetic spectrum.

The abundantly available solar energy can be utilised either directly as *heat* or indirectly by converting it into electrical power using **photovoltaic cells**. Accordingly solar energy utility systems are broadly divided into two categories:

(*i*) Active systems, and

(*ii*) Passive systems.

Green-houses and solariams are the most common examples of the direct use of solar energy with glass windows concentrating the visible light from the sun but restricting the heat from escaping. Flat plate collectors are another direct method and mounted on roof tops. They can provide one third of the energy required for space heating. Hence windows and collectors are considered as *passive systems*. Most popular and widely used solar water heaters, solar cookers also belong to this category.

Active solar systems differ from passive systems in that they operate machineries such as pump or electric fans, electronic devices and so many. A huge or small panel of photovoltaic cells, connected in series and framed on a rigid platform. The photovoltaic cells are made of semiconductor materials mainly silicon with small amount of *gallium arsenide* or *cadmium sulfide* added so that the cell emits electrons, when exposed to sunlight. These are connected in series, as per power requirements. These cells are capable of absorbing part of the solar flux to produce a direct electric current with about 14% efficiency. A thin film technology is being perfected for the production of these cells and the cost per unit watt will eventually be reduced because less material will be required. These modules based on photovoltaic cells are used economically to charge storage batteries aboard boats, operate light houses and supply power for emergency telephones on highways. They are also of immense use in remote areas, rural villages which are not connected to a power supply grid for pumping water for irrigation. Larger solar systems have been most effective using trackers that follow the sun or mirror reflectors that concentrate the rays.

Both passive and active solar systems can be installed very easily. Neither produces air pollution and both have a very low environmental impact.

Worldwide efforts are being made to exploit the solar energy as a major non-conventional energy source. The Solar Energy Co-ordination and Management Programme and the Solar Energy Research Institutes Colorado are putting their efforts to achieve this goal.

8.33 THE GREEN HOUSE EFFECT

A green house effect is important in heating the earth's atmosphere. Much of the short-wavelengths visible light from the sun that reaches the earth's surface is reradiated as long-wavelength infrared light that is readily absorbed by CO_2 and H_2O in the atmosphere. This means that the atmosphere is heated mainly from below by the earth rather than from above by the sun. The total energy that the earth and its atmosphere radiate into space on the average equals the total energy that they receive from the sun.

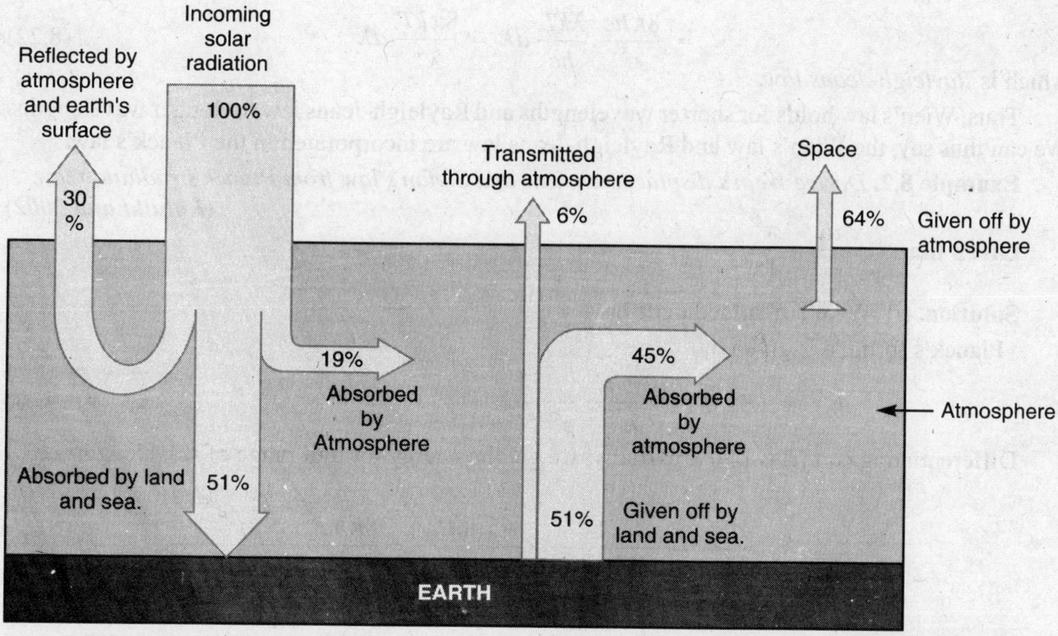

Fig. 8.24

It is worth mentioning that everybody of condensed matter radiates according to Planck's radiation law, regardless of its temperature. An object need not be so hot that it glows conspicuously in the visible region in order to be radiating. The radiation from an object at room temperature is chiefly in the infrared part of the spectrum to which the eye is not sensitive. Therefore, the interior of a green house is warmer than the out side air because sunlight can enter through its windows but the infrared radiation given off by the interior cannot escape through them as shown in Fig. 8.24.

SOLVED EXAMPLES

Example 8.1. *Show that Planck's law reduces to Wien's law for shorter wavelengths and Rayleigh-Jeans law for longer wavelengths.* ***(OSM. U. 2004, Madurai U. 2003, Luck U. 2002, Aadh 2000; Meerut U. 2007, 2006, 2005)***

Solution. Planck's radiation law is

$$E_\lambda\, d\lambda = \frac{8\pi hc}{\lambda^5} \cdot \frac{d\lambda}{e^{hc/\lambda kT} - 1}$$

For *Shorter wavelengths,* $e^{hc/\lambda kT}$ becomes large as compared to unity and hence Planck's law reduces to

$$E_\lambda\, d\lambda = \frac{8\pi hc}{\lambda^5} \cdot \frac{d\lambda}{e^{hc/\lambda kT}}$$

$$= \frac{8\pi hc}{\lambda^5} \cdot e^{-hc/\lambda kT} . d\lambda \qquad \text{... (8.21)}$$

which is *Wien's law.*

For longer wavelengths, $e^{hc/\lambda kT}$ may be approximated to $\left(1 + \frac{hc}{\lambda kT}\right)$ and hence Planck's law reduces to

$$E_\lambda\, d\lambda = \frac{8\pi hc}{\lambda^5} \cdot \frac{d\lambda}{\left(1 + \frac{hc}{\lambda kT} - 1\right)}$$

$$= \frac{8\pi hc}{\lambda^5} \cdot \frac{\lambda kT}{hc} d\lambda = \frac{8\pi kT}{\lambda^4} d\lambda \qquad \text{... (8.22)}$$

which is *Rayleigh-Jeans law.*

Thus, Wien's law holds for shorter wavelengths and Rayleigh-Jeans law for longer wavelengths. We can thus say, the Wien's law and Rayleigh-Jeans law are incorporated in the Planck's law.

Example 8.2. *Derive Wein's displacement law and Stefan's law from Planck's radiation law.*

(*Rohilkhand 2002*)

Given that $\int_0^\infty \frac{x^3 dx}{e^x - 1} = \frac{\pi^4}{15}$

Solution. (*i*) Wein's displacement law,

Planck's formula is given by

$$E_\lambda d\lambda = \frac{8\pi hc}{\lambda^5} \cdot \frac{1}{e^{hc/\lambda kT} - 1} \qquad \text{... (i)}$$

Differentiating eqn (*i*) w.r.t. to λ partially, we get the energy for unit range of wavelength, *i.e.*,

$$\frac{\partial E_\lambda}{\partial \lambda} = \frac{1}{e^{hc/\lambda kT} - 1} \times \frac{-5(8\pi hc)}{\lambda^6} + \frac{8\pi hc}{\lambda^5} \times \frac{\frac{hc}{\lambda^2 kT} e^{hc/\lambda kT}}{(e^{hc/\lambda kT} - 1)^2}$$

$$= \frac{-40\pi hc}{\lambda^6} \times \frac{1}{e^{hc/\lambda kT} - 1} + \frac{8\pi hc}{\lambda^5} \times \frac{hc}{\lambda^2 kT} \times \frac{e^{hc/\lambda kT}}{(e^{hc/\lambda kT} - 1)^2}.$$

For maximum energy of emission, *i.e.*, for maximum value of E_λ, $\frac{\partial E_\lambda}{\partial \lambda}$ must be zero. Hence

$$\frac{-40\pi hc}{\lambda^6} \cdot \frac{1}{e^{hc/\lambda kT} - 1} + \frac{8\pi hc}{\lambda^5} \times \frac{hc}{\lambda^2 kT} \times \frac{e^{hc/\lambda kT}}{(e^{hc/\lambda kT} - 1)^2} = 0$$

or

$$\frac{8\pi hc}{(e^{hc/\lambda kT} - 1)\lambda^6} \left[-5 + \frac{hc}{\lambda kT} \times \frac{e^{hc/\lambda kT}}{(e^{hc/\lambda kT} - 1)} \right] = 0$$

or

$$-5 + \frac{hc}{\lambda kT} \times \frac{e^{hc/\lambda kT}}{(e^{hc/\lambda kT} - 1)} = 0 \qquad \text{... (ii)}$$

Putting $\frac{hc}{\lambda kT} = x$, say in equation (*ii*), we get

$$-5 + \frac{xe^x}{e^x - 1} = 0$$

or

$$\frac{xe^x}{e^x - 1} = 5 \qquad \text{... (iii)}$$

It suggests that there must be a root in the neighbourhood of 5. By applying the method of approximation, the exact value of x is found to be equal to 4.965.

$$\therefore \qquad x = \frac{hc}{\lambda kT} = 4.965 \qquad \text{... (iv)}$$

Here λ represents the wavelength at which energy per unit range of wavelength is maximum. Representing this particular wavelength by λ_m, equation (*iv*), may be written as

$$\frac{hc}{\lambda_m kT} = 4.965$$

or

$$\lambda_m . T = \frac{hc}{4.965 k} = \text{constant} \qquad \text{... (v)}$$

This is **Wein's displacement law.**

***(ii)* Stefan's law from Planck's formula**

The Planck's radiation formula in terms of frequency is given by eq. (8.14) as

$$E_\nu \, d\nu = \frac{8\pi h\nu^3}{c^3} \cdot \frac{1}{e^{h\nu/kT} - 1} d\nu$$

Therefore, the total radiant energy over all frequencies will be

$$E = \int_0^\infty E_\nu d\nu = \frac{8\pi h}{c^3} \int_0^\infty \frac{\nu^3 d\nu}{e^{h\nu/kT} - 1} \quad \text{... } (vi)$$

Putting $\frac{h\nu}{kT} = x$, so that $\nu = \frac{kT}{h} x$ and $d\nu = \frac{kT}{h} dx$

Substituting in equation (*vi*), we get

$$E = \frac{8\pi k^4 T^4}{c^3 h^3} \int_0^\infty \frac{x^3 \, dx}{e^x - 1}$$

But $$\int_0^\infty \frac{x^3 \, dx}{e^x - 1} = \frac{\pi^4}{15}$$

$\therefore$ $$E = \frac{8\pi^5 k^4 T^4}{15 c^3 h^3}$$

or $$E = AT^4 \quad \text{... } (vii)$$

where $$A = \frac{8\pi^5 k^4}{15 c^3 h^3}$$

But $\frac{Ac}{4} = \sigma$, the Stefan's constant.

$\therefore$ $$E = \sigma T^4 \quad \text{... } (viii)$$

This is **Stefan's law**, where $\sigma = \frac{2\pi^5 k^4}{15 h^3 c^2}$ gives the Stefan's constant.

Example 8.3. *Two large closely spaced concentric spheres (both are black body radiators) are maintained at temperatures 200 K and 300 K respectively. The space in between the two spheres is evacuated. Calculate the net rate of energy transfer between the two spheres.* [$\sigma = 5.672 \times 10^{-8}$ *M.K.S. units*].

Solution. Here, $T_1 = 300$ K, $T_2 = 200$ K, $\sigma = 5.672 \times 10^{-8}$ M.K.S. units

$$E = \sigma [T_1^4 - T_2^4]$$

$$= 5.672 \times 10^{-8} [(300)^4 - (200)^4]$$

$$E = \mathbf{368.68 \text{ watts/m}^2}$$

Example 8.4. *Calculate the radiant emittance of a black body at a temperature of (i) 400 K (ii) 4000 K.* [σ *is* 5.672×10^{-8} *M.K.S. units*]. ***(Purvanchal 2007, Bhopal 2000)***

Solution. According to Stefan's law,

$$E = e \sigma T^4$$

$$= \sigma T^4 \quad (\because e = 1, \text{ for a black body})$$

(*i*) For $T = 400$ K, $$E = 5.672 \times 10^{-8} \times [400]^4$$

$$= \mathbf{1452 \text{ watts/m}^2}$$

(*ii*) For $T = 4000$ K, $E = 5.672 \times 10^{-8} \times [4000]^4$

$= 1452 \times 10^4$ Watts/m^2

$=$ **14520 Kilo-watts/m²**

Example 8.5. *An aluminum foil of relative emittance 0.1 is placed in between two concentric spheres at temperatures 300 K and 200 K respectively. Calculate the temperature of the foil after the steady state is reached. Assume that the spheres are perfect black body radiators. Also calculate the rate of energy transfer between one of the spheres and the foil.* [$\sigma = 5.672 \times 10^{-8}$ *M.K.S. units*].

(Agra 2000)

Solution. Here, $T_1 = 300$ K, $T_2 = 200$ K, $e = 0.1$.

(*i*) Let x be the temperature of the foil, after the steady state is reached, then

$$e\sigma\,(T_1^4 - x^4) = e\sigma\,(x^4 - T_2^4)$$

$$[(300)^4 - x^4] = [x^4 - (200)^4]$$

$$x^4 = 48.5 \times 10^8$$

$$\therefore \quad \mathbf{x = 263.8\ K}$$

(*ii*) $E = e\sigma\,(T_1^4 - x^4)$

$= 0.1 \times 5.672 \times 10^{-8}\,[(300)^4 - (263.8)^4]$

$=$ **18.5 Watts/m²**

Example 8.6. *Calculate the energy radiated per minute from the filament of an incandescent lamp at 2000 K, if the surface area is 5.0 × 10⁻⁵ sq. metres and its relative emittance is 0.85.*

(Lucknow 2000)

Solution. Here, $A = 5 \times 10^{-5}$ m^2

$e = 0.85$

$\sigma = 5.672 \times 10^{-8}$ M.K.S. units

$t = 60$ sec.

$T = 2000$ K

$\therefore$ $E = A\,e\sigma\,T^4 \times t$

$E = 5 \times 10^{-5} \times 0.85 \times 5.672 \times 10^{-8} \times (2000)^4 \times 60$

$E =$ **2315 Joules**

Example 8.7. *An iron furnace radiates 1.53 × 10⁵ calories per hour through an opening of cross-section 10⁻⁴ sq. metre. If the relative emittance of the furnace is 0.80, calculate the temperature of the furnace. Given : σ = 1.36 × 10⁻⁸ cal/m²–S–K⁴.* ***(Kanpur 2001)***

Solution. $E = A\,e\sigma t\,(T^4)$

Here, $A = 10^{-4}\ m^2$, $e = 0.80$, $t = 3600$ sec., $T = ?$

Thus, $$T^4 = \frac{E}{Ae\sigma t}$$

$$\therefore \quad T = \left(\frac{E}{Ae\sigma t}\right)^{\frac{1}{4}} = \left[\frac{1.53\times10^5}{10^{-4}\times0.80\times1.36\times10^{-8}\times3600}\right]^{\frac{1}{4}}$$

$$\mathbf{T = 2500\ K}$$

Example 8.8. *A black body has its cavity in the shape of a cube. Determine the number of modes of vibration per unit volume in the wavelength region 4990 Å to 5010 Å.* ***(Osmania 2004)***

Solution. The number of modes per c.c. in the wavelength range λ to $\lambda + d\lambda$ in a chamber (or cavity) is

$$N = \frac{8\pi}{\lambda^4}\,d\lambda$$

Here, $d\lambda = (5010 - 4990)$ Å = 20 Å = 20×10^{-8} cm. and λ = 5000 Å = 5000×10^{-8} cm

$$\therefore \qquad N = \frac{8 \times 3.14 \times 20 \times 10^{-8}}{(5000 \times 10^{-8})^4}$$

$$= \mathbf{8.038 \times 10^{11}}$$

Example 8.9. *Calculate the surface temperature of the sun and moon, given that λ_m = 4753 Å and 14μ respectively, λ_m being wavelength of maximum intensity of emission.* **(Meerut 2000)**

Solution. We know from Wien's displacement law,

$$\lambda_m T = \text{constant} = 0.2898$$

(*i*) For the sun,

given $\qquad \lambda_m = 4753$ Å $= 4753 \times 10^{-8}$ cm

∴ the temperature of the sun

$$T_s = \frac{0.2898}{4753 \times 10^{-8}} = \mathbf{6097\ K.}$$

(*ii*) For the moon,

given $\qquad \lambda_m = 14\mu = 14 \times 10^{-4}$ cm $\qquad$ [∵ 1μ = 1 micron = 10^{-4} cm]

∴ the temperature of the moon,

$$T_m = \frac{0.2898}{14 \times 10^{-4}} = \mathbf{207\ K}$$

Example 8.10. *A body at 1500 K emits maximum energy of wavelength 2000 nm. If the sun emits maximum energy of wavelength 550 nm, what would be the temperature of the sun?*

(Nagpur 2010, 2007, 2004/s)

Solution. Temperature of the body = 1500 K

Wavelength for maximum energy emission = 2000×10^{-9} m

Wavelength for maximum energy emission from sun = 550×10^{-9} m.

If T is the temperature of the sun, then according to Wien's displacement law

$$\lambda_m T = \text{constant}$$

$$550 \times 10^{-9} \times T = 2000 \times 10^{-9} \times 1500$$

$$\therefore \qquad T = \frac{2000 \times 10^{-9} \times 1500}{550 \times 10^{-9}} = \mathbf{5454\ K.}$$

Example 8.11. *What is the wavelength at which human body radiates maximum energy? Temperature of human body is 37°C. Wien's constant is 2.898×10^{-3} mK.*

(Meerut 2004/s, Nagpur 2009)

Solution. According to Wien's displacement law,

$$\lambda_m T = \text{constant} = 2.898 \times 10^{-3} \text{ m K.}$$

where λ_m is the wavelength at which black body radiates maximum energy at temperature T.

Temperature of human body = T = 37° C = 273 + 37 = 310 K.

Considering the human body as a black body, the wavelength at which it radiates maximum energy = $\lambda_m \times 310 = 2.898 \times 10^{-3}$ m

$$\therefore \qquad \lambda_m = \frac{2.898 \times 10^{-3}}{310} = \mathbf{9.35 \times 10^{-6}\ m.}$$

Example 8.12. *A black body with temperature 127°C is put in an evacuated enclosure whose walls are blackened and maintained at 27°C. What is net amount of energy lost by body per second per unit area.*

Stefan's constants $\qquad \sigma = 5.672 \times 10^{-8}$ S.I. units

Solution. Here,

$$T_1 = 273 + 127 = 400 \text{ K}$$
$$T_2 = 273 + 27 = 300 \text{ K}$$
$$\sigma = 5.672 \times 10^{-8} \text{ S.I. units}$$
$$R = \sigma [T_1^4 - T_2^4]$$
$$R = 5.672 \times 10^{-8} [(400)^4 - (300)^4]$$
$$R = \mathbf{992.6 \text{ watts/m}^2}$$

Example 8.13. *If Wien's constant b = 0.3 cm-K°, calculate the temperature of the sun whose radiation has maximum energy at wavelength λ = 5500 Å.* **(*Meerut 2004*)**

Solution. According to Wien's displacement law

$$\lambda_m T = \text{constant} = b$$
$$T = \frac{b}{\lambda_m}$$

Here

$$b = 0.3 \text{ cm-K}$$
$$\lambda_m = 5500 \text{ Å} = 5500 \times 10^{-8} \text{ cm}$$
$$\therefore \quad T = \frac{0.3}{5500 \times 10^{-8}}$$
$$T = \mathbf{5455 \text{ K}}$$

Example 8.14. *Deduce the temperature at which a black body loses thermal energy at the rate of 1 watt/cm².* **(*Bhopal 2000*)**

Solution.

$$R = 1 \text{ watt/cm}^2 = 10^4 \text{ watts/m}^2$$
$$R = \sigma T^4$$
$$T^4 = \left(\frac{R}{\sigma}\right)$$
$$T = \left(\frac{R}{\sigma}\right)^{\frac{1}{4}}$$
$$T = \left[\frac{10^4}{5.672 \times 10^{-8}}\right]^{\frac{1}{4}}$$
$$T = \mathbf{648 \text{ K.}}$$

Example 8.15. *If a black body at a temperature 6174 K emits 4700 Å with maximum energy; calculate the temperature at which it will emit a wavelength of 1.4 × 10⁻⁵ m with maximum energy.* **(*Mumbai 2005*)**

Solution. According to Wien's displacement law

$$\lambda_m T = \text{constant}$$
$$\lambda_m T = \lambda_m' T'$$
$$T' = \frac{\lambda_m T}{\lambda_{m'}} = \frac{4700 \times 10^{-10} \times 6174}{1.4 \times 10^{-5}}$$
$$T' = \mathbf{207.27 \text{ K}}$$

Example 8.16. *Assume the sun to be a black body at temperature 5800 K. Use Stefan's law to show that the total radiant energy emitted by sun per second is 3.95 × 10²⁶ J. Also show that the rate at which energy is reaching the top of earth's atmosphere is 1.4 k W-m⁻².* **(*Rohilkhand 2000*)**

Solution. Here radius of the sun, $r = 7 \times 10^8$ m

Surface area of the sun, $A = 4\pi r^2$

$$= 4 \times 3.142 \times (7 \times 10^8)^2 \text{ m}^2$$

$$\sigma = 5.672 \times 10^{-8} \text{ S.I. units}$$

$$T = 5800 \text{ K}$$

Total energy emitted by sun per second

$$U = A\,\sigma\,T^4$$

$$U = 4 \times 3.142 \times (7 \times 10^8)^2 \times 5.672 \times 10^{-8} \times (5800)^4$$

$$U = \mathbf{3.95 \times 10^{26}\ J}$$

Distance of the earth's atmosphere from the sun

$$= r_1 = 1.5 \times 10^{11} \text{ m}$$

Energy reaching per unit area per second.

$$R = \frac{U}{4\pi r_1^2}$$

$$R = \frac{3.95 \times 10^{26}}{4 \times 3.142 \times (1.5 \times 10^{11})^2}$$

$$R = 1.3968 \times 10^3 \text{ W/m}^2$$

$$R = 1.3968 \text{ kW/m}^2$$

$$\mathbf{R = 1.4\ kW/m^2}$$

Example 8.17. *Calculate the number of modes in a chamber of volume 50 c.c. in the frequency range 4×10^{14} and 4.01×10^{14} sec^{-1}.* **(Indore 2000)**

Solution. The no. of modes per c.c. in the frequency range v and $v + dv$ is

$$= \frac{8\pi v^2}{c^3} dv$$

$$= \frac{8 \times 3.14 \times (4 \times 10^{14})^2}{(3 \times 10^{10})^3} \times (4.01 \times 10^{14} - 4 \times 10^{14}) = 1.5 \times 10^{11}$$

$\therefore$ The number of modes in the whole chamber of volume $V = 50$ c.c. is

$$= 1.5 \times 10^{11} \times 50$$

$$= \mathbf{7.5 \times 10^{12}}$$

Example 8.18. *A solid copper sphere (density ρ and specific heat C) of radius r at initial temperature 200 K is suspended inside a chamber whose walls are at almost 0K. The time required for the temperature of the sphere to fall to 100 K is ......* **(Purvanchal 2000)**

Solution. Here $\frac{dQ}{dt} = e\,\sigma\,A\,(T^4 - T_0^4)$

Here $e = 1$

and $A = 4\pi r^2$

Also $\frac{dQ}{dt} = mc\left(\frac{dT}{dt}\right)$

Also $m = \left(\frac{4}{3}\pi r^3\right)\rho$

$\therefore$ $mc\left(\frac{dT}{dt}\right) = \sigma\,A\,(T^4 - T_0^4)$

$$T_0 = 0$$

$$dt = \left(\frac{m}{A}\right)\left(\frac{c}{\sigma}\right)\frac{dT}{T^4}$$

Integrating

$$\int dt = \left(\frac{\frac{4}{3}\pi r^3 \rho}{4\pi r^2}\right)\left(\frac{c}{\sigma}\right)\int_{200}^{100}\frac{dT}{T^4}$$

$$t = \left(\frac{r\rho c}{3\sigma}\right)\left(\frac{1}{3}\right)\left[\left(\frac{1}{200}\right)^3 - \left(\frac{1}{100}\right)^3\right]$$

$$t = \left[\frac{7\, r\rho c}{(72\times 10^6)\,\sigma}\right]$$

Example 8.19. *Compare the radiant emittance of a black body at 200 K and 2000 K. Given* $\sigma = 5.672 \times 10^{-8}$ *M.K.S. units.* **(Delhi 2000)**

Solution. Here,

$$T_1 = 200 \text{ K}$$

$$T_2 = 2000 \text{ K}$$

$$R_1 = \sigma T_1^4$$

$$R_2 = \sigma T_2^4$$

$$\left(\frac{R_1}{R_2}\right) = \left(\frac{T_1}{T_2}\right)^4$$

$$= \left(\frac{200}{2000}\right)^4$$

$$= \mathbf{10^{-4}}$$

Example 8.20. *A black sphere of diameter 8 cm. is heated to 500 K when the surrounding temperature is 300 K. What is the rate at which energy is radiated? Stefan's constant is* 6×10^{-8} $W\,m^{-2}\,K^{-4}$. **(Nagpur 2000)**

Solution. Here,

$$\sigma = 6 \times 10^{-8} \text{ W m}^{-2} \text{ K}^{-4}$$

$$t = 1\, s$$

$$T = 500 \text{ K}$$

$$T_0 = 300 \text{ K}$$

$$d = 8 \text{ cm} = 0.08 \text{ m}$$

$$A = \pi d^2$$

$$= 3.14 \times (0.08)^2$$

$$= 0.02 \text{ m}^2$$

$$e = 1$$

$$R = \sigma A e\,[T^4 - T_0^4]$$

$$= 6 \times 10^{-8} \times 0.02 \times 1\,[(500)^4 - (300)^4]$$

$$= 65.28 \text{ J/s}$$

$$= \mathbf{65.28 \text{ watts}}$$

Example 8.21. *An unclothed boy is in a room at 20°C. If the skin temperature of the boy is 37°C, how much heat is lost from his body in 10 minutes. The surface area of the boy is 3 m² and emissivity is 0.90 and*

$\sigma = 5.67 \times 10^{-8}\ Wm^{-2}\ k^{-4}$ **(Mumbai 2000)**

Solution. Here

$$e = 0.9$$
$$A = 3\ \text{m}^2$$
$$\sigma = 5.67 \times 10^{-8}\ \text{W/m}^2 - \text{k}^4$$
$$\theta_1 = 273 + 37 = 310\ \text{K}$$
$$\theta_2 = 273 + 20 = 293\ \text{K}$$
$$R = eA\,\sigma\,(\theta_1^4 - \theta_2^4)$$
$$R = 0.9 \times 3 \times 5.67 \times 10^{-8}\ [(310)^4 - (293)^4]$$
$$R = 286\ \text{J/s}$$
$$t = 10\ \text{minutes} = 600\ \text{s}$$

Total heat lost by the skin in 10 minutes.

$$Q = Rt = 286 \times 600$$
$$= \mathbf{1.716 \times 10^5\ J}$$

Example 8.22. *The relative emittance of tungsten is approximately 0.35. A tungsten sphere of surface area 10^{-3} sq metres is suspended inside a large evacuated enclosure whose walls are at 300 K. What power input is required to maintain the sphere at a temperature of 3000 K? The conduction of heat along the supports can be neglected.* **(Mysore 2000)**

$$[\sigma = 5.672 \times 10^{-8}\ \text{M.K.S. units}]$$
$$R = \sigma A\, e[T^4 - T_0^4]$$

Solution. Here

$$A = 10^{-3}\ \text{sq metres}$$
$$e = 0.35$$
$$\sigma = 5.672 \times 10^{-8}\ \text{M.K.S. units}$$
$$R = 5.672 \times 10^{-8} \times 10^{-3} \times 0.35\ [(3000)^4 - (300)^4]$$
$$= 1609\ \text{watts}$$

The power input $= \mathbf{1609\ watts.}$

Example 8.23. *The order of magnitude of the energy received from sun at earth's surface is 10^{-1} Joule cm^{-2} sec^{-1}. Calculate the order of magnitude of the total force due to solar radiation on the earth (assumed perfectly absorbing). Given: earth's diameter = 10^7 metre.*

(Lucknow 2001, Purvanchal 2006)

Solution. The energy received from the sun at earth's surface per unit area per second is $E = 10^{-1}$ Joule/cm^2 sec $= 10^{-3}$ Joule m^{-2} sec^{-1}.

∴ Radiation pressure

$$P = \frac{E}{C} = \frac{10^{-3}}{3 \times 10^8}\ N/\text{m}^2$$
$$= 3.33 \times 10^{-6}\ \text{N/m}^2$$

∴ Total force due to solar radiation on the earth

$$= p \times \text{surface area}$$
$$= 3.33 \times 10^{-6} \times 4\pi r^2$$
$$= 3.33 \times 10^{-6} \times 4 \times 3.14 \times \left(\frac{10^7}{2}\right)^2$$
$$= \mathbf{1.052 \times 10^9\ N}$$

Example 8.24. *Calculate the average energy $\bar{\varepsilon}$ of an oscillator of frequency 0.60×10^{14} sec^{-1} at T = 1800 K treating it as (i) classical oscillator, (ii) Planck's oscillator.*

Solution. (*i*) The average energy of a classical oscillator

$$\bar{\varepsilon} = kT$$
$$= (1.38 \times 10^{-23} \text{ Joule/K}) \times 1800 \text{ K}$$
$$= \mathbf{2.484 \times 10^{-20} \text{ Joule}}$$

(*ii*) The average energy of a Planck's oscillator

$$\bar{\varepsilon} = \frac{h\nu}{e^{h\nu/kT} - 1} = \frac{h\nu / kT}{e^{h\nu/kT} - 1} \times kT$$

Let us calculate $h\nu/kT$ first.

$$\frac{h\nu}{kT} = \frac{(6.6\times10^{-34})(0.60\times10^{14})}{1.38\times10^{-23}\times1800} = \frac{3.96\times10^{-23}}{2.484\times10^{-20}}$$
$$= 1.59$$

$$\therefore \quad \bar{\varepsilon} = \frac{h\nu / kT}{e^{h\nu/kT} - 1} kT = \frac{1.59\times2.484\times10^{-20}}{e^{1.59} - 1}$$

$$= \frac{1.59}{(2.72)^{1.59} - 1} \times 2.484\times10^{-20} \qquad (\because e = 2.72)$$

$$= \frac{1.59\times2.484\times10^{-20}}{3.909} = \mathbf{1.01 \times 10^{-20} \text{ Joule}}$$

Example 8.25. *Calculate the maximum amount of heat which may be lost per second by radiation from a sphere of 5 cm. in diameter at a temperature of 600 K when placed in an encloser at a temperature of 300 K. [$\sigma = 5.7 \times 10^{-12}$ Watts × cm^{-2} (°C)$^{-4}$]* **(*Garhwal 2005*)**

Solution. The rate of loss of thermal energy will be maximum, if the sphere radiates as black body. Applying Stefan Boltzmann's law, the net energy lost per unit area of the sphere per second is given by

$$E = \sigma (T^4 - T_0^4)$$

where T and T_o are the absolute temperatures of sphere and the surroundings respectively, and σ is Stefan's constant. Total surface area of the sphere = $4\pi r^2$

∴ The net loss of heat from the total surface of the sphere

$$Q = E \times 4\pi r^2$$
$$= \sigma (T^4 - T_0^4) \times 4\pi r^2$$
$$= \frac{\sigma (T^4 - T_0^4) \times 4\pi r^2}{J} \text{ calories/second}$$

Given:

$$r = 2.5 \text{ cm} = 2.5 \times 10^{-2} \text{ m}$$
$$T = 600 \text{ K}, \; T_o = 300 \text{ K}.$$
$$\sigma = 5.7 \times 10^{-12} \text{ watts} \times \text{cm}^{-2} (°\text{C})^{-4}$$
$$= 5.7 \times 10^{-8} \text{ Joule m}^{-2} \text{ sec}^{-1} \text{ K}^{-4}$$
$$= 5.7 \times 10^{-8} \text{ Watt m}^{-2} (\text{K})^{-4}$$

and

$$J = 4.2 \text{ Joule/cal}$$

$$\therefore \quad Q = \frac{5.7\times10^{-8}\times[(600)^4 - (300)^4]\times4\times3.14\times(2.5\times10^{-2})^2}{4.2}$$

$$= \frac{5.7\times10^{-8}\times1215\times10^{8}\times4\times3.14\times6.25\times10^{-4}}{4.2}$$

$$= \frac{543651.75 \times 10^{-4}}{4.2}$$
$$= \mathbf{12.944\ cal/s}$$

Example 8.26. *Filament of a tungsten bulb is emitting radiation of maximum energy at wavelength $\lambda_m = 2.16 \times 10^{-5}$ cm. Calculate filamant temperature.* **(H.P.U. 2001)**

Solution. Maximum energy wavelength $\lambda_m = 2.16 \times 10^{-5}$ cm $= 2.16 \times 10^{-7}$ m

According to Wien's displacement law

$$\lambda_m T = \text{a constant} = b = 2.898 \times 10^{-3} \text{ m K}$$

$$T = \frac{2.898 \times 10^{-3}}{\lambda_m} = \frac{2.898 \times 10^{-3}}{2.16 \times 10^{-7}} = 1.34 \times 10^4 = 13400 \text{ K}.$$

Example 8.27. *If the Sun emits maximum energy at wavelength 4753 A, calculate the temperature of its surface. Wien's constant b = 0.288 cm°C.* **(Nagpur U., 2001; Luck U., 1991)**

Solution. Wavelength for maximum energy emission $\lambda_m = 4753$ A $= 4753 \times 10^{-10}$ m

Wien's constant $b = 0.288$ cm °C $= 2.88 \times 10^{-3}$ mK

According to Wien's displacement law, $\lambda_m T = b$ (a constant)

$\therefore$ Temperature of the surface of Sun $T = \dfrac{2.88 \times 10^{-3}}{4753 \times 10^{-10}} = 6.06 \times 10^3 = 6060 \text{ K}$

Example 8.28. *Two stars radiate maximum energy at wavelength 3.6×10^{-7} m and 4.8×10^{-7} m respectively. What is the ratio of their temperature ?* **(Pbi. U. 2008)**

Solution. According to Wien's displacement law $\lambda_m T$ = constant

$$\therefore \quad \lambda_{m_1} T_1 = \lambda_{m_2} T_2 \text{ or } 3.6 \times 10^{-7}\, T_1 = 4.8 \times 10^{-7}\, T_2 \text{ or } \frac{T_1}{T_2} = \frac{4.8 \times 10^{-7}}{3.6 \times 10^{-7}} = \frac{4}{3}.$$

Example 8.29. *If the filament of a 100 W bulb has an area 0.25 cm² and behaves as a perfect black body, find the wavelength corresponding to the maximum in its energy distribution. Stefan's constant = 5.67×10^{-5} erg/cm² sec K⁴.* **(G.N.D.U., 1991)**

Solution. Energy radiated by the filament per second per sq. m. $= \dfrac{100}{0.25 \times 10^{-4}} = 4 \times 10^6 \text{ J/m}^2$

Stefan's constant $\sigma = 5.67 \times 10^{-5}$ erg/cm² sec K⁴ $= 5.67 \times 10^{-8}$ J/m² sec K⁴

According to Stefan's law $E = \sigma T^4$

$$\therefore \quad T^4 = \frac{E}{\sigma} = \frac{4 \times 10^6}{5.67 \times 10^{-8}} = 0.7054 \times 10^{14} \text{ or } T = 2898 \text{ K}$$

According to Wien's displacement law

$$\lambda_m . T = \text{a constant} = 2.898 \times 10^{-3} \text{ mK}$$

$$\therefore \quad \lambda_m = \frac{2.898 \times 10^{-3}}{2898} = 10^{-6}\ m = 10^4 \text{ Å}$$

Example 8.30. *What is the wavelength of maximum intensity radiation radiated from a source at temperature 3000°C ? Wien's constant = 2.898×10^{-3} mK.* **(Kan. U., 1994)**

Solution. Temperature of the source $T = 3000$ °C $= 3000 + 273 = 3273$ K

Wien's constant $b = 2.898 \times 10^{-3}$ mK

According to Wien's displacement law $\lambda_m . T = b$ (a constant)

$\therefore$ Wavelength for maximum intensity radiation $\lambda_m = \dfrac{b}{T}$

$$= \frac{2.898 \times 10^{-3}}{3273} = 8.854 \times 10^{-7} \text{ m} = 8854 \times 10^{-10} \text{ m} = 8854 \text{ Å}$$

Example 8.31. *A piece of copper sheet is blackened on both sides and placed normally to sun rays just outside the earth's atmosphere. After some time its temperature is found to have become constant. Calculate this temperature if solar energy incident on the sheet is 1.9 cal/cm². min, Stefan's constant = 5.67 × 10⁻⁵ erg/cm² sec. K⁴.* **(G.N.D.U., 1992)**

Solution. Solar energy incident on the blackened copper sheet per second per square cm.

$$E = \frac{1.9}{60} \text{ cal} = \frac{1.9 \times 4.2 \times 10^7}{60} = 1.33 \times 10^6 \text{ ergs}$$

Stefan's constant $\sigma = 5.67 \times 10^{-5}$ ergs/cm² sec K⁴

The temperature of copper sheet will become constant when the heat energy received by it per sq. cm per second is equal to the heat energy radiated by it per sq. cm per second. Let this temperature be T K.

Heat energy radiated by the blackened copper sheet per second per sq. cm according to Stefan's law

$$E = \sigma T^4 = 5.67 \times 10^{-5} T^4 \text{ ergs} \quad \therefore 5.67 \times 10^{-5} T^4 = 1.33 \times 10^6$$

or $$T^4 = \frac{1.33}{5.67} \times 10^{11} = \frac{.133}{5.67} \times 10^{12}$$

or $$T = \left(\frac{.133}{5.67}\right)^{\frac{1}{4}} \times 10^3 = 0.3913 \times 10^3 \text{ K} = 391.3 \text{ K} = 118.3\ °\text{C}$$

Example 8.32. *Estimate the temperature of the sun from the knowledge of solar constant and Stefen's Law* **(Meerut U. 2007, 2005, 2003, 2001, 2000; Kerala U. 2000; Nagpur U. 2002)**

Solution. The radius of the sun = R cms.

$\therefore$ Surface area of sun = $4\pi R^2$ sq. cms.

Stefan's constant $\sigma = 1.37 \times 10^{-12}$ cals. cm⁻² sec.⁻¹ deg⁻⁴

Suppose the temperature of the sun's surface is TK. If the sun is considered as perfectly black body the radiations emitted by it obey Stefen's law.

$\therefore$ Radiations emitted by the sun = σT^4 cals. cm⁻² sec⁻¹

Total radiations emitted by sun's surface in one minute = $60 \times 4\pi R^2 \sigma T^4$ cals.

These total radiations spread uniformly in all directions. If d is mean distance of the earth from the sun, then these radiations will spread over an area $4\pi d^2$ which is the area of the sphere of radius d. Radiations falling on a unit area of this sphere per minute = $\dfrac{60 \times 4\pi R^2 \sigma T^4}{4\pi d^2}$

This by definition is equal to the solar constant S in calories of heat per minute per sq. cm.

$\therefore$ Solar constant $S = \dfrac{60 \times 4\pi R^2 \sigma T^4}{4\pi d^2} = \dfrac{60 R^2 \sigma T^4}{d^2}$

and Temperature of sun, $T = \left(\dfrac{Sd^2}{60 R^2 \sigma}\right)$

Example 8.33. *Earth receives 2.0 cal cm^{-2} min^{-1} from sun. If angular diameter of the sun is 32 minutes and it is treated as black body, calculate temperature of the sun.*

Solution. Angular diameter $= \dfrac{\text{Diameter of the sun}}{\text{Distance of sun from earth}} = \dfrac{2R}{d}$

$$\therefore \quad \frac{2R}{d} = \frac{32}{60} \times \frac{\pi}{180} \quad \text{or} \quad \frac{R}{d} = \frac{16}{60} \times \frac{\pi}{180}$$

Solar constant = Energy received by the earth per cm^2 per minute

$$S = 2.0 \text{ cal/minute} = 2 \times 4.2 \times 10^7 \text{ ergs/mt}$$

$$\sigma = 5.7 \times 10^{-5} \text{ erg sec}^{-1} \text{ cm}^{-2} \text{ deg}^{-4}$$

$\therefore$ Temperature of sun

$$T = \left[\frac{S}{60\sigma}\left(\frac{d}{R}\right)^2\right]^{\frac{1}{4}} = \left[\frac{2 \times 4.2 \times 10^7}{60 \times 5.7 \times 10^{-5}} \times \left(\frac{16}{60} \times \frac{\pi}{180}\right)^2\right]^{\frac{1}{4}} = 5802.4 \text{ K}$$

Example 8.34. *Sun light arrives at the earth at the rate of about 1.4 KWm^{-2} when the Sun is exactly overhead. The average radius of the earth's orbit is 1.5 × 10^{11} m and radius of the Sun is 7 × 10^8 m. Find the surface temperature of the Sun assuming it radiates like a black body.*

Solution. In such a case the temperature of the Sun is given by

$$T = \left[\frac{Sd^2}{R^2\sigma}\right]^{1/4}$$ [For proof See ...]

where S is solar constant, d the mean distance of the earth from the Sun, R the radius of the Sun and σ Stefen's constant.

Here $S = 1.4$ k Wm^{-2} = 1.4×10^3 Js^{-1} m^{-2}; R = 7×10^8 m; $d = 1.5 \times 10^{11}$ m

and $\sigma = 5.67 \times 10^{-8}$ Js^{-1} m^{-2} T^{-4}

$$T = \left[\frac{Sd^2}{R^2\sigma}\right]^{1/4} = \left[\frac{1.4 \times 10^3 \times 1.5 \times 1.5 \times 10^{22}}{7 \times 7 \times 10^{16} \times 5.67 \times 10^{-8}}\right]^{1/4} = 5803 \text{ K}$$

Example 8.35. *Luminosity of Rigel star in Orion constellation is 17,000 times that of our sun. If the surface temperature of the sun is 6,000 K, calculate the temperature of the star.* **(A.U. 1994)**

Solution. Consider the sun and the star to be perfect black bodies. Let the temperature of the star be T K, then

Heat radiated by the sun per second per unit area = σ (6000)^4

Heat radiated by the star per second per unit area = σT^4

Since the luminosity of the star is 17,000 times that of the sun

$$\therefore \quad \sigma T^4 = 17{,}000 \times \sigma(6000)^4 \quad \text{or} \quad T = 6000 \times (17000)^{\frac{1}{4}} = 68520 \text{ K}$$

Example 8.36. *To what temperature a black body be raised to double its total radiation if its original temperature is 727°C.*

Solution. Total radiation per second per unit area at 727°C *i.e.*, at (727 + 273) = 1000 K, $E = \sigma(1000)^4$

If T is the temperature at which total radiation is doubled, then

$$2E = \sigma T^4$$

$$\therefore \qquad 2 = \frac{T^4}{(1000)^4} \text{ or } T^4 = 2\times(1000)^4 \text{ or } T = 2^{1/4}\times 1000$$

$$= 1189\text{ K} = 1189 - 273 = 916°\text{C}$$

Example 8.37. *A uniformally heated enclosure is maintained at 2500 K and has a cavity of diameter 5 × 10^{-3} m which acts like a black body. Calculate the heat energy radiated per minute from the cavity. Given Stefen's constant = 5.7 × 10^{-8} Wm^{-2} K^{-4}.*

Solution. Heat energy radiated per second per unit area $E = \sigma T^4$

$$= 5.7 \times 10^{-8} \times (2500)^4 \text{ Jm}^{-2}$$

Area of the cavity $= \pi r^2 = 3.14 \times 2.5 \times 2.5 \times 10^{-6}$

∴ Heat energy radiated per minute from the cavity

$$= 3.14 \times 2.5 \times 2.5 \times 10^{-6} \times 60 \times 5.7 \times 10^{-8} \times (2500)^4 \text{ Joules.}$$

$$= 2620 \text{ Joules.}$$

Example 8.38. *A black body with an initial temperature 300°C is cooled inside an evacuated chamber surrounded by melting ice at a rate 0.35°C per sec. If the mass of the body is 32 gm., specific heat 0.10, surface area 8 cm², then find Stefan's constant.* ***(A.U., 1995)***

Solution. Temperature of the black body $T_1 = 300°\text{ C} = 573\text{ K}$

Temperature of enclosure $T_0 = 0°\text{C} = 273\text{ K}$

∴ Rate of loss of heat $\propto (T_1 - T_0)$

Surface area of the body $A = 8$ sq cm

Mass of the body $m = 32$ gm

Specific heat $s = 0.10$

Rate of fall of temperature $\frac{dT}{dt} = 0.35°\text{C per sec.}$

∴ Rate of loss of heat energy $= ms\frac{dT}{dt}$

Rate of loss of heat energy is ergs per unit area $= \frac{Jms}{A}\frac{dT}{dt}$

$$e = \frac{4.18\times10^7 \times 32\times0.1\times0.35}{8}$$

According to Stefan's law $e = (T_1^{\,4} - T_0^{\,4})$

$$\therefore \qquad \sigma = \frac{e}{(T_1^4 - T_0^4)} = \frac{4.18\times10^7 \times 32\times0.1\times0.35}{8[(573)^4 - (273)^4]} = 5.73\times10^{-5} \text{ erg/ sq cm/ K}^4$$

EXERCISE

8.1 Define thermal radiation. ***(Rohilkhand 2001 Improvement)***

8.2 Explain the meaning of emissive power and absorptive power. ***(Kanpur 2000, Agra 2000, Purvanchal 2002, Calicut 2003)***

8.3 Explain the characteristics of radiation in a hollow encloser at constant temperature. ***(Rohilkhand 2001, Kanpur 2002)***

8.4 Give a brief account of Prevost's theory of exchanges.

8.5 What is perfectly black body? Describe how has the idea of a black body been achieved in practice? ***(Nagpur 2009, 2008, 2007, Meerut 2001, 2003, Agra 2000)***

8.6 What is black-body radiation? Explain its temperature dependence.
(Nagpur 2009, 2008, 2007, 2005/s, 2005/w, Ranchi 2005, Madras 2005, Rohilkhand 2001)

8.7 Explain the terms emissive power and absorptive power. Prove that at any temperature the ratio of the emissive power to the absorptive power of a substance is constant and is equal to the emissive power of a perfectly black body. ***(Purvanchal 2007, Agra 2003, Calcutta 2003)***

8.8 State and explain Kirchhoff's law. Give its application. ***(Lucknow 2004, Meerut 2003, Purvanchal 2005, 2002, 2003, Bundelkhand 2001, Madras 2005, Rohilkhand 2001)***

8.9 Prove that the pressure due to diffuse radiation on a surface is $u/3$, where u is energy density of radiation. ***(Purvanchal 2002, Ranchi 2005)***

8.10 Show that the diffuse radiation exert pressure on the wall of the encloser and derive a relation between energy and pressure of diffuse radiation. ***(Nagpur 2005/w, Calcutta 2004)***

8.11 Why polished reflectors are used in electric heaters? ***(Lucknow 2004)***

8.12 Describe the main characteristics of radiation emitted by a perfectly black body. Show that the diffused radiation exerts a pressure on the walls of the container equal to one-third of its energy density. ***(Garhwal 2005, Purvanchal 2003, 2004, 2005, Rohilkhand 2001)***

8.13 State Stefan-Boltzmann law of radiation. Deduce this law on thermodynamic considerations. ***(Nagpur, 2009, 2007, 2005/w, Meerut 2003, Lucknow 2003, Osmania 2005, Bangalore 2002, Purvanchal 2006)***

8.14 Explain the distribution of energy of a black body at different temperatures by drawing the graphs. Discuss briefly the different laws which explain the above energy spectrum. ***(Nagpur, 2007, 2005/s, 2004/s, Madras 2004, Bangalore 2002, 2005, Meerut 2001, Alagappa 2004, Osmania 2004)***

8.15 What is a perfectly black body? Draw curves for the distribution of energy in the spectrum of a black body for two different temperatures. Explain the important results obtained from these curves. ***(Nagpur, 2010, 2007, 2009, Lucknow 2004, Meerut 2003)***

8.16 State and explain the laws relating to radiation and temperature of a radiating body. What do you mean by
(*a*) perfectly black body,
(*b*) coefficient of absorption and
(*c*) coefficient of emission?

8.17 Deduce Wien's displacement law for the distribution of energy in black body spectrum. Explain its validity in terms of experimental results. ***(Agra 2003, Bangalore 2005, Gauhati 2000, Rohilkhand 2002 Imp. G.N.D.U. 2004, 2003, Kerala 2001, Osmania 2004, Nagpur 2002)***

8.18 State and explain Rayleigh-Jeans law and its failures.
(Nagpur 2010, Agra 2008, Rohilkhand 2002, Imp.)

8.19 State Planck's radiation law. Give the Planck's quantum postulates.
(Nagpur 2005/s, Ranchi 2004)

8.20 Derive Planck's radiation formula. Show that Rayleigh-Jeans law and Wien's law are special cases of Planck's law. ***(Nagpur 2005/s, Pune 2010, Meerut 2007, 2003, 2005, Ranchi 2004, Mumbai 2005, G.N.D.U. 2001, Punjab 2002, Agra 2006, 2002, Rohilkhand 2002 Imp., M.D.U. 2003, Madurai 2003, Purvanchal 2007, 2005)***

8.21 Explain the ultraviolet catastrophe according to Rayleigh–Jeans distribution law.
(Agra 2006, Nagpur 2009, 2007, 2005/s, 2004/s, 2004/w)

8.22 Assuming Planck's Law of black body spectrum, obtain Wien's law of distribution for short wavelength. ***(Nagpur 2004/s, Bangalore 2004, Alagappa 2004, Mysore 2004/s, 2004/w)***

8.23 Derive Wein's displacement law and Stefan's law from Planck's law of black body radiation. ***(Pbi U. 2005, 2003; Kerala U. 2001; Nagpur 2007, Punjab 2008, 2006, 2005, Alagappa 2004, Madras 2004)***

8.24 Describe Wien's displacement law. Using this law, calculate the temperature of the sun. ***(Purvanchal 2007, Agra 2005)***

8.25 Give Planck's theory of black body radiation. Show this reduces to Wien's formula for small wavelengths and Rayleigh-Jeans law for large wavelength. ***(Madras 2004, Purvanchal 2003, Rohilkhand 2000, Rajasthan 1996, 1997, 2000, Kumaun 1998, 2003)***

8.26 Derive Rayleigh-Jeans law from Planck's law. ***(Agra 2004, Rohilkhand 2001 Imp., Bangalore 2004, 2005, Alagappa 2004, Mysore 2004)***

8.27 Discuss Planck's hypothesis of quantum theory of radiation and obtain an expression for $\vec{\varepsilon} = \frac{h\nu}{e^{h\nu/kT} - 1}$ mean kinetic energy of Planck's oscillator. ***(Purvanchal 2006, Meerut 2004, Sagar 2000)***

8.28 What is Planck's law of radiation? Derive Planck's formula for energy distribution in black body spectrum. ***(Nagpur 2007, Meerut 2004, M.S. Uni. Tamil Nadu 2007)***

8.29 Explain what do you understand by a black body. State Stefan's law of radiation and prove it from thermodynamical considerations. Indicate how it can be verified. How is this law distinguished from Newton's law of cooling ? ***(Madurai Kamraj 2003, Mysore 2004, Purvanchal 2002, Agra 2002, Kanpur 1996)***

8.30 Show that Planck's law and Rayleigh-Jeans law become identical if the size of the quantum is allowed to vanish or if the temperature becomes very high. ***(Mumbai 2005)***

8.31 State Stefan's law and write its equation. Explain how would you calculate surface temperature of the sun using the law. ***(Alagappa 2004, Purvanchal 2003)***

8.32 What do you understand by a black body? State and explain Stefan's law of heat radiation. ***(Madras 2005)***

8.33 In what way the Planck's formula overcomes the deficiencies in both Wien's law and Rayleigh-Jeans formula for black body energy distribution? ***(Purvanchal 2004)***

8.34 Prove that the energy density of a black body radiation in a cavity is proportional to the fourth power of absolute temperature. ***(Calcutta 2005)***

8.35 What do you mean by black body? Discuss the distribution of energy in the spectrum of black body on the basis of the spectrum obtained in the experiment performed by Lummer and Pringsheim. ***(Pune 2010, M.S. Uni. Tamil Nadu 2007, Agra 2007, 2005, Rohilkhand 2000)***

8.36 Describe an experiment to verify Stefan's law and the determination of Stefan's constant. How is the temperature of sun estimated from the data of the solar constant? ***(Madras 2004, 2005, Rohilkhand 2000, Meerut 2000, Kanpur 2002, Gorakhpur 2000)***

8.37 Define solar constant. How do you find the temperature of the sun from a knowledge of solar constant and Stefan's law. ***(Purvanchal 2005, Meerut 2003, Madurai 2003)***

8.38 Define solar constant. How is it experimentally determined? Comment on the source of energy in the sun. ***(Purvanchal 2007, Madras 2004, Osmania 2005, Bangalore 2005, Lucknow 2004)***

8.39 Derive Planck's law of black body radiation and discuss its experimental verification. ***(Patna 2005)***

8.40 Deduce Kirchhoff's law of radiation and describe an experiment for its verification. ***(Patnu 2004)***

8.41 What is the principle of optical pyrometer? With the help of a diagram, explain the construction and working of an optical Pyrometer. ***(Osmania 2005, Madras 2005)***

8.42 Define solar constant. Describe water flow pyrheliometer and explain how solar constant is determined from it. ***(Alagappa 2004)***

8.43 Give an account of the methods for measuring high temperatures by radiation pyrometry. ***(Madras 2005, 2004)***

8.44 Prove that for long wavelength the black body radiation obeys Rayleigh-Jeans formula $E_\lambda d\lambda = \dfrac{8\pi kT}{\lambda^4} d\lambda$ and is a special case of Plancks law of radiation.

(Osm. U. 2004, Madurai U. 2003, Luck U. 2002, Meerut U. 2007, 2006, 2005)

8.45 Prove that for short wavelength the black body rediation obeys Wien's radiation law $E_\lambda d\lambda = A\lambda^{-5} e^{-a/\lambda T}.d\lambda$ and is a special case of Planck's law of radiation.

(P.U. 2000; Meerut U. 2006, 2005, 2004; G.N.D. U. 2001; Pbi. U. 2005; M.D.U. 2003)

8.46 Obtain Stefan's law of radiation from Planck's law given that $\int_0^\infty x^3 e^{-ax} dx = \dfrac{6}{a^4}$.

(Pbi. U. 2008, 2007, 2003; P.U. 2006, 2005, 2003, 2002; G.N.D.U. 2008, 2003, 2000; Utkal U. 2003; M.D.U. 2003)

8.47 Obtain Stefan's law of radiation from Planck's law given that $\int_0^\infty \dfrac{x^3 dx}{e^x - 1} = \dfrac{\pi^4}{15}$

(Rohilkhand 2002)

8.48 Write notes on :

(*i*) Solar spectrum ***(Alagappa 2004)***

(*ii*) Green house effect ***(Alagappa 2004)***

(*iii*) Wien's displacement law ***(Kumaun 2000)***

(*iv*) Solar constant and temperature of sun ***(Meerut 2001)***

(*v*) Optical pyrometer ***(Osmania 2005)***

(*vi*) Planck's law of radiation ***(Meerut 2001)***

(*vii*) Kirchhoff's law of black body radiation ***(Patna 2005, Rohilkhand 2000)***

(*viii*) Radiation pyrometry ***(Madras 2004)***

(*ix*) Sources of solar energy ***(Madras 200)***

8.49 A metallic ball at a temperature of 527 °C is placed in an encloser at 27 °C. If emissivity of the body is 0.4, find the rate of loss of energy per unit area by the body. Given: $\sigma = 5.67 \times 10^{-8}$ J/m²/s/deg⁴. [**Ans.** 9106.02 J/s] ***(Nagpur 2005/s)***

8.50 Spectra of three stars A, B and C show maximum brightness for the respective wavelengths of 6500 Å, 4800 Å and 5600 Å. Compare their temperatures T_A, T_B and T_C.

[**Ans.** $T_A = 4459$ K, $T_B = 6038$ K, $T_C = 5175$ K] ***(Nagpur 2004/s)***

8.51 In an atomic explosion, the maximum temperature reached was of the order of 10^7 K. Calculate the wavelength of maximum energy. Given: Wien's constant = 0.2898 cm K.

[**Ans.** 2896 Å] ***(Osmania 2004)***

8.52 What is wavelength of maximum intensity radiation radiated from a source at temperature 3000°C? Wien's constant = 2.898×10^{-3} m K. [**Ans.** 8854 Å] ***(Nagpur 2001)***

8.53 If the sun emits maximum energy at wavelength 4753 Å; calculate the temperature of its surface. Given: Wien's constant $b = 0.288$ cm °C. [**Ans.** 6060 K] ***(Nagpur 2001, 2004/w, 2005/w)***

8.54 Luminosity of rigid star in Orion constellation is 17000 times that of our sun. If the surface temperature of the sun is 6000 K; calculate the temperature of the star.

[**Ans.** 68520 K] ***(Agra 2008)***

8.55 Sun light arrives at the earth at the rate of about 1.4 kWm^{-2}, when the sun is exactly overhead. The average radius of the earth's orbit is 1.5×10^{11} m and the radius of the sun is 7×10^8 m. Find the surface temperature of the sun, assuming it radiates like a black body.

[**Ans.** 5803 K] ***(Rohilkhand 2001)***

8.56 Two large closely spaced concentric spheres (both of which are black body radiators) are maintained at temperatures of 400 K and 600 K respectively. The space in between the two spheres is evacuated. Calculate net rate of energy transfer between the two temperatures.

$\sigma = 5.672 \times 10^{-8}$ M.K.S. units

[**Ans.** 5.90 kilowatts/m^2] ***(Gorakhpur 2000)***

8.57 Calculate the radiant emittance of a black body at a temperature of (*i*) 200 K and (*ii*) 2000 K.

$\sigma = 5.672 \times 10^{-8}$ M.K.S. units

[**Ans.** (*i*) 90.75 watts/m^2 (*ii*) 907.5 kilowatts/m^2] ***(Madras 2005, Kanpur 2000)***

8.58 The relative emittance of tungsten is approximately 0.70. A tungsten sphere of surface area 5×10^4 sq metres is suspended inside a large evacuated enclosure whose walls are at 300 K. What power input is required to maintain the sphere at a temperature of 3000 K. The conduction of heat along the supports can be neglected.

$\sigma = 5.672 \times 10^{-8}$ M.K.S. units

[**Ans.** 1609 watts] ***(Kanpur 1999, 2001, Meerut 2000)***

8.59 An aluminium foil of relative emittance 0.2 is placed in between two concentric spheres at temperatures 300 K and 200 K respectively. Calculate the temperature of the foil after the steady state is reached. Assume the spheres are perfect black body radiators. Also calculate the rate of energy transfer between one of the spheres and the foil.

$\sigma = 5.672 \times 10^{-8}$ M.K.S. units

[**Ans.** (*i*) 263.8 K, (*ii*) 37 watts/m^2] ***(Nagpur 2000, Indore 2000)***

8.60 Calculate the energy radiated per minute from the filament of an incandescent lamp at 3000 K, if the surface area is 10^{-4} sq metres and its relative emittance is 0.425.

[**Ans.** 11710 joules] ***(Allahabad 2001, Kumaun 2000)***

8.61 What is the wavelength of the maximum intensity radiation, radiated from source having temperature 3000 K, the Wien's constant is 0.29 cm-K. ***(Osmania 2005, Bangalore 2004)***

8.62 If the average energy radiated per unit area of the surface of the sun is 7.452×10^6 kW. Estimate the surface temperature of the sun assuming it to be a black body. Stefan's constant is 5.67×10^{-8} W m^{-2} K^{-4}. [**Ans.** 19790 K] ***(Bangalore 2002, Punjab 2001)***

8.63 Calculate the number of modes of vibration in a 100 C.C. chamber in the wavelength range (*i*) 5000 to 5002 Å and (*ii*) 8000 to 8005 Å. [**Ans.** $N_1 = 8.03 \times 10^{12}$, $N_2 = 3.066 \times 10^{12}$]

8.64 Calculate the energy radiated by unit area of a black body in one second when its temperature is 1000 K. ($\sigma = 5.672 \times 10^{-8}$ J/s/m^2) ***(Alagappa 2004)***

8.65 An iron furnace radiates 7.65×10^4 calories per hour through an opening of cross-section 10^{-4} sq metres. If the relative emittance of the furnace is 0.40, calculate the temperature of the furnace. [**Ans.** 2500 K] ***(Indore 2001)***

8.65 Two large closely spaced concentric spheres (both of which are black body radiators) are maintained at temperatures of 400 K and 600 K respectively. The space in between the two spheres is evacuated. Calculate net rate of energy transfer between the two temperatures.

$\sigma = 5.672 \times 10^{-8}$ M.K.S. units. [**Ans.** 5.90 kilowatts/m^2] ***(Jiwaji 2000)***

8.66 Two ideal black bodies A and B at temperatures 227 °C and 327 °C respectively are placed in an evacuated encloser whose walls are blackened and kept at 27 °C. Compare their rates of loss of heat. [**Ans.** $A : B = 544 : 1215$]

8.67 A black body with an initial temperature of 300 °C is allowed to cool inside an evacuated encloser surrounded by melting ice at the rate of 0.35 °C per sec. If the mass, specific heat and surface area of the body are 32 g; 0.10 cal/g/°C and 8 cm^2 respectively. Calculate the Stefan's constant. [**Ans.** 5.7 × 10^{-8} Joule m^{-2} sec^{-1} (K)$^{-4}$] ***(Agra 2000)***

8.68 Calculate the number of modes of vibration per unit volume in a black body cavity having frequencies between 5.5 × 10^{12} Hz and 5.55 × 10^{12} Hz. [**Ans.** 1.40718 × 10^{12}] ***(Mumbai 2005)***

MULTIPLE CHOICE QUESTIONS

8.1 A perfectly black body is that which:

(*a*) is totally black in colour
(*b*) can radiate all its energy
(*c*) is made of ideal gas names Krishnika
(*d*) absorbs all the radiations incident on it ***(Garhwal 2005)***

8.2 Average energy of a Planck's oscillator is:

(*a*) $E = h\nu$
(*b*) $E = nh\nu$
(*c*) $E = \dfrac{h\nu}{e^{h\nu/kT} - 1}$
(*d*) $E = mc^2$ ***(Garhwal 2005)***

8.3 Rayleigh–Jeans law of radiation:

(*a*) applies to smaller wavelengths
(*b*) applies to longer wavelengths
(*c*) applies to all wavelengths
(*d*) does not apply to any wavelength ***(Garhwal 2005)***

8.4 According to Stefan-Boltzmann's law:

(*a*) $E = \sigma T_0^4$
(*b*) $E = \sigma T^2$
(*c*) $E = \sigma (T^2 - T_0^2)$
(*d*) $E = \sigma (T^4 - T_0^4)$ ***(Agra 2003)***

8.5 Wien's displacement law is expressed by the equation:

(*a*) $\lambda_m T$ = constant
(*b*) $\lambda_m \sqrt{T}$ = constant
(*c*) $\lambda_m T^2$ = constant
(*d*) $\lambda_m T^3$ = constant ***(Ranchi 2005)***

8.6 'Good absorbers are good emitters' – The statement is called:

(*a*) Prevost's law
(*b*) Kirchhoff's law
(*c*) Stefan's law
(*d*) Wien's law ***(Rohilkhand 2002 Imp.)***

8.7 Electromagnetic radiation is emitted by

(*a*) all bodies at all temperatures
(*b*) all bodies at 100 °C
(*c*) all bodies at absolute zero
(*d*) only a few bodies at all temperatures. ***(Rohilkhand 2003)***

8.8 For a perfectly black body, the absorptive power is

(*a*) 1
(*b*) 0.5
(*c*) 0
(*d*) ∞ ***(Rohilkhand 2002, Imp.)***

8.9 The absolute temperature of a perfectly black body is increased to twice its value. The rate of emission of energy per unit area will be:

(*a*) 2 times
(*b*) 4 times
(*c*) 8 times
(*d*) 16 times ***(Kanpur 2002)***

8.10 Plots of intensity verses wavelength for three bodies at temperatures T_1, T_2 and T_3 respectively are shown in Fig. Their temperatures are such that:

(*a*) $T_1 > T_2 > T_3$

(*b*) $T_1 > T_3 > T_2$

(*c*) $T_2 > T_3 > T_1$

(*d*) $T_3 > T_2 > T_1$ ***(I.I.T. 2000)***

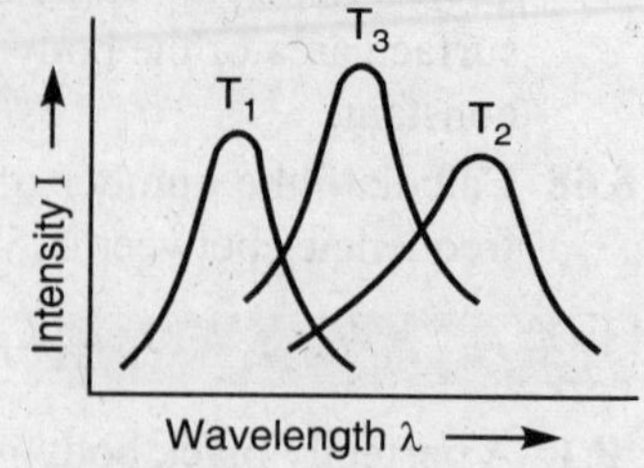

8.11 Wien's displacement law is a special case of

(*a*) Newton's law (*b*) Planck's law

(*c*) Stefan's law

8.12 The wavelength λ at which a black body emits maximum amount of rediation is proportional to

(*a*) T (*b*) $\frac{1}{T}$

(*c*) T^5 (*d*) T^4 ***(Agra 2001)***

ANSWERS

8.1 (*d*)	**8.2** (*c*)	**8.3** (*b*)	**8.4** (*d*)	**8.5** (*a*)	**8.6** (*b*)
8.7 (*a*)	**8.8** (*a*)	**8.9** (*d*)	**8.10** (*b*)	**8.11** (*b*)	**8.12** (*b*)

SECTION III

Statistical Physics

STATISTICAL PHYSICS

Statistical physics is the branch of physics that uses methods of probability theory and statistics, and particularly the mathematical tools for dealing with large populations and approximations, in solving physical problems. Its applications include many problems in the fields of physics, biology, chemistry, neurology, and so many. A statistical approach can work well in classical systems when the number of degrees of freedom (and so the number of variables) is so large that exact solution is not possible, or not really useful. Statistical mechanics can also describe work in non-linear dynamics, chaos theory, thermal physics, fluid dynamics, or plasma physics successfully.

Statistical mechanics provided a molecular-level interpretation of macroscopic thermodynamic quantities such as work, heat, free energy, and entropy. It enables the thermodynamic properties of bulk material to be related to the spectroscopic data of individual molecules. This ability to make macroscopic predictions based on microscopic properties is the main advantage of statistical mechanics over classical thermodynamics. Both theories are governed by the second law of thermodynamics through the medium of entropy. However, entropy in thermodynamics can only be known empirically, whereas in **statistical** mechanics, it is a function of the distribution of the system on its micro-states.

Statistical mechanics was initiated in 1870 with the work of Austrian physicist Ludwig Boltzmann, much of which was collectively published in Boltzmann's 1896 Lectures on Gas Theory. Boltzmann's original papers on the statistical interpretation of thermodynamics, the H-theorem, transport theory, thermal equilibrium, the equation of state of gases, and similar subjects, occupy about 2,000 pages in the proceedings of the Vienna Academy and other societies. The term "statistical thermodynamics" was proposed for use by the American thermodynamicist and physical chemist J.Willard Gibbs in 1902. According to Gibbs, the term " **Statistical**", in the context of mechanics. *i.e.* statistical mechanics, was first used by the Scottish physicist James Clerk Maxwell in 1871. "Probabilistic mechanics" might today seem a more appropriate term, but " **Statistical mechanics**" is firmly entrenched.

The study of statistical physics is mainly classified in two categories. These are:

1. Classical statistics or Maxwell-Boltzmann statistics.
2. Quantum statistics or B.E. and F.D. statistics.

The classical statistics employs the classical results of Maxwell's laws of distribution of molecular velocities and Boltzmann theorem, relating entropy and probability. Hence classical statistics is known as Maxwell-Boltzmann statistics.

On the other hand, the quantum statistics was developed by Bose, Einstein, Fermi and Dirac scientists. The essential difference between these two B.E. and F.D. statistics is due to the consideration of particles having different spins. B.E. statistics deals the case of identical, indistinguishable particles having zero or integral spin, while F.D. statistics is used in case of identical, indistinguishable particles having half-integral spin. Classical statistics is only a limiting case of quantum statistics. Four chapters are devoted for gradual understanding of the subject.

STATISTICAL BASIS OF THERMODYNAMICS

9

Chapter

9.1 STATISTICAL BASIS

Statistics is that branch of science which deals with the collection, classification and tabulation of numerical data as the basis of explanation, description and comparison of various phenomenon. When statistical concepts are applied to physics, the new branch that emerges is called *statistical physics.*

Statistical physics deals with *macroscopic systems. i.e.* the system consisting of a large number of individual particles such as atoms, molecules etc. It is not concerned with the behavior of the individual particles of the system but takes into consideration the average or *most probable* properties of the system. As a system is a collection of a large number of particles of one type, we may have a system composed of atoms, molecules, protons, neutron or electrons.

A study of thermodynamics gives us various macroscopic properties that are related through an equation of state having only two independent parameters. However, the equation of state can not be deduced from the laws of thermodynamics. It has to be obtained experimentally. The ordinary laws of mechanics were the only tool to explain physical phenomena, up to the end of 17th century. In certain cases, particularly where the system contains a large number of particles, ordinary laws of mechanics could not be used, as it is impossible to follow the motion of each particle. For example, a pinch of matter contains a very large number of atoms or molecules. (Avogadro's number $N_o = 6 \times 10^{26}$ per kg mole). Therefore, it is impossible to apply the ordinary laws of mechanics to a physical system containing large number of particles, particularly that of electrons. Such problems are however, successfully solved by *statistical mechanics*. The larger is the number of particles in the physical system considered, the more nearly correct are the statistical predictions. Smaller the number of particles in a mechanical system, the statistical mechanics goes on becoming meaningless. Before the advent of quantum theory, Maxwell, Boltzmann, Gibbs etc. applied the statistical methods making the use of classical physics. These methods are known as **classical statistics or Maxwell-Boltzmann statistics**. The classical statistics successfully explained the phenomenon like temperature, pressure, energy etc., but failed to explain other observed phenomenon like black body radiation, specific heat at low temperature etc. For this, new approach was introduced by Bose, Einstein, Fermi and Dirac. The Planck's quantum concept of discrete exchange of energy between systems was used. The new statistics was subdivided into two categories (*i*) Bose-Einstein statistics and (*ii*) Fermi-Dirac statistics. Thus, there are three statistics depending upon three different kinds of particles :

1. **Maxwell-Boltzmann statistics:** This is applicable to the identical, distinguishable particles of any spin. The molecules of a gas are the particles of this kind.
2. **Bose-Einstein statistics:** This is applicable to the identical, indistinguishable particles of zero or integral spin. These particles are called *bosons*. The examples of bosons are helium atoms at low temperature and the photons.
3. **Fermi-Dirac statistics:** This is applicable to the identical, indistinguishable particles of half-integral spin. These particles obey Pauli exclusion principle and are called *fermions*. The examples of fermions are electrons, protons, neutrons etc.

It's worth mentioning here that the classical Maxwell Boltzmann statistics is only the limiting case of the other two quantum statistics. Let us consider first some of the basic concepts of classical statistics.

9.2 PROBABILITY

The probability of an event may be defined as the ratio of the number of cases in which the event occurs to the total number of cases.

$$\text{Thus, the probability of an event} = \frac{\text{Number of cases in which the event occurs}}{\text{Total number of cases}} \qquad \text{... (9.1)}$$

Suppose an event can happen in a ways and fails to happen in b ways, then the probability of happening of the event $= \dfrac{a}{a+b}$ and the probability of failing of the event $= \dfrac{b}{a+b}$. Here $(a + b)$ represents the total number of equally likely possible ways. It should be noted that the sum of these two probabilities is always 1; since the event must either occur or fail.

To evaluate the probability of random event, we consider a unit of measurement. An event is called a 'sure' event if it occurs in an experiment. Thus, the probability of a 'sure' event is assumed to be equal to 1 and that of an impossible event to be equal to zero. Thus, the probability P of a random event lies between 0 and 1, *i.e.*,

$$0 \le P \le 1 \qquad \text{... (9.2)}$$

This is further explained by the following experiments:

1. **Throwing a coin:** Suppose we toss a coin. Either the 'head' can come upward or the 'tail' *i.e.* an event can occur in a total number of *two* equally likely ways. The number of ways in which the 'head' can come up is only *one*. Therefore, the probability that the 'head' may come up is $\frac{1}{2}$. Similarly, the probability that the 'tail' may come up is also $\frac{1}{2}$.
2. **Throwing a die:** A die is a homogeneous, regular and balanced cube with six faces marked respectively with number of dots from 1 to 6 engraved on them. It is supposed that the die is symmetrical and cannot fall on its edges. When the die is thrown it falls on one of its faces upward *i.e.*, yields one of the six results and no other. All outcomes are equally likely in a single throw of dice. Out of six possible results only one is favourable for the appearance of six spots. In other words, the probability of any one face (say with a number 3) to come up is $\frac{1}{6}$. The set of all possible outcomes can be written as S (1, 2, 3, 4, 5, 6). The probability of the die coming up with an even number is $\frac{3}{6}$ as there are only three even number on the die 2, 4 and 6.

$$\therefore \qquad P\,(\text{even}) = \frac{3}{6} = \frac{1}{2}$$

Likewise, the probability of the die coming up with odd number is $\frac{3}{6}$ or $\frac{1}{2}$.

The probability of the die coming up with any number less than 6 is given by

$$P\,(\text{a number} < 6) = \frac{5}{6}$$

as there are five numbers 1, 2, 3, 4, 5 which are less than 6.

Zero Probability. Suppose we want to know the probability of the die coming up with a face marked with a number 7. The dic has only six faces marked serially from 1 to 6. There is no face marked as 7. Therefore, probability of appearing of a number 7 is zero, *i.e.*,

$$\therefore \qquad P\,(\text{number } 7) = \frac{0}{6} = 0$$

In other words, impossible event is always zero.

Probability one. The probability of appearing any number less than 7 is one. This is because all the six faces of the die are marked from 1 to 6, *i.e.*, the numbers less than 7.

$$P\,(\text{number} < 7) = \frac{6}{6} = 1$$

Total probability. If a is the number of cases in which an event occurs and b the number of cases in which an event fails, then

$$\text{Probability of occurrence of the event} = \frac{a}{a+b}$$

$$\text{and Probability of failing of the event} = \frac{b}{a+b}$$

The sum *i.e.*, the total probability is always one, since the event may either occur or fail.

9.3 PRINCIPLE OF EQUAL A PRIORI PROBABILITY

Suppose we toss a coin. It is clear in our mind that the coin will fall either with its 'Head' up or 'Tail' up. Similarly if a six faced cubical dice is thrown, it is sure that the dice will fall with one of its six faces upward. In the same way, if we have an open box divided into two equal sized compartments X and Y (Fig. 9.1), and a small particle is thrown from a large distance in such a way that it must fall in either of the two compartments, then the probability of the particle to fall in the compartment marked X is equal to the probability that it may fall into the compartment marked Y. Again there is an equal probability.

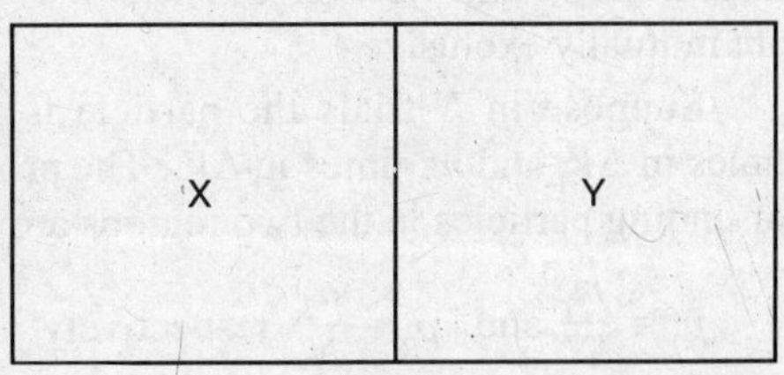

Fig. 9.1

This principle of assuming equal probability for events which are equally likely to occur is known as the principle of equal a priori probability.

A priori means something which exists in our mind prior to and independently of the observations we are going to make.

This principle of equal a priori probability will not hold good, if in the above example, the two compartments are of unequal size.

The principle of equal a priori probability is the fundamental basis of statistical mechanics.

9.4 PROBABILITY AND FREQUENCY

All the possible outcomes, in the examples discussed above, are assumed to be known. This is not always possible in practice and hence formula given by equation (9.1) is not a general formula. It can be used only in experiments that possess symmetry. Suppose, the die is asymmetric by adding a little load (by wax) to one of its faces. Now all the outcomes are not equally likely and we cannot say that the probability of appearance of six spots is $\frac{1}{6}$. For all such cases, we adopt a technique as under:

Suppose we toss a coin, say N times and we find that 'Head' appears H times. Here we introduce a term frequency of an event F as

$$F = \frac{\text{No. of trials in which head occurs}}{\text{Total number of trials}} = \frac{\text{H}}{\text{N}} \qquad \text{... (9.3)}$$

Thus, if a coin is tossed 50 time and in 10 of them the coin shows Heads, the frequency of this event is $\frac{10}{50} = 0.2$. From the classical definitions of probability, the probability of occurrence of 'Head' is 0.5 or 50%. Hence, we conclude that frequency is not the same as probability. There must be a relationship between frequency and probability. As the number of trials is increased, the frequency of the event progressively tends to stabilize and gradually approaches a constant value, known as the probability of the event. We define probability in terms of frequency as

$$P = \lim_{N \to \infty} \frac{M}{N} \quad \text{... (9.4)}$$

Thus, in order to obtain sufficiently accurate result, trials must be conducted until the ratios $\frac{M}{N}$ differ from one another by a very small value.

9.5 SOME BASIC RULES OF PROBABILITY THEORY

Additive Law of Probability

This law is applicable to mutually exclusive events. Two or more events are said to be mutually exclusive if the occurrence of any one of them prevents the occurrence of others. Such events never occurs simultaneously.

For example, consider two small non-overlapping region ΔV_1 and ΔV_2 in a box of volume V as shown in (Fig. 9.2). A particle in ΔV_1 rules out the possibility of its being present, at the same instant in ΔV_2 and vice-versa. The two events, thus, are mutually exclusive.

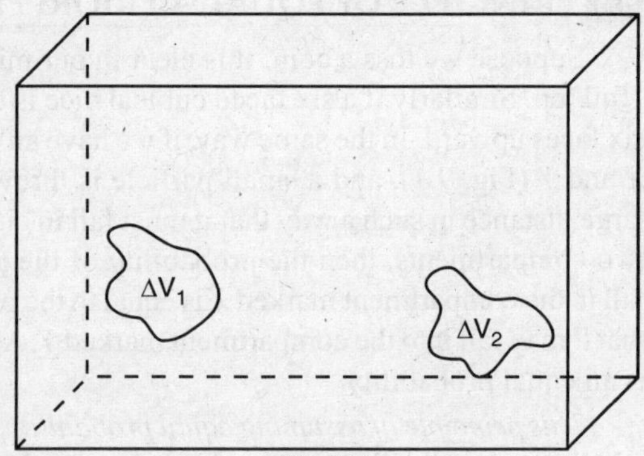

Fig 9.2

Suppose in N trials the particle is found m_1 times in ΔV_1 and m_2 times in ΔV_2. The probabilities of finding particles in the two regions are

$$p_1 = \frac{m_1}{N} \text{ and } p_2 = \frac{m_2}{N} \text{ respectively}$$

The number of times that the particle will be found at least in one of the two regions in N trials is $(m_1 + m_2)$. Hence, the probability will be

$$P\,(\Delta V_1 \text{ or } \Delta V_2) = \frac{m_1 + m_2}{N} = \frac{m_1}{N} + \frac{m_2}{N} = p_1 + p_2$$

This law can be generalized to any number of mutually exclusive events, giving probabilities, say, p_1, p_2, p_3 ...., p_n then the probability that any one of them occurs is the sum of the probability of these events, *i.e.*,

$$P = p_1 + p_2 + p_3 + \text{........} + p_N = \sum_{i=1}^{N} p_i$$

This is known as additive law of probability.

Example 9.1 *A card is drawn from a well shuffled pack of 52 cards. Calculate the probability for this card to be either a king or a queen.* ***(Kerala 1995)***

Solution. We are taking out one card out of 52 cards of pack. So the total number of ways in which the event can occur is, $N = 52$.

But we want to draw a specific card *i.e.* king. There are 4 kings in a pack of cards. Therefore, the number of ways favourable to the first event, $m_1 = 4$

$\therefore$ Probability of drawing a king, $p_1 = \frac{m_1}{N} = \frac{4}{52} = \frac{1}{13}$

The number of ways in which second event *i.e.*, drawing a queen may happen, $m_2 = 4$, as there are four queens too.

$\therefore$ Probability of drawing a queen, $p_2 = \frac{m_2}{N} = \frac{4}{52} = \frac{1}{13}$

Both the events are mutually exclusive. Therefore the probability that the card drawn is either a king or a queen is

$$P = p_1 + p_2 = \frac{1}{13} + \frac{1}{13} = \frac{2}{13}$$

Multiplication Rule: Joint Probability

In the calculations of probabilities, we sometimes come across random events; such that the probability of occurrence of one does not affect the probability of occurrence of the other. For example in Fig. 9.2, the probability that a molecule A gets into ΔV_1, at a particular instant is $p_1 = \frac{\Delta V_1}{V}$. The probability that another molecule gets into volume ΔV_2 at the same instant is $p_2 = \frac{\Delta V_2}{V}$, regardless of whether or not the molecule A gets into ΔV_1. We want to calculate the probability of joint occurrence of these events, *i.e.*, the probability of A getting into ΔV_1 and B getting into ΔV_2 *at the same instant*. Suppose in N trials, the molecule A is found m times in ΔV_1. If p_2 is the probability that B gets into ΔV_2, irrespective of the presence of A in ΔV_1, the number of times the two events will occur simultaneously is $m p_2$. Thus, the joint probability of occurrence of these two events, is

$$P = \frac{m\,p_2}{N}$$

$$= \frac{m}{N} \times p_2$$

$$P = p_1 \times p_2 \qquad \left(\because p_1 = \frac{m}{N}\right)$$

Thus, probability of joint occurence of two independent events is equal to the product of the probabilities of each of these independent events.

Example 9.2 *We throw a die twice and obtain two numbers. What is the probability that these numbers are 6 and 4 precisely in that order?*

Solution. The probability that the first throw gives a 6 is $\frac{1}{6}$. Similarly, the probability that the second throw gives a 4 is also $\frac{1}{6}$. These two events are independent.

$\therefore$ Required probability $= \frac{1}{6} \times \frac{1}{6} = \frac{1}{36}$

Example 9.3 *Two six faced dice are thrown simultaneously list the various ways in which event can happen. What conclusion do you draw from it.*

Solution. Throwing of two dice. If we throw two dice simultaneously the first die can fall in 6 different ways. Corresponding to *each* of these ways the second die can also fall in 6 different ways. Hence the total number of different ways in which the two dice can fall simultaneously is 6 × 6= 36.

These ways are

[1,1] [2,1] [3,1] [4,1] [5,1] [6,1]
[1,2] [2,2] [3,2] [4,2] [5,2] [6,2]
[1,3] [2,3] [3,3] [4,3] [5,3] [6,3]
[1,4] [2,4] [3,4] [4,4] [5,4] [6,4]
[1,5] [2,5] [3,5] [4,5] [5,5] [6,5]
[1,6] [2,6] [3,6] [4,6] [5,6] [6,6]

Conclusion. From the above example, we conclude that "*If one event can occur in n_1, different ways and another independent event can occur in n_2 different ways. then both the events can occur togehter simultaneouly in $n_1 \times n_2$ different ways*"

Example 9.4 *Two six faced dice, each marked 1 to 6 are thrown. Calculate the probability that one of the dice shows 6 and the other shows 5.* ***(G.N.D.U.2001)***

Solution. The first dice can fall in *six* different ways with any of its faces marked 1 to 6 upward. Out of these only two ways are favourable to the event *i.e.* when the dice falls with the face marked 5 or 6 upward.

∴ Probability that the first dice shows 5 or 6

$$p_1 = \frac{2}{6}$$

The second dice can also fall in *six* different ways but out of these only one way is now favourable to the event. If the first dice falls with 5 upward, the second dice must show 6 and if the first dice falls with 6 upward, the second dice must show 5.

∴ Probability that the second dice must shows one out of 5 or 6 which the first dice has not shown;

$$p_2 = \frac{1}{6}$$

∴ Probability that one of the dice shows 5 and the other shows 6 is,

$$P = p_1 \times p_2 = \frac{2}{6} \times \frac{1}{6} = \frac{1}{18}$$

Example 9.5 *A coin and a six faced dice are thrown. What is the probability that the coin falls with its head up and the dice with its face 6 up?* ***(P.U. 2006)***

Solution. Probability that the coin falls with its head up $= \frac{1}{2}$

Probability that the dice falls with its face 6 up $= \frac{1}{6}$

Since the events are independent, the probability that the two events take place simultaneously

$$= \frac{1}{2} \times \frac{1}{6} = \frac{1}{12}$$

Conditional Probability

The probability for an event A to occur the condition that event B has occurred is called the *conditional probability* and is denoted by $P\left(\frac{A}{B}\right)$.

This can be better understood by the following numerical.

Example 9.6 *A bag contains 4 black and 3 white balls. What is the probability that on two successive draws, the balls drawn are both black?* ***(Allahabad 2002)***

Solution. Suppose the first ball drawn is black. Let the event be called as A. Its probability is $P(A) = \frac{4}{4+3} = \frac{4}{7}$. This event is followed by second event B in which the second ball drawn is also black (conditional). Since an event A is already realized, the balls remained are 3 black and 3 white, out of which the event B is taking place, *i.e.* second ball out of 6 balls. Hence, the probability of the event B to occur under the condition that A has already realized is $P\left(\frac{B}{A}\right) = \frac{3}{3+3} = \frac{3}{6} = \frac{1}{2}$. The first event thus has an effect on the probability of the second event.

∴ The probability of the composite event $(A + B)$ is

$$P(A+B) = P(A) \cdot P\left(\frac{B}{A}\right) = \frac{4}{7} \times \frac{1}{2} = \frac{2}{7}$$

9.6 PERMUTATIONS AND COMBINATIONS

The word permutation means arrangement and combination means formation of groups.

Permutations: To understand the meaning of permutation *i.e.* arrangement, let us consider an example of four distinguishable objects marked *a, b, c* and *d*. Taking any two objects at a time, the possible arrangements are

ab, ba, ac, ca, ad, da, bc, cb, bd, db, cd, dc.

There are total 12 arrangements possible. In arranging these objects the *order* of their placing is also taken into consideration. Thus, *ab* and *ba* are two different permutations. Thus, 4 objects can be arranged in 12 ways by taking 2 objects at a time, *i.e.* the number of permutations is 12.

Symbolically, $^4p_2 = 12$

Instead, if 3 objects are taken at a time out of 4 objects *a, b, c* and *d* the various arrangements are

abc	*abd*	*acd*	*bcd*
acb	*adb*	*adc*	*bdc*
bca	*bad*	*cda*	*cbd*
bac	*bda*	*cad*	*cdb*
cab	*dab*	*dac*	*dbc*
cba	*dba*	*dca*	*dcb*

Thus, 4 objects can be arranged in 24 different ways by taking 3 at a time out of 4 objects. Symbolically,

$$^4P_3 = 24$$

In general, the number of arrangements of *n* distinguishable objects by taking *r* at a time is given by

$$^nP_r = \frac{n!}{(n-r)!} \qquad \text{... (9.5)}$$

Thus, $$^4P_2 = \frac{4!}{(4-2)!} = \frac{4!}{2!} = \frac{4\times3\times2\times1}{2\times1} = 12$$

and $$^4P_3 = \frac{4!}{(4-3)!} = \frac{4!}{1!} = \frac{4\times3\times2\times1}{1} = 24$$

This includes all types of arrangements, meaningful as well as meaningless.

Combinations: Here we consider only combinations *i.e.* groups without considering the order of their placement *i.e.* only meaningful combinations. Thus, the combinations of 4 objects *a, b, c* and *d* taking two at a time are

ab, ac, ad, bc, bd, cd.

i.e., only 6 combinations. Symbolically, $^4C_2 = 6$

Similarly, combinations of 4 objects *a, b, c* and *d* taking three at a time are

abc, abd, acd, bcd

i.e., only 4 combinations. Symbolically, $^4C_3 = 4$

In general, the number of groups *i.e.* meaningful combinations of *n* distinguishable objects by taking *r* at a time is given by

$$^nC_r = \frac{n!}{r!(n-r)!} \qquad \text{... (9.6)}$$

But $$^nP_r = \frac{n!}{(n-r)!}$$

$$\therefore \qquad {}^nC_r = \frac{{}^nP_r}{r!}$$

or $$ {}^nP_r = r!\,{}^nC_r \qquad \text{... (9.7)}$$

From equation 9.5
$$ {}^nP_n = \frac{n!}{(n-n)!} = \frac{n!}{o!} = n! \qquad (\because 0! = 1)$$

and from equation 9.6
$$ {}^nC_n = \frac{n!}{n!\,(n-n)!} = \frac{n!}{n!\,o!} = \frac{n!}{n!} = 1 \qquad (\because 0! = 1)$$

9.7 MACROSTATE AND MICROSTATE

Macrostate: Consider 4 distinguishable particles and we wish to distribute them into two exactly similar compartments in an open box. Let the particles be *a, b, c* and *d*. When any particle is thrown into the box, it must fall into one of the two compartments. Since the compartments are alike, the particles have the same a priori probability of going into either of them and will be $\frac{1}{2}$. The possible ways in which 4 particles can be distributed in two compartments are shown in table 9.1.

Table 9.1

Compartment	*No. of particles*				
	Distribution 1	*Distribution 2*	*Distribution 3*	*Distribution 4*	*Distribution 5*
1	0	1	2	3	4
2	4	3	2	1	0

Thus, there are 5 different distributions (0, 4), (1, 3), (2, 2), (3, 1) and (4, 0). Each compartmentwise distribution of a system of particles is known as a *macrostate*. In general, for a system of *n* particles to be distributed in two similar compartments, the various macrostates are (0, *n*), (1, *n* – 1), (2, *n* – 2), ... (*n* – 1, 1) and (*n*, 0). Therefore, the total no. of macrostates for *n* particles is (*n* + 1).

Microstates: Since the particles are *distinguishable*, the number of different possible arrangements in each compartment is shown below :

Table 9.2

Macrostate	*Possible arrangements*		*No. of microstates W (Thermodynamic probability)*
	Compartment 1	*Compartment 2*	
0,4	0	*abcd*	1
1,3	*a* *b* *c* *d*	*bcd* *cda* *dab* *abc*	4
2,2	*ab* *ac* *ad* *bc* *bd* *cd*	*cd* *bd* *bc* *ad* *ac* *ab*	6

3,1	*bcd* *cda* *dab* *abc*	*a* *b* *c* *d*	4
4,0	*abcd*	0	1

Each distinct arrangement is known as the *microstate* of the system. The distribution (0, 4) can have only one arrangement, distribution (1, 3) can have 4 distinct arrangements, distribution (2, 2) can have 6, distribution (3, 1) have 4 and distribution (4, 0) have only one. Thus, a given macrostate may consists of a number of microstates. In the above example of 4 particles, the total number of microstates are $16 = 2^4$. In general, for a system of n particles, the total number of microstates are 2^n.

Suppose we consider a thermodynamic system of N identical, non-interacting particles occupying a volume V. Let $E_1, E_2, E_3, \ldots\ldots\ldots E_i$ be the energy of particles $n_1, n_2, n_3 \ldots\ldots\ldots n_i$ respectively, then total energy of the system, $E = n_1 E_1 + n_2 E_2 + \ldots\ldots\ldots + n_i E_i$

$$= \Sigma n_i E_i \text{ and } N = \Sigma n_i$$

According to quantum theory, the possible values of energy of each particle are discrete and hence the total energy E also has a discrete value. But when the volume V is very large, the spacings of different energy levels are so small in comparison with total energy of the system that E may be regarded as almost *continuous*.

The actual values of parameters N, V and E defines a particular *macrostate* (N, V, E) of the given system.

There are a large number of different ways in which the individual Ei's can be chosen, so as to make total energy equal to E.

Each of the different ways in which the total energy E of the system can be distributed among N particles constituting it specifies a *microstate*. Thus, *a given macrostate corresponds to many microstates* and it is very natural to assume that at any time the system is equally likely to be in any of these *microstates*.

9.8 THERMODYNAMIC PROBABILITY

The number of microstates corresponding to any given macrostate is called its thermodynamic probability. In other words, the thermodynamic probability of a particular macrostate is defined as the number of microstates corresponding to that macrostate. In general, this is a very large number and is represented by W (or Ω).

The number of microstates corresponding to a given macrostate is equal to the number of meaningful arrangements or permutations of various particles in the macrostate excluding those permutations which are meaningless *i.e.* merely interchange the order of particles in a particular compartment (or cell). For the case of n particles and two compartments (or cells), if r is the number of particles in the compartment No. 1 and the remaining $(n - r)$ are in compartment No. 2, then

$$\text{No. of meaningful arrangements} = \frac{n!}{r!(n-r)!}$$

$$= {}^nC_r$$

Therefore, the number of microstates in a macrostate (r, $n - r$) or thermodynamic probability

$$W_{(r,n-r)} = \frac{n!}{r!(n-r)!} = {}^nC_r \qquad \ldots (9.8)$$

Applying this to a system of 4 distinguishable particles, for a macrostate (1, 3), $r = 1$, $(n - r) = 3$ and $n = 4$ the number of microstates,

$\Omega(1, 3)$, or

$$W_{(1,3)} = \frac{4!}{1!\,3!} = 4$$

Similarly, for a macrostate (2, 2), $r = 2$, $(n - r) = 2$ and $n = 4$ the thermodynamic probability *i.e.* no. of microstates will be

$$W_{(2,2)} = \frac{4!}{2!\,2!} = 6$$

and for macrostate (0, 4) or (4, 0) will be

$$W_{(0,4)} = W_{(4,0)} = \frac{4!}{0!\,4!} = \frac{4!}{4!} = 1 \qquad (\because 0! = 1)$$

This agrees with the table 9.2.

Most probable macrostate. *The probability (or occurrence) of a macrostate is defined as the ratio of the number of microstates (i.e. thermodynamic probability W) in it to the total number of possible microstates of the system.*

Thus,
$$P_{macro} = \frac{\text{No. of microstates in the macrostate}}{\text{Total no.of microstates of the system}} \qquad \text{... (9.9)}$$

The total number of ways of arranging n distinguishable particles in c numbered compartments $= c^n$

If there are only 2 compartments or cells, then the total no. of microstates of the system $= 2^n$. For 4 particles system, total no. of microstates will be $2^4 = 16$. From equation (9.9), we get.

$$P_{macro} = \frac{W}{c^n} = \frac{W}{2^n} \qquad \text{... (9.10)}$$

Substituting for W from equation (9.8),

$$P_{(r,\,n-r)} = \frac{n!}{r!\,(n-r)!} \times \frac{1}{2^n} = \frac{1}{2^n}\,{}^nC_r \qquad \text{... (9.11)}$$

This gives the probability (of occurrence) of macrostate (r, n–r).

Combination Possessing Maximum Probability

We know from elementary algebra that if n is even the value of ${}^nC_r$ is maximum (greatest) when $r = \frac{n}{2}$, refer table 9.2. Hence the maximum value of probability is given by

$$P_{max} = \frac{1}{2^n}\,{}^nC_{n/2} = \frac{n!}{\frac{n}{2}!\,\frac{n}{2}!} \times \frac{1}{2^n}$$

$$= \frac{n!}{\left(\frac{n}{2}!\right)^2} \times \frac{1}{2^n} \qquad \text{... (9.11A)}$$

Combination Possessing Minimum Probability

We know from elementary algebra that ${}^nC_r$ is minimum, if r is equal to either zero or n.

If $r = 0$, the probability is

$$P_{\min} = \frac{n!}{r!(n-r)!} \times \frac{1}{2^n}$$

$$= \frac{n!}{o!(n-o)!} \times \frac{1}{2n} \qquad (\because o! = 1)$$

$$= \frac{1}{2^n} \qquad ...(i)$$

If $r = n$, the probability is

$$P_{min} = \frac{n!}{r!(n-r)!} \times \frac{1}{2^n}$$

$$= \frac{n!}{r!(n-r)!} \times \frac{1}{2^n}$$

$$= \frac{n!}{n!o!} \times \frac{1}{2^n} \qquad (\because o! = 1)$$

$$= \frac{1}{2^n} \qquad ...(ii)$$

Thus, from (*i*) and (*ii*), we conclude that the minimum probability is given by

$$P_{min} = \frac{1}{2^n} \qquad ... (9.12)$$

Relation between *P* and *W*

The probability of occurrence of a particular microstate is given by

$$p = \frac{1}{2^n}$$

because 2^n gives the total no. of microstates of a system of n particles to be arranged in two compartments (or cells) and all the microstates of a system have *a priori* probability. Thus, from equation (9.10), we have

$$P = W \times p$$

= No. of microstates in the macrostate × probability of occurrence of a microstate

or $\quad P \propto W$

Therefore, the probability (of occurrence) of a macrostate (P) is directly proportional to the thermodynamic probability (W).

Example 9.7 *Calculate the probability that in tossing a coin 5 times, we get 3 heads and 2 tails.* ***(Punjab 1998)***

Solution. Tossing a coin 5 times is equivalent to tossing 5 similar coins simultaneously. Hence $n = 5$. We wish to get 3 heads and 2 tails i.e.,

$$r = 3 \text{ and } (n - r) = 2$$

The probability of distribution $(r, n - r)$ is given by

$$P_{(r,\, n-r)} = \frac{1}{2^n}\,{}^nC_r = \frac{1}{2^n} \cdot \frac{n!}{r!(n-r)!}$$

$$= \frac{1}{2^5} \times \frac{5!}{3!\,2!} = \frac{5 \times 4 \times 3 \times 2 \times 1}{32 \times 3 \times 2 \times 1 \times 2 \times 1}$$

$$= \frac{10}{32} = \mathbf{\frac{5}{16}}$$

Example 9.8 Three coins are tossed simultaneously. Find the probability of getting one head and two tails. ***(PU 1999, Pbi. U. 2002)***

Solution. Here $n = 3$; $r = 1$; $n - r = 2$

$$\therefore \qquad P = \frac{n!}{r!(n-r)!} \times \frac{1}{2^n} = \frac{3!}{1!\,2!}\,\frac{1}{2^3}$$

$$= \frac{3}{8} = 0.375$$

Example 9.9 *Calculate the probability that in tossing a coin 10 times, we get (i) all heads (ii) 5 heads & 5 tails and (iii) 3 heads & 7 tails. (iv) 7 heads & 3 tails.* ***(Meerut 2004)***

Solution. The probability of distribution of $(r, n-r)$ is given by

$$P_{(r,\, n-r)} = \frac{1}{2^n} \cdot \frac{n!}{r!(r-r)!}$$

(*i*) Tossing a coin 10 times is equivalent to tossing 10 similar coins simultaneously

Here, $n = 10$

We wish to have all heads

$\therefore$ $r = 10$ and $(n-r) = 0$

Hence, $$P_{(10,\,0)} = \frac{1}{2^{10}} \cdot {}^{10}C_{10} = \frac{1}{2^{10}} \times \frac{10!}{10!\,0!} = \frac{10!}{10!} \qquad (\because 0! = 1)$$

$$\therefore \quad P_{(10,\,0)} = \frac{1}{2^{10}} = \frac{1}{1024} = \mathbf{0.00098}$$

(*ii*) We wish to have 5 heads and 5 tails

$\therefore$ $r = 5$ and $(n-r) = 5$

Hence $$P_{(5,\,5)} = \frac{1}{2^{10}}\,{}^{10}C_5 = \frac{1}{2^{10}} \frac{10!}{5!\,5!}$$

$$= \frac{10\times9\times8\times7\times6\times5\times4\times3\times2\times1}{5\times4\times3\times2\times1\times5\times4\times3\times2\times1} \times \frac{1}{1024}$$

$$= \frac{252}{1024} = \mathbf{0.24609}$$

(*iii*) We wish to have 3 heads and 7 tails

$\therefore$ $r = 3$ and $(n-r) = 7$

Hence, $$P_{(3,\,7)} = \frac{1}{2^{10}}\,{}^{10}C_3 = \frac{1}{1024} \times \frac{10!}{3!\,7!}$$

$$= \frac{10\times9\times8\times7\times6\times5\times4\times3\times2\times1}{3\times2\times1\times7\times8\times6\times5\times4\times3\times2\times1} \times \frac{1}{1024} = \frac{120}{1024} = 0.11719$$

(*iv*) We wish to have 7 heads and 3 tails

$\therefore$ $r = 7$ and $(n-r) = 3$

Hence, $$P_{(7,\,3)} = \frac{{}^{10}C_7}{2^{10}} = \frac{10!}{7!\,3!} \frac{1}{2^{10}}$$

$$= \frac{10\times9\times8\times7\times6\times5\times4\times3\times2\times1}{7\times6\times5\times4\times3\times2\times1\times3\times2\times1} \times \frac{1}{1024} = \frac{120}{1024}$$

Thus, the probability of distribution (7, 3) is equal to the probability of distribution (3, 7).

Example 9.10 *Calculate the probability that in tossing a coin 5 times, we get 3 heads and two tails.* ***(Pbi U. 2008)***

Solution: Here $n = 5$, we wish to have 3 heads and 2 tails

$\therefore$ $r = 3$, $n - r = 2$

Hence $$P(3,2) = \frac{5!}{3!\,2!} \times \frac{1}{2^5} = \frac{5\times4\times3\times2\times1}{3\times2\times1\times2\times1} \times \frac{1}{32} = \frac{10}{32} = 0.3125$$

9.9 FLUCTUATIONS AND THEIR DEPENDENCE ON *N*. (NARROWING OF PROBABILITY DISTRIBUTION WITH INCREASING *N*)

Let us consider n molecules of an ideal gas enclosed in a container or a box. Let us investigate the positions and distribution of these particles in space. For this purpose, imagine that the box is divided into two equal parts by some not actual partition Fig. 9.3. Let n_1 and n_2 be the number of molecules in A and B compartments *i.e.*, in equal halves.

Fig. 9.3

Thus,

$$n_1 + n_2 = n$$

If n is large, it would be ordinarily expected that

$$n_1 \cong n_2.$$

i.e approximately half of the molecules are in each half of the box.

Due to intermolecular interactions, the molecules collide with each other and on the walls of the container; as a result some of the molecules enter left half of the box while some leave it. Thus the number n_1 of the molecules actually located in the left half fluctuates constantly with time. For large n, these fluctuations are so small that n does not differ too mcuh from $\frac{n}{2}$. The state $\left(\frac{n}{2}, \frac{n}{2}\right)$ is the state of maximum probability and is called **equilibrium state**. We shall show now that the probability distribution narrows rapidly as n increases. For simplicity, again we consider two similar compartments. From a priori probability, we expect $p = p' = \frac{1}{2}$

Consider $n = 40$. The most probable distribution is (20, 20). Using equation (9.11), its probability is

$$P_{max} = P_{(20,\,20)} = \frac{1}{2^{40}} \cdot \frac{40!}{20!\,20!} \qquad \text{... }(i)$$

A distribution 25% off from this is with $r = 15$ (or 25). Its probability is

$$P_{(15,\,25)} = \frac{1}{2^{40}} \cdot \frac{40!}{15!\,25!} \qquad \text{... }(ii)$$

The ratio R of the two probabilities is obtained by dividing (*ii*) by (*i*), and we get

$$R\,(25\%\text{ off}) = \frac{20!\,20!}{15!\,25!} = \frac{20\times19\times18\times17\times16}{25\times24\times23\times22\times21}$$

$$\cong \left(\frac{4}{5}\right)^5 \cong 0.3 \qquad \text{... }(iii)$$

With $n = 400$, a distribution 25% off from the most probable one will have the relative probability

$$R\,(25\%\text{ off}) = \frac{200!\,200!}{150!\,250!} = \frac{200\times199\times\ldots\times151}{250\times249\times\ldots\times201}$$

$$\cong \left(\frac{4}{5}\right)^{50} \cong 10^{-5} \qquad \text{... }(iv)$$

Thus, as n increases the probability of distributions appreciably far from the most probable distribution becomes negligibly small.

General Formula: In order to derive general formula, let us consider $2n$ particles distributed between two compartments with equal probability, the most probable distribution is (n, n). The probability $P_{(n,\,n)}$ may be compared with the probability ($P_{n+s,\,n-s}$) where $\frac{s}{n} << 1$. From equation (9.11), the ratio of probabilities for these distributions are given by

$$R = \frac{P_{(n+s,\,n-s)}}{P_{(n,\,n)}} = \frac{^{2n}C_{n+s}}{^{2n}C_n}$$

$$= \frac{2n\times(2n-1)\times.......\times(n-s+1)}{1\times2\times3.......\times(n+s)}$$

$$\times\frac{1\times2\times3\times.......\times n}{2n\times(2n-1)\times.......\times(n+1)}$$

$$= \frac{(n-s+1)\times(n-s+2)\times.......\times n}{(n+1)\times(n+2)\times.......(n+s)}$$

$$\simeq \left(1-\frac{s}{n}\right)^s, \qquad \text{Since } s/n \ll 1$$

If we write $\frac{s}{n} = f$ to represent the fractional deviation from the most probable distribution, we get (for large n values)

$$\log_e R = s\log_e\left(1-\frac{s}{n}\right) = s\left[-\frac{s}{n}-\frac{s^2}{n^2}-\frac{s^3}{n^3}-.....\right] = s\left(-\frac{s}{n}\right) \text{ for } s \ll n = -\frac{s^2}{n^2}.n$$

$$= -f^2 n$$

$$\therefore \qquad R = e^{-f^2 n} \qquad ...(9.13)$$

Let $P(f)\,df$ represents the probability that the distribution is between fractions f and $(f+df)$ away from the most probable distribution (n, n). Then

$$P(f)\,df = A\,.\,e^{-f^2 n}df$$

where A is a constant determined by the condition

$$\int_{-1}^{+1} P(f)\,df = 1 \qquad \text{(normalization condition)}$$

This gives $A = \sqrt{\frac{n}{\pi}}$, so that

$$P(f)\,df = \sqrt{\frac{n}{\pi}}\,.\,e^{-f^2 n}df \qquad ...(9.14)$$

Thus, as the fractional deviation f from most probable distribution increases, the probability falls exponentially. Further, with larger n, the fall will be more rapid (Fig. 9.4). The probability falls to $\frac{1}{e}$ times the maximum, when

$$f^2 n = 1 \quad \text{or} \quad f = \frac{1}{\sqrt{n}} \qquad ...(9.15)$$

A graph is plotted between the relative probability R against the fractional departure $f\left(=\frac{s}{n}\right)$ from equal distribution for different values of $2n$. There is a rapid fall of R and the curve becomes narrower with increasing $2n$ or n.

Alternative method: The probability distribution narrows rapidly as n increases. The variation of probability due to a small deviation from the state of maximum probability is calculated as under:

Suppose we have n number of particles distributed in two compartments. Total number of microstates = 2^n. The microstate of maximum probability is represented by $\left(\frac{n}{2}, \frac{n}{2}\right)$ (Refer table 9.2) and is given by equation (9.11A).

$$P_{\max} = \frac{{}^{n}C_{n/2}}{2^n} = \frac{n!}{\left(\frac{n}{2}!\right)^2} \times \frac{1}{2^n}$$

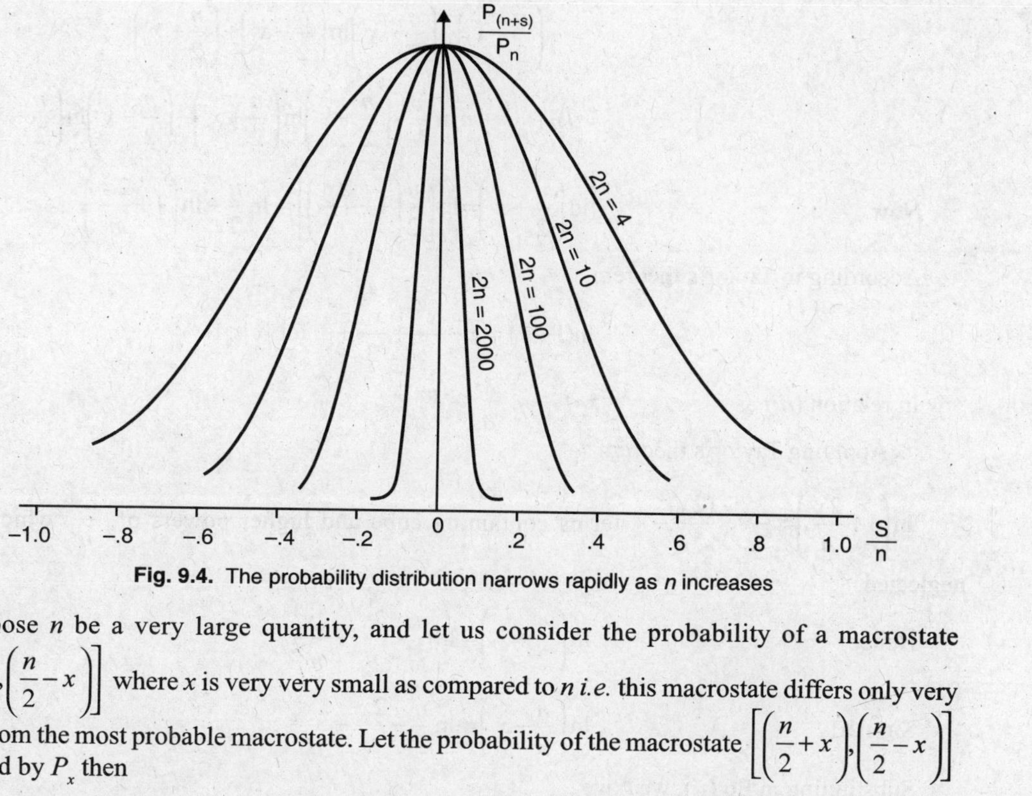

Fig. 9.4. The probability distribution narrows rapidly as n increases

Suppose n be a very large quantity, and let us consider the probability of a macrostate $\left[\left(\frac{n}{2}+x\right), \left(\frac{n}{2}-x\right)\right]$ where x is very very small as compared to n *i.e.* this macrostate differs only very slightly from the most probable macrostate. Let the probability of the macrostate $\left[\left(\frac{n}{2}+x\right), \left(\frac{n}{2}-x\right)\right]$ be denoted by P_x then

$$P_x = \frac{{}^{n}C_{\left(\frac{n}{2}+x\right)}}{2^n} = \frac{n!}{\left(\frac{n}{2}+x\right)!\left(\frac{n}{2}-x\right)!} \frac{1}{2^n}$$

$$= \frac{n!\frac{n}{2}!\frac{n}{2}!}{\frac{n}{2}!\frac{n}{2}!\left(\frac{n}{2}+x\right)!\left(\frac{n}{2}-x\right)!} \frac{1}{2^n}$$

$$\therefore \qquad P_x = p_{\max} \frac{\left(\frac{n}{2}!\right)^2}{\left(\frac{n}{2}+x\right)!\left(\frac{n}{2}-x\right)!}$$

Taking natural logarithms on both sides and representing $\log_e = \ln$; we have

$$\ln P_x = \ln P_{\max} + 2\ln\left(\frac{n}{2}\right)! - \ln\left(\frac{n}{2}+x\right)! - \ln\left(\frac{n}{2}-x\right)! \qquad \text{... (i)}$$

According to Stirling's formula

$$\ln (n!) = n \ln n - n$$

Applying Stirling's formula to relation (*i*), we have

$$\ln P_x = \ln P_{\max} + 2\frac{n}{2}\ln\frac{n}{2} - 2\frac{n}{2} - \left(\frac{n}{2}+x\right)\ln\left(\frac{n}{2}+x\right)$$

$$+\left(\frac{n}{2}+x\right) - \left(\frac{n}{2}-x\right)\ln\left(\frac{n}{2}-x\right) + \left(\frac{n}{2}-x\right)$$

$$= \ln P_{max} + n\ln\frac{n}{2} - \left(\frac{n}{2}+x\right)\ln\left(\frac{n}{2}+x\right) - \left(\frac{n}{2}-x\right)\ln\left(\frac{n}{2}-x\right) \qquad \text{... (ii)}$$

Now $$\ln\left(\frac{n}{2}+x\right) = \ln\left[\frac{n}{2}\left(1+\frac{2x}{n}\right)\right] = \ln\frac{n}{2} + \ln\left(1+\frac{2x}{n}\right) \qquad \text{... (iii)}$$

According to Taylor's theorem

$$\ln(1+x) = x - \frac{x^2}{2} + \frac{x^3}{3} + \text{....for } |x| < 1$$

In relation (*iii*) as $x << n, \ \frac{x}{n} << 1$

∴ Applying Taylor's theorem

$\ln\left(1+\frac{2x}{n}\right) = \frac{2x}{n} - \frac{(2x)^2}{2n^2}$ + terms containing cube and higher powers of $\frac{2x}{n}$ which can be neglected

Hence $$\ln\left(\frac{n}{2}+x\right) = \ln\frac{n}{2} + \frac{2x}{n} - \frac{2x^2}{n^2}$$

Similarly $$\ln\left(\frac{n}{2}-x\right) = \ln\frac{n}{2} - \frac{2x}{n} - \frac{2x^2}{n^2}$$

Substituting in Eq (*ii*), we have

$$\ln P_x = \ln P_{\max} + n\ln\frac{n}{2} - \left(\frac{n}{2}+x\right)\left(\ln\frac{n}{2} + \frac{2x}{n} - \frac{2x^2}{n^2}\right) - \left(\frac{n}{2}-x\right)\left(\ln\frac{n}{2} - \frac{2x}{n} - \frac{2x^2}{n^2}\right)$$

$$= \ln P_{\max} + n\ln\frac{n}{2} - \frac{n}{2}\ln\frac{n}{2} - x + \frac{x^2}{n} - x\ln\frac{n}{2} - \frac{2x^2}{n} + \frac{2x^2}{n^2}$$

$$-\frac{n}{2}\ln\frac{n}{2} + x + \frac{x^2}{n} + x\ln\frac{n}{2} - \frac{2x^2}{n} - \frac{2x^3}{n^2} = \ln P_{\max} - \frac{2x^2}{n}$$

$$\therefore \quad \ln\frac{p_x}{P_{\max}} = -\frac{2x^2}{n}$$

or $$\frac{P_x}{P_{\max}} = e^{\left(\frac{-2x^2}{n}\right)} = e^{\left(-n\frac{f^2}{2}\right)} \qquad \text{.... (iv)}$$

where $f = \frac{2x}{n} = \frac{x}{n/2}$ and represents the *fractional deviation* from the most probable number of particles in a compartment.

Half width. From Eq. (iv) we find that for $f = 0$

$e^{\left(-n\frac{f^2}{2}\right)} = e^0 = 1$ and $P_x = P_{\max}$ *i.e.* the probability has a maximum value for $f = 0$. This probability goes on decreasing as f increases. A curve showing the variation of $\frac{P_x}{P_{\max}}$ with $f = \frac{x}{n/2}$ is given in Fig. 9.5. The curve is symmetrical about the point $f = 0$.

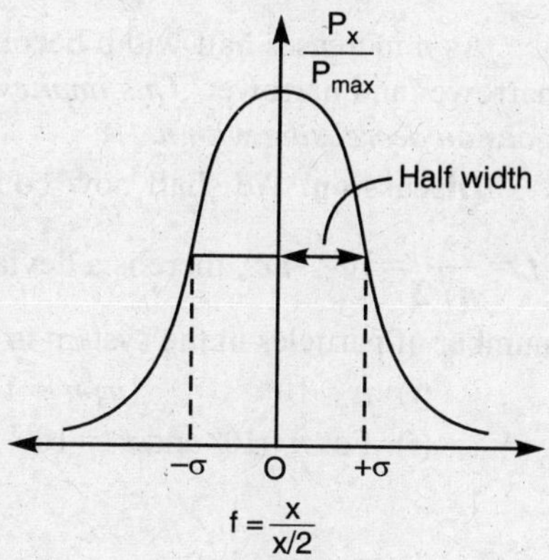

Fig. 9.5

Now suppose the value of $\frac{P_x}{P_{\max}}$ falls to $\frac{1}{e}$ for fractional deviation $f = \sigma$

$$\therefore \qquad \frac{P_a}{P_{\max}} = \frac{1}{e}$$

or

$$P_\sigma = \frac{P_{\max}}{e} = P_{\max} e^{-1} \qquad \text{... } (v)$$

Putting $f = \sigma$ in Eq. (*iv*), we have

$$P_\sigma = P_{\max} e^{\left(\frac{-n\sigma^2}{2}\right)} \qquad \text{... } (vi)$$

Comparing Eqs. (*v*) and (*vi*), we get

$$\frac{n\sigma^2}{2} = 1$$

or

$$\sigma = \sqrt{\frac{2}{n}} \qquad \text{... (9.16)}$$

σ is known as *half width* of the curve. The smaller the value of σ, the narrower is the curve. Thus σ is a measure of the sharpness of the curve

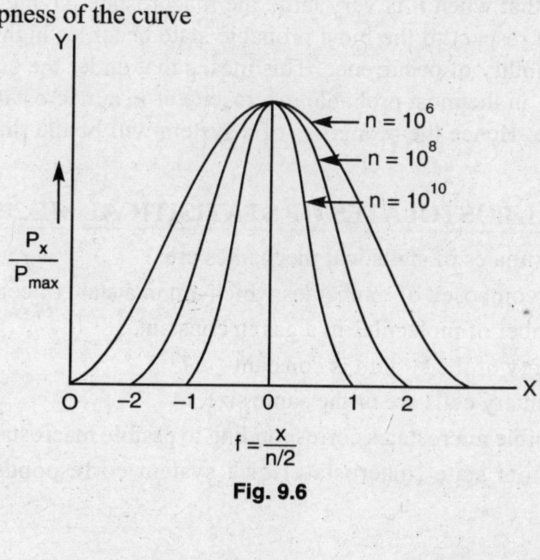

Fig. 9.6

Eq. 9.16 shows that half width is inversely proportional to the square root of n

or
$$\sigma \propto \frac{1}{\sqrt{n}} \qquad ... (9.17)$$

As n increases half width becomes smaller and smaller. Hence the probability curve becomes narrower and narrower. *This implies that the probability of deviation from the most probable state goes on decreasing with n.*

Discussion. We shall now consider the result of equation (*iv*) for a macrostate for which $f = \frac{x}{n/2} = 10^{-3}$ *i.e.*, there is a deviation of only 0.1% from the most probable state when the total number of particles in the system is

(*i*) $n = 10^6$ (*ii*) $n = 10^8$ and (*iii*) $n = 10^{10}$

(*i*) For $n = 10^6$ and $f = 10^{-3}$

$$\frac{P_x}{P_{max}} = e^{-(10^{-6} \times 10^6)} = e^{-1} = \frac{1}{e} = \frac{1}{2.718}$$

(*ii*) For $n = 10^8$

$$\frac{P_x}{P_{max}} = e^{-(10^{-6} \times 10^8)} = e^{-10^2} = \frac{1}{e^{100}}$$

(*iii*) For $n = 10^{10}$

$$\frac{P_x}{P_{max}} = e^{-(10^{-6} \times 10^{10})} = e^{-10^4} = \frac{1}{e^{10^4}}$$

We, therefore, find that for $n = 10^6$ when a macrostate has a deviation of only 0.1% from the most probable state the probability of this state is $\frac{1}{e}$ times the probability of the most probable state; it is $\frac{1}{e^{100}}$ when the number is increased to 10^8 and becomes $\frac{1}{e^{10^4}}$ when the number rises to 10^{10}. In other words as n goes on increasing the probability of the state goes on falling rapidly and becomes almost negligible. A graph between $\frac{P_x}{P_{max}}$ taken along Y- axis and f as a percentage taken along the X-axis is shown in Fig. 9.6.

Hence we conclude that when n is very large the macorostate having even a very very small amount of deviation with respect to the most probable state becomes highly **improbable** *i.e.* has an extremely small probability of occurrence. This means that under the circumstances the system can be supposed to exist in the most probable macrostate or in a macrostate very very close to the most probable macrostate. Hence the properties of a system will be the same as deduced from the most probable state.

9.10 FUNDAMENTAL POSTULATES OF STATISITICAL MECHANICS

The fundamental postulates of statistical mechanics are:

(*i*) All gases are composed of molecules which are in a state of constant random motion.

(*ii*) The total number of molecules in a gas in constant.

(*iii*) The total energy of the system is constant

(*iv*) All the elementary cells are of the same size.

(*v*) All the accessible microstates corresponding to pssible macrostates are equally probable.

(*vi*) The equilibrium state (macrostate) of a system corresponds to state of maximum probability.

9.11 CONSTRAINTS ON A SYSTEM

A set of conditions or restrictions that must be obeyed by a system are known as constraints.

Let us take an example of the distribution of 3 particles in two compartments, the system must obey the constraint that total number of particles in the two compartments must be 3. In general, if there are N particles to be distributed in two compartments and there are n_1 particles in compartment No. 1 and n_2 particles in compartment No. 2, then we must have

$$n_1 + n_2 = N$$

This relation is known as *equation of constraint* on the system.

A typical set of constraints is

$$\Sigma n_i = \text{constant} = N$$

$$\Sigma E_i n_i = \text{constant} = E$$

where n_i is the number of particles (or molecules) in the ith compartment (or ith cell), E_i is the energy of each particle (or molecule) in the ith compartment (or ith cell).

9.12 ACCESSIBLE STATES

Accessible states are the states consistent with the given constraints of the system.

Suppose we wish to distribute 3 particles in two compartments, then the system will have following macro and microstate (Table 9.3) provided the manner of relative placement of particles is immaterial i.e. only meaningful are considered. The system has four macro and 8 microstates when particles are distinguishable and only four microstates when they are indistinguishable.

Table 9.3

MACROSTATE	MICROSTATES			
	Particles distinguishable		*Particles indistinguishable*	
	Description	*Number*	*Description*	*Number*
3,0	*(abc, 0)*	1	*(aaa,0)*	1
2,1	*(ab, c)(bc, a) (ca, b)*	3	*(aa,a)*	1
1,2	*(c,ab), (a,bc), (b, ac)*	3	*(a, aa)*	1
(0,3)	*(0,abc)*	1	*(0, aaa)*	1

If we put the constraint that no compartment should remain empty, then the macrostate (3, 0) and (0, 3) can not exist. The system will then have only two macrostates (2, 1) and (1, 2) with their corresponding microstates. Thus, there will be only two macrostates and six microstates, when the particles are distinguishable, and only two microstates when they are indistinguishable.

Suppose again we put the constraint that the particles are indistinguishable, then we shall have four macrostates (3, 0), (2, 1), (1, 2) and (0, 3) but each macrostate will have only one microstate as shown in table 9.3. The total number of microstates will also be four.

Thus, the constraints decrease the number of macro as well as microstates of the system.

Accessible macrostates: ***The macrostates which are allowed under a constraint are called accessible macrostates.***

For example, in distributing 3 particles in 2 compartments under the constraint that no compartment will remain empty, the only accessible macrostates are (2, 1) and (1, 2).

Inaccessible macrostates: ***The macrostates which are not allowed under a constraint are called inaccessible macrostates.***

In the above example, the macrostates (3, 0) and (0, 3) are inaccessible macrostates.

Sometimes we have constraints like this; 'energy of the system in between E and $E + \delta E$'. In this case, all microstates which correspond to energy less than E or greater than $E + \delta E$ are considered inaccessible.

9.13 STATIC AND DYNAMIC SYSTEMS

Static System : ***A system is said to be static if the constituent particles of the system remain at rest in a particular microstate.***

In static system, the particles do not change their relative positions in the compartment in which they are located. They do not move from one compartment to even the neighbouring compartment. In other words, the static systems do not change of their own due to internal forces or the internal conditions. However, the change in possible due to external forces or causes.

Whenever the change in static system takes place due to external causes or forces, the system goes from one macrostate to another macrostate. The system now stays in the new macrostate even after the external cause of force is withdrawn.

Example: A system of tossed coin is an example of static system. After every toss, the coins remain permanently latched in the last state.

Dynamic system: ***A system is said to be dynamic if the constituent particles of the system can move so that the system goes from one microstate to another.***

The particles of the dynamic system change their relative positions within the compartment in which they are. They also go from one compartment to another. Thus, both the micro as well as macrostates of the system change with time. Dynamic systems can change of their own due to internal causes or forces. They can also undergo changes due to external causes or forces affecting them.

Example: Any gas is an example of dynamic system (except at absolute zero of temperature). The molecules of gas are always in constant, random (haphazard) motion, continuously colliding with one another, following Brownian motion. During this, they change their position, momentum and energy continuously.

If the dynamic system is disturbed by some external causes or forces and then left to itself the system tends to approach the macrostate with maximum thermodynamic probability and remains there or stays very close to the most probable macrostate.

9.14 MOST PROBABLE STATE (EQUILIBRIUM STATE)

The most probable state of a system is that macrostate which has the maximum probability of occurrence. In the case of N particles distributed in two compartments, the probability of occurrence of the most probable state is given by

$$P_{max} = \frac{{}^N C_{N/2}}{2^N}$$

In a **static system**, the particles remain latched *i.e.*, at rest when put in a compartment or placed in a cell.

The thermodynamic probability of the macrostate (n_1, n_2) in such a system is given by

$$W_{(n_1, n_2)} = \frac{N!}{n_1!\, n_2!} = \frac{N!}{n_1!(N - n_1)!} = {}^N C_{n_1} = {}^N C_{n_2}$$

But in **dynamic system**, the particles are in a constant state of motion and are not stationary. At N.T.P, the molecules of 1 cm^3 volume of air undergoes 6.6×10^{28} collisions per second *i.e.*, one collision in 1.5×10^{-29} second, molecules are moving with very high speeds. Suppose, such molecules are enclosed in a vessel having two compartments, using a permeable partition. Due to collisions, the number of molecules in the two compartments go on changing and the system moves from one macrostate to another very very large number of times per second. This number is generally of the order of millions of billions. Some macrostates are more probable than the others depending upon frequency of the microstate. *When the number of particles, in a dynamic system is very large, the system exists practically all the time in the most probable macrostate.* This is because the probabilities of macrostates fall extremely rapidly, even for a very small deviation (Refer Fig. 9.4) from the most probable state. This most probable state is known as the **equilibrium state** of the system. A dynamic system always tends to go from a state of lower probability towards the state of maximum probability *i.e.*, the equilibrium state. Even it is disturbed by some external causes of force, again it quickly comes back to its equilibrium state.

Importance in Statistical Physics

In statistical physics, particularly in statistical thermodynamics, we handle the systems having a very large number of particles and assume that the system exists in the most probable state *i.e.*, equilibrium state. Since n is very large, no significant error is caused by making this assumption. In other words, the importance of most probable state in determining the behaviour of a macroscopic system in equilibrium lies in the fact that all the properties of the system can be deduced by knowing the most probable or equilibrium state of the system.

9.15 LIFE-TIME OF A MICROSTATE AND MACROSTATE

Life-time of a Microstate: The life-time of a microstate is the average time taken by a particle of a system to move from one phase space* compartment to an adjascent phase space compartment.

Let us consider a particle in a phase-plane (x, p_x), moving from compartment 1 to compartment 2, with a change in position coordinate from (x, p_x) to $(x + \Delta x, p_x)$ in time T_{micro}, keeping its momentum p_x constant (Fig. 9.7). If m is the mass of the particle, then its velocity

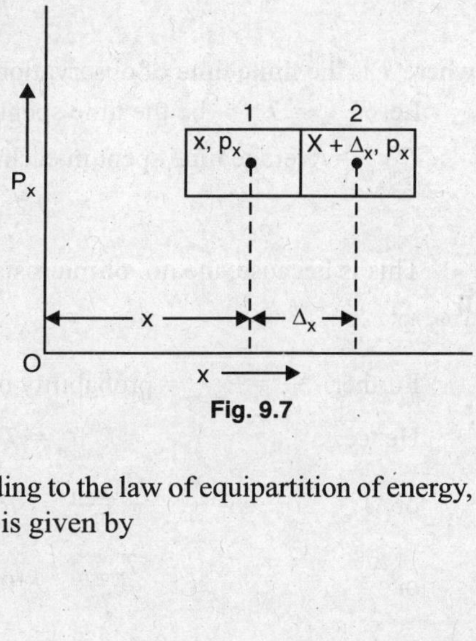

Fig. 9.7

$$v_x = \frac{\Delta x}{T_{micro}} = \frac{p_x}{m}$$

or

$$T_{micro} = \frac{\Delta x}{V_x} \qquad \text{... } (i)$$

This T_{micro} is called the life-time of the microstate. According to the law of equipartition of energy, the average K.E. of a gas molecule at temperature T Kelvin is given by

$$\frac{1}{2} m v_x^2 = \frac{1}{2} kT$$

where k is Boltzmann constant (1.38×10^{-16} erg K^{-1}).

$$\therefore \qquad v_x = \sqrt{\frac{kT}{m}} \qquad \text{... } (ii)$$

Substituting in equation (i), we get

$$T_{micro} = \frac{(\Delta x)}{\sqrt{\frac{kT}{m}}} = \Delta x . \sqrt{\frac{m}{kT}} \qquad \text{... } (iii)$$

For Hydrogen gas: Mass of hydrogen molecule $m = 3.2 \times 10^{-24}$ gm.

At room temperature, $T = 300$ K, and assuming $\Delta x = 10^{-8}$ cm,

$$T_{micro} = 10^{-8} \sqrt{\frac{3.2 \times 10^{-24}}{1.38 \times 10^{-16} \times 300}} = 10^{-13} \text{ second (approx.)}$$

This gives the average life-time of a microstate.

* For phase space, refer next chapter

Life-time of a Macrostate

If we try to examine the state of the system as shown in Fig. 9.7 i.e. the number of particles in each compartment, we will need some **finite time**. This finite time would be of the order of one-millionth (10^{-6}) of a second, even if we use a superfast camera. This finite time T would be very large of the order of 10^7 as compared to the life-time of a microstate, as calculated below :

$$\frac{T}{T_{micro}} = \frac{10^{-6}}{10^{-13}} = 10^7$$

During such a single experimental observation, the system passes through 10 million (10^7) microstates.

Further, in a dynamic system, the change from one microstate to another microstate takes place even at very fast rate, the system will spend equal amounts of time in all the microstates. This is because, all the microstates of the system have the *same a priori probability*.

Therefore, the average time spent in each microstate

$$T_{micro} = \frac{T}{\text{Total no. of microstates } (N)}$$

$$= \frac{T}{N} \quad \text{... } (i)$$

where T is the finite time of observation.

Let $T_{macro} = T_{(n_1, n_2)}$ be the time spent in the macrostate (n_1, n_2) then

T_{macro} = Average time spent in each microstate × No. of microstates in a given macrostate

$$= \frac{T}{N} \times W_{(n_1, n_2)} \quad \text{... } (ii)$$

This is because, the no. of microstates in a given macrostate is the thermodynamic probability $W_{(n_1, n_2)}$.

Further, $\frac{1}{N} = P_{micro}$ = probability of microstate.

Hence,
$$T_{macro} = T.P_{micro} W_{(n_1, n_2)} \quad \text{... } (iii)$$

or
$$\frac{T_{macro}}{T} = P_{micro} W_{(n_1, n_2)} \quad \text{... } (iv)$$

or
$$\frac{T_{macro}}{T} \propto W_{(n_1, n_2)} \quad \text{... (9.18)}$$

Thus, the fraction of the time spent by a system in a macrostate is proportional to the thermodynamic probability of the state.

9.16 CONCEPT OF A CELL IN A COMPARTMENT

In practice, if we wish to distribute n distinguishable particles in k compartments which are of unequal size, we find that the a *priori* probability of a particle going into a particular compartment is not the same. *Large probability is in a bigger compartment*. This is because of unequal size of the compartment. This difficulty is overcomed by introducing the concept of a cell.

Each compartment is divided into a very large number of sub-compartments, called as cells, in such a way that each cell is of the same size. Therefore, all the cells have the same a priori probability. The size of the cell is very very small and hence the number of cells in each compartment is exceedingly large. For a given size of the cell, the number of cells in each compartment is fixed.

Distribution of particles
(In compartments of unequal size)

The concept of a cell is well understood by considering the distribution of particles in different compartments marked as 1, 2, 3, ... , *i*..., $k-1$, k of unequal size (Fig. 9.8). We wish to distribute n distinguishable particles in these k compartments. As the compartments are of unequal size, a priori probability of a particle going into particular compatment is not the same. To overcome this difficulty, we introduce the **concept of a cell**. For this purpose we divide each compartment, from compartment 1 to k, into cells of **equal sizes**, so that compartment 1 contains g_1 cells, compartment 2 contains g_2 cells, compartment 3 contains g_3 cells and so on, the ith compartment containing g_i cells and lastly k^{th} compartment has g_k cells.

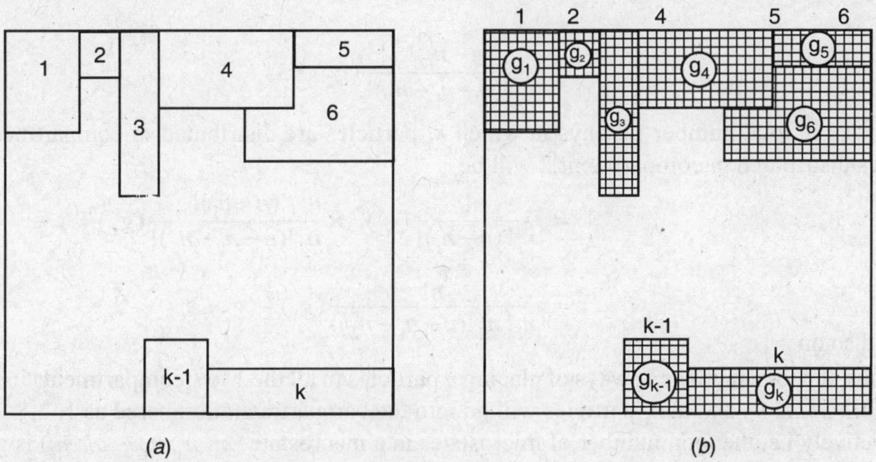

Fig. 9.8

As the compartments are of unequal size, the number of cells in each compartment will be different.

The total number of cells, each of same size $= g_1 + g_2 + \dots\dots\dots + g_i + \dots\dots\dots g_k = G$

Hence

(*i*) Since all the cells are of same size, there is an equal a priori probability *i.e.* a particle in a given compartment can go into any cell within the compartment with equal probability.

(*ii*) Moreover, since the size of the compartment is big, the number of cells in any compartment are exceedingly large.

Let us find number of ways of placing n_1 particles in g_1 cells for compartment 1. The first particle can go into any one of the g_1 cells of compartment 1. Thus, the number of ways the first particle can be arranged in g_1 cells is $= g_1$.

In M.B. statistics*, as there is no restriction on the number of particles that can be placed in a single cell, the 2nd particle can also go into any one of the g_1 cells of compartment 1.

$\therefore$ Total number of ways in which two particles can be arranged in g_1 cells of compartment 1 is given by

$$= g_1 \times g_1 = g_1^{\,2}$$

Proceeding in the same manner, the total number of ways in which n_1 particles can be distributed in g_1 cells of compartment 1 is

$$= g_1 \times g_1 \times g_1 \times \dots\dots\dots \times g_1 \ (n_1 \text{ times}) = g_1^{\,n_1}$$

* for this refer next chapter

The number of ways in which n_1 particles can be chosen out of n particles

$$= {}^{n}C_{n_1} = \frac{n!}{n_1!(n-n_1)!}$$

Therefore, the total number of ways in which n_1 particles can be chosen out of n particles and placed in g_1 cells of compartment 1 will be

$$= \frac{n!}{n_1!(n-n_1)!}(g_1)^{n_1} \qquad \text{... (9.19)}$$

Next, let us find the number of ways of placing n_2 particles in g_2 cells of compartment 2 as under: The number of particles remained after placing n_1 particles in compartment 1 are = $(n - n_1)$. Out of these n_2 particles are to be placed in g_2 cells of compartment 2. The number of ways this can be done is

$$= \frac{(n-n_1)!}{n_2!(n-n_1-n_2)!}(g_2)^{n_2}$$

Thus, the total number of ways in which n_1 particles are distributed in compartment 1 and n_2 particles distributed in compartment 2 will be

$$= \frac{n!}{n_1!(n-n_1)!}(g_1)^{n_1} \times \frac{(n-n_1)!}{n_2!(n-n_1-n_2)!}(g_2)^{n_2}$$

$$= \frac{n!}{n_1!\,n_2!(n-n_1-n_2)!}(g_1)^{n_1}(g_2)^{n_2} \qquad \text{...(9.20)}$$

and so on.

Thus, the total number of ways of placing n particles in all the 1 to k compartments in such a way that $n_1, n_2, \ldots\ldots\ldots\, n_i, \ldots\ldots\ldots\, n_k$ particles will go into the compartment numbered as 1, 2, ... i .... k respectively i.e. the total number of microstates in a macrostate $(n_1, n_2, n_3 \ldots\ldots\ldots\, n_k)$ is given by

$$W_{(n_1,n_2,n_3,\ldots\ldots\ldots n_k)} = \frac{n!}{n_1!\,n_2!\,n_3!\ldots\ldots.n_k!(n-n_1-n_2-n_3\ldots\ldots.-n_k)!}$$

$$\times (g_1)^{n_1}(g_2)^{n_2}(g_3)^{n_3}\ldots\ldots\ldots(g_k)^{n_k}$$

But $(n - n_1 - n_2 - n_3 \ldots.. - n_k) = n - (n_1 + n_2 + n_3 + \ldots\ldots\, n_k) = n - n = 0$

$\therefore \qquad (n - n_1 - n_2 - n_3 \ldots.. - n_k)! \;= 0! = 1$

Therefore $\qquad W_{(n_1, n_2 \ldots\ldots n_k)} = \dfrac{n!}{n_1!\,n_2!\ldots\ldots\, n_k!} \times (g_1)^{n_1}(g_2)^{n_2}(g_3)^{n_3}\ldots\ldots(g_k)^{n_k}$

or $$W = n!\prod_{i=1}^{k}\frac{(g_i)^{n_i}}{n_i!} \qquad \text{...(9.21)}$$

where $\prod$ denotes multiplication of terms stated above for various values of i.

Thus, the concept of a cell plays an important role in overcoming the difficulty in the distribution of particles in different compartments of unequal sizes, keeping the consistency of the principle of a priori probability.

Distribution of particles in compartments of equal size

The number of different meaningful ways of arranging n distinguishable particles in k compartments of equal size such that n_1 are placed in compartment no. 1, n_2 in compartment no. 2 and so on is given by

$$W = \frac{n!}{n_1!\,n_2!\,n_3!\ldots..n_k!} = \frac{n!}{\prod n_i!}$$

The probability of the macrostate $(n_1, n_2, n_3, \ldots n_k)$ is given by

$$P_{(n_1, n_2, n_3, \ldots nk)} = \frac{n!}{n_1!n_2!n_3!\ldots\ldots n_k!} \times \frac{1}{k^n}$$

9.17 ENSEMBLE AND AVERAGE PROPERTIES

The usual meaning of ensemble* is the whole or all the parts taken together. In statistical mechanics an ensemble is a collection of identical systems with all accessible microstates represented in it with equal frequency. Consider a system of 4 particles in two similar compartments. The system, under constraint $\Sigma n_i = 4$, has 16 different microstates (table 9.2). We can draw all the sixteen microstates as 16 identical systems (Fig. 9.9). Consider just one system, which at time $t = 0$ has a particular microstate, say (*ab*, *cd*). Then due to interactions among the particles, the system will pass through all the accessible microstates as time passes. In a short time, we may find that same microstates repeat more frequently than the other. But if we observe for long enough time, then we should expect all the microstates are repeated with equal frequency. In the present case of 16 accessible states, suppose we make 48 observations in all, then we expect that each microstate should appear just 3 times. But in practice, some microstates may occur even 7 times and some microstates may not occur at all in these 48 observations! On the other hand, if we make 48×10^{20} observations in all, then the observed frequency of each of the 16 microstates would not be different from 3×10^{20} by more than (say) 1 part in 10^{12} or so.

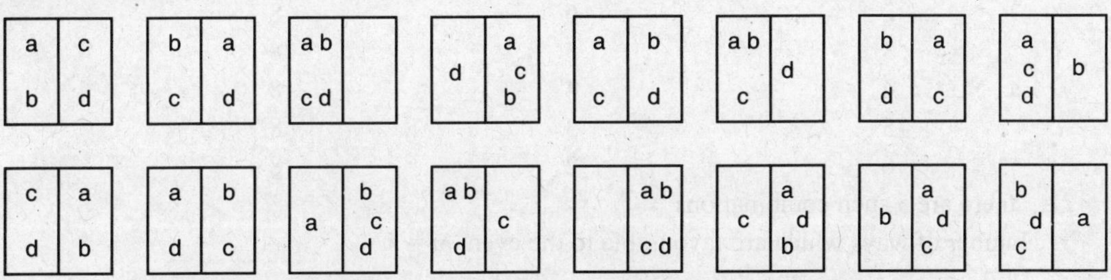

Fig. 9.9 An Ensemble for a system of 4 particles in two equal compartments

We may either make these 48×10^{20} times observations on a single particle system or at a single time on a collection of 48×10^{20} identical systems. The conclusion would be the same. Here we take into consideration *the postulate of equal a priori probabilities*. It is a fundamental postulate of statistical mechanics and it says that a system can exist in each of the accessible microstates with equal probability. In other words, no preference is to be given to any one or more microscopic states over others.

The probability of particular macrostate is given by the number of microstates falling under it divided by the total number of accessible microstates. Let a macrostate be designated by suffix j and let W_j represent the frequency with which it occurs in the ensemble (*i.e.*, the number of microstates falling under jth macrostate. Then, the probability of occurrence of the jth macrostate) is given by

$$P_j = \frac{W_j}{\Sigma W_j} \qquad \ldots(9.22)$$

If any property x (may be energy, one component of momentum or any thing) has value x_j in the jth macrostate then the average value of this property over the ensemble, denoted by $\bar{x}$ or $<x>$, is given by

$$\bar{x} = <x> = \sum_j x_j p_j = \frac{\Sigma x_j W_j}{\Sigma W_j} \qquad \ldots(9.23)$$

It may be noted that we have no means of knowing how the system really passes through the various macrostates, much less about the further details of the microstates.

* for this, please refer next chapter

Example 9.11 *What is the probabiloity of drawing (a) a queen (b) a queen of diamond from a well shuffled pack of cards* **(Pbi .U. 2007)**

Solution: (*a*) Probability of drawing a queen is

$$P = \frac{4}{52} = \frac{1}{13}$$

(*b*) Probability of drawing a queen of diamond is

$$P = \frac{1}{52}$$

Example 9.12 *If a pair of six faced dice with faces marked 1 to 6 is thrown, what is the probability that the sum of the numbers which shown up is 8* **(P.U. 2003 ; H.P.U. 1991)**

Solution: When we throw two dice simultaneously, the first dice can fall in $n_1 = 6$ ways with any of its faces marked 1 to 6 upward. Corresponding to **each** of these ways, the second dice can fall in $n_2 = 6$ ways. Hence the total number of different ways in which the two dice can fall will be

$$n = n_1 \times n_2 = 6 \times 6 = 36$$

∴ The number of equally likely possible ways $n = 36$

The sum of the numbers which show up will be 8 for the following combinations

First dice	*Second dice*	*Total*
2	6	8
3	5	8
4	4	8
5	3	8
6	2	8

i.e., there are 5 such combinations

∴ Number of ways which are favourable to the event $m = 5$

Hence probability $P = \frac{m}{n} = \frac{5}{36}$

Example 9.13 *In a system of 14 distinguishable particles distributed in two equally probable halves of a box, determine the probability of distribution (10,4), (14,0), (7,7)* **(Agra. U. 1995)**

Solution: Total number of particles $n = 14$

Total number of compartment $C = 2$

∴ Total number of microstates $= C^n = 2^{14} = 16384$

(*i*) Number of microstates in the macrostate (10, 4)

$$W_{(10,4)} = {}^nC_{n_1} = {}^nC_{n_2} = {}^{14}C_{10} = {}^{14}C_4 = \frac{14!}{10!\,4!}$$

$$= \frac{14\times13\times12\times11}{4\times3\times2\times1} = 1001$$

∴ Probability of the macrostate (10, 4)

$$P_{(10,4)} = \frac{W_{(10,4)}}{C^n} = \frac{1001}{2^{14}} = \frac{1001}{16384}$$

(*ii*) Number of microstates in the macrostate (7, 7)

$$W_{(7,7)} = {}^{14}C_7 = \frac{14!}{7!7!} = \frac{14\times13\times12\times11\times10\times9\times8}{7\times6\times5\times4\times3\times2\times1}$$

∴ Probability of the macrostate (7,7)

$$P_{(7,7)} = \frac{W_{(7,7)}}{C^n} = \frac{3432}{2^{14}} = \frac{3432}{16384}$$

(*iii*) Number of microstates in the macorstate (14,0)

$$W_{(14,0)} = {}^{14}C_{14} = {}^{14}C_0 = \frac{14!}{14!\,0!} = 1$$

∴ Probability of the macrostate (14, 0)

$$P_{(14,0)} = \frac{W_{(14,0)}}{C_n} = \frac{1}{2^{14}} = \frac{1}{16384}$$

Example 9.14 *A box is divided into 9 equal sized compartments. A ball thrown in the box is equally likely to go into any one of these compartments. If 9 balls are thrown, calculate the probability for a particular compartment to contain.*

(i) one paprticular ball only (ii) all the nine balls (iii) none of the balls. State the principle used.

(Pbi. U. 1993)

Solution If p_1 is the probability that a first event will occur and p_2 and probability that a second independent event will occur, then the probability that both the events will occur together is $p_1 \times p_2$.

The result can be extended to more than two events and can be stated as follows:

If p_1 is probability that a first event will occur, p_2 the probability that a second independent event will occur and so on, p_i the probability that the ith independent event will occur, then the probability that all events will occur together is

$$P = p_1 \times p_2 \,....\, \times p_i$$

(*i*) In the present problem as the ball is equally likely to go into any one of the 9 equal sized compartments, the probability of a ball to go into particular compartment.

$$p_1 = \frac{1}{9}$$

The probability of the next ball *not to* go into this compartment *i.e.*, its probability to go into the remaining 8 compartments.

$$p_2 = \frac{8}{9}$$

∴ The probability that one ball may go into a particular compartment and remaining 8 balls go into other compartments is

$$P = p_1 \times p_2 \times p_3 \times p_4 \times p_5 \times p_6 \times p_7 \times p_8 \times p_9$$

$$= \frac{1}{9} \times \left(\frac{8}{9} \times \frac{8}{9} \times \frac{8}{9} \times \frac{8}{9} \times \frac{8}{9} \times \frac{8}{9} \times \frac{8}{9} \times \frac{8}{9}\right) = \frac{1}{9} \times \left(\frac{8}{9}\right)^8 = \frac{8^8}{9^9}$$

(*ii*) The probability of each ball to go into a particular compartment is $\frac{1}{9}$

∴ Probability of all the 9 balls to go into the same compartment is $\left(\frac{1}{9}\right)^9 = \frac{1}{9^9}$

(*iii*) The probability of each ball not to go into a particular compartment *i.e.* to go into the remaining 8 compartments is $\left(\frac{8}{9}\right)$. Therefore, the probability that none of the nine balls may go into a particular compartment is $\left(\frac{8}{9}\right)^9 = \frac{8^9}{9^9}$.

Example 9.15. *What is the probability of drawing two kings in succession from a pack of cards?* **(*Punjab 2005*)**

Solution: The no. of ways in which the first card can be drawn $n_1 = 52$.

As there are 4 kings, the number of ways favourable to the first event, *i.e.*, the first card drawn may be king, $m_1 = 4$

$\therefore$ Prabability of getting a king in the first draw,

$$P_1 = \frac{m_1}{n_1} = \frac{4}{52} = \frac{1}{13} \quad \text{...} (i)$$

The number of ways in which the second card can be drawn, $n_2 = 51$, because one king is removed from the pack, the cards left are only 51. The no. of kings left in the pack are only 3. Therefore, the no. of ways favourable to the second event, *i.e.* the second card drawn may also be, king, $m_2 = 3$.

$\therefore$ Probability of getting a king in the second draw,

$$P_2 = \frac{m_2}{n_2} = \frac{3}{51} = \frac{1}{17}. \quad \text{...}(ii)$$

Since the events are independent, the total probability of drawing two kings,

$$P = p_1 \times p_2 = \frac{1}{13} \times \frac{1}{17}$$

$$= \frac{1}{221}$$

Exercise 9. 16 *From a well shiffted deck of cards what is the probability that three cards drawn in succession are all aces?*

Solution: The probability will be

$$P = p_1 \times p_2 \times p_3$$

Here $\quad p_1 = \frac{m_1}{n_1} = \frac{4}{52} = \frac{1}{13} \quad$ and $\quad p_3 = \frac{m_3}{n_3} = \frac{2}{50} = \frac{1}{25}$

$$p_2 = \frac{m_2}{n_2} = \frac{3}{51} = \frac{1}{17}$$

$$\therefore \quad P = p_1 \times p_2 \times p_3 = \frac{1}{13} \times \frac{1}{17} \times \frac{1}{25}$$

$$= \frac{1}{5525}$$

Example 9.17 *A system consists of 6000 particles distributed in three energy states with equal spacing. The energy of the three states are $E_1 = 0$, $E_2 = x$ and $E_3 = 2x$. All the three states have the same intrinsic probability g. At a certain instant there are 3000 particles in the lower level, 2500 in the middle level and 500 in the upper level. Compare the relative probabilities with the distribution obtained by the transfer of one particle from the middle to the lower level and one particle from the middle to the upper level and the original distribution.*

Solution : Let P_1 and P_2 be the probabilities in the two cases

In the first case,

$$n = 6000, n_1 = 3000, n_2 = 2500, n_3 = 500$$

$$P_1 = \frac{g^n}{n_1 ! n_2 ! n_3 !}$$

$$P_1 = \frac{g^{6000}}{3000!\,2500!\,500!} \quad \text{... }(i)$$

In the second case

$$n = 6000$$

$$n_1 = 3001,\ n_2 = 2498,\ n_3 = 501$$

$$\therefore \quad P_2 = \frac{g^{6000}}{3001!\,2498!\,501!} \quad \text{... }(ii)$$

Dividing (*ii*) by (*i*)

$$\frac{P_2}{P_1} = \frac{3000!\,2500!\,500!}{3001!\,2498!\,501!}$$

$$\frac{P_2}{P_1} = \frac{2500 \times 2499}{3001 \times 501} \quad \text{... }(iii)$$

$$\frac{P_2}{P_1} = 4.155 \quad \text{... }(iv)$$

$$\approx \mathbf{4.2}$$

It means the transfer of one particle from the middle to the upper and one particle from the middle to the lower state has changed the probability by a factor 4.2. This shows that both these distributions are not near the equilibrium state.

Example 9.18 *Three similar dice A, B and C each having six equally likely faces marked as 1, 2, 3, 4, 5, 6 are thrown simultaneously and in a toss all the faces have equal probability of appearing up. Calculate the probability of getting the faces of all the dice up marked with 1 number.*

(*Agra 2003*)

Solution Probability of dice *A* to show face marked 1 up, $p_1 = \frac{1}{6}$

Probability of dice *B* to show face marked 1 up, $p_2 = \frac{1}{6}$

Probability of dice *C* to show face marked 1 up, $p_3 = \frac{1}{6}$

∴ Probability of all the three dice *A*, *B* and *C* to show face marked 1 up,

$$P = p_1 \times p_2 \times p_3 = \frac{1}{6} \times \frac{1}{6} \times \frac{1}{6}$$

$$= \mathbf{\frac{1}{216}}$$

Example 9.19 *Three particles are distributed in three compartment of equal size. Find the number of microstates in (i) macrostate (0, 0, 3) and (ii) macrostate (1, 2, 0)*

(*G.N.D.U. 2007*)

Solution:

(*i*) For macrostate (0, 0, 3)

number of micr states $= \frac{3!}{0!\,0!\,3!} = 1$

(*ii*) For macrostate (1, 2, 0)

number of microstates $= \frac{3!}{1!2!0!} = 3$

Example 9.20 *Calculate the thermodynamic probability of various macrostates corresponding to distribution of 5 particles in two compartments.* **(*P.U. 2001*)**

Soltuion. Number of particles $n = 5$

Number of compartments $C = 2$

Total number of macrostates = $n + 1 = 6$

The various macrostates are (0, 5), (1, 4), (2, 3), (3, 2), (4, 1) and (5, 0)

Thermodynamic probability $W_{(n_1,n_2)} = \dfrac{n!}{n_1!n_2!}$

(*i*) ∴ For the macrotstate (0, 5) thermodynamic probability

$$W_{(0,5)} = \frac{5!}{n_1!\,5!} = 1$$

(*ii*) For the macrostate (1, 4). thermodynamic probability

$$W_{(1,4)} = \frac{5!}{1!4!} = 5$$

(*iii*) For the macrostate (2, 3), thermodynamic probability

$$W_{(2,3)} = \frac{5!}{2!3!} = 10$$

Similarly for the macrostates (3, 2), (4, 1) and (5, 0), the thermodynamic probabilities respectively are:

$$W_{(3,2)} = 10\ ;\ W_{(4,1)} = 5 \text{ and } W_{(5,0)} = 1$$

Example 9.21 *In a system of 8 distinguishable particles distributed in two equal sized compartments, calculate the probability of the macrostate (3,5), (4,4), (2,6)*

(Pbi. U. 2001; P.U. 2007. 2000 K.U. 2002)

Solution: Total number of particles $n = 8$; Total number of compartments $C = 2$

∴ Total number of microstates $C^n = 2^8 = 256$

(*i*) Number of microstates in the macrostate (3, 5) $= \dfrac{8!}{3!5!} = \dfrac{8\times7\times6}{3\times2\times1} = 56$

∴ Probability of the macrostate $(3,5) = \dfrac{56}{256} = \dfrac{7}{32}$

(*ii*) Number of microstates in the macrostate (4, 4) $= \dfrac{8!}{4!4!} = \dfrac{8\times7\times6\times5}{4\times3\times2\times1} = 70$

∴ Probability of the macrostate (4, 4) $= \dfrac{70}{256} = \dfrac{35}{128}$

(*iii*) Number of microstates in the macrostate (2,6) $= \dfrac{8!}{2!6!} = \dfrac{8\times7}{2\times1} = 28$

∴ Probability of the macrostate (2, 6) $= \dfrac{28}{256} = \dfrac{7}{64}$

Example 9.22 *Eight distinguishable particles are distributed among three compartments of equal size. Find the probability of the macrostate (i) (4, 3, 1) and (ii) (3, 3, 2).*

(Nagpur 2010, Punjab 2001)

Solution. The probability of occurrence of the macrostate ($n_1, n_2, n_3, \dots n_k$) corresponding to the distribution of n distinguishable particles into k compartments of equal size is given by

$$P_{(n_1, n_2, n_3, \ldots nk)} = \frac{n!}{n_1!n_2!\ldots n_k!} \times \frac{1}{k^n}$$

Here, total no of particles, $n = 8$

and the total no. of compartments, $k = 3$

(*i*) For the macrostate (4, 3, 1) ; $n_1 = 4, n_2 = 3, n_1 = 1$

$$\therefore \qquad P_{(4, 3, 1)} = \frac{8!}{4!3!1!} \times \frac{1}{3^8}$$

$$= \frac{8 \times 7 \times 6 \times 5}{3 \times 2} \times \frac{1}{3^8} = \frac{280}{6581}$$

$$= \mathbf{0.0427}$$

(*ii*) For the macrostate (3, 3, 2); $n_1 = 3, n_2 = 3, n_3 = 2$

$$\therefore \qquad P_{(3, 3, 2} = \frac{8!}{3!3!2!} \times \frac{1}{3^8}$$

$$= \frac{8 \times 7 \times 6 \times 5 \times 4}{3 \times 2 \times 2} \times \frac{1}{3^8} = \frac{560}{6561}$$

$$= 0.0854$$

Example 9.23 *What is the maximum and minimum value of the probability of occurrence of a macrostate for the case of n distinguishable particles distributed in two compartments ? Prove.*

(*Punjab 2003, Kerala 2002*)

Solution. The distribution of *n* distinguishable particles between two halves of a box with equal probability is similar to the tossing of *n* coins. As in the case of tossing of a coin, either head or tail may be uppermost, in the case of distribution of molecules in two halves of a box, the molecule may either go in first half or in the other half. If *a* and *b* denote that a molecule will be in the first and second half of a box respectively, then the possible combinations in the distribution of *n* molecules will be

$$(a + b)^n = a^n + n_{c_1} a^{n-1} b + n_{c_2} a^{n-2} b^2 + \ldots + n_{c_r} a^r b^{n-r} + \ldots + n_{c_n} b^n$$

Total number of terms representing all possible combinations is

$$\Sigma n_{cr} = 2^n$$

Therefore, the probability of distribution ($r, n - r$), is given by

$$P_{(r, n-r)} = \frac{n!}{r!(n-r)!} \times \frac{1}{2^n} \qquad \ldots (i)$$

The probability of most probable distribution is

$$p_{max} = \frac{n!}{\left(\frac{n}{2}\right)!\left(\frac{n}{2}\right)!} \times \frac{1}{2^n}$$

$$= \frac{\boldsymbol{n!}}{\left(\frac{\boldsymbol{n}}{\boldsymbol{2}}\boldsymbol{!}\right)^{\boldsymbol{2}}} \times \frac{\boldsymbol{1}}{\boldsymbol{2^n}} \qquad \ldots (ii)$$

The probability of least or minimum probable distribution is

$$P_{min} = \frac{n!}{n!(n-n)!} \times \frac{1}{2^n} = \frac{\mathbf{1}}{\mathbf{2^n}} \qquad \ldots (iii)$$

Example 9.24 *A system of n particles has only two allowed states A and B. The probability for state A is p and that for state B is (1 – p). What is the probability for system to be in macrostate defined by distribution (r, n – r).* **(*Rohilkhand 2003*)**

Solution The probability of state A is p and that for B is $(1-p)$. This means the weight factors a and b for states A and B are p and $(1-p)$ respectively. Therefore, the probability for the distribution of n particles in states A and B in macrostate $(r, n-r)$, is given by

$$P_{(r,\,n-r)} = \frac{n!}{r!\,(n-r)!}\,p^r\,(1-p)^{n-r}$$

Example 9.25 *Eight distinguishable particles are distributed in two compartments of unequal sizes. The first compartment is further subdivided into 6 cells and second into 2 cells of equal sizes. Calculate the probability of macrostate (6, 2) and (4, 4).* **(*Punjab 2000*)**

Solution. The probability of occurrence of a macrostate $(n_1, n_2 \ldots n_k)$ for the distribution of n distinguishable particles in k compartments in given by

$$P_{(n_1,\,n_2,\ldots nk)} = \frac{n!}{n_1!\,n_2!\,\ldots n_k!} \times \frac{(g_1)^{n_1}\,(g_2)^{n_2}\,(g_3)^{n_3}\ldots(g_k)^{n_k}}{(g_1+g_2+\ldots g_k)^n}$$

where $g_1, g_2 \ldots g_k$ are the no. of cells belonging to compartments marked 1, 2, ... k respectively.

In the given problem,

Total no. of particles, $n = 8$

Total no. of compartments, $k = 2$

No. of cells in the compartment No. 1, $g_1 = 6$

No. of cells in the compartment No. 2, $g_2 = 2$

For the macrostate (6, 2), $n_1 = 6$ and $n_2 = 2$

$$\therefore \text{ Probability of occurrence, } P_{(6,\,2)} = \frac{8!}{6!\,2!}\times\frac{(6)^6\,(2)^2}{(6+2)^8} = 28\times\frac{729}{65536}$$

$$= \mathbf{0.311}$$

For the macrostate (4, 4), $n_1 = 4$ and $n_2 = 4$

$$\therefore \text{ Probability of occurrence, } P_{(4,\,4)} = \frac{8!}{4!\,4!}\times\frac{(6)^4\,(2)^4}{(6+2)^6}$$

$$= \frac{70\times 8}{65536}$$

$$= \mathbf{0.0865}$$

Exercise 9.26 *In the above problem, calculate the probabilty of the most probable macrostate.* **(*Punjab 2000*)**

Solution Total number of particles, $n = 8$

Total number of cells in the two compartments

$$G = g_1 + g_2 = 6 + 2 = 8$$

For the most probable macrostate, the number of particles in compartment No. 1 $= \frac{n}{G}g_1 = \frac{8}{8} \times 6 = 6$ and the number of particles in compartment No. 2

$$= \frac{n}{G}g_2 = \frac{8}{8}\times 2 = 2$$

$\therefore$ Most probable macrostate is (6, 2) for which the probability as calculated above is 0.311.

Example 9.27 *Calculate the percentage error made in Stirling's formula.*

$\log_e (n!) = n \log_e n - n$ when (*i*) $n = 5$ and (*ii*) $n = 10$

(*Punjab 2000, 2001, 2005, G.N.D.U. 2004*)

Solution

(*i*) $n = 5$; $\log_e (n!) = \log_e (5 \times 4 \times 3 \times 2 \times 1) = \log_e 120$

$$= \log_e 1.20 + \log_e 10^2 = 0.1823 + 2 \times 2.3026 = 4.7875$$

$$n \log_e n - n = 5 \log_e (5) - 5 = 5 \times 1.6094 - 5 = 3.047$$

Difference $= 4.7875 - 3.047 = 1.7405$

$\therefore$ Percentage error $= \dfrac{1.7405}{4.7875} \times 100 = \mathbf{36.38\%}$

(*ii*) $n = 10$; $\log_e (10!) = \log_e (10 \times 9 \times 8 \times 7 \times 6 \times 5 \times 4 \times 3 \times 2 \times 1) = \log_e 3628800$

$$= 15.1044$$

$$n \log_e n - n = 10 \log_e (10) - n = 10 \times 2.3026 - 10$$

$$= 13.026$$

Difference $= 15.1044 - 13.026 = 2.0784$

$\therefore$ Percentage error $= \dfrac{2.0784}{15.1044} \times 100 = \mathbf{13.76\%}$

Example 9.28 *A system 5 microstates, under which there are 6, 20, 42, 12 and 2 microstates. A property x associated with the system has values 4, 4, 2, 6 and 10 respectively for the 5 macrostates. Calculate*

(*i*) *the probabilities for the different macrostates, and*

(*ii*) *average values of x and of* $\sqrt{x^2}$

Solution: The sum of thermodynamic probabilities of total no. of microstates are

$$\Sigma W_i = 6 + 20 + 42 + 12 + 2 = 82$$

$$\therefore \quad P_j = \frac{6}{82}, \frac{20}{82}, \frac{42}{82}, \frac{12}{82}, \frac{2}{82} \quad \text{[This gives } \Sigma P_j = 1]$$

(*i*) $$<x> = \frac{1}{82} (6 \times 4 + 20 \times 4 + 42 \times 2 + 12 \times 6 + 2 \times 10) = 3.41 \quad \text{and}$$

(*ii*) $$<x^2>^{\frac{1}{2}} = \left[\frac{1}{82}(6\times16+20\times16+42\times4+12\times36+2\times100)\right]^{\frac{1}{2}}$$

$$= \mathbf{3.85}$$

Example 9.29 *100 molecules of a gas are enclosed in a cubical volume. Imagine the volume to be divided into two equal halves. Calculate the ratio of the time spent by the system in the most probable macrostate and the macrostate (i) (40 : 60) (ii) (45: 55)*

(P.U. 2006; Pbi. U. 2005; H.P.U. 1992)

Solution: Number of particles $n = 100$

Number of compartment $C = 2$

As the number of particles is 100, the most probable macrostate is (50:50).

$\therefore$ For the macrostate (50 ; 50), $\dfrac{T_{(50,50)}}{T} = P_{(50,50)} = \dfrac{^{100}C_{50}}{2^{100}}$

(*i*) For the macrostate (40: 60), $\dfrac{T_{(40,60)}}{T} = P_{(40,60)} = \dfrac{^{100}C_{40}}{2^{100}}$

$$\therefore \quad \frac{T_{(50,50)}}{T_{(40,60)}} = \frac{^{100}C_{50}}{^{100}C_{40}} = \frac{\dfrac{100!}{50!\,50!}}{\dfrac{100!}{40!\,60!}} = \frac{40!60!}{50!50!}$$

$$= \frac{60\times59\times58\times57\times56\times55\times54\times53\times52\times51}{50\times49\times48\times47\times46\times45\times44\times43\times42\times41} = 7.3$$

(*ii*) For the macrostate (45, 55), $\frac{T_{(45,55)}}{T} = P_{(45,55)} = \frac{{}^{100}C_{45}}{2^{100}}$

$$\frac{T_{(50,50)}}{T_{(45,55)}} = \frac{{}^{100}C_{50}}{{}^{100}C_{45}} = \frac{45!55!}{50!50!}$$

$$= \frac{55\times54\times53\times52\times51}{50\times49\times48\times47\times46} = 1.64$$

Example 9.30 *80 molecules of a gas are enclosed in a cubical volume. Let this volume be divided into two equal halves by means of an imaginary partition. Calculate the ratio of the time spent by the system in the most probable macrostate (40, 40) and (i) the macrostate (38,42) and (ii) the macrostate (35,45)* **(P.U. 2007, 1999)**

Soltuion : $n = 80$; $c = 2$; As the number of particles is 80, the most probable macrostate is (40,40)

(*i*) $\therefore \quad \frac{T_{(40,40)}}{T_{(38,42)}} = \frac{{}^{80}C_{40}}{{}^{80}C_{38}} = \frac{\frac{80!}{40!\,40!}}{\frac{80!}{38!42!}} = \frac{38!42!}{40!40!} = \frac{42\times41}{40\times39} = \frac{1722}{1560} = 1.104$

(*i*) $\therefore \quad \frac{T_{(40,40)}}{T_{(35,45)}} = \frac{\frac{80\,!}{40!\,40!}}{\frac{80\,!}{45\,!35\,!}} = \frac{45!\,35!}{40!40!} = \frac{45\times44\times43\times42\times41}{40\times39\times38\times37\times36} = 1.8567$

EXERCISE

9.1 Define probability. When will the probability be (*i*) zero (*ii*) one? What is total probability? **(H.P. 2003, Meerut 2000, Pb. 2002, Purvanchal 2001)**

9.2 State and explain the principle of equal a priori probability. **(Nagpur U. 2010, Pune 2010, 2007, Purvanchal 2001)**

9.3 Prove that the probability of a composite event is equal to the product of the probability of the individual independent events.

9.4 Discuss additive law and multiplicative law of probability. **(Meerut 2004)**

9.5 Write a short note on permutations and combinations, giving examples.

9.6 n similar coins are tossed simultaneously for a large number of times. Prove that the probability of getting r heads upper most is

$$\frac{n!}{r!\,(n-r)!}\cdot\frac{1}{2^n}$$ **(Purvanchal 2007)**

9.7 Clearly distinguish between probability and frequency.

9.8 Define and explain the terms Macrostate and Microstate with the help of an example. **(Nagpur 2011/s, 2008, 2007, 2006, 2005/winter, 2006/s, Punjab 2001; Pbi 2000, 2001, 2002, H.P. 2001, G.N.D. 2000, Purvanchal 2005, 2002, 2001, Agra 2007)**

9.9 What is meant by the term thermodynamic probability of a macrostate? How it is related to probability of occurrence of that state. How does it differ from mathematical probability.
(Punjab 2000, 2001, M.D. 2001, 2003, H.P. 2001, Pune 2010, Kerala 2000; Nagpur 2011/s, 2009, 2008, 2007, 2004, 2005, Agra 2003)

9.10 For n distinguishable particles to be distributed in two compartments, prove that the thermodynamic probability

$$W_{(n_1,n_2)} = \frac{n!}{n_1!\,n_2!} \quad \text{where } n_1 + n_2 = n$$

9.11 Explain the term constraints on a system. Hence define accessible and inaccessible states.
(Punjab 2000, 2001, 2003; Purvanchal 2001, Nagpur 2004, Agra 2003)

9.12 Derive the general formula to explain that the probability distribution narrows rapidly as n increases.

9.13 Define an ensemble and average properties. Considering the postulate of equal a priori probability, deduce the formula for the average value over the ensemble. ***(Meerut 2005)***

9.14 Explain in brief static and dynamic systems. ***(Kerala 2001, G.N.D. 2004)***

9.15 Find the expression for the probability of a macrostate corresponding to the distribution of n particles in k compartments of unequal sizes. ***(Punjab 2001, 2002, 2005 Pbi. 2000, 2001)***

9.16 What is the role played by the most probable states in determining the behaviour of a system in equilibrium? What is its importance in statistical physics. ***(Kerala 2002)***

9.17 Show with the help of an equation that on the average the fraction of the time spent by a dynamic system in any particular macrostate is proportional to the thermodynamic probability.
(Punjab 2001, 2003, 2005)

9.18 Distinguish between Accessible and Inaccessible Macrostates. ***(Punjab 2005)***

9.19 Define the terms with examples:

(*i*) Most probable macrostates. ***(Rohilkhand 2002 Improvement)***

(*ii*) Thermodynamic probability. ***(Rohilkhand 2001, Nagpur 2004)***

(*iii*) Accessible and inaccessible microstates.
(Purvanchal 2001, Meerut 2001, Nagpur 2004)

(*iv*) Postulate of equal a priori probability. ***(Nagpur 2004)***

(*v*) Macrostate and microstate. ***(Nagpur 2006, 2004/s, 2004/w, Punjab 2005)***

(*vi*) Probability. ***(Nagpur 2004)***

9.20 If a pair of 6 faced dice with faces marked 1 to 6 is thrown, what is the probability that the sum of the numbers which shows up is 8? **[Ans.** $\frac{5}{36}$**]** ***(Punjab 2003)***

9.21 Two Six faced dice, each marked 1 to 6 are thrown. Calculate the probability that one of the dice shows 6 and the other shows 5? **[Ans.** $\frac{1}{18}$**]**

9.22 In a system of 14 distinguishable particles distributed in two equally probable halves of a box, determine the probability of distribution (10, 4), (14, 0) and (7, 7).

$$\left[\textbf{Ans. } \frac{1001}{16384}, \frac{3432}{16384}, \frac{1}{16384}\right]$$

9.23 A six faced dice with faces marked 1 to 6 and in any throw each face is equally likely to land uppermost. If the dice is tossed 5 times, calculate the probability of the following: (*a*) number 1 appearing just once, (*b*) number 2 and 5 appearing just once each. (*c*) Number 4 appearing in all the five tosses.

9.24 A box is divided into 9 equal sized compartments. A ball thrown in the box is equally likely to go into any one of these compartments. If 9 balls are thrown, calculate the probability for a particular compartment to contain (*i*) one particular ball only (*ii*) all the nine balls (*iii*) none of the balls. State the principle used. $\left[\textbf{Ans. } \frac{8^9}{9^9}, \frac{1}{9^9}, \frac{8^9}{9^9}\right]$

9.25 Calculate the thermodynamic probability of various macrostates corresponding to distribution of 5 particles into two compartments. **[Ans.** 1, 5, 10, 10, 5, 1**]** ***(Punjab 2001)***

9.26 In a system of 8 distinguishable particles distributed in two equal sized compartments, calculate the probability of the macrostate (3, 5), (4, 4), (2, 6).

(Agra 2007, Punjab 2000, Kerala 2002, Pbi 2001)

$\left[\textbf{Ans. } \frac{7}{32}, \frac{35}{128}, \frac{7}{64}\right]$

9.27 Ten particles are distributed in two equal sized cells. Find the number of macrostates and microstates.

[**Hint :** Number of macrostates = $n + 1 = 10 + 1 = 11$ ***(Punjab 2002)***

Number of microstates = $2^n = 2^{10}$]

9.28 4 molecules are to be distributed in 2 cells. Find possible no. of macrostates and corresponding no. of microstates. ***(Nagpur 2010, 2004)***

[**Hint :** No. of macrostates = $n + 1 = 4 + 1 = 5$

No. of microstates = $c^n = 2^4 = 16$

(Here $n = 4$ and $c = 2$)

9.29 5 distinguishable particles are to be arranged in 3 compartments of a box. (*i*) calculate total no. of ways in which this can be done; assuming that there is no restriction on the no. of particles that can go in any compartment (*ii*) Also calculate the probability of macrostate (2, 1, 2). **[Ans.** (*i*) 243 (*ii*) $\frac{10}{81}$**]** ***(Punjab 2005)***

MULTIPLE CHOICE QUESTIONS

9.1 A card is drawn from a pack of 52. The probability of its being an ace or king is

(*a*) $\frac{2}{52}$ (*b*) $\frac{8}{52}$ (*c*) $\frac{4}{52}$ (*d*) $\frac{4\times3}{52\times51}$

(P.U. 1993)

9.2 A coin and a six faced dice are thrown. Probability that coin shows tail and dice shows five is

(*a*) $\frac{7}{12}$ (*b*) $\frac{1}{8}$ (*c*) $\frac{2}{12}$ (*d*) $\frac{1}{12}$ ***(P.U. 1993)***

9.3 Constraints imposed on a system

(*a*) increase number of inaccessible microstates

(*b*) decrease the number of inaccessible microstates

(*c*) have no effect

9.4 $\frac{P_x}{P_{max}}$ falls rapidly with deviation x. The ratio is maximum when ***(P.U. 1993)***

(*a*) $n = 10^2$ (*b*) $n = 10^4$ (*c*) $n = 10^6$ (*d*) $n = 10^8$

9.5 Statistical methods give greater accuracy when the number of observation is

(*a*) very small (*b*) very large (*c*) neither very small nor very large

(H.P.U., 1992, 1991)

9.6 The probability of drawing two kings in succession from a well shuffled pack of cards is **(*H.P.U. 1993*)**

(*a*) $\frac{1}{52}$ (*b*) $\frac{1}{144}$ (*c*) $\frac{1}{221}$ (*d*) none of these

9.7 The value of probability of an event can not be **(*H.P.U. 1996, 1994*)**

(*a*) Zero (*b*) 1 (*c*) $\frac{1}{2}$ (*d*) negative

9.8 The thermodynamic probability of a system in equilibrium is **(*H.P.U. 1996*)**

(*a*) maximum (*b*) minimum but not 1

(*c*) 1 (*d*) Zero

9.9 The number of most probable macrostates for a system having odd number of particles is

(*a*) 1 (*b*) 2 (*c*) 3 (*d*) 4

9.10 The probability of occurrence of two independent events is equal to their

(*a*) sum (*b*) difference (*c*) product (*d*) ratio

9.11 If 12 particles are distributed randomly between two boxes A and B with equal probability, what is the probability of the distribution (8, 4)?

(*a*) 4.95×2^{-12} (*b*) 495×2^{-12} (*c*) 49.5×2^{-12} (*d*) 495×2^{-11}

(*Rohilkhand 2003*)

9.12 The relative probability between two different energy states having difference 1.1×10^{-20} joules at 400 K temperature is:

(*a*) e^{-1} (*b*) e^{-2} (*c*) e (*d*) e^{2}

(*Rohilkhand 2003 Improvement*)

ANSWERS

9.1 (*b*)	**9.2** (*d*)	**9.3** (*a*)	**9.4** (*a*)	**9.5** (*b*)	**9.6** (*c*)
9.7 (*d*)	**9.8** (*a*)	**9.9** (*b*)	**9.10** (*c*)	**9.11** (*c*)	**9.12** (*d*)

SOME UNIVERSAL LAWS IN STATISTICAL MECHANICS

10

Chapter

INTRODUCTION

The distribution of particles with a given total energy into a discrete set of energy states becomes complicated when we deal with very large value of n and possesses a quite complex energy pattern. A special procedure **for handling such huge number of molecules** has therefore to be developed. This is governed by some universal laws in statistical mechanics. Statistical mechanics is concerned with the properties of matter in equilibrium in the empirical sense used in thermodynamics.

10.1 DEGREES OF FREEDOM

From the statistical point of view, a monatomic gas constitutes the simplest system. The energy expression for a system involves f co-ordinates of *position* and f co-ordinates of *momentum.* We then say that a system comprising one molecule has f degrees of freedom. If there are N molecules in the system, the energy expression for the system of N molecules has Nf co-ordinates of position and Nf co-ordinates of momentum. Thus, the system has Nf degrees of freedom.

10.2 POSITION SPACE

Consider a system consisting of n particles distributed in a given volume V. If the system is *static* all the particles will remain fixed at various points in space. To completely specify the location of any particle in the three dimensional space we must know the value of its three co-ordinates x, y and z which are mutually perpendicular to each other. As there are n particles, a knowledge of $3n$ co-ordinates given complete information about such a *static system.*

The three dimensional space in which the location of a particle is completely given by the three positions co-ordinates, is known as **position space.**

A small element in position space is denoted by dV and is given by

$$dV = dx\,dy\,dz \qquad \text{... }(i)$$

10.3 MOMENTUM SPACE

If the system is *dynamic,* its particles move about with various velocities and hence possess momenta. A complete specification of such a system cannot be described in terms of position co-ordinates only. For a *dynamic* system, we must specify the three components of momentum* with the help of three velocity components v_x, v_y and v_z in addition to the three position co-ordinates x, y and z.

If m is the mass of the particle moving with a velocity v, its momentum $p = mv$.

The three components of momentum are

$$p_x = mv_x, p_v = mv_y, p_z = mv_z$$

Just as the position of a particle is completely given by the three position co-ordinates x, y, z *in three dimensional position space,* the momentum of the particle is completely specified by the three

* The state of motion of a particle is described in terms of velocity components $\dot{x}, \dot{y}, \dot{z}$ (dot means derivative with respect to time). Sometimes it is more convenient to use momentum coordinates instead of velocity coordinates.

mutually perpendicular momentum co-ordinates p_x, p_y and p_z in three dimensional space known as **momentum space.**

A small volume element in momentum space is denoted by Γ and is given by

$$d\Gamma = dp_x\, dp_y\, dp_z \qquad \text{... (ii)}$$

10.4 PHASE SPACE

A combination of the position space and momentum space is known as phase space. Thus phase space has six dimensions. A point in phase space is, therefore, completely specified by six co-ordinates x, y, z, p_x, p_y, p_z. Complete information about any particle in a dynamic system can be obtained from a knowledge of these six co-ordinates which completely determine its position as well as momentum. As there are n particles a knowledge of $6n$ co-ordinates gives complete information regarding position and momentum of all the n particles in the phase space for a dynamic system. The concept of phase space is very useful while dealing with dynamical systems actually existing in nature.

A small volume element in phase space is denoted by $d\tau$ and is given by

$$d\tau = dx\, dy\, dz\, dp_x\, dp_y\, dp_z \qquad \text{... (iii)}$$

Relation. A small volume element in phase space is given by

$$d\tau = dx\, dy\, dz\, dp_x\, dp_y\, dp_z$$

Now according to relation (*i*) $dx\, dy\, dz$ gives a small volume element dV in position space and according to relation (*ii*) $dp_x\, dp_y\, dp_z$ gives a small volume element $d\Gamma$ in momentum space

$$\therefore \quad d\tau = dV\, d\Gamma$$

Thus a volume element $d\tau$ in phase space is the product of a volume element dV in position space and volume element $d\Gamma$ in momentum space.

Dimensionality of phase space. Phase space has *six* dimensions. A point in phase space is completely specified by six mutually perpendicular co-ordinates $x, y, z; p_x, p_y, p_z$. The first three co-ordinates are *space* co-ordiantes, and the last three co-ordinates are *momentum* co-ordinates. Complete information about any particle in a dynamic system can be obtained from a knowledge of these six co-ordinates which determine its position as well as momentum.

10.5 AVAILABLE VOLUME

Suppose the particles of a dynamical system occupy a given volume (V) in ordinary space. The geometry of this volume determines the minimum and maximum possible values x, y and z for the particles of the system. Let the particles of the system possess momenta with magnitude ranging from zero to some maximum possible value say p_{max}. Just as the limits on position co-ordinates define a particular volume (V) available to the system in position space, the limits on momentum values p_x, p_y and p_z define a volume (Γ) available to the system in momentum space.

The product of these two volumes is known as the available volume in phase space (τ).

A small volume element in phase space is given by

$$d\tau = dx\, dy\, dz\, dp_x\, dp_y\, sp_z \qquad \text{... (i)}$$

The total phase space volume accessible to the system is obtained by integrating equation (*i*) over the allowed values of x, y, z, p_x, p_y, p_z.

$$\therefore \quad \text{Phase space volume available to the system } \tau = \iiiint\!\!\iint dx\, dy\, dz\, dp_x\, dp_y\, dp_z$$

$$= \iiint dx\, dy\, dz \iiint dp_x dp_y dp_z = \iiint dV \iiint d\Gamma$$

or

$$\tau = V\Gamma$$

where $V = \iiint dV$ is the physical volume occupied by all the particles of the system and $\Gamma = \iiint d\Gamma$ is the momentum space volume occupied by all the particles. Thus Γ will depend upon the momentum of the particles forming the system.

Phase space volume. Suppose p_m is the maximum value of momentum of the particles in the system. Let p_x, p_y, p_z represent three mutually perpendicular axes in momentum space (fig. 10.1). With the origin O as centre and radius p_m draw a sphere, then all the points lying in this sphere will have their momenta lying between 0 and p_m.

The momentum volume accessible to the system

$$= \text{Volume of the sphere of radius } p_m$$

or $$\Gamma = \frac{4}{3}\pi p_m^3$$

or $$\tau = \frac{4}{3}\pi p_m^3 V$$

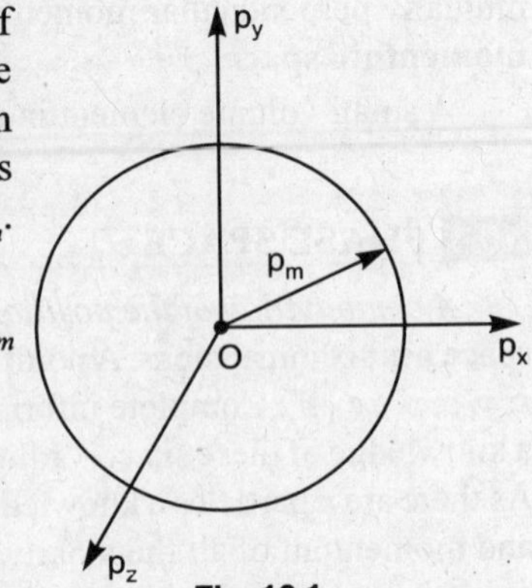

Fig. 10.1

This equation gives the total phase space volume available to the system occupying a volume V and having a momentum range 0 to p_m.

10.6 THE μ-SPACE AND GAMMA SPACE

Let us limit ourselves to a system comprising of 1 molecule. Moreover, consider it to be mono-atomic as the simplest system. The state of the gas is completely known if the position and momentum of each atom of the gas is specified. The position of an atom can be specified in terms of its rectangular coordinates *x*, *y* and *z*. This is known as *position space*. The momentum of an atom is completely known if its rectangular momentum components p_x, p_y and p_z are specified. This is called *momentum space*. These six quantities x, y, z, p_x, p_y, p_z for each atom when known, the state of the gas can be determined. The space covered by the infinite set of these six coordinates is called the *phase-space* for a single monatomic molecule.

Since the energy of a molecule is given by $(p_x^2 + p_y^2 + p_z^2)/2m$ which is equal to $p^2/2m$, the momentums space representation also gives representation of energy of the state.

Such a six-dimensional space for a *single particle* is called *molecular phase space* or *mu-space* or *μ-space*. Since 6-dimensional diagram can not be drawn, phase space is purely a mathematical concept.

Suppose that position or momentum of one of the molecules is changed slightly. The *phase point* of this molecule will undergo a displacement in the phase space; and the microscopic state of the gas will be modified. To explain further, let us consider that the μ-phase space is divided into two-dimensional *energy sheet*, having position coordinate as *x*-axis and momentum coordinate as p_x-axis. Let us subdivide the range of variables *x* and p_x into arbitrary small discrete intervals called as *phase cells* of arbitrary size (Fig. 10.2). The minimum size of phase space cell is explained as under:

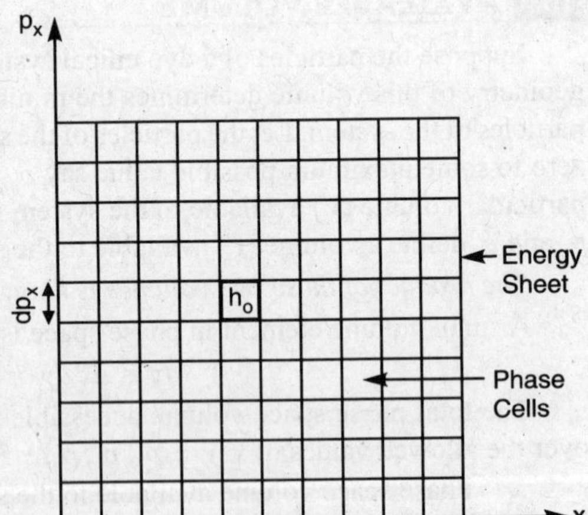

Fig. 10.2

(i) Minimum size in classical statistics

The meaning of a point in phase space is well understood with the help of **Heisenberg uncertainty principle.**

Let *dx* be the size of the interval on the *x*-axis and dp_x, the corresponding length on p_x-axis. The phase space is divided into energy sheets of smaller and smaller energy ranges. The energy sheet thus consists of small cells of equal area $= dx \,.\, dp_x = h_o$. The state of the system can then be specified by locating a phase point within a particular cell of phase space. In classical description h_o can be chosen arbitrarily as small as possible and we may set it 10^{-30} erg. sec or 10^{-35} erg. sec. or even smaller value. It may be noted that h_o has dimensions of *energy* × *time*. The lengths dx and dp_x are assumed to be so small that if a phase point within a cell is slightly displaced to any point within the cell, the change in the position and the momentum will be so minute that even the most accurate experiment would be unable to detect the changes occurred. Such an infinisimally small sized phase cell is considered. If the particle lies within the cell, its state is represented by the pair of coordinates (x, p_x). If there is a change in the state of the particle, it will move from this cell to another appropriate cell. The dimension h_o = length × momentum = (metre × Newton. sec) = (Joule. sec).

(*ii*) Minimum size in Quantum Statistics

In Boltzmann's time, there was no reason to ascribe any specific value to h_o. All the results of classical statistics are independent of the size of h_o. Therefore, we permit h_o to go over to zero size ($h_o \to 0$). But the nature itself has provided a finite and constant volume for the unit phase cell with the discovery of Planck's constant and the advent of quantum theory discussed under quantum statistics. We see that the assumption $dx.\ dp_x = h$ is now consistent with the Heisenberg uncertainty principle given by $(dx.dp_x)_{min} = h = 6.625 \times 10^{-27}$ erg. sec. $= 6.6 \times 10^{-34}$ J-sec. According to this relation,

$$dx \cdot dp_x \simeq h$$

$$dy \cdot dp_y \simeq h$$

$$dz \cdot dp_z = h$$

$$\therefore \quad dx \cdot dp_x\, dy \cdot dp_y \cdot dz \cdot dp_z = h^3$$

or
$$dx\, dy\, dz\, dp_x\, dp_y\, dp_z = h^3 \qquad \text{... (10.1)}$$

Here $dx\, dy\, dz\, dp_x\, dp_y\, dp_z$ represents the volume of each state $d\tau$.

$$\therefore \quad d\tau = h^3$$

Therefore, according to uncertainty principle, in phase space (mu), the coördinates of a particle can be specified only to the extent that lying within an element of phase space of volume h^3. (Planck's constant $h = 6.625 \times 10^{-34}$ J – sec). The unit of phase space volume is (Joule – s)^3.

The Γ – Space

If we are to deal with a system of N molecules, there are two alternative methods. One is to have the *6-dimensional* (*3 position* + *3 momentum*) *phase space* representation, and we take N points to represent the positions and momenta of the N molecules at a given instant. This is called μ-space or mu-space representation.

The other alternative method is Γ –space (*i.e.*, gamma space) representation. It is a 6N–dimensional phase-space representation in which a single point represents a state of entire N–particle system and is referred as the representative point. This is in contrast to the μ-space, which refers to only one particle, *i.e.*, one point for one particle. The Γ –space representation is mathematically more difficult to handle. But it has one great advantage. If an *ensemble* (*i.e.*, assembly of all accessible microstates of the system corresponding to one macrostate) is to be represented, this 6N–dimensional representation alone is applicable. Different points will then represent the corresponding all microstates of another macrostate. The Γ –space is considered to be a conceptual Euclidean space having $2fN$ rectangular axes.

For a general case of f degrees of freedom per molecule, we shall have:

(*i*) position space	f – dimensional
(*ii*) momentum space	f – dimensional
(*iii*) μ – space	$2f$ – dimensional
(*iv*) Γ – space	$2fN$ – dimensional.

10.7 DIVISION OF PHASE SPACE INTO CELLS

Let us consider a mu-space representation. It is characterised by $2f$ dimensional phase space configuration. For a $2f$ dimensional phase space (μ–space) defined by position co-ordinates $q_1, q_2, \ldots\ldots q_i \ldots\ldots q_f$ and momentum coordinates $p_1, p_2, \ldots\ldots p_i \ldots\ldots p_f$, an element of phase volume may be represented by

$$d\tau = (dq_1\, dq_2 \ldots\ldots dq_i \ldots\ldots dq_f) \times (dp_1\, dp_2 \ldots\ldots dp_i \ldots\ldots dp_f). \quad \ldots(10.2)$$

$$= \Pi^f_{i=1}\, dq_i\, dp_i$$

This quantity has the dimension of (length × momentum)^f or (metre × Newton. sec)^f or (Joule–s)^f.

Now suppose p_i and q_i are allowed to take all possible continuously variable values, then for a finite volume of phase space, there can be infinite number of representation points and hence infinite number of possible microstates. In statistics, however we deal with finite number of microstates, howsoever large. Hence phase space is divided into "Cells", each cell of a certain **fixed** volume $\boldsymbol{h_0^f}$. Therefore, in the element of volume of phase space given by equation (10.2) will contain the number of microstates

$$\frac{d\tau}{h^f} = \frac{(dq_1\, dq_2 \ldots\ldots dq_i \ldots\ldots dq_f)\,(dp_1\, dp_2 \ldots\ldots dp_i \ldots\ldots dp_f)}{h_0^f} \quad \ldots (10.3)$$

here h_0 is an arbitrary constant with the dimension of (erg–sec) or (Joule–sec). This value of h_0 is taken to be as small as possible so the number of microstates obtained by using relation (10.3) are large enough to handle with approximations.

For Γ –space representation having $2fN$–dimensional phase space, which is useful to describe a microstate of the system of N particles with a single point, an element of phase volume is given by

$$(dq_1\, dq_2 \ldots\ldots dq_i \ldots\ldots dq_{Nf})\,(dp_1\, dp_2 \ldots\ldots dp_i \ldots\ldots dp_{Nf}) \quad \ldots (10.4)$$

and the number of phase cells in this volume element is given by

$$\frac{d\tau}{h^{Nf}} = \frac{(dq_1\, dq_2 \ldots\ldots dq_i \ldots\ldots dq_{Nf})\,(dp_1\, dp_2 \ldots\ldots dp_i \ldots\ldots dp_{Nf})}{h_0^{Nf}} \quad \ldots (10.5)$$

Purpose of dividing phase space into cells

The phase space available to a dynamic system of particles is given by $\tau = \Gamma V$ where V is the physical volume occupied by the particles in position space and Γ is the available volume in momentum space. The value of Γ is given by $\Gamma = \frac{4}{3}\pi\, p_m^3$, where p_m is the maximum value of the momentum of the particles in the system.

The kinetic energy of a particle of momentum p is given by

$$E = \frac{p^2}{2m}$$

Thus the kinetic energy of a particle depends only on the values of its momentum co-ordinates.

The volume in momentum space available to the system is equal to the volume of the sphere of radius p_m as shown in Fig. 10.3.

Divide this sphere into a large number of spherical shells such that the width of each shell is the same. Let k be the number of shells, then

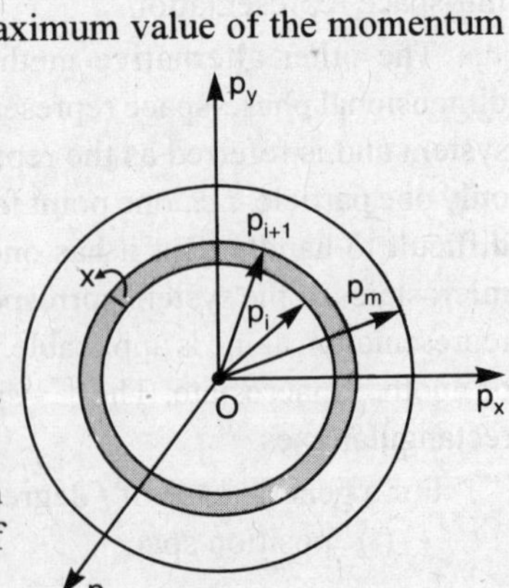

Fig. 10.3

$$\text{Width of each shell } x = \frac{P_m}{k}$$

Each spherical shell is called an energy (or momentum) compartment and has different outer and inner radii.

Now, consider the ith compartment. The inner radius of ith compartment in momentum space $p_i = xi$.

The outer radius of ith compartment = The inner radius of $(i+1)$th compartment

$$= p_{i+1} = x(i+1)$$

$\therefore$ Volume of the ith compartment in momentum space

$$\Gamma_i = \frac{4}{3}\pi[(p_{i+1})^3 - (p_i)^3] = \frac{4}{3}\pi x^3 [(i+1)^3 - i^3] = \frac{4}{3}\pi x^3 (3i^2 + 3i + 1)$$

As i is very large, $(3i+1)$ is very small quantity as compared to $3i^2$ and can be neglected

$$\therefore \qquad \Gamma_i = \frac{4}{3}\pi x^3 \times 3i^2 = 4\pi i^2 x^3 = 4\pi i^2 \left(\frac{p_m}{k}\right)^3 = 4\pi\left(\frac{p_m\, i}{k}\right)^2\left(\frac{p_m}{k}\right)$$

This gives the volume of the ith compartment in momentum space.

$$\therefore \quad \text{Volume of } i\text{th compartment in phase space } \tau_i = V\Gamma_i = 4\pi V\left(\frac{p_m\, i}{k}\right)^2\left(\frac{p_m}{k}\right) \qquad \text{... (10.6)}$$

Thus we find that the volume of the ith compartment in phase space is proportional to i^2. In other words the phase volume of each compartment is different *i.e.*, the compartments are of unequal size and hence do not have the same a priori probability. *It is therefore necessary to divide each compartment into cells of equal size so that each cell has the same a priori probability.*

The total number of cells in phase space is given by

$$\frac{\text{Total available volume in phase space}}{\text{Volume of one cell}} = \frac{V\Gamma}{d\tau}$$

where $d\tau = h_0^3$ in classical mechanics which may even approach zero and $d\tau = h^3$ in quantum mechanics where h is Planck's constant.

The number of cells in the phase space in the ith compartment

$$g_i = \frac{\tau_i}{h_0^3} = \frac{4\pi V}{h_0^3}\left(\frac{p_m\, i}{k}\right)^2\left(\frac{p_m}{k}\right) \qquad \text{... (10.7)}$$

Thus the phase space is first divided into compartments representing various intervals of momentum (or energy). These compartments are marked 1, 2, 3 ... k. Each compartment is further divided into elementary cells of equal size. Thus the compartment contain $g_1, g_2, g_3 \ldots g_k$ elementary cells respectively. If the comportments contain $n_1, n_2, n_3, \ldots n_k$ particles respectively then the distribution of particles in various compartments represents a *macrostate* of the dynamical system. The arrangement of the particles in various cells of a particular compartment represents the various microstates in it.

Again, the purpose of dividing the phase space into cells is to study the distribution of particles of a dynamical system among these compartments and cells and therefrom to develop various relations in statistical thermodynamics for the most probable macrostate of the system when the system is in equilibrium.

10.8 APPLICATIONS

(i) One–Dimensional Harmonic Oscillator

Consider a linear simple harmonic oscillator of mass m and oscillating freely with frequency ν. Let us calculate the number of phase cells in the energy range from 0 to E. The phase diagram contains one position coordinate x and one momentum coordinate p. So it is a p, x plane. At any instant, the total energy E of the oscillator will comprise of potential energy and kinetic energy.

$\therefore$ Total energy, $\qquad E = P.E + K.E.$

$$= \frac{1}{2}kx^2 + \frac{1}{2}mv^2$$

$$= \frac{1}{2}kx^2 + \frac{1}{2}\frac{m^2v^2}{m}$$

$$\therefore \qquad E = \frac{1}{2}kx^2 + \frac{1}{2}\frac{p^2}{m}$$

$$2E = kx^2 + \frac{p^2}{m}$$

where k is called force constant.

Dividing by $2E$,

$$1 = \frac{x^2}{2E/k} + \frac{p^2}{2mE} \qquad \text{... (10.8)}$$

$$\therefore \qquad \frac{x^2}{a^2} + \frac{y^2}{b^2} = 1$$

where $a^2 = \frac{2E}{k}$ and $b^2 = 2m\,E$

Thus, Equation (10.8) represents an ellipse in the x, y phase diagram (Fig. 10.4) and the points of constant E lie on this ellipse, having major and minor axes as $\sqrt{\frac{2E}{k}}$ and $\sqrt{2mE}$ respectively. For energy range from O to E, the phase area is $A = \pi\,ab$ *i.e.*, $\pi\left(\frac{2E}{k} \times 2mE\right)^{\frac{1}{2}}$ or $2\pi E\left(\frac{m}{k}\right)^{\frac{1}{2}}$. Each phase cell has an area $q_i \times p_i = h_o$. Hence, the number of phase cells in energy range from O to E will be

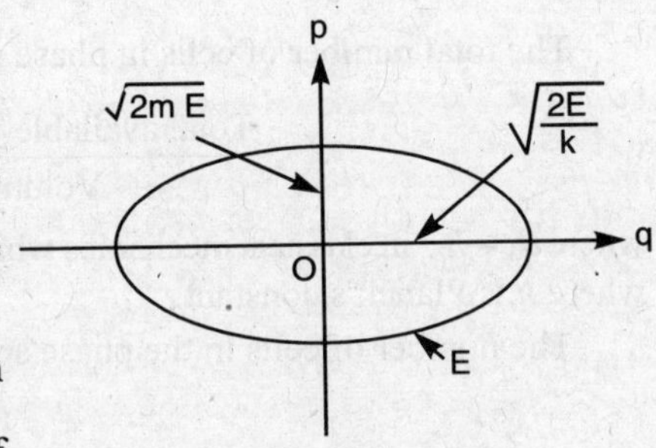

Fig. 10.4

$$\phi\,(E) = \frac{\pi}{h_o}\,2E\left(\frac{m}{k}\right)^{\frac{1}{2}}$$

But the frequency ν is given by

$$\nu = \frac{1}{2\pi}\sqrt{\frac{k}{m}}$$

$$\therefore \qquad \phi\,(E) = \frac{E}{h_o\nu} \qquad \text{... (10.9)}$$

or Energy will be

$$E = h_o\nu \,.\, \phi\,(E) \qquad \text{... (10.10)}$$

Thus, if the oscillator is so governed that the number of phase cells can take only integral values *i.e.*, $\phi\,(E) = 0, 1, 2, \ldots\ldots n$, then we get

$$E = n\,h_o\nu \qquad \text{... (10.11)}$$

This means that energy can occur in discrete states and has values integral multiple of $h_o\nu$ only.

(*ii*) A Free Particle

For simplicity, consider a phase space representing a single particle of mass m in a volume V. Let us calculate the number of phase cells in energy range from 0 to E. The volume of each phase cell is say, $h_o^{\,3}$. An elemental volume in phase space is

$$dp_x\, dp_y\, dp_z\, dq_x\, dq_y\, dq_z = dp_x\, dq_z\, dp_y\, dq_y\, dp_z\, dq_z$$
$$= h_0^3$$

Assuming only 3–translational degrees of freedom, the phase space volume is

$$V_{phase} = \iiiiiint dp_x\, dp_y\, dp_z\, dq_x\, dq_y\, dq_z$$

Here the integration over all position coordinates $q^{'s}$ gives the total physical volume V and integration over all linear momenta $p^{'s}$ is limited to $(p_x^{\,2} + p_y^{\,2} + p_z^{\,2})_{max} = 2m\,E = (p_{max})^2$. So the phase space volume is

$$V_{phase} = V \times \frac{4\pi}{3}(p_{max})^3$$

$$= \frac{4\pi V}{3}(2m\,E)^{\frac{3}{2}} \qquad [\because\ (p_{max})^2 = 2m\,E]$$

The number of phase cells will be

$$\phi\,(E) = \frac{\text{Volume of Phase Space}}{\text{Volume of One Cell}}$$

or

$$\phi\,(E) = \frac{V_{Phase}}{h_o^3}$$

For a single particle, the number of accessible microstates will be equal to the number of accessible cells in the phase space. Therefore, the number of microstates in the energy range 0 to E is given by

$$\phi\,(E) = \frac{4\pi V}{3h_o^3}(2mE)^{\frac{3}{2}} \qquad \text{... (10.12)}$$

Example 10.1. *For a single particle of mass m enclosed in volume V, show that the number of accessible microstates in the energy range E to E + dE is given by*

$$\Omega\,(E) = \frac{4\sqrt{2}.\pi V}{h^3}\, m^{\frac{3}{2}}\, E^{\frac{1}{2}}\, \delta E$$

where phase space is divided by the rule $\delta p_i\, \delta q_i = h$ ***(Purvanchal 2005, 2001, Mumbai 2005)***

Solution. The number of microstates in the phase space having energies between E and $E + dE$ is obtained by differentiating equation (10.12) with respect to E. Thus we have

$$\Omega\,(E) = \frac{\partial \phi\,(E)}{\partial E}$$

$$= \frac{3}{2}\cdot\frac{4\pi V}{3h^3}(2m\,E)^{\frac{1}{2}}\; 2m\,\delta E$$

$$= \frac{4\sqrt{2}.\pi V}{h^3}\, m^{\frac{3}{2}}\, E^{\frac{1}{2}}\, \delta E$$

or,

$$\Omega\,(E) = \frac{2\pi V}{h^3}(2m)^{\frac{3}{2}}\, E^{\frac{1}{2}}\, dE$$

10.9 FUNDAMENTAL POSTULATES OF STATISTICAL MECHANICS

The application of statistical mechanics to gases, certain fundamental postulates are made. These are:

1. *Any gas under consideration may be considered to be composed of large number of molecules which are constantly in motion and behave like very small elastic spheres.*

2. *All the cells in phase space are of equal size.*

3. *All accessible microstates corresponding to possible macrostates are equally probable.* This is the most fundamental and hence important postulate of statistical mechanics. For example, if we consider as earlier, the distribution of 4 distinguishable particles *a, b, c, d* in two cells *i* and *j*, then

$$\text{Probability of microstate } (a\ b\ c\ d,\ \text{o}) = \frac{1}{2^4} = \frac{1}{16}$$

$$\text{Probability of microstate } (a\ b\ c,\ d) = \left(\frac{1}{2^3}\right)\left(\frac{1}{2}\right) = \frac{1}{16}$$

$$\text{Probability of microstate } (ab,\ cd) = \left(\frac{1}{2^2}\right)\left(\frac{1}{2^2}\right) = \frac{1}{16}$$

$$\text{Probability of microstate } (a,\ bcd) = \left(\frac{1}{2}\right)\left(\frac{1}{2^3}\right) = \frac{1}{16}$$

Thus, we can conclude that the probability of all accessible microstates of the system is equal. From this postulate it follows that the probability of occurrence of a given macrostate is proportional to the number of microstates, corresponding to the given macrostate. Thus the probability $P(E)$ of system possessing energy E is proportional to the thermodynamic probability $W(E)$. *i.e.*,

$$P(E) \propto W(E)$$

or

$$P(E) = cW(E)$$

where c is a constant of proportionality.

4. *The equilibrium state of a gas corresponds to the macrostate of maximum probability.*

5. *The total number of molecules is constant.* This is in accordance with the **principle of conservation of matter.** If n_1 is the number of molecules in cell 1, n_2 in cell 2, n_3 in cell 3 etc. If N is the total number of molecules, then

$$n_1 + n_2 + n_3 + \ldots\ldots + n_i + \ldots\ldots\ n_k = \Sigma n_i = N = a \text{ constant.} \qquad \ldots (10.11)$$

6. *The total energy of the system is constant.*

This is again in accordance with the **conservation of energy** of the system. If E_1 is the energy of each particle in cell 1, E_2 the energy of each particle in cell 2 etc. and E the total energy, then

$$E_1 n_1 + E_2 n_2 + E_3 n_3 + \ldots = \Sigma E_i n_i = E = \text{a constant.} \qquad \ldots (10.14)$$

10.10 DENSITY OF QUANTUM STATES OF ENERGY OF A PARTICLE

According to quantum mechanics, the solution of a 3-dimensional time independent Schrodinger's equation

$$\left(\frac{\partial^2}{\partial x^2} + \frac{\partial^2}{\partial y^2} + \frac{\partial^2}{\partial z^2}\right)\psi_k(r) + \frac{2mE}{\hbar^2}\psi_k(r) = 0$$

the allowed energy levels *i.e.* the eigenvalues of energy of a particle of mass m in a 3-dimensional cubic potential box of each side L are given by

$$E_{n_x n_y n_z} = \frac{\pi^2 \hbar^2 n^2}{2mL^2} = \frac{\pi^2 \hbar^2}{2mL^2}(n_x^2 + n_y^2 + n_z^2) \qquad \ldots (i)$$

where,

$$n_x = 1, 2, 3, \ldots\ldots$$
$$n_y = 1, 2, 3, \ldots\ldots$$
$$n_z = 1, 2, 3, \ldots\ldots$$

corresponding to each value of the integer, there is a single quantum state of the particle. The lowest energy level *i.e.* the energy in the ground state, occurs for

$$n_x = n_y = n_z = 1$$

leads to the same value of $n^2 = 6$ for all the three quantum states.

In other words, the particle having the same energy in the first excited state can be in any one of the three quantum states. Such an energy level (with three wave functions ψ_1, ψ_2, ψ_3) is said to be three-fold degenerate. One energy level corresponds to three states. The lowest energy level ($n_x = n_y = n_z = 1$) is however non-degenerate. If the energy levels are to degenerate, there will be a very large number* of quantum states contained in dE. It is represented in terms of density of quantum states, denoted by $g(E)$. These are the available quantum states per unit energy range at the given energy E.

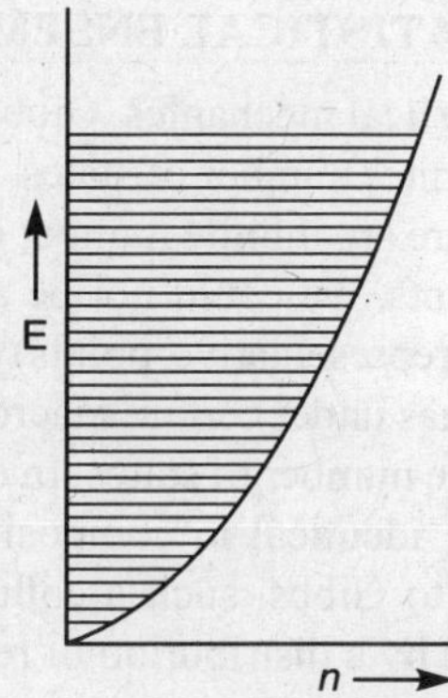

Fig. 10.5. A parabolic curve (energy versus n)

Definition. *The density of quantum states g (E) is defined as the number of quantum states per unit energy range at the given energy E.*

The quantity $g(E)\,dE$ represents the number of quantum states in the energy range between E and $E + dE$ and is given by

$$g(E) = \frac{2\pi}{h^3}(2m)^{\frac{3}{2}} E^{\frac{1}{2}} \quad \text{... (10.15)}$$

Here the spin of the particle is not taken into consideration. For particles of spin $\pm\frac{1}{2}$ like electrons, there will be two allowed quantum states for each representative point corresponding to the two possible spin orientations of the particle. Taking spin in to account, the number of quantum states with energy between E and $E + dE$, is given by

$$g(E)\,dE = 2 \times \frac{2\pi}{h^3}(2m)^{\frac{3}{2}} E^{\frac{1}{2}}\,dE$$

or

$$g(E)\,dE = \frac{4\pi}{h^3}(2m)^{\frac{3}{2}} E^{\frac{1}{2}}\,dE \quad \text{... (10.16)}$$

Distribution Function

In order to determine average properties of a system of particles, it is necessary to find the number of particles actually occupying quantum states in the energy range between E and $E + dE$. This number is denoted by $n(E)\,dE$ and is expressed as

$$n(E)\,dE = g(E)\,f(E)\,dE \quad \text{... (10.17)}$$

where $g(E)$ is the density of (available) quantum states at energy E, $g(E)\,dE$ is the number of quantum states (available) in the energy range between E and $F + dE$ and $f(E)$ is the distribution function of the system, depending upon which statistics (M.B., B.E. or F.D.) we use corresponding to the spin of the particle under consideration.

From equation (10.17), the distribution function $f(E)$ is defined as the average number of particles of the system, which occupy a single quantum state of energy E. The unit of $n(E)$ is particle/joule.

---

* The maximum value of the energy difference between consecutive energy levels for a free electron in a cube of metal of side 10^{-2} m (1 cm) is less than 10^{-6} eV. Therefore, energy levels may be regarded as continuous and we can use calculus method to define the **density of energy states**.

Any small interval dE of energy contains a large number of quantum states. For example, if $dE = 10^{-3}$ eV = 0.001 eV.

Min. number of energy levels contained in dE

$$= \frac{dE}{\text{Energy difference between two consecutive energy levels}} = \frac{10^{-3}}{10^{-6}} = 1000.$$

Further, each energy level degenerates (except the ground level) and hence, the number of quantum states are very large *i.e.* quantum states are out numbering the energy levels.

10.11 STATISTICAL ENSEMBLES

In statistical mechanics, Gibbs introduced the concept of ensemble. It is obvious that a very large (in fact infinite) number of states of the gas corresponds to a given macroscopic condition of the gas because there are infinite number of ways to distribute the molecules in space. Through macroscopic measurements, we could not be able to distinguish between two gases existing in different states (2 distinct representative points) but satisfying the same macroscopic conditions. Thus, when we speak of a gas under certain macroscopic conditions, we are in fact referring not to a single state, but to an infinite number of states. In other words, we are referring not a single system but to a collection of systems, identical in composition and macroscopic conditions but existing in different states. According to Gibbs, such a collection of systems is called **an ensemble**, which is geometrically represented by a distribution of representative points in Γ (gamma) space.

Thus an ensemble is defined as a *collection of large number macroscopically identical but essentially* independent systems. By the term, *macroscopically identical*, we mean that each of the systems constituting an ensemble satisfies the same macroscopic conditions, *e.g.*, volume, energy, pressure, temperature, total number of particles etc. By the term independent systems, we mean that the system constituting an ensemble are mutually non-interacting, just like the role of non-interacting molecules do in a gas.

Kinds of Ensembles

According to Gibbs, there are three standard ensembles to which real experiments could be approximated. These are :

1. The Microcanonical ensemble,

2. The Canonical ensemble, and

3. The Grand-canonical ensemble.

This classification depends on the manner in which their systems interact. It has its own characteristic distribution. Physical systems can remain perfectly insulated from each other or can interact in a variety of ways; in particular, they may exchange energy or matter or both with each other.

10.11.1 Microcanonical Ensemble

The microcanonical ensemble is a collection of essentially independent systems having the ***same energy E, volume V and number of particles N***. For simplicity, we assume that all the particles are identical. The individual systems of a microcanonical ensemble are separated by rigid, impermeable and well insulated walls such that the values of E, V and N for a particular system are not affected at all by the presence of other systems (Fig. 10.6).

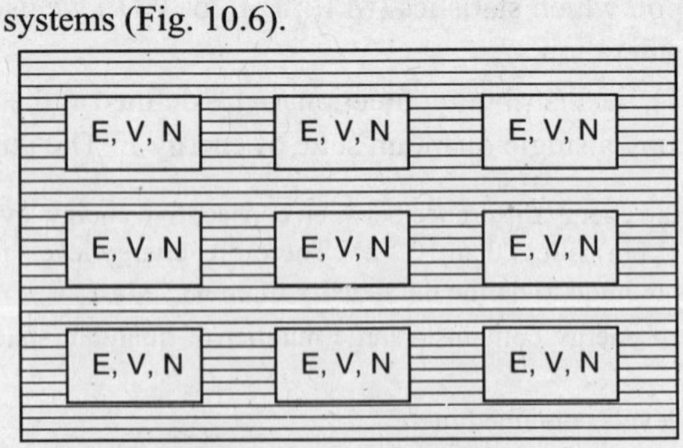

Fig. 10.6. Microcanonical ensemble

10.11.2 Canonical Ensemble

The canonical ensemble is a collection of essentially independent systems having the **same *temperature T, volume V and number of identical particles N.*** This is achieved by bringing all the systems in thermal contact with a large heat reservoir at constant temperature T or by simply bringing all *of the systems in thermal contact with each other.* Fig. 10.7 represents symbolically a canonical ensemble. The individual systems of the canonical ensemble are separated by *rigid, impermeable but diathermic walls i.e., conducting walls.* As the separating walls are conducting, heat can be exchanged between the systems; till they reach a common temperature T. Thus, in canonical ensemble, systems can exchange energy but not the matter *i.e.*, particles.

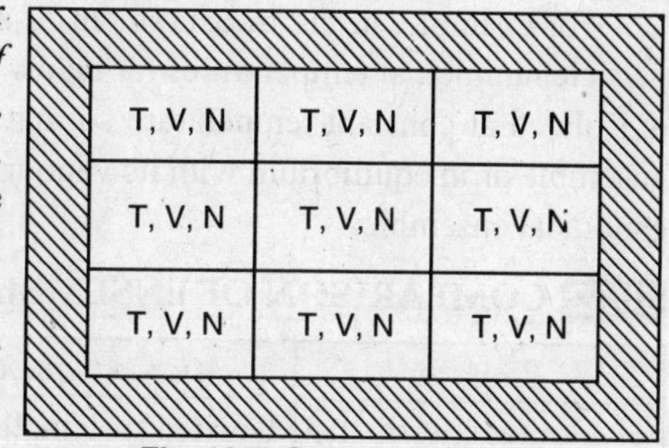

Fig. 10.7. Canonical ensemble

10.11.3 Grand-Canonical Ensemble

In going from microcanonical ensemble (in which independent systems having the same energy E, volume V and number of particle N) to canonical ensemble (in which independent systems have the same temperature T, volume V and number of particles N), the condition of constant energy has been relaxed. This simplifies the calculations in thermodynamics where the exchange of energy takes place. Now the next logical step is to abandon the condition of same number of particles. Actually, in chemical process this number N varies in various physical problems, *e.g.*, radio active decay. Thus, the grand-canonical ensemble is a collection of large number of essentially independent system having the ***same temperature T, volume V and chemical potential*** μ. The individual systems of a grand-canonical ensemble are separated by rigid, permeable and conducting walls, as shown in Fig. 10.8. As the separating walls are conducting and permeable, the exchange of heat energy as well as that of particles between the systems take place in such a way that all the systems arrive at common temperature T and the chemical potential μ.

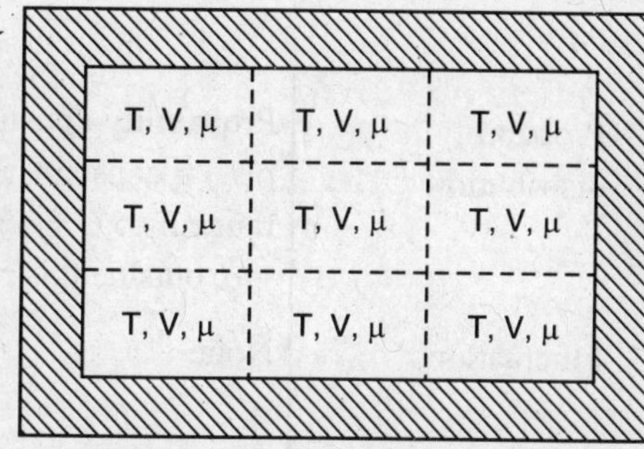

Fig. 10.8. Grand-canonical ensemble

Thus, an ensemble in which the systems exchange energy but not matter is called a *canonical ensemble*, that in which both energy and matter are exchanged between the systems is called a *grand canonical ensemble*; and that in which neither energy nor matter is exchanged is called a *microcanonical ensemble*. A specific choice of ensembles may be thought of as corresponding to a particular physical situation. The microcanonical ensemble is fundamental and refers to the simple situation in which constituent systems are isolated and not influenced in any way by external disturbances. Therefore, it is best suited for isolated systems. However, there is always some energy exchange with the surroundings. Hence canonical ensembles have been found to be more appropriate for dealing with such systems. Besides, canonical approach gives results applicable even when the constituents interact strongly. We will examine critically the Boltzmann's canonical distribution law.

Uses of Ensemble: In practice, we consider the three main types of ensembles and each ensemble corresponds to a different experimental situation. In fact, the three ensembles, which we have introduced and discussed above are only examples of the infinite number of ensembles that can be considered. These are particularly useful for two main reasons:

(*a*) Firstly, they correspond approximately to the types of thermodynamic measurements most frequently made in practices.

(*b*) Secondly, in large number of systems, it is useful to find that the values of thermodynamic quantities that are not very sensitive to the method of measurement.

For example, in the measurement of specific heat of a liquid of known mass and thus known particle number, at temperatures far below the boiling point, it matters very little whether the liquid is isolated at constant temperature so that the number of particles is fixed as in a microcanonical ensemble or in equilibrium with its vapour so that the number of particles can fluctuate as in grand-canonical ensemble.

10.12 COMPARISON OF ENSEMBLES

Property	*Microcanonical*	*Canonical*	*Grand-canonical*
1. Contact with the environment	Isolated system, that is, energy and number of particles of the system are constant.	Systems are in thermal contact with heat reservoir. Hence system can exchange energy but no matter *i.e.*, particles	Systems can exchange both energy and matter *i.e.*, particles.
2. Probability distribution	Probability density $\rho(E)$ = constant, in the range E to $E + dE$ $= o$, outside this range	$\rho(E) = Ae^{E/kT}$	$\rho(E) = e^{[(\Omega + n\mu - E)/KT]}$
3. Fluctuations	None	Fluctuations in energy only	Fluctuations in both energy and matter *i.e.*, particles.
4. Partition function Z	$Z = \Delta\Gamma$	$Z = \int e^{-\beta E}\, d\Gamma$	$Z = \sum_{n=0}^{\infty} e^{\mu n/T} Z_n$

10.13 EQUILIBRIUM BETWEEN TWO SYSTEMS IN THERMAL CONTACT

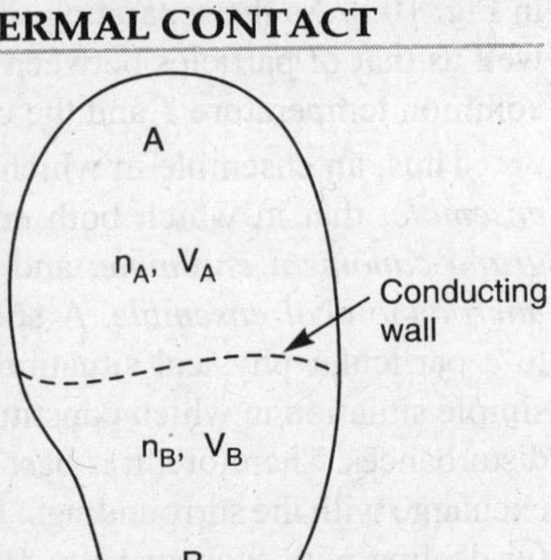

Fig. 10.9.

Two systems are said to be in thermal contact when they are capable of exchanging energy of thermal agitation between one another.

Consider two systems A and B with specified volumes V_A and V_B and having number of particles n_A and n_B respectively as shown in Fig. 10.9. Total energy E is also specified. Let us find how this energy E gets shared between two systems in the equilibrium state. This is obtained by deriving the condition of maximum probability under the given constraints.

Out of the total energy, let E_A be the energy shared by system A. Applying the principle of conservation of energy, the energy shared by system B will be

$$E_B = E - E_A. \quad \text{... (10.18)}$$

The probability that system A will possess energy E_A is given by

$$P(E_A) = C\, W(E_A) \quad \text{... (10.19)}$$

where $W(E_A)$ gives the number of microstates for a system A in an energy range δE at E and C is an arbitrary constant. Similarly for system B, we can write

$$P(E_B) = C'\, W'(E_B) \quad \text{... (10.20)}$$

where C' is another arbitrary constant.

Now applying the multiplication rule the joint probability (Ref. 9.5) of the system, will be

$$P(E_A, E_B) = CC'\,W(E_A)\,W'(E_B) \qquad \text{... (10.21)}$$

The two systems *A* and *B* are in perfect thermal contact, and hence they will exchange energy amongst them and an equilibrium state is attained for which the probability $P(E_A, E_B)$ will have a maximum value.

In order to incorporate this fact, take natural logarithms of both sides,

$$\log P(E_A, E_B) = \log CC' + \log W(E_A) + \log W'(E_B) \qquad \text{... (10.22)}$$

Differentiating *w.r.* to E_A, and equating to zero,

$$0 = \frac{\partial}{\partial E_A}\log W(E_A) + \frac{\partial}{\partial E_B}\log W'(E_B)\,.\,\frac{\partial E_B}{\partial E_A} \qquad \text{... (10.23)}$$

But $\dfrac{\partial E_B}{\partial E_A} = -1.$ (∵ Eqn. 10.18)

This means, whatever energy is gained by system *A* is equal and opposite (–ve sign) to the energy gained by system *B*. This can be seen from equation 10.18.

Hence, Equation 10.23 leads to

$$\frac{\partial}{\partial E_A}\log W(E_A) = \frac{\partial}{\partial E_B}\log W'(E_B) \qquad \text{... (10.24)}$$

Thus, according to statistical mechanics, the two systems in thermal contact will be in equilibrium if their functions $\left[\dfrac{\partial}{\partial E}\log W\right]$ are equal.

This is true for all systems, however complex they may be.

10.14 BRIDGE WITH MACROSCOPIC PHYSICS

The functions $\dfrac{\partial}{\partial E}\log W$ is generally represented by parameter β. Using this, we have

$$\beta = \frac{\partial}{\partial E_A}\log W = \frac{1}{W}\frac{\partial W}{\partial E_A} \qquad \text{... (10.25)}$$

and $$\beta' = \frac{\partial}{\partial E_B}\log W' = \frac{1}{W'}\frac{\partial W'}{\partial E_B} \qquad \text{... (10.26)}$$

β′ has dimensions of reciprocal of energy.

Now thermal equilibrium is specified by the condition given by equation (10.24). This means β = β′. In terms of temperature, thermal equilibrium is specified by $T = T'$ *i.e.* the macroscopic parameter temperature is reached to a constant and maximum value. In other words, *equality of temperature between the two systems in thermal contact means the equality of the* β *parameters of the two systems.* This fact is used to define temperature, a macroscopic parameter as

$$\frac{1}{\beta} = kT \text{ and } \frac{1}{\beta'} = kT' \qquad \text{... (10.27)}$$

where *k* is a constant. The choice of *k* will depend upon the scale we select. The dimensions of this constant *k* are erg, $\deg^{-1}$. When this scale matches with the perfect gas scale temperature then this *k* must be set equal to the Boltzmann's constant which is defined by R/N_o where *R* is the universal gas constant and N_o is Avogadro's number.

10.15 THEORIES BASED ON STATISTICAL MECHANICS

The entropy and the distribution of molecular speeds, these two theories in physics are based on the statistical concepts. The thermodynamic property entropy (*S*) is explained on the basis of statistical mechanics. According to statistical considerations equilibrium state of an isolated system is the macrostate of maximum thermodynamic probability. Hence in equilibrium state, both entropy and the thermodynamic probability have the maximum value. On the other hand, with the help of Maxwell Boltzmann statistics, the number of molecules per unit volume for a gas having speeds between v and $(v + dv)$ has been calculated and hence the value of most probable speed.

10.16 ENTROPY AND PROBABILITY

Statististical Definition of entropy.

Consider a system consisting of μ gram molecules (moles) having a volume V and total energy U. The total number of molecules (n) in the system is given by the value of μ and $N = \mu \times 6.02 \times 10^{23}$ molecules. The thermodynamic probability W for any macrostate of the system, in which n particles can be arranged in k compartments having $g_1, g_2, ... g_k$ cells, belonging to compartments 1, 2, ..., k respectively, such that $n_1, n_2, ..., n_k$ particles go into respective compartments *i.e.*, the total number of microstates in the macrostate, $n_1, n_2, ..., n_k$ is given by

$$W(n_1, n_2, ..., n_k) = \frac{N!}{n_1!n_2!... n_k!} \times (g_1)^{n_1}(g_2)^{n_2}...(g_k)^{n_k}$$

The above expression is the product of a number of quantities and, in general is an inconveniently large number. We, therefore, deal with *ln W*, which is comparatively a smaller number, consisting of the sum of the natural logarithms of these quantities.

We know that when an isolated system (*i.e.,* a system which does not interact with its surroundings so that its energy is constant) undergoes an irreversible process, there is a net increase of the entropy of the system. Spontaneous processes, such as spontaneous expansion of a gas into an evacuated space, diffusion of one gas into another, etc., are all irreversible processes. Hence, in all irreversible processes, there is a net increase of entropy, and an increase of randomness.

In general, all spontaneous processes represent changes from a less probable to a more probable state, and since in such processes the entropy increases, there should be a relation between the entropy S of the system in a given state and the probability of that state.

Boltzmann, in 1896 derived a relation between entropy (a thermodynamical quantity) and probability (a statistical quantity). Boltzmann started thinking from a simple idea in mind that the equilibrium state of any system is the state of its maximum probability, *i.e.*, the probability of the system in equilibrium state is maximum. But from the thermodynamical point of view, the equilibrium state of a system is the state of its maximum entropy. If the system is not in equilibrium, then changes take place within the system until the equilibrium state is reached and the system attains a maximum entropy. Thus, in the equilibrium state, both the entropy and the thermodynamical probability have their maximum value. This observation inspired Boltzmann to think that there must be corelation between these two quantities. The relation between them is known as Boltzmann's entropy relation.

Boltzmann's entropy probability relation. In dealing with a macroscopic system we assume a new quantity (or parameter) $S = k \ln W$ where k is Boltzman's constant = 1.38×10^{-16} ergs/K or 1.38×10^{-23} J/K. This quantity is known as the *entropy* of the system.

$$\therefore \qquad S = k \ln W$$

Hence, according to statistical definition, **entropy** *is the product of Boltzmann's constant k and natural log of thermodynamic probability W.*

Thus the concept of entropy is applicable to any macrostate whether most probable or not. In this respect entropy differs from other state variables like pressure, volume and temperature which only apply to equilibrium state.

If the system is in a state of equilibrium, it is in the most probable macrostate so that $W = W_{max}$.

$\therefore$ For equilibrium state $S = k \ln W_{max}$

Units of entropy. As ln W is a mere number having no dimensions, the dimensions of entropy S are the same as that of k *i.e.*, ergs/K on the C.G.S. system and Joules/K on the M.K.S. or S.I. system.

Proof of Boltzmann's Entropy Relation

According to thermodynamics, the function entropy S of a system is related with temperature T by the relation

$$\frac{1}{T} = \frac{\partial S}{\partial E} \qquad ... (i)$$

According to the approach of statistical mechanics, using equations (10.25), (10.26) and (10.27), we write

$$\frac{1}{T} = k\beta = k\frac{\partial}{\partial E}\log W \qquad ... (ii)$$

Equating equations (*i*) and (*ii*), we have

$$\frac{\partial S}{\partial E} = k\frac{\partial}{\partial E}\log W$$

Integrating, we get

$$S = k \log W \qquad ... (10.28)$$

Aliter. Boltzmann assumed that entropy S of a physical system in a definite state is a function of the maximum probability W of that state. Thus

$$S = f(W) \qquad ... (i)$$

where S is entropy and W is the thermodynamical probability of the state.

Let us consider two completely independent systems A and B having entropies S_1 and S_2 respectively. Since entropy is an extensive (*i.e.* additive) quantity, the entropy of the two systems together will be

$$S = S_1 + S_2 \qquad ... (ii)$$

If W_1 is the thermodynamic probability of system A and W_2 that of system B, then the probability W will be the product of their probabilities, *i.e.*,

$$W = W_1 \times W_2 \qquad ... (iii)$$

Thus, $$S = f(W) = f(W_1 \times W_2) \qquad ... (iv)$$

$$S_1 = f(W_1) \qquad ... (v)$$

$$S_2 = f(W_2) \qquad ... (vi)$$

Substituting (*iv*), (*v*) and (*vi*) in eqn. (*ii*), we get

$$f(W_1 \times W_2) = f(W_1) + f(W_2) \qquad ... (vii)$$

Differentiating partially w.r. to W_1 and W_2, we get

$$W_2 f'(W_1 \times W_2) = f'(W_1) \qquad ... (viii)$$

$$W_1 f'(W_1 \times W_2) = f'(W_2) \qquad ... (ix)$$

Dividing eqn. (*viii*) by eqn. (*ix*), we get

$$\frac{f'(W_1)}{f'(W_2)} = \frac{W_2}{W_1}$$

or, $$W_1 f'(W_1) = W_2 f'(W_2) = ... = k$$

k being any constant.

Thus, $$f'(W_1) = \frac{k}{W_1}, f'(W_2) = \frac{k}{W_2} \text{ and so on.}$$

Integrating, we get

$$f(W_1) = k \log (W_1) + C_1$$

and $$f(W_2) = k \log (W_2) + C_2$$

In general,

$$f(W) = k \log (W) + C$$

where C is a constant of integration.

Thus, $$S = k \log (W) + C \quad \text{... (∵ Eqn. } i)$$

The constant of integration C is evaluated by putting the condition that at absolute zero ($T = 0K$), the entropy $S = O$, and $W = 1$.

$$\therefore \quad C = 0$$

Putting in equation (x), we get

$$S = k \log (W) \quad \text{... (10.26)}$$

This is the required relation between entropy and thermodynamic probability and is called the *Boltzmann's entropy relation.* Here k is universal constant known as Boltzmann's constant. The constant of integration is treated as zero because we are interested in changes in entropy and not in its absolute values. Thus, *the Boltzmann's entropy relation states that the entropy of a system is proportional to the logarithm of thermodynamic probability of that system.*

The law that the entropy of the universe always tends to increase (refer article 5.5) now means that W always tends to increase, which in turn means simply that the universe always changes towards statistically more probable states.

Example 10.2 *Calculate the number of accesible microstates W of a system having entropy 10 cal K^{-1} in the equilibrium state given 1 cal = 4.2×10^7 ergs, and Boltzmann's constant $k = 1.38 \times 10^{-16}$ ergs^{-1}.* ***(M.D.U. 1999)***

Solution. Entropy $S = 10$ Cal $K^{-1} = 10 \times 4.2 \times 10^7 = 42 \times 10^7$ ergs K^{-1}

Boltzmann's constant $k = 1.38 \times 10^{-16}$ ergs K^{-1}

Now $S = k \ln W$ or $42 \times 10^7 = 1.38 \times 10^{-16} \ln W$

or $$\ln W = \frac{42 \times 10^7}{1.38 \times 10^{-16}} = 30.4 \times 10^{23} = 3.04 \times 10^{24}$$

$$\therefore \quad W = e^{3.04 \times 10^{24}}$$

10.17 RELATION OF ENTROPY WITH TEMPERATURE

The value of thermodynamic probability W, according to the three types of statistics, is as under.

M.B. statistics $$W = N! \Pi \frac{(g_i)^{n_i}}{n_i!}$$

B.E. statistics $$W = \Pi \frac{(n_i + g_i)!}{n_i!\, g_i!}$$

F.D. statistics $$W = \Pi \frac{g_i!}{n_i!\,(g_i - n_i)!}$$

W is therefore, a function of two variables

(1) Number of particles in the system N

(2) Total number of available phase space cells g

The number of particles in a system $N = \mu \times 6.02 \times 10^{23}$ where μ is the number of moles in the system.

The number of phase space cells depends upon the total volume V and the total energy U of the system. Hence the value of W depends upon μ the number of the moles constituting the system, its volume V and total energy U. Any change in any one of these three quantities will bring about a change in W and thereby a change in entropy S. If we keep m and V constant and supply to the system in equilibrium an infinitesimally small amount of heat δQ so as to change the value of U, by dU then from the first law of thermodynamics, we have

$$\delta Q = dU + P\,dV$$

But $\quad dV = 0 \qquad [\because V = \text{a constant}]$

$\therefore \quad \delta Q = dU$

Since the system is in equilibrium, an infinitesimally small amount of heat given to the system will bring about only an infinitesimally small change in the state of equilibrium, so that we can still suppose the system to exist in its most probable or equilibrium state.

For the most probable state, the thermodynamic probability W and hence $\ln W$ has a maximum value

$$\therefore \quad d(\ln W) = 0 \qquad \text{... } (i)$$

In a system composed of a given number of particles (N) having total energy U

$$N = n_1 + n_2 + \ldots = \sum n_i$$

$$\therefore \quad dN = \sum dn_i = 0 \qquad \text{... } (ii)$$

and

$$U = n_1 u_1 + n_2 u_2 + \ldots = \sum n_i u_i$$

$$\therefore \quad dU = \sum dn_i u_i = 0$$

where u_i's are fixed quantities representing the mean energy values of intervals.

Multiplying equation (*ii*) by $(-\alpha)$, equation (*iii*) by $(-\beta)$ where α and β are arbitrary constants and adding to equation (*i*), we get

$$d(\ln W) - \alpha\, dN - \beta\, dU = 0 \qquad \text{... } (iv)$$

Now $\beta = \dfrac{1}{kT}$ and it is given that N the number of particles is a constant

$$\therefore \quad dN = 0 \quad \text{or} \quad \alpha\, dN = 0$$

Hence $\quad d \ln W - \dfrac{dU}{kT} = 0$

or $\quad d \ln W = \dfrac{dU}{kT}$

or $\quad kd \ln W = \dfrac{dU}{T} \qquad \text{... } (v)$

Also $\delta Q = dU$

$$\therefore \quad \frac{\delta Q}{T} = \frac{dU}{T} \qquad \text{... } (vi)$$

According to the statistical definition of entropy, $S = k \ln W$

$$\therefore \quad dS = k\, d \ln W$$

Hence $\quad dS = k\, d\ln W = \dfrac{\delta Q}{T} \qquad \text{... } (vii)$

Hence change in entropy is defined as the ratio of the infinitesimally small amount of heat supplied to the system divided by the absolute temperature keeping μ *and* V *constant.* The relation $k\, d \ln W = dS = \dfrac{\delta Q}{T}$ gives the relation between probability a statistical quantity and entropy a thermodynamic quantity.

Note. It should be noted that small changes in S and U are denoted an dS and dU but a small change in Q is denoted as δQ and not as dQ. The reason for the same is that S and U are both continuous single valued and differentiable variables. But this is not the case with the quantity of heat given to or taken away from a system. Therefore, δQ is not a perfect differential and should not be written as dQ.

10.18 LAW OF INCREASE OF ENTROPY (On statistical basis)

Consider a system the particles of which can exchange energy and interact only with one another. Such a system neither loses nor gains energy from outside and is, therefore, known as an *isolated system.* If all the parts of the system are at the same temperature, it is said to be in thermal equilibrium *i.e.*, it is in the state of maximum thermodynamic probability. The probability of the system existing in a macrostate having even a small deviation from the state of maximum probability is negligible. Hence if the system initially exists in a state different from the state of maximum probability, it very quickly comes to the equilibrium state, due to mutual collisions between the particles constituting the system. As the entropy S of a system is proportional to the thermodynamic probability W, the system, in moving from a state of lower probability to the state of maximum probability also moves from the state of lower entropy to a state of maximum entropy. It is not however possible for the system to itself move in the opposite direction *i.e.*, from the state of maximum probability (or entropy) to a state of lower probability (or entropy). Since, when the system is in equilibrium entropy S has its maximum value.

∴ For the state of equilibrium, change in entropy $\Delta S = 0$

In other words, the change in entropy of a body and its surroundings in thermal equilibrium with each other is zero.

When the system moves from a state of lower probability (state of inequilibrium) to the state of maximum probability (equilibrium state) there is an increase in entropy. As this process is irreversible, we have

For an irreversible process, the change in entropy,

$$\Delta S > 0$$

Hence *the entropy of an isolated system remains constant or increases according as the changes it undergoes are reversible or irreversible.*

The tendency of an isolated system to shift towards a state of higher entropy in irreversible processes is called the law of increase of entropy.

Entropy during natural process. According to second law of thermodynamics, the natural direction of heat flow is from hot regions to cold regions. Heat cannot flow from cold regions to hot regions unless work is done by an external agency. Thus all natural processes are irreversible. *This is why entropy increases during natural processes.* Entropy is also called '*thermal inertia*'. It bears to *heat motion* the same relation as mass does to linear motion and moment of interia to rotational motion.

Increase in entropy of the universe. The entropy of an isolated system remains constant or increases according as the changes it undergoes are reversible or irreversible *i.e.*, $dS \geq 0$.

The sign of equality applies to a reversible process and that of inequality to an irreversible process. All actual processes are inherently irreversible. Physical processes like conduction, radiation etc. are continuously going on in nature. For all these physical processes the final entropy is always higher than the initial entropy. Hence all natural processes take place in such a direction as to increase the entropy of the system as well as that of the environment.

The entropy of the universe, therefore, always increases.

Loss of available energy. According to second law of thermodynamics no device can convert all the heat absorbed from the source into useful mechanical work. It must reject some heat energy to the sink and this heat energy rejected to the sink is not available for conversion into useful work.

In a Carnot's cycle, suppose Q_1 is the quantity of heat absorbed from the source at temperature T_1 and a quantity of heat Q_2 is rejected to the sink at a temperature T_2, *then* Q_2 *is the unavailable portion of the energy.* According to Carnot's principle

$$\frac{Q_1}{T_1} = \frac{Q_2}{T_2}$$

$\therefore$ Unavailable heat energy $Q_2 = \frac{Q_1}{T_1} T_2$

Thus for a given value of T_2 the unavailable energy is determined by the increase in entropy during the isothermal process at T_1. Hence entropy is a measure of unavailability of useful energy. *In other words, an increase in entropy is always followed by loss of available energy.*

Increase of entropy in a process of equalisation of temperature. In a process which tends to equalise temperature of the parts of a system, heat flows from parts at higher temperature to parts at a lower temperature. According to second law of thermodynamics heat cannot flow from a lower temperature to a higher temperature unless work is done by an external agency. The process of equalisation of temperature is, therefore irreversible. *Hence there is an increase of entropy.*

The entropy of the system increases even when it loses heat. The entropy of an isolated system remains constant or increases according as the changes it undergoes are reversible or irreversible *i.e.*, $dS \geq 0$. For all physical processes like conduction, radiation etc. the final entropy is always higher than initial entropy.

10.19 ENTROPY: A MEASURE OF DISORDER

Order and disorder. Consider a system in a macrostate, having thermodynamic probability $W = 1$. Such a system will then exist only in one microstate. In other words, it is possible to arrange the particles of the system in *one* and *only one way.* We therefore, have complete and definite information about all the particles constituting such a system.

When complete and definite information is known for all the particles of a system it is said to be in perfect order.

Now consider a macrostate of the system in which $W = 2$. This means that the system has two microstates and it can exist in any one of them with equal probability. It is, therefore, not possible for us to say with certainty in which of the two states the system exists at any particular instant of time.

The probability that it may exist in any one of the two state is $\frac{1}{2}$.

Similarly, when the system exists in a macrostate with $W = 3$, it may be in any one of the three states at any particular instant of time. The probability that it may be in any particular microstate is $\frac{1}{3}$.

Thus when the number of microstates in a macrostate is more than one we do not have complete and definite information about the particles of the system. The uncertainty becomes greater and greater as the value of W increases. *The uncertainty or loss of information about the state of a system is called disorder.* The disorder is proportional to W.

Entropy a measure of disorder. As stated above disoder is proportional to W, the thermodynamic probability.

Now entropy $S = k \ln W$.

$\therefore$ Entropy, being proportional to W is also related to disorder in the system. Entropy is taken to be a measure of disorder. According to the law of increase of entropy, the entropy of every natural system tends to increase. The same can be stated as follows.

Every system tends to proceed towards the state of maximum disorder.

Crystal-low entropy system. A crystal is a solid of definite shape with its atoms, ions or molecules arranged in some regular repititious three dimensional pattern.

Thus a crystal presents *spatial regularity* extending over a large volume *i.e.*, it has a *long range order* in its structure.

As entropy is a measure of disorder, a crystal having long range order has very little disorder and is, therefore, an example of a low entropy system.

Entropy-physical significance. The statistical definition of entropy is

$$S = k \ln W$$

where W is the thermodynamic probability and k Boltzmann's constant. Entropy being proportioanl to W is also related to disorder.

Entropy, therefore, represents the degree of disorder in a system and is a state variable.

10.20 ENTROPY OF A PERFECT GAS

Suppose we have a vessel containing a volume V of a perfect gas. The gas has N molecules and is in a state of equilibrium at a temperature T. As the gas is in equilibrium it is in the most probable macrostate.

Let the N molecules be divided into k compartments (energy intervals) numbered $(1, 2, ... k)$ having $g_1, g_2, ... g_k$ equal sized cells and $n_1, n_2 ... n_k$ molecules respectively. As a perfect gas obeys Maxwell Boltzmann statistics, the thermodynamic probability of the macrostate $(n_1, n_2, ... n_k)$ is given by

$$W(n_1, n_2, ... n_k) = \frac{N!(g_1)^{n_1}(g_2)^{n_2} \text{........} (g_k)^{n_k}}{n_1!\, n_2! .... n_k!}$$

or $$W(n_1, n_2 ... n_k) = N! \prod_{i=1}^{k} \frac{(g_i)^{n_i}}{n_i!} \qquad ...(i)$$

where $$n_i = g_i\, e^{-\alpha}\, e^{-\beta u_i} = g_i\, e^{-\alpha}\, e^{-u_i/kT}$$

α being an arbitrary constant, $\beta = \frac{1}{kT}$ and u_i the mean energy of the molecules in the ith compartment (energy interval)

Taking natural logarithms of both sides of equation (i), we have

$$\ln W = \ln N! + \sum_{i=1}^{k} n_i \ln g_i - \sum_{i=1}^{k} \ln n_i!$$

Using Sterling's formula

$$\ln N! = N \ln N - N, \text{ we have}$$

$$\ln W = N \ln N - N + \sum_{i=1}^{k} n_i \ln g_i - \sum_{i=1}^{k} n_i \ln n_i + \sum_{i=1}^{k} n_i$$

But $\sum_{i=1}^{k} n_i = N$ total number of molecules

$$\therefore \quad \ln W = N \ln N + \sum_{i=1}^{k} n_i \ln g_i - \sum_{i=1}^{k} n_i \ln n_i = N \log N - \sum_{i=1}^{k} n_i \ln \frac{n_i}{g_i} \qquad ...(ii)$$

Substituting $n_i = g_i\, e^{-\alpha} e^{-u_i/kT}$ in equation (ii), we get

$$\ln W = N \ln N - \sum_{i=1}^{k} n_i \ln (e^{-\alpha}\, e^{-u_i/kT}) = N \ln N - \sum_{i=1}^{k} n_i \ln e^{-\alpha} - \sum_{i=1}^{k} n_i \ln e^{-u_i/kT}$$

$[\because \ln mn = \ln m + \ln n]$

or $$\ln W = N \ln N - N \ln e^{-\alpha} + \sum_{i=1}^{k} \frac{n_i u_i}{kT} \ln e$$

$$= N \ln N - N \ln e^{-\alpha} + \frac{1}{kT} \sum_{i=1}^{k} n_i u_i \qquad [\because \ln e = 1]$$

But $\sum_{i=1}^{k} n_i u_i = U$ where U is the total energy of the system

$$\therefore \qquad \ln W = N \ln N - N \ln e^{-\alpha} + \frac{U}{kT} = N \ln \frac{N}{e^{-\alpha}} + \frac{U}{kT} = N \ln N e^{\alpha} + \frac{U}{kT} \qquad \text{... (iii)}$$

For a system obeying Maxwell Boltzmann statistics

$$e^{-\alpha} = \frac{Nh^3}{V}\left(\frac{\beta}{2\pi m}\right)^{3/2} = \frac{Nh^3}{V}\left(\frac{1}{2\pi mkT}\right)^{3/2} \qquad \text{... (iv)}$$

$$\therefore \qquad Ne^{\alpha} = \frac{V}{h^3}(2\pi\, mkT)^{3/2}$$

Substituting the above value of Ne^{α} in equation (*iii*), we have

$$\ln W = N \ln\left[\frac{V}{h^3}(2\pi\, mkT)^{3/2}\right] + \frac{U}{kT}$$

For a mono-atomic gas $U = \frac{3}{2} NkT$ [$\because$ energy per molecule = 3/2 kT]

$$\therefore \qquad \ln W = N \ln\left[\frac{V}{h^3}(2\pi\, mkT)^{3/2}\right] + \frac{3}{2} N$$

According to statistical definition of entropy $S = k \ln W$

$$\therefore \qquad S = \frac{3}{2} Nk + Nk \ln\left[\frac{V}{h^3}(2\pi\, mkT)^{3/2}\right] \qquad \text{... (10.29)}$$

Equation (10.29) gives the entropy of a perfect gas. According to this equation *entropy only depends upon two thermodynamic variables V (volume) and T (temperature). But V and T define uniquely the state of the system. Hence entropy is a function of state of the system. The entropy S will change if either V or T changes.*

Entropy for one mole. If the vessel contains one mole of perfect gas, then the total number of molecules S = Avogadro's number = N and $Nk = R$. Also V is gram molecular volume

$$\therefore \qquad S = \frac{3}{2} R + R \ln\left[VT^{3/2}\left(\frac{2\pi mk}{h^2}\right)^{3/2}\right] \qquad \text{... (10.29A)}$$

10.21 BOLTZMANN'S CANONICAL DISTRIBUTION LAW

In equilibrium state, let $n_1, n_2, n_3, \ldots\ldots n_i$ be the number of gas molecules in different cells, say in cell no. 1, 2, 3, ... i respectively. As the gas molecules are continuously moving, n_i's will change continuously in many different ways, but will always keep value very close to that for the state of equilibrium, *i.e.*, the state of maximum probability. At any instant, the gas molecules obey the following three fundamental postulates of statistical mechanics. There are :

I. The total number of molecules in constant, *i.e.*,

$$N = n_1 + n_2 + n_3 + \ldots\ldots\ldots\ldots + n_i + \ldots\ldots\ldots\ldots = \text{constant}$$

or

$$\delta N = \delta n_1 + \delta n_2 + \delta n_3 + \ldots\ldots\ldots\ldots + \delta n_i + \ldots\ldots\ldots\ldots = 0 \qquad \text{... (i)}$$

II. The total energy of the system is constant, *i.e.*,

$$E = E_1 n_1 + E_2 n_2 + E_3 n_3 + \ldots\ldots\ldots\ldots + E_i n_i + \ldots\ldots\ldots\ldots = \text{constant}$$

$$\delta E = E_1 \delta n_1 + E_2 \delta n_2 + E_3 \delta n_3 + \ldots\ldots\ldots\ldots + E_i \delta n_i + \ldots\ldots\ldots\ldots = 0 \qquad \text{... (ii)}$$

III. In equilibriunm state, the thermodynamical probability (W) is maximum, *i.e.*,

$$\delta W = 0 \qquad \text{... (iii)}$$

We know that $\log_e W$ is more convenient than W for mathematical treatment. When W is maximum, $\log W$ is also maximum. Therefore

$$\delta (\log W) = 0 \qquad \text{... } (iv)$$

The thermodynamical probability W is expressed as

$$W = \frac{N!}{n_1!\,n_2!\,n_3!\,........n_i!........} \qquad \text{... } (v)$$

Taking logarithm of both sides, we get

$$\log W = \log N! - \log n_1! - \log n_2! - \log n_3! \text{ .... } - \log n_i! \text{ ....} \qquad \text{... } (vi)$$

As number of molecules (n_i's) are large in number, we can apply Stirling's theorem*. Applying we get

$$\log W = N \log N - N - (n_1 \log n_1 - n_1) - (n_2 \log n_2 - n_2) \text{ ... } - (n_i \log n_i - n_i) \text{ ...}$$

$$= N \log N - N - \sum_i (n_i \log n_i - n_i) \qquad \text{... } (vii)$$

Differentiating equation (*vii*) with respect to $n_1, n_2, \text{........} n_i \text{.......}$ and keeping in mind that N is constant, we have

$$\delta (\log W) = -\sum_i [n_i (\delta \log n_i) + \log n_i \; \delta n_i - \delta n_i] \qquad \text{... } (viii)$$

or $$\sum_i \left[n_i \left(\frac{1}{n_i} \right) \delta n_i + \log n_i \; \delta n_i - \delta n_i \right] = 0 \qquad (\because \; iv) \qquad \text{... } (ix)$$

or $$\sum \log n_i \; \delta n_i = 0$$

or $$\log n_1 \; \delta n_1 + \log n_2 \; \delta n_2 + \log n_3 \; \delta n_3 + \text{.........} + \log n_i \; \delta n_i + \text{.........} = 0 \qquad \text{... } (x)$$

It should be noted that the equation (*x*), (*i*) and (*ii*) are independent of one another and they must be satisfied at any instant. To maintain their independence, let us multiply each by a separate arbitrary constants α_1, α_2 and α_3 respectively and adding them, we get

$$\alpha_1 \{\log n_1 \; \delta n_1 + \log n_2 \; \delta n_2 + \log n_3 \; \delta n_3 + \text{.....} + \log n_i \; \delta n_i + \text{.....}\} + \alpha_2 \{\delta n_1 + \delta n_2 + \delta n_3 + \text{.....} + \delta n_i + \text{......}\} + \alpha_3 \{E_1 \; \delta n_1 + E_2 \; \delta n_2 + E_3 \; \delta n_3 + \text{......} + E_i \; \delta n_i + \text{......}\} = 0$$

Dividing by α_1, we get

$$\{\log n_1 \; \delta n_1 + \log n_2 \; \delta n_2 + \text{....} + \log n_i \; \delta n_i + \text{.....}\} + \frac{\alpha_2}{\alpha_1} \{\delta n_1 + \delta n_2 + \text{......} + \delta n_i + \text{.....}\} + \frac{\alpha_3}{\alpha_1} \{E_1 \; \delta n_1 + E_2 \; \delta n_2 + \text{......} + E_i \; \delta n_i + \text{......}\} = 0 \qquad \text{... } (xi)$$

Let $\frac{\alpha_2}{\alpha_1} = \alpha$ and $\frac{\alpha_3}{\alpha_1} = \beta$ where α and β are new constants. Substituting in equation (*xi*), we have

$$\{\log n_1 \; \delta n_1 + \log n_2 \; \delta n_2 + \text{.....} + \log n_i \; \delta n_i + \text{......}\} + \alpha \{\delta n_1 + \delta n_2 + \text{.....} + \delta n_i + \text{.....}\} + \beta \{E_1 \; \delta n_1 + E_2 \; \delta n_2 + \text{......} + E_i \; \delta n_i + \text{......}\} = 0$$

Rearranging, we get

$$(\log n_1 + \alpha + \beta E_1) \; \delta n_1 + (\log n_2 + \alpha + \beta E_2) \; \delta n_2 + \text{......} + (\log n_i + \alpha + \beta E_i) \; \delta n_i + \text{......} = 0 \qquad \text{... } (xii)$$

In this equation, all δn_i's are all independent of each other. Therefore, if equation (*xii*) is to be satisfied, the coefficient of each term must vanish. Thus

$$\log n_i + \alpha + \beta E_i = 0 \quad \text{for all values of } i. \qquad \text{... } (xiii)$$

$$\therefore \qquad \log n_i = -\alpha - \beta E_i$$

---

* When x is large, according to Sterling's theorem loge $x! = [x \log_e x] - x$

or $$n_i = e^{-(\alpha + \beta E_i)} = e^{-\alpha} . e^{-\beta E_i}$$

or $$n_i = A\, e^{-\beta Ei} \qquad \text{... (10.30)}$$

where $A = e^{-\alpha}$, a new constant.

Equation (10.30) gives the number of molecules in each cell as a function of energy associated with each particle in that cell. This is known as ***Boltzmann's canonical distribution law***.

Using equation (10.30), if a graph is plotted between the number of molecules per cell n_i and energy E_i associated with the cell, an exponential decrease nature is observed as shown in (Fig. 10.10). From the graph, we conclude that low energy cells will contain more particles than high energy cells, the number of particles being maximum in zero-energy cell. It does not mean that the number of molecules having zero-energy is greater than any other energy because there are more cells corresponding to high energy.

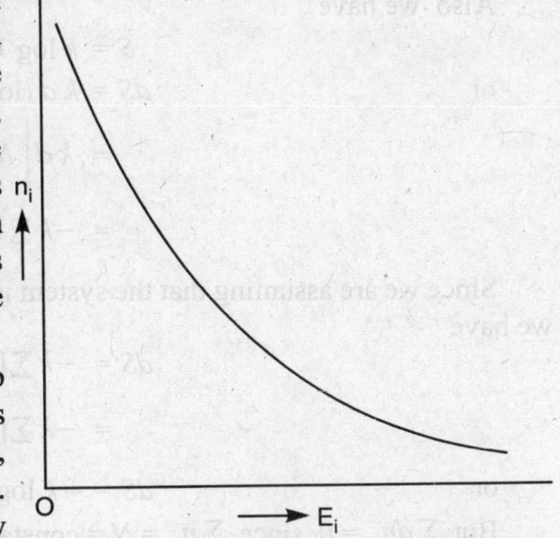

Fig. 10.10

It is also convenient sometimes to consider a group of adjacent cells in the phase space. Such a group of cells is called a **zone of cells**. As all the cells are of equal size, the size of zone is proportional to the number of cells in it. If g_i is the number of cells in a zone, then the probability of finding a molecule in the zone will be proportional to g_i. This g_i is called then a **priori probability** for the zone which solely depends on the size of the zone.

If we consider a zone of cells, all corresponding to about the same energy E_i, then the number of molecules in this zone will be

$$n_i = A\, g_i\, e^{-\beta E_i} \qquad \text{... (10.31)}$$

or $$n_i = e^{-\alpha} g_i e^{-\beta E_i} \qquad \text{... (1031A)}$$

Evaluation of Constant A

Since the sum of all n_i's must be equal to the total number of molecules, say N, then

$$\sum_i n_i = N = \text{A}\, \Sigma e^{-\beta E_i} \quad \text{(From Eqn. 10.30)} \qquad \text{...(10.32)}$$

The summation term $\Sigma e^{-\beta Ei}$ is called the **partition function or sum of states** and is represented by a symbol Z, *i.e.*,

Partition function, $$Z = \Sigma e^{-\beta Ei} \qquad \text{... (10.33)}$$

Substituting in equation 10.32, we get

$$N = \text{AZ}$$

or $$A = \frac{N}{Z} \qquad \text{... (10.34)}$$

Substituting this value of A in equation 10.30, the number of molecules in the ith cell in its state of maximum thermodynamic probability will be

$$n_i = \frac{N}{Z} e^{-\beta E_i} \qquad \text{... (10.35)}$$

Evaluation of Constant β

Suppose a small amount of heat dQ is supplied to a gas at constant volume so that some of the molecules move from cell of lower energy to cells of higher energy. Let dn_1, dn_2, dn_3, ... etc. be the change in the number of molecules in cell 1, cell 2, cell 3, ... etc., respectively.

According to thermodynamics, the increase in internal energy at constant volume is given by

$$dE = \delta Q = TdS \qquad \text{... (10.36)}$$

Using equ. (*ii*),

$$dE = E_1\, dn_1 + E_2\, dn_2 + E_3\, dn_3 + ... + E_i\, dn_i + ...$$

$$= \sum_i E_i\, dn_i \qquad \text{... (10.37)}$$

Also, we have

$$S = k \log W$$

or

$$dS = k\, d\,(\log W)$$

$$= k\, d\left[N \log N - \sum_i \{n_i \log n_i - n_i\}\right] \qquad \text{[using eqn. } viii]$$

$$= -k \sum_i \log n_i\, dn_i \qquad (\because N \text{ is a constant}) \qquad \text{... (10.38)}$$

Since we are assuming that the system is in thermal equilibrium, we may use $n_i = Ae^{-\beta E_i}$, so that we have

$$dS = -k \sum_i [\log (Ae^{-\beta E_i})\, dn_i]$$

$$= -k \sum_i [\log A - \beta E_i)\, dn_i]$$

or

$$dS = -k \log A\, \Sigma dn_i + k\beta\, \Sigma E_i\, dn_i \qquad \text{... (10.39)}$$

But $\sum_i dn_i = 0$, since $\sum_i n_i = N =$ constant.

and

$$dE = \sum_i E_i\, dn_i \qquad (\because \text{Eq. } 10.37)$$

Therefore, eqn. (10.39) becomes

$$dS = k\beta\, dE \qquad \text{... (10.40)}$$

Comparing (10.36) and (10.40), we have

$$k\beta = \frac{1}{T}$$

or

$$\beta = \frac{1}{kT} \qquad \text{... (10.41)}$$

Here k is a Boltzmann constant and T is absolute temperature.

Therefore, the number of molecules in a cell having energy E_i is given by

$$n_i = A\, e^{-E_i/kT}$$

Hence, the probability that a molecule will possess energy E_i will be

$$P_{(E_i)} = \frac{n_i}{N} = \frac{A e^{-E_i/kT}}{N}$$

where N is the total number of molecules.

$$\therefore \qquad P_{(E_i)} = C e^{-E_i/kT} \qquad \text{... (10.42)}$$

where $C = \dfrac{A}{N}$ and can be evaluated by using the fact that the probability of a molecule having energy E_i is definite, i.e.,

$$\Sigma P_{(E_i)} = 1$$

$\therefore$ Equation (10.42) gives

$$\Sigma P_{(E_i)} = C \Sigma e^{-E_i/kT}$$

or

$$1 = C \Sigma e^{\ E_i/kT}$$

or
$$C = \frac{1}{\sum e^{-E_i/kT}}$$

Substituting for C in equation (10.42), the probability $P_{(Ei)}$ is given by

$$P_{(E_i)} = \frac{e^{-E_i/kT}}{\sum e^{-E_i/kT}} \quad \text{... (10.43)}$$

10.22 APPLICATION OF BOLTZMANN'S CANONICAL DISTRIBUTION LAW

The most important application of Boltzmann's distribution law expressed in the form given by eqn. (10.43) is to evaluate the probability that a given molecule assembly of N molecules in thermal equilibrium that exist in a state of energy E_i. The molecule under consideration is referred as the '***system***' and remaining (N-1) molecules as the '***reservoir***' of temperature T. The equation (10.43)

$$P_{(E_i)} = \frac{e^{-E_i/kT}}{\sum e^{-E_i/kT}} \quad \text{... (10.44)}$$

is then applied to each one of the N molecules in turn by this procedure.

The Boltzmann's canonical distribution $P_i \propto \exp(-E_i/kT)$ plays a very crucial role while deriving the theorem of equipartition of energy (article 10.23).

If the system deviates and does not follow the equipartition theorem for particular energy terms which are known to be quadratic type, there is a 'break-down of classical-mechanics'. The assumption of parameters as continuously variable need to be modified. Thus, when the quantum steps (of energy) are much greater than kT, the equipartition theorem suffers total break-down. This is the field of 'quantum mechanics'.

10.23 THE EQUIPARTITION OF ENERGY

The theorem of equipartition of energy is usually stated as:

"For every degree of freedom in a system in thermal equilibrium at temperature T, the average energy of the system is $\frac{1}{2}kT$, where k is Boltzmann's constant."

For simplicity, let us consider a system of N monoatomic molecules. To specify the microstate of the system uniquely, we need 3N position coordinates and $3N$ momentum coordinates. Thus, in phase space representation, in all we need $6N$ coordinates. However, the energy E is determined by $3N$ momentum coordinates only. Thus

$$E = \sum_{i=1}^{3N} \frac{p_i^2}{2m} \quad \text{... (10.45)}$$

where p_i is the ith momentum coordinate. The number of such coordinate is $3N$, hence the number of degrees of freedom = $3N$

$$\therefore \quad \text{Total energy} = 3N \times \frac{1}{2} kT \quad \text{... (10.46)}$$

This is kinetic energy due to translational motion.

A system consisting of N harmonic oscillators about mean positions (in case of solid), we need $3N$ displacement coordinates and $3N$ momentum coordinates to specify the microstate uniquely. The total energy is comprised of potential energy and kinetic energy. Thus

$$\text{Total Energy} = E = \text{P.E.} + \text{k.E.}$$

$$= \sum_{i=1}^{N} \frac{p_i^2}{2m} + \sum_{i=1}^{N} \frac{1}{2} c_i\, q_i^2 \quad \text{... (10.47)}$$

where c_i is the ith force constant and q_i is the ith displacement coordinate. Thus, there are ($6N$) squared terms in the energy expression. The number of degrees of freedom is now to be considered as $6N$, so that

$$\text{Total energy} = 6N \times \frac{1}{2} kT \quad \text{... (10.48)}$$

In case of Van der Waals gas, where interaction energy terms are also present, is much more complex. Here the total energy is given by

$$E = \sum_{i=1}^{N} \frac{p_i^2}{2m} + \sum_a \sum_b \phi(r_{ab}) \quad \text{... (10.49)}$$

where r_{ab} is the distance between the ath and b^{th} molecule and $\phi\,(r_{ab})$ is a function of r_{ab}, which is not of the quadratic form. The number of these interaction terms is $\frac{1}{2} N\,(N-1)$. The average energy corresponding to these in **not** given by $\frac{1}{2} kT$. In this case the degrees of freedom is 3N but the total energy of the system is **not** given by $3N \times \frac{1}{2} kT$. Therefore, the theorem of equipartition of energy stated above is modified and restated as:

'For every independent quadratic term appearing in the energy expression of a system, the average energy corresponding to temperature T of the system is $\frac{1}{2} kT$.'

Again let us consider a monatomic gas with N molecules forming a system. We need f position coordinates and f momentum coordinates to specify the system uniquely. The total energy can be expressed as

$$E = E\,(q_1, q_2, \ldots\ldots\ldots\, q_f, p_1, p_2, \ldots\ldots\ldots\, p_f) \quad \text{... (10.50)}$$

Now imagine that the energy expression is so arranged that there is a term $E_i\,(p_i)$ which depends on the coordinate p_i alone and this term does not appear in any other energy term. Then we may write

$$E = E_i\,(p_i) + E'\,(q_1, q_2, \ldots\ldots\, q_f, p_1, p_2, \ldots\ldots\, p_{i-1}, p_{i+1} \ldots\ldots\, p_f) \quad \text{... (10.51)}$$

where E' is not a function of p_i. The mean value of E_i i.e., $\overline{E_i}$ may now be obtained by using formula (10.43) as

$$\overline{E_i} = \frac{\sum E_i\, e^{-\beta E}}{\sum e^{-\beta E}} \quad \text{where } \beta = \frac{1}{kT}$$

and the summation extends to all the microstates which pertains to total energy E.

If the q and p coordinates are treated as continuously variable, the summation may be changed into integration as

$$\overline{E_i} = \frac{\iint \ldots\ldots \int E_i - e^{-\beta(E_i+E')} dq_1\, dq_2 \ldots\ldots dq_f . dp_1\, dp_2 \ldots\ldots dp_f}{\iint \ldots\ldots \int e^{-\beta(E_i+E')} dq_1\, dq_2 \ldots\ldots dq_f . dp_1\, dp_2 \ldots\ldots dp_f}$$

since E_i is a function of p_i only and E' is not a function of p_i, we get

$$\overline{E_i} = \frac{\left[\int E_i e^{-\beta E_i} dp_i\right]}{\left[\int e^{-\beta E_i} dp_i\right]} \times \frac{\left[\iint \ldots\ldots \int e^{-\beta E'} dq_1\, dq_2 \ldots\ldots dq_f\, dp_1 \ldots\ldots dp_{i-1} dp_{i+1} \ldots\ldots dp_f\right]}{\left[\iint \ldots\ldots \int e^{-\beta E'} dq_1\, dq_2 \ldots\ldots dq_f\, dp_1 \ldots\ldots dp_{i-1}\, dp_{i+1} \ldots\ldots dp_f\right]}$$

$$= \frac{\int E_i\, e^{-\beta E_i} dp_i}{\int e^{-\beta E_i} dp_i} \quad \text{... (10.52)}$$

Thus, all variables other than p_i become irrelevant in deducing $\overline{E_i}$.

Now we introduce the statement that E_i is a quadratic function of p_i. If we put

$$E_i = bp_i^2$$

then using standard integrals, we have

$$\overline{E_i} = \frac{\int b\, p_i^2\, e^{-b\beta p_i^2}\, dp_i}{\int e^{-b\beta p_i^2}\, dp_i} = b \cdot \frac{1}{2b\beta}$$

or $$\overline{E_i} = \frac{1}{2\beta} = \frac{1}{2} kT \qquad \left[\text{Since } \beta = \frac{1}{kT}\right] \qquad \ldots (10.53)$$

Here the constant b also does not appear in $\overline{E}_i$. Equation (10.53) represents the equipartition theorem. Thus, the mean energy corresponding to each quadratic term in the energy expression depends on temperature of the system alone and is equal to $\frac{1}{2} kT$.

The meritorious point to be noted here is that we have considered the position and momentum coordinates as continuously variable. That is the fundamental basis of all classical mechanics. Moreover, we have not laid down any other specification about the system. Therefore, the theorem has a general validity, *viz.* whether the particles have mass large or small, the harmonic oscillator experiences force constant large or small so long as we consider quadratic energy terms and continuously variable coordinates. Again it should be noted that the Boltzmann's canonical distribution $p_i \; \alpha \; \exp\left(-\frac{E_i}{kT}\right)$ has a crucial role in deriving the above for formula.

10.24 STATISTICAL INTERPRETATION OF SECOND LAW OF THERMODYNAMICS

The second law of thermodynamics is a generalization of experiences and observations and is concerned with the direction in which energy transfer takes place. In practice, there is no engine which can convert heat from the single source in to useful work without rejecting some heat to a heat sink at a lower temperature. In other words, it is impossible to have perpetual motion of the second kind. This is the basis of the second law of thermodynamics.

In statistical mechanics, the basis of the second law is our experience that natural phenomenon must proceed from states of less probability to larger probability. Here, it is a general rule that more ordered states have smaller probability. As a result, we may enunciate that natural processes proceed from more ordered states to less ordered states. This is the statistical interpretation of second law of thermodynamics.

10.25 GIBB'S PARADOX

According to Boltzmann's relation, the entropy S of an ideal gas is related with the maximum thermodynamic probability W by the relation

$$S = k \log W \qquad \ldots (i)$$

From the Maxwell - Boltzmann count [Refer eqn 11.11]

$$\log W = \log N! + \sum_i^k (n_i \log g_i - \log n_i!)$$

$$\log W = \log N! + \sum_i n_i \log g_i - n_i \log n_i + n_i$$

$$\log W = \log N! + \sum_i^k n_i (\log g_i - \log n_i + 1)$$

$$= \sum_i^k n_i \left(\log \frac{g_i}{n_i} + 1\right) + \log N! \qquad \ldots (ii)$$

For maximum value of W or log W, we have

$$n_i = e^{-\alpha} g_i\, e^{\frac{-E_i}{kT}} \qquad \ldots (iii)$$

or
$$\frac{g_i}{n_i} = e^{\alpha+\left(\frac{E_i}{kT}\right)}$$

Taking logarithm of both sides, we have

$$\log \frac{g_i}{n_i} = \alpha+\left(\frac{E_i}{kT}\right)$$

$$= \frac{E_i}{kT}+\alpha \qquad \text{... (iv)}$$

Substituting this value in equation (*ii*), we get the maximum value log W as

$$\log W = \left[\sum_i n_i\left(\frac{E_i}{kT}+\alpha+1\right)\right]+\log N!$$

$$= \frac{1}{kT}\sum_i n_i E_i+\alpha\sum_i n_i+\sum_i n_i+\log N!$$

But
$$\sum_i n_i E_i = \text{total energy } E = \frac{3}{2}NkT$$

and
$$\sum_i n_i = N$$

$\therefore$
$$\log W = \frac{E}{kT}+\alpha N+N+\log N!$$

$$= \frac{1}{kT}\left(\frac{3}{2}NkT\right)+N+\alpha N+\log N!$$

$$= \frac{3}{2}N+N+\alpha N+\log N!$$

$$= \frac{5}{2}N+\alpha N+\log N! \qquad \text{... (iv)}$$

But
$$\alpha = -\log\left[\frac{N}{V}\left(\frac{h^2}{2\pi mkT}\right)^{\frac{3}{2}}\right] \qquad [\because 10.20\ iv]$$

Substituting, we get

$$\log W = \frac{5}{2}N - N\log\left[\frac{N}{V}\left(\frac{h^2}{2\pi mkT}\right)^{\frac{3}{2}}\right]+\log N! \qquad \text{... (v)}$$

Now substituting this value in eqn. (*i*), we get

$$S = \frac{5}{2}Nk - Nk\log\left[\frac{N}{V}\left(\frac{h^2}{2\pi mkT}\right)^{\frac{3}{2}}\right]+k\log N!$$

or
$$S = \frac{5}{2}Nk + Nk\log\left[\frac{N}{V}\left(\frac{2\pi mkT}{h^2}\right)^{\frac{3}{2}}\right]+k\log N! \qquad \text{...(10.54)}$$

When two different ideal gases (distinguishable) with number of particles N_1 in compartment A of volume V_1 and N_2 that in comportment B of volume V_2, both the gases are at same pressure and temperature and are separated by a partition (Fig. 10.11). When the partition is removed and gases are allowed to mix in a volume $V = V_1 + V_2$ at the same temperature, then since the process is irreversible, the entropy of the mixture should be greater than the sum of the entropies before mixing.

If the gases are of the same kind (indistinguishable), the change in entropy on mixing should be zero.

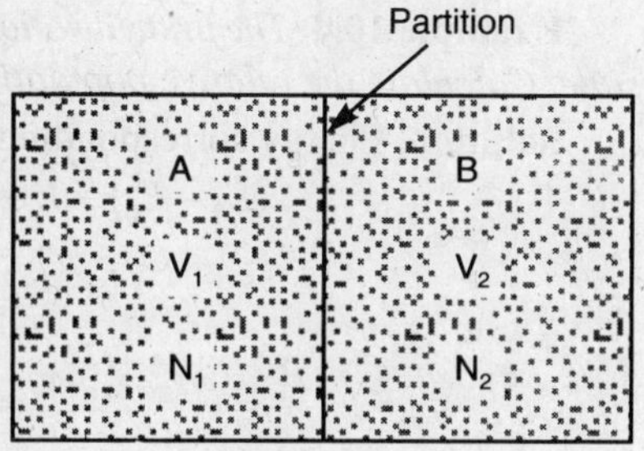

Fig. 10.11

If Eqn. (10.54) is applied to the process, then for mixing of two different gases it gives the correct result. But for mixing of two identical gases (indistinguishable), the equation gives increase in entropy. This is in contradiction. This contradiction or conclusion drawn from this equation is called the **Gibbs paradox.**

Removal of paradox: To resolve this paradox, the expression for the Maxwell-Boltzmann count must be divided by N!. In other words, we should subtract the term $k \log N!$ from equation (10.54), so that

$$S = \frac{5}{2} Nk + Nk \log \left[\frac{N}{V} \left(\frac{2\pi mkT}{h^2} \right)^{\frac{3}{2}} \right] \quad ... (10.55)$$

This has been verified experimentally for a gas even at high temperatures. It is known as **sackur-Tetrode equation.**

Eqn. (10.55) can be written in the form

$$S = Nk \log \left[\frac{V}{N} \left(\frac{2\pi mkT}{h^2} \right)^{\frac{3}{2}} . e^{\frac{5}{2}} \right] \quad ... (10.56)$$

where $e = 2.718$ is the base of Napierian logarithm.

SOLVED EXAMPLES

Example 10.3 *In a system in thermal equilibrium at temperature T, two states with energy difference 4.8×10^{-14} erg occur with relative probability e^2 erg deg^{-1}. Calculate the temperature. Given $k = 1.38 \times 10^{-16}$ erg. deg^{-1}*

Solution. The probabilities P_1 and P_2 of the state of energies E_1 and E_2 are given by

$$P_1 = C\, e^{-E_1/kT}$$

and

$$P_2 = C\, e^{-E_2/kT}$$

$$\therefore \quad \frac{P_1}{P_2} = \frac{C\, e^{-E_1/kT}}{C\, e^{-E_2/kT}}$$

or

$$\log_e \left(\frac{P_1}{P_2} \right) = \frac{E_2 - E_1}{kT}$$

or

$$T = \frac{E_2 - E_1}{k \log_e \left(\frac{P_1}{P_2} \right)}$$

Here $E_2 - E_1 = 4.8 \times 10^{-14}$ erg and $\frac{P_1}{P_2} = e^2$

$$T = \frac{4.8 \times 10^{-14}}{k \log_e (e^2)} = \frac{4.8 \times 10^{-14}}{k \times 2}$$

$$= \frac{4.8 \times 10^{-14}}{2 \times 1.38 \times 10^{-16}}$$

$$= \mathbf{173.913\ °K}$$

Example 10.4 *The first vibrational energy of a diatomic molecule is 600 cm^{-1} above the ground state. Calculate the relative population of molecules in these two levels at T = 400°K.*

Solution. Energy corresponding to 600 cm^{-1} is given by

$$\begin{aligned}(E_1 - E_0) = E &= h\nu = hc/\lambda = hc\,\bar{\nu} \\ &= 6.6 \times 10^{-27} \times 3 \times 10^{10} \times 600 \\ &= 1.2 \times 10^{-13} \text{ erg}\end{aligned}$$

$$\begin{aligned}\therefore \quad \frac{P_1}{P_2} = e^{(E_2 - E_1)/kT} &= e^{-(E_1 - E_2)/kT} \\ &= e^{-\left(1.2\times10^{-13}\right)/\left(1.38\times10^{-16}\times 400\right)} \\ &\cong \mathbf{0.1}\end{aligned}$$

Example 10.5 *Using statistical definition of entropy, show that the difference in entropy between a state of volume V_f and a state of volume V_i (temperature and number of molecules remaining constant) is equal to nR log $\frac{V_f}{V_i}$.*

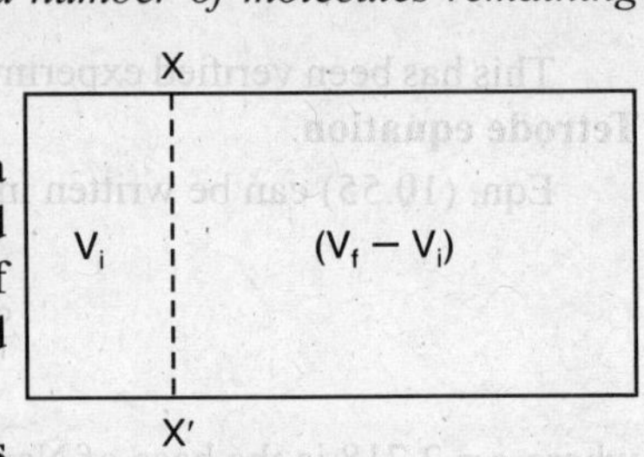

Fig. 10.12

Solution. Consider a rectangular-box of volume V_f; with a partition wall XX', which may be moved anywhere inside it, and the box gets divided into two compartments. If V_i is volume of one part, then remaining $(V_f - V_i)$ will be the volume of the second part (Fig. 10.12).

Let V_i represents volume in initial state, and V_f represents volume in the final state.

Imagine that there is a single molecule of an ideal gas present in the box. The probability of the molecule to be found in volume V_i at any given instant is obviously $\frac{V_i}{V_f}$.

If the box contains two molecules instead one, then the probability of finding both the molecules in the volume V_i would be $\left(\frac{V_i}{V_f}\right)^2$ and if there be p molecules in the box, the probability of finding all p molecules in volume V_i would be $\left(\frac{V_i}{V_f}\right)^P$.

Instead, suppose the box contains n gram-moles of a perfect gas, then probability of finding all the gas molecules in volume V_i will be

$$W_i = \left(\frac{V_i}{V_f}\right)^{nN}$$

where N is Avogadro's number. If now the partition is removed, the probability of finding all the molecules in volume V_f would be unity, *i.e.*, $W_f = 1$.

Therefore, dividing, we get

$$\frac{W_i}{W_f} = \left(\frac{V_i}{V_f}\right)^{nN}$$

Taking logarithm of both sides, we get

$$\log \frac{W_i}{W_f} = nN \log \frac{V_i}{V_f}$$

Multiplying both sides by Boltzmann constant k, we have

$$k \log \frac{W_i}{W_f} = nkN \log \frac{V_i}{V_f}$$

But $kN = R$, universal gas constant,

$$\therefore \quad k \log \frac{W_i}{W_f} = nR \log \frac{V_i}{V_f} \quad \text{... } (i)$$

If S_i and S_f represent the entropy of the gas in volume V_i and V_f respectively, then using the relation $S = k \log W$, we get

$$\Delta S = S_f - S_i = k \log W_f - k \log W_i \quad \text{(By definition of entropy S)}$$

$$= k \log \frac{W_f}{W_i}$$

$$\Delta S = nR \log \frac{V_f}{V_i} \quad [\because \text{ equation } (i)] \text{ ... } (ii)$$

This equation relates the change of entropy for the gas with the change in volume from V_i to V_f

Example 10.6. *Consider 100 molecules and 10 cells of equal energy. Find log Ω for (i) the most probable distribution, (ii) the lesast probable distribution.*

Solution. (*i*) For most probable dissipation,

$$N = 100$$

and

$$n_1 = n_2 = ... = n_{10} = \frac{100}{10} = 10$$

$$\therefore \quad \log \Omega = \log \frac{100!}{(10!)^{10}} = \log 100! - 10 \log 10!$$

$$= 100 \log_e 100 - 100 - 10\,[10 \log_e 10 - 10] \quad \text{...}(i)$$

By using Sterling theorem, $\log_e x! = [x \log_e x] - x$ solving (*i*), we get

$$\log \Omega = 100 \times 2.303 \times 2.00 - 100 - 10\,[10 \times 2.303 \times 1.00 - 10]$$

$$= 460 - 100 - 230 + 100$$

$$= \mathbf{230}$$

(*ii*) For least probable distribution,

$$N = 100,\ n_1 = 100 \text{ and } n_2 = n_3 = n_4 = ... n_{10} = 0$$

$$\therefore \quad \log \Omega = \log \frac{100!}{100!\,(0!)^9} = \frac{100!}{100!\,(1)^9} \quad (\because 0! = 1)$$

$$= \log 1$$

$$= \mathbf{0}$$

EXERCISE

10.1 Explain the terms

(*i*) mu-space. ***(Nagpur 2011/s, 2005)***

(*ii*) Γ-space. ***[Nagpur 2004 (s) 2005 (w) Gujarat 2005]***

10.2 Give two reasons for keeping the size of phase cell h_o^{Nf} small. In classical statistics h_o is an arbitrary choice, but in quantum statistics the corresponding constant h is a fixed natural constant. Why?

10.3 Calculate the number of phase cells in energy range 0 to E for a linear simple harmonic oscillator of mass m and frequency ν. ***(Meerut 2004, Agra 2002)***

10.4 For the phase space representing a single particle of mass m in a volume V, calculate the number of phase cells in energy range 0 to E, given that volume of each phase cell is h_o^3.

10.5 What is the minimum size of phase space in classical and Quantum Mechanics. ***(Purvanchal 2006, Punjab 2005)***

10.6 Explain the term phase space. ***(Purvanchal 2006, Nagpur 2009, Rohilkhand 2003, Agra 2004)***

10.7 Give the concept of ensemble and phase space. Calculate the number of states per unit volume of the phase space. ***(Purvanchal 2001, 2003, Rohilkhand 2000)***

10.8 Define phase space and find volume of a unit cell. ***(Magadh 2001, Purvanchal 2003, Nagpur 2009, 2005 Summer)***

10.9 Show that the volume of μ-space corresponding to a single quantum state for particles of no spin is h^3, where h is Planck's constant. ***(Nagpur 2004/w, 2004/s)***

10.10 Explain the statistical meaning of entropy. ***(Nagpur 2004(s), 2005/w)***

10.11 What do your mean by canonical ensemble? For what type of system is it suitable. ***(Lucknow 2000, Pune 2010)***

10.12 Explain concept of ensemble. Differentiate between microcanonical, canonical and grand-canonical ensembles. Where are they used? ***(Magadh 2002, Rohilkhand 2000)***

10.13 What do you understand by a priori probability and thermodynamical probability of distribution? ***(Meerut 2000)***

10.14 Give the statistical definition of entropy. Use this definition to obtain the entropy of a perfect gas. ***(Pbi. U. 2008, 2003, 2000, 1999; G.N.D.U. 2007, 2006, 2000, 1999 K.U. 2000)***

10.15 (*a*) What is the purpose of dividing phase space into cells ? ***(P.U. 2006, 2001, H.P.U. 2001; M.D.U. 2008, 2007, 2006, 2005, G.N.D.U. 2001)***

(*b*) How is the phase space related to elementary cell? ***(G.N.D.U. 2008)***

10.16 (*a*) Explain the terms position space, momentum space and phase space. Obtain relationship between volume in phase space and volume in position and momentum space. ***(H.P.U. 2001; P.U. 2006, 2003, 2001; Luck. U. 2002; K.U. 2000; M.D.U. 2008, 2007, 2006, 2005, 2000, G.N.D.U. 2006, 2002; Meerut. U. 2007)***

(*b*) What is dimensionality of phase space?

10.17 What do you understand by thermodynamical probability ? How is it related with the entropy of the system ? Establish the necessary relation.

10.18 Deduce Boltzmann's entropy probability relation

$$S = k \log \Omega\,(E)$$

where S is entropy, $\Omega\,(E)$ is the number of microstates in the energy interval between E and $E{+}dE$ and k is a Boltzmann's constant. ***(Meertu 2004, 2000, Rohilkhand 2000, 2003, Nagpur 2004/s, 2005/s, 2005/w, 2006/s, 2007 2008, Agra 2004, 2007, Purvanchal 2005, 2007)***

10.19 Define entropy and probability. Show that the entropy of the system is proportional to the logarithm of probability of that system. ***(Rohilkhand 2001 Improvement)***

10.20 What are the postulates of statistical mechanics. ***(Nagpur 2009, Purvanchal 2006)***

10.21 Derive the condition of equilibrium between two systems in thermal contact and explain how it plays a role of bridge with macroscopic physics. ***(Agra 2004, Nagpur 2004/s)***

10.22 Derive the condition $\beta_1 = \beta_2$ for equilibrium of two systems in thermal contact. ***(Meerut 2002, Agra 2000)***

10.23 What is Gibb's Paradox ? How this paradox is solved ? Discuss it by deriving necessary equations. Obtain Suckur Tetrode equation from it. ***(Magadh 2003, Gujrat 2004, 2005)***

10.24 Define thermodynamic probability and state its minimum value. ***(Nagpur 2005 summer)***

10.25 Obtain an expression for the number of microstates accessible to a free particle confined in a box of length L. ***(Nagpur 2005 winter)***

10.26 State and prove the Boltzmann's theorem between entropy and probability. ***(Rohilkhand 2001, Agra 2001, 2004)***

10.27 Explain the statistical interpretation of second law of thermodynamics.

10.28 Show that the number of molecules in a cell of energy E_i in the equilibrium state of temperature T is given by

$$n_i = Ae^{-E_i/kT}$$

where A is any constant.

10.29 Derive Boltzmann's canonical distribution law. From this calculate the probability for the system. ***(Agra 2008, Rohilkhand 2002, Meerut 2002, Gujarat 2005)***

10.30 Derive the relation for probability

$$P_{(Ei)} = \frac{e^{-E_i/kT}}{\sum e^{-E_i/kT}}$$ ***(Agra 2002)***

10.31 Discuss the applications of Boltzmann's canonical distribution law. ***(Meerut 2000, Agra 2000, Nagpur 2005)***

10.32 State and prove the theorem of equipartition of energy. Give merits of this theorem. Discuss the condition when the theorem of equipartition of energy fails to explain the classical mechanics. ***(Purvanchal 2007, Nagpur 2006, Gujarat 2005)***

10.33 What do you mean by breakdown of equipartition theorem? When it occurs?

10.34 Show that when two systems are placed in thermal contact

$$\beta(E) = \beta'(E) = \frac{1}{kT} = \frac{\partial\Omega(E)}{\Omega(E).\partial E}.$$ ***(Meerut 2002)***

where $\Omega(E)$ represents the number of microstates of energy E.

10.35 Show that in a general interaction between two systems, mechanical equilibrium is reached when the temperatures of the two systems become equal and the pressures of the two becomes equal. ***(Mumbai 2005)***

10.36 What is meant by available volume ? show that the phase space accessible to a system occupying a volume V and having a momentum range 0 to p_m is given by $4\pi/3p_m^3 V$. ***(G.N.D.U. 2003)***

10.37 Derive an expression for phase space volume. ***(M.D.U. 2008, 2007, 2006, 2005)***

10.38 (*a*) Give statistical definition of entropy and deduce Boltzmann's entropy probability relation $S = k \ln W$. What are the units of entropy in C.G.S. and S.I. system ? ***(M.D.U. 2001, 2000; P.U. 2007, 2006, 2005, 2003, 2002, 2001, 2000; Pbi. U. 2003, 2001; Meerut, U. 2000; K.U. 2000; Gauhati. U. 2000; G.N.D.U. 2008, 2007, 1993)***

10.39 Show that entropy increases when two gases at the same temperature and pressure diffuse into each other. Discuss Gibbs paradox in this connection. ***(Calcutta 2003)***

10.40 Obtain an expression for Maxwell-Boltzmann distribution $n_i = g_i Ae^{-\beta E_i}$. Evaluate the values of constants β and A. ***(Nagpur 2008)***

10.41 In M–B statistics, prove that the constant β is equal to $\frac{1}{kT}$ from thermodynamic considerations. ***(Nagpur 2010)***

10.42 Obtain partition function for canonical ensemble ? What is the effect of shifting the uro level of energy on the partition function ? ***(Magadh 2001)***

10.43 What is meant by microcanonical ensemble ? Establish Sackur-Tetrode formula for the entropy of a perfect gas. ***(Magade 2001)***

10.44 What do you understand by microcanonical ensemble and to what kind of systems does it apply ? Use the ensemble derive an expression for the entropy of classical idea gas. ***(Purvanchal 2006)***

10.45 Derive the relation $S = Nk \log_e Z + 3/2\, Nk$, where N = Avogadro's number, Z = Partition function, k = Boltzmann constant, S = entropy. ***(Purvanchal 2007)***

10.46 (*a*) What is order and disorder ? Entropy of a substance is a measure of disorder prevailing among its molecules. Discuss. Why does a natural system always tend to a change in the direction of increasing disorder ? ***(Pbi. U. 2003; P.U. 2008, 2007, 2006, 2003, 2001, 2000; H.P.U. 2003, 2001; K.U. 2000; Kan. U. 2000; G.N.D.U. 2008)***

(*b*) A crystal is an example of low entropy system. Why? Explain.

(*c*) What is statistical definition of enetropy and its physical significance. ***(H.P.U. 2003; P.U. 2006, Meerut. U. 2005; G.N.D.U. 2007, 2006)***

10.47 (*a*) Discuss the law of increase of entropy on the basis of statistical Physics. Discuss the applications of the law. ***(G.N.D.U. 2002; Nag. U. 2001; K.U. 2001, 2000; Kan. U., 2002,) P.U. 2001, 2000; H.P.U., Pbi. U. 2001)***

(*b*) Entropg increases during natural process Explain ***(Bang. U. 2007)***

10.48 Discuss the law of entropy and its importance in natural processes. ***(M.D.U. 2003)***

10.49 Starting from the statistical definition of entropy, show that when an infinitesimally small amount of heat δQ is added to a system, keeping its volume and number of particles constant, the change of entropy is $\dfrac{\delta Q}{T}$.
(P.U. 2007, 2001, K.U. 2002, 2000; Meertu. U. 2001, Pbi. U. 2007; H.P.U. 2003 G.N.D.U. 2006, 2002)

10.50 Discuss the concept of entropy and derive its relation with temperature. ***(M.D.U. 2003)***

10.51 (*a*) Entropy of universe always increases. Why ? ***(Meerut. U. 2005; P.U. 2000; Calicut. U. 2004; K.U. 2000)***

(*b*) Show that an increase in entropy is always followed by the loss of available energy. ***(Kerala. U. 2001; A.U., 2002)***

(*c*) Show that the process which tends to equalise temperature of the parts of a system increases entropy. ***(Kan. U. 1996)***

(*d*) Can entropy of a system increase when it loses heat ? ***(P.U. 2006)***

10.52 Write notes on the followings:

(*i*) Density of States. ***(Purvanchal 2006, 2000)***

(*ii*) Statistical ensembles. ***(Purvanchal 2002, Meerut 2000)***

(*iii*) Canonical Ensemble. ***(Rohilkhand 2001)***

(*iv*) Microcanonical Ensemble. ***(Purvanchal 2006)***

(*v*) Grand-Canonical Ensemble. ***(Magadh 2001)***

(*vi*) Partition function ***(Purvanchal 2006, 2002, 2000)***

(*vii*) Phase space. ***(Purvanchal 2004, Agra 2000, Rohilkhand 2000, 2002)***

(*viii*) Thermodynamic Probability. ***(Agra 2004)***

(*ix*) Entropy and Probability. ***(Agra 2000)***

(*x*) Fundamental postulates of statistical mechanics.

(*xi*) Statistics and Entropy.

(*xii*) Boltzmann's factor. ***(Agra 2002, Rohilkhand 2001)***

(*xiii*) Gibbs paradox. ***(Magadh 2001, Gujarat 2005)***

10.53 Consider 100 molecules and 10 cells of equal energy. Find $\log_e W$ for (*i*) the most probable distribution and (*ii*) the least probable distribution. **[Ans. (*i*) 230 (*ii*) 0]**

10.54 Given 3 cells and 5 particles. Calculate the thermodynamical probability for

(*i*) $n_1 = 5, n_2 = 0, n_3 = 0$

(*ii*) $n_1 = 4, n_2 = 1, n_3 = 0$

(*iii*) $n_1 = 3, n_2 = 2, n_3 = 0$ **[Ans. (*i*) = 1, (*ii*) 5, (*iii*) 10]**

10.55 The first vibrational energy level of a particular diatomic molecule is 600 cm^{-1}, above the ground state. Calculate the relative population of the molecules in these levels at 127°C.

[Ans. 0.1]

MULTIPLE CHOICE QUESTIONS

10.1 According to Boltzmann canonical distribution law, the number of molecules per cell

(*a*) increases linearly with energy associated with the cell,

(*b*) increases exponentially with energy associated with the cell,

(*c*) decreases linearly with energy associated with the cell,

(*d*) decreases exponentially with energy associated with the cell.

10.2 In the equilibrium state;

(*a*) probability is maximum,

(*b*) β parameters of two systems are equal,

(*c*) both (*a*) and (*b*),

(*d*) number of particles is maximum.

10.3 Five particles are distributed in two phase cells. Then number of macrostates is;

(*a*) 6 (*b*) 10

(*c*) 32 (*d*) $\frac{5}{2}$

10.4 According to Boltzmann canonical distribution law :

(*a*) low energy cells contain more particles,

(*b*) high energy cells contain more particles,

(*c*) number of molecules having zero energy is greater than any other energy,

(*d*) zero energy molecules are zero.

10.5 The thermodynamic probability of a system in equilibrium is

(*a*) maximum (*b*) minimum but not 1

(*c*) 1 (*d*) zero

10.6 Partition function is

(*a*) $Z = \sum e^{-\beta E_i}$ (*b*) $Z = \sum_i e^{-2\beta E_i}$

(*c*) $Z = \sum e^{-\beta/E_i}$ (*d*) none of these **(*Agra 2004*)**

ANSWERS

10.1 (*d*)	**10.2** (*c*)	**10.3** (*a*)	**10.4** (*a*)	**10.5** (*a*)	**10.6** (*a*)

MAXWELL-BOLTZMANN STATISTICS

11

Chapter

INTRODUCTION

On the basis of assumptions of the kinetic theory of gases and by applying the laws of probability, James Clerk Maxwell in 1859 derived the law of distribution of speeds among the molecules for an ideal gas in thermal equilibrium. Later, in 1872, Boltzmann derived his fundamental non-linear integro-differential relation (known as Boltzmann transport equation) from which he derived the velocity distribution function for the molecules of an ideal gas in thermal equilibrium. From this equation, speed and energy distributions among the molecules of ideal gas can be obtained easily. These distributions are known as Maxwell or Boltzmann or Maxwell-Boltzmann distribution, after the founders of the theory.

In this chapter, we will derive the laws of distribution of energy, speed and velocity among the gas molecules of an ideal gas, followed by the experimental evidences.

11.1 THREE KINDS OF PARTICLES

The assemblies consist of, in general, three kinds of particles :

1. *Identical but distinguishable particles without any spin*. The molecules of a gas are the particles of this kind. In quantum terms, the wave functions of the particles overlap to a negligible extent. The **Maxwell-Boltzmann distribution function** holds for such particles.
2. *Identical and indistinguishable particles of zero or integral spin*. They cannot be distinguished one from another because their wave functions overlap. Such particles are called **bosons** (refer next chapter), and do not obey the Pauli exclusion principle. The **Bose-Einstein distribution function** holds for them. The helium atoms at low temperature and the photons are the particles of this kind and we shall use Bose-Einstein statistics to account for the spectrum of radiation from a black body.
3. *Identical and indistinguishable particles with odd half-integral spin* $\left(\frac{1}{2}, \frac{3}{2}, \frac{5}{2}, ...\right)$. Their wave functions overlap on each other. Such particles are called **fermions** or **fermi particles**, obeying the Pauli exclusion principle. The **Fermi-Dirac distribution function** holds for them. Electrons, protons, neutrons etc. are the examples of this kinds. We shall use F.D. statistics to study the behaviour of the free electrons in a metal that are responsible for its ability to conduct electric current.

11.2 M.B. STATISTICS APPLICABLE TO IDEAL GAS

A system consisting of molecules of an ideal gas under ordinary conditions of temperature and pressure is governed by the laws of classical *Maxwell-Bcltzmann Statitics*. Here the gas molecules are treated as *distinguishable* particles. In the classical statistics, we can choose the size of the cell in the phase space as small as we please. We, therefore, take the size of the cell so small ($h \to 0$) that the total number of cells available are very large as compound to the number of particles, with the result that the probability of a cell containing more than one particle is almost negligible. In other words, *it means that a cell can contain either one or no particle.* In this way, every particle of the system will occupy a different numbered cell. It will, therefore, be possible to *distinguish* one molecule from the other by the number of cell which it occupies. However, there is *no restriction on the number of particles which can occupy a phase space cell.* This is because, *there is a little chance of two molecules occupying the same cell, the molecules of the gas can be treated as distinguishable particles.*

Statistical Equilibrium

Consider a gas having N molecules at a certain temperature and pressure. Suppose its volume, temperature and pressure are kept constant *i.e.*, it is a thermodynamically isolated system of N particles. The total energy U of this system remains constant. The energy states available to the particles are E_1, E_2, E_3, ... etc. These energy states may be *quantized* or may be *continuous* and are due to the vibrational and rotational energy of the particles.

Suppose, that at any given instant of time, n_1 particles are in state of energy E_1, n_2 particles in energy state E_2 and so on.

The total number of particles in the system is,

$$N = n_1 + n_2 + n_3 + \ldots = \sum_i n_i \qquad \ldots(i)$$

Here $\qquad i = 1, 2, 3, \ldots$ etc.

The total energy of the system,

$$U = n_1E_1 + n_2E_2 + n_3E_3 + \ldots$$

or

$$U = \sum_i n_iE_i \qquad \ldots(ii)$$

Equation (*ii*) refers to the total energy of a system in which the particles are *non-interacting.* Here the energy of each particle depends only on the coordinates of the particle in the system. For an **isolated system,** the total energy U is constant. Thus, for an isolated system.

$$U = \sum_i n_iE_i = \text{constant} \qquad \ldots(iii)$$

Since *the system is isolated*, the total energy of the system remains constant. But, the molecules of the gas collide with each other and also with the walls of the container. Consequently, the number of molecules change from one energy state to the other energy state. It means that the values of n_1, n_2, n_3 etc. continuously change. It can be reasonably assumed that for each microscopic state of a system of particles, there is a particular most favoured distribution. When this distribution or partition is reached, the system attains *statistical equilibrium.* For an isolated system, the values of n_1, n_2, n_3 etc. vary only near the value corresponding to the most probable distribution. Hence, the basic problem in statistical thermodynamics is to obtain the most probable distribution law for a given composition of the system.

11.3 MAXWELL-BOLTZMANN ENERGY DISTRIBUTION LAW

Suppose the particles are distributed among k energy levels E_1, E_2, ... E_i, ... E_k. Let n_1, n_2, ... n_i, ... n_k be the number of independent quantum states associated with the energy levels. According to the statistical basic postulates.

$$N = n_1 + n_2 + ... + n_i + n_k = \sum_i n_i \quad ... (11.1)$$

and since the force of interaction between particles are negligible, we have

$$U = n_1E_1 + n_2E_2 + ... + n_iE_i + ... + n_kE_k \quad ... (11.2)$$

Taking differential of equation (11.1) and (11.2), we have

$$dN = \sum_i dn_i = 0 \quad ... (11.3)$$

and

$$dU = \sum_i E_i dn_i = 0 \quad ... (11.4)$$

The set of occupation numbers n_1, n_2, ... is a possible mode of distribution of the particles among the energy levels, and it determines a macrostate of the system. According to the fundamental postulate of *equal a priori probabilities,* all the quantum states of energy of the particles in the system in equilibrium corresponding to the constant values of U, V and N are equally probable. It means that in the equilibrium state the distribution of the particles among various quantum states has the maximum probability of occurrence. This distribution can be obtained in a maximum number of statistically independent ways. Therefore, let us count (*i.e.*, determine) the total number of independent ways W of distributing the particles among the quantum states corresponding to the set of number n_1, n_2, ... and then determine the set of number of which W is maximum.

Maxwell-Boltzmann Statistical Count

Out of N total number of particles, n_1 are in energy level E_1, n_2 are in energy level E, and so on as shown in Fig. 11.1. For the first energy level E_1, we have to find the number of ways in which n_1 particles are chosen from among N distinguishable particles.

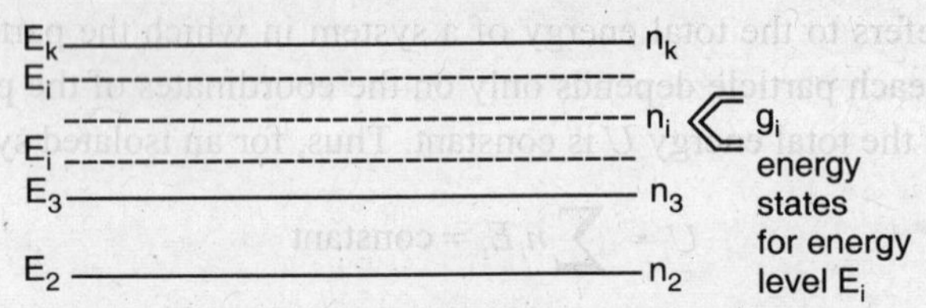

Fig. 11.1

The number of ways of choosing the first particle = N. When the first particle has been chosen, the number of ways of choosing the second particle = $N - 1$.

Similarly, the number of ways of choosing the third, fourth, ... n_1th particles are respectively $(N - 2)$, $(N - 3)$, ... $(N - n_1 + 1)$.

Therefore, the number of ways of choosing n_1 particles from N distinguishable particles is

$$= N(N-1)(N-2) ... (N - n_1 + 1)$$

$$= \frac{N(N-1)(N-2)...(N-n_1+1)(N-n_1)!}{(N-n_1)!}$$

$$= \frac{N!}{(N-n_1)!}$$

But this number also includes the $n_1!$ arrangement of n_1 particles among themselves*. This is because any sequence of the particles in the first energy level gives the same distribution of n_1 particles in the energy level, the number of independent (*i.e.*, meaningful) ways of choosing n_1 particles for the first energy level is, thus, given by

$$x_1 = \frac{N!}{n_1!(N-n_1)!} = {}^{N}C_{n_1}$$

For the second state E_2, only $(N - n_1)$ particles are available and n_2 particles are in state E_2. The number of independent ways of choosing n_2 particles for the second energy level is similarly given by

$$x_2 = \frac{(N-n_1)!}{n_2!(N-n_1-n_2)!} = {}^{N-n_1}C_{n_2}$$

Proceeding in the same way, the number of independent ways of choosing $n_3, n_4, \ldots n_i \ldots n_k$ particles for the third, fourth, ... i^{th} and k^{th} energy levels are given by

$$x_3 = \frac{(N-n_1-n_2)!}{n_3!(N-n_1-n_2-n_3)!} = {}^{(N-n_1-n_2)}C_{n_3}$$

$$\ldots\ldots\ldots \quad \ldots \quad \ldots \quad \ldots$$

$$\ldots\ldots\ldots \quad \ldots \quad \ldots \quad \ldots$$

$$x_k = \frac{(N-n_1-n_2\ldots n_{k-1})!}{n_k!(N-n_1-n_2\ldots n_k)!} = {}^{(N-n_1-n_2\ldots n_{k-1})}C_{n_k}$$

Since these numbers are independent, the number of distinct ways of obtaining the distribution of the particles among the energy levels is given by

$$x = x_1 x_2 x_3 \ldots x_i \ldots x_k$$

$$= \frac{N!}{n_1!n_2!n_3\ldots n_k!} \qquad \ldots (11.5)$$

Note that the denominator will contain $[N - (n_1 + n_2 + \ldots\ldots n_k)]! = 0! = 1$

Now our next job is to find the number of independent ways in which the particles in the various energy level can be distributed among the *quantum states* in the respective levels. For example, in the ith energy level, containing n_i particles. there are g_i quantum states. Each of the n_i particles can be in any one of the quantum states. Thus, there are g_i independent choices for the first particle to go in any

* To understand this, let us consider some finite number, of particles (say *a*, *b*, *c*, and *d*) in the first energy level E_1. Hence $n_1 = 4$. These four particles are *identical and distinguishable.* The identical particles refer to the particles having the same structure. The only distinction between the particles is due to their energy states at a given instant.

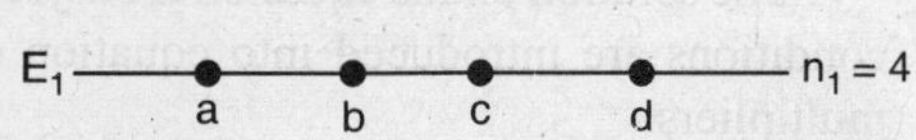

The first particle a in state E_1 can be chosen in N ways. The second particle *b* in state E_1 can be selected in (N – 1) ways and so on. Therefore, the total number of ways in which the first four particles in state E_1 can be selected is given by

$$N(N-1)(N-2)(N-3) = \frac{N!}{(N-4)!}$$

Moreover the four particles in state E_1 can be arranged in 4 different orders *i.e.*, in all 24 ways. For example: *abcd*, *bcda*, *cdab* and so on. But it is immaterial for these particles to be arranged in any particular order in state E_1, because they are all identical. Thus, the total number of *distinguishable* and meaningful different ways are obtained by dividing the above value by 4!. Hence, we have

$$x_1' = \frac{N!}{4!(N-4)!}$$

This can also be simplified by considering example 11.1

one of the quantum states. For each of these choices, these are g_i choices for the second particle and so on as there is no restriction of number of particles to be occupied in any quantum state, unlike to Pauli-exclusion principle in quantium statistics.

Hence there are g_i^{ni} ways of arranging n_i particles among g_i quantum states available in the *i*th level. Each of these arrangements is a separate independent way of distributing the particles among the quantum states. Thus, all the quantum states belonging to all the energy levels are considered, the total number of ways of arranging all the particles in the given distribution is given by

$$W = \frac{N!}{n_1!n_2!.....n_k!} \times g_1^{n_1} \times g_2^{n_2} \times ......g_k^{n_k}$$

$$= N!\prod_{i=1}^{k}\left[\frac{g_i^{n_i}}{n_i!}\right] \qquad ... (11.6)$$

where the symbol Π represents the product of the quantities $g_i^{ni}/n_i!$ for all values of *i* from 1 to *k*.

Condition for most Probable Distribution

The most probable distribution of the particles among the energy states in equilibrium is that for which the probability of occurrence is maximum, i.e. for which ***W is maximum***. For mathematical convenience, we consider the condition for maximum value of log *W*. The condition for maximum value of log *W* is

$$d\,(\log W) = 0 \qquad ... (11.7)$$

where *W* is a function of $n_1, n_2, ...n_k$.

Differentiating equation (11.6), we get

$$d \log (W) = \frac{\partial \log W}{\partial n_1} dn_1 + \frac{\partial \log W}{\partial n_2} dn_2 + ... + \frac{\partial \log W}{\partial n_k} dn_k$$

$$= \sum_{i=1}^{k} \frac{\partial \log W}{\partial n_i} dn_i$$

Applying the condition from eqn. (11.7), we get

$$\sum_{i=1}^{k} \frac{\partial \log W}{\partial n_i} dn_i = 0 \qquad ... (11.8)$$

The solution of this equation is subject to the condition laid down by eqn.(11.3) and (11.4). These conditions are introduced into equation (11.8) by using the method of Lagrange's undetermined multipliers.

Let α and β be these multipliers independent of n_i's. We multiply eq. (11.3) by $-\alpha$ and eq. (11.4) by -- β and adding these equations to eq. (11.8), we have

$$\sum_{i=1}^{k} \frac{\partial \log W}{\partial n_i} - \alpha - \beta E_i dn_i = 0 \qquad ...(11.9)$$

For maximum value of log *W,* eq. (11.9) must be true for the magnitudes of the individual increments dn_i. Hence, for all values of *i*, the term in the bracket must be zero.

$$\therefore \qquad \frac{\partial \log W}{\partial n_i} - \alpha - \beta E_i = 0 \qquad ... (11.10)$$

M-B Energy Distribution Law in General Form

Taking logarithms of Eq. (11.6), we have

$$\log W = \log N! + \log \left[\prod_i \left(\frac{g_i^{ni}}{n_i!}\right)\right]$$

$$= \log N! + \sum_{i=1}^{k} (n_i \log g_i - \log n_i!) \qquad ...(11.11)$$

Since the number of particles is very large, $N!$ and $n_i!$ may be approximated by Sterling's formula*, according to which

$$\log N! = N \log N - N \qquad ...(11.12\,(a))$$

and

$$\log n_i! = n_i \log n_i - n_i \qquad ...(11.12\,(b))$$

$$\therefore \quad \log W = N \log N - N - \sum_{i=1}^{k} (n_i \log g_i - n_i \log n_i + n_i) \qquad ...(11.13)$$

Differentiating Eq. (11.13) partially with respect to n_i and remembering that the partial derivatives of the terms, except the ith term, are zero. Thus,

$$\frac{d(\log W)}{dn_i} = \sum \left[\log g_i - \log n_i + \frac{n_i}{n_i} - 1 \right] \delta n_i$$

$$= \sum [\log g_i - \log n_i] \delta n_i$$

$$0 = \sum \left[-\log \left(\frac{n_i}{g_i} \right) \right] \delta n_i$$

Substituting this equation into (11.10), we have

$$\sum \left[-\log \left(\frac{n_i}{g_i} \right) - \alpha - \beta E_i \right] \delta n_i = 0 \quad \therefore -\log \left(\frac{n_i}{g_i} \right) - \alpha - \beta E_i$$

or

$$\log \left(\frac{n_i}{g_i} \right) = -\alpha - \beta E_i$$

or

$$n_i = g_i e^{(-\alpha - \beta E_i)}$$

or

$$n_i = g_i e^{-\alpha} e^{-\beta E_i}$$

$$(\text{where } i = 1, 2, 3, \ldots k) \qquad ...(11.15)$$

This equation is known as the **Maxwell – Boltzmann energy distribution law** in the general form.

The value of undetermined multiplier β is evaluated (refer article of chapter 10) as

$$\beta = \frac{1}{kT} \qquad ...(11.16)$$

Substituting in eqn. (11.15), we get

$$n_i = g_i e^{-\alpha} e^{-Ei/kT} \qquad ...(11.17)$$

where k is Boltzmann constant and T its absolute temperature. The quantity $e^{-Ei/kT}$ is known as the **Boltzmann factor.**

From the fundamental postulates of statistical mechanics, we write, N, the total number of particles as

$$N = \sum_i n_i$$

$$= \sum_i g_i e^{-\alpha} e^{-E_i/kT}$$

* Where x is large, according to Sterling's theorem, $\log_e x! = [x \log_e x] - x$

$$= e^{-\alpha} g_i e^{-E_i/kT}$$

$$\therefore \quad e^{-\alpha} = \frac{N}{\sum_i g_i e^{-E_i/kT}} \quad \text{... (11.18)}$$

Substituting this value in equation (11.17), we get

$$n_i = \frac{N g_i e^{-E_i/kT}}{\sum_i g_i e^{-E_i/kT}} \quad \text{... (11.19)}$$

The quantity $\sum_i g_i e^{-E_i/kT}$ is the sum over all the states of the system and is called the **partition function** of the system devoted by Z. Thus

$$Z = \sum_i g_i e^{-E_i/kT} \quad \text{... (11.20)}$$

Substituting in equation (11.19), we have

$$n_i = \frac{N}{Z} g_i e^{-E_i/kT} \quad \text{... (11.21)}$$

M. B. Energy Distribution Function

The general energy distribution function $f(E_i)$ is the average number of particles per quantum state in the ith energy level E_i. It is given by

$$f(E_i) = \frac{n_i}{g_i}$$

Substituting for $\frac{n_i}{g_i}$ from equation (11.17), we have

$$f(E_i) = e^{-\alpha} e^{-E_i/kT} \quad \text{... (11.22)}$$

Therefore, energy E, the M.B. energy distribution function may be expressed as

$$f(E_i) = e^{-\alpha} e^{-E/kT} \quad \text{... (11.23)}$$

M. B. Energy Distribution Function for an Ideal Gas

If the energy levels of the system are very close to each other, then the number of particles $n(E)\,dE$, whose energies lie in the range between E and $E + dE$ is given by

$$n(E)\,dE = f(E)\,g(E)\,dE \quad \text{... (11.24)}$$

where $g(E)\,dE$ represents the number of quantum states available between the energy E and $(E + dE)$. For an ideal gas, $e^{-\alpha}$ is given by (11.30) as

$$e^{-\alpha} = \frac{N}{V}\left(\frac{h^2}{2\pi mkT}\right)^{3/2} \quad \text{... (11.25)}$$

Substituting this value in eqn. (11.23), we get M.B. energy distribution function for an ideal gas as

$$f(E) = \frac{N}{V}\left(\frac{h^2}{2\pi mkT}\right)^{3/2} e^{-E/kT} \quad \text{... (11.26A)}$$

But for an ideal gas consisting of free particles with no spin $g(E)\,dE$ is given by (Example 10.1), as

$$g(E)\,dE = 2\pi V\left(\frac{2m}{h^2}\right)^{3/2} E^{\frac{1}{2}}\,dE \quad \text{... (11.26B)}$$

Substituting the value of $f(E)$ in equation (11.24), we get

$$n(E)\,dE = \frac{N}{V}\left(\frac{h^2}{2\pi mkT}\right)^{3/2} e^{-E/kT} \times 2\pi V\left(\frac{2m}{h^2}\right)^{3/2} E^{\frac{1}{2}}\,dE$$

or
$$n(E)\,dE = \frac{2\pi N}{(\pi kT)^{3/2}} E^{\frac{1}{2}} e^{-E/kT}\,dE \qquad \text{... (11.27A)}$$

or
$$n(E)\,dE = \frac{2N}{\sqrt{\pi}}\left(\frac{1}{kT}\right)^{3/2} E^{\frac{1}{2}} e^{-E/kT}\,dE \qquad \text{...(11.27B)}$$

This equation is the **Maxwell-Boltzmann energy distribution law** for the particles of an ideal gas.

The equation (11.27) gives the number of particles with energies between E and $E + dE$ of an ideal gas containing the total number of particles N (or molecules N) at absolute temperature T.

Evaluation of the Multiplier α.

From eqn. (11.18), we write

$$e^{-\alpha} = \frac{N}{\sum_i g_i\, e^{-Ei/kT}}$$

Since the variation of energy of free particles of an ideal gas is continuous, we will replace g_i by $g(E)dE$ and E_i by E and the sign of summation is replaced by the sign of integration. Thus, we get

$$e^{-\alpha} = \frac{N}{\int_0^\infty g(E)dE.\, e^{-E/kT}}$$

$$= \frac{N}{\int_0^\infty g(E)\, e^{-E/kT}\,dE} \qquad \text{... (11.28)}$$

Here the limit of integration is taken from 0 to ∞, because energy of the particles of an ideal gas is entirely kinetic and so they can have any kinetic energy. The value of $g(E)dE$ for particles with **no spin** is given by

$$g(E)dE = 2\pi V\left(\frac{2m}{h^2}\right)^{3/2} E^{\frac{1}{2}}\,dE$$

Substituting in eqn. (11.28), we get

$$e^{-\alpha} = \frac{N}{2\pi V\left(\frac{2m}{h^2}\right)^{3/2} \int_0^\infty E^{\frac{1}{2}} e^{\frac{-E}{kT}}\,dE} \qquad \text{... (11.29)}$$

Let us evaluate the definite integral as under :

For this, let $\frac{E}{kT} = x$, so that $dE = kTdx$

$$\therefore \quad \int_0^\infty E^{\frac{1}{2}} e^{-E/kT}\,dE = \int_0^\infty (kTx)^{\frac{1}{2}} e^{-x} kTdx$$

$$= (kT)^{3/2} \int_0^\infty x^{\frac{1}{2}} e^{-x}\,dx$$

$$= (kT)^{3/2} \int_0^\infty x^{\frac{3}{2}-1} e^{-x}\,dx$$

The value of the integral on the R.H.S. is the Γ (*i.e.* gamma) function,

$$\Gamma(3/2) = \frac{\sqrt{\pi}}{2}$$

$$\therefore \quad \int_0^\infty E^{\frac{1}{2}} e^{-E/kT} dE = (kT)^{\frac{3}{2}} \frac{\sqrt{\pi}}{2}$$

Substituting in eqn. (11.29), we get

$$e^{-\alpha} = \frac{N}{V\left(\frac{2\pi mkT}{h^2}\right)^{\frac{3}{2}}}$$

$$= \frac{N}{V}\left(\frac{h^2}{2\pi mkT}\right)^{\frac{3}{2}} \qquad \text{... (11.30)}$$

or
$$e^{\alpha} = \frac{V}{N}\left(\frac{2\pi mkT}{h^2}\right)^{\frac{3}{2}} \qquad \text{... (11.31)}$$

Taking logarithms of both sides, we get

$$\alpha = -\log\left[\frac{N}{V}\left(\frac{h^2}{2\pi mkT}\right)^{\frac{3}{2}}\right]$$

or
$$\alpha = \log\left[\frac{V}{N}\left(\frac{2\pi mkT}{h^2}\right)^{\frac{3}{2}}\right] \qquad \text{... (11.32)}$$

Here e^{α} is called the **degeneracy parameter.**

Condition for Application of M.B. Statistics.

The M.B. Statistics is applicable to a system of particles for which the mean distance between the particles is *greater than* the thermal De-Broglie wavelength of the particles.

The volume per particle = $\left(\frac{V}{N}\right)$

$\therefore$ The mean distance between the particles = $\left(\frac{V}{N}\right)^{\frac{1}{3}}$

The De-Broglie wavelength associated is given by

$$\lambda = \left(\frac{h^2}{2\pi mkT}\right)^{\frac{1}{2}}$$

Thus, the condition for M.B. statistics to became applicable is

$$\left(\frac{V}{N}\right)^{\frac{1}{3}} \geq \left(\frac{h^2}{2\pi mkT}\right)^{\frac{1}{2}}$$

or,
$$\left(\frac{V}{N}\right) \geq \left(\frac{h^2}{2\pi mkT}\right)^{\frac{3}{2}}$$

or, $$\frac{V}{N}\left(\frac{2\pi mkT}{h^2}\right)^{\frac{3}{2}} \geq 1$$

or, $$e^{\alpha} \geq 1 \qquad (\because \text{Eqn. 11.31}) \qquad \ldots (11.33)$$

Three cases :

Case (*i*) : When the degeneracy parameter e^{α} satisfies this condition, the gas is said to be **nondegenerate.** *i.e.*

$$e^{\alpha} \geq 1$$

Case (*ii*) : When $e^{\alpha} > 1$, but not too large, the gas is said to be **weakly degenerate.**

Case (*iii*) : When $e^{\alpha} < 1$, the gas is said to be **strongly degenerate.**

From inequality (11.33), we infer that the M.B. statistics is valid for systems at high temperature. *i.e.,* at low densities $\frac{N}{V}$ has a low value.

EVALUATION OF g_i

To evaluate the value of g_i, let us consider a *continuous distribution* of molecular energies and within this continuous distribution imagine a very small interval of energy lying between E and $(E+dE)$. As dE finally tends to zero, the energy of any molecule lying within this energy range can be taken to be equal to E. Now consider a unit interval of energy within mean value E *i.e.,* the interval between $(E - ½)$ and $(E + ½)$. If $n(E)$ denotes the number of molecules lying in this energy interval, then eqn. (11.15), the Maxwell-Boltzmann (general) energy distribution law becomes

$$n(E) = g(E)\, e^{-\alpha}\, e^{-\beta E} \qquad \ldots (11.34)$$

where $g(E)$ gives the number of cells in the phase space corresponding to *unit* energy interval with E as its mean value.

If M is the mass of the molecule and v its velocity, then its momentum $= p = mv$ and its energy $E = \frac{1}{2}mv^2$ or $v = \frac{p^2}{2m}$. Substituting this in equation (11.34), we have

$$n(p) = g(p)e^{-\alpha}e^{-\beta\left(\frac{p^2}{2m}\right)} \qquad \ldots (11.35)$$

where $n(p)$ gives the number of molecules in *unit momentum interval* between $(p - 1/2)$ and $(p + 1/2)$ and $g(p)$ gives the number of cells in the phase space corresponding to this *unit momentum interval.*

The number of molecules in the momentum interval p and $(p + dp)$ is obtained by multiplying (11.35) by dp,

$$n(p)dp = g(p)\, e^{-\alpha}\, e^{-\beta\frac{p^2}{2m}} dp \qquad \ldots (11.36)$$

Here, $g(p)dp$ represents the number of cells in the phase space corresponding to the momentum interval between p and $(p + dp)$. If h^3 is the volume of a single cell in the phase space, then

$$g(p)dp = \frac{\text{Volume in phase space corresponding to momentum interval between } p \text{ and } (p+dp)}{h^3}$$

$$= \frac{\iiint dxdydz \iiint dp_x\, dp_y\, dp_z}{h^3}$$

But $\iiint dx\, dy\, dz = V$, the volume of the system in ordinary *position space i.e.* the volume occupied by gas molecules.

$$\therefore \qquad g(p)dp = \frac{V}{h^3}\iiint dp_x\, dp_y\, dp_z \qquad \text{... (11.37)}$$

The quantity $\iiint dp_x\, dp_y\, dp_z$ represents the volume of the system in *momentum space* corresponding to momentum interval between p and $(p+dp)$.

If we draw a sphere with the origin 0 as centre and radius equal to p, then all the points lying on this sphere will have a momentum value p, given by $p^2 = p_x^2 + p_y^2 + p_z^2$. If we draw another concentric sphere of radius $(p + dp)$, then all the points lying on this sphere will have momentum value $p+dp$ as shown in (Fig. 11.2).

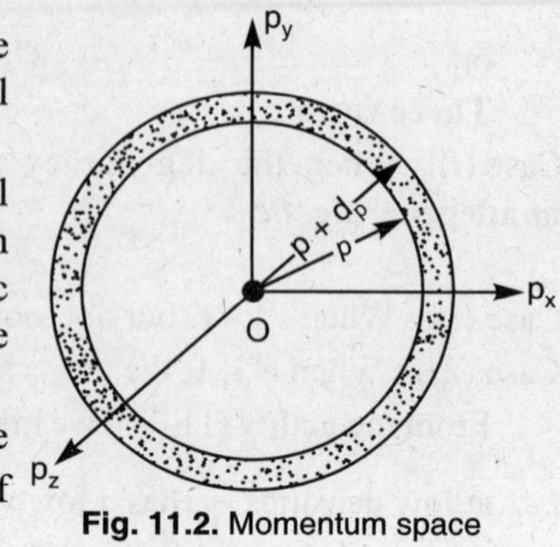

Fig. 11.2. Momentum space

The volume in between these two concentric spheres will be = (Surface area of the inner sphere of radius p) × thickness of the shell.

$$\therefore \qquad \iiint dp_x\, dp_y\, dp_z = 4\pi p^2\, dp \text{ [See details in eqn. (vi) of 11.4]} \qquad \text{...(11.38)}$$

Substituting in eqn. (11.37), we have

$$g(p)dp = \frac{V}{h^3}.\, 4\pi p^2\, dp \qquad \text{... (11.39)}$$

(Note that here $g(p)$ corresponds to g_i).

Substituting eqn. (11.39) in eqn. (11.36), we have

$$n(p)dp = \frac{4\pi V p^2}{h^3}\; e^{-\alpha}\; e^{-\beta p^2/2m}\, dp \qquad \text{... (11.40)}$$

Substituting the value of $\beta = \frac{1}{kT}$ in eqn. (11.40), we get

$$n(p)dp = \frac{4\pi V p^2}{h^3}.\frac{N}{V}\left(\frac{h^2}{2\pi mkT}\right)^{\frac{3}{2}}.\; e^{-p^2/2mkT}\, dp$$

$$\text{or} \qquad n(p)dp = 4\pi N\left(\frac{1}{2\pi mkT}\right)^{\frac{3}{2}}.\; p^2.\, e^{-p^2/2mkT}\, dp \qquad \text{... (11.41)}$$

This relation gives the number of molecules having momentum between p and $(p + dp)$ and is known as **Maxwell-Boltzmann law of distribution of momentum.**

11.4 APPLICATIONS OF M.B. DISTRIBUTION LAW

1. Total internal energy and specific heat at constant volume of an ideal gas.

The total internal energy U of an ideal gas is given by

$$U = \int_0^\infty E\; n(E)\, dE \qquad \text{... (11.42)}$$

$$= \int_0^\infty E\left[\frac{2\pi N}{(\pi kT)^{\frac{3}{2}}}\, E^{\frac{1}{2}}\, e^{-E/kT}\right] dE$$

$$= \frac{2\pi N}{(\pi kT)^{\frac{3}{2}}}\int_0^\infty E^{\frac{3}{2}}\, e^{-E/kT}\, dE$$

Suppose $\frac{E}{kT} = x$, so that $E = kTx$ and $dE = kTdx$

$$\therefore \quad U = \frac{2\pi N}{(\pi kT)^{\frac{3}{2}}} \int_0^\infty (kTx)^{\frac{3}{2}} e^{-x}\, kTdx$$

$$= \frac{2\pi N}{(\pi kT)^{\frac{3}{2}}} (kT)^{\frac{3}{2}}\, kT \int_0^\infty x^{\frac{3}{2}} e^{-x}\, dx$$

$$= \frac{2\pi NkT}{\pi\sqrt{\pi}} \int_0^\infty x^{\frac{5}{2}-1} e^{-x}\, dx$$

The integral is the gamma function

$$\Gamma\frac{5}{2} = \frac{3}{2}\Gamma\frac{3}{2} = \frac{3}{2}.\frac{1}{2}\sqrt{\pi} = \frac{3\sqrt{\pi}}{4}$$

Substituting the value of the integral, we have

$$U = \frac{2NkT}{\sqrt{\pi}}.\frac{3\sqrt{\pi}}{4}$$

$$U = \frac{3}{2}NkT. \qquad \text{... (11.43)}$$

Therefore, internal energy per molecule *i.e.* average internal energy $\overline{U}$ is given by

$$\overline{U} = \frac{U}{N} = \frac{3}{2}kT \qquad \text{... (11.44)}$$

Thus the average internal energy of an ideal monatomic gas depends only on its temperature, as k is constant.

For 1 mole of an ideal gas, the quantity N in equation (11.43) is nothing but Avogadro's number and hence, the value of U given by equation (11.43) is the total **internal energy of 1 mole of an ideal gas.**

The molar specific heat of the gas at constant volume is given by

$$C_V = \left(\frac{\partial U}{\partial T}\right)_V$$

$$C_V = \frac{\partial}{\partial T}\left(\frac{3}{2}NkT\right) \qquad \text{... } (\because \text{Eqn. 11.43})$$

$$= \frac{3}{2}Nk \qquad \text{... (11.45)}$$

or, $$C_V = \frac{3}{2}N\left(\frac{R}{N}\right) \qquad \left(\because k = \frac{R}{N}\right)$$

or, $$C_V = \frac{3}{2}R \qquad \text{... (11.46)}$$

where ***R* is** the universal gas constant for 1 mole.

2. Maxwell - Boltzmann Speed Distribution Law

The Maxwell-Boltzmann equation for distribution of energy among the molecules of an ideal gas is given by

$$n(E)\,dE = \frac{2\pi N}{(\pi kT)^{\frac{3}{2}}} E^{\frac{1}{2}} e^{-E/kT} dE \qquad \text{... }(i)$$

A classical ideal gas is defined as an assembly of non-interacting molecules, each distinguishable from one another. Therefore, the molecules have no internal degrees of freedom, all the energy of the gas molecules is completely kinetic in nature. Thus,

$$E = \frac{1}{2}mv^2, \text{ Hence } dE = mvdv$$

Substituting these values of E and dE in equation (*i*), the number of molecules $n(v)dv$ whose speeds lie between v and $(v + dv)$ is given by

$$n(v)dv = \frac{2\pi N}{(\pi kT)^{\frac{3}{2}}}.\left(\frac{1}{2}mv^2\right)^{\frac{1}{2}} e^{-mv^2/2kT}\; mvdv$$

$$= \frac{2\pi N}{(\pi kT)^{\frac{3}{2}}}.2\left(\frac{m}{2}\right)^{\frac{3}{2}} v^2 e^{-mv^2/2kT}\; dv$$

or,

$$n(v)dv = 4\pi N\left(\frac{m}{2\pi kT}\right)^{\frac{3}{2}} v^2 e^{-mv^2/kT}\; dv \qquad \text{... (11.47)}$$

This equation is known as the **Maxwell or Maxwell-Boltzmann law of distribution of speeds** among the molecules of a gas. In this $n(v)$ is the number of molecules per unit speed range.

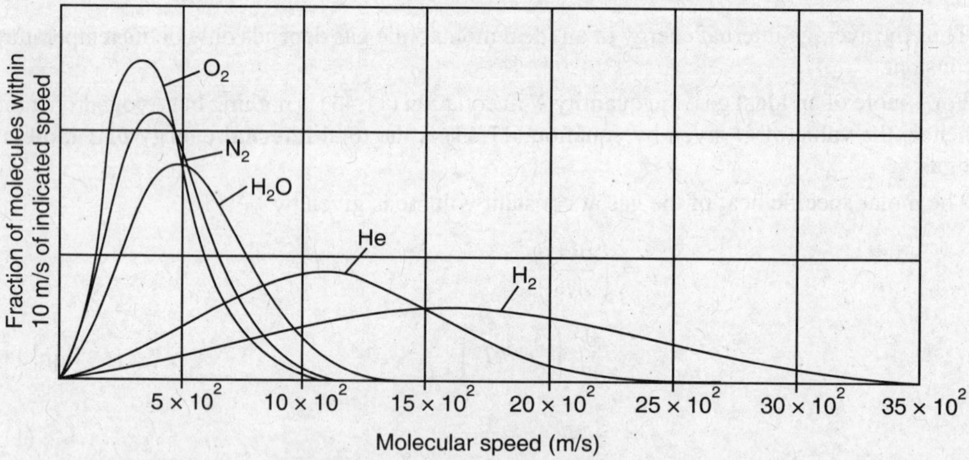

Fig. 11.3

Aliter : Let us consider an ideal gas in a vessel of volume V. In order to derive an expression for the number of molecules of the gas whose speeds lie between v and $v + dv$.

For this, let us consider a zone of cells, *i.e.* a group of cells in the *phase space* corresponding to the range in speed from v to $v + dv$. If g_i represents the number of cells in the zone, then the number of molecules in the *i*th zone, each having energy close to $E_i = \frac{1}{2}mv^2$, is given by Maxwell-Boltzmann law as

$$n_i = Ag_i e^{-\beta E_i} \qquad \text{(From eqn. 10.31)} \qquad \text{... }(i)$$

where $\beta = \dfrac{1}{kT}$ and $A = \dfrac{Z}{N}$

Here k is Boltzmann constant and T the absolute temperature.

As the number of molecules having speed ranging between v and $v + dv$ will be proportional to dv, we may write

$$n_i = N(v)dv \qquad \text{... (ii)}$$

where $N(v)$ is known as the *distribution function* pertaining to speeds. Substituting for n_i in equation (*i*), we get

$$N(v)dv = Ag_i\, e^{-E_i/kT} \qquad \text{... (iii)}$$

To evaluate g_i

Let us now express g_i in terms of v and dv. We have

$$g_i = \text{Number of cells in the zone under consideration.}$$

$$= \frac{\text{Volume of the zone in the phase space}}{\text{Volume of one cell}}$$

i.e.,

$$g_i = \frac{\iiint\iiint dx\, dy\, dz\, dp_x\, dp_y\, dp_z}{h^3} \qquad \text{... (iv)}$$

If V is the volume occupied by the gas; then it is

$$V = \iiint dx\, dy\, dz \qquad \text{... (v)}$$

Now we have to integrate with respect to p_x, p_y and p_z. Here we will consider only the resultant momentum p and do not take care of the momentum components. For this let us imagine that p_x, p_y and p_z are along three mutually perpendicular axes and draw a sphere of radius p about the origin, in the momentum space (refer Fig. 11.4). Then the surface of the sphere will contain all the points for which $p_x^2 + p_y^2 + p_z^2 = p^2$. Now draw another concentric sphere of radius $(p + dp)$. The volume between these two concentric spheres of radius p and $p + dp$ will contain space having momentum between p and $p + dp$.

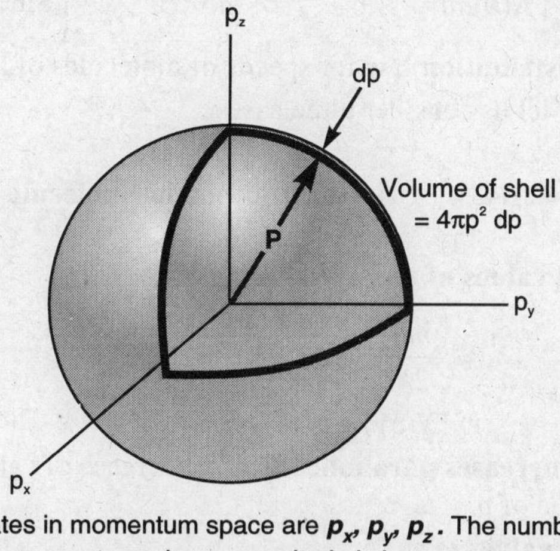

Fig. 11.4. The coordinates in momentum space are p_x, p_y, p_z. The number of momentum states available to a particle with a momentum whose magnitude is between p and $p + dp$ is proportional to the volume of a spherical shell in momentum space of radius p and thickness dp.

The volume between two concentric spheres will be

$$\iiint dp_x\, dp_y\, dp_z = \left(\frac{4}{3}\right)\pi\,(p + dp)^3 - \left(\frac{4}{3}\right)\pi\, p^3$$

$$= \left(\frac{4}{3}\right)\pi\left[(p+dp)^3 - p^3\right]$$

$$= \left(\frac{4}{3}\right)\pi\left[p^3 + 3p^2dp + 3p(dp)^2 + dp^3 - p^3\right]$$

$$= \left(\frac{4}{3}\right)\pi\left(3p^2dp\right) \text{ neglecting higher terms } dp^2 \text{ and } dp^3.$$

$$= 4\pi p^2 dp \qquad \text{... (vi)}$$

Now substituting eqn. (*v*) and (*vi*) in (*iv*), we get

$$g_i = \frac{V\, 4\pi p^2 dp}{h^3}$$

or,
$$g_i = \frac{4\pi\, Vp^2 dp}{h^3} \qquad \text{... (vii)}$$

Substituting for g_i in eqn. (*iii*), and using $E_i = \frac{1}{2}mv^2$, we get

$$N(v)dv = \frac{4\pi\, AV}{h^3} p^2\, e^{-\beta mv^2/2} dp$$

putting
$$p = mv \text{ and } dp = mdv$$

$$N(v)dv = \frac{4\pi\, AV\, m^3}{h^3} v^2\, e^{-\beta mv^2/2} dv$$

We may put $\frac{4\pi AVm^3}{h^3}$ = B, where B is any constant, we have

$$N(v)dv = Bv^2\, e^{-\beta mv^2/2}\, dv$$

or,
$$N(v)dv = Bv^2\, e^{-mv^2/2kT}\, dv \qquad \left(\text{Since } \beta = \frac{1}{kT}\right) \qquad \text{... (11.48)}$$

This is **Maxwell's distribution law** for speeds of molecules of a perfect gas.

Discussion. For this, let us consider three cases :

Case (*a*) : If $v = 0$

Then from equation (11.48), $N(v) = 0$. It means **no molecule has a zero speed.** This is an important conclusion.

Case (*b*) : For small values of v

In this case,
$$\frac{mv^2}{2kT} \ll 1 \text{ and } e^{\frac{-mv^2}{2kT}} = 1$$

$$\therefore \qquad N(v) \propto v^2$$

Thus, value of $N(v)$ **increases parabolically** with increase of value of v and attains a maximum value for a particular value of v.

Case (c) : For large values of v

In this case, $\frac{mv^2}{2kT} >> 1$ and the exponential term $e^{-mv^2/2kT}$ dominates over v^2. Hence

$$N(v) \propto e^{-mv^2/2kT}$$

Thus, value of $N(v)$ **decreases exponentially** with increase of value of v.

This behaviour is explained by Fig. 11.5. If a graph is plotted $N(v)$ against v, the curve first rises parabolically, reaches a maximum and then falls exponentially to zero as shown in Fig. 11.5.

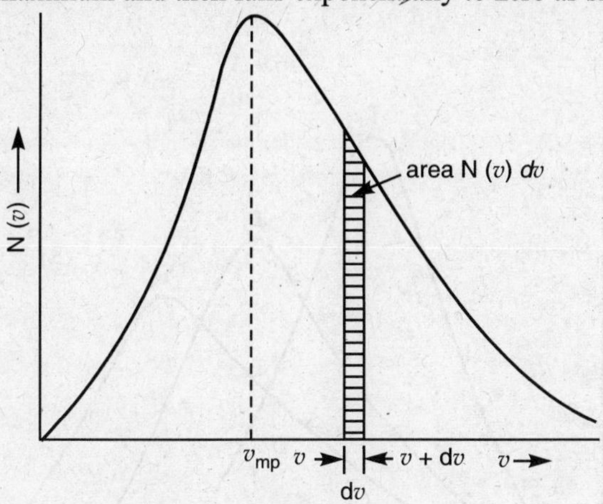

Fig. 11.5.

The maximum corresponds to the most probable speed v_{mp}. As the area under the curve towards right of maximum is greater than to the left, it means that the mean speed v is greater than v_{mp}. The area of the shaded portion gives the number of molecules lying in the speed range v to $v + dv$.

Evaluation of constant B

The quantity $N(v)\,dv$ represents the number of gas molecules having speed in the range of v to $(v + dv)$ and is given by equation (11.48). In fact, the molecules may have speed ranging from 0 to ∞, hence we may write

$$\int N(v)dv = N$$

where N is total number of molecules in the given state.

Thus,

$$\int Bv^2 e^{-\frac{mv^2}{2kT}}\, dv = N$$

or
$$B\int v^2\, e^{-\frac{mv^2}{2kT}}\, dv = N$$

or
$$B\left[\frac{1}{4}\sqrt{\frac{8\pi k^3T^3}{m^3}}\right] = N$$

or
$$B = 4N\sqrt{\frac{m^3}{8\pi k^3T^3}}$$

$$= \frac{4N}{\sqrt{\pi}}\left(\frac{m}{2kT}\right)^{3/2} = 4\pi N\left(\frac{m}{2\pi kT}\right)^{3/2} \quad \text{... (11.49)}$$

Substituting for B in equation (11.48), the **Maxwell's distribution law of speed** becomes

$$N(v)dv = 4\pi N\left(\frac{m}{2\pi kT}\right)^{3/2} v^2\, e^{-\frac{mv^2}{2kT}}\, dv \quad \text{... (11.50)}$$

or
$$N(v) = 4\pi N\left(\frac{m}{2\pi kT}\right)^{3/2} v^2\, e^{-\frac{mv^2}{2kT}} \quad \text{... (11.51)}$$

The effect of temperature on the distribution of speed of a given number of molecules can be understood by using this relation. The graphs $N(v)$ against v are plotted using equation (11.51) at different temperatures.

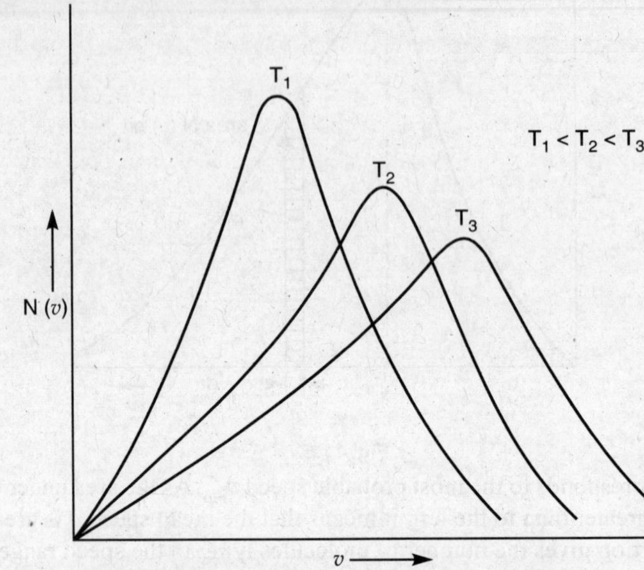

Fig. 11.6. Distribution function at different temperatures

Temperatures T_1, T_2 and T_3 such that $T_1 < T_2 < T_3$ are drawn in Fig. 11.6. From the graphs, we conclude that higher the temperature, higher are the speeds possessed by the gas molecules. However, the area under all the three curves is the same.

The fraction of molecules having speed ranging between v and $v + dv$ is given by

$$\frac{N(v)dv}{N} = 4\pi\left(\frac{m}{2\pi kT}\right)^{3/2} v^2 e^{-\frac{mv^2}{2kT}} dv \quad \text{... (11.52)}$$

But according to the definition of probability, this ratio represents the probability of molecules to possess the speed between v and $(v + dv)$.

$$\therefore \quad P(v)dv = 4\pi\left(\frac{m}{2\pi kT}\right)^{3/2} v^2 e^{-\frac{mv^2}{2kT}} dv$$

The probability function $P(v)$ is given by

$$P(v) = 4\pi\left(\frac{m}{2\pi kT}\right)^{3/2} v^2 e^{-\frac{mv^2}{2kT}} \quad \text{... (11.53)}$$

11.5 MEAN, RMS AND MOST PROBABLE SPEEDS

Mean or Average Speed $(\overline{v})$:

Let there be N_i molecules each having speed v_i, then mean or average speed $(\overline{v})$ is defined as

$$(\overline{v}) = \frac{\sum_i N_i v_i}{\sum_i N_i}$$

But $\sum_i N_i = N$, the total number of molecules,

$$\therefore \quad \overline{v} = \frac{1}{N}\sum N_i v_i$$

As the speed of molecules is approximately continuous, we may replace summation sign by integration, *i.e.*,

$$\bar{v} = \frac{1}{N}\int_0^\infty v\, N(v)\, dv$$

$$= \frac{1}{N}\int_0^\infty v\,.\,4\pi N\left(\frac{m}{2\pi kT}\right)^{3/2}.\, v^2\,.\, e^{-\frac{mv^2}{2kT}}\, dv$$

$$= \int_0^\infty 4\pi\left(\frac{m}{2\pi kT}\right)^{3/2} v^3\, e^{-\frac{mv^2}{2kT}}\, dv$$

$$= 4\pi\left(\frac{m}{2\pi kT}\right)^{3/2}\int_0^\infty v^3\, e^{-\frac{mv^2}{2kT}}\, dv$$

$$= 4\pi\left(\frac{m}{2\pi kT}\right)^{3/2}\left[\frac{2(kT)^2}{m^2}\right]$$

$$\bar{v} = \sqrt{\frac{8kT}{m\pi}} = 1.59\sqrt{\frac{kT}{m}} \qquad \text{... (11.54)}$$

This equation gives the mean or average speed of a gas molecule.

Root Mean Square Speed (v_{rms})

The mean square speed $\left(\overline{v^2}\right)$ is given by

$$\left(\overline{v^2}\right) = \frac{1}{N}\int_0^\infty v^2 N(v)\, dv$$

$$= \frac{1}{N}\int_0^\infty v^2\,.\,4\pi N\left(\frac{m}{2\pi kT}\right)^{3/2} v^2\, e^{-\frac{mv^2}{2kT}}\, dv$$

$$= 4\pi\left(\frac{m}{2\pi kT}\right)^{3/2}\int_0^\infty v^4\, e^{-\frac{mv^2}{2kT}}\, dv$$

$$= 4\pi\left(\frac{m}{2\pi kT}\right)^{3/2}\left[\frac{3}{8}\sqrt{\left\{\frac{32\pi\,(kT)^5}{m^5}\right\}}\right]$$

$$= \frac{3kT}{m}$$

Taking square root of both sides,

$$v_{rms} = \sqrt{\left(\overline{v^2}\right)} = \sqrt{\frac{3kT}{m}} \qquad \text{... (11.55)}$$

The quantity $\sqrt{\left(\overline{v^2}\right)}$ is called **root mean square speed** of the gas molecules.

Most Probable Speed (v_{mp})

For the most probable speed the probability $P(v)$ will be maximum. Therefore, to obtain the most probable speed, let us differentiate equation (11.53) with respect to v and equate it to zero. Thus

$$4\pi\left(\frac{m}{2\pi kT}\right)^{3/2}\left[2v\, e^{\left(-\frac{mv^2}{2kT}\right)} - \frac{mv^3}{kT}\, e^{\left(-\frac{mv^2}{2kT}\right)}\right] = 0$$

$$2v\, e^{-\frac{mv^2}{2kT}} = \frac{mv^3}{kT} e^{-\frac{mv^2}{2kT}}$$

or
$$2v = \frac{mv^3}{kT}$$

or
$$v^2 = \frac{2kT}{m}$$

or
$$v = \sqrt{\frac{2kT}{m}}$$

This v is the most probable speed, therefore putting it equal to v_{mp}, we have

$$v_{mp} = \sqrt{\frac{2kT}{m}} \qquad \text{... (11.56)}$$

Thus, from equations (11.54), (11.55) and (11.56), it is clear that $v_{rms} > \bar{v} > v_{mp}$.

This ratio of two speeds $\frac{V_{rms}}{V_{mp}} = \sqrt{\frac{3kT}{m}} \times \sqrt{\frac{m}{2kT}} = \sqrt{\frac{3}{2}}$.

11.6 MAXWELL'S DISTRIBUTION LAW OF VELOCITIES

Assumptions for velocity distribution of gas

To derive the velocity distribution of gas molecules we use Maxwell Boltzmann statistics. The main assumptions of Maxwell Boltzmann statistics are:

(*i*) The particles of the system are *identical* and *distinguishable.*

(*ii*) The available *volume* of the phase space cell can be *as small as we please* and may even approach zero.

(*iii*) The phase space can be divided into a *very large number of cells.*

(*iv*) The state of each particle is specified by giving its *instantaneous position and momentum* co-ordinates or by giving its *cell number* in phase space

(*v*) There is *no restriction on the number of particles* which can occupy a phase space cell.

(*vi*) The system of particles contained in a vessel form an **isolated system** *i.e.,* the system satisfies two conditions.

(*a*) *The total number of particles in the system remains constant* and

(*b*) *The total energy of the system remains constant.*

Let us consider an ideal gas contained in a container of volume *V*. Then applying the Maxwell-Boltzmann's canonical distribution law, in equilibrium, the number of molecules in a cell of energy E_i will be

$$n_i = A\, e^{-\beta E_i}$$

The number of molecules having position coordinates in the range x to $x + dx$, y to $y + dy$, z to $z + dz$ and the velocity components in ranges v_x to $v_x + dv_x$, v_y to $v_y + dv_y$, v_z to $v_z + dv_z$ will be proportional to the element of phase of volume $dx\, dy\, dz\, dv_x\, dv_y\, dv_z$.

Therefore, according to Maxwell-Boltzmann's canonical distribution law, the number of molecules having energy E_i and having position coordinates between x and $x + dx$, y and $y + dy$, z and $z + dz$ and velocity components between v_x and $v_x + dv_x$, v_y and $v_y + dv_y$, v_z and $v_z + dv_z$ is given by

$$n_i\, dx\, dy\, dz\, dv_x\, dv_y\, dv_z = A\, e^{-\beta E_i}\, dx\, dy\, dz\, dv_x\, dv_y\, dv_z \qquad \text{... (11.57)}$$

But E_i = energy of the particle = $\frac{1}{2}mv^2 = \frac{1}{2}m\left(v_x^2 + v_y^2 + v_z^2\right)$

Substituting for E_i in equation (11.57), we get

$$n_i\, dx\, dy\, dz\, dv_x\, dv_y\, dv_z = A\, e^{-m\beta\left[\frac{1}{2}\left(v_x^2+v_y^2+v_z^2\right)\right]}\, dx\, dy\, dz\, dv_x\, dv_y\, dv_z \quad \text{... (11.58)}$$

The constant A is evaluated by integrating equation (11.58) over all available volume and all ranges of velocities, *i.e.*,

$$\begin{aligned} N &= \iiiiiint A\, e^{-m\beta\,\frac{1}{2}(v_x^2+v_y^2+v_z^2)}\, dx\, dy\, dz\, dv_x\, dv_y\, dv_z \\ &= \iiint dx\, dy\, dz \iiint A\, e^{-m\beta\frac{1}{2}(v_x^2+v_y^2+v_z^2)}\, dv_x\, dv_y\, dv_z \end{aligned}$$

But $\iiint dx\, dy\, dz = V$ = Volume of vessel containing gas.

$$\therefore \quad \begin{aligned} N &= AV \iiint e^{-m\beta\left[\frac{1}{2}(v_x^2+v_y^2+v_z^2)\right]}\, dv_x\, dv_y\, dv_z \\ &= AV \int_{-\infty}^{+\infty} e^{-\frac{\beta m v_x^2}{2}}\, dv_x \int_{-\infty}^{+\infty} e^{-\frac{\beta m v_y^2}{2}}\, dv_y \int_{-\infty}^{+\infty} e^{-\frac{\beta m v_z^2}{2}}\, dv_z \quad \text{...(11.59)} \end{aligned}$$

Using standard definite integrals, we have

$$\int_{-\infty}^{+\infty} e^{-\frac{\beta m v_x^2}{2}}\, dv_x = \sqrt{\frac{2\pi}{m\beta}}$$

$$\int_{-\infty}^{+\infty} e^{-\frac{\beta m v_y^2}{2}}\, dv_y = \sqrt{\frac{2\pi}{m\beta}}$$

and

$$\int_{-\infty}^{+\infty} e^{-\frac{\beta m v_z^2}{2}}\, dv_z = \sqrt{\frac{2\pi}{m\beta}}$$

Substituting these values in equation (11.59), we get

$$N = AV\left(\frac{2\pi}{m\beta}\right)^{3/2}$$

or

$$A = \frac{N}{V}\left(\frac{m\beta}{2\pi}\right)^{3/2}$$

or

$$A = \frac{N}{V}\left(\frac{m}{2\pi\, kT}\right)^{3/2} \qquad \left(\because\ \beta = \frac{1}{kT}\right) \quad \text{... (11.60)}$$

Now substituting values of A and β in equation (11.57), we get

$$n_i\, dx\, dy\, dz\, dv_x\, dv_y\, dv_z = \frac{N}{V}\left(\frac{m}{2\pi kT}\right)^{3/2} e^{-\frac{m(v_x^2+v_y^2+v_z^2)}{2kT}}\, dx\, dy\, dz\, dv_x\, dv_y\, dv_z \quad \text{... (11.61)}$$

The number of molecules having velocity coordinates in the range v_x to $v_x + dv_x$, v_y to $v_y + dv_y$, v_z to $v_z + dv_z$ irrespective of the position coordinates, can be found by integrating equation (11.61) with respect to position coordinates, which gives

$$n_i\, dv_x\, dv_y\, dv_z = \frac{N}{V}\left(\frac{m}{2\pi kT}\right)^{3/2} .\, e^{-\frac{m(v_x^2+v_y^2+v_z^2)}{2kT}}\, .\, dv_x\, dv_y\, dv_z\, .\, V$$

$$= N\left(\frac{m}{2\pi kT}\right)^{3/2} . e^{-\frac{m(v_x^2+v_y^2+v_z^2)}{2kT}} . dv_x\, dv_y\, dv_z \qquad ...(11.62)$$

Lastly let us find the number of molecules having velocity components in the range v_x to $v_x + dv_x$ irrespective of v_y. This will be obtained by integrating eqn. (11.62) with respect to v_y and v_z, *i.e.*,

$$n(v_x)\, dv_x = N\left(\frac{m}{2\pi kT}\right)^{3/2} dv_x \iint e^{-m\left(\frac{v_x^2+v_y^2+v_z^2}{2kT}\right)} dv_y\, dv_z$$

$$= N\left(\frac{m}{2\pi kT}\right)^{3/2} . e^{-mv_x^2/2kT}\, dv_x \int_{-\infty}^{+\infty} e^{-mv_y^2/2kT} . dv_y \int_{-\infty}^{+\infty} e^{-mv_z^2/2kT}\, dv_z$$

$$= N\left(\frac{m}{2\pi kT}\right)^{3/2} e^{-mv_x^2/2kT}\, dv_x \left[\sqrt{\frac{2\pi kT}{m}}\right] \times \left[\sqrt{\frac{2\pi kT}{m}}\right]$$

$$\therefore \quad n(v_x)dv_x = N\left(\frac{m}{2\pi kT}\right)^{1/2} e^{-mv_x^2/2kT}\, dv_x \qquad ...(11.63)$$

This equation (11.63) gives the number of molecules having *x*-component of velocity in the range between v_x and $v_x + dv_x$. Eq. (11.63) shows that the velocity component v_x is distributed symmetrically about the value $v_x = 0$.

The probability that a molecule will have *x*-component of velocity in the range v_x to $v_x + dv_x$ is given by

$$P(v_x)\, dv_x = \frac{n(v_x)}{N} dv_x = \left(\frac{m.}{2\pi kT}\right)^{\frac{1}{2}} e^{-\frac{mv_x^2}{2kT}}\, dv_x \qquad ...(11.64)$$

Equations (11.63) and (11.64) represent **Maxwell's distribution law of velocities.**

The probability function

$P(v_x)$ is given by $\dfrac{P(v_x)dv_x}{dv_x}$

$$\therefore \quad P(v_x) = \left(\frac{m}{2\pi kT}\right)^{\frac{1}{2}} . e^{-mv_x^2/2kT} \qquad ...(11.65)$$

Equation (11.65) can also be expressed in terms of momentum component p_x. Substituting $mv_x = p_x$ and $mdv_x = dp_x$ on R.H.S. of equation (11.64) and dividing by dp_x, we get

$$P(p_x) = \left(\frac{1}{2\pi mkT}\right)^{\frac{1}{2}} . e^{-\frac{p_x^2}{2mkT}} \qquad ...(11.66)$$

Discussion: Consider the following three situations :

The probability that a molecule will have velocity component between v_x and $v_x + dv_x$ is given by

$$P(v_x)dv_x = \left(\frac{m}{2\pi kT}\right)^{\frac{1}{2}} . e^{-mv_x^2/2kT}\, dv_x \qquad ...(11.67)$$

(*i*) This equation indicates that the function $P(v_x)$ is symmetrically distributed about the value $v_x = 0$, since average value of v_x for Maxwellian distribution is zero, *i.e.*,

$$\bar{v}_x = \int_{-\infty}^{+\infty} v_x\, P(v_x)\, dv_x$$

$$= \int_{-\infty}^{+\infty} \left(\frac{m}{2\pi kT} \right)^{\frac{1}{2}} v_x \,.\, e^{-mv_x^2/2kT} \, dv_x = 0$$

$$\left[\text{Since} \int_{-\infty}^{+\infty} v_x \,.\, e^{-mv_x^2/2kT} \, dv_x = 0 \right]$$

Also from equation (11.67), it is clear that

$$P(-v_x) = P(v_x) \qquad \text{... (11.68)}$$

This again indicates that $P(v_x)$ is symmetrical about $v_x = 0$.

(*ii*) The probability distribution function is maximum for that value of v_x for which

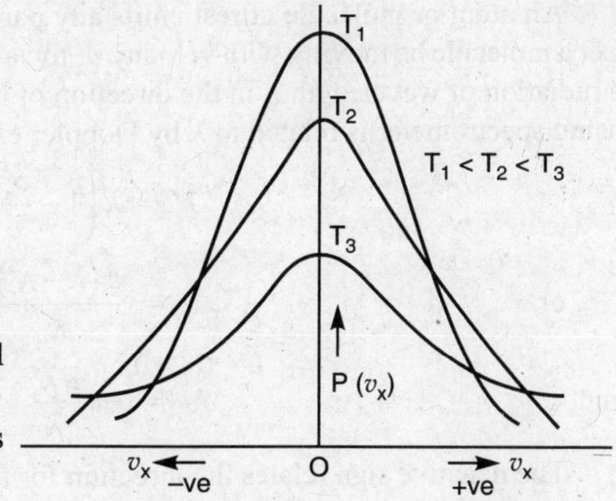

Fig. 11.7

$$\frac{\partial P(v_x)}{\partial v_x} = 0$$

or $$\frac{\partial}{\partial v_x} \left[\left(\frac{m}{2\pi kT} \right)^{\frac{1}{2}} e^{-mv_x^2/2kT} \right] = 0$$

This gives $v_x = 0$ and the maximum value is

$$P_{max} = \left(\frac{m}{2\pi kT} \right)^{\frac{1}{2}} \qquad \text{... (11.69)}$$

This shows that the P_{max} increases as m increases and T decreases.

(*iii*) P falls to $\frac{1}{e}$ times P_{max} at value of v_x, given by

$$P(v_x) = \frac{1}{e} P_{max} = \frac{1}{e} \left(\frac{m}{2\pi kT} \right)^{\frac{1}{2}}$$

or $$\left(\frac{m}{2\pi kT} \right)^{\frac{1}{2}} e^{-mv_x{}^2/2kT} = \frac{1}{e} \left(\frac{m}{2\pi kT} \right)^{\frac{1}{2}}$$

or $$e^{-mv_x^2/2kT} = \frac{1}{e} = e^{-1}$$

This gives $$\frac{mv_x^2}{2kT} = 1 \text{ or } v_x = \sqrt{\frac{2kT}{m}} \qquad \text{... (11.70)}$$

The probability function $P(v_x)$ is plotted against v_x for three different temperatures T_1, T_2 and T_3 such that $T_1 < T_2 < T_3$ as shown in Fig. 11.7. As T increases the peak at $v_x = 0$ lowers down and the distribution spreads out. However, the area under the curve is always *unity.*

11.7 DOPPLER BROADENING OF SPECTRAL LINES

The Maxwellian distribution law of velocities has been verified quantitatively and proved to be the best from Doppler effect in light. The experimental verification of Maxwell distribution function is obtained by measuring the width of a spectral line emitted by hot gas molecules at low densities. The reason for broadening is Doppler's effect. The broadening (*i.e.* spread) arises from the distribution of velocities of the molecules in a gas.

An atom or molecule at rest emits any particular spectral line with a fairly sharp frequency v. Let a molecule be moving with velocity v_x towards an observer. Further, let the molecule be emitting a radiation of wavelength λ in the direction of motion. The wavelength recorded λ' by the observer using spectrometer is related to λ by Doppler expression as

$$\lambda' = \lambda\left(1 - \frac{v_x}{c}\right)$$

or,
$$v_x = \frac{c(\lambda - \lambda')}{\lambda} \quad \text{... (i)}$$

and
$$dv_x = -c\frac{d\lambda'}{\lambda} \quad \text{... (ii)}$$

The negative sign relates the direction for v_x.

The probability that a molecule has component velocity in the range from v_x to $v_x + dv_x$ is given by

$$\left(\frac{m}{2\pi kT}\right)^{\frac{1}{2}} e^{\left\{\frac{-mv_x^2}{2kT}\right\}} dv_x \quad \text{... (iii)}$$

Substituting the value of v_x and dv_x from Eqn. (*i*) and (*ii*), in Eqn. (*iii*), we get

$$\left(\frac{m}{2\pi kT}\right)^{\frac{1}{2}} e^{\left\{-\frac{mc^2}{2kT}\left(\frac{\lambda-\lambda'}{\lambda}\right)^2\right\}} \cdot \frac{c}{\lambda} d\lambda'$$

or
$$A\, e^{\left\{-\frac{mc^2}{2kT}\left(\frac{\lambda-\lambda'}{\lambda}\right)^2\right\}} d\lambda' \quad \text{... (iv)}$$

where
$$A = \left(\frac{m}{2\pi kT}\right)^{\frac{1}{2}} \cdot \frac{c}{\lambda} \quad \text{... (v)}$$

Expression (*iv*) represents the intensity of radiation $I_{\lambda'} d_{\lambda'}$ received by the observer. Hence

$$I_{\lambda'} d_{\lambda'} = A\, e^{\left\{\frac{-mc^2}{2kT}\left(\frac{\lambda-\lambda'}{\lambda}\right)^2\right\}} d\lambda' \quad \text{... (vi)}$$

when
$$\lambda' = \lambda, \text{ then } A = I_\lambda$$

$\therefore$
$$I_{\lambda'} d_{\lambda'} = I_\lambda \exp\left\{-\frac{mc^2}{2kT}\left(\frac{\lambda-\lambda'}{\lambda}\right)^2\right\} d\lambda' \quad \text{... (vii)}$$

A graph is plotted for intensity (I) against frequency (v), we get an intensity distribution curve, which has a Gaussian line shape as shown in Fig. 11.8. The width of the spectral line is usually measured at the half-intensity k points as shown in figure. If eqn. (*vii*) is expressed in terms of frequency, then half-frequency width Δv_d can be expressed as

$$\frac{\Delta v_d}{v} = 2\left(\frac{2kT \log_e(2)}{mc^2}\right)^{\frac{3}{2}} \quad \text{... (11.71)}$$

The theoretical and experimental agreement was found to be excellent.

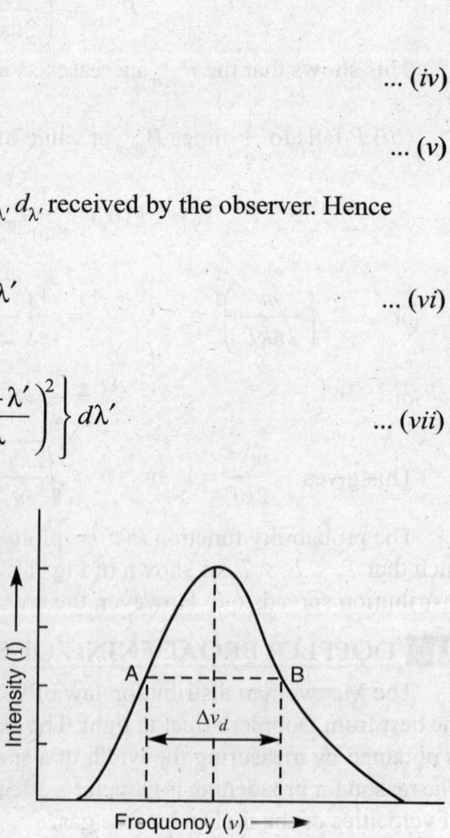

Fig. 11.8

If we put $\lambda - \lambda' = s$ = the shift in wavelength, then from equation (*vii*), we write

$$I_{\lambda'} = I_{\lambda} \exp\left\{-\frac{mc^2 s^2}{2 kT \lambda^2}\right\}$$

$$= I_{\lambda} \exp\{-bs^2\} \quad \text{... (11.72)}$$

where $$b = \frac{mc^2}{2kT\lambda^2}$$

Equation (11.72) may be written as,

$$I_{\lambda'} = I_{\lambda} e^{-bs^2}$$

or, $$\frac{I_{\lambda'}}{I_{\lambda}} = e^{-bs^2} \quad \text{... (11.73)}$$

$\therefore$ $$\frac{I_{\lambda}}{I_{\lambda'}} = e^{bs^2}$$

or $$\log\left(\frac{I_{\lambda}}{I_{\lambda'}}\right) = bs^2$$

or $$s^2 = \frac{1}{b}\log\left(\frac{I_{\lambda}}{I_{\lambda'}}\right)$$

or $$s = \left[\frac{1}{b}\log\left(\frac{I_{\lambda}}{I_{\lambda'}}\right)\right]^{\frac{1}{2}} \quad \text{... (11.74)}$$

11.8 LIMITATIONS OF MAXWELL-BOLTZMANN'S METHOD

The method has lot of limitations. Some of them are discussed below:

1. It is applicable only to an isolated gas of identical molecules in equilibrium for which the following conditions are satisfied :

(*i*) The mean potential energy due to mutual interaction between the molecules is very small compared to their mean kinetic energy. (*Under this condition the gas is said to be* ***ideal.***)

(*ii*) The gas is dilute, *i.e.* the number of molecules per unit volume is small, so that the average separation between the molecules is large and hence individual molecules can be distinguished. (*Under this condition the gas is said to be non-degenerate*).

Therefore, the important results, such as expressions for E, C_v and P, obtained by this method are the same as those derived by applying simple kinetic theory.

2. The expression for the Maxwell-Boltzmann count does not give correct expression for the entropy S of an ideal gas, and thus leads to the Gibbs paradox. To resolve the paradox, the expression must be divided by $N!$

3. In the expression for entropy of an ideal gas

$$S = Nk \log\left[\frac{2\pi mkT}{h^2}\right]^{\frac{3}{2}} e^{\frac{5}{2}}$$

if we put, $T = 0$, we get

$$S = Nk \log 0$$

$$= \text{negative quantity}$$

Thus, the expression for S does not satisfy the third law of thermodynamics. According to this law, as the temperature of the system approaches the absolute zero, the entropy aproaches a constant value S_0, independent of the system.

4. It cannot be applied to a system of indistinguishable particles. If we apply the Maxwell-Boltzmann distribution law to the thermionic emission, we get the expression for the emission current density

$$J = A_0 T^{\frac{1}{2}} e^{-\phi/kT}$$

which is not correct. The correct expression which has been varified experimentally is

$$J = A_0\, e^2\, e^{-\phi\,/kT}.$$

11.9 EXPERIMENTAL VERIFICATION OF MAXWELLIAN DISTRIBUTION OF MOLECULAR SPEEDS

O'Stern tried for the first time to prove experimentally the Maxwell's distribution law of molecular speeds. Eldrige improved this method as the results were not satisfactory. Later a more sophisticated method was suggested by *I.F.* Zartman and C.C. *KO*. Lammert used another method based on Fizeau's toothed wheel model for finding the velocity of light. The basic principle employed is the same in all these methods.

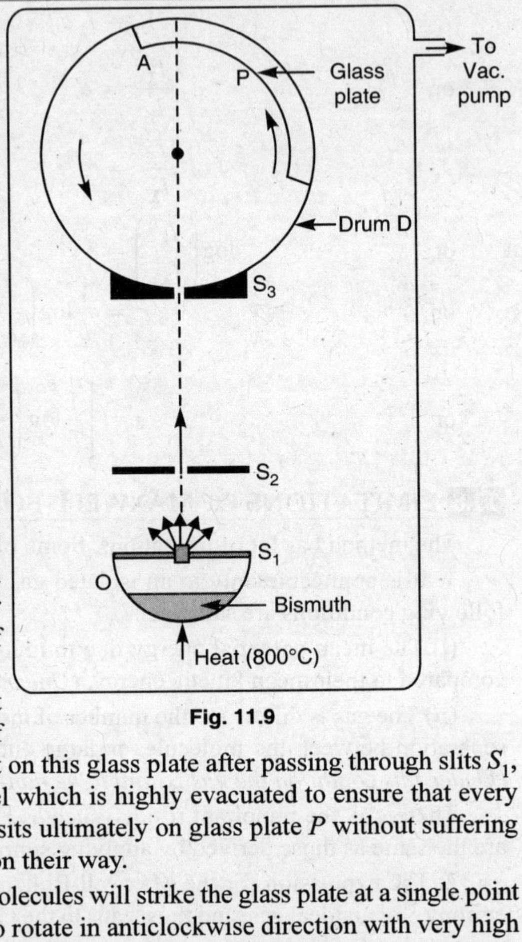

Fig. 11.9

Zartman and C.C. KO Experiment. Fig. 11.9 describes the experimental arrangement used by *I.F.* Zartman and C.C. *KO*. Bismuth is taken in the oven O and heated upto 800°C. Bismuth melts and vaporises. The bismuth molecules in vapour form come out through a slit S_1 and the beam of bismuth molecules is collimated by another slit S_2, kept close to S_1. A hollow cylindrical drum D is kept exactly vertically up that of slit S_2, capable of rotating with a very high speed, nearly 1000 – 15000 rotations per second about an axis perpendicular to the plane of the paper in anticlockwise direction as shown in fig. 11.9. The drum has a narrow slit S_3 in line with that of slit S_2. A glass plate P is fixed on the inner side of the drum so that the bismuth molecules can deposit on this glass plate after passing through slits S_1, S_2 and S_3. The whole system is enclosed in a vessel which is highly evacuated to ensure that every bismuth molecule goes from $S_1 - S_2 - S_3$ and deposits ultimately on glass plate P without suffering any scattering due to collision with air molecules on their way.

If the drum is stationary, then all the bismuth molecules will strike the glass plate at a single point A just opposite to S_3. But when the drum is made to rotate in anticlockwise direction with very high and constant speed, the glass plate P moves towards left and bismuth molecules get deposited on the glass plate, at different points depending upon their speeds. The more rapidly moving bismuth molecules reach the glass plate earlier and condense on it at a point nearer to A than the slowly moving ones which take a longer time to reach the glass plate. After some time, when sufficient amount of bismuth is deposited, the glass plate P is removed and the relative intensity of bismuth molecules at various points is measured with the help of micro-photometer. The relative intensity of bismuth molecules is proportional to the number of bismuth molecules striking at that point and gives the relative number of molecules having different velocities. This gives directly the *velocity spectrum* of bismuth molecules in terms of distance from point A. The distance from A is a measure of distribution of molecular speeds. This distribution obtained is found to agree well with the Maxwellian distribution of molecular speeds.

The results obtained from Zartman and C.C. Ko experiment are shown in Fig. 11.10 where the intensity of deposit is plotted on ordinate (Y – axis) and the displacement from the zero-mark or the velocity of the corresponding molecule on the abscissa (X – axis). The points drawn in the diagram are the observed densities of the deposit and the line as calculated on the assumption that the velocities of the molecules are distributed according to Maxwell's law. There is an excellent agreement between the theoretical curve and the observed point. This experiment suggested by Zartman and C.C. Ko proved to be an accurate test of Maxwell's law of distribution of velocities employing classical statistics.

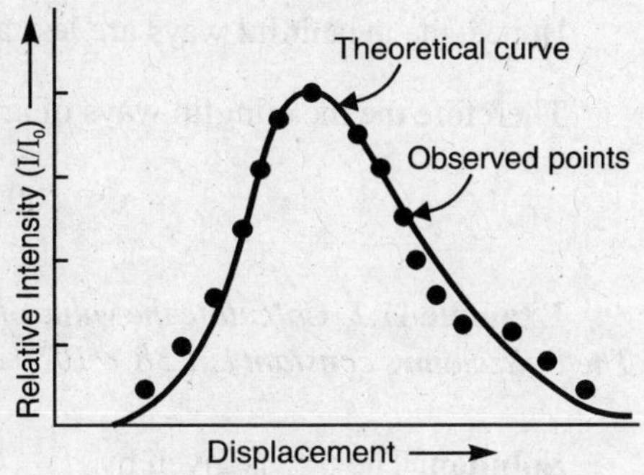

Fig. 11.10

11.10 SOME USEFUL STANDARD DEFINITE INTEGRALS

If $$I_n = \int_0^\infty v^n e^{-\alpha v^2} dv,$$

then the value of I_n for various value of n, which may be used directly, are given below

$$I_0 = \int_0^\infty e^{-\alpha v^2} dv = \frac{1}{2}\sqrt{\left(\frac{\pi}{\alpha}\right)} \quad ...(1)$$

$$I_1 = \int_0^\infty ve^{-\alpha v^2} dv = \frac{1}{2\alpha} \quad ...(2)$$

$$I_2 = \int_0^\infty v^2 e^{-\alpha v^2} dv = \frac{1}{4}\sqrt{\frac{\pi}{\alpha^3}} \quad ...(3)$$

$$I_3 = \int_0^\infty v^3 e^{-\alpha v^2} dv = \frac{1}{2\alpha^2} \quad ...(4)$$

$$I_4 = \int_0^\infty v^4 e^{-\alpha v^2} dv = \frac{3}{8}\sqrt{\left(\frac{\pi}{\alpha^5}\right)} \quad ...(5)$$

$$I_5 = \int_0^\infty v^5 e^{-\alpha v^2} dv = \frac{1}{\alpha^3} \quad ...(6)$$

If the limits of integration are $-\alpha$ to $+\alpha$ in place of 0 to ∞, then the integrals I_0', I_2', I_4' will be double while the integrals I_1', I_3', I_5' will vanish.

NUMERCIAL EXAMPLES

Example 11.1 *Calculate the meaningful ways in which 4 particles out of 100 be arranged in first energy level.*

Solution. No. of ways 1st particle out of 100 is = 100

No. of ways 2nd particle out of 99 is = 100 – 1 = 99

No. of ways 3rd particle out of 98 is = 100 – 2 = 98

No. of ways 4th particle out of 97 is = 100 – 3 = 97

$$\text{Total no. of ways} = 100 \times 99 \times 98 \times 97 = \frac{100 \times 99 \times 98 \times 97 \times 96 \times 95 \times ... \times 2 \times 1}{96\,!}$$

$$= \frac{100!}{96\,!}$$

The four particles are also arranged in themself in 24 ways (*i.e.*, 4!).

Hence, the meaninful ways are less and are obtained by dividing the above equation by 4!.

Therefore the meaningful ways of arangement $= \dfrac{100!}{4!96!}$

$$= \frac{100 \times 99 \times 98 \times 97}{24} = 3921225$$

Example 11.2 *Calculate the value of root mean square speed of a molecule of hydrogen at N.T.P. The Boltzmann's constant is 1.38 × 10^{-16} erg per degree and Avogadro's number is 6 × 10^{23}/gm. mol.*

(*Nagpur 2006, Rajasthan 2001*)

Solution. The v_{rms} is given by

$$v_{rms} = \sqrt{\frac{3kT}{m}}$$

Here $k = 1.38 \times 10^{-16}$ erg per degree

$T = 273$ K

m = mass of hydrogen molecule

$= 2$ a.m.u. $= 2/6 \times 10^{23}$ gm

$$\therefore \quad v_{rms} = \sqrt{\frac{3kT}{m}}$$

$$= \sqrt{\frac{3 \times 1.38 \times 10^{-16} \times 273}{(2/6 \times 10^{23})}}$$

$$= \sqrt{\frac{3 \times 1.38 \times 10^{-16} \times 273 \times 6 \times 10^{23}}{2}}$$

$$= \mathbf{1.838 \times 10^5 \text{ cm/sec}}$$

Example 11.3 *You are given the following group of particles (n_i represents the number of particles with speed v_i)*

n_i	v_i *(m/sec)*
2	1.0
4	2.0
8	3.0
6	4.0
3	5.0

(i) Compute the average speed ($\overline{v}$),

(ii) Compute the root mean square speed (v_{rms})

(iii) Which one is the most probable speed (v_{mp}) among the entire group.

(*Nagpur 2011/s; Rohilkhand 2000*)

Solution. (*i*) The average speed ($\overline{v}$) is given by

$$\overline{v} = \frac{\sum_i N_i v_i}{\sum_i N_i}$$

$$= \frac{2 \times 1 + 4 \times 2 + 8 \times 3 + 6 \times 4 + 3 \times 5}{2 + 4 + 8 + 6 + 3}$$

$$= \frac{2+8+24+24+15}{23} = \frac{73}{23}$$

$$= \mathbf{6.17\ m/sec.}$$

(*ii*) The Mean square speed $\left(\overline{v^2}\right)$ is given by

$$\overline{v^2} = \frac{\sum_i N_i v_i^2}{\sum_i N_i} = \frac{2\times1^2 + 4\times2^2 + 8\times3^2 + 6\times4^2 + 3\times5^2}{2+4+8+6+3}$$

$$= \frac{2+16+72+96+75}{23} = \frac{281}{23}$$

∴ The root mean square speed $\left(\sqrt{\overline{v^2}}\right)$ will be

$$v_{rms} = \sqrt{\frac{281}{23}}$$

$$= \mathbf{3.36\ m/sec.}$$

(*iii*) Out of 23 particles 8 have speed of 3 m/sec., 6 have speed of 4 m/sec., 4 have speed 2 m/sec., while remaining 5 particles have different speeds, therefore, most probable speed v_{mp} is **3 m/sec.**

Example 11.4 *Find the value of v_x for which the probability falls to (i) $\frac{1}{e}$ times (ii) $\frac{1}{10}$ times, the maximum value.* **(*Rohilkhand 2002 Imp*)**

Solution. The probability of a molecule having *x* component of velocity between v_x and $v_x + dv_x$

$$P(v_x) = \left(\frac{m}{2\pi kT}\right)^{\frac{1}{2}} e^{-mv_x^2/2kT}\, dv_x \quad \text{... } (i)$$

The probability will be maximum, when $v_x = 0$, *i.e.*,

$$P_{max}(v_x) = \left(\frac{m}{2\pi kT}\right)^{\frac{1}{2}}$$

(*i*) $\frac{1}{e}$ times:

Given $$P(v_x) = \frac{1}{e}(P_{max}) = \frac{1}{e}\left(\frac{m}{2\pi kT}\right)^{\frac{1}{2}}$$

Substituting this value in equation (*i*), we get

$$\frac{1}{e}\left(\frac{m}{2\pi kT}\right)^{\frac{1}{2}} = \left(\frac{m}{2\pi kT}\right)^{\frac{1}{2}} e^{-mv_x^2/2kT}$$

or $$\frac{1}{e} = e^{-mv_x^2/2kT}$$

Taking $\log_e$ of both sides,

$$\log_e e = -\frac{mv_x^2}{2kT}$$

$$\therefore \qquad v_x = \sqrt{\frac{2kT}{m}} \qquad (\because \log_e e = 1)$$

(*ii*) $\frac{1}{10}$ times :

Given $\qquad P(v_x) = \frac{1}{10} P_{max} = \frac{1}{10}\left(\frac{m}{2\pi kT}\right)^{\frac{1}{2}}$

$$\text{or} \qquad \frac{1}{10}\left(\frac{m}{2\pi kT}\right)^{\frac{1}{2}} = \left(\frac{m}{2\pi kT}\right)^{\frac{1}{2}} e^{-mv_x^2/2kT}$$

$$\text{or} \qquad \frac{1}{10} = e^{-mv_x^2/2kT}$$

Taking $\log_e$ of both sides,

$$-\log_e 10 = -\frac{mv_x^2}{2kT}$$

$$\text{or} \qquad v_x = \sqrt{\frac{2kT}{m}\log_e 10}$$

$$= \sqrt{\frac{2kT}{m} \times 2.303}$$

$$= \sqrt{\frac{4.606\, kT}{m}}$$

Example 11.5 *Deduce an expression for Doppler half-width of a spectral line.*

(Rohilkhand 2002, Jiwaji 2000)

Solution. Doppler half-width is that value of shift s at which intensity falls to half the maximum value. Recalling equation (11.73), and putting

$$\frac{I_{\lambda'}}{I_\lambda} = \frac{1}{2} \quad \text{or} \quad \frac{I_\lambda}{I_{\lambda'}} = 2$$

from equ. (11.73), we have $\qquad I_{\lambda'} = I_\lambda\, e^{-bs^2}$

$$\therefore \qquad \frac{I_{\lambda'}}{I_\lambda} = e^{-bs^2}$$

Taking $\log_e$, we get, $\log_e\left(\frac{I_{\lambda'}}{I_\lambda}\right) = -bs^2$

$$\therefore \qquad s = \left[\frac{\log_e \frac{I_\lambda}{I_{\lambda'}}}{b}\right]^{\frac{1}{2}} = \left[\frac{\log_e 2}{b}\right]^{\frac{1}{2}}$$

$$\text{or} \qquad (\Delta\lambda)_{half} = s = \left(\frac{0.69 \times 2kT}{m}\right)^{\frac{1}{2}} \cdot \frac{\lambda_o}{c} \quad \textbf{Ans.}$$

Example 11.6 *Calculate the Doppler broadening in hydrogen line 4861 Å for T = 400 K. Given k = 1.380 × 10^{-16} erg per degree.*

Solution. Broadening is expressed by half intensity shift, *i.e.*, S_{half} which is expre ssed as $(\Delta v)_{half}$ This may also be expressed in terms of wavelength shift $(\Delta\lambda)_{half}$

Thus,

$$\left|\frac{(\Delta\lambda)_{half}}{\lambda_0}\right| = \left(\frac{0.69\times 2kT}{m}\right)^{\frac{1}{2}} \cdot \frac{1}{c}$$

$$\therefore \qquad (\Delta\lambda)_{half} = \left(\frac{0.69\times 2kT}{m}\right)^{\frac{1}{2}} \times \frac{\lambda_o}{c}$$

Given:

k = Boltzmann's constant = 1.38×10^{-16} erg/degree

T = 400 K

λ_0 = 4861 A

m = mass of H atom (and not of H_2 molecule)

= mass of proton (rest mass)

= 1.6725×10^{-24} gram.

$$\therefore \qquad (\Delta\lambda)_{half} = \left(\frac{0.69\times 2kT}{m}\right)^{\frac{1}{2}} \cdot \frac{\lambda_0}{c}$$

$$= \left(\frac{0.69\times 2\times 1.38\times 10^{-16}\times 400}{1.6725\times 10^{-24}}\right)^{\frac{1}{2}} \times \frac{4861}{3\times 10^{10}} A^\circ$$

$$= 0.034577\ \text{A}^\circ$$

Example 11.7 *Calculate the probability that the speed of oxygen molecule lies between 99.5 and 100.5 metre/sec. at 200k.* ***(Rohilkhand 2001)***

Solution. The probability of a molecule with its speed between v to $v + dv$ is given by

$$P(v)dv = 4\pi\left(\frac{m}{2\pi kT}\right)^{\frac{3}{2}} v^2\, e^{-mv^2/2kT}\, dv$$

Given :

m = 32 a.m.u. = $\frac{32}{6\times 10^{26}}$ kg

v = 100 m/s

dv = 100.5 – 99.5 = 1 metre/sec.

k = 1.38×10^{-23} Joule/kelvin

T = 200 K

Substituting, we get

$$P(v) = 4\times 3.14\left[\frac{\frac{32}{6}\times 10^{26}}{2\times 3.14\times 1.38\times 10^{-23}\times 200}\right]^{\frac{3}{2}} \times 100^2$$

$$\left[\exp -\left\{\frac{32\times 100^2 / 6\times 10^{26}}{2\times 1.38\times 10^{-23}\times 200}\right\}\right]\times 1$$

$$= \frac{4}{\sqrt{3.14}}\left[\left\{\frac{32}{2\times1.38\times10^{-23}\times200\times6\times10^{26}}\right\}^{\frac{3}{2}}\times10^{4}\right.$$

$$\left.\left[\exp-\left\{\frac{32}{2\times1.38\times10^{-23}\times200\times6\times10^{26}}\right\}^{\frac{3}{2}}\right]\right]$$

$$= \frac{4}{1.77}\left(0.96\times10^{-5}\right)^{\frac{3}{2}}\times10^{4}\times e^{-9.6\times10^{-2}}$$

$$P(v) = \mathbf{6.11\times10^{-4}}.$$

Example 11.8 *The first excited state of hydrogen atom is 10.2 eV above its ground state. What temperature is needed to excite hydrogen atoms to the first excited level ?* **(Gorakhpur 2002)**

Solution. The kinetic interpretation of temperature suggests the average kinetic energy per molecule at absolute temperature T is given by

$$K.E. = \frac{1}{2}mv^2 = \frac{3}{2}kT$$

as it has 3 degrees of freedom.

Given that, $$\frac{1}{2}mv^2 = 10.2\ eV = 10.2\times1.6\times10^{-19}\text{ Joule}$$

$\therefore$ $$\frac{3}{2}kT = 10.2\times1.6\times10^{-19}\text{ Joule}$$

or, $$T = \frac{2}{3}\times\frac{10.2\times1.6\times10^{-19}}{k}$$

or, $$T = \frac{2\times10.2\times1.6\times10^{-19}}{3\times1.38\times10^{-23}}$$ (Since $k = 1.38\times10^{-23}$ Joule / K)

$$= \mathbf{7.888\times10^{4}\ Kelvin}$$

Example 11.9 *Calculate the r.m.s. velocity of H_2 at 27°C. Given $k = 1.38\times10^{-23}$ J/deg and mass of hydrogen molecule = 3.34×10^{-27} kg.* **(M.D.U. 2003, Agra 2000, Nagpur 2006/s)**

Solution. The r.m.s. velocity is given by

$$v_{rms} = 1.73\sqrt{\frac{kT}{m}}$$

$$= 1.73\times\sqrt{\frac{1.38\times10^{-23}\times(273+27)}{3.34\times10^{-27}}}$$

$$= \mathbf{1.926\ km/s}$$

Example 11.10 *At what temperature will the average speed of molecules of hydrogen gas be double the average speed of oxygen at 300 K ?*

Solution. Average speed $\bar{v} = 1.59\sqrt{\frac{kT}{m}}$

Average speed of oxygen molecule at 300 K is

$$\bar{v}_{oxy} = 1.59\sqrt{\frac{k\ 300}{m_{oxy.}}}$$

Let T be the temperature at which speed of hydrogen molecule be double the average speed of oxygen at 300 K.

$\therefore$ Average speed of hydrogen molecule at $T\,K$ will be

$$\bar{v}_{Hydro.} = 1.59\sqrt{\frac{kT}{m_{Hydro.}}}$$

Given : $$\bar{v}_{Hydro.} = 2\,\bar{v}_{oxy.}$$

$$\therefore \quad 1.59\sqrt{\frac{kT}{m_{Hydro}}} = 2\times 1.59\sqrt{\frac{300k}{m_{oxy}}}$$

or $$\frac{T}{m_{Hydro}} = 4\times\frac{300}{m_{oxy.}}$$

or $$T = 4\times 300\times\frac{m_{Hydro}}{m_{oxy.}} = \frac{4\times 300}{16}$$

$$= \mathbf{75\ K.}$$

Example 11.11 *The average kinetic energy of a gas molecule at a certain temperature is 6.21 × 10^{-21} joule. Find the temperature. Boltzmann's constant K = 1.38 × 10^{-23} joule K^{-1}.*

(*Guwahati 2000, Nagpur 2003/s, Jiwaji 2000*)

Solution. Average kinetic energy $= \frac{3}{2}kT = 6.21\times 10^{-21}$ Joule

Given : $$k = 1.38\times 10^{-23}\ JK^{-1}$$

$$\therefore \quad T = \frac{2\times 6.21\times 10^{-21}}{3\times 1.38\times 10^{-23}}$$

$$= 300\ K$$

$$= \mathbf{27^{\circ}C}$$

Example 11.12 *At what temperature will the mean speed of hydrogen molecules be the same as that of Nitrogen molecules at 35 °C. Molecular weight of N_2 = 28 and that of H_2 = 2.*

(*Nagpur 2007, 2005, winter 2004/s, Rajasthan 2000, Agra 2003*)

Solution. The mean or average speed $\bar{v}$ is given by

$$\bar{v} = \sqrt{\frac{8kT}{\pi m}}$$

For Hydrogen molecules,

$$\bar{v}_{H_2} = \sqrt{\frac{8kT}{\pi m_{H_2}}} = \sqrt{\frac{8kT}{\pi\times 2}} \quad \text{... (i)}$$

For Nitrogen molecules,

$$\bar{v}_{N_2} = \sqrt{\frac{8k\,(273+35)}{\pi\, m_{N_2}}} = \sqrt{\frac{8k\times 308}{\pi\times 28}} \quad \text{... (ii)}$$

But $$\bar{v}_{H_2} = \bar{v}_{N_2}$$

$$\therefore \quad \sqrt{\frac{8kT}{2\pi}} = \sqrt{\frac{8k\times 308}{28\pi}}$$

or, $$\frac{T}{2} = \frac{308}{28}$$

or, $$T = \frac{308 \times 2}{28}$$
$$= 22K$$

Example 11.13 *Calculate the most probable velocity of hydrogen at 27°C. Given k = 1.38 × 10^{-23} J/degree and mass of hydrogen molecule 3.34 × 10^{-27} kg.* **(M.D.U. 2005)**

Solution. $$v_{mp} = 1.41\sqrt{\frac{kT}{m}} = 1.57 \times 10^3 \text{ ms}^{-1}.$$

Example 11.14 *For O_2 gas at N.T.P. calculate (i) $v_{m.p.}$ (ii) $\overline{v}$ (iii) $v_{rms.p}$ Given Boltzmann's constant k = 1.38 × 10^{-16} ergs/K. Avogadro's number = 6.02 × 10^{25}/mole.* **(H.P.U. 2003, Nagpur 2010/w)**

Solution. Mass of an oxygen molecule $= \dfrac{32}{6.02 \times 10^{23}} = 5.31 \times 10^{-23}$ gm $= 5.31 \times 10^{-26}$ kg

Boltzmann's constant $k = 1.38 \times 10^{-16}$ ergs/K = 1.38×10^{-23} Joule/K

Thus $$\sqrt{\frac{kT}{m}} = \sqrt{\frac{1.38 \times 10^{-23} \times 273}{5.31 \times 10^{-26}}} = 2.66 \times 10^2 \text{ ms}^{-1}$$

$\therefore$ Average speed $$\overline{v} = 1.59\sqrt{\frac{kT}{m}} = 1.59 \times 2.66 \times 10^2 = 4.23 \times 10^2 \text{ ms}^{-1}$$

Root mean square speed $$v_{rms} = 1.73\sqrt{\frac{kT}{m}} = 1.73 \times 2.66 \times 10^2 = 4.6 \times 10^2 \text{ ms}^{-1}$$

Most probable velocity $$v_{m.p.} = 1.41\sqrt{\frac{kT}{m}} = 1.41 \times 2.66 \times 10^2 = 3.76 \times 10^2 \text{ ms}^{-1}.$$

Example 11.15 *Calculate the r.m.s velocity and average speed of H_2 at 27°C. Given k = 1.38 × 10^{-23} J/deg. and mass of hydrogen molecule = 3.34 × 10^{-27} kg.* **(P.U. 2008, 2007; M.D.U. 2002)**

Solution. (*i*) $$v_{r.m.s} = 1.73\sqrt{\frac{kT}{m}} = 1.73\sqrt{\frac{1.38 \times 10^{-23} \times (273+27)}{3.34 \times 10^{-27}}} = 1.73 \times 1.113 \times 10^3$$

$$= 1.926 \text{ km/s}$$

(*ii*) $$v_{av} = 1.59\sqrt{\frac{kT}{m}} = 1.59\sqrt{\frac{1.38 \times 10^{-23} \times (273+27)}{3.34 \times 10^{-27}}} = 1.59 \times 1.113 \times 10^3 = 1.77 \text{ km/s}$$

Example 11.16 *Find the temperature at which the average speed of hydrogen gas molecule is 7 times the average speed of nitrogen gas molecules at 300 K.* **(P.U. 2001)**

Solution. Let T be the temperature at which average speed of hydrogen molecule be 7 times the average speed of nitrogen at 300 K, then $$\sqrt{\frac{8}{\pi}\frac{kT}{m_H}} = 7 \times \sqrt{\frac{8}{\pi}\frac{k\,300}{m_N}}$$

or $$T = 49 \times 300 \times \frac{m_H}{m_N} = \frac{49 \times 300}{14} = 1050 \text{ K}$$

Example 11.17 *Find the temperature at which the average speed of molecules of nitrogen is* $\frac{1}{\sqrt{8}}$ *times the average speed of oxygen gas molecules at 300 K.* **(G.N.D.U. 2006)**

Solution. Average speed of oxygen molecules at 300 $K = \bar{v}_{ox} = \sqrt{\frac{8}{\pi} k \frac{300}{m_{ox}}}$.

Let T be the temperature at which the average speed of molecules of nitrogen is $\frac{1}{\sqrt{8}}$ times, the average speed of oxygen molecules at 300 K, then

Average speed of nitrogen molecules at $TK = \bar{v}_N = \sqrt{\frac{8}{\pi} k \frac{T}{m_N}}$

$$\therefore \qquad \bar{v}_N = \frac{1}{\sqrt{8}} \bar{v}_{0x.} \text{ or } \sqrt{\frac{8}{\pi} \frac{kT}{m_N}} = \frac{1}{\sqrt{8}} \sqrt{\frac{8}{\pi} k \frac{300}{m_{ox}}}$$

$$\therefore \qquad \frac{T}{m_N} = \frac{1}{8} \frac{300}{m_{ox}} \text{ or } T = \frac{300}{8} \frac{m_N}{m_{ox}} = \frac{300}{8} \times \frac{28}{32} = 32.81 \text{ K}$$

Example 11.18 *Calculate the fraction of molecules of a gas possessing speeds within 1% of the most probable speed.* **(Luck U., 2001)**

Solution. The most probable velocity $v_{mp} = \sqrt{\frac{2kT}{m}}$

The fraction of molecules having speeds (or velocities) lying between v and $(v + dv)$ is given by

$$\frac{n(v)\,dv}{n} = 4\pi \left(\frac{m}{2\pi kT} \right)^{3/2} e^{\frac{-mv^2}{2kT}} v^2 \, dv$$

Substituting $v = v_{mp} = \left(\frac{2kT}{m} \right)^{1/2}$, we have

$$\frac{n\,(v)\,dv}{n} = 4\pi \left(\frac{m}{2\pi\, kT} \right)^{3/2} e^{\frac{-m}{2kT} \cdot \frac{2kT}{m}} \frac{2kT}{m} \, dv$$

As v varies within 1% of the most probable velocity, it lies between 0.99 v_{mp} and 1.01 v_{mp}

$$\therefore \quad dv = (1.01 - 0.09)\, v_{mp} = 0.02\, v_{mp} = 0.02 \left(\frac{2kT}{m} \right)^{1/2}$$

$$\therefore \qquad \frac{n(v)\,dv}{n} = \frac{4}{\sqrt{\pi}} \left(\frac{m}{2\,kT} \right)^{3/2} e^{-1} \frac{2kT}{m} \left(\frac{2kT}{m} \right)^{1/2} \times 0.02$$

$$= \frac{4 \times 0.02}{\sqrt{\pi}} e^{-1} = 0.045\, e^{-1} = 0.016$$

Example 11.19 *A gas molecule of mass* 3×10^{-24} *gm escapes from a furnance having temperature 4000 K with root mean square speed. Calculate the speed of escaping molecule.* **(Cal. U., 2003)**

Solution. Here $m = 3 \times 10^{-24}$ gm $= 3 \times 10^{-27}$ kg; $T = 4000$ K; $k = 1.38 \times 10^{-23}$ JK^{-1}

Root mean square speed $v_{rms} = \sqrt{\frac{3kT}{m}} = \sqrt{\frac{3 \times 1.3810^{-23} \times 4000}{3 \times 10^{-27}}} = 7.43 \times 10^{-3} \text{ ms}^{-1}$

This gives the root mean square speed of the escaping molecule.

Example 11.20 *If the most probable speed of a molecule of hydrogen gas is 100 ms^{-1} at temperature T K, calculate the most probable speed of an oxygen molecule at temperature 2T K.* **(Luck. U., 2002)**

Solution. The most probable velocity $v_{mp} = \sqrt{\dfrac{2kT}{m}}$

$$\therefore \quad v_{mp}(H) = 100 \text{ ms}^{-1} = \sqrt{\frac{2kT}{m_H}} \text{ and } v_{mp}(Ox) \text{ at } 2T\text{K} = \sqrt{\frac{2k2T}{m_{ox}}}$$

$$\therefore \quad \frac{v_{mp}(Ox)}{100} = \sqrt{\frac{\dfrac{2k \times 2T}{m_{ox}}}{\dfrac{2kT}{m_H}}} = \sqrt{\frac{2T}{T}\frac{m_H}{m_{ox}}} = \sqrt{2 \times \frac{1}{16}} = \sqrt{\frac{1}{8}}$$

$$\therefore \quad v_{mp}(Ox) = \frac{100}{\sqrt{8}} = 35.36 \text{ ms}^{-1}$$

Example 11.21 *If root mean square velocity of molecules of hydrogen at N.T.P. is 1.84 km per sec; calculate the root mean square velocity of oxygen molecules at N.T.P. Given the molecular weight of hydrogen and oxygen are two and thirty two respectively.* ***(Pbi. U. 2007; M.D.U. 2000)***

Solution.
$$v_{rms} = \sqrt{\frac{3\,kT}{m}}$$

$$\therefore \quad v_{rms}(H) = \sqrt{\frac{3\,kT}{m_H}} \text{ and } v_{rms}(Ox) = \sqrt{\frac{3\,kT}{m_{ox}}}$$

or
$$\frac{v_{rms}(Ox)}{v_{rms}(H)} = \sqrt{\frac{m_H}{m_{ox}}} = \sqrt{\frac{2}{32}} = \sqrt{\frac{1}{16}} = \frac{1}{4}$$

$\therefore$ Root mean square velocity of oxygen $v_{rms}(Ox) = \dfrac{1}{4} \times v_{rms}(H) = \dfrac{1}{4} \times 1.84 = 0.46\ km/\text{sec}.$

Example 11.22 *Calculate the most probable speed of nitrogen at 27°C. Given N = 6 × 10^{23} molecules/ mole; k = 1.38 × 10^{-16} erg K^{-1}.* ***(H.P.U., 2000; M.D.U. 2006)***

Solution. Most probable speed $v_{mp} = \sqrt{\dfrac{2\,kT}{m}}$

Mass of nitrogen molecule $m_N = \dfrac{\text{Molecular weight}}{\text{Avogadro's number}} = \dfrac{28}{6 \times 10^{23}} = 4.66 \times 10^{-23}$ gm

$$\therefore \quad v_{mp} = \sqrt{\frac{2 \times 1.38 \times 10^{-16} \times (273+27)}{4.66 \times 10^{-23}}} = \left[\frac{6 \times 1.38}{4.66} \times 10^9\right]^{1/2} = 4.2 \times 10^4 \text{ cm/s}$$

EXERCISE

11.1 Derive Maxwell-Boltzmann distribution law and hence show that

$$n(E)dE = \frac{2\pi N}{(\pi kT)^{\frac{3}{2}}} E^{\frac{1}{2}} \exp(-E/kT)\,dE,$$

when $n(E)dE$ is the number of molecules with energies E and $E + dE$.

(*Agra 2006, 2000, Madras 2001, Rohilkhand 2001, Nagpur 2010, Mumbai 2005, Punjab 2001, Purvanchal 2002, M.D.U. 2003, Magadh 2002, Pbi.U. 2000, 2003, Kerala U. 2000, 2002, Meerut 2000, 2002*)

11.2 Obtain Maxwell-Boltzmann's distribution of speeds for gas molecules and explain the variation of number of molecules per unit speed range with the speed of molecules.

(*Nagpur 2009, 2005, Kolkata 2004, Agra 2004*)

11.3 Using M-B distribution obtain relation between the average energy of the particle and its temperature in equilibrium. **(*Purvanchal 2007*)**

11.4 Using Maxwell's law of distribution of speeds of molecules in a gas obtain expressions for most probable speed, average speed and root-mean square speed.
Describe a method for the verification of Maxwell's law.

(*Nagpur Uni. 2010/w, 2008, Purvanchal 2003, 2005, 2007, Meerut 2004, 2005, Agra 2002*)

11.5 Define root mean square speed of the gas molecules. Using Maxwell's distribution law of molecular speeds, show that r.m.s. speed is given by $\sqrt{\frac{3\,kT}{m}}$.

(*Nagpur 2004/w, 2006/s, 2007, G.N.D.U. 2002, Kerala 2001, 2004/W, M.S. Uni. Tamil Nadu 2007, Kanpur 2004/w, 2006/s, Garhwal 2002*)

11.6 Assuming *M.B.* distribution of molecular speeds, show that the most probable speed is given by $v_{mp} = \sqrt{\frac{2\,kT}{m}}$.

(*Nagpur 2011/s, 2005/w, 2005'/s 20007; Punjab 2000, 2001, 2002; Pune 2010, Manonmanian S. Uni. 2007, Garhwal 2000, Kerala 2001*)

11.7 Assuming M.B. distribution of molecular speeds, show that mean speed for the molecules of an ideal gas is given by $\bar{v} = \sqrt{\frac{8\,kT}{\pi\,m}}$.

(*Nagpur 2011/s, 2005, 2009, winter; Punjab 2005, MDU. 2001, 2003, GNDU. 2002, Punjab 2001, Pune 2010, Kerala 2001, Calcutta 2001, H.P.U. 2003*)

11.8 Give the experimental verification of Maxwell-Boltzmann's law of distribution of molecular speeds.

(*Alagappa U. 2004, Kerala 2001, Ranchi 2004, Rohilkhand 2002 Improvement, M.S. Uni. Tamil Nadu 2007, Agra 2006, Lucknow 2002, Punjab 2001*)

11.9 Show that the probability of a molecule to have its velocity component between v_x and $v_x + dv_x$ is given by

$$P(v_x)\,dv_x = \left(\frac{m}{2\pi kT}\right)^{\frac{1}{2}} e^{-\frac{mv_x^2}{2kT}}\,dv_x$$

(*Nagpur 2008*)

11.10 Explain the variation of probability function $P(v_x)$ with temperature using Maxwell's distribution law of molecular velocities. **(*Rohilkhand 2003*)**

11.11 Explain clearly by plotting graphs, the variation of distribution function $N(v)$ with temperature using Maxwell's distribution law of molecular speeds.

(*GNDU 2001, Punjab 2001, Garhwal 2000, Nagpur 2006/s*)

11.12 What do you mean by Doppler broadening of spectral lines. Derive the relation for the wavelength shift $s = (\lambda - \lambda')$ given by ***(Nagpur 2011/s, 2010, 2009, 2007, 2006/s)***

$$I_{\lambda'} = I_{\lambda}\, e^{-bs^2}$$

where $$b = \frac{mc^2}{2kT\lambda^2}$$

11.13 State the assumptions made by M–B for the distribution of velocities of the molecules of a ideal gas and hence derive Maxwell's law of distribution of velocities of the molecules of an ideal gas using Maxwell-Boltzmann's energy distribution formula.
(Nagpur 2011/s, 2008, 2007, 2006 /s, 2005/w winter, 2004/W, Alagappa U.2004, Patna 2004, Ranchi 2004, Agra 2001, 2003, Kanpur 2000, Kumaun 2001,Rajasthan 2001, Rohilkhand 2002, Imp 2003 M.S. Uni. Tamil Nadu 2007)

11.14 It v_p, $\overline{v}$ and v_{rms} stand respectively for the most probable, average and root-mean square speeds, show that for the Maxwellian distribution.

$$\frac{v_{rms}}{v_{mp}} = \sqrt{\frac{3}{2}}$$ ***(GNDU. 2004)***

11.15 What are the limitations of Maxwell-Boltzmann method ? ***(Nagpur, 2010 2011/s, 2005/s)***

11.16 Applying Maxwell-Boltzmann distribution show that the internal energy of an ideal monatomic gas depends only on its temperature. ***(Madras 2005)***

11.17 A perfect gas of molecules of mass '*m*' each is at absolute temperature *T.* The velocity of a molecule is denoted by $\overline{v}$. Stating the assumptions clearly, find an expression for the average number of molecules per unit volume with *x*-component of velocity lying between v_x and $v_x + dv_x$ irrespective of the values of the other two components. Find also the expression for the average number of molecules per unit volume having speed between v and $v + dv$.
(Calcutta 2005)

11.18 Draw the graph of the Maxwell-Boltzmann velocity distribution curve and state any two features of the distribution curve. ***(Nagpur 2007)***

11.19 For O_2 gas at N.T.P. calculate (*i*) $\overline{v}$ (*ii*) v_{mp} and (*iii*) v_{rms}. Given Boltzmann's constant 1.38×10^{-16} erg/degree, Avogadro's number = 6.02×10^{23}/mole. ***(Nagpur 2009, 2006)***
[**Ans.** 4.23×10^2 ms^{-1}, 4.61×10^2 ms^{-1}, 3.76×10^2 ms^{-1}]

11.20 At what temperature will the average speed of molecules of hydrogen gas is seven times the average speed of nitrogen at 300 K. [**Ans.** 1050 K]

11.21 Calculate the probability that the speed of Oxygen molecule lies between 100 and 101 ms^{-1} at 200 K. [**Ans.** 6.13×10^{-4}] ***(Rohilkhand 2003)***

11.22 Calculate the fraction of molecules of a gas possessing speeds within 1% of the most probable speed. [**Ans.** 0.016]

11.23 If the most probable speed of a molecule of Hydrogen gas is 100 ms^{-1} at temperature T *K*; calculate the most probable speed of an Oxygen molecule at temperature 2T *K*.
(Chennai 2001)

11.24 Calculate average translational kinetic energy per molecule of a gas at 27°*C* in joules. Given $k = 1.38 \times 10^{-23}$ Joule/mole – *K*. [**Ans.** 6.21×10^{-21} Joule/mole] ***(Nagpur 2005 winter)***

11.25 The speed of 10 particles in *m/s* is 0, 1, 2, 3, 3, 3, 4, 4, 5 and 6, Calculate the average speed, the root mean square speed and the most probable speed. **Ans.** 3.1, $5\sqrt{5}, 3$

11.26 Find the Number of possible arrangements of three particles in two cells assuming the particles obey M.B. statistics. ***(Nagpur 2009)***

11.27 Calculate the rms speed of hydrogen molecule at NTP. Given R = 8.31 J/k^{-mole} K.
(Nagpur 2011/s)

MULTIPLE CHOICE QUESTIONS

11.1 The formula for the most probable speed of the molecules in a gas is:

(*a*) $C_m = \sqrt{\frac{3kT}{m}}$ (*b*) $C_m = \sqrt{\frac{m}{2kT}}$

(*c*) $C_m = \sqrt{\frac{2kT}{m}}$ (*d*) $C_m = \sqrt{\frac{m}{2kT}}$ **(*Agra 2003*)**

11.2 The root mean square speed of molecules of mass '*m*' at temperature *T* is:

(*a*) $\sqrt{\frac{2kT}{m}}$ (*b*) $\sqrt{\frac{3kT}{\pi m}}$

(*c*) $\sqrt{\frac{3kT}{m}}$ (*d*) $\sqrt{\frac{8kT}{m}}$ **(*Ranchi 2005*)**

11.3 RMS speed of an ideal gas is inversely proportional to the square-root of its :

(*a*) mass (*b*) temperature

(*c*) none of these (*d*) both of these **(*Meerut 2003*)**

11.4 For a gas at *N.T.P.* which shall be maximum?

(*a*) V_{av} (*b*) $V_{r.m.s.}$

(*c*) $V_{m.p.}$ (*d*) none

11.5 Out of *n* particles in a gas, the number of particles having exactly the most probable velocity is :

(*a*) zero (*b*) *n*

(*c*) $\frac{n}{2}$ (*d*) 1

11.6 In Zartman Ko experiment distance from the plate is a measure of

(*a*) electron speed (*b*) molecular speed

(*c*) density of molecules (*d*) electron density

11.7 The root mean square speed of a gas molecules of mass *m* at a given temperature *T* is proportional to

(*a*) m^o (*b*) *m*

(*c*) $m^{-1/2}$ (*d*) $m^{1/2}$

11.8 Four particles have speeds 2, 3, 6, and 5 cm/s respectively. Their average speed is :

(*a*) 4 cm/s (*b*) 2 cm/s

(*c*) 0.25 cm/s (*d*) $\sqrt{2}$ cm/s

11.9 The *rms* speed of hydrogen gas molecules at *N.T.P.* is *m/s.* The gas is heated at constant volume till the pressure becomes 9 times. The final *rms* speed will be :

(*a*) 3 v (*b*) 9 v

(*c*) 18 v (*d*) v/3

$$\left[\textbf{Hint}: v_{rms} = \sqrt{\frac{3kT}{m}} \text{ put } PV = RT \text{ i.e. } T = \frac{PV}{R},\ v_{rms} = \sqrt{\frac{3kPV}{Rm}}\right]$$

11.10 The ratio of most probable speed and average speed of a gas enclosed in a vessel is :

(*a*) $\frac{\sqrt{\pi}}{4}$ (*b*) 1 (*c*) $\frac{\sqrt{\pi}}{2}$ (*d*) $\frac{2}{\sqrt{\pi}}$

11.11 The speeds of 10 particles in *m/s* are 0, 1.0, 2.0, 3.0, 3.0, 4.0, 4.0, 4.0, 5.0 and 6.0. The most probable speed is :

(*a*) 3 *m/s* (*b*) 4 *m/s*
(*c*) zero (*d*) none of these

11.12 Four particles have speeds 1, 2, 3 and 4 *cm/s* respectively. Their *rms* speed is :

(*a*) 2.5 (*b*) 10
(*c*) $\frac{\sqrt{30}}{2}$ (*d*) $\sqrt{30}$

11.13 A gas having *rms* speed at 400 *K* is :

(*a*) twice the value at 100 *K* (*b*) four times the value at 100 *K*
(*c*) half the value at 100 *K* (*d*) same as at 100 *K*

11.14 The most probable speed of molecules varies with temperature T as $v_{mp} \propto T^n$, the value of n is :

(*a*) 0 (*b*) $\frac{1}{2}$
(*c*) 2 (*d*) $\frac{1}{3}$

11.15 The particles obeying Maxwell-Boltzmann statistics are :

(*a*) identical (*b*) identical and indistinguishable
(*c*) distinguishable (*d*) photons

ANSWERS

11.1 (*c*)	**11.2** (*b*)	**11.3** (*a*)	**11.4** (*b*)	**11.5** (*a*)	**11.6** (*b*)
11.7 (*c*)	**11.8** (*a*)	**11.9** (*a*)	**11.10** (*c*)	**11.11** (*b*)	**11.12** (*c*)
11.13 (*a*)	**11.14** (*b*)	**11.15** (*c*)			

QUANTUM STATISTICS

12 Chapter

12.1 NEED OF QUANTUM STATISTICS

Classical statistics known as Maxwell-Boltzmann statistics explained the energy and velocity distribution of the molecules of an ideal gas to a fair degree of accuracy but it failed to explain several observed phenomena such as black body radiation, photoelectric effect, specific heat at low temperatures etc.

There is ample evidence to show that within a metallic conductor there are *free* electrons confined within the volume of the metal in the same manner as the molecules of a gas are confined to the volume of the containing vessel. But when Maxwell-Boltzmann statistics is applied to the '*electron gas*' it is unable to explain the observed facts. Similarly a hollow enclosure at a constant temperature T is supposed to be filled with radiations characteristic of that temperature and known as black body radiation. The energy of radiation is proportional to its frequency ν and supposed to consist of '*photons*' each of energy $h\nu$ where h is Plack's constant. These *photons* move about, collide with one another and with the walls of the enclosure. Thus these behave like gas molecules and exert pressure. But when M.B. statistics is applied to '*photon gas*' it predicts a continuouly increasing number of photons per unit range of frequency as the frequency increases whereas the actual distribution as given by Planck's law shows a maximum and falls off rapidly to zero on either side.

In classical statistics we are free to make the cell size *as small as we please*. In other words, the number of cells g_i in a compartment of the phase space can be made *as large as we like* so that it can be much greater than the number of particles n_i available to the ith compartment of the phase space.

$\therefore$ Occupation index $\dfrac{n_i}{g_i} << 1$

In such a case when the occupation index << 1 the probability of a cell containing two or more particles is *negligible* and the particles can, therefore, be treated as distinguishable so that classical statistics is applicable to them.

But all natural system obey uncertainty principle and we are *not free* to choose the cell size as small as we like. The minimum size of a cell in phase space in such a case is h^3 so that number of cells g_i is limited by the value of h^3 and $\dfrac{n_i}{g_i}$ connot be made as small as we please. When the occupation index n_i/g_i is of the order of unity, the particles cannot be treated as distinguishable. Classical statistics, therefore, cannot be applied.

Altogether new quantum cancept of discrete exchange of energy between systems and along with its a new statistics, known as *quantum statistics* has resolved all the above difficulties, and hence the need.

In 1924, the quantum statistics was first formulated by Bose in the deduction of Planck's radiation law. Einstein in the same year employed it in kinetic theory of gases as a substitute for the classical Boltzmann statistics, known as Bose-Einstein (*B-E*) statistics. Two years later, in 1926, Fermi and Dirac quite independently, modified *B-E* statistics in certain cases, on the basis of Pauli's exclusion principle, known as Fermi-Dirac (*F-D*) statistics. Thus, the quantum statistics can be put in to two sub-classes as

(*i*) Bose-Einstein statistics
(*ii*) Fermi-Dirac statistics

depending upon symmetric or antisymmetric wavefunctions.

Concept of energy compartment. According to classical statistics, a classical particle can have any arbitrary value of energy. But according to quantum statistics the energy of a quantum particle is quantised. As a result, the *quantum particle has a set of discrete energy levels.* In such a case each energy level defines an *energy compartment.*

When, the successive energy levels are very close, *i.e.*, the separation between the adjacent energy levels is extremely small, then the energy levels form a *quasi-continuous spectrum.* In such a case, *the energy interval between two adjacent energy levels having values E and E + dE define an energy compartment.*

12.2 DEVELOPMENT OF QUANTUM STATISTICS

For developing a quantum statistical theory for a system containing a large number of identical and indistinguishable particles, it is necessary to consider the wavefunction for many body state as

$$\Psi(\vec{r}_1, \vec{r}_2, \vec{r}_3, \ldots\ldots\ldots\ldots)$$

where $\vec{r}_1, \vec{r}_2, \vec{r}_3, \ldots$ are position vectors of the first, second, third ... particles respectively.

Symmetric wavefunction : If on interchanging any pair of identical particles, there is no change of the sign of the wave function, *i.e.*, if

$$\Psi(\vec{r}_2, \vec{r}_1, \vec{r}_3, \ldots\ldots\ldots\ldots) = +\Psi(\vec{r}_1, \vec{r}_2, \vec{r}_3, \ldots\ldots\ldots\ldots)$$

the wavefunction is said to be **symmetric**.

Anti-symmetric wavefunction: If on interchanging any pair of identical particles, the sign of the wavefunction changes, *i.e.*, if

$$\Psi(\vec{r}_2, \vec{r}_1, \vec{r}_3, \ldots\ldots\ldots\ldots) = -\Psi(\vec{r}_1, \vec{r}_2, \vec{r}_3, \ldots\ldots\ldots\ldots)$$

the wave function is said to be **anti-symmetric.**

For particles having spin angular momenta that are zero or integral multiples of the unit $\hbar \left(= \frac{h}{2\pi}\right)$, the wave-functions are symmetric, and for particles with spin angular momenta that are odd half-integral multiples of $\hbar$, the wave-functions are antisymmetric. Based on this requirement the following two quantum statistics have been developed.

1. Bose-Einstein Statistics

This B.E. statistics was developed by S.N. Bose for light quanta, called **photons** and generalised by A. Einstein to describe the energy distribution among particles, whose spin angular momenta are $n\hbar$ where $n = 0, 1, 2, 3, \ldots$. Such particles are called **bosons**. Wave functions are symmetric on interchanging two bosons. The Pauli exclusion principle which states that in each quantum state there cannot simultaneously be more than one particle is not applicable to bosons, so that there is no upper limit to the number of bosons in a given quantum state.

Examples of bosons are α particles, each having spin quantum number $s = 0$, photon for which $s = 1$, deutrons for which $s = 1$, π mesons for which $s = 0$, etc.

2. Fermi-Dirac Statistics

This statistics was developed by E. Fermi for electrons and its relation to quantum mechanics was established by P.A.M. Dirac. It describes the energy distribution among particles whose spin angular momenta are $\left(n+\frac{1}{2}\right)\hbar$ where $n = 0, 1, 2, 3, ...$ In other words, they are $\frac{1}{2}\hbar, \frac{3}{2}\hbar, \frac{5}{2}\hbar, ...$ Such particles are called **fermions**. Wave functions are antisymmetric on interchanging two fermions. The Pauli exclusion principle is applicable to fermions, so that there can be a maximum of one fermion in a given quantum state.

Examples of fermions are electrons for which $s = \frac{1}{2}$, positrons for which $s = \frac{1}{2}$, protons $\left(s = \frac{1}{2}\right)$, neutrons $\left(s = \frac{1}{2}\right)$, μ-mesons $\left(s = \frac{1}{2}\right)$, etc.

12.3 'h' AS A NATURAL CONSTANT

In classical mechanics h_o is an arbitrary quantity and we may set it as 10^{-30} erg. sec or 10^{-35} erg. sec or even smaller value. Here we are free to take the cell size *as small as we please*. In other words, the number of cells g_i in a compartment of the phase space can be made as large as we like so that it can be much greater than the number of particles n_i available to the *i*th compartment of the phase space. Therefore, occupation index $\frac{n_i}{g_i} << 1$. In such a case, the probability of a cell containing two or more particles is negligible and particles can, therefore, be treated as distinguishable so that classical mechanics is applicable to them.

However, all natural systems obey Heisenberg's uncertainty principle and we are not free to choose the cell size as small as we like. The size of the phase cell is fixed by nature through this uncertainty principle as

$$(\Delta p \,.\, \Delta x)_{min} = h = 6.6 \times 10^{-27} \text{ erg.sec}$$

The minimum size of a cell in phase space in such a case is h^3 as given by

$$dx\, dy\, dz\, dp_x\, dp_y\, dp_z \geq h^3$$

i.e., the volume of the phase cell in phase space cannot be less than h^3, where h is known as Planck's constant. In contrast to classical mechanics, since the size of the cell is restricted to h, the number of cell g_i is limited and hence $\frac{n_i}{g_i}$ cannot be made as small as we please. When the occupation index $\frac{n_i}{g_i}$ is of the order of unity, the particles cannot be treated as distinguishable. *i.e.* become indistinguishable. Classical statistics cannot, therefore, be applied. Here lies the necessity of introduction of quantum statistics.

Two Cases :

(*i*) Particle in a One Dimensional Box

Consider a particle of mass m enclosed in a one dimensional box of length L. This case is useful for understanding the role of size of a phase cell which we have taken to be constant h_o and energy range δE_0 in statistical mechanics.

In one-dimensional box, the *x-p* plane is the phase space. The 'phase volume' is given by

$$V_{phase} = \iint dx\, dp = \int dx \int dp = L \int dp$$

Where L is the length within which the particle is constrained in its motion, and the energy range considered is 0 to E. The number of phase celis corresponding to this energy range is

$$g_i = \phi(E) = \frac{L\sqrt{2mE}}{h_o}$$

where unit phase cell of phase volume h_o (and not h^3_0 as in 6 dimensional phase space.)

As an example, if we consider a particle of mass $m = 2 \times 10^{-24}$ gm in a 'box' of length $L = 1 \times 10^{-3}$ cm. If h_o is set as 1×10^{-26} erg. sec, then in energy range 0 to 10^{-12} erg the number of phase cells would be

$$g_i = \phi(E) = 2 \times 10^5$$

This means that the variation of energy within one phase cell would be nearly 1 part in 2×10^5 of the total energy.

Moreover, the number of microstates in energy range δE_o at energy E is, for this case,

$$W(E) = \frac{L\sqrt{2m}}{h_o} \cdot \frac{1}{2} E^{-\frac{1}{2}} \delta E_o \qquad \text{... (12.2)}$$

For the same values of L, m and h_0 considered above, let us consider an energy sheet of $\delta E_o = 10^{-14}$ erg at $E = 10^{-12}$ erg. This gives

$$W(E) = 10^3$$

In classical mechanics, we are free to set h_0 being an arbitrary quantity. We may set it 10^{-30} erg. sec or 10^{-34} erg. sec or even smaller value. That would make $W(E)$ larger and larger, or it would permit us to divide the phase space into energy sheets of smaller and smaller energy ranges.

But in quantum mechanics, the size of the phase cell is fixed by nature through the uncertainty principle, as we know, to be $h = 6.6 \times 10^{-27}$ erg. sec.

Using this value of h and the values taken for L, m and E in the above problem, would make g_i and $W(E)$ some what larger, but small enough to retain the validity of the statistical method. But for the particles of smaller mass, like that of electron ($m = 9.11 \times 10^{-28} g$), the situation would be worse, but with larger L, the validity would improve. Thus, in order to incorporate the particles of smaller mass, particle in a one-dimensional box fully supports to naturally set value of constant h through the uncertainty principle.

***(ii)* One-Dimensional Harmonic Oscillator.**

The thermal radiation are electromagnetic waves is well established. In a black body encloser, the equilibrium between the walls and the radiation can be obtained only if the walls contain electrical oscillators of some kind. Therefore, one must conceive electrical oscillators (atoms of the spherical encloser black body or cavity) of corresponding frequencies in the walls of the chamber and lay down that in thermal equilibrium the average energy of these oscillators equals the average energy of the corresponding mode in the waves. For simplicity, let us consider a mechanical linear one-dimensional harmonic oscillator. Let m be its reduced mass and force constant c. Let p be momentum at any instant, then

$$p = mv = m\frac{dx}{dt}$$

potential energy of the oscillator $= \frac{1}{2}cx^2$

and, kinetic energy of the oscillator $= \frac{1}{2}m\left(\frac{dx}{dt}\right)^2 = \frac{p^2}{2m}$

$\therefore$ Total energy $\qquad E = \frac{1}{2}cx^2 + \frac{p^2}{2m} \qquad \text{... }(i)$

The frequency of oscillation is given by

$$\nu = \frac{1}{2\pi}\sqrt{\frac{c}{m}} \qquad \text{... } (ii)$$

According to classical theory, radiation of a given frequency would be affected only by oscillators of the *same frequency*. Therefore, in equilibrium the radiation density at a frequency *v* and the average energy of the oscillators of this frequency in the walls must bear a definite ratio. The 'oscillators' are thus these electrical oscillators in the walls and we are concerned with the average energy they possess, when the walls are at temperature *T*.

From the classical theory, which gives the equipartition law, this average energy comes out to be *kT*. Planck realised the failure of Rayleigh–Jeans law, and therefore, *discarded the equipartition principle*. Thus, the classical theory fails to explain black body radiation.

To explain, a new concept of quantum theory was proposed by Planck. To understand this, let us again examine the mechanical case of one-dimensional harmonic oscillator. Consider the *p, x* phase space for such an oscillator as shown in (Fig. 12.1). Let us draw ellipses with origin *O* as centre so as to divide the *x, p* plane into rings of equal area *h*. The first ellipse is numbered 0 and others are numbered as 1, 2, 3, ... *r* .... *n*.

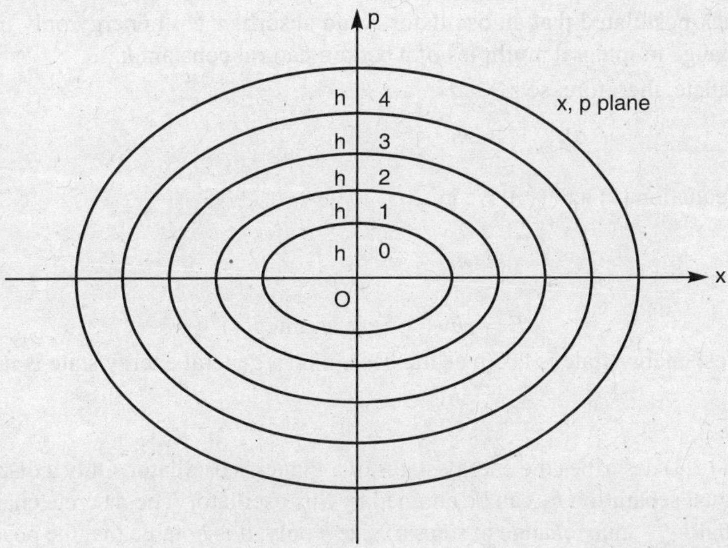

Fig. 12.1

From equation (*i*) it is obvious that for a given energy *E*, the *p, x* relation gives a parabola in which

$$p_{max} = (2mE)^{\frac{1}{2}}$$

and
$$x_{max} = \left(\frac{2E}{c}\right)^{\frac{1}{2}}$$

where *m* is reduced mass and *c* is a force constant for the oscillator.

The phase volume (V_{phase}) (for the 2*f* dimensional phase space, the term phase volume is adopted for $\iint dp\, dx$ for uniformity, here *f* = 1. Hence V_{phase} here means an area in *p, x* plane) corresponding to energy range 0 to *E* is therefore, given by area of the ellipse

$$V_{phase} = \pi\ P_{max}\ x_{max}$$

$$= \pi(2mE)^{\frac{1}{2}}\left(\frac{2E}{c}\right)^{\frac{1}{2}}$$

$$= 2\pi\sqrt{\frac{m}{c}}\,E \qquad \text{... (iii)}$$

Eqn. (*ii*) gives $\quad 2\pi\sqrt{\frac{m}{c}} = \frac{1}{\nu}$

substituting in equation (*iii*), we get

$$V_{phase} = \frac{E}{\nu} \qquad \text{... (iv)}$$

From this equation we predict that if an oscillator of a given frequency ν gains energy ΔE, the 'phase volume' change is given by

$$\Delta V_{phase} = \frac{\Delta E}{\nu} \qquad \text{... (v)}$$

Now Planck postulated that an oscillator could absorb or emit energy only in such a way that V_{phase} could change in integral multiples of a certain natural constant h.

This postulate, therefore, sets

$$\Delta V_{phase} = rh \qquad \text{... (vi)}$$

Equating equation (*v*) and (*vi*), we get

$$\frac{\Delta E}{\nu} = rh$$

$$\Delta E = rh\nu \qquad (r \text{ is an integer}) \qquad \text{... (vii)}$$

If the lowest energy state is taken as the base, then a general energy state is now given by

$$E_n = nh\nu \qquad \text{... (12.3)}$$

where $n = 0, 1, 2, ...$

Equation (12.3) describes the energy states of a Planck's oscillator. Only a discrete set of energy values, with equal separation $h\nu$ can be attained by this oscillator. The discrete character arises from the postulate that V_{phase} may change in steps of size h only. It is notable that the postulate is universal so far as h is the same for all cases. Thus, in energy expression, the quantum comes out as $h\nu$. Hence h has a universal natural constant, known as Planck's constant and has set its value as 6.6×10^{-27} erg. sec.

12.4 INDISTINGUISHABILITY OF PARTICLES AND ITS CONSEQUENCES

The consequences of indistinguishability of particles in quantum statistics may be explained in the following manner :

Let there be two points of collection and only two cells to be occupied. In classical statistics, each particle has a recognizable individuality, hence the particles are distinguishable and each particle is as likely to be in one cell as in the other. This means that in classical statistics, each or both of the particles can occupy any one of the two cells. Thus, in this case, we have four possible arrangements, each of which is counted for assigning a probability to the distribution. This is as shown in Fig. 12.2 (*a*).

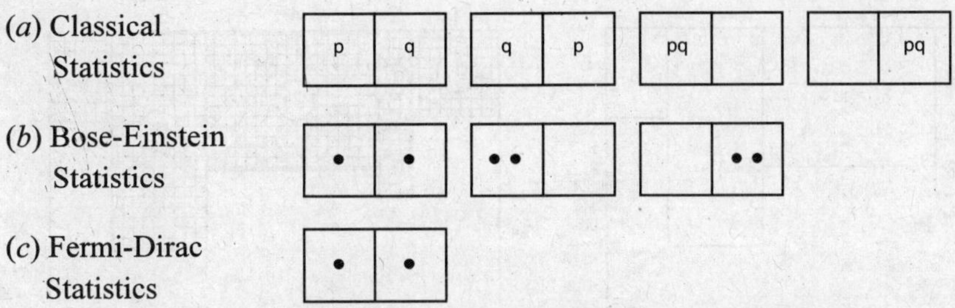

Fig. 12.2. Essential difference in three statistics

In Bose-Einstein statistics, the particles loose their individuality, hence they are indistinguishable. So we must concentrate our attention on the cell rather than the individual particles. Each cell may have any number of particles, so that in this case, we have only three possible arrangements, as shown in Fig. 12.2 (*b*).

In Fermi-Dirac statistics, since the particles are indistinguishable, again we have to concentrate our attention on the cells rather than on individual particles. But in this case, the particles obey Pauli's exclusion principle according to which it is not possible to have more than one representative point in any one cell. Obviously, each cell will contain at the most one particle. Therefore, in this case, only one arrangement is possible, as shown in Fig. 12.2 (*c*).

12.5 BOSE-EINSTEIN DISTRIBUTION LAW

Bose used Planck's hypothesis, according to which radiation in a temperature encloser are composed of light quantas known as photons each of energy, $E = h\nu$. These photons in the encloser are indistinguishable.

Basic Postulates :

In Bose-Einstein statistics, the conditions are:

1. The particles of the system are identical and indistinguishable.

2. Any number of particles can occupy a single cell in the phase space.

3. The size of the cell cannot be less than h^3, where h is Planck's constant ($h = 6.63 \times 10^{-34}$ Joule. sec).

4. The number of phase space cells is comparable with the number of particles, *i.e.,* the occupation index $f(E_i)$ is $\frac{n_i}{g_i} = 1$.

5. Bose-Einstein statistics is applicable to particles with integral spin angular momentum in units of $\frac{h}{2\pi}$. All particles which obey *B.E.* statistics are known as Bosons.

Consider a system consisting of n independent and indistinguishable particles. The particles have definite energies (or momenta) and occupy definite positions. Therefore, they can be represented by phase points in the phase space. In order to determine the energy (or momenta) distribution of these particles in the most probable state known as equilibrium state, we divide the available volume in the phase space into large number (say k) of compartments (quantum groups or energy levels), each compartment representing a small interval of energy (or momentum) as shown in Fig. 12.3 (*a*). Further we divide each compartment into elementary cells, each of size h^3 where h is Planck's constants Fig. 12.3(*b*). We suppose that the size of the compartment is very large as compared to the size of the cell so that each compartment contains a very large number of elementary cells.

Let there be $n_1, n_2, \ldots n_i \ldots n_k$ particles having mean energy values $E_1, E_2, \ldots E_i \ldots E_k$ respectively in compartments numbered as 1, 2, ... i ... k containing $g_1, g_2, \ldots g_i \ldots g_k$ cells respectively in them. The total number of particles in the system is

$$n = n_1 + n_2 + \ldots + n_i + \ldots + n_k$$

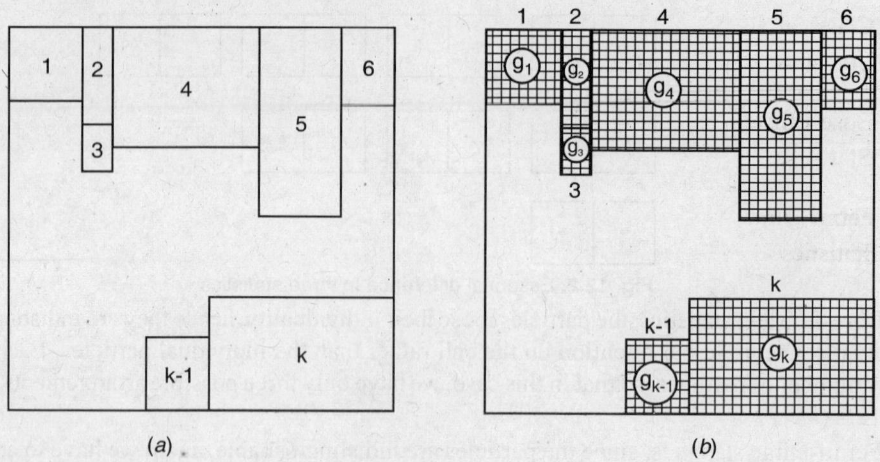

Fig. 12.3

Now consider ith compartment. It has n_i indistinguishable particles distributed among its g_i cells. Suppose, that n_i particles are arranged in a row and distributed among g_i quantum states with (g_i-1) partitions in between. The total number of possible arrangements of particles and partitions is equal to the total number of permutations of $(n_i + g_i - 1)$ objects in a row. Therefore, the total possible ways of arranging n_i particles with $g_i - 1$ partitions

$$= (n_i + g_i - 1)!$$

As the particles are indistinguishable, the rearrangement of particles among themselves will not give rise to any new distribution. The number of such meaningless permutations is n_i ! Hence above result would be divided by n_i !. Moreover, this also includes permutations of $(g_i - 1)$ partitions among themselves. These permutations also do not produce different states and hence are meaningless. The above result is further divided by $(g_i - 1)$!.

Hence, the actual number of ways in which n_i particles are to be distributed in g_i cells in the ith compartment is given by

$$W_i = \frac{(n_i + g_i - 1)!}{n_i\,!(g_i - 1)!} \qquad \text{... (12.4)}$$

Similar expressions will be obtained for other compartments. Therefore, the total number of different arrangements for all the n particles of the system gives thermodynamic probability.

$$W_{(n_1, n_2, \ldots\ldots n_i \ldots\ldots n_k)} = \frac{(n_1 + g_1 - 1)!}{n_1\,!(g_1 - 1)!} \times \frac{(n_2 + g_2 - 1)!}{n_2\,!(g_2 - 1)!} \times \ldots \times \frac{(n_i + g_i - 1)!}{n_i\,!(g_i - 1)!} \times \ldots \frac{(n_k + g_k - 1)!}{n_k\,!(g_k - 1)!}$$

$$= \prod_{i=1}^{k} \frac{(n_i + g_i - 1)!}{n_i\,!(g_i - 1)!} \qquad \text{... (12.5)}$$

where $\prod$ denotes multiplication of terms stated above for various values of i from $i = 1$ to k.

Most Probable Microstate

The most probable microstate corresponds to the state of maximum thermodynamic probability. In equation (12.5), n_i and g_i both are very large numbers. Hence we may neglect 1 in the above expression

$$\therefore \qquad W_{(n_1, n_2, \ldots\ldots n_i \ldots\ldots n_k)} = \Pi \frac{(n_i + g_i)!}{n_i \, ! \, g_i \, !} \qquad \ldots (12.6)$$

Taking natural logarithm on both sides, we have

$$\ln W = \sum_{i=1}^{k} \left[\ln (n_i + g_i)! - \ln n_i ! - \ln g_i ! \right]$$

As n_i and g_i are very large numbers, we can use Sterling approximation.

$$\ln n! = (n \ln n) - n$$

Applying Sterling approximation, we get

$$\ln W = \sum_{i=1}^{k} \left[(n_i + g_i) \ln (n_i + g_i) - (n_i + g_i) - \right.$$

$$\left. -n_i \ln n_i + n_i - g_i \ln g_i + g_i \right]$$

$$= \sum_{i=1}^{k} \left[(n_i + g_i) \ln (n_i + g_i) - n_i \ln n_i - g_i \ln g_i \right] \qquad \ldots (12.7)$$

Here g_i is not subject to variation whereas n_i varies continuously.

To get the state of maximum thermodynamic probability, we differentiate equation (12.7) and equate it to zero; *i.e.*,

$$\delta (\ln W) = 0$$

$$\therefore \qquad \delta (\ln W) = \sum_{i=1}^{k} \left[\delta n_i \ln (n_i + g_i) + (n_i + g_i) \frac{1}{(n_i + g_i)} \delta n_i \right.$$

$$\left. - \delta n_i \ln n_i - n_i \frac{1}{n_i} \delta n_i \right]$$

$$= \sum_{i=1}^{k} \delta n_i \ln (n_i + g_i) - \delta n_i \ln n_i$$

$$(\because g_i \text{ is a mere number } \delta g_i = 0)$$

$$\text{or} \qquad \delta (\ln W) = \sum_{i=1}^{k} \left[\ln \frac{(n_i + g_i)}{n_i} \right] \delta n_i = - \sum_{i=1}^{k} \left[\ln \frac{n_i}{(n_i + g_i)} \right] \delta n_i = 0$$

$$\text{or} \qquad \sum_{i=1}^{k} \left[\ln \frac{n_i}{(n_i + g_i)} \right] \delta n_i = 0 \qquad \ldots (12.8)$$

In addition, our system must satisfy two subsidiary or auxillary conditions:

(*i*) Conservation of total number of particles, *i.e.*, n = a constant.

$$\therefore \qquad n = \sum_i n_i = \text{constant.}$$

$$\textit{i.e.,} \qquad \delta n = \sum_i \delta n_i = 0 \qquad \ldots (12.9)$$

(*ii*) Conservation of total energy of the system, *i.e.*, E = a constant.

$$\therefore \qquad E = \sum_i n_i \, E_i = \text{constant.}$$

$$\textit{i.e.,} \qquad \delta E = \sum_i E_i \delta n_i = 0 \qquad \ldots (12.10)$$

Now we shall apply the Langrangian method of undetermined multipliers. For this let us multiply eqn. (12.9) by α and eqn. (12.10) by β, and add the resulting expression to eq. (12.8) so that we get

$$= \sum_{i=1}^{k}\left[\ln\left(\frac{n_i}{n_i+g_i}\right)+\alpha+\beta E_i\right]\delta n_i = 0 \qquad \text{... (12.11)}$$

The variations δn_i are independent of each other. Hence we get

$$\ln\left(\frac{n_i}{n_i+g_i}\right)+\alpha+\beta E_i = 0$$

or $$\ln\left(\frac{n_i}{n_i+g_i}\right) = -\alpha-\beta E_i$$

or $$\frac{n_i}{n_i+g_i} = e^{-\alpha-\beta E_i} = e^{-(\alpha+\beta E_i)}$$

or $$\frac{n_i+g_i}{n_i} = e^{\alpha+\beta E_i}$$

or $$1+\frac{g_i}{n_i} = e^{\alpha+\beta E_i}$$

or $$\frac{g_i}{n_i} = e^{\alpha+\beta E_i}-1$$

or $$n_i = \frac{g_i}{e^{\alpha+\beta E_i}-1} \qquad \text{... (12.12)}$$

This equation represents the most probable distribution of the particles among various energy levels for a system obeying Bose-Einstein statistics and is therefore, known as **Bose-Einstein's Distribution Law** for an assembly of **bosons**.

Substituting for $\beta = \frac{1}{kT}$, the equation (12.12) becomes

$$n_i = \frac{g_i}{e^{\alpha}.e^{\beta E_i}-1} = \frac{g_i}{e^{\alpha}.e^{E_i/kT}-1} \qquad \text{... (12.13)}$$

B-E energy distribution function.

The energy distribution function $f(E_i)$ is the average number of particles per quantum state in the energy level E_i. It is given by

$$f(E_i) = \frac{n_i}{g_i}$$

Substituting the value of n_i/g_i in equation (12.13), we get

$$f(E_i) = \frac{1}{e^{\alpha}e^{E_i/kT}-1} \qquad \text{... (12.14)}$$

For continuous distribution of energy, the distribution function is written as

$$f(E) = \frac{1}{e^{\alpha}e^{E/kT}-1} \qquad \text{... (12.15)}$$

B-E energy distribution law for continuous variation of energy.

When the energy levels of the system are very closely packed, they form a *quasi-continuous* spectrum. In such a case, if $g(E)\,dE$ is the number of energy states between the energy range E and $(E + dE)$, then

$$g_i = g(E)\,dE$$

If $n(E)\,dE$ is the number of particles whose energy lies between E and $(E + dE)$, then $n_i = n(E)\,dE$. Substituting in eqn. (12.13), we get

$$n(E)dE = \frac{g(E)\,dE}{e^{\alpha}.e^{E_i/kT} - 1} \qquad \text{... (12.16)}$$

The quantity $g(E)$ denotes the *density of states.*

For the system consisting of free particles with no spin $g(E)dE$ is given by (See Eqn. 11.26B)

$$g(E)dE = 2\pi V\left(\frac{2m}{h^2}\right)^{3/2} E^{\frac{1}{2}}dE \qquad \text{... (12.17)}$$

Substituting this value in eqn. (12.14), we get

$$n(E)dE = 2\pi V\left(\frac{2m}{h^2}\right)^{3/2} \frac{E^{\frac{1}{2}}dE}{e^{\alpha}.e^{E_i/kT} - 1} \qquad \text{... (12.18)}$$

This is **Bose-Einstein energy distribution law** for continuous distribution of energy among free particles with no spin.

12.6 PHOTON GAS

S.N. Bose, in 1924, introduced the concept, considering thermal radiation as a photon gas and obtained Planck's formula. Subsequently Einstein developed this idea further into what is now known as Bose-Einstein statistics.

A hollow encloser, say a double walled hollow sphere having a very narrow opening at one point is a perfect absorber as its inner surface is coated with lamp black and it has a sharp projection opposite to the opening. If the radiation enters through the narrow opening, it is absorbed completely inside by successive reflections on the inner walls, as designed by Fery. If the encloser is maintained at a constant temperature T, then atoms of the walls of the encloser emit electromagnetic radiation and at the same time these radiations are absorbed by the atoms in the walls. Thus, the atoms of the walls will emit and reabsorb photons continuously. After a certain lapse of time, a situation will be established when emission and absorption of photons attains thermodynamic equilibrium. At this stage, the amount of energy emitted by the atoms per unit time is equal to the amount of energy absorbed by atoms per unit time.

Thus, the interaction of electromagnetic radiation with matter led to the idea that electromagnetic radiation are composed of discrete energy particles called photons. Each photon has an energy $h\nu$ and momentum $\frac{h}{\lambda}$. Hence ν is the frequency and λ is the wavelength of the radiation. The electromagnetic radiation trapped in a cavity and in thermal equilibrium with the walls of the cavity are termed as *black body radiation.* In the equilibrium condition, the black body radiation can be considered as the photon gas.

12.7 PLANCK'S RADIATION LAW

Statement: According to the law, radiation of wavelength lying between λ and $\lambda + d\lambda$ emitted per unit volume by a perfectly black body at temperature T is given by

$$E_{\lambda}d\lambda = \frac{8\pi hc}{\lambda^5}\,\frac{d\lambda}{e^{hc/\lambda kT} - 1} \qquad \text{... (12.19)}$$

Derivation of the law

Some Important Postulates: It is assumed that the photons do not interact among themselves. The photons interact only with the atoms of the walls of the cavity. It is further assumed that the photons are indistinguishable and many photons can have the same energy. Photons are considered as bosons and they obey Bose-Einstein statistics.

The most probable distribution of the particles among various energy levels for a system obeying Bose-Einstein's distribution law is

$$n_i = \frac{g_i}{e^{(\alpha+\beta E_i)}-1} = \frac{g_i}{e^{\alpha}.e^{E_i/kT}-1} \quad \text{where } \beta = \frac{1}{kT}$$

Most important point to be noted here is that photon gas differs from the ideal gas in one respect. Although the total energy of photons inside the hollow encloser at a particular temperature T remains constant, the total number of photons may be completely absorbed on striking the wall or the hot wall may emit a new photon of energy $h\nu$, *i.e.,* the photons may be created or destroyed. In other words, for a photon gas system, $\sum_i \delta n_i = 0$ is no longer valid. Hence $\delta n = \sum_i \delta n_i \neq 0$.

Due to this reason, the value of α is equal to zero in the Bose-Einstein distribution law. Thus

$$n_i = \frac{g_i}{e^{E_i/kT}-1} \qquad (\because \alpha = 0) \qquad \text{...(12.20)}$$

In this case, the cavity is large as compared to the wavelength of radiation, the energy spectrum of photons is taken to be continuous. The energy difference between successive allowed energy values is very small. Thus, replacing g_i by $g(E)\,dE$ and n_i by $n(E)\,dE$, we get

$$n(E)\,dE = \frac{g(E)\,dE}{e^{E/kT}-1} \qquad \text{... (12.21)}$$

Since $E = h\nu$, the value of $g(E)\,dE$ can be written as $g(\nu)\,d\nu$. The quantity $g(\nu)\,d\nu$ corresponds to the number of oscillatory modes in the frequency range ν and $(\nu + d\nu)$ in the energy range E and $(E + dE)$.

The number of states in a black body radiation in the frequency range ν and $(\nu + d\nu)$ can be obtained by calculating the spherical volume bounded by two spheres of radius $\frac{h(\nu+d\nu)}{c}$ and $\frac{h\nu}{c}$.

The volume of the spherical shell

$$= \frac{4}{3}\pi\frac{h^3}{c^3}(\nu+d\nu)^3 - \frac{4}{3}\pi\frac{h^3}{c^3}\nu^3$$

$$= \frac{4}{3}\pi\frac{h^3}{c^3}\left[\nu^3 + 3\nu^2\,d\nu + ... - \nu^3\right]$$

$$= \frac{4}{3}\pi\frac{h^3}{c^3}\times 3\nu^2\,d\nu \qquad \text{(higher terms neglected)}$$

$$= 4\pi\frac{h^3}{c^3}\nu^2\,d\nu \qquad \text{... (12.22)}$$

The phase space has volume $V = h^3$ and there are two states of polarization (left hand and right hand) for the radiation. The number of states in the black body radiation in the frequency range ν and $(\nu + d\nu)$ is given by

$$g(\nu)d\nu = 2\times 4\pi\frac{V}{c^3}\nu^2\,d\nu \qquad \text{... (12.23)}$$

$$= \frac{8\pi V}{c^3}\nu^2\,d\nu$$

Substituting in equation (12.21), we get the number of photons in the frequency range ν and (ν + dν) given by

$$n(\nu)\,d\nu = \frac{\left(8\pi V\,\nu^2 d\nu\right)}{c^3} \times \frac{1}{e^{h\nu/kT}-1} \quad \text{... (12.24)}$$

Since each photon has energy $h\nu$, the energy density $E(\nu)\,d\nu$ defined as the amount of energy per unit volume lying between the frequency range ν and (ν + dν) is given by

$$E(\nu)\,d\nu = \frac{h\nu . n(\nu)\,d\nu}{V}$$

$$\therefore \qquad E(\nu)\,d\nu = \frac{\left(8\pi h\,\nu^3 d\nu\right)}{c^3} \times \frac{1}{e^{h\nu/kT}-1} \quad \text{... (12.25)}$$

Equation (12.25) represents **Planck's radiation law for black body radiation.**

In terms of wavelength: To express the law in terms of wavelength (λ), substituting $\nu = \frac{c}{\lambda}$ and $d(\nu) = -\frac{c}{\lambda^2}$ in equation (12.25), we get

$$E_\lambda d_\lambda = \frac{8\pi h}{c^3} \cdot \frac{c^3}{\lambda^3} \cdot \frac{c d\lambda}{\lambda^2} \cdot \frac{1}{e^{hc/\lambda kT}-1} = \frac{8\pi hc}{\lambda^5}\,\frac{d\lambda}{e^{hc/\lambda kT}-1}$$

or
$$E_\lambda = \frac{8\pi hc\lambda^{-5}}{e^{hc/\lambda kT}-1} \quad \text{... (12.26)}$$

This equation (12.26) also represents **Planck's law for black body radiation in terms of wavelength.**

12.8 FERMI-DIRAC DISTRIBUTION LAW

Consider a system consisting of n independent and indistinguishable particles. The particles have definite energies (or momenta) and occupy definite positions. Therefore, they can be represented by phase points in the phase space. In order to determine the energy (or momenta) distribution of these particles in the most probable state known as equilibrium state; we divide the available volume in the phase space into large numbers (say k) of compartments (quantum groups or energy levels), each compartment representing a small interval of energy (or momentum). Further each compartment is divided into elementary cells; each of size h^3, where h is a Planck's constant.

Further, we suppose that the size of the compartment is very large as compared to the size of the cell so that each compartment contains a very large number of elementary cells. Let the compartments be marked 1, 2, ... i, ... k and their mean energy values be represented by E_1, E_2, ... E_i .... E_k containing g_1, g_2, ... g_i .... g_k cells respectively in them. The total number of particles in the system is

$$n = n_1 + n_2 + \text{........} + n_i \text{........} + n_k$$

Our aim is to find the Fermi-Dirac distribution law *i.e.*, the distribution of n_i Fermions out of the total n Fermions in g_i cells of the ith compartment. For this, the basic postulates are:

Basic Postulates

(*i*) The particles are indistinguishable so that there is no distinction between the various ways in which n_i particle are chosen.

(*ii*) The particles obey Pauli's exclusion principle according to which there can be either no particle or only one particle in a given cell. Therefore, the number of cells must be much greater than the number of particles, *i.e.*, $g_i >> n_i$.

Out of n_i number of particles in the ith compartment with mean value E_i, the first particle can be placed in any one of the available g_i states, *i.e.,* this particle can be assigned to any of the g_i sets of quantum numbers. Thus, the first particle can be distributed in g_i different ways, in accordance with Pauli's exclusion principle and the remaining $(g_i - 1)$ will remain vacant. The second particle can be arranged in $(g_i - 1)$ different ways and the process continues.

Thus, the total number of different ways of arranging n_i particles among available g_i states with energy level E_i is given by

$$= g_i (g_i - 1)(g_i - 2)(g_i - 3) \,........\, [g_i - (n_i - 1)]$$

$$= \frac{g_i\,!}{(g_i - n_i)!} \qquad ...(12.27)$$

Further, since the particles are indistinguishable, it would not be possible to detect any difference when n_i particles are reshuffled into different ways occupied by them in the energy level E_i. Thus, out of these n_i permutations of the indistinguishable particles among themselves will be meaningless. Therefore, the total number of different and distinguishable ways is

$$= \frac{g_i\,!}{n_i\,!(g_i - n_i)!} \qquad ...(12.28)$$

Therefore, the thermodynamic probability for the microstate $(n_1, n_2, \,.....\, n_i, \,.....\, n_k)$ of the system is given by

$$W(n_1, n_2, ... i, ... k) = \frac{g_1\,!}{n_1\,!(g_1 - n_1)!} \times \frac{g_2\,!}{n_2\,!(g_2 - n_2)!} \times \,.........\, \times \frac{g_i\,!}{n_i\,!(g_i - n_i)!}$$

$$.........\times \frac{g_k\,!}{n_k\,!(g_k - n_k)!}$$

$$= \prod_{i=1}^{k} \frac{g_i\,!}{n_i\,!(g_i - n_i)!} \qquad ...(12.29)$$

where $\prod$ denotes multiplication of terms for various values of i from 1 to k.

Most Probable Microstate

Taking natural logarithm on both sides of eqn. (12.29),

$$\ln W = \sum_{i=1}^{k} [\ln g_i\,! - \ln n_i\,! - \ln (g_i - n_i)\,!]$$

As n_i and g_i are very large numbers, we can use Sterling approximation

$\ln n! = (n \ln n) - n$

Applying Sterling's approximation, we have

$$\ln W = \sum_{i=1}^{k} \big[g_i \ln g_i - g_i - n_i \ln n_i + n_i - (g_i - n_i) \ln (g_i - n_i) + (g_i - n_i) \big]$$

$$= \sum_{i=1}^{k} \big[g_i \ln g_i - n_i \ln n_i - (g_i - n_i) \ln (g_i - n_i) \big]$$

Here g_i is not subject to variation whereas n_i varies continuously. Differentiating both sides, we have

$$\delta(\ln W) = \sum_{i=1}^{k}\left[-n_i \frac{1}{n_i}\delta n_i - \ln n_i\, \delta n_i + (g_i - n_i)\frac{1}{(g_i - n_i)}\delta n_i + \ln(g_i - n_i)\,\delta n_i\right]$$

$$= \sum_{i=1}^{k}\left[\ln(g_i - n_i) - \ln n_i\right]\delta n_i$$

To get the state of *maximum thermodynamic probability*

$$\delta(\ln W) = 0$$

$\therefore$
$$\sum_{i=1}^{k}\left[\ln(g_i - n_i) - \ln n_i\right]\delta n_i = 0$$

or
$$\sum_{i=1}^{k}\left[\ln\frac{g_i - n_i}{n_i}\right]\delta n_i = 0$$

or
$$-\sum_{i=1}^{k}\left[\ln\frac{n_i}{g_i - n_i}\right]\delta n_i = 0$$

or
$$\sum_{i=1}^{k}\left[\ln\frac{n_i}{g_i - n_i}\right]\delta n_i = 0 \qquad \text{... (12.30)}$$

In addition, our system must satisfy the two auxillary conditions:

(*i*) Conservation of total number of particles, *i.e.*, $n = a$ constant

$\therefore$
$$n = \sum_i n_i = \text{constant}$$

i.e.,
$$\delta n = \sum_i \delta n_i = 0 \qquad \text{... (12.31)}$$

(*ii*) Conservation of total energy of the system, *i.e.*, E = a constant.

$\therefore$
$$E = \sum_i n_i\, E_i \text{ constant}$$

i.e.,
$$\delta E = \sum_i E_i\, \delta n_i = 0 \qquad \text{... (12.32)}$$

We shall now apply Langrangian method of undetermined multipliers. For this, let us multiply eqn. (12.31) by α and eqn. (12.32) by β and add the resulting expression to eqn. (12.30),

$$\sum_{i=1}\left[\ln\left(\frac{n_i}{g_i - n_i}\right) + \alpha + \beta E_i\right]\delta n_i = 0$$

As the variations δn_i are independent of each other, we get

$$\ln\left(\frac{n_i}{g_i - n_i}\right) + \alpha + \beta E_i = 0$$

or
$$\frac{n_i}{g_i - n_i} = e^{-(\alpha + \beta E_i)}$$

or
$$\frac{g_i - n_i}{n_i} = e^{\alpha + \beta E_i}$$

or
$$\frac{g_i}{n_i} - 1 = e^{\alpha + \beta E_i}$$

or $$\frac{g_i}{n_i} = e^{\alpha+\beta E_i} + 1$$

or $$n_i = \frac{g_i}{e^{\alpha+\beta E_i} + 1} \qquad \text{... (12.33)}$$

This equation represents the most probable distribution of the particles among various energy levels for a system obeying Fermi-Dirac statistics and is therefore known as **Fermi-Dirac distribution law**, for an assembly of *fermions*.

The parameter $\beta = \frac{1}{kT}$, where k is Boltzmann's constant, has the same role as in case of *M-B* distribution law. Substituting in equation (12.33), we have

$$n_i = \frac{g_i}{e^{\alpha + E_i/kT} + 1} = \frac{g_i}{e^{\alpha}.e^{E_i/kT} + 1} \qquad \text{... (12.34)}$$

Fermi-Dirac energy distribution function.

Since there can be a *maximum of one particle per quantum state*, the function $f(E_i)$ is the ratio of the number of quantum states of energy E_i occupied by electrons to the total number of quantum states available in the energy level E_i. Therefore, the value of $f(E_i)$ for the Fermi-Dirac distribution at a paricular energy E_i is the probability that under equilibrium a quantum state of that energy is occupied by a particle.

From equation (12.34), $f(E_i)$ is given by

$$f(E_i) = \frac{n_i}{g_i}$$

$$= \frac{1}{e^{\alpha} e^{E_i/kT} + 1} \qquad \text{... (12.35)}$$

For continuous distribution of energy at energy E, the distribution function is written as

$$f(E) = \frac{1}{e^{\alpha} e^{E/kT} + 1} \qquad \text{... (12.36)}$$

F-D energy distribution law for continuous variation of energy

If the energy levels are very close together, then the distribution of energy of the particles may be considered continuous. For this distribution the number of particles $n(E)dE$ whose energies lie between E and $E + dE$, then $n_i = n(E)dE$ and is given by

$$n(E)dE = f(E)\, g(E)dE \qquad \text{... (12.37)}$$

where $g(E)dE$ is the number of quantum states of energy between E and $E + dE$.

Substituting the value of $f(E)$ in eqn. (12.37), we have

$$n(E)dE = \frac{g(E)dE}{e^{\alpha}.e^{E/kT} + 1} \qquad \text{... (12.38)}$$

For particles like electrons of spin angular momentum $\pm \frac{1}{2}\hbar$, there are two possible spin orientations. For a system of such particles $g(E)dE$ is given by (see eqn. 11.26B). Substituting, we get

$$n(E)\, dE = 2 \times 2\pi V \left(\frac{2m}{h^2}\right)^{3/2} \frac{E^{\frac{1}{2}} dE}{e^{\alpha}.e^{E/kT} + 1} \qquad \text{... (12.39)}$$

This is the Fermi-Dirac energy distribution law for continuous variation of energy among free particles with spin $\frac{1}{2}$.

12.9 PHOTON GAS AND IDEAL GAS

An ideal gas differs from a photon gas in many respects.

(*i*) The rest mass of the photon is zero. A gas molecule has a finite rest mass.

(*ii*) The molecules of a perfect gas are distinguishable. The photons in a photon gas are indistinguishable from each other.

(*iii*) Photon gas obeys Bose-Einstein statistics whereas an ideal gas obeys Maxwell Boltzmann statistics.

(*iv*) Ideal gas molecules collide with each other and also with the walls of the containing vessel but no molecule is either absorbed or emitted by the walls. *Thus the total number of molecules in the gas remains constant.*

In case of photon gas, a photon can be absorbed or emitted by the walls of the containing vessel. *Thus total number of photons does not remain constant.*

(*v*) The total energy of an ideal gas as well as of a photon gas remains constant if the temperature remains constant.

12.10 FREE ELECTRONS IN METAL: ELECTRON GAS

Metals are very good conductors. The high conductivity of metals is due to the presence of free electrons. These free electrons inside a metallic conductor move freely inside the metal. They continuously collide with the fixed (Localized) atoms and thus behave like an *electron gas*. These free electrons in a metal belong to a most characteristic system of fermions because electrons obey Pauli's exclusion principle, *i.e.*, the electrons have half integral spin angular momentum in unit of $\frac{h}{2\pi}$ or $\hbar$. They are Fermi particles and obey Fermi-Dirac statistics. The electrons in metal have their energy quantised. Fermi-Dirac derived an expression for the distribution of energies among electrons in the *electron gas* in a metal on the basis of these assumptions.

Energy Distribution among Electrons in a Metal

Consider an electron gas having n free electrons in a metallic conductor whose volume is V. Let the total energy of electrons be E. Practically at all temperatures, the lower level energy bands are filled with electrons. The upper level energy bands are only partially filled with electrons. The distribution of electrons is to be considered only in the upper bands called the *conduction band*. The zero energy level is taken at the lowest level of conduction band. It is also assumed the electrons have free movement within the conductor, provided the energy associated with the electrons is of the order of upper level energy bands.

As the energy of the electron in the conduction band is continuous, the term g_i replaced by $g(E)$ dE. Hence $n(E)\,dE$ electrons have energy in the range E and $E + dE$. According to Fermi-Dirac distribution law.

$$n_i = \frac{g_i}{e^{\alpha}.e^{E_i/kT}+1}$$

Substituting for $g_i = g(E)\,dE$, in above equation and replacing n_i by $n(E)\,dE$ we have

$$n(E)\,dE = \frac{g(E)\,dE}{e^{\alpha}.e^{E/kT}+1}$$

As an electron has spin $\pm\frac{1}{2}$, the total number of states in the sphere is twice. The quantity $\frac{V}{(2\pi)^3}$ refers to the translational states per unit volume in the Fermi space. Fermi sphere of radius k_F has the total number of particles accommodated

$$n(E) = \frac{2\left[\frac{V}{(2\pi)^3}\right]\left(\frac{4}{3}\pi k_F{}^3\right)}{e^{\alpha}.e^{E/kT}+1}$$

Here $$E = \frac{p^2}{2m} = \frac{\left(\frac{h}{2\pi}\right)^2 k_F^2}{2m}$$

$$\therefore \quad k_F = \frac{p}{h/2\pi} = \frac{2\pi p}{h}$$

substituting, we get

$$n(E) = \frac{2\left[\frac{V}{(2\pi)^3}\right]\left(\frac{4}{3}\pi\left(\frac{2\pi p}{h}\right)^3\right)}{e^{\alpha}.e^{E/kT}+1}$$

But $$p = (2mE)^{\frac{1}{2}}$$

$$\therefore \quad n(E) = \frac{2\left[\frac{V}{(2\pi)^3}\right]\left(\frac{4}{3}\pi\left(\frac{2\pi}{h}\right)^3 (2mE)^{\frac{3}{2}}\right)}{e^{\alpha}.e^{E/kT}+1}$$

$$= \frac{\left[\frac{8\pi V}{3h^3}\right](2m)(2m)^{\frac{1}{2}} E^{\frac{3}{2}}}{e^{\alpha}.e^{E/kT}+1}$$

Differentiating,

$$n(E)\,dE = \frac{\left[\frac{8\pi V}{3h^3}\right](2m)(2m)^{\frac{1}{2}} \times \frac{3}{2} E^{\frac{1}{2}}\,dE}{e^{\alpha}.e^{E/kT}+1}$$

$$n(E)\,dE = \frac{\left[\frac{8\pi V\left(2m^3\right)^{\frac{1}{2}}}{h^3}\right] E^{\frac{1}{2}}}{e^{\alpha}.e^{E/kT}+1}\,dE = \frac{8\sqrt{2}\,\pi V}{h^3}\frac{m^{\frac{3}{2}} E^{\frac{1}{2}}\,dE}{e^{\alpha}.e^{E/kT}+1} \quad \text{... (12.40)}$$

Equation 12.40 represents the distribution of energy among free electrons. The value of α is still unknown. To evaluate, let us define first the Fermi energy.

12.11 FERMI LEVEL AND FERMI ENERGY

The equation (12.40) reveals that as the value of T decreases, the energy of the electrons goes on decreasing, but at 0 K, the energy of electrons is not zero. Moreover according to Pauli's exclusion principle, only one electron can occupy one energy level. Therefore, there will be one electron in the ground state having zero energy. If n is the total number of electrons at 0 K, then total number of energy states which will be occupied will also be n; each state with one electron. Let E_F be the energy of highest state occupied by the electrons at 0 K, then all the energy states having energy values greater than E_F will be empty and all the energy states having energy values less than E_F will be fully occupied, each with one electron.

Definition: Thus, the energy value up to which all the energy states are completely filled by electrons at absolute zero (*i.e.* 0 K) of temperature and above which all the energy states are completely empty is known as **Fermi energy**. It is denoted by E_F.

Thus, the occupation index, $f(E) = \dfrac{n(E)}{g(E)} = 0$ for $E > E_F$

and the occupation index, $f(E) = \dfrac{n(E)}{g(E)} = 1$ for $E \leq E_F$

From these two conditions, we can prove that $\alpha = -\dfrac{E_F}{kT}$ as under:

For values $E \leq E_F$, $T = 0\ K$ and assuming $\alpha = -\dfrac{E_F}{kT}$

$$\text{Occupation Index} = \frac{n(E)}{g(E)} = \frac{1}{\left(e^{-E_F/kT}\,.\,e^{E/kT}+1\right)} = \frac{1}{e^{-\infty}+1} \cong 1 \qquad \text{... (12.41)}$$

In other words, all energy states having $E \leq E_F$, each is occupied by a single electron.

For values $E > E_F$, at absolute zero, $T = 0\ K$ and assuming $\alpha = -\dfrac{E_F}{kT}$.

$$\text{Occupation Index,} \quad \frac{n(E)}{g(E)} = \frac{1}{e^{(E-E_F)/kT}+1} = \frac{1}{e^{\infty}+1} \cong 0 \qquad \text{...(12.42)}$$

i.e., This means that at absolute zero all quantum states of energy more than E_F are empty. (At T = 0 K, E_F is expressed as E_{FO}).

This discussion gives us sufficient proof to assign the value of $\alpha = -\dfrac{E_F}{kT}$. Hence substituting for α in equation (12.40), we get

$$n(E)dE = \frac{8\sqrt{2}\,\pi V}{h^3} \cdot \frac{m^{3/2} E^{\frac{1}{2}} dE}{e^{(E-E_F)/kT}+1} \qquad \text{... (12.42A)}$$

This is the final expression for Fermi-Dirac law of energy distribution of free electrons in a metal.

Discussion. The physical interpretation of the Fermi energy at absolute zero is that it represents highest energy level at $T = 0\ K$ in which all the quantum states are occupied.

Eqn. (12.41) and (12.42) show that at absolute zero, the function $f(E)$ is constant equal to 1 for all values of energies up to E_{FO} and at E_{FO}, it suddenly falls to zero. Thus, at absolute zero of temperature, $f(E)$ is a *step function,* as shown in Fig. 12.4.

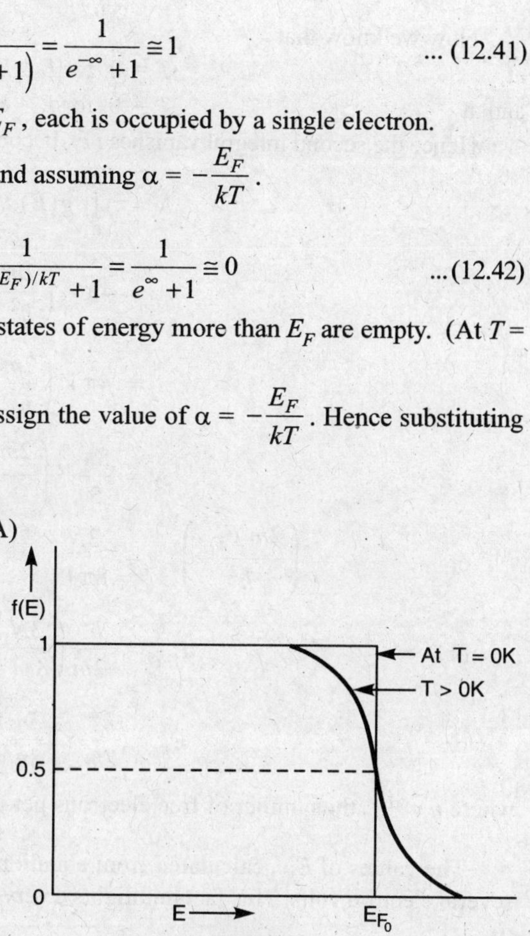

Fig. 12.4

Fermi energy at $T > 0\,K$

For $T > 0\,K$ at $E = E_F$, we have

$$f(E) = \frac{1}{e^{0/kT}+1} = \frac{1}{e^0+1} = \frac{1}{1+1} = \frac{1}{2}$$

This gives the physical interpretation of the Fermi energy at a finite temperature ($T > 0\,K$). Thus, the Fermi energy E_F at $T > 0\,K$ is that energy level in which 50% of the quantum states are occupied and 50% are empty.

It can be shown that as T increases, E_F decreases slightly. Therefore, for practical purposes E_F at temperature T may be considered constant and is nearly equal to E_{FO}.

12.12 E_{FO} FOR ELECTRONS IN A METAL

The total number of free electrons in a metal is given by

$$N = \int_0^\infty n(E)\,dE$$

$$= \int_0^\infty f(E)g(E)\,dE$$

$$= \int_0^{E_{FO}} f(E)g(E)\,dE + \int_{E_{FO}}^\infty f(E)g(E)\,dE$$

Now we know that

at $T = 0$, for $E \le E_{FO}$, $f(E) = 1$

and at $T = 0$, for $E > E_{FO}$, $f(E) = 0$

Hence the second integral vanishes *i.e.*, becomes zero and in first integral we put $f(E) = 1$.

$$\therefore \quad N = \int_0^{E_{FO}} g(E)\,dE$$

$$= 4\pi V\left(\frac{2m}{h^2}\right)^{3/2}\int_0^{E_{FO}} E^{\frac{1}{2}}\,dE$$

$$= 4\pi V\left(\frac{2m}{h^2}\right)^{3/2}\frac{2}{3}E_{FO}^{\;3/2}$$

$$N = \frac{8\pi}{3}V\left(\frac{2m\,E_{FO}}{h^2}\right)^{3/2}$$

or

$$\left(\frac{2m\,E_{FO}}{h^2}\right)^{3/2} = \frac{3N}{8\pi V}$$

or

$$E_{FO} = \frac{h^2}{2m}\left(\frac{3N}{8\pi V}\right)^{2/3}$$

or

$$E_{FO} = \frac{h^2}{2m}\times\left(\frac{3n}{8\pi}\right)^{2/3}, \qquad \text{... (12.43)}$$

where $n = \dfrac{N}{V}$, the number of free electrons per unit volume, *i.e.*, electron density.

The values of E_{FO} calculated from equation (12.43) for a number of metals are of the order of several electron volts. This fact highlights a very important difference between the classical statistics and Fermi-Dirac statistics. According to classical statistics all electrons in a metal at absolute zero would have zero energy.

12.13 FERMI GAS AND ORDINARY GAS

Electronic gas (or Fermi gas). Metals are very good conductors. The high conductivity of metals is due to the presence of free electrons. The free electrons inside a metallic conductor move about freely within the metal. They continuously collide with the fixed atoms and thus behave like an *electron gas*. In other words a metallic solid consists of a set of fixed positive ions immersed in a free electron gas. Since electrons have *half integral* spin angular momentum in units of $\frac{h}{2\pi}$, they are Fermi particles (*or Fermions*) and obey Fermi-Dirac statistics. The electrons in a metal have their energies quantised and being Fermi particles obey *Pauli's exclusion principle.* Such a system of Fermions confined in a volume is known as Fermi gas. Fermi-Dirac derived an expression for the distribution of energies among the electrons in the electron gas in a metal on the basis of these assumptions.

Example. A familiar example of Fermi gas is the conduction electrons in a metal.

Difference between free electron gas and ordinary gas. Free electron gas differs from ordinary gas in the following respects:

(*i*) The free electron gas consists of electrically charged particles whereas the molecules of an ordinary gas are neutral (uncharged)

(*ii*) The number of electrons per unit volume in free electron gas is greater than the number of molecules per unit volume in an ordinary gas being 10^{29} electrons per m^3 as compared to 10^{25} molecules per m^3.

(*iii*) Electrons in free electron gas obey Pauli's exclusion principle whereas molecules in an ordinary gas do not.

(*iv*) The molecules of an ordinary gas are considered *distinguishable* whereas free electrons are *indistinguishable*.

(*v*) The free electron gas obeys Fermi-Dirac statistics whereas ordinary gas obeys Maxwell Boltzmann statistics.

12.14 FERMI-DIRAC VELOCITY DISTRIBUTION IN GIVEN DIRECTION

The Fermi-Dirac energy distribution function for free electrons in a metal at temperature TK is given by

$$f(E) = \frac{1}{e^{(E-E_F)/kT}+1}$$

for energies $E > E_F$, such that $(E - E_F) \gg kT$. Under this condition, we may omit 1 from the denominator.

Thus,
$$f(E) = \frac{1}{e^{(E-E_F)/kT}}$$
$$= e^{E_F/kT} \cdot e^{-E/kT} \qquad \text{... (i)}$$

In a metal if all the electrons are assumed to be free, then the whole energy is kinetic in nature.

$$E = \frac{1}{2}mv^2$$
$$= \frac{1}{2}m\,(v_x^2 + v_y^2 + v_z^2) \qquad \text{... (ii)}$$

where the symbols have their usual meanings.

The function of E in terms of the function of the velocity components (v_x, v_y, v_z) in the form

$$f(v_x, v_y, v_z) = e^{E_F/kT} \cdot e^{-m(v_x^2+v_y^2+v_z^2)/2kT} \qquad \text{... (iii)}$$

The number of electrons with velocity components between v_x and $v_x + dv_x$; v_y and $v_y + dv_y$; v_z and $z_z + dv_z$ is given by

$$n(v_x, v_y, v_z)\, dv_x\, dv_y\, dv_z = f(v_x, v_y, v_z)\, g(v_x, v_y, v_z)\, dv_x\, dv_y\, dv_z \qquad \text{... (iv)}$$

where $g(v_x, v_y, v_z)\, dv_x\, dv_y\, dv_z$ is the number of quantum states within the velocity space element $dv_x dv_y dv_z$.

The volume of momentum space corresponding to one quantum state is (taking into account spin $\frac{1}{2}$ of electrons) $\frac{h^3}{2V}$.

Therefore, the number of quantum states in the volume $dp_x\, dp_y\, dp_z$ of momentum space.

$$= \frac{2V}{h^3}\, dp_x\, dp_y dp_z$$

$$= \frac{2m^3 V}{h^3}\, dv_x\, dp_y dp_z$$

$$\text{Hence, } g(v_x, v_y, v_z)\, dv_x\, dv_y\, dv_z = \frac{2m^3 V}{h^3}\, dv_x\, dv_y dv_z \qquad \text{... (v)}$$

Substituting Eqns. (*iii*) and (*v*) into Eqn. (*iv*), we get

$$n(v_x, v_y, v_z)\, dv_x\, dv_y\, dv_z = \frac{2m^3 V}{h^3}\, e^{E_F/kT} \times e^{-m(v_x^2+v_y^2+v_z^2)/2kT} \times dv_x dv_y dv_z \qquad \text{... (vi)}$$

Now the number $n(v_x)dv_x$ of electrons is obtained by integrating Eqn. (*vi*) over all possible values of v_y and v_z. Thus, we get

$$n(v_x)\, dv_x = \int_{-\infty}^{+\infty}\int_{-\infty}^{+\infty} [n(v_x, v_y, v_z)] dv_y\, dv_z$$

$$= \frac{2m^3\, V}{h^3}\, e^{E_F/kT} \times e^{-mv_x^2/2kT} dv_x \times \int_{-\infty}^{\infty} e^{-mv_y^2/2kT}\, dv_y \times \int_{-\infty}^{\infty} e^{-mv_z^2/2kT}\, dv_z$$

The integrals are standard integrals, the value of each being $\left(\frac{2\pi\, kT}{m}\right)^{\frac{1}{2}}$.

Substituting this value in the above equation

$$n(v_x)dv_x = \frac{2m^3\, V}{h^3}\, e^{E_F/kT} \times e^{-mv_x^2/2kT}\, dv_x \left(\frac{2\pi\, kT}{m}\right)$$

$$= V\left(\frac{4\pi\, m^2\, kT}{h^3}\right) e^{E_F/kT} \times e^{-mv_x^2/2kT}\, dv_x \qquad \text{... (12.44)}$$

Eqn. (12.44) gives the Fermi-Dirac velocity distribution law for electrons in a metal, whose x-component of velocity lies between v_x and $v_x + dv_x$.

12.15 MAXWELL-BOLTZMANN DISTRIBUTION AS A LIMITING CASE OF BOSE-EINSTEIN AND FERMI-DIRAC DISTRIBUTIONS

The general form of the Bose-Einstein and Fermi-Dirac distribution laws given by

$$n_i = \frac{g_i}{e^{\alpha} e^{E_i/kT} \pm 1} \qquad \text{... (i)}$$

where the negative sign is for the B-E distribution and the positive sign for the F-D distribution.

If for any value of E_i, the condition

$$e^{\alpha}\, e^{Ei/kT} \gg 1 \qquad \text{... } (ii)$$

is satisfied, then 1 in the denominator of Eq. (*i*) can be neglected. This gives

$$n_i = e^{-\alpha} .\, g_i\, e^{-Ei/kT} \qquad \text{... } (iii)$$

This is the general form of the Maxwell-Boltzmann distribution law.

Moreover, we can show that the M.B. distribution is applicable to rarefied gases. For this, Equation (*iii*) can be written as

$$\frac{n_i}{g_i} = e^{-\alpha}\, e^{-Ei/kT}$$

From the inequality (*ii*), we get the inequality as

$$\frac{n_i}{g_i} << 1 \qquad \text{... } (iv)$$

The ratio gives the average number of particles per quantum state. Thus, the inequality (*iv*) shows that the Maxwell-Boltzmann distribution law is true in the limiting case of small number of particles per quantum state, *i.e.*, the distribution law is true for rarefied gases.

12.16 APPLICATION TO LIQUID HELIUM

Helium has proven to be one of the most remarkable elements in nature, particularly from the standpoint of statistical physics. It is the only substance which remains liquid under ordinary pressure at absolute zero; all others freeze to the solid state if sufficiently cooled. Solidification of helium takes place under external pressure exceeding 25 atmospheres. According to classical concepts, at absolute zero, all the atoms of a substance cease to move and occupy fixed positions within the body, causing it to become a solid. The fact that at this low temperature *i.e.* 0 *K*, helium remains in liquid state, indicates that entirely different concepts of *quantum mechanics* are required for its understanding.

Helium atoms occur in nature in two stable isotopic forms, ^{3}He and ^{4}He. The former is rarer of these two isotopes. Chemically both these are virtually identical. The only difference between them is a difference of mass and that ^{3}He has a nuclear spin $\frac{1}{2}$ and ^{4}He has spin zero. Therefore, ^{3}He obeys Fermi-Dirac statistics while ^{4}He obeys Bose-Einstein statistics.

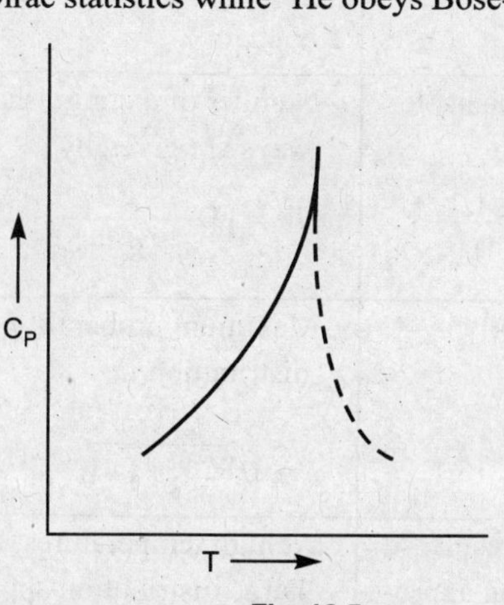

Fig. 12.5

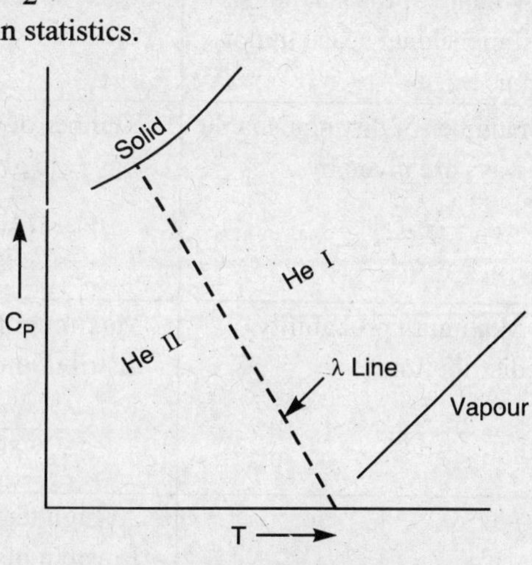

Fig. 12.6

Due to the difference in their statistics, the two systems exhibit very different behaviour upon cooling. ^{4}He which is boson liquid exhibits a second order transition to a superfluid state. This was first observed as an abrupt jump in liquid specific heat Fig. 12.5. ^{3}He also undergoes a transition to superfluid state, but at much lower temperature nearly 2.7×10^{-3} K. The mechanisms of the two transitions, however, are different. (Fig. 12.6) shows the co-existence curves for ^{4}He. At low temperature, ^{4}He has four phases : a solid phase, liquid phases and a vapour phase. As the normal liquid is cooled, a line of λ-point occurs near 2K, the exact value of temperature depends upon pressure. The λ-line indicates that a continuous phase transition has occurred. Liquid helium above the λ-line is called Helium I (or He I) and that below it Helium II (or He II).

The specific heat-temperature graph resembles λ and hence this temperature at which the specific heat changes abruptly (2.19 K) is called the λ-point. Liquid helium above 2.19 K is called He I because it behaves in normal way and below 2.19K, it is called He II because of its abnormal properties. Viscosity of liquid He I decreases with decrease in temperature and this property is contrary to the property of normal liquids but it resembles that of a gas. The viscosity of He II is practically almost zero and it can flow rapidly through narrow capillary tubes. It has zero entropy and no viscosity.

The thermal conductivity of He II has an abnormally high value and it is many times more than that of copper and silver.

These properties of He II are described well in quantum statistics by a two-fluid model. It is assumed that He II is a mixture of two fluids: the normal fluid and the superfluid and that there is no viscous interaction between them. While the normal fluid has all the usual properties of a fluid, superfluid has some curious properties *viz.* zero entropy and no viscosity.

12.17 COMPARISON OF THE THREE STATISTICS

Maxwell-Boltzmann	*Bose-Einstein*	*Fermi-Dirac*
1. Particles are distinguishable	Particles are indistinguishable	Particles are indistinguishable
2. Only particles are taken into consideration.	Only quantum states are taken into consideration.	Only quantum states are taken into consideration.
3. There is no restriction on the number of particles in a given state.	There is no restriction on the number of particles in a given quantum state.	The is restriction on the number of particles in a given quantum state
4. Volume of the state in six dimensional space in not known	Phase space is known $V = h^3$	Phase space is known $V = h^3$
5. Number of distinguishable ways are given by $W = \prod_i^N \frac{g_i^{n_i}}{n_i!}$	Number of distinguishable ways are given by $W = \prod_i \frac{(n_i + g_i - 1)!}{n_i!(g_i - 1)!}$	Number of distinguishable ways are given by $W = \prod_i \frac{g_i!}{n_i!(g_i - n_i)!}$
6. Maximum probability distribution $\propto \frac{1}{e^{(\alpha+\beta E_i)}}$	Maximum probability distribution $\propto \frac{1}{[e^{(\alpha+\beta E_i)}]-1}$	Maximum probability distribution $\propto \frac{1}{[e^{(\alpha+\beta E_i)}]+1}$
7. —	At high temperatures, Bose-Einstein distribution approaches Maxwell- Boltzmann distribution.	At high temperatures, Fermi-Dirac distribution approaches Maxwell-Boltzmann distribution.

8. Applicable to ideal gas molecules	Applicable to photons and symmetrical particles known as bosons.	Applicable to electrons and antisymmetric particles known as fermions.
9. Internal energy of an ideal monatomic gas depends only on its temperature. At absolute zero, energy is zero.	The energy at absolute zero is taken to be zero.	Due to Pauli's exclusion principle, all the electrons cannot occupy the lowest energy level. Even at absolute zero temperature, some of the electrons are distributed at higher energy levels. Hence, at absolute zero, the energy is not zero. Therefore, at the complete degenerate state, the energy is independent of temprature.
10. $n_i = \dfrac{g_i}{e^{(\alpha+\beta_i)}}$	$n_i = \dfrac{g_i}{[e^{(\alpha+\beta E_i)}-1]}$	$n_i = \dfrac{g_i}{[e^{(\alpha+\beta E_i)}+1]}$

11. All the three distributions can be represented by a single equation

$$\frac{g_i}{n_i} = [e^{(\alpha+\beta E_i)} + \delta]$$

Here $\delta = 0$ | Here $\delta = -1$ | Here $\delta = +1$

If $\dfrac{g_i}{n_i} \geq 1$ or $n_i \leq g_i$, *i.e.*, δ can be neglected. In the case of diluted systems, the three statistics lead practically to similar results. It means that at high temperatures, the three statistics give practically the same results.

12.18 DIFFERENCE BETWEEN CLASSICAL AND QUANTUM STATISTICS

The main points of difference between classical and quantum statistics are :

(*i*) Classical statistics is applied to a system of identical particles which obey the laws of classical mechanics, whereas quantum statistics is applied to a system of identical particles which obey the laws of quantum mechanics.

(*ii*) In classical statistics the particles of the system are considered *distinguishable*, whereas in quantum statistics these are considered *indistinguishable.*

(*iii*) In classical statistics the *volume* of the elementary cell in the phase space can be made as small as we please whereas in quantum statistics its value cannot be less than h^3, where h is Planck's constant, given by 6.63×10^{-34} Joules second. In other words, the number of cells in the phase space can be made as large as we like in classical statistics but this is not so in quantum statistics.

(*iv*) As the number of cells available in phase space in classical statistics is very large as compared to the number of particles, the occupation index $\dfrac{n_i}{g_i} \ll 1$. In quantum mechanics, the number of cells available is approximately equal to the number of particles. The occupation index $\dfrac{n_i}{g_i} = 1$.

SOLVED EXAMPLES

Example 12.1 *Calculate the average energy of Planck's Oscillator of frequency ν in thermal equilibrium at temperature T.* ***(Kolkata 2000)***

Solution. The probability that the oscillator may be in nth state is (by Boltzmann's canonical theorem) proportional to $e^{-E_i/kT}$, the phase volume available being equal in all cases. Hence

$\bar{E} = \frac{\sum E_n e^{-E_n/kT}}{\sum e^{-E_n/kT}}$. The summation extends from $n = 0$ to $n = \infty$. Substituting for $E_n = nh\nu$ we get

$$\bar{E} = \frac{\sum nh\nu\, e^{-nh\nu/kT}}{\sum e^{-nh\nu/kT}} = \frac{h\nu}{e^{h\nu/kT} - 1}$$

This gives the average energy of Planck's oscillator of frequency ν in thermal equilibrium at temp. T.

Example 12.2 *Show that if h is treated as arbitrary and allowed to tend to zero, the average energy expressed by* ***(Chennai 2001)***

$$\bar{E} = \frac{h\nu}{e^{h\nu/kT} - 1}$$

would be that given by the equipartition theorem.

Solution. Let $\frac{h\nu}{kT} = x$ then $h\nu = kT.\ x$

substituting, we get

$$\bar{E} = kT \frac{x}{e^x - 1}$$

For $x \to 0$, $e^x - 1 = x + \frac{x^2}{2!} + ... = x$

$$\therefore \quad \bar{E} = kT$$

This is the average energy given by equipartition theorem.

Example 12.3 *A system consists of 5 particles arranged in two compartments. The first compartment is divided into 6 cells and the second into 8 cells. The cells are of equal size. Calculate the number of microstates in the macrostate (2, 3), if the particles obey Fermi-Dirac statistics.* ***(Purvanchal 2006)***

Solution. As the particles obey F-D statistics, the number of microstates in the macrostates $(n_1, n_2, ... n_k)$ or thermodynamic probability of the macrostate is given by

$$W_{(n_1, n_2, ... n_i, ... n_k)} = \prod_{i=1}^{k} \frac{g_i!}{n_i!(g_i - n_i)!}$$

Given : $k = 2$

$$\therefore \quad W_{(n_1, n_2)} = \frac{g_1!}{n_1!(g_1 - n_1)!} \times \frac{g_2!}{n_2!(g_2 - n_2)!}$$

Now $g_1 = 6$ and $g_2 = 8$

$$\therefore \quad W_{(2,3)} = \frac{6!}{2!(6-2)!} \times \frac{8!}{3!(8-3)!}$$

$$= \frac{6!}{2!4!} \times \frac{8!}{3!5!} = \frac{6 \times 5}{2 \times 1} \times \frac{8 \times 7 \times 6}{3 \times 2 \times 1}$$

$$= 15 \times 56$$

$$= \mathbf{840}$$

Example 12.4 *Calculate the value of Fermi-energy at absolute zero temperature.*

(Nagpur 2000)

Solution. *At absolute zero temperature, i.e., T = 0 K, the number of electrons is equal to the total number of energy states occupied by the electrons i.e., Fermions ranging from zero to E_F since each energy state can occupy only one electron; i.e.,*

$$n = \int_o^{E_F} g(E)\,dE = \frac{8\sqrt{2}\pi V}{h^3} m^{\frac{3}{2}} \int_o^{E_F} E^{\frac{1}{2}} dE \left(\because e^{(E-E_F)/kT} = 0 \text{ as } T = 0\right)$$

$$= \frac{8\sqrt{2}\,\pi V}{h^3} m^{\frac{3}{2}} \left[\frac{2}{3} E^{\frac{3}{2}}\right]_0^{E_F}$$

$$= 16\sqrt{2}\,\pi V \frac{m^{\frac{3}{2}}}{3h^3} E_F^{3/2}$$

$$\therefore \qquad E_F^{3/2} = \frac{3nh^3}{16\sqrt{2}\,\pi V\, m^{\frac{3}{2}}}$$

or
$$E_F = \frac{h^2}{2m}\left(\frac{3n}{8\pi V}\right)^{\frac{2}{3}}$$

Example 12.5 *The number of conduction electrons per c.c. is 24.2×10^{22} in beryllium and 0.91×10^{22} in cesium. If the fermi energy of conduction electrons in Be is 14.44 eV. Calculate that is cesium.*

Solution.

Electrons per *c.c.* in Be $= \frac{n}{V} = 24.2\times10^{22}$

Electrons per *c.c.* in $Cs = \frac{n}{V} = 0.91\times10^{22}$

Fermi energy, $E_F = \frac{h^2}{2m}\left(\frac{3}{8\pi}\right)^{\frac{2}{3}}\left(\frac{n}{V}\right)^{\frac{2}{3}} = k\left(\frac{n}{V}\right)^{\frac{2}{3}}$

$$\therefore \quad \frac{\text{Fermi energy for } Be}{\text{Fermi energy for } Cs} = \frac{k\times\left(24.2\times10^{22}\right)^{\frac{2}{3}}}{k\times\left(0.91\times10^{22}\right)^{\frac{2}{3}}} = \left(\frac{24.2}{0.91}\right)^{\frac{2}{3}}$$

or Fermi energy for $Cs = 14.44\times\left(\frac{0.91}{24.2}\right)^{\frac{2}{3}}$

$= \mathbf{1.587\ eV}$

Example 12.6 *Fermi energy of conduction electrons in silver is 5.48 eV. Calculate the number of such electrons per cm^3 given that $h = 6.62\times10^{-27}$ erg sec. and 1 eV= 1.62×10^{-12} erg.*

Solution. Fermi energy is given by

$$E_F = \left(\frac{h^2}{2m}\right)\left(\frac{3n}{8\pi V}\right)^{\frac{2}{3}}$$

$$E_F^{3/2} = \frac{h^3}{(2m)^{\frac{3}{2}}} \cdot \frac{3n}{8\pi V}$$

$\therefore$ Number of electrons per cm^3, $\dfrac{n}{V} = \dfrac{(2m)^{\frac{3}{2}} E_F^{\frac{3}{2}} .8\pi}{3h^3}$

$$= \frac{2^{\frac{3}{2}} \times \left(9.1\times10^{-28}\right)^{\frac{3}{2}} \times 8\pi \times \left(5.48\times1.62\times10^{-12}\right)^{\frac{3}{2}}}{3\times\left(6.62\times10^{-27}\right)^3}$$

$$= \mathbf{5.9 \times 10^{22}\ cm^{-3}}$$

Example 12.7 *How many photons are present in 1 cm^3 of radiation at 727 °C? What is their average energy? Given that*

$$\int_0^\infty \frac{x^2 dx}{e^x - 1} = 2.405.$$

Solution. The number of photons per unit volume having frequencies between ν and (ν + *d*ν) is given by

$$n(\nu)d\nu = \frac{8\pi}{c^3} \times \frac{\nu^2\, d\nu}{e^{h\nu/kT} - 1} \qquad \text{[Equation 12.24 with } V = 1]$$

$\therefore$ Total no. of photons in a unit volume having all the frequencies is given by

$$N = \int_o^\infty n(\nu)\, d\nu = \frac{8\pi}{c^3} \int_o^\infty \frac{\nu^2 d\nu}{e^{h\nu/kT} - 1}$$

Substituting $\dfrac{h\nu}{kT} = x$, we have $\nu = \left(\dfrac{kT}{h}\right)x$ and $d\nu = \left(\dfrac{kT}{h}\right)dx$

$$\therefore \qquad N = 8\pi \left(\frac{kT}{hc}\right)^3 \int_o^\infty \frac{x^2\, dx}{e^x - 1} = 2.405 \times 8\pi \left(\frac{kT}{hc}\right)^3$$

$$= 8 \times 3.14 \times 2.405 \times \left[\frac{1.38\times10^{-16}\times(273+727)}{6.63\times10^{-27}\times3\times10^{10}}\right]^3$$

$$= 2.027 \times 10^{10} \text{ photons.}$$

Total radiation energy per unit volume is given by

$$E_S = \left(\frac{8\pi^5\, k^4}{15\, c^3 h^3}\right) T^4$$

$$\text{Average energy per photon} = \frac{E_s}{N}$$

$$= \frac{8\pi^5 k^4 T^4 h^3\, c^3}{15\, c^3 h^3\, k^3\, T^3 \times 8\pi \times 2.405}$$

$$= \frac{\pi^4}{15} \cdot \frac{kT}{2.405}$$

$$= \frac{(3.14)^4 \times 1.38\times10^{-16} \times 1000}{15\times2.405}$$

$$= 3.73\times10^{-13} \text{ erg}$$

$$= \frac{3.73\times10^{-13}}{1.6\times10^{-12}}$$

$$= \mathbf{0.233\ eV.}$$

Example 12.8 *A gas has only two particles a and b. Show with the help of diagrams how these two particles can be arranged in three quantum series 1, 2, 3 using (i) Maxwell-Boltzmann (ii) Fermi-Dirac and (iii) Bose-Einstein statistics.* **(*Chennai 2004*)**

Solution. **(*i*) Maxwell Boltzmann statistics**

In this case, the particles are distinguishable and there is no limit to the number of particles in any one state. The total number of ways is $3^2 = 9$

States	*Possible Distribution in various States*								
1.	*a*	*b*	–	–	*a*	*b*	*ab*	–	–
2.	*b*	*a*	*a*	*b*	–	–	–	*ab*	–
3.	–	–	*b*	*a*	*b*	*a*	–	–	*ab*

(*ii*) Fermi-Dirac statistics

In this case, the particles are indistinguishable and not more than one particle can be in any one state. Here $a = b$. The total number of ways = 3.

States	*Possible Distribution in various States*		
1.	*a*	*a*	–
2.	*a*	–	*a*
3.	–	*a*	*a*

(*iii*) Bose-Einstein statistics

In this case, the particles are indistinguishable and there is no limit to the number of particles in any one state. Here $a = b$. The total number of ways = 6.

States	*Possible Distribution in various States*					
1.	*a*	*a*	–	*aa*	–	–
2.	*a*	–	*a*	–	*aa*	–
3.	–	*a*	*a*	–	–	*aa*

Example 12.9 *Which of the statistics will you use for the systems having : (i) electrons, (ii) photons, (iii) mesons, (iv) oxygen molecules, (v) phonons, (vi) holes, (vii) neutrons, (viii) protons, (ix) He^4 atom at low temperature, (x) α–particle, (xi) positron.* **(*Purvanchal 2007, Agra 2000*)**

Solution. (*i*) **Electron, Proton, Neutron, Positron and Holes** are spin half particles. They are therefore, *fermions* and hence obey *Fermi-Dirac statistics.*

(*ii*) **Photons, Phonons, Mesons, α-particles and Helium atom at low temperature** are particles of integral spin. Therefore, they are *Bosons* and obey *Bose-Einstein Statistics.*

(*iii*) **Oxygen molecules** are classical particles. Therefore, they obey *Maxwell-Boltzmann statistics.*

Example 12.10 *Classify the following particles according to B-E and F-D statistics : proton, neutron, electron, photon, α-particle, hydrogen atom, hydrogen molecule, positron, lithium ion ($^6_3Li^+$).* **(*Purvanchal 2006, Madras 2005*)**

Solution. The atoms or molecules having an *even number of fundamental particles have integral spin* value are called *Bosons.*

Similarly, the atoms or molecules having *odd number* of fundamental particles have *half integral spin* value are called *Fermions.* Accordingly, the particles are classified as under :

Proton, Neutron, Electron and Positron are fundamental particles having half spin value. Hence, they all are *Fermions.* **Photon** has spin 1 and hence it is a *Boson.*

α-particle ($_2He^4$) = $2p + 2n + 2e \rightarrow$ Even $\rightarrow$ *Boson*

Hydrogen atom = $1p + 1n + 1e \rightarrow$ Odd $\rightarrow$ *Fermion*

Hydrogen molecule = $2p + 2n + 2e \rightarrow$ Even $\rightarrow$ *Boson*

Positron = $e^+ \rightarrow$ odd $\rightarrow$ *Fermion*

Lithium ion ($^6_3Li^+$) = $3p + 3n + 2e \rightarrow$ Even $\rightarrow$ *Boson.*

Example 12.11. *Three particles are to be distributed in four energy levels a, b, c and d. Calculate all possible ways of this distribution when particles are (i) Fermions, (ii) Bosons, (iii) Classical particles.*

Solution. (*i*) Fermions: These particles are identical, indistinguishable having half integral spin. They obey F-D statistics, according to which only one particle can occupy one cell. Accordingly, different ways of distribution of $n_i = 3$ particles in $g_i = 4$ cells.

$$= \frac{g_i\,!}{n_i\,!\,(g_i - n_i)!} = \frac{4!}{3!\,(4-3)!} = 4$$

(*ii*) Bosons: These are identical, indistinguishable particles which do not obey Pauli's exclusion principle. They obey B-E statistics, according to which any number of particles can be in any cell (quantum state) and that all quantum states are equally probable. Thus, the number of ways in which $n_i = 3$ indistinguishable particles can be distributed among $g_i = 4$ cells.

$$= \frac{(n_i + g_i - 1)!}{n_i\,!\,(g_i - 1)!} = \frac{(3+4-1)!}{3!\,(4-1)!} = \frac{6!}{3!\,3!} = \mathbf{20}$$

(*iii*) Classical Particles: These are identical particles of any spin and are distinguishable from one another. They obey M-B statistics, according to which any number of particles can be in any cell. The number of ways in which $n_i = 3$ distinguishable particles can be arranged among $g_i = 4$ energy levels.

$$= g_i^{ni} = (4)^3 = \mathbf{64.}$$

Example 12.12. *Find the number of ways in which three bosons may be distributed in four cells.* ***(P.U. 1999)***

Solution. The number of different arrangements of distributing n_i distinguishable particles or bosons among g_i cells in the compartment is given by

$$W_i = \frac{(n_i + g_i - 1)!}{n_i\,!\,(g_i - 1)!}$$

Here $n_i = 3$ and $g_i = 4$

$$W_i = \frac{(3+4-1)!}{3!(4-1)!} = \frac{6!}{3!3!} = \frac{6\times5\times4}{3\times2\times1} = 20$$

Example 12.13. *Five bosons are distributed in two compartments the first having 3 cells and the second 4. Find the thermodynamic probability for the macrostate (i) (5, 0) (ii) (4, 2)*

Solution. Number of compartments $k = 2$

Number of cells in compartments 1, $g_1 = 3$

Number of cells in compartment 2, $g_2 = 4$

Thermodynamic probability $W(n_1, n_2 \ldots n_i \ldots n_k) = \prod_{i=1}^{k} \frac{(n_i + g_i - 1)!}{n_i\,!\,(g_i - 1)\,!}$

or

$$W(n_1, n_2) = \prod_{i=1}^{2} \frac{(n_i + g_i - 1)!}{n_i\,!\,(g_i - 1)!} = \frac{(n_1 + g_1 - 1)!}{n_1\,!(g_1 - 1)!} \times \frac{(n_2 + g_2 - 1)!}{n_2\,!\,(g_2 - 1)!}$$

(*i*) For the macrostate (5, 0); $n_1 = 5, n_2 = 0, g_1 = 3, g_2 = 4$

$$\therefore \quad W_{(5,\,0)} = \frac{(5+3-1)}{5!(3-1)!} \times \frac{(0+4-1)!}{0!(4-1)!} = \frac{7!}{5!2!} \times \frac{3!}{3!} = \frac{7\times6}{2\times1} = 21$$

(*ii*) For the macrostate (4, 2); $n_1 = 4$, $n_2 = 2$, $g_1 = 3$, $g_2 = 4$

$$\therefore \quad W_{(4,\,2)} = \frac{(4+3-1)\,!}{4\,!(3-1)\,!} \times \frac{(2+4-1)\,!}{2\,!(4-1)\,!} = \frac{6\,!}{4!2\,!} \times \frac{5\,!}{2\,!3\,!}$$

$$= \frac{6\times5}{2\times1} \times \frac{5\times4}{2\times1} = 15\times10 = 150$$

Example 12.14. *Calculate the average energy of a Planck's oscillator of frequency 1.5 × 10*^{14} *at T = 1800 K, h = 6.62 × 10*$^{-34}$ *Js and k = 1.38 × 10*$^{-23}$ *JK*$^{-1}$. **(Kan. U., 1996, 1993)**

Solution. Frequency $\quad v = 1.5 \times 10^{14}\ s^{-1}$; $h = 6.62 \times 10^{-34}$ Js

$k = 1.38 \times 10^{-23}$ JK^{-1}; $T = 1800$ K

Average energy of a Planck's oscillator

$$\bar{E} = \frac{hv}{e^{hv/kT} - 1} = \frac{6.62\times10^{-34}\times1.5\times10^{14}}{e^{6.62\times10^{-34}\times1.5\times10^{14}/1.38\times10^{-23}\times1800} - 1}$$

Now $$\frac{6.62\times10^{-34}\times1.5\times10^{14}}{1.38\times10^{-23}\times1800} = 3.997 \text{ and } e^{3.997} = 54.46$$

and $6.62 \times 10^{-34} \times 1.5 \times 10^{14} = 9.93 \times 10^{-20}$

$$\therefore \quad E = \frac{9.93\times10^{-20}}{54.46-1} = \frac{9.93}{53.46}\times10^{-20} = 0.186\times10^{-20}\text{ J}$$

$$= 1.86 \times 10^{-21}\text{ J}$$

Example 12.15. *Calculate the number of different arrangements of 10 indistinguishable particles in 15 cells of equal a priori probability considering that one cell contains only one particle.* **(K.U. 2002)**

Solution. As the particles are indistinguishable and one cell contains only one particle, the particles are Fermions. The number of meaningful arrangements is, therefore given by

$$\frac{g_i\,!}{n_i\,!(g_i - n_i)\,!}$$

where g_i is the number of cells and n_i the number of particles.

Here $\quad g_i = 15$; $n_i = 10$

$$\therefore \quad \text{Number of different arrangements } W = \frac{15\,!}{10\,!(15-10)\,!} = \frac{15\,!}{10\,!5\,!}$$

$$= \frac{15\times14\times13\times12\times11\times10!}{10\,!\times5\times4\times3\times2\times1} = 21\times13\times11 = 3003$$

Example 12.16. *Calculate the number of ways of arranging 8 Fermions in 12 Phase space cells.* **(Nagpur 2011/s, P.U. 2005)**

Solution. Number of Fermions $n_i = 8$, Number of Cells $g_i = 12$ phase space cells

$$\therefore \quad \frac{g_i\,!}{n_i\,!(g_i - n_i)!} = \frac{12!}{8!(12-8)!} = \frac{12!}{8!4!} = \frac{12\times11\times10\times9}{4\times3\times2\times1} = 495$$

Example 12.17. *There are two compartments a and b. The compartment a has three cells whereas b has four cells. Five fermions are distributed in these compartments. Find the thermodynamic probabily for macrostawte (i) (3, 2) (ii) (1, 4).* **(G.N.D.U. 2004)**

Solution. As the particles are Fermions, they obey Fermi-Dirac statistics. Therefore, the number of microstate in the macrostate $W(n_1, n_2, ... n_k)$ is given by

$$W(n_1 . n_2 . ... n_k) = \prod_{i=1}^{k} \frac{g_i !}{n_i ! (g_i - n_i)!}$$

Here $k = 2$

$$\therefore \qquad W(n_1 . n_2) = \frac{g_1 !}{n_1 ! (g_1 - n_1)!} \times \frac{g_2 !}{n_2 ! (g_2 - n_2)!}$$

(*i*) For macrostate (3, 2)

$$g_1 = 3, \qquad g_2 = 4$$
$$g_1 = 3, \qquad n_2 = 4$$

$\therefore$ Thermodynamic probability

$$W_{(3,\,2)} = \frac{3!}{3!(3-3)!} \times \frac{4!}{2!(4-2)!} = \frac{1}{0!} \times \frac{4!}{2! \times 2!}$$

$$= 1 \times \frac{4 \times 3 \times 2 \times 1}{2 \times 1 \times 2 \times 1} = 1 \times 6 = 6 \qquad [\because 0! = 1]$$

$$\therefore \qquad W_{(3,\,2)} = 6$$

(*ii*) For macrostate (1, 4)

$$g_1 = 3, \qquad g_2 = 4$$
$$n_1 = 1, \qquad n_2 = 4$$

$$\therefore \qquad W_{(1,\,4)} = \frac{3!}{1!(3-1)!} \times \frac{4!}{4!(4-4)!} = \frac{6}{2} \times \frac{1}{0!}$$

$$= 3 \qquad [\because 0! = 1]$$

or $\qquad W_{(1,\,4)} = 3.$

Example 12.18. *What are the number of ways to arrange two particles a and b in three energy states according to MB, BE and FD stawtistics ? Write in tabular form and explain.* **(K.U. 2000)**

Solution. The distribution of particles and hence the number of macrostates in the three types of statistics will be different on account of different basic assumptions.

Suppose we have two particles and three cells. The number of ways in which these two particles can be arranged in three cells depends upon the type of statistics applicable to these particles. If the particles are distinguishable let them be denoted by *a* and *b*. There will be *nine* possible ways in which these *two* particles can be arranged in *three* cells as shown in Table (*i*). This is M.B. statistics.

M.B. Statistics		
1	2	3
ab	×	×
×	*ab*	×
×	×	*ab*
a	*b*	×
b	*a*	×
a	×	*b*
b	×	*a*
×	*a*	*b*
×	*b*	*a*

(*i*)

B.E. Statistics		
1	2	3
aa	×	×
×	*aa*	×
×	×	*aa*
a	*a*	×
a	×	*a*
×	*a*	*a*

(*ii*)

F.D. Statistics		
1	2	3
a	*a*	×
a	×	*a*
×	*a*	*a*

(*iii*)

If these particles are indistinguishable let them be denoted by *a, a* then there will be *six* ways in which these two particles can be arranged in three cells as shown in Table (*ii*). This is B.E. statistics. If not more than one particle can be placed in any one cell then there are only *three* ways in which these two particles can be arranged in 3 cells as shown in Table (*iii*). This is Fermi-Dirac statistics.

Example 12.19. *Do electrons have zero energy at 0K ? If not why ?*

(*P.U. 2001, G.N.D.U. 2006, 2004; Pb. U. 2005*)

Solution. Energy of electrons at 0K. The electrons at 0K do not have zero energy. The energy of a system of *n* electrons at 0 K is given by $U = \frac{3}{5} n u_f$

The electrons at 0 K do not have zero energy because according to Pauli's exclusion principle only one electron can occupy the zero energy level. All other electrons will occupy higher energy levels and will not, thereofe, have zero energy.

Example 12.20. *Calculate the Fermi energy of free electrons in copper assuming that there is one free electron per atom. Hence find the average energy per electron at 0 K and show that it is about 100 times the average energy per gas molecule at room temperature (say 27°C). Given Avogadro's number = 6.02×10^{23} atoms/mole; atomic wt. of Cu = 0.0635 Kg. mole^{-1}, density = 8.94 gm/cm^3, h = 6.63×10^{-27} erg. sec and mass of electron m = 9.1×10^{-28} gm, Boltzmann's constant k = 1.38×10^{-16} erg/°C.*

(*H.P.U. 2000; P.U. 2001*)

Solution. Fermi energy $\qquad u_f = \frac{h^2}{2m}\left(\frac{3n}{8\pi V}\right)^{2/3}$

$$\text{Atomic weight} = 0.0635 \text{ kg mole}^{-1} = 63.5 \text{ gm mole}^{-1}$$

But $\qquad V = \frac{\text{Atomic weight}}{\text{density}} = \frac{63.5}{8.94}$ and $n = 6.02 \times 10^{23}$ atoms/ mole

$$\therefore \qquad \text{Electron density} = \frac{n}{V} = \frac{6.02 \times 10^{23} \times 8.94}{63.5} = 8.48 \times 10^2 \text{ electrons/ cm}^3$$

$$\therefore \qquad \text{Electron energy} = u_f = \frac{(6.63 \times 10^{-27})^2}{2 \times 9.1 \times 10^{-28}}\left[\frac{3}{8\pi} \times \frac{6.02 \times 10^{23}}{63.5} \times 8.94\right]^{2/3}$$

$$= 1.13 \times 10^{-11} \text{ erg.} = \frac{1.13 \times 10^{-11}}{1.6 \times 10^{-12}} = 7.06 \text{ eV}$$

In other words at absolute zero there will be electrons with energies upto 7.06 eV only in copper. The students should note that all the molecules of an ideal gas have *zero* energy at 0 K.

Average energy at 0K = $\bar{u} = \frac{3}{5} u_f = 4.23$ eV

Average energy per gas molecule at room temperature (27°C = 300 K)

$$= \frac{3}{2} kT = \frac{3}{2} \times 1.38 \times 10^{-16} \times 300 \text{ erg.}$$

$$= \frac{3}{2} \times \frac{1.38 \times 10^{-16} \times 300}{1.6 \times 10^{12}} = 3.83 \times 10^{-2} \text{ eV}$$

$$\therefore \qquad \frac{\bar{u}}{\text{Average energy per gas molecule}} = \frac{4.23}{3.83} \times 100 = 1.1 \times 100$$

Thus, we find that the electrons possess much more (about 100 times) energy at room temperature than that predicted by Maxwell Boltzmann's classical statistics.

Example 12.21. *The number of conduction electrons per c.c. is 24.2 × 10^{22} in beryllium and 0.91 × 10^{22} in Cesium, If the Fermi energy of conduction electrons in Be is 14.44 eV calculate that in Cesium.* **(G.N.D.U. 2007)**

Solution. Electrons per c.c. in Be $= \frac{n}{V} = 24.2 \times 10^{22}$

$$\text{Electrons per c.c. in Cs} = \frac{n}{V} = 0.91 \times 10^{22}$$

$$\text{Fermi energy } u_f = \frac{h^2}{2m}\left(\frac{3}{8\pi}\right)^{2/3}\left(\frac{n}{V}\right)^{2/3} = K\left(\frac{n}{V}\right)^{2/3}$$

$$\therefore \quad \frac{\text{Fermi energy for Be}}{\text{Fermi energy in Cs}} = \frac{K \times (24.2 \times 10^{22})^{2/3}}{K \times (0.91 \times 10^{22})^{2/3}} = \left(\frac{24.2}{0.91}\right)^{2/3}$$

$$\text{or} \quad \text{Fermi energy in Cs} = 14.44 \times \left(\frac{0.91}{24.2}\right)^{2/3} = 1.587 \text{ eV}$$

Example 12.22. *Assume that in Tungsten (At. wt. = 183.8 and density = 19.3 gm/c.c.) there are two free electrons per atom. Calculate the Fermi energy and electron density. Mass of electron = 9.05 × 10^{-28} gm; Avogadro's number = 6.02 × 10^{23} and h = 6.62 × 10^{-27} erg sec.* **(G.N.D.U. 2000)**

Solution. Volume per gram atom of Tungsten $V = \frac{\text{Atomic weight}}{\text{density}} = \frac{183.8}{19.3}$

As there are two free electrons per atom, the total number of electrons in one gm. atom of Tungsten

$$n = 2 \times 6.02 \times 10^{23}$$

$$\therefore \quad \text{Electron density} = \frac{\text{Total number of electrons in one gram atom}}{\text{Volume per gram atom}}$$

$$= \frac{n}{V} = \frac{2 \times 6.02 \times 10^{23} \times 19.3}{183.8} = 1.264 \times 10^{23} \text{ electrons per c.c.}$$

$$\text{Fermi energy} \quad u_f = \frac{h^2}{2m}\left(\frac{3n}{8\pi V}\right)^{2/3}$$

$$= \frac{h^2}{2m}\left(\frac{3}{8\pi}\right)^{2/3}\left(\frac{n}{V}\right)^{2/3} = \frac{(6.62 \times 10^{-27})^2}{2 \times 9.05 \times 10^{-28}} \times \left(\frac{3}{8\pi}\right)^{2/3} \times (1.264 \times 10^{23})^{2/3}$$

$$= 1.479 \times 10^{-11} \text{ erg.} = \frac{1.479 \times 10^{-11}}{1.6 \times 10^{-12}} = 9.24 \text{ eV}$$

Example 12.23. *Fermi energy of conduction electrons in Silver is 5.48 eV. Calculate the number of such electrons per cm³ given that h = 6.62 × 10^{-27} erg sec. and 1 eV = 1.62 × 10^{-12} erg.* **(G.N.D.U. 2001)**

Solution. Fermi energy $\quad u_f = \frac{h^2}{2m}\left(\frac{3n}{8\pi V}\right)^{2/3}$

$$\therefore \quad u_f^{3/2} = \frac{h^3}{(2m)^{3/2}} \frac{3}{8\pi} \frac{n}{V}$$

$$\therefore \text{ Number of electrons per cm}^3 \quad \frac{n}{V} = \frac{(2m)^{3/2}\, u_f^{3/2}\, 8\pi}{3h^3}$$

$$= \frac{2^{3/2} \times (9.1 \times 10^{-28})^{3/2} \; 8\pi \times (5.48 \times 1.62 \times 10^{-12})^{3/2}}{3 \times (6.62 \times 10^{-27})^3} = 5.9 \times 10^{22} \text{ cm}^{-3}$$

Example 12.24. *The Fermi energy of silver at 0K is 5.5 eV. Find (i) Average energy per electron in silver. (ii) Average speed of the electron.* **(G.N.D.U. 2007, 2006)**

Solution. Fermi energy of silver at 0K = u_f = 5.5 eV

$\therefore$ Average energy per electron in silver $\bar{u} = u_{av} = \frac{3}{5} u_f$

$$= \frac{3}{5} \times 5.5 = 3.3 \text{ eV}$$

$$= 3.3 \times 1.6 \times 10^{-19} = 5.28 \times 10^{-19} \text{ Joule}$$

Now, average energy $u_{av} = \frac{1}{2} m\, a_v^2$

$$\therefore \qquad a_v = \sqrt{\frac{2u_{av}}{m}} = \sqrt{\frac{2 \times 5.28 \times 10^{-19}}{9.11 \times 10^{-31}}} = 1.07 \times 10^6 \text{ ms}^{-1}$$

EXERCISE

12.1 Explain the need for introducing quantum statistics.
(M.D.U. 2006; H.P.U. 2008; P.U. 2001, 2000; GNDU 2008, 2004; Nagpur 2009, 2004/W)

12.2 'In quantum statistics, 'h' is selected as a constant by nature', explain by using the case of particle in a one-dimensional box. ***(Nagpur 2011/s)***

12.3 'One-dimensional harmonic oscillator employing quantum concept, has set a choice to select h as a natural constant, explain. What are its implications? ***(Nagpur 2005/W)***

12.4 What is indistinguishability of particles? What role it plays in quantum statistics? What are its consequences? ***(Nagpur 2005/W)***

12.5 What is Bose-Einstein statistics? What are the basic postulates used? Derive an expression

$$n_i = \frac{g_i}{e^{\alpha + \beta E_i} - 1}$$ for the most probable distribution of the particles of a system obeying *B.E.*

statistics and hence deduce Planck's black-body radiation formula.
(M.S. Uni. Tamil nadu 2007; Magadh 2001, 2003, Purvanchal 2006, 2000 Nagpur 2006/S)

12.6 What are bosons? Which statistics is used to study them?
(Agra 2007, 2002, 2004, Nagpur 2005/W, 2005/s, Madras 2004, Purvanchal 2005, Pune 2010, Madras 2004/S)

12.7 What are Fermions? Write down the postulates of Fermi-Dirac statistics. Derive an expression for the probability distribution of particles governed by Fermi-Dirac statistics.
(Madras 2004/W, 2005/S; Agra 2004, Nagpur 2009, 2004 (W), 2005/W, Pune 2010, Purvanchal 2005, 2001, Mumbai 2005, M.S. Uni. Tamil Nadu 2007)

12.8 Applying Fermi-Dirac distribution to an electron gas show that the distribution of energy among free electrons of a conductor is

$$n(E)\, dE = \frac{8\sqrt{2}\,\pi V}{h^3} \cdot \frac{m^{3/2}\, E^{\frac{1}{2}}\, dE}{e^{(\alpha + \beta E)} + 1}$$

12.9 What is electron gas? Starting from Fermi-Dirac distribution law, drive the expression for energy distribution of free electrons in a metal.

12.10 Starting from Bose-Einstein energy distribution law, derive Planck's law of black body radiation. ***(Purvanchal 2007, Garhwal 2005, Meerut 2004, Nagpur 2004/W, 2005/S, 2005/W, 2006/S, 2007, 2008, 2011/s, Purvanchal 2003)***

12.11 What is meant by Fermi-energy of conduction electrons. Calculate the value of Fermi-energy. ***(Purvanchal 2002, 2003, Nagpur, 2010, 2004/S)***

12.12 What is occupation index. Under what conditions do Bose-Einstein & Fermi-Dirac statistics yield classical statistics. ***(Nagpur 2009, Purvanchal 2001)***

12.13 Explain why behaviour of liquid helium cannot be explained by classical statistics. How it is over-comed by quantum statistics.

12.14 Describe the peculiar behaviour of liquid helium.

12.15 What is the difference between classical and quantum statistics. ***(Agra 2005, 2002, Purvanchal 2003, Rohilkhand 2002, Nagpur 2011/S 2006/S)***

12.16 Using the expression for Fermi-Dirac distribution of energy among electrons within a metal, prove that at absolute zero, the average *K.E.* is $\overline{E} = \frac{3}{5} E_F$. ***(Purvanchal 2007)***

12.17 Do electrons have zero energy at absolute zero ? If not why? Explain.

12.18 Define Fermi energy and Fermi temperature. Explain the significance of Fermi energy. ***(Agra 2005, Nagpur 2007)***

12.19 Discuss the free electron gas model for metals. Obtain the distribution law for the electron gas in a metal at absolute zero. ***(Nagpur 2005/W)***

12.20 Distinguish between Classical Statistics, Fermi-Dirac Statistics and Bose-Einstein Statistics. ***(Nagpur 2009, Purvanchal 2002, 2003, Rohilkhand 2001, 2002, Imp., Agra 2006, 2000, 2002, 2004, Meerut 2003, Avadh 2000, Madras 2005)***

12.21 Discuss the thermodynamic probability in B-E statistics. Obtain an expression for Bose-Einstein distribution law and derive Planck's radiation formula from the above law. ***(Purvanchal 2005, 2007, Nagpur 2007, Agra 2000, 2004, Madras 2005)***

12.22 Is neutron a boson or a fermion? What are bosons? Give example for boson. ***(Madras 2005/S)***

12.23 Discuss the postulates of quantum statistics. ***(Purvanchal 2006, Madras 2004/W)***

12.24 Compare the basic postulates of M-B, B-E and F-D statistics. ***(Napur 2011; Purvanchal 2001, Pune 2010)***

12.25 Derive Fermi-Dirac distribution formula. How does this distribution tend to classical. ***(Wagpur 2011/s, 2010; Purvanchal 2002, 2003, Agra 2003)***

12.26 How does F-D statistics differ from B.E. statistics ? ***(Magadh 2003, Nagpur 2005/S)***

12.27 What do you mean by distinguishable and indistinguishable particles. ***(Nagpur 2005/S, 2005/W)***

12.28 Explain the concepts of quantum mechanics which have influenced the transition from classical to quantum mechanics. ***(Nagpur 2004/S, 2004/W)***

12.29 What is Fermi function ? Give its significance. ***(Nagpur 2007, 2004/S)***

12.30 What are symmetric and antisymmetric wave functions ? ***(Pune2010; Madras 2004)***

12.31 In Fermi-Dirac statistics, particles are not obeying Pauli's exclusion principle, is it true ? What is the spin value of Fermions ? ***(Madras 2005)***

12.32 What are classical and quantum particles. ***(Madras 2004)***

12.33 (*i*) State two conditions for applicability of Bose-Einstein statistics.

(*ii*) Using Bose-Einstein statistics, derive Planck's formula for energy distribution in the black body spectrum. ***(Nagpur U. 2011/s)***

(*iii*) Draw the energy distribution curve of a black body spectrum. ***(Nagpur Uni. 2007)***

12.34 Derive the relation for Fermi energy of electron in metal at $T = 0K$.

$$E_{F_0} = \frac{h^2}{2m} \times \left(\frac{3n}{3\pi}\right)^{2/3}$$

where letters have their usual meaning. ***(Nagpur 2011/s, 2007)***

12.35 Write short notes on:

(*i*) Statistical equilibrium
(*ii*) Electron gas
(*iii*) Photon gas
(*iv*) Phase gas
(*v*) Quantum statistics
(*vi*) Fermi temperature ***(Purvanchal 2007)***
(*vii*) Fermions and Bosons ***(Nagpur 2000, Madras 2004/W)***
(*viii*) Planck's radiation law ***(Meerut 2001)***
(*ix*) Statistical thermodynamics
(*x*) Maxwell-Boltzmann distribution law
(*xi*) Fermi-Dirac distribution law ***(Magadh 2001)***
(*xii*) Bose-Einstein distribution law

12.36 What is electron gas (or Fermi gas) ? Give an example. How does free electron gas differ from an ordinary gas ?
(P.U., 2007, 2003, 2001; G.N.D.U., 2007, 2006, 2004, 2000; M.D.U. 2006, 2005)

12.37 Five bosons are distributed in two compartments the first having 3 cells and the second 4. Find the thermodynamic probability for the macrostate (*i*) (5, 0) and (*ii*) (4, 2)
[Ans. 21,150**]** ***(Purvanchal 2006)***

12.38 Calculate the average energy of a Planck's oscillator of frequency 1.5×10^{14} at $T = 1800K$, $h = 6.62 \times 10^{-34}$ *Js* and $k = 1.38 \times 10^{-23}$ Jk^{-1}. **[Ans.** 1.86×10^{-21} J**]**

12.39 A system has 7 particles arranged in two compartments. The first compartment has 8 cells and the second has 10 cells. All cells are of equal size. Calculate the number of microstates in the macrostate (3,4) when the particles obey *F.D.* statistics. **[Ans.** 11760**]**

12.40 Find the average velocity of electrons at $0°C$ in a metal having 3×10^{22} electrons per cm^3.
[Ans. 0.835×10^8 cm^{-1}**]**

12.41 A system consists of 8000 particles distributed in three energy states with equal spacing. The energy of the three states are $E_1 = 0$, $E_2 = x$, $E_3 = 2x$. All the three states have the same intrinsic probability *g*. At a certain instant, there are 4000 particles in the lower level, 3000 particles in the middle level and 1000 particles in the upper level. Compare the relative probabilities with the distribution obtained by the transfer of one particle from middle to the lower level and one particle from the middle to the upper level and the original distribution. **[Ans.** 2.247**]**

12.42 Calculate the number of ways of arranging 8 fermions in 12-phase space cells.
[Ans. 495**]** ***(Punjab 2005)***

MULTIPLE CHOICE QUESTIONS

12.1 Average energy of a Planck's oscillator is : ***(Garhwal 2005)***

(*a*) $E = h\nu$ (*b*) $E = nh\nu$ (*c*) $E = \dfrac{h\nu}{(e^{h\nu/kT} - 1)}$ (*d*) $E = mc^2$

12.2 From Fermi-Dirac statistics $n_i = ?$ ***(Agra 2004)***

(*a*) $\dfrac{g_i}{e^{\alpha+\beta E_i}+1}$ (*b*) $\dfrac{2g_i}{e^{\alpha+\beta E_i}+1}$ (*c*) $\dfrac{2g_i}{e^{\alpha+\beta E_i}+1}$ (*d*) $\dfrac{g_i}{e^{\alpha+\beta E_i}+1}$.

12.3 Deduction of Planck's law is possible on the basis of ***(Agra 2003)***
(*a*) Fermi-Dirac (F.D) statistics (*b*) Classical statistics
(*c*) Maxwell-Boltzmann (M.B) statistics (*d*) Bose-Einstein (B-E) statistics.

12.4 Boson particles obey Pauli's exclusion principle : ***(Agra 2003)***
(*a*) True (*b*) False
(*c*) Can't say (*d*) Sometimes true sometimes false.

12.5 According to which statistics, the energy at absolute zero cannot be zero.
(*a*) M-B (*b*) B-E (*c*) F-D (*d*) None of these.

12.6 Fermions have spin value :
(*a*) $\frac{1}{2}$ (*b*) 1 (*c*) Zero (*d*) Any one.

12.7 Boson have spin value :
(*a*) 0 (*b*) 1 (*c*) $\frac{1}{2}$ (*d*) 0 or 1.

12.8 Photons obey the statistics :
(*a*) M-B (*b*) B-E (*c*) F-D (*d*) None of these.

12.9 In how many ways two particles can be arranged in three phase cells according to B-E statistics :
(*a*) 6 (*b*) 9 (*c*) 3 (*d*) 27.

12.10 According to F-D statistics, 3 particles can be distributed in 4 energy states, in number of ways :
(*a*) 16 (*b*) 4 (*c*) 9 (*d*) 12

12.11 The spin of photon is :
(*a*) Zero (*b*) $\frac{1}{2}\hbar$ (*c*) $\hbar$ (*d*) $\frac{3}{2}\hbar$

12.12 Pauli's exclusion principle applies to : ***(H.P.V 2001)***
(*a*) M.B. Statistics (*b*) B.E. Statistics (*c*) F.D. Statistics (*d*) None.

12.13 In which the probability of occupation index can tend to zero :
(*a*) Maxwell-Boltzmann (*b*) Fermi-Dirac
(*c*) Bose-Einstein (*d*) None of these.

12.14 The number of meaningful ways 4 Fermions can be arranged in 5 compartments :
(*a*) 1 (*b*) 4 (*c*) 5 (*d*) 9

12.15 The number of different arrangements of 6 indistinguishable particles among 4 cells of equal a priori probability, there being no restriction of number of particles entering into a cell is
(*a*) 24 (*b*) 30 (*c*) 84 (*d*) 210

12.16 The average energy of an electron in Fermi gas at 0K is
(*a*) 0.2 u_f (*b*) 0.4 u_f (*c*) 0.6 u_f (*d*) 0.8 u_f

ANSWERS

12.1 (*c*)	**12.2** (*a*)	**12.3** (*d*)	**12.4** (*b*)	**12.5** (*c*)	**12.6** (*a*)
12.7 (*d*)	**12.8** (*b*)	**12.9** (*a*)	**12.10** (*b*)	**12.11** (*c*)	**12.12** (*c*)
12.13 (*d*)	**12.14** (*c*)	**12.15** (*c*)	**12.16** (*c*)		

SECTION IV

Waves

WAVES IN MEDIA

13

Chapter

INTRODUCTION

The propagation of a wave through any medium (solid, liquid or gas) is associated with disturbance and results into wave motion. The motion of a single particle is simple to understand under different conditions. But when we have large number of particles linked (i.e. Connected) together by elastic chains, it is not possible to displace any one of them without disturbing the other. The motion imparted to one is communicated through the chains to others and they in turn are set into *motion*. An elastic medium consists of very fine particles with elastic connection between them. If we impart periodic motion to any particle, its energy of motion will be communicated to the adjacent particles surrounding it, which will thus execute similar periodic motions and in doing so will in their turn transfer their energy to the neighbouring particles surrounding them and so on. The motion imparted to the particle will move uniformly to all directions from particle to particle with a definite velocity.

13.1 WAVE MOTION

The propagation of a disturbance due to transfer of energy from particle to particle results into wave motion. Thus, wave motion is defined as under:

Wave motion is a form of disturbance which travels through a medium due to the repeated periodic motion of the particles of the medium about their mean positions, the disturbance being handed on from one particle to the next particle of the medium.

It may be noted that there is no bodily transfer of the medium through which the waves propagates, but it is only the disturbance which travels forward.

Examples. When a stone is dropped into a calm pond containing water, waves are produced at the point where the stone strikes the water in the pond. The waves travel outward, the particles of water (*i.e.* Medium) vibrate only up and down about their mean positions. This can be seen by placing a light piece of paper or a wooden cork. The piece of paper only moves up and down about its mean position but does not move forward. In other words, water particles do not travel along with the wave.

Then-what is propagated in a wave motion? The answer to this question is that the physical condition due to a disturbance generated at some point in the medium is propagated to other points in the medium. *In all the waves, the particles of the medium vibrate about their mean positions*. Hence, in case of wave motion, it is not matter that is propagated but it is only **state of motion of the matter** that is propagated from one point to the other point in the medium. It is a form of dynamic condition that is propagated from one point to the other point in the medium.

Another example that can be cited is that of light. As we know, when the bulb is switched 'ON' the room is filled with light. Light waves have been generated and propagated. Light waves also carry energy. It is possible to transmit an electric signal (or a message) in the form of electromagnetic (E-M) wave from one place to another. Although these various processes of transport of energy are different yet they have a common feature which we shall call wave motion.

The concept of wave motion in a medium is essential during the study of various branches in physics, where it deals with the propagation of various forms of energy like sound, X- rays, γ-rays, radio waves etc. in the form of wave motion. For transfer of energy through medium, the medium must possess the properties of **elasticity, inertia and negligible frictional resistance.**

As mentioned earlier, it is the dynamic condition that is propagated in the medium from one point to the other. According to the laws of physics, any dynamic condition is related to transfer of *momentum* and *energy*. To conclude, it may be said that in wave motion momentum and energy are transferred or propagated and not the propagation of matter of medium.

13.2 DIFFERENT TYPES OF PROGRESSIVE WAVES

There are two different types of progressive waves :

1. Progressive transverse wave, and

2. Progressive longitudinal wave

Progressive transverse wave in which every particle vibrates in a direction perpendicular to the direction of propagation of the wave. They have the same amplitude but differ in phase with distance.

Water waves (ripples) and rope waves are transverse waves. Alternate dipping and rising of a pin at any point on the surface of still water will generate ripples which propagate outwards uniformly in all directions in the form of circles of alternate crests and troughs of gradually increasing radius. Every particle of water executes S.H.M. about their mean position of rest at right angles to the line of advance *i.e.* propagation. A cork floating on water which was originally at rest, will rise and fall as wave advance through it.

A thick rope say a rubber tubing suspended from one end, the other end of which is moved a little at right angles to the length with given horizontal periodic motion will cause the disturbance in the form of sinusoidal wave to travel along the length upwards. Every particle of the rope will execute to and fro motion perpendicular to the length of the rope without advancing in the forward direction.

The motion of stringed (wires used) musical instruments are the examples of transverse wave. The total length of a crest and a trough, is called the wavelength. The distance between two consecutive crests or consecutive troughs is also a measure of wavelength of the transverse wave.

Progressive longitudinal wave in which every particle vibrates along the direction of propagation of the wave. They are same in amplitude but differ in phase with distance.

Consider the example of a spring suspended horizontally from a rigid support by a bifilar suspension as shown in fig. 13.1. Whe a small push is applied at one end of the spring, progressive logitudinal waves are formed in it. The rims move to and fro along the axis of the spring producing compressions (C) and rare-factions (R) that propogates along the length of the spring.

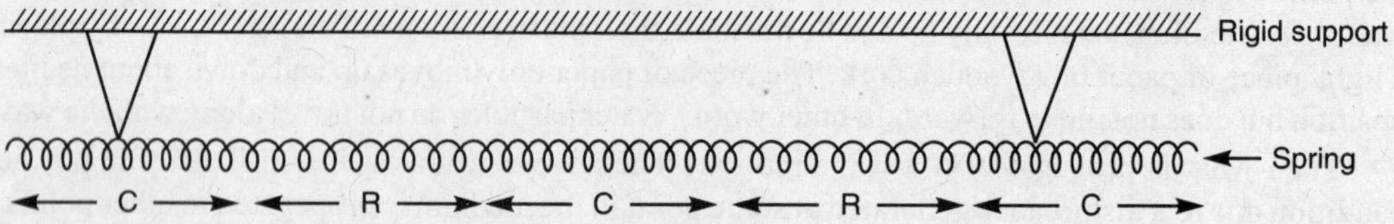

Fig. 13.1

Here the motion of individual turn of the spring is along the direction of propagatio. This is an example of longitudinal wave.

13.3 CHARACTERISTICS OF WAVE MOTION

1. Wave motion is a disturbance produced in the medium by the repeated periodic motion of the particles of the medium.
2. Only the wave travels forward whereas the particles of the medium vibrate about their mean positions.
3. There is a regular phase change between the various particles of the medium. The particle ahead starts vibrating a little later than a particle just preceding it.

4. The velocity of the wave is different from the velocity of the particles. The wave travels with a **uniform** velocity whereas the velocity of the particles is **different** at different positions. It is maximum at the mean position and zero at the extreme position of the particles.
5. Wave motion is possible only in media which have the property of inertia and elasticity.
6. The energy of the vibrating particle at the extreme positions is wholly potential and in the mean position wholly kinetic.
7. When a wave travels in a medium, there is a flow of energy without any transfer of matter of the medium.

Vector relation. Mathematical vector relation between unit vector along the displacement of the vibrating particles about its mean position and the unit vector along the direction of propagation of the wave can be put as under.

If the displacement vector of the vibrating particles of the medium is denoted by $\vec{y}$ and the vectorial distance along the direction of prapogation of the wave by $\vec{x}$, then unit vectors $\hat{x}$ and $\hat{y}$ are related as:

For transverse wave $\hat{x}.\hat{y}=0$ but $\hat{x}\times\hat{y}\neq 0$ and

For longitudinal wave $\hat{x}.\hat{y}\neq 0$ but $\hat{x}\times\hat{y}=0$

13.4 DEFINITIONS

Wavelength. ***It is distance travelled by the wave in the time in which the particle of the medium completes one vibration.***

It is also defined as the minimum distance along the line of propagation between two particles of the medium vibrating in the same phase. The phase difference between two particles separated by a wavelength λ is 2π. The distance *AB* (Fig.13.2) is equal to the wavelength λ.

Frequency. ***It is the number of vibrations made by a particle in one second and is usually denoted as n.***

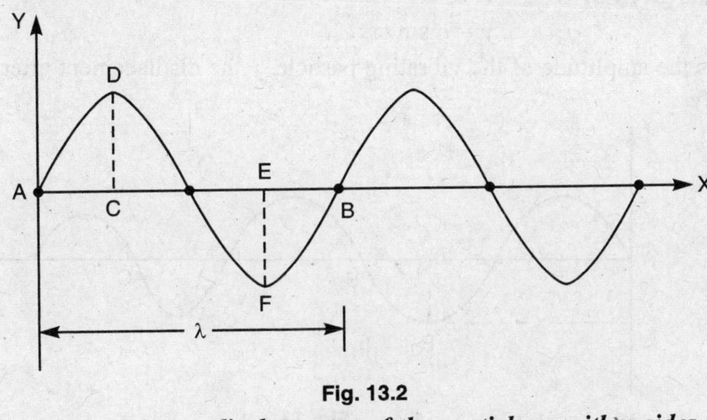

Fig. 13.2

Amplitude. ***It is the maximum displacement of the particle on either sides from its mean position of rest.*** In Fig. 13.2, *CD* (or *EF*) is the amplitude.

Time period. ***It is the time taken by a particle to complete one vibration.***

Suppose the frequency is *n*. Time taken to complete *n* vibrations = 1 sec. Hence, time taken to complete one vibration = (1/*n*) second. Thus, from the definition of time period, time taken to complete one vibration is the time-period (*T*)

$$\therefore \quad T=\frac{1}{n} \quad \text{or} \quad nT=1$$

or, Frequency (*n*) = 1/ Time period

Wave number. ***The number of wavelengths in a unit distance is called wave number. It is denoted by*** $\bar{\nu}$

$$\therefore \quad \bar{\nu} = \frac{1}{\lambda}$$

Vibration. ***It is the to and from motion of a particle from one extreme position to the other and back again.*** It is also equal to the motion of a particle from the mean position to one extreme position, then to the other extreme position and finally back to the mean position.

Phase. ***It is defined as the ratio of the displacement of the vibrating particle at any instant to the amplitude of the vibrating particle.*** It is also defined as the fraction of the time interval that has elapsed since the particle crossed the mean position of rest in the positive direction or it is also equal to the angle swept by the radius vector since the vibrating particle last crossed its mean position of rest.

Relation between frequency and wavelength

If the distance through which the energy is propogated *i.e.*, wave is propagated is unit time is v, the velocity of propagation of the wave and n be the frequency of vibration, then the wavelength $\lambda = v/n$ or $v = n\lambda$. The S.I. unit of wavelength is metre, frequency Hertz (Hz) and time period second.

13.5 GENERAL EQUATION OF A PROGRESSIVE WAVE

In a wave motion, the particles of the medium execute simple harmonic vibrations about their mean position along or perpendicular to the direction of propagation of the wave according as it is a longitudinal or a transverse wave. The disturbance is handed on from particle to particle after a definite time and there is a gradual fall of phase in the direction of motion.

Consider a transverse wave in a string propagating from left to right along the positive X-direction starting from A as shown in Fig.13.3. A particle on the right will begin its vibrations a certain time later than the one on the left.

As each particle of the medium executes simple harmonic motion the equation of motion of any particle, say A, is given by

$$y = a \sin \omega t$$

Where a is the amplitude of the vibrating particle, y the displacement after a time t and ω the angular velocity.

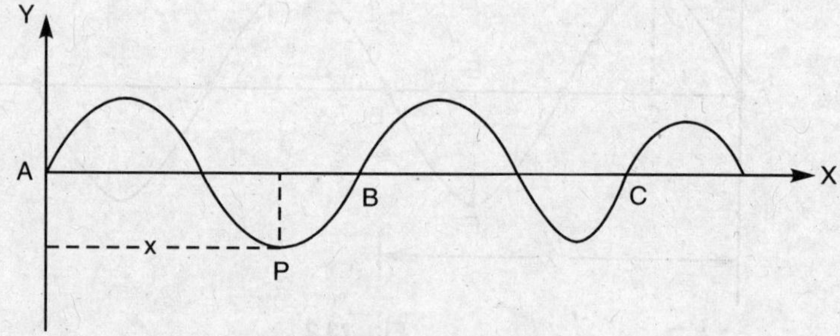

Fig. 13.3

If n the frequency of vibration, then $\omega = 2\pi n$

$$\therefore \quad y = a \sin 2\pi nt$$

When the particle A passes through its mean position, the particles like B, C, etc., also pass through their mean postions in the same direction.

Hence the particles A, B, C... etc. are in the *same phase*.

The distance AB between the two consecutive particles in the same phase is called the wavelength and denoted by λ. The phase changes by 2π in going from A to B.

Therefore, in going from the point A to any point P at a distance x from A the phase changes by ϕ and is given by

$$\phi = \frac{2\pi}{\lambda}x$$

Hence the displacement of P is given by

$$y = a \sin(\omega t - \phi)$$

$$= a\sin\left(2\pi nt - \frac{2\pi}{\lambda}x\right)$$

$$= a\sin\left(2\pi\frac{v}{\lambda}t - \frac{2\pi}{\lambda}x\right)$$

or

$$y = a\sin\frac{2\pi}{\lambda}(vt - x) \qquad ...(i)$$

The equation is quite general and gives the displacement of any particles whose distance x from the fixed point A at any time is known.

Hence this is the equation of a progressive simple harmonic wave.

The number of wavelengths in a unit distance is called *wave number*. It is denoted by $\overline{v}$

$$\therefore \qquad \overline{v} = \frac{1}{\lambda} \qquad ...(ii)$$

The quantity $\frac{2\pi}{\lambda} = k$ is called *propagation constant*

Other forms: Equation (i) can also be written as

$$y = a\sin\left(\frac{2\pi v}{\lambda}t - \frac{2\pi}{\lambda}x\right)$$

$$= a \sin(\omega t - kx) \qquad ... (iii) \qquad \left[\because \frac{2\pi v}{\lambda} = 2\pi n = \omega\right]$$

In exponential form

$$y = ae^{i(\omega t - kt)} \qquad ... (iv)$$

The equations $y = a \sin \frac{2\pi}{\lambda}(vt - x)$, $y = a \sin(\omega t - kx)$ and $y = ae^{i(\omega t - kx)}$ denote a wave moving to the right along $+X$ axis.

If the wave is moving to the left (along $-X$–direction) the sign of ϕ is changed because the oscillations at x begin before that at $x = 0$ and the equations are represented as

$$y = a\sin\frac{2\pi}{\lambda}(vt + x), y = a\sin(\omega t + kx) \qquad ... (v)$$

and

$$y = ae^{i(\omega t + kx)} \qquad ... (vi)$$

13.6 VOLUMETRIC STRAIN

Consider a displacement curve $ABCDE$ of a wave set up in a cylinder having unit area of cross-section and its axis coincident with the direction of propagation of the wave ACE. Consider two neighbouring points P and Q distance dx apart. The displacement at P is PR and that at Q is QS. The volume of the gas between the points P and Q is given by

$$PQ \times 1 = PQ$$

When the wave is in the position shown, the particle P is displaced by PR and the particle Q is displaced by QS. Hence the change in volume between P and Q is given by

$$QS - PR = ST = dy$$

$$\therefore \quad \text{Volumetric strain} = \frac{\text{change in volume}}{\text{original volume}}$$

$$= \frac{ST}{PQ} = \frac{ST}{RT} = \frac{dy}{dx}$$

= the slope of the displacement curve

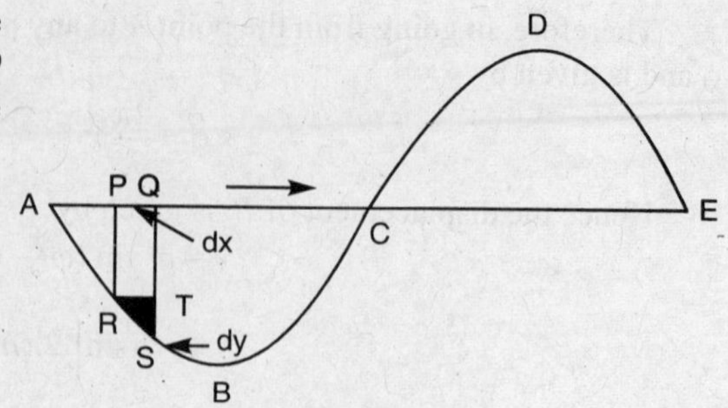

Fig. 13.4

Thus the slope of the displacement curve at any point measures the volumetric strain at that point.

13.7 VELOCITY AND ACCELERATION OF A PARTICLE IN A PROGRESSIVE WAVE

The displacement of a particle in a medium is given by $y = a\sin\frac{2\pi}{\lambda}(vt - x)$ ... (*i*)
where v is the velocity of the progressive wave.

The velocity of the particle at a place at an instant t is obtained by differenting equation (*i*) with respect to time, we get

$$\text{Velocity} = \frac{dy}{dt} = \frac{2\pi v}{\lambda} a\cos\frac{2\pi}{\lambda}(vt - x) \quad \text{... (ii)}$$

But

$$\frac{dy}{dx} = -\frac{2\pi}{\lambda} a\cos\frac{2\pi}{\lambda}(vt - x) \quad \text{... (iii)}$$

Now $\frac{dy}{dx}$ represents the strain or compression in the medium.

When $\frac{dy}{dx}$ is positive a rarefaction takes place and when it is negative, a compression takes place.

From equation (*ii*) and (*iii*), we have

$$\frac{dy}{dt} = -v\frac{dy}{dx} \quad \text{... (iv)}$$

Particle velocity = Wave velocity × slope of displacement curve or strain.

Thus, the velocity of the particle in the wave is not the same that of velocity of the wave. At any point x, at an instant t, the velocity of the particle (dy/dt) is $-v$ times the slope of the wave displacement curve at the point. (Explained in next article). Differentiating velocity dy/dt in equation (*ii*), with respect to time, we get acceleration as,

$$\frac{d^2y}{dt^2} = -\left(\frac{2\pi v}{\lambda}\right)^2 a\sin\frac{2\pi}{\lambda}(vt - x)$$

$$= -\left(\frac{2\pi v}{\lambda}\right)^2 y$$

$$= -\omega^2 y \quad \text{... (v)}$$

As

$$\omega = 2\pi n = \frac{2\pi}{T} = \frac{2\pi v}{\lambda}$$

where n is the frequency, T the time period. From equation (v), it is also implied that the particle executes S.H.M.

Differentiating equation (*iii*) w.r.to x, we get

$$\frac{d^2y}{dx^2} = -\left(\frac{2\pi}{\lambda}\right)^2 a\sin\frac{2\pi}{\lambda}(vt - x)$$

or

$$\frac{d^2y}{dx^2} = -\frac{\omega^2}{v^2}.y \qquad \text{... (vi)}$$

Comparing equation (v) and (vi), we get

$$\frac{d^2y}{dt^2} = v^2\frac{d^2y}{dx^2} \qquad \text{... (vii)}$$

Thus the acceleration of the particle is equal to v^2 times the curvature of the wave displacement curve. Equation (*vii*) represents the differential equation of the wave motion.

The general solution of equation (*vii*) is

$$y = f(vt - x) + g(vt + x)$$

Where f and g are any two functions. f represents plane wave moving in the positive x-direction with velocity v i.e., forward wave and g represent a plane wave moving in the negative x-direction *i.e., backward wave.*

Propogation of sound waves through air

A source of sound is always a vibrating body or vibrating air column. All vibrating bodies do not perform a single S.H.M. but a complex note. However, a tuning fork is ordinarily the only source which produces sound by performing a single S.H.M. of a fixed frequency. Hence the tuning forks producing different frequencies, say 308 Hz, 438 Hz, 512 Hz etc are available. This is due to the structure of a tuning fork. Rayleigh regarded the tuning fork as two clamped free bars mounted parallel to each other on a heavy rigid block of metal. The two bars form a U elongated. This peculiar structure of tuning fork produces a single S.H.M.

When a tning fork vibrates and motion of one of the prongs is outwardly for the first time at the beginning will compress the layer of air immediately in front of it. When the prong moves back to the extreme opposite position the compressed layer expands and in doing so, pushes the layer in front of it and comes backwards in the space in which pressure has been released due to motion backwards. The compression will successively travel outward from layer to layer from the prong. By this time, the prong has swung back and starts a journey leftward. As such, it creates a partial vacuum behind causing a rarefaction of the layer in contact with it. This rarefied layer tends to return to its normal condision and causes a rarefaction of the layer in contact with it. This rarefied layer tends to return to its normal condition and causes a rarefaction of the adjoining layer. Thus a pulse of rarefaction travels outward from layer to layer from the prong. During the vibration of the prong, alternate compression (condensation) and rarefaction are produced in the medium which travels outwards from the source. A compression and rarefaction constitute a longitudinal wave progressing in the outward direction. The distance between successive compression or rarefaction is called wavelength of longitudinal waves.

The frequency of sound that an human ear responds is from 20 Hz to 20 KHz. Hence it is called audio frequency (A.F) range. (Musical instruments produce the sound waves in this range. Electronic instruments like audio frequency oscillator (A.F.O) or A.F. geneators can also be employed to produce sound of desired frequency using a loud-speaker.

Here, again the question comes in mind. Do the molecules of the medium flow from the speaker to the listener or audiance? The answer is No.

The molecules of the medium do not actually flow from the speaker to the listener. The molecules only oscillate about their mean positions. Each molecule receives evergy from the neighbouring molecule immediately preceeding it and is itself set into oscillation. In the process it further transfers energy to its immediate next neighbouring moleculule. In this way it is the energy that flows from the speaker to the audience in the form of longitudinal sound waves due to the repeated periodic motion of the molecules of the medium about their mean positions.

We will discuss first the transverse waves and then longitudinal waves.

I. TRANSVERSE WAVES

13.8 TRANSVERSE WAVES IN A STRING

A string is a cord whose length is very large as compared to its diameter and which is perfectly uniform and flexible. When a string is streched between two points and is plucked in a direction at right angles to its length transverse vibrations are produced in it. The particles vibrate perpendicular to its length and movement is handed on from particle to particle. A transverse wave travels along the string with a velocity depending upon certain constants of the string.

Velocity of transverse waves. Consider a part of the wave *AB* travelling in the string from left to right with a velocity *c*. If the string is pulled from right to left with the same velocity the wave *AB* will remain stationary with respect to the paper in space. A small past PQ of the wave *AB* can be considered to be the arc of a circle. As the string moves along the circular arc *PQ* a centripetal force acts on it along the radius towards the centre O. (fig.13.5)

A uniform tension or stretching force *T* acts in the string throughout. This tension *T* acts along *CP* at *P* and along *CQ* at *Q*, where *CP* and *CQ* are tangents to the arc *PQ* at *P* and *Q* respectively. Draw *PO* and *QO* perpendiculars to the arc at *P* and *Q* meeting at *O* the centre of the circular arc having a radius *r*.

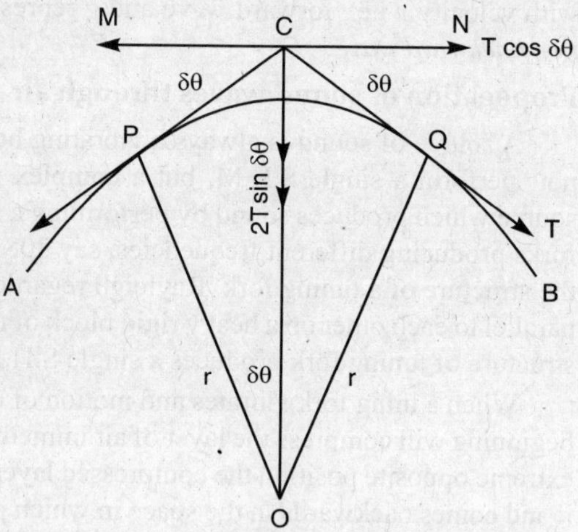

Fig 13.5

Let the angle *POQ* be denoted by 2 δθ. Join *OC*; then *OC* is the bisector of the angle *POQ*.

$\because$ Arc $PQ = r \times 2\,\delta\theta$

Let ρ be the mass per unit length of the string, then

Mass of the part $PQ = \rho \times PQ = 2\,\rho r\delta\theta$

The centripetal force acting on $PQ = \dfrac{2\rho r\delta\theta c^2}{r} = 2\rho c^2\delta\theta$ ...(*i*)

The tension *T* can be resolved into two rectangular components. The horizontal component $T\cos\delta\theta$ along *CM* and $T\cos\delta\theta$ alongt *CN*, cancel each other being equal and opposite. The vertical components of tension act in the same direction and total force along *CO*.

$$= 2\,T\sin\delta\theta = 2\,T\,\delta\theta \qquad [\because \delta\theta \text{ is small}] \qquad \text{...}(ii)$$

This provides the necessary centripetal force. Equating (*i*) and (*ii*), we have

$$2T\delta\theta = 2\,\rho c^2\,\delta\theta$$

or
$$T = \rho c^2$$

$\therefore$ Velocity of the transverse wave $c = \sqrt{\dfrac{T}{\rho}}$

Frequency of the fundamental note. When a string is fixed at both ends, it is said to produce a *fundamental note*, if it vibrates as a whole in one segment. In such a case the length *l* of the string is equal to $\dfrac{\lambda}{2}$ where λ is the wavelength of the note

$$\therefore \qquad \lambda = 2\,l$$

Now frequency $$n = \frac{c}{\lambda} = \frac{l}{2l}\sqrt{\frac{T}{\rho}}$$

This is the frequency of the fundamental note.

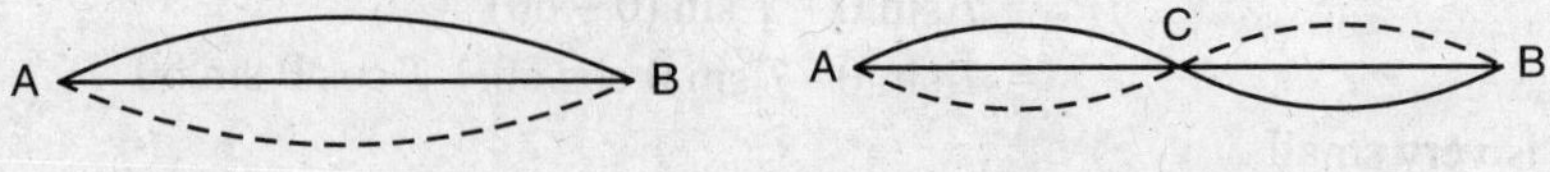

Fig 13.6

The two fixed ends *A* and *B* of the string act as nodes.

Frequency of first octave. When the string vibrates in two segments, it is said to produce *first octave.* In such a case $\lambda = l$

$$\therefore \text{Frequency of the first octave} = n_1 = \frac{c}{\lambda} = \frac{1}{l}\sqrt{\frac{T}{\rho}}$$

In addition to the two ends *A* and *B,* there is a node at the mid point *C.*

13.9 WAVE EQUATION FOR TRANSVERSE WAVE IN A STRING

Consider a string *AB* plucked in the middle so that when left to itself it begins to vibrate. The components of the tension *T* at right angles to the length of the string tend to bring the string back to equilibrium position. Consider a small element *PQ* of the string, as shown in Fig 13.7.

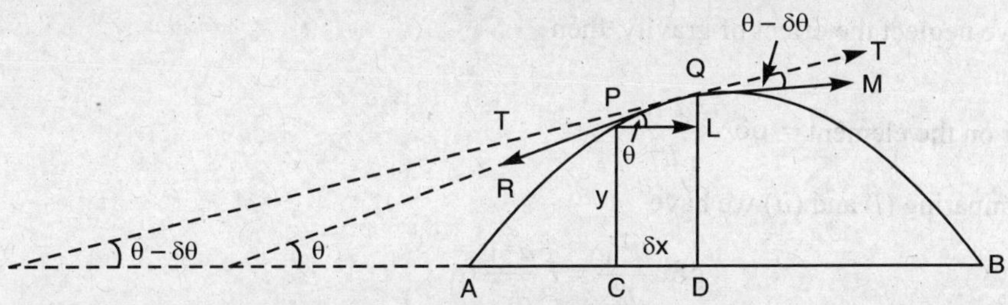

Fig. 13.7

The tension *T* at *P* acts along the tangent *PR* and at *Q* along the tangent *QT*. It is supposed that the gradient of the curve formed by the string in the displaced position is very small so that the tension *T* acts uniformly along the string and the normal position of the string coincides with the *X*-axis of the co-ordinate system.

The small element *PQ* of the string executes simple harmonic motion and undergoes a vertical displacement which varies with the position of the element as well as with time.

From *P* and *Q* draw *PL* and *QM* respectively parallel to *AB.* The tangent at *P* is inclined to the *X*-axis and the line *PL* at an angle θ and the tangent at *Q* is inclined to *X*- axis and the line *QM* at an angle θ – δθ. The angle δθ is very small as the amplitude of string is small.

The components of tension *T* acting at *P* are

$T\cos\theta$ in the horizontal direction

and $T\sin\theta$ in the vertically downward direction.

Similarly, the components of tension *T* acting at *Q* are

$T\cos(\theta - \delta\theta)$ in the horizontal direction

and $T\sin(\theta - \delta\theta)$ in the vertically upward direction

As θ and $(\theta - \delta\theta)$ differ by a very small value $\delta\theta$

$$T\cos\theta = T\cos(\theta - \delta\theta)$$

Hence the two horizontal components cance each other as they act almost in the same line in opposite directions. The resultant force F acting in the vertically downward direction.

$$= T\sin\theta - T\sin(\theta - \delta\theta)$$

$$= T\sin\theta - T\sin\theta\cos\delta\theta + T\cos\theta\sin\delta\theta$$

But $\delta\theta$ is very small

$$\therefore \quad \cos\delta\theta = 1 \text{ and } \sin\delta\theta = \delta\theta$$

Hence the resultant force $F = T\sin\theta - T\sin\theta + T\cos\theta\delta\theta$

$$= T\delta(\sin\theta)$$

As θ is small, $\sin\theta = \tan\theta = \dfrac{dy}{dx}$

where $\dfrac{dy}{dx}$ is the slope of the curve

$$\therefore \quad F = T\delta\left(\frac{dy}{dx}\right) = T\left[\frac{d}{dx}\left(\frac{dy}{dx}\right)\right]\delta x$$

$$= T\frac{d^2y}{dx^2}\delta x \qquad \text{... (i)}$$

If ρ is the mass per unit length of the string, then mass of the element PQ of length $\delta x = \rho\delta x$.

The accleration of the element in the direction of Y-axis $= \dfrac{d^2y}{dt^2}$

If we neglect the effect of gravity, then

For on the element $= \rho\delta x \times \dfrac{d^2y}{dt^2}$ ... (ii)

Comparing (i) and (ii) we have

$$\rho\delta x\frac{d^2y}{dt^2} = T\frac{d^2y}{dx^2}\delta x$$

or

$$\frac{d^2y}{dt^2} = \frac{T}{\rho}\frac{d^2y}{dx^2}$$

or

$$\frac{d^2y}{dt^2} = c^2\frac{d^2y}{dx^2} \qquad \text{... (iii)}$$

Where $\sqrt{\dfrac{T}{\rho}} = c = a$ `constant. The dimensions of c are those of velocity $[L^1 T^{-1}]$

The equation (iii) is the differential equation of wave motion and can be put in the form.

$$\frac{d^2y}{dx^2} = \frac{1}{c^2}\frac{d^2y}{dt^2}$$

The wave equation $\dfrac{d^2y}{dt^2} = c^2\dfrac{d^2y}{dx^2}$ has been derived by considering the vibrating string as a proto type. Any partial differential equation of the form.

$$\frac{\partial^2 \psi}{\partial t^2} = c^2 \frac{\partial^2 \psi}{\partial x^2}$$

or
$$\frac{\partial^2 \psi}{\partial x^2} = \frac{1}{c^2}\frac{\partial^2 \psi}{\partial t^2}$$

represents a general wave equation where ψ is a function of space and time co-ordinates and c the velocity with which the wave travels without any change of form. The wave variable ψ represents the transverse displacement y in the case of a string, pressure in the case of sound waves in a gas and electric field vector $\vec{E}$ in the case of electro-magnetic waves and so on. For plane waves in one dimension only it is a function of x but in the case of waves in two or three dimensions it is a function of two or three space co-ordinates.

Solution of the wave equation. We shall consider the wave equation in the form

$$\frac{d^2 y}{dt^2} = c^2 \frac{d^2 y}{dx^2}$$

It is clear from the form of the wave equation that the solution of this equation must be a linear function of the variables x and t. A function of the form

$$y = f_1(ct - x) \quad \text{or} \quad y = f_2(ct + x)$$

is one of its solutions. Therefore, a general solution is given by

$$y = f_1(ct - x) + f_2(ct + x)$$

To show that $y = f_1(ct - x)$ is a solution of the wave equation, we have

$$\frac{dy}{dt} = cf_1'(ct - x)$$

and
$$\frac{d^2 y}{dt^2} = c^2 f_1''(ct - x)$$

where f_1' and f_1'' are the first and second derivatives of f_1 respectively with respect to $(ct - x)$. Similarly

$$\frac{dy}{dx} = -f_1'(ct - x)$$

and
$$\frac{d^2 y}{dx^2} = f_1''(ct - x)$$

Hence
$$\frac{d^2 y}{dt^2} = c^2 \frac{d^2 y}{dx^2}$$

Similarly it can be shown that $y = f_2(ct + x)$ as well as $y = f_1(ct - x) + f_2(ct + x)$ is a solution of the wave equation.

In a harmonic progressive wave each particle of the medium executes simple harmonic motion which may be represented by a sine or a cosine function.

For a wave travelling along the positive x direction, we have

For sine function $y = f(ct - x) = a \sin k(ct - x)$ ... (*i*)

where a is the amplitude and k a constant. The wave profile repeats itself regularly at a distance $\frac{2\pi}{k}$ because $a \sin k\left(ct - x + \frac{2\pi}{k}\right) = a \sin k(ct - x)$. The distance $2\pi/k$, therefore, gives the wavelength λ.

Hence $\lambda = \frac{2\pi}{k}$ or $k = \frac{2\pi}{\lambda}$

Substituting in (*i*), we have

$$y = a\sin\frac{2\pi}{\lambda}(ct - x) = a\sin\left(\frac{2\pi ct}{\lambda} - \frac{2\pi}{\lambda}x\right)$$

$$= a\sin 2\pi\left(vt - \frac{x}{\lambda}\right) = a\sin(\omega t - kx)$$

where $\omega = 2\pi v$ (v = frequency) and $k = \frac{2\pi}{\lambda} = 2\pi\bar{v}$ and $\bar{v}$ is the wave number *i.e.* the reciprocal of the wavelength or the *number of waves contained in a unit length.* The constant k is known as *propagation constant.*

For cosine function $y = a\cos\frac{2\pi}{\lambda}(ct - x) = a\cos 2\pi\left(vt - \frac{x}{\lambda}\right)$

$$= a\cos(\omega t - kx)$$

Exponential form: The exponential form of the equation $y = ae^{i(\omega t - kx)}$ is equally valid solution of the differential equation as it contains both the sine and cosine functions. The real part of the expression is the cosine function and the part multiplied by ($i = \sqrt{-1}$) is the sine function.

$$y = a\sin\frac{2\pi}{\lambda}(ct - x) \text{ or } y = a\cos\frac{2\pi}{\lambda}(ct - x) \text{ or } y = ae^{i(\omega t - kx)}$$

represents a wave moving to the right whereas

$$y = a\sin\frac{2\pi}{\lambda}(ct + x) \text{ or } y = a\cos\frac{2\pi}{\lambda}(ct + x) \text{ or } y = ae^{i(\omega t + kx)}$$

represents a wave moving to the left in the negative ($-x$) direction.

13.10 ENERGY OF A VIBRATING STRING

A vibrating string possesses both kinetic as well as potential energy.

Kinetic energy. Consider a small element of length dx of a stretched vibrating string of total length l and mass per unit length ρ, If $\dot{y}$ is the velocity of the element, then

$$\text{K.E. of the element} = \frac{1}{2}\rho\, dx\, \dot{y}^2$$

$$\therefore \text{ Total K.E. of the string} = \frac{1}{2}\int_0^l \rho\dot{y}^2\, dx$$

Potential energy. It is the work done by tension T in extending an element of length dx in the normal position of the string to a new element of length ds in the vibrating position, fig (13,8).

$$\therefore \text{ Potential energy of the element} = T(ds - dx)$$

$$= T[(dx^2 + dy^2)^{1/2} - dx]$$

$$= T\left\{\left[1 + \left(\frac{dy}{dx}\right)^2\right]^{1/2} - 1\right\}dx$$

$$= \frac{1}{2}T\left(\frac{dy}{dx}\right)^2 dx$$

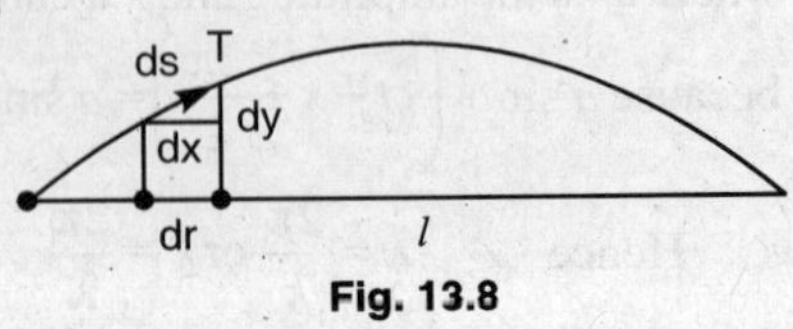

Fig. 13.8

as
$$\left[1+\left(\frac{dy}{dx}\right)^2\right]^{\frac{1}{2}}=1+\frac{1}{2}\left(\frac{dy}{dx}\right)^2$$

neglecting higher power of $\frac{dy}{dx}$ and also assuming. T to be a constant.

$\therefore$ Total potential energy of the string $=\frac{1}{2}T\int_0^e\left(\frac{dy}{dx}\right)^2 dx$

Energy in each normal mode. When a string vibrates in a pure normal mode, every point in the string vibrates with the same frequency and at any moment all parts move either in the same or the opposite phase. In a normal mode the total displacement y in the string is the superposition of the displacements y_n of the individual harmonics and the total energy of the resultant vibration is obtained by adding the energy of each mode as if it existed independently of other modes. Thus, the kinetic energy of the nth mode.

$$E_n\text{ (kinetic)}=\frac{1}{2}\int_0^l \rho\dot{y}_n^2 dx$$

and potential energy E_n (potential) $=\frac{1}{2}T\int_0^l\left(\frac{dy_n}{dx}\right)^2 dx$

The displacement for the nth harmonic is given by

$$y_n=(A_n\cos\omega_n t+B_n\sin\omega_n t)\sin\frac{\omega_n}{c}x \quad \text{[Refer Eq. (}vii\text{) article 15.9]}$$

where $A_n=-2\,ai$ and $B_n=2a$

$$\therefore \quad \dot{y}_n=(-A_n\omega_n\sin\omega_n t+B_n\omega_n\cos\omega_n t)\sin\frac{\omega_n}{c}x$$

and
$$\frac{dy_n}{dx}=\frac{\omega_n}{c}(A_n\cos\omega_n t+B_n\sin\omega_n t)\cos\frac{\omega_n}{c}x$$

$\therefore$ Total Energy $=E_n$ (kinetic) $+E_n$ (potential)

$$=\frac{1}{2}\rho\,\omega_n^2(-A_n\sin\omega_n t+B_n\cos\omega_n t)^2\int_0^l\sin^2\frac{\omega_n}{c}x\,dx$$

$$+\frac{1}{2}T\frac{\omega_n^2}{c^2}(A_n\cos\omega_n t+B_n\sin\omega_n t)^2\int_0^l\cos^2\frac{\omega_n}{c}x\,dx$$

Now
$$\frac{T}{c^2}=\rho,\ \int_0^l\sin^2\frac{\omega_n}{c}x\,dx=\frac{1}{2}l,\ \int_0^l\cos^2\frac{\omega_n}{c}x\,dx=\frac{1}{2}l$$

and
$$(-A_n\sin\omega_n t+B_n\cos\omega_n t)^2+(A_n\cos\omega_n t+B_n\sin\omega_n t)^2=(A_n^2+B_n^2)$$

$$\therefore \quad \text{Total energy}=\frac{1}{4}\rho l\omega_n^2(A_n^2+B_n^2)=\frac{1}{4}m\omega_n^2(A_n^2+B_n^2)$$

where $m=\rho l=$ mass of the whole string.

As mass m, normal mode frequency ω_n, and the normal mode amplitudes A_n and B_n are constants, the total energy in the normal mode vibration of the string remains constant.

13.11 ENERGY FLOW ALONG A STRETCHED STRING

When a transverse wave is propagated through a string, the displacement of the particles of the string is given by $y = Ae^{i(\omega t - kx)}$

We shall, however, assume that the displacement varies as a sine function and is given by the imaginary part of the above equation *i.e.*

$$y = A \sin(\omega t - kx)$$

where $\omega = 2\pi\nu$, ν being the frequency and $k = \frac{2\pi}{\lambda}$, λ being the wavelength of the wave travelling through the string.

$\therefore$ Particle velocity $v = \frac{dy}{dt} = A\omega \cos(\omega t - kx)$

and acceleration $a = \frac{d^2 y}{dt^2} = -A\omega^2 \sin(\omega t - kx) = -\omega^2 y$

Consider a very small element of length dx of the string of mass m, then

Potential energy of the element $E_p = -\int_0^y m \frac{d^2 y}{dt^2} dy$

$$= m\omega^2 \int_0^y y\,dy = \frac{1}{2} m\omega^2 y^2$$

$$= \frac{1}{2} m\omega^2 A^2 \sin^2(\omega t - kx) \quad \text{... (i)}$$

and kinetic energy of the element $E_k = \frac{1}{2} mv^2 = \frac{1}{2} m \left(\frac{dy}{dt}\right)^2$

$$= \frac{1}{2} m\omega^2 A^2 \cos^2(\omega t - kx) \quad \text{... (ii)}$$

$\therefore$ Total energy of the element $E = E_p + E_k$

$$= \frac{1}{2} m\omega^2 A^2 \left[\sin^2(\omega t - kx) + \cos^2(\omega t - kx)\right]$$

$$= \frac{1}{2} m\omega^2 A^2 \quad \text{... (iii)}$$

If ρ is the mass per unit length (or linear density) of the string, then $m = \rho dx$

$$\therefore \quad E = \frac{1}{2} \rho\omega^2 A^2 dx \quad \text{... (iv)}$$

Hence, total energy per unit length, $E_l = \frac{1}{2} \rho\omega^2 A^2$ ... (v)

Rate of flow of energy. The rate of flow of energy across any cross-section of the string is given by

$$I = \frac{dE}{dt} = \frac{1}{2} \rho\omega^2 A^2 \frac{dx}{dt} \quad (\because \text{Eq. } iv)$$

But $\frac{dx}{dt} = c$, the velocity of propagation of the wave in the string.

$$\therefore \qquad I = \frac{1}{2}\omega^2 A^2 (\rho c)$$

$$= \frac{1}{2}\omega^2 A^2 Z \qquad (\because \; Z = \rho c) \qquad ...(vi)$$

From relation (*vi*), we find that

(*i*) The rate of the flow of energy through the string when a wave propagates through it is proportional to *Z*, the characteristic impedance of the string.

(*ii*) It is independent of *x* and *t*.

(*iii*) The rate of flow of energy is the same everywhere in the string and at all instants of time.

13.12 CHARACTERISTIC IMPEDANCE OF A STRING

A wave of any type-mechanical, electrical or electro-magnetic-experiences an impedance while propagating through a medium. When there is no resistive or dissipating mechanism, the impedance is determined by two energy storing parameters *i.e.*, inertia and elasticity for mechanical waves and inductance and capacitance in the case of electrical waves. The impedance in such a case is a real quantity. The presence of a loss mechanism introduces a complex term into the impedance.

Characteristic impedance. *It is the impedance offered by a string to transverse progressive waves and is defined as the ratio of transverse force to the transverse velocity.*

It is also called as transverse impedance

$$\therefore \qquad \text{Impedance, } Z = \frac{\text{Transverse force}}{\text{Transverse velocity}} = \frac{F}{v}$$

Suppose we have a continuous string stretching from left to right with the left end at $x = 0$. Consider a progressive wave on the string which is generated by an impressed oscillating force $F = F_0 e^{i\omega t}$ acting at this end. The force acts only in the plane of the paper in a *transverse* direction *i.e.*, perpendicular to the equilibrium position of the string.

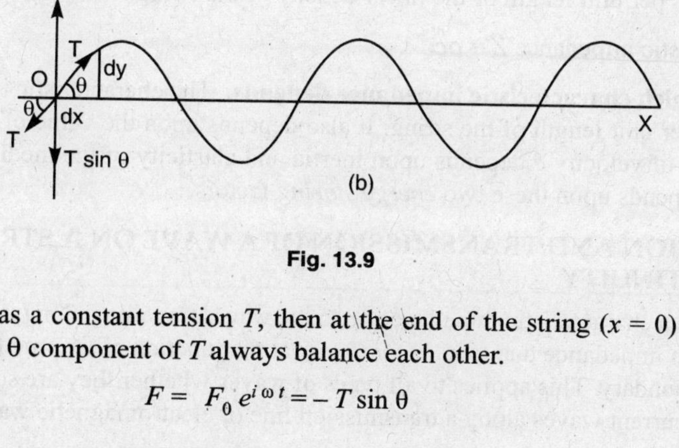

Fig. 13.9

If the string has a constant tension *T*, then at the end of the string ($x = 0$) the applied force $F_0 e^{i\omega t}$ and the $T \sin\theta$ component of *T* always balance each other.

$$\therefore \qquad F = F_0 e^{i\omega t} = -T\sin\theta$$

As θ is small, then as shown in Fig. 13.9 (*b*)

$$\sin\theta = \tan\theta = \frac{dy}{dx}$$

$$\therefore \quad F_0 e^{i\omega t} = -T\left(\frac{dy}{dx}\right)_{x=0}$$

The displacement of the progressive wave is represented in the exponential form by the equation

$$\vec{y} = \vec{A} e^{i(\omega t - kx)}$$

where $\vec{A}$ is the amplitude and may be a *complex* quantity due to its phase relation with F and $k = \frac{2\pi}{\lambda} = \frac{\omega}{c}$, ω being the angular frequency of the applied periodic force and c the wave velocity.

Hence at the end of the string where $x = 0$

$$F_0 e^{i\omega t} = -T\left(\frac{dy}{dx}\right)_{x=0} = ik\,T\vec{A} e^{i\omega t}$$

$$\therefore \quad \vec{A} = \frac{F_0}{ikT} \quad \text{and} \quad \vec{y} = \frac{F_0}{iT} \times \frac{c}{\omega} e^{i(\omega t - kx)}$$

$$\therefore \text{ Transverse velocity} \quad \vec{v} = \frac{d\vec{y}}{dt}$$

$$\therefore \quad = F_0 \frac{c}{T} e^{i(\omega t - kx)} = ve^{i(\omega t - kx)}$$

where v is the velocity amplitude $= \frac{F_0 c}{T}$

According to definition, impedance $Z = \frac{F_0}{v}$

$$\therefore \quad Z = \frac{F_0}{v} = \frac{T}{c}$$

Now for a string $\quad c = \sqrt{\frac{T}{\rho}}$ or $T = \rho c^2$

where ρ is the mass per unit length or the linear density of the string.

∴ Characteristic impedance $Z = \rho c$.

Factors on which characteristic impedance depends. The characteristic impedance depends upon ρ the mass per unit length of the string. It also depends upon the value of the velocity of the wave c. The value of velocity c depends upon inertia and elasticity of the medium. Therefore, the impedance also depends upon these two *energy storing* factors.

13.13 REFLECTION AND TRANSMISSION OF A WAVE ON A STRING AT DISCONTINUITY

A string offers a characteristic impedance ρc to the waves travelling along it. Whenever there is a *sudden* change in impedance due to a *discontinuity* at the junction of two media the wave suffers reflection at the boundary. This applies to all types of waves whether they are sound waves along a string, voltage or current waves along a transmission line or electromagnetic waves in a medium.

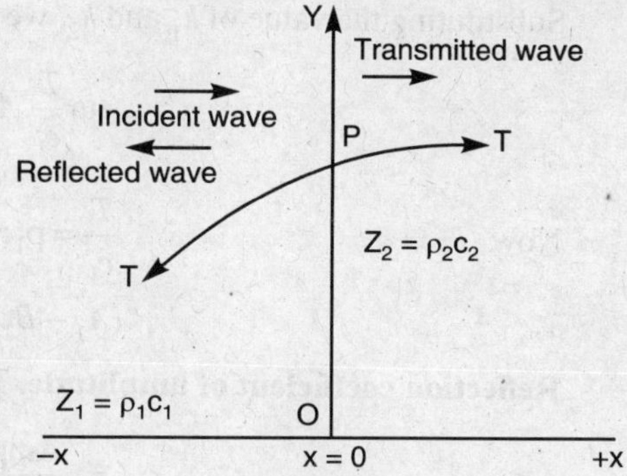

Fig. 13.10

Let us suppose that the string consists of two parts smoothly joined at the point P, $x = 0$ and has a constant tension T along the whole string. The two sections have different linear densities ρ_1 and ρ_2 and, therefore, different wave velocities c_1 and c_2 given by

$$c_1 = \sqrt{\frac{T}{\rho_1}} \text{ and } c_2 = \sqrt{\frac{T}{\rho_2}}$$

The specific or characteristic impedances respectively are $Z_1 = \rho_1 c_1$ and $Z_2 = \rho_2 c_2$. If the incident wave is represented by the real part of the displacement equation.

$$y_1 = A_1 e^{i(\omega t - k_1 x)}$$

where A_1 is the amplitude of the incident wave travelling in the *positive X-direction* with velocity c_1 and $k_1 = \frac{\omega}{c_1}$, then the reflected and transmitted waves are represented by the real part of the equations.

$$y_r = B_1 e^{i(\omega t + k_1 x)}$$

and $$y_t = A_2 e^{i(\omega t - k_2 x)}$$

where B_1 is the amplitude of the reflected wave travelling in the *negative X-direction* with velocity c_1 and A_2 is the amplitude of the transmitted wave travelling in the *positive* X-direction with velocity c_2 and $k_2 = \frac{\omega}{c_2}$. The value of B_1 and A_2 can be determined with the help of two boundary condition which must be satisfied at the point of impedance discontinuity at $x = 0$. The boundary conditions are:—

(1) The displacement y is the same immediately to the left and to the right of $x = 0$ for all times so that there is *no discontinuity of displacement.* This is the *geometrical condition* and is stated as

$$y_i + y_r = y_t$$

or $$A_1 e^{i(\omega t - k_1 x)} + B_1 e^{i(\omega t + k_1 x)} = A_2 e^{i(\omega t - k_2 x)}$$

At $$x = 0$$

$$A_1 e^{i\omega t} + B_1 e^{i\omega t} = A_2 e^{i\omega t}$$

or $$A_1 + B_1 = A_2 \quad \text{... } (i)$$

(2) The transverse force $T\left(\frac{dy}{dx}\right)$ at $x = 0$ is *continuous across the boundary.* This means that *the gradient* $\frac{dy}{dx}$ *must have the same value on both sides.* If it is not so a finite transverse force will act on an infinitesimally small mass of the string producing an infinite acceleration which is physically an absurd statement. This is the *dynamical condition* and is stated as

$$T\frac{dy_i}{dx} + T\frac{dy_r}{dx} = T\frac{dy_t}{dx}$$

or $$-Tk_1 A_1 e^{i(\omega t - k_1 x)} + Tk_1 B_1 e^{i(\omega t + k_1 x)} = -T k_2 A_2 e^{i(\omega t - k_2 x)}$$

At $$x = 0$$

$$-Tk_1 A_1 e^{i\omega t} + Tk_1 B_1 e^{i\omega t} = -Tk_2 A_2 e^{i\omega t}$$

or $$-Tk_1 A_1 + Tk_1 B_1 = -TK_2 A_2$$

Substituting the value of k_1 and k_2, we have

$$-\omega\frac{T}{c_1}A_1 + \omega\frac{T}{c_1}B_1 = -\omega\frac{T}{c_2}A_2$$

Now $$\frac{T}{c_1} = \rho_1 c_1 = Z_1 \text{ and } \frac{T}{c_2} = \rho_2 c_2 = Z_2$$

$\therefore$ $$(A_1 - B_1)\,Z_1 = A_2\,Z_2 \qquad \text{... (ii)}$$

Reflection coefficient of amplitude. The reflection coefficient

$$= \frac{\text{Amplitude of reflected wave}}{\text{Amplitude of incident wave}} = \frac{B_1}{A_1}$$

Now from (*ii*)

$$\frac{A_2}{A_1 - B_1} = \frac{A_1 + B_1}{A_1 - B_1} = \frac{Z_1}{Z_2}$$

$\therefore$ Reflection co-effiieint $$\frac{B_1}{A_1} = \frac{Z_1 - Z_2}{Z_1 + Z_2}$$

Transmission coefficient of amplitude. The transmission coefficient

$$= \frac{\text{Amplitude of transmitted wave}}{\text{Amplitude of incident wave}} = \frac{A_2}{A_1}$$

Substituting $B_1 = A_2 - A_1$ in (*ii*), we have

$$(A_1 - A_2 + A_1)\,Z_1 = A_2\,Z_2$$

or $$2A_1\,Z_1 = A_2(Z_1 + Z_2)$$

$\therefore$ $$\frac{A_2}{A_1} = \frac{2\,Z_1}{Z_1 + Z_2}$$

It is clear from the above expression that both these coefficients are independent of ω i.e., these hold for waves of all the frequencies. These are real quantities and, therefore, are free from phase changes except the phase change of π when reflection takes place at a fixed end.

Amplitude after reflection at dense medium ($Z_2 \cong \infty$)

If $Z_2 = \infty$ *i.e.*, $x = 0$ is a *fixed* end, then $A_2 = 0$ and there is no transmitted wave. This also gives

$$\frac{B_1}{A_1} = \frac{Z_1/Z_2 - 1}{Z_1/Z_2 + 1} = -1$$

so that the incident wave is completely reflected with a phase change of π i.e. the phase is reversed. This condition is necessary for standing waves to exist. A group of waves having a number of component frequencies will *retain* its shape on reflection.

Amplitude after reflection at rarer medium ($Z_2 \cong 0$)

If $Z_2 = 0$ *i.e.*, $x = 0$ is a *free* end, then

$$\frac{B_1}{A_1} = 1 \text{ and } \frac{A_2}{A_1} = 2$$

i.e., the amplitude of the transmitted wave A_2 is twice that of the incident wave A_1. It is due to this reason that there is a 'flick' at the end of a whip or the free end of a string when a wave reaches it.

As $B_1 = A_1$, the incident wave is completely reflected without any phse change.

13.14 FLICK

There is a '*flick*' at the end of a whip or the free end of a string because when a wave travelling through the string reaches the free end it meets a highly rare medium for which the characteristic impedance $Z_2 \cong 0$. The transmission coefficient of amplitude A_2 is given by

$$\frac{A_2}{A_1} = \frac{2Z_1}{Z_1 + Z_2}$$

when
$$Z_2 = 0, \frac{A_2}{A_1} = 2 \quad \text{or} \quad A_2 = 2A_1$$

Then amplitude at the free end is thus doubled and hence the intensity of the wave becomes four times. The sound becomes very loud and is called '*flick*'.

13.15 REFLECTION AND TRANSMISSION OF ENERGY

When a wave meets a boundary between two media having different impedance values a part of it is reflected and a part transmitted. If A_1 is the amplitude of the incident wave in the medium of impedance Z_1 (linear density ρ_1, wave velocity c_1), B_1 the amplitude of the reflected wave and A_2 the amplitude of the transmitted wave in the medium of impedance Z_2 (linear density ρ_2, wave velocity c_2), then

The reflection coefficient $$\frac{B_1}{A_1} = \frac{Z_1 - Z_2}{Z_1 + Z_2} \quad \text{... } (i)$$

and transmission coefficient $$\frac{A_2}{A_1} = \frac{2Z_1}{Z_1 + Z_2} \quad \text{... } (ii)$$

In a general case, the total energy per unit length of a string of linear density ρ considering it a simple harmonic oscillator of maximum amplitude A and angular frequency ω is given by

$$E = \frac{1}{2} \rho \omega^2 A^2$$

As the wave travels along the string with a velocity c each unit length of the string take up its oscillation with the passage of the wave

∴ Rate of flow of energy along the string = energy × velocity = $\frac{1}{2} \rho \omega^2 A^2 c$

If the point at which reflection takes place is at $x = 0$, then the rate of energy arriving at the boundary $x = 0$, with the incident wave

$$= \frac{1}{2}\rho_1\omega^2 A_1^2 c_1 = \frac{1}{2} Z_1\omega^2 A_1^2 \qquad [\because \rho_1 c_1 = Z_1]$$

The energy leaves the boundary *via* the reflected and the transmitted waves.

∴ Rate at which energy leaves the boundary

$$= \frac{1}{2} \rho_1 \omega^2 B_1^2 c_1 + \frac{1}{2} \rho_2 \omega^2 A_2^2 c_2$$

$$= \frac{1}{2} Z_1 \omega^2 B_1^2 + \frac{1}{2} Z_2 \omega^2 A_2^2$$

Substituting the values of B_1 and A_2 from (*i*) and (*ii*) the above expression

$$= \frac{1}{2} \omega^2 A_1^2 \frac{Z_1(Z_1 - Z_2)^2 + 4Z_1^2 Z_2}{(Z_1 + Z_2)^2} = \frac{1}{2} Z_1 \omega^2 A_1^2$$

which is equal to the energy arriving with the incident wave.

This shows that all the energy arriving at the boundary in the incident wave leaves the boundary in the reflected and transmitted wave.

Reflected and transmitted energy co-efficients. The reflected and transmitted energy co-efficients are defined as

$$\text{Reflected energy co-efficient} = \frac{\text{Reflected energy}}{\text{Incident energy}}$$

$$= \frac{\frac{1}{2} Z_1 \omega^2 B_1^2}{\frac{1}{2} Z_1 \omega^2 A_1^2} = \frac{B_1^2}{A_1^2} = \left(\frac{Z_1 - Z_2}{Z_1 + Z_2}\right)^2 \quad \text{... (iii)}$$

$$\text{Transmitted energy coefficient} = \frac{\text{Tansmitted energy}}{\text{Incident energy}}$$

$$= \frac{\frac{1}{2} Z_2 \omega^2 A_2^2}{\frac{1}{2} Z_1 \omega^2 A_1^2} = \frac{Z_2 A_2^2}{Z_1 A_1^2} = \frac{4Z_1 Z_2}{(Z_1 + Z_2)^2} \quad \text{... (iv)}$$

∴ Reflected energy coefficient + transmitted energy coefficient

$$= \left(\frac{Z_1 - Z_2}{Z_1 + Z_2}\right)^2 + \left(\frac{4Z_1 Z_2}{(Z_1 + Z_2)}\right) = 1$$

II. LONGITUDINAL WAVES

13.16 VELOCITY OF LONGITUDINAL WAVE IN A SOLID

Sound is a form of longitudinal wave motion. Longitudinal vibrations can be set up in a solid medium in the form of a rod whose thickness is small as compared to its length. To set up longitudinal vibrations in a rod it is either stroked at one end with a hammer or rubbed along its length by a resined cloth. The longitudinal vibrations travel along the rod with a velocity which depends upon the Young's modulus Y and density ρ of the material of the rod.

Consider a rod of uniform area of cross-section a. Let the axis of the rod be along the X-axis, Fig. (13.11) Suppose longitudinal sound waves are travelling in the solid medium of the rod from left to right along the X- axis OX. Take two planes perpendicular to OX at A and B a small distance δx apart. When sound waves travel through the medium let at any instant the plane A be displaced to A' and plane B to B'. Suppose the displacement of the plane at $A = AA' = y$ and the rate of change of displacement $= \frac{dy}{dx}$

∴ Displacement of the plane $B = BB' = y + \left(\frac{dy}{dx}\right)\delta x$

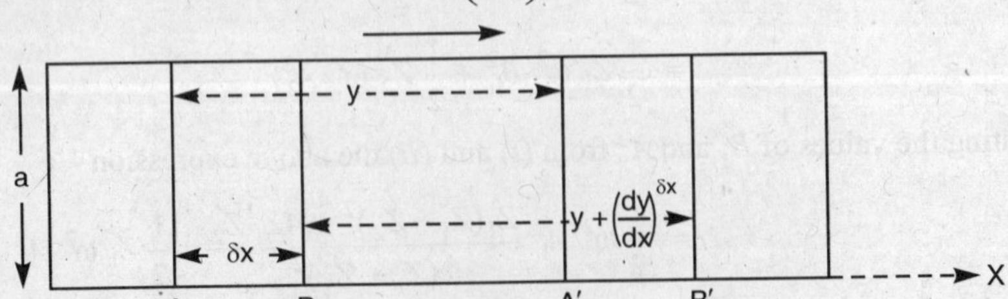

Fig 13.11

As the displacement of the plane B is greater than the displacement of the plane A, the distance $B'A'$ is greater than BA by an amount given by

$$y+\left(\frac{dy}{dx}\right)\delta x-y=\left(\frac{dy}{dx}\right)\delta x$$

$\therefore$ Change of distance between the planes A and $B = \left(\frac{dy}{dx}\right)\delta x$

Original distance between the planes A and $B = \delta x$

$$\therefore \quad \text{Longitudinal strain} = \frac{\text{Change in length}}{\text{original length}}$$

$$= \frac{\text{Change of distance between the planes } A \text{ and } B}{\text{Original distance between the planes } A \text{ and } B} = \frac{\left(\frac{dy}{dx}\right)\delta x}{\delta x}$$

or Longitudinal strain $= \frac{dy}{dx}$

When the rod is under strain elastic forces automatically come into play.

Let F be the force per unit area on the plane A and $\frac{dF}{dx}$ the rate of change of force, then

Fore on the plane $A = aF$

and Force on the plane $B = aF + a\left(\frac{dF}{dx}\right)\delta x$

Hence net force on the layer $AB = a\left(\frac{dF}{dx}\right)\delta x$

Mass of the medium in the layer $AB = a\,\delta x\,\rho$ where ρ is the density of the medium.

If $\frac{d^2y}{dt^2}$ represents the acceleration of the particles in the layer, then

$$a\left(\frac{dF}{dx}\right)\delta x = a\,\delta x\,\rho\frac{d^2y}{dt^2} \text{ or } \frac{dF}{dx}=\rho\frac{d^2y}{dt^2} \qquad \text{... }(i)$$

Now, Young's modulus (or Longitudinal elasticity) of the medium

$$Y=\frac{\text{Stress (Force per unit area)}}{\text{Longitudinal strain}}=\frac{F}{\frac{dy}{dx}}$$

$$\therefore \qquad F = Y\frac{dy}{dx}$$

or

$$\frac{dF}{dx}=\frac{d}{dx}\left(Y\frac{dy}{dx}\right)=Y\frac{d^2y}{dx^2}$$

Comparing (i) and (ii), we have

$$\rho\frac{d^2y}{dt^2}=Y\frac{d^2y}{dx^2}$$

or
$$\frac{d^2y}{dt^2}=\frac{Y}{\rho}\frac{d^2y}{dx^2}$$

This expression represents the differential equation of longitudinal wave propagation along the rod. Comparing it with standard equation of wave motion

$$\frac{d^2y}{dt^2}=v^2\frac{d^2y}{dx^2}$$

we get the velocity of the wave

$$v^2=\frac{Y}{\rho}$$

or
$$v=\sqrt{\frac{Y}{\rho}}$$

This shows that the velocity of longitudinal waves travelling through a rod depends upon Young's modulus Y and density ρ. It is independent of the applied force and area of cross-section.

13.17 VELOCITY OF LONGITUDINAL WAVES IN A GAS

The velocity of longitudinal waves in a fluid is given by the relation.

$$v=\sqrt{\frac{E}{\rho}}$$

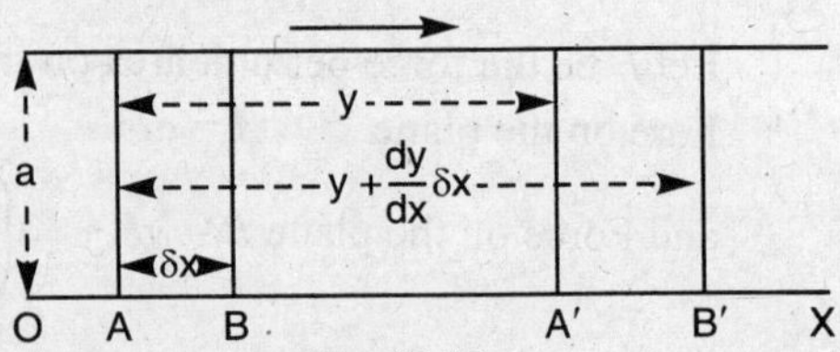

Fig. 13.12

Where E is the coefficient of volume elasticity (bulk modulus) of the medium and ρ its density.

To prove this suppose longitudinal sound waves are travelling in a gaseous medium from *left* to *right*. Due to these longitudinal waves *rarefactions* and *condensations* will be formed in the medium. Let these rarefractions and condensations travel from left to right with a velocity v along the X-axis OX in a tube of area of cross-section a.

Consider two planes perpendicular to OX at A and B a small distance δx apart. When sound waves travel through the medium let at any instant the plane A be displaced to the position A' and plane B to the position B' (Fig. 13.12).

Suppose the displacement of the plane at $A = AA' = y$ and the rate of change of displacement = $\frac{dy}{dx}$

$$\therefore \qquad \text{Displacement at } B = BB' = y+\left(\frac{dy}{dx}\right)\delta x$$

As the displacement of the plane B is greater than the displacement of the plane A, the distance $B'A'$ is greater than BA by an amount given by

$$y+\left(\frac{dy}{dx}\right)\delta x-y=\left(\frac{dy}{dx}\right)\delta x$$

$$\therefore \qquad \text{Change in volume} = a\left(\frac{dy}{dx}\right)\delta x$$

Original volume of the layer $AB = a\ \delta x$

$$\therefore \quad \text{Volumetric strain} = \frac{a\left(\frac{dy}{dx}\right)\delta x}{a\,\delta x} = \frac{dy}{dx}$$

Let P be the excess of pressure over the atmosphere at the plane A. Suppose the excess pressure varies with distance x at the rate $\frac{dP}{dx}$, then

$$\text{Excess pressure at the plane } B = P + \left(\frac{dP}{dx}\right)\delta x$$

The excess of pressure at A and B act in *opposite* directions

$\therefore$ The resultant pressure acting on the element AB

$$= P + \left(\frac{dP}{dx}\right)\delta x - P = \left(\frac{dP}{dx}\right)\delta x$$

$$\text{Hence force acting on the element } AB = a\left(\frac{dP}{dx}\right)\delta x$$

As ρ is the density of the gaseous medium, the mass of the element $AB = a\,\delta x\,\rho$

If $\frac{d^2y}{dt^2}$ is the acceleration of the mass of the element AB, then according to Newton's second law of motion.

$$a\,\delta x\,\rho\frac{d^2y}{dt^2} = a\left(\frac{dP}{dx}\right)\delta x$$

or
$$\rho\frac{d^2y}{dt^2} = \frac{dP}{dx} \qquad \text{... }(i)$$

Now Bulk modulus (or volume elasticity) of the gas

$$E = \frac{\text{Stress (force per unit area or pressure)}}{\text{Volumetric strain}}$$

$$= -\frac{P}{\frac{dy}{dx}}$$

or
$$P = E\frac{dy}{dx} \text{ (numerically)} \qquad \text{... }(ii)$$

$$\therefore \quad \frac{dP}{dx} = \frac{d}{dx}\left(E\frac{dy}{dx}\right) = E\frac{d^2y}{dx^2} \qquad \text{... }(ii)\,(a)$$

Comparing (i) and (ii) (a) we have

$$\rho\frac{d^2y}{dt^2} = E\frac{d^2y}{dx^2}$$

or
$$\frac{d^2y}{dt^2} = \frac{E}{\rho}\frac{d^2y}{dx^2}$$

This represents the differential equation of a longitudinal (or sound) wave propagating in a gaseous (fluid) medium. Comparing it with standard equation of wave motion.

$$\frac{d^2y}{dt^2} = v^2 \frac{d^2y}{dx^2}, \text{ we get}$$

$$v^2 = \frac{E}{\rho}$$

or $$v = \sqrt{\frac{E}{\rho}}$$

This shows that the velocity of sound waves through a gas depends upon volume elasticity and density of the gas.

13.18 ENERGY STORED IN A MEDIUM DUE TO LONGITUDINAL PROGRESSIVE WAVE

When a wave travels through a medium, it means transfer of energy from one place to another in the direction of propagation of the wave. If v is the velocity of the wave, then energy transferred per second is equal to the energy possessed by the particles of the medium in a length v. The energy of the progressive wave at any instant is partially potential and partially kinetic. As the particles of the medium execute simple harmonic motion, the kinetic energy is maximum at the mean position where the displacement is zero and particle velocity is maximum. The potential energy is maximum at the extreme position where the displacement is maximum and particle velocity is zero. It means that at the mean position entire energy of the wave is kinetic (potential energy is zero) and at the extreme position entire energy of the wave is potential (kinetic energy is zero).

It should be clearly noted that there is no transfer of the medium in the direction of propagation of the progressive wave but there is only a transfer of energy in the direction of propagation of the wave.

A progressive wave of wavelength λ travelling with a velocity v along the positive direction of X-axis is represented by the expression.

$$y = a \sin \frac{2\pi}{\lambda}(vt - x)$$

where y is the displacement at a point distant x from the origin at any time t and a the amplitude. The velocity of the particle

$$\frac{dy}{dt} = a\, 2\pi \frac{v}{\lambda} \cos \frac{2\pi}{\lambda}(vt - x)$$

and aceleration $$\frac{d^2y}{dt^2} = -\frac{4\pi^2 v^2}{\lambda^2} \sin \frac{2\pi}{\lambda}(vt - x)$$

The negative sign suggests that the acceleration of the particles is always directed towards the mean position.

Kinetic energy. Consider a unit volume of the medium of mass ρ equal to the density of the medium, then

$$\text{K.E. per unit volume} = \frac{1}{2}\text{mass} \times (\text{velocity})^2$$

$$= \frac{1}{2}\rho . \left\{ a \frac{2\pi v}{\lambda} \cos \frac{2\pi}{\lambda}(vt - x) \right\}^2$$

$$= \frac{2\pi^2 v^2 \rho}{\lambda^2} a^2 \cos^2\left[\frac{2\pi}{\lambda}(vt - x)\right]$$

$$= 2\pi^2 a^2 n^2 \rho \cos^2\left[\frac{2\pi}{\lambda}(vt - x)\right]$$

where $$n = \frac{v}{\lambda}$$

Potential energy. To find the potential energy per unit volume of the medium, we have Force on a unit volume = mass × acceleration

$$= -\rho a \frac{4\pi^2 v^2}{\lambda^2} \sin\frac{2\pi}{\lambda}(vt - x)$$

$$= -\rho \frac{4\pi^2 v^2}{\lambda^2} y$$

The negative sign only indicates that the force is directed towards the mean position opposite to the displacement. Hence work done to produce a small displacement dy

$$= \rho \frac{4\pi^2 v^2}{\lambda^2} y dy$$

∴ Work done during the whole displacement from 0 to y

$$= \int_0^y \rho \frac{4\pi^2 v^2}{\lambda^2} y dy = \frac{4\pi^2 v^2 \rho}{\lambda^2} \frac{y^2}{2}$$

$$= \frac{2\pi^2 v^2 \rho}{\lambda^2} a^2 \sin^2\left[\frac{2\pi}{\lambda}(vt - x)\right]$$

$$= 2\pi^2 n^2 a^2 \rho \sin^2\left[\frac{2\pi}{\lambda}(vt - x)\right]$$

This work is stored in a unit volume of the medium as potential energy of the wave.

∴ Potential energy per unit volume $= 2\pi^2 a^2 n^2 \rho \sin^2\left[\frac{2\pi}{\lambda}(vt - x)\right]$

Total Energy. Total energy per unit volume

= Kinetic energy per unit volume + Potential energy per unit volume

$$= 2\pi^2 n^2 a^2 \rho \left\{\sin^2\left[\frac{2\pi}{\lambda}(vt - x)\right] + \cos^2\left[\frac{2\pi}{\lambda}(vt - x\right]\right\} = 2\pi^2 n^2 a^2 \rho$$

or Total energy per unit volume $E = 2\pi^2 n^2 a^2 \rho$ ... (i)

Average kinetic energy density. Now consider a cylindrical column of the medium of unit area of cross-section with its axis parallel to the direction of propagation. If we take an element of length dx, then its volume = $dx \times 1 = dx$.

The K.E. of this volume $= 2\pi^2 a^2 n^2 \rho \cos^2\left[\frac{2\pi}{\lambda}(vt - x)\right] dx$ ... (ii)

The different elements of the column will have different kinetic energies. The total kinetic energy of a column of length λ equal to the wavelength of the wave can be calculaved by integrating equation (*ii*) between $x = 0$ and $x = \lambda$. Dividing the result by λ we get the **average kinetic energy per unit volume or average kinetic energy density.**

∴ Average K.E. per unit volume.

$$= 2\pi^2 a^2 n^2 \rho \frac{1}{\lambda}\int_0^\lambda \cos^2 \frac{2\pi}{\lambda}(vt - x)dx$$

$$= 2\pi^2 a^2 n^2 \rho \frac{1}{\lambda}\frac{\lambda}{2} = \pi^2 a^2 n^2 \rho \qquad \left[\because \int_0^\lambda \cos^2 \frac{2\pi}{\lambda}(vt - x)dx = \frac{\lambda}{2}\right]$$

Again $2\pi n = \omega$

∴ $\pi^2 n^2 = \frac{1}{4}\omega^2$

Hence average kinetic energy per unit volume $= \frac{1}{4}\rho a^2 \omega^2$

Average potential energy density. The potential energy per unit volume

$$= 2\pi^2 a^2 n^2 \rho \sin^2\left[\frac{2\pi}{\lambda}(vt - x)\right]$$

∴ Potential energy for a volume dx

$$= 2\pi^2 a^2 n^2 \rho \sin^2\left[\frac{2\pi}{\lambda}(vt - x)\right]dx$$

and the **average potential energy per unit volume or Average potential energy density.**

$$= 2\pi^2 a^2 n^2 \rho \frac{1}{\lambda}\int_0^\lambda \sin^2 \frac{2\pi}{\pi}(vt - x)dx$$

$$= 2\pi^2 a^2 n^2 \rho \frac{1}{\lambda}\frac{\lambda}{2} \qquad \left[\because \int_0^\lambda \sin^2 \frac{2\pi}{\lambda}(vt - x)dx = \frac{\lambda}{2}\right]$$

$$= \pi^2 a^2 n^2 \rho = \frac{1}{4}\rho a^2 \omega^2$$

∴ **Average total energy per unit volume or Average total energy density**

$$E = \pi^2 a^2 n^2 \rho + \pi^2 a^2 n^2 \rho = 2\pi^2 a^2 n^2 \rho$$

$$= \frac{1}{2}\rho a^2 \omega^2$$

∴ Average kinetic energy per unit volume $(K.E.)_{av}$

= Average potential energy per unit volume $(P.E)_{av}$

$= \frac{1}{2} \times$ Average total energy per unit volume.

Hence average kinetic energy per unit volume is equal to the average potential energy per unit volume and is equal to half the total energy per unit volume of the progressive wave.

Maximum potential energy. It should be noted that both potential and kinetic energies of the wave depend upon the position and time *i.e.*, the values of x and t, but the total energy is independent of both these quantities.

The potential energy at a point is **maximum** when

$$\sin\frac{2\pi}{\lambda}(vt - x) = 1$$

or $$y = a \qquad \left[\because y = a\sin\frac{2\pi}{\lambda}(vt - x)\right]$$

i.e. when the point is at extreme end of its displacement on either side of its mean position. The value of maximum potential energy.

$$= 2\pi^2 a^2 n^2 \rho = \frac{1}{2}\rho a^2 \omega^2$$

Maximum kinetic energy. The kinetic energy at a point is maximum when

$$\cos\frac{2\pi}{\lambda}(vt - x) = 1$$

or $$\sin\frac{2\pi}{\lambda}(vt - x) = 0$$

or $$y = 0$$

i.e. when the point is passing through its mean position.

The value of maximum kinetic energy

$$= 2\pi^2 n^2 a^2 \rho = \frac{1}{2}\rho a^2 \omega^2$$

Thus the maximum value of the potential energy is the same as the maximum value of kinetic energy and is equal to the average total energy per unit volume of the medium. Further at a time, when the potential energy at a point is maximum, the kinetic energy is zero and *vice versa*.

The kinetic energy is maximum when the particle is at its mean position, and is gradually converted into potential energy as the particle moves towards the extreme position. At the extreme position the energy is wholly potential. When the particle again moves backward to the mean position the potential energy is gradually converted into kinetic energy till the particle reaches the mean position where the energy is wholly kinetic.

13.19 INTENSITY OF ACOUSTIC WAVE

The intensity of an acoustic wave is defined as the amount of energy flowing through the medium per unit area per unit time. It is denoted by I.

In an advancing wave train a new length v equal to the velocity of the wave is set in motion every second. Hence the energy transferred per unit area per unit time is equal to the energy contained in a length v.

Now the total energy contained in a unit volume of the medium or energy density $E = 2\pi^2 n^2 a^2 \rho$ (For proof see article 13.18. Eq.(*i*) where ρ is the density of the medium, n the frequency and a the amplitude of the wave.

$\therefore$ Intensity of the wave $I = Ev = 2\pi^2 n^2 a^2 \rho\, v$

Hence intensity of the wave is equal to the product of the energy density and velocity of the wave. It is, therefore, known as *energy current*.

Now, $$I = 2\pi^2 n^2 a^2 \rho v = \frac{1}{2}\rho a^2 \omega^2 v \qquad \text{... }(i) \qquad [\because \omega = 2\pi n]$$

This shows that intensity of a wave is

(*i*) proportional to the density of the medium

(*ii*) velocity of the wave

(*iii*) square of frequency of the wave and

(*iv*) square of amplitude of the wave

Total energy and intensity independent of space and time co-ordinates. The average total energy per unit volume of a progressive wave is given by

$$E = \frac{1}{2}\rho a^2\omega^2 = 2\pi^2 n^2 a^2 \rho \qquad \text{... }(ii)$$

[For proof see article 13.18. Eq. (*i*)].

The intensity of the wave is given by

$$I = 2\pi^2 n^2 a^2 \rho v = \frac{1}{2}\rho a^2\omega^2 v \qquad [\because \text{ Eq. } (i)] \qquad \text{... }(iii)$$

From (*ii*) and (*iii*) it is clear that average total energy per unit volume and intensity of a progressive wave are independent of space and time co-ordinates (x and t).

13.20 SPECIFIC ACOUSTIC IMPEDANCE

It is defined as the ratio of excess of pressure to the particle velocity in the wave motion. It is denoted by Z.

$$\therefore \qquad Z = \frac{P}{\frac{dy}{dt}}$$

But $$P = -E\frac{dy}{dx} \qquad [\text{Eq. } (ii)]$$

and $$y = a\sin\frac{2\pi}{\lambda}(vt - x)$$

$$\therefore \qquad \frac{dy}{dt} = a\frac{2\pi v}{\lambda}\cos\frac{2\pi}{\lambda}(vt - x)$$

and $$\frac{dy}{dx} = -a\frac{2\pi}{\lambda}\cos\frac{2\pi}{\lambda}(vt - x)$$

$$\therefore \qquad Z = \frac{Ea\frac{2\pi}{\lambda}\cos\frac{2\pi}{\lambda}(vt - x)}{a\frac{2\pi}{\lambda}v\cos\frac{2\pi}{\lambda}(vt - x)} = \frac{E}{v}$$

$\therefore$ *Specific acoustic impedance may also be defined as the ratio of the elasticity of the medium to the velocity of the wave.*

Units. The unit of Z are that of $\frac{P}{\frac{dy}{dx}} = \frac{\text{Force}}{\text{area}} \times \frac{\text{time}}{\text{length}} = \text{Newton sec/m}^3$

13.21 WAVE VELOCITY AND GROUP VELOCITY

Wave motion is a form of disturbance which travels through a medium due to the repeated periodic motion of the particles of the medium about their mean positions, the motion being handed over from one particle to the next. The individual oscillators which make up the medium only execute simple harmonic motion about their mean positions and do not themselves travel through the medium with the wave. Every particle begins its vibrations a little later than its predecessor and there is a progressive change of phase as we travel from one particle to the next. *It is the phase relationship of these particles that we observe as a wave and not their progressive motion through the medium.* There are three distinct velocities connected with the wave motion.

(*i*) **The particle velocity.** It is the velocity of the simple harmonic motion of the oscillating particle about its equilibrium position.

(*ii*) **The wave or phase velocity.** It is the velocity with which the plane of equal phase (crest or trough) travels through the medium.

The equation of a plane progressive wave is given by

$$y = a\sin\frac{2\pi}{\lambda}(vt - x) = a\sin(\omega t - kx)$$

where $\omega = 2\,\pi\,n$ and $k = \dfrac{2\pi}{\lambda}$

$$\therefore \qquad \frac{\omega}{k} = \frac{2\pi n}{2\pi/\lambda} = n\lambda = v$$

This v is the phase *velocity or the wave velocity*. Hence "*phase velocity is the velocity with which a plane progressive wave front travels forward and which has a constant phase given by* $(\omega t - kx)$'

Thus $\qquad \omega t - kx =$ a constant

Differentiating with respect to t, we get

$$\omega - k\frac{dx}{dt} = 0$$

or $$\frac{dx}{dt} = \frac{\omega}{k} = v \text{ (phase velocity)} \qquad \text{... } (i)$$

The group velocity. A group consists of a number of waves of different frequencies superimposed upon each other. For example, white light consists of a continuous visible wave length spectrum ranging from about 3000Å in the violet to about 7000 Å in the red region. The motion of such a pulse is described by its group velocity.

The wave velocity of each components of such a group would be different in all media except vacuum. Hence such a group of waves will disperse with time when it is travelling through a medium.

The importance of the group velocity lies in the fact that *this is the velocity with which the maximum amplitude and thus the energy of the wave group is transmitted.*

For *monochromatic* wave their group velocity and wave velocity are identical.

Wave groups. Let us consider a group of waves consisting of only two components of equal amplitudes and having frequencies ω_1 and ω_2, differing by a *small* amount and represented by the equation.

$$y_1 = a\cos(\omega_1 t - k_1 x)$$

and $$y_2 = a\cos(\omega_2 t - k_2 x)$$

where $\dfrac{\omega_1}{k_1}$ and $\dfrac{\omega_2}{k_2}$ represent their respective velocities, k is known as wave number and is $= 2\pi/\lambda$. It is further supposed that $\dfrac{\omega_1}{k_1} \neq \dfrac{\omega_2}{k_1}$ i.e., it is a *dispersive medium.*

The resultant amplitude is given by

$$y = y_1 + y_2 = a\,[\cos(\omega_1 t - k_1 x) + \cos(\omega_2 t - k_2 x)]$$

$$= 2a\cos\left[\frac{(\omega_1 + \omega_2)}{2}t - \frac{(k_1 + k_2)}{2}x\right]\cos\left[\frac{(\omega_1 - \omega_2)}{2}t - \frac{(k_1 - k_2)}{2}x\right]$$

$$\therefore \qquad y = 2\,a\cos(\omega t - kx)\cos\left(\frac{\Delta\omega t}{2} - \frac{\Delta kx}{2}\right) \qquad \text{... } (ii)$$

where $$\omega = \frac{\omega_1 + \omega_2}{2} \text{ and } k = \frac{k_1 + k_2}{2}$$

and $$\Delta\omega = (\omega_1 - \omega_2) \text{ and } \Delta k = (k_1 - k_2).$$

The resultant wave has two parts.

Part (*a*) : A wave of angular frequency ω, propagation constant k and velocity.

$$v = \frac{\omega}{k}$$

Part (*b*) : A second wave of angular frequency $\frac{\Delta\omega}{2}$, propagation constant $\frac{\Delta k}{2}$ and velocity $v_g = \frac{\Delta\omega}{\Delta k}$

The first type of wave has a maximum amplitude $2a$. It is modulated in space and time by the second wave which consists of a group of waves of the first type and is a very slowly varying envelope of frequency $\frac{\Delta\omega}{2}$ and propagation constant $\frac{\Delta k}{2}$. The modulated pattern moves with a velocity $v_g = \frac{\Delta\omega}{\Delta k}$ known as the group velocity.

$$\therefore \qquad \text{Group velocity } v_g = \frac{\Delta\omega}{\Delta k} = \frac{d\omega}{dk} \qquad \text{... (iii)}$$

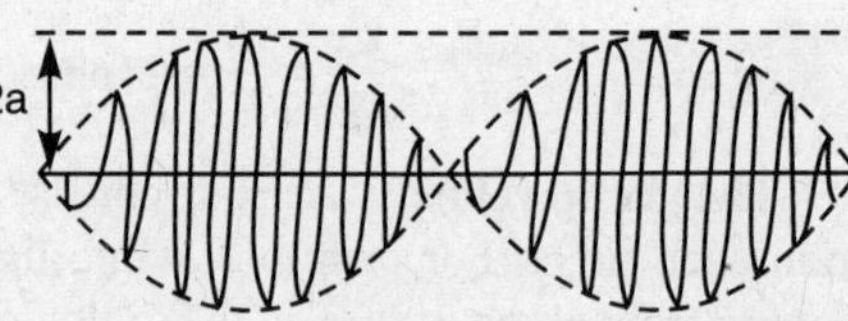

Fig. 13.13

If the waves are sound waves, the intensity is maximum whenever the amplitude is maximum = $2a$. This occurs twice in every period of the modulating frequency *i.e.*, at a frequency.

$$2\frac{\Delta\omega}{2.2\pi} = \frac{\omega_1 - \omega_2}{2\pi} = v_1 - v_2$$

where v_1 and v_2 are the frequencies corresponding to ω_1 and ω_2. Thus the *beats* of maximum intensity fluctuations have a frequency equal to the difference $v_1 - v_2$ of the two components. If the two components have equal amplitudes, the amplitude of the envelope will vary between 0 and $2a$. In such a case the modulation is said to be complete or 100%.

Relation between group velocity and wave velocity. Suppose the two frequency components have different phase velocities $\frac{\omega_1}{k_1}$ and $\frac{\omega_2}{k_2}$ so that $\frac{\omega_1}{k_1} \neq \frac{\omega_2}{k_2}$, then the velocity $v_g = \frac{\omega_1 - \omega_2}{k_1 - k_2} = \frac{\Delta\omega}{\Delta k}$ will be different from each of the phase velocities. In such a case the superposition of the two waves will no longer remain constant and hence the group profile will change with time. Such a medium in which the phase velocity depends upon frequency is called a *dispersive medium*. The variation of ω as a function of k represents a *dispersion relation.*

If the group of waves consists of a number of component velocities very close to each other, we can write

$$\frac{\Delta\omega}{\Delta k} = \frac{d\omega}{dk} = v_g \qquad \text{... (iv)}$$

If the phase velocity is represented by v, then

$$v = \frac{\omega}{k} \text{ or } \omega = kv$$

$$\therefore \qquad \frac{d\omega}{dk} = v + k\frac{dv}{dk}$$

But $$\frac{d\omega}{dk} = v_g$$

$\therefore$ $$v_g = v + \frac{k}{dk} dv \qquad \text{... } (v)$$

Also $$k = \frac{\omega}{v} = \frac{2\pi n}{v} = \frac{2\pi}{\lambda}$$

$$\frac{dK}{d\lambda} = -\frac{2\pi}{\lambda^2} = -\frac{1}{\lambda} K$$

or $$\frac{k}{dk} = -\frac{\lambda}{d\lambda}$$

Substituting the above value of $\frac{k}{dk}$ in (*v*), we have

$$v_g = v - \lambda \frac{dv}{d\lambda} \qquad \text{... } (vi)$$

The group velocity v_g may be < = > *v*, where v is the wave or phase velocity.

Now three cases arise:

(*i*) $\frac{dv}{d\lambda}$ is a positive quantity. In such a case $v_g < v$ *i.e.*, group velocity is less than wave velocity.

(*ii*) $\frac{dv}{d\lambda} = 0$. In such a case $v_g = v$ *i.e.*, group velocity is equal to wave velocity or phase velocity. In other words if the phase velocity is the same for all the component waves, the group velocity of the wave obtained by their superposition is equal to wave (or phase) velocity.

(*iii*) $\frac{dv}{d\lambda}$ is a negative quantity. In such a case

$$v_g = v + \lambda \frac{dv}{d\lambda}$$

and $v_g > v$ *i.e.*, group velocity is greater than wave velocity.

Hence group velocity is less than, equal to or greater than wave velocity according as $\frac{dv}{d\lambda}$ is positive, zero or negative.

It should be noted that when light travels through vacuum, all the waves of different wavelengths travel with the same velocity *i.e.* there is no change of wave velocity with wavelength or $\frac{dv}{d\lambda} = 0$. Hence from Eq.(*vi*) $v_g = v$. This explains why there is no difference in wave velocity and group velocity when light travels in vacuum.

13.22 DISPERSION OF WAVES (NORMAL AND ANOMALOUS)

Group velocity and wave velocity of a group of waves is connected by the relation.

$$v_g = v - \lambda \frac{dv}{d\lambda} \qquad \text{... } (i)$$

where the symbols have their usual meaning. When $\frac{dv}{d\lambda}$ is zero i.e., the wave velocity does not very with wavelength group velocity v_g = wave velocity *v*. i.e. all the waves in the group even of different wavelengths (or frequencies) travel with the same velocity and remain together. The most familiar example is that of white light travelling through vacuum. Such a medium is non-dispersive medium.

A medium in which $\frac{dv}{d\lambda} \neq 0$, the group velocity depends upon wavelength (or frequencies). As a result waves of different wavelengths travel with different velocities and separate out. *The process of separation of waves of different wavelengths (or frequencies) while passing through a medium is known as dispersion. Such a medium is called a dispersive medium.*

The dispersion of white light into its constituent colours while passing through a prism is a familiar example.

Dispersion relation

(*i*) **Normal dispersion.** If $\frac{dv}{d\lambda}$ is a positive quantity, in relation (*i*) $v_g < v$. *i.e.*, group velocity is less than wave velocity. This is the case of *normal dispersion.* In this case $v_g = v - \lambda \frac{dv}{d\lambda}$ gives the dispersions relation.

(*ii*) **Non-dispersion.** If $\frac{dv}{d\lambda} = 0$, $v_g = v$. i.e., group velocity is equal to the wave velocity. This is the case of *non-dispersive medium*. In such a case $v_g = v$ gives the dispersion relation. Hence group velocity is equal to wave velocity in a non-dispersive medium.

(*iii*) **Anomalous dispersion.** If $\frac{dv}{d\lambda}$ is a negative quantity $v_g > v$ *i.e.* group velocity is greater than wave velocity. This is the case of anomalous dispersion. In this case $v_g = v + \lambda \frac{dv}{d\lambda}$ gives the dispersion relaltion.

Examples. Electromagnetic wave in vacuum are non-dispersive. In dielectric media (glass, water etc.) there is normal dispersion except at the natural resonant frequency of its atoms and $v_g < v$. In electrical conductors $v_g > v$ and there is *anomalous dispersion*.

Dispersive medium is distinguished from a non-dispersive medium as under:

In a non-dispersive medium group velocity v_g is equal to wave velocity v.

According to the dispersion relation

$$v_g = v - \lambda \frac{dv}{d\lambda}$$

For a non-dispersive medium $\frac{dv}{d\lambda} = 0$

$$\therefore \qquad v_g = v$$

For a dispersive medium $\frac{dv}{d\lambda} \neq 0$. If $\frac{dv}{d\lambda}$ is positive, $v_g < v$

If $\frac{dv}{d\lambda}$ is negative, $v_g > v$

$\therefore$ For a dispersive medium $v_g < v$ or $v_g > v$

SOLVED EXAMPLES

Example 13.1 *If a wave of frequency 500 Hz is travelling with a velocity of 200 ms^{-1}, then find the change in phase at a given point in space between a time interval of 10^{-3} sec. Also find the path difference between two points which differ in phase by $\frac{\pi}{2}$ radian.* **(N.U. 2001)**

Solution. At a given point in space, the change in phase is 2π during a time interval T, the time period of wave motion.

$\therefore$ For a time interval t, phases change $\phi = \frac{2\pi}{T}t$

Here frequency of the wave $n = 500$ Hz

$\therefore$ Time period $\qquad T = \frac{1}{n} = \frac{1}{500}$ sec

Hence phase change in a time interval 10^{-3} second; $\phi = \frac{2\pi}{T}t$

$$= 2\pi \times 10^{-3} \times 500 = \pi \text{ radian}$$

At a given time, the phase difference between two points separated by a distance λ, the wavelength of the wave $= 2\pi$.

$\therefore$ For two points separated by a distance x, phase difference

$$\phi = \frac{2\pi}{\lambda}x$$

Here $\qquad \phi = \frac{\pi}{2}; \lambda = \frac{v}{n} = \frac{200\,\text{ms}^{-1}}{500\text{s}^{-1}} = 0.4\text{m}$

$\therefore \qquad \frac{\pi}{2} = \frac{2\pi}{0.4}x$ or path difference $x = 0.1$ m

Example 13.2 *A simple harmonic wave travelling along X-axis is given by*

$$y = 5 \sin 2\pi\,(0.2\,t - 0.5\,x)$$

Calculate the amplitude, frequency, wavelength, wave velocity, particle velocity, velocity amplitude, particle acceleration and acceleration amplitude (x is in metres t in seconds).

(Kerala U. 2000)

Solution. The given equation is

$$y = 5 \sin 2\,\pi\,(0.2\,t - 0.5\,x) \text{ m}$$

Comparing with equation, $y = a \sin \frac{2\pi}{\lambda}(vt - x)$, we get

$$y = a \sin 2\pi\left(\frac{vt}{\lambda} - \frac{x}{\lambda}\right) = a \sin 2\pi\left(nt - \frac{x}{\lambda}\right)$$

We have

Amplitude $\qquad a = 5$ m

Frequency $\qquad n = 0.2\ \text{s}^{-1}$

Now $\qquad \frac{x}{\lambda} = 0.5x$

$\therefore \qquad$ Wave length $\lambda = \frac{1}{0.5} = 2\text{m}$

Wave velocity $\qquad v = n\lambda = 0.2 \times 2 = 0.4\,\text{ms}^{-1}$

Particle velocity $\qquad = \frac{dy}{dt} = 5 \cos 2\pi(0.2t - 0.5x) \times 0.2$

$$= \cos 2\pi\,(0.2\,t - 0.5\,x)\text{ms}^{-1}$$

Particle velocity amplitude $= 1 \text{ ms}^{-1}$

Particle acceleration $= \dfrac{d^2y}{dt^2} = -\sin 2\pi(0.2t - 0.5x)0.2$

$= -0.2 \sin 2\pi\,(0.2\,t - 0.5\,x)\ \text{ms}^{-2}$

Particle acceleration amplitude $= -0.2 \text{ ms}^{-2}$

Example 13.3 *Find the amplitude, frequency, velocity and wavelength of the transverse wave in a string represented by*

$y = 5.\sin 2\pi\,(t - 0.04\,x)$ *in SI units* **(P.U. 2001)**

Solution Comparing with equation,

$$y = a \sin 2\pi\,(nt - \frac{x}{\lambda})\text{ ,we get}$$

Amplitude, $a = 5$ m

Frequency, $n = 1$ Hz

Wavelength, $\lambda = \dfrac{1}{0.04} = 25\text{m}$

Velocity, $v = n\lambda = 25 \text{ ms}^{-1}$

Example 13.4 *Write down the equation of wave travelling in the negative direction along x-axis and having an amplitude 0.01m, a frequency 550 Hz and speed 330 m/s.* **(Luck U. 2001)**

Solution. The equation of a wave travelling in the negative X-direction is

$$y = a \sin \frac{2\pi}{\lambda}(vt + x)$$

Now $a = 0.01$ m, $v = 330 \text{ ms}^{-1}$, $n = 550$ Hz

$$\therefore \quad \lambda = \frac{v}{n} = \frac{330}{550} = \frac{3}{5}\text{m}$$

Substituting, we get

$$y = 0.01 \sin\left[2\pi \times \frac{5}{3}(330t + x)\right]$$

or

$$y = 0.01 \sin 2\pi\left(550t + \frac{5}{3}x\right)$$

Example 13.5 *This displacement equation for a transverse plane wave at any instant is y (x,t) = 0.03 sin (3πt – 0.03 πx) where x and t are in metres and seconds. Calculate wavelength, frequency and velocity of the wave. Also calculate phase difference between two particles 0.05 metre apart at same instant.* **(Agra U. 2001)**

Solution. The general displacement equation of a transverse plane wave is given by

$$y = a \sin \frac{2\pi}{\lambda}(vt - x) = a \sin\left(2\pi \frac{v}{\lambda} t - \frac{2\pi}{\lambda} x\right)$$

$$= a \sin\left(2\pi nt - \frac{2\pi}{\lambda} x\right)$$

Comparing with the given equation

$$y\,(x, t) = 0.03 \sin\,(3\pi\,t - 0.03\,\pi x)$$

We have $2\pi n = 3\pi$

or $$n = \frac{3}{2} = 1.5 \text{ Hertz}$$

$$\frac{2\pi}{\lambda} = 0.03\pi$$

or $$\lambda = \frac{2}{0.03} = 66\frac{2}{3}\text{m}$$

velocity $$v = n\lambda = \frac{3}{2} \times \frac{200}{3} = 100 \text{ ms}^{-1}$$

Phase difference between two particles $x = 0.05$m apart

$$\phi = \frac{2\pi}{\lambda}x = 0.03\pi \times 0.05 = .0015\pi$$

Example 13.6 *Equation of plane progressive wave is given below*

$$y = 10 \sin \pi\, [0.01x - 2.00\, t]$$

where y and x are expressed in cm and t in sec. Determine (i) amplitude of the wave, (ii) frequency of the wave (iii) phase difference at an instant between two points 40 cm apart **(*Indore U. 2001*)**

Solution. The given equation can be written as

$$y = -10 \sin 2\pi\, [1.00\, t - 0.005x)$$

Hence, Amplitude, $a = 10$ cm

Frequency, $n = 1$ Hz

and phase difference, $\phi = \frac{2\pi}{\lambda}x = \frac{2\pi}{200} \times 40 = \frac{2}{5}\pi$ rad

Example 13.7 *A wave of frequency 400 Hz is travelling with a velocity 800 m/s. How far are two points situated whose displacements differ in phase by* $\frac{\pi}{4}$? **(*Calcutta 2001*)**

Solution. Frequency of the wave $n = 400$ Hz

Velocity of the wave $v = 800 \text{ ms}^{-1}$

∴ Wavelength $$\lambda = \frac{v}{n} = \frac{800}{400} = 2\text{m}$$

Phase difference between two points a distance x apart

$$\phi = \frac{2\pi}{\lambda}x$$

∴ For the displacements to differ in phase by $\frac{\pi}{4}$

$$\phi = \frac{\pi}{4} = \frac{2\pi}{2}x$$

or $$x = \frac{1}{4}\text{m} = 0.25\,\text{m} = 25\,\text{cm}$$

Example 13.8 *A wave has frequency of 400 Hz and velocity 320 ms*$^{-1}$. *Find the distance between points which are 45° out of phase.* **(*Pbi. U. 2008*)**

Solution. Wavelength, $$\lambda = \frac{V}{n} = \frac{320}{400} = 0.8 \text{ m}$$

Phase difference, $\phi = \frac{2\pi}{\lambda}x$ or $45° = \frac{\pi}{4} = \frac{2\pi}{0.8}x$

$\therefore$ $x = \frac{0.8}{8} = 0.1\text{m} = 10 \text{ cm}$

Example 13.9 *Find the speed of propagation of transverse wave on a 0.8 mm wire which is under a tension of 700 N. The density of steel is 7.9×10^3 kg/m³* **(Pbi. U. 2008)**

Solution. Diameter of the wire $d = 0.8$ mm $= 0.8 \times 10^{-3}$m $= 8 \times 10^{-4}$ m

Density of steel $D = 7.9 \times 10^3$ kg /m³

$\therefore$ Mass per unit length of the wire, $\rho = \frac{\pi d^2}{4} \times D = \frac{\pi}{4} \times 8 \times 8 \times 10^{-8} \times 7.9 \times 10^3$

$= 397 \times 10^{-5}$ kg/m

$= 3.97 \times 10^{-3}$ kg/m

Tension $T = 700$ N (Given)

$\therefore$ Speed (velocity) of propagation of transverse wave in the string

$$c = \sqrt{\frac{T}{\rho}} = \sqrt{\frac{700}{3.97 \times 10^{-3}}} = 4.2 \times 10^2 = 420 \text{ m/s}$$

Example 13.10 *A plane wave travelling in a string along X-axis is represented by*

$$y = 5 \cos \frac{\pi}{2}\left(\frac{x}{60} - \frac{t}{0.03}\right)$$

where x and y are in metres and t in sec. Find the speed of the wave. **(P.U. 2008)**

Solution. Given $y = 5 \cos \frac{\pi}{2}\left(\frac{x}{60} - \frac{t}{0.03}\right)$

$$= 5 \cos \frac{\pi}{120}\left(x - \frac{60t}{0.03}\right)$$

$$= 5 \cos \frac{2\pi}{240}\left(x - \frac{60t}{0.03}\right)$$

$$= 5 \cos \left[\frac{-2\pi}{240}\right](2000t - x)$$

Comparing this with standard equation for a progressive wave in cosine form as

$$y = a \cos \frac{2\pi}{\lambda}(vt - x) \text{, we have}$$

$$v = 200 \text{ m/s}$$

as x and y are in metres and t is in seconds.

Example 13.11 *Prove that $\psi = Ae^{i(\omega t - kx)}$ is the solution of wave equation* **(G.N.D.U. 2001)**

Solution: $\psi = Ae^{i(\omega t - kx)}$ where ω is the angular frequency $= 2\pi n$, k the propagation constant

$$= \frac{2\pi}{\lambda}$$

$$\therefore \quad \frac{d\psi}{dt} = i\omega A e^{i(\omega t - kx)} \text{ and } \frac{d^2\psi}{dt^2} = -\omega^2 A e^{i(\omega t - kx)}$$

$$= -\omega^2 \psi$$

$$\frac{d\psi}{dx} = -ik A e^{i(\omega t - kx)} \text{ and } \frac{d^2\psi}{dx} = k^2 A e^{i(\omega t - kx)} = -\omega^2 \psi$$

$$\therefore \quad \frac{d^2\psi}{dt^2} = \frac{\omega^2}{k^2}\frac{d^2\psi}{dx^2}$$

Now $\quad \dfrac{\omega}{k} = \dfrac{2\pi n\lambda}{2\pi} = n\lambda = c$ the velocity of the wave

$$\therefore \quad \frac{\omega^2}{k^2} = c^2$$

and $\quad \dfrac{d^2\psi}{dt^2} = c^2 \dfrac{d^2\psi}{dx^2}$

or $\quad \dfrac{d^2\psi}{dx^2} = \dfrac{1}{c^2}\dfrac{d^2\psi}{dt^2}$

which represents a general wave equation where ψ is a function of space and time co-ordinates (x and t) and c the velocity with which the wave travels without any change in form

Hence $\psi = Ae^{i(\omega t - kx)}$ is the solution of the wave equation.

Example 13.12 *If a wave is represented by $y = a \sin k(vt - x)$*

Prove that $\quad \dfrac{d^2y}{dt^2} = v^2 \dfrac{d^2y}{dx^2}$ **(P.U. 2000)**

Solution : $y = a \sin k(vt - x)$

$$\therefore \quad \frac{dy}{dt} = kva \cos k(vt - x) \text{ and } \frac{d^2y}{dt^2} = -k^2v^2 a \sin k(vt - x)$$

$$\therefore \quad \frac{dy}{dx} = -k\,a \cos k(vt - x) \text{ and } \frac{d^2y}{dx^2} = -k^2 a \sin k(vt - x)$$

Hence $\quad \dfrac{d^2y}{dt^2} = v^2 \dfrac{d^2y}{dx^2}$

Example 13.13 *Show that $y = x^2 + c^2t^2$ is a solution of one dimensional wave equation* **(P.U. 2008)**

Solution $\quad y = x^2 + c^2 t^2$

$$\therefore \quad \frac{dy}{dt} = 2c^2 t \text{ and } \frac{d^2y}{dt^2} = 2c^2$$

Again $\quad \dfrac{dy}{dx} = 2x \text{ and } \dfrac{d^2y}{dx^2} = 2$

Hence $\quad \dfrac{d^2y}{dt^2} = c^2 \dfrac{d^2y}{dx^2}$

Which is the one dimensional wave equation for a wave travelling along x-axis with a velocity c. Hence $y = x^2 + c^2 t^2$ is a solution of one dimensional wave equation.

Example 13.14 *Prove that 10 sin kx cos ct can not be a solution to the one dimensional wave equation (k and c are constants)* **(P.U. 2007, 2001, 2000)**

Solution. Here y = 10 sin kx cos ct

$$\therefore \quad \frac{dy}{dt} = -10c \sin kx . \sin ct \text{ and } \frac{d^2y}{dt^2} = -10c^2 \sin kx . \cos ct = -c^2 y$$

Again $$\frac{dy}{dx} = 10k \cos kx . \cos ct \quad \text{and} \quad \frac{d^2y}{dx^2} = -10k^2 \sin kx . \cos ct = -k^2 y$$

$$\therefore \quad \frac{d^2y}{dt^2} = \frac{c^2}{k^2}\frac{d^2y}{dx^2}$$

This does not satisfy the one dimensional wave equation given by

$$\frac{d^2y}{dt^2} = c^2 \frac{d^2y}{dx^2}$$

Hence, y = 10 sinkx cos ct can not be a solution to one dimensional wave equation.

Example 13.15 *A string is stretched by suspending a load of 5 kg. The mass per unit length of the string is 5 gm m^{-1}. The travelling waves are sent through the string by oscillating one of the ends with a frequency of 250 Hz and amplitude 5mm. Calculate the velocity and the wave length of the wave. Also write the equation of the travelling wave. Take g = 10 ms^{-2}.*

(Nagpur U. 2008, CPU. 2002)

Solution. Load suspended = 5 kg; g = 10 ms^{-2}

$\therefore$ Tension $T = mg = 5 \times 10 = 50$ N

Mass per unit length, ρ = 5 gm m^{-1} = 5×10^{-3} kg m^{-1}

$$\therefore \quad \text{Velocity } c = \sqrt{\frac{T}{\rho}} = \sqrt{\frac{50}{5 \times 10^{-3}}} = 100 \, \text{ms}^{-1}$$

Frequency n = 250 Hz

$$\therefore \text{ Wavelength} \quad \lambda = \frac{c}{n} = \frac{100}{250} = 0.4 \, \text{m}$$

Amplitude a = 5 mm = 5×10^{-3} m

$\therefore$ Wave equation is given by

$$y = a \sin \frac{2\pi}{x}(ct - x)$$

$$= 5 \times 10^{-3} \sin \frac{2\pi}{0.4}(100t - x)$$

$$= 5 \times 10^{-3} \sin 5\pi (100t - x)$$

Example 13.16 *Consider a rope 10 metre long with a mass of 0.05 kg. One end of the rope is given a transverse motion of small amplitude with a frequency of 10 Hz. If the tension of the rope is 200 N, what is the wavelength of the wave created in the rope.* **(Meerut U. 2006)**

Solution. Mass per unit length

$$p = \frac{0.05}{10} = 0.005\,\text{kg/m}$$
$$T = 200\text{ N}$$

$\therefore$ $$\text{Velocity } c = \sqrt{\frac{T}{\rho}} = \sqrt{\frac{200}{0.005}} = 200\,\text{m/s}$$

Frequency $$n = 10\text{ Hz}$$

Wavelength $$\lambda = \frac{v}{n} = \frac{200}{10}$$
$$= 20\text{ m}$$

Example 13.17 *A string 1.3 metre in length is divided into three segments such that their frequencies are in the ratio 2:3:4. Find the length of each segment.*

Solution. The frequency $n = \dfrac{1}{2l}\sqrt{\dfrac{T}{\rho}}$

As T and ρ are constant, the frequency $n \propto \dfrac{1}{l}$

As the frequencies are in the ratio 2: 3: 4, the lengths of the segments, l_1, l_2, l_3 are in the ratio $\dfrac{1}{2}:\dfrac{1}{3}:\dfrac{1}{4}$ or $6:4:3$

Also total length $l = l_1 + l_2 + l_3 = 1.3$ m

$\therefore$ $$l_1 = \frac{1.3}{6+4+3} \times 6 = 0.6\text{m}$$
$$l_2 = \frac{1.3}{6+4+3} \times 4 = 0.4\text{m}$$
$$l_3 = \frac{1.3}{6+4+3} \times 3 = 0.3\text{m}$$

Example 13.18 *A string is streched with a force of 1000 N. Its linear mass density is 0.1 kg m^{-1}. One end of the string (x = 0) is oscillating with an amplitude of 0.05 m and frequency 400 Hertz so that the travelling waves are set up in the positive x-direction.*

(*i*) *Calculate the velocity and wave-length of the wave.*

(*ii*) *Write the equation of the travelling wave*

Solution. Tension $T = 1000$ N

Mass per unit length $\rho = 0.1$ kg m^{-1}

$\therefore$ Velocity $$c = \sqrt{\frac{T}{\rho}} = \sqrt{\frac{1000}{0.1}} = 100\,\text{ms}^{-1}$$

Frequency $$v = 400\text{ Hertz}$$

$\therefore$ Wavelength $$\lambda = \frac{c}{v} = \frac{100}{400} = 0.25\,\text{m}$$

Amplitude a = 0.05 m

∴ Wave equation is given by

$$y = a\sin\frac{2\pi}{\lambda}(ct - x) = 0.05\sin\frac{2\pi}{0.25}(100t - x)$$

or $$y = 0.05 \sin 8\pi\,(100\,t - x)$$

Example 13.19. *Two wires of steel of the same length are stretched on a sonometer. The tension of the first and second are 8 kg wt and 2 kg wt respectively. Find the ratio of the fundamental notes emitted by the two wires when the diameter of cross section of the first wire is half that of the second.* **(Cal. U. 2000)**

Solution. Let d be the density of steel

Diameter of the second wire = D

∴ Diameter of the first wire = $\dfrac{D}{2}$

Mass per unit length of the first wire $\rho_1 = \dfrac{\pi}{4}\left(\dfrac{D}{2}\right)^2 d = \dfrac{\pi}{16}D^2 d$

Mass per unit length of the second wire $\rho_2 = \dfrac{\pi}{4}D^2 d = \dfrac{\pi}{4}D^2 d$

∴ $$\frac{\rho_2}{\rho_1} = 4$$

Tension for first wire T_1 = 8 kg wt = 8 g Newton

Tension for the second wire T_2 = 2 kg wt = 2 g Newton

Let l be the length of each wire, then

Frequency of the fundamental note of first wire $n_1 = \dfrac{1}{2l}\sqrt{\dfrac{T_1}{\rho_1}}$

Frequency of the fundamental note of second wire $n_2 = \dfrac{1}{2l}\sqrt{\dfrac{T_2}{\rho_2}}$

∴ $$\frac{n_1}{n_2} = \sqrt{\frac{T_1}{\rho_1}.\frac{\rho_2}{T_2}} = \sqrt{\frac{8}{2}.4} = 4$$

Example 13.20 *A string AB of length 100 cm is kept under uniform tension. A bridge kept at C such that BC = 60 cm. When the string is subjected to vibration BC part vibrates with frequency 252 Hz. Find the frequency of both parts of the string when the bridge is moved through 10 cm towards A.* **(Vid S.U. 2001)**

Solution. The position of the bridge in the first and second case is shown in Fig. 13 (*a*) and (*b*) respectively.

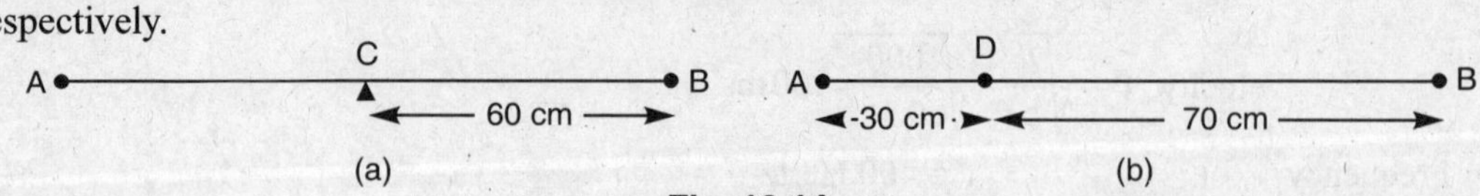

Fig. 13.14

When BC = 60 cm, frequency of BC part, $n = \dfrac{1}{2l}\sqrt{\dfrac{T}{\rho}}$

or $$252 = \frac{1}{120}\sqrt{\frac{T}{\rho}}$$

$$\therefore \qquad \sqrt{\frac{T}{\rho}} = 252 \times 120$$

When the bridge is moved 10 cm towards A from C to D

$$BD = 70 \text{ cm and } DA = 30 \text{ cm}$$

$$\therefore \text{ Frequency of part } BD,\ n_1 = \frac{1}{2 \times 70}\sqrt{\frac{T}{\rho}} = \frac{1}{140} \times 252 \times 120 = 216 \text{ Hz}$$

$$\text{Frequency of part } DA,\ n_2 = \frac{1}{2 \times 30}\sqrt{\frac{T}{\rho}} = \frac{1}{60} \times 252 \times 120 = 504 \text{ Hz}$$

Example 13.21 *A stretched string is observed to vibrate with a frequency 30 c.p.s. in the fundamental note when the length of the string is 60 cm. The string has a mass of 0.5 gm/cm. Find the velocity of propagation of the transverse wave and compute the tension of the string.*

(*Nagpur U. 2011, 2010; Burd U. 2003*)

Solution. Frequency of the fundamental note $n = 30$ c.p.s

Length of the string $l = 60$ cm

Mass per unit length $\rho = 0.5$ gm/cm

Let T be the tension of the string, then

$$n = \frac{1}{2l}\sqrt{\frac{T}{\rho}}$$

or

$$30 = \frac{1}{2 \times 60} \times \sqrt{\frac{T}{0.5}}$$

$\therefore\ T = 30 \times 30 \times 2 \times 2 \times 60 \times 60 \times 0.5 = 6.48 \times 10^6$ dynes/cm

$$\text{Velocity of propagation } v = \sqrt{\frac{T}{\rho}} = n \times 2l = 30 \times 2 \times 60$$

$$= 3600 \text{ cm/sec}$$

Example 13.22 *A plane progressive wave of frequency 256 Hz and amplitude 1/1000 mm is propagating through air of density 1.29 kg/m³. Calculate the energy density in the wave*

(*Kerala U. 2001*)

Solution. Here $\rho = 1.29$ kg/m³; $a = 1/1000$ mm $= 10^{-6}$ m; $\omega = 2\pi \times 256$ rad s^{-1}

$$\text{Average energy density} = \frac{1}{2}\rho a^2 \omega^2 = \frac{1}{2} \times 1.29 \times 10^{-12} \times 4\pi^2 \times 256 \times 256$$

$$= 1.67 \times 10^{-6} \text{ Jm}^{-3}$$

Example 13.23 *Calculate the increase in pressure when a sound wave produces an energy 10^{-7} Watt/cm². Velocity of sound =340 m/sec and density of air = 0.001293 gm/cc.*

Solution. Energy flow or intensity $I = 10^{-7}$ Watt/cm²

$= 10^{-7}$ Joule /sec/cm²

$= 10^{-3}$ Joule /sec/m²

Velocity of sound $\qquad v = 340$ m /sec

Density of air $\qquad \rho = 0.001293$ gm/c.c

$= 0.001293 \times 10^3$ kg m²

∴ Excess pressure $p_{max} = \sqrt{2I\rho v}$

$$= \sqrt{2\times10^{-3}\times0.001293\times10^{3}\times340}$$

$$= 0.9376 \text{ Newton/sq m.}$$

Example 13.24 *A tuning fork has a frequency of 256 and an amplitude of 0.5 cm. If the velocity of sound is 340 m/s, calculate the energy current. Density of air is 0.001293 gm/c.c*

(*Kan. U. 1990*)

Solution. Given $n = 256,\ a = 0.5 \text{ cm} = 0.5\times10^{-2}\text{ m}$

$v = 340 \text{ m/s},\ \rho = 0.001293 \text{ gm/c.c}$

$= 0.001293\times10^3 \text{ kg/m}^3$

Energy current = Intensity $I = 2\pi^2 n^2 a^2 \rho v$

$$= \left[2\times\frac{22}{7}\times\frac{22}{7}\times256\times256\times0.5\times0.5\times10^{-4}\times0.001293\times10^{3}\times340\right]$$

$$= 119 \text{ J/s/m}^2 = 119 \text{ watt/m}^2$$

Example 13.25 *The wave velocity v of a transverse wave in a crystal of atomic separation 'a' is given by* $v = c\left[\dfrac{\sin\frac{ka}{2}}{k\frac{a}{2}}\right]$ *where k is the wave number and c is a constant. Show that the value of group velocity is* $c\left[\dfrac{\cos ka}{2}\right]$

(*G.N.D.U 2008*)

Solution. Group velocity $v_g = v + k\dfrac{dv}{dk}$

where v is the waves velocity and k the wave number

Now $$v = c\left[\frac{\sin\frac{ka}{2}}{\frac{ka}{2}}\right] = c\left[\frac{ka}{2}\right]^{-1}\sin\frac{ka}{2}$$

∴ $$\frac{dv}{dk} = c\left[\frac{ka}{2}\right]^{-1}\frac{a}{2}\cos\frac{ka}{2} - \sin\frac{ka}{2}\,c\left[\frac{ka}{2}\right]^{2}\frac{a}{2}$$

$$= \frac{c}{k}\cos\frac{ka}{2} - \frac{c}{k^2\frac{a}{2}}\sin\frac{ka}{2}$$

or $$k\frac{dv}{dk} = c\cos\frac{ka}{2} - c\frac{\sin\frac{ka}{2}}{k\frac{a}{2}} = c\cos\frac{ka}{2} - v$$

∴ group velocity, $v_g = v + k\dfrac{dv}{dk} = c\left[\cos\dfrac{ka}{2}\right]$

Example 13.26 *The phase velocity of waves under the combined effect of surface tension and gravity is given by* $v = \sqrt{\frac{1}{k}\left(g + \frac{k^2 T}{\rho}\right)}$ *where k is the wave number, g the acceleration due to gravity, T the surface tension and ρ the density. Calculate the group velocity* **(G.N.D.U. 2008)**

Solution. Wave velocity $v = \frac{\omega}{k} = \sqrt{\frac{1}{k}(g + k^2 \frac{T}{\rho})} = \frac{1}{k}\sqrt{kg + \frac{k^3 T}{\rho}}$

$$\therefore \qquad \omega = \sqrt{kg + \frac{k^3 T}{\rho}} = \left(kg + \frac{k^3 T}{\rho}\right)^{\frac{1}{2}}$$

Group velocity
$$V_g = \frac{d\omega}{dk} = \frac{d}{dk}\left(kg + \frac{{}^3T}{\rho}\right)^{\frac{1}{2}}$$

$$= \frac{1}{2}\left(kg + \frac{k^3 T}{\rho}\right)^{-\frac{1}{2}}\left(g + \frac{3k^2 T}{\rho}\right)$$

$$= \frac{1}{2}\frac{\left(g + \frac{3k^2 T}{\rho}\right)}{\omega}$$

Example 13.27 *Two sinusoidal waves*

$$y_1 = 0.03 \cos (7t - 10\,x) \text{ m and}$$
$$y_2 = 0.03 \cos (5t - 8x) \text{ m}$$

were superimposed. Calculate the group velocity. **(G.N.D.U. 2001)**

Solution. Comparing with the wave equation $y = a \cos (\omega t - kx)$ we have,

Amplitude $a = 0.03$ m

Angular frequency $\omega_1 = 7$; $\omega_2 = 5$

Propagation constant $k_1 = 10$; $k_2 = 8$

$$\therefore \text{ Group velocity } v_g = \frac{\Delta\omega}{\Delta k} = \frac{\omega_1 - \omega_2}{k_1 - k_2} = \frac{7-5}{10-8} = 1 \text{ ms}^{-1}$$

Example 13.28 *Calculate the group velocity when the two waves*

$$y_1 = 10 \sin (2\pi t - 5\,x)$$
$$y_2 = 15 \sin (5\pi t - 5x)$$

superimpose. y_1 *and* y_2 *are in metres* **(G.N.D.U. 2001)**

Solution. Comparing with the wave equation $y = a \sin (\omega t - kx)$ we have,

Angular frequency $\omega_1 = 2\pi$, $\omega_2 = 5\pi$

Propagation number (constant) $k_1 = +5$: $k_2 = -5$

$$\therefore \text{ Group velocity } v_g = \frac{\Delta\omega}{\Delta k} = \frac{\omega_1 - \omega_2}{k_1 - k_2} = \frac{2\pi - 5\pi}{+5-(-5)}$$

$$-\frac{3}{10}\pi = -0.3\,\pi \text{ ms}^{-1}$$

Example 13.29 *Show that in a conducting medium*

$$v_g = v\left[1 + \frac{\lambda}{2\varepsilon_r} - \frac{d\varepsilon_r}{d\lambda}\right]$$

where ε_r *= relative permittivity* $\frac{\varepsilon}{\varepsilon_0}$ *and* $\mu = 1$. *Comment on the type of dispersion*

Solution. The wave velocity is given by

$$v = \frac{1}{\sqrt{\varepsilon_r \mu}} = \frac{1}{\sqrt{\varepsilon_r}} = \varepsilon_r^{-1/2} \qquad [\Theta\ \mu = 1]$$

$$\therefore \quad \frac{dv}{d\lambda} = -\frac{1}{2}\varepsilon_r^{-3/2}\frac{d\varepsilon_r}{d\lambda} = -\frac{1}{2\varepsilon_r}\varepsilon_r^{-1/2}\frac{d\varepsilon_r}{d\lambda}$$

$$= -v\frac{1}{2\varepsilon_r}\frac{d\varepsilon_r}{d\lambda}$$

Now $$v_g = v - \lambda\frac{dv}{d\lambda}$$

$$= v + \lambda v\frac{1}{2\varepsilon_r}\frac{d\varepsilon_r}{d\lambda} = v\left[1 + \frac{\lambda}{2\varepsilon_r}\frac{d\varepsilon_r}{d\lambda}\right]$$

As the quantity within brackets is greater than one, $v_g > v$ *i.e.*, it is the case of *anomalous dispersion.*

Example 13.30 *Calculate the frequency of fundamental note of a string 1 metre long and mass 2 gm when stretched by a weight of 400 kg.* ***(Nagpur U. 2011)***

Solution. Length of string $l = 1$ m

Mass per unit length $\rho = \frac{2\times10^{-3}}{1} = 2\times10^{-3}\,\text{kg/m}$

Tension $T = 400 \times 9.8 = 3920$ N

Fundamental frequency, $n = \frac{1}{2l}\sqrt{\frac{T}{\rho}}$

$$= \frac{1}{2\times1}\sqrt{\frac{3920}{2\times10^{-3}}} = \frac{1}{2}\times14\times10^2$$

$$= 700 \text{ Hz.}$$

EXERCISE

1. What is a wave motion? What are transverse and longitudinal waves? Give the condition regarding the direction of oscillation and direction of propagation of longitudinal and transverse waves. ***(Nagpur U.2010, 2009, 2008)***
2. State the characteristics of progressive wave. Derive an equation for the progressive wave of wavelength 0 and amplitude *a* moving with a velocity v in the positive *x*-direction.
   (*a*) What is a plane wave ***(G.N.D U.2007, Pbi U.2005, 2006)***
3. (*a*) Show that the slope of the displacement curve for a wave gives the volumetric strain in the medium in which it travels.
   (*b*) Derive a relation between v (Wave velocity), $\frac{dy}{dt}$ (particle velocity) and $\frac{dy}{dx}$ (volumetric strain)
   (Mithila U. 2001; Kan.U.2002; Vid. S.U. 2003)

4. Show that for a transverse wave in a string $T/p = c^2$ where T is the tension with which the string is stretched, ρ the mass per unit length, and c the velocity of the wave produced in it. Hence find the frequency of the fundamental note and the first octave of the vibrating string fixed at both end. ***(Meerut U.2006, 2005; Indore U.2001, Pbi U.2000, G.N.D.U.2006, 2000)***
5. The wave velocity in a string is a function of elasticity and inertia of the medium. Comment. ***(P.U. 2000)***
6. Prove that the wave equation for a transverse wave in a string is given by

$$\frac{\delta^2 y}{\delta x^2} = \frac{1}{c^2}\frac{\delta^2 y}{\delta t^2}$$

Where $c = \sqrt{\frac{T}{\rho}}$, T being the tension and ρ the linear density of the string. What are its possible solutions ***(H.P.U. 2001, G.N.D.U.2003, P.U. 2004)***
7. Deduce an expression for the velocity of transverse waves in a string. Hence find the frequency of the fundamental note and the first octave for a vibrating string fixed at both ends. ***(Nagpur U. 2011, 2010)***
8. (*a*) Derive the relation for the characteristic impedance of a string. Explain the factors on which it depends. ***(G.N. D.U. 2008, 2007, 2004, P.U. 2009, 2006, 2005, 2001, 2000, Pbi.U. 2008)***
(*b*) Explain why we hear a 'flick' at the end of a string. ***(G.N. D.U. 2000)***
9. What happens if incident waves fall at a common boundary of two strings having different impedances but joined in such a way that tension is the same throughout?
10. Define reflection coefficient and transmission coefficient of amplitude. What is the amplitude of transverse waves reflected at the denser medium (i.e. $Z_2 \cong \infty$) ?
11. Show that all energy arriving at the boundary in the incident wave leaves the boundary in the reflected and transmitted wave. Define reflection and transmission coefficients of energy.
12. Show that the velocity of sound in a solid medium is $\sqrt{\frac{Y}{\rho}}$ where Y and ρ denote respectively the Young's modulus and density of the solid. ***(Luke U.2001, Gauhati, U., 2000; Cal U. 2003)***
13. Derive an expression for the velocity of longitudinal waves in a gaseous medium. ***(Luck U.2001, Nagpur U. 2003)***
14. What is longitudinal wave? Do the molecules actually flow from speaker to the audience? Discuss what happens in reality.
15. Obtain the expression for total mechanical (kinetic and potential) energy per unit volume of the medium when a wave is travelling through it. Prove that in a progressive wave half the energy is kinetic and half potential. Also show that maximum kinetic energy is equal to maximum potential energy. ***(Bang U. 2004; Pbi. U. 2007)***
16. Explain the term intensity of an acoustic wave. Obtain expression for intensity of the wave using expression for total energy possessed by a wave in terms of density of the medium, frequency, amplitude and velocity of the wave.
17. Show that the total energy and intensity of a progressive wave are independent of space and time co-ordinates (x and t). ***(G.N.D.U.2000)***
18. (*a*) Show that the intensity of the wave at a point is given by

$$I = \frac{P_{rms}^2}{\rho v}$$

where P_{rms} is the root mean square value of the excess pressure.

(*b*) Calculate the increase in pressure when a sound wave produces an energy 10^{-7} watt/cm^2. Velocity of sound =340 m/sec. and density of air = 0.001293 gm/c.c.

[**Ans.** 0.9376 N/m^2] ***(Kan. U., 1990)***

19. What is specific acoustic impedance? State its units. ***(OS U. 2004)***

20. Define the terms wave velocity and group velocity. Find a relation connecting the two

(Nagpur U. 2011, 2010, G.N.D.U. 2004, 2002, 2000, Bhopal U. 2004; P.U. 2002, 2001, 2000, Pbi. U. 2000, Luck U. 2001)

21. Is group velocity less than wave velocity or greater than wave velocity? Comment. Prove that if the phase velocity is the same for all the component waves, group, velocity of the wave obtained by their superposition is equal to the wave velocity. ***(Luck 2001, Nagpur 2003)***

22. What is dispersion of waves? Discuss and distinguish between normal dispersions, non-disperions and anomalous dispersion. Give dispersion relations for them. Also give examples.

(P.U. 2004, 2001, 1998, 1992, Pbi. U. 2000, G.N.D.U. 1995)

23. Explain why there is no difference in wave velocity and group velocity when light travels in vacuum.

24. How dispersive medium is distinguished from non-dispersive medium?

(G.N.D.U. 2003, 2001, P.U. 2002, Pbi U. 2003, 2002, 2001)

25. Equation of a progressive wave is given by

$$y = \sin[500\pi t - \frac{\pi}{70}x]\text{ units}$$

What are the amplitude, frequency and wavelength of the wave.

[Ans. Amplitude = 1 unit, Frequency = 250 units, Wavelength = 140 units]

26. Write down the equation of a simple harmonic progressive wave of amplitude 0.001 m and frequency 5 Hz travelling in the negative *x*- direction with a velocity 3000 m/s.

[**Ans.** $y = 0.001 \sin 2\pi\ (5t + \frac{x}{600})$]

27. Two waves $y = 0.07 \cos(3t - 4x)$ m and $y_2 = 0.07 \cos(t - 2x)$ m are superimposed. Calculate the group velocity. ***(P.U. 2000)*** [**Ans.**1 ms^{-1}]

MULTIPLE CHOICE QUESTIONS

13.1 In a non dispersive medium the group velocity is

(*a*) greater than phase velocity
(*b*) less than phase velocity
(*c*) independent of phase velocity
(*d*) equal to phase velocity ***(H.P.U. 2003)***

13.2 In a normal dispersive medium the group velocity c_g and phase velocity c are related as

(*a*) $c_g > c$
(*b*) $c_g = c$
(*c*) $c_g = c^{-1}$
(*d*) $c_g < c$ ***(H.P.U. 2000)***

13.3 The phase velocity *c* and group velocity c_g are related as

(*a*) $c_g = c - \lambda\frac{dc}{d\lambda}$
(*b*) $c_g = c + \lambda\frac{d\lambda}{dc}$
(*c*) $c = c_g + \lambda\frac{dc_g}{d\lambda}$
(*d*) $c = c_g - \lambda\frac{dc_g}{d\lambda}$

ANSWERS

13.1 (*d*)	13.2 (*d*)	13.3 (*a*)

SUPERPOSITION OF WAVES

14
Chapter

INTRODUCTION

In the previous chapter, we have studied the propagation of a wave, longitudinal and transverse, through any medium (solid, liquid or gas). Here, we have discussed the simplest and smoothest, type of oscillatory motion, namely, simple harmonic motion (S.H.M.) of systems having one degree of freedom. In the present chapter, we will discuss the physical situations that involve the simultaneous application (or superposition) of two or more simple harmonic oscillations on the same system. Such situations are very common in acoustics. The most illustrative example is the working of microphone diaphragm and that of human ear drum. When both of them are subjected simultaneously to various vibrations, the displacement of vibration corresponds to the *resultant effect* due to superposition of various vibrations. The resultant of two or more harmonic oscillations is calculated by making use of a very important principle, known as the *superposition principle.*

14.1 THE SUPERPOSITION PRINCIPLE

The superposition principle states that ***the resultant of two or more harmonic displacements is simply the algebraic sum of the individual displacements.***

If y_1 and y_2 are the displacements of a particle due to two individual simple harmonic motions, when subjected simultaneously, then the resultant displacement of the particle is given by $y = (y_1 + y_2)$.

14.2 LINEAR AND NON-LINEAR EQUATIONS OF MOTION

A particle (or a body) executing simple harmonic motion (S.H.M.) has only one degree of freedom (displacement only along one-axis, either X, Y or Z axis). Moreover, the moving part always stays close to the equilibrium (*i.e.*, mean) position. The differential equation of S.H.M. is given by

$$\frac{d^2y}{dt^2} + \omega^2 y = 0 \quad \text{or} \quad \frac{d^2y}{dt^2} = -\omega^2 y \qquad \text{... (14.1)}$$

In this equation, d^2y/dt^2 is proportional to $-y$. This differential equation is the basis to prove any oscillatory motion is S.H.M. or not and is used to find the periods for a simple pendulum, compound penclulum, mass and spring system, torsion pendulum, bifilar oscillations, $L - C$ circuit etc. In all these cases, the restoring force F is proportional to its displacement y from the mean position *i.e.*, $F = -ky$. In other words, the restoring force depends only upon the term y. The equation does not contain terms like y^2, y^3 etc. The differential equations which do not contain higher powers of y, such as y^2, y^3 ..., etc. terms are called *linear equations.* Moreover, this equation does not contain any term independent of y.

Therefore, Eq. 14.1 is called a *linear homogeneous equation.*

Non-linear equation of motion. If, in a particular differential equation, higher powers of y *i.e.*, y^2, y^3 .... etc. terms are present, the equation is said to be *non-linear equation* of motion. Moreover, if it contains terms independent of y also, it is said to be non-homogeneous and non-linear equation. Non-linear equations are generally difficult to solve. Often one has to take recourse to numerical methods.

The equation for a simple pendulum is

$$M\frac{d^2y}{dt^2} = -Mg\sin\theta \qquad \text{... (14.2)}$$

It is assumed that amplitude *i.e.*, θ is small, then sin θ ≅ θ, and the equation

$$M\frac{d^2y}{dt^2} = -Mg\theta \text{ is linear}$$

However, $\sin\theta = \theta - \frac{\theta^3}{3!} + \frac{\theta^5}{5!} \cdots$ (Maclaurin's theorem)

Substituting this value of sinθ, we get

$$M\frac{d^2y}{dt^2} = -Mg\left(\theta - \frac{\theta^3}{3!} + \frac{\theta^5}{5!} - \ldots\right) \qquad \text{... (14.3)}$$

Eq. (14.3) is non-linear because it contains terms θ^3, θ^5, ... etc.

Linear homogeneous equations. One of the important properties of linear homogeneous equations is that *the sum of any two solutions is a solution by itself.* This property is not true in case of non-linear equations. This can be proved as under:

Consider the differential equation,

$$\frac{d^2y}{dt^2} = -\omega^2 y + Ay^2 + By^3 + Cy^4 \text{ etc.} \qquad \text{... (14.4)}$$

where *A*, *B*, *C* ... are constants. If the values of these constants are zero *i.e.*, *A* = *B* = *C* = ... = 0 or they can take sufficiently nearer value to zero, then Eq. (14.4) becomes

$$\frac{d^2y}{dt^2} = -\omega^2 y \qquad \text{... (14.5)}$$

which is *linear* and *homogenous* differential equation.

However, in Eq. 14.4, the constants *A, B, C* ... etc. are not zero, then the equation is *non-linear.*

Let us examine the important criterion mentioned above. Suppose, y_1 is the first solution of Eq. 14.5 at some other instant of time t_1 and y_2 is the second solution of Eq. 14.5 at some other instant of time t_2 (different displacement and velocity). The equations of motions will be

$$\frac{d^2y_1}{dt^2} = \omega^2 y_1 + Ay_1^2 + By_1^3 + Cy_1^4 + \ldots \qquad \text{... (14.6)}$$

and

$$\frac{d^2y_2}{dt^2} = -\omega^2 y_2 + Ay_2^2 + By_2^3 + Cy_2^4 + \ldots \qquad \text{... (14.7)}$$

It is interesting to find out whether or not the superposition of y_1 and y_2 defined by the sum $y = y_1 + y_2$, satisfies the equation of motion (14.4). We shall assume that it does and then arrive at contradiction. If $y = y_1 + y_2$ is also a solution of Eq. (14.4), we must have replacing *y* by $(y_1 + y_2)$

$$\frac{d^2y}{dt^2} = \frac{d^2}{dt^2}(y_1+y_2) = -\omega^2(y_1+y_2) + A(y_1+y_2)^2 + B(y_1+y_2)^3 + C(y_1+y_2)^4 + \ldots \qquad \text{... (14.8)}$$

Adding Eq. 14.6 and 14.7,

$$\frac{d^2y_1}{dt^2} + \frac{d^2y_2}{dt^2} = -\omega^2(y_1+y_2) + A(y_1^2+y_2^2) + B(y_1^3+y_2^3) + C(y_1^4+y_2^4) + \ldots \qquad \text{... (14.9)}$$

The equations 14.8 and 14.9 are identical, only if

$$\frac{d^2}{dt^2}(y_1+y_2) = \frac{d^2y_1}{dt^2} + \frac{d^2y_2}{dt^2} \qquad \text{... (14.10)}$$

$$-\omega^2(y_1+y_2) = -\omega^2 y_1 - \omega^2 y_2 \quad \text{... (14.11)}$$

$$A(y_1+y_2)^2 = A(y_1^2+y_2^2) \quad \text{... (14.12)}$$

$$B(y_1+y_2)^3 = B(y_1^3+y_2^3) \quad \text{... (14.13)}$$

$$C(y_1+y_2)^4 = C(y_1^4+y_2^4) \quad \text{... (14.14)}$$

14.10 and Eq. 14.11 are true. But Eqns. 14.12, 14.13 and 14.14 are true only if $A = 0$, $B = 0$, $C = 0$.

When constants A, B, C etc are zero, the equations become linear. In other words, our assumption that $(y_1 + y_2)$ is a solution of Eq. 14.4 can be realized if and only if the non-linear terms (*i.e.*, those with constants A, B, C, etc.) are absent. Thus, we conclude that the superposition of two solutions is itself a solution, if and only if the equation is linear. The superposition principle does not hold for non-linear equations.

14.3 THE IMPORTANCE OF SUPERPOSITION PRINCIPLE

To understand the importance of superposition principle, let us take the example of a simple pendulum. For small oscillations it executes linear homogeneous simple harmonic motion and is given by Eq. 14.1.

i.e., $$\frac{d^2y}{dt^2} = -\omega^2 y \quad (\because \text{ Eq. 14.1})$$

Suppose y_1 is the displacement at some instant of time t. Under a given set of initial conditions, let the solution y_1 be

$$y_1 = A_1 \cos(\omega t + \phi_1) \quad \text{... } (i)$$

where A_1 and ϕ_1 are determined from the initial conditions. Remember that the angular frequency ω does not depend upon the initial conditions.

Under another set of initial conditions, the displacement y_2 is given by

$$y_2 = A_2 \cos(\omega t + \phi_2) \quad \text{... } (ii)$$

where A_2 and ϕ_2 are determined from a new set of initial conditions.

Now, suppose we establish a third set of initial conditions in which the displacement is exactly equal to $(y_1 + y_2)$ *i.e.*, the sum of displacements of previous two individual initial conditions. Then from the superposition principle, the new motion described by the solution y_3 is simply given by the superposition of y_1 and y_2

In other words, $$y_3 = y_1 + y_2 \quad \text{... } (iii)$$

It should be noted that there is no need to solve the equation of motion to find the new position. Thus, *the resultant of two or more simple harmonic displacements is given by the algebraic sum of the individual displacements.*

As discussed in article 14.2, this superposition principle is valid only if the equation of motion is linear.

Consequences. The superposition of two linear harmonic oscillations results in to various phenomena, depending upon their relative frequencies, and phases and the way they are subjected *i.e.*, either parallel or perpendicular. The most commonly observed phenomena are production of beats, Lissajous figures etc. These consequences are explained below.

14.4 SUPERPOSITION OF TWO COLINEAR HARMONIC OSCILLATIONS

14.4.1 Oscillations having equal frequencies

Suppose we have two SHMs of equal frequencies beet of two different amplitudes and phase constants acting on a particle (or a system) in the x direction. Let ω be the same angular frequency of both the SHMs acting having initial phases ϕ_1 and ϕ_2 and amplitudes a_1 and a_2. The displacements x_1 and x_2 of the two harmonic motions are given by

$$x_1 = a_1 \sin(\omega t + \phi_1) \quad \ldots (i)$$

and

$$x_2 = a_2 \sin(\omega t + \phi_2) \quad \ldots (ii)$$

The resultant can be calculated analytically as well as geometrically.

(a) Analytical method. The resultant displacement due to the two simple harmonic vibrations is obtained by using *the superposition principle* and is given by

$$\begin{aligned} x &= x_1 + x_2 \\ &= a_1 \sin(\omega t + \phi_1) + a_2 \sin(\omega t + \phi_2) \\ &= \sin\omega t\,(a_1 \cos\phi_1 + a_2 \cos\phi_2) + \cos\omega t\,(a_1 \sin\phi_1 + a_2 \sin\phi_2) \end{aligned}$$

Put $a_1 \cos\phi_1 + a_2 \cos\phi_2 = R\cos\theta$

and $a_1 \sin\phi_1 + a_2 \sin\phi_2 = R\sin\theta$

Then, $x = R\sin\omega t.\cos\theta + R\sin\omega t.\sin\theta = R\sin(\omega t + \theta)$

which gives the equation of resultant simple harmonic vibration of amplitude R and initial phase θ where

$$\begin{aligned} R^2 = R^2\cos^2\theta + R^2\sin^2\theta &= a_1^2 + a_2^2 + 2a_1a_2(\sin\phi_1 \sin\phi_2 + \cos\phi_1 \cos\phi_2) \\ &= a_1^2 + a_2^2 + 2a_1a_2\cos(\phi_2 - \phi_1) \\ &= a_1^2 + a_2^2 + 2a_1\,a_2\cos\phi \quad \ldots (iii) \end{aligned}$$

where the phase difference

$$\phi = \phi_2 - \phi_1$$

and $\tan\theta = \dfrac{R\sin\theta}{R\cos\theta} = \dfrac{a_1\sin\phi_1 + a_2\sin\phi_2}{a_1\cos\phi_1 + a_2\cos\phi_2}$

(b) Geometrical method. Let us represent the vibration

$$x_1 = a_1 \sin(\omega t + \phi_1)$$

as the vector $\overrightarrow{OA}$ where the vector $\overrightarrow{OA} = a_1$ makes an angle ϕ_1 with the X-axis and the vibration

$$x_2 = a_2 \sin(\omega t + \phi_2)$$

as the vector $\overrightarrow{OB}$ where the vector $\overrightarrow{OB} = a_2$ makes an angle ϕ_2 with the X-axis as shown in Fig. 14.1.

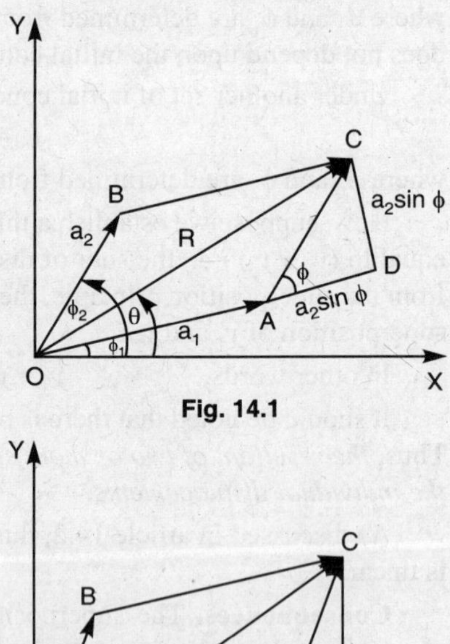

Fig. 14.1

The resultant is given by the vector $\vec{R} = \overrightarrow{OC}$.

and $\angle BOA = \phi_2 - \phi_1 = \phi$

$$\therefore \quad R^2 = (a_1 + a_2\cos\phi)^2 + (a_2\sin\phi)^2$$
$$= a_1^2 + a_2^2 + 2a_1\,a_2\cos\phi$$

which is the same result as obtained in (*iii*)

If θ is the angle that the resultant R makes with the x-axis then referring to Fig. 14.2, we have

$$\tan\theta = \frac{CD}{OD} = \frac{FD + CF}{OE + ED}$$

$$= \frac{AF + CF}{OE + AF}$$

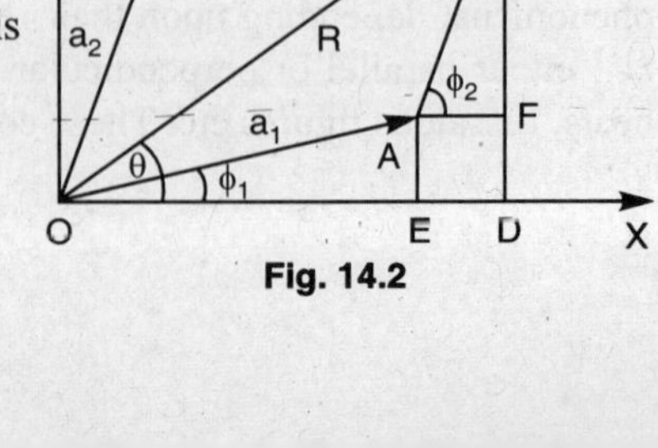

Fig. 14.2

$$= \frac{a_1 \sin\phi_1 + a_2 \sin\phi_2}{a_1 \cos\phi_1 + a_2 \cos\phi_2}$$

The resulting vibration is, therefore given by

$$x = R \sin(\omega t + \theta) \qquad \ldots (v)$$

(*i*) **Phase difference, (0).** When the phase difference

$$\phi = (\phi_2 - \phi_1) = 0$$

$$R^2 = a_1^2 + a_2^2 + 2a_1 a_2 \cos(0)$$

$$= a_1^2 + a_2^2 + 2a_1 a_2 = (a_1 + a_2)^2$$

or amplitude, $R = a_1 + a_2$ ... (*vi*)

Let $\phi_1 = 0$ and $\phi_2 = 0$, then

$$\tan\theta = \frac{a_1 \sin\phi_1 + a_2 \sin\phi_2}{a_1 \cos\phi_1 + a_2 \cos\phi_2} \qquad \ldots (vii)$$

$$= 0$$

(*ii*) **Phase difference, $\pi/2$.** When the phase difference

$$\phi = (\phi_2 - \phi_1) = \pi/2$$

$$R^2 = a_1^2 + a_2^2 + 2a_1 a_2 \cos\pi/2 = a_1^2 + a_2^2$$

or amplitude $R = \sqrt{a_1^2 + a_2^2}$ ... (*viii*)

Let $\phi_1 = 0$ and $\phi_2 = \pi/2$, then

$$\tan\theta = \frac{a_1 \sin\phi_1 + a_2 \sin\phi_2}{a_1 \cos\phi_1 + a_2 \cos\phi_2} = \frac{a_1 \sin 0 + a_2 \sin\pi/2}{a_1 \cos 0 + a_2 \cos\pi/2} = \frac{a_2}{a_1} \qquad \ldots (ix)$$

(*iii*) **Phase difference, π.** When the phase difference

$$\phi = (\phi_2 - \phi_1) = \pi$$

$$R^2 = a_1^2 + a_2^2 + 2a_1 a_2 \cos\pi$$

$$= a_1^2 + a_2^2 - 2a_1 a_2 = (a_1 - a_2)^2$$

or amplitude, $R = (a_1 - a_2)$ ... (*x*)

Again, $\tan\theta = 0$ ... (*xi*)

14.4.2 Oscillations having different frequencies : Beats

In practice, we come across many times the phenomenon in which the moving part of a system (or a particle) is subjected simultaneously to two harmonic oscillations of different frequencies. The resultant displacement is again calculated by using the superposition principle. When the frequency difference between two S.H.M.s is small, we observe the waxing and vaning of sound at regular intervals. This gives the phenomenon of beats.

Beats. When two wave-trains of nearly the same frequency travel along the same straight line in the same direction the resultant displacement at a point alternately waxes and wanes in amplitude as many times per second as the difference in their frequencies. This phenomenon is known as **beats.** The explanation is as follows:

The displacement *y* of a particle (along *y*-axis) at any time *t* when a harmonic wave of frequency *n* is moving with a velocity *v* is given by

$$y = a \sin 2\pi nt.$$

where *a* is the amplitude of the vibration.

If we consider two wave motions of slightly different frequencies of values *n* and *m* per second, then the displacement of a particle due to these waves is given by

$$y_1 = a \sin 2\,\pi\, n\, t \quad \text{and} \quad y_2 = a \sin 2\,\pi\, m\, t$$

When the two waves superimpose the resultant displacement Y is obtained by applying superposition principle, and is given by

$$Y = y_1 + y_2 = a \sin 2\,\pi\, n\, t + a \sin 2\, m\, t$$

$$= 2\, a \cos 2\pi\left(\frac{n-m}{2}\right)t \sin 2\,\pi\left(\frac{n+m}{2}\right)t$$

$$= A \sin 2\pi\left(\frac{n+m}{2}\right)t$$

This equation represents a periodic vibration of amplitude $A = 2a \cos 2\,\pi\left(\frac{n-m}{2}\right)t$ and frequency $\frac{n+m}{2}$, which is the arithmetic mean of the individual frequencies. The amplitude is, however, not constant as it changes with time.

The maximum value of the amplitude $A = \pm\, 2\, a$ when

$$\cos 2\pi\left(\frac{n-m}{2}\right)t = \pm 1$$

or $$\pi(n-m)\, t = k\,\pi \text{ where } k = 0, 1, 2, 3, \text{ etc.}$$

or $$t = \frac{k}{n-m} = 0, \frac{1}{n-m}, \frac{2}{n-m}, \frac{3}{n-m} \ldots$$

and so on *i.e.*, t is an integral multiple of $\frac{1}{n-m}$.

$\therefore$ Time interval between two consecutive maxima

$$= \frac{1}{n-m}$$

or The frequency of maxima $= n - m$.

The minimum value of the amplitude $A = 0$ when

$$\cos 2\pi\left(\frac{n-m}{2}\right)t = 0$$

or $$\pi(n-m)\, t = k\,\pi + \pi/2 \text{ where } k = 0, 1, 2, 3, \text{ etc.}$$

or $$t = \frac{k}{n-m} + \frac{1}{2(n-m)}$$

$$= \frac{1}{2(n-m)}, \frac{3}{2(n-m)}, \frac{5}{2(n-m)} \ldots \text{ for } k = 0, 1, 2, 3 \ldots$$

and so on, *i.e.*, t is an odd multiple of $\frac{1}{2(n-m)}$.

The minima are, therefore, regularly timed between the maxima. The time interval between two consecutive minima

$$= \frac{1}{n-m}$$

or The frequency of minima $= n - m$.

Since one maximum and one minimum sound constitutes a beat, the number of beats $= n - m$.

The intensity of the resultant sound rises and falls $n - m$ times per second.

The resultant wave is thus a simple harmonic wave whose frequency is equal to the arithmetic mean of the component frequencies and whose amplitude varies alternately from $2a$ or the sum of the amplitudes of the component waves, to zero or their difference. The vibration has a frequency equal to the difference of the component frequencies.

Determination of frequency. The phenomenon of beats is used to determine the frequency of a tuning fork. The number of beats produced by two tuning forks of nearly the same frequencies is equal to the difference in frequencies of the two forks. Thus if n is the frequency of a tuning fork A and b the number of beats produced, then the unknown frequency m of the tuning fork B is given by

$$m = n \pm b$$

To decide about the *positive* or *negative* sign proceed as follows:

Attach a little of wax to the prong of the tuning fork B of unknown frequency and again find out the number of beats produced. *If the number of beats decreases, then take* **positive** *sign and if the number of beats increases take* **negative** *sign.*

The effect of attaching a little of wax to the prongs of B is to make it heavy and so its frequency will decrease. Thus if the frequency of this tuning fork of unknown frequency was already greater, the difference between the two frequencies will decrease and so the number of beats will also decrease, hence *positive sign should be taken.* If the frequency of the unknown tuning fork, *i.e.*, of B had been less than that of A, the effect of loading the prongs of B is to increase the frequency difference and so the number of beats will increase, hence *negative sign* should be taken.

Example 14.1 *A fork of unknown frequency gives 4 beats per second when sounded with another of frequency 256. The fork is now loaded with a piece of wax and again 4 beats per second are heard. Calculate the frequency of the unknown fork.*

Solution. Number of beats produced = 4

Frequency of one tuning fork = 256

$\therefore$ Frequency of the other $256 \pm 4 = 252$ or 260

When the tuning fork is loaded with wax its frequency decreases. If the frequency is taken to be 252 it should decrease further and the number of beats should always increase.

If the frequency is taken to be 260 the number of beats should decrease when the tuning fork is loaded with a very small quantity of wax. If the quantity used is large the value of frequency will fall below 256. In the present case it falls to 252 thus making the difference again 4 and producing 4 beats per second. Hence

The frequency before loading = 260.

Applications of beats. The phenomenon of beats is of great practical importance. Beats can be used to determine the small difference between frequencies of two sounding bodies. If the frequency of one sounding body is known, the frequency of another (unknown) sounding body (source) can be determined. Another important application of phenomenon of beats is used in tuning the musical instruments. Musicians often make use of beats in tuning their instruments. A piano tuner uses beats to tell whether his standard tuning fork has the same frequency as the string of his instrument. If the two differ in frequency, *i.e.*, are out of tune, he will hear beats. He adjusts the tension in the string and thus changes the frequency of the note emitted by the string and matches it with his fork, by observing no beats at all. In Orchestra, sometimes beats are deliberately produced in a particular section, to yield a pleasing tone.

14.4.3 Superposition of two perpendicular harmonic oscillations: Lissajous Figures

(*a*) Analytical Method

The equation of a simple harmonic motion is given by

$$x = a \sin (\omega t + \phi)$$

where x is the displacement of the vibrating particle at any instant t, a the amplitude of vibration, ω the angular velocity and ϕ the initial phase.

Let the displacements of two perpendicular simple harmonic motions of the same period (or frequency ω) taking place along the *X*-axis and *Y*-axis respectively be represented by

$$x = a \sin(\omega t + \phi_1) \quad \text{... } (i)$$

and

$$y = b \sin(\omega t + \phi_2) \quad \text{... } (ii)$$

where *a* is the amplitude of the vibration along the *X*-axis, *b* is the amplitude of the vibration along the *Y*-axis, ϕ_1 and ϕ_2 the initial phases of *X* and *Y* vibrations respectively. The phase difference between the two vibration, then

$$\phi = \phi_2 - \phi_1$$

From (*i*) we have

$$\frac{x}{a} = \sin \omega t \cos \phi_1 + \cos \omega t \sin \phi_1 \quad \text{... } (iii)$$

From (*ii*) we have

$$\frac{y}{b} = \sin \omega t \cos \phi_2 + \cos \omega t \sin \phi_2 \quad \text{... } (iv)$$

Multiplying (*iii*) by $\sin \phi_2$ (*iv*) by $\sin \phi_1$ and subtracting, we have

$$\left(\frac{x}{a}\sin\phi_2 - \frac{y}{b}\sin\phi_1\right) = \sin \omega t (\cos \phi_1 \sin \phi_2 - \cos \phi_2 \sin \phi_1)$$

$$= \sin \omega t \sin(\phi_2 - \phi_1) \quad \text{... } (v)$$

Similarly multiplying (*iv*) by $\cos \phi_1$ and (*iii*) by $\cos \phi_2$ and subtracting, we have

$$\left(\frac{y}{b}\cos\phi_1 - \frac{x}{a}\cos\phi_2\right) = \cos \omega t (\sin \phi_2 \cos \phi_1 - \sin \phi_1 \cos \phi_2)$$

$$= \cos \omega t \sin(\phi_2 - \phi_1) \quad \text{... } (vi)$$

Squaring (*v*) and (*vi*) and adding, we get

$$\left(\frac{x}{a}\sin\phi_2 - \frac{y}{b}\sin\phi_1\right)^2 + \left(\frac{y}{b}\cos\phi_1 - \frac{x}{a}\cos\phi_2\right)^2$$

$$= \sin^2(\phi_2 - \phi_1)\,[\sin^2 \omega t + \cos^2 \omega t]$$

$$= \sin^2(\phi_2 - \phi_1)$$

or

$$\frac{x^2}{a^2} + \frac{y^2}{b^2} - \frac{2\,xy}{ab}(\sin\phi_2 \sin\phi_1 + \cos\phi_2 \cos\phi_1) = \sin^2(\phi_2 - \phi_1)$$

or

$$\frac{x^2}{a^2} + \frac{y^2}{b^2} - \frac{2xy}{ab}\cos(\phi_2 - \phi_1) = \sin^2(\phi_2 - \phi_1)$$

Substituting $\phi_2 - \phi_1 = \phi$ we get

$$\frac{x^2}{a^2} + \frac{y^2}{b^2} - \frac{2xy}{ab}\cos\phi = \sin^2\phi \quad \text{... } (vii)$$

This is the equation giving the expression for resultant motion and represents the equations of an ellipse whose major and minor axes are inclined to the co-ordinate axes. This ellipse can be inscribed in a rectangle whose sides are $2a$ and $2b$.

Important cases. (*i*) Straight line. When $\phi = 0$, $\cos \phi = 1$ and $\sin \phi = 0$

The relation (*vii*) is, then reduced to

$$\frac{x^2}{a^2} + \frac{y^2}{b^2} - \frac{2\,xy}{ab} = 0$$

or

$$\left(\frac{x}{a} - \frac{y}{b}\right)^2 = 0$$

This represents, a pair of coincident straight lines lying in the *first* and *third* quadrants as shown in Fig. 14.3 (*i*). The straight line is inclined to the *X*-axis at angle θ given by $\theta = \tan^{-1}\left(\frac{b}{a}\right)$

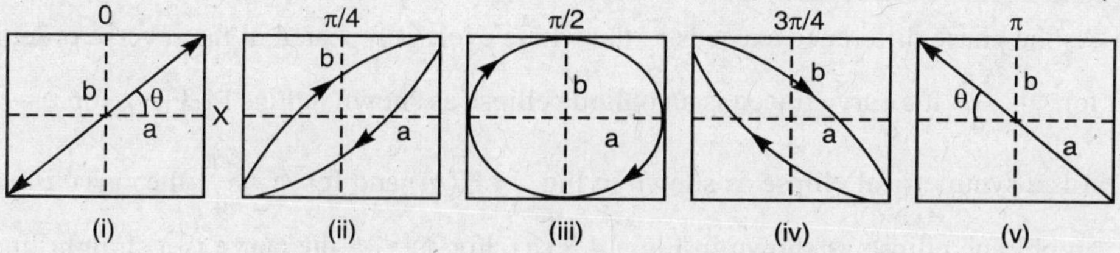

Fig. 14.3

(*ii*) **Oblique ellipse.** When $\phi = \frac{\pi}{4}$, $\cos\frac{\pi}{4} = \frac{1}{\sqrt{2}}$ and $\sin\frac{\pi}{4} = \frac{1}{\sqrt{2}}$

The relation (*vii*) is, then reduced to

$$\frac{x^2}{a^2} + \frac{y^2}{b^2} - \frac{\sqrt{2}\,xy}{ab} = \frac{1}{2}$$

This represents an *oblique ellipse* lying in *first* and *third* quadrant as shown in Fig. 14.3 (*ii*).

(*iii*) (***a***) **Symmetrical ellipse.** When $\phi = \frac{\pi}{2}$, $\cos\frac{\pi}{2} = 0$ and $\sin\frac{\pi}{2} = 1$

The relation (*vii*) is reduced to

$$\frac{x^2}{a^2} + \frac{y^2}{b^2} = 1$$

This represents a symmetrical ellipse whose major and minor axes coincide with the co-ordinate axes as shown in Fig. 14.3 (*iii*).

(***b***) **Circle.** If $a = b$ the resultant vibration is circular and is represented by

$$x^2 + y^2 = a^2$$

(*iv*) **Oblique ellipse.** When $\phi = \frac{3\pi}{4}$, $\cos\frac{3\pi}{4} = -\frac{1}{\sqrt{2}}$ and $\sin\frac{3\pi}{4} = \frac{1}{\sqrt{2}}$.

The relation (*vii*) is reduced to

$$\frac{x^2}{a^2} + \frac{y^2}{b^2} + \frac{\sqrt{2}\,xy}{ab} = \frac{1}{2}$$

This again represents an oblique ellipse lying in *second* and *fourth* quadrants as shown in Fig. 14.3 (*iv*).

(*v*) **Straight line.** When $\phi = \pi$, $\cos\phi = -1$ and $\sin\phi = 0$

The relation (*vii*) is reduced to

$$\frac{x^2}{a^2} + \frac{y^2}{b^2} + \frac{2xy}{ab} = 0$$

$$\left(\frac{x}{a} + \frac{y}{b}\right)^2 = 0$$

This represents a pair of coincident straight lines lying in the *second* and *fourth* quadrants as shown in Fig. 14.3 (*v*). The straight lines is inclined to the *X*-axis at an angle θ which is given by

$$\theta = \tan^{-1}\left(-\frac{b}{a}\right)$$

(*vi*) When ϕ lies between π and 2π.

As the phase difference exceeds π. the whole cycle is repeated in the reverse order *i.e.*, for $\phi = \frac{5\pi}{4}$ the curve traced is an oblique ellipse as shown in Fig. 14.3 (*iv*), for $\phi = \frac{3\pi}{2}$, it is a symmetrical ellipse as shown in Fig. 14.3 (*iii*) and for $\phi = \frac{7\pi}{4}$ the curve is again an oblique ellipse as shown in Fig. 14.3 (*ii*). For $\phi = 2\pi$ the curve is a straight line as shown in Fig. 14.3 (*i*), *i.e.*, for $\phi = 0$.

These figures traced are known as Lissajous figures and are as shown in fig. 14.3.

Lissajous figures. *When a particle is acted upon simultaneously by two simple harmonic motions at right angles to each other, the resultant path traced out by the particle is called* **Lissajous figures.**

The nature of the resultant path or the curve traced out depends upon

(*i*) the amplitudes,

(*ii*) the periods (or frequencies), and

(*iii*) the phase difference between the two component vibrations.

(*b*) Graphical method of tracing Lissajous' figures

(*i*) **Amplitudes different, periods same and phase difference** $\frac{\pi}{4}$ (45°). Let the simple harmonic vibration along the *X*-axis have an amplitude *a* and the vibration along the *Y*-axis an amplitude *b*.

Draw two circles of reference of radii *a* and *b* equal to the amplitudes of the corresponding simple harmonic motions taking place along the *X*-axis and the *Y*-axis respectively. In the circle of radius *a* draw a diameter *XX′* parallel to the *X*-axis and in the circle of radius *b* draw a diameter *YY′* parallel to the *Y*-axis. As the *periods* are *equal* divide both the circles into an equal number of parts, say 8, so that each part is travelled in the *same* time. Draw lines through these points perpendicular to the lines *XX′* and *YY′* respectively, so as to enclose a rectangle *PQRS* as shown in Fig. 14.4. It is supposed that the particle O_1 vibrating along *XX′* (*X*-vibration) first begins its journey *towards the right* and the particle O_2 vibrating along *YY′* (*Y*-vibration) first begins its journey in the *upward direction.*

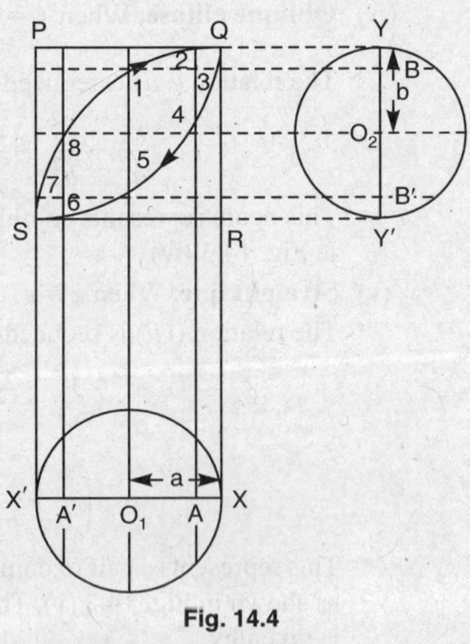

Fig. 14.4

Let the *Y*-vibration be ahead of the *X*-vibration by an angle $\pi/4$. It means that when the particle O_1 starts from its mean position along the *X*-axis, the particle O_2 has already completed 1/8th of its vibration along the *Y*-axis, *i.e.*, it is at *B*. The resultant position of a particle when both the vibrations act on it is represented by the point marked 1.

When the particle O_1 reaches *A* the particle O_2 reaches *Y* and the resultant position is represented by the point marked 2. Proceeding in this manner the positions as indicated by the points 3, 4 .... 8 are marked. Joining all these points by a free hand curve an oblique ellipse as shown in Fig. 14.4 is obtained.

Thus, we can apply graphical method to trace Lissajous figures for different input conditions. Fig. 14.5 shows the formation of another 8 shaped Lissajous figure corresponding to two S.H.M.s of different amplitudes, periods in the ratio of 1 : 2 and phase difference zero. (for more details, refer article 6.7 of physics for degree students B.Sc. I by the same authors).

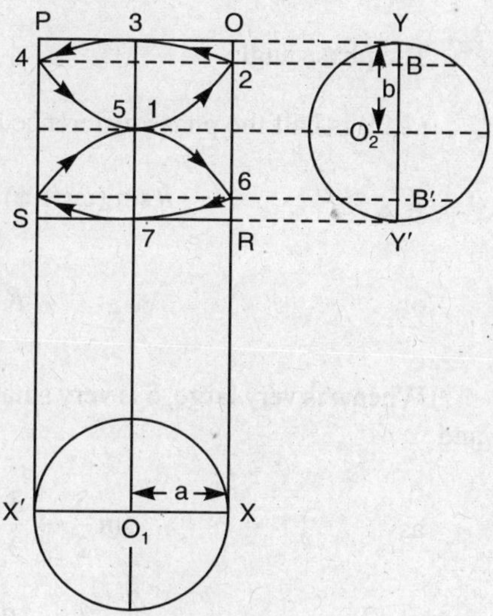

Fig. 14.5

14.5 RESULTANT OF LARGE NUMBER OF S.H.M.s

To find the resultant of a large number of simple harmonic vibrations having the same amplitude 'a', but constant phase difference δ with its predecessor, we represent the various simple harmonic vibrations as $a \sin \omega t$, $a \sin (\omega t + \delta)$, $a \sin (\omega t + 2\delta)$, $a \sin [(\omega t + (n-1)\delta]$. If R is the amplitude of the resultant vibration and ϕ its phase difference with respect to the first component $a \sin \omega t$, then

$$R \sin (\omega t + \phi) = a \sin \omega t + a \sin (\omega t + \delta) + a \sin (\omega t + 2\delta) \ldots + a \sin [\omega t + (n-1)\delta]$$

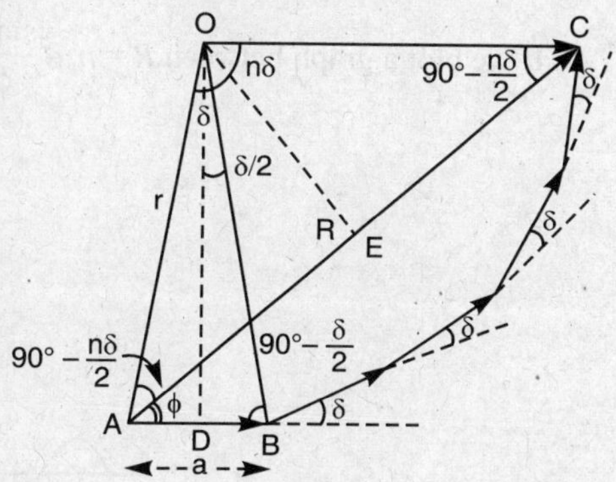

Fig. 14.6

To find the value of R and ϕ, the various amplitudes are represented in magnitude and direction by different sides of the polygon. The resultant amplitude is represented by the closing side of the polygon.

If O is the centre of the circle enclosing the (incomplete) polygon of n sides each of magnitude a and external angle δ, then

Radius of the circle $r = OA = OC$

In triangle AOB, $\angle AOB = \delta$ and $\angle OAB = \angle OBA = 90 - \frac{\delta}{2}$

Draw $OD \perp AB$, then

$$a = AB = AD + DB = r \sin \frac{\delta}{2} + r \sin \frac{\delta}{2}$$

$$= 2r \sin \frac{\delta}{2}$$

or $$2r = a / \sin \frac{\delta}{2}$$

In ΔAOC, $\angle AOC = n\delta$ and $\angle OAC = \angle OCA = 90 - \frac{n\delta}{2}$

Draw $OE \perp AC$, then

$$R = AC = AE + EC = r \sin \frac{n\delta}{2} + r \sin \frac{n\delta}{2} = 2r \sin \frac{n\delta}{2}$$

or $$R = a \frac{\sin n\,\delta/2}{\sin \delta/2}$$

The phase angle $$\phi = \left(90 - \frac{\delta}{2}\right) - \left(90 - \frac{n\delta}{2}\right) = (n-1)\frac{\delta}{2}$$

i.e., it is half the phase difference between the first and the last vibration.

$$\therefore \qquad R \sin(\omega t + \phi) = a\frac{\sin n\,\delta/2}{\sin \delta/2} \sin\left\{\omega t + (n-1)\frac{\delta}{2}\right\}$$

or $$R = a\frac{\sin n\,\delta/2}{\sin \delta/2}$$

When n is very large, δ is very small and in such a case ϕ is very nearly equal to $n\,\delta/2$ or $(n-1)\frac{\delta}{2}$ and

as $$\sin\frac{\delta}{2} \to \frac{\delta}{2} = \frac{\phi}{n}$$

$$\therefore \qquad R = a\frac{\sin n\,\delta/2}{\sin \delta/2} = \frac{a \sin\phi}{\phi/n} = n\,a\frac{\sin\phi}{\phi}$$

If we plot a graph between $R = n\,a\frac{\sin\phi}{\phi}$ and ϕ, it is as shown in Fig. 14.7.

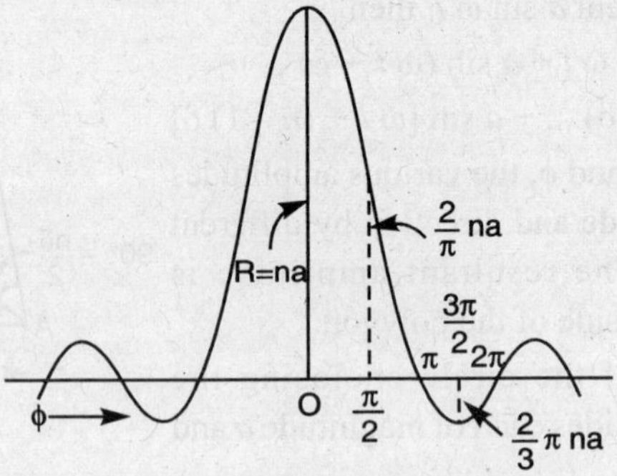

Fig. 14.7

The pattern obtained is symmetrical about $\phi = 0$. The value of R is maximum when $\phi = 0$ so that $\frac{\sin\phi}{\phi} = 1$ and $R = na$.

When $\phi = \frac{\pi}{2}$, $\sin\phi = 1$ and $R = \frac{2}{\pi}\,na$

When $\phi = \pi$, $\sin\phi = 0$ and $R = 0$

When $\phi = \frac{3\pi}{2}$, $\sin\phi = -1$ and $R = -\frac{2}{3\pi}\,na$

Aliter using exponential series:

Let us represent the successive simple harmonic motions by exponential functions

$$ae^{i\omega t},\ ae^{i(\omega t + \delta)},\ ae^{i(\omega t + 2\delta)} \ldots\ldots\ ae^{i[\omega t + (n-1)\delta]}$$

Let the resultant simple harmonic motion be represented by

$$Re^{i(\omega t + \phi)} = ae^{i\omega t} + ae^{i(\omega t + \delta)} + ae^{i(\omega i + 2\delta)} + \ldots..$$

$$..... + ae^{i[\omega t + (n-1)\delta]}$$

$$= ae^{i\omega t}[1 + e^{i\delta} + e^{i2\delta} + ... + e^{i(n-1)\delta}]$$

The expression within brackets is a geometrical series with common ratio $e^{i\delta}$ and

$$= \frac{1 - e^{in\delta}}{1 - e^{i\delta}}$$

$$\therefore \quad Re^{(\omega t + \phi)} = ae^{i\omega t}\frac{1 - e^{in\delta}}{1 - e^{i\delta}}$$

$$= ae^{i\omega t}\frac{e^{in\delta/2}}{e^{i\delta/2}}\left[\frac{e^{-in\delta/2} - e^{in\delta/2}}{e^{-i\delta/2} - e^{i\delta/2}}\right]$$

$$= ae^{i[\omega t + (n-1)\delta/2]}\frac{\sin n\delta/2}{\sin \delta/2}$$

It is a complex solution of the addition of a large number of displacements. Taking only the real part of the solution, we have

$$R \sin(\omega t + \phi) = a\frac{\sin n\delta/2}{\sin \delta/2}\sin\left[\omega t + (n-1)\frac{\delta}{2}\right]$$

$$\therefore \quad R = a\frac{\sin n\delta/2}{\sin \delta/2} \text{ and } \phi = (n-1)\frac{\delta}{2}$$

EXERCISE

1. State the principle of superposition and prove that it holds only for linear differential equations.
2. Two collinear simple harmonic motions acting simultaneously on a particle are given by
$$x_1 = A_1 \cos \omega t$$
$$x_2 = A_2 \cos(\omega t + \phi)$$
show that the resultant motion of the particle is simple harmonic. Also obtain the expression for the amplitude and phase constant of the resultant motion in terms of A_1, A_2 and ϕ.
3. Write the non-linear equation of motion and from this prove that the superposition principle does not hold for non-linear equations.
4. Enunciate the importance of superposition principle.
5. What are the consequences of superposition principle applied in case of two S.H.M.S. acting simultaneously.
6. Calculate the resultant of two simple harmonic vibrations of the same frequency acting along the same line but differing in phase, by applying superposition principle. What is the amplitude when the phase difference is 0, $\frac{\pi}{2}$ and π? ***(P.U. 2005; Bhopal U. 2004)***
7. A particle is subjected simultaneously to two S.H.M. of the same period but of different amplitudes and phases in perpendicular directions. Find the expression for the resultant motion. For what condition the path may be a straight line, ellipse or circle ? Discuss the different important cases. ***(G.N.D.U. 2007; Calicut U. 2003; Meerut U. 2003, 2002; Nagpur U. 2003, 2001; C.U. 2002)***

8. Show that the resultant of two simple harmonic motions at right angles to each other and having equal periods and amplitudes but phase difference 90° is a circle.

(Nagpur U. 2003, 2002)

9. What are Lissajous figures? How will you trace graphically the Lissajous figures ? When (*i*) the periods are equal and the phase difference is $\frac{\pi}{4}$ and (*ii*) the periods are in the ratio of 2: 1 and phase difference is (*a*) zero and (*b*) a quarter of the smaller period.

10. What are beats ? ***(Meerut U. 2007, G.N.D.U. 2006)***

Given an analytical treatment using superposition principle. Show that the beat frequency is equal to the difference between the frequencies of the component oscillations.

11. Discuss the superposition of a large number of simple harmonic vibrations of equal amplitude and equal successive phase difference.

12. By using exponential series, find the resultant of '*n*' number of S.H. vibrations of equal amplitude but successive phase angle δ.

13. Two organ pipes when sounded together produce 4 beats per second. Find their frequencies if one of them is 75 cm long and the other is 80 cm long. [**Ans.** 64 Hz and 60 Hz.]

14. Discuss linear superposition and non-linear superposition with suitable examples.

(Nagpur U. 2009)

MULTIPLE CHOICE QUESTIONS

14.1 The superposition principle holds only in case of

(*a*) linear equations of motion (*b*) non-linear equations of motion
(*c*) non of these

14.2 The resultant of two S.H.M. using superposition principle is given by

(*a*) $y = y_1 - y_2$ (*b*) $y = y_1 + y_2$ (*c*) $y = y_1^2 + y_2^2$ (*d*) $y = y_1^2 - y_1^2$

14.3 Beats are produced when the difference in frequencies of two S.H.M.s acting simultaneously is

(*a*) very large (*b*) very small (*c*) zero
(*d*) does not depend upon frequencies

14.4 Phenomena of beats and Lisajous figures are the consequences of superposition principle of

(*a*) linear equations of motion (*b*) non-linear equations of motion
(*c*) none of these

ANSWERS

14.1 (*a*) **14.2** (*b*) **14.3** (*b*) **14.4** (*a*)

STANDING WAVES

15
Chapter

INTRODUCTION

In the chapters 13 and 14, we have discussed waves which travel uninterrupted in open or unbounded systems. We assumed that the medium was homogeneous and infinite in extent. In this chapter we will discuss closed or bounded systems. We will describe some of the effects when a wave travelling in medium meets the boundary of another medium.

Sound wave is a longitudinal wave. It can be reflected, refracted, interfered and diffracted under appropriate conditions. Polarisation in longitudinal is out of question as it is possible only in transverse mode. Sound Waves can form stationary (standing) waves and beats in appropriate conditions.

15.1 REFLECTION AND TRANSMISSION

When a wave travelling in medium 1 meets the boundary of a medium 2, it is partly reflected and partly transmitted at the boundary. The reflected wave travels in medium 1 and the transmitted wave in medium 2. In the previous chapter 13, we have seen that a medium offers a characteristic impedance to the waves travelling through it. The characteristic impedance depends on the properties of the medium. Since the two media offer different impedances, it will be interesting to find out how waves will respond to an abrupt change of independence at the boundary separating the two media. We will analyse this problem for the waves described in chapter 13; transverse waves on a string and longitudinal (sound) waves in a medium.

15.2 REFLECTION OF SOUND WAVES

Echo, reverberation are some of the known and important phenomena caused due to reflection of sound waves in buildings. When a sound wave (longitudinal wave) meets an obstacle its behaviour is determined by the dimensions of the obstacle compared to the wavelength of the wave.

When the size of the obstacle is large compared with the wavelength of the wave, the phenomenon of reflection occurs, and the wave is reflected from the obstacle in the same medium. Since sound waves are generally of large wavelength, so very large obstacle is required to produce reflection and to produce shadow behind. Sound waves, moreover, obey the laws of reflection and refraction and Huygen's principle as is given in the case of light waves.

When the size of the obstacle is comparable with the wavelength of wave, the wave bends round the corner and encroaches into shadow appreciably. This results into the phenomenon of *diffraction.*

When the size of the obstacle is sufficiently smaller compared to the wavelength of the wave, the particle behaves as a source of new waves spreading out in all direction. This is known as *scattering* of waves.

15.3 CHANGE OF PHASE ON REFLECTION

When a wave (either longitudinal or transverse) suffers reflection at the boundary of a rigid wall or a denser medium, or any reflecting surface, the velocities of vibrating particles are reversed and suffer a change of phase of π.

As the displacement of the vibrating particles are reversed in direction, the crest is reflected as trough and trough as crest in case of transverse wave. Similar phenomenon happens in longitudinal waves.

However, the pressure waves are reflected back without change of phase at the surface of reflection. This will be found in case of closed organ pipes. The stale of compression or rarefaction will remain unchanged in the form in case of longitudinal waves in such cases.

15.4 FORMATION OF STATIONARY WAVES

When two simple harmonic waves of the same amplitude, frequency and time period travel in *opposite directions* in a straight line, the resultant wave obtained is called a stationary or a standing wave. The formation of stationary waves is due to the superposition of the two waves on the particles of the medium.

The formation of stationary waves can be represented graphically as follows :

Consider two wave trains A and B of the same amplitude, frequency and wavelength travelling in opposite directions. At an instant of time $t = 0$, the waves are as shown in Fig. 15.1. The resultant displacement curve is a straight line. All the particles of the medium are at their mean positions.

At time $t = T/4$, the wave A will advance through a distance $\lambda/4$ towards right, and the wave B will advance through a distance $\lambda/4$ towards left. The resultant displacement pattern is shown in Fig. 15.2.

The particles 1, 3, 5, and 7 are at their extreme positions and particles 2, 4 and 6 are at their mean positions.

At time $t = T/2$, the wave A will advance through a distance $\lambda/2$ towards right and the

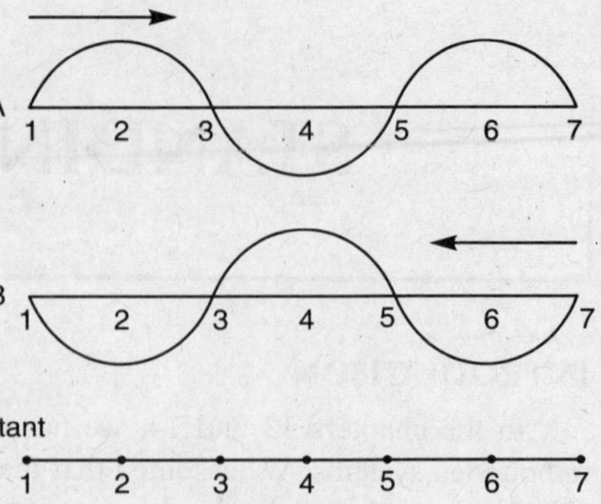

Fig. 15.1

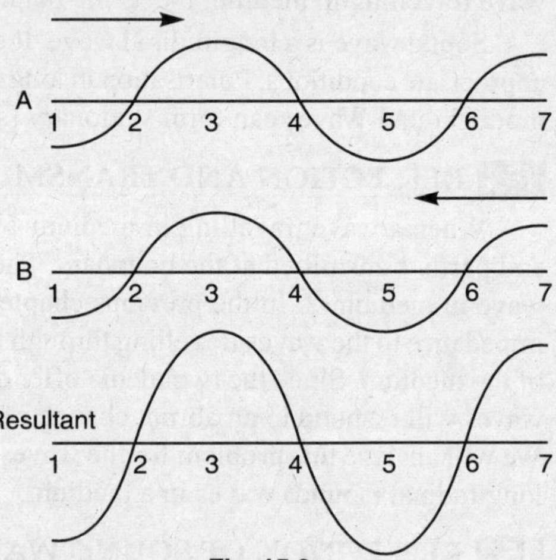

Fig. 15.2

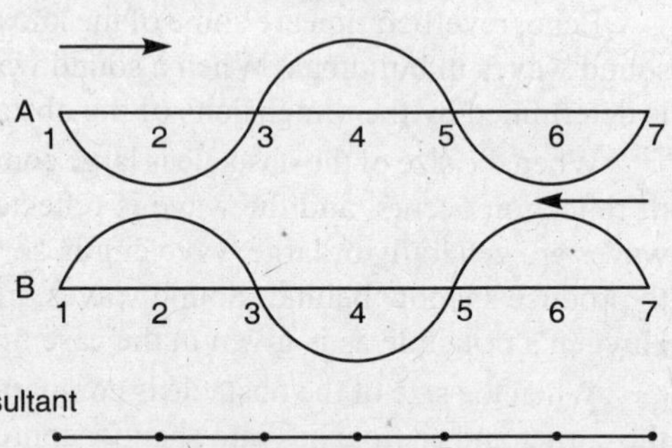

Fig. 15.3

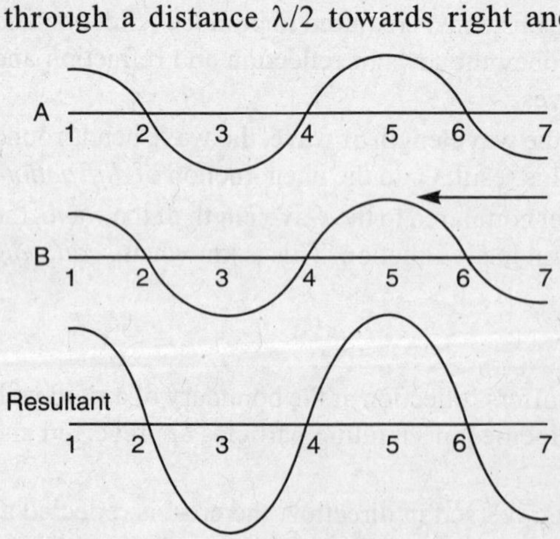

Fig. 15.4

wave B will advance through a distance $\lambda/2$ towards left (with reference to zero time).

The resultant displacement pattern is shown in Fig. 15.3.

All the particles of the medium are at their mean positions.

At time $t = 3T/4$, the wave A will advance through a distance $3\lambda/4$ towards right and the wave B will advance through a distance $3\lambda/4$ towards left (with reference to zero time).

The resultant displacement pattern is shown in Fig. 15.4.

The particles 1, 3, 5 and 7 are at their extreme positions and 2, 4, 6 are at their mean positions.

At time $t = T$, the wave A will advance through a distance λ towards right and the wave B will advance through a distance λ towards left (with reference to zero time). The resultant displacement pattern is shown in Fig. 15.5.

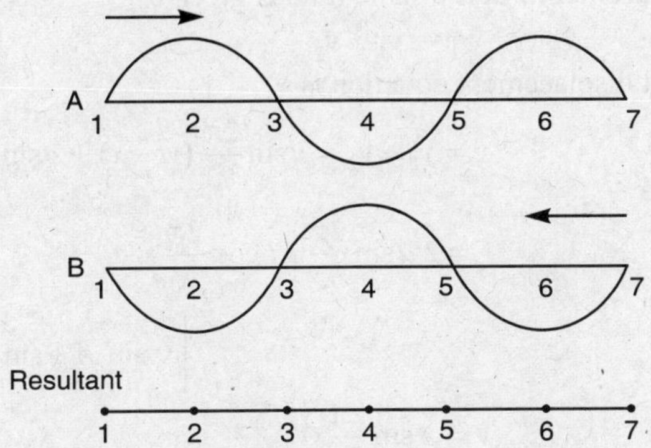

Fig. 15.5

All the particles are at their mean positions

From the patterns discussed above it is clear that the particles of the medium such as 2, 4, 6 etc. always remain at their mean positions. The particles such as 1, 3, 5, 7 etc. continue to vibrate simple harmonically about their mean positions with double the amplitude of each wave. It appears as though the wave pattern is stationary in space. The resultant displacement patterns at intervals of time, 0, $\frac{T}{4}, \frac{T}{2}, \frac{3T}{4}$ and T are shown in fig. 15.6.

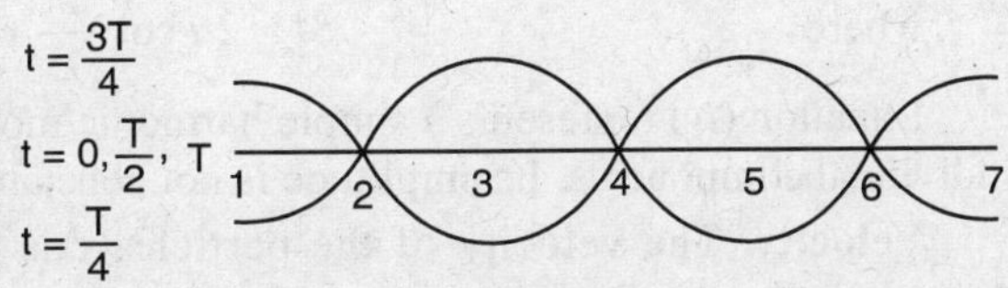

Fig. 15.6

The positions of the particles 2, 4, 6 etc. which always remain at their mean positions are called *nodes.* Node is a position of zero displacement and maximum strain.

The positions of the particles 1, 3, 5, 7 etc. which vibrate simple harmonically with maximum amplitude (twice the amplitude of each wave) are called antinodes. At the antinodes, the strain is minimum. The distance between any two consecutive nodes or antinodes is equal to $\lambda/2$. Between a node and an antinode, the amplitude gradually increases from zero to maximum.

15.5 STATIONARY WAVES (STANDING WAVES)

Analytical Treatment

Stationary waves are produced when two wave trains of the same period, wavelength amplitude and velocity travelling along the same straight line in opposite directions superimpose upon each other.

The analytical expression for a progressive wave of wavelength λ, amplitude a travelling with a velocity v along the positive direction of X-axis is given by

$$y_1 = a \sin \frac{2\pi}{\lambda}(vt - x) \qquad \text{... } (i)$$

where y_1 is the displacement at any time t and x is the distance of the particle from the origin ($x = 0$).

For a wave travelling in the opposite direction with the same velocity the distance of a particle in the negative direction of X-axis is $-x$, hence the displacement y_2 of this particle at the same time

t is given by

$$y_2 = a \sin \frac{2\pi}{\lambda}(vt + x) \quad \text{... (ii)}$$

These displacement equations actually refer to the incident and the reflected waves in the case of an open pipe or a string free at the other end, when the boundary where reflection taken place is taken as the origin $x = 0$.

The resultant displacement y at the same time is given by

$$y = y_1 + y_2$$

Hence the resultant displacement equation is

$$y = y_1 + y_2 = a \sin \frac{2\pi}{\lambda}(vt - x) + a \sin \frac{2\pi}{\lambda}(vt + x)$$

$$= 2\, a \sin \frac{2\pi}{\lambda} v\, t \cos \frac{2\pi}{\lambda} x \quad \text{... (iii)}$$

$$\left[\because \sin A + \sin B = 2 \sin \frac{A+B}{2} \cos \frac{A-B}{2}\right]$$

or
$$y = A \sin \frac{2\pi}{\lambda} vt \quad \text{... (iv)}$$

where
$$A = 2a \cos \frac{2\pi}{\lambda} x$$

Equation (*iv*) represents a simple harmonic motion of the same wavelength and velocity as the individual components. Its amplitude is not constant but varies with x the position of the particle.

Velocity. The velocity of the particles due to the resultant wave motion is obtained by differentiating the displacement y with respect to time t (keeping x constant) in equation (*iii*).

$$\therefore \quad \text{Velocity } \frac{dy}{dt} = \frac{4a\pi v}{\lambda} \cos \frac{2\pi}{\lambda} x \cos \frac{2\pi}{\lambda} vt \quad \text{... (v)}$$

Compression or strain. The compression or strain of the resultant vibration at any point is obtained by differentiating the displacement y with respect to x (keeping t constant) in equation (*iii*).

$$\therefore \quad \text{Strain } \frac{dy}{dx} = -\frac{4\pi a}{\lambda} \sin \frac{2\pi}{\lambda} x \sin \frac{2\pi}{\lambda} vt \quad \text{... (vi)}$$

The equations (*iii*), (*v*) and (*vi*) show that the displacement, velocity and strain at a given point change with time and at a given time are different at different points.

Changes with respect to position

At positions where $\frac{2\pi}{\lambda} x = 0, \pi, 2\pi \,....,\, m\pi$

or
$$x = 0, \frac{\lambda}{2}, ...,\, m\frac{\lambda}{2}$$

$$\sin \frac{2\pi}{\lambda} x = 0 \text{ and } \cos \frac{2\pi}{\lambda} x = \pm 1$$

At positions where $\frac{2\pi}{\lambda} x = \frac{\pi}{2}, \frac{3\pi}{2}, ... (2m+1)\frac{\pi}{2}$

or
$$x = \frac{\lambda}{4}, \frac{3\lambda}{4}, ... (2m+1)\frac{\lambda}{4}$$

$$\sin \frac{2\pi}{\lambda} x = \pm 1 \text{ and } \cos \frac{2\pi}{\lambda} x = 0$$

(i) **Displacement.** The displacement is given by

$$y = 2a \sin \frac{2\pi}{\lambda} vt \cos \frac{2\pi}{\lambda} x$$

This is maximum when $\cos \frac{2\pi}{\lambda} x = \pm 1$

or At points where $\quad x = 0, \frac{\lambda}{2}, \lambda \ldots. \frac{m\lambda}{2}$

the displacement is maximum. These points are called the **antinodes.**

The displacement is minimum when $\cos \frac{2\pi}{\lambda} = 0$

or $\quad$ At points where $x = \frac{\lambda}{4}, \frac{3y}{4}, \ldots. (2m+1)\frac{\lambda}{4}$

The points are called **nodes.**

Hence the interface $x = 0$ is an *antinode* and other antinodes lie at distance $\frac{\lambda}{2}, \lambda, \frac{3\lambda}{2}, \ldots..,$ from this point

The *nodes* lie in between the *antinodes* at distances $\frac{\lambda}{4}, \frac{3\lambda}{4}$ etc.

At points other than nodes and antinodes the displacement decreases gradually from their maximum value at the antinode to zero value at the node.

(ii) **Velocity.** From (*v*) we find that velocity is maximum when

$$\cos \frac{2\pi}{\lambda} x = \pm 1$$

or At points where $\quad x = 0, \frac{\lambda}{2}, \lambda \ldots., \frac{m\lambda}{2}$

i.e., the velocity is maximum at the antinodes where the displacement is also maximum.

The velocity is minimum (zero) when $\cos \frac{2\pi}{\lambda} x = 0$

or At points where $x = \frac{\lambda}{4}, \frac{3\lambda}{4} \ldots., (2m+1)\frac{\lambda}{4}$

i.e., the velocity is minimum (zero) at the nodes where the displacement is also minimum.

(iii) **Strain.** From (*vi*), we find that strain is maximum when

$$\sin \frac{2\pi}{\lambda} x = \pm 1$$

or At points where $x = \frac{\lambda}{4}, \frac{3\lambda}{4}, \ldots. \quad (2m+1)\frac{\lambda}{4}$

i.e., the strain is maximum at the nodes where displacement and velocity are zero.

It is minimum when $\sin \dfrac{2\pi}{\lambda} x = 0$

or At points where $x = 0, \dfrac{\lambda}{2}, \lambda \ldots., m\dfrac{\lambda}{2}$

i.e., the strain is minimum at the antinodes where the displacement and velocity are maximum.

Changes with respect to time. If T is the period of vibration of a particle due to either wave and v its velocity, then the distance travelled in time T is given by

$$\lambda = v\,T$$

or
$$\frac{v}{\lambda} = \frac{1}{T}$$

Substituting the value of $\dfrac{v}{\lambda}$ in (*iii*), (*v*) and (*vi*), we have

Displacement
$$y = 2a \sin \frac{2\pi t}{T} \cos \frac{2\pi}{\lambda} x \qquad \ldots (vii)$$

Velocity
$$\frac{dy}{dt} = \frac{4a\pi v}{\lambda} \cos \frac{2\pi}{\lambda} x \cos \frac{2\pi t}{T} \qquad \ldots (viii)$$

and Strain
$$\frac{dy}{dx} = -\frac{4\pi a}{\lambda} \sin \frac{2\pi}{\lambda} x \sin \frac{2\pi t}{T} \qquad \ldots (ix)$$

At times when
$$\frac{2\pi t}{T} = 0, \pi, 2\pi \ldots., m\pi$$

or when
$$t = 0, \frac{T}{2}, T, \ldots., m\frac{T}{2}$$

$$\sin \frac{2\pi t}{T} 0 \text{ and } \cos \frac{2\pi t}{T} = \pm 1$$

At times when
$$\frac{2\pi t}{T} = \frac{\pi}{2}, \frac{3\pi}{2}, \ldots., (2m+1)\frac{\pi}{2}$$

or
$$t = \frac{T}{4}, \frac{3T}{4} \ldots, (2m+1)\frac{T}{4}$$

$$\sin \frac{2\pi t}{T} = \pm 1 \text{ and } \cos \frac{2\pi t}{T} = 0$$

(*i*) **Displacement.** From (*vii*) displacement is maximum when

$$\sin \frac{2\pi t}{T} = \pm 1$$

or when
$$t = \frac{T}{4}, \frac{3T}{4}, \ldots.., (2m+1)\frac{T}{4}$$

Being alternately maximum positive or negative.

The displacement is minimum, when

$$\sin \frac{2\pi t}{T} = 0$$

or when $t = 0, \frac{T}{2}, T, ..., m\frac{T}{2}$

(ii) **Velocity.** From (*viii*) the velocity is maximum, when

$$\sin\frac{2\pi t}{T} = \pm 1$$

or when $$t = 0, \frac{T}{2}, T, ... m\frac{T}{2}$$

The velocity is minimum, when

$$\cos\frac{2\pi t}{T} = 0$$

or when $$t = \frac{T}{4}, \frac{3T}{4}, ..., (2m+1)\frac{T}{4}$$

Hence when the displacement is maximum, *i.e.*, the particles are at their extreme positions, the velocity is zero and when the displacement is zero, *i.e.*, the particles are passing through their mean positions, the velocity is maximum.

(iii) **Strain.** From (*ix*) the stain is maximum, when

$$\sin\frac{2\pi t}{T} = \pm 1$$

or when $$t = \frac{T}{4}, \frac{3T}{4}, ..., (2m+1)\frac{T}{4}$$

The strain is minimum, when

$$\sin\frac{2\pi t}{T} = 0$$

or when $t = 0, \frac{T}{2}, T, ... m\frac{T}{2}$

Hence the strain is maximum at times when the displacement is maximum and the velocity is zero. The strain is minimum when the displacement is zero and velocity is maximum.

15.6 CHARACTERISTIC PROPERTIES OF STANDING WAVES

1. The stationary waves are not progressive, *i.e.,* the condensation (crests) or rarefactions (troughs) do not travel forward or backward. Hence *there is no transfer of energy from one particle to the other.* The displacement and velocity at a point continuously change from maximum positive to maximum negative value and *vice versa.*
2. Every particle except at the nodes, executes simple harmonic motion with the same period as the component waves. The particles *at the nodes always have zero displacement, zero velocity and undergo the maximum changes in strain.* The displacement increases as we proceed from the nodes on either side and becomes maximum at the antinodes which is midway between the two nodes. *At the antinodes the displacement and velocity are maximum and the strain is zero.*
3. The distance between two consecutive nodes or antinodes is equal to *half the wave length* of the stationary wave, which is the same as that of component waves.
4. Twice in one vibration at intervals of $\frac{T}{2}$ all the particles have maximum displacement and are momentarily at rest having zero velocity. After a time $\frac{T}{2}$ the particles having maximum positive displacement have maximum negative displacement and *vice versa.*

5. Twice in one vibration all the particles pass through their mean positions and have maximum velocity. After a time $T/2$ they again pass through their mean positions in the opposite directions with maximum velocity.

The medium splits up into segments. For half a time period all the particles in a segment between any two consecutive nodes move in one direction and in the next half in the opposite direction. At any particular instant the direction of motion of the particles in one segment is opposite to that of the particles in the segment next to it or just behind it.

15.7 WAVE FUNCTION OF STATIONARY WAVES

The equation of a progressive wave is $y = a \sin \frac{2\pi}{\lambda}(vt - x)$ where λ is the wavelength and v the velocity of the wave.

The velocity of stationary wave is $y = a \sin \frac{2\pi}{\lambda} vt \cdot \cos \frac{2\pi}{\lambda} x$. Thus, the wave function of progressive wave is composite function of x and t. *i.e.*, $y = f(x, t)$. On the other hand, that for standing wave it is a product of two separate functions of x and t, one function of time t containing $\sin \frac{2\pi}{\lambda} vt$ and the second of distance x containing $\cos \frac{2\pi}{\lambda} x$ *i.e.*, $y = f_1(x) f_2(t)$.

15.8 COMPARISON BETWEEN STATIONARY AND PROGRESSIVE WAVES

The distinction between progressive and stationary waves is given below in a tabular form.

Progressive waves	*Stationary waves*
1. The disturbance travels forward, *i.e.*, it is handed over from one particle to the next after some time.	The distrubance is fixed, *i.e.*, no particle transfers its motion to the next at any time.
2. Each particle has the same constant amplitude. The phase varies along the wave, *i.e.*, a particle reaches any stage of its displacement at a time different from that of the last and the next particle.	The amplitude of each particle is not the same. It is maximum at an antinode and decreases gradually from an antinode to the node according to the cosine law. All the particles between two consecutive nodes are in the same phase, *i.e.*, they reach their maximum or minimum displacement at the same time. The phase of particles in one segment is opposite to that of the particles in the next or the last segment.
3. No particle is permanently at rest. Every particle is momentarily at rest at the extreme positions of its displacement. Different particles reach this position at different times.	Particles at the nodes are permanently at rest. Other particles are also momentarily at rest at the extreme position of their displacements. All the particles reach this position at the same time. This condition is repeated after half a period.
4. All the particles have the same maximum velocity which they have on passing through their mean positions one after the other.	The velocity at the nodes is always zero. It increases gradually as we go to the antinodes and is maximum at the antinodes. All the particles have their maximum velocity when they pass through their mean positions at the same time.

(Contd.)

5. Every region passes successively through conditions of compression, normal density and rarefaction and these conditions travel forward.	The condensation, regions of normal density and rarefactions are fixed. In any region the same condition appears and disappears alternately.
6. The displacement, velocity and strain equations are represented by $y = a\sin\frac{2\lambda}{\lambda}(vt - x)$ $\frac{dy}{dt} = \frac{2\pi av}{\lambda}\cos\frac{2\pi}{\lambda}(vt - x)$ $\frac{dy}{dx} = \frac{-2\pi a}{\lambda}\cos\frac{2\pi}{\lambda}(vt - x)$ A particle has its maximum velocity and maximum strain (change of pressure) at the same time and has its maximum displacement a quarter period later. The velocity and the pressure curves, therefore, agree and are $\frac{T}{4}$ ahead of the displacement curve.	The displacement, velocity and strain equation are represented by $y = 2a\cos\frac{2\pi}{\lambda}x\sin\frac{2\pi}{\lambda}vt$ $\frac{dy}{dt} = \frac{4\pi av}{\lambda}\cos\frac{2\pi}{\lambda}x\cos\frac{2\pi}{\lambda}vt$ $\frac{dy}{dx} = \frac{-4\pi a}{\lambda}\sin\frac{2\pi}{\lambda}x\sin\frac{2\pi}{\lambda}vt$ A particle has its maximum displacement and maximum strain at the same time but has its maximum velocity a quarter period later. No two curves agree in this case.
7. There is a transmission of energy across every plane.	There is no flow of energy across any plane. This is due to the fact that condensation and velocity curves differ in phase by $\frac{T}{4}$.

15.9 NORMAL MODES OF BOUNDED SYSTEM

Let us consider the formation of stationary (standing) waves in a string of fixed length clamped rigidly at both ends. The nodes are formed at both the fixed ends *as the system is bounded at either ends*. The phenomenon occurs due to the superposition of progressive wave travelling in one direction and on reflection (reversal of phase by an angle π), the reflected wave in opposite direction. This results into the formation of loops. The waves are localised along the length of the string fixed at both ends. The waves formed are named as *standing waves*.

Standing waves in a string of fixed length.

Consider a string of fixed length clamped rigidly at both ends. Consider a monochromatic wave *i.e.*, a wave of only one (angular) frequency ω and amplitude a travelling in a positive X-direction, then the wave equation is given by

$$y_1 = a\,e^{i(\omega t - kx)}$$

where $$k = \frac{2\pi}{\lambda} = \frac{\omega}{c}$$

This wave is reflected from the fixed end and if the amplitude of the reflected wave travelling along rhe *negative* X-direction is b, then the equation of the reflected wave is given by

$$y_2 = b\,e^{i(\omega t + kx)}$$

The displacement of a point on the string due to both these waves is given by

$$y = y_1 + y_2 = a\,e^{t(\omega t - kx)} + b\,e^{i(\omega t + kx)} \quad \ldots (i)$$

The boundary condition to be satisfied at all times are

(*a*) At $x = 0$, displacement $y = 0$

(*b*) At $x = l$, displacement $y = 0$

Applying the boundary condition (*a*) we have from (*i*)

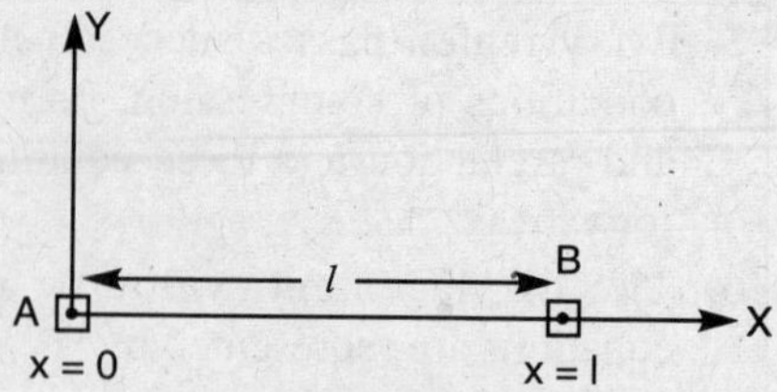

Fig. 15.7

$$0 = a\,e^{i\omega t} + b\,e^{i\omega t}$$

or
$$a = -b$$

This condition implies that the wave is completely reflected at either end which offers infinite impedance with a *reversal of phase* or phase change of π.

Substituting $b = -a$ in (*i*), we have

$$y = a\,e^{i\omega t}\,[e^{-ikx} - e^{+ikx}]$$

$$= (-2i)\,a\,e^{i\omega t}\frac{e^{-ikx} - e^{+ikx}}{-2i}$$

or
$$y = -2\,i\,a\,e^{i\omega t}\sin kx \qquad \text{... (ii)}$$

This equation satisfies the *standing wave time independent* form of the wave equation

$$\frac{d^2y}{dt^2} + k^2 y = 0, \text{ because}$$

$$\frac{dy}{dt} = (-2i)\,a(i\,\omega)\,e^{i\omega t}\sin kx$$

and
$$\frac{d^2y}{dt^2} = (-2i)\,a\,(-\omega^2)\,e^{i\omega t}\sin kx = -\omega^2 y$$

$$\therefore \quad \frac{1}{c^2}\cdot\frac{d^2y}{dt^2} = -\frac{\omega^2}{c^2}y = -k^2 y$$

Substituting in the differential form of wave equation

$$\frac{d^2y}{dx^2} = \frac{1}{c^2}\frac{d^2y}{dt^2}, \text{ we have}$$

$$\frac{d^2y}{dx^2} + k^2 y = 0$$

Thus equation (*ii*) gives a general result for all wave shapes and frequencies.

Applying the boundary condition (*b*), we have from (*ii*)

$$0 = -2\,ia\,e^{i\omega t}\sin kl$$

or
$$\sin kl = \sin\frac{\omega}{c}l = 0 \qquad \left[\because k = \frac{\omega}{c}\right]$$

or
$$\frac{\omega l}{c} = n\pi \qquad \text{... (iii)}$$

where *n* is an integer.

This condition limits the value of *allowed* frequencies. If we denote the value of ω corresponding to integer *n* by ω_n, then

$$\frac{\omega_n l}{c} = n\pi$$

or
$$\frac{\omega_n}{c} = \frac{n\pi}{l} \qquad \text{... (iv)}$$

from which
$$\sin kx = \sin\frac{\omega_n}{c}x = \sin\frac{n\pi}{l}x \qquad \text{... (iv) (a)}$$

Substituting $\omega_n = 2\pi\nu_n$ in (*iv*), we have

$$\frac{2\pi\nu_n}{c} = \frac{n\pi}{l}$$

or $$\nu_n = \frac{nc}{2l} \qquad \text{... (v)}$$

If λ_n is the corresponding wavelength, then

$$\lambda_n = \frac{c}{\nu_n} = \frac{2l}{n}$$

or $$l = \frac{n\lambda_n}{2} \qquad \text{... (vi)}$$

Hence we find that *the allowed frequencies define the length of the string as an exact number of half wavelengths.*

Eigen frequencies. The frequencies given by equation (*v*) for various integral values of n are the *normal frequencies* or *normal modes of vibration* and are often called *Eigen frequencies.* The term Eigen frequencies is commonly used in *wave-mechanics.*

The lowest frequency corresponding to $n = 1$ is called the *fundamental* and those corresponding to $n > 1$ are known as *harmonics.* The first three harmonics for $n = 2, 3, 4$ are shown in Fig. 15.8. In general all normal modes may be present at the same time. The general displacement is the superposition of all the displacements due to different allowed frequencies.

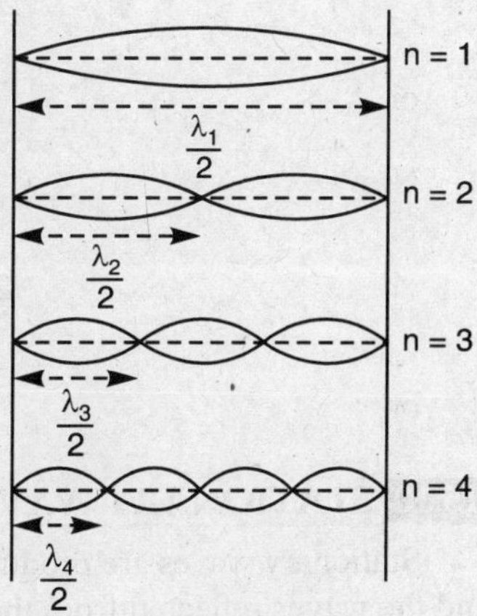

Fig. 15.8

Nodes and antinodes. From equation (*iv*) (*a*), we find that for values of n greater than 1, there are a number of positions along the string which are always at rest. These position are given by

$$\sin\frac{\omega_n}{c}x = \sin\frac{n\pi}{l}x = 0$$

or $$\frac{n\pi x}{l} = r\pi$$

where $$r = 0, 1, 2, 3, ... n$$

The position of the two fixed ends is given by $r = 0$ and $r = n$ because $r = 0$ gives $x = 0$ and $r = n$ gives $x = l$. Between these two positions there are $n - 1$ positions for the nth harmonic which are equally spaced along the string where the positions of zero for tall times. These positions are known as nodes or nodal points and give the positions of zero motion in a *system of standing waves.* These sanding waves are produced by the superposition of the incident and the reflected waves travelling in the opposite directions. The nodal points will exist if the reflection is complete so that the amplitude of the incident and the reflected progressive waves are equal and opposite. For $n = 4$, the number of nodes $= n - 1 = 3$. In between two consecutive nodes and equally spaced are the positions having maximum displacement. These points are known as *antinodes.*

Equations for standing waves. Equation (*ii*) for nth harmonic can be expressed as

$$y_n = -2aie^{i\omega_n t}\sin\frac{\omega_n}{c}x$$

$$= 2a(-i)(\cos\omega_n t + i\sin\omega_n t)\sin\frac{\omega_n}{c}x$$

$$= (A_n \cos \omega_n t + B_n \sin \omega_n t) \sin \frac{\omega_n}{c} x \qquad \text{... (vii)}$$

where $A_n = -2\,ai$ and $B_n = 2a$

Equation (*vii*) gives the equation of standing wave, along the string.

Example 15.1 *A wire of linear density 0.01 gm/cm is stretched with a tension of 10N. It resonates at a frequency of 200 Hz. The next higher frequency at which it resonates is 240 Hz. Find the length of the wire.*

Solution. Let the string be vibrating in p segments corresponding to 200 Hz frequency and $(p + 1)$ segments corresponding to the frequency 240 Hz. If T is the tension and ρ the density, then

$$200 = \frac{p}{2l}\sqrt{\frac{T}{\rho}} \qquad \text{... (i)}$$

and

$$240 = \frac{p+1}{2l}\sqrt{\frac{T}{\rho}} \qquad \text{... (ii)}$$

From (*i*) and (*ii*), we get

$$\frac{240}{200} = \frac{p+1}{p}$$

or

$$p = \frac{200}{240 - 200} = 5$$

Now $T = 10$ N, $\rho = 0.01$ gm / cm = 0.001 kg/m

$$\therefore \quad 200 = \frac{5}{2l}\sqrt{\frac{10}{0.001}} = \frac{500}{2l}$$

or

$$l = \frac{500}{400} = 1.25 \text{ m}$$

15.10 STATIONARY WAVES IN A CLOSED ORGAN PIPE

Stationary waves are produced in closed organ pipe due to the superposition of the direct waves and the waves reflected from the closed end.

The analytical expression for the direct wave is given by

$$y_1 = a \sin \frac{2\pi}{\lambda}(vt - x)$$

where λ is the wavelength, a the amplitude and v the velocity of sound.

The closed end of the organ pipe must remain a *node* hence on reflection the amplitude must change from $+ a$ to $- a$ and the wave must travel in the backward direction, *i.e.*, along the negative direction of the X-axis.

Hence the equation of the reflected wave is given by

or

$$y_2 = -a \sin \frac{2\pi}{\lambda}(vt + x)\,*$$

* *Note.* When reflection takes place at the open end of an open organ pipe, the end must ramain an antinode. The equation is then represented by

$$y_2 = a \sin \frac{2\pi}{\lambda}(vt + x)$$

The resultant displacement y at the some time is given by

$$y = y_1 + y_2$$

or
$$y = a \sin\frac{2\pi}{\lambda}(vt - x) - a\sin\frac{2\pi}{\lambda}(vt + x)$$

or
$$y = -2\, a \cos\frac{2\pi}{\lambda}\, vt \sin\frac{2\pi}{\lambda}\, x \quad \text{... (i)}$$

The resultant vibration of the particle represents a simple harmonic wave of the same period. The amplitude of the resultant wave is given by

$$A = -2a \sin\frac{2\pi}{\lambda} x$$

This shows that the resultant amplitude is a function of x and hence varies from point to point along the length of the pipe.

The velocity of the particle is obtained by differentiating (*i*) with respect to time

$$\frac{dy}{dt} = \frac{4\pi av}{\lambda} \sin\frac{2\pi x}{\lambda} \sin\frac{2\pi vt}{\lambda} \quad \text{... (ii)}$$

And acceleration
$$\frac{d^2y}{dt^2} = \frac{8\pi^2 av^2}{\lambda^2} \sin\frac{2\pi x}{\lambda} \cos\frac{2\pi vt}{\lambda} \quad \text{... (iii)}$$

The strain or compression at any point of the resultant vibration is obtained by differentiating (*i*) with respect to x.

$$\therefore \quad \frac{dy}{dx} = -\frac{4\pi a}{\lambda} \cos\frac{2\pi x}{\lambda} \cos\frac{2\pi vt}{\lambda} \quad \text{... (iv)}$$

These equations show that *the amplitude, velocity, acceleration and strain or compression vary with position and time.*

Position of nodes. Consider the points where

$$\sin\frac{2\pi}{\lambda} x = 0 \text{ and } \cos\frac{2\pi}{\lambda} x = \pm 1$$

From (*i*) we have

Displacement $y = 0$ and amplitude $A = 0$

From (*ii*) and (*iii*), we have

Velocity
$$\frac{dy}{dt} = 0 \text{ and Acceleration } \frac{d^2y}{dt^2} = 0$$

From (*iv*), we find that

Strain
$$\frac{dy}{dx} = \mp\frac{4\pi a}{\lambda} \cos\frac{2\pi\, yt}{\lambda}$$

Thus we find that at these position, displacement, amplitude, velocity and acceleration are all zero, but the strain is maximum. The positions correspond to **nodes.**

Since
$$\sin\frac{2\pi x}{\lambda} = 0$$

But $\sin m\,\pi = 0$ when $m = 0, 1, 2, \ldots$ etc.

$$\therefore \quad \frac{2\pi x}{\lambda} = m\pi \text{ or } x = m\frac{\lambda}{2}$$

Hence $$x = 0, \frac{\lambda}{2}, \lambda, \frac{3\lambda}{2} \text{ ...etc.}$$

Thus the nodes are equidistant and are separated by $\frac{\lambda}{2}$. At $x = 0$, the position of closed end is, therefore, a **node**.

Position of antinodes. Considering the position, where

$$\sin \frac{2\pi x}{\lambda} = \pm 1 \text{ and } \cos \frac{2\pi x}{\lambda} = 0, \text{ we find}$$

Displacement $$y = \pm 2a \cos \frac{2\pi vt}{\lambda}$$

Amplitude $$A = \pm 2a$$

Velocity $$\frac{dy}{dt} = \pm \frac{4\pi av}{\lambda} \sin \frac{2\pi vt}{\lambda}$$

Acceleration $$\frac{d^2y}{dt^2} = \mp \frac{8\pi^2 av^2}{\lambda^2} \cos \frac{2\pi vt}{\lambda}$$

and Strain $$\frac{dy}{dx} = 0$$

Such points where the amplitude is maximum and strain is zero are called **antinodes.**

Since $$\sin \frac{2\pi x}{\lambda} = \pm 1$$

But $$\sin \frac{(2m+1)\lambda}{2} = \pm 1 \text{ when } m = 0, 1, 2, \text{ ... etc.}$$

$$\therefore \quad \frac{2\pi x}{\lambda} = \frac{(2m+1)\pi}{2}$$

or $$x = \frac{(2m+1)\lambda}{4}$$

or $$x = \frac{\lambda}{4}, \frac{3\lambda}{4}, \frac{5\lambda}{4} \text{ .... etc.}$$

Thus we find that the antinodes are also equidistant and separated by $\frac{\lambda}{2}$. The first antinode is at a distance $\frac{\lambda}{4}$ from the closed end of the organ pipe and the second antinode at a distance $\frac{3\lambda}{4}$.

Henced an antimoode lies between two consecutive nodes.

15.11 MODES OF VIBRATION OF CLOSED AND OPEN ORGAN PIPE

At the end of a closed pipe the layer in contact with the closed end cannot move freely as the closed end acts like a rigid wall, therefore a node is formed at the closed end.

The simplest case is that of a pipe having an antinode at the open end and a node at the closed end as shown in Fig. 15.9 (*a*) if l is the length of the pipe it is evident that $l = \frac{\lambda_1}{4}$ for the first note where λ_1 is the wavelength.

Now as $$V = n\lambda$$

$$\therefore \quad n = \frac{V}{4l} \qquad \text{... (i)}$$

The note is called the **fundamental.**

The next possible mode of vibration is shown in Fig. 15.9(*b*). Here the length of the pipe contains two nodes and two antinodes and so the length of the pipe corresponds to $\frac{3}{4}\lambda_2$ where λ_2 is the new wave length.

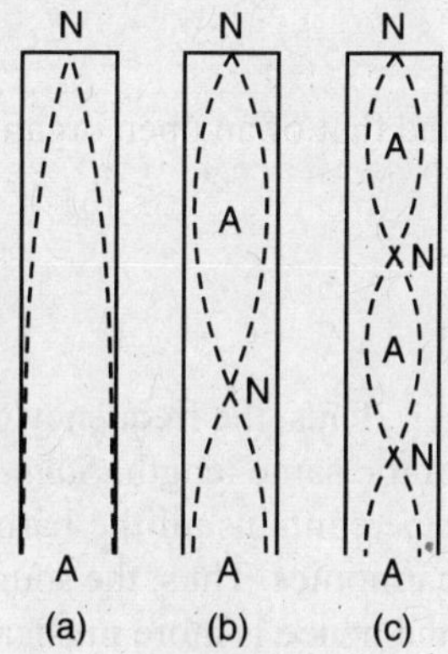

Fig. 15.9

$$\therefore \qquad l = \frac{3}{4}\lambda_2 \text{ or } \lambda_2 = \frac{4}{3}l$$

$$\therefore \qquad n_1 = \frac{V}{\lambda_2} = \frac{3V}{4l} = 3n$$

This note corresponds to three times the frequency of the fundamental.

The next mode of vibration is represented in Fig. 15.9. (*c*). Here the pipe contains three nodes and three antinodes and so the length of the pipe corresponds to $\frac{5}{4}\lambda_3$ where λ_3 is the new wave length.

$$\therefore \qquad l = \frac{5}{4}\lambda_3 \text{ or } \lambda_3 = \frac{4}{5}l$$

Hence
$$n_2 = \frac{V}{\lambda_3} = \frac{5V}{4l} = 5n$$

This note corresponds to *five* times the frequency of the fundamental note. Proceeding in this manner we get the various *odd harmonics* along with the fundamental of the following frequencies, *n*, 3*n*, 5*n*, 7*n*, ... etc.

Mode of vibration or overtones in an open pipe. The open end of the organ pipe is free to communicate with the outside air and thus can vibrate freely, hence the open end will be an *antinode.* Now two antinodes must have a node in-between them as shown in Fig. 15.10 (*a*). It is evident that if l is the length of the pipe and λ_1 the wavelength, then

$$l = \lambda_1/2$$

$$N = \frac{V}{\lambda_1} = \frac{V}{2l} \qquad \text{... (ii)}$$

where N is the frequency of the fundamental note.

The next possible mode of vibration is shown in Fig. 15.10 (*b*). Here the same length of the pipe contains *three* consecutive antinodes instead of two and so the length of the pipe corresponds to λ_2 the new wave length.

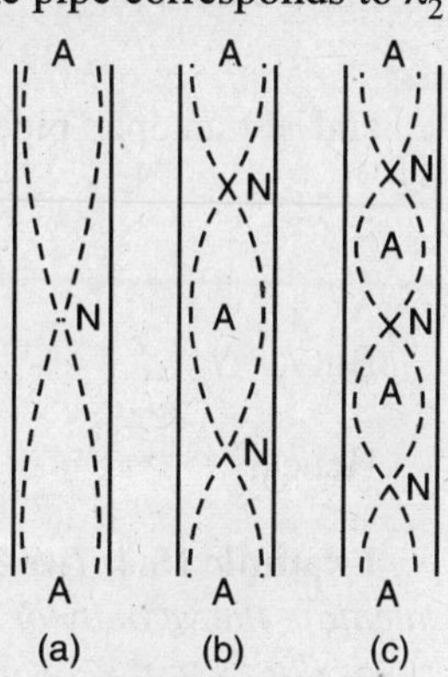

Fig. 15.10

$$\therefore \qquad l = \lambda_2$$

or
$$N_1 = \frac{V}{\lambda_2} = \frac{2V}{2l} = 2N$$

The next overtone occurs when the pipe contains *four* consecutive antinodes as shown in Fig. 15.10 (*c*) and so the length of the pipe corresponds to $\frac{3}{2}\lambda_3$ where λ_3 is the new wave length.

$$\therefore \qquad l = \frac{3}{2}\lambda_3 \text{ or } \lambda_3 = \frac{2}{3}l$$

Hence
$$N_2 = \frac{V}{\lambda_3} = \frac{3V}{2l} = 3N$$

This shows that the open pipe emits notes of frequencies N, $2N$, $3N$, $4N$, etc. Thus all the harmonics will be present when air vibrates in an open pipe.

Comparing equations (*i*) and (*ii*) we find that frequency of the fundamental note of a closed organ pipe.

$$n = \frac{V}{4l}$$

and that of an open organ pipe

$$N = \frac{V}{2l}$$

$$\therefore \quad N = 2n$$

Thus, the frequency of the fundamental note of an open organ pipe is twice that of a closed pipe of the same length. *Sound of an open pipe is more musical.* The sound produced by an open organ pipe contains all the harmonics where as that produced by a closed organ pipe contains only *odd* harmonics. Thus, the sound produced from an open organ pipe is richer in harmonics (or overtones) and hence is more musical.

Example 15.2 *Compare the lengths of an open end and closed end pipes which emit the same fundamental note.*

Solution. For a closed pipe

$$n_c = \frac{V}{4l_c} \qquad [\because \text{Eq. 15.11 } (i)]$$

For an open pipe

$$n_o = \frac{V}{2l_o} \qquad [\because \text{Eq. 15.11 } (ii)]$$

Here, $\quad n_o = n_c$

$$\therefore \quad \frac{V}{4l_c} = \frac{V}{l_o} \quad \text{or} \quad \frac{l_o}{l_c} = 2$$

Example 15.3 *Compare the fundamental frequencies of an open end and closed end pipes of the same length.*

Solution

For a closed pipe, $\quad n_c = \dfrac{V}{4l_c}$

and for an open pipe, $\quad n_o = \dfrac{V}{2l_o}$

$$\therefore \quad \frac{n_o}{n_c} = \frac{4l_c}{2l_o} = \frac{2l_c}{l_o}$$

But $l_o = l_c$

Hence, $\quad \dfrac{n_o}{n_c} = 2$

Example 15.4 *Two open organ pipes of lengths 50 cm and 50.5 cm produce 3 beats per second Calculate the velocity of sound in air.*

Solution. Given: $l_1 = 50$ cm and $l_2 = 50.5$ cm

Let V be the velocity of sound in air.

$$\therefore \qquad n_1 = \frac{V}{2l_1} = \frac{V}{2\times 50} = \frac{V}{100}$$

and
$$n_2 = \frac{V}{2l_2} = \frac{V}{2\times 50.5} = \frac{V}{101}$$

Beats per second = 3.

Hence $n_1 - n_2 = 3$

$$\therefore \qquad \frac{V}{100} - \frac{V}{101} = 3 \quad \text{or} \quad V = 3 \times 10100 = 30300 \text{ cm/sec.} = 303 \text{ m/sec.}$$

Example 15.5 *An air column is a tube closed at one end and open at the* ***other end*** *has a length of 42.5 cm. The velocity of sound in air is 340 m/sec. Neglecting the end correction, determine the fundamental frequency and frequency of second orertone.* **(Nagpur U. 2008, 2010)**

Solution. Given: $l = 42.5$ cm = 0.425 m; $V = 340$ m/sec.

Fundamental frequency, $n = \frac{V}{4l} = \frac{340}{4\times 0.425} = \frac{340}{1.7} = 200$ Hz

Second overtone, $n_1 = \frac{V}{\lambda_2} = \frac{3V}{4l} = 3n = 3\times 200 = 600$ Hz $\qquad \left(\because l = \frac{3}{4}\lambda_2\right)$

15.12 VIBRATIONS IN RODS

We know that sound waves travel through all the three media, solid, liquid and gases. It has been established that velocity of sound waves in an isotropic medium is given by

$$V = \sqrt{\frac{E}{\rho}}$$

The elasticity of a solid is higher as compared to liquids and gases. It is also found that velocity of sound through iron is higher than the velocity of sound through water or air.

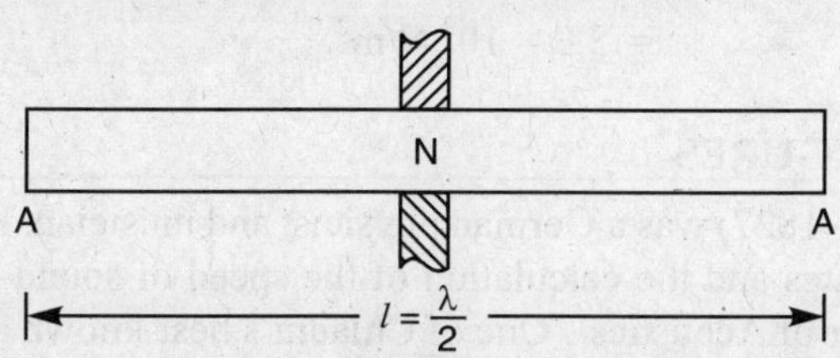

Fig. 15.11

Fig. 15.11 shows a metal rod clamped firmly at its midpoint. If one end of the rod is gently rubbed with a refined cloth, *longitudinal stationary vibrations* are set up in the rod. At the clamped position, there is a node (compression). At both the free ends, there are antinodes (rarefactions). The distance between the two ends of the rod is equal to half the wavelength of sound through the rod.

Hence, $\qquad l = \frac{\lambda}{2} \quad \text{or} \quad \lambda = 2l$

The longitudinal vibrations in rods has been used in the experiments used for the determination of sound through iron, glass, wood, brass, copper etc.

Example 15.6 *The frequency of the note next higher to the fundamental, as given by a rod of an alloy, 1 m long and clamped at its midpoint is 1000 Hz. If the density of the alloy be 7.5 × 10^3 kg-m^3, calculate the value of Young's modulus of the rod.*

Solution.

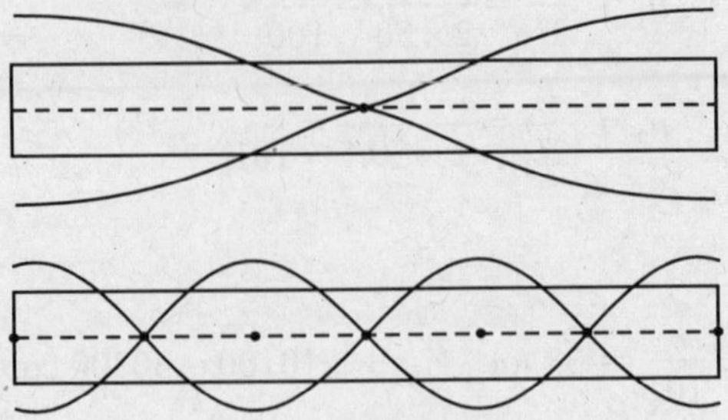

Fig. 15.12

Here

$$n_2 = 3n_1 \quad \text{and} \quad n_2 = 1000 \text{ Hz}$$

$$\therefore \quad n_1 = \frac{n_2}{3} = \frac{1000}{3} \text{ Hz}$$

Also, $l = 1$m

Hence,

$$V = n_1\lambda_1 = 2n_1 l$$

$$= 2 \times \frac{1000}{3} \times 1 = \left(\frac{2000}{3}\right) \text{ m/s}$$

and ρ = density of the alloy = 7.5×10^3 kg-m^3

The velocity of sound is given by

$$V = \sqrt{\frac{E}{\rho}}$$

or

$$E = \rho V^2 = 7.5 \times 10^3 \times \left(\frac{2000}{3}\right)^2$$

$$= 3.3 \times 10^9 \text{ N/m}^2.$$

15.13 CHLADNI'S FIGURES

Ernst Chladni (1756–1827) was a German physicist and musician. His important works include research on vibrating plates and the calculation of the speed of sound for different gases. For this some call him the 'Father of Acoustics'. One of Chladni's best-known achievements was inventing a technique to show the various modes of vibration on a mechanical surface.

Chladni's technique, in his book, 'Discoveries in the Theory of Sound', published in 1787, consisted of drawing a bow over a piece of metal whose surface was lightly covered with sand. The plate was bowed until it reached resonance and the sand formed a pattern showing the nodal regions. Since the 20th century it has become more common to place a loudspeaker driven by an electronic signal generator over or under the plate to achieve a more accurate adjustable frequency.

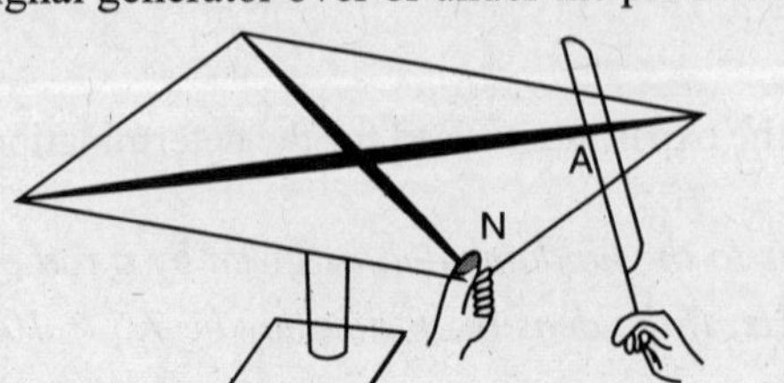

Fig. 15.13. Formation of Node (N) and Antinode (A)

Fig. 15.14. Bow at different points

Experimental demonstration. The vibrations in a plate and formation of Chladni's figures can be studied by the following experiment. The formation of Chladni's figures is demonstrated in Fig. 15.13. Fix a glass plate on a stand, sprinkle sand on the plate to form a uniform thin layer. Hold the plate at one corner and bow the middle of one side. It is found that the sand particles arrange themselves into a symmetrical pattern as shown in Fig. 15.14. By holding the plate firmly at other points and bowing at different positions, different symmetrical patterns are obtained as shown in Fig. 15.15. The position of the plate where it is firmly held becomes the node (N) and the point where the plate is bowed becomes antinode (A). The patterns obtained are known as Chladni's figures. Each pattern is specific and different and corresponds to a particular mode of vibration. However, it should be noted that these modes are not harmonically related in frequency. Various Chladni's patterns are shown in Fig. 15.15.

Fig. 15.15. Various Chladni's patterns

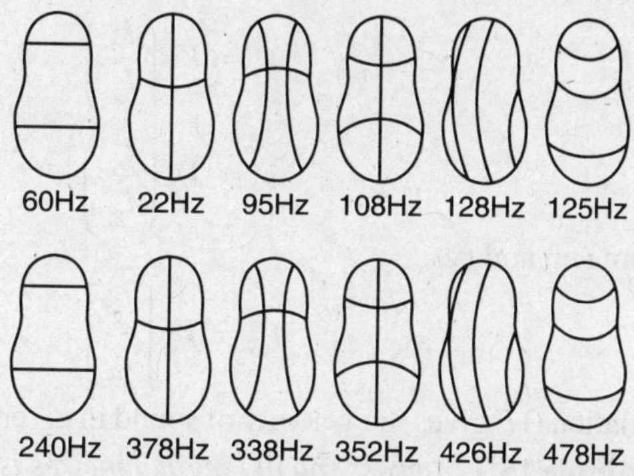

Fig. 15.16. Chladni's modes of guitar plates

Even today, variations of this technique are commonly used in the design and construction of acoustic instruments such as violins, guitars, and cellos. Fig. 15.16 shows the different modes of vibrations of guitar plates.

15.14 KUNDT'S TUBE

It consists of a glass tube. A cork B is attached to a wooden handle and can be moved inside the tube. The rod of the required material for which the velocity of sound is to be determined, is clamped in the middle at C. The front end of the rod has an aluminium disc A attached to it. The tube contains air at room temperature and a little lycopodium powder is put along the length of the tube (Fig. 15.17).

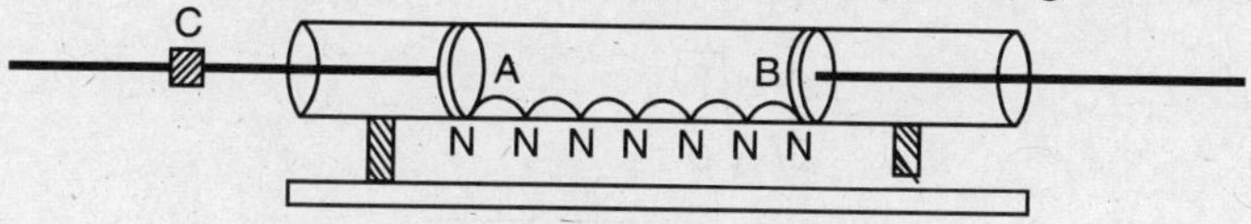

Fig. 15.17

Velocity of sound in solids. When the rod is rubbed with a resined cloth gently, longitudinal vibrations are set up in the rod. The free ends of the rod behave as antinodes and the clamped point C acts as a node. These vibrations are transferred to the air column and the air column vibrates. The cork B is adjusted so that the air column resonates with the vibrations of the rod. At this stage, the lycopodium powder sets itself in the form of heaps at regular intervals of distance.

Let the length of the rod be l_r and the mean distance between two consecutive heaps be l_a.

$$\therefore \quad l_r = \frac{\lambda_r}{2} \text{ and } l_a = \frac{\lambda_a}{2}$$

For the rod,

$$n = \frac{V_r}{\lambda_r} = \frac{V_r}{2l_r} \qquad \text{... } (i)$$

For the air column $$n = \frac{V_a}{\lambda_a} = \frac{V_a}{2l_a} \quad \text{... (ii)}$$

From (*i*) and (*ii*)

$$V_r = V_a\left[\frac{l_r}{l_a}\right] \quad \text{... (iii)}$$

Velocity of sound in gases. Here the experiment is performed first with air and then with the required gas, using the same rod.

Suppose, the distance between two consecutive heaps in the air column = l_a and the mean distance between two consecutive heaps in the gas column = l_g.

From equation (*iii*)

$$V_r = V_a\left(\frac{l_r}{l_a}\right) \quad \text{... (iv)}$$

and

$$V_r = V_g\left(\frac{l_r}{l_g}\right) \quad \text{... (v)}$$

From (*iv*) and (*v*)

$$V_g = V_a\left[\frac{l_g}{l_a}\right] \quad \text{... (vi)}$$

Equation (*vi*) gives the velocity of sound in given gas (V_g) in terms of velocity of sound in air (V_a).

Example 15.7 *A brass rod of length 3 metres is clomped in the middle. Calculate the frequency of the fundamental note emitted by it, when it vibrates longitudinally. Density of brass = 8.3 g/cm*^{3} *and Young's modulus of brass = 10.76 × 10*^{11} *dynes/cm*^{2}.

Solution. Given

$$l = 3 \text{ metres}$$

$$\rho = 8.8 \text{ gm/cm}^2 = 8800 \text{ kg/m}^3$$

$$y = 10.76 \times 10^{11} \text{ dynes/cm}^2$$

$$= 10.76 \times 10^{10} \text{ newtons/m}^2$$

Here, $$V = \sqrt{\frac{Y}{\rho}} \quad \text{and} \quad n = \frac{v}{2l}$$

$$\therefore \quad n = \frac{1}{2l}\sqrt{\frac{Y}{\rho}}$$

$$= \frac{1}{2\times 3}\sqrt{\frac{10.76\times 10^{10}}{8300}}$$

$$= 600 \text{ Hz}$$

15.15 ULTRASONIC WAVES

Human ear is sensitive to sounds of frequency lying between 20 Hertz (Hz) and 20 KHz. The ear is unable to hear sounds of frequency less than 20 Hz and more than 20 KHz. The sounds of frequency less than 20 Hz are called *infrasonics* whereas those of frequency more than 20K are called *ultrasonics.* In other words ultrasonics are longitudinal mechanical waves of frequency beyond the highest audible frequency (*i.e.*, 20 KHz). As their frequency is very high, the wave length is very low (about 1.6 cm). (Note that 1 KHz means 1000 Hz). Hence, both ultrasonic and infrasonic waves can not be heard by humans. The speed of ultrasound does not depend on its frequency, but depends on what material or tissue it is travelling in. They travel faster in dense materials and slower in compressible material. In soft tissues sound travels at 1500 m/s, in bones about 3400 m/s, and in air 330 ms/. When the ultrasound 'hits' different structures in the body of different density, it sends back echoes of varying strength.

15.16 PRODUCTION OF ULTRASONIC WAVES

The important methods of producing ultrasonic waves are

(*i*) ***Mechanical method.*** The most commonly used mechanical generator of ultrasonics is Galton's whistle. It produces frequencies upto 100 kilo-cycles with a constant amplitude.

(*ii*) ***Magnetostriction oscillator*** and

(*iii*) ***Piezo-electric oscillator.*** We shall describe in detail the last two methods.

15.16.1 Magnetostriction oscillator

When a bar of a ferro magnetic material like iron or Nickel is suddenly magnetised with the help of a strong magnetic field applied parallel to its length it undergoes a slight change in length. This is known as *magnetic-striction effect.*

If a rod of *invar* (36% Ni + 64% Fe) or *monel* (65% Ni + 31% Cu + 4% Fe, Mn and C) is placed inside a coil parallel to its axis and a high frequency current is passed through the coil, the rod is magnetised and demagnetised with the current. The length of the rod changes accordingly and its free ends produce high frequency vibrations or ultrasonics. If the length of the rod is so adjusted that its natural frequency of vibration is the same as the frequency of the applied alternating current, *resonance* takes place and the amplitude of vibrations becomes very large.

As the rod vibrates longitudinally its fundamental frequency of vibration is given by

$$n = \frac{1}{2l}\sqrt{\frac{Y}{\rho}}$$

where l is the length, Y the Young's modulus of the material and ρ the density.

A circuit showing the working of a magnetostriction oscillator is shown in Fig. 15.18. The high frequency current is obtained from a valve oscillator. The best results are, however, obtained if the rod *AB* is previously magnetised by means of a second coil wound over it and carrying a direct current.

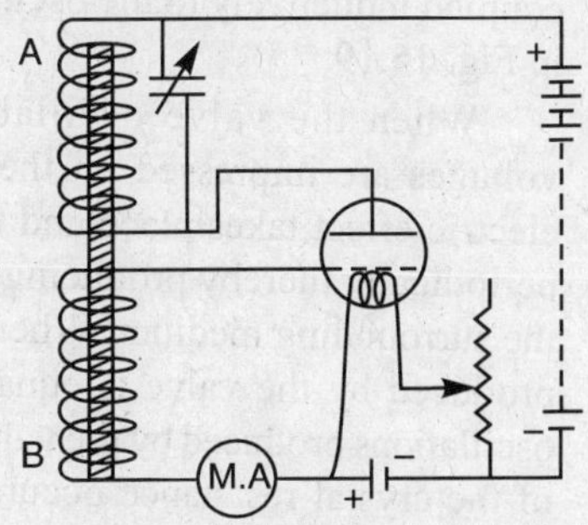

Fig. 15.18

15.16.2 Piezo electric oscillator

When a slab of quartz, tourmaline, Zinc blende or Rochelle salt cut with its faces perpendicular the optic axis is subjected to a mechanical pressure on one pair of parallel faces, electric charges are developed on the other pair of parallel faces perpendicular to the first pair. The quantity of charge developed is proportional to the applied pressure. The sign of charge is reversed when the pressure is changed into tension.

This phenomenon is known as *piezo-electric effect.*

On the other hand when an electric potential difference is applied across the faces of a crystal, the crystal undergoes a strain. If the direction of applied potential difference is reversed the direction of strain is also reversed *i.e.*, if in the first case it is a compression strain now it will be an extension strain and vice-versa. Hence if a *high frequency alternating voltage* is applied to the crystal slab rapid alternation of compression and extension takes place in two perpendicular directions resulting in the forced vibrations of the crystal.

This phenomenon is known as *inverse piezo-electric effect.*

When the frequency of these forced vibrations becomes equal to the natural frequency of the *slab, resonance* takes place and the amplitude of vibration becomes sufficiently large.

When an alternating voltage is applied to the slab along its *thickness* there are alternating changes in its length and thickness *i.e.*, when the length increases, the thickness decreases and *vice-versa* with the same frequency as that of the applied A.C. voltage. When the frequency of the applied A.C. voltage becomes equal to the natural frequency of the slab, resonant vibrations take place in the slab.

The natural frequency of vibration of the slab is given by

$$n = \frac{1}{2d}\sqrt{\frac{E}{\rho}}$$

where d = distance between the opposite faces of the crystal slab

E = Bulk modulus of quartz (material of the slab)
$= 8 \times 10^{10}$ N/m^2

ρ = density of quartz = 2.65×10^3 kg/m^3

$$\therefore \quad n = \frac{1}{2d}\sqrt{\frac{8\times10^{10}}{2.65\times10^3}} = \frac{2.75\times10^3}{d} \text{ Hz}$$

If $\quad d = 1$ mm $= 10^{-3}$ m, then

$$n = \frac{2.75\times10^3}{10^{-3}} = 2.75\times10^6 \text{ Hz} = 2.75 \text{ MHz}$$

The crystal is placed between two metal plates A and B so as to form a parallel plate condenser with crystal C as the dielectric. The plates are connected to the primary of a transformer which is coupled inductively to the oscillatory circuit of a valve as shown in Fig. 15.19.

When the valve oscillates high frequency alternating voltages are impressed on the plates A and B. Inverse Piezo-electric effect takes place and the crystal contracts and expands periodically thereby producing compressions and rarefactions in the surrounding medium. When the frequency of the oscillations produced by the valve is equal to the natural frequency of the oscillations produced by the valve is equal to the natural frequency of the crystal resonance occurs and the amplitude of vibrations becomes very large.

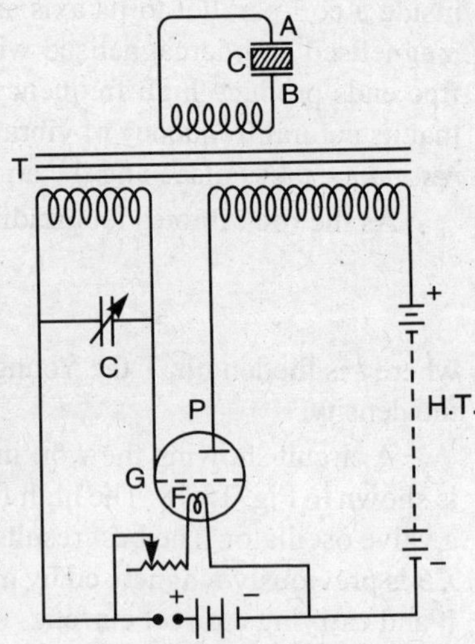

Fig. 15.19

15.17 DETECTION OF ULTRASONIC WAVES

The ultrasonic waves are detected using the following methods:

(*i*) **Kundt's tube method.** Kundt's tube can be used to detect ultrasonic waves in a similar way as for ordinary sound waves.

(*ii*) **Sensitive flame.** Since very high frequency sound waves change the intensity of the flame, a flame can be used to detect the presence of ultrasonics.

(*iii*) **Thermal detectors.** A *probe* of fine platinum wire is placed in the region to be tested for ultra-sonic waves. At a node compression and rarefactions occur very rapidly and cause adiabatic changes. The platinum probe is, therefore, alternately heated or cooled and the resistance changes accordingly which can be detected by suitable methods. No such changes occur at an antinode.

(*iv*) **Piezo electric detectors.** When one pair of faces of a piezo-electric crystal like quartz is subjected to ultrasonic waves opposite charges are developed on the other pair perpendicular to the first. These changes are amplified by an electronic amplifier and then detected by suitable means.

15.18 APPLICATIONS OF ULTRASONIC WAVES

(*i*) **Velocity of sound in liquids and gases.** Ultrasonics have been used to find velocity of sound in liquids and gases, particularly when available in small quantities. This determination reveals many physical and chemical properties of the substances.

To find the velocity stationary waves are set up in the liquid or gas similar to the Kundt's tube method and the wavelength of the wave λ is determined. The velocity $V = n\lambda$ where n is the frequency of oscillations. Changes in velocity with frequency and temperature can also be determined by this method.

The velocity of ultrasonic waves can also be determined by using an *acoustic grating.* For details refer next article.

(*ii*) **Elastic symmetries of crystals.** When ultrasonic waves are passed through a solid or a liquid and a beam of light is made to travel in a direction at right angles the ultrasonic wave system acts like a grating. The compressions act as lines of greater density or *opacities* and rarefactions like lines of lesser density or *transparencies.*

If ultrasonic waves are applied to an isotropic crystal they give rise to interference fringes from which the elastic symmetry of the crystal can be determined.

(*iii*) **Depth sounding.** Ultrasonic waves are used for signalling, for finding the depth of sea and to detect the positions of submerged rocks, submarines and ice bergs.

A strong beam of ultrasonic waves is transmitted and received, after reflection from the submerged object. By noting the time taken the distance can be calculated.

(*iv*) **Chemical effect.** Ultrasonic waves have been used to form stable emulsions of immiscible liquids like water and oil; to liquify *gells* like aluminium hydroxide in the same manner as they are liquified by shaking; to coagulate *aerosols i.e.*, dispersed fine particles of a solid or a liquid in a gas, *e.g.*, dust, smoke, mist, etc. to accelerate crystallisation and to produce oxidation.

(*v*) **Sound signalling.** Ultrasonic waves have been used for directional signalling. Because of their high frequency these waves can be sent out in the form of a narrow beam and therefore have a great advantage over ordinary sound waves in the audible range which spread out in all directions. It is thus possible to radiate out a large amount of energy without making the amplitude of the vibrating crystal very large. To radiate out energy at the same rate using ordinary audible sound frequency a very much larger amplitude will be required.

It has been found experimentally that if a vibrating crystal in the form of a circular disc of radius r is used for transmission of signal, the angle of the core containing the waves is given by $\sin\theta = \frac{0.61\lambda}{r}$. This has a very small value for short wavelength.

(*vi*) **Thermal effects.** A fluid subjected to ultrasonic waves shows a slightly higher temperature than that of its surroundings due to the absorption of waves.

(*vii*) **Biological effects.** Small animals like frog, fish, etc., are maimed or killed when exposed to these waves. Micro-organisms like bacteria are torn to pieces or burnt or otherwise destroyed and yeast loses its power of reproduction.

(*viii*) **Medical effects.** Ultrasonics are now widely used for medical purposes. The ultrasonic vibrations have a good massaging effect and can be used for treatment of muscular pains. These waves are also helpful for treatment of neuro-patients where other conventional method like psychotherapy, drug treatment or electric shock treatment fail. These waves have been used for location of the position of eye tumour and for extraction of broken teeth.

Ultrasound technique is widely and safely used in sonography. Ultrasound waves images are produced by a transducer. These waves can be bounced of tissues using special devices. The echoes are then converted into a picture called *sonogram.* Ultrasound imaging, or ultrasonography, allows physicians and patients to get an inside view of soft tissues and body cavities. Ultrasound is often used to examine a foetus during pregnancy, or a fatal heart. Ultrasound usually ranges from one MHz to 20 MHz. This is because tissues absorb higher frequency energy readily, hence producing fainter images. There are 3 modes of the pulse echo visualization (*i.e.*, A-mode, M-Mode, and B-mode.

Ultrasound based diagnostic imaging technique can be enhanced with Doppler measurements, which employ the *Doppler effect** to assess whether structures (usually blood) are moving towards or away from the probe, and its relative velocity. The principal applications of Doppler effect are in cardiac and peripheral vesicular diagnostic specialities.

(*ix*) **Other effects.** The ultrasonic waves are used to detect crocks or other defects in homogeneity noticeable by reflection or absorption and washing of silk and delicate fabrics.

Example 15.8. *Calculate the fundamental frequency of ultrasonics produced by a quartz crystal of thickness 0.5 mm. The value of Young's modulus for quartz is 8 × 10[11] dynes/cm[2] and density 2.65 gm/cm[3].* **(Nagpur U., 2010, 2008, 2002)**

Solution. Here thickness of quartz crystal $d = 0.5$ mm = 0.05 cm

Density of quartz $\rho = 2.65$ gm/cm^3

Young's modulus of quartz $Y = 8 \times 10^{11}$ dynes/cm^2

∴ Frequency of fundamental note produced

$$n = \frac{1}{2d}\sqrt{\frac{Y}{\rho}} = \frac{1}{2 \times 0.05}\sqrt{\frac{8 \times 10^{11}}{2.65}}$$

$$= 5.49 \times 10^6 \text{ Hz} = 5.49 \text{ MHz}.$$

15.19 MEASUREMENT OF VELOCITY OF ULTRASONIC WAVES

The measure the velocity of ultrasonic waves we use a piezo electric oscillator which consists of a quartz crystal set into high frequency resonant vibrations with the help of a variable radio frequency electronic oscillator.

Radio frequency oscillator. A radio frequency oscillator produce *A.C. voltages* of frequency of the order of a few megacycles per second. Its input is connected to the *A.C.* mains and the output to the quartz slab as shown in Fig. 15.20. The frequency of the *R.F.* out put can be varied by rotating a knob which in turn changes the capacitance of the oscillating LC circuit. The value of output frequency is read on a calibrated dial fixed on the panel known as *Frequency meter.* It has normally a range from 0 – 10 MHz.

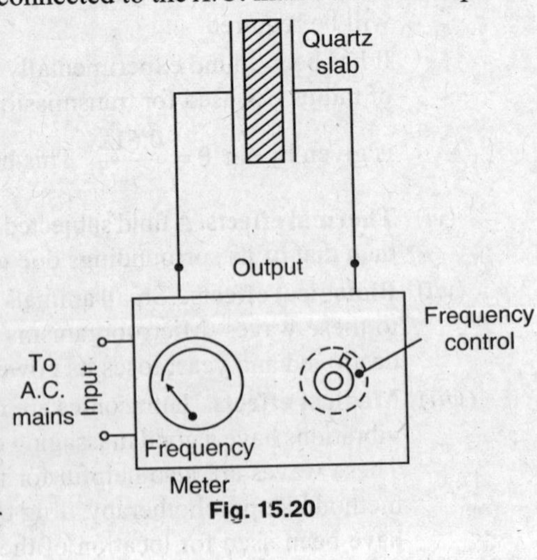

Fig. 15.20

Acoustic grating. The velocity of ultrasonic waves set up in a liquid can be determined with the help of the *acoustic grating* formed by them. When plane ultrasonic waves are produced in a liquid the pressure varies periodically with the distance from the ultrasonic source being higher at a compression and lower at a rarefaction. The density of the medium and the refractive index also show a periodic variation with distance from the source along the direction of

* Pl. refer article 15.21

propagation of the wave. Under such a condition, if monochromatic light from a slit is passed through the liquid in a direction at right angles to the wave propagation the liquid behaves as a diffraction grating with grating element (width of opacity + width of transparency) equal to the wave length of ultrasonic waves, λ_u. Such a grating is known as *acoustic grating.* The light which suffers diffraction at the grating shows a number of diffraction images of the slit when seen through a telescope eye-piece. The arrangement of the apparatus is shown in Fig. 15.21.

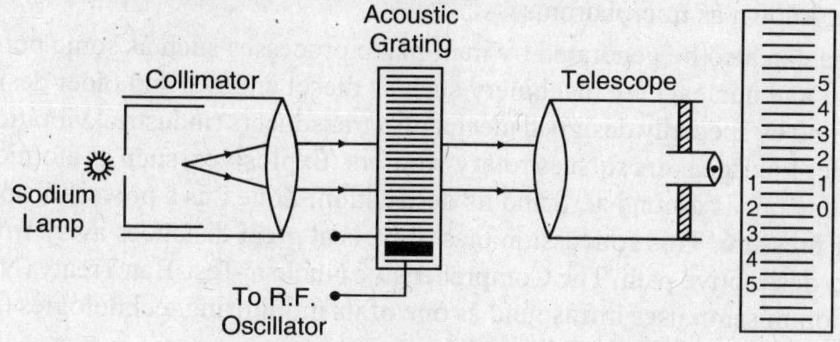

Fig. 15.21

The acoustic grating is mounted on the prism table of a spectrometer. A parallel beam of monochromatic sodium light rendered parallel by the collimator is incident normally on the acoustic grating. The diffracted light is viewed through the telescope. The diffraction pattern consists of a central maxima (0-order) and a number of principal maxima on either side (1, 2, 3, 4, 5 ...).

If λ is the wavelength of sodium light which suffers diffraction at the acoustic grating of grating element $(a + b) = \lambda_u$ and θ_n is the angle of diffraction for the nth order maxima, then

$$\lambda_u \sin\theta_n = n\lambda$$

$$\therefore \text{ Wavelength of ultrasonic wave } \lambda_u = \frac{n\lambda}{\sin\theta_n}$$

If D is the distance between the telescope objective and the eye piece cross wire (which is equal to the focal length of the objective), and d_n the distance of the nth order diffraction maxima from the central zero order diffraction maxima, then

$$\sin\theta_n = \tan\theta_n = \frac{d_n}{D}$$

Hence
$$\lambda_u = \frac{n\lambda D}{d_n}$$

If v is the frequency of ultrasonic waves as given by the frequency meter of the R.F. oscillator,

then Velocity of ultrasonic waves $= v\lambda_u = \dfrac{vn\lambda D}{d_n}$

15.20 INFRASONIC WAVES

Infrasound refers to extreme bass waves or vibrations, having frequency below the audibility range of the human ear (*i.e.*, less than 20 Hz). Hence these waves can not be heard by humans. Hearing becomes gradually less sensitive as frequency decreases, so for humans to perceive infrasound, the *sound pressure* must be sufficiently high. Under normal conditions, although infrasound is inaudible for human ear, but at higher levels it is possible to feel infrasound vibrations in various parts of the body. The study of such sound waves is referred to as *infrasonics,* covering sounds beneath 20 Hz down to 0.001 Hz. This frequency range is utilized in monitoring earthquakes, charting rock and petroleum formations below the earth, and also in *ballistocardiography* and *seismocardiography* to study the mechanics of heart. Infrasound is characterized by an ability to cover long distances and get around obstacles with little dissipation.

Infrasound was first used by allied forces in World war I to locate artillery. French scientist Vladimir Gaureau noticed surprisely for the first time in his lab during 1960s, when he experienced pain in the ear drums and shaking lab equipments, *but no audible sound was* picked up on his microphone.

Infrasound sometimes results naturally from severe weather, surf, lcc waves, avalanches, earthquakes, volcanoes, waterfalls, calving of icebergs, aurorae, lightning and upper-atmospheric lighting. Nonlinear ocean wave interactions in ocean strorms produce pervasive infrasound vibrations around 0.2 Hz, known as microbaroms.

Infrasound can also be generated by man-made processes such as sonic booms and explosions (both chemical and nuclear), by machinery such as diesel engines and older designs of down tower wind turbines and by specially designed mechanical transducers (industrial vibration tables) and large-scale subwoofer loudspeakers such as rotary woofers. Explosives, such as atomic weapons, produce infrasound. Zone one is ground zero and its destruction. Zone 2 is a powerful, speeding, sonic wave of reduced air pressure. This concussion blast travels at great distances away from ground zero and few survive its destructive path. The Comprehensive Nuclear-Test-Ban Treaty (NTBT) Organization Preparatory Commission uses infrasound as one of its monitoring technologies (along with seismic, hydro-acoustic, and atmospheric radionuclide monitoring).

Animals have been known to perceive the infrasonic waves going through the earth by natural disasters and can use these as an early warning. A recent example of this is the 2004 Indian Ocean earthquake and tsunami. Animals were reported to flee the area long before the actual tsunami hit the shores of Asia.

Whales, elephants, hippopotamuses, rhinoceros, giraffes, okapi, and alligators are known to use infrasound to communicate over distances—up to hundreds of kilometres in the case of whales. It has also been suggested that migrating birds use naturally generated infrasound, from sources such as turbulent airflow over mountain ranges, as a navigational aid. Infrasonic waves can carry over long distances [thousands of kilometers] and are less susceptible to disturbance or interference than waves of higher frequencies. Elephants, in particular, produce infrasound waves (12 – 35 Hz) that travel through solid ground and are sensed by other heards using their feet, although they may be separated by hundreds of kilometres. Even tigers emit infrasound.

Infrasound has been known to cause feelings of awe or fear in humans. They can be felt and, as studies have shown, they produce a range of effects in some people including anxiety, uneasiness, extreme sorrow, nervous feelings of revulsion or fear, and chills down the spine and feelings of pressure on the chest.

Applications of infrasound

1. Medical:

(*a*) **(Therapeutic devices):** Several studies conducted in Russia and Europe reported that infrasound has therapeutic effects.

(*b*) **Infrasound peumomassage:** At 4 Hz, the progression of myopia in school children can be stabilized.

Infrasound phonophersis in antibacterial drugs: In treatment of patients which bacterial keratitis, it is as effective as local instillations of the same drugs.

2. Monitoring activities of the atmosphere: Infrasonic waves will be influenced by the atmosphere during its propagation, which is closely related with the distribution of temperature and wind in the atmosphere. By measuring the propagation properties of infrasonic waves generated by natural sources, one can detect some characteristics and rules of the large scale meteorological motions.

3. Forecasting natural disasters: Many disasters, such as volcanic eruptions, earthquakes, land-slides and clear-air turbulences, radiate infrasound in advance. By monitoring the infrasound waves, we can forecast these disasters.

15.21 DOPPLER EFFECT

When a sounding body produces a note of frequency n, a listener at a distance receives n waves per second if he is at rest. If there is a relative motion between them due to the motion of the source or the listener or both the number of waves received by the listener per second or the apparent frequency of the sound changes and is not the same as that of the source. The pitch of the note heard appears to rise if the two are approaching each other and appears to fall if the two are receding away from each other.

The apparent change in frequency due to a relative motion between the source and the listener is known as **Doppler's principle.**

Example 1. If a railway engine travelling with a high speed with its whistle blowing is approaching an observer, the pitch appears to rise. The pitch of the note appears to fall just as the engine passes the observer on the platform.

Example 2. If a tuning fork mounted on a resonance box is moved in a circle, then to a listener who places himself in the plane of the fork, the pitch seems to be higher as the fork is moving towards him than when it is moving away from him.

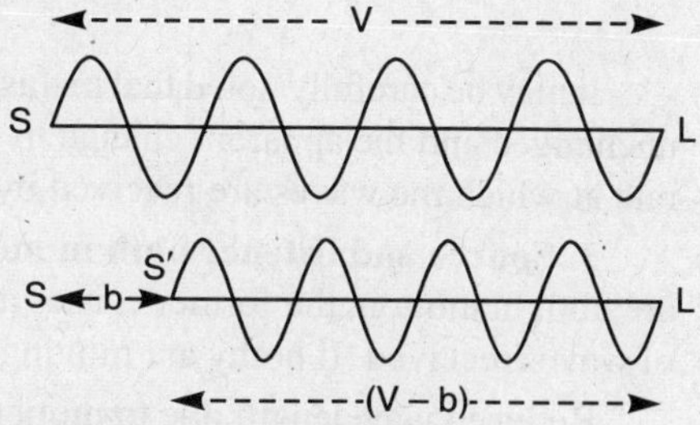

Fig. 15.22

Expression for change in frequency. 1. Source in motion, listener and medium at rest. Let the frequency of the note produced by the source be n and V be the velocity of sound. The number of waves emitted by the source per second is n and these would occupy a length V if the source is stationary at S. If the *source now moves towards the listener* (in the same direction in which the sound is travelling) with a velocity b, then after one second it will be at S'. The n waves produced per second will now be crowded in the space $S'L$ instead of SL and occupy a length $(V-b)$ as shown in Fig. 15.22.

$$\therefore \quad \text{Reduced wavelength} = \frac{V-b}{n}$$

Remember: *The motion of the source causes a change in wavelength.*

Hence the apparent pitch of the sound heard by the listener.

$$n' = \frac{\text{Velocity}}{\text{Wavelength}} = \frac{V}{(V-b)}n$$

Since $(V-b) < V$, hence $n' > n$. This shows that the apparent frequency, as heard by the listener, is **higher** than the actual frequency of the note when the source moves towards the listener.

If the source instead of moving towards the listener *moves away from him* (in a direction opposite to that of the sound), then b is negative and the new frequency n' is given by

$$n' = \frac{V}{V+b}\cdot n. \qquad \text{... (ii)}$$

Since $(V+b) > V$, hence $n' < n$. This shows that the apparent frequency as heard by the listener, is **lower** than the actual frequency of the note when the source moves away from the listener.

2. Listener in motion, source and medium at rest. When the listener is at rest the length of the block of waves passing him per second is V containing n waves. Hence the wavelength of sound.

$$\lambda = \frac{V}{n}.$$

If the listener moves away from the source with a velocity c, then it will be at L' after one second. The length of the block of waves which has now passed the observer is $(V-c)$ as shown in Fig. 15.23.

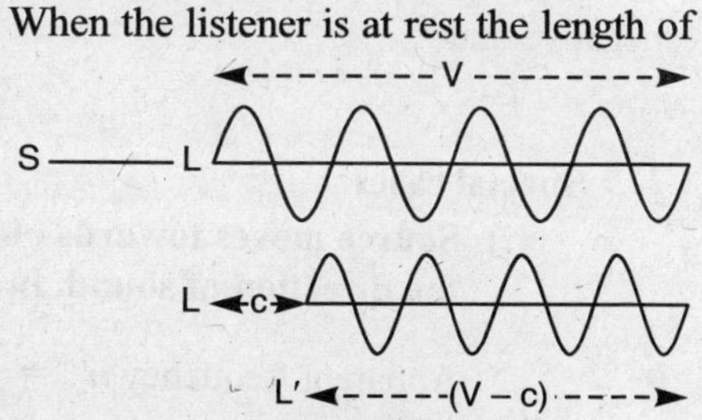

Fig. 15.23

Remember: *The motion of the listener causes a change in the number of waves received per second.*

$$\therefore \quad \text{Apparent frequency } n'' = \frac{\text{length of block of waves}}{\text{wavelength}}$$

$$= \frac{V - c}{\lambda} = \frac{V - c}{V - b} n \qquad \text{... (iii)}$$

When the listener is moving towards the source, c is negative and the apparent frequency is given by

$$n'' = \frac{V + c}{V} n \qquad \text{... (iv)}$$

It may be carefully noted that in this case the wavelength of the note emitted by the source remains unchanged and the apparent change in the frequency or pitch of the note is due to the change in the rate at which the waves are received by the listener.

3. Source and listener both in motion and medium at rest. When the source and the listener are both in motion, the former causes a change in wavelength and the latter a change in the number of waves received. If bothy are moving in the **same** direction in which the sound is travelling, thehn

Reduced wavelength due to motion of the source

$$= \frac{V - b}{n}$$

Length of the block of waves received due to motion of the listener

$$= V - c$$

$$\therefore \quad \text{Apparant frequency } n'' = \frac{\text{Length of the block of waves}}{\text{Reduced wavelength}}$$

$$= \frac{V - c}{V - b} n$$

This relation covers all possible cases, with appropriate signs given to c and b. Thus if the listener is at rest $c = 0$ and we get the relation as in (1). If the source is at rest $b = 0$ and we get the same relation as in (2) for the apparent frequencies.

If $c = b$ *i.e.*, the source and the listener move with the same velocity in the same direction, then

$$n'' = n$$

i.e., there is no change in the frequency of the note, in such a case there is no relative motion between the source and the listener. A familiar example of this case is that of the whistle of the engine as listened by the observer sitting in the train carried by the engine.

4. Effect of motion of the medium. When the medium is in motion *i.e.*, there is a wind blowing from S to L with velocity w, the apparent velocity of sound increases from V to $(V + w)$.

$$\therefore \quad n'' = \frac{(V + w) - c}{(V + w) - b} n$$

If the wind is blowing with the same velocity in the opposite direction, *i.e.*, from L to S, w is negative and

$$n'' = \frac{(V - w) - c}{(V - w) - b} n$$

Special cases

(*i*) **Source moves towards observer and observer moves away from the source both in the direction of sound.** In this case as proved in relation (*v*),

$$\text{Apparent frequency } n'' = \frac{V - c}{V - b} n \qquad \text{... (vi)}$$

(*ii*) **Source and observer move towards each other.** In this case c is negative and relation (*vi*) becomes

$$\text{Apparent frequency } n'' = \frac{V-(-c)}{V-b}n = \frac{V+c}{V-b}n \quad \text{... (vii)}$$

(*iii*) **Source and observer move away from each other.** In this case b is negative and relation (*vi*) becomes

$$\text{Apparent frequency } n'' = \frac{V-c}{V-(-b)}n = \frac{V-c}{V+b}n \quad \text{... (viii)}$$

(*iv*) **Source moves away from observer and observer moves towards the source.** In this case both b and c are negative and relation (*vi*) becomes

$$\text{Apparent frequency } n'' = \frac{V-(-c)}{V-(-b)}n = \frac{V+c}{V+b}n \quad \text{... (ix)}$$

Example 15.9. *Two trains are approaching each other with speeds of 60 Km/hr. and 45 Km/hr. A whistle of frequency 512 Hz is sounded in the first train. Calculate the frequency of the note heard by a listener in the second train (i) before and (ii) after the trains pass each other. (speed or sound 332 m/s)*

Solution. Velocity of sound $V = 332$ m/s.

$$\text{Velocity of first train (source) } b = 60 \text{ Km/hr} = \frac{60\times1000}{60\times60}$$

$$= \frac{100}{6} = 16.67 \text{ m/sec.}$$

$$\text{Velocity of second train (listener) } c = 45 \text{ Km/hr} = \frac{45\times1000}{60\times60}$$

$$= \frac{50}{4} = 12.5 \text{ m/sec.}$$

***(i)* Before the trains pass each other.** In this case source and the listener are moving towards each other.

$$\therefore \text{ Apparent frequency} \quad n' = \frac{V+c}{V-b}\times n$$

$$= \frac{332+12.5}{332-16.67}\times512 = \frac{344.5}{315.33}\times512 = 559.3 \text{ Hz.}$$

***(ii)* After the trains pass each other.** In this case source and listener are moving away from each other.

$$\text{Apparent frequency} \quad n' = \frac{V-c}{V+b}\times n$$

$$= \frac{332-12.5}{332+16.67}\times512 = \frac{319.5}{348.67}\times512 = 469.1 \text{ Hz}$$

Example 15.10. *Calculate the velocity with which a source of frequency 10,000 Hz should approach an observer at rest in order to produce a Doppler shift of 200 Hz. Velocity of sound 332 m/s.* ***(Nagpur 2002)***

Solution. Velocity of sound $V = 332$ m/s; Frequency of source $n = 10{,}000$ Hz.

Doppler shift = change in frequency = 200 Hz.

As the source is in motion and listener is at rest,

Apparent frequency $n' = \frac{V}{V-b}n$

As $(V-b) < b$, hence $n' > n$

$\therefore$ $n' = 10{,}000 + 200$ Hz

or $10{,}200 = \frac{332}{332-b} \times 10{,}000$

or $10{,}200 \times 332 - 10200\,b = 332 \times 10{,}000$

or $(10{,}200 - 10{,}000)\,332 = 10200\,b$

or $b = \frac{200 \times 332}{10200} = 6.5$ m/s

$= \frac{6.5 \times 60 \times 60}{1000} = 23.4$ Km/hour.

Example 15.11. *A train sounding a whistle of frequency 500 approaches a stationary observer at a speed of 72 Km/hr. Find the frequency of sound heard by the listener. (velocity of sound = 340 m/s)*
(Nagpur 2003, 2001)

Solution. $72 \text{ Km/hr} = \frac{72 \times 1000}{60 \times 60} = 20 \text{ ms}^{-1}$

Apparent frequency $n' = \frac{V}{V-b}n = \frac{332}{332-20} \times 500$

$= 532$ Hz.

EXERCISE

1. Explain the term standing waves. Investigate theoretically the formation of standing (stationary) waves and discuss their characteristic properties.
(G.N.D.U. 2008, 2006; Cal. U. 2001; Lucknow U. 2003)
2. How does the wave function of a progressive wave differ from that of a stationary wave ?
(Pbi. U. 2006)
3. What is the effect of reflection on the phase of a wave ?
4. Distinguish between progressive and stationary waves.
(Cal. U. 2001; Phi. U. 2002; P.U. 2009; Nagpur U. 2011)
5. What do you mean by nodes and antinodes in a stationary wave ? ***(P.U. 2008)***
6. What do you understand by normal modes of a bounded system ?
7. Discuss analytically the formation of standing waves in a string of fixed length. What are its eigen frequencies ? How does the position of nodal points change with eigen frequencies ?
(G.N.D.U. 2007; 2006; Pbi. U. 2006; P.U. 2000)
8. Derive an expression for the position of nodes and antinodes of stationary waves formed in a closed organ pipe.
9. Discuss the different modes of vibrations in a
(*i*) closed and (*ii*) open organ pipe.
Which sound is more musical and why ? ***(Nagpur 2010)***
10. Discuss the production of harmonics in an organ pipe.
11. Explain the formation of stationary waves and positions of nodes and antinodes when a metallic rod is firmly fixed at its middle and one end is gently rubbed with resined cloth.

12. What are Chladni's figures ? What is the role of Ernst Chladni in the discovery of musical instrument ? Give the experimental demonstration of formation of different Chladni's patterns?

13. Explain the use of Kundt's apparatus for different acoustical determinations.

14. Describe the method of determining the velocity of sound by using a Kundt's tube.

15. What are ultrasonic waves? Describe two methods of producing ultrasonic waves and briefly describe their detection and uses.

(Meerut U. 2007, 2006, 2003, 2001, 2000, Nagpur, 2010, 2009)

16. Describe a method of measurement of velocity of ultrasonic waves.

17. What are infrasonic waves ? Distinguish between ultrasonic and infrasonic waves ?

18. Write notes on:

(*i*) Ultrasonic waves (*ii*) Infrasonic waves

19. Give applications of ultrasonic waves in medical therapeutic devices commonly used

(Nagpur U. 2008)

20. Give applications of infrasonic waves in different fields.

21. What is meant by Doppler effect ? Derive an expression for the change in frequency of a note due to relative motion of the observer, source and medium. Hence give the relation for the following special cases :

(*i*) Source moves towards the observer and observer moves away from the source both in the direction of sound.

(*ii*) Source and observer move towards each other.

(*iii*) Source and observer move away from each other.

(*iv*) Source moves away from observer and the observer moves towards the source.

(Cal. U. 2003; Nagpur U., 2003, 2002, 2001)

22. Explain the production of ultra-sonic waves using piezo-electirc crystals. ***(Nagpur U. 2011)***

23. Explain the formation of standing waves in case of pipe open at both ends and show that all harmonics are produced. ***(Nagpur U. 2011)***

24. Compare formation of standing waves in case of a pipe open at both ends and a pipe closed at one end. ***(Nagpur U. 2010)***

MULTIPLE CHOICE QUESTIONS

15.1 The distance between two nearest antinodes is

(*a*) λ (*b*) $\frac{\lambda}{2}$ (*c*) $\frac{\lambda}{4}$ (*d*) zero

15.2 The amplitude of stationary wave is zero at

(*a*) an antinode (*b*) node

(*c*) a point midway between the node and antinode

(*d*) no where

15.3 Two wave trains reinforce each other when the phase difference between them is

(*a*) zero (*b*) $\frac{\pi}{2}$ (*c*) π (*d*) any value

15.4 When a wave suffers reflection at the boundary of a rigid wall (or a denser medium), there is a change of phase of

(*a*) 0 (*b*) $\frac{\pi}{2}$ (*c*) $\frac{\pi}{4}$ (*d*) π

15.5 Velocity of ultrasonic waves is given by

(*a*) $\frac{v\lambda D}{ndn}$ (*b*) $\frac{vn\lambda d}{dn}$ (*c*) $\frac{vn\lambda d^2}{dn}$ (*d*) $\frac{vnd^2}{\lambda dn}$

15.6 Ultrasonic waves have frequency

(*a*) < 20 Hz (*b*) > 20 Hz (*c*) < 20 KHz (*d*) > 20 KHz

15.7 Infrasonic waves have frequency

(*a*) < 20Hz (*b*) > 20Hz (*c*) < 20 KHz (*d*) > 20 KHz

15.8 Speed of ultrasonic waves depends upon

(*a*) nature of material through which it is passing

(*b*) its frequency

(*c*) its wavelength

(*d*) none of these

ANSWERS

15.1 (*b*)	**15.2** (*b*)	**15.3** (*a*)	**15.4** (*d*)	**15.5** (*b*)
15.6 (*d*)	**15.7** (*a*)	**15.8** (*b*)		

SECTION V

Acoustics

16. NOISE AND MUSIC

17. REFLECTION, REFRACTION AND DIFFRACTION OF SOUND

18. APPLIED ACOUSTICS

ACOUSTICS

Acoustics is the branch of physics that deals with the process of generation, reception and propagation of sound. This may be described by the following steps

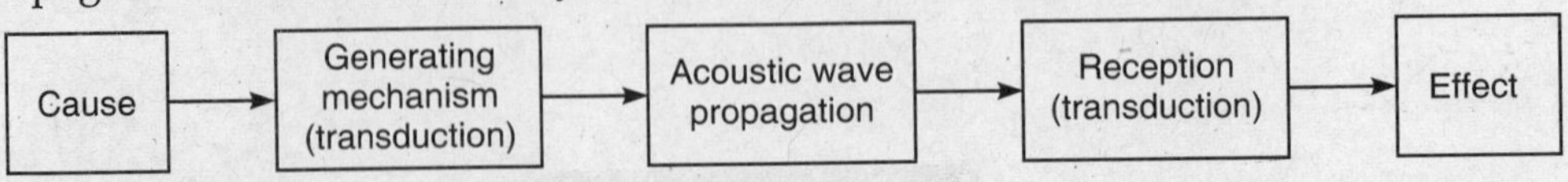

The steps shown in the above diagram can be found in any acoustical event or process. There are many kinds of cause, both natural and artificial. There are many kinds of transduction process that convert energy from some other form into acoustic energy, producing the acoustic wave. This branch, acoustics in fact covers many fields and is closely related to various branches of science and technology. Some of them are (i) musical acoustics that deals with the design of acoustical instruments, (ii) electro-acoustics viz., recording of sound and sound production (microphones, loudspeakers, amplifiers etc.), (iii) architectural acoustics, that deals with the design and construction of auditoriums, cinema theaters, operas, music recording rooms, radio & T.V. program recording rooms so that the system is acoustical defects (viz., reverberation, noise, echelon effect, etc.) free.

In a broad sense, acoustics is the interdisciplinary science that deals with the study of all mechanical waves in gases, liquids, and solids including vibration, sound, ultrasound and infrasound. The Latin synonym is "sonic", after which the term sonics used to be a synonym for acoustics and later a branch of acoustics. After acousticians had extended their studies to frequencies above and below the audible range, it became conventional to identify these frequency ranges as "ultrasonic" and "infrasonic" respectively, while letting the word "acoustics" refer to the entire frequency range without limit.

NOISE AND MUSIC

16

Chapter

INTRODUCTION

Hearing is one of the most crucial means of survival of human being, and speech is one of the most distinctive characteristics of human development and culture. The science of acoustics has widen its impact, through communication on social and economical aspects of our society, particularly music, architecture, industrial production, warfare, medicine and more. Art, craft, science and technology have forced one another to advance, as in many other fields of knowledge. The sound that produces a pleasing effect or sensation on the ear is called a *musical sound* while sound that produces harsh (jarring) effect or displeasing effect on the ear is called a *noise.*

16.1 HUMAN VOICE

Human voice is a natural source of sound. The human voice is generated due to mainly four organs: (*i*) Lungs, as a power generator, (*ii*) vocal cords in the larynx as vibrators (*iii*) cavities formed by the larynx, mouth and nose as a resonator, and (*iv*) lips, tongue and teeth etc. as a articulator. During the process of production of sound, an expansion and contraction of lungs takes place, due to the effect of which air pushed out through a narrow slit of the larynx. The vocal cord is opened by the membranous reed under tension and vibrates with the desired frequency. These vibrations are communicated to the cavities of mouth and nose which serve as resonators. The volume of resonator, mouth and the larynx are adjustable to produce sound of different pitch. Loudness or intensity of sound in a human voice primarily depends upon the force with which the air from the lungs is allowed to pass through the larynx. The frequency of the human voice is dependent upon the elasticity and vibrations of the vocal cords in the larynx.

The shape of the mouth is changed to produce different vowels. We can whisper without forcing air through vocal cord simply by motion of the lips and tongue. However, without the movement of lips and tongue we can sing by changing the frequency of the vocal cord where it is not converted into speech. This is just like the musical instruments which can produce tones of different frequencies but not the speech. In human sound, tongue plays a crucial role in producing speech. Keeping the pitch (frequency) of the voice constant, the loudness of that particular pitch can be varied. In other words, it is possible to keep the frequency of vibrations the same but to increase the amplitude of the vibration by forcing more air through the larynx. This means that the resultant speech sound wave produces greater motion in the air molecules. These, therefore, strike the ear drum of the listener more forcefully and the sound is perceived to be louder.

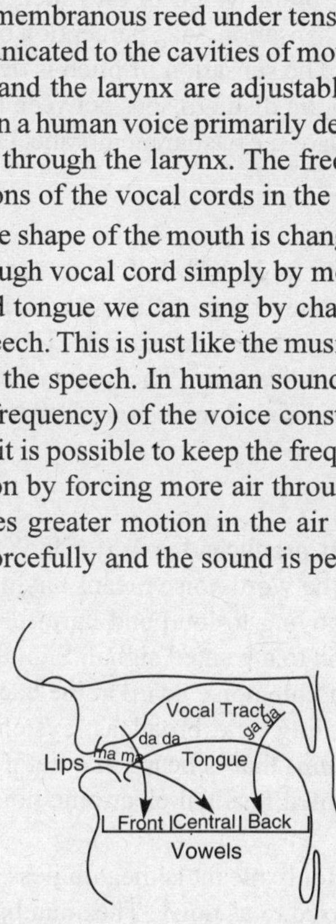

Fig. 16.1. Structure of Human mouth

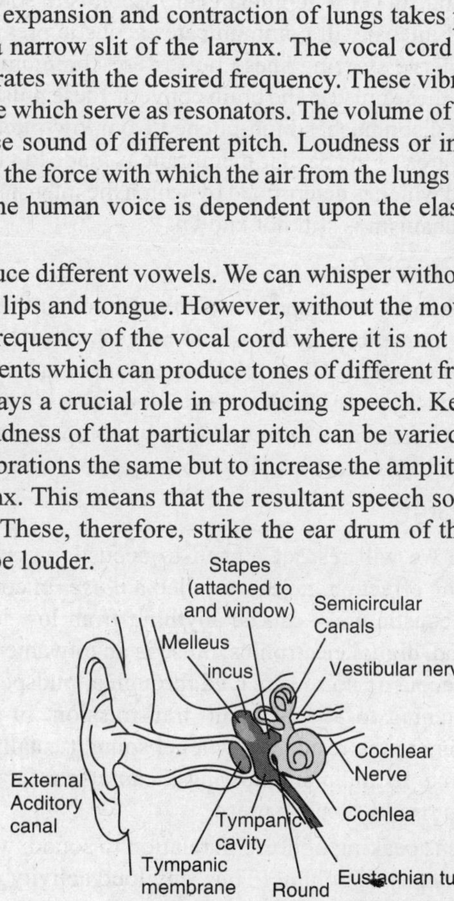

Fig. 16.2. Structure of Human ear

16.2 HUMAN EAR

Ear structure

The human ear consists of outer, middle, and inner parts. Thc outer ear is the visible portion; it includes the skin-covered flap of cartilage known as the auricle and the opening (auditory canal) leading to the eardrum (tympanic membrane).The middle ear, separated from the outer ear by the eardrum, contains three small bones, or ossicles. Because of their shapes, these bones are known as the hammer (malleus), anvil (incus), and stirrup (stapes). Air reaches the middle ear through the Eustachian tube, or auditory tube, which connects it to the throat.

The inner ear, or labyrinth, contains the cochlea, which houses the sound-analyzing cells of the ear, and the vestibule, which houses the organs of equilibrium. The cochlea is a coiled, fluid-filled tube divided into the three canals: the vestibular, tympanic, and cochlear canals. The basilar membrane forms a partition between the cochlear canal and the tympanic canal and houses the organ of Corti. Anchored in the Corti structure are some 20,000 hair cells, with filaments varying in length in a manner somewhat analogous to harp strings. These are the sensory hearing cells, connected at their base with the auditory nerve.

The Hearing Process

Human ear is a natural sound receiving organ. The human voice and the human ear together form a fundamental and natural sound system. The hearing mechanism is a highly sensitive electro acoustic transducer. The human ear responds to sound waves of a wide range of frequencies, wave forms and intensity. It communicates acoustic pressure variation of the ear drum into pulses in the auditory nerve system. These pulses are communicated to the brain which in turn identifies and interprets these pulses. The brain converts these pulses into aural sensation, viz., perception of sound. The brain distinguishes high-pitched from low-pitched sounds. The sensation of pitch is dependent on which area of the basilar membrane is made to vibrate. The brain distinguishes between loud and soft sound which is determined by which the intensity of vibration of the basilar membrane. However, exact mechanism is still not known.

Ear Responses

The human ear responds to frequencies in the range 20 Hz to 20 KHz*. It does not respond to sound frequencies in ultrasonic range (> 20 KHz) and infrasonic range (< 20 KHz). Hence the frequency range 20 Hz to 20 KHz is known as audio frequency range (AF range). The range of sound intensity over which the ear is sensitive is 1 watt/m^2 to 10-12 watt/m^2. The human ear is more sensitive to variations in frequencies as compared to variations in sound intensities. The human ear is comparatively more sensitive to sounds of low intensity.

16.3 NOISE

Here we will restrict ourself to acoustic noise. Sound that produces harsh (jarring) effect or displeasing effect on the ear is called a *noise.* In common use, the word noise means any unwanted sound. Acoustic noise can be anything from low level but annoying to loud and harmful. In both analog and digital electronics, noise is an unwanted perturbation to a wanted signal. Signal noise is heard as acoustic noise if played through a loudspeaker. The audible noise heard at the background when listening to a weak radio transmission. In audio, recording, and broadcast systems audio noise refers to the residual low level sound (usually hiss and hum) that is heard in quiet periods of programme. In audio engineering it can also refer to the unwanted residual electronic noise signal that gives rise to acoustic noise.

When speaking of noise in relation to sound, what is commonly meant is meaningless sound of greater than usual volume. Thus, any loud activity may be referred to as noisy. The sound produced

* Frequency 1 KHz is equal to 1000 Hz *i.e.*, 1000 cycles per second.

by a bullet fired from gun, a plate falling on ground, bomb explosions etc. are examples of noise. Conversations of other people may be called noise for people not involved in any of them, and noise can be any unwanted sound such as the noise of dogs barking, neighbours playing loud music, road traffic sounds, chainsaws, or aircraft, etc. can cause permanent irreversible hearing damage and also spoils the quiet environment. The activity results into what we call as "sound pollution," which is detrimental (hazardous) to human as well as plant life.

16.4 MUSICAL SOUND

Musical sound of any tone has characteristics such as controlled pitch and quality (timbre). The sounds are produced by instruments in which the periodic vibrations can be controlled by the performer. Some sounds are intrinsically musical, while others are not. From the tinkle of a bell to the slam of a door, any sound is a potential ingredient for the kinds of sound organization called music. The choices of sounds for music making have been severely limited in all places and periods by a diversity of physical, aesthetic, and cultural considerations.

The fundamental distinction usually made has been between tone and noise, a distinction best clarified by referring to the physical characteristics of sound. Tone differs from noise mainly in that it possesses features that enable it to be regarded as autonomous. Noises are most readily identified, not by their character but by their sources; *e.g.*, the noise of the dripping faucet, the grating chalk, or the squeaking gate. Although tones too are commonly linked with their sources (violin tone, flute tone, etc.,) they more readily achieve autonomy because they possess controlled pitch, loudness, timbre, and duration, attributes that make them amenable to musical organization. Instruments that yield musical sounds, or tones, are those that produce periodic vibrations. Their periodicity is their controllable (*i.e.* Musical) basis. The instruments that produce musical sounds are shown in Fig. 16.3.

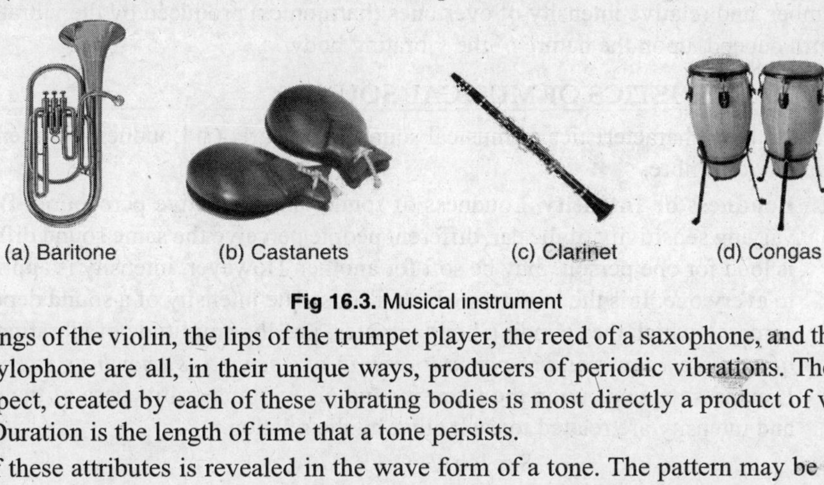

(a) Baritone (b) Castanets (c) Clarinet (d) Congas

Fig 16.3. Musical instrument

The strings of the violin, the lips of the trumpet player, the reed of a saxophone, and the wooden slabs of a xylophone are all, in their unique ways, producers of periodic vibrations. The pitch, or high-low aspect, created by each of these vibrating bodies is most directly a product of vibrational frequency. Duration is the length of time that a tone persists.

Each of these attributes is revealed in the wave form of a tone. The pattern may be visualized as an elastic reed-like that of a clarinet—fixed at one end, moving like a pendulum in a to-and-fro pattern when set into motion. Clearly, this reed's motion reed will be in proportion to the applied force. Its arc of movement will be lesser or greater depending upon the degree of pressure used to set it into motion. Once moving, it will oscillate until friction and its own inertia cause it to return to its original state of rest. As it moves through its arc the reed passes through a periodic number of cycles per time unit, although its speed is not constant. With these conditions prevailing, its motion through time could be charted by placing a carbon stylus on its moving head, then pulling a strip of paper beneath it at a uniform rate. The reed's displacement to-and-fro diminishes in a smooth fashion as time passes (decreasing intensity). Each cycle of its arc is equally spaced (uniform frequency). Each period of the motion forms the same arc pattern (uniform wave content). If this vibratory motion were audible, it could be described as follows: it grows weaker from the beginning (diminishing loudness) until it becomes inaudible; it remains at a stable level of highness (steady pitch); and it is of unvarying tonal quality (uniform timbre). If the reed were a part of a clarinet and the player continued blowing it with unvaried pressure, loudness, pitch, and timbre would appear as constants.

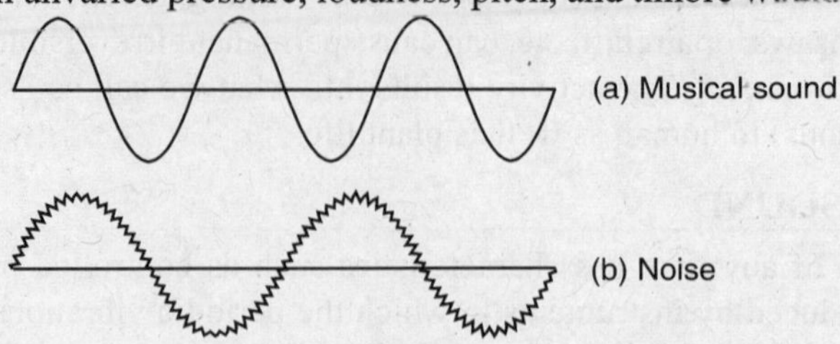

Fig. 16.4

Musical sounds are distinguished from noises in that they are composed of regular, uniform vibrations Fig 16.4 (a), while noises are irregular and disordered vibrations Fig 16. 4 (b). Composers, however, frequently use noises as well as musical sounds. One musical tone is distinguished from another on the basis of pitch,intensity, or loudness, and quality, or timbre. Pitch describes how high or low a tone is and depends upon the rapidity with which a sounding body vibrates, i.e., upon the frequency of vibrations. The higher the frequency of vibration, the higher the tone; the pitch of a siren gets higher and higher as the frequency of vibration increases. The apparent change in the pitch of a sound as a source approaches or moves away from an observer is described by the Doppler effect. The sound waves given off by different vibrating bodies differ in quality, or timbre. A note from a saxophone, for instance, differs from a note of the same pitch and intensity produced by a violin or a xylophone; similarly vibrating reeds, columns of air, and strings all differ. Quality is dependent on the number and relative intensity of overtones (harmonics) produced by the vibrating body, and these in turn depend upon the nature of the vibrating body.

16.5 CHARACTERISTICS OF MUSICAL SOUND

There are three characteristics of musical sound. These are: (*i*) Loudness or intensity (*ii*) pitch and (*iii*) quality or timbre.

1. **Loudness or Intensity.** Loudness of sound is a subjective perception. Because of the varying sensitivity of the ear, different people perceive the same sound differently. What is loud for one person, may be soft for another. However, intensity is applicable equally to every one. It is the loudness which differs. The intensity of a sound depends upon the extent to which the sounding body vibrates, i.e., the amplitude of vibration. *The amount of sound energy crossing per unit area in one second is known as intensity of sound.* Loudness depends upon the intensity and also upon sensitiveness of the ear. Loudness and intensity are related to each other by the relation.

$$S \propto \log_e I \qquad \text{...(16.1)}$$

where S represents the sensation of loudness and I, the intensity of sound. Loudness or intensity depends upon the following factors.

(*i*) **Amplitude.** Loudness is directly proportional to the square of the amplitude of the sounding body. A sound is louder as the amplitude of vibration is greater, and the intensity decreases as the distance from the source increases. The amplitude of sound produced by men is large and hence loud sound is produced. The amplitude of sound produced by ladies or children is small, therefore a feeble sound is produced. Loudness is measured in units called decibels.

(*ii*) **Surface area** Loudness is directly proportional to the surface area of sounding body. A tuning fork of large size produces a loud sound as compared to a tuning fork of small size. Beating drums with large surfaces produce a loud sound as compared to drums with small surface area. In sonometer experiment using tuning fork, sound produced at resonance is feeble that of a tuning fork. But the moment its stem is pressed against the sonometer, the loudness increases. An air inside the hollow sonometer is set in to resonance. The apparent surface area increases, hence a loud sound is heard.

(*iii*) **Distance between the source and listener.** The intensity of sound is inversely proportional to the square of the distance between the source and the listener. Therefore, the sound loudness becomes weaker and weaker as the distance between the source and listener goes on increasing.

(*iv*) **Density of the medium.** The intensity of sound is directly proportional to the density of the medium. Hence loud sound is generated in a denser medium as compared to a rarer medium, through which the sound propagates.

(*v*) **Motion of air.** Loudness increases in the direction of flow of air. If the sound is travelling in opposite direction of motion of air (air flow), the loudness becomes feeble.

2. **Pitch.** A sound which has a well defined frequency is called a tone. Thus, every sinusoidal sound wave is a pure tone. In general, sound sources do not produce a single frequency. Musical notes, the pitch of the note being that corresponding to the lowest tone it contains. The tone having frequency n is called the **fundamental tone.** The additional tones of frequency 2n, 3n, 4n, ...are called **overtones** or **harmonics.** Pitch is a subjective sensation (to the ear) perceived when a tone of a given frequency is sounded. It enables us to classify a tone as high or low and to distinguish a shrill (harshcy sound from a flat sound of the same intensity sounded on the same musical instrument. Thus, pitch is referred to the tone frequency. Higher the pitch means higher the frequency of oscillation of the sounding body.

3. **Quality or Timbre.** Timbre is another subjective sensation It is the quality of sound that corresponds to the quality of wave form produced by a source. Since the **relative intensity of the harmonics** produced by different types of sound sources is different, each one has a characteristic timbre distinguishable from the timbre of the other sources, even if all sources produce the same fundamental frequency (n). Therefore, memory of timbre helps us identifying sounds such as instruments being played or the person who is speaking or singing. The quality of sound enables us to distinguish two sounds having the same pitch and loudness.

   Pitch does not depend upon **loudness** or **quality.** The voice produced by ladies and children has high pitch because the frequency is high. The voice of old man has low pitch and is hoarse because the frequency of sound is low. Pitch of a male voice ranges from 82 Hz to 140 Hz (average = 106 Hz) and that of female from 170 Hz to 333 Hz (average = 202 Hz).

   The frequency of the sound produced by a mosquito is of high pitch due to high frequency. The sound is not heard when the mosquito is even a few metres away from the ear but it is a shrill sound. This is because the sound produced has high frequency (pitch) but low amplitude. The roaring of a lion has high amplitude but low frequency. Therefore, the roaring of a lion can be heard even at far distant places, but it is not a shrill (harsh) sound.

16.6 INTENSITY OF SOUND

As defined earlier, the intensity of sound I is the energy transported by sound waves in unit time across a unit cross-sectional area normal to the direction of wave propagation. Similar to light waves the sound intensity is proportional to the square of the wave amplitude. Thus,

$$I \propto p^2$$

or $$I = \frac{p_{\max}^2}{2\rho v} \quad \text{... (16.2)}$$

Where $p_{\max}$ stands for the maximum pressure amplitude

Proof : The amount of sound energy transfer per unit area per second

$$I = 2\pi^2 \rho n^2 a^2 \nu$$

where ρ is the density of the medium, n is the frequency a is the amplitude and ν is the velocity of sound. The velocity of sound is given by

$$\nu = \sqrt{\frac{E}{\rho}} \quad \text{and} \quad E = -\frac{p}{dV/V}$$

dV is the change in volume, V is the original volume and P is the excess of pressure.

$$\therefore \quad v = \sqrt{\frac{-p}{\left(\frac{dV}{V}\right)\rho}}$$

Taking $\frac{dV}{V} = \frac{dy}{dx}$ and simplifying

$$p = -v^2\rho\frac{dy}{dx} \quad \text{... (ii)}$$

A simple harmonic wave is represented by equation

$$y = a\sin\frac{2\pi}{\lambda}(\nu t - x)$$

$$\frac{dy}{dx} = -\frac{2\pi a}{\lambda}\cos\frac{2\pi}{\lambda}(\nu t - x)$$

Substituting this value of dy/dx in Eq (ii), we get

$$p = \frac{2\pi a v^2 \rho}{\lambda}\cos\frac{2\pi}{\lambda}(vt - x) \quad \text{... (iii)}$$

The maximum excess of pressure

$$p_{\max} = \frac{2\pi a v^2 \rho}{\lambda} \quad \text{... (iv)}$$

and in general,

$$p = p_{\max}\cos\frac{2\pi}{\lambda}(\nu t - x) \quad \text{... (v)}$$

Rearranging Eq.(iv), $$p_{\max} = 2\pi a\rho\nu\left(\frac{\nu}{\lambda}\right)$$

$$= 2\pi a\rho\nu n \quad \text{... (vi)}$$

Eq. (i) and (iv) gives,

$$I = 2\pi^2\rho n^2 a^2 \nu$$

$$= \frac{(2\pi\rho n a \nu)^2}{2\rho\nu}$$

$$I = \frac{p_{\max}^2}{2\rho\nu} \quad \text{... (16.3)}$$

Thus, the intensity of sound heard varies directly as the square of excess of pressure. In acoustics, it is important to measure the excess of pressure to obtain the value of intensity of sound. For ordinary conversation $p_{max} = 0.1$. The intensity of the faintest sound wave that can be heard is about 10^{-16} Watt/cm^2 or 10^{-12} W/m^2.

In case of spherical waves propagating from a source, the intensity is inversely proportional to the square of the distance from the source to the receiver. The sound gets weaker as the distance from the source is increased as shown in Fig.16.5.

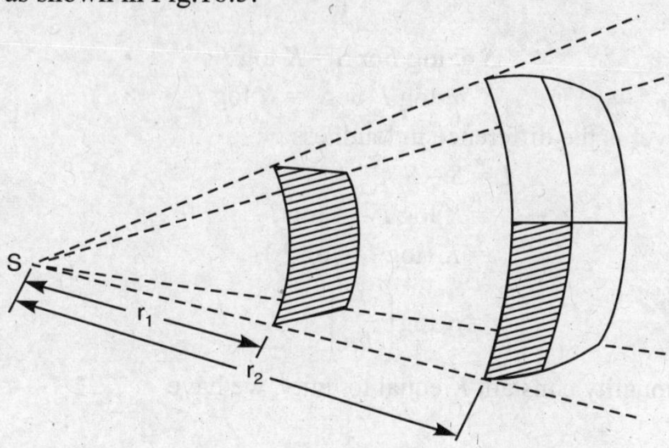

Fig. 16.5

Thus, $I \alpha \frac{1}{r^2}$

If I_1 and I_2 are intensities at distances r_1 and r_2 respectively, then

$$\frac{I_1}{I_2} = \frac{r_2^2}{r_1^2} \quad \text{... (16.4)}$$

16.7 MEASUREMENT OF INTENSITY OF SOUND

The intensity of sound is defined as the quantity of energy propagating through a unit area per unit time, the direction of propagation being perpendicular to the area. The loudness of a sound is just an aural sensation and is a physiological phenomenon rather than a physical phenomenon. In other words, intensity of sound is a definite physical quantity and loudness is merely a degree of sensation. Loudness of sound increases with the intensity of sound, governed by Weber-Fecher law in physiology. According to this law, the loudness produced is proportional to the logarithm of intensity. This law holds good within the limits of audibility. According to this law,

$$S \propto \log I$$

or

$$S = K \log I \quad \text{... }(i)$$

where K is a constant

or

$$\frac{dS}{dI} = \frac{K}{I} \quad \text{... }(ii)$$

The quantity dS/dI is called the **sensitiveness** of the ear. The sensitiveness of the ear decreases with increase in intensity of sound and vice-versa.

For all practical purposes, it is the relative intensity that is important and not the absolute value. Hence, the intensity of sound is often measured as its ratio to a standard intensity I_0. Hence, the intensity level is I/I_o. The standard intensity I_o is taken as 0.01 watt/m^2. The faintest sound that can be heard depends upon the frequency of the note. For a frequency of 440 Hz, 10^{-10} I_0 is the threshold of audibility for an average person. Thus, the range of hearing for a human ear is from 10^{-10} Io up to 100 I_0. Hence the human ear has a wide range of 10^{12} in intensity.

16.8 INTENSITY LEVEL (UNIT) OF SOUND: DECIBEL

As discussed earlier, the range of variation of intensity is very large. The loudness of a sound as sensed by the ear is proportional to the logarithm of intensity ($S \alpha \log I$) If I and I_0 represent the intensities of two sounds of a particular frequency and S and S_0 are their corresponding measures of loudness, then

$$S \propto \log I \text{ or } S = K \log I$$

and
$$S_0 \propto \log I_0 \text{ or } S_o = K \log I_0$$

The intensity level is the difference in loudness

$$\begin{aligned} L &= S - S_0 \\ &= K \log I - K \log I_0 \\ &= K (\log I - \log I_0) \\ &= K \log\left(\frac{I}{I_o}\right) \end{aligned}$$

Taking proportionality constant K equal to unity, we have

$$L = \log\left(\frac{I}{I_o}\right) \text{ in bel} \qquad \text{... (16.5)}$$

Here L is called the **intensity level** and is expressed in bels, a unit named after Alexander Graham Bell, the inventor of telephone. The unit of bel is large and hence in practice a smaller unit *decibel* (abbreviated as dB) is used

$$1 \text{ decibel} = \frac{1}{10} \text{bel}$$

Accordingly, the intensity level of a sound wave is defined as

$$L = 10 \log\left(\frac{I}{I_o}\right) \text{ in dB} \qquad \text{... (16.5)}$$

To construct a scale of loudness, a reference is to be chosen. The sound intensity corresponding to the threshold of hearing is chosen as the reference. $I_o = 10^{-16}$ watt/cm^2 is taken as the reference.

Now, if the intensity of the sound wave is I_o, its intensity level is 0 dB.

When $I = I_o$, then $L = 10 \log\left(\frac{I}{I_o}\right) = 10 \log 1 = 0$. This represents the **threshold of audibility**

If $L = 1$ dB, $\log (I/I_o) = 1/10$

$$\therefore \qquad \frac{I}{I_o} = 1.26 \qquad \text{... (16.7)}$$

This implies that a change in intensity by 26% enhances the intensity level by 1 dB. This is the smallest change in intensity level that the human ear can ordinarily detect.

Moreover, if $I = 100\, I_o$, then $L = 10 \log 10^2 = 20$ dB

if $I = 1000\, I_o$, then $L = 10 \log 10^3 = 30$ dB

if $I = 10^4 I_o$, then $L = 40$ log dB and so on.

Thus, when two sounds differ by 20 dB, the louder of them is 100 times more intense and when they differ by 40 dB, the louder one is 10,000 times more intense. The loudesl-sound the can be heard without pain is about 130 dB. The lowest change in intensity level that can be detected by the human ear is 1 dB.

Example 16.1 Calculate the change in intensity level when the intensity of sound increases by 10^6 times its original intensity.

Solution. Let I_o be the intial intensity and I, the final intensity

$$\therefore \quad \frac{I}{I_o} = 10^6$$

$$\text{Increase in intensity level,} \quad L = 10 \log_{10}\left(\frac{I}{I_o}\right)$$

$$= 10 \log_{10}(10^6)$$

$$= 60 \text{ decibels}$$

16.9 LIMITS OF AUDIBILITY

The limits of audibility of sound depend upon (*i*) the intensity and (*ii*) the frequency of sound. In order that a song to be audible, it must have a certain minimum intensity and a certain minimum frequency. The minimum intensity of sound necessary for the sound to be audible, is called **threshold intensity** of audibility. The minimum audible frequency is called **lower pitch limit** of audibility.

Moreover, there is also a maximum intensity limit, beyond which the sound produces a sensation of pain on the ear. Similarly, there is a maximum limit of frequency, beyond which the sound is not audible. The maximum audible intensity limit is called **threshold intensity of feeling.** The maximum audible frequency limit is the upper pitch limit of audibility.

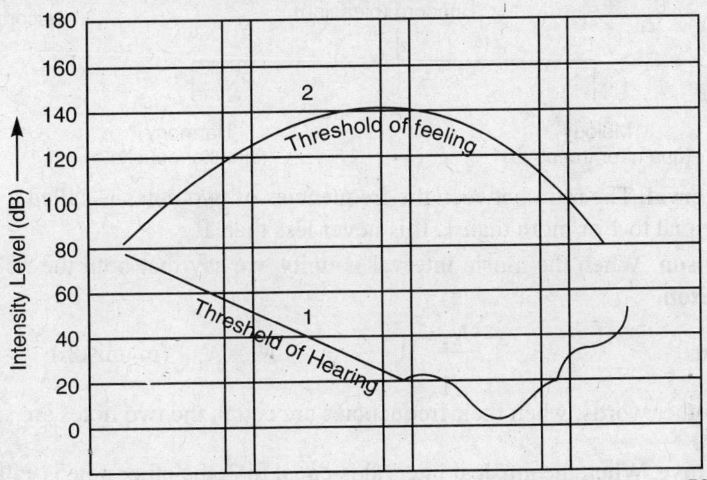

Fig. 16.6. Human Audiogram

Wegel (1922) obtained an interesting diagram, showing the range of response of average human ear with respect to intensity and frequency, known as audiogram as shown in Fig. 16.6. The variation of intensity level (in dB) is taken along linear axis (Y-axis) and frequency (in Hz) along logarithmic axis (X-axis) of a semi-log paper. In the figure, curve 1 represents the threshold of audibility (i.e. hearing) and curve 2 represents the threshold of feeling. The area enclosed between these two curves is called the *auditory sensation area.* This audiogram represents the auditory sensation area for a normal human ear. If the pitch and the intensity of any sound are beyond the limits set by the auditory sensation area, the sound is not heard. The sound wave may be felt by an ear as pressure but not as an audible sound. The frequency range of audibility for a normal human ear extends from 20 Hz to 20 KHz. Some persons can hear sound of vibration of frequency lower than 20 Hz even. Young persons can hear frequency upto 20 KHz but with age audibility reduces to a frequency of 12 KHz or less.

16.10 MUSICAL SCALE

In music, a scale is a sequence of musical notes in ascending and descending order. Most commonly, the notes of a scale belong to a single key of a keyboard instrument such as piano. Let us define some terms popularily used. These are:

(*i*) **Chord.** When two or more notes are sounded together the combined note produced is called **chord.** If the combined note produces a pleasing effect on the ear, it is called **concord.** If the combined note produces displeasing effect on the ear, it is called **discord.**

(*ii*) **Harmony.** When two or more notes are sounded simultaneously, the combined note, producing a pleasant effect on the ear, is called **harmony.**

Example. In English music, where a number of instruments play simultaneously, harmony is produced.

(*iii*) **Melody.** When two or more notes are sounded one after the other, the combined note producing pleasant effect on the ear, is called **melody.**

Example: In Indian folk music, melody is produced

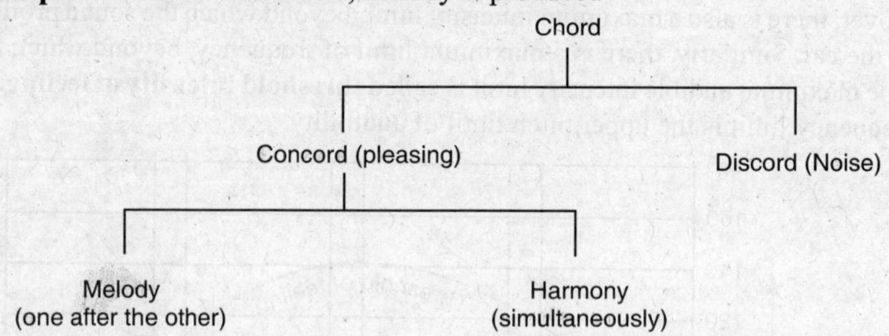

(*iv*) **Interval.** The ratio between the frequencies of two notes is called a musical interval. It is equal to 1 or more than 1. It is never less than 1.

(*v*) **Unison.** When the music interval is unity, we say that both the notes are vibrating in **unison.**

Thus, $$\frac{N_2}{N_1} = 1 \quad \text{or} \quad N_2 = N_1 \quad \text{(in unison)}$$

In other words, when their frequencies are equal, the two notes are said to be in unison.

(*vi*) **Octave.** When the musical interval is equal to 2, the other note is called as an **octave** of the first note.

$$\therefore \quad \frac{N_2}{N_1} = 2 \quad \text{or} \quad N_2 = 2N_1$$

Here N_2 is the octave of N_1

Major tone. When the musical interval $= \frac{9}{8}$

Minor tone When the musical interval = $\frac{10}{9}$

Semi tone. When the musical interval = $\frac{16}{15}$

Fifth tone : When the musical interval = $\frac{3}{2}$

(*vii*) **Consonant interval.** The degree of minimum discord was estimated by Helmholtz. The order of consonance is given below with their intervals and names.

Name	*Interval*
Unison	1.1
Octave	2:1
Fifth	3:2
Fourth	4:3
Major third	5:4
Minor third	6:5
Major sixth	5:3
Minor sixth	6:3

(*viii*) **Diatonic musical scale:** This musical scale has eight keys and seven intervals. This is described in fig.16.7

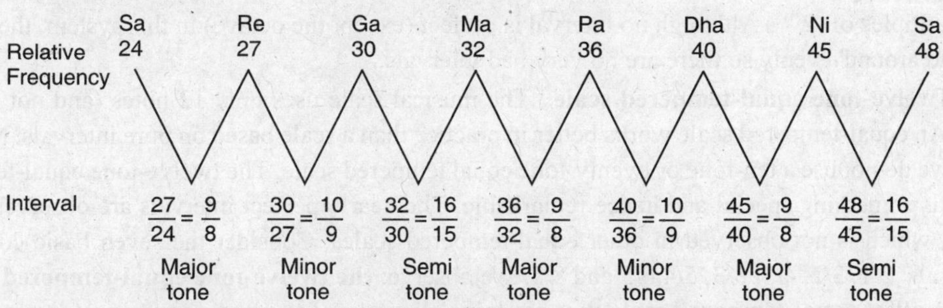

Fig. 16.7

The frequencies of diatonic musical scales are as shown in following table

Symbol	*Western name*	*Indian name*	*Proportional Frequency (in Hz)*	*Frequency in the basis of 256 Hz*	*Interval*
C	DO	Sa	24	256	$\frac{9}{8}$ Major tone
D	RE	Re	27	288	$\frac{10}{9}$ Minor tone
E	MI	Ga	30	320	$\frac{16}{15}$ Semitone
F	FA	Ma	32	341.3	$\frac{9}{8}$ Major tone

G	SOL	Pa	36	384	$\frac{10}{9}$ Minor tone
A	LA	Dha	40	326.7	$\frac{9}{8}$ Major tone
B	SI	Ni	45	480	$\frac{16}{15}$ Semitone
Si	Do	Sa	48	512	

16.11 TEMPERAMENT

One key instruments, tuning pure fifth intervals causes octaves to be out-of-tune. A temperament is a system of tuning in which the tuning of the fifth is altered to keep the octaves in-tune, and to allow all notes of the musical scale to be usable. Hence equal tempered scale is used.

In the past (before 1500 AD), designers of instruments were using scales based on pure or natural ratios of small integrers. However this method runs into serious problems. Although, some of the intervals are perfect, other combination of notes sound very bad, hence named "Wolf intervals". The modern **equal temperament** system was invented (in the 1500 ADs) to solve this problem . (Galileo's father, a music theorist, was one early proponent of equal temperament). The octave is divided into twelve exactly equal intervals. In this system, the smallest interval, the semitone, is not a simple integer ratio, but is the twelfth root of two ($2^{1/12}$) or approximately 1.059. Larger intervals are multiples of ($2^{1/12}$) Although no interval is perfect (except the octave) in this system, the error is 'spread around' evenly so there are no very bad intervals.

Twelve-tone equal-tempered scale. The musical scale uses only 12 notes (and not more or less) An equal-tempered scale works better in practice than a scale based on pure intervals. Not only this, we do not use a ten-tone or twenty-tone equal tempered scale. The twelve-tone equal-tempered scale is something special and hence remarkable. The nearly perfect intervals are observed in this scale, which is not observed in other equal-tempered scales. Consider the seven basic consonant intervals: 2/1, 3/2, 4/3, 5/3, 5/4,6/5 and 8/5. We observe: **the twelve-tone equal-tempered scale is the smallest equal-tempered scale that contains all seven of the basic consonant intervals to a good approximation-within one percent.**

Furthermore, for the most important intervals, the fifth (3/2) and fourth (4/3), the approximations are better-within one tenth of one percent.

These are reasons, why twelve-tone equal-tempered scale is something special as compared to other scales.

In an **equally tempered scale,** there are 13 keys and 12 intervals. The intervals are equal and each interval = $2^{\frac{1}{12}}$. This is shown in Fig. 16.8

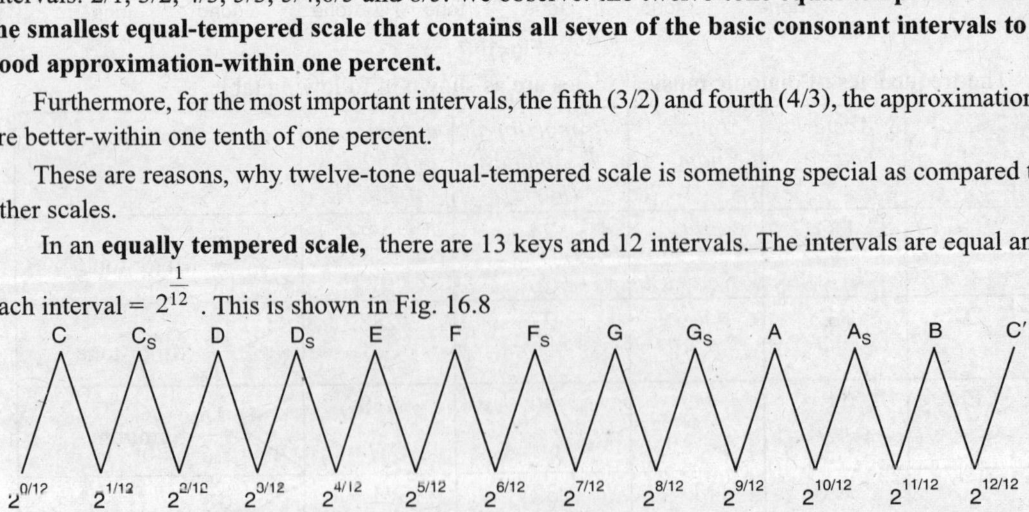

Fig. 16.8

In this scale five notes have been introduced. They are C_s, D_s, F_s, G_s and A_s in diatonic musical scale. The main advantage of this scale is that the interval is the same between the consecutive notes and singer can conveniently use any key as his fundamental.

For example, suppose C = 256 Hz

$$D = 256 \times \frac{9}{8} = 288 \text{ Hz}$$

$$E = 288 \times \frac{10}{9} = 320 \text{ Hz}$$

$$F = 320 \times \frac{16}{15} = \frac{1024}{3} = 341.33 \text{ Hz}$$

$$G = \frac{1024}{3} \times \frac{9}{8} = 384 \text{ Hz}$$

$$A = 384 \times \frac{10}{9} = \frac{1280}{3} = 426.66 \text{ Hz}$$

$$B = \frac{1280}{3} \times \frac{9}{8} = 480 \text{Hz}$$

$$C' = 480 \times \frac{16}{15} = 512 \text{ Hz}$$

Example 16.2 The frequency of the key note C= 512 Hz. Find the frequency of the note G.

Solution

$$C = 512 \text{ Hz}$$

$$\therefore \quad G = 512 \times \frac{36}{24} = 768 \text{ (Because, if } C = 24,\ G = 36)$$

EXERCISE

1. Describe the structure of human ear and the mechanism of hearing *(Nagpur U. 2011)*
2. Discuss the range of response of human ear.
3. Discuss 'Human voice is a natural source of sound'.
4. Distinguish between noise and music. *(Nagpur U. 2010)*
5. Give the characteristics of musical sound.
6. Is there a difference between loudness and intensity of sound. Explain *(Nagpur U. 2008)*
7. On what factors the intensity of sound produced depend?
8. What do you mean by pitch of a sound? What are tones and overtones?
9. Explain quality of a sound? How it plays an important role to distinguish two sounds of same pitch and loudness?
10. Does pitch depend upon loudness? Explain.
11. Define intensity of sound produced?
12. Derive the relation for relative intensities $\frac{I_1}{I_2} = \frac{r_2^2}{r_1^2}$ for a spherical wave propagation.
13. Prove the relation $I = \frac{p^2_{\max}}{2\rho v}$ for the intensity of sound.

14. Define sensitiveness of human ear.

15. Show that the sensitivity of human ear decreases with increase in intensity.

16. Define threshold of audibility. What is the audible range of intensity for a human ear?

17. Define the unit of measuring intensity level of sound.

18. What is bell and decibel? ***(Nagapur U. 2009, 2008)***

19. What is the lowest change in intensity level than a normal human ear can detect?

20. What is Weber-Fetchner law?

21. Describe human audiogram in terms of threshold of hearing and threshold of feeling.

22. What is musical scale? Why is it necessary? Discuss diatonic musical scale?

23. Define the terms

(*i*) Chord

(*ii*) Harmony

(*iii*) Melody

(*iv*) Interval

(*v*) Unison

(*vi*) Octave

24. (*a*) What do you mean by temperament of a musical scale?

(*b*) Explain why tempered scale is used in keyed musical instrument.

(*c*) Discuss the superiority of 12 tone equal-tempered scale over other scales

(Nagpur U. 2008)

25. Calculate the change in intensity level when the intensity of sound increases 100 times its original intensity. **[Ans 20 dB]**

26. Explain the term limit of human audibility. ***(Nagpur U. 2011, 2008)***

MULTIPLE CHOICE QUESTIONS

16.1 Human ear responds to frequencies in the range

(*a*) 100 Hz to 40 KHz (*b*) 20 Hz to 20 KHz

(*c*) All frequencies below 20 Hz (*d*) All frequencies above 20 kHz

16.2 The main feature of musical sound is that the waves are

(*a*) regular and uniform (*b*) irregular but uniform

(*c*) regular but not uniform (*d*) irregular and not uniform

16.3 The intensity of sound is

(*a*) directly proportional to the distance between the source and listener

(*b*) inversely proportional to the distance between source and listener.

(*c*) directly proportional to the square of the distance between source and listener.

(*d*) inversely proportional to the square of the distance between source and listener.

16.4 Pitch of a sound refers to its

(*a*) amplitude (*b*) phase

(*c*) frequency (*d*) loudness

16.5 Intensity of sound heard varies

(*a*) directly as the square of pressure (*b*) inversely as the pressure

(*c*) directly as the pressure (*d*) inversely as the square of the pressure

16.6 Sensitiveness of the human ear

(*a*) decreases with decrease in intensity of sound

(*b*) increases with increase in intensity of sound

(*c*) decreases with increase in intensity of sound

(*d*) does not depend upon the intensity of sound

16.7 The lowest change in intensity level that can be detected by normal human ear is

(*a*) 100 dB (*b*) 40 dB

(*c*) 10 dB (*d*) 1 dB

16.8 The most remarkable equal-tempered musical scale is

(*a*) 6-tone (*b*) 10-tone

(*c*) 12-tone (*d*) 20-tone

ANSWERS

16.1 (*b*)	**16.2** (*a*)	**16.3** (*d*)	**16.4** (*c*)	**16.5** (*a*)	**16.6** (*c*)	**16.7** (*d*)	**16.8** (*c*)

REFLECTION, REFRACTION AND DIFFRACTION OF SOUND

17

Chapter

INTRODUCTION

Sound wave is a longitudinal wave. It can be reflected, refracted, interfered and diffracted under appropriate conditions. The question of polarization of sound wave does not arise because polarization is a phenomena that takes place in case of transverse wave, where the particles vibrate in all possible directions perpendicular to the direction of propogation. In longitudinal wave, the particles of the medium vibrate in one direction along the line of propogation of wave. Sound waves can form stationary waves, beats depending upon the conditions imposed.

17.1 REFLECTION OF SOUND

In every day life, we come across many acoustic phenomena, e.g. echos, roaring of thunder, reverberation in halls etc. All these phenomena are based on reflection of sound. Sound waves play an important role in acoustic applications like speaking tubes, horns, trumpets, sounding boards etc.

Formation of stationary waves in open and closed organ pipes are also based on the principle of reflection of sound. Sound waves follow the law of reflection and refraction and Huygen's principle as is given in the case of light waves.

17.2 REFLECTION AND TRANSMISSION OF SOUND WAVES AT A BOUNDARY BETWEEN TWO MEDIA

When a sound wave meets a boundary separating two media of different acoustic impedances* it is partly reflected and partly transmitted at the boundary. Consider a plane sound wave travelling in a medium 1 of density ρ_1 and incident normally on a plane boundary at $x = 0$ separating medium 1 from another medium 2 of density ρ_2 as shown in fig 17.1.

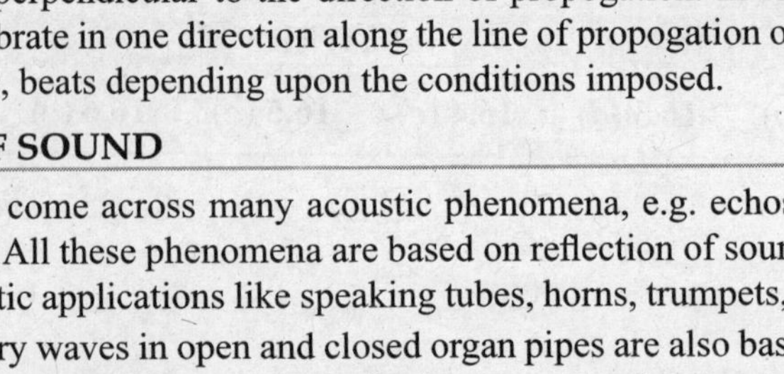

Fig. 17.1

The acoustic impedances of the two media respectively are

$$Z_1 = \rho_1 v_1 \quad \text{... (i)}$$

$$Z_2 = \rho_2 v_2 \quad \text{... (ii)}$$

where v_1 and v_2 are the sound speeds in medium 1 and medium 2 respectively. The incident, reflected and transmitted waves are respectively given by

$$\psi_i (x, t) = A_i \sin (\omega t - k_1 x) \quad \text{... (iii)}$$

$$\psi_r (x, t) = A_r \sin (\omega t - k_1 x) \quad \text{... (iv)}$$

$$\psi_t (x, t) = A_t \sin (\omega t - k_2 x) \quad \text{... (v)}$$

* Refer article 13.20

The boundary conditions are

(1) The particle displacement $\psi(x,t)$ is continuous across the boundary. Hence the particle velocity $\frac{\partial \psi}{\partial t}$ is also continuous.

(2) The excess pressure $(p = \Delta p = -E\frac{\partial \psi}{\partial x} = -\gamma p_o \frac{\partial \psi}{\partial x})$ is continuous across the boundary [see article 13.17 Eq. (*ii*) replacing ψ by y]. Here $\gamma = C_p/C_v$ and p_o is the equilibrium pressure Thus, the two boundary conditions to be satisfied are (at $x = 0$)

$$\psi_i(x,t) + \psi_r(x,t) = \psi_t(x,t) \qquad \text{... (vi)}$$

and
$$p_i + p_r = p_t$$

or
$$+\frac{\partial \psi_i}{\partial x} - \frac{\partial \psi_r}{\partial x} = \frac{\partial \psi_t}{\partial x}$$

The two boundary conditions give

$$A_i + A_r = A_t$$

and
$$k_l(A_r - A_i) = k_2 A_t \qquad \text{...(vii)}$$

But
$$k_1 = \frac{\omega}{v_1} = \frac{\omega}{\rho_1 v_1^2}\rho_1 v_1 = \frac{\omega Z_1}{\gamma p_0} \qquad (\because \text{Eq. } (i) \text{ and } v_1 = \sqrt{\frac{\gamma p_o}{\rho_1}})$$

and
$$k_2 = \frac{\omega}{\gamma P_o} Z_2$$

Therefore, we have

$$Z_1(A_r - A_t) = Z_2 A_t \qquad \text{... (viii)}$$

Equations (*vii*) and (*viii*) gives,

$$r_{12} = \frac{A_r}{A_i} = \frac{Z_1 - Z_2}{Z_1 + Z_2} \qquad \text{... (17.1)}$$

$$t_{12} = \frac{A_t}{A_i} = \frac{2Z_1}{Z_1 + Z_2} \qquad \text{... (17.2)}$$

These equations give the reflection coefficient (r_{12}) and transmitted coefficient (t_{12}).

Case 1: It is clear from Eq. 17.1 that if $Z_1 > Z_2$; A_r/A_i is positive indicating that the incident and reflected displacements are in phase.

Case 2: If $Z_1 < Z_2$; A_r/A_i is negative, showing that the reflected wave undergoes a phase change of π with respect to the incident wave.

Case 3: From Eq. 17.2, it is also clear that A_t/A_i remains positive independent of whether Z_1 is less or more than Z_2. This means that the transmitted wave does not undergo any phase change.

At a rigid wall where $Z_2 = \infty$ (i.e. Infinity), $A_r = -A_i$, showing that the wave is completely reflected (no transmission at all).

17.3 REFLECTION AND TRANSMISSION OF SOUND ENERGY

The intensity of a sound wave of amplitude A and frequency n travelling with a speed V in a medium of density ρ is given by [Article 13.19, Eq. (*i*)]

$$I = 2\pi^2 n^2 A^2 \rho V$$
$$= 2\pi^2 V^2 A^2 Z$$

where Z is the characteristic acoustic impedance offered by the medium. The intensity coefficients of reflection and transmission are, therefore, given by

$$\text{Reflection coefficient} = \frac{I_r}{I_i} = \frac{2\pi^2 V^2 A_r^2 Z_1}{2\pi^2 V^2 A_i^2 Z_1} = \frac{A_r^2}{A_i^2} = \left(\frac{Z_1 - Z_2}{Z_1 + Z_2}\right)^2 \quad \text{... (17.3)}$$

$$\text{Transmission coefficient} = \frac{I_t}{I_i} = \frac{2\pi^2 v^2 A_t^2 Z_2}{2\pi^2 v^2 A_i^2 Z_1} = \frac{Z_2 A_t^2}{Z_1 A_i^2} = \frac{4Z_1 Z_2}{(Z_1 + Z_2)^2} \quad \text{... (17.4)}$$

Note that,
$$\frac{I_r}{I_i} + \frac{I_t}{I_i} = \frac{(Z_1 - Z_2)^2}{(Z_1 + Z_2)} + \frac{4Z_1 Z_2}{(Z_1 + Z_2)^2} = 1 \quad \text{... (17.5)}$$

or,
$$I_i = I_r + I_t \quad \text{... (17.6)}$$

This by itself proves that the energy is conserved

Experiments show that there is almost total reflection of sound energy at the air-water interface, where as, in the case of steel-water interface, the reflection coefficient is about 0.85 or 85%. Thus , only about 15% of sound energy in transmitted at a steel-water interface. This badly limits the transmission and detection devices used in submarines using ultrasonic waves.

Example 17.1 *A plane sound wave in air of density 1.29 kg m^{-3} falls on a water surface at normal incidence. The speed of sound in air is 334 ms^{-1} and in water the speed of sound is 1480 ms^{-1}. (a) what is the ratio of the amplitude of sound wave that enters water to that of the incident wave? (b) What fraction of the incident energy flux enters the water?*

Solution. The characteristic impedances of air and water respectively are

$$Z_1 = \rho_1 v_1$$
$$Z_2 = \rho_2 v_2$$
$$\therefore \quad \frac{Z_1}{Z_2} = \frac{\rho_1 v_1}{\rho_2 v_2} = \frac{1.29 \times 334}{1000 \times 1480} = 0.000291$$

(*a*)
$$\frac{A_t}{A_t} = \frac{2Z_1}{Z_1 + Z_2} = \frac{2Z_1 / Z_2}{Z_1 / Z_2 + 1}$$

Substituting the value of Z_1 / Z_2 we get

$$\frac{A_t}{A_i} = 5.82 \times 10^{-4}$$

(*b*) The fraction of the incident energy flux entering water is

$$\frac{I_t}{I_i} = \frac{4Z_1 Z_2}{(Z_1 + Z_2)^2} = \frac{4Z_1 / Z_2}{(Z_1 / Z_2 + 1)^2} = 1.16 \times 10^{-3}$$

Example 17.2 *Sound waves are incident normally on water-steel interface. If the characteristic impedances of water and steel are 1.43 × 10^6 kg m^{-2} s^{-1} and 3.90 × 10^7 kg m^{-2} s^{-1} respectively, what percentage of the incident energy is reflected at the interface?*

Solution.

$$Z_1 = 1.43 \times 10^6 \text{ kg m}^{-2} \text{ s}^{-1}$$
$$Z_2 = 3.90 \times 10^7 \text{ kg m}^{-2} \text{ s}^{-1}$$

$$\text{Energy reflection coefficient} = \left(\frac{Z_1 - Z_2}{Z_1 + Z_2}\right)^2 \simeq 0.86$$

Thus 86% of the incident energy is reflected at the interface.

17.4 IMPEDANCE MATCHING

When a wave meets a boundary between two media having different impedance values, a part of it is reflected and a part transmitted. The reflection energy coefficient is given by

$$\frac{\text{Reflected enengy}}{\text{Incident energy}} = \left(\frac{Z_1 - Z_2}{Z_1 + Z_2}\right)^2 \quad \text{... (17.7)}$$

If $Z_1 = Z_2$, the reflected energy = 0

The transmitted coefficient of energy is given by

$$\frac{\text{Transmitted energy}}{\text{Incident energy}} = \frac{4Z_1 Z_2}{(Z_1 + Z_2)^2} \quad \text{... (17.8)}$$

If $Z_1 = Z_2$, the transmitted coefficient of energy = 1. In such a case

Transmitted energy = Incident energy

This shown that when $Z_1 = Z_2$, no energy is reflected and all energy is transmitted.

In such a case the impedances are said to be perfectly matched.

This can be achieved by supressing the refrection .

Reflections are often undesirable, Reflections are often undesirable and need to be suppressed in many situations. Suppose we wish to transmit energy from one medium into another by means of tranvelling waves. When the wave reaches the boundary between the two media, it is partly reflected. Thus some energy is lost by reflection. If somehow the reflection could be prevented or suppressed, the energy would have been transmitted in the second medium without any loss. Look at the following examples.

(1) Suppose we wish to study some characteristics of a travelling wave on a string. The string must end somewhere. At the free ends reflections will occur giving rise to a standing wave pattern superposed on travelling waves on the string, thus altering the travelling wave pattern.

(2) In a loudspeaker the sound energy has to be transferred from air in the loudspeaker to air in a room or hall without generating reflections. The reflections are undesiruble because they represent a loss and hence decrease the power output of an amplifier.

(3) Skindivers have a breathing apparatus and have their faces covered with masks made of glass. Suppose a skindiver wishes to talk to another skindiver under water, the sound of a diver is reflected at the glass of the mask and very little sound is transmitted in water. The reason is that the characteristic impedances of air, glass, and water are very different. If the impedances were equal, no reflections would occur and the divers could easily talk to each other.

(4) Long distance cables carry electrical energy from a power house to distant places. The cables have joints where reflections occur. The wastage of energy due to reflections must be avoided or minimised so that the power transfer from the generator is maximum.

(5) Suppose we wish to send travelling waves of visible light from air into a lens or a slab without generating a reflection. The reflections are undesirable because they decrease the intensity of the transmitted beam and also cause glare in some other part of the apparatus.

Transferring energy by travelling waves from one medium to another without reflection is called *impedance matching.* We have seen that if a wave of any kind (transverse wave on a string, sound wave in a medium, voltage and current wave in a transmission line and electromagnetic wave in a medium) travelling in medium 1 of characteristic impedance Z_1 meets the boundary of medium 2 of characteristic impedance Z_2, the amplitude reflection and transmission coefficients are given by

$$r_{12} = \frac{A_r}{A_i} = \frac{Z_1 - Z_2}{Z_1 + Z_2} \quad \text{... (17.9)}$$

$$t_{12} = \frac{A_t}{A_i} = \frac{2Z_1}{Z_1 + Z_2} \quad \text{.. (17.10)}$$

On the other hand, if a wave travelling in medium 2 meets the boundary of medium 1, these coefficients become

$$r_{21} = \frac{A_r}{A_i} = \frac{Z_2 - Z_1}{Z_1 + Z_2} \quad \text{... (17.11)}$$

$$t_{21} = \frac{A_t}{A_i} = \frac{2Z_2}{Z_1 + Z_2} \quad \text{... (17.12)}$$

It is clear from these relations that if $Z_1 = Z_2$, no reflection takes place and the wave is completely transmitted. The impedances of the two media are said to be *matched.*

17.5 IMPEDANCE OF THE COUPLING ELEMENT

Two media of characteristic impedance Z_1 and Z_2 ($Z_1 \neq Z_2$) coupled through an element of impedance Z and length l is shown in Fig 17.2. The boundaries of the coupling element are located at $x = 0$ and $x = l$. The wave enters the coupling medium at $x = 0$ and leaves it at $x = l$. The value of wave numbers for the three media are k_1, k and k_2 respectively.

The incident reflected and transmitted waves in all the three media are as shown in the figure.

For maximum transfer of power the two boundary conditions that must be satisfied are

(*i*) The displacement y is continuous at the interfaces $x = 0$ and $x = l$ and

(*ii*) The transverse force $T\frac{dy}{dx}$ is also continuous at these two points.

At $x = 0$. According to first condition at $x = 0$

$$y_{il} + y_{rl} = y_i + y_r$$

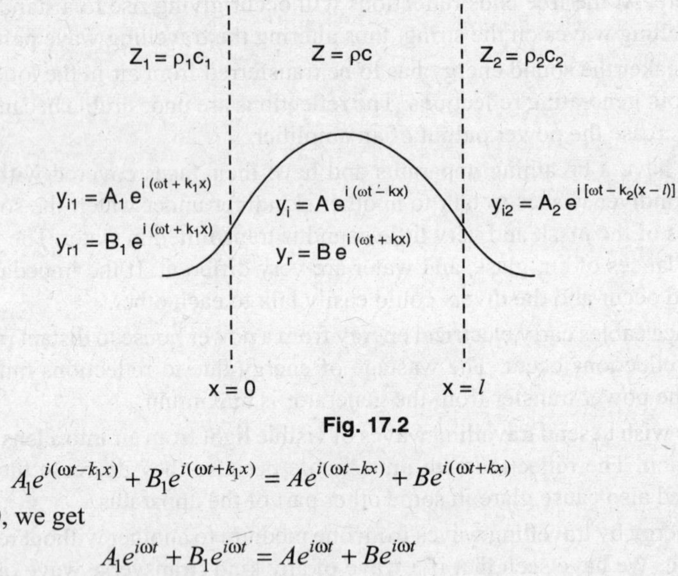

Fig. 17.2

or $$A_1 e^{i(\omega t - k_1 x)} + B_1 e^{i(\omega t + k_1 x)} = A e^{i(\omega t - kx)} + B e^{i(\omega t + kx)}$$

Putting $x = 0$, we get

$$A_1 e^{i\omega t} + B_1 e^{i\omega t} = A e^{i\omega t} + B e^{i\omega t}$$

or $$A_1 + B_1 = A + B \quad \text{... (i)}$$

According to the second condition at $x = 0$

$$T[-ik_1 A_1 + ik_1 B_1] = T[-ikA + ikB]$$

or $$-i[A_1 Tk_1 - B_1 Tk_1] = i[ATk - BTk]$$

Now $$Tk = T\frac{2\pi}{\lambda} = T\frac{2\pi\nu}{c} = \frac{T}{c}\omega = Z\omega \qquad \text{since } \frac{T}{c} = Z$$

$\therefore$ Substituting $Tk_1 = Z_1\omega$ and $Tk = Z\omega$ in the above equation we get

$$Z_1\omega(A_1 - B_1) = Z\omega(A - B)$$

or $$Z_1 (A_1 - B_1) = Z (A - B) \qquad \text{... (ii)}$$

At $x = l$ At the boundary of the coupling medium and the medium into which the wave is finally transmitted, we have, according to the first condition.

$$y_i + y_r = y_{i2}$$

or $$A\, e^{i(\omega t - kx)} + B\, e^{i\,(\omega t + kx)} = A_2\, e^{i\,[\omega t - k_2 (x - l)]}$$

At $x = l$, we have

$$Ae^{i(\omega t - kl)} + Be^{i(\omega t + kl)} = A_2 e^{i\omega t}$$

or $$e^{i\omega t} [Ae^{ikl} + B\, e^{ikl}] = A_2 e^{i\omega t}$$

or $$Ae^{-ikl} + Be^{+ikl} = A_2 \qquad \text{... (iii)}$$

Similarly according to second condition at $x = l$

$$-iATke^{i(\omega t - kl)} + i\,BTke^{i(\omega t + kl)} = -i\, A_2 T\, k_2 e^{i\omega t}$$

or $$-ATke^{-ikl} + BTke^{++ikl} = -A_2 Tk_2$$

Substituting $Tk = Z\omega$ and $Tk_2 = Z_2\omega$, we get

$$\omega[AZe^{-ikl} - BZe^{+ikl}] = \omega\, A_2 Z_2$$

or $$Z[Ae^{-ikl} - Be^{+ikl}] = A_2 Z_2 \qquad \text{... (iv)}$$

To find the value of $\dfrac{\text{transmitted energy}}{\text{Incident energy}} = \dfrac{Z_2 A_2^2}{Z_1 A_1^2}$, we first find the value of $\dfrac{A_2}{A_1}$

From Eq (*i*) $$B_1 = A + B - A_1$$

Substituting in Eq. (*ii*) we get

$$Z_1 (A_1 - A - B + A_1) = Z (A - B)$$

or $$2\, Z_1\, A_1 = Z_1\, (A + B) + Z\, (A - B)$$

or $$2A_1 = (A + B) + \frac{Z}{Z_1}(A - B)$$

$$= A\left[1 + \frac{Z}{Z_1}\right] + B\left[1 - \frac{Z}{Z_1}\right]$$

or $$A_1 = \frac{1}{2} A\left[1 + \frac{Z}{Z_1}\right] + \frac{1}{2} B\left[1 - \frac{Z}{Z_1}\right] \qquad \text{... (v)}$$

To find the value of A and B in terms of A_2 we have, from Eq. (*iv*)

$$Ae^{-ikl} - Be^{+\,ikl} = \frac{Z_2}{Z} A_2 \qquad \text{... (vi)}$$

Adding Eqs. (*iii*) and (*vi*), we get

$$2\, Ae^{-ikl} = A_2\left(1 + \frac{Z_2}{Z}\right)$$

or $$A = \frac{A_2}{2}\left(1 + \frac{Z_2}{Z}\right)e^{+ikl}$$

Subtracting (*vi*) from (*iii*), we get

$$2\, Be^{+ikl} = A_2\left(1 - \frac{Z_2}{Z}\right)$$

or $$B = \frac{A_2}{2}\left(1 - \frac{Z_2}{Z}\right)e^{-ikl}$$

Substituting these values of A and B in eq. (v), we get

$$A_1 = \frac{A_2}{4}\left(1+\frac{Z}{Z_1}\right)\left(1+\frac{Z_2}{Z}\right)e^{+ikl} + \frac{A_2}{4}\left(1-\frac{Z}{Z_1}\right)\left(1-\frac{Z_2}{Z}\right)e^{-ikl}$$

or
$$\frac{A_1}{A_2} = \frac{1}{4}\left(1+\frac{Z}{Z_1}+\frac{Z_2}{Z}+\frac{Z_2}{Z_1}\right)e^{+ikl} + \frac{1}{4}\left(1-\frac{Z}{Z_1}-\frac{Z_2}{Z}+\frac{Z_2}{Z_1}\right)E^{-ikl}$$

$$= \frac{1}{2}\left(1+\frac{Z_2}{Z_1}\right)\frac{e^{ikl}+e^{-ikl}}{2} + \frac{1}{2}\left(\frac{Z}{Z_1}+\frac{Z_2}{Z}\right)\frac{e^{ikl}-e^{-ikl}}{2}$$

$$= \frac{1}{2}\left(1+\frac{Z_2}{Z_1}\right)\cos kl + i\frac{1}{2}\left(\frac{Z}{Z_1}+\frac{Z_2}{2}\right)\sin kl$$

$$\therefore \quad \frac{\text{Transmitted energy}}{\text{Incident energy}} = \frac{E_T}{E_i} = \frac{Z_2(A_2)^2}{Z_1(A_1)^2} = \frac{Z_2}{Z_1}\left(\frac{A_2}{A_1}\right)\left(\frac{A_2}{A_1}\right)$$

$$= \frac{Z_2}{Z_1}\frac{4}{\left(1+\frac{Z_2}{Z_1}\right)^2\cos^2 kl + \left(\frac{Z}{Z_1}+\frac{Z_2}{Z}\right)^2\sin^2 kl}$$

If $l = \frac{\lambda}{4}$, then $\cos kl = \cos\frac{k\lambda}{4} = \cos\frac{2\pi}{\lambda}.\frac{\lambda}{4} = \cos\frac{\pi}{2} = 0$ and $\sin kl = \sin\frac{\pi}{2} = 1$

$$\therefore \quad \frac{E_T}{E_i} = \frac{Z_2}{Z_1}\frac{4}{\left(\frac{Z}{Z_1}+\frac{Z_2}{Z}\right)^2}$$

For maximum transfer of power $\frac{E_T}{E_i} = 1$

$$\therefore \quad \frac{Z_2}{Z_1}\times\frac{4}{\left(\frac{Z}{Z_1}+\frac{Z_2}{Z}\right)^2} = 1$$

or
$$4\frac{Z_2}{Z_1} = \left(\frac{Z}{Z_1}+\frac{Z_2}{Z}\right)^2 = \left(\frac{Z}{Z_1}\right)^2 + \left(\frac{Z_2}{Z}\right)^2 + 2\frac{Z_2}{Z_1}$$

or
$$\left(\frac{Z}{Z_1}\right)^2 + \left(\frac{Z_2}{Z}\right)^2 - 2\frac{Z_2}{Z_1} = \left(\frac{Z}{Z_1}-\frac{Z_2}{Z}\right)^2 = 1$$

or
$$\frac{Z}{Z_1} - \frac{Z_2}{Z} = 1 \qquad \text{... (17.13)}$$

or
$$Z^2 = Z_1 Z_2 \quad \text{or} \quad Z = \sqrt{Z_1 Z_2} \qquad \text{... (17.14)}$$

Thus perfect matching is achieved and all the incident energy is transmitted when the following two conditions are satisfied.

1. The impedance of the coupling element is given by $Z = \sqrt{Z_1 Z_2}$ *i.e.* it is the geometric mean of the impedances of the media to be matched.
2. The coupling element has a length $l = \frac{\lambda}{4}$ where λ is the wavelength of the incident wave in the medium of the coupling element.

17.6 IMPEDANCE MATCHING FOR TRANSDUCERS

A transducer is a device for converting a physical quantity viz. sound, temperature, pressure, light. etc into its equivalent electrical quantity, say electric current and vice-versa. Microphones and Loud speakers are employed for this purpose in acoustics.

Impedance matching is a very important practical problem in many branches of optics and engineering where transfer of energy from one medium into another is desired. Electrical energy is carried over long distance cables which have joints. The impedances across the joints must be accurately matched to avoid wastage of energy by reflection. A generator delivers power across a load. For maximum power transfer the impedance of the load must match that of the generator. For the same reason the impedance of a loudspeaker (sound transducer) has to be matched with the impedance of the power output of an amplifier.

However, impedance matching problem exists when transferring sound energy from one medium to another. If the acoustic impedance of the two media are very different, then most of the sound energy will be reflected or absorbed; rather than transferred across the border. Most of the Loudspeaker systems themselves contain impedance matching mechanism.

Acoustic impedance matching (or the lack of it) affects the operation of a megaphone, an echo, reverberation and sound proofing.

The gel used in medical ultrasonography helps to transfer acoustic energy from the transducer to the body and back. Without the gel, the "impedance mismatch" **in** the transducer-to-air and the air-to-body discontinuity reflects almost all the energy, leaving very little to go into the body.

Horns are used like transformers, matching the impedance of the transducer to the impedance of the air. This principle is used in both horn loudspeakers and musical instruments.

17.7 DIFFRACTION OF SOUND

If the diamension of the obstacle is comparable with the wavelength of the wave, the waves bend round the corner and encroaches into shadow appreciably. This phenomenon is called **diffraction.**

It is observed that sound bends appreciably round any obstacle, while light appears to be linear for that. In case of obstacle such as the edge of a razor blade or at the end of a needle, light bends round the appropriate obstacle. It is due to the fact that wavelength of light is very short while that of sound is more than million times larger.

Similar to diffraction in light, diffraction takes place in sound also, provided the size of diffracting obstacle is comparable to the wavelength of sound. **A region behind the obstacle placed before the sound source where sound is absent is known as acoustic shadow,** Similar to optical analogy. This bending of waves round the obstacle and their encroachment within the geometrical shadow is known as **diffraction** of sound wave.

The acoustical shadow is not so sharp as the optical shadow. In order to produce sharp acoustic shadow, the wavelength of sound wave should be short, that is, its frequency should be high and the obstacle be of moderate size.

The sound wave can be diffracted by suitable acoustic grating similar to optical grating. Grating for this purpose has been designed and constructed by using glass rods spaced at about one centimetre apart. Source of sound generated by spark discharge of a condenser, which produces sound wave of a few millimetre in length.

17.8 ACOUSTIC GRATING

A quartz crystal Q placed between two metal plates is set into vibration by connecting audio frequency oscillator. The frequency of AFO is adjusted such that the crystal vibrates with its natural frequency and resonance takes place.

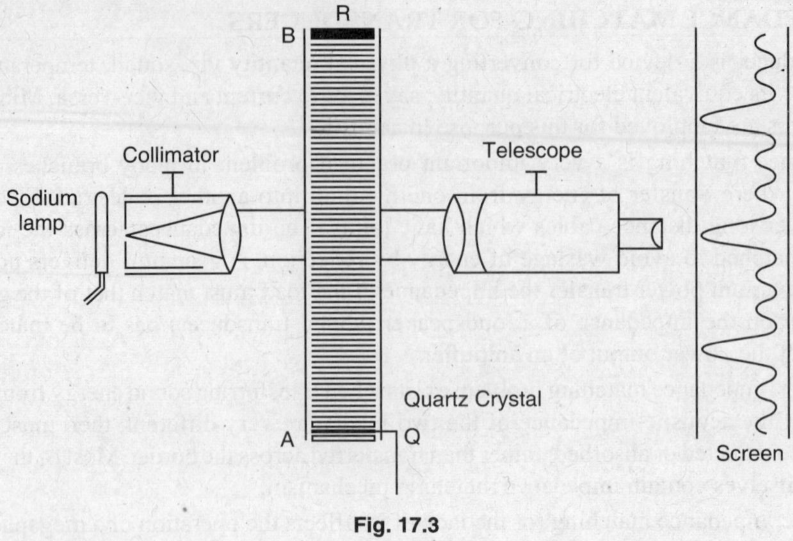

Fig. 17.3

Due to longitudinal compression and expansion of the crystal, ultra sonic waves are produced. These waves are reflected by reflector R (fig. 17.3) The direct and reflected waves superpose on each other and produce stationary waves in the medium.

17.9 SONAR

Although some animals (dolphins and bats) have used sound for communication and object detection for millions of years, use by humans in the water is initially recorded by *Leonardo Da Vinci in* 1490: a tube inserted into the water was said to be used to detect vessels by placing an ear to the tube. Sonar is an underwater echo ranging device seems to have been prompted by the Tinanic disaster of 1912.

Sonar (acronym for Sound Navigation and Ranging) is a technique that uses sound propagation (usually underwater, as in submarine navigation) to navigate, communicate with or detect other vessels. The use of sound to 'echo locate' underwater in the same way as bats use sound for aerial navigation.

Principle of Sonar system: ***When ultrasonic waves are transmitted through water, it is reflected by the object in the water and will produce an echo signal. By noting the time interval between the generation of the pulse and its reception, the exact location of the object from the source*** **of signal generator can be determined.**

Sonar may be used as a means of acoustic location and of measurement of the echo characteristics of "targets' in the water. The term *sonar* is also used for the equipment used to generate and receive the sound. The acoustic frequencies used in sonar systems vary from very low (infrasonic) to extremely high (ultrasonic). The study of underwater sound is known as underwater acoustics or **hydro-acoustics.**

British ASDIC research technology (first developed by the British in 1920) was transferred for free to the United States and then Americans began to use the term SONAR for their systems in World War II, in coined as the equivalent of RADAR.

Sonar used a sound transmitter and a receiver as shown in Fig.17.4. When the two are in the same place it is monostatic operation. When the transmitter and receiver are separated it is bistatic operation.When more transmitters (or more receivers) are used, again spatially separated, it is multistatic operation. Most sonars are used monastically with the same array often being used for transmission and reception.

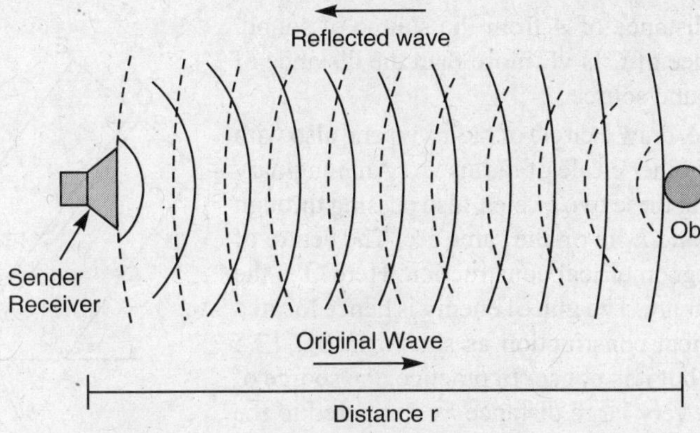

Fig. 17.4

Sonar creates a pulse of sound, often called a "ping' and then listens for reflections (echo) of the pulse. This pulse of sound is generally created electronically using a sonar projector consisting of a signal generator, power amplifier and electro-acoustic transducer. A beam former is usually employed to concentrate the acoustic power into a beam, which may be swept to cover the required search angles. Generally, the electro-acoustic transducers are designed and optimised to achieve maximum efficiency over the widest bandwidth, in order to optimise performance of the overall system, Occasionally, the acoustic pulse may be created by other means e.g (1) chemically using explosive, or (2) airguns or (3) plasma sound sources.

To measure the distance to an object, the time from transmission of a pulse to reception is measured and converted into a range by knowing the speed of sound, To measure the bearing, several hydrophones are used, and the set measures the relative arrival time to each, or with an array of hydrophones, by measuring the relative amplitude in beams formed through a process called beam-forming. Use of an array reduces the spatial response so that to provide wide cover multibean systems are used. Further process may be carried out to classify the target and localise it, as well as measuring its velocity.

17.10 SOUND RANGING

The position of enemy gun can be located by observing the time of the reception of sound at three different positions. This technique is known as **Sound ranging.**

Suppose a gun is fired from some unknown position *O*. Let *A, B* and *C* be the three different positions of observation as shown in Fig. 17.5. Microphones are fitted at each position.The sound received by each microphone is converted into an electrical pulse and fed to time-base of a cathode ray oscilloscope (C.R.O) at some common control position. As the distances of *A, B* and *C* are different from the source of sound, sound will be received at different instants.

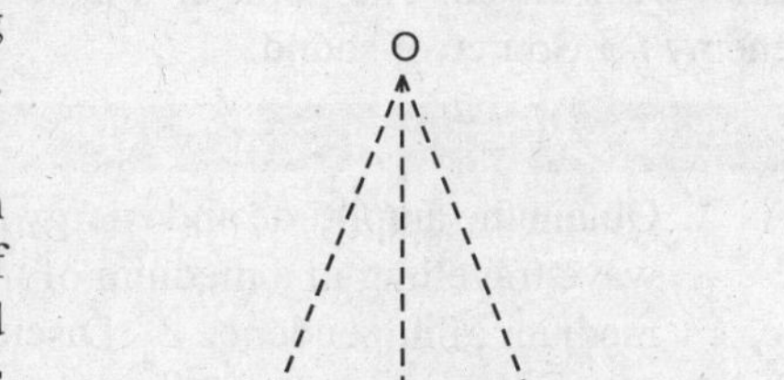

Fig. 17.5

From the signals obtained on the C.R.O., screen, It is possible to find the difference in time intervals between various signals. Let these time intervals be (*i*) t_1 between *A* and *B*,and (*ii*) t_2 between *A* and C.

Let v be the velocity of sound in air. It means sound reaches B, t_1 seconds later than *A*. The distance of *B* is

vt_1 more than the distance of A from the source of sound. Similarly the distance of C is vt_2 more than the distance of A from the same sound source.

With B as centre, draw a circle of radius vt_1 and also with C as centre, draw another circle of radius vt_2. An imaginary arc is drawn touching these two circles, also passing through A. The points D, A and E lie on the same arc. The centre of the arc is found by geometrical construction. Here O is the centre of the circular arc. The gun of enemy is hence located at O. The geometrical construction as shown in fig. 17.5 seems to be simple but it is not so. In practice, the source of sound *i.e.* gun is at very large distance as compared to the distances between A, B and C.

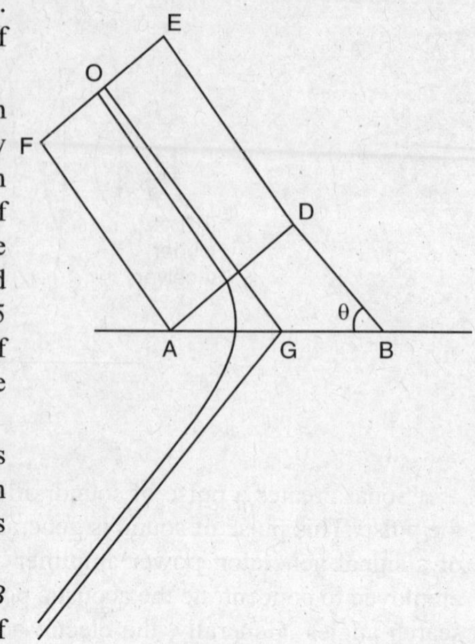

Fig. 17.6

Analytical treatment. Let A and B be the two positions at which observations are made (fig 17.6). The enemy's gun is located at O. The sound from O reaches B, after t seconds later than A. Therefore, distance $OB - OA = vt$.

The point O lies on the parabola drawn with A and B as its focus. O is at a large distance from G- (the centre of AB). EOF and AD are perpendiculars on OG, then O lies on the asymtote OG such that

$$OB - OA = EB - FA = BD = vt$$

Taking $AB = l$ and $\angle ABD = \angle\theta = \angle OGA$

$$\cos\theta = \frac{BD}{AB} = \frac{vt}{l}$$

or $$\theta = \cos^{-1}\left(\frac{vt}{l}\right) \quad \text{... (17.15)}$$

Knowing v,t and l, the angle θ can be calculated Here θ gives the direction of source of sound with respect to the positions A and B.

Using the same procedure, the angle θ corresponding to other positions say B and C, C and A can be calculated. The point of intersection of these three directions, gives the position of gun of enemy *i.e.* Source of sound.

EXERCISE

1. Obtain the amplitude and energy reflection and transmission coefficients when a plane sound wave travelling in a medium of impedance Z_1 is incident normally on a plane boundary of a medium of impendence Z_2. Discuss the phase change when $Z_1 > Z_2$ and $Z_1 < Z_2$.
2. Show that energy is conserved in reflection and transmission of sound energy
3. What is impedance matching? Why is it necessary?
4. State the conditions for perfect impedance matching between two media. What are the uses of impedance matching? Give application for anti-reflection films. ***(P.U. 2002 G.N.D.U.2000)***
5. What does the term 'the matching of impedance' imply? Show that for perfect matching the impedance of the coupling element is the geometric mean of the impedances of the media to be matched when the impedances of the two media are not equal. ***(G.N.D.U. 2006; P.U. 2003)***
6. Enunciate the practical importance of 'impedance matching'.
7. Discuss impedance matching for transducers.
8. 'Reflations are often undesirable' explain giving few examples.
9. Explain the use of gel in ultrasonography.

10. What should be the range of wavelength of sound to produce diffraction phenomenon at the obstacle?

11. What is acoustic grating? Explain how velocity of ultrasonic waves is determined?

12. What is 'acoustic' shadow'

13. What is SONAR? Give its applications (***Nagpur 2008***)

14. Explain the principle involved in SONAR system

15. What is sound ranging? Explain clearly how sound ranging is used in locating the position of source of sound.

MULTIPLE CHOICE QUESTIONS

17.1 A sound is completely reflected from a wall when the impedances are

(*a*) $Z_1 < Z_2$ (*b*) $Z_1 > Z_2$ (*c*) $Z_2 = \infty$ (*d*) $Z_2 = 0$

17.2 The sound energy is conserved when

(*a*) $I_i = I_r - I_t$ (*b*) $I_i = I_r + I_t$ (*c*) $I_i^2 = I_r^2 - I_t^2$ (*d*) $I_i^2 = I_r^2 + I_t^2$

17.3 Impedances are said to be completely matched when transmitted coefficient of energy is equal to

(*a*) 0 (*b*) <1 (*c*) > 1 (*d*) 1

17.4 Gel is used in medical ultrasonography because it

(*a*) reduces absorption of sound waves

(*b*) increases absorption of sound waves

(*c*) avoids impedance mismatch at air to body discontinuity

(*d*) increases reflection of sound waves

17.5 Diffraction of sound occurs when its wavelength is

(*a*) not comparable with the size of obstacle

(*b*) comparable with the size of the obstacle

(*c*) for any size obstacle

(*d*) none of the above

17.6 'Acoustic shadow' is a region where there

(*a*) sound is present

(*b*) sound is absent

(*c*) partial sound is present

ANSWERS

17.1 (*c*)	17.2 (*b*)	17.3 (*d*)	17.4 (*c*)	17.5 (*b*)	17.6 (*b*)

APPLIED ACOUSTICS

18
Chapter

18.1 TRANSDUCERS

A transducer is a device for converting one form of energy into another. In an acoustical context, this usually means converting sound energy into electrical energy (or vice versa). For nearly all acoustic applications, same type of acoustic transducer is necessary. Acoustic transducers include loudspeakers, microphones, hydrophones and sonar projectors. These devices convert an electric signal to or from a sound pressure wave. The most widely used transduction principles are electromagnetism, electrostatics and piezo-electricity.

The transducers in most common loudspeakers (e.g. woofers and tweeters) are electromagnetic devices that generate waves using a suspended diaphragm driven by an electromagnetic voice coil, sending off pressure waves. Condenser microphones employ electrostatics. As the sound wave strikes the microphone's diaphragm, it moves back and forth, and induces a voltage change. The ultrasonic systems used in medical ultrasonography employ piezo-electric transducers. These are made from special ceramics in which elastic vibrations and electrical fields are interlinked through a property of the material itself. The acoustics have been grouped in three domains: physical acoustics, biological acoustics include aeroacoustics, general linear acoustics, non-linear acoustics and structural acoustics. Biological acoustics and acoustical engineering. Physical acoustics include bioacoustics, musical acoustics, physiological acoustics, psycho-acoustics and speech acoustics. Acoustical engineering includes acoustic measurements and instrumentation, acoustic signal processing, architectural acoustics, environmental acoustics, transduction, ultrasonics and room acoustics.

18.2 CHARACTERISTICS OF SENSORS AND TRANSDUCERS

Sensor or transducer is a device which converts one type of energy to another, usually electric. They are used for various purposes or information transfers. Generally speaking sensor or transducer is a device that converts a signal from one form to another. The characteristics of a sensor or transducer describes its performance.

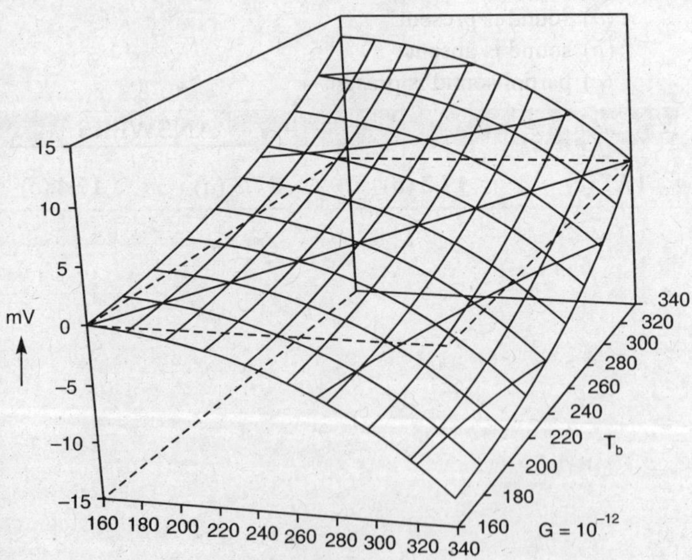

Fig. 18.1

One of the characteristics, that describes the functional relationship between physical input and electrical output i.e., **transfer function.** Transfer function is a graph that shows relationship between the input and output signal. Depending upon the details of this characteristic, it may be a complete description of sensor characteristics. Transfer function may be used as a **calibration curve**. For instance let us take an example of infrared temperature sensor. Its transfer function can be represented as 3D function of a thermal radiation. It is because sensor is affected by two temperatures: T_b, the absolute temperature of the object and T_s, the absolute temperature of transducer sensor surface. Then output voltage $V = G\,(T_b^4, T_s^4)$, where G is constant, then we can represent transfer function in SD space as shown in Fig. 18.1.

Second characteristic is **Dynamic range** of sensor. This term defines the input signal that can be converted to electrical signal. Signals outside the range may cause large inaccuracy. Dynamic range of transducer is typically defined by manufacturer with other characteristics as well.

Third actual parameter is **sensitivity** of a transducer. This describes the relationship between input and output signals. Sensitivity is defined as the ratio between change in electrical signal to a change of input physical quantity (signal). As mentioned above, temperature sensitivity of the sensor would be described in units VOHS/kelvin. High sensitivity would be when small temperature change would result in a large voltage change.

Other characteristics that are used are:

(*i*) **Accuracy** as the largest expected error between actual and ideal output signals. Unit is in percentage (%), of full scale output (FSO), and

(*ii*) **Hysteresis** originates when sensors do not return to the same value during cyclic changes of input. Again unit is percentage (%), of full scale output (FSO). This is shown in Fig. 18.2.

(*iii*) **Linearity**, as a deviation from linear transfer function over dynamic range;

(*iv*) **Noise,** as output noise that adds signal at the output;

(*v*) **Resolution**, as minimal detectable signal fluctuations.

(*vi*) **Band width**, that defines a response time to a instant change of physical signal.

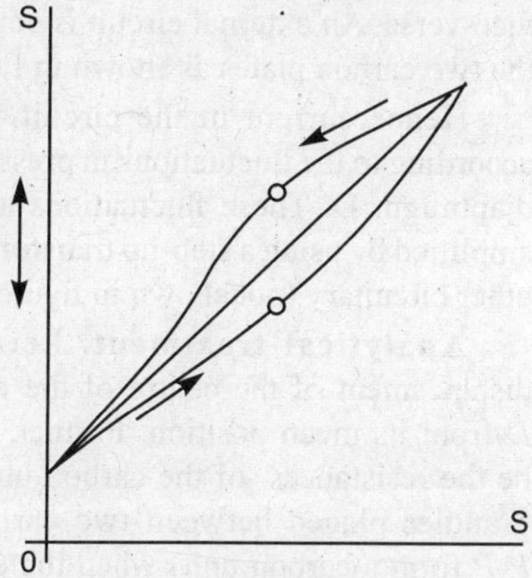

Fig. 18.2

18.3 RECORDING DEVICES OF SOUND

Microphone converts acoustical energy into equivalent electrical energy called as transducer. Hence, it is called electro-acoustic transducer. The transducers used in air are known as microphones, and those operating in water are hydrophones.

The microphones which are very widely used are classified into two groups: (1) Pressure microphones and (2) Velocity microphones (also known as pressure gradient microphones). Some commonly used pressure microphones are:

(*i*) Carbon microphones,

(*ii*) Condenser microphones,

(*iii*) Crystal (Piezo-electric) microphones,

(*iv*) Moving coil electro-dynamic microphones,

(*v*) Hot wire microphones,

The essential characteristics that measures the quality or performance of the microphones are:

(*a*) **Sensitivity:** The sensitivity of a microphone must be very high *i.e.*, open circuit voltage developed per unit sound pressure, at a given frequency, should be high.

(*b*) **Frequency-Response:** The frequency response should be uniform throughout the whole range (20Hz to 20KHz) of audio-frequency.

(*c*) **Noise-Free:** The output must be free from any self-noise, generally known as background noise.

18.4 CARBON MICROPHONE

This type of microphones are usually used in telephones. The operation of carbon microphone depends on the variation of resistance of carbon in contact.

Construction. Fine carbon granules are closely packed between two thin polished carbon plates *PP*. The first plate is connected to a boss *S* which is connected to the centre of thin metal diaphragm *D*. The pressure variation in air produced by sound wave causes the metal diaphragm to vibrate, forces the front carbon plate to vibrate. The pressure change on front carbon plate will produce the *change in its resistance* of carbon grains. It is found that when the pressure increases, the resistance of carbon grains decreases and vice-versa. An external circuit is connected to the two carbon plates as shown in Fig. 18.3.

Hence current in the circuit fluctuates according to the fluctuations in pressure on the diaphragm D. These fluctuations are further amplified by using a step-up transformer *T* and other circuitory (not shown in figure).

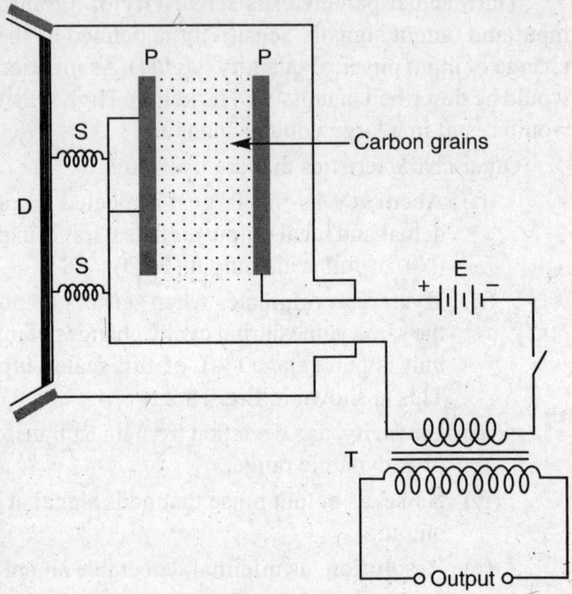

Fig. 18.3

Analytical treatment. Let *x* be the displacement of the centre of the diaphragm *D* from its mean position. Further, R_x and R_o be the resistances of the carbon unit (carbon granules placed between two carbon plates *P-P* form a carbon unit) when the diaphragm *D* is displaced through *x* and zero (*i.e.*, mean position) respectively, then

$$R_x = R_o + \rho_x \quad \text{... (1)}$$

where ρ is the resistance per unit displacement.

The amplitude of oscillation of the diaphragm is very small at the frequencies other than the fundamental resonant frequency. The motion of the diaphragm is *stiffness controlled.* Therefore,

The force exerted on the diaphragm will be

$$F = kx = AP_o \sin\omega t \quad \text{... (ii)}$$

where P_0 is pressure amplitude,

A is area of the diaphragm, and ω is angular frequency (ω = 2 π *n*, where *n* is the frequency) of the incident sound wave.

From Eq. (*ii*), we have

$$x = \frac{AP_o}{k} \sin \omega t \quad \text{... (iii)}$$

or,

$$x = x_0 \sin \omega t \quad \text{... (iv)}$$

where

$$x_0 = \frac{AP_o}{k},$$

Substituting in Eq. (*i*), we get

$$R_x = R_o + \rho x_o \sin\omega t$$

or

$$R_x = R_o + \frac{\rho AP_o}{k} \sin \omega t \quad \text{... (v)}$$

The equivalent fluctuating (*a.c.*) current is given by

$$I_x = \frac{E_o}{R_x} = \frac{E_o}{R_o + \dfrac{\rho A P_o}{k}\sin\omega t}$$

$$= \frac{E_o}{R_o + \rho x_o \sin\omega t} \qquad (\because \text{Eq. } iv)$$

$$= \frac{E_o}{R_o}\left(1 + \frac{\rho x_o}{R_o}\sin\omega t\right)^{-1} \qquad \text{... } (vi)$$

$$= \frac{E_o}{R_o}\left\{1 - \frac{\rho x_o}{R_o}\sin\omega t + \left(\frac{\rho x_o}{R_o}\right)^2 \sin^2\omega t \text{ ...}\right\} \qquad \text{... } (vii)$$

$$= \frac{E_o}{R_o}\left\{\left(1 + \frac{\rho^2 x_0^2}{2R_0^2}\right) - \frac{\rho x_0}{R_0}\sin\omega t - \frac{\rho^2 x_0^2}{2R_0^2}\cos 2\omega t \text{ ...}\right\}$$

The first harmonic of alternating current is

$$= -\frac{E_o \rho x_o}{R_o^2}\sin\omega t \qquad \text{... } (viii)$$

The quantity $\dfrac{E_o}{R_o}\left(1 + \dfrac{\rho^2 x_0^2}{2R_0^2}\right) \cong \dfrac{E_0}{R_0}$ as $\rho x_0 << E$ which represents steady current. The E.M.F. of the alternating current is

$$E_x = I_x R_o = -\frac{E_0 \rho x_0}{R_0}\sin\omega t \qquad \text{... (18.1)}$$

The amplitude of the voltage developed is given by

$$E_a = \frac{E_o \rho A P_o}{k R_o} \qquad \text{... (18.2)}$$

Hence, $$\frac{E_a}{P_o} = \frac{E_o \rho A}{k R_o} \qquad \text{... (18.3)}$$

The quantity E_o/ρ_o is a measure of *sensitivity* of the microphone and is known as *open circuit voltage response'*.

Frequency response: The frequency response curve is given in Fig. 18.4. The straight portion represents the region of uniform response. The response increases with the polarizing battery voltage E_o, area of the diaphragm A, and decreases with the increase of circuit resistance R_o and stiffness k.

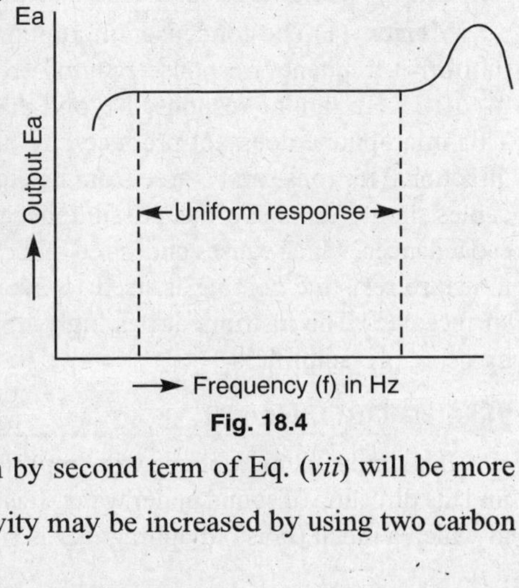

Fig. 18.4

Drawbacks. (*i*) The increase of E_o gives rise to *heating effect* which causes *interval noise.*

(*ii*) Sensitivity cannot be increased by increasing the ratio $\dfrac{\rho}{R_o}$ in which case the second harmonics (as given by second term of Eq. (*vii*) will be more pronounced to produce distortion). However, the sensitivity may be increased by using two carbon packs one on each side of the diaphragm. The diaphragm while moves the resistance of one increases and that of other decreases. In such type, the even harmonics generated in microphone cancel out. This type of microphone is known as *double-button push-pull type microphone*. The main disadvantage is its 'self-noise' generation.

18.5 CONDENSER MICROPHONE

Principle. The capacitance of a parallel-plate capacitor is given by

$$C_o = \frac{KA\epsilon_o}{d} \text{ (farad)} \quad \ldots (18.4)$$

where A = area of each plate (m^2)

d = distance of seperation between two plates (m)

ϵ_o = permittivity of free space = 9.85×10^{12} (F/m)

K = dielectric constant.

Since the capacitance is inversely proportional to the spacing of the parallel plates, any variation in d causes a corresponding variation in the capacitance. This principle is applied in the condenser microphone, the capacitive transducer as shown in Fig. 18.5.

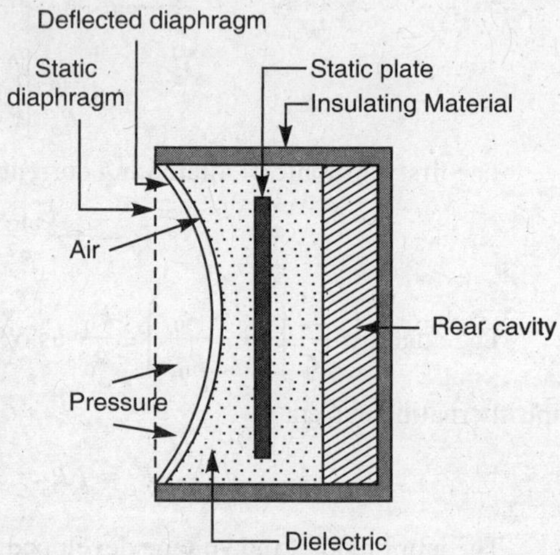

Fig. 18.5

Construction and Working. The construction of a condenser microphone is schematically shown in Fig. 18.5. When the sound waves strike on the diaphragm that functions as one plate of a simple capacitor, the varying air pressure changes the distance between the diaphragm and the static plate. The resultant change in capacitance could be measured with an a.c. bridge, but it is usually measured with an oscillator circuit. Usually this capacitor is connected parallel to the tank circuit of an oscillator. The change in capacitance, changes the frequency of the oscillator as given by

$$f_o = \frac{1}{2\pi\sqrt{L(C+C_o)}} \quad \ldots (18.5)$$

where C_o is the capacity of the condenser microphone in equilibrium state. Thus, the variation in frequency f_o will be a direct measure of the variation in C_o.

Merits. (1) The condenser microphone has excellent *i.e.* wide and flat (uniform) frequency response ranging from very low frequency (say $f \geq 0$) to 10 KHz and the response beyond that falls rapidly to a low value (2) This microphone does not produce any hissing sound or noise (3). It has directional response and is free from background noise. (4) The connection cables should be as short as possible because they contribute to the 'stray capacitances' which causes undesired reduction in its sensitivity, (4) A special moisture resistive coating is used to avoid leakage across the insulating surfaces. (5) The instrument has high input impedance, which is reduced by using pre-amplifiers.

18.6 HYDROPHONE

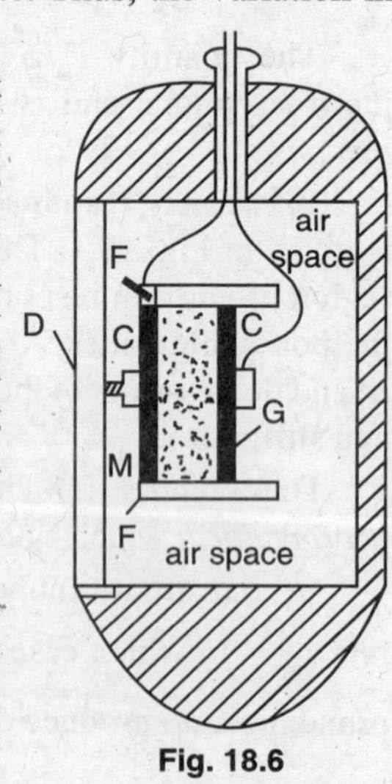

Fig. 18.6

It is similar to microphone with important changes in design. It is used to find the direction of sound under water. It is a massive lens-shaped metal disc to which a metal (brass) diaphragm D is rigidly fixed making it watertight.

There is a small carbon microphone *M*. It has two parallel carbon discs *CC*, supported in a small brass capsule, the other end is soldered to the brass body as shown in Fig. 18.6. A soft ring of iron surrounds the carbon discs and presses tightly. Two-third of the cavity is filled with carbon granules. The varying pressure of sound wave in water causes the diaphragm to vibrate which produces the change of resistance between carbon plates, which produces the change of polarising current. The hydrophone kept under water from a ship is connected to a battery of cells, transformer and telephone cables which are kept at the test centre.

18.7 CRYSTAL MICROPHONE

Principle. Crystal microphone is a electromechanical device based upon the principle of photo-electric effect. According to this effect, when crystals like quartz, Barium Titanate, Rochelle salt, tourmline etc. are stretched or compressed along certain axis known as mechanical axis then an electrical potential difference is developed along a perpendicular axis known as electrical axis. Quartz crystal is prefered as it produces large E.M.F. for the same stress. Moreover quartz is inexpensive and easily available and hence widely used as piezo-electric crystal.

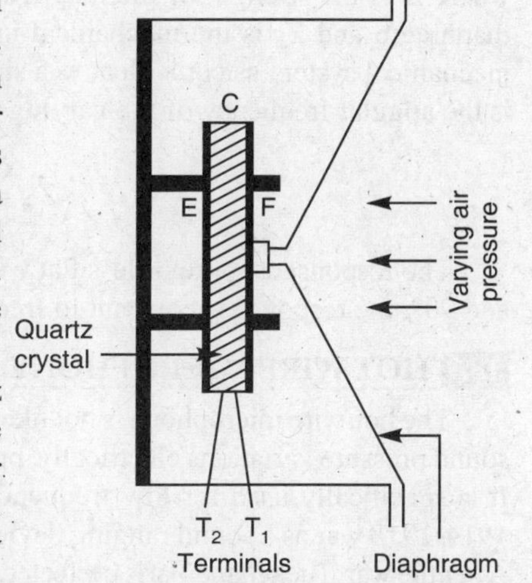

Fig. 18.7

Construction and Working. A specially cut quartz crystal *C* is fitted between two parallel plates *E* and *F*. One plate *E* is fixed while the diaphragm is attached to the other plate *F*. When sound vibrations strike, diaphragm vibrates. These vibration force the quartz crystal to vibrate, accordingly and thereby produces electric signal of varying magnitude. This output of quartz crystal is very weak and hence the signal obtained is voltage amplified by using a voltage amplifier and then the power amplifier before feeding to loudspeaker. The structural design of diaphragm type crystal microphone is shown in Fig. 18.7.

Drawbacks. Though the crystal microphone has an excellent wide frequency response, the drawback is that the piezo-electric property decreases rapidly at high temperatures, say above 100°C. Hence, it is not successful in hot countries like India as it limits its performance.

18.8 MOVING COIL ELECTRO-DYNAMIC MICROPHONE

This microphone is based upon the principle of electromagnetic induction in a coil, moving in strong magnetic field.

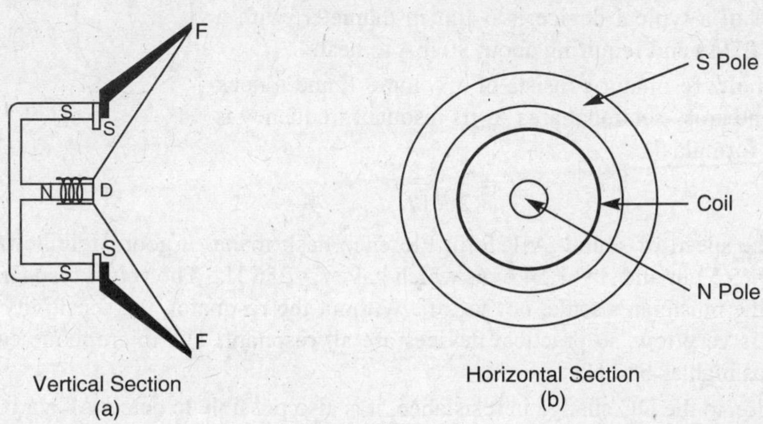

Fig. 18.8

The instrument consists of a E shaped permanent strong magnet as shown in Fig. 18.8. A coil of insulated copper wire is wound on light cylindrical frame which is rigidly attached to diaphragm D of waxed paper or very thin aluminium sheet. The coil can move backward and forward freely into the annular space between north pole and south pole, but it is not capable of movement in any other direction and the movement of the coil generates an induced E.M.F., which is amplified by valve amplifier. The open circuit response

$$\frac{E}{P} = \frac{BlA}{Z_m} \qquad \text{... (18.6)}$$

where B is the strength of the magnetic filed, l is the total length of the coil, A is the area of the diaphragm and Z_m is the mechanical impedance of the system. Z_m is obtained by considering the mechanical system as equivalent to a simple oscillator of mass m stiffness μ, and resistance R and p is the angular frequency of the varying force acting, given by

$$Z_m = \sqrt{\left\{R^2 + \left(\frac{\mu}{p} - mp\right)^2\right\}} \qquad \text{... (18.7)}$$

The response is more or less flat within a wide range of frequency 50 Hz to 7 KHz. Between 0° and 90°, the response is constant to frequency of about 2KHz.

18.9 HOT WIRE MICROPHONE

The hot-wire microphone is not like the other microphones we have studied. It does not reproduce sound pressure variations electrically, but is more of a detector of sound and an indicator of its energy. It is specifically used for low frequencies and for infrasonic signals. It was developed during the 1914-1918 war as a sound ranging device, for acoustic location of artillery to aid counter battery fire. After the war, Tucker and Paris perfected the hot-wire microphone for infrasonic detection.

An example of a hot-wire microphone is shown in Fig. 18.9. It consists of a very fine platinum wire placed over the neck of a Helmholtz resonator and heated by a current passed through it. The wire is supported by a thin glass rod and a disc of mica. The disc is clamped between silver rings that make the contacts. When a sound wave of the resonant frequency arrives, air rushes in and out of the neck of the resonator at that frequency. This air flow cools the wire by forced convection, so its resistance decreases. The resistance decrease is easily detected by a Wheatstone bridge. The hot wire of a typical device is 6 μm in diameter, with a resistance of 350Ω and requiring about 30 mA to heat.

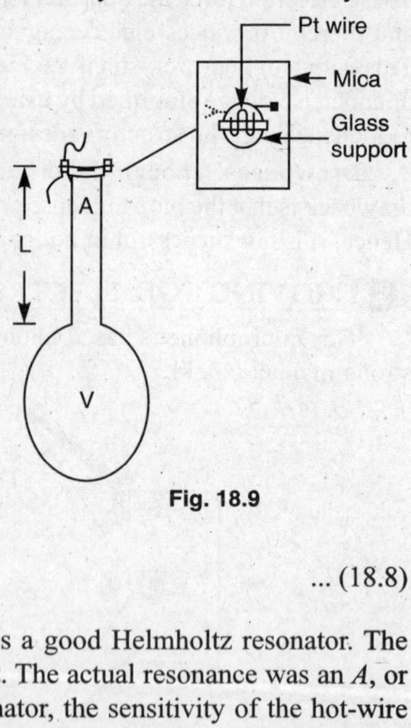

Fig. 18.9

A Helmholtz resonator consists of a volume V and a neck of length L and cross-sectional area A. Its resonant frequency is given by the formula

$$f = \frac{c}{2}\sqrt{\frac{S}{VL}} \qquad \text{... (18.8)}$$

where c is the speed of sound. A 125-ml Florence flask makes a good Helmholtz resonator. The measured $L = 5.5$ cm and $A = 1.54$ cm^2, which gave $f = 256$ Hz. The actual resonance was an A, or 220 Hz, on the musician's scale, not far off. Without the resonator, the sensitivity of the hot-wire microphone is very low, so practical devices are all resonant. The microphone can be applied to frequencies as high as 512 Hz.

In addition to the DC change in resistance, it is also possible to detect AC variations in the hot-wire resistance. These variations are at twice the sound frequency, since the air blows alternately in and out, and the cooling does not depend on the direction of the air velocity. The hot-wire microphone is, accordingly, not applicable to speech or music. As its use in acoustic ranging indicates, it has a rather quick response. It is useful in a frequency range where other microphones are unresponsive.

18.10 LOUD SPEAKER

It is a device to convert the fluctuating electric current produced in a microphone by varying pressure of sound wave, into sound energy for radiation in a room or open space. Thus a loud speaker is a microphone worked back.

Loud speakers are mainly of two type

(*i*) Fixed coil or moving iron type

(*ii*) Moving coil type

(*i*) **Fixed coil or moving iron type.** It is similar to telephone receivers. It consists of a permanent horseshoe magnet with two pole pieces. A steel diaphragm is fixed in the their horn in front to the pole pieces. The fluctuating current corresponding to the original is made to pass through the coil, wound round the pole pieces. It produces corresponding motion of the diaphragm and sound is thus produced.

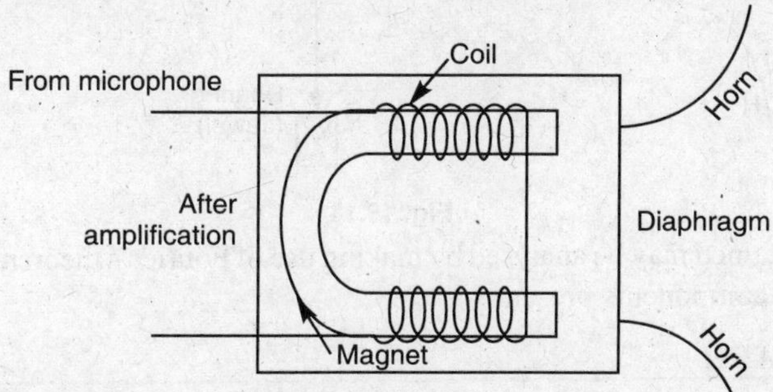

Fig. 18.10

(*ii*) **Moving coil type loud speaker.** The construction is similar to moving coil microphone. In such a microphone voltage induced in the voice coil is proportional to the velocity of the coil which must, in turn, be proportional to the sound pressure, whereas in the loud speaker it is proportional to the amplitude and hence pressure of sound is being reproduced.

Description of loud speaker is same as moving coil microphone shown in Fig. 18.10. When fluctuating current flows through the coil, it with its diaphragm, vibrates and hence sound is produced.

18.11 RECORDING OF SOUND

Principle. *The sound, which is a function of the variation of air pressure, is transformed in to movements of a cutting stylus and engraved in to the surface of a rotating medium. This was previously done purely mechanically.*

Miller's Phonodeik. The simplest technique to understand the method of sound recording is the Miller's phonodeik. The principle involved can be explained by the schematic diagram as shown in Fig. 18.11.

H is a horn, neck of which is fitted with thin glass or mica diaphragm of thickness 0.00762 cm which is held tightly between soft rubber rings and closes the neck of the horn. One end of a thin platinum wire *W* (of 0.00127 cm thick) is attached perpendicularly to the centre of the diaphragm. The wire wraps on the small pulley *P* fixed with a small steel spindle, mounted on jewelled bearing, and is connected to a spring *S*. The spindle carries a very tiny mirror *M* and the plane of the mirror *M* is kept vertical. Light from a pin hole *L* is focussed by lens (not shown) fall upon the mirror and is reflected to a moving film *F* or a rotating mirror. As sound waves produced before the horn, which fall on diaphragm, it vibrates. This will cause the platinum wire resting upon the pulley to move in the forward and backward direction. This will give rise to rotation of the spindle and mirror. The spot of light reflected from the mirror, falling on a distant screen performs a horizontal oscillation when the diaphragm is disturbed. The film running in the vertical direction will record the wave form of sound. The natural frequency of the diaphragm should be high so as to eliminate resonance effect but difficulty arises to eliminate the resonance effect of the horn.

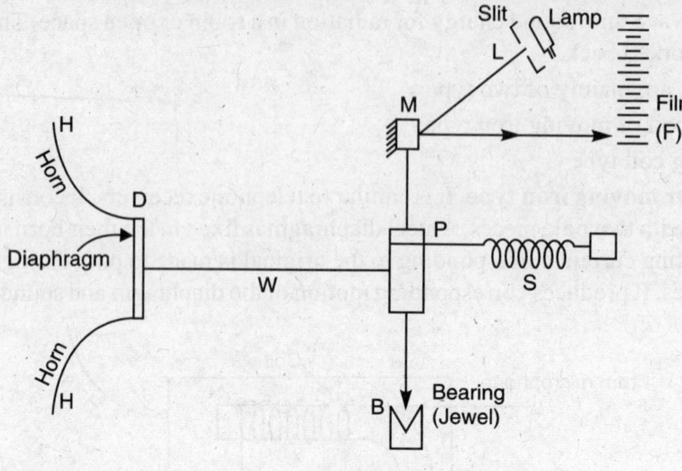

Fig. 18.11

The curves, so obtained may be analysed by making use of Fourier's theorem to find the intensities of different harmonic components present.

18.12 PHONOGRAPH

Recording. Another method employed for recording and reproduction of sound is a *phonograph.* It consists of a cylinder *C* which is capable of rotating about its axis mechanically by a handle *H* as shown in Fig. 18.12. A thick layer of wax is coated on the cylinder *C*. The axle is fixed on two vertical stands S_1 and S_2 through which the axle can be turned end-wise by the use of pinion and screw arrangement. The cylinder can be turned by handle *H* as shown.

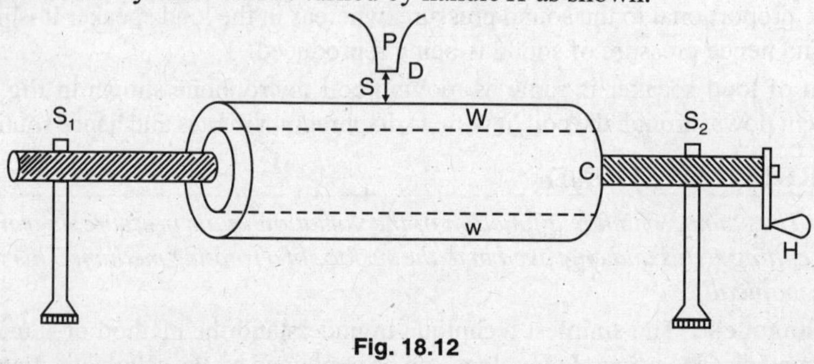

Fig. 18.12

The sound to be recorded is converged through the funnel *P* upon a thin glass diaphragm *D*. A small stylus *S* with a sharp cutting edge is fixed to the bottom of diaphragm *D*, touching the waxed surface of the cylinder. The vibration of diaphragm causes the cutting stylus to vibrate longitudinally at right angles to the surface of the cylinder and cut groove of variable depth on the wax of the rotating cylinder. This groove of variable depth ("*hill and dale*") has the information to be recorded (*i.e.,* sound) as per the vibrations of the diaphragm, which is in accordance with the sound.

Reproduction. To reproduce the sound that is recorded, the cylinder is rotated back to the original position and the cutting stylus is replaced by a blunt point. As the cylinder rotates in the same direction and with the same speed (in practice, handle is replaced by speed controlled electric motor), the blunt point passes along the grooves and vibrates exactly in the same way as the recording stylus. The grooves formed on the cylinder surface is the exact record of the original sound.

18.13 GRAMOPHONE

Emile Berliner invented the gramophone in 1887. A disc type record player was developed usually named as gramophone in which grooves are arranged as a spiral on the surface of the disc. The horn was used, in place of todays loud-speaker for reproduction of sound. For recording, a circular disc was used which has highly polished wax uniform coating on its surface. The diaphragm of a sound collecting horn is connected to a *L*-shaped sharp stylus *S* which touches to the rotating disc. The stylus *S*, moves gradually in a spiral path on the rotating disc. The forward and backward motion of the stylus to the diaphragm owing to varying pressure of sound is converted into lateral displacements and cuts the record on the revolving disc in wavy spiral groove.

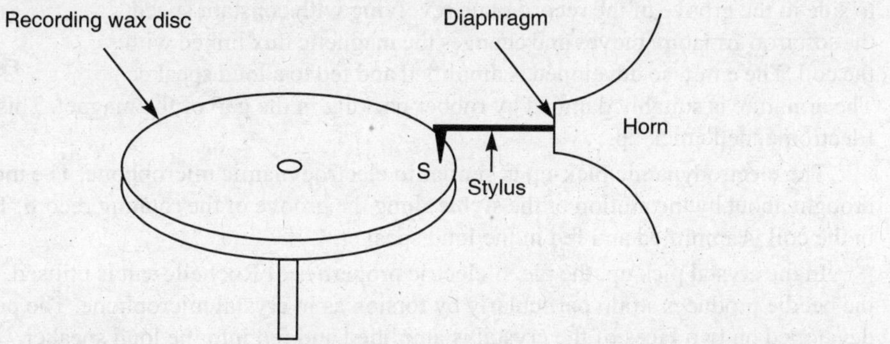

Fig. 18.13

As wax is soft, the surface is dusted with graphite powder and copper is deposited electrolytically. The copper plate with projections corresponding to depression in the wax disc is, called 'negative' record or master shell. Hot disc of plastic material when pressed upon the master shell will form a 'positive' record with depression similar to wax disc. This becomes hard on cooling. Any number of such records on plastic disc can be obtained, as replica.

18.14 IMPROVED TECHNIQUES OF RECORDING OF SOUND

Around 1925, this acousto-mechanical process was replaced by an electrically amplified system. In this system, the sound was transformed by a microphone into an electrieal single, which moved in electrically driven cutting stylus. The reproduction was also improved by the use of electrical pick-up system, the signals of which, suitably amplified, were reconverted into mechanical movements by loudspeakers or headphones. Recently, system for optical, contact-less replay of mechanical carriers have been developed.

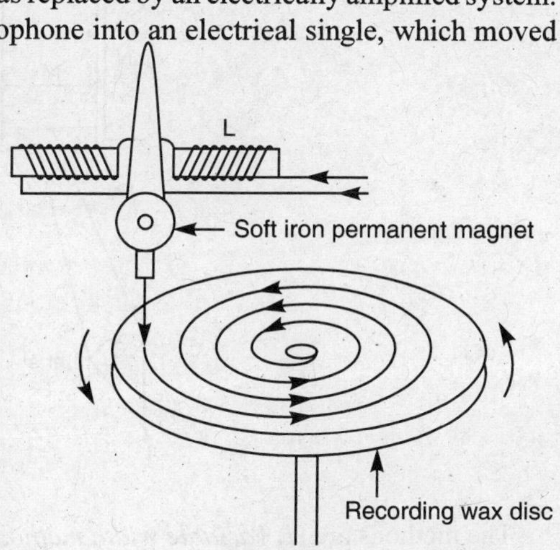

Fig. 18.14

For recording a (sound) signal obtained from a microphone is passed through a coil *L* kept perpendicular to the magnetic field produced at the centre of the armature, at the end of which the sharp stylus is fixed. Depending upon the varying current through the coil, the sharp stylus vibrates due to electro-magnetic effect. The vibrating sharp stylus cuts a wavy groove in a spiral path on the rotating disc.

Large number of copies of records are obtained by the method described above.

Reproduction of Sound. The same system is used for reproducing the sound but in reverse order. The sharp cutting stylus is replaced by a blunt stylus which moves through the spiral grooves on the record, whose angular speed in maintained (equal to that of recording process) by an external electrical motor (not shown in figure). The vibrations produced in the stylus are converted into equivalent electrical current by an arrangement shown in Fig. 18.15.

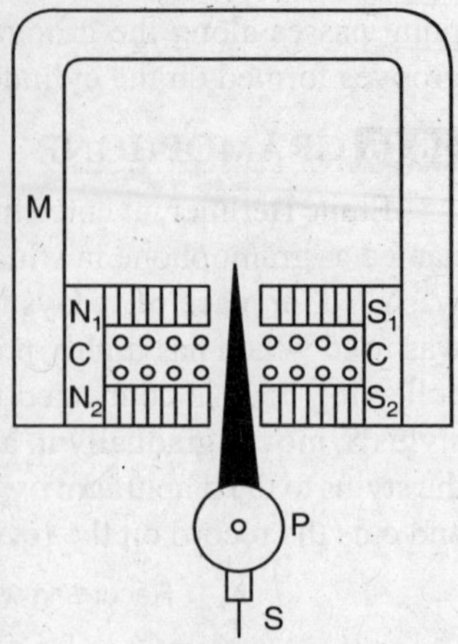

Fig. 18.15

It consists of a small permanent horseshoe magnet M with four pole pieces $N_1N_2S_1S_2$. A soft iron armature in the form of magnetic needle pivoted at the centre P, is placed inside the pole pieces of the magnet M. This is fitted with a stylus N. A fixed coil of insulated copper wire C is placed inside the gap of the magnet pole. As the needle N moves side to side in the groove of the record plate revolving with constant speed, the soft iron armature moves and changes the magnetic flux linked with the coil. The e.m.f. so developed is amplified and fed to a loud speaker. The armature is suitably damped by rubber packing in the gap of the magnet. This type is known as Electromagnetic pick up.

The electrodynamic pick-up is similar to electrodynamic microphone. The motion of the coil is brought about by the motion of the stylus along the groove of the rotating record. The induced e.m.f. in the coil is amplified and fed in the loud speaker.

In the crystal pick up, the piezo-electric properties of Rochelle salt is utilised. The movement of the needle produces strain particularly by torsion as in crystal microphone. The potential difference developed on two faces of the crystal is amplified and fed into the loud speaker.

18.15 FILM RECORDING AND REPRODUCTION

In film recording, there are two different methods of recording and reproduction of sound. The sound track occupies a space of width 2.5 mm, in a film and video accupies a space of 35 mm width

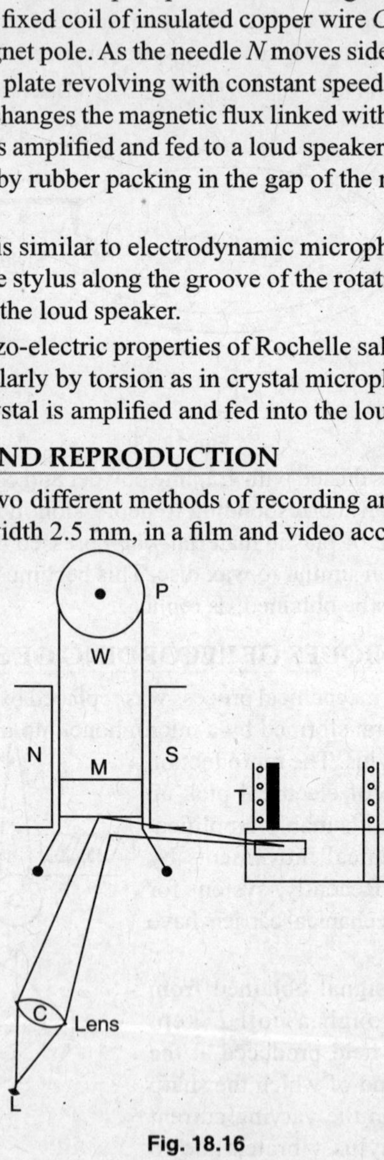

Fig. 18.16

The methods are (*i*) *Variable width method* and (*ii*) *Variable density method* (*i*) **Variable Width Method.** The amplified current from the microphone is passed through a loop of wire *W* passing through a small pulley *P* placed in a strong magnetic field transversely.

The current enters through one arm of the loop and goes out through the other arm. A small mirror *M* is attached to the loop. The beam of light from a powerful recording lamp *L* falls on the mirror after concentration by a lens *C*. The light is reflected from the mirror *M* falls on the slit *T* on the track of the film moving with uniform speed of 0.4572 metre per second. The mirror *M* vibrates as the loop vibrates due to the passage of fluctuating current from the microphone. The vibrating mirror illuminates a length of the slit which varies with the tilt of the mirror. The tilt varies with strength of the current from the microphone in the loop placed in the magnetic field. Thus a variable width on the film represents the amplitude of the sound vibration along its length.

(*ii*) **Variable Density Method.** The sound record in this type consists of series of parallel lines of different transparency. These are the images of a narrow slit illuminated by a source of variable intensities by amplified microphone current. The intensities are varied either (*i*) by modifying the intensity of the source or (*ii*) by modifying the width of the slit through which passes the light beam.

(*a*) **By the modification of intensity.** Fig 18:17. The amplified microphone current is applied to a gas discharge lamp *G*. The charging current produces the change in brightness of the lamp light through S_1 and lens *L* illuminates the longitudinal slit S_2 in contact with the film moving in the vertical direction with constant speed. The film will be most opaque in places of maximum brightness of the lamp. The situation will be reversed when positive will be made for reproduction.

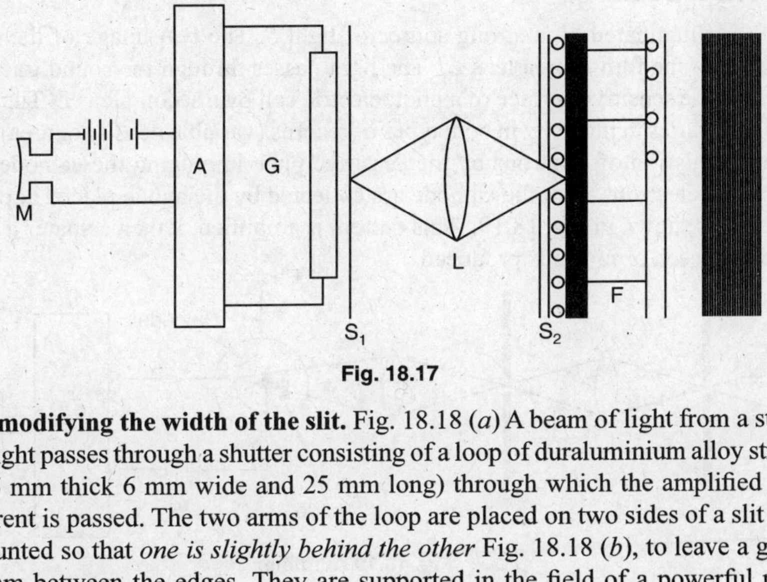

Fig. 18.17

(*b*) **By modifying the width of the slit.** Fig. 18.18 (*a*) A beam of light from a strong source of light passes through a shutter consisting of a loop of duraluminium alloy strip or ribbon (0.5 mm thick 6 mm wide and 25 mm long) through which the amplified microphone current is passed. The two arms of the loop are placed on two sides of a slit S_1. They are mounted so that *one is slightly behind the other* Fig. 18.18 (*b*), to leave a gap normally 1 mm between the edges. They are supported in the field of a powerful magnet. The force is exerted by the magnetic field on the current from microphone passing through the ribbons which set in vibration, that is, they either come closer or diverge depending on the direction and magnitude of the current flowing. The movement of the two parts of the strips causes the change of the width of the slit. Thus it acts as a light valve. Light through various widths of the slit S_1 passes through another slit S_2 of constant width. The image of the slit falls on a film running at a constant speed. A continuous series of parallel lines of various opacity and number per unit length appear on the sound track of the film. The number of lines per unit length depends on frequency, and opacity depends on intensity.

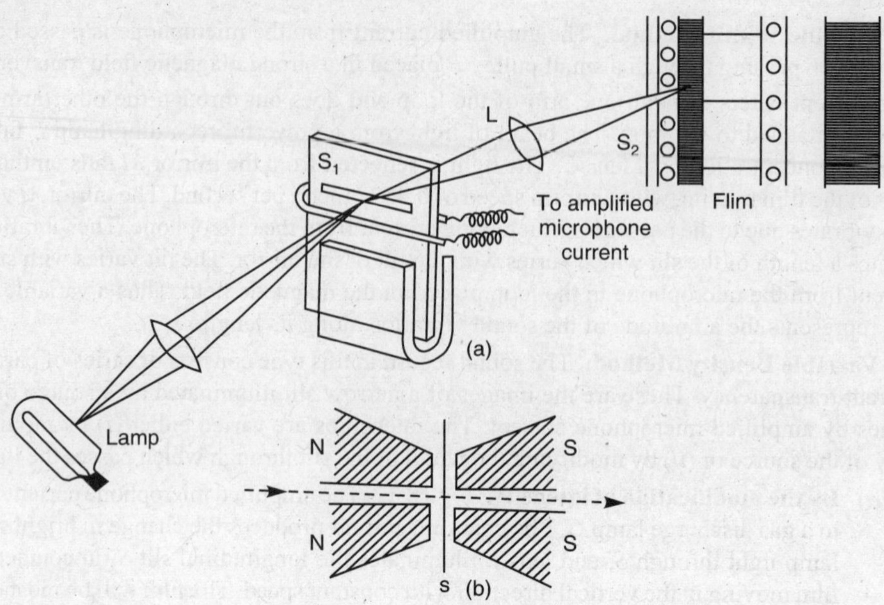

Fig. 18.18

Reproduction of Sound

A slit S_1 is illuminated by a strong source of light S. The real image of the slit is formed on the sound track of the film F by a lens L_1. The light passes through the sound track of the film and converges upon the sensitive surface of a photoelectric cell by a second lens L. The transmitted light through the film varies in intensity in both types of records (variable density or area), produces feeble current due to emission of electrons by the sensitive plate impinging the cathode of photoelectric cell. The emitted electrons from the cathode are collected by the anode placed at positive potential, using battery B as shown in Fig. 18.19. This current is amplified, it then actuates a loud speaker and corresponding speech or music is produced.

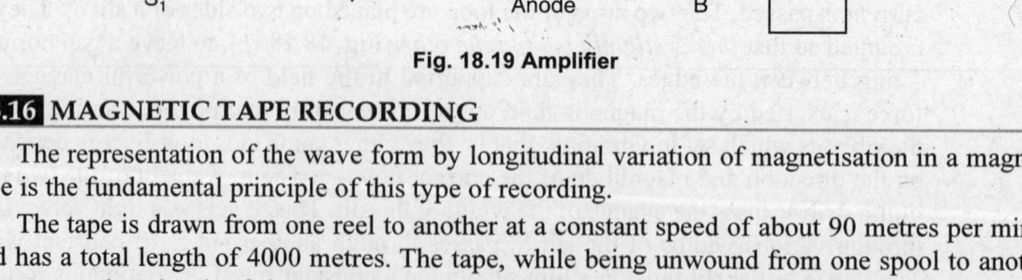

Fig. 18.19 Amplifier

18.16 MAGNETIC TAPE RECORDING

The representation of the wave form by longitudinal variation of magnetisation in a magnetic tape is the fundamental principle of this type of recording.

The tape is drawn from one reel to another at a constant speed of about 90 metres per minute and has a total length of 4000 metres. The tape, while being unwound from one spool to another, moves over three heads of electromagnets, successively each having chisel-shaped poles as shown in Fig. 18.20. The first one is known as washing or erasing head. Here the tape is subjected to a strong high frequency (ultrasonic frequency of 100 K Hz). magnetic field which wipes out any residual field in the tape and thereafter becomes suitable for fresh record.

The second electromagnet is known as recording head. The amplified current from microphone superimposed upon high frequency current is allowed to pass through the coil of the electromagnet. The tape while passing over the chisel-ended poles of this magnet with uniform speed is magnetised transversely varying along the length which fluctuates like the microphone current and are used in impressing the magnetic record.

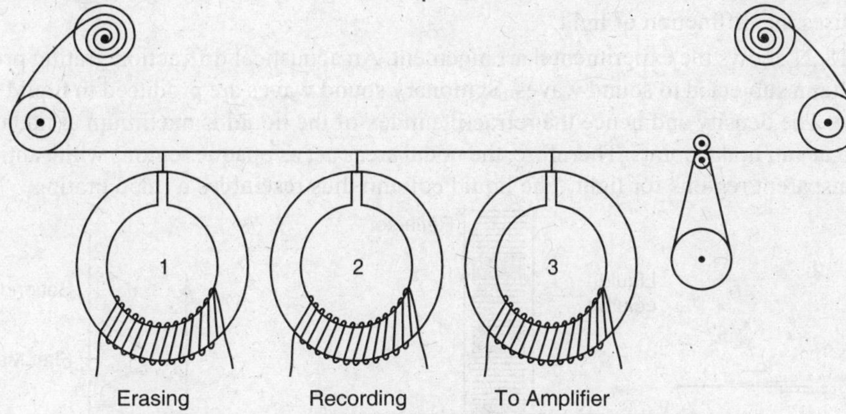

Fig. 18.20

Such a magnetised tape is passed between two chisel-shaped poles of the third electromagnet. The flux variation induces corresponding current variation in the coil of the electromagnet. This current is amplified and is fed to a loud speaker which reproduces the original sound.

18.17 MODERN METHODS OF AUDIO AND VIDEO RECORDING

The most wide-spread type of instantaneous discs are the 'lacquer' or 'acetate' disc. A lacquer coating, consisting mainly of cellulose nitrate carries the information. The substrate is generally a glass. However lacquer coating become brittle with age. The data should, therefore, be transformed into digital before sound gets erased. From late 1940s onwards, a new material, a co-polymer of polyvinyl chloride (PVC) and polyvinyl acetate (PVA) was introduced. PVC/PVA copolymer is chemically very stable and not vulnerable to mechanical damage, e.g. to scratches.

Magnetic recording was invented in the 19th century. It was used in small scale parallel to cylinders and gramophones. After world-war second it spread world-wide. In the 1960 s, several cassette formats were developed. After number of experiments, digital audio recording magnetic tape was introduced in the 1980s. In 1987, R-DAT, a digital recording on cassette format was developed and gained popularity. By 1991, however, it was also obsolete. Today, magnetic tape was replaced by compact disc. (CDs) and CD-ROM. By 1991, both recordable CDs (CD–Rs) and Rewritable CDs (CD-RW) were developed. In order to increase the capacity of the optical disks, mainly to make suitable to store video films the DVD (Digital Video Disc) was introduced from 1995 onwards, using the same principle as CDs. By 2005/06 to accomodate high definition video programmes, the storage capacity was increased further by the use of **lasers** of shorter wavelength. The technology has led to two formats, the HD DVD (High density DVD) and the Blu-ray Disc (BD). Both take another step towards smaller signal representation and thereby increased data density. Finally, magneto-optical disks, originally used in computer world for storing data with huge capacity of hard disk drives (HDD) at ever lower prices are of prime importance and popularity worldwide.

Magneto-optical recording is based on the phenomenon that magnetic orientation influences reflection of light (Kerr effect).The recording is achieved by heating the magnetic information layer with a laser beam beyond its Curie point which allows magnetic (re-) orientation by applying very low magnetic fields. Replay i.e. Reproduction works in a similar way as other optical disks.

18.18 DETERMINATION OF WAVELENGTH AND VELOCITY

When sound waves (particularly ultrasonic) propagate in a medium say liquid, the alternating compressions and rarefactions change the density of the medium. It leads to a periodic variation of refractive index (μ) of the liquid. Such a liquid column subjected to sound waves constitute an acoustical grating. If monochromatic light is passed through the liquid at right angles to the waves,the liquid causes the diffraction of light.

Fig 18.21 shows the experimental arrangement. An acoustical diffraction grating produced by a liquid column subjected to sound waves. Stationary sound waves are produced in liquid column in a glass tube. The density and hence the refractive index of the liquid is maximum at nodal points and minimum at anti nodal points. Therefore, the nodal areas act as opaque regions while anti nodal areas act as transparent regions for light. The liquid column thus resembles a ruled grating.

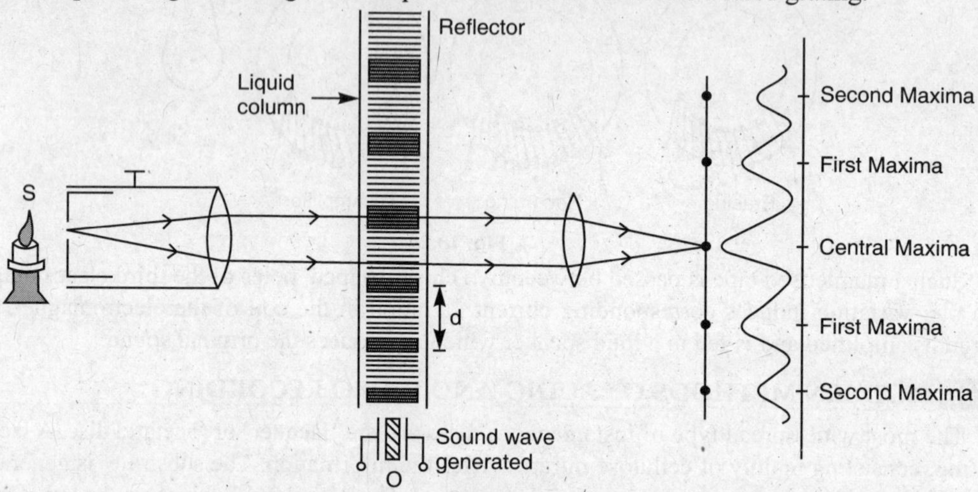

Fig. 18.21

When the crystal is at rest, a single image of the slit is formed on the screen. When the crystal is excited a diffraction pattern is obtained. It consists of a central maxima followed by first order, second order maxima and minima etc. The grating period d equals to $\frac{\lambda_s}{2}$ and is given by

$$\frac{\lambda_s}{2}\sin\theta = m\lambda \qquad \text{... (18.9)}$$

Where λ_s is the wavelength of sound waves,
λ is the wavelength of monochromatic light beam
and m is the order of the maxima

$$\therefore \qquad \lambda_s = \frac{2m\lambda}{\sin\theta} \qquad \text{... (18.10)}$$

Knowing λ , and m and by measuring θ, λ_s can be determined can be determined using relation 18.10. The frequency f of the waves is known from the frequency of the oscillator. The velocity of waves in the liquid can be found from the relation

$$\text{Velocity, } v = f\lambda_s \qquad \text{... (18.11)}$$

Thus, by using acoustical diffraction method, the wavelength, frequency and velocity of ultrasonic sound waves can be determined.

18.19 DEFECTS DUE TO REFLECTED SOUND

The sound of foot steps is heard louder while walking in narrow passages or corridors, owing to reflection of sound from walls and ceilings. The whispering galleries such as famous St.Paul's cathedral due to acoustical properties of sound relfections from the walls. The reflection of sound in an enclosed space may lead to two important defects, namely *echo* and reverberation.

Echoes and reverberation are both reflections of sound. A reflection is called an **echo**, if the time between the original sound and its reflection is long enough that both sounds can be heard distinctly. If a room has lots of echoes and they are closely spaced in time so that they are not discriminable, then this large number of echoes is known as reverberation.

18.20 ECHOES

An echo is produced when the sound reflected from obstacle reaches the ear after the sound from the source has already been heard. Thus, there is a repetition of the sound in this case. The sensation of sound persists for about 100 ms after the source stopped producing sound. Hence, in order to hear a distinct echo, the reflected sound must reach the ear 100 ms later than the direct sound. When the distance of the obstacle from the source is 17m or more, echo will be heard distinctly. When the reflection arrives within 60 ms or less after the original sound, the listener will not hear the reflection as distinct echo. When sound is reflected from a number of reflecting surfaces, multiple echoes are heard. Thus, the heavy rolling sound of a thunder is a result of successive reflections from clouds, mountains, rocks etc. reflecting surfaces.

18.21 REVERBERATION

Sound produced in a closed hall (i.e encloser) continues to be heard for same time even after the source stops producing sound. A sound produced in a hall undergoes multiple reflections from the walls, floor and ceiling before it becomes inaudible. A person in the hall continues to receive the successive reflections of progressively diminishing intensity as shown in Fig. 18.22. Therefore, the sound lasts for some time even after the source ceases emitting sound. **The persistence of sound in a hall due to multiple reflections, even after the source stops emitting sound, is known as reverberation.** Reverberation is familiar phenomenon in halls without furniture of a newly built building.

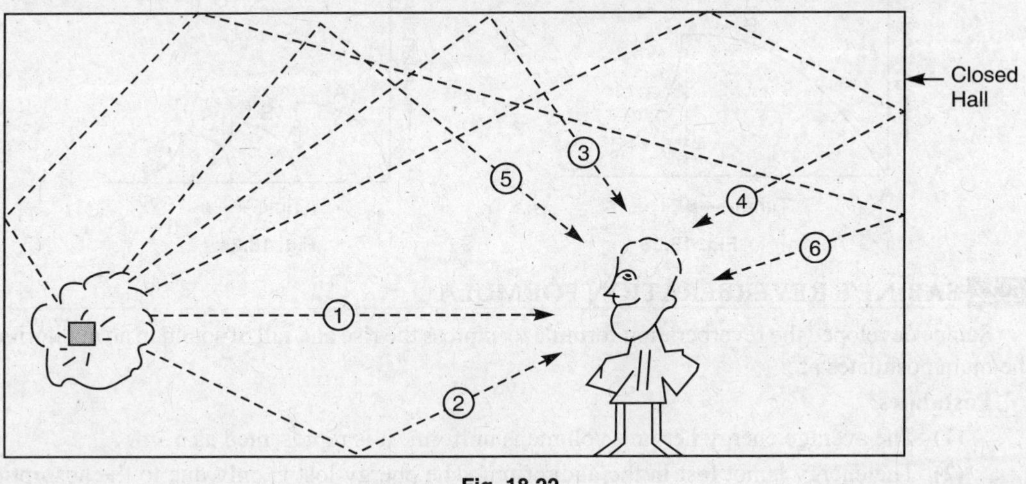

Fig. 18.22

Reverberation time. The time taken by a sound in a room (or hall) to fall from its average intensity to audibility is called the reverberation time of the room (or hall). In practice, the reverberation time is defined as the time during which the sound energy density falls from its steady state value to its one-millionth value after the source is shut off. If $Ii/I_f = 10^6$, $L = 60\ d$B. Thus, the reverberation time is the period of time in seconds, which is required for sound energy to diminish by 60 dB after the sound source has stopped emitting sound. The reverberation time will depend on the size of the room or the auditorium the nature of reflecting material on the wall and the ceiling and the area of the reflecting surfaces.

However, small reverberation is sometime desirable, especially in a hall used for a musical performance. A small amount of reverberation, improves the original sound and aids the ability to hear and understand the source. Too much reverberation makes speech unintelligible and causes muddy and boomy sound quality in a musical performance, and hence a hall is constructed reverberation free.

In a good auditorium it is necessary to keep the reverberation time negligibly small. The intensity of sound as received by the listener, is shown graphically in Fig.18.23. When a source emits sound, the waves spread out and the listener is aware of the commencement of sound when the direct waves reach his ears. Subsequently the listener receives sound energy due to reflected waves also. If the note is continuously sounded. The intensity of sound at the listener's ears gradually increases. After sometime a balance is reached between the energy emitted per second by the source and the energy lost or dissipated by walls or other materials. The resultant energy attains an average steady value, and to the listener the intensity of sound appears to be steady and constant. This is represented by the portion BC of the curve $ABCD$. If at C, the source stops emitting sound, the intensity of sound falls exponentially as shown by the curve CD. When the intensity of sound falls below the minimum audibility level, the listener will not hear the sound.

When a series of notes are produced in an auditorium (say speech or music) each note will give rise to its own intensity curve with respect to time. The curves for these notes are shown in Fig. 18.24.

For clear audibility of speech or music, it is necessary that (*i*) each separate note should give sufficient intensity of sound in every part of the auditorium, and (*ii*) each note should die down rapidly before the maximum average intensity due to the next note is heard by the listener. **(Fig** 18.24) This is particularly important with speech. In the case of music *comparatively more* reverberation can be tolerated.

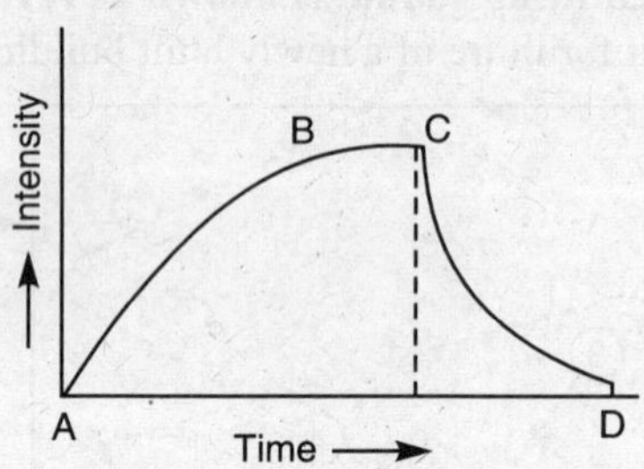

Fig. 18.23

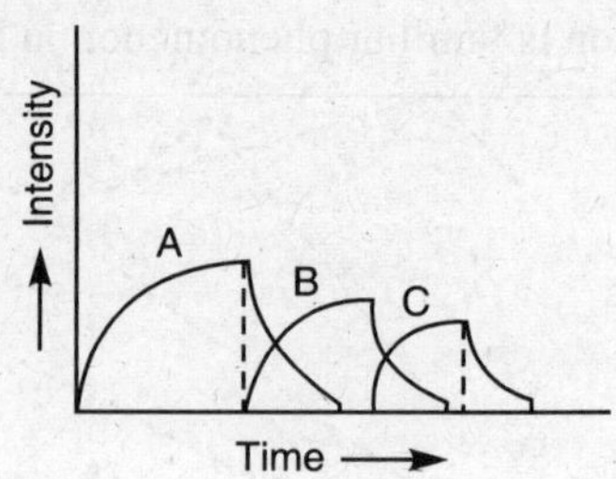

Fig. 18.24

18.22 SABINE'S REVERBERATION FORMULA

Sabine developed the reverberation formula to express the rise and fall of sound in an auditorium. The main postulates are:

Postulates:

(1) The average energy per unit volume is uniform. It is represented as σ.

(2) The energy is not lost in the auditorium. The energy lost is only due to the absorption of the material of the walls and ceiling and also due to escape through the windows and ventilators. Both these factors are included in the term **'absorption'** of energy.

Theory: Suppose a source is producing sound continuously. This sound energy is propagated in all directions. Let σ be the energy contained in a unit volume. The energy that is contained in a solid angle $d\phi$ is

$$= \frac{\sigma . d\phi}{4\pi}$$

Let this energy be incident on a unit surface area of the wall at an angle θ (Fig.18.25). If the velocity of sound is v, then the total energy falling per second on a unit surface area of the wall.

$$= \left(\frac{\sigma . d\phi}{4\pi}\right)(\cos\theta).v$$

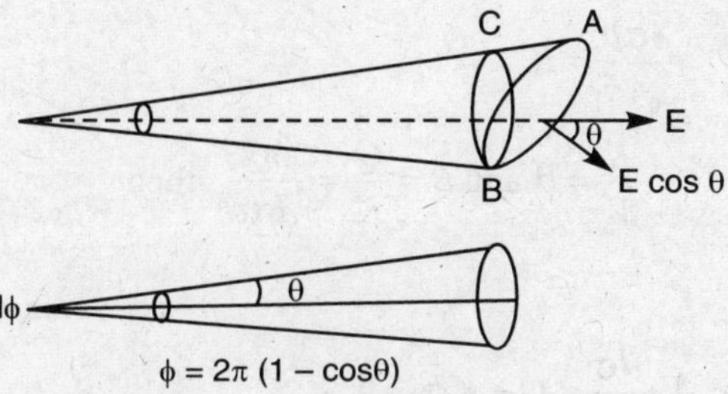

Fig. 18.25

The total energy falling per second within a hemisphere

$$= \frac{\sigma v}{4\pi}\int \cos\theta . d\phi$$

But $\qquad \phi = 2\pi\,(1 - \cos\theta)$

or $\qquad d\phi = 2\pi \sin\theta .\, d\theta$

Substituting this value of $d\phi$,

The total energy falling per second within a hemisphere

$$= \frac{\sigma v}{4\pi}\int_0^{\pi/2} 2\pi \sin\theta . \cos\theta . d\theta$$

$$= \frac{\sigma v}{2}\left[-\frac{\cos^2\theta}{2}\right]_0^{\pi/2}$$

$$= \frac{\sigma v}{4}$$

Suppose α is the absorption coefficient of the walls that refers to the fraction of the incident energy not reflected from the walls. The amount of energy absorbed per second per unit area = $\frac{\alpha\sigma v}{4}$

If A is the area of the walls and the other absorbing materials including ceiling, windows and ventilators etc., the amount of energy absorbed per second.

$$= \frac{A\alpha\sigma v}{4}$$

Let V be the volume of the auditorium, then the total energy = $V\sigma$. The rate of increase of energy

$$= \frac{d}{dt}(V\sigma)$$

$$= V\frac{d\sigma}{dt} \qquad \text{... (i)}$$

Suppose, the source supplies energy at the rate of Q units per second

Then, the rate of increase of energy

$$= Q - \frac{A\alpha\sigma v}{4} \qquad \text{... (ii)}$$

Equating (*i*) and (*ii*)

$$V.\frac{d\sigma}{dt} = Q - \frac{A\alpha\sigma v}{4} \quad \text{... (iii)}$$

Take $\frac{A\alpha v}{4} = K$

and $\frac{K}{V} = \beta$ and $B = \frac{Q}{K} = \frac{4Q}{A\alpha v}$, then

From equation (*iii*)

$$V.\frac{d\sigma}{dt} = Q - K\sigma$$

$$\frac{d\sigma}{dt} = \frac{Q}{V} - \frac{K}{V}.\sigma \quad \text{... (iv)}$$

The general solution of this equation is

$$\sigma = B + be^{-\beta t} \quad \text{... (v)}$$

When $t = 0$ $\sigma = 0$

∴ From equation (*v*)

$$0 = B + b$$

or $b = -B$

$$\sigma = B - Be^{-\beta t}$$

$$\sigma = B\,[1 - e^{-\beta t}]$$

Substituting the values of B and β

$$\sigma = \frac{4Q}{A\alpha v}\left[1 - e^{-\frac{(A\alpha v).t}{4V}}\right] \quad \text{... (vi)}$$

Equation (*vi*) represents the rise of average sound energy per unit time from the time source commences to produce sound.

The maximum value of average energy per unit volume

$$\sigma_{max} = \frac{4Q}{A\alpha v} \quad \text{... (vii)}$$

Similarly, after the source ceases to emit sound, the decay of the average energy per unit volume is given by

$$\sigma = \frac{4Q}{A\alpha v}e^{-\frac{(A\alpha v.t)}{4V}} \quad \text{... (viii)}$$

or

$$\sigma = \sigma_{max}e^{-\frac{(A\alpha v.t)}{4V}} \quad \text{... (ix)}$$

The factor $\frac{A\alpha v}{4V}$ gives the reverberation time in the auditorium. If σ_0 represents the minimum audible intensity after a time t_1, then from equation (*ix*).

$$\sigma_0 = \sigma_{max}e^{-\frac{(A\alpha v),t_1}{4V}} \quad \text{... (x)}$$

Here t_1 is the time interval between the cutting off the sound and the time at which intersity falls below the minimum audible level.

From equation (*X*)

$$\sigma_{max} = \sigma_0 . e^{+\frac{(A\alpha v.t_1)}{4V}}$$

Taking logarithms

$$\log_e \left(\frac{\sigma_{max}}{\sigma_0} \right) = \frac{A\alpha v}{4V} . t_1 \quad \text{... (xi)}$$

Here α and σ_0 change with the frequency of sound.

For calculating the reverberation time, a standard steady intensity is required. Sabine took the value of $\frac{\sigma_{max}}{\sigma_0} = 10^6$

From equation (*xi*)

$$\log_e (10^6) = \frac{A\alpha v}{4V} t_1$$

$$2.303 \times 6 = \frac{A\alpha v}{4V} . t_1$$

Taking velocity of sound approximately at room temperature as 350 m/s,

$$2.303 \times 6 = \frac{A\alpha \times 350}{4V} . t_1$$

or

$$t_1 = \frac{2.303 \times 24V}{350 A\alpha}$$

$$t_1 = \frac{0.158V}{A\alpha} \quad \text{... (18.12)}$$

In general

$$t_1 = \frac{0.158V}{\Sigma A\alpha} \quad \text{... (18.13)}$$

In equation (18.12) the quantities are represented in M.K. S. units.

[Note. In F.P.S. units, taking v =1120 ft/s.]

Equation (18.12) represents the Sabine's reverberation time formula.

$$t_1 = \frac{0.05V}{A\alpha} \quad \text{... (18.14)}$$

According to equation 18.12 the reverberation time is (*i*) directly proportional to the volume of the auditorium, (*ii*) inversely proportional to the area of the ceiling, walls, etc. and (*iii*) inversely proportional to the total absorption plus transmission through open surfaces. It has been experimentally found that a reverberation time of 1.03 seconds is most suitable for all rooms having approximately a volume of less than 350 cubic metres.

To decrease the reverberation time, the walls of the auditorium are usually covered with materials having large absorption coefficients. The area of the surfaces of the walls is also increased in good cinema halls to decrease the reverberation time.

Example 18.1 *A hall of volume 5500 metre³ is found to have a reverberation time of 2.3 seconds. The sound absorbing surface of the hall has an area of 750 metre². Calculate the average absorption coefficients.*

Solution. Here $V = 5500\ \text{m}^3$

$t = 2.3$ s

$A = 750\ \text{m}^2$

Absorption coefficient $= \alpha$

Reverberation time,

$$t = \frac{0.158V}{A\alpha}$$

$$\alpha = \frac{0.158V}{At} = \frac{0.158 \times 5500}{750 \times 2.3}$$

$$= 0.504.$$

18.23 DETERMINATION OF ABSORPTION COEFFICIENT

A source of frequency 512 hertz is taken. The time of reverberation in a hall is determined using a chronograph, (*i*) without the absorbing materials in the hall, and (*ii*) with the absorbing materials in the hall. Let these two times be t_1 and t_2 respectively. According to Sabine's formula

$$t_1 = \frac{0.158V}{\Sigma\alpha_1 A_1} \qquad \text{... (i)}$$

and

$$t_2 = \frac{0.158V}{\Sigma\alpha_1 A_1 + \alpha_2 A_2} \qquad \text{... (ii)}$$

Here α_2 is the absorption coefficient of the material of area A_2.

$$\therefore \quad \frac{1}{t_1} = \frac{\Sigma\alpha_1 A_1}{0.158V} \qquad \text{... (iii)}$$

$$\frac{1}{t_2} = \frac{(\Sigma\alpha_1 A_1) + \alpha_2 A_2}{0.158V} \qquad \text{... (iv)}$$

$$\therefore \quad \frac{1}{t_2} - \frac{1}{t_1} = \frac{\alpha_2 A_2}{0.158V}$$

$$\therefore \quad \alpha_2 = \frac{0.158V}{A_2}\left[\frac{t_1 - t_2}{-t_1 t_2}\right] \qquad (18.15)$$

From equation (18.15) knowing the values of t_1, t_2, A_2 and V, the value of α_2 can be calculated.

The absorption coefficients of some common materials calculated with a source of frequency 512 hertz are given in the following table.

The absorption coefficient of a given material is different at different frequencies. It is generally higher at higher frequencies.

Table 18.1

No.	*Material*	*Absorpiton Coefficient*
1.	Marble	0.01
2.	Common plaster	0.03
3.	Glass	0.027
4.	Concrete	0.17
5.	Cork	0.23
6.	Asbestos	0.26

7.	Carpet	0.30
8.	Acoustic Plaster	0.30
9.	Acoustic Felt	0.45
10.	Fibre Board	0.50
11.	Heavy Curtains	0.50
12.	Hair or Felt	0.58
13.	Fibre Glass	0.75
14.	Perforated Cellulose Fibre Tiles	0.85

18.24 REQUIREMENTS FOR GOOD ACOUSTICS

The main requirements in the design of a good auditorium are smooth decay and growth of sound. In an auditorium, the sound must be distributed equally and evenly over the whole area particularly main floor and balcony. This is clear from Fig. 18.25

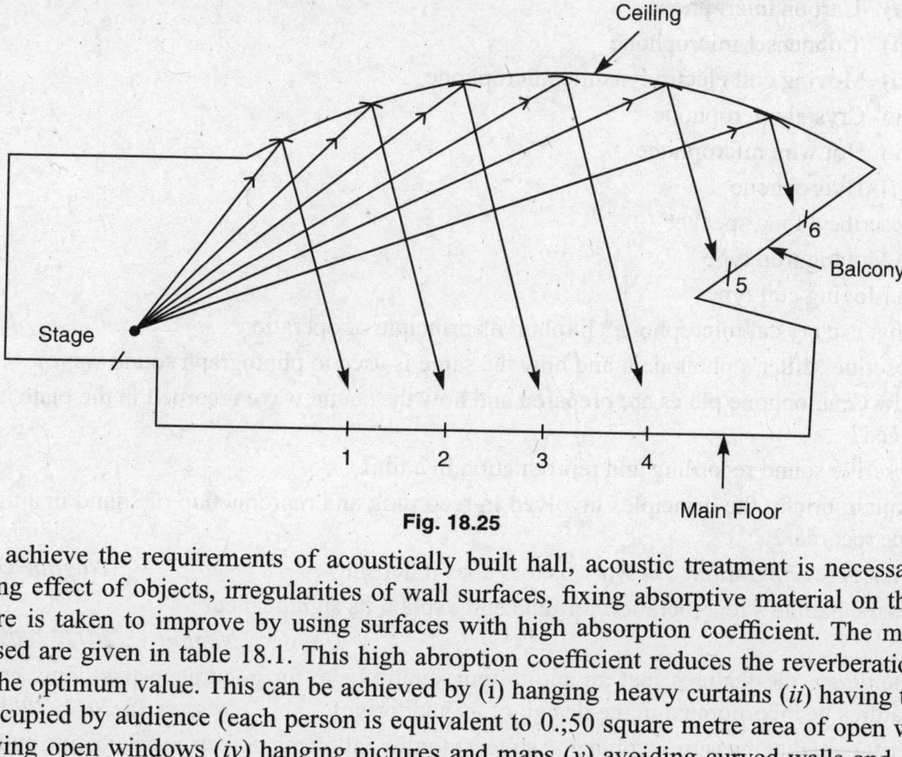

Fig. 18.25

To achieve the requirements of acoustically built hall, acoustic treatment is necessary, *vis*, scattering effect of objects, irregularities of wall surfaces, fixing absorptive material on the wall, etc. Care is taken to improve by using surfaces with high absorption coefficient. The materials to be used are given in table 18.1. This high abroption coefficient reduces the reverberation time below the optimum value. This can be achieved by (i) hanging heavy curtains (*ii*) having the hall fully occupied by audience (each person is equivalent to 0.;50 square metre area of open window (*iii*) having open windows (*iv*) hanging pictures and maps (*v*) avoiding curved walls and corners bounded by two walls (*vi*) upholstered seats should be provided (*vi*) walls and ceiling covered with material of high absorption co-efficient (*vii*) having walls engraved and rough so as to increase their absorption

Example 18.2 *A broadcasting studio measuring 25×12×7 metres has a reverberation time of 0.90 sec when empty. What will be the reverberation time when an audience of 250 persons is present? (Given absorption coefficient of a person = 0.4)*

Solution Given: Absorption coefficient of each person = 0.4

Average absorbing area of each person = 0.6 sq.m

Volume of the studio = $25 \times 12 \times 7 = 2100$ m^3

Initially if A be total absorption then

$$0.9 = \frac{0.158V}{A} = \frac{0.158 \times 2100}{A}$$

$$\therefore \quad A = 368.67$$

Due to audience A increases by 250 × 0.4 × 0.6 Sabins = 60 sabins

$$\therefore \text{ New reverberation time} = \frac{0.158 \times 2100}{(368.67 + 60)}$$

$$= 0.774 \text{ sec}$$

EXERCISE

1. What are transducers? Discuss the characteristics of sensors and transducers?
2. What are different parameters which characterize the performance of a transducer?
3. What is a microphone? Explain any one type of microphone? ***(Nagpur U. 2009)***
4. Describe the different parts and explain the action of
   (*i*) Carbon microphone
   (*ii*) Condenser microphone
   (*iii*) Moving coil electrodynamic microphone
   (*iv*) Crystal microphone
   (*v*) Hot wire microphone
   (*vi*) Hydrophone
5. Describe a loud speaker
   (*a*) Moving iron type
   (*b*) Moving coil type
6. What is a crystal microphone? Explain its principle of operation
7. Describe Miller's phonodeik and how the same is used to photograph sound wave?
8. How Gramophone plates are prepared and how the sound wave recorded in the plate is reproduced?
9. Describe sound recording and reproduction in a film.
10. Explain briefly the principles involved in recording and reproduction of sound in a magnetic tape recorder?
11. What is reverberation? On what factors does it depend? ***(Nagpur U. 2009)***
12. Derive Sabine's reverberation formula and explain its significance? ***(Nagpur U. 2011, 2009, 2008)***
13. Enumerate the features that an auditorium should have for good acoustics. How can these features be incorporated in the design of an auditorium ***(Nagpur U. 2011, 2009, 2008)***
14. What remedial measures would you suggest for the following defects in an existing auditorium
    (*i*) Uneven distribution of Sound,
    (*ii*) Unintelligible audibility of speech. ***(Nagpur U. 2011)***
15. Explain what causes reverberation in a hall and how it can be reduced? ***(Kanpur 2002)***
16. Describe any modern method of recording and reproducing the sound ***(Nagpur 2009, 2007, 2006)***
17. The volume of the hall is 5900 m^3 and its total surface area of absorption of sound is 200 m^2 and has absorption coefficient 0.4. Calculate the reverberation period of the hall. [**Ans.** 11.6525 sec.]***(Nagpur U. 2009)***

18. Derive an expression for the velocity of sound in a fluid. ***(Nagpur 2010, 2008, 2007)***

19. Calculate the time of reverberation of the empty hall of volume 80,000 cu.ft. and total absorption without audience 775 sabines. [**Ans.** $t = \frac{0.05\,V}{A} = 5.161\,\text{sec}.$] ***(Nagpur U. 2010)***

20. What is reverberation time? ***(Nagpur U. 2011, 2010)***

MULTIPLE CHOICE QUESTIONS

18.1 For a good auditorium, the reverberation time should be

(*a*) moderate (*b*) large (*c*) negligibly small (*d*) very large

18.2 For the reverberation free hall, the walls and ceiling should be covered by materials having absorption coefficient:

(*a*) small (*b*) zero (*c*) high (*d*) moderate

18.3 At nodal points, the refractive index of the medium is :

(*a*) maximum (*b*) minimum (*c*) does not depend upon R.I

ANSWERS

18.1 (*c*) **18.2** (*c*) **18.3** (*a*)

SECTION VI

Geometrical Optics

19. FERMAT'S PRINCIPLE
20. THEORY OF IMAGE FORMATION
21. ABERRATION IN IMAGES
22. OPTICAL INSTRUMENTS

REFRACTION AT SPHERICAL SURFACES FERMAT'S PRINCIPLE

19

Chapter

19.1 SIGN CONVENTIONS

In order to study the phenomena in optics, it is necessary to consider the sign conventions and assumptions which are essentially required in deriving various relations for refraction at spherical surfaces. The sign conventions to be followed are:

(*i*) All figures are drawn with incident light travelling from left to right.

(*ii*) All distances are measured from the optical centre of a lens or from the pole of a reflecting or a refracting surface.

(*iii*) Distances measured in the same direction in which light is travelling, are counted as **positive,** while those measured in the opposite direction are taken as ***negative.***

(*iv*) Transverse distances are taken as positive when measured upward from the principal axis and **negative** when measured below it.

Assumptions. In deriving various relations, for the refraction at spherical surfaces following assumptions are made:

(*i*) The aperture of the surface is small.

(*ii*) The object is a point object and lies on the principal axis.

(*iii*) The incident ray from the object strikes the surface at a point very close to the pole, so that the angles of incidence and refraction are very small.

(*iv*) The incident ray, the refracted ray and the normal make small angles with the principal axis. In other words the rays striking the surface have oblique incidence and oblique refraction.

19.2 OPTICAL PATH AND GEOMETRICAL PATH

Before discussing the optical and geometrical paths, let us define first the refractive index (μ) of the medium. Different ways of defining refractive index are:

Refractive index

(*i*) When a ray of light travels from one medium '*a*' into another medium '*b*' its suffers refraction. If *i* is the angle of incidence (in the medium '*a*') and *r* the corresponding angle of refraction (in the medium '*b*'), then

$$\frac{\sin i}{\sin r} = {}^{a}\mu_{b} \qquad \text{... (19.1)}$$

where ${}^{a}\mu_{b}$ is the refractive index of the medium '*b*' with respect to the medium '*a*'.

(*ii*) According to the wave theory of light, if *c* is the velocity of light in vacuum or free space and *v* that in a medium, then

Refractive index of the medium $\mu = \dfrac{c}{v}$

The velocity of light in air is taken to be very nearly the same as in vacuum.

$$\therefore \qquad \mu = \frac{\text{velocity of light in air or vacuum}}{\text{velocity of light in the medium}}$$

Thus, the refractive index of a medium is defined as the ratio of the velocity of light in vacuum *(or air or free space) to the velocity of light in the medium.*

(*iii*) Velocity of light in a medium $v = n\lambda$, where n is the frequency of light and λ the wave-length in the medium.

Therefore, when monochromatic light travels from one medium to the other, its velocity and hence the wave length changes but the frequency remains constant.

$$\text{Hence} \quad {}^a\mu_b = \frac{\text{velocity of light in medium } a}{\text{velocity of light in medium } b}$$

$$= \frac{n\lambda_a}{n\lambda_b} = \frac{\lambda_a}{\lambda_b} \qquad \text{... (19.2)}$$

Optical path. The geometrical path between two points is the actual distance travelled by light in going from one point to the other.

The optical path is defined as the product of the geometrical path and the refractive index of the medium.

If a ray of light travels a geometrical path x in a medium of refractive index μ, then

Optical path = μx = Refractive index × geometrical path. In a given time, light travels the same optical path in different media. If in time t light travels a distance x_1 in a medium of refractive index μ_1, and a distance x_2 in a medium of refractive index μ_2. then

$$\mu_1 x_1 = \mu_2 x_2 \qquad \text{... (19.3)}$$

The wave length of light in vacuum λ will chang to λ/μ in a medium of refractive index μ; as explained below:

Refractive index of a medium $\mu = \dfrac{\text{velocity of light and vacuum}}{\text{velocity of light in medium}} = \dfrac{c}{v}$. When monochromatic light travels from vacuum (or air) into a medium its velocity and hence the wavelength λ changes but the frequency n remains unchanged.

As $v = n\lambda$, we have

$$\mu = \frac{c}{v} = \frac{n\lambda}{n\lambda_m} = \frac{\lambda}{\lambda_m}$$

where λ is the wave length of monochromatic light in vacuum and λ_m that in the medium.

$$\text{Hence} \quad \lambda_m = \frac{\lambda}{\mu} \qquad \text{... (19.4)}$$

19.3 FERMAT'S PRINCIPLE

As stated by Fermat in the year 1658, the principle is as follows:

When a ray of light goes from one point to another through a set of media, it always follows a path along which the time taken is minimum.

It should be note that this statement of Fermat's principle is not complete, for there are a number of cases in which the optical path is *maximum* or else neither a maximum nor a minimum but *stationary*.

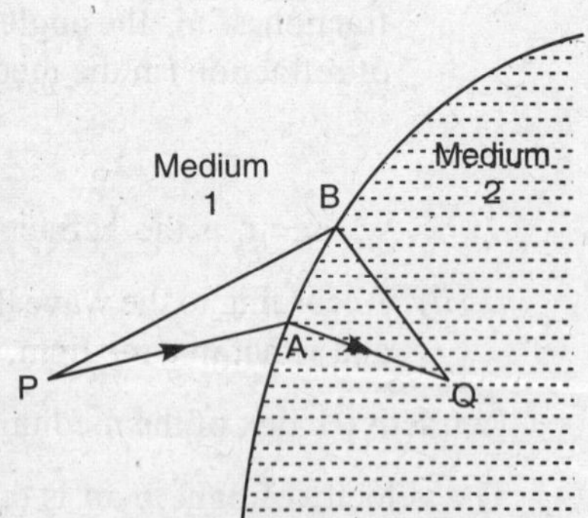

Fig. 19.1

In Fig.19.1, P and Q are two points in two different media i.e. medium 1 and 2 respectively. Let PAQ be a real path of light.

The time taken by light in reaching from P to Q along this path is given by

$$t = \frac{PA}{V_1} + \frac{AQ}{V_2} \qquad \text{... } (i)$$

Where V_1 and V_2 are the speed of light in the two media.

A neighbouring path is PBQ, **and this may** be expressed in terms of parameter S, which could be either the lateral distance AB or angle APB or anything which shows a **lateral shift** of path. Fermat's principle now says that

$$\frac{dt}{ds} = 0 \qquad \text{... } (ii)$$

In practice, *i.e.* in 3-dimensional space, a lateral shift of path can always be expressed in terms of two independent parameters S_1 and S_2. Therefore, Eq. (*ii*) is equivalent to two independent conditions:

$$\left(\frac{dt}{ds_1}\right)_{s_2} = 0 \text{ and } \left(\frac{dt}{ds_2}\right)_{s_1} = 0 \qquad \text{... } (iii)$$

In a medium of refractive index μ, the speed of light is given by $v = c/\mu$, where c is the speed of light in vacuum. Therefore Eq. (*i*) can be expressed as

$$t = \frac{1}{c}(\mu_1 PA + \mu_2 AQ) \qquad \text{... } (iv)$$

The quantity ($\mu_1 PA + \mu_2 AQ$) is often called as the optical path. If we represent it by p, then since c is constant, Eq. (*ii*) is equivalent to

$$\frac{dp}{ds} = 0 \qquad \text{... } (19.5)$$

This states Fermat's principle in terms of optical path, instead of time t.

19.4 LAWS OF REFLECTION FROM FERMAT'S PRINCIPLE

Consider a plane mirror $M_1 M_2$ and a plane $abcd$ normal to it. A ray of light AO incident at O is reflected along OB (Fig 19.2). The rays AO, OB and normal ON lie in the plane $abcd$. Consider another ray AO' reflected from the point O' where OO' is normal to the plane $abcd$. The incident ray AO' is reflected along $O'B$. Now $AO' > AO$ and $O'B > OB$. Thus the path $(AO' + O'B) > (AO + OB)$, which is against Fermat's principle. Hence O' must coincide with O for the path to be minimum. Hence AO, OB and ON all lie in the same plane.

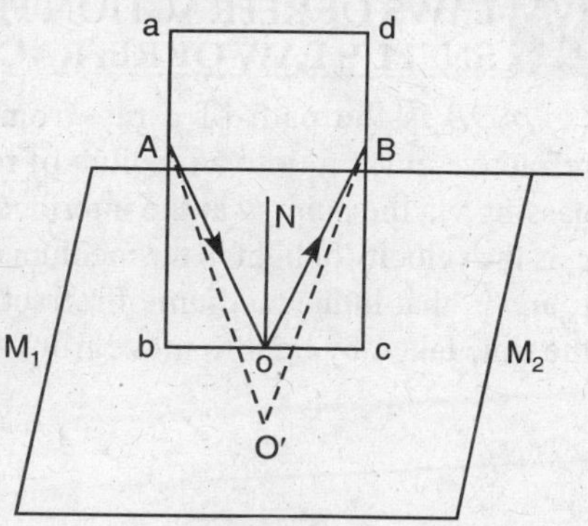

Fig. 19.2

To derive the *second part of the law of reflection*, consider the point of incidence O on the mirror $M_1 M_2$ so that the time taken by light to travel from A to B via O is minimum. Let i be the angle of incidence and r that of reflection (Fig.19.3). If c is the velocity of light, then time t to travel the optical path $(AO + OB)$ is given by

$$t = \frac{AO + OB}{c}$$

Now $$AO = \sqrt{a^2 + x^2}$$

and $$OB = \sqrt{b^2 + (d - x)^2}$$

$$t = \frac{\sqrt{a^2 + x^2} + \sqrt{b^2 + (d-x)^2}}{c} \qquad \text{... } (i)$$

If the point O is displaced slightly along $M_1 M_2$, x will change by dx and slightly different path will be obtained for light and time will be different. As A and B are fixed d will remain the same for the new path.

According to Fermat's principle the time of actual traverse should be *extremum i.e.* minimum, maximum or stationary or

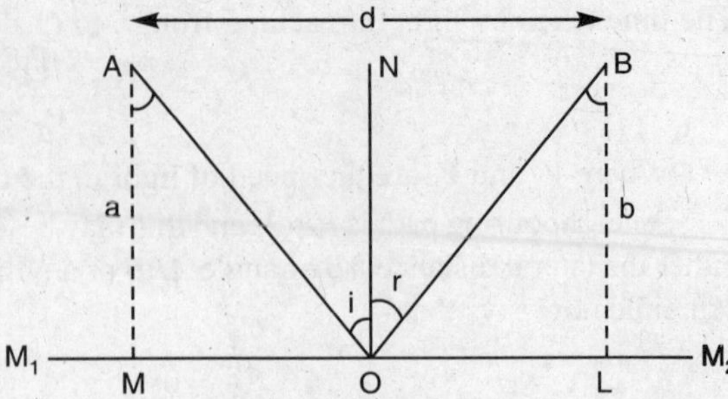

Fig. 19.3

$$\frac{dt}{dx} = 0$$

$$\therefore \quad \frac{dt}{dx} = \frac{1}{c}\left[\frac{x}{\sqrt{a^2+x^2}} - \frac{d-x}{\sqrt{b^2+(d-x)^2}}\right] = 0$$

or $$\frac{x}{\sqrt{a^2+x^2}} = \frac{d-x}{\sqrt{b^2+(d-x)^2}}$$

or $$\sin i = \sin r$$

or $$i = r \qquad \text{... (19.6)}$$

Thus the path of light from A to B via the plane mirror traversed in the least time is such that the angle of incidence is equal to the angle of reflection.

The second derivative of (*i*) is

$$\frac{d^2t}{dx^2} = \frac{1}{c}\left\{\frac{a^2}{(a^2+x^2)^{3/2}} + \frac{d^2}{\{b^2+(d-x)^2\}^{3/2}}\right\} \qquad \text{... (19.7)}$$

This is a positive quantity. Hence under the condition $i = r$, the time along the path AOB is minimum.

Thus, Fermat's principle leads to both parts of the law of reflection.

19.5 LAWS OF REFRACTION FROM FERMAT'S PRINCIPLE SNELL'S LAW OF REFRACTION

AOB is the path of a ray from the medium of refractive index μ_1 into a medium of refractive index μ_2 passing via the point O at the interface where $\mu_2 > \mu_1$. If c_1 is the velocity of light in the medium of refractive index μ_1 and c_2 that in the medium of refractive index μ_2, then the time taken by light to move from A to B is given by

$$t = \frac{AO}{c_1} + \frac{OB}{c_2}$$

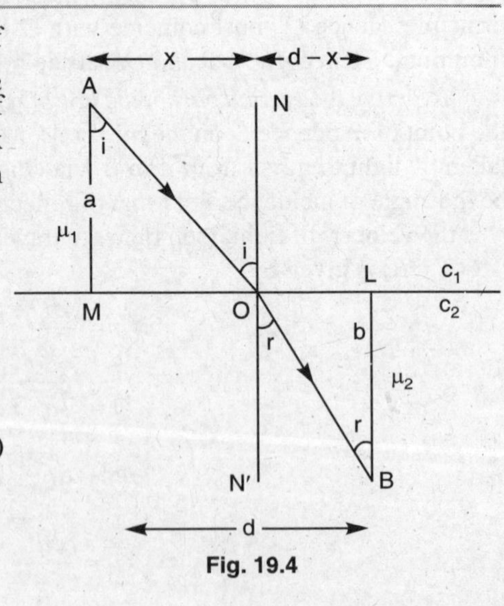

Fig. 19.4

Now $AO = \sqrt{a^2+x^2}$ and $OB = \sqrt{b^2+(d-x)^2}$

$$\therefore \quad t = \frac{\sqrt{a^2+x^2}}{c_1} + \frac{\sqrt{b^2+(d-x)^2}}{c_2} \qquad \text{... } (i)$$

For t to be minimum, maximum or stationary

$$\frac{dx}{dt} = \frac{1}{c_1}\frac{x}{\sqrt{a^2+x^2}} + \frac{1}{c_2}\frac{(d-x)}{\sqrt{b^2+(d-x)^2}} = 0$$

or $$\frac{\sin i}{c_1} = \frac{\sin r}{c_2}$$

Now $\mu_1 = \frac{c}{c_1}$ and $\mu_2 = \frac{c}{c_2}$

where c is the velocity of light in vacuum.

$\therefore$ $$\mu_1 \sin i = \mu_2 \sin r \quad \text{... (19.8)}$$

Thus Snell's law is proved. It determines the actual path of the ray.

It is seen that the second derivative of (*i*) is positive and hence the actual path of the ray is minimum.

19.6 REFRACTION AT A SINGLE CONVEX SPHERICAL SURFACE TO PROVE RELATION $\frac{\mu}{v} - \frac{1}{u} = \frac{\mu - 1}{R}$

When refraction takes place at a convex spherical surface, the image formed may be *real* or *virtual*.

Case 1: When refraction takes place at a convex spherical surface and a real image is formed. Consider a convex spherical surface of radius of curvature R separating a less dense medium of refractive index μ_1 on the left from a more dense medium of refractive index μ_2 ($\mu_2 > \mu_1$). Let C be the centre of curvature and P the pole of the spherical surface. Consider a point object O lying on the principal axis in a rarer medium of refractive index μ_1 at a distance u from the pole P in front of the convex spherical surface, the distance u being greater than the radius of curvature R of the surface. A ray of light OD incident at D after refraction bends towards the normal and gives rise to a real image at I at a distance v from the pole P. If i and r are the angles of incidence and refraction respectively and α, β and γ the angles which the incident ray, the refracted ray and the normal at D make with the principal axis, then according to Snell's law.

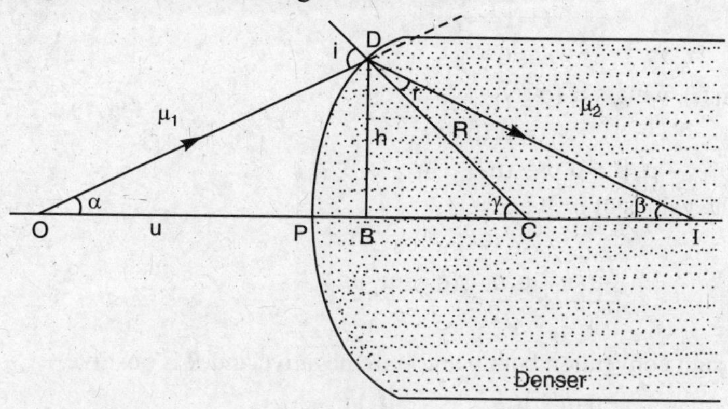

Fig. 19.5

$$\mu_1 \sin i = \mu_2 \sin r$$

As the aperture of the spherical surface is small, the angles i and r are also very small and

$$OD = OP = u, \; ID = IP = v \text{ and } CD = R$$

$$\mu_1 \, i = \mu_2 \, r \quad \text{... } (i)$$

Referring to Fig 19.5 in Δ ODC, i is the exterior angle so that

$$i = \alpha + \gamma$$

In triangle IDC $\quad \gamma = r + \beta$ or $r = \gamma - \beta$

Substituting the values of i and r in (*i*), we have

$$\mu_1(\alpha + \gamma) = \mu_2 (\gamma - \beta)$$

or $$\mu_2\beta + \mu_1\alpha = (\mu_2 - \mu_1)\gamma \qquad \text{... (ii)}$$

Denoting *DB* the perpendicular on the axis by *h*, we get

$$\alpha = \frac{h}{u}, \beta = \frac{h}{v} \text{ and } \gamma = \frac{h}{R}$$

According to the sign convention *u* is negative and both *v* and *R* are positive. Substituting the values of α, β, γ, in (*ii*) and applying the sign convention, we get

$$\frac{\mu_2 h}{v} - \frac{\mu_1 h}{u} = \frac{(\mu_2 - \mu_1)h}{R}$$

or $$\frac{\mu_2}{v} - \frac{\mu_1}{u} = \frac{\mu_2 - \mu_1}{R} \qquad \text{... (iii)}$$

When the rarer medium is air and the denser medium has a refractive index μ with respect to it, then $\mu_1 = 1$ and $\mu_2 = \mu$ (say).

Substituting these values in (*iii*), we have

$$\frac{\mu - 1}{R} = \frac{\mu}{v} - \frac{1}{u} \qquad \text{... (19.9)}$$

Generally the refracting surface lies in air, hence relation (19.9) is of great significance in practice.

Case 2: When refraction takes place at a convex spherical surface and virtual image is formed. If *O* is a point object and *I* its virtual image formed after refraction at a convex spherical surface, then in Δ*ODC* in Fig **19.6.**

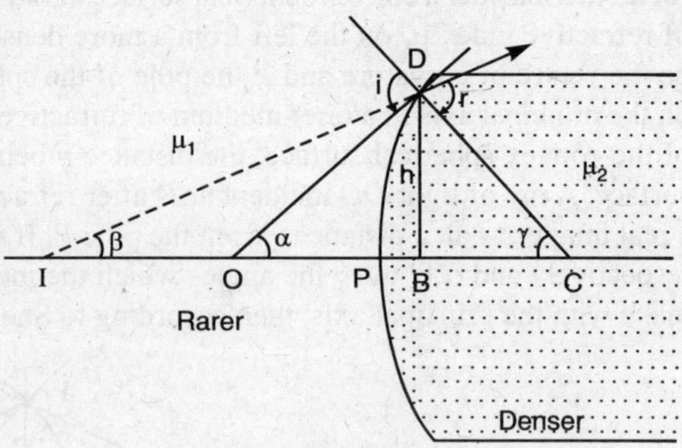

Fig. 19.6

$$i = \alpha + \gamma$$

and in Δ *IDC* $$r = \beta + \gamma$$

Substituting in (*i*), we get $\mu_1(\alpha + \gamma)$

$$= \mu_2(\beta + \gamma)$$

or $$-\mu_2\beta + \mu_1\alpha = (\mu_2 - \mu_1)\gamma$$

Substituting the values of α, β and γ, we have

$$\frac{-\mu_2 h}{v} + \frac{\mu_1 h}{u} = \frac{(\mu_2 - \mu_1)h}{R}$$

According to sign convention both *u* and *v* are negative and *R* is positive.

$$\frac{-\mu_2 h}{-v} + \frac{\mu_1 h}{-u} = \frac{(\mu_2 - \mu_1)h}{R}$$

or $$\frac{\mu_2}{v} - \frac{\mu_1}{u} = \frac{\mu_2 - \mu_1}{R} \qquad \text{... (19.10)}$$

It is same relation as proved in case 1 (*iii*)

19.7 REFRACTION AT A SINGLE CONCAVE SPHERICAL SURFACE

In this case the incident ray *OD* after refraction at *D* bends towards the normal *CD* and meets the principal axis at *I* when produced backward. The image formed is thus virtual, as shown in Fig. 19.7. Referring to Δ *ODC*, we have

$$\alpha + i = \gamma \quad \text{or} \quad i = \gamma - \alpha$$

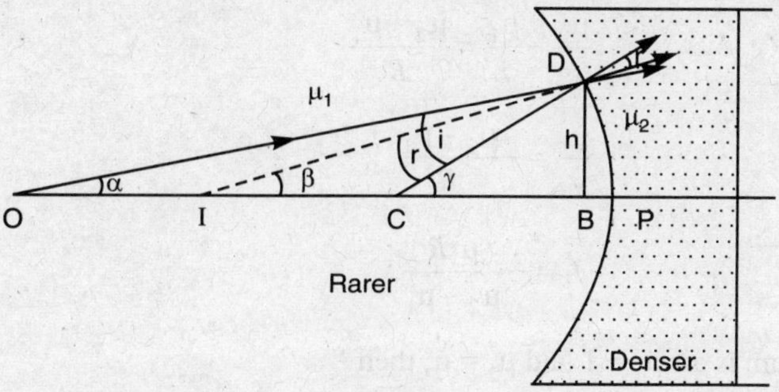

Fig. 19.7

Referring to ΔIDC, we have

$\beta + r = \gamma$ or $r = \gamma - \beta$

Substituting the values of i and r in the relation $\mu_1 i = \mu_2 r$, we get

$$\mu_1(\gamma - \alpha) = \mu_2(\gamma - \beta)$$

or $$\mu_2 \beta - \mu_1 \alpha = (\mu_2 - \mu_1)\gamma$$

Substituting the values of α, β and γ, we get

$$\frac{\mu_2 h}{v} - \frac{\mu_1 h}{u} = \frac{(\mu_2 - \mu_1)h}{R}$$

According to sign convention u, v and R are all negative

$$\therefore \quad \frac{\mu_2}{v} - \frac{\mu_1}{u} = \frac{\mu_2 - \mu_1}{R} \qquad \text{... (19.11)}$$

19.8 SOME DEFINITIONS

(*i*) **Power of a spherical surface**

If a surface of refractive index μ_2 lies in a medium of refractive index μ_1 and has a radius of curvature R, then the expression $\frac{(\mu_2 - \mu_1)}{R}$ is called the **power** of the refracting surface. According to the sign convention the radius of curvature of a convex surface is positive and since $(\mu_2 - \mu_1)$ is also positive, the power of a convex spherical surface is **positive.** For a concave surface $(\mu_2 - \mu_1)$ is positive and R is negative, hence its power is **negative.** If we denote this power by P, then

$$P = \frac{\mu_2}{v} - \frac{\mu_1}{u} \qquad \text{... } (i)$$

If the first medium is air then $\mu_1 = 1$ and $\mu_2 = \mu$.

$$\therefore \quad P = \frac{\mu}{v} - \frac{1}{u}$$

If R, v and u are expressed in metres, the power is expressed in **diopters.**

(*ii*) **Principal foci.** The equation $\frac{\mu_2}{v} - \frac{\mu_1}{u} = \frac{\mu_2 - \mu_1}{R}$ gives two important points on the principal axis, called **the first and the second principal foci** F_1 and F_2 respectively. F_1 is the position of object for which the image lies at infinity, while F_2 is the position of the image for which the object lies at infinity.

First principal focus. If f_1 denotes the distance of the object for which the image lies at infinity *i.e.* $v = \infty$, then

$$\frac{\mu_2}{\infty}-\frac{\mu_1}{f_1}=\frac{\mu_2-\mu_1}{R}$$

or $$\frac{\mu_1}{f_1}=-\frac{\mu_2-\mu_1}{R}$$

$$f_1=-\frac{\mu_1 R}{\mu_2-\mu_1} \quad \text{... (ii)}$$

If the first medium is air $\mu_1 = 1$ and $\mu_2 = \mu$, then

$$f_1=-\frac{R}{\mu-1} \quad \text{... (19.13)}$$

Second principal focus. If the object is at ∞, the position of image is represented by F_2. If f_2 is the distance of F_2 from the pole, then

$$\frac{\mu_2}{f_2}-\frac{\mu_1}{\infty}=\frac{\mu_2-\mu_1}{R}$$

or $$\frac{\mu_2}{f_2}=\frac{\mu_2-\mu_1}{R}$$

∴ $$f_2=\frac{\mu_2 R}{\mu_2-\mu_1} \quad \text{... (iii)}$$

If the first medium is air then $\mu_1 = 1$ and $\mu_2 = \mu$

$$f_2=\frac{\mu R}{\mu-1} \quad \text{... (19.14)}$$

(*iii*) **Principal planes.** Planes through F_1 and F_2 perpendicular to the principal axis are called the **first and the second principal planes** respectively of the refracting surface.

From (*ii*) and (*iii*), we have

$$\frac{f_1}{f_2}=-\frac{\mu_1}{\mu_2} \text{ or } \frac{f_1}{\mu_1}+\frac{f_2}{\mu_2}=0$$

or $$\mu_1 f_2+\mu_2 f_1=0$$

If the first medium is air, then $\mu_1 = 1$ and $\mu_2 = \mu$

∴ $$f_2+\mu f_1=0 \quad \text{... (19.15)}$$

This gives the distinction between the two focal distances.

Example 19.1. *Show that* $\frac{f_1}{u}+\frac{f_2}{v}=1$ *where f_1 and f_2 are the principal focal lengths of a spherical refracting surface, u and v having their usual meaning.*

Solution. If an object is placed at a distance u in a medium of refractive index μ_1 from a convex spherical surface of radius of curvature R and image is formed at a distance v in the medium of refractive index μ_2, then

$$\frac{\mu_2}{v}-\frac{\mu_1}{u}=\frac{\mu_2-\mu_1}{R} \quad \text{... (i)}$$

If the object is placed at infinity and f_2 is the corresponding value of v, then

$$\frac{\mu_2}{f_2} = \frac{\mu_2 - \mu_1}{R}$$

or

$$v = f_2 = \frac{\mu_2 R}{\mu_2 - \mu_1} \quad \text{... (ii)}$$

This value of f_2 is called the **second principal focal length.**

Again for $v = \infty$ *i.e.*, the object is so placed that the rays after refraction become parallel to the principal axis, then

$$u = f_1 = \frac{-\mu_1 R}{\mu_2 - \mu_1} \quad \text{... (iii)}$$

This value of f_1 is called the **first principal focal length** of the refracting surface.

From (*ii*) and (*iii*), we get

$$\frac{f_1}{\mu_2} = \frac{R}{\mu_2 - \mu_1} \text{ and } \frac{f_1}{\mu_1} = -\frac{R}{\mu_2 - \mu_1}$$

Adding, we gevt

$$\frac{f_1}{\mu_1} + \frac{f_2}{\mu_2} = 0 \text{ or } \frac{f_1}{f_2} = -\frac{\mu_1}{\mu_2}$$

As μ_1 and μ_2 are always positive and unequal; f_1 and f_2 are always unequal and opposite in sign.

From (*i*), we get

$$\frac{\mu_2 R}{v(\mu_2 - \mu_1)} - \frac{\mu_1 R}{u(\mu_2 - \mu_1)} = 1$$

or

$$\frac{f_2}{v} + \frac{f_1}{u} = 1$$

Example 19.2 *A convex surface of radius of curvature 30 cm separates two media of refractive indices 4/3 and 1.5. An object is kept at a distance of 20 cm from the surface in the first medium. Calculate the position of the image.* **(Nagpur U. 2008)**

Solution. Here $\mu_1 = \frac{4}{3}$, $\mu_2 = 1.5$, $R = 30$ cm $= 0.3$ m

$$u = -20 \text{ cm} = -0.2 \text{ m}, v = ?$$

Substituting these in the relation $\frac{\mu_2}{v} - \frac{\mu_1}{u} = \frac{\mu_2 - \mu_1}{R}$, we get

$$\frac{1.50}{v} + \frac{\frac{4}{3}}{0.2} = \frac{1.5 - \frac{4}{3}}{0.3} = \frac{5}{9}$$

or

$$\frac{1.50}{v} = \frac{5}{9} - \frac{20}{3} = -\frac{55}{9}$$

$\therefore$

$$v = \frac{1.50 \times 9}{-55} = -0.2455\text{m}$$

Example 19.3 *The incident face of a glass block is bounded by a concave of radius 0.03 m. A small object 2× 10^{-2} m high is situated in air at a distance 0.1 m from the vertex. Find (a) the first and the second focal lengths, (b) the image distance, (c) the lateral magnification and (d) the power of the surface. Give μ glass = 1.50*

Solution Here $\mu_2 = 1.50$, $\mu_1 = 1.0$, $u = -0.1$ m, $R = -0.03$ m

(*a*) Now
$$f_2 = \frac{\mu_2 R}{\mu_2 - \mu_1} = \frac{-1.5 \times 0.03}{1.50 - 1.00} = -0.09\text{m}$$

$$f_1 = \frac{-\mu_1 R}{\mu_2 - \mu_1} = \frac{-1 \times (-0.03)}{1.5 - 1.0} = -0.06\text{m}$$

(*b*) $\dfrac{\mu_2}{v} - \dfrac{\mu_1}{u} = \dfrac{\mu_2 - \mu_1}{R}$

∴
$$\frac{1.5}{v} - \frac{1}{(-0.1)} = \frac{1.5 - 1.0}{-0.03} = -16.667$$

or
$$\frac{1.5}{v} = -16.667 - 10 = -26.667$$

or
$$v = \frac{-1.5}{26.667} = -0.05625\text{m}$$

(*c*) Magnification $m = \dfrac{\mu_1}{\mu_2} \times \dfrac{v}{u} = \dfrac{1}{1.50}\left(\dfrac{-0.05625}{-0.1}\right) = 0.375$

(*d*) Power $P = \dfrac{\mu_2 - \mu_1}{\mu_1 R} = \dfrac{1.5 - 1.0}{0.03} = 16.67\,D$

Example 19.4 *An object is situated at the optical axis of a convex spherical surface. A ray of light starting from the object subtends an angle of 5° with the optical axis and intersects the spherical surface at a height of 2.5 cm. If the radius of curvature of the surface is +5 cm and refractive index of the material 1.5, find the angle which the refracted ray makes with the optical axis.*

Solution Let u be the distance of the object, v that of the image, R the radius of curvature and μ the refractive index of the material of the surface.

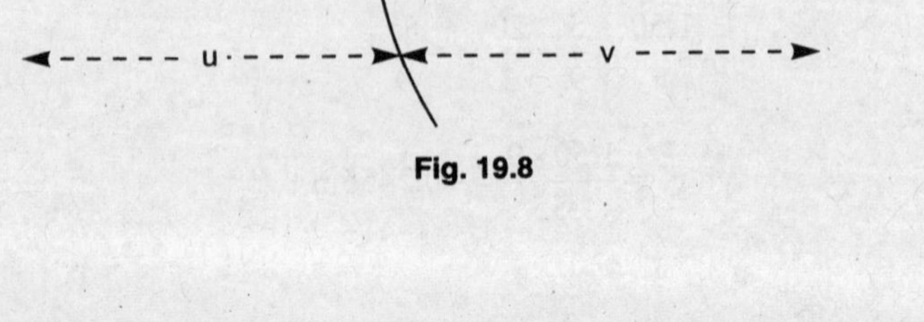

Fig. 19.8

Here $AB = 2.5$ cm, $R = 5.0$ cm, $\alpha = 5°$, $\mu = 1.5$

$$OB = OP = -u,\ IB = IP = v$$

Now $$\tan\alpha = \frac{AB}{OB} = \frac{AB}{OP} = \frac{2.5}{-u}$$

$\therefore$ $$u = \frac{-2.5}{\tan 5} = \frac{-2.5}{0.0875} = -28.57 \text{ cm}$$

As refraction is taking place from a rarer medium into a denser medium

$\therefore$ $$\frac{\mu}{v} - \frac{1}{u} = \frac{\mu - 1}{R}$$

or $$\frac{\mu}{v} = \frac{1.5-1}{5.0} - \frac{1}{28.57}$$

$$= 0.1 - 0.035 = 0.065$$

$$v = \frac{1.5}{0.065} = 23.08 \text{ cm}$$

The angle β which the refracted ray makes with the optical axis is given by

$$\tan\beta = \frac{2.5}{23.08} = 0.1083 \text{ or } \beta = 6.2°$$

19.9 LATERAL, LONGITUDINAL AND ANGULAR MAGNIFICATION

(*i*) **Lateral (Transverse) magnification:** The section of a spherical surface *MPN* separating the two media of refractive index μ_1 and μ_2 having its centre of curvature at *C* and pole at *P* is shown in Fig. 19.9. A small object *AB* is situated on the axis of the refracting surface far away from *P* in the medium of refractive index μ_1.

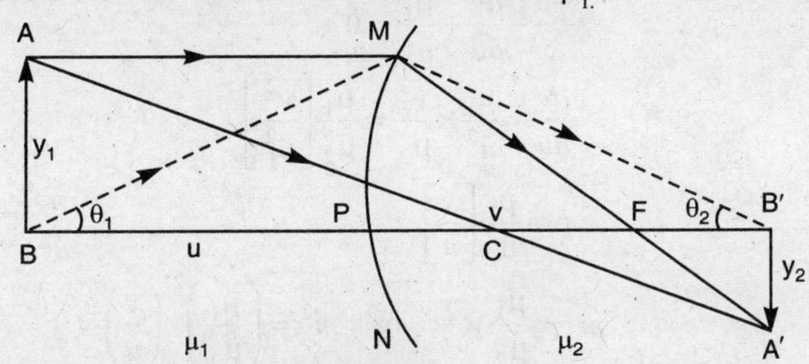

Fig. 19.9

A ray of light parallel to the principal axis starting from *A* passes through the second focus *F* after refraction. A second ray from *A* passing through *C* the centre of curvature goes undeflected. The two rays meet at *A'* which is the image of *A*. Also *B'* is the image of *B*. Hence *A'B'* is the linear image of the object *AB*.

$\therefore$ $PB = -u$, $PB' = +v$, $PC = R$, $AB = y_1$ and $A'B' = -y_2$ according to sign convention.

Lateral (or linear) magnification is defined as the size of the image divided by the size of the object.

$\therefore$ Lateral (or linear) magnification $m = \dfrac{A'B'}{AB} = \dfrac{y_2}{y_1}$

Now Δ S *ABC* and *A'B'C* are similar

$\therefore$ $$\frac{A'B'}{AB} = \frac{CB'}{CB} = \frac{PB' - PC}{PB + PC} = \frac{+v - (+R)}{-u + (+R)} = \frac{v - R}{-u + R}$$

or $$\frac{-y_2}{y_1} = \frac{v-R}{-u+R} \text{ or } \frac{y_2}{y_1} = \frac{v-R}{u-R} \quad \text{...(i)}$$

For refraction at a curved surface

$$\frac{\mu_2}{v} - \frac{\mu_1}{u} = \frac{\mu_2 - \mu_1}{R} = \frac{\mu_2}{R} - \frac{\mu_1}{R}$$

or $$\frac{\mu_2}{v} - \frac{\mu_2}{R} = \frac{\mu_1}{u} - \frac{\mu_1}{R} \text{ or } \frac{\mu_2(R-v)}{vR} = \frac{\mu_1(R-u)}{uR}$$

or $$\frac{\mu_2(R-v)}{v} = \frac{\mu_1(R-u)}{u} \text{ or } \frac{R-v}{R-u} = \frac{\mu_1}{u} \times \frac{v}{\mu_2}$$

or $$\frac{v-R}{u-R} = \frac{\mu_1}{\mu_2} \times \frac{v}{u} \quad \text{... (ii)}$$

Comparing (*i*) and (*ii*), we get

Lateral (or Linear) magnification $m = \dfrac{y_2}{y_1} = \dfrac{\mu_1}{\mu_2} \times \dfrac{v}{u}$ ... (19.16)

(*ii*) **Longitudinal magnification.** Longitudinal magnification is defined as $L = \dfrac{dv}{du}$ where dv is a small change in v corresponding to a small change du in u. For refraction at a curved surface.

$$\frac{\mu_2}{v} - \frac{\mu_1}{u} = \frac{\mu_2 - \mu_1}{R}$$

Differentiating, we have

$$\frac{-\mu_2}{v^2}\frac{dv}{du} + \frac{\mu_1}{u^2} = 0$$

$$\frac{dv}{du} = \frac{\mu_1}{u^2} \times \frac{v^2}{\mu_2} = \frac{\mu_1}{\mu_2}\left[\frac{v^2}{u^2}\right]$$

or $$L = \frac{\mu_1}{\mu_2}\left[\frac{v}{u}\right]^2$$

But $$m = \frac{\mu_1}{\mu_2} \times \frac{v}{u} \quad \text{or} \quad m^2 = \left(\frac{\mu_1}{\mu_2}\right)^2 \left(\frac{v}{u}\right)^2$$

$\therefore$ $$L = \left(\frac{\mu_1}{\mu_2}\right)^2 \left(\frac{v}{u}\right)^2 \left(\frac{\mu_2}{\mu_1}\right) = m^2 \frac{\mu_2}{\mu_1} \quad \text{... (19.17)}$$

(*iii*) **Angular magnification (Lagrange's Law)**

Now join *BM* and *MB'*, then *BM* is the incident ray starting from the point *B* of the object which after refraction at *M* meets the principal axis at *B'*. Let θ_1 be the angle that the ray *BM* starting from the object makes with the axis and θ_2 the corresponding angle that the refracted ray meeting the image at *B'* makes with the axis, then,

Angular magnification $\alpha = \dfrac{\theta_2}{\theta_1} = \dfrac{\tan\theta_2}{\tan\theta_1}$ as both θ_1 and θ_2 are small angles.

Now $$\frac{\tan\theta_2}{\tan\theta_1} = \frac{MP/v}{MP/u} = \frac{u}{v}$$

But linear magnification $m = \dfrac{y_2}{y_1} = \dfrac{\mu_1}{\mu_2}\dfrac{v}{u}$

$$\therefore \quad \frac{y_2}{y_1} \times \frac{\tan\theta_2}{\tan\theta_1} = \frac{\mu_1}{\mu_2} \times \frac{v}{u} \times \frac{u}{v} = \frac{\mu_1}{\mu_2}$$

or $$\mu_1 y_1 \tan\theta_1 = \mu_2 y_2 \tan\theta_2 = \text{a constant.} \qquad \text{... (19.18)}$$

This is known as Lagrange's law or Helmholtz relation.

Lagrange's law can be used for refraction at a number of co-axial refracting surfaces. In such a case $\mu_1 y_1 \tan\theta_1 = \mu_2 y_2 \tan\theta_2 = \mu_3 y_3 \tan\theta_3 .... = \mu_n y_n \tan\theta_n$. As θ is small, we have

$$\mu_1 y_1 \theta_1 = \mu_2 y_2 \theta_2 = \mu_3 y_3 \theta_3 ... = \mu_n y_n \theta_n \qquad \text{... (19.19)}$$

where y_n is the size of the final image formed by refraction at the $(n-1)$th refracting surface situated in a medium of refractive index μ_n

Example 19.5 *When seen normally through a flat surface the greatest thickness of a hemispherical glass block appears to be 4.8 cm but when seen normally from the curved surface it appears to be 6 cm thick. The actual thickness is 7.2 cm. Find (a) the refractive index of glass and (b) radius of curvature of the surface.*

Solution. The two cases are shown in Fig 19.10 (*a*) and (*b*).

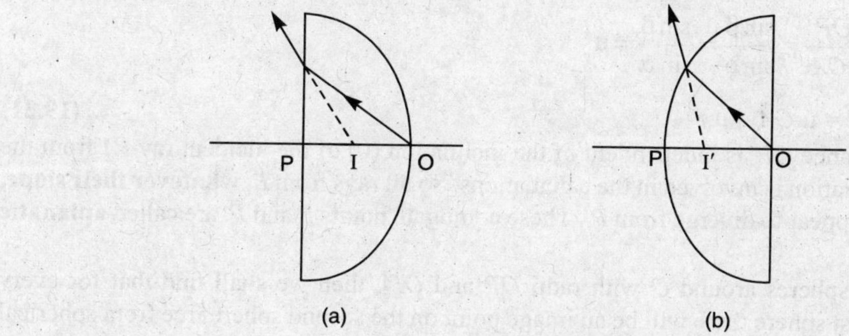

Fig. 19.10

1st Case $\mu_1 = ?, \mu_2 = 1, R = \infty, u = OP = 7.2$ cm

and $v = IP = 4.8$ cm

$$\frac{\mu_2 - \mu_1}{R} = \frac{1}{v} - \frac{\mu_1}{u} \text{ or } 0 = \frac{1}{4.8} - \frac{\mu_1}{7.2} \text{ or } \mu_1 = \frac{7.2}{4.8} = 1.5$$

2nd Case. $u = OP = 7.2$ cm, $v = I'P = 6$ cm and $R = ?$

$$\therefore \quad \frac{1-1.5}{R} = \frac{1}{6} - \frac{1.5}{7.2} \text{ or } R = 24 \times 0.5 = 12.0 \text{ cm}$$

19.10 APLANATIC POINTS OF A SPHERE

Consider a spherical medium with centre O, radius r and refractive index μ as shown in Fig. 19.11. The medium surrounding the sphere is air. Consider a object point P at distance $OP = r/\mu$. Consider any ray PA from P. It is refracted along AB. Let BA produced meet the line OP at P'. We have to calculate the distance OP'

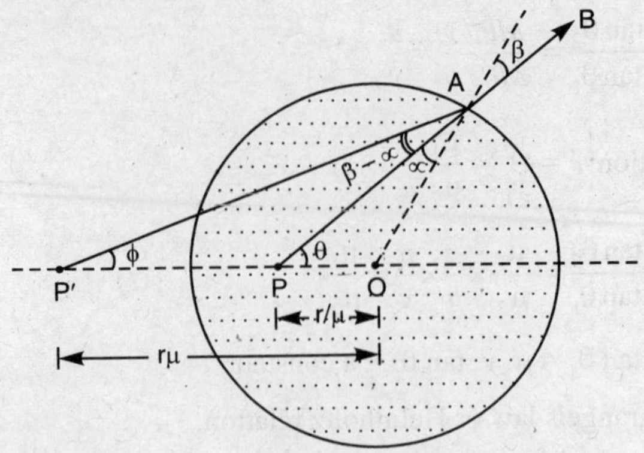

Fig. 19.11

In Δ *OPA*,

$$\frac{\sin\theta}{\sin\alpha} = \frac{OA}{OP} = \frac{r}{r/\mu} = \mu \qquad \text{... (i)}$$

But by Snell's law, $\mu = \dfrac{\sin\beta}{\sin\alpha}$ ... (*ii*)

Hence, it follows that $\beta = \theta$ [∴ using Eq. (*i*) and (*ii*)]

In Δ *AP′P*,

$$\phi = \theta - (\beta - \alpha) = \beta - (\beta - \alpha) = \alpha$$

$$\therefore \qquad \frac{OP'}{OA} = \frac{\sin\beta}{\sin\phi} = \frac{\sin\beta}{\sin\alpha} = \mu$$

This implies, $OP' = \mu.OA = \mu.r$ **... (19.21)**

We note that distance *OP′* is independent of the inclination (θ) of the incident ray *PA* from the axis *PO*. No approximation is involved in the calculations. So all rays from *P*, **whatever their slope,** will after refraction, appear to diverge from *P′*. These conjugate points *P* and *P′* are called **aplanatic points** of the sphere.

If we draw two spheres around *O* with radii *OP* and *OP′*, then we shall find that for every object point on the first sphere there will be an image point on the second sphere free from spherical aberation. These two surfaces are called **aplanatic surfaces.** The linear magnification is $r\mu \div r/\mu = \mu^2$, and the surface magnification is μ^4.

Practical use of aplanatic points of a sphere is made in the construction of (*i*) an oil immersion lens, and (*ii*) a 'meniscus' lens (Refer article 21.8, 21.9, 21.10).

EXERCISE

1. State the sign convention and assumptions in deriving various relations for refraction at spherical surfaces.
2. (*a*) Defined refractive index
   (*b*) Define optical path. What is the difference between optical path and geometrical path of a ray? How are they related? ***(Meerut U.2003, 2002, G.N.D.U. 2000)***
3. (*a*) The wave length of light in vacuum is λ. What will be its wave length in a medium of refractive index μ? ***(Pbi. U. 2001)***
   (*b*) The wave length λ of light in vacuum changes to $\dfrac{\lambda}{\mu}$ in a medium of refractive index μ Comment. ***(Pbi. U. 2004)***

4. State and explain Fermat's principle of least time (or stationary time)

(Nagpur U.2011, 2008, Meerut U. 2002, 2001, Kerala U. 2001)

5. Using Fermat's principle establish the laws of reflection of light.

(Nagpur W/2008, Bangalore 2002)

6. Using Fermat's principle establish the laws of refraction.

(Gujrat U. 2004, Nagpur S/2006)

7. Derive the relation $\frac{\mu}{v} - \frac{1}{u} = \frac{\mu - 1}{R}$ for refraction at a single convex spherical surface.

8. Derive a relation connecting *u, v, R* and μ when refraction takes place at a concave spherical surface.

9. What do you understand by (*i*) the power (*ii*) principal foci and (*iii*) Principal planes of a spherical surface?

10. Define (*i*) lateral magnification and (*ii*) longitudinal magnification and calculate their values.

11. Establish Helmholtz Lagrange formula for angular magnification for a single curved surface. ***(Gauhati U.2000 Nagpur U.2008, Gujrath 2000)***

12. Explain what do you understand by "aplanatic point'? What are aplanatic points of a refracting sphere? Discuss the aplanatic points in case of two concentric refracting spheres.

10. A convex surface of radius of curvature 0.2m separates two media of refractive indices 1.33 and 1.50. An object is kept in the first medium at a distance of 2.4 m from the refracting surface. Calculate the position and nature of image.

[**Ans.** The image is formed at a distance of 5.07m from the refracting surface and is real]

THEORY OF IMAGE FORMATION

20

Chapter

INTRODUCTION

In this chapter we shall deal with another important generalization in the laws of optics. This pertains to the theory of image formation by any complex optical system. For this we will discuss first the cardinal points of an optical system and then we will apply this concept to thick lens and lens combination. At the end, we will study the eyepieces which are used in telescopes and microscopes.

20.1 THIN CONVEX LENS

Suppose A and B are the two curved surfaces of a thin double convex lens of refractive index μ_2 and separating it from a medium of refractive index μ_1 on either side. Let R_1 and R_2 be the radii of curvature of the surfaces A and B respectively. Consider a point-object O placed in front of the face A on the principal axis as shown in fig. 20.1. A ray of light starting from O and passing through the optical centre C suffers no deviation. An oblique ray OD meeting the convex surface A becomes convergent and will meet the axis at I_1 if the lens material were continuous. Thus I_1 is the real image of O formed by refraction at the surface A at a distance v_1, then

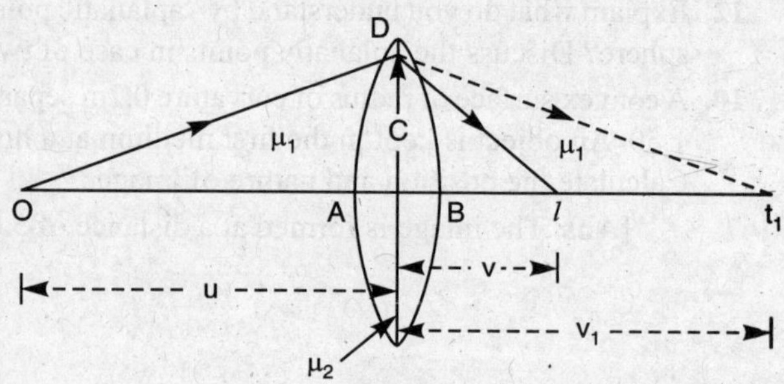

Fig. 20.1

$$\frac{\mu_2}{v_1} - \frac{\mu_1}{u} = \frac{\mu_2 - \mu_1}{R_1} \qquad \text{... (i)}$$

The rays are now refracted from the second surface B of radius of curvature R_2 of the lens. The virtual image I_1 is regarded as the object for this surface and the final image is formed at I at a distance v and lies in the medium of refractive index μ_1. As the rays are now passing from the medium of refractive index μ_2 to the medium of refractive index μ_1, the equation for refraction can be written as

$$\frac{\mu_1}{v} - \frac{\mu_2}{v_1} = \frac{\mu_1 - \mu_2}{R_2} = \frac{\mu_2 - \mu_1}{-R_2} \qquad \text{... (ii)}$$

Adding (*i*) and (*ii*), we get

$$\frac{\mu_1}{v} - \frac{\mu_1}{u} = (\mu_2 - \mu_1)\left(\frac{1}{R_1} - \frac{1}{R_2}\right)$$

or

$$\frac{1}{v} - \frac{1}{u} = \left(\frac{\mu_2 - \mu_1}{\mu_1}\right)\left(\frac{1}{R_1} - \frac{1}{R_2}\right)$$

If the object lies at infinity, then $\frac{1}{u} = 0$ and $v = f$

$$\therefore \qquad \frac{1}{f} = \left(\frac{\mu_2 - \mu_1}{\mu_1}\right)\left(\frac{1}{R_1} - \frac{1}{R_2}\right) \qquad \text{... (iii)}$$

If the lens is placed in air, then $\mu_1 = 1$ and $\frac{\mu_2}{\mu_1} = \mu$

$$\frac{1}{f} = (\mu - 1)\left(\frac{1}{R_1} - \frac{1}{R_2}\right) \qquad \text{...(20.1)}$$

Approximations. (*i*) The aperture is small and hence the angles made by the rays with the optical axis are also small.

(*ii*) The lens is very thin so that the distances measured from the surfaces can be taken equal to those measured from the optical centre.

Limitations. This formula is applicable to a very thin lens of a very small aperture.

20.2 THIN CONCAVE LENS

Consider a thin concave lens bounded by surfaces A and B of refractive index μ_2 placed in a medium of refractive index μ_1 on either side.

Let R_1 and R_2 be the radii of curvature of the two surfaces A and B respectively and O be a point object placed in front of the face A on the principal axis. A ray of light OD suffers refraction at the concave surface and diverges along DE and appears to come from I_1. Hence I_1 is the virtual image formed by refraction at a distance $CI_1 = v_1$, then

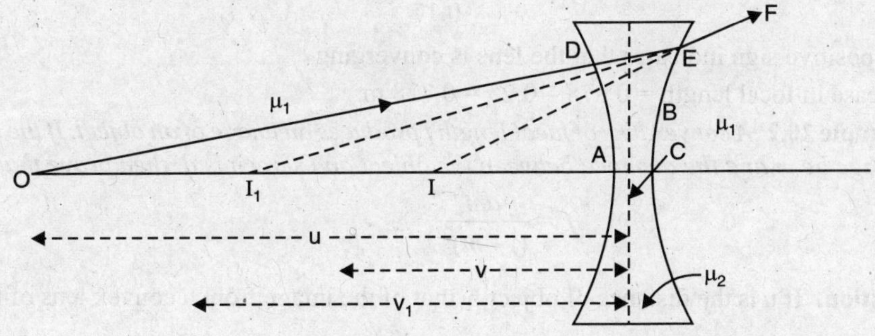

Fig. 20.2

$$\frac{\mu_2}{v_1} - \frac{\mu_1}{u} = \frac{\mu_2 - \mu_1}{R_1} \qquad \text{... (i)}$$

The rays on striking the second face B of the lens are deviated along EF. The virtual image I_1 is regarded as object for the face B and final image is formed at I at a distance v and lies in the medium of refractive index μ_1. As the rays are now passing from a denser medium of refractive index μ_2 to the medium of refractive index μ_1, the equation for refraction can be written as

$$\frac{\mu_1}{v} - \frac{\mu_2}{v_1} = \frac{\mu_1 - \mu_2}{R_2} = \frac{\mu_2 - \mu_1}{-R_2} \qquad \text{... (ii)}$$

Adding (*i*) and (*ii*), we get

$$\frac{\mu_1}{v} - \frac{\mu_1}{u} = (\mu_2 - \mu_1)\left(\frac{1}{R_1} - \frac{1}{R_2}\right)$$

or $$\frac{1}{v}-\frac{1}{u}=\left(\frac{\mu_2-\mu_1}{\mu_1}\right)\left(\frac{1}{R_1}-\frac{1}{R_2}\right)$$

If the lens is placed in air, then $\mu_1 = 1$ and $\mu_2 = \mu$

$$\therefore \quad \frac{1}{v}-\frac{1}{u}=(\mu-1)\left(\frac{1}{R_1}-\frac{1}{R_2}\right) \quad \text{... (20.2)}$$

An equation similar to the one obtained for a thin convex lens.

Example 20.1 *A convex lens of focal length 0.25 m and refractive index 1.5 is immersed in water (μ=1.33). Find the change in the focal length of the lens.*

Solution. If f is the focal length of glass lens of $\mu_2 = 1.5$, then

$$\frac{1}{f}=(\mu-1)\left(\frac{1}{R_1}-\frac{1}{R_2}\right)=0.5\left(\frac{1}{R_1}-\frac{1}{R_2}\right) \quad \text{... (i)}$$

If f_1 is the focal length of the lens when immersed in water of refractive index 1.33, then

$$\frac{1}{f_1}=\left(\frac{1.50-1.33}{1.33}\right)\left(\frac{1}{R_1}-\frac{1}{R_2}\right)=\frac{0.17}{1.33}\left(\frac{1}{R_1}-\frac{1}{R_2}\right) \quad \text{... (ii)}$$

Dividing (*i*) by (*ii*), we get

$$\frac{f_1}{f}=\frac{0.5\times1.33}{0.17}$$

$$\therefore \quad f_1=\frac{0.5\times1.33\times0.25}{0.17}=+0.978 \text{ m} \quad (\because f=0.25 \text{ m})$$

The positive sign indicates that the lens is converging

Increase in focal length = 0.978 – 0.25 = 0.728 m.

Example 20.2 *A convex lens of focal length f produces an image of an object. If the magnification of the image be m and the distance between the object and screen is d, then prove that*

$$f=\frac{dm}{(1+m)^2}$$

Solution. If u is the distance of object, v that of the image from a convex lens of focal length f, then

$$\frac{1}{v}-\frac{1}{u}=\frac{1}{f} \quad \text{... (i)}$$

Multiplying (*i*) by v, we get

$$1-\frac{v}{u}=\frac{v}{f}$$

or $$1+m=\frac{v}{f} \quad \text{... (ii)}$$

as $\frac{v}{u}=m$ and is negative.

Multiplying (*i*) by u, we get

$$\frac{u}{v}-1=\frac{u}{f}$$

or $$-\frac{1}{m}-1=\frac{u}{f} \qquad \text{... } (iii)$$

Subtracting (*iii*) from (*ii*), we get

$$1+m+\frac{1}{m}+1=\frac{v-u}{f}$$

or $$\frac{m^2+2m+1}{m}=\frac{d}{f} \qquad (\because d=v-u)$$

or $$\frac{(1+m)^2}{m}=\frac{d}{f}$$

or $$f=\frac{dm}{(1+m)^2}$$

Example 20.3 *Show that minimum distance between an object and its real image in a convex lens is four times the focal length of the lens.*

Solution If u the distance of object and v that of the image from the optical centre of the lens, then the distance x between the object O and its real image I is given by

$$x=-u+v \qquad \text{... } (i)$$

As u is negative according to sign convention. The lens formula is

$$\frac{1}{v}-\frac{1}{u}=\frac{1}{f} \qquad \text{... } (ii)$$

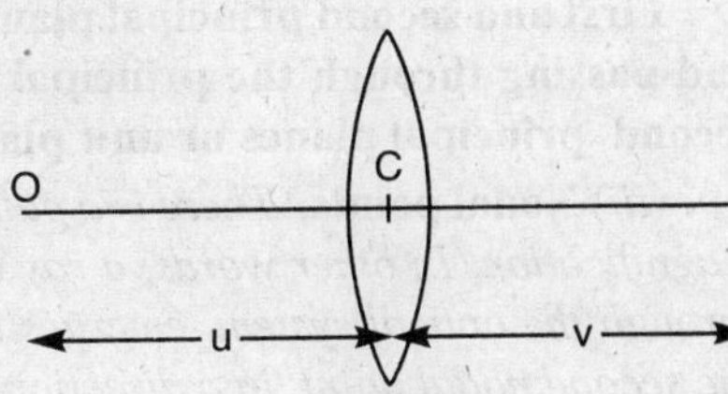

Fig 20.3

Eliminating u from (*i*) and (*ii*), we get

$$\frac{1}{v}-\frac{1}{v-x}=\frac{1}{f} \text{ or } \frac{v-x-v}{v(v-x)}=\frac{1}{f}$$

or $$v(v-x)=-xf \qquad \text{... } (iii)$$

or $$v^2-vx+xf=0$$

Solving equation (*iii*), we get

$$v=\frac{x\pm\sqrt{x^2-4xf}}{2}$$

For the real value of v

$$x^2-4xf=+\text{ve or zero}$$

$$\therefore \quad x^2\geq 4xf \text{ or } x\geq 4f$$

Hence the minimum value of x is $4f$

20.3 CARDINAL POINTS

In a system of lenses or mirrors or both, which forms ideal images of objects, there are six cardinal points. These are:

(*i*) Two focal points (*ii*) Two principal (or unit) points and (*iii*) Two nodal points.

(*i*) **Two focal points.** A ray of light incident on the lens in a direction parallel to the principal axis on the object side is bent on emergence in such a way that it passes through F_2 the point on the axis on the image side. Similarly a ray of light passing through the point F_1 on the axis, on the object side is rendered parallel to the axis after emergence through the lens. The points F_1 and F_2 as shown in Fig. 20.4 are called the first and the second focal points respectively.

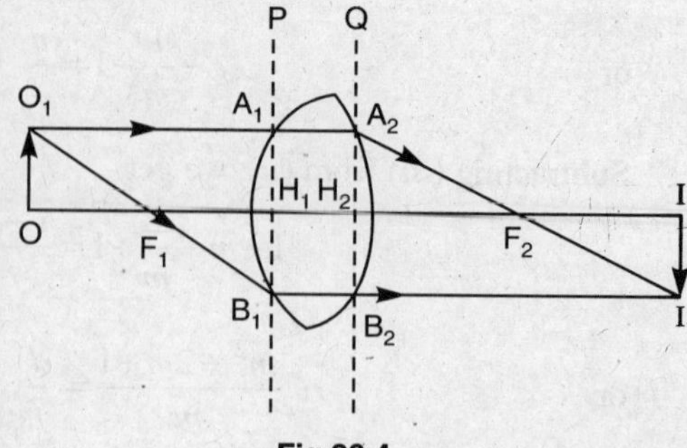

Fig 20.4

Hence the *first focal point* is a point on the principal axis of a lens such that a ray of light passing through it is rendered parallel to the principal axis after refraction.

The *second focal point* is a point on the principal axis such that a ray of light parallel to the principal axis after refraction through the lens passes through it.

Focal planes. The planes perpendicular to the principal axis and passing through F_1 and F_2 are called the first and the second focal planes respectively.

(*ii*) **Two principal points.** These are two points on the principal axis for which the lens has *a positive magnification of unity.* In other words, *an object placed at one principal point gives rise to an erect image of the same size on the other principal point.* Thus in Fig. 20.4 H_1 and H_2 are the first and second principal points respectively.

Focal length. The distance of the first focal point from the first principal point, *i.e.* $F_1 H_1$ is called the **first focal length** of the lens and the distance of the second focal point from the second principal point, *i.e.* F_2H_2 is called the *second focal length.*

If the medium on the two sides of the lens, having the same radii of curvature for the two surfaces is the same, then the first focal length is equal to the second focal length.

First and second principal planes. The planes P and Q, **perpendicular to the principal axis and passing through the principal points H_1 and H_2 are respectively called the first and the second principal planes or unit planes.**

(*iii*) **Nodal points.** *There is a pair of conjugate points on the axis, having unit positive angular magnification. In other words, a ray of light directed towards one of these points after refraction through the optical system, emerges out from the second nodal point, in a direction parallel to the incident ray.*

Fig 20.5

Let P and Q be the first and the second principal planes of an optical system and AF_1 and BF_2 the first and the second focal planes respectively. A ray of light AA_1 from a point A on the first focal plane parallel to the axis will proceed towards the second focal point F_2 from point A_2 on the second principal plane Q (Fig. 20.5)

Another ray AT_1 parallel to the emergent ray A_2F_2 striking the first principal plane at T_1 emerges from a point T_2 on the second principal plane, such that $T_1 H_1 = T_2 H_2$ .This ray T_2R will proceed parallel to A_2F_2. The point of intersection N_1 of the incident ray AT_1 and N_2 of the conjugate ray T_2R with the axis gives the positions of the two nodal points.

Distance between two nodal points is equal to distance between two principal points.

The right angled triangles $T_1H_1N_1$ and $T_2 H_2N_2$ are congruent as $T_1H_1= T_2 H_2$ and $\theta_1= \theta_2$

$$\therefore \quad H_1N_1 = H_2N_2$$

$$\text{or} \quad H_1N_1+ N_2H_2 = H_2N_2 + N_1H_2$$

$$\therefore \quad H_1H_2 = N_1N_2 \qquad ... (20.3)$$

This shows that the distance between two nodal points is equal to the distance between the two principal points.

Principal and nodal points coincide if medium on both sides same.

The right angled triangles AF_1N_1 and $A_2H_2F_2$ are congruent as

$$\angle AN_1F_1 = \angle A_2F_2H_2 \text{ and } AF_1 = A_2H_2$$

$$\therefore \quad F_1N_1 = H_2F_2$$

$$\text{or} \quad F_1H_1 + H_1N_1 = H_2F_2$$

$$\text{or} \quad H_1N_1 = H_2F_2 - F_1H_1$$

$$\text{Now} \quad F_1H_1 = -f_1 \text{ and } H_2F_2 = +f_2$$

$$\therefore \quad H_1N_1 = -H_2N_2 = f_2 + f_1 \quad \text{... (20.4)}$$

If the medium on both sides of the system is the same, then

$$f_2 = -f_1$$

$$\text{Hence} \quad H_1N_1 = H_2N_2 = 0$$

This show that the principal points will coincide with the nodal points if the medium on both sides of the system is the same.

Nodal planes. The planes passing through the nodal points and perpendicular to the principal axis are called **the nodal planes.**

20.4 COMBINATION OF TWO THIN LENSES (OBJECT AT INFINITY)

Equivalent lens. If we consider two thin lenses placed on a common axis and separated by some finite distance, then such a combination can be replaced by a single thin lens which when placed at a suitable fixed point, will produce an image of the *same size but not generally in the same position* as that produced by the combination. It is in this restricted sense, that it produces the same transverse *magnification* as does the combination for small objects placed normal to its axis. Such a lens is said to be an **equivalent lens** and its focal length is said to be the **equivalent focal length.**

Equivalent focal length. Consider two thin lenses L_1 and L_2 of focal lengths f_1 and f_2 placed on a common axis separated by a distance d, as shown in Fig. 20.6.

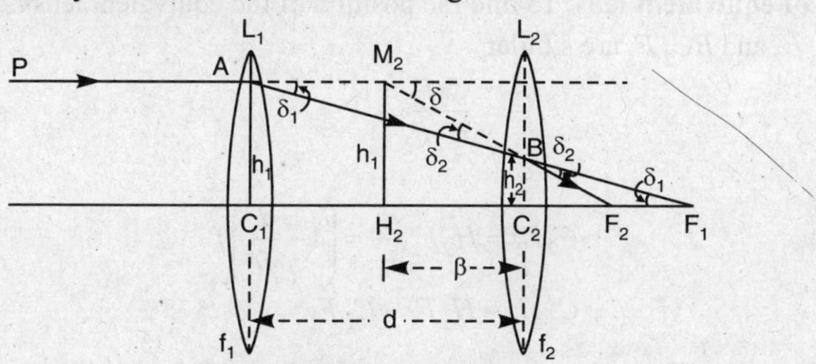

Fig 20.6

Consider a ray PA parallel to the axis incident at A on the lens L_1 at a height $C_1A = h_1$. It suffers deviation through an angle δ_1 at the first lens and proceeds towards the principal focus F_1. It meets the second lens at a point B at a height h_2 from the axis. On refraction it is further deviated through an angle δ_2 and meets finally the axis at F_2.

As the incident parallel ray PA after refraction through both the lenses meets the axis at F_2, the point F_2 must be the second principal focus of the combination.

The ray F_2B produced backward meets the ray PA produced forward at M_2. Therefore M_2H_2 is the *principal plane* and H_2 the *principal point*. Hence $F_2H_2 = f$ is the required focal length of the combination. The lens of focal length f placed at H_2 is called the **equivalent lens** which can replace the combination.

Deviation produced by the equivalent lens $\delta = \frac{h_1}{f}$

Deviation produced by the first lens $\delta_1 = \frac{h_1}{f_1}$

Deviation produced by the second lens $\delta_2 = \frac{h_2}{f_2}$

Now $\delta = \delta_1 + \delta_2 \therefore \frac{h_1}{f} = \frac{h_1}{f_1} + \frac{h_2}{f_2}$ ... (*i*)

Referring to Fig. 20.6,

$$h_2 = C_2B = C_2K - BK = h_1 - \frac{dh_1}{f_1} = h_1\left(1 - \frac{d}{f_1}\right) \qquad \left[\because \delta_1 = \frac{BK}{d} = \frac{h_1}{f_1}\right] \quad ...(ii)$$

Substituting the value of h_2 in (*i*), we get

$$\frac{h_1}{f} = \frac{h_1}{f_1} + \frac{h_1}{f_2}\left(1 - \frac{d}{f_1}\right)$$

or $$\frac{1}{f} = \frac{1}{f_1} + \frac{1}{f_2} - \frac{d}{f_1 f_2} \quad ... (20.5)$$

or $$\frac{1}{f} = \frac{f_1 + f_2 - d}{f_1 f_2}$$

$\therefore$ $$f = \frac{f_1 f_2}{f_1 + f_2 - d} = -\frac{f_1 f_2}{\Delta} \quad ... (20.6)$$

where $\Delta = d - (f_1 + f_2)$ is known as the **optical separation** or **optical interval** between the two lenses.

Position of equivalent lens. To find the position of the equivalent lens *i.e.,* $C_2 H_2$ or $C_1 H_2$ $\Delta s\ M_2 H_2 F_2$ and $BC_2 F_2$ are similar.

$\therefore$ $$\frac{h_2}{h_1} = \frac{C_2F_2}{H_2F_2}$$

or $$C_2F_2 = H_2F_2 \frac{h_2}{h_1} = \left(1 - \frac{d}{f_1}\right)f$$

Now $$C_2H_2 = H_2F_2 - C_2F_2$$

$$= f - \left(1 - \frac{d}{f_1}\right)f = f\frac{d}{f_1}$$

If we donote $C_2 H_2$ by β, then conventionally β is negative.

$\therefore$ $$-\beta = \frac{fd}{f_1} \text{ or } \beta = -\frac{fd}{f_1} \quad ... (20.7)$$

Thus the equivalent lens must be placed at a distance $-\frac{f.d}{f_1}$ in front of the second lens.

Second principal plane and second principal point. The plane $M_2 H_2$ where the deviation of the incident parallel ray coming from the left appears to occur is called the **second principal plane**

and the point H_2 on the axis is called the **second principal point.** The **second principal focus** F_2 is situated to the right of this point at a distance f.

First principal plane and first principal point. To find the position of **first principal focus** of the lens system and the **first principal plane** and **first principal point** consider a ray F_1A starting from F_1, the first principal focus of the equivalent lens (Fig. 20.7). The ray $F_1 A$ after refraction at the first lens L_1 goes along AB and after refraction at the second lens L_2 is rendered parallel to the axis and goes along BQ. The ray BQ when produced backward meets the ray F_1A produced at M_1, then M_1H_1 is the **first principal plane** and H_1 the **first principal point.** Referring to Fig. 20.7 and proceeding as above, we find that

$$F = +\frac{f_1 f_2}{f_1 + f_2 - d}$$

and
$$C_1H_1 = \frac{fd}{f_2} = \frac{df_1}{f_1 + f_2 - d}$$

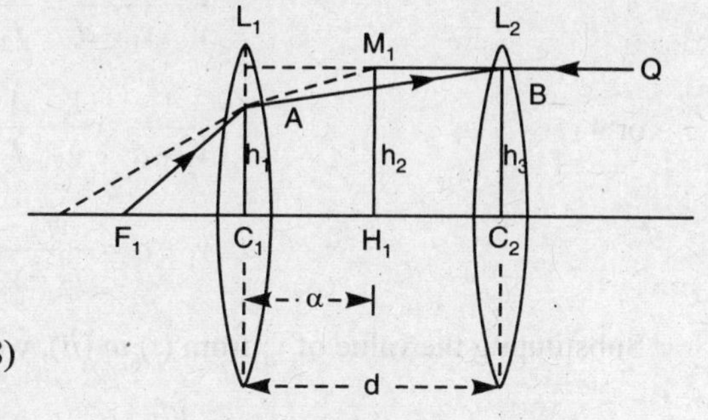

Fig 20.7

If we denote this distance by α , then

$$\alpha = \frac{df_1}{f_1 + f_2 - d} = +\frac{fd}{f_2} \quad \text{... (20.8)}$$

and
$$C_2 H_1 = d\left(1 - \frac{f}{f_2}\right)$$

The **first principal focus** F_1 is situated to the left of the point H_1 at a distance f.

The behaviour of the system cannot be reproduced in all respects by the equivalent thin lens placed in a fixed position. It will have to be placed in the position $M_2 H_2$ (Fig. 20.6) when light enters from left and in position $M_1 H_1$ (Fig. 20.7) when light enters from the right.

20.5 COMBINATION OF TWO THIN LENSES (OBJECT AT FINITE DISTANCE)

Focal length of an equivalent lens. Consider a point object O lying on the principal axis at a distance u from the first lens of focal length f_1. A ray of light OA incident on the surface of this lens after suffering refraction meets the axis at I_1 at a distance v_1. Thus, I_1 is the image formed by the first lens as shown in Fig. 20.8.

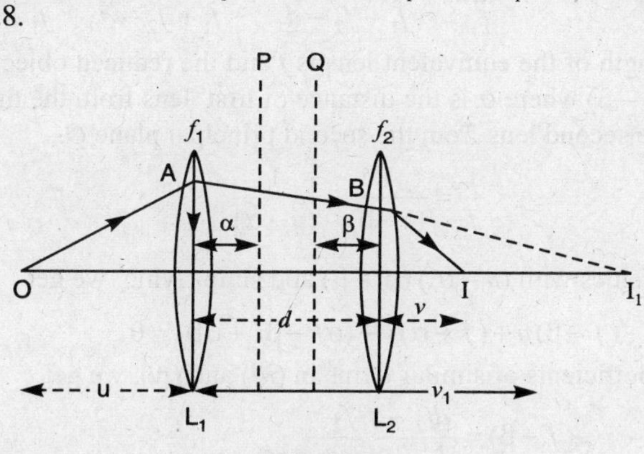

Fig. 20.8

The relation connecting the distance of the image, the distance of the object and the focal length f_1 for the first lens is given by

$$\frac{1}{v_1} - \frac{1}{u} = \frac{1}{f_1} \quad \text{or} \quad \frac{1}{v_1} = \frac{1}{f_1} + \frac{1}{u}$$

$$\therefore \qquad v_1 = \frac{uf_1}{u+f_1} \qquad \text{... } (i)$$

If a second lens of focal length f_2 is placed at a distance d from the first lens, the ray AB suffers a further deviation at B and hence the final *real image* is formed at I at a distance v from the second lens. In other words I_1 behaves as an object at a distance $v_1 - d$ from the second lens forming the final image at I. The relation connecting the distance of the object, the distance of image and the focal length f_2 for the second lens is given by

$$\frac{1}{v} - \frac{1}{v_1 - d} = \frac{1}{f_2}$$

or

$$\frac{1}{v_1 - d} = \frac{1}{v} - \frac{1}{f_2} = \frac{f_2 - v}{vf_2}$$

$$\therefore \qquad v_1 - d = \frac{vf_2}{f_2 - v} \qquad \text{... } (ii)$$

Substituting the value of v_1 from (*i*) in (*ii*), we get

$$\frac{uf_1}{u+f_1} - d = \frac{vf_2}{f_2 - v}$$

Multiplying both sides by $(u+f_1)(f_2 - v)$, we get

$$u f_1(f_2 - v) - d(u + f_1)(f_2 - v) = vf_2(u + f_1)$$

or

$$uf_1f_2 - uvf_1 - duf_2 + duv - df_1f_2 + dvf_1 - uvf_2 - vf_1f_2 = 0$$

or

$$-uv(f_1 + f_2 - d) + u(f_1f_2 - df_2) + v(df_1 - f_1f_2) - df_1f_2 = 0$$

or

$$uv(f_1 + f_2 - d) + u(df_2 - f_1f_2) + v(f_1f_2 - df_1) + df_1f_2 = 0$$

Dividing by $(f_1 + f_2 - d)$, we have

$$uv + \frac{u(df_2 - f_1f_2)}{f_1 + f_2 - d} + \frac{v(f_1f_2 - df_1)}{f_1 + f_2 - d} + \frac{df_1f_2}{f_1 + f_2 - d} = 0 \qquad \text{... } (iii)$$

Suppose focal length of the equivalent lens is f and the reduced object distance is $(u - \alpha)$ and that of the image is $(v - \beta)$ where α is the distance of first lens from the first principal plane P and β is the distance of the second lens from the second principal plane Q.

Hence

$$\frac{1}{f} = \frac{1}{(v-\beta)} - \frac{1}{(u-\alpha)}$$

Multiplying both sides with $(u - \alpha)(v - \beta)$ and simplifying, we get

$$uv - (f + \beta)u + (f - \alpha)v + (\alpha f - \beta f + \alpha\beta) = 0 \qquad \text{... } (iv)$$

Comparing the coefficients of similar terms in (*iii*) and (*iv*), we get

$$-(f + \beta) = \frac{df_2 - f_1f_2}{f_1 + f_2 - d} \qquad \text{... } (v)$$

$$f - \alpha = \frac{f_1f_2 - d\,f_1}{f_1 + f_2 - d} \qquad \text{... } (vi)$$

and

$$\alpha f - \beta f + \alpha\beta = \frac{df_1f_2}{f_1 + f_2 - d} \qquad \text{... } (vii)$$

Multiplying (v) and (vi), we get

$$-(f-\alpha)(f-\beta)=\frac{(f_1 f_2 - df_1)(df_2 - f_1 f_2)}{(f_1+f_2-d)^2}$$

or $$-f^2+\alpha f-\beta f+\alpha\beta=\frac{(f_1 f_2 - df_1)(df_2 - f_1 f_2)}{(f_1-f_2-d)^2}$$

or $$f^2=\frac{df_1 f_2}{f_1+f_2-d}-\frac{(f_1 f_2 - df_1)(df_2 - f_1 f_2)}{(f_1+f_2-d)^2}$$

$$=\frac{(df_1^2 f_2 + df_1 f_2^2 - d^2 f_1 f_2 - df_1 f_2^2 + f_1^2 f_2^2 + d^2 f_1 f_2 - df_1^2 f_2)}{(f_1+f_2-d)^2}$$

$$=\frac{f_1^2 f_2^2}{(f_1+f_2-d)^2}$$

or $$f=\frac{f_1 f_2}{(f_1+f_2-d)} \qquad ...(viii)$$

$$\frac{1}{f}=\frac{1}{f_1}+\frac{1}{f_2}-\frac{d}{f_1 f_2} \qquad ...(ix)$$

or $$f=\frac{f_1 f_2}{f_1+f_2-d} \qquad ...(20.9)$$

Positions of principal points (planes). From equation (vi) the position of the first principal plane at a distance α is given by

$$\alpha=f-\frac{f_1 f_2 - df_1}{f_1+f_2-d}=\frac{f_1 f_2}{f_1+f_2-d}-\frac{f_1 f_2 - df_1}{f_1+f_2-d}$$

$$=\frac{df_1}{f_1+f_2-d}=\frac{df}{f_2} \qquad ...(20.10)$$

Similarly from (v), we have the position of the second principal plane given by

$$-\beta=f+\frac{df_2 - f_1 f_2}{f_1+f_2-d}=\frac{f_1 f_2}{f_1+f_2-d}+\frac{df_2 - f_1 f_2}{f_1+f_2-d}$$

$$=\frac{df_2}{f_1+f_2-d}=\frac{df}{f_1}$$

or $$\beta=-\frac{df}{f_1} \qquad ...(20.11)$$

Hence $$\alpha f_2=-\beta f_1 \qquad (20.12)$$

Here α is the distance of first principal plane from the first lens and β is the distance of second principal plane from the second lens.

Positions of principal foci. The **first principal focus** is situated to the left of the first principal plane P at a distance f.

The **second principal focus** is situated to the right of the second principal plane Q at a distance f.

NOTE. In article 20.4 the relation $\frac{1}{f}=\frac{1}{f_1}+\frac{1}{f_2}-\frac{d}{f_1 f_2}$ and position of principal points and focal points has been derived by considering the object and the image at infinity using the deviation method whereas in article 20.5 the same relations have been proved by considering the object and the image at finite distances from the lens system. The latter is a more general method.

20.6 SIGNIFICANCE OF PRINCIPAL POINTS AND PRINCIPAL FOCI

The principal points H_1 and H_2 for a co-axial system are *conjugate points* for which transverse magnification is plus unity. To prove this, consider the relation

$$\frac{1}{v-\beta}-\frac{1}{u-\alpha}=\frac{1}{f}$$

or $$\frac{1}{v-\beta}=\frac{1}{f}+\frac{1}{u-\alpha}=\frac{(u-\alpha)+f}{f(u-\alpha)}$$

or $$v-\beta=\frac{f(u-\alpha)}{(u-\alpha)+f}$$

If $u=\alpha$, then $v-\beta=0$ or $v=\beta$

This is true for all values of *f*.

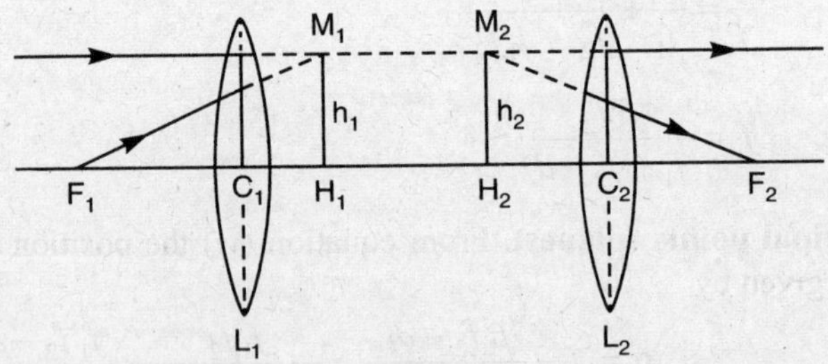

Fig. 20.9

Hence object at the first principal point will produce an image at the second principal point. Thus the first and the second principal points are conjugate points.

If we consider two rays, one originally parallel to the principal axis and the other through F_1 the first principal focus of the system drawn **in such** a manner that the height of the incident parallel ray and that of the ray which passes on emergence through F_2, the second principal focus is the same, then the pair of rays converging to M_1 before entering the system becomes the pair of rays diverging from M_2 after emergence. Thus M_1 and M_2 are also conjugate points of the system and the transverse planes M_1H_1 and M_2H_2 are also conjugate planes and transverse magnification is unity. In other words, a small object placed normal to the axis at H_1 will have an equal size image at H_2. Thus the magnification is unity. The planes M_1H_1 and M_2H_2 are called **unit planes** or **principal planes** of the co-axial system.

(*iii*) The position of the principal points of course depend upon the particular set of lenses.

(*iv*) The distance of the object is measured from the first principal point and that of the image from the second principal point.

(*v*) When the lens system is situated in a uniform medium, a ray directed towards the first point will, after refraction through the lens system, will appear to move from the second point and parallel to the incident ray.

The co-axial lens system also has **principal foci** and **principal focal planes**

(*i*) The **first principal focus** lies at distance *f* towards the left from the first principal point H_1 and the second to the right from the second principal point H_2 at the same distance.

(*ii*) A ray of light directed to pass through F_1 and incident on the first lens, after refraction will emerge parallel to the axis and a ray incident parallel to the axis, after refraction will emerge from the second lens in a direction passing through the second principal focus F_2.

(*iii*) Numerically the first focal length $H_1 F_1$ is equal to the second focal length $H_2 F_2$. But as they are opposite in sign,

$$H_2 F_2 = -H_1 F_1 = f \quad \text{... (20.13)}$$

20.7 CONSTRUCTION OF IMAGE USING CARDINAL POINTS

Knowing the cardinal points of the optical system, the image corresponding to any object placed on the principal axis can be constructed. It is, however, not necessary to know the **position** and **curvatures** of the refracting surfaces or the nature of the **intermediate media**. Only knowledge of cardinal points and cardinal planes is sufficient.

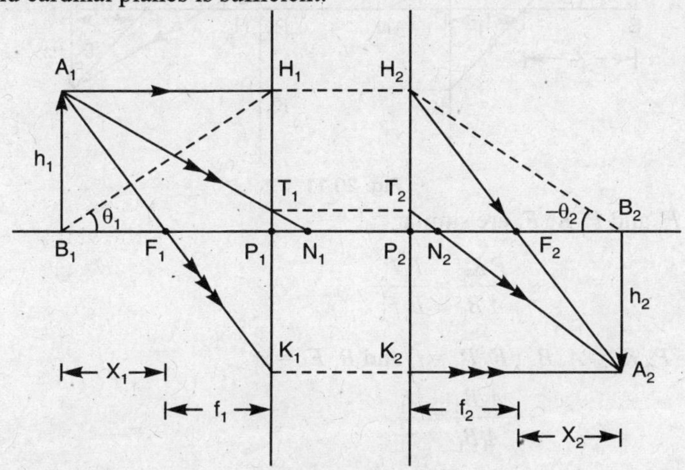

Fig. 20.10

Let F_1, F_2 be the principal foci, P_1, P_2 the principal points and N_1, N_2 the nodal points of the optical system, shown in Fig. 20.10. A_1B_1 is a linear object placed on the axis. In order to find image of point A_1, follow the following procedure: (*i*) A ray A_1H_1 is drawn parallel to the axis touching the first principal plane at H_1. The conjugate ray will proceed from H_2 a point on the second principal plane such that $H_2P_2 = H_1 P_1$ and will pass through the second principal focus F_2.

(*ii*) A second ray $A_1T_1N_1$ is drawn which is directed towards the first nodal point N_1. This ray passes after refraction through N_2, in a direction parallel to A_1N_1.

(*iii*) A third ray $A_1F_1K_1$ is drawn which is directed towards first principal focus F_1 and touching the first principal plane at K_1. Its conjugate ray will proceed from K_2 such that $K_2P_2 = K_1 P_1$ and will be parallel to the axis.

A point of intersection of any of the above two refracted rays will give the image of A_1. Let it be A_2. From A_2, if a perpendicular is drawn on to the axis, it gives the image A_2B_2 of the object A_1B_1.

20.8 NEWTON'S FORMULA FOR COAXIAL SYSTEM OF TWO LENSES

Newton's formula states $x_1 x_2 = f_1 f_2$

Where f_1 and f_2 are the first and second principal focal length , x_1 the distance of the object from the first principal focus and x_2 that of the image from the second principal focus.

Proof. A knowledge of six cardinal points is sufficient to construct the image of an object placed on the principal axis of an optical system. The principal points P_1 and P_2, the focal points F_1 and F_2 and the nodal points N_1 and N_2 of an optical system are shown in Fig. 20.11.

To find the image of an object $A_1 B_1$ draw a ray $A_1 H_1$ parallel to the principal axis touching the first principal plane at H_1. The conjugate ray will proceed from H_2 a point on the second principal plane such that $H_2 P_2 = H_1 P_1$. It will pass through the second principal focus F_2.

A secondray $A_1 T_1 N_1$ directed towards the first nodal point N_1 after refraction will pass through the second nodal point N_2 such that the ray $T_2 N_2 A_2$ is parallel to $A_1 T_1 N_1$.

Another ray $A_1 F_1 K_1$ passing through the first principal focus F_1 touches the first principal plane at K_1, emerges from K_2 such that $K_2 P_2 = K_1 P_1$ and proceeds parallel to the principal axis.

All these rays meet at A_2 which is the image of A_1. Hence $A_2 B_2$ is the image of $A_1 B_1$.

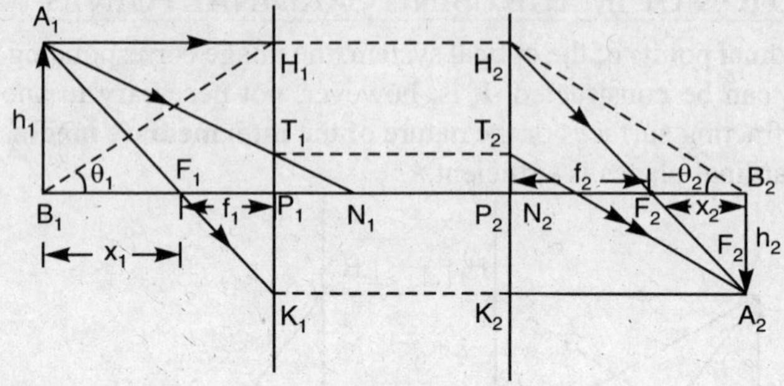

Fig. 20.11

Now $\Delta S\ A_1 B_1 F_1$ and $P_1 K_1 F_1$ are similar

$$\therefore \qquad \frac{P_1K_1}{A_1B_1} = \frac{F_1P_1}{B_1F_1}$$

But $\quad P_1 K_1 = P_2 K_2 = A_2 B_2$; $F_1 P_1 = f_1$ and $B_1 F_1 = x_1$

$$\therefore \qquad \frac{A_2B_2}{A_1B_1} = \frac{f_1}{x_1} \qquad \text{... } (i)$$

Also $\Delta s\ A_2 B_2 F_2$ and $H_2 P_2 F_2$ are similar.

$$\therefore \qquad \frac{A_2B_2}{H_2P_2} = \frac{B_2 F_2}{P_2 F_2}$$

But $H_2 P_2 = H_1 P_1 = A_1 B_1$; $F_2 P_2 = f_2$ and $B_2 F_2 = x_2$

$$\frac{A_2B_2}{A_1B_1} = \frac{x_2}{f_2} = \frac{h_2}{h_1} \qquad \text{... (20.14)}$$

Using eq. (i), we have

$$\frac{f_1}{x_1} = \frac{x_2}{f_2} \qquad \text{... (20.14A)}$$

or $$x_1 x_2 = f_1 f_2 \qquad \text{... (20.15)}$$

This is Newton's formula.

Note: The distances of the image and the object have been measured from their respective foci. However, it is sometimes more convenient to measure the conjugate distances from the principal points.

20.9 RELATION SHIP BETWEEN f_1 and f_2.

Consider Fig. 20.10 and using the sign convention, $P_1 B_1 = H_1 A_1 = -u$, and $P_2 B_2 = K_2 A_2 = +v$

Also, $A_1 B_1 = P_1 H_1 = P_2 H_2 = + h_1$ and $A_2 B_2 = K_1 P_1 = K_2 P_2 = - h_2$

$$K_1H_1 = K_1P_1 + P_1H_1 = -h_2 + h_1 \text{ and } K_2H_2 = K_2P_2 + P_2H_2 = -h_2 + h_1$$

Both $\Delta s\ K_1F_1P_1$ and $\Delta\ K_1A_1H_1$ are similar, hence we write

$$\frac{P_1F_1}{H_1A_1} = \frac{K_1P_1}{K_1H_1}$$

$$\frac{f_1}{-u} = \frac{h_2}{-h_2 + h_1}$$

or,

$$\frac{f_1}{u} = \frac{-h_2}{-h_2 + h_1} \qquad \text{... (20.16)}$$

Moreover, $\Delta s\ H_2P_2F_2$ and $H_2K_2A_2$ are also similar

$$\therefore \quad \frac{P_2F_2}{K_2A_2} = \frac{P_2H_2}{K_2H_2}$$

$$\therefore \quad \frac{f_2}{v} = \frac{h_2}{-h_2 + h_1} \qquad \text{... (20.17)}$$

Adding Eq. (20.16) and Eq. (20.17), we get

$$\frac{f_1}{u} + \frac{f_2}{v} = \frac{-h_2 + h_1}{-h_2 + h_1}$$

$$\therefore \quad \frac{f_1}{u} + \frac{f_2}{v} = 1 \qquad \text{... (20.18)}$$

Eq.(20.18) can be rewritten as

$$\frac{v}{u} = \frac{(v - f_2)}{f_1} \qquad \text{... (20.19)}$$

When the system is situated in air, $f_2 = -f_1 = f$.

Hence, $f_1 = -f$ and $f_2 = f$ ... (20.20)

20.10 RELATIONSHIP BETWEEN f_1 and f_2 AND μ_1 and μ_2

In fig. 20.10, the conjugate rays B_1H_1 and H_2B_2 make angles θ_1 and θ_2 respectively with the principal axis.

$$\therefore \quad \frac{\tan\theta_1}{\tan(-\theta_2)} = \frac{h_1/(-u)}{h_1/v} = -\frac{v}{u} \qquad \text{... (20.21)}$$

From Eq. 20.14, we have

$$\frac{h_2}{h_1} = \frac{x_2}{f_2} = \frac{v - f_2}{f_2} \quad (\because \text{ from figure } x_2 = v - f_2) \qquad \text{... (20.22)}$$

Using Eq. (20.19) into Eq. (20.22), we get

$$\frac{\tan\theta_1}{\tan\theta_2} = -\frac{h_2}{h_1}\cdot\frac{f_2}{f_1}$$

In a paraxial approximation, the above expression reduces to

$$\frac{\theta_1}{\theta_2} = -\frac{h_2}{h_1}\cdot\frac{f_2}{f_1} \qquad \text{... (20.23)}$$

According to the Lagrange law, we have

$$\mu_1 h_1 \theta_1 = \mu_2 h_2 \theta_2$$

or,
$$\frac{\theta_1}{\theta_2} = \frac{\mu_2 h_2}{\mu_1 h_1} \qquad \text{... (20.24)}$$

Comparing Eq. (20.23) and Eq. (20.24), we obtain

$$\frac{f_1}{f_2} = -\frac{\mu_1}{\mu_2} \qquad \text{... (20.25)}$$

When the system is situated in air, $f_2 = -f_1 = f$

$\therefore$
$$f_1 = -f \text{ and } f_2 = f \qquad \text{... (20.26)}$$

20.11 GAUSSIAN FORMULA

Using the result given by Eq. (20.26) into Eq. (20.18), we get

$$-\frac{f}{u} + \frac{f}{v} = 1$$

$\therefore$
$$\frac{1}{v} - \frac{1}{u} = \frac{1}{f} \qquad \text{... (20.27)}$$

This is Gauss' formula:

Thus, the simple formula for a thin lens can also be used for a thick lens or for a compound lens system, provided the distances are measured from the corresponding points. A similar equation can be derived for a divergent system.

20.12 THE THREE MAGNIFICATIONS AND THEIR RELATIONSHIP

(1) Lateral magnification

The lateral (or transverse) magnification is given by

$$m = \frac{y_2}{y_1} = \frac{h_2}{h_1} \qquad \text{... (20.28)}$$

where h_1 and h_2 are the heights of image and object respectively.

(2) Longitudinal magnification. The longitudinal (or axial) magnification is given by

$$m_L = \frac{dv}{du} = \frac{\Delta x_2}{\Delta x_1} \qquad \text{... (20.29)}$$

Differentiating Newton's formula (Eq. 20.15), we get

$$x_1 \Delta x_2 + \Delta x_1 . x_2 = 0$$

$\therefore$
$$m_L = \frac{\Delta x_2}{\Delta x_1} = -\frac{x_2}{x_1}$$

Using Eq. (20.14A) and Eq. (20.28) in the above equation, we have

$$m_L = -\frac{mf_2}{f_1 / m} = m^2 \frac{f_2}{f_1} \qquad \text{... (20.30)}$$

Using Eq. (20.25) and Eq. (20.30), we write

$$m_L = m^2 . \frac{\mu_2}{\mu_1} \qquad \text{... (20.31)}$$

(3) **Angular magnification.** The angular magnification is given by

$$\alpha = \frac{\theta_2}{\theta_1}$$

In case of paraxial approximation, the Lagrange law gives

$$\frac{\theta_2}{\theta_1} = \frac{\mu_1 h_1}{\mu_2 h_2} = \frac{\mu_1}{\mu_2}.\frac{1}{m}$$

$$\therefore \qquad \alpha.m = \frac{\mu_1}{\mu_2} \qquad \text{... (20.32)}$$

Combining Eq. (20.31) and Eq. (20.32), we obtain the relationship between the linear, axial and angular magnifications.

$$\alpha.m_L = m \qquad \text{... (20.33)}$$

When $\mu_1 = \mu_2$, Eq. (20.31) gives that $m_L = m^2$.

Therefore, $\alpha \, . \, m = 1$ ... (20.34)

Eq (20.34) shows that the angular magnification and lateral magnification are inversely proportional to each other, when $\mu_1 = \mu_2$.

Example 20.4 *Two thin convex lenses of focal lengths 20 cm and 5 cm are kept coaxially separated by a distance of 10cm. Plot the positions of the cardinal points for the combination.*

(Ngapur 2010, 2008, Agra 2001)

Solution. Given: $f_1 = 20$ cm, $f_2 = 5$ cm and $d = 10$ cm

Equivalent focal length, $$f = \frac{f_1 f_2}{f_1 + f_2 - d} = \frac{20 \times 5\,\text{cm}}{20\,\text{cm} + 5\,\text{cm} - 10\,\text{cm}} = 6.67$$

First Principal Point

$$\alpha = \frac{fd}{f_2} = \frac{10\text{cm} \times 6.67 \text{ cm}}{5 \text{ cm}} = 13.33 \text{ cm}$$

Second Principal Point

$$\beta = -\frac{fd}{f_1} = -\frac{10 \text{ cm} \times 6.67 \text{ cm}}{20 \text{ cm}} = -3.33 \text{ cm}$$

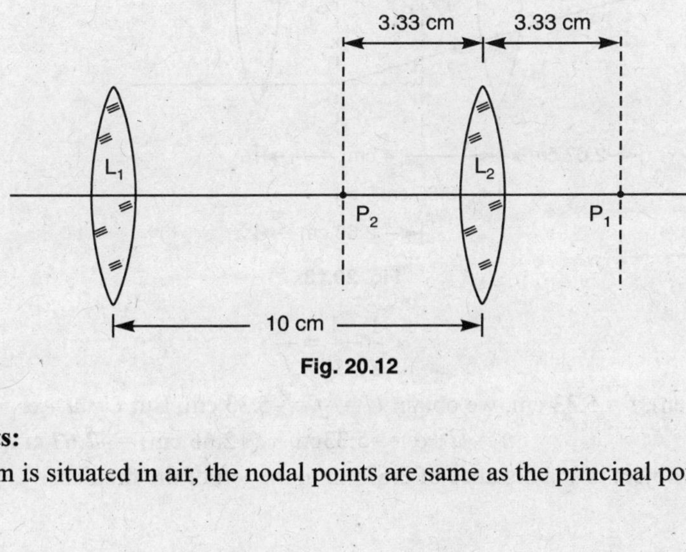

Fig. 20.12

Nodal Points:

As the system is situated in air, the nodal points are same as the principal points.

First Focal Point:

The distance of F_1 from the lens L_1 is

$$L_1F_1 = -f\left[1-\frac{d}{f_2}\right] = -6.67\text{cm}\left[1-\frac{10\text{cm}}{5\text{ cm}}\right] = 13.33\text{cm}$$

Second Focal Point

The distance of F_2 from the lens L_2 is

$$L_2F_2 = +f\left[1-\frac{d}{f_1}\right] = +6.67\text{ cm}\left[1-\frac{10\text{cm}}{20\text{cm}}\right] = +3.33\text{ cm}$$

The first principal point P_1 is to the right of the first lens and is at a distance of 13.33 cm from it. The second principal point P_2 is to the left of the second lens and is at a distance of 3.33 cm from it. The cardinal points are plotted Fig. 5.12.

Example 20.5 *Two identical thin convex lenses of focal lengths 8 cm each are coaxial and 4 cm apart. Find the equivalent focal length and the positions of the principal points. Also, find the position of the object for which the image is formed at infinity.*

Solution Here, $f_1 = f_2 = 8$ cm, $d = 4$ cm

$$f = \frac{f_1f_2}{f_1+f_2-d} = \frac{8\text{cm}\times 8\text{cm}}{8\text{cm}+8\text{ cm}-4\text{ cm}} = \frac{64}{12}\text{cm} = 5.33\text{ cm}$$

$$\alpha = +\frac{df}{f_2} = \frac{4\text{ cm}\times 5.33\text{ cm}}{8\text{ cm}} = 2.66\text{ cm}$$

$$\beta = -\frac{df}{f_1} = -\frac{4\text{cm}\times 5.33\text{ cm}}{8\text{ cm}} = -2.66\text{cm}$$

The points are plotted in Fig. 20.13. For the final image to be formed at infinity.

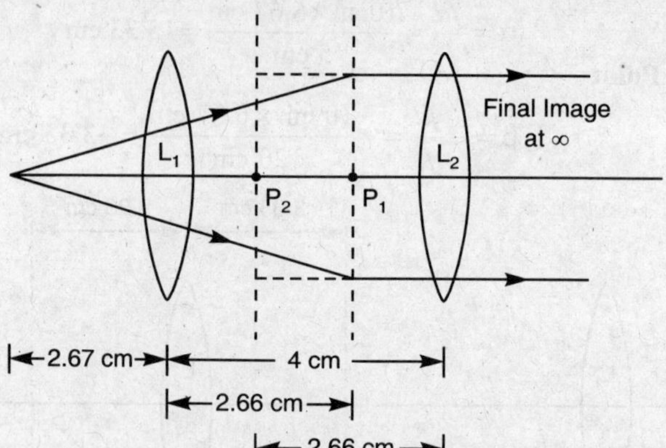

Fig. 20.13

$$\frac{1}{v}-\frac{1}{u}=\frac{1}{f}$$

As $V = \infty$ (given), $f = 5.33$ cm, we obtain $U = -f = -5.33$ cm. But $U = u - \alpha$

$\therefore$ $u = U + \alpha = -5.33\text{cm} + (+2.66\text{ cm}) = \mathbf{-2.67\text{ cm.}}$

Therefore, the object is at a distance of 2.67 cm to the left of the first lens.

Example 20.6 *Two thin convergent lenses each of 20 cm focal length are set coaxially 5 cm apart. An image of upright pole 200 m distant and 10m high is formed by the combination. Find the position of the unit and focal planes and the image. Also find the size of the image.*

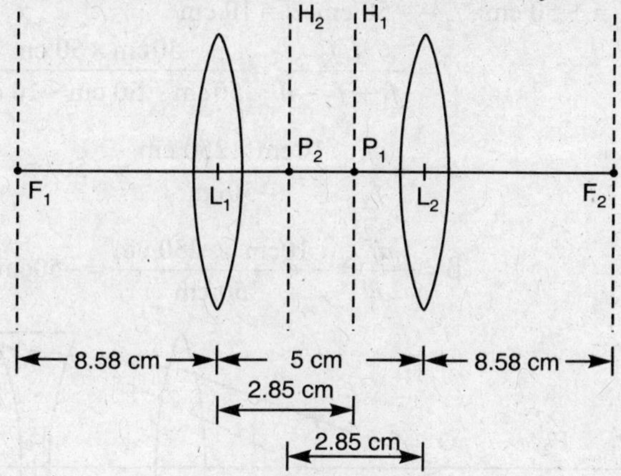

Fig. 20.14

Solution. Here, $f_1 = f_2 = 20$ cm, $d = 5$ cm

$$f = \frac{f_1 f_2}{f_1 + f_2 - d} = \frac{20\,\text{cm} \times 20\,\text{cm}}{20\,\text{cm} + 20\text{ cm} - 5\,\text{cm}} = \frac{80}{7}\text{cm} = 11.43 \text{ cm}$$

$$\alpha = +\frac{df}{f_2} = \frac{5\,\text{cm} \times 11.43\,\text{cm}}{20\,\text{cm}} = 2.85\,\text{cm}$$

$$\beta = -\frac{df}{f_1} = \frac{5\,\text{cm} \times 11.43 \text{ cm}}{20\,\text{cm}} = -2.85\,\text{cm}$$

We have $u = -20000$ cm, $\alpha = 2.85$ cm.

$$U = u - \alpha = -20000 \text{ cm} - 2.85 \text{ cm} = -20002.85 \text{ cm}$$

$$\frac{1}{V} - \frac{1}{U} = \frac{1}{f}$$

$\therefore$
$$\frac{1}{V} = \frac{1}{f} + \frac{1}{U} = \frac{1}{11.43\,\text{cm}} + \frac{1}{-20002.85 \text{ cm}}$$

or $V = 11.4$ cm

But $V = \upsilon - \beta$

$\therefore$ $\upsilon = V + \beta = 11.4 \text{ cm} - 2.85 \text{ cm} =$ **8.55 cm**

Therefore, the first unit plane P_1 is at a distance of 2.85 cm to the right of the lens L_1 and the second unit plane P_2 is at a distance of 2.85 cm to the left of the lens L_2.

The first focal plane is at a distance of 11.43 cm – 2.85 cm = 8.58 cm to the left of the lens L_1 and the second focal plane is at a distance of 8.58 cm to the right of the lens L_2.

The final image is formed at a distance of 8.55 cm to the right of the lens L_2.

Magnification, $m = \frac{I}{O} = \frac{V}{U}$

As $O = 1000$ cm, $\frac{I}{1000\,\text{cm}} = -\frac{11.4\,\text{cm}}{20002.85\,\text{cm}}$

$\therefore$ $I = -0.57$ cm

Example 20.7 *A thin convex and a thin concave lens each of 50 cm focal length are coaxially situated and separated by 10 cm. Find the position and nature of the final image formed of an object placed 20 cm from the convex lens (i.e. 30 cm from the concave lens).* **(Nagpur 2011, 2010)**

Solution. Here, $f_1 = +50$ cm, $f_2 = -50$ cm, $d = 10$ cm

$$f = \frac{f_1 f_2}{f_1 + f_2 - d} = \frac{-50\,\text{cm} \times 50\text{ cm}}{50\,\text{cm} - 50\text{ cm} - 10\text{ cm}} = +250\text{cm}$$

$$\alpha = +\frac{df}{f_2} = \frac{10\,\text{cm} \times 250\text{ cm}}{-50\,\text{cm}} = -50\text{ cm}$$

$$\beta = -\frac{df}{f_1} = -\frac{10\text{cm} \times 250\text{ cm}}{50\text{ cm}} = -50\text{cm}$$

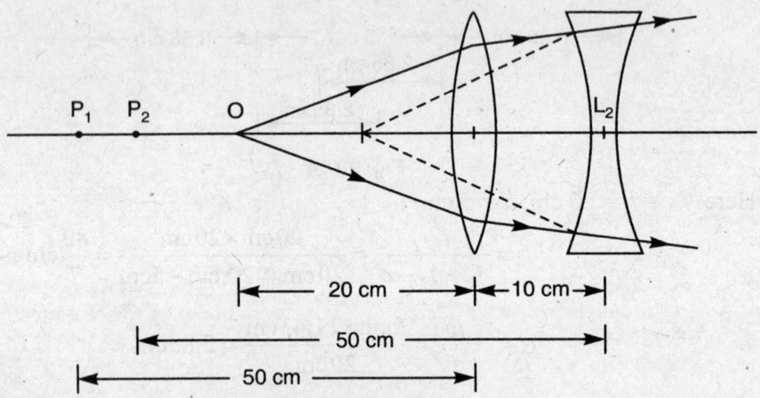

Fig . 20.15

$$U = 30\text{ cm}, f = +250\text{ cm}$$

$$\frac{1}{V} - \frac{1}{U} = \frac{1}{f}$$

$$\therefore \quad \frac{1}{V} = \frac{1}{f} + \frac{1}{U} = \frac{1}{30\text{cm}} + \frac{1}{250\text{cm}} = 26.8\text{cm}$$

The image is formed at a distance 26.8 cm from P_2. Therefore, the image is at a distance 13.21 cm to the left of the convex lens. The image is virtual and erect.

Example 20.8 *A concave-convex lens has an index of refraction 1.5 and the radii of curvature of its surface are 10 cm and 20cm. The concave surface is upwards and it is filled with an oil of refractive index 1.6. Calculate the focal length of the oil-glass combination.*

Solution. For the glass lens

$$\frac{1}{f_1} = (\mu - 1)\left[\frac{1}{R_1} + \frac{1}{R_2}\right]$$

Here $R_1 = 10$ cm, $R_2 = 20$ cm and $\mu = 1.5$

$$\frac{1}{f_1} = (1.5 - 1)\left[\frac{1}{10\,\text{cm}} + \frac{1}{20\text{cm}}\right] = \frac{1}{40\,\text{cm}}$$

For the oils lens

$$\frac{1}{f_2} = (\mu - 1)\left[\frac{1}{R_1} + \frac{1}{R_2}\right]$$

Here $R_1 = 20$ cm, $R_2 = \infty$ and $\mu = 1.6$

$$\frac{1}{f_2} = (1.6-1)\left[\frac{1}{20\,\text{cm}} - \frac{1}{\infty}\right] = \frac{3}{100\,\text{cm}}$$

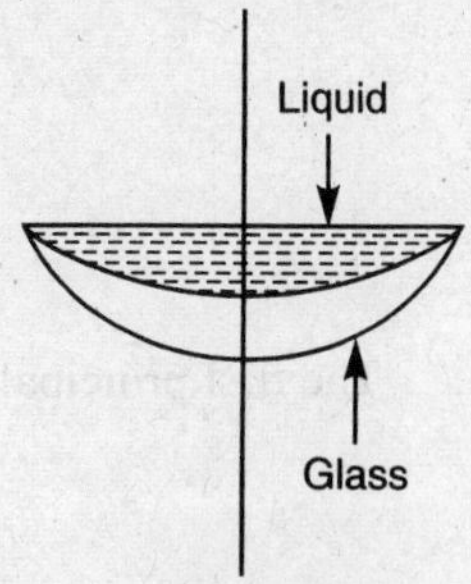

Fig 20.16

For the oil-glass combination

$$\frac{1}{f} = \frac{1}{f_1} + \frac{1}{f_2} = \frac{1}{40\,\text{cm}} + \frac{3}{100\,\text{cm}} = \frac{11}{200\text{cm}}$$

$\therefore$ $\quad f = 18.2$ **cm**

The equivalent focal length is measured from the respective principal planes. Often, we do not know where the principal planes are, or they often are inaccessible within the lens or lens system. But we need easily accessible reference points from which to measure. These reference points are the vertices of the lens.

Example 20.9 *Two thin converging lenses L_1 and L_2 of powers 5 D and 4 D are placed coaxially 10 cm apart. Find the focal length f of the combination and positions of principal points P_1 and P_2.* **(*Purvanchal 2000*)**

Solution: The focal length of the first lens, $f_1 = \frac{1}{P} = \frac{1}{5D} = \frac{1\text{m}}{5} = 20\text{cm}$

The focal length of the second lens, $f_2 = \frac{1}{P} = \frac{1}{4D} = \frac{1\text{m}}{4} = 25\,\text{cm}$

The equivalent focal length of the coaxial system is given by

$$f = \frac{f_1 f_2}{f_1 + f_2 - d}$$

$$\therefore \quad f = \frac{20\text{ cm} \times 25\,\text{cm}}{20\,\text{cm} + 25\text{ cm} - 10\text{ cm}} = 14.29\text{cm}$$

$$\alpha = +\frac{df}{f_2} = -\frac{10\,\text{cm} \times (14.29\text{ cm})}{25\,\text{cm}} = 5.7\text{cm}$$

The first principal point P_1 is at a distance 5.7 cm to the right of the first lens.

$$\beta = -\frac{df}{f_1} = -\frac{-10\,\text{cm} \times 14.29\,\text{cm}}{20\,\text{cm}} = -7.14\text{ cm}$$

The second principal point P_2 is at a distance of 7.14 cm to the left of the second lens.

Example 20.10 *Two lenses of powers + 2.5 D and – 5 D are kept coaxially separated by 30 cm. Find the position of cardinal points* **(*Allahabad 2002*)**

Solution. The focal length of the first lens, $f_1 = \frac{1}{P} = \frac{1}{2.5D} = \frac{1\text{m}}{2.5} = 40\,\text{cm}$

The focal length of the second lens,

$$f_2 = \frac{1}{P} = -\frac{1}{5D} = -\frac{1\text{m}}{5} = -20\,\text{cm}$$

The equivalent focal length of the coaxial system is given by

$$f = \frac{f_1 f_2}{f_1 + f_2 - d}$$

$$\therefore \quad f = \frac{40\text{cm} \times (-20\text{ cm})}{40\text{cm} - 20\text{ cm} - 30\text{ cm}} = +80\text{cm}$$

$$\alpha = +\frac{df}{f_2} = -\frac{(30\text{cm}) \times (80\text{cm})}{-20\text{ cm}} = -120\text{cm}$$

The first principal point P1 is at a distance 120 cm to the left of the first lens.

$$\beta = -\frac{df}{f_1} = -\frac{-30\text{cm} \times 80\text{cm}}{40\text{cm}} = -60\text{cm}$$

The second principal point P_2 is at a distance of 60 cm to the lef of the second lens.

First Focal point:

The distance of F_1 from the lens L_1 is

$$L_1F_1 = -f\left[1 - \frac{d}{f_2}\right] = -80\text{cm}\left[1 - \frac{30\text{ cm}}{(-20\text{cm})}\right] = -200\text{cm}$$

Second Focal Point

The distance of F_2 from the lens L_2 is

$$L_2F_2 = +f\left[1 - \frac{d}{f_1}\right] = 80\text{cm}\left[1 - \frac{30\text{cm}}{40\text{cm}}\right] = +20\text{cm}$$

The first focal point F_1 is to the left of the first lens and is at a distance of 200 cm from it. The second focal point F_2 is to the right of the second lens and is at a distance of 20 cm from it. The cardinal points are plotted in Fig. 20.17.

Nodal Points

As the system is situated in air, the nodal points are same as the principal points.

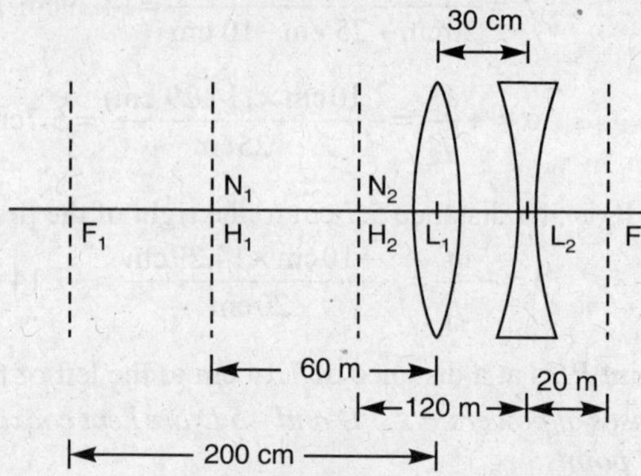

Fig. 20.17

Example 20.11 *A lens combination consists of two lenses of focal length +20 cm and –10 cm separated by a distance of 8 cm. An object 2 cm high is placed at a distance of 40 cm from the convex lens. Find the position and size of the image.* ***(M.D.U. 2002)***

Solution Here $f_1 = +20$ cm; $f_2 = -10$ cm ; and $d = 8$ cm

$$\therefore \quad \text{Equivalent focal length } f = \frac{f_1 f_2}{f_1 + f_2 - d} = \frac{20 \times (-10)}{20 - 10 - 8} = -100\text{ cm}$$

Taking the lenses as thin, the distance α of the first principal plane from the optical centre of the first lens (of focal length + 20 cm) is given by

$$\alpha = \frac{df}{f_2} = \frac{8\times(-100)}{(-10)} = 80 \text{ cm}$$

As α is positive the first principal plane lies to the right of the convex lens of focal length + 20 cm. The distance β of the second principal plane from the optical centre of the second lens (of focal length –10 cm) is given by

$$\beta = \frac{-df}{f_1} = -\frac{8\times(-100)}{20} = +40 \text{ cm}$$

Now if the final image is formed at a distance v from the second lens, and $u = -40$ cm, then

$$\frac{1}{(v-\beta)} - \frac{1}{(u-\alpha)} = \frac{1}{f}$$

or
$$\frac{1}{(v-\beta)} = \frac{1}{f} + \frac{1}{u-\alpha} = -\frac{1}{100} + \frac{1}{-40-80}$$

$$= -\left[\frac{1}{100} + \frac{1}{120}\right] = -\frac{22}{1200} = -\frac{11}{600}$$

or
$$v - \beta = -\frac{600}{11} \text{ or } v = -\frac{600}{11} + 40 = -\frac{160}{11} \text{ cm}$$

i.e. the final image lies to the left of the second lens of focal length – 10 cm

$$\text{Now magnification } m = \frac{\text{size of the image } I}{\text{size of the object } O} = \frac{v}{u}$$

$$= \frac{\frac{-160}{11}}{-40} = \frac{4}{11}$$

size of the object $O = 2$ cm

$$\therefore \text{ Size of the image } I = 2\times\frac{4}{11} = \frac{8}{11} \text{ cm}$$

Note. For the guidance of the students we shall also calculate the value of v by considering first the refraction at the first (convex) lens of focal length + 20 cm and again further refraction of the image formed at the second (concave) lens of focal length –10 cm

For the first lens. $u = -40$ cm, $f_1 = +20$ cm , $v_1 = ?$

$$\frac{1}{v_1} - \frac{1}{u} = \frac{1}{f_1} \text{ or } \frac{1}{v_1} = \frac{1}{f_1} + \frac{1}{u} = \frac{1}{20} - \frac{1}{40} = \frac{1}{40}$$

$$\therefore \quad v_1 = 40 \text{ cm}$$

As v_1 is positive the image lies at a distance of 40 cm towards the right of the convex lens *i.e.* at a distance of (40 – 8) = 32 cm to the right of the second concave lens.

For the second lens this serves as an object. Therefore, for the second lens

$u = +32$ cm; $v = ?$ $f_2 = -10$ cm

$$\frac{1}{v} - \frac{1}{u} = \frac{1}{f_2} \text{ or } \frac{1}{v} = \frac{1}{f_2} + \frac{1}{u} = -\frac{1}{10} + \frac{1}{32} = -\frac{11}{160}$$

or $$v = -\frac{160}{11}$$ i.e., the final image lies to the left of the second (concave) lens

Example 20.12 *Two thin convex lenses having focal lengths 0.05 m and 0.02 m are coaxial and separated by a distance of 0.03 m. Find the equivalent focal length and the positions of principal points.*

(Purvanchal 2001, Nagapur U. 2009)

Solution. Here, $f_1 = 0.05$ m, $f_2 = 0.02$ m, $d = 0.03$ m

$$\therefore \quad f = \frac{f_1 f_2}{f_1 + f_2 - d}$$

$$= \frac{0.05 \times 0.02}{0.05 + 0.02 - 0.03}$$

$$= 0.025 \text{ m}$$

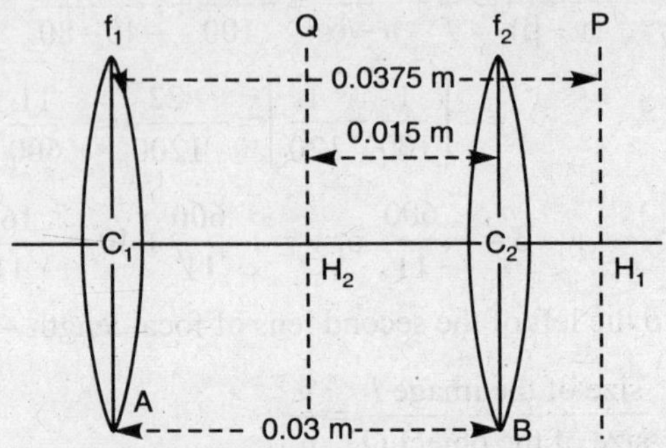

Fig. 20.18

If α is the distance of the first principal point H_1, from the lens A,

then $$\alpha = +\frac{df}{f_2} = \frac{0.03 \times 0.025}{0.02} = 0.0375 \text{m}$$

The distance of second principal point H_2 from the lens B

$$\beta = -\frac{df}{f_1} = -\frac{0.03 \times 0.025}{0.05} = -0.015 \text{ m}$$

The negative sign shows that the second principal point H_2 is to the left of lens B at a distance of 0.015 m as shown in Fig. 20.18.

Example 20.13 *Two convex lenses of focal lengths 10 cm and 20 cm are placed 5 cm apart in air. Find the cardinal points of the system.*

Solution. Here f_1 = 10 cm = 0.1 m, f_2 = 20 cm = 0.2 m and d = 5 cm = 0.05 m

$$f = \frac{f_1 f_2}{f_1 + f_2 - d} = \frac{0.1 \times 0.2}{0.1 + 0.2 - 0.005} = +0.08\text{m}$$

Principal points. If α is the distance of first principal point H_1 from the lens A of focal length f_1, then

$$C_1H_1 = \alpha = \frac{df}{f_2} = \frac{0.05 \times 0.08}{0.2} = 0.02$$

If β is the distance of second principal point H_2 from the second lens B of focal length f_2, then

$$C_2 H_2 = \beta = -\frac{df}{f_1} = \frac{-0.05\times0.08}{0.1} = -0.04 \text{ m}$$

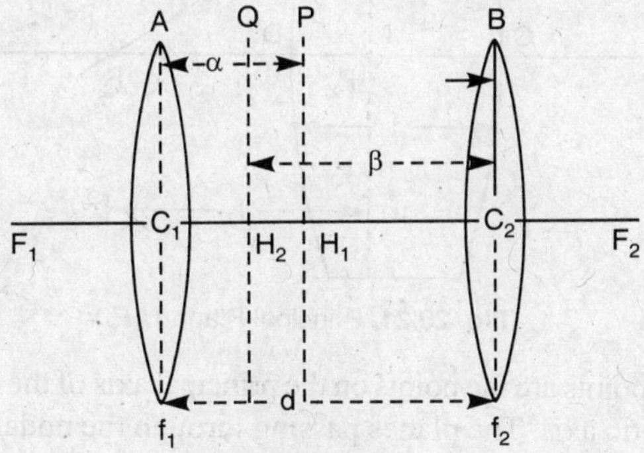

Fig. 20.19

Negative sign indicates that H_2 lies to the left of lens B

Nodal points. As the system lies in air the nodal points are the same as the principal points.

Focal points. The first focal point F_1 lies to the left of the lens A at a distance

$$= (F_1 H_1 - \alpha) = 0.08 - 0.02 = 0.06 \text{ m}$$

The second focal point F_2 lies to the right of the lens B at a distance

$$= (F_2 H_2 - \beta) = 0.08 - 0.04 = 0.04 \text{ m.}$$

20.13 CARDINAL POINTS OF A THICK LENS

As compared to a thin lens, a thick lens has larger physical separation between two spherical surfaces. The separation is appreciable in comparison with the radii of curvature of the spherical surfaces. In other words, a thick lens whose thickness can not be treated negligible as compared to its focal length. We will study the size and position of the image formed due to such a thick lens by considering refraction at each surface successively. The procedure is, however, complicated. The knowledge of cardinal points makes the problem simpler in case of thick lens or system of lenses.

Like thin lens optical system, a thick lens optical system is also characterized by six cardinal points, namely two principal points, two focal points and two nodal points. Fig 20.20 and Fig. 20.21 shows the formation of two principal planes. $H_1 P_1$ is the *first principal plane* and $H_2 P_2$ the **second principal plane.**

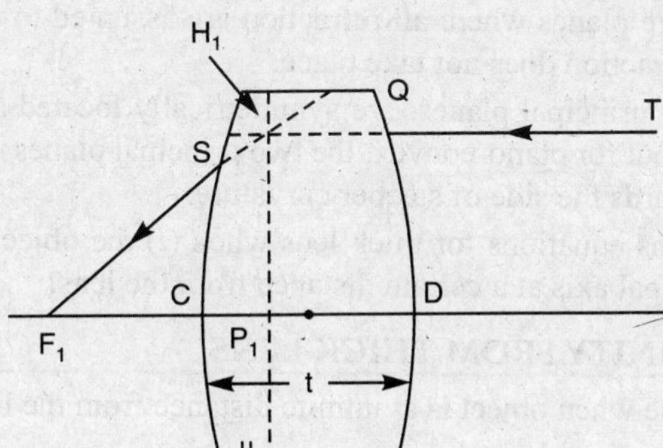

Fig. 20.20. Principal Plane H_1P_1

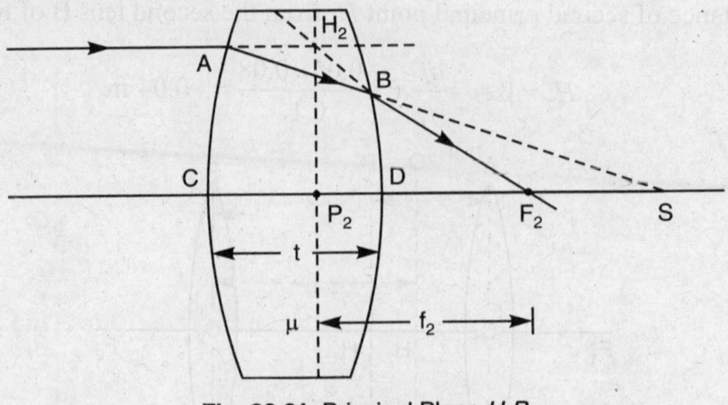

Fig. 20.21. Principal Plane H_2P_2.

Nodal Points. Nodal points are the points on the principal axis of the lens where light rays, without refraction, intersect the optic axis. The planes passing through the nodal points and perpendicular to the principal axis are called the **nodal planes.** The formation of image of an object OO' is shown in Fig. 20.22. The paraxial incident ray $O'A$ travels parallel to the axis and is incident on the lens at A. The ray after refraction passes through second focal point F_2. Another parpaxial ray $O'F_1Q$ passes through the first focal point F_1 and after refraction through lens emerges out at E parallel to the principal axis. Another paraxial ray $O'N_1$, passes through first nodal point N_1 and emerges out along N_2D after refraction through the lens. All the emergent rays intersect at I', which is the image of O'. Thus, an image II' is formed that of object OO'.

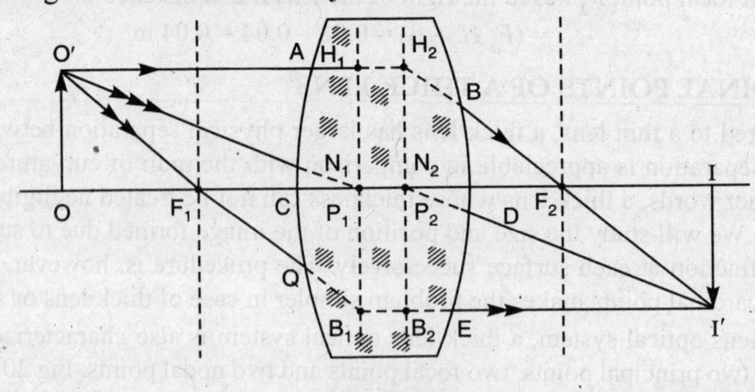

Fig. 20.22

The principal planes are planes where all refraction are assumed to occur, where as the nodal planes are planes where refraction does not take place.

It is not necessary that principal planes are symmetrically located and lie within the lens. In symmetrical lens, it is true but for plano-convex, the two principal planes move to one side and even move outside the lens, towards the side of steeper curvature.

We shall derive the lens equations for thick lens when (*i*) the object is at infinity and (*ii*) the object located on the principal axis at a certain distance from the lens.

20.14 OBJECT AT INFINITY FROM THICK LENS

Let us consider the case when object is at infinite distance from the lens.

20.14.1 Focal Length

Since the object is at infinity, a parallel ray AB to axis of the thick lens strikes it at B. It refracts along BG through the lens and comes to focus at F_2. Let H_1P_1 be the first principal plane and H_2P_2

the second principal plane. Let R_1 be the radius of curvature of the first surface and R_2 that of second surface. Let μ be the refractive index of the lens material and t be its thickness. Let f_1 and f_2 be the first and second focal lengths of the lens respectively. Let the medium outside the lens be air.

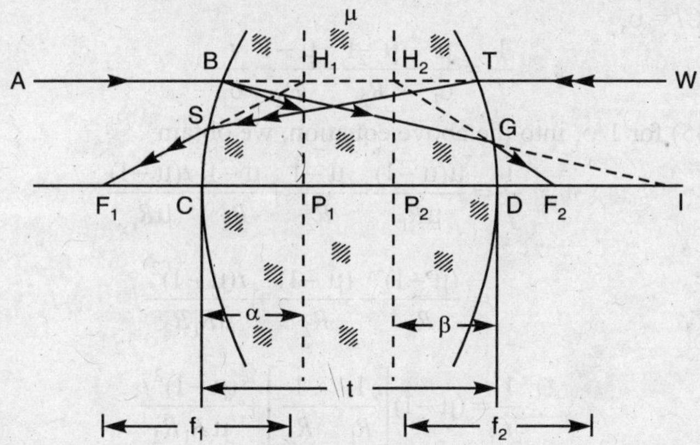

Fig. 20.23

Let $P_1F_1 = -f_1$ and $P_2F_2 = f_2$. But $f_2 = -f_1$. We designate the numerical value by f.

Let I be the position of the image formed by refraction at the first surface BC when the object is lying at infinity. Thus,

$$CI = \upsilon_1$$

Since $u = \infty$, we can write for refraction at the first surface

$$\frac{\mu}{\upsilon_1} - \frac{1}{\infty} = \frac{\mu-1}{R_1} \quad \text{or} \quad \frac{1}{\upsilon_1} = \frac{\mu-1}{\mu R_1} \qquad \text{... (20.35)}$$

As the second surface refracted the ray BG along F_2 and formed the final image at F_2, we can write for refraction at the second surface

$$\frac{1}{DF_2} - \frac{\mu}{DI} = \frac{\mu-1}{-R_2} \qquad \text{... (20.36)}$$

The Δs $H_2P_2F_2$, GDF_2 and BCI, GDI are similar. Therefore,

$$\frac{P_2F_2}{DF_2} = \frac{H_2P_2}{GD} = \frac{BC}{GD} = \frac{CI}{DI}$$

$$\therefore \qquad \frac{DF_2}{P_2F_2} = \frac{DI}{CI} \qquad \text{... (20.37)}$$

or

$$\frac{1}{f} = \frac{1}{DF_2} \cdot \frac{DI}{\upsilon_1} \qquad \text{... (20.38)}$$

Multiplying Eq. (20.36) by DI, we get

$$\frac{DI}{DF_2} - \mu = \frac{\mu-1}{-R_2} \cdot DI$$

$$\therefore \qquad \frac{DI}{DF_2} = \mu + \frac{\mu-1}{-R_2} \cdot DI$$

Substituting the above expression in Eq. (20.38) we obtain

$$\frac{1}{f} = \frac{1}{\upsilon_1}\left[\mu - \frac{\mu-1}{R_2} \cdot DI\right]$$

$$= \frac{1}{\upsilon_1}\left[\mu - \frac{\mu-1}{R_2}(CI - CD)\right]$$

As $CD = t$ and $CI = \upsilon_1$

$$\frac{1}{f} = \frac{\mu}{\upsilon_1} - \frac{\mu-1}{R_2} - \frac{\mu-1}{R_2}\cdot\frac{t}{\upsilon_1} \qquad \text{... (20.39)}$$

Using Eq. (20.35) for $1/\upsilon_1$ into the above equation, we obtain

$$\frac{1}{f} = \frac{\mu(\mu-1)}{\mu R_1} - \frac{\mu-1}{R_2} + \frac{\mu-1}{R_2}\cdot\frac{t(\mu-1)}{\mu R_1}$$

$$= \frac{(\mu-1)}{R_1} - \frac{(\mu-1)}{R_2} + \frac{t(\mu-1)^2}{\mu R_1 R_2}$$

$$\frac{1}{f} = (\mu-1)\left[\frac{1}{R_1} - \frac{1}{R_2}\right] + \frac{(\mu-1)^2 t}{\mu R_1 R_2}$$

or

$$\frac{1}{f} = (\mu-1)\left[\left(\frac{1}{R_1} - \frac{1}{R_2}\right) + \frac{(\mu-1)t}{\mu R_1 R_2}\right] \qquad \text{... (20.40)}$$

On putting $t = 0$ (Eq. 20.40) reduces to equation for a thin lens

$$\frac{1}{f_t} = (\mu-1)\left[\frac{1}{R_1} - \frac{1}{R_2}\right]$$

Therefore, we can write Eq. (20.40) as also

$$\frac{1}{f} = \frac{1}{f_t} + \frac{(\mu-1)^2 t}{\mu R_1 R_2} \qquad \text{... (20.41)}$$

20.14.2 Cardinal Points of Thick Lens

When a beam of light passes through a thick lens or a combination of lenses, we encounter problems in measuring the object distance, the image distance and focal length. These distances *cannot* be measured from the centre of the lens system. They are therefore measured from the *principal planes.*

***(i)* Second Focal Point.** The distance of the second focal point F_2 from the second surface of the lens is DF_2. From the relation (20.37) we have

$$DF_2 = P_2F_2\frac{DI}{CI} = f.\frac{DI}{CI}$$

$$= f.\frac{CI - CD}{CI} = f.\frac{CI - t}{CI}$$

$$\therefore \qquad DF_2 = f\left(1 - \frac{t}{CI}\right) = f\left(1 - \frac{t}{\upsilon_1}\right)$$

Using the value of υ_1 from (Eq. 20.35) into the above equation, we get

$$DF_2 = f\left[1 - \frac{(\mu-1)t}{\mu R_1}\right] \qquad \text{... (20.42)}$$

***(ii)* Second Principal Plane:** The second principal plane is located at a distance $P_2 D = \beta$ from the second refracting surface. From the Fig. 20.23 it is seen that the pairs of Δs $H_2 P_2 F_2$, GDF_2 and BCI, GDI are similar

$$\therefore \quad \frac{P_2D}{P_2F_2} = \frac{H_2G}{H_2F_2} = \frac{BG}{GI} = \frac{CD}{CI} = \frac{t}{\upsilon_1} \quad \text{... (20.43)}$$

$$\therefore \quad P_2D = P_2F_2 \cdot \frac{t}{\upsilon_1}$$

$$\therefore \quad -\beta = f_2 . t . \frac{(\mu - 1)}{\mu R_1} \qquad \text{But } f_2 = f$$

$$\therefore \quad \beta = -f . \frac{(\mu - 1)t}{\mu R_1} \quad \text{... (20.44)}$$

(iii) First Focal Point: If we consider the incident ray *WT* coming from the right, then the first focal point F_1 from the first surface *C* is CF_1. Obviously, the positions of R_1 and R_2 interchange and all the signs become opposite. The distance of the first focal point is then given by

$$CF_1 = -f\left[1 + \frac{(\mu - 1)t}{\mu R_2}\right] \quad \text{... (20.45)}$$

(iv) First Principal Plane: Th first principal plane is located at a distance $CP_1 = \alpha$ from the first refracting surface. Consider the ray WT coming from the right side parallel to the axis and incident on the second surfaces. The surface TGD and BSC successively refract it. The final ray emerges at the first surface and goes through the first focal point F_1. Following arguments similar to those used for deriving an expression for β, we obtain for the distance of the first principal plane.

$$\alpha = -f . t . \frac{(\mu - 1)}{\mu R_2} \quad \text{... (20.46)}$$

(*v*) and (*vi*) Nodal Points: As the medium on either side of the lens is air, the nodal points N_1 and N_2 will coincide with the principal points P_1 and P_2 respectively.

20.14.3 Power of Thick Lens

The power of a thick lens is given by

$$P = \frac{1}{f} = \frac{(\mu - 1)}{R_1} - \frac{(\mu - 1)}{R_2} + \frac{(\mu - 1)^2}{R_1 R_2} \cdot \frac{t}{\mu}$$

The power of the first refracting surface is

$$P_1 = \frac{(\mu - 1)}{R_1}$$

and the power of the second refracting surface is

$$P_2 = -\frac{(\mu - 1)}{R_2}$$

$$\therefore \quad P = P_1 + P_2 - P_1 P_2 \cdot \frac{t}{\mu} \quad \text{... (20.47)}$$

20.15 OBJECT NEAR THE THICK LENS

Let us consider the case when the point object is located on the axis of a thick lens. We will find its focal length and position of cardinal points.

20.15.1 Focal Length

Consider a lens of thickness t and of refractive index μ, placed in air. The radii of curvature are R_1 and R_2. A point object O is situated on the axis at a distance u from the first refracting surface and forms an image I at a distance υ_1 from P.

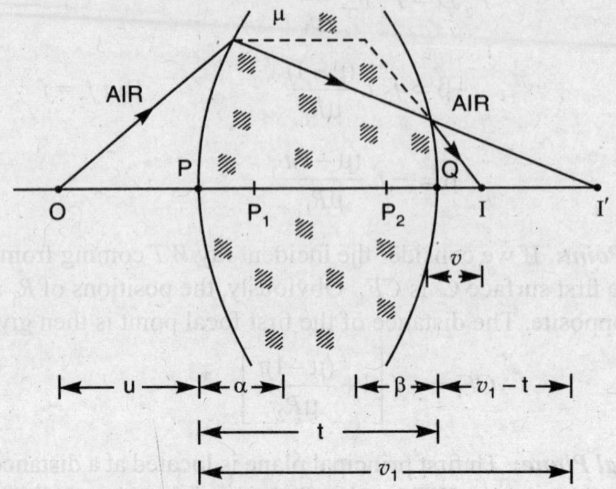

Fig. 20.24

$$\frac{\mu}{\upsilon_1} - \frac{1}{u} = \frac{\mu - 1}{R_1}$$

$$\therefore \qquad \upsilon_1 = \frac{R_1 u \mu}{R_1 + u(\mu - 1)} \qquad \text{... (20.48)}$$

The image formed by the first surface acts as the object for the second surface and the final image is formed at I.

$$\frac{1/\mu}{\upsilon} - \frac{1}{(\upsilon_1 - t)} = \frac{1/\mu - 1}{R_2}$$

$$(\upsilon_1 - t) = \frac{\mu \upsilon R_2}{R_2 + \upsilon(\mu - 1)} \qquad \text{... (20.49)}$$

Substituting the value of υ_1 from Eq. (20.48) in Eq. (20.49), we get

$$\frac{R_1 u \mu}{R_1 + u(\mu - 1)} - t = \frac{\mu \upsilon R_2}{R_2 + \upsilon(\mu - 1)}$$

On simplification of the above expression, we obtain

$$u\upsilon\left[\mu(\mu-1)(R_1 - R_2) - (\mu-1)^2 t\right] + u[\mu R_1 R_2 - tR_2(\mu - 1)]$$

$$+ \upsilon[-\mu R_1 R_2 - tR_1(\mu-1)] - tR_1 R_2 = 0 \qquad \text{... (20.50)}$$

The equation is of the form

$$u\upsilon A + uB + \upsilon C + D = 0$$

or
$$u\upsilon + u\frac{B}{A} + \upsilon\frac{C}{A} + \frac{D}{A} = 0 \qquad \text{... (20.51)}$$

where A, B, C and D are coefficients

Let us take $V = \upsilon - \beta$ and $U = u - \alpha$ and the focal length $= f$

$$\frac{1}{\upsilon-\beta}-\frac{1}{u-\alpha}=\frac{1}{f} \quad \text{... (20.52)}$$

Simplifying and rearranging the terms, we obtain

$$u\upsilon+(-\beta-f)u+(-\alpha+f)\upsilon+(-\beta f+\alpha f+\alpha\beta)=0 \quad \text{... (20.53)}$$

Comparing (Eq. 20.51) and (Eq. 20.53), we have

$$-\beta-f=\frac{B}{A} \quad \text{... (20.54)}$$

$$-\alpha+f=\frac{C}{A} \quad \text{... (20.55)}$$

and

$$-\beta f+\alpha f+\alpha\beta=\frac{D}{A} \quad \text{... (20.56)}$$

From Eq. (20.54) to (Eq. 20.56)

$$f^2=\frac{D}{A}-\frac{BC}{A^2}=\frac{AD-BC}{A^2}$$

Substituting the values of *A, B, C* and *D* in the above expression and after simplification, we get

$$f^2=\frac{\mu^2 R_1^2 R_2^2}{\left[\mu(\mu-1)(R_1-R_2)-(\mu-1)^2 t\right]^2}$$

$$f=\pm\frac{\mu R_1 R_2}{\left[\mu(\mu-1)(R_1-R_2)-(\mu-1)^2 t\right]}$$

$$\therefore \quad \frac{1}{f}=(\mu-1)\left[\frac{1}{R_1}-\frac{1}{R_2}+\frac{(\mu-1)t}{\mu R_1 R_2}\right] \quad \text{... (20.57)}$$

For a thin lens, $t = 0$ and from Eq. (20.57) we see that

$$\therefore \quad \frac{1}{f}=(\mu-1)\left[\frac{1}{R_1}-\frac{1}{R_2}\right] \quad \text{... (20.58)}$$

which is the same as Eq. (20.1)

20.15.2 Position of the Principal Points

The first principal point P_1 is at a distance α from *P*. From Eq. (20.55)

$$-\alpha+f=\frac{C}{A}$$

$$\therefore \quad \alpha=f-\frac{C}{A}$$

Substituting the value of *C, A* and *f* and after simplification, we have

$$\alpha=\frac{+tR_1}{\mu(R_1-R_2)-t(\mu-1)} \quad \text{... (20.59)}$$

The second principal point P_2 at a distance β from *Q*. We have

$$-\beta-f=\frac{B}{A}$$

$$\therefore \qquad \beta = -\left(\frac{B}{A} + f\right)$$

Substituting the values of B, A, and f and after simplification we get

$$\beta = \frac{+tR_2}{\mu(R_1 - R_2) - t(\mu - 1)} \qquad \text{... (20.60)}$$

It is easy to see that

$$\frac{\alpha}{\beta} = \frac{R_1}{R_2}$$

Eq. (20.52) may now be written as

$$\frac{1}{V} - \frac{1}{U} = \frac{1}{f} \qquad \text{... (20.61)}$$

This is a relation similar to that of a thin lens; with the difference that the object distance is measured from the first principal plane and the image distance is measured from the second principal plane.

20.15.3 Critical Thickness

In case of a double convex lens, R_1 is positive and R_2 is negative. When we take these signs into consideration, the thick lens equation may be rewritten as

$$\frac{1}{f} = (\mu - 1)\left[\frac{1}{R'} + \frac{1}{R''} - \frac{(\mu - 1)t}{\mu R'R''}\right] \qquad \text{... (20.62)}$$

Eq. (20.62) indicates that the value of $1/f$ decreases with the increasing thickness. As t increases, a stage will be reached when $1/f$ approaches zero and beyond that it becomes negative. It then acts as a divergent lens. The **critical thickness, t_c,** beyond which a thick convergent lens changes into a divergent lens is given by

$$\left[\frac{1}{R'} + \frac{1}{R''} - \frac{(\mu - 1)t_c}{\mu R'R''}\right] = 0$$

$$\therefore \qquad t_c = \frac{\mu(R' + R'')}{\mu - 1} \qquad \text{... (20.63)}$$

For a lens in which $R' = R''$ $(= R)$, the above equation reduces to

$$t_c = \frac{2\mu R}{\mu - 1} \qquad \text{... (20.64)}$$

Eq. (20.64) gives the critical thickness of a thick lens.

20.16 GLASS SPHERE AS A LENS

(a) Focal length

Consider a glass sphere of radius R and refractive index, μ placed in air as shown in Fig. 20.25. It may be treated as a *thick lens* with O_1 and O_2 as poles of two refracting surfaces. The focal length of a thick lens in air is given by

$$\frac{1}{f} = (\mu - 1)\left[\frac{1}{R_1} - \frac{1}{R_2} + \frac{(\mu - 1)t}{\mu R_1 R_2}\right]$$

In the case of a sphere-lens, we have $R_1 = +R$, $R_2 = -R$ and $t = 2R$.

$$\therefore \qquad \frac{1}{f} = (\mu - 1)\left[\frac{1}{R} + \frac{1}{R} - \frac{(\mu - 1)(2R)}{\mu R^2}\right]$$

$$= (\mu - 1)\frac{2}{R}\left[1 - \frac{\mu - 1}{\mu}\right]$$

$$= \frac{2(\mu - 1)}{\mu R}$$

$$\therefore \qquad f = \frac{\mu R}{2(\mu - 1)} \qquad ...(20.65)$$

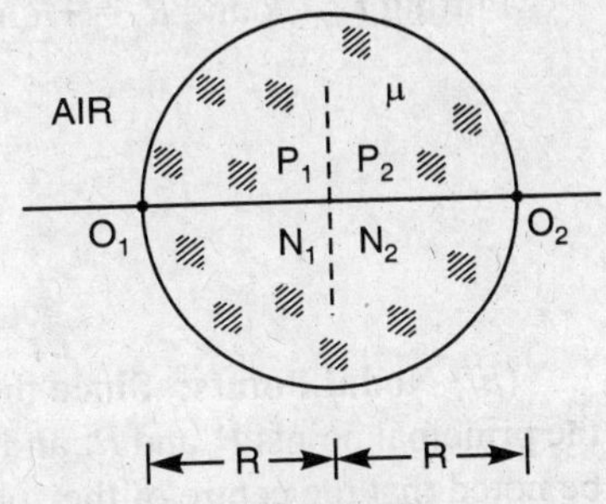

Fig. 20.25

The relation (20.65) gives the focal length of a sphere-lens

***(b)* Cardinal Points**

(i) Principal Points: The distance of the first principal point P_1 from the first surface of the lens is given by

$$O_1 P_1 = -f\frac{(\mu - 1)t}{\mu R_2}$$

Using $t = 2R$ and $R_2 = -R$ and the relation (20.65) for f into the above equation, we get

$$O_1 P_1 = \frac{\mu R}{2(\mu - 1)} \cdot \frac{(\mu - 1)(2R)}{\mu(-R)} = +R \qquad ... (20.66)$$

It means that the first principal point P_1 is at a distance on the right of O_1, that is, at the centre of the sphere.

The distance of the second principal point P_2 from the rear surface is given by

$$O_2 P_2 = -f\frac{(\mu - 1)t}{\mu R_2}$$

Again putting $t = 2R$ and $R_1 = +R$ and the relation 20.66 for f into the above equation, we get

$$O_2 P_2 = -\frac{\mu R}{2(\mu - 1)} \cdot \frac{(\mu - 1)(2R)}{\mu(R)} = -R \qquad ...(20.67)$$

It means that the second principal point P_2 is at a distance on the left of O_2, that is, at the centre of the sphere.

(*ii*) **Focal Points:** The distance of the *first focal point* F_1 from the first surface is given by

$$O_1 F_1 = -f\left[1 + \frac{(\mu - 1)t}{\mu R_2}\right]$$

Putting $t = 2R$ and $R_2 = -R$ and the relation (20.63) for f into the above equation, we get

$$O_1 F_1 = -\frac{\mu R}{2(\mu - 1)}\left[1 + \frac{(\mu - 1)2R}{\mu(-R)}\right] = -\frac{\mu R}{2(\mu - 1)} + R$$

$$\therefore \qquad O_1 F_1 = -\frac{R(2 - \mu)}{2(\mu - 1)} \qquad ... (20.68)$$

The distance of the *second focal point* F_2 from the second surface is given by

$$O_2 F_2 = +f\left[1 - \frac{(\mu - 1)t}{\mu R_1}\right]$$

Putting $t = 2R$ and $R_1 = +R$ and the relation (20.58) for f into the above equation, we get

$$O_2F_2 = \frac{\mu R}{2(\mu-1)}\left[1-\frac{(\mu-1)2R}{\mu(R)}\right] = \frac{\mu R}{2(\mu-1)} - R$$

$$\therefore \quad O_2F_2 = \frac{R(2-\mu)}{2(\mu-1)} \qquad \text{... (20.69)}$$

(*iii*) **Nodal Points:** Since the sphere is located in air, the nodal points N_1 and N_2 coincide with the principal points P_1 and P_2 and thus, lie at the centre of the sphere, as shown in Fig. 20.25. It may be noted that the centre of the sphere is also the optical centre of the sphere-lens because the nodal points are coinciding with the centre.

20.17 COMBINATION OF TWO THICK LENSES

Object at infinity

As already mentioned in article 20.4, the image formed by the first optical system becomes the object for a second system and the two systems acts as one forming the final image from the original object. We have to find the cardinal points of the combined system.

Principal Points P_{11} and P_{12} and the focal points F_1 and F_1' represent the first system. Principal points P_{21} and P_{22} and the focal points F_2 and F_2' represent the second system. The separation of the two systems is specified by the separation of their adjacent principal planes P_{12} and P_{21}, which is say d. d is positive when measured to the right from P_{12} to P_{21}. Principal points P_1 (not shown in diagram) and P_2 and the focal points F (again, not shown in Fig. 20.26 and F' represent the combined system.

Consider a ray AB' parallel to the axis and at height h_1. It means the P_{11} plane, at B'. It emerges from the first lens at B at the same height h_1 and would have followed the path BF_1. Thus, BF_1' is the emergent ray from the first system. Ray BF_1' meets the plane P_{21} at D, which has height h_2. F_1' forms the object point for the second system. The second lens deviates the ray BF_1 along EF' which is the final emergent ray. The F' will be the *second focal point* of the combined system.

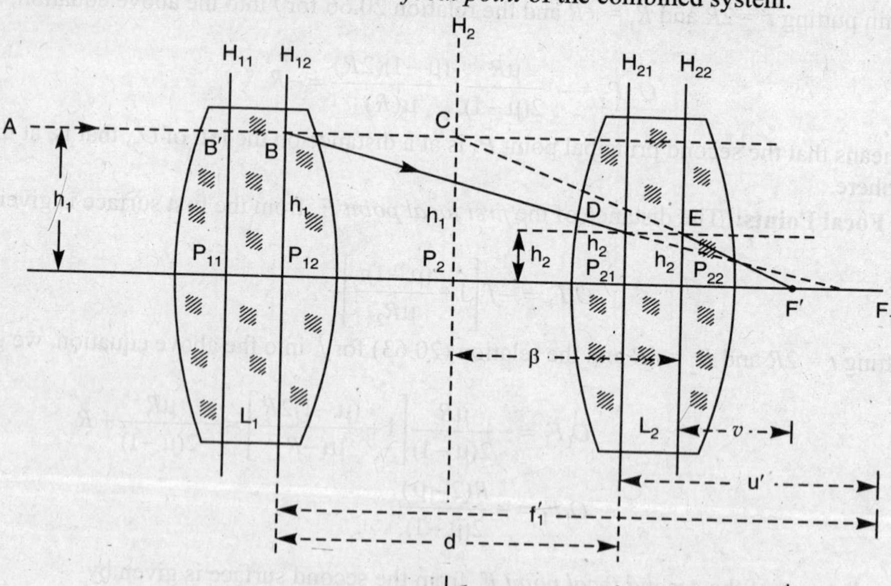

Fig. 20.26

If the ray AB is produced forward and the ray EF' backward, they intersect at C. The plane CP_2 is *the principal plane of the equivalent lens*. $P_2F' = f$ is the *focal length of the equivalent lens*. Δs $BF_1'P_{12}$ and $EF_1'P_{22}$ are similar we have

$$\frac{BP_{12}}{EP_{22}} = \frac{P_{12}F_1'}{P_{22}F_1'} \quad \therefore \quad \frac{h_1}{h_2} = \frac{f_1'}{f_1' - d} \tag{20.70}$$

Also Δs CP_2F' and $EP_{22}F'$ are similar

$$\frac{CP_2}{EP_{22}} = \frac{P_2F'}{P_{22}F'} \quad \therefore \quad \frac{h_1}{h_2} = \frac{f}{f - \beta} \tag{20.71}$$

From the Equ. 20.70 and 20.71, we obtain

$$\frac{f}{f-\beta} = \frac{f_1'}{f_1' - d} \quad \text{or} \quad \frac{1}{f-\beta} = \frac{f_1'}{f(f_1' - d)} \tag{20.72}$$

Further, $\dfrac{1}{\upsilon} - \dfrac{1}{u'} = \dfrac{1}{P_{22}F'} - \dfrac{1}{P_{21}F_1'} = \dfrac{1}{f_2}$ or $\dfrac{1}{f-\beta} - \dfrac{1}{f_1' - d} = \dfrac{1}{f_2'}$

$$\therefore \quad \frac{1}{f-\beta} = \frac{1}{f_1' - d} + \frac{1}{f_2'} \tag{20.73}$$

Equating 20.72 and 20.73, we get

$$\frac{f_1'}{f(f_1' - d)} = \frac{1}{f_1' - d} + \frac{1}{f_2'}$$

$$\therefore \quad f_1' = f\left[\frac{f_1' - d}{f_2'} + 1\right]$$

or

$$f = \frac{f_1' f_2'}{f_1' + f_2' - d} \tag{20.74}$$

$$\therefore \quad \frac{1}{f} = \frac{1}{f_1'} + \frac{1}{f_2'} - \frac{d}{f_1' f_2'} \tag{20.75}$$

Eq. (20.75) gives the equivalent focal lens of combination of two thick lenses-seperated by a distance d.

Determination of '*f*' by constant deviation method.

We shall now apply the method of deviation to determine the equivalent focal length of the combination of two coaxial thick lenses. For the sake of simplicity, we represent the lens system by its two principal planes.

Referring to the Fig. 20.27, we see that there is a deviation of path of the incident ray OA due to the presence of optical system and the ray subsequently follows the path $BA'C$. The angular deviation is δ, which is given by

$$\delta = \angle BAA' + \angle BA'A = \angle BOO' + \angle BO'O$$

or,

$$\delta = \frac{h}{-u} + \frac{h}{\upsilon} = h\left(\frac{1}{\upsilon} - \frac{1}{u}\right) = \frac{1}{f_2}$$

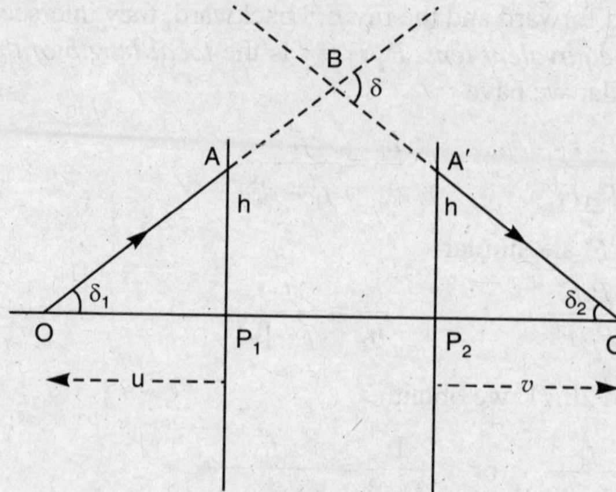

Fig. 20.27

It is easy to see that the deviation is independent of u; it is proportional to the height where the ray meets the principal plane and inversely proportional to the focal length f_2.

We can now apply the above result to the combination of lenses. Referring to the Fig. 20.28, the ray AB is parallel to the principal axis, and the first lens causes a deviations of $\delta_1 = \dfrac{h_1}{f_1'}$ and the second lens causes a deviation of $\delta_2 = \dfrac{h_2}{f_2'}$.

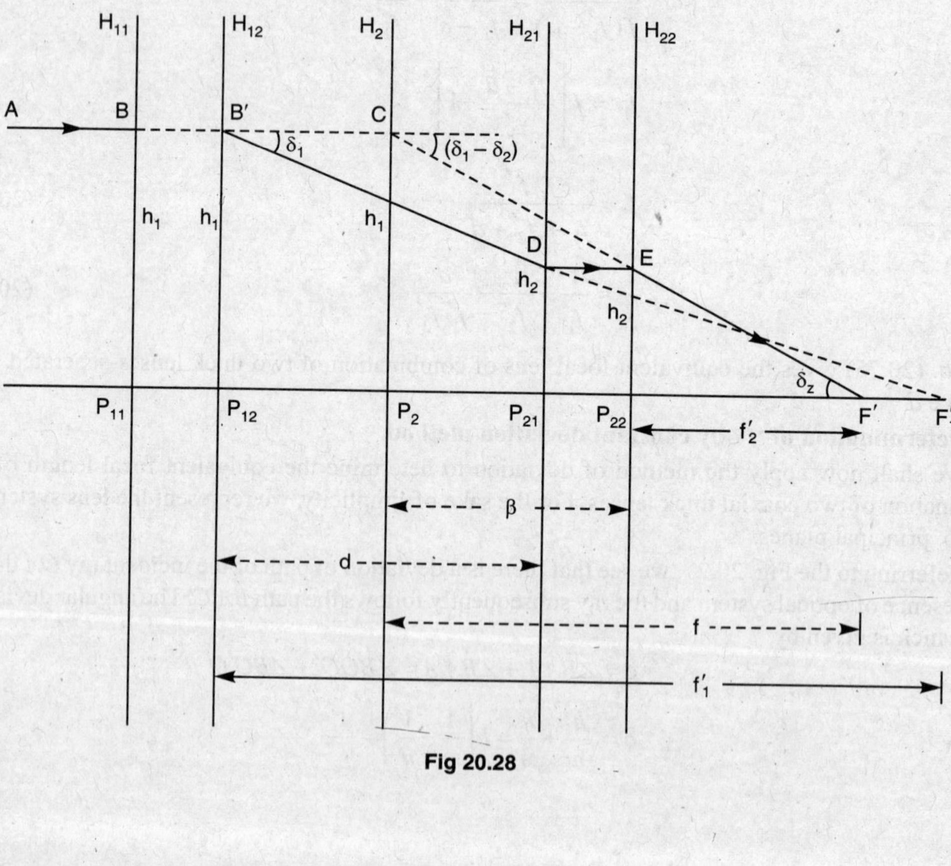

Fig 20.28

The deviation due to the total system is $\delta = \delta_1 + \delta_2 = \frac{h_1}{f}$

$$\therefore \quad \frac{h_1}{f} = \frac{h_1}{f_1'} + \frac{h_2}{f_2'} \quad \text{or} \quad h_1\left(\frac{1}{f} - \frac{1}{f_1'}\right) = \frac{h_2}{f_2'}$$

$$\therefore \quad \frac{h_1}{h_2} = \frac{1}{f_2'\left(\frac{1}{f} - \frac{1}{f_1'}\right)} \tag{20.76}$$

The Δs $B' P_{12} F_1'$ and $DP_{21} F_1''$ are similar. Therefore,

$$\frac{B'P_{12}}{DP_{21}} = \frac{P_{12}F_1'}{P_{21}F_1'} \quad \text{or} \quad \frac{h_1}{h_2} = \frac{f_1'}{(f_1' - d)} \tag{20.77}$$

Equating (20.76) and (20.77), we find that

$$\frac{f_1'}{(f_1' - d)} = \frac{1}{f_2'\left(\frac{1}{f} - \frac{1}{f_1'}\right)}$$

On simplifying the above expression, we get

$$\frac{1}{f} = \frac{f_1' + f_2' - d}{f_1' f_2'} \tag{20.78}$$

Eq. (20.78) is same as eq. (20.75)

Considering the similar Δs $CP_2 F'$ and $EP_{22} F'$ we get

$$\frac{h_1}{h_2} = \frac{f}{f - \beta} \tag{20.79}$$

Equating 20.77 and (20.79), we find that

$$\frac{f}{f - \beta} = \frac{f_1'}{f_1' - d}$$

We get the following result after simplification of the above equation

$$\beta = +\frac{fd}{f_1'} \tag{20.80}$$

Note: Principal planes change their positions when two positive lenses of equal power are first in contact and then are gradually separated. When the lenses are in contact the two principal planes are close together within the system. As the distance of separation is increased, the principal planes also separate. When the distance between the lenses to equal to their equivalent focal length, the principal planes coincide and it is interesting to note that when the separation of two lenses is $2f$, the principal planes moves out to infinity.

20.18 APPLICATIONS OF LENS COMBINATIONS

Although there is a very large variety of lens combinations of lenses in optical instruments as per desired functions, we shall study only two typical applications, telescopic combinations and telephoto lens. We will study some other combinations, eyepieces in chapter 22.

20.18.1 Telescopic Combinations

Telescope performs two main functions. One is to produce an enlarged image of a distant object and the other is to gather more light from the object so that it is possible to view with the unaided eye at ease.

In telescopic combination, the spacing (d) between the lenses is made equal to $(f_1 + f_2)$ As a result $(f_1 + f_2 - d) = \Delta = 0$, so that the equivalent focal length becomes infinity $(f = \frac{f_1 f_2}{\Delta}$ ∞ further $\alpha = \frac{fd}{f_1} = \infty$ and $\beta = -\frac{fd}{f_1} = -\infty$ and the cardinal points (focal and principal) lie at infinity. When image is formed at ∞, it can be seen by the relaxed eye. A simple astronomical telescopic system uses this principle in which images are formed at infinity for distant objects. This is achieved in simple astronomical telescopic system by using combination of two single convex lenses, such that back focal plane of the first lens coincides with the front focal plane of the second lens. This makes the separation *d* between the two lenses equal to $(f_1 + f_2)$.

Fig. 20.29 shows the telescopic lens combinations. The linear magnification h'/h is constant for all object and image positions, which is the desired requirement.

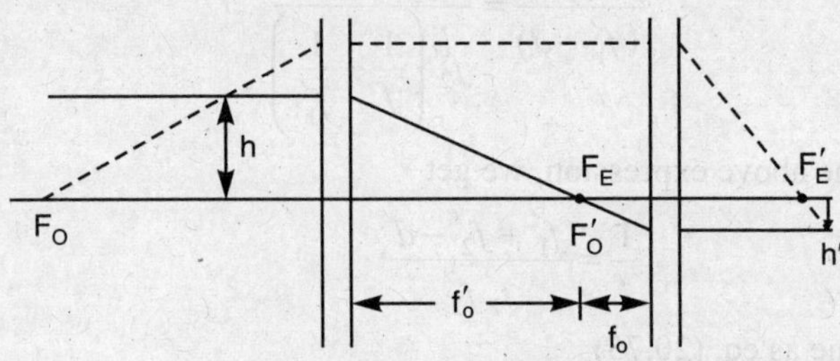

Fig. 20.29. Telescopic lens combination

20.18.2 Telephoto Lens

Another important application of combination of lenses is telephoto lens in camera. To photograph a distant object, a camera should have a convex lens of long focal length. The image size for distant object is directly proportional to the focal length of the lens. Therefore, to produce a large image on the film surface, the distance between the lens and photographic film should be large. Thus, a magnified image requires a long camera. On the other hand, the overall size of the camera should be reasonable and manageable. We can achieve this *only* if the image plane is close to the lens. These two contradictory requirements, of long focal length and short lens-to-focal-plane distance, can be resolved if we use a lens combination in stead of a single lens. Such a lens combination used in photographing distant objects is known as a *telephoto lens*.

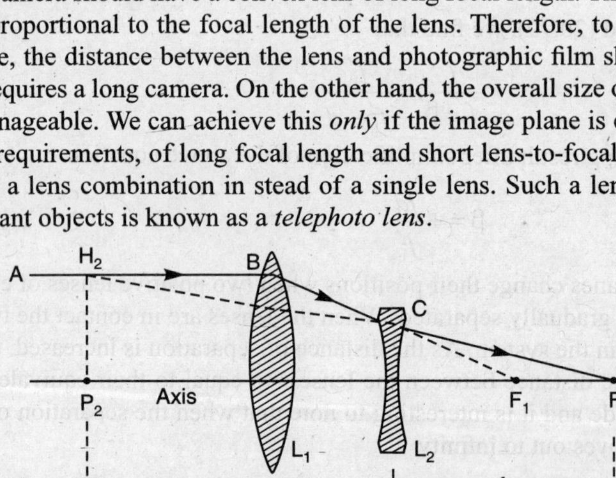

Fig. 20.30. Lens combination in camera

Theory. A telephoto lens consists of a convex lens of focal length f_1 and a concave lens of focal length f_2 separated by a distance slightly greater than $f_1 + f_2$. The concave lens L_2 is placed in the original position of the lens in the camera whereas the convex lens L_1 is in front of L_2, as shown in Fig. 20.30. The combination of these two lenses acts as single convex lens of large focal length. The equivalent focal length of the combination is given by

$$\frac{1}{f} = \frac{1}{f_1} + \frac{1}{f_2} - \frac{d}{f_1 f_2}$$

The focal length f of the system is measured from the second principal plane $H_2 P_2$ and the lens to focal plane distance, $L_2 F_2$ is measured from L_2 to the photographic film. The distance from the first lens to the photographic film, $L_1 F_2$, determines the overall length of the camera. What we require is shorter $L_1 F_2$ and longer $P_2 F_2$. The position to the second principal plane is determined by the intersection of a ray, AB, parallel to the axis in the object space with its conjugate ray, CF_2, in the final image space. It is seen that the second principal plane $H_2 P_2$ lies well in front of the first lens L_2. Therefore, the equivalent focal length of the system, $f = P_2 F_2$, is quite large. The lens-to-focal-plane distance is reduced because the back focal length, $f_b = L_2 F_2$, is shorter. Thus, the overall length of the photo lens plus camera is made shorter.

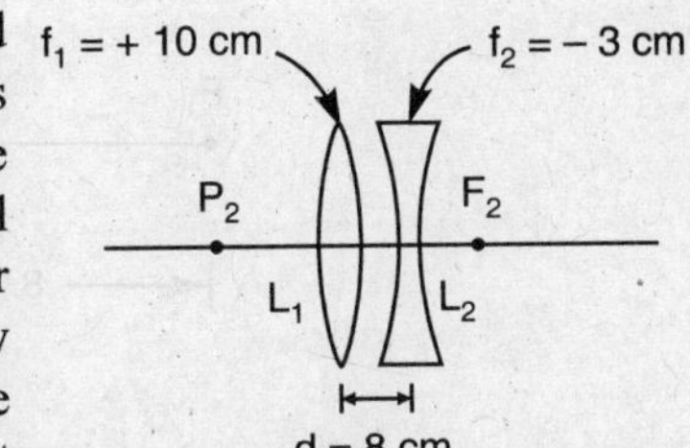

Fig. 20.31

To fully appreciate the function of the lens combination in the telephoto lens, let us consider the following example. Fig. 20.31 shows a telephoto lens consisting of a convex lens of focal length 10 cm and a concave lens of focal length (–3 cm), separated by a distance 8 cm.

The focal length of the combination, $f = \dfrac{f_1 f_2}{f_1 + f_2 - d} = \dfrac{10\text{cm} \times (-3\,\text{cm})}{10\text{ cm} - 3\text{cm} - 8\text{cm}} = +30\text{cm}$

$\therefore \quad f = P_2 F_2 = 30$ cm

$$\beta = -\frac{fd}{f_1} = -\frac{30\,\text{cm} \times 8\text{ cm}}{10\text{cm}} = -24\text{cm}$$

$\therefore \quad L_2 P_2 = -\beta = 24$ cm

$\therefore$ Lens-to-focal plane distance $L_2 F_2 = P_2 F_2 - P_2 L_2 = (30 - 24)\text{cm} = 6$ cm

$$\alpha = \frac{fd}{f_2} = -\frac{30\,\text{cm} \times 8\text{cm}}{-3\text{cm}} = -80\text{ cm}$$

$\therefore \quad L_1 P_1 = -80$ cm

Overall length of the camera $L_1 F_2 = d + L_2 F_2 = (8 + 6)\text{cm} =$ **14 cm**

However, if a single convex lens is used the length of the camera would have been 30 cm. The lens combination discussed above reduces it to only 14 cm, even less than half. This is the advantage of lens combinations with adequate separation.

Example 20.14 *A convex lens of thickness 4 cm has radii of curvature 6 cm and 10 cm. Find the focal length and the positions of the focal points and the principal points. The refractive index of lens material is* $\mu = 1.5$.

Solution. Here $R_1 = 6$ cm, $R_2 = -10$ cm and $t = 4$ cm

$$f = \frac{-\mu R_1 R_2}{\mu(\mu-1)(R_1 - R_2) - (\mu-1)^2 t}$$

$$= \frac{-1.5 \times 6\text{cm} \times (-10\text{ cm})}{1.5 - (1.5-1)\,(6\text{cm} + 10\text{ cm}) - (1.5\,-1)^2 \times 4\text{cm}}$$

$$f = 8.18 \text{ cm}$$

$$\alpha = \frac{R_1 t}{\mu(R_1 - R_2) - t(\mu - 1)}$$

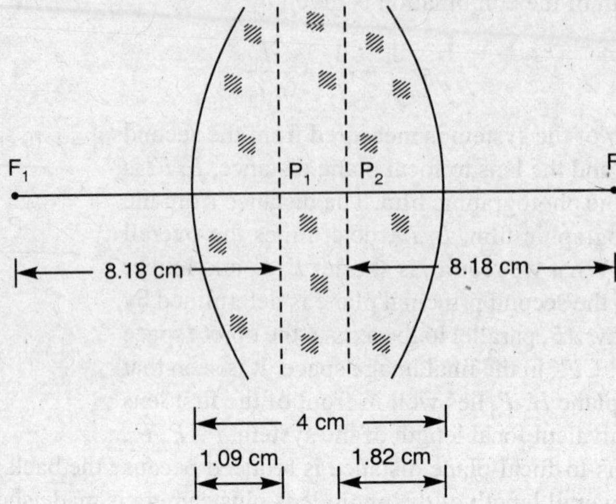

Fig. 20.32

$$\frac{6\text{cm} \times 4\,\text{cm}}{1.5\,[6\text{cm} - (-10 \text{ cm})] - 4\text{cm}\,(1.5 - 1)}$$

$$= 1.09 \text{ cm}$$

$$\beta = \frac{R_2 t}{\mu(R_1 - R_2) - t(\mu - 1)}$$

$$= \frac{4\text{cm}\,(-10 \text{ cm})}{1.5\,[6\text{cm} - (-10 \text{ cm})] - 4 \text{ cm}\,(1.5 - 1)} = -1.82 \text{ cm}$$

In Fig. 20.32 P_1 and P_2 show the positions of the principal points and F_1 and F_2 show the positions of the focal points. As the medium on the two sides of the lens is the same, the nodal points coincide with the principal points.

Example 20.15 *Determine the positions of the focal points, principal points and nodal points in the case of a sphere of radius 10 cm and μ = 1.5. Indicate their positions in a diagram.*

Solution. Here $R_1 = 10$ cm, $R_2 = -10$ cm and $t = 20$ cm

$$f = \frac{-\mu R_1 R_2}{\mu(\mu - 1)(R_1 - R_2) - (\mu - 1)^2 t}$$

$$= \frac{-1.5 \times 10\text{cm} \times (-10 \text{ cm})}{1.5(1.5 - 1)(10\text{cm} + 10 \text{ cm}) - (1.5 - 1)^2 \times 20\text{cm}}$$

$$f = \mathbf{15 \text{ cm}}$$

$$\alpha = \frac{R_1 t}{\mu(R_1 - R_2) - t(\mu - 1)}$$

$$= \frac{10\,\text{cm} \times 20\text{cm}}{1.5\,[10\text{cm} - (-10\text{ cm})] - 20\text{ cm}(1.5-1)}$$

$$= +\,10\text{ cm}$$

$$\beta = \frac{R_{2}t}{\mu(R_1 - R_2) - t(\mu - 1)}$$

$$= \frac{20\text{cm}\,(-10\text{ cm})}{1.5[10\text{cm} - (-10\text{ cm})] - 20\text{ cm}(1.5-1)}$$

$$= -\,10\text{ cm}$$

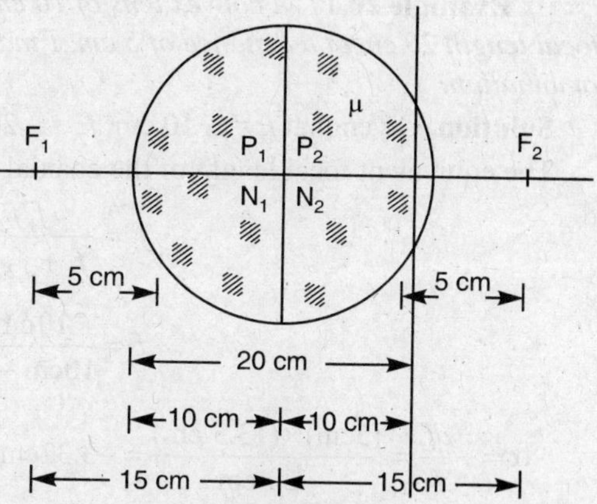

Fig. 20.33

Example 20.16 *Find the focal length and the positions of cardinal points of a plano-convex lens of refractive index 1.5, the radius of curvature of the curved surface being 20 cm and thickness 1.0 cm.*

(*Kurukshetra 2001*)

Solution. Given $R_1 = \infty$, $R_2 = -20$ cm, $t = 1$ cm and $\mu = 1.5$

(*i*) **Focal length** : The focal length of a lens of thickness t is given by

$$\frac{1}{f} = (\mu - 1)\left[\frac{1}{R_1} - \frac{1}{R_2} + \frac{(\mu-1)}{\mu}, \frac{t}{R_1 R_2}\right]$$

$$\frac{1}{f} = (1.5-1)\left[\frac{1}{\infty} - \frac{1}{-20\text{cm}} + \frac{(1.5-1)}{1.5}.\frac{1\text{cm}}{\infty(-20\text{cm})}\right]$$

$$\frac{1}{f} = (1.5-1)\left[\frac{1}{20\,\text{cm}}\right] = \frac{0.5}{20\text{ cm}}$$

$$f = +\,\mathbf{40\text{ cm}}$$

(*ii*) **Second Focal Point**

$$DF_2 = f\left[1 - \frac{(\mu-1)t}{\mu R_1}\right] = 40\text{cm}\left[1 - \frac{(1.5-1)1\text{cm}}{1.5\,(\infty)}\right] = \mathbf{+40}\text{ cm}$$

(*iii*) **Second Principal Plane**

$$\beta = -f.\frac{(\mu-1)t}{\mu R_1} = -40\text{cm}\;\frac{1.5-1}{1.5\,(\infty)} = 0$$

(*iv*) **First Focal Point**

$$CF_1 = -f\left[1 + \frac{(\mu-1)t}{\mu R_2}\right] = -40\,\text{cm}\left[1 + \frac{(1.5-1)1\text{cm}}{1.5\,(-20\text{cm})}\right] = \mathbf{-39.3\,cm}$$

(*v*) **First Principal Plane**

$$\alpha = -f\,t.\frac{(\mu-1)}{\mu R_2} = -40\text{cm}.\;1\text{ cm}\;\frac{(1.5-1)}{1.5(-20\,\text{cm})} = \mathbf{0.66\,cm}$$

(*vi*) **and** (*vii*) **Nodal Points:** The nodal points, N_1 and N_2 coincide with the principal points P_1 and P_2 respectively.

Example 20.17 *A convex lens of 10 cm focal length is placed in air from a concave lens of focal length 20 cm at a distance of 5 cm. Find the distance between the two principal points of the combination.* ***(Garhwal 2002)***

Solution. Given that $f_1 = +\ 10$ cm, $f_2 = -\ 20$ cm and $d = 5$ cm

The equivalent focal length of the coaxial system is given by

$$f = \frac{f_1 f_2}{f_1 + f_2 - d}$$

$$\therefore \qquad f = \frac{10\,\text{cm} \times (-20\ \text{cm})}{10\text{cm} - 20\ \text{cm} - 5\text{cm}} = 13.3\ \text{cm}$$

$$\alpha = +\frac{df}{f_2} = \frac{(5\text{cm})\times(13.3\ \text{cm})}{-20\ \text{cm}} = -3.32\text{cm}$$

The first principal point P_1 is at a distance 3.32 cm to the left of the first lens.

$$\beta = -\frac{df}{f_1} = \frac{(5\text{cm}) \times (13.3\,\text{cm})}{10\,\text{cm}} = -6.65\text{cm}$$

The second principal point P_2 is at a distance of 6.65 cm to the left of the second lens.

The distance between the two principal points of the combination

$$= \beta - \alpha - d = [-6.65 - (-3.32) - (-5)]\,\text{cm} = 1.67\ \text{cm}$$

EXERCISE

1. Define principal foci, principal points and nodal points of a coaxial lens system. ***(Nagpur U. 2008)***
2. For a co-axial lens system, show that the distance between two principal points (p_1, p_2) is equal to the distance between two nodal points (N_1, N_2). ***(Nagpur U. 2008)***
3. (*a*) Deduce for a thin convex lens an equation connecting the focal length, the radii of curvature of the surfaces and the refractive indices of the material of the lens and the surrounding medium.

   (*b*) What approximations have been used and what limitations do they impose on the use of this formula ?
4. Obtain the relation between the distance of an object and its image formed by a thin concave lens in terms of the radii of curvature of the lens.
5. (*a*) What are cardinal points of a co-axial (optical) lense system ? What are their properties ?

   (*b*) Show that (*i*) Distance between two principal points is the same as the distance between two nodal points (*ii*) If the medium on both sides of the lens system is the same the principal (or unit) points and nodal points coincide.

   (Gharwal U. 2004; Meerut. U., 2003, 2002, 2000; M.D.U. 2000; Osm. U. 2004, Nagpur U. 2009, 2008)
6. (*a*) What is an equivalent lens ? Show that the equivalent focal length of two thin lenses with focal lengths f_1 and f_2 separated by a distance d is given by

   $$\frac{1}{f} = \frac{1}{f_1} + \frac{1}{f_2} - \frac{d}{f_1 f_2}$$ ***(Guahati. U., 2000)***

   (*b*) Derive expression for the position of principal points and focal points for the system. ***(Utkal. U. 2003; Meerut. U. 2001)***

7. Derive an expression for equivalent focal length and position of principal points and focal points of a co-axial system of two thin lenses separated by a distance.

(Bhopal, U. 2004; Luck. U. 2002; Meerut. U. 2002, Nagpur U. 2011, 2010, 2008)

8. Explain the significance of principal points and principal foci.

9. Explain the term cardinal points with reference to a coaxial system. ***(Nagpur U. 2011)***

10. What are the principal points and principal planes ? Show that the principal planes are the planes of unit linear magnification.

11. What are nodal points and nodal planes ? Give their properties. Show that the nodal planes are planes of unit angular magnification.

12. Derive Netwton's formula for a convergent system of lenses forming a real image.

13. Show that for a co-axial lens system, $xx' = ff'$ where x and x' are the respective distances of the object and the image from the first and the second focal points and f and f' are the two focal lengths. What form does the expression take when the media on the two sides of the system are the same?

14. Define cardinal points of a system of co-axial lenses. Describe how you would determine experimentally the principal planes of a combination of two thin lenses separated by a distance.

15. What do you understand by the term power of a lens ?

16. Define cardinal points of a coaxial lens system. Show how the principal planes of such a system can be located using the theory of the method of deviation or otherwise. State the sign convention you use.

17. Derive an expression for the equivalent focal length of a system of two thin lenses separated by a finite distance when the space between them is filled with a medium of refractive index $\mu(\mu > 1)$.

18. Derive expressions for the equivalent focal length and the positions of principal points and focal points of a coaxial system of two thin lenses separated by a finite distance.

(Kurukshetra-2002)

19. Find the position of principal points for a convex lens of thickness 3 cm, refractive index 1.5 and radius of curvature of each surface 10 cm.

[**Ans.** 1.05 cm inside from each surface] ***(Nagpur U. 2008)***

20. Two thin converging lenses each of 30 cm focal length are set coaxially 10 cm apart. An image of an upright pole 100 metres distant and 5 metre high is formed by the combination. Find the position of the unit and focal planes and the image. Also find the size of the image.

[**Ans.** f = 18 cm; α = + 6 cm; β = –6 cm; υ = 12.03 cm, h_2 = 0.9 cm]

21. Determine the positions of the focal points, principal points and nodal points in the case of sphere of radius 12 cm (μ = 1.5).

[**Ans.** f = 18 cm; α = + 12 cm; β = –12 cm. The nodal points and the principal points are at the centre of the sphere.]

22. Two thin converging lenses of focal lengths 20 cm and 40 cm are placed coaxially 20 cm apart. An object is placed at a distance of 50 cm from the first lens. Calculate the positions of the principal points, focal points and the position of the image.

[**Ans.** α = + 12 cm; β = –16 cm. P_1F_1 = – 24 cm, P_2F_2 = + 24 cm; Image is formed at a distance of 23.16 cm to the right of the second lens.]

23. The focal length of a convex lens for a certain monochromatic light is 0.3 m and refractive index of the material is 1.5. What will be the focal length for another monochromatic light whose refractive index is higher than the first 2%? [**Ans.** 0.283]

24. Two thin converging lenses of focal lengths 30 cm and 40 cm are placed co-axially 20 cm apart. An object is located at a distance of 48 cm from the first lens. Find (*i*) the position of image (*ii*) the position of principal points.

[**Ans.** = + 12 cm, = – 16 cm, the final image is formed at a distance of 24 cm to the right of second lens.]

25. Two similar thin convex lenses of focal length 10 cm each are placed co-axilly 5 cm apart. Find the equivalent focal length and the position of principal points. Also find the position of the object for which the image is formed at infinity.

[**Ans.** $F = 6.67$ cm, $\alpha = +3.33$ cm, $\beta = -3.33$ cm, $u = -3.33$ cm]

26. What are cardinal points of a thick lens or a coaxial combination of lenses ? Define them, state their characteristics and show them diagrammatically for a lens system. Explain image formation with the help of these points in the case of a coaxial lens system. ***(Delhi, 2000)***

27. Calculate the focal length of a thick convex lens of radii of curvature R_1 and R_2 and thickness t. Calculate the position of the principal points.

28. Define cardinal points of a system of coaxial lenses. Show that the principal planes are conjugate planes of +1 magnification. ***(Udaipur, 2000)***

29. Show that the focal length of a thick lens is given by

$$\frac{1}{f} = (\mu - 1)\left[\frac{1}{R_1} - \frac{1}{R_2} + \frac{(\mu - 1)}{\mu} \cdot \frac{d}{R_1 R_2}\right]$$

30. Describe nodal assembly method to locate cardinal points of a lens system experimentally. Explain its working with particular reference to the characteristics of nodal points. ***(Mysore, 2001)***

31. What are cardinal points of a thick lens ? How they differ from that of thin lens ?

32. Establish the relations for equivalent focal length and cardinal points of a thick lens when the object is at infinity.

33. Derive the relations for equivalent focal length and positions of principal points of the thick lens when the object is near to it.

34. Derive the relations for focal length and cardinal points of a glass sphere as a lens.

35. State the applications of combination of lenses ? How they are used in optical instrument ?

36. Write notes on:

(*i*) Telescopic combination of lenses.

(*ii*) Importance of Telephoto lenses ***(Nagpur U. 2008)***

37. (*a*) What is meant by optical separation of two lenses ? Two lenses of focal length, 20 and 30 cm are separated by a distance of 20 cm. Find the optical separation. **(Ans. Δ = 40 cm)**

(*b*) What is telescopic combination of lenses. **(P.U. 2003)**

38. Explain the construction and working of a telephoto lens. ***(Nagpur U. 2011, 2010)***

39. An optical system consists of two thin lenses, a convex lens of focal length 20 cm and a concave lens of focal length 10 cm and are separated by a distance of 8 cm. An object 1 cm in height is placed at a distance 50 cm from the convex lens. Find the position and size of the image.

[**Ans.** $f = -100$ cm, $\alpha = +80$ cm, $\beta = +40$ cm, $I = 0.435$ cm, virtual, erect.]

40. The radii of curvature of a convex meniscus lens are 15 cm and 10 cm and its thickness is 2 cm. Calculate its focal length and the positions of cardinal points. If an object is placed at a distance of 100 cm from this lens, calculate the position and magnification of the image.

(**Ans.** $f = 52.9$ cm, $\alpha = -3.53$ cm, $\beta = 2.35$ cm magnification $= -1.08$)

41. A convex lens of thickness 4 cm has radii of curvature 6 cm and 10 cm. Find the focal length and the position of focal points. The R.I. of lens material is $\mu = 1.5$.
[**Ans.** $f = 8.18$ cm, $\alpha = 1.09$ cm, $\beta = -1.82$ cm; Nodal points coincide with principal points]

42. A convex lens of thickness 6 cm has radii of curvature 8 cm and 10 cm. Find the focal length and position of the focal points and the principal points ($\mu = 1.5$).
[**Ans.** $f = 10$ cm; $\alpha = +2$ cm; $\beta = -2.5$ cm]

43. Determine the position of the focal points, principal points and nodal points in the case of sphere of radius 12 cm. ($\mu = 1.5$). [**Ans.** $f = 18$ cm; $\alpha = +2$ cm; $\beta = -12$ cm. The nodal points and the principal points are at the centre of the sphere.]

44. A concave-convex lens has a refractive index of 1.5 and the radii of curvature of its surfaces are 15 cm and 30 cm. The concave surface is upwards and it is filled with a liquid of refractive index 1.6. Calculate the focal length of the liquid-glass combination.
[**Ans.** $f = 27.27$ cm]

45. Two convex lenses of focal lengths 20 and 5 are 10 cm apart. Calculate the power of such combination. [**Ans.** $P = 15$ diopters]

46. Two thin convex lenses having focal lengths 6 cm and 2 cm are co-axial and separated by a distance of 4 cm. Calculate the combined focal length and the positions of the principal planes.
[**Ans.** $f = 3$ cm; $\alpha = +6$ cm; $\beta = -2$ cm]

47. What will be the focal length of a plano-convex lens of refractive index 1.5 and radius of curvature of curved surface equal to 50 cm? [**Hint.** $t = (\mu - 1)\left(\frac{1}{R_1} - \frac{1}{\infty}\right)$[**Ans.** 100 cm]
(***Nagpur, 2009***)

MULTIPLE CHOICE QUESTIONS

20.1 In a lens or lens system, the number of cardinal points are
(*a*) 2 (*b*) 4 (*c*) 6 (*d*) 8

20.2 Optical separation Δ of a lens is given by
(*a*) $f_1 - f_2 - d$ (*b*) $f_1 + f_2 - d$ (*c*) $f_1 - f_2 + d$ (*d*) $f_1 + f_2 + d$

20.3 In astronomical telescope, the object and the image both lie at
(*a*) infinity (*b*) near the objective
(*c*) near the eyepiece (*d*) in between the objective and eyepiece.

20.4 Gaussian formula for a lens is given by
(*a*) $-\frac{1}{v} + \frac{1}{u} + \frac{1}{f}$ (*b*) $\frac{1}{v} - \frac{1}{u} = \frac{1}{f}$
(*c*) $-\frac{1}{v} - \frac{1}{u} = \frac{1}{f}$ (*d*) $\frac{1}{v} - \frac{1}{u} = \frac{1}{f}$

20.5 Equivalent focal length of two thin lenses is
(*a*) $\frac{1}{f} = \frac{1}{f_1} + \frac{1}{f_2} + \frac{d}{f_1 f_2}$ (*b*) $\frac{1}{f} = \frac{1}{f_1} - \frac{1}{f_2} + \frac{d}{f_1 f_2}$
(*c*) $f = f_1 + f_2 + \frac{f_1 f_2}{d}$ (*d*) $\frac{1}{f} = \frac{1}{f_1} + \frac{1}{f_2} - \frac{d}{f_1 f_2}$

20.6 Newton's formula for coaxial system of two lenses is

(*a*) $x_1 f_2 = x_2 f_1$ (*b*) $x_1 x_2 = f_1 f_2$

(*c*) $x_1 f_1 = x_2 f_2$ (*d*) none of these

20.7 The relationship between f_1 and f_2 is

(*a*) $\frac{f_1}{u}+\frac{f_2}{v}=0$ (*b*) $\frac{f_1}{u}+\frac{f_2}{v}=-1$

(*c*) $\frac{f_1}{u}+\frac{f_2}{v}=1$ (*d*) $\frac{f_1}{u}-\frac{f_2}{v}=1$

20.8 The power (P) of a thick lens is

(*a*) $P_1 + P_2 + P_1 P_2 \frac{t}{\mu}$ (*b*) $P_1 + P_2 - P_1 P_2 \frac{t}{\mu}$

(*c*) $P_1 + P_2 + P_1 P_2 \frac{\mu}{t}$ (*d*) $P_1 + P_2 - P_1 P_2 \frac{\mu}{t}$

ANSWERS

20.1 (*c*)	**20.2** (*b*)	**20.3** (*a*)	**20.4** (*b*)	**20.5** (*d*)
20.6 (*b*)	**20.7** (*c*)	**20.8** (*b*)		

21 Chapter

ABERRATION IN IMAGES

21.1 ABERRATIONS

One of the basic problems of lenses is the imperfect quality of images. The deviations in size, shape, position and colour in the actual images produced by a lens in comparison to the object are called aberrations produced by the lens.

There are two type of aberrations : (*i*) chromatic aberrations and (*ii*) monochromatic aberrations.

Chromatic aberrations are distortions of the image due to the dispersion* of light in the lenses of an optical system, when white light is used. *The defect of coloured image formed by a lens with white light is called chromatic aberration.*

If monochromatic light is used then such defects are automatically removed. Besides these defects, there are defects which are present even when monochromatic light is used. Such defects are called *monochromatic aberrations*. These aberrations are the results of (*i*) the large aperture of the optical system, (*ii*) the large angle subtended by the rays with the principal axis, and (*iii*) the large size of the object.

The following are the monochromatic aberrations:

(*i*) Spherical aberration (*ii*) Astigmatism (*iii*) Coma, (*iv*) Curvature of field and (*v*) Distortion

21.2 CHROMATIC ABERRATIONS OF A LENS

Light waves of different colours have different velocities in a refracting medium. Every refracting medium, therefore has a different refractive index for each colour or wavelength. The refractive index is least for red colour and maximum for violet colour in the visible region. As the focal length of the lens depends upon the refractive index, it is different for different colours or wavelengths. As $\mu_v > \mu_r$, the focal length $f_v < f_r$. A single lens will, therefore, not form one image of an object point, but a series of coloured images at different distances from the lens, corresponding to each colour constituting the incident beam are formed. The variation of the image distance from the lens, with refractive index measures the **axial or longitudinal chromatic aberration.**

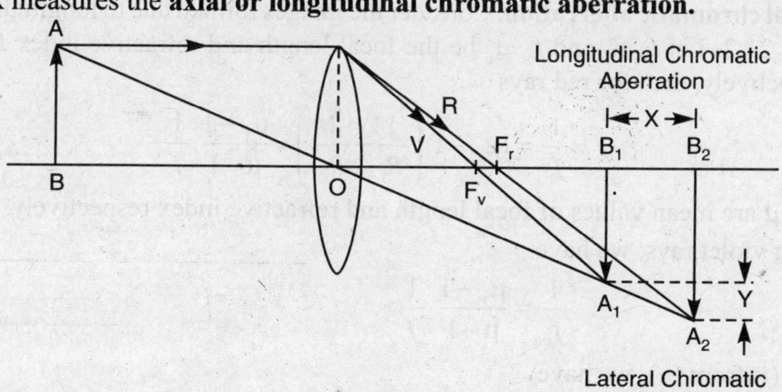

Fig. 21.1

---

* **Dispersion** means splitting of light into its constituents. A ray of white light after passing through a prism, is split up into its constituent colours *i.e.*, from red through violet, all seven colours arranged in bands. This is because each colour has different angular dispersion.

Again the magnification of the image depends upon the focal length of the lens. The size of the image is different for different colours or wavelengths. The variation in the size of the image measures the **lateral chromatic aberration.** As shown in Fig. 21.1, the object AB gives the image $A_1 B_1$ corresponding to violet light and $A_2 B_2$ corresponding to the red light. The violet image is formed nearer to the lens than the red image. The distance $B_1 B_2 = x$ measures the longitudinal chromatic aberration and the distance $y = A_2 B_2 - A_1 B_1$ measures the lateral chromatic aberration. The images of intervening colours between violet and red lie between the images $A_1 B_1$ and $A_2 B_2$ and their size increases from violet to red.

It is clear that no one position the images are in sharp focus. The effect due to both the aberrations is known as **chromatic aberration.** The chromatic aberration of a lens for an object at infinity can be calculated as given below:

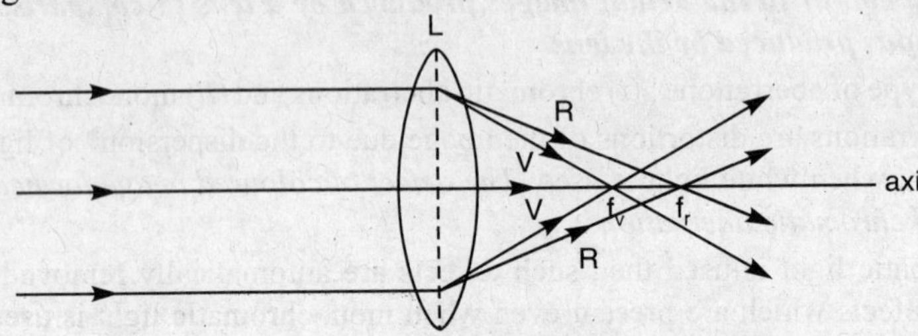

Fig. 21.2(a). Longitudinal Aberration due to Convex Lens

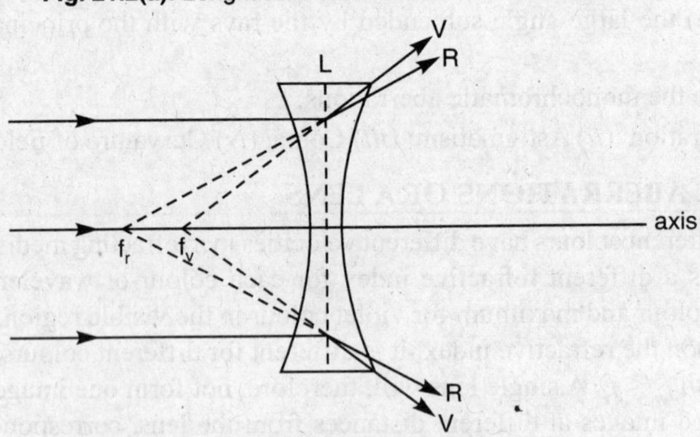

Fig 21.2 (b). Longitudinal Aberration due to Concave Lens

Longitudinal chromatic aberration. Consider the images formed due to longitudinal aberration, as shown in Fig. 21.2. Let f_r, μ_r and f_v, μ_v be the focal length and refractive index for the red and violet rays respectively, then for red rays

$$\frac{1}{f_r} = (\mu_r - 1)\left(\frac{1}{R_1} - \frac{1}{R_2}\right) = \frac{\mu_r - 1}{\mu - 1} \cdot \frac{1}{f} \qquad \dots (i)$$

where f and μ are mean values of focal length and refractive index respectively.

Similarly for violet rays, we have

$$\frac{1}{f_v} = \frac{\mu_v - 1}{\mu - 1} \cdot \frac{1}{f} \qquad \dots (ii)$$

Subtracting (*i*) from (*ii*), we have

$$\frac{1}{f_v} - \frac{1}{f_r} = \frac{\mu_v - \mu_r}{\mu - 1} \cdot \frac{1}{f} = \frac{\omega}{f}$$ where $\omega = \left[\frac{\mu_v - \mu_r}{\mu - 1}\right]$ and is known as dispersive power of the material of the lens.

$$\therefore \qquad \frac{f_r - f_v}{f_v f_r} = \frac{\omega}{f}$$

or $$f_r - f_v = \frac{\omega}{f} f^2 = \omega f. \qquad (\because f_r f_v = f^2) \qquad ...(21.1)$$

Thus longitudinal chromatic aberration ($f_r - f_v$) is equal to the product of dispersive power and mean focal length for yellow light. It is also clear that a single lens can not form an image free from chromatic aberration.

21.3 ACHROMATISM OF A LENS

The term achromatism of a lens means that all the light rays even of different colours are brought to focus by the lens at one and the same point.

Longitudinal Chromic Aberration

Let us consider a point object illuminated by white light and situated on the axis of the lens as shown in Fig. 21.3. Coloured images are formed along the axis. The violet image is nearest the lens and the red image is the farthest. In between these two images, if a screen is placed at the position XY, the image of *least chromatic aberration* is formed *i.e.*, image *AB*.

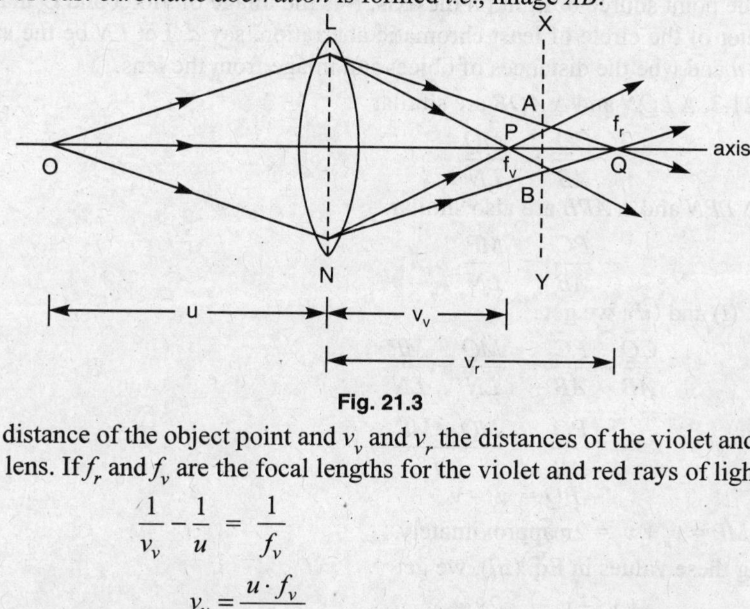

Fig. 21.3

If u is the distance of the object point and v_v and v_r the distances of the violet and red images on the axis of the lens. If f_r and f_v are the focal lengths for the violet and red rays of light then,

$$\frac{1}{v_v} - \frac{1}{u} = \frac{1}{f_v}$$

or $$v_v = \frac{u \cdot f_v}{u + f_v} \qquad ...(i)$$

and $$\frac{1}{v_r} - \frac{1}{u} = \frac{1}{f_r}$$

or $$v_r = \frac{u.f_r}{u + f_r} \qquad ...(ii)$$

Subtracting Eq. (*ii*) from Eq. (*i*), we get

$$\frac{1}{v_v} - \frac{1}{v_r} = \frac{1}{f_v} - \frac{1}{f_r}$$

$$\frac{v_r - v_v}{v_r v_v} = \frac{f_r - f_v}{f_r\, f_v}$$

Taking $v_r \cdot v_v = v^2$ and $f_r \cdot f_v = f^2$

$$\frac{v_r - v_v}{v^2} = \frac{f_r - f_v}{f^2}$$

But $f_r - f_v = \omega f$ $(\because \text{Eq. } 21.1)$

$$\frac{v_r - v_v}{v^2} = \frac{\omega f}{f^2} = \frac{\omega}{f}$$

$$\therefore \quad v_r - v_v = \frac{\omega v^2}{f} \qquad \text{... (21.2)}$$

Thus, it is clear that the longitudinal chromatic aberration depends on the distance of the image and hence on the distance of the object from the lens, in addition to it dependence on the dispersive power (ω) and the focal length (*f*) of the lens.

21.4 CIRCLE OF LEAST CHROMATIC ABERRATION

The diameter of circle of least chromatic aberration is calculated as under. For this consider Fig. 21.3. Let *O* be the point source of light on the axis. *P* is the image of violet and *Q* that for red rays. *AB* is the diameter of the circle of least chromatic aberration, say *d*. Let *LN* be the aperture of the lens, say *D*. Let *u* and *v* be the distances of object and image from the lens.

From Fig. 21.3, Δ *LQN* and Δ *AQB* are similar

$$\therefore \quad \frac{CQ}{AB} = \frac{MQ}{LN} \qquad \text{... (i)}$$

Similarly, Δ *LPN* and Δ *APB* are also similar.

$$\therefore \quad \frac{PC}{AB} = \frac{MP}{LN} \qquad \text{... (ii)}$$

Adding Eq. (*i*) and (*ii*), we get

$$\frac{CQ}{AB} + \frac{PC}{AB} = \frac{MQ}{LN} + \frac{MP}{LN}$$

$$\therefore \quad \frac{PQ}{AB} = \frac{MQ + MP}{LN}$$

But $PQ = v_r - v_v$ ... (*iii*)

and $MQ + MP = v_r + v_v = 2v$ **approximately.**

Substituting these values in Eq. (*iii*), we get

$$\frac{v_r - v_v}{d} = \frac{2v}{D} \qquad (\because AB = d \text{ and } LN = D)$$

$$\therefore \quad d = D \cdot \frac{v_r - v_v}{2v}$$

Using Eq. (21.2), we have $v_r - v_v = \dfrac{\omega v^2}{f}$

$$\therefore \quad d = D \cdot \frac{\omega v^2}{2v \cdot f} = \frac{1}{2} D\omega \cdot \frac{v}{f} \qquad \text{... (21.3)}$$

If incident light is parallel beam then,

$$v = f \text{ and } d = \frac{1}{2} D\omega \cdot \frac{f}{f}$$

or $$d = \frac{1}{2} D\omega \quad \text{... (21.4)}$$

Thus, the diameter of the circle of least chromatic aberration depends on the diameter of the lens aperture and the dispersive power of the material, but is *independent* of the focal length of the lens.

Example 21.1 *A convex lens has a focal length for red colour 15.5 × 10⁻² m. Its focal length for violet colour is 14.45 × 10⁻² m. If an object is kept at a distance of 0.4 m from the lens, calculate the longitudinal chromatic aberration produced by the lens.*

Solution. Here $f_r = 15.5 \times 10^{-2}$ m, $f_v = 14.45 \times 10^{-2}$ m, $u = -0.4$ m

If v_r and v_v are the distance of the images for red and violet lights from the lens respectively, then

$$v_r = \frac{uf_r}{u + f_r} = \frac{-0.4 \times 15.5 \times 10^{-2}}{-0.4 + 15.5 \times 10^{-2}} = 0.25306 \text{ m}$$

and $$v_v = \frac{-0.4 \times 14.5 \times 10^{-2}}{-0.4 + 0.1445} = 0.22622 \text{ m}$$

∴ Longitudinal chromatic aberration = 0.25306 – 0.22622 = 0.02684 m.

Example 21.2 *A crown glass convex lens has radii of curvature 10 cm and 20 cm. If $\mu_b = 1.523$ and μ_r 1.513, find chromatic aberration produced by the lens.* **(M.D.U. 2003; K.U. 2002)**

Solution. Here $R_1 = 10$ cm, $R_2 = 20$ cm, $\mu_b = 1.523$, $\mu_r = 1.513$

∴ Mean $$\mu = \frac{\mu_b + \mu_r}{2} = \frac{1.523 + 1.513}{2} = 1.518$$

Mean focal length f is given by $$\frac{1}{f} = (\mu - 1)\left(\frac{1}{R_1} - \frac{1}{R_2}\right)$$

$$= (1.518 - 1)\left(\frac{1}{10} + \frac{1}{20}\right) \quad (\because R_2 \text{ is negative.})$$

$$= .518 \times \frac{3}{20} = \frac{1.554}{20}$$

∴ $$f = \frac{20}{1.554} = 12.87 \text{ cm}$$

Dispersive power $$\omega = \frac{\mu_b - \mu_r}{\mu - 1} = \frac{1.523 - 1.513}{1.518 - 1}$$

$$= \frac{.01}{.518} = 0.0193$$

Chromatic aberration = $f_r - f_b = \omega_f = 0.0193 \times 12.87$

$$= 0.2484 \text{ cm.}$$

Example 21.3 *The refractive indices of hard crown glass for red and violet are 1.553 and 1.567 respectively. If a beam of white light falls on a convex lens of the same glass in a direction parallel to the axis and if the radii of curvature of the two surfaces of the lens are 30 cm and 20 cm, calculate the axial chromatic aberration of the lens for the above colours.* **(Gauhati U. 2000)**

Solution. Mean $$\mu = \frac{1.553 + 1.567}{2} = 1.560$$

Mean f is given by $$\frac{1}{f} = (\mu - 1)\left(\frac{1}{R_1} + \frac{1}{R_2}\right) = (1.56 - 1)\left(\frac{1}{20} + \frac{1}{30}\right)$$

$$= 0.560 \times \frac{1}{12}$$

or $$f = \frac{12}{.560} = 21.43 \text{ cm}$$

Dispersive power $$\omega = \frac{\mu_v - \mu_r}{\mu - 1} = \frac{1.567 - 1.563}{1.56 - 1} = 0.025$$

$\therefore$ Chromatic aberration $= \omega f = 0.025 \times 21.43$

$$= 0.536 \text{ cm.}$$

21.5 ACHROMATIC COMBINATION OF LENSES

Chromatic aberration is eliminated by:

(*i*) Keeping two lenses in contact with each other, and

(*ii*) Keeping two lenses out of contact.

When two or more lenses are combined together in such a way that the combination is free of chromatic aberration, then such a combination is called achromatic combination of lenses.

The term achromatism (or sometimes achromatisation) of a lens means that all the light rays of different colour are brought to focus by the lens at one and the same point.

In practice, however, chromatic aberration can not be removed completely. Usually achromatism is achieved for two prominent colours.

21.5.1 Condition for Achromatism of Two Thin Lenses Placed in Contact

Conditions for achromatism. The achromatic combination is made by placing in contact lenses of different materials and of suitable focal lengths so that the focal length of the combination is the same for both the extreme colours; violet and red. The refraction formula for a lens is

$$\frac{1}{f} = (\mu - 1)\left[\frac{1}{R_1} - \frac{1}{R_2}\right].$$

Differentiating, we get

$$-\frac{df}{f^2} = d\mu\left(\frac{1}{R_1} - \frac{1}{R_2}\right) = \frac{d\mu}{\mu - 1} \cdot \frac{1}{f}$$

Now if μ_v and μ_r are the refractive indices for the violet and the red rays, then $d\mu$ is the change in refractive index, *i.e.*, $(\mu_v - \mu_r)$. Hence,

Dispersive power $$\omega = \frac{d\mu}{\mu - 1}$$

$$\therefore \quad -\frac{df}{f^2} = \frac{\omega}{f} \qquad \text{... } (i)$$

When two lenses of focal lengths f_1 and f_2 are placed in contact to form an achromatic combination, then focal length f of the combination is given by

$$\frac{1}{f} = \frac{1}{f_1} + \frac{1}{f_2}$$

$$\therefore \quad -\frac{df}{f^2} = -\frac{df_1}{f_1^2} - \frac{df_2}{f_2^2}$$

In order that rays of different colours are brought to focus at one point, dF the change in focal length of the combination should be zero.

$$\therefore \qquad \frac{df_1}{f_1^2} + \frac{df_2}{f_2^2} = 0$$

or
$$\frac{\omega_1}{f_1} + \frac{\omega_2}{f_2} = 0 \qquad ...(21.5)$$

where ω_1 and ω_2 are the dispersive powers for the materials of the two lenses and are positive. Equation (*ii*) indicates that f_1 and f_2 must have opposite signs, *i.e., the combination should consist of one concave and the other convex lens of different materials.* Such a combination is called an **achromatic doublet.**

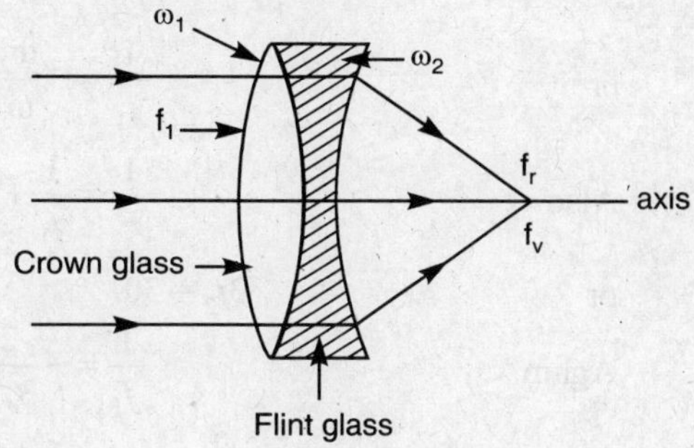

Fig. 21.4

Achromatic doublet. Relation (21.5) can be put in the form

$$\frac{\omega_1}{f_1} = \frac{\omega_2}{f_2}$$

or
$$-\frac{f_2}{f_1} = \frac{\omega_2}{\omega_1} \qquad ...(21.6)$$

Thus an achromatic combination of two thin lenses or an achromatic doublet can be prepared by taking two lenses of different materials one convex and the other concave, such that their focal lengths are in the ratio of their dispersive powers.

In the achromatic *convex* combination of two lenses of focal length $+f$, one lens is convex having a focal length $+f_1$ and the other is concave having a focal length $-f_2$. As $\frac{1}{f} = \frac{1}{f_1} - \frac{1}{f_2}$, for $\frac{1}{f}$ to be positive, $\frac{1}{f_1} > \frac{1}{f_2}$ or $f_2 > f_1$ i.e., the focal length of the concave lens should be greater than that of the convex lens. The converging lens is, therefore, made of crown glass (smaller dispersive power) and the diverging lens of flint glass (larger dispersive power). This is shown in fig. 21.4.

Example 21.4 *Calculate the focal length of a lens of dispersive power 0.031 which should be placed in contact with a convex lens of focal length 0.88 m and dispersive power 0.022 to make the combination achromatic.* **(*M.D.U. 2002*)**

Solution. Here $f_1 = 0.88$ m, $\omega_1 = 0.022$, $\omega_2 = 0.031$, $f_2 = ?$

For achromatism
$$\frac{\omega_1}{f_1} + \frac{\omega_2}{f_2} = 0 \text{ or } \frac{\omega_1}{f_1} = \frac{\omega_2}{f_2}$$

or
$$\frac{f_2}{f_1} = \frac{\omega_2}{\omega_1} = \frac{0.031}{0.022}$$

or
$$f_2 = -0.88 \times \frac{0.031}{0.022} = -1.24 \text{ m}$$

Example 21.5 *Calculate the focal length of a lens of flint glass of dispersive power 0.4 which will make an achromatic doublet with a converging lens of crown glass of focal length 0.4 m and dispersive power 0.2.* **(*K.U. 2001*)**

Solution.
$$f_2 = -\frac{\omega_2}{\omega_1} f_1 = \frac{0.4}{0.2} \times 0.4 = -0.8 \text{ m.}$$

Example 21.6 *An achromatic doublet of a convex lens and a concave lens of focal length 0.30 m is to be formed out of flint and crown glass whose dispersive powers are 0.03 and 0.02 respectively. Calculate the focal length of the two lenses.* **(*G.N.D.U 2003*)**

Solution. Let f_1 and f_2 be the focal lengths and ω_1 and ω_2 the dispersive powers of the two lenses respectively forming the achromatic doublet of focal length $F = 0.30 = 30$ cm.

Now $\omega_1 = 0.03$, $\omega_2 = 0.02$, $f_1 = ?$ and $f_2 = ?$

For achromatism $\quad \dfrac{\omega_1}{f_1} + \dfrac{\omega_2}{f_2} = 0$ or $\dfrac{\omega_1}{f_1} = -\dfrac{\omega_2}{f_2}$

or $\quad \dfrac{1}{f_1} = -\dfrac{\omega_2}{\omega_1} \times \dfrac{1}{f_2} = -\dfrac{0.02}{0.03}\dfrac{1}{f_2} = -\dfrac{2}{3f_2}$

Also $\quad \dfrac{1}{F} = \dfrac{1}{f_1} + \dfrac{1}{f_2}$ or $\dfrac{1}{30} = \dfrac{-2}{3f_2} + \dfrac{1}{f_2} = \dfrac{1}{f_2}\left[1 - \dfrac{2}{3}\right] = \dfrac{1}{3f_2}$

or $\quad 3f_2 = 30 \quad \therefore \quad f_2 = 10 \text{ cm} = 0.10 \text{ m}$

Again $\quad \dfrac{1}{f_1} = -\dfrac{2}{3f_2} = -\dfrac{2}{3 \times 10} = -\dfrac{1}{15}$

$\therefore \quad f_1 = -15 \text{ cm} = -0.15 \text{ m}.$

Example 21.7 *The focal length of an achromatic combination of two lenses is 1.5 m. If the dispersive powers of the materials of the two lenses are 0.018 and 0.027, calculate the focal length of the two lenses.* **(G.N.D.U. 2001; Cal. U. 2002)**

Solution. $\quad \dfrac{1}{f_1} = \dfrac{\omega_2}{\omega_1}\dfrac{1}{f_2} = \dfrac{0.027}{0.018} \times \dfrac{1}{f_2} = -\dfrac{3}{2f_2}$

$\dfrac{1}{f_1} + \dfrac{1}{f_2} = \dfrac{1}{15}$ or $-\dfrac{3}{2f_2} + \dfrac{1}{f_2} = \dfrac{1}{1.5}$ or $-\dfrac{1}{2f_2} = \dfrac{1}{1.5}$

or $\quad f_2 = -0.75 \text{ m}$ and $\dfrac{1}{f_1} = +\dfrac{3}{2 \times 0.75 \text{ m}}$ or $f_1 = 0.5 \text{ m}$

Example 21.8 *Dispersive powers for crown and flint glass are 0.25 and 0.50 respectively. Determine the focal lengths of an achromatic combination of two lenses in contact having equivalent focal length 30 cm.* **(Nagpur U. 2011, P.U. 2002)**

Solution. Let f_1 and f_2 be the focal lengths are ω_1 and ω_2 the dispersive powers of the two lenses respectively forming an achromatic combination of focal length $F = 30$ cm.

Now $\omega_1 = 0.25$; $\quad \omega_2 = 0.50$; $\quad f_1 = ?$; $\quad f_2 = ?$

For achromatism $\quad \dfrac{\omega_1}{f_1} + \dfrac{\omega_2}{f_2} = 0$ or $\dfrac{\omega_1}{f_1} = -\dfrac{\omega_2}{f_2}$

or $\quad \dfrac{1}{f_1} = -\dfrac{\omega_2}{\omega_1} \times \dfrac{1}{f_2} = -\dfrac{0.50}{0.25} \times \dfrac{1}{f_2} = -\dfrac{2}{f_2}$

Also $\quad \dfrac{1}{f_1} = \dfrac{1}{f_1} + \dfrac{1}{f_2}$ or $\dfrac{1}{30} = -\dfrac{2}{f_2} + \dfrac{1}{f_2} = -\dfrac{1}{f_2}$

$\therefore \quad f_2 = -30 \text{ cm}$

Again $\quad \dfrac{1}{f_1} = -\dfrac{2}{f_2} = +\dfrac{2}{30} = \dfrac{1}{15}$ or $f_1 = +15 \text{ cm}.$

Example 21.9 *Calculate the focal length of a convex lens of crown glass of dispersive power 0.12 and concave lens of flint glass of dispersive power 0.20 such that when placed in contact they form achromatic combination of focal length 30 cm.* **(P.U. 2004; Gauhati. U. 2000)**

[Ans. $f_1 = 12$ cm; $f_2 = -20$ cm. (The solution is left as an exercise for students)**]**

Example 21.10 *This dispersive power of crown and flint glass are 0.015 and 0.030 respectively. The refractive index for the mean ray are 1.52 and 1.65. If one of the surfaces of the flint glass is plane, Calculate the radii of curvature of the other two surfaces of the two lenses which form an achromatic combination of focal length 30 cm.* ***(M.D.U. 2001)***

Solution. Let f_1 and f_2 be the focal lengths of the two lenses of crown glass and flint glass respectively placed in contact.

Given: $\omega_1 = 0.015$, $\omega_2 = 0.030$, $f = 30$ cm.

For achromatism $\frac{\omega_1}{f_1} + \frac{\omega_2}{f_2} = 0$ or $\frac{0.015}{f_1} + \frac{0.030}{f_2} = 0$

$\therefore$ $\frac{1}{f_1} = -\frac{0.030}{0.015} \cdot \frac{1}{f_2} = -\frac{2}{f_2}$

Also $\frac{1}{f} = \frac{1}{f_1} + \frac{1}{f_2}$ or $\frac{1}{f_1} + \frac{1}{f_2} = \frac{1}{30}$

or $\frac{-2}{f_2} + \frac{1}{f_2} = \frac{1}{30}$ or $-\frac{1}{f_2} = \frac{1}{30}$ $\therefore f_2 = -30$ cm.

Again $\frac{1}{f_1} = -\frac{2}{f_2} = +\frac{2}{30} = \frac{1}{15}$ or $f_1 = +15$ cm

Thus the focal length of the double convex lens of crown glass is 15 cm.

For plano – concave flint glass lens, $\mu_2 = 1.65$, $R_2 = \infty$ and $f_2 = -30$ cm.

$$\therefore \quad \frac{1}{f_2} = (\mu_2 - 1)\left(\frac{1}{R_1} - \frac{1}{R_2}\right)$$

or $$-\frac{1}{30} = (1.65 - 1)\frac{1}{R_1} = \frac{0.65}{R_1}$$

or $$R_1 = -30 \times 0.65 = -19.5 \text{ cm.}$$

Since the cemented surfaces of the concave and convex lens have got the same radius of curvature. Therefore, for the convex lens we have

$R_2 = -19.5$ cm; $f_1 = 15$ cm; $R_1 = ?$, $\mu_1 = 1.52$

Now $\frac{1}{f_1} = (\mu_1 - 1)\left(\frac{1}{R_1} - \frac{1}{R_2}\right)$ or $\frac{1}{15} = (1.52 - 1)\left(\frac{1}{R_1} + \frac{1}{19.5}\right)$

$$\frac{1}{15 \times 0.52} - \frac{1}{19.5} = \frac{1}{R_1} \quad \text{or} \quad \frac{1}{R_1} = \frac{1}{7.8} - \frac{1}{19.5} = \frac{11.7}{7.8 \times 19.5}$$

$$\therefore \quad R_1 = \frac{7.8 \times 19.5}{11.7} = +13 \text{ cm.}$$

Hence for the convex lens $R_1 = 13$ cm and $R_2 = -19.5$ cm.

Example 21.11 *A telescope objective is made by cementing together two lenses, the cemented surface having the same radii of curvature. The combination is plano-convex and is of focal length 1.0 m. Calculate the radii of curvature of the component lenses : given that refractive index and dispersive power of crown glass are 1.50 and 0.02 respectively and that of flint glass are 1.60 and 0.03.*

Solution. $\frac{\omega_1}{f_1} + \frac{\omega_2}{f_2} = 0$ or $\frac{0.02}{f_1} + \frac{0.03}{f_2} = 0$ or $\frac{1}{f_1} = -\frac{3}{2f_2}$

$\frac{1}{f_1} + \frac{1}{f_2} = 1$ or $\frac{-3}{2f_2} + \frac{1}{f_2} = 1$ $\therefore f_2 = -0.5$ m

$$\frac{1}{f_1} = -\left[\frac{3}{2\times(-0.5)}\right] = 3 \qquad \text{or} \qquad f_1 = \frac{1}{3}\text{ m.}$$

For plano-concave lens, $-\dfrac{1}{0.5} = (1.6-1)\left(\dfrac{1}{R_1}\right)$ $\qquad \therefore \quad R_1 = -0.3$ cm.

For double convex lens $\quad R_2 = -0.3$ m

and $\dfrac{3}{1} = (1.5-1)\left(\dfrac{1}{R_1} + \dfrac{1}{0.3}\right)$ $\qquad \therefore \quad R_1 = +0.375$ cm.

Example 21.12 *It is desired to make an achromatic lens by placing a thin convex lens in contact with a thin concave lens. The covex and concave lenses are made up of materials of dispersive power in the ratio of 1 : 2. If the focal length of the convex lens is 20 cm, determine the focal length of the achromatic combination. How does the combination behave ? Convergent or divergent.* **(K.U. 2000)**

Solution. Focal length of the convex lens $f_1 = 20$ cm.

Ratio of dispersive powers $\dfrac{\omega_2}{\omega_1} = \dfrac{2}{1}$

Focal length of the other lens $f_2 = ?$

Focal length of the combination $F = ?$

Now $\qquad \dfrac{\omega_1}{f_1} + \dfrac{\omega_2}{f_2} = 0 \quad \text{or} \quad \dfrac{\omega_1}{f_1} = -\dfrac{\omega_2}{f_2}$

or $\qquad f_2 = \dfrac{-\omega_2}{\omega_1} f_1 = -2\times 20 = -40$ cm.

$\therefore$ Focal length of the combination is given by

$$\frac{1}{f} = \frac{1}{f_1} + \frac{1}{f_2} = \frac{1}{20} - \frac{1}{40} = \frac{1}{40}$$

or $\qquad f = +40$ cm.

The combination behaves as a convergent lens.

Example 21.13 *The mean refractive indices of two specimens of glass are 1.52 and 1.66 respectively; the difference in indices for the same two lines of the spectrum are 0.013 for the first lens and 0.022 for the second. Find the focal length of a lens of the second glass which when combined with a convex lens of 0.5 m focal length will make an object glass achromatic for these two lines.*

Solution. Let the dispersive powers of the two specimens of glass be ω_1 and ω_2 and the two lines be violet and red, then

$$\omega_1 = \frac{\mu_v - \mu_r}{\mu - 1} \quad \text{and} \quad \omega_2 = \frac{\mu_v' - \mu_r'}{\mu' - 1}.$$

Now for the first lens, we have

$$\mu_v - \mu_r = 0.013 \text{ and } \mu' = 1.52$$

and for the second lens, we have

$$\mu_v' - \mu_v' = 0.022 \text{ and } \mu' = 1.66$$

$$\therefore \qquad \omega_1 = \frac{0.013}{1.52-1} = \frac{0.013}{0.52} \quad \text{and} \quad \omega_2 = \frac{0.022}{1.66-1} = \frac{0.022}{0.66}$$

Let f_1 and f_2 be the focal lengths of two lenses, then for achromatism.

$$\frac{\omega_1}{f_1} + \frac{\omega_2}{f_2} = 0.$$

Here $f_1 = 0.5$ m

$$\therefore \quad \frac{0.013}{0.52}\times\frac{1}{0.5} = -\frac{0.022}{0.66}\times\frac{1}{f_2}$$

or $$f_2 = -\frac{0.022}{0.66}\times\frac{0.52}{0.013}\times 0.5 = 0.667 \text{ m}$$

21.5.2 Condition for Acromatism of two thin Lenses Separated by a Finite Distance

Let f_1 and f_2 be the focal lengths of the two lenses placed at a distance d apart as shown in Fig. 21.5. Let f be the focal length of the combination, then

$$\frac{1}{f} = \frac{1}{f_1} + \frac{1}{f_2} - \frac{d}{f_1 f_2}$$

Differentiating, we get

$$\frac{df}{f^2} = -\frac{df_1}{f_1^2} - \frac{df_2}{f_2^2} + d\left[\frac{df_1}{f_1^2 f_2} + \frac{df_2}{f_1 f_2^2}\right] \quad \dots (i)$$

Now for a single lens, we know

$$\frac{1}{f} = (\mu - 1)\left(\frac{1}{R_1} - \frac{1}{R_2}\right)$$

Differentiating, we get

$$-\frac{df}{f^2} = d\mu\left(\frac{1}{R_1} - \frac{1}{R_2}\right) = \frac{d\mu}{\mu - 1}\cdot\frac{1}{f} = \frac{\omega}{f}$$

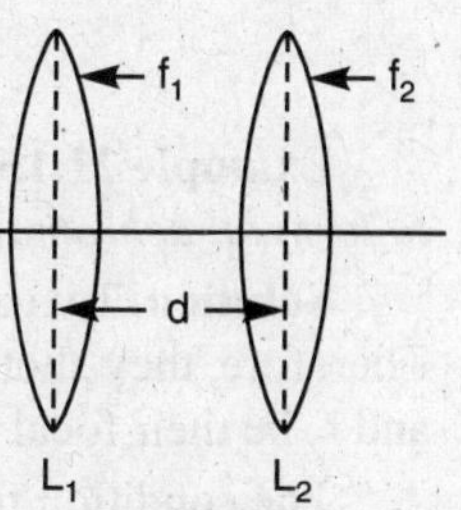

Fig. 21.5

Thus $$-\frac{df}{f^2} = \frac{\text{dsipersive power}}{\text{focal length}} \quad \dots (ii)$$

If ω_1 and ω_2 are the dispersive powers for the two lenses of focal length f_1 and f_2 respectively and ω that of the combination, then from (*i*), we have

$$\frac{\omega}{f} = \frac{\omega_1}{f_1} + \frac{\omega_2}{f_2} - d\left\{\frac{\omega_1}{f_1 f_2} + \frac{\omega_2}{f_1 f_2}\right\}$$

In order that rays of different colours are brought to focus at a point, the dispersive power of the combination ω should be zero. In other words,

$$\frac{\omega_1}{f_1} + \frac{\omega_2}{f_2} - d\left\{\frac{\omega_1}{f_1 f_2} + \frac{\omega_2}{f_1 f_2}\right\} = 0$$

or $$\frac{\omega_1}{f_1} + \frac{\omega_1}{f_2} = \frac{d(\omega_1 + \omega_2)}{f_1 f_2}$$

If the lenses are in contact, then $d = 0$

$$\therefore \quad \frac{\omega_1}{f_1} + \frac{\omega_2}{f_2} = 0 \quad \dots (21.7)$$

(*b*) If the two lenses are of the *same material* and are separated by a distance d then $\omega_1 = \omega_2 = \omega$.

$$\therefore \quad \frac{\omega}{f_1} + \frac{\omega}{f_2} = \frac{2d\cdot\omega}{f_1 f_2}$$

or $$\omega\times\frac{f_1 + f_2}{f_1 f_2} = \frac{2d\cdot\omega}{f_1 f_2}$$

or $$f_1 + f_2 = 2d$$

$$\therefore \qquad d = \frac{f_1 + f_2}{2} \qquad \text{... (21.8)}$$

This shows that two lenses of the *same material* when separated by a distance equal to half the sum of the focal lengths of the two lenses, form an *achromatic combination.*

Advantage. Chromatic aberration is completely eliminated for all colours as the expression does not contain the dispersive power of the material of the lenses. Further since d is to be positive, either both the lenses should be convex or the lens of higher focal length should be convex.

Example 21.14 *Two convex lenses made from same material and of focal lengths 0.20 m and 0.30 m are to be placed a certain distance apart so that they form an image free from chromatic aberration. Find the distance between them.* **(Nagpur 2009, 2008; G.N.D.U. 2002)**

Solution. Here $f_1 = 0.2$ m; $f_2 = 0.3$ m

For achromatism the distance d between the two lenses

$$d = \frac{f_1 + f_2}{2}$$

$$= \frac{0.2 + 0.3}{2} = 0.25 \text{ m}$$

Example 21.15 *Two thin converging lenses of focal lengths 10 cm and 20 cm respectively are to form an achromatic combination. Calculate the effective focal length of the combination.*

Solution. Two converging lenses can not form an achromatic combination if they are in cantact. Therefore, they should be seperated by a distance d. Both the lenses are of the same material. Let f_1 and f_2 be their focal lengths respectively.

The condition for the combination to be achromatic is

$$d = (f_1 + f_1)/2$$

The effective focal length f of the combination is

$$\frac{1}{f} = \frac{1}{f_1} + \frac{1}{f_2} - \frac{d}{f_1 f_2} = \frac{1}{f_1} + \frac{1}{f_2} - \frac{(f_1 + f_2)}{2 f_1 f_2} = \frac{f_1 + f_2}{2 f_1 f_2}$$

$$\therefore \qquad f = \frac{2 f_1 f_2}{f_1 + f_2} = \frac{2 \times 10 \times 20}{10 + 20} = 13.33 \text{ cm}$$

Example 21.16 *A system consisting of two crown glass lenses is to be achromatic for the focal length which is 0.4 m. The focal length of one lens being 0.25 m, find the focal length of the other and the required distance between the two lenses.*

Solution. Here, $f = 0.4$ m, $f_1 = 0.25$ m

$$\frac{1}{f} = \frac{1}{f_1} + \frac{1}{f_2} - \frac{d}{f_1 f_2}$$

$$\text{or} \qquad \frac{1}{0.4} = \frac{1}{0.25} + \frac{1}{f_2} - \frac{d}{0.25 f_2} \qquad \text{... }(i)$$

The condition for achromatism is $d = (f_1 + f_2)/2$ ... (*ii*)

$$\therefore \qquad d = (0.25 + f_2)/2$$

Substituting value of d is Eq. (*i*), we get

$$\frac{1}{0.4} = \frac{1}{0.25} + \frac{1}{f_2} - \frac{(0.25 + f_2)}{2 \times 0.25 f_2}$$

$$\therefore \qquad f_2 = 1 \text{ m}$$

From Eq. (*ii*), $d = (0.25 + 1)/2 = 0.625$ m.

21.6 MONOCHROMATIC ABERRATION

Monochromatic aberration takes place even when the incident light is monochromatic, *i.e.*, composed of rays of a single wavelength.

Types of monochromatic aberrations. Various types of monochromatic aberrations are (*i*) Spherical aberration (*ii*) Astigmatism (*iii*) Coma (*iv*) Curvature of the field and (*v*) distortion

21.6.1 Spherical Aberration in Lenses

If a point object is placed on the axis of a large lens, then the rays incident further away from the axis the brought to focus at I_p which is nearer to the lens than the point I_c where the rays from the central portion of the lens are brought to focus, as shown in Fig. 21.6.

It is clear from the figure that the image formed by the central rays lies at a longer distance than the image formed by the marginal rays. The image formed is, therefore, not sharp at any point on the axis. If a screen is placed perpendicular to the axis at I_P, the outer portions of the object are in focus, while at I_c the central portions are in focus. If the screen is placed at AB, the image appears to be a circular patch of diameter AB. This patch of diameter AB is called the **circle of least confusion** and corresponds to the position of best image.

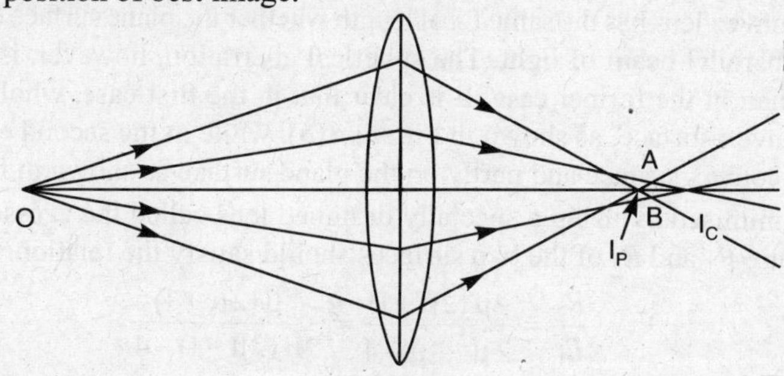

Fig. 21.6

The distance $I_c I_p$ measures *axial* or *longitudinal spherical aberration*.

The radius of the circle of least confusion lying between I_c and I_p is a measure of the *lateral spherical aberration.*

It is clear that for an object point O on the axis, the image extends over the length $I_p I_c$. This defect is called spherical aberration and is due to the fact that different zones have different focal lengths.

The spherical aberration produced by a concave lens is shown in Fig. 21.7. It is *positive* for a convex lens and is negative for a concave lens. When a parallel beam of light is incident on the surface of a concave lens, the central rays, after refraction appear to diverge from the point F_c which lies away from the lens than the point F_p where the paraxial rays come to focus. Hence the spherical aberration is negative.

Removal of spherical aberration in lenses: (*a*) It can be minimized if the total deviation of a ray is equally shared by all the refracting surfaces. For a single lens, spherical aberration is minimised under the following conditions :

(*i*) The spherical aberration depends upon the aperture of the lens. It can, therefore, be reduced by reducing the lens aperture with the help of a co-axial aperture stop, thus using only the central portion of the lens.

(*ii*) *By using a plano-convex lens with its convex surface towards the incident or the emergent beam, whichever is more parallel to the axis.*

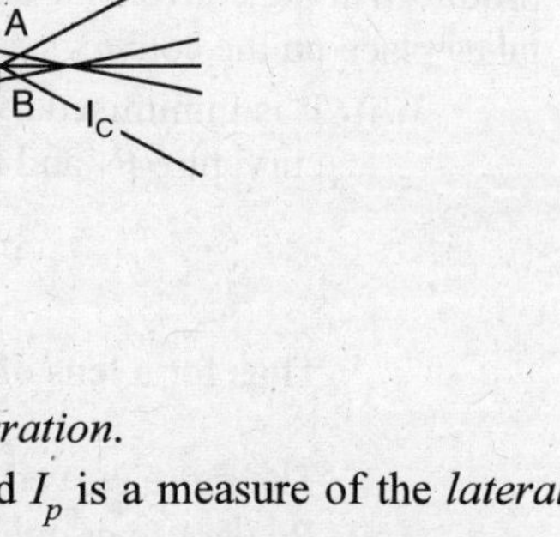

Fig. 21.7

We have seen that spherical aberration is due to greater deviation of the marginal rays as compared to that of axial rays. If the deviation of the marginal rays is made minimum, the focal point F_p will be nearest to the focal point F_c for central rays and the spherical aberration will be the least. Just as in the case of a prism, the deviation is minimum when the incident and the emergent rays make equal angles with its faces; similarly the deviation of marginal rays falling on a lens will be minimum when these enter the first lens surface and leave the second lens surface at more or less equal angles. Hence the rule is that the spherical aberration is minimum when a lens is so designed or used that *the total deviation of a given ray is divided equally between the two refractions.*

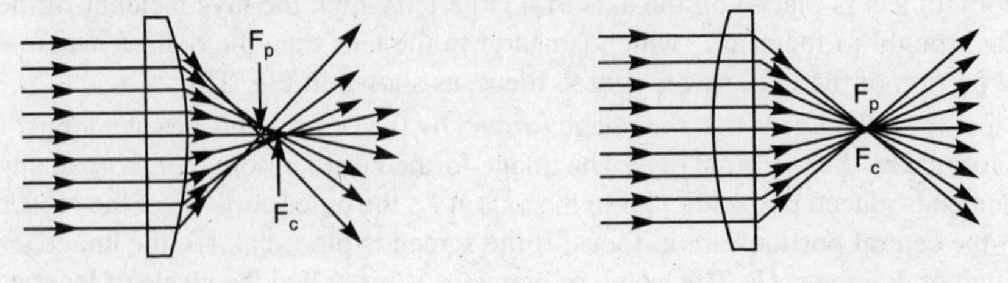

Fig. 21.8

A thin plano-convex lens has the same focal length whether the plane surface or the convex surface faces the incident parallel beam of light. The spherical aberration, however, is very much reduced in the latter case than in the former case. It is clear that in the first case, whole of the deviation is produced at the convex surface, as shown in Fig. 21.8(*a*), while in the second case, deviation partly takes place on the convex surface and partly on the plane surface as shown in Fig. 21.8(*b*).

(*iii*) It is minimised by using a specially designed lens called the **crossed lens.** The radii of curvature R_1 and R_2 of the two surfaces should satisfy the relation.

$$\frac{R_2}{R_1} = \frac{-\mu(2\mu+1)}{2\mu^2-\mu-4} = -\frac{\mu(2\mu+1)}{\mu(2\mu-1)-4}$$

Thus for a lens of refractive index 1.5, we have

$$R_2 : R_1 :: 6 : 1.$$

The more curved surface has to face the more parallel beam of light.

(*iv*) *By the use of aplanatic surface* (see article 19.10).

(*b*) For a combination of **two or more lenses,** the spherical aberration is minimised under the following conditions.

(*i*) When two convergent lenses, separated by a distance are used, the refraction takes place at four surfaces. The spherical aberration is least when there is an equal deviation at all the surfaces.

This is achieved, if $f_2 - f_1 = d$

where f_2 and f_1 are the focal lengths of the two lenses and d is the distance between them. This arrangement is used in **eye-pieces.**

(*ii*) The spherical aberration can be minimised by using a *stop* at a suitable position, so as to restrict the incidence of rays by reducing the effective aperture of the lens. This, however, reduces the intensity of the image formed, as the amount of light passing through the lens is decreased.

(*iii*) It is also minimised by combining a suitable concave lens with a convex lens, so as to shift the focus F_p towards F_c. This method is, however, rarely used, as such a combination works only for a particular pair of object and image, for which it is constructed.

21.6.2 Astigmatism

This defect is due to the aberration of an oblique pencil of rays, coming from the object points lying far away from the axis and passing through the centre of the lens. Consider a point source of light O at some distance off the axis of a convex lens and a screen placed on the other side, as shown in Fig. 21.9. It will be found that in no position of the screen the rays are brought to a sharp point focus. In the position T all the rays contained in the vertical plane RS of the lens, are brought to focus in a *short line lying in the plane of the paper.* If the screen is drawn back into the position S, all the rays from the horizontal plane PQ are brought to focus on the screen in a *short line in a plane perpendicular to the plane of the paper.* The nearest approach to a point image is somewhere in between the two focal lines as at L where a circular patch of light is obtained. This is known as the *circle of least confusion.* This defect is known as *astigmatism* and the distance between T and S is known as **astigmatic difference.** It increases with the obliquity of the rays.

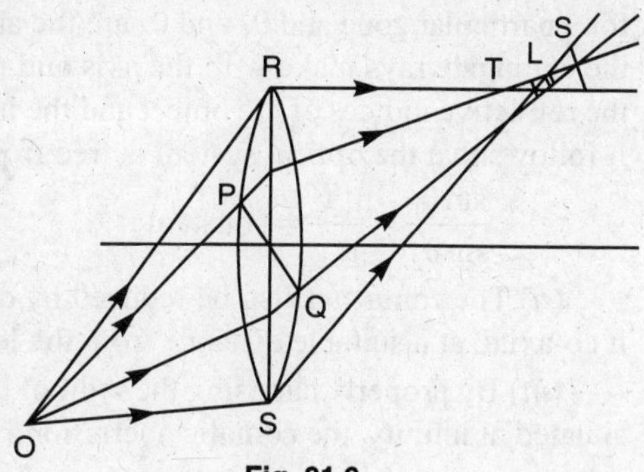

Fig. 21.9

Removal. (*i*) The astigmatic difference is in opposite direction for a convex and a concave lens. Hence by suitably combining the two lenses to form a compound lens, astigmatism can be corrected to a large extent. Such a combination is known as an astigmat.

(*ii*) It can also be removed by using two plano-convex lenses (spherical or cylindrical) separated by a suitable distance.

(*iii*) It is also absent if the lens has surfaces of different horizontal and vertical curvatures. Such a lens is called a **toric lens.**

21.6.3 Coma

The aberration known as *coma* occurs for the rays which come from object points that lie off the lens axis. It is because for these non-axial point objects, there is either a decrease or increase of lateral magnification with the height of the narrow circular zone through which refraction takes place. Comatic aberration is similar to spherical aberration in that both are due to failure of the lens to bring all rays from a point object to focus at the same point. In spherical aberration, the image is a circle of varying diameter belong the axis but in case of comatic aberration, the image is comet-shaped (see Fig. 21.10) and hence the name coma.

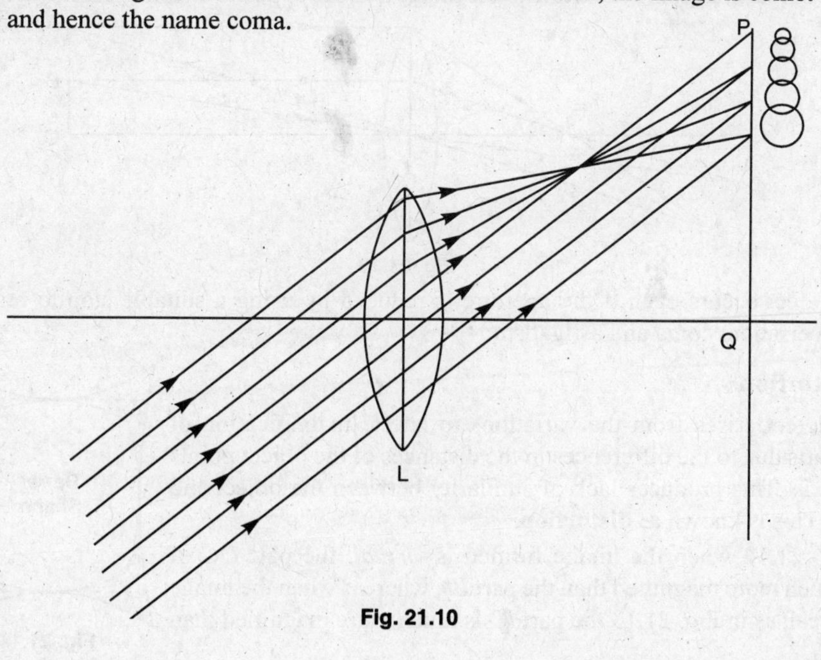

Fig. 21.10

Reduction of coma (*i*) To avoid coma, it is necessary that for all rays from the axial points of the object, the ratio of the sines of the angles of the slopes of the object and the image rays for all zones must be constant. This is known as **Abbe sine condition:** according to which

$$\mu y \sin\theta_1 = \mu' y' \sin\theta_2$$

where y and y' are the lengths of the object and the image for a particular zone and θ_1 and θ_2 are the angles which the conjugate rays make with the axis and μ and μ' are the refractive indices of the object and the image space. It follows that the optical system is free from coma, if

$$\frac{\sin\theta_1}{\sin\theta_2} = \frac{\mu' y'}{\mu\, y} = \text{constant}.$$

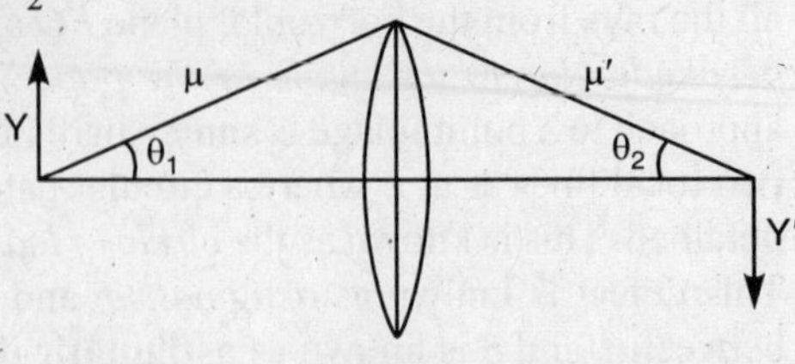

Fig. 21.11

(*ii*) The coma can also be reduced by decreasing the radius of the stop sufficiently, and placing it co-axial, at a suitable distance from the lens.

(*iii*) By properly choosing the radii of curvature of the lens surfaces. For example, for an object situated at infinity, the comatic aberration may be minimised by taking a lens of $\mu = 1.5$ and

$$K = \frac{R_1}{R_2} = -\frac{1}{9}$$

21.6.4 Curvature of the Field

If an extended plane object is viewed through a convex lens, the image formed is not a flat one. Generally the central part of the image is in sharp focus, but the edges will be quite blurred. It is because the focal length of the lens for a paraxial beam F_c is greater than the focal length F_p for the oblique pencil. In the case of a *virtual image* the image points situated off the axis is drawn farther from the lens than those situated in the central paraxial part. The image observed is curved outward as shown in Fig. 21.12. This defect is known as the curvature of the field.

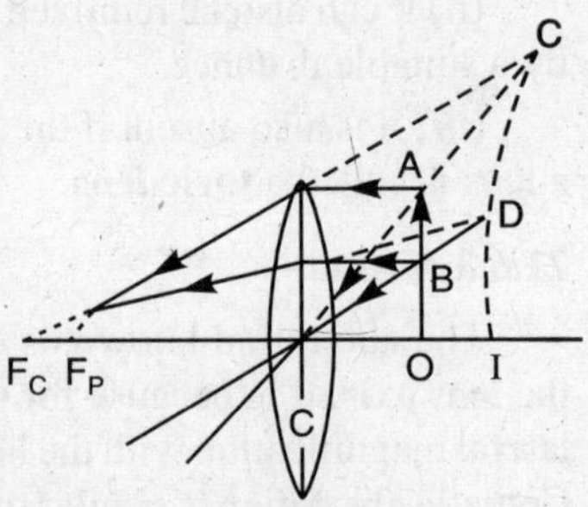

Fig. 21.12

In case of real image, the curvature of the image is in the opposite direction as shown in Fig. 21.13.

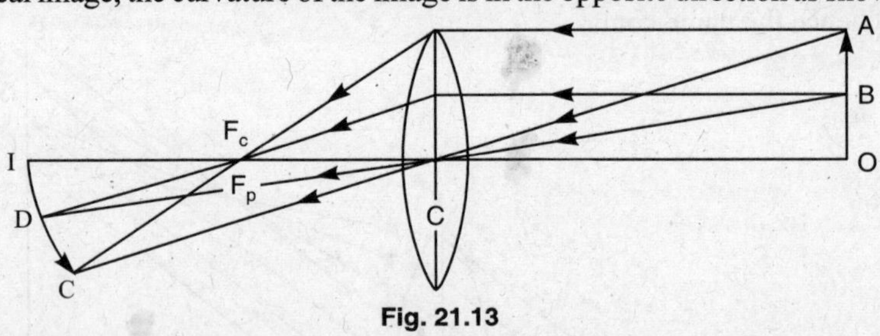

Fig. 21.13

This defect occurs even if the aperture is reduced by using a suitable stop to reduce central spherical aberration, coma and astigatism.

21.6.5 Distortion

This defect arises from the variations in lateral magnification of different parts due to the differences in the distances of the object points from the axis. This produces lack of similarity between the object and the image. This is known as **distortion.**

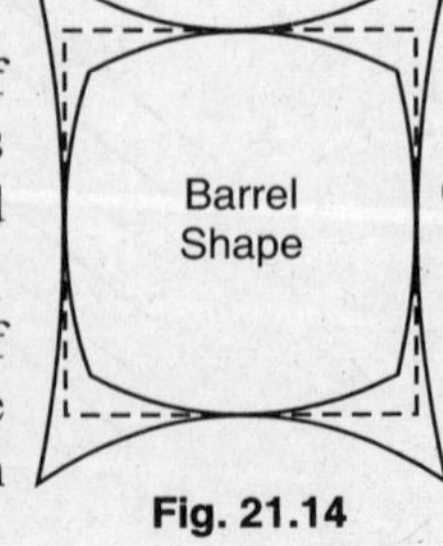

Fig. 21.14

In Fig. 21.12 when the image formed is *virtual,* the part *DC* of image is much more magnified than the part *ID*, whereas when the image formed is real, as in Fig. 21.13 the part *ID* is much more magnified than the part *DC*.

(*i*) If the image of a rectangular wire gauze is viewed through a thick convex lens, the appearance is *barrel-shaped* when the image formed is *real* and it is *pin-cushion shaped* when the image formed is *virtual.* It is clear that in the first case the magnification of the central part is much more than that of the edges. In the second case the magnification at the edges is much more than that of the central part of the image.

(*ii*) The types of distortion, even for a real image, changes with the position of the stop. If a stop is placed between the object and the lens the distortion is *barrel shaped* or *negative,* whereas it is *cushion-shaped* or *positive* when the stop is placed between the image and the lens.

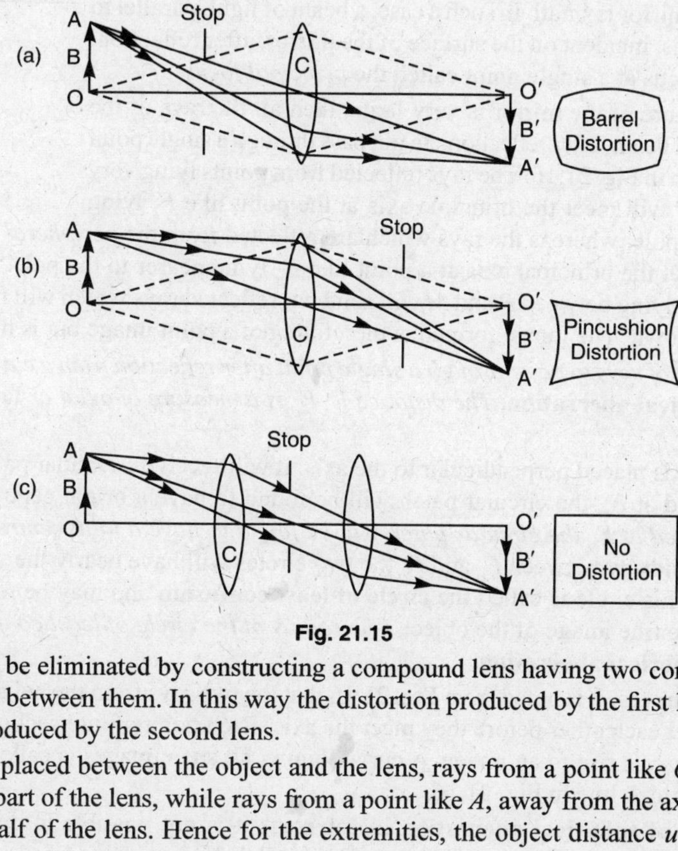

Fig. 21.15

This defect can be eliminated by constructing a compound lens having two components with a stop placed suitably between them. In this way the distortion produced by the first lens is cancelled by the distortion produced by the second lens.

When a stop is placed between the object and the lens, rays from a point like O on the axis go through the central part of the lens, while rays from a point like A, away from the axis, are refracted only by the lower half of the lens. Hence for the extremities, the object distance u will exceed the paraxial value of the object distance and the magnification $m = \dfrac{f}{u+f}$ will decrease. The image is thus more pulled in at the extremities than elsewhere. In other words, the outermost parts of the object, as the corners of a square, are less magnified, thereby giving rise to a barrel-shaped distortion, as shown in Fig. 21.15(*a*). When the stop is placed between the lens and the image, the rays passing through the upper half of the lens take part in the formation of the image. This results in the increase of the image distance V and hence the magnification $m = \dfrac{f-v}{f}$ will be greater than that for the paraxial value. In other words, the outermost parts of the object are more magnified, thereby giving rise to a pin cushion-shaped distortion, as shown in Fig. 21.15(*b*).

By combining two identical lenses with a stop midway between them, as shown in Fig. 21.15(*c*), the *pin-cushion* distortion produced by the first lens is compensated by the *barrel shaped* distortion produced by the second lens. Such an arrangement is commonly used in the construction of a lens system for a photographic camera and the projection lantern.

21.7 SPHERICAL MIRRORS

The common defects observed in images produced by reflection at curved surfaces are:

(1) Spherical aberration (2) Curvature and distortion (3) Astigmatism.

1. Spherical aberration

The mirror formula has been derived on the assumption that the aperture of the mirror is small. In such a case, a beam of light, parallel to the principal axis, incident on the surface of the mirror, after reflection, is brought to focus at a single point called the *principal focus.*

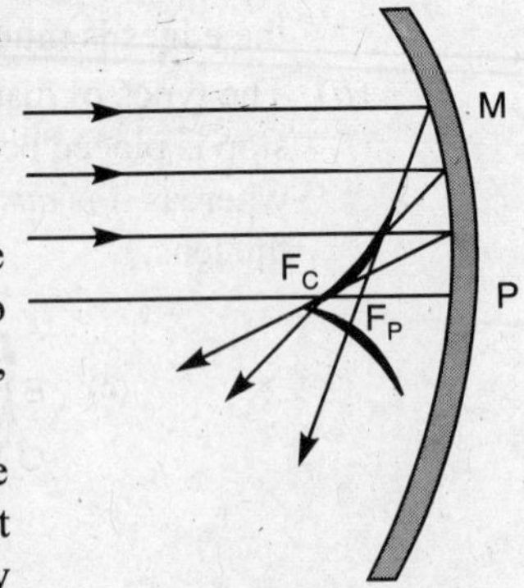

Fig. 21.16

If the aperture of the mirror is very large, then all the rays of the incident parallel beam, after reflection, do not pass through a single point like F_c as shown in Fig. 21.16. The rays reflected from points lying very near the pole P will meet the principle axis at the point like F_c lying away from the pole; whereas the rays which are reflected from the *peripheral portion* of the mirror like M will meet the principal axis at a point like F_p lying nearer to the pole. All the rays reflected from the points lying between P and M will send out reflected rays which will meet the axis between F_c and F_p as shown. The image formed is therefore, not a point image but is blurred.

The failure of rays to pass through a single point after reflection from a curved spherical surface is called **spherical aberration.** *The distance $F_c F_p$ is a measure of axial or* **longitudinal spherical aberration.**

If a screen is placed perpendicular to the axis, it will receive a circular patch of light. When the screen is placed at F_c, the circular patch will be found to have a bright centre and a faint rim; but when *it is placed at F_p the circular patch will be found to have a faint centre and a bright rim.* At a point lying midway between F_c and F_p the two circles will have nearly the same illumination and will overlap. This circle is called the **circle of least confusion** and may be regarded as the nearest approach to the true image of the object. *The radius of the circle of least confusion is a measure of the* **lateral spherical aberration.**

Caustic curves. It is noted from Fig. 21.16 that the two rays from the neighbouring points of the mirror intersect each other before they meet the axis. Each point of intersection of the neighbouring reflected rays gives rise to an image. A curve joining all such images is called a **caustic curve.** It is shown by a thick line in Fig. 21.16.

Removal of spherical aberration in mirrors. It is not possible to design a reflecting or a refracting surface of wide aperture, which can bring all the rays from any point on the axis to a point focus. But it is possible to construct such surfaces which will produce a point image of a point object placed only at a particular point on the axis, however, wide the incident pencil of light may be. Such a reflecting surface is called **an aplanatic surface** and the points of particular positions of the object and the image are termed **aplanatic foci.**

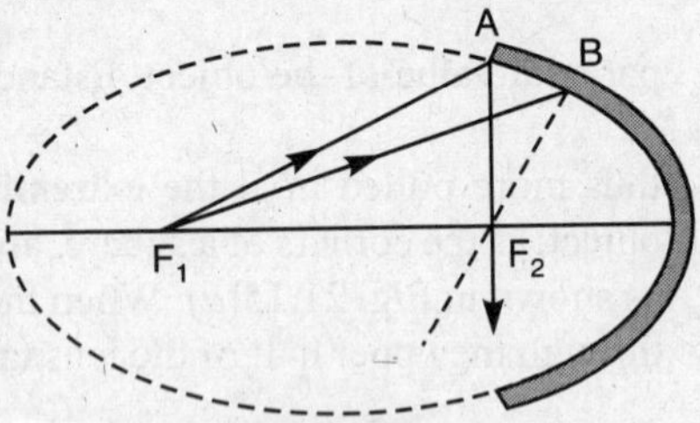

Fig. 21.17

Examples of aplanatic surfaces are :

(*i*) **Ellipsoidal mirror.** It is the part of an ellipsoid having a reflecting surface as shown in Fig. 21.17. In this mirror all the rays proceeding from one focus F_1 after reflection will pass through the other focus F_2. The total lenth of the path of all the rays starting from F_1 and meeting at the point F_2 after reflection is the same, *i.e.*,

$$F_1 A + AF_2 = F_1 B + BF_2$$

Hence the points F_1 and F_2 give rise to the image, which is free from spherical aberration. The points F_1 and F_2 are called the **aplanatic points** and the surface is called **an aplanatic surface.**

(*ii*) **Paraboloidal mirrors.** If one of the foci of the ellipsoid say F_1 moves to infinity, the ellipsoid becomes a *paraboloid* and has a reflecting surface as shown in Fig. 21.18. Light rays incident parallel to the axis, after reflection, converge to a point F called the focus. Conversely light rays starting from F, after reflection become parallel to the axis. Hence the focus and infinity are **aplanatic points.** Such mirrors of wide aperture are used in reflecting telescopes, as these bring all rays coming from infinity to a point focus.

2. Curvature and distortion. Due to the large aperture of mirrors the images obtained are usually *curved* and *distorted.* It is because (*i*) the various parts of an extended object situated close to the mirror cannot be brought to focus at the same distance. Thus, the image formed is curved. (*ii*) The magnification of the ends of the object is different from the magnification of the middle part and, therefore, the *image becomes distorted.*

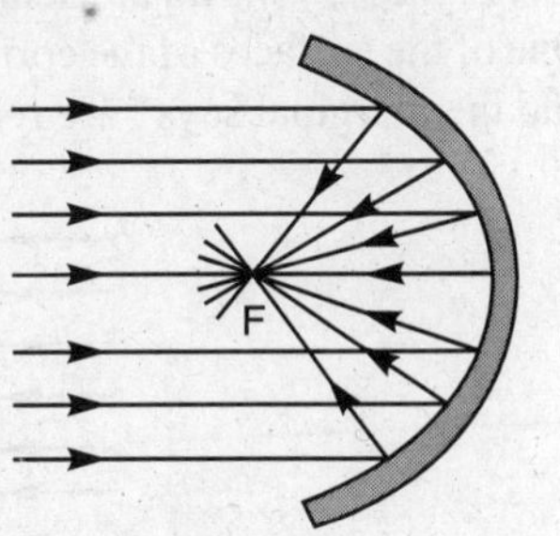

Fig. 21.18

These two defects are eliminated by using a stop at the right position, *i.e.*, by allowing only a narrow pencil of rays to fall centrally on the mirror.

3. Astigmatism. If an oblique concentric pencil OXY is incident on a hemispherical surface of which C is the centre of curvature and P the pole, then the rays after reflection, intersect with each other at F_3 and cross the axis at different points like F_1 and F_2. If the figure is supposed to rotate through a small angle about the axis OP, then the curved line XY of the reflecting surface will describe a small regular element and the triangle XOY will describe a conical pencil diverging from O. The point F_3 will trace out a short, approximately straight line, perpendicular to the plane of the paper. Every ray in the reflected pencil will pass through this line and will meet the axis between the points F_1 and F_2. (Fig. 21.19).

The line which F_3 traces out is called the **first principal focal line** of the reflected pencil. The short line $F_1 F_2$ in the plane of the paper is termed the **second principal focal line** of the reflected pencil.

It is clear that the section of the reflected pencil changes from a straight line perpendicular to the plane of the paper at F_3 to a straight line $F_1 F_2$ in the plane of the paper. Consequently somewhere between F_3 and $F_1 F_2$, the pencil must be equally wide in both directions. This approximately circular section of the pencil is called the **circle of least confusion** and is the nearest approach to a point focus of the reflected pencil. Since the reflected pencil in the above case does not pass through a single point, it is termed as the **astigmatic** pencil, which means without a spot or a point.

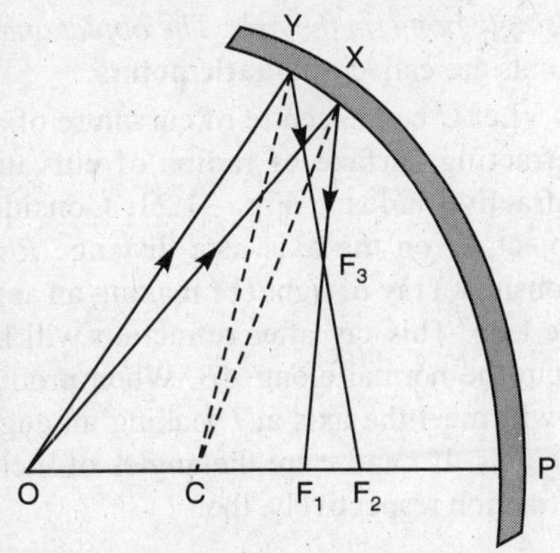

Fig. 21.19

Removal of astigmatism. The astigmatism can be minimised by placing a stop at the centre of curvature of the mirror. Light from each part of the object reaches the mirror in the form of a narrow paraxial pencil. Each point in the object forms a point image and astigmatism is thus overcome.

The paraboloidal mirrors show a large astigmatic difference if the object is placed off the axis. This is why paraboloidal reflectors are limited in their use in devices which require only a small angular spread, such as telescopes and search-lights.

21.8 SCHMIDT CORRECTOR PLATE

In large telescopes, large spherical mirrors are used to form images and the images are recorded on a curved film. The systems are free from astigmatism and coma etc. However, because of large aperture spherical aberration occurs.

In 1929 Schmidt designed a system that eliminates spherical aberration in such large spherical mirrors. He showed that a thin glass plate having a shallow toroidal surface and held before the spherical mirror could help minimizing the spherial aberration in a mirror. Such plates are known as **corrector plates.** Light rays passing through the outer regions of the corrector plate would be deviated by just the amount needed to be sharply focused on the curved film.

Fig. 21.20 shows the Schmidt corrector plate P used used with large aperture reflector telescopes. One of the surfaces of the corrector plate P is plane and the second surface is convex towards the axis and the marginal says* are reflected outwards. All the refracted rays come to focus at F.

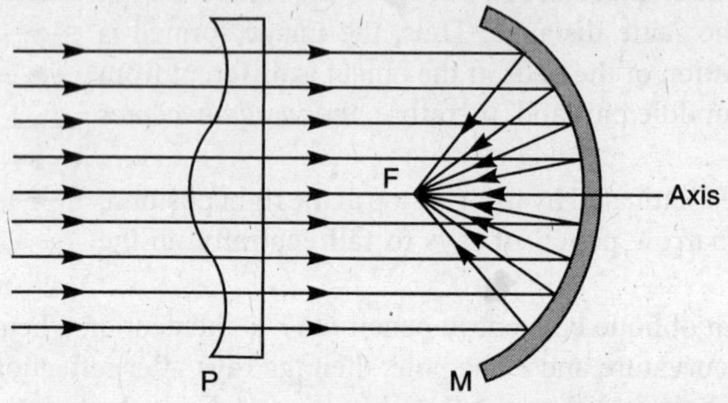

Fig. 21.20

21.9 APLANATIC SURFACE AND APLANATIC POINTS

Aplanatic surface. *This suface has the property by virtue of which all the rays starting from a point object lying on its axis, after refraction at the surface, coverage to or appear to diverge from a single point on the axis. The object and the image* points are called **aplanatic points.**

Let C be the centre of curvature of a spherical refracting surface of radius of curvature R and refractive index μ Fig. 21.21. Consider a point object, O on the axis at a distance R/μ from C. Consider a ray of light OA making an angle θ_1 with the axis. This ray after refraction will bend away from the normal along AB. When produced back, it will meet the axis at I making an angle θ_2 with the axis. If i and r are the angles of incidence and refraction respectively, then

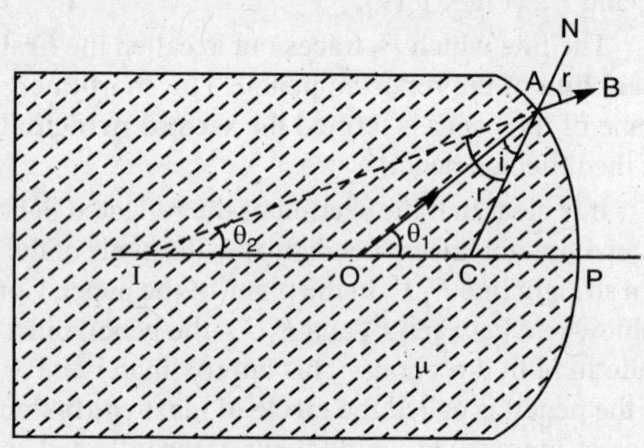

Fig. 21.21

$$\frac{\sin i}{\sin r} = \frac{1}{\mu}$$

---

* While deriving the different relations such as lens-maker's relations, lens formula etc. it was assumed that the objects are small in size and situated close to the axis, known as *paraxial approximations,* so that the different rays make very small angles with the axis of the lens. In acutal practice, the objects are bigger. Hence the rays falling on lens are wide angle rays and falling on top and bottom part of the lens. These rays are known as *peripheral* or *marginal rays.*

In $\Delta\,AOC$, we have

$$\frac{OC}{\sin i} = \frac{AC}{\sin \theta_1}$$

or
$$\frac{\sin i}{\sin \theta_1} = \frac{OC}{AC} = \frac{R}{\mu} \times \frac{1}{R} = \frac{1}{\mu}$$

∴
$$\frac{\sin i}{\sin r} = \frac{\sin i}{\sin \theta_1}$$

or
$$\angle r = \angle \theta_1$$

In $\Delta\,AOI$, we have

$$\theta_2 + r - i = \theta_1 \qquad (\because r = \theta_1)$$

or
$$\theta_2 = i$$

In $\Delta\,AOI$, we have
$$\frac{IC}{\sin r} = \frac{CA}{\sin \theta_2}$$

or
$$\frac{\sin r}{\sin \theta_2} = \frac{IC}{CA} = \frac{\sin r}{\sin i} = \mu$$

∴
$$IC = \mu CA = \mu R \qquad \text{... (19.20)}$$

This shows that if the object is placed at a distance $\frac{R}{\mu}$ from the centre of curvature of the spherical refracting surface then the image is formed at a distance μR irrespective of the slopes θ_1 and θ_2.

Such a surface eliminates both the spherical aberration and coma.

Applications. It is usefully employed in microscope objectives where it is essential that as wide a pencil of light as possible from the object enter the objective to have sufficient brightness of the image.

21.10 OIL IMMERSION LENS

This is used in microscope objectives. The objective is a hemispherical lens of radius r and refractive index μ. The object is placed on a platform at depth r/μ, and a drop of oil of refractive index μ fills the space between the platform and the lens. The arrangement is shown in Fig. 21.22. As discussed above, there is no spherical aberation* for object point at P.

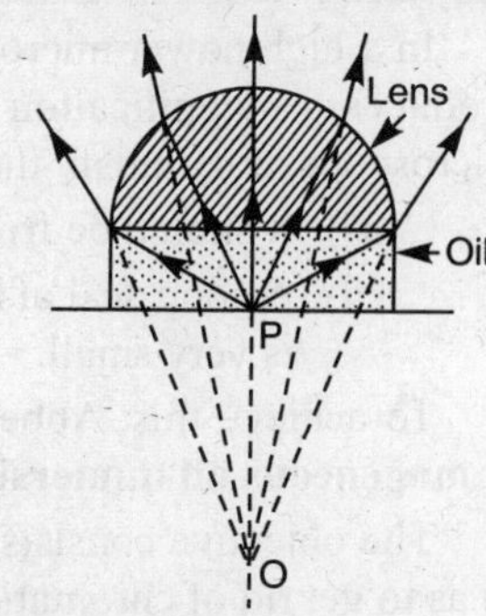

Fig. 21.22. Oil immersion objective

21.11 APLANATIC MENISCUS LENS

Fig. 19.11 requires the object point P to be placed 'inside' the sphere. Meniscus lens [Fig. 21.23(a)] is a way of achieving this. Face II has radius r and its centre is at 0. Face I has its centre at P, with distance $OP = r/\mu$. All rays from P are thus normal to face I and suffer refraction at face II. They form image at P'. Thus, P, P' form an aplanatic pair of points. Often the meniscus lens is used together with an oil immersion lens in high power microscope objectives as shown in Fig. 21.23 (*b*).

Example 21.17 *Calculate the radii of aplanatic surfaces for a refracting sphere of* $\mu = 1.60$, $R = 3.2$ *cm. Calculate also the magnification of the image in area.*

Solution. r (internal) $= R \div \mu = 3.2/1.6 = 2.0$ cm

r (external) $= R \times \mu = 3.2/1.6 = 5.1$ cm

Surface magnification $= \mu^4 = (1.6)^4 = 6.5$.

* For spherical aberation, refer article 21.6.1.

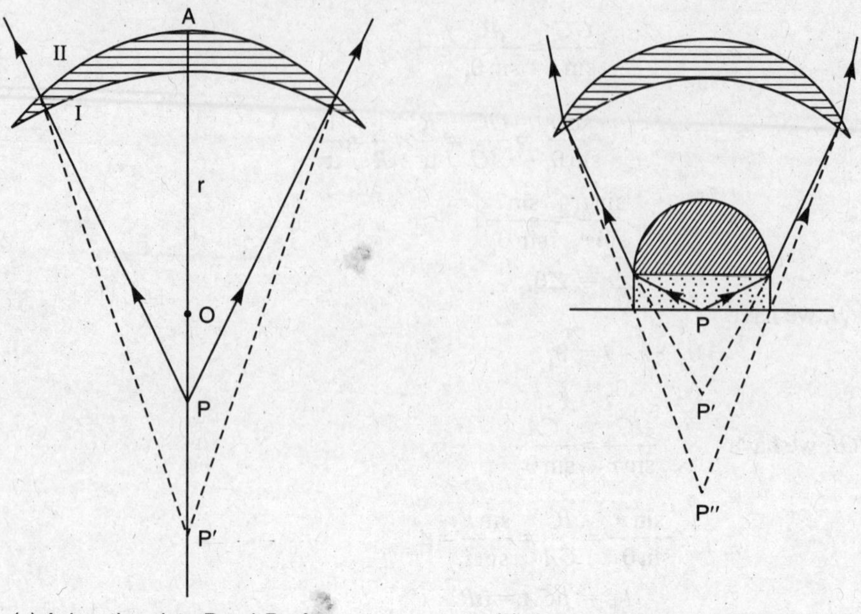

(*a*) Aplanatic points *P* and *P′* of a meniscus lens

(*b*) A meniscus lens in tandem with an oil immersion lens

Fig. 21.23

21.12 OIL IMMERSION OBJECTIVE MICROSCOPE

In a high power microscope having an overall magnification of about 500, the objective alone produces a magnificaiton of the order of 50. Hence the objective of such a microscope must satisfy the following conditions:

(*i*) It should be free from chromatic and spherical aberrations

(*ii*) A good deal of light should enter through it, as the object to be viewed is very small.

To achieve this, Abbe devised a complicated objective, known as **Abbe's homogeneous oil immersion objective.**

The objective consists of combination of a large number of lenses arranged so as to get rid of chromatic and spherical aberrations. The lower most lens L_1 is a plano-hemispherical convex lens. The space between this lens and the object at *O* is filled with **cedar wood oil,** which has the same refractive index as that of the material of the lens. The rays of light starting from *O* thus suffer no refraction till they reach the spherical surface of the lens L_1 (21. 24).

The point *O* where the object is placed, is an *aplanatic point* for the lens L_1. It thus forms a geometrical point image, at the other aplanatic point I_1, however wide the incident pencil of light may be. The conditions for aplanatism are as derived below :

Let the distance of the object *O* from *C* the centre of curvature of the hemispherical surface of radius *R* and refractive index μ, be *R*/μ. The ray *OA* incident at *A* at an angle *i* after refraction appears to come from I_1 thus making angle $CAI_1 = r$.

$$\text{Now } \frac{CA}{CO} = \frac{R}{R/\mu} = \mu = \frac{\sin r}{\sin i} \quad \ldots (i)$$

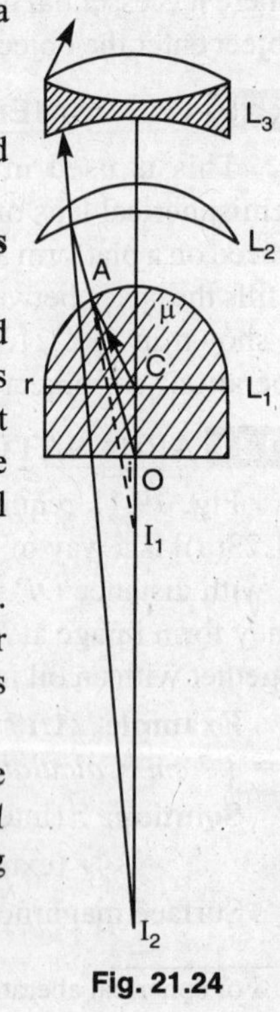

Fig. 21.24

In triangle CAO, $$\frac{CA}{CO} = \frac{\sin \angle COA}{\sin i} \quad \text{... (ii)}$$

From relations (*i*) and (*ii*), we have

$$\angle COA = r$$

Now $$\angle COA = r = \angle AI_1O + \angle I_1AO$$

and $$\angle r = \angle i + \angle I_1AO$$

$\therefore$ $$\angle AI_1O = i$$

In ΔI_1AC

$$\frac{CI_1}{CA} = \frac{\sin \angle CAI_1}{\sin \angle AI_1O} = \frac{\sin r}{\sin i} = \mu$$

$\therefore$ $$CI_1 = \mu \times CA = \mu R.$$

This result is independent of the angle of incidence. Hence all the rays starting from O, after refraction, appear to come from I_1 whatever may be the value of angle of incidence.

Now the position of the second lens L_2 is adjusted in such a manner that the point I_1 lies at the centre of curvature of lower surface of L_2, so that the beam emerging from L_1 suffers no refraction at the lower surface. The points I_1 and I_2 are aplanatic points for the second surface of less L_2. Thus, a geometrical point image of I_1 is formed at I_2. This arragement changes the wide angle beam of light, starting from O, into a narrow angle pencil, originating from I_2. This diverging beam is brought to a good focus by means of various other lenses, which minimise the spherical as well as the choromatic aberrations.

A really good microscope objective consists of such 10 lenses. Such a combination corrects spherical abrraction for two colours and chromatic aberration for three colours.

EXERCISE

1. What is meant by chromatic aberration of a lens ? What causes chromatic aberration ? ***(Osm. U. 2004 Nagpur U. 2003; M.D.U. 2005, 2003, K.U. 2000; Meerut. U. 2003, 2002, 2000)***
2. Show that the chromatic aberration in lenses is equal to the product of mean focal length and dispersive power of the material ***(M.D.U. 2008)***
3. (*a*) What do you understand by the term achromatism ? Find the condition of achromatism for two thin lenses in contact. How is achromatic doublet of two thin lenses in contact prepared ? ***(G.N.D.U. 2005; M.D.U. 2006, 2003, 2000; P.U. 2003; K.U. 2002; Gauhati U. 2000; Nagpur U. 2010)***

   (*b*) In the achromatic convex combination of two lenses in contact which lens should have more focal lenth and why ? ***(Pbi. U. 2003)***
4. What is achromatism ? Explain the principle and use of achromatic doublet. Derive the condition for achromatic combination of two lenses in contact ***(Nagpur U. 2008)***
5. What is longitudinal chromatic aberration ? Prove the relationship $V_r - V_v = \dfrac{\omega V^2}{f}$
6. Explain circle of least chromatic aberration? Derive the relation for its diameter.
7. How chromatic aberration in lenses can be eliminated ? Explain.
8. (*a*) Obtain the condition that must be satisfied in order that a combination of two lenses of focal lengths f_1 and f_2 at a distance d apart be achromatic. What will be the condition when the two lenses are of the same material ? Is such a combination of lenses truly achromatic ? ***(G.N.D.U. 2005, Osm. U. 2004; Meerut. U. 2001, Nagpur U. 2009, M.D.U. 2007, 2005, Gauhati U. 2000; P.U. 2000)***

9. What is aberration ? What are different types of monochromatic aberrations in a lens ? Name at least four types ? **(P.U. 2002, K.U. 2001, Nagpur U. 2001)**
10. What is meant by spherical aberration ? Explain the cause of spherical aberration fully. Describe without mathematical details how this defect can be reduced. **(Osm. U. 2004, Meerut U. 2003; G. N. D. U. 2001, P.U. 2002)**
11. Derive the condition for minimum spherical aberration by using two plano convex lenses seperated by a distance. **(Nagpur U. 2011, 2009)**
12. What is spherical aberration ? Give the methods to reduce spherical aberration. **(Nagpur U. 2010, 2008)**
13. (*a*) What are aberrations in images ?
    (*b*) What is spherical aberration and why does it occur ?
    Explain how it can be removed by using a plano-convex lens. **(Nagpur U. 2009)**
14. Write notes on:
    (*i*) Coma (*ii*) Astigmatism
    (*iii*) Distortion (*iv*) Spherical mirrors
15. Explain what do you understand by curvature of field and distortion. What is the difference between pin cushion distortion and barrel shaped distortion ? How can these defects be removed ? **(M.D.U. 2002)**
16. What are the common defects produced in images formed by reflection at curved surfaces ? Explain how these defects can be removed. **(Utkal. U. 2003; Bhopal U. 2004)**
17. Define spherical aberration. What is lateral, longitudinal, spherical aberration and circle of least confusion. Explain the formation of a caustic curve due to reflection of parrallel rays incident on a large concave mirror. **(Nagpur. U., 2002)**
18. What is meant by an 'aplanatic surface' ? Show that a spherical refracting surface is aplantic with respect to a certain position of the object. Give practical applications of the same. **(G.N.D.U. 2004; Madurai U. 2003)**
19. What is meant by a 'aplanatic surface' ? What are aplanatic points of a refracting surface ? Show that a spherical refracting surface is aplanatic with respect to a certain position of the object ? Give practical applications of the same. **(Meerut U. 2007, G.N.D.U. 2004, Modurai U. 2003)**
20. What are conjugate points in a spherical medium ? How do they serve as aplanatic points ?
21. Write notes on:
    (*i*) oil imersion lens
    (*ii*) aplanatic meniscus lens
22. Describe the high power microscope using oil immersion objective. Explain how it is based upon the aplanatic points ?
23. Explain the construction and working at an oil immersion objective **(Madurai U. 2003)**
24. A telescope objective is made by cemeting together two lenses, the cemented surfaces having the same radii of curvatures. The combination is plano-convex and is of focal length 100 cm. Calculate the radii of curvatures of the component lenses given that

	Crown glass	*Flint glass*
Refractive index	1.50	1.60
Dispersive power	0.02	0.03

(Hint. It is a chromatic aberation) **Ans.** (*i*) *Radii of curvature 37.45 cm, – 30 cm of convex lens*; (*ii*) *Concave lens are – 30 cm and* ∞)

25. Compound achromatic lens of focal length of 40 cm is to be made from two different glasses. Find the focal length of the component lenses if dispersive power of crown glass is 0.21 and that of flint glass is 0.46. [**Ans.** + 21.71 (crown) and – 47.62 (flint)]

26. An achromatic objective of focal length 30 cm, is to comprise of two thin lenses cemented together with canada balsam and are made from hard crown and dense flint glasses respectively. The free surface of the divergent lens is plain. Calculate the radii of curvature of the surfaces of both the lenses. Given.

For crown glass	For flint glass
$\mu_v = 1.5231$	$\mu'_v = 1.6347$
$\mu_r = 1.5171$	$\mu'_r = 1.6224$

[**Ans.** Concave lens of flint: $R_1 = -13.28$ cm, $R_2 = \infty$, Convex lens of crown: $R_1 = +13.28$ cm, $R_2 = -12.65$ cm]

MULTIPLE CHOICE QUESTIONS

21.1 The longitudinal chromatic aberration of a lens $(f_r - f_v)$ is give by

(*a*) ω/f (*b*) f/ω (*c*) $\omega^2 f^2$ (*d*) ωf

21.2 The longitudinal chromatic aberration the object and image distances are related by

(*a*) $V_r - V_v = \dfrac{\omega V^2}{f}$ (*b*) $V_r - V_v = \dfrac{\omega V^2}{f^2}$ (*c*) $V_r - V_v = \dfrac{\omega^2 V}{f}$ (*d*) $V_r - V_v = \dfrac{\omega^2 V}{f^2}$

21.3 The diameter (*d*) of a circle of least chromatic aberration in terms of diameter of a lens (D) used is given by

(*a*) $\dfrac{1}{4} D\omega$ (*b*) $\dfrac{3}{4} D\omega$ (*c*) $\dfrac{1}{2} D\omega$ (*d*) $D^2\omega^2$

21.4 Condition for achromatism of two thin lenses placed in contact is

(*a*) $\dfrac{\omega_1}{f_1} + \dfrac{\omega_2}{f_2} = 1$ (*b*) $\dfrac{\omega_1}{f_1} + \dfrac{\omega_2}{f_2} = -1$ (*c*) $\dfrac{\omega_1}{f_1} + \dfrac{\omega_2}{f_2} = 0$ (*d*) $\dfrac{\omega_1}{f_1} - \dfrac{\omega_2}{f_2} = 0$

21.5 Condition for achromatism of two thin lenses seperated by a finite distance *d* is given by

(*a*) $d = \dfrac{f_1 + f_2}{4}$ (*b*) $d = \dfrac{f_1 + f_2}{3}$ (*c*) $d - \dfrac{f_1 + f_2}{2}$ (*d*) $d = f_1 + \dfrac{1}{2}$

21.6 To reduce coma in monochromatic aberration, the Abbe sine condition used is

(*a*) $\mu y \sin\theta_1 = \mu' y' \sin\theta_2$ (*b*) $\mu^2 y^2 \sin\theta_1 = \mu' y' \sin\theta_2$

(*c*) $\mu y \sin^2\theta_1 = \mu' y' \sin^2\theta_2$ (*d*) $\mu y \sin\theta_1 = -\mu' y' \sin\theta_2$

ANSWERS

21.1 (*d*)	**21.2** (*a*)	**21.3** (*c*)	**21.4** (*c*)	**21.5** (*c*)	**21.6** (*a*)

OPTICAL INSTRUMENTS

22

Chapter

INTRODUCTION

The optical instruments can be broadly divided into two groups. One group forms the *real images* of an object while the other forms *virtual images.* The group which produces real images can be projected on the screen or photographic plates and can be seen simultaneously by many observers. Projectors used in cenema hall belongs to this category. The other group which forms virtual images of an object can be viewed only by one observer. The virtual image formed by the instrument is transformed by the eye into a real image which our brain interprets it. Therefore, the optical instruments forming a virtual image are often called as *optical aids.* The spectacles, microscopes and telescopes belong this group. Eyepieces of different types play an important role in optical instruments.

22.1 HUMAN EYE

The observer's eye is an essential part of all optical instruments. A sectional view of the human eye is shown in Fig. 22.1(*a*). It is nearly spherical in shape and about 2.5 cm in diameter.

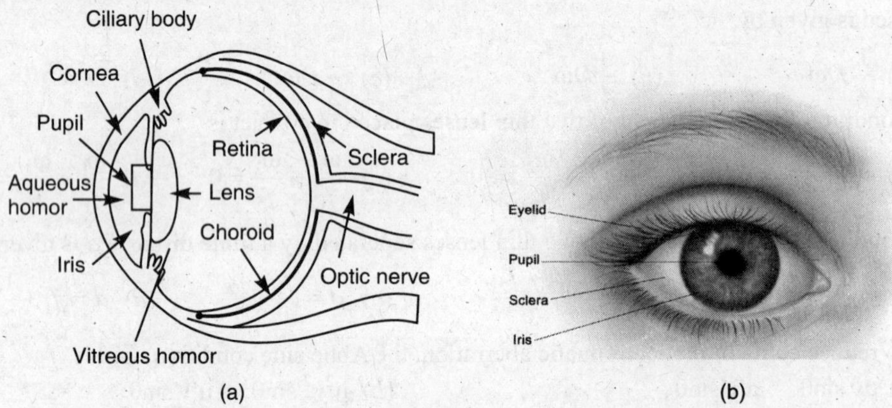

Fig. 22.1

Eye is like a camera. Human eye function is more sophisticated than any man-made optical device. The fundamental task of the eye is to focus light originating from sources outside of the eye on to the retina, which is inside and at the back of the eye. At the front of the eye, a tough white sheet called sclera covers the outside of the eye that extends into a thin convex transparent membrane called the *cornea,* which acts as an entrance lens for the eye. Back of the cornea is the *iris,* which is pigmented muscular ring. The iris can be of different colours, which explains why people have blue, green or black eyes. The external object is seen like the camera takes the picture of any object. Light enters the eye through a small hole called the pupil and is focused on the retina, which is like a camera film. Eye also has a focusing lens, which focuses images from different distances on the retina. The colored ring of the eye, the iris, controls the amount of light entering the eye. Iris closes when light is bright and opens when light is dim. Ciliary muscles in ciliary body control the focusing of lens automatically. Image formed on the retina is transmitted to brain by optic nerve. The image is finally perceived by the brain.

Function of Lens. The job of the lens is to make sure that the rays refracted by the cornea come to a sharp focus on the retina. It accomplishes this by adjusting its shape, an action that is analogous to the auto-focus function on cameras, but markedly more sophisticated. This elegant seamless auto-focus function is made possible by the ability of the lens to change its refractive power by altering its form and thickness. This eye function is called **accommodation**.

The optical power of a lens is measured in dioptre, with increasing numbers representing stronger optical power. In a camera, optical power is achieved by combining the optical power of individual lenses. This is efficient, because two lenses of different dioptre ratings can be used individually or combined to yield three different levels of optical power (3 lenses can produce 6 unique levels). Because the eye has just one lens, it uses a different strategy to adjust its optical power. Through accommodation, the lens of the eye can change its shape (either thinner or thicker) by tensing or relaxing the muscles of the eye. This unique property is continuous and makes eye function unsurpassed by any other optical device in its focusing ability.

Function of Cornea: Refraction at the cornea and surfaces of the lens produce a real and inverted image of the object on the retina. The job of the lens is to make sure that the rays refracted by the cornea come to a sharp focus on the retina. It accomplishes this by adjusting its shape, an action that is analogous to the auto*focus function on cameras, but markedly more sophisticated. This elegant seamless auto-focus function is made possible by the ability of the lens to change its refractive power by altering its form and thickness. This eye function is called *accommodation*.

Function of Retina and Image Processing. Once light rays are bent at the lens, they pass through the vitreous humor, a jelly-like substance filling the back part of the eye (vitreous chamber), and arrive at the retina. The *retina* is a complex multi-layered structure consisting of three types of light-sensitive photoreceptors: rods, cones, and photosensitive ganglion cells.

The human eye has a total of 125 million rods and 6.5 million cones. They sense the image and transmit it via the *optic nerve* to the brain. These photoreceptors are responsible for converting light rays (called photons) into electrical impulses through a process called ***signal transduction***. These impulses are processed and transmitted through the *optic nerve* to the visual cortex in the brain, where the image is processed and perceived.

Both Eyes Works Together. Humans have binocular vision, which means that our eyes work together rather than separately like those of many birds and reptiles. To accomplish this, both eyes normally aim at the same spot. The brain then combines the two images into a single three-dimensional representation. This three-dimensional image gives us depth perception, a very important aspect of eye function.

We describe the limited ability of the eye in terms of the *far point* and the *near point* of the eye. The far point of a normal eye is infinity. The ciliary muscle is fully relaxed in this position. The nearest point on which the eye is focused with the ciliary muscle fully contracted is the near point of the eye. The minimum distance at which an eye can see objects distinctly and without getting tired is called the *Normal Distance of Distinct Vision* (NDDV) and is 25 cm for a normal eye. The range of accommodation gradually diminishes with age because the crystalline lens grows through a person's life and ciliary muscles are less able to distort a larger lens. This is the reason why the near point gradually recedes with age.

22.2 PERCEPTION OF SIZE OF AN OBJECT

Our perception about the *size of an object* is judged by its *linear dimensions i.e.*, height of the object. The feeling of the size of an object is actually derived from the size of its image cast on the retina. The size of the retinal image depends upon the *visual angle,* which is the angle subtended by the object at the eye. Therefore, it is known as the *angular size.*

Consider two objects A_1 and A_2 placed in front of the eye as shown in Fig. 22.2. Both the objects are of different sizes; but they subtend equal angles *i.e.*, θ_1 at the eye and the images formed by them on the retina are of same size. If d is the distance between retina and the lens, then

$$\frac{h_1}{d} = \theta_1 \text{ or } h_1 = d \cdot \theta_1 \qquad \text{... (22.1)}$$

$$\therefore \qquad h_1 \propto \theta_1 \qquad (\because d \text{ is constant})$$

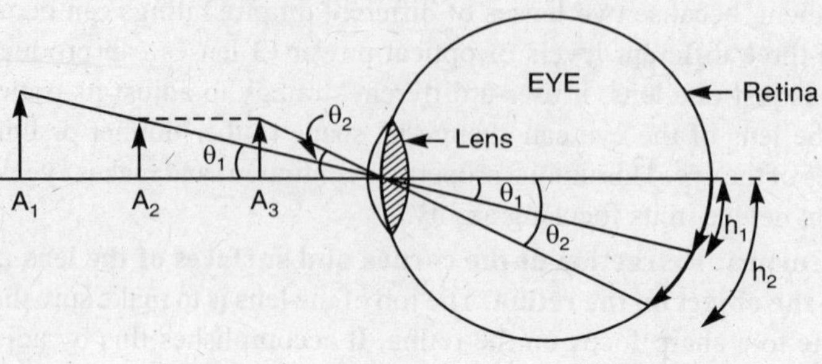

Fig. 22.2

Thus, the size of the image on the retina is proportional to the angle subtended θ_1. In practice, we bring the object closer to the eye to see more details of the object. This action makes the subtended angle θ_1, consequently the retinal image as large as possible. However, if the object is brought closer to the eye, say to position A_3, the image on the retina becomes larger and larger on account of increase in visual angle θ_2, we feel strain on the retina. The object can not be brought nearer to the eye beyond the limit, known as least distance of distinct vision (NDDV) which is 25 cm for a normal human eye. If we try to view the object within this limit, the eye can not accommodate and as a result, the eye experiences strain on it.

22.3 THE SIMPLE MAGNIFIER (SIMPLE MICROSCOPE)

The image formed on the retina can be increased in size by using a simple optical instrument like a magnifying glass. Thus, we see small objects which can not be seen by the naked eye. Such instruments (artificially) increase the virtual angle and are known as *magnifying instruments.* A converging lens is used to form a virtual image that is larger and farther from the eye than the object itself as shown in Fig. 22.3. Then the object may be moved closer to the eye and the *angular size* of the image may be substantially larger than the angular size of the object held at 25 cm (*i.e.*, NDDV) without the lens. The lens used is called a *magnifier or a simple microscope.* The virtual image is most comfortable to view when it is placed at infinity, as the muscle of the eye is relaxed.

In Fig. 22.3(*a*), object is shown situated at *near point* (*i.e.*, NDDV) where it subends an angle θ at the eye. In fig. 22.3(*b*), a magnifier is held in front of the eye forming an image at infinity and the angle θ' subtended at the magnifier.

Angular magnification, $M = \frac{\theta}{\theta'}$. Consider an object of length $AB = h$ placed at the near point A and viewed by naked eye [Fig. 22.4(*a*)]. Then the visual angle is α.

$$\therefore \qquad \tan\alpha = \frac{h}{D} \qquad \text{... (i)}$$

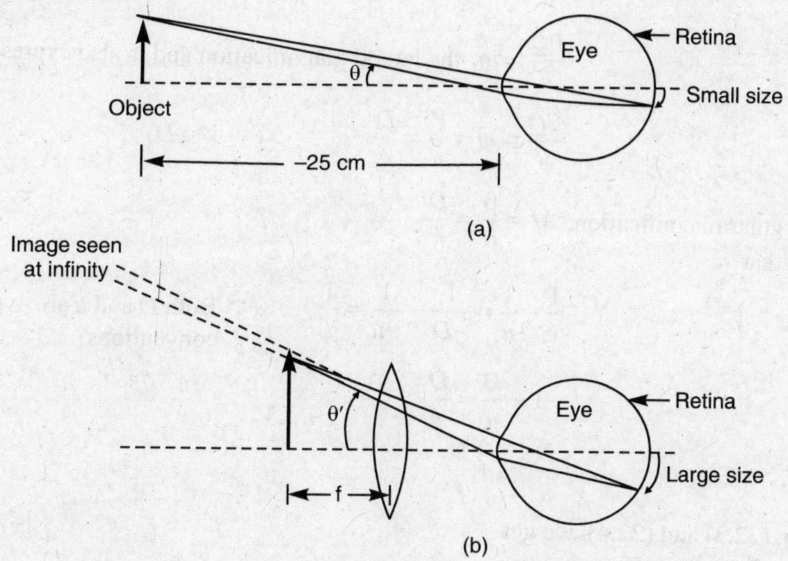

Fig. 22.3

Suppose that a converging lens (the magnifier) L of focal length f is used as a magnifying glass and the object is placed between O and F, such that the magnified erect image PQ is formed. If the observer's eye is close to the lens, then the distance OP is the least distance of distinct vision. In this condition, the visual angle is given by

$$\tan\beta = \frac{PQ}{D} \qquad \text{... } (ii)$$

Dividing Eq. (*ii*) by (*i*), we get

$$\frac{\tan\beta}{\tan\alpha} \equiv \frac{\beta}{\alpha} = \frac{PQ}{D} \times \frac{D}{h} = \frac{PQ}{h}$$

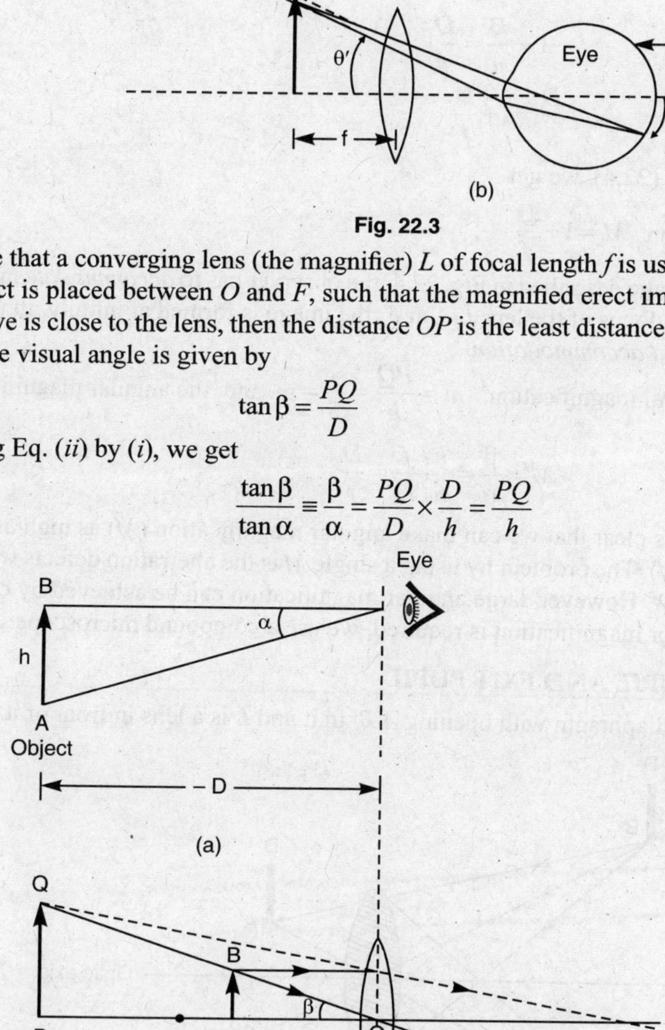

Fig. 22.4

But $$\frac{PQ}{h} = m, \text{ the lateral magnification and is also expressed as}$$

$$\frac{PQ}{h} = m = \frac{V}{u} = \frac{D}{u} \qquad (\because V = D)$$

Hence, angular magnification, $$M = \frac{\beta}{\alpha} = \frac{D}{u} \qquad \text{... (22.2)}$$

We know that

$$\frac{1}{v} - \frac{1}{u} = \frac{1}{-D} - \frac{1}{-u} = \frac{1}{f} \qquad (\because \text{ Both } D \text{ and } u \text{ are –ve as per sign conventions})$$

$$\therefore \qquad -1 + \frac{D}{u} = \frac{D}{f}$$

or $$\frac{D}{u} = 1 + \frac{D}{f} \qquad \text{... (22.3)}$$

Using Eq. (22.3) and (22.4), we get

Angular magnification, $$M = 1 + \frac{D}{f} \qquad \text{... (22.4)}$$

In order to see the image described in Fig. 22.4, the observer has to *accommodate* his eye When the object is placed at the focus of the lens *i.e.*, at F, the image is formed at infinity. In this case, the image can be seen *without accommodation.*

In this case, the lateral magnification, $m = \frac{PQ}{h} = \frac{\infty}{h} = \infty$ and, the angular magnification,

$$M = \frac{\beta}{\alpha} = \frac{h/f}{h/D} = \frac{D}{f} \qquad \text{... (22.5)}$$

From Eq. (22.5), it is clear that we can make angular magnification (M) as high as we wish by decreasing focal length (f). The problem by using a single M is the aberration defects which puts the limit to M about $3X$ to $4X$. However, large angular magnification can be achieved by correcting the aberrations. When greater magnification is required, we use a compound microscope.

22.4 ENTRANCE PUPIL AND EXIT PUPIL

In fig. 22.5, D' is a diaphragm with opening $A'B'$ in it and L is a lens in front of it.

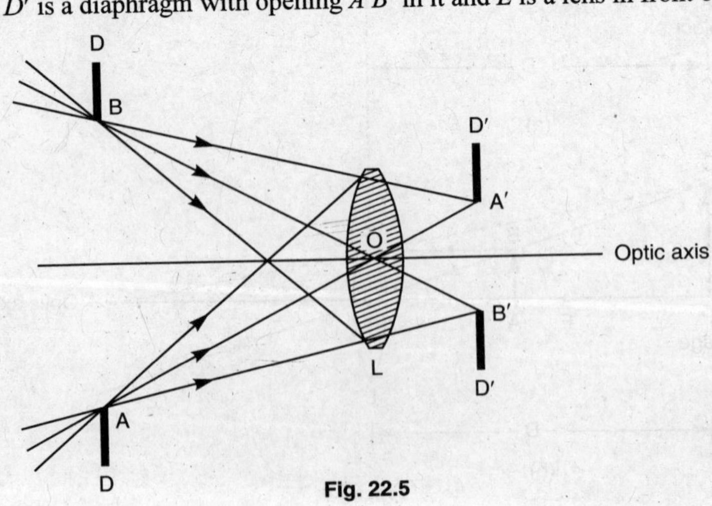

Fig. 22.5

By backward drawing we obtain the points A, B in object space conjugate to A', B'. If an eye or any instrument is placed to the right of D', the width of the exit beam is limited to $A'B'$ which is called the *exit pupil* for this highly simple single lens viewer. The width of the entrance beam is AB which is called the *entrance pupil.*

If, however, we introduce a diaphragm in plane D, and its opening is smaller than AB, the situation changes. Imagine that the opening is $\left(\frac{1}{3}\right)$ AB. Then the entrance pupil is 1/3 (AB) and that limit the exit pupil to $\frac{1}{3}$ ($A'B'$).

In this discussion, it does not matter for us what is placed at the plane D or at plane D'– it may be a simple opening, or a convex lens, or a concave lens.* All what we assumed is that D is the *first element* on the entrance side, and D' is the *last element* on the exit side.

In the intervening space between D and D' there may be any number of elements–convex lenses, concave lenses, or both. All that we assume is that $A'B'$ is the image of AB by all the *intervening* elements. In other words, the *exit pupil* is the image of the *entrance pupil.*

What we place in the opening AB is very important. If it is a simple opening like a slit, we see, objects in the projection cone AOB. If there is a convex lens placed (near AB) there, we see objects in a small cone (field of view being small in this case). If there is a plane mirror, there, we see objects in a cone behind ourselves. (see article 22.5).

In most situations the eye itself is the exit pupil. A diaphragm with an aperture is provided in the instrument in the plane where the image of the front lens (not of the object) is made by the intervening lenses. This aperture is slightly larger than the aperture of human eye (~ 2 mm); therefore, the human eye sets the limit at this exit end.

To understand this, let us take the example of a telescope. In a telescope the optimum ratio of exit pupil to entrance pupil is equal to 1/m; where m is the lateral magnification. If a 40 cm diameter objective is used and magnification is 100 (m = 100), the exit pupil to use full objective aperture should be 40 / 100 = 0.4 cm = 4 mm. The eye would provide only about half the required pupil. Thus, there is a waste of the outer half of the objective. For full use of 40 cm diameter of the objective, the magnification should be 200. For a camera or an epidiascope no such limitation exists.

22.5 FIELD OF VIEW

Consider the fig. 22.6 in which light enters through the aperture AB and reaches the eye E. The eye can view objects in the angle range PEQ. This is known as the *field of view*.

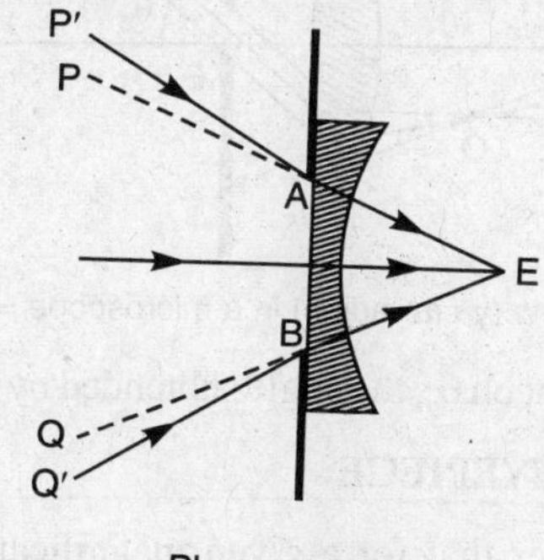

Plano-concave lens

Fig. 22.6

* Note that image of AB by a lens at the same place is AB itself. Similarly, the image of $A'B'$ by a lens at the same place is $A'B'$ itself.

If a concave lens (plano-concave) is placed in plane AB, light from a wider field $P'Q'$ can reach the eye. We say that the *field of fiew is enlarged.*

On the otherhand, if a convex lens (plano-convex) is placed in plane AB, it will *narrow the field of view,* for compound systems like telescope, microscopes, the situation is some-what different.

Telescope. Fig. 22.7 shows the field of view for a telescope. Let O be the centre of the objective, L_2 is the eye lens and the image O by L_2 is formed at centre E of the eyering. The outermost ray AE subtends an angle θ_2 with the optic axis. So the eye placed at the exit pupil E receives a cone of rays of semi-angle θ_2. in Fig. 22.7, the field of view is θ_1 given by

$$\theta_1 = \theta_2/m \qquad \text{... (22.6)}$$

where m is magnifying power of the telescope, and θ_2 is determined by the eye lens aperture. The aperture of the objective has no role in this.

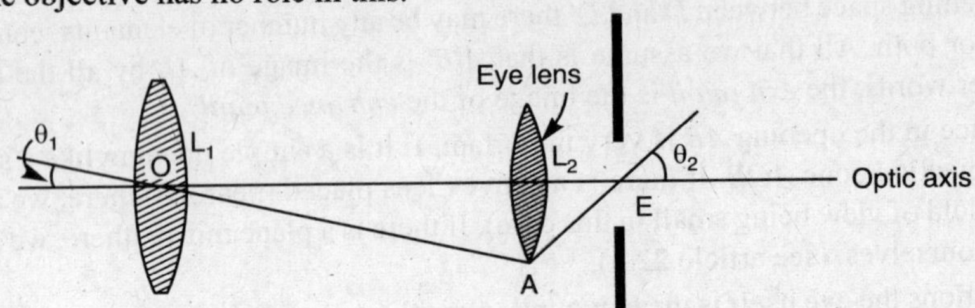

Fig. 22.7. Field of view θ_1 given by $\frac{1}{m}$ times angles AEL_2

Microscope. Fig. 22.8 shows the field of view for a microscope. The angle θ_2 is related to object height by the ratio of θ_2 to y/D, where D is the least distance of distinct of distinct vision (25 cm). We see here

$$m = \frac{\theta_2}{y/D} \quad or \quad y = \frac{\theta_2}{m}.D. \qquad \text{... (22.7)}$$

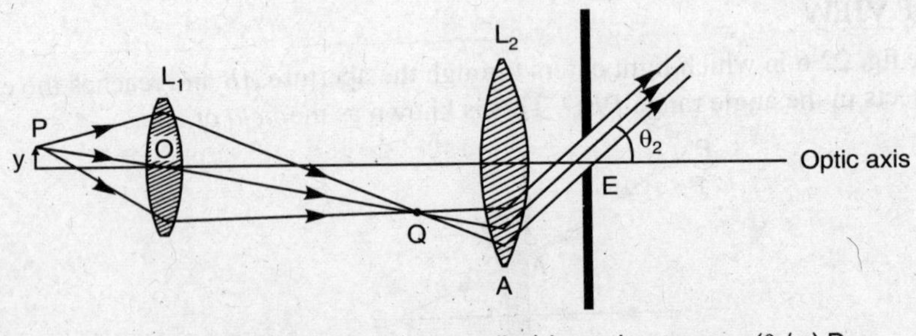

Fig. 22.8. Field of view (y cm radius) in a microscope = $(\theta_2/m).D$

The field of view of y is again dependent on θ_2, the angle subtended by the eye lens at the exit pupil.

22.6 NEED FOR MULTIPLE LENS EYEPIECE

From Eq. (22.6) and (22.7), it is clear that for a given magnification m, the field of view is proportional to the angle θ_2 subtended by the eye lens at the exit pupil. Using only single eye lens, the distance L_2E is close to focal length of L_2, so that θ_2 is approximately the ratio of the radial aperture to focal length of L_2. For a single lens this can not be made large without serious aberrations. Hence, adequate care should be taken while constructing eye-pieces with multiple lenses so as to produce a *large field of view without much aberrations:*

(*a*) **Enlarging the filed of view.** The field lens of the eye-piece effectively pushes the eye lens closer to the exit pupil, so as to enlarge the angle subtended.

(*b*) **Minimising aberrations.** Firstly, the deviations of rays are divided over a larger number of surfaces, so that spherical aberration is minimised. Secondly, for a separated doublet chromatic aberration can be reduced to zero or considerably minimised without involving different materials for the lenses.

The above two arrangements helps us to minimise the aberrations. Fig. 22.9 shows the general function of an eye piece (for simplicity we have taken the Ramsden's eye piece). In the two-lens eyepieces the one nearer the eye is called *eye lens,* the other is called *field lens.*

First Condition (*a*). To impliment the functions mentioned in (*a*) and (*b*) above, let us take the case of Ramsden's eyepiece (see article 22.8). Fig. 22.9 shows this case. If $f'_1 = f'_2 = 4.5$ cm, the focal length of the combination becomes 3.4 cm and the distance of *F*′ from the eye lens is 1.1 cm. This means that for a given *aperture* of eye lens, the angle θ_2 obtained in the two-lens eyepiece is 3 times that obtainable by using a single lens of equivalent focal length (3.4 cm) and same aperture. Thus the field of view is considerably enlarged.

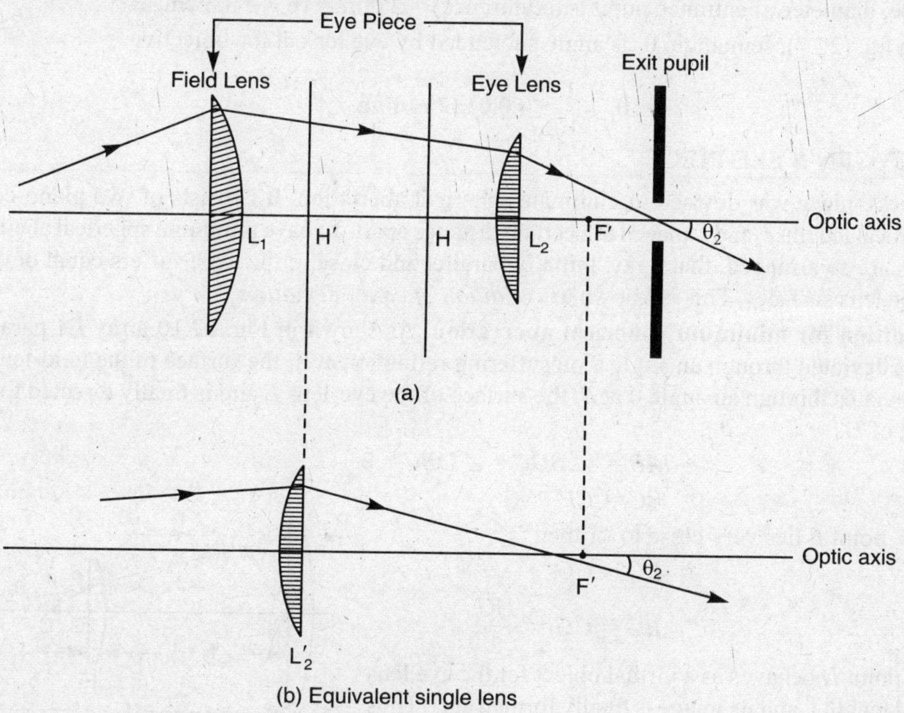

Fig. 22.9

Second Condition (*b*). For the same aperture, a single lens of $f' = 3.4$ cm will have more aberration than the eye lens of $f'_2 = 4.5$ cm used in the two-lens eyepiece. The use of two separated lenses also reduces the chromatic and spherical aberrations as compared with a single lens.

Functions of field lens and eye lens in an eyepiece. The *field-lens* which is relatively of a large aperture as compared to the eye-lens turns the rays of light passing through the outer portion towards the principal axis, so that these are made to pass through the centre of eye-lens. Thus rays from different parts enter the eye of the observer and field of view is increased.

If single eye-lens of small focal length is used, then magnified image suffers from spherical and chromatic aberrations. By using two lenses field-lens and the eye-lens, the total deviation of rays is spread over four surfaces to reduce spherical aberration. The distance between the lenses is so adjusted that chromatic aberration is also minimum.

Example 22.1. *In a telescope with a single lens eyepiece of $f = 6$ cm the objective is 106 cm from the eyepiece. Deduce the position of the exit pupil, magnification of the telescope; and size of entrance pupil if exit pupil is taken as 2 mm diameter. If the diameter of the eye lens is 3 cm, deduce the semi angle of the field of view.*

Solution. Exit pupil is at the image of the objective by the eye lens (Fig. 22.5). Given : $u = -106$ cm, $f = 6$ cm, we get the exit pupil position

$$v = \frac{106 \times 6}{100} \text{ cm} = 6.4 \text{ cm from the eye lens.}$$

For a telescope, magnification = angle subtended by eye lens at exit pupil ÷ angle subtended by eye lens at the objective (fig. 22.7). Hence

$$m = 106 \div 6.4 = 16.7$$

Size of entrance pupil ÷ size of exit pupil = magnification

Hence, diameter of entrance pupil is d (entrance) = 2 mm × 16.7 = 3.3 cm

From fig. (22.7), semiangle θ_1 is angle subtended by eye lens at the objective

$$\theta_1 = \frac{1.5}{106} = \mathbf{0.0142 \text{ radian}}$$

22.7 HUYGEN'S EYE-PIECE

This eye-piece was devised to eliminate spherical aberration. It consists of two plano-convex lenses of focal lengths f_1 and f_2 placed at a certain distance apart. To have minimum spherical aberration, the lenses are so arranged, that a ray, initially parallel and close to the axis, suffers equal deviation at both the lens surfaces. This is known as *condition of equal deviation.*

Condition for minimum spherical aberration. As shown in Fig. 22.10 a ray PA parallel to the axis is deviated through an angle θ on suffering refraction at A, the surface of the field-lens F. It further deviates through an angle θ at B, the surface of the eye-lens E and is finally directed towards C instead of D.

Now $$\angle IAB = \angle BDC = \angle DBC = \theta$$

$\therefore$ $$BC = CD$$

If the point B lies very close to G, then

$$BC = CG$$

$\therefore$ $$BC = CG = \frac{DG}{2}$$

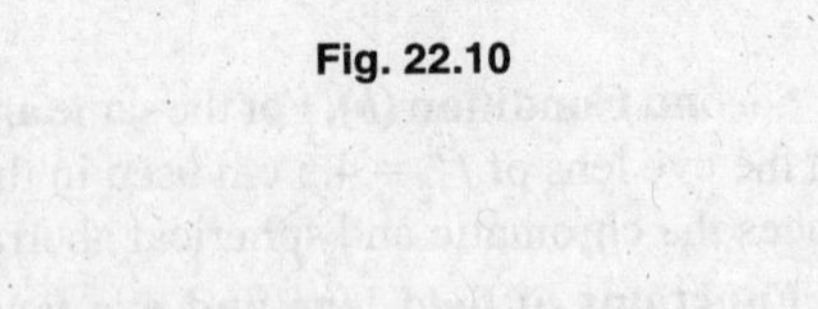

Fig. 22.10

The point D behaves as a virtual object for the eye lens E of focal length f_1 and its image is finally formed at C. Thus

$$DG = +u;\ CG = v = \frac{u}{2}$$

$\therefore$ $$\frac{1}{f_1} = \frac{1}{v} - \frac{1}{u} = \frac{2}{u} - \frac{1}{u} = \frac{1}{u}$$

or $$f_1 = u = DG$$

Since the ray PA is parallel to the axis and would have converged to D if the eye-lens E was not present.

$\therefore$ $$HD = f_2$$

or $$HG = f_2 - f_1 = d \qquad \text{... (i)}$$

Thus, we see that equal deviation is produced at the surface of the field as well as the eye-lens when the distance between the lenses is equal to the difference between their focal lengths. The field and the eye-lenses used are plano-convex and are placed with their convex surfaces towards the incident ray. In this way, the total deviation due to the combination is divided into four parts which makes the combination to have minimum spherical aberration.

Simultaneous minimum chromatic and spherical aberration. For minimum chromatic aberration, we know that

$$\frac{f_2 + f_1}{2} = d \quad \text{... } (ii)$$

If spherical and chromatic aberrations are to be simultaneously minimised, then form (*i*) and (*ii*), we have

$$\frac{f_2 + f_1}{2} = f_2 - f_1$$

or $\quad f_2 = 3f_1$

$\therefore \quad d = 3f_1 - f_1 = 2f_1$

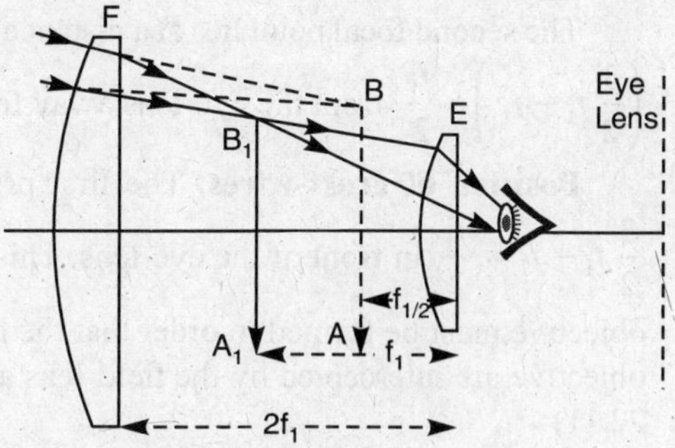

Fig. 22.11

Construction. To satisfy the condition for minimum chromatic and spherical aberrations, the focal length of the field-lens should be three times the focal length of the eye-lens and the distance between them should be equal to twice the focal length of the eye-lens. On this principle, Huygen constructed the eye-piece. The path of rays is shown in the ray diagram (Fig. 22.11).

Focal length of the equivalent lens. If f_1 is the focal length of the eye-lens, $3f_1$ is the focal length of the field lens and $2f_1$ the distance between them, then

The focal length F of the equivalent lens is given by

$$\frac{1}{F} = \frac{1}{f_1} + \frac{1}{f_2} - \frac{d}{f_1 f_2}$$

$$= \frac{1}{f_1} + \frac{1}{3f_1} - \frac{2f_1}{f_1 3f_1} = \frac{2}{3f_1}$$

$$\therefore \quad F = \frac{3}{2} f_2$$

Position of principal points. The equivalent lens should be placed at a distance α from the field lens which is given by

$$\alpha = \frac{d \times F}{f_1} = \frac{2f_1 \times \frac{3}{2} f_1}{f_1} = 3f_1$$

Since the distance d between the field-lens and the eye-lens is $2f_1$, the position of the equivalent lens is $3f_1 - 2f_1 = +f_1$ *i.e.*, it should be placed away from the eye-lens, as shown by dotted line in Fig. 22.12.

The first principal point P_1 lies at a distance $\alpha = 3f_1$ from the field lens. The second principal point P_2 lies at a distance β from the eye lens towards the field lens and is given by

$$\beta = -\frac{d \times f}{f_2} = \frac{-2f_1 \times \frac{3}{2} f_1}{3f_1} = -f_1$$

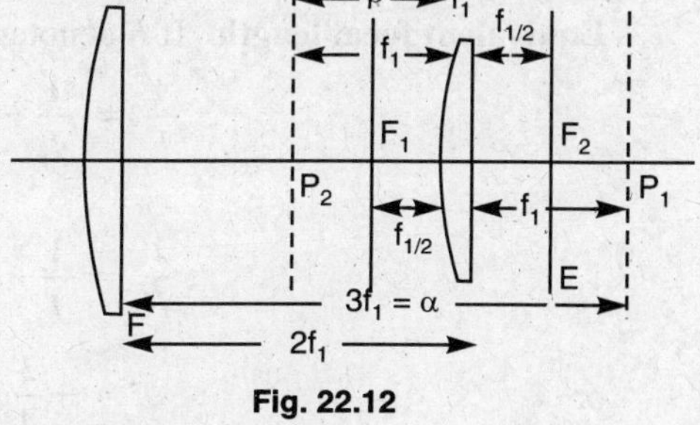

Fig. 22.12

Position of the focal points. The first focal point F_1 lies at a distance $\frac{3}{2}f_1$ from the first principal point P_1 i.e., at a distance $\left(\frac{3}{2}f_1 - f_1\right) = \frac{f_1}{2}$ from the eye-lens on the side of the field-lens.

The second focal point lies at a distance $\frac{3}{2}f_1$ from the second principal point P_2 *i.e.*, at a distance $\left(\frac{3}{2}f_1 - f_1\right) = \frac{f_1}{2}$ from the eye-lens away from the field-lens.

Position of cross-wires. The first principal focus of the equivalent lens lies at a distance $\frac{3}{2}f_1 - f_1 = \frac{f_1}{2}$ in front of the eye-lens. This point gives the position where the image *AB* due to the objective must be formed in order that the final image may lie at infinity. The rays coming from the objective are intercepted by the field-lens and the image *AB* is displaced to the position A_1B_1. (Fig. 22.11)

Now for the field-lens of focal length $3f_1$, *AB* lying at a distance $\frac{3}{2}f_1$ serves as a virtual object. Therefore, the distance *v* of the real image A_1B_1 is given by

$$\frac{1}{v} = \frac{1}{3f_1} + \frac{2}{3f_1} = \frac{1}{f_1} \quad \text{or} \quad v = f_1$$

Hence A_1B_1 lies midway between the two lenses. This fixes the position of the cross-wires.

Negative eye-piece. *As the image due to the objective is formed on the negative side (behind the field lens) the eye-piece is termed* as **negative eye-piece.** The image seen by the eye will be distinct but slightly curved without any distortion. The measurements, however, are not reliable as the cross-wires are seen only through the eye-lens and hence suffer from defects of spherical and chromatic aberrations, its outer portions being disproportionately magnified.

22.8 RAMSDEN'S EYEPIECE

It consists of two plano-convex lenses each of focal length *f*, separated by a distance equal to $\frac{2}{3}f$. The lenses are placed with their curved surfaces facing each other as shown in Fig. 22.13 and the eye-piece is placed beyond the image formed by the objective. This eye-piece is provided with cross-wires and is used in all optical instruments where accurate quantitative measurements are required.

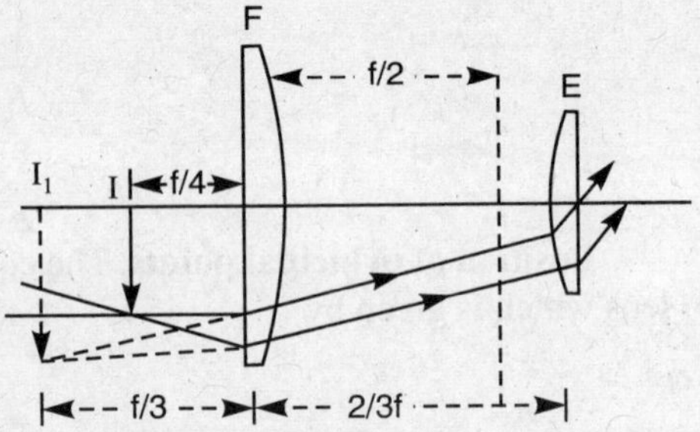

Fig. 22.13

Ray diagram. The objective forms an image of a distant object at *I*. The eye piece further forms its final image at I_1. The course of rays through the eye-piece is shown in the ray diagram (Fig. 22.13).

Equivalent focal length. If *F* denotes the focal length of the equivalent lens, then

$$\frac{1}{F} = \frac{1}{f_1} + \frac{1}{f_2} - \frac{d}{f_1 f_2}$$

$$\therefore \quad \frac{1}{F} = \frac{1}{f} + \frac{1}{f} - \frac{\frac{2}{3}f}{f^2} = +\frac{4}{3f}$$

or

$$F = +\frac{3}{4}f.$$

Position of equivalent lens. The equivalent lens of focal length $\frac{3}{4}f$ must be placed behind the field lens at a distance.

$$\alpha = \frac{F \times d}{f} = \frac{\frac{3}{4}f \times \frac{2}{3}f}{f} = \frac{f}{2}$$

Thus the equivalent lens lies between the field and the eye-lens.

Position of cross-wire. As the focal length of the eye-piece (Equivalent focal length) is $\frac{3}{4}f$, the image of the object due to objective must be formed at a distance $\frac{3}{4}f - \frac{1}{2}f = \frac{1}{4}f$ in front of the field lens at *I*. This image will act as an object for the eye-piece and the final image will be formed at infinity. Thus the cross-wires must be placed at the position of the image *I*, *i.e.*, at a distance of $\frac{1}{4}f$ in front of the field lens *F*. This is the advantage of Ramsden's eye-piece.

Advantages. The chromatic aberration is small as both the lenses are made of a combination of a crown and flint glass to eliminate chromatic aberration. As both the lenses are plano-convex with their convex surfaces facing each other, the spherical aberration produced is also small.

Position of the principal planes. The first principal plane of the equivalent lens lies at P_1 behind the field-lens at a distance α given by the relation.

$$\alpha = \frac{d \times F}{f_2}$$

where *d* is the distance between the two lenses and f_2 is the focal length of the eye-lens. Hence

$$\alpha = \frac{\frac{2}{3}f \times \frac{3}{4}f}{f} = +\frac{1}{2}f$$

Thus the equivalent lens is placed at P_1 in between the field lens *F* and the eye-lens E at a distance $\frac{1}{2}f$ behind the field-lens, as shown in Fig. 22.14. The point H_1 where the principal plane P_1 cuts the axis is the **first principal point.**

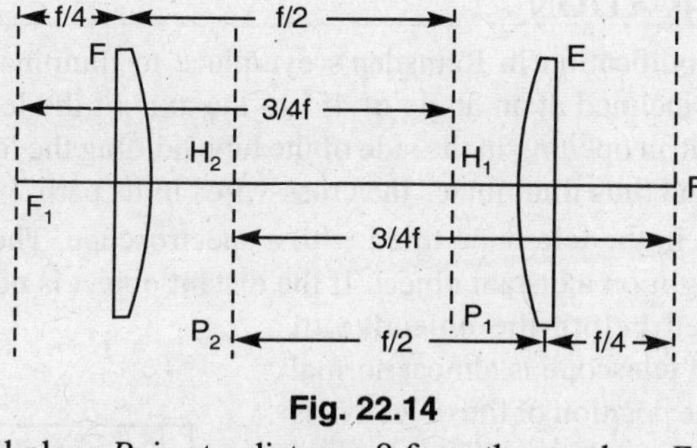

Fig. 22.14

The second principal plane P_2 is at a distance β from the eye-lens *E*, given by

$$\beta = -\frac{d \times F}{f_1} = \frac{-\frac{2}{3}f \times \frac{3}{4}f}{f} = -\frac{f}{2}$$

The second principal plane P_2 lies behind the eye-lens E at a distance $\frac{1}{2}f$. The point H_2 where the principal plane P_2 cuts the axis is the **second principal point.**

Nodal point. As the system lies in air the two nodal points coincide with the principal points.

Position of focal points. The equivalent focal length of the eye-piece is $\frac{3}{4}f$ Hence the **first focal point** F_1 lies at a distance $\frac{3}{4}f$, from the first principal point H_1 *i.e.*, at a distance $\left(\frac{3}{4}f-\frac{1}{2}f\right)=\frac{1}{4}f$ to the left of the field lens F.

The **second focal point** F_2 lies at a distance $\frac{3}{4}f$ from the second principal point H_2 *i.e.*, at a distance $\frac{1}{4}f$ behind the eye-lens E away from the field-lens.

Merits and Demerits of Ramsden's eye-piece

(*i*) In this arrangement the field-lens is not very efficient but the field of view is fairly wide.

(*ii*) The arrangement is not entirely achromatic. The chromatic effects are minimised by using achromatic combinations for both the field and the eye-lenses.

(*iii*) The spherical aberration is minimised as the image is seen through two lenses and hence the deviation is small, as it takes place at four different surfaces. It is further minimised by making both the lenses plano-convex, their convex surfaces facing each other. The coma and distortion are also sufficiently reduced.

(*iv*) This eye-piece is very commonly used in all the optical instruments, where measurements are to be made, as it is a positive eye-piece and the cross-wires are placed outside in front of the field lens F.

(*v*) It can be used as a simple-mciroscope, because the first focal plane lies to the left of the field-lens and is therefore real.

Ramsden's eye-piece is positive eye-piece. The image formed by the objective acts as a real object for the eye-lens and the final image is formed at infinity. The Ramsden's eye-piece is, therefore, called a **positive eye-piece.** This is an advantage of Ramsden's eye-piece over Huygen's eye-piece, as the cross-wires can be replaced by the measuring scales and the magnification produced by the eye-piece in respect of the cross-wires and the image is the same.

22.9 GAUSS'S MODIFICATION

Gauss introduced a modification in Ramsden's eye-piece to illuminate the field of view. He introduced a glass plate G inclined at an angle of 45° to the axis of the lens system. Light from a source S is admitted through an opening in the side of the tube holding the lens system. It is reflected along the axis of the tube and thus illuminates the cross-wires in its path.

This eye-piece is used in the telescope fitted with a spectroscope. The telescope has to be set for parallel rays by focusing it on a distant object. If the distant object is not available, then a plane reflecting surface is placed before the objective of telescope so that the axis of telescope is almost normal to the reflecting surface. The position of the objective is adjusted till the image of the cross-wires after reflection coincide with the cross-wires. In this way the axis of telescope is exactly normal to the reflecting surface placed on the prism table. When the cross-wires and their image are both in sharp focus, the telescope is correctly adjusted for parallel rays.

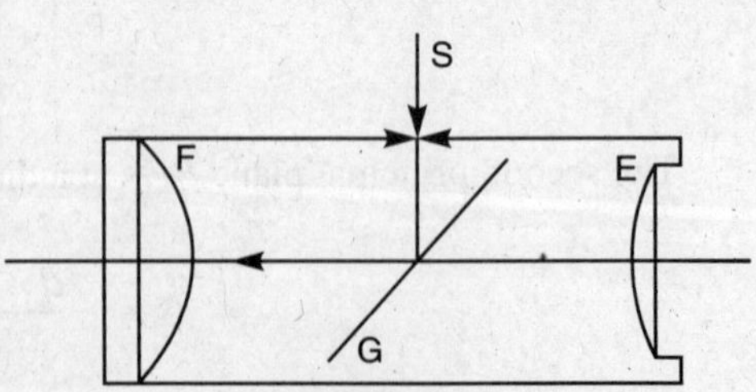

Fig. 22.15

22.10 COMPARISON OF HUYGEN'S EYEPIECE AND RAMSDEN'S EYEPIECE

Relative merits and demerits

(*i*) *Field of view.* The field of view in both these eye-pieces is fairly large in the image space.

(*ii*) *Achromatism and spherical aberration.* Huygen's eye-piece is constructed so that it has minimum spherical aberration. Ramsden's eye-piece has also minimum spherical aberration as in this case the deviation of rays is shared by all the four surfaces of the two lenses comprising it.

The Huygen's eye-piece is free from chromatic aberration as

$$d = \frac{f_1 + f_2}{2}.$$

Ramsden's eye-piece does not satisfy this condition and shows lateral colours. This effect can be reduced by using achromats in place of field and eye-lenses.

Huygen's eye-piece is achromatic for all colours whereas Ramsden's eye-piece is ahcromatic for only the selected colours.

(*iii*) *Position of cross-wires.* In Ramsden's eye-piece a cross-wire or a scale is placed in front of the eye-piece for making measurements. The image formed by the objective along with the cross-wires or transparent scale are equally magnified by the eye-piece and hence the measurements are trustworthy.

In Huygen's eye-piece the cross-wires or the scale has to be placed midway between the field-lens and the eye-lens. This does not give trustworthy measurements because the image undergoes refraction in both the lenses of the eye-piece while the light from the cross-wires suffers refraction only at the surface of the eye-lens. This gives rise to unequal magnification of the image and the crosss-wires.

(*iv*) *Cross-wire facility.* In Huygen's eye-piece the cross-wires are placed in between the field and the eye-lens at a definite position. This involves mechanical difficulty. Again these are adjusted permanently for persons of normal vision only.

In Ramsden's eye-piece the cross wires are placed outside in front of the field-lens and hence neither there is the difficulty of placing them nor in adjusting to the vision of varying visions.

(*v*) In Huygen's eye-piece, if the crosss-wires are replaced by measuring scales, the measurements are not trustworthy due to unequal magnification. In Ramsden's eye-piece, this difficulty is not present and hence the measurements are trustworthy.

In Ramsden's eye-piece the field of view is much better than that in the case of Huygen's eye-piece.

(*vi*) The distance between the eye-lens and the eye-ring is greater in Ramsden's eye-piece than in Huygens' eye-piece. Hence eye-relief is better in Ramsden's eye-piece.

In Ramsden's eye-piece the cross wires are placed in front of the eye-piece and hence magnification of cross wire and image is the same, but in Huygen's eye-piece the cross wires are placed in between the field-lens and the eye-lens, hence magnification of image and cross wires is not the same. The measurements with Huygen's eye-piece are not trustworthy. Hence Ramsden's eyepiece are prefered over Huygen's eyepiece.

Example 22.2. *Two thin lenses of focal length 10 cm and 8 cm are kept in position of least spherical aberration. Find the distance of principal points for the system.* **(Gharwal U. 2000)**

Solution. If f_1 is the focal length of one lens and f_2 that of the other, then if they are kept in the position of least spherical aberration, the distance d between them is given by

$$d = f_1 - f_2$$

Here $f_1 = 10$ cm; $f_2 = 8$ cm; $\therefore d = 10 - 8 = 2$ cm.

∴ Equivalent focal length F is given by

$$\frac{1}{F} = \frac{1}{f_1} + \frac{1}{f_2} - \frac{d}{f_1 f_2} = \frac{1}{10} + \frac{1}{8} - \frac{2}{10 \times 8} = \frac{1}{5}$$

or $F = 5$ cm

Position of first principal point $\alpha = \frac{dF}{f_2} = \frac{2 \times 5}{8} = 1.25$ cm

Position of second principal point $\beta = \frac{-dF}{f_1} = -\frac{2 \times 5}{10} = -1$ cm

Example 22.3 *Two lenses of focal length 10 cm and 30 cm are separated by 20 cm in an eye-piece. Show that both spherical and chromatic aberration are removed. Name the eye-piece.*

(H.P.U. 2001)

Solution. If f_1 and f_2 are the focal lengths of the two lenses forming an eye piece and d is the distance between them, then for

Removal of spherical aberration $d = f_1 - f_2$ and for

Removal of chromatic aberration $d = \frac{f_1 + f_2}{2}$

As in this case $f_1 - f_2 = 30 - 10 = 20$ cm $= d$ and $\frac{f_1 + f_2}{2} = \frac{30 + 10}{2} = 20$ cm $= d$

The eye piece satisfied both the conditions. Hence both spherical and chromatic aberrations are removed.

It is a Huygen's eye-piece.

Example 22.4 *Light from the sun is falling upon the Ramsden's eye-piece. Locate the position of the image thus formed and also find the point from which the distance of the image is to be measured. The focal length of each lens of the eye-piece is 3.0 cm.*

Solution. Focal length of each lens $f = 0.03$ m

Distance between the lenses $d = \frac{2}{3} f = \frac{2}{3} \times 0.03 = 0.02$ m

Focal length of equivalent lens $F = \frac{0.03 \times 0.03}{0.03 + 0.03 - 0.02} = 0.0225$ m

Since the object, *i.e.*, the sun lies at infinity, the image is formed at the second focal plane of the eye-piece, *i.e.*, at a distance 0.0225 m or 2.25 cm from the second principal plane.

Distance of the second principal plane from the eye-lens E

$$= \frac{-dF}{f_1} = \frac{-2 \times 0.0225}{3} = -0.015 \text{ m}$$

The negative sign indicates that the second principal plane lies behind the eye-lens E.

Example 22.5 *Find how much longer a telescope will have to be when it is fitted with Ramsden's eye-piece than when it is fitted with the simple equivalent lens when a distant object is viewed.*

(P.U. 1999)

Solution. Let focal length of the telescope objective = F. If f is the focal length of the field lens as well as the eye lens of Ramsden's eye piece, then its

Equivalent focal length $= \frac{3}{4} f$.

∴ Length of the telescope with simple equivalent lens = Focal length of the objective + Focal length of the eyepiece $= F + \frac{3}{4} f$.

The distance of the second principal point from the eye lens $= \frac{1}{2} f$ *i.e.*, second principal point lies to the left of the eye lens.

$\therefore$ Length of telescope with Ramsden's eye piece

$$= F + \left(\frac{3}{4} f + \frac{2}{3} f - \frac{f}{2} \right) = F + \frac{11}{12} f$$

Hence the telescope will be longer by $\left(F + \frac{11}{12} f \right) - \left(F + \frac{3}{4} f \right) = \frac{1}{6} f.$

EXERCISE

1. Explain the function of human eye and the image processing.
2. What do you understand by visual angle and angular magnification ? Give the construction and working of a simple microscope. Calculate its magnifying power.
3. What are the advantages of a compound microscope over a simple microscope ? Calculate its magnifying power.
4. Explain what are entrance pupil and exit pupil. State their importance. ***(Nagpur U. 2008)***
5. What are the functions of field lens used in an eyepiece ? Give the construction of a Huygen's eyepiece and calculate the positions of the cardinal points.
6. What is an eyepiece and what is its advantage over a single lens. What is the need for a multiple lens eye-piece ? ***(Nagpur U. 2008)***
7. Calculate and show in a diagram the positions of principal points, the focal and nodal points of Ramsden's eye piece. ***(Meerut U. 2001; Gharwal U. 2004)***
8. Give the construction, ray diagram and working of Ramsden's eyepiece. Prove that the equivalent focal length is equal to $\frac{3}{4} f$ where f is the focal length of each of the field lens and eye lens of the eyepiece. Mark the position of the cross-wire in it. ***(G.N.D.U. 2005; Pbi. U. 2003, 2001; H.P.U. 2003, 2002; P.U. 2003, 2002; Meerut U. 2001; Nagpur U. 2002, 2001; Gauhati 2000)***
9. Give the merits and demerits of Ramsden's eye piece. Why is it called a positive eye-piece ? ***(H.P.U., 2003, 2002, P.U. 2004; 2003 Pbi. U. 2002, 2000, Nag. U. 2002; Guahati. U. 2000; G.N.D.U., 2000)***
10. What modification was introduced in Ramsden's eye-piece by Gauss and why ?
11. With the help of a neat diagram give the construction and working of Huygen's eye piece. Deduce the condition for minimum spherical aberration for two thin lenses separated by a finite distance. Give its merits and demerits. ***(P.U. 2004; 2002, 2000; Bhopal. U. 2004; G.N.D.U. 2002, 2001; Meerut 2002, 2000; H.P.U. 2001; Pbi. U. 2002; K.U. 2001, Gharwal U. 2004; Nagpur U. 2010)***
12. Calculate the equivalent focal length, position of cross-wires and the position of cardinal points of a Huygen's eye-piece and show them by a diagram. Why is it called a negative eye piece ? ***(Nag. U. 2010, 2001)***
13. Compare Ramsden's eye-piece with Huygen's eye-piece. Give their relative merits and demertis. ***(H.P.U. 2003, Nagpur U. 2008, 2002; Madurai U. 2003)***
14. Why we can not use Huygen's eyepiece for quantitative measurements ? ***(P.U. 2003, Nagpur U. 2009)***

15. Why do we call Huygen's eyepiece theoretically perfect and negative ?
Hint. It is called theoretically perfect as both the chromatic and spherical aberrations are minimised simultaneously by taking the focal length of field lens $3f$ and eye lens f. These two are separated by $2f$. It is called a negative eye-piece as the image formed by the objective lies on the negative side *i.e.,* behind the field lens. ***(G.N.D.U. 2003; Pbi. U. 2001)***

16. What are the functions of field lens and the eye lens in an eyepiece ? ***(G.N.D.U. 2002)***

17. In Ramsden's eyepiece why the two lenses of equal focal lengths are separated by $\frac{2}{3}f$ instead of f which will make it free from chromatic aberration ?
[**Hint.** To avoid magnification of dust particles on the surface of field lens which will make the image blurred.]

18. Why we prefer Ramsden's eyepiece over Huygen's eyepiece ?

19. Explain how spherical and chromatic aberration are overcome in Huygen's eyepiece ***(Nagpur U. 2011)***

20. Two thin lenses of focal lengths 6 cm and 2 cm are placed 4 cm apart to form an eye-piece. Calculate the focal length of the eye-piece and locate the positions of the principal planes and focal planes. [**Ans.** f = 3 cm, α = 6 cm and β = –2 cm]

21. Two thin lenses of focal lengths f_1 and f_2 separated by a distance d have an equivalent focal length 50 cm. The combination satisfies the condition of no chromatic aberration and minimum spherical aberration. If the lenses are made of the same material, find the values of f_1, f_2 and d.
[**Ans.** f_1 = 100 cm, f_2 = 33.2 cm, d = 66.67 cm] ***(Madurai U. 2003)***

22. Two lenses of focal lengths 6 cm and 2 cm are placed co-axially at a distance apart such that spherical aberration is minimum. What is the distance between the lenses ? **[Ans. 4 cm]**

MULTIPLE CHOICE QUESTIONS

22.1 Huygen's eyepiece consists of:
(*a*) two plano-convex lenses of focal lengths $3f$ and f separated by $2f$.
(*b*) two plano-concave lenses of focal lengths $3f$ and f separated by $2f$.
(*c*) two plano-convex lenses of focal lengths $3f$ and f separated by $\frac{2}{3}f$.
(*d*) two plano-convex lenses of focal lengths $3f$ and f separated by $\frac{2}{3}f$.

22.2 In Huygen's eyepiece focal length of lenses are $3f$ and f and the distance between them is
(*a*) f (*b*) $2f$ (*c*) $3f$ (*d*) $\frac{3}{2}f$

22.3 Huygen's eyepiece is free from spherical and chromatic aberrations:
(*a*) true (*b*) false (*c*) can't say

22.4 Angular magnification M of a simple microscope is given by :
(*a*) $1+\frac{D}{f}$ (*b*) $1-\frac{D}{f}$ (*c*) $1+\frac{f}{D}$ (*d*) $1-\frac{f}{D}$

22.5 Huygen's eyepiece is free from chromatic aberration since :
(*a*) $d=\frac{f_1-f_2}{2}$ (*b*) $d=\frac{f_1^2-f_2^2}{2}$ (*c*) $\frac{f_1+f_2}{2}$ (*d*) $\frac{f_1^2+f_2^2}{2}$

ANSWERS

22.1 (*a*)	22.2 (*b*)	22.3 (*a*)	22.4 (*a*)	22.5 (*c*)

SECTION VII

Physical Optics

INTERFERENCE OF LIGHT

23

Chapter

INTRODUCTION

Light is a form of wave motion. The waves are transverse in nature. A single source of light gives out energy which is distributed uniformly in the surrounding medium. If two independent sources of light which are capable of giving out *continuous waves of the same amplitude; same wavelength and same phase difference or constant phase difference* are held close to each other, the distribution of energy in the surrounding medium is not uniform, but bright and dark regions are observed.

At places where the *crest* of one wave falls on the *crest* of the other or a *trough* of one falls on the *trough* of the other, the amplitude increases and hence the energy or intensity of light is increased. At places where the *trough* of one falls on the *crest* of the other, the resultant amplitude is zero, which results in the decrease of intensity. This gives rise to a phenomenon of interference. This is studied by two different ways: (*i*) by considering division of wave front and (*ii*) by division of amplitude. In this chapter we will focus our attention on principle of superpositions, optical path retardation, localized fringes, thin films and some important interferometers.

23.1 INTERFERENCE

The *modification in the distribution of light energy due to superposition of two or more waves is called* interference.

The phenomenon may be better understood by considering two point sources A and B which produce exactly similar waves. The crests of the waves are represented by complete arcs and the troughs by dotted arcs. At any particular instant the displacement at any point on the surface of water will be the resultant of the displacements produced by the two waves. Constructive interference will take place at points where the crest of one falls over the crest of other or the trough of one falls over the trough of the other. Such places are represented by points marked with crosses. The total displacement at such points is double the displacement due to a single wave and the intensity will, therefore, be four times. The full lines passing through C, O or E give regions of maximum intensity.

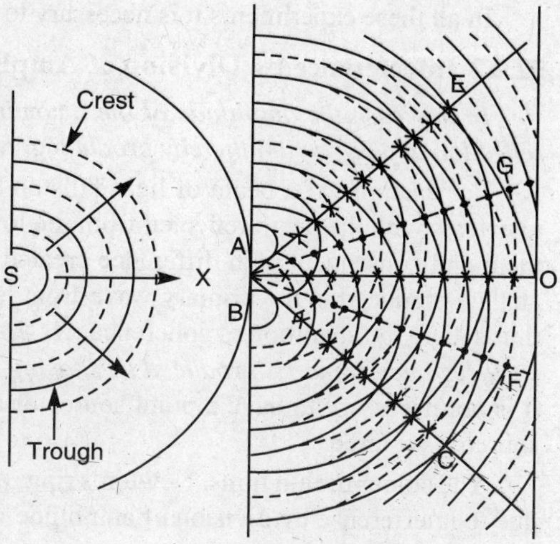

Fig. 23.1

Destructive interference takes place at points where the trough of one falls over the crest of the other. Such places are represented by points marked with small circles. The total displacement at such points is zero or a minimum and hence the intensity is minimum. The dotted lines passing through F or G give the regions of minimum intensity. No flow of energy takes place along these regions.

The two sources which vibrate with a fixed phase difference between them are said to be **coherent**. If on the other hand the two sources are vibrating with varying phase difference *i.e.*, the phase changes rapidly, then stationary interference can not be observed. In such a case, the source's are said to be **incoherent.**

23.2 METHODS OF PRODUCING INTERFERENCE

23.2.1 Interference by Division of Wave Front

Two slits illuminated by light coming from a single slit, a narrow source and its virtual image or two virtual images of a narrow source can be used as coherent sources of light to produce interference. In these cases *a wave front is divided into two parts by utilising the phenomenon of reflection; refraction or diffraction in such a way that after traversing slightly different optical paths the light from the two coherent sources so produced finally unites to produce interference bands.*

For example, in Young's double slit experiment two slits illuminated by light coming through a single slit serve as two coherent sources. Young allowed sunlight to fall on a pin hole S in a darkened room and then at some distance away from S on two pin holes A and B about 0.5 mm apart, as shown in Fig. 23.1. The pin holes A and B are equidistant from S. Spherical waves spread out from A and B. These waves are of the same amplitude and wavelength, as they are derived from the same wave front. The two sets of spherical waves coming out of A and B interfere and form a symmetrical pattern of varying intensity on the screen.

If the pin holes are replaced by narrow slits held perpendicular to the plane of paper and are illuminated by monochromatic light, cylindrical wavefronts, instead of spherical wavefronts, come our and *evenly spread bright and dark* fringes parallel to the length of the slit in a direction at right angle to AB are produced on the screen. When the distance between the slits A and B is increased the bands become narrow and finally disappear.

In the case of Fresnel's bi-prism two virtual images of single point source produced by refraction at a bi-prism serve as two coherent sources. In the case of Lloyd's single mirror a source and its virtual image formed by reflection in a mirror and in Fresnel's double mirror two virtual images formed by reflection in the two mirrors serve as two coherent sources of light.

In all these experiments it is necessary to use point source or a line source of light.

22.2.2 Interference by Division of Amplitude

In this case the amplitude of the incoming wave is divided into two or more parts by partial reflection and refraction thereby producing two or more beams which produce interference.

For example, if a beam of light falls on a thin film the beam reflected from the upper surface interferes with the beam reflected from the lower surface due to the different distances travelled by them and consequent path difference created between the secondary wavelets. Any random phase change produced in the primary wave front is equally passed on to the two beams. The two beams thus satisfy the condition of coherence. *As the same beam of light is partially reflected from different surfaces interference is produced by sharing of light intensity or division of amplitude.* In this case it is not necessary to have a point source or a narrow line source. On the other hand an extended source is required.

The colour of thin films, Newton's rings and fringes produced in Michelson's interferometer are due to interference by division of amplitude.

23.3 PRINCIPAL OF SUPERPOSITION

When two wave trains travel in the same direction they superimpose upon each other and the resultant displacement at any point is given by the vector sum of the displacement of the two individual waves.

Superposition of two waves. Light consists of transverse electro-magnetic wave. A light wave travelling along the X- axis is represented by the equation $y = a \sin (\omega t - \phi)$ where y is the instantaeous value of the *electric field vector* or the *displacement* of the transverse light wave along the Y-axis in a direction perpendicular to the direction of propagation, a the amplitude of electric field vector, ω

the angular frequency = $2\pi n$, n being the frequency of the electromagnetic wave corresponding to wavelength λ and ϕ the initial phase.

Consider two light waves of the same wavelength *i.e.* the same value of ω travelling in the same direction given by

$$y_1 = a_1 \sin(\omega t - \phi_1)$$

and

$$y_2 = a_2 \sin(\omega t - \phi_2)$$

There two waves super impose upon each other and the value of resultant electric field vector or displacement of the light wave is given by

$$\begin{aligned} y = y_1 + y_2 &= a_1 \sin(\omega t - \phi_1) + a_2 \sin(\omega t - \phi_2) \\ &= a_1 [\sin \omega t \cos \phi_1 - \cos \omega t \sin \phi_1) + a_2[\sin \omega t \cos \phi_2 - \cos \omega t \sin \phi_2] \\ &= \sin \omega t [a_1 \cos \phi_1 + a_2 \cos \phi_2] - \cos \omega t [a_1 \sin \phi_1 + a_2 \sin \phi_2] \end{aligned}$$

Substituting $a_1 \cos \phi_1 + a_2 \cos \phi_2 = A \cos \theta$

and $a_1 \sin \phi_1 + a_2 \sin \phi_2 = A \sin \theta$

we get

$$\begin{aligned} y &= A \sin \omega t \cos \theta - A \cos \omega t \sin \theta \\ &= A \sin(\omega t - \theta) \end{aligned}$$

Thus the resultant wave has the same value of ω (or wave length λ), amplitude A and initial phase ϕ given by

$$\begin{aligned} A^2 &= A^2 \cos^2 \theta + A^2 \sin^2 \theta \\ &= (a_1 \cos \phi_1 + a_2 \cos \phi_2)^2 + (a_1 \sin \phi_1 + a_2 \sin \phi_2)^2 \\ &= a_1^2 + a_2^2 + 2a_1 a_2 (\cos \phi_1 \cos \phi_2 + \sin \phi_1 \sin \phi_2) \\ &= a_1^2 + a_2^2 + 2a_1 a_2 \cos(\phi_2 - \phi_1) \\ &= a_1^2 + a_2^2 + 2 a_1 a_2 \cos \phi \qquad \text{... (23.1)} \end{aligned}$$

where $\phi = \phi_2 - \phi_1$

and

$$\tan \theta = \frac{a_1 \sin \phi_1 + a_2 \sin \phi_2}{a_1 \cos \phi_1 + a_2 \cos \phi_2} \qquad \text{... (23.2)}$$

(*i*) **Intensity.** Intensity of the resultant light wave is given by (from relation 23.1)

$$I = A^2 = a_1^2 + a_2^2 + 2a_1 a_2 \cos \phi$$

(*ii*) **Condition for $I_{max.}$** The amplitude and therefore the intensity of light wave is maximum when $\cos \phi = +1$ *i.e.*,

$$\phi = 2n\pi \qquad \text{where } n = 0, 1, 2 \text{ etc}$$

and

$$I_{max} = a_1^2 + a_2^2 + 2a_1 a_2 = (a_1 + a_2)^2$$

(*iii*) **Condition for $I_{min.}$** The amplitude and therefore the intensity of light is minimum when $\cos \phi = -1$ *i.e.*,

$$\phi = (2n+1)\pi \text{ where } n = 0, 1, 2 \text{ etc}$$

and

$$I_{min} = a_1^2 + a_2^2 - 2a_1 a_2 = (a_1 - a_2)^2$$

(*iv*) **When amplitudes are equal** *i.e.*, $a_1 = a_2 = a$

$$I_{max} = 4a^2$$

and

$$I_{min} = 0$$

(*v*) **Interference fringes.** Thus we find that due to superposition of two light waves we get alternate regions of maximum and minimum light intensity. These are known as **interference fringes.**

(*vi*) **Visibility of fringes or fringe contrast (F.C).** The visibility of fringes or fringe contrast is measured by

$$\text{Fringe Contrast (F.C)} = \frac{I_{max} - I_{min}}{I_{max} + I_{min}}$$

When $a_1 \neq a_2$ $\quad I_{max} - I_{min} = (a_1^2 + a_2^2 + 2a_1a_2) - (a_1^2 + a_2^2 - 2a_1a_2)$

$$= 4a_1a_2$$

$$I_{max} + I_{min} = (a_1^2 + a_2^2 + 2a_1a_2) + (a_1^2 + a_2^2 - 2a_1a_2)$$

$$= 2(a_1^2 + a_2^2)$$

and Visibility $= F.C = \dfrac{I_{max} - I_{min}}{I_{max} + I_{min}} = \dfrac{4a_1a_2}{2(a_1^2 + a_2^2)} = \dfrac{2a_1 a_2}{a_1^2 + a_2^2}$

When $a_1 = a_2$: Fringe contrast $F.C = \dfrac{2a^2}{2a^2} = 1$

Let I_1 be the intensity of one source having amplitude a_1 and I_2 of the other source having amplitude a_2 then $\dfrac{I_2}{I_1} = \dfrac{a_2^2}{a_1^2} = \alpha$ $\quad$ or $\quad \dfrac{a_2}{a_1} = \sqrt{\alpha}$ $\quad$ or $\quad a_2 = a_1\sqrt{\alpha}$

$$\therefore \quad \frac{I_{max} - I_{min}}{I_{max} + I_{min}} = \frac{2a_1a_2}{a_1^2 + a_2^2} = \frac{2a_1a_1\sqrt{\alpha}}{a_1^2 + \alpha a_1^2} = \frac{2\sqrt{\alpha}}{1+\alpha} \qquad \dots (23.4)$$

Example 23.1 *The amplitude of light waves emerging from two slits in Young's experiment is in the ration of 1:2. Find the intensity ratio of the interference pattern.* **(*Pbi. U. 2002*)**

Solution At the bright band intensity is maximum give by

$$I_{max} = a_1^2 + a_2^2 + 2\,a_1\,a_2$$
$$= a_1^2 + (2a_1)^2 + 2\,a_1\,(2a_1) = 9a_1^2$$

At the dark band intensity is minimum given by

$$I_{min} = a_1^2 + a_2^2 - 2\,a_1\,a_2$$
$$= a_1^2 + (2\,a_1)^2 - 2\,a_1\,(2a_1) = 9a_1^2$$

$$\therefore \quad \frac{\text{Intensity of bright band}}{\text{Intensity of dark band}} = \frac{I_{max}}{I_{min}} = \frac{9a_1^2}{a_1^2} = 9$$

$\therefore$ Intensity ratio of the interference pattern is 9 : 1.

Example 23.2. *Two coherent sources of intensity ratio 100:1 interfere. Deduce the ratio of intensity between maxima and minima.* **(*M.D.U. 2008*)**

Solution. $\dfrac{I_1}{I_2} = \dfrac{100}{1}$ $\therefore \dfrac{a_1}{a_2} = \dfrac{10}{1}$ or $a_1 = 10a_2$

$$\frac{I_{max}}{I_{min}} = \frac{a_1^2 + a_2^2 + 2a_1 a_2}{a_1^2 + a_2^2 - 2a_1 a_2} = \frac{100a_2^2 + a_2^2 + 20a_2^2}{100a_2^2 + a_2^2 - 20a_2^2} = \frac{121}{81}$$

23.4 COHERENT SOURCES

Two sources are said to be coherent if they emit continuous light waves of the same frequency or wavelength, nearly of the same amplitude (preferably equal) and should either be in phase with each other or have a constant phase difference.

Condition for two sources to be coherent. The two sources should have the same wavelength

(*i*) There should be zero or constant phase difference between the light waves from the two sources which should not change with time.

(*ii*) The amplitude of light coming from the two sources should be equal or nearly equal.

Two independent point sources cannot be coherent. Light from a source is emitted when the orbital electrons of the atom jump from a higher energy orbit E_2 to a lower energy orbit E_1 thereby releasing the difference in energy of the two levels in the form of light of wave length λ or frequency $\nu = \frac{c}{\lambda}$ given by $E_2 - E_1 = h\nu = \frac{hc}{\lambda}$. The emission of light pulse from different atoms is a random process and so the phase of the light pulses varies randomly and independently of each other. Thus it is not possible to maintain a constant phase difference between the light waves from two different independent light sources.

The phase relation between the waves at the time of emission rapidly changes with time, not only in different sources but even the different parts of the same source. As a result there are rapid changes of brightness and darkness, which produce general illumination. It is thus not possible to produce interference with two independent sources which can not be coherent.

Production of coherent sources. In practice, two coherent sources are realized from a single source by the following methods.

(*i*) *A real narrow source and its virtual image produced by reflection, as in* **Lloyd's single mirror.**

(*ii*) *Two virtual images of the same source produced by reflection,as in* **Fresnel's double mirror.**

(*iii*) *Two virtual images of the same source produced by refraction, as in* **Fresnel's biprism.**

(*iv*) *Two real images of the same source produced by refraction*, as in **Billet's split lens.**

(*v*) *By dividing the amplitude of a portion of the wave front into two parts by reflection or refraction or both.* This type of interference is produced in Newton's rings, Michelson's interferometer, and thin films.

23.5 CONDITIONS FOR GOOD INTERFERENCE PATTERN

The interference pattern must satisfy the following three conditions to produce good interference pattern (*i.e.*, fringes): These are:

(*a*) Observance

(*b*) Good contrast of fringes

(*c*) Stationary interference pattern

All the above three are explained below:

Condition I: Observance

The interference fringes are observed when obtained from two coherent sources. If d is the distance between the two sources, D the distance of screen from the source, then for the given wavelength λ, the fringe-width β is given by

$$\beta = \frac{D}{d}\lambda \qquad \ldots (23.5)$$

(*i*) To get a reasonable fringe-width, the *distance d between the two sources should be small.* If d is large, β will be small and due to limited resolving power, the eye will not be able to see the fringes as separate.

(*ii*) The distance D should be large for the same reason.

(*iii*) To have good visibility the background should be dark. If this is not so, then due to low intensity of fringes, these will not be visible.

Condition II: Good contrast

(*i*) *The amplitudes of the interfering waves should be preferably equal.* If a_1 and a_2 are the amplitudes of the interfering beams, then

Intensity of maxima ∝ $(a_1 + a_2)^2$ and

Intensity of minima ∝ $(a_1 - a_2)^2$

If the difference between the amplitudes a_1 and a_2 is large, then the intensity of the minima will be practically the same as that of maxima and hence the contrast will be poor. For a good contrast $a_1 = a_2$ so that the minima has a low intensity.

(*ii*) The *two sources should be very narrow*. A broad source of light is equivalent to a large number of narrow sources lying side by side. Each set of such two sources will give rise to its own pattern of interference fringes. The overlapping of a number of such patterns will result in general illumination and hence the fringes will vanish.

(*iii*) *The sources should be monochromatic.* The fringe-width $\left(\beta = \frac{D}{d}\lambda\right)$ depends upon the wavelength of light. If the source is monochromatic, β will be constant and hence good intensity fringes will be observed.

If the source used is emitting white light, it is equivalent to an infinite number of monochromatic sources. Each monochromatic pair gives its own set of fringes and the fringe-width, therefore, will be different for each pair of sources. As a result of this, the zero order maxima of all wavelengths will lie in the same position, while the other order maxima will lie in different positions for each wave length. This results in overlapping of fringes due to different wave lengths, and thus only a few coloured fringes with poor contrast will be visible. When the path-difference is large, it results in uniform illumination.

Conditions III: Stationary interference. (*i*) *The sources should be coherent i.e., they should emit continuous waves of the same wavelength and should have either the same phase or a constant difference in phase.* The phase difference rapidly changes not only in different sources but also in the different part of the same source. The intensity depends upon the amplitude and the phase-difference. Hence in order to have constant intensity, there should be a constant phase-difference or the waves should be in the same phase.

(*ii*) The *source should emit waves of the same wavelength and periodic time.* When the periodic time of the two waves in different, the intensity of light at a point will alternate between maximum and minimum and thus will be continuously changing. The resultant effect, which is due to a large number of waves is, therefore, not constant and the phenomenon will be similar to be phenomenon of beats, as in sound.

23.6 ANALYTICAL TREATMENT OF INTERFERENCE (ON THE BASIS OF WAVE THEORY)

Consider a narrow slit *S* illuminated with monochromatic source emitting wave of wave length λ. Let *A* and *B* be two parallel slits lying very close to each other and equidistant from the source *S*. The waves arriving at *A* and *B* will be in phase. On emergence from the slits *A* and *B*, the waves proceed as if they have started from *A* and *B*. To find the intensity at any point *P* on the screen *XY* placed parallel to *A* and *B*, let the general equation of the wave reaching *P* from *A* be.

$$y_1 = a \sin \frac{2\pi}{\lambda} vt$$

where y_1 is the displacement of particle from its mean position at any time *t*, *v* the velocity of propagation of the wave of wave length λ and *a* the amplitude.

If y_2 is the displacement of the wave reaching *P* from *B* and *x* the path-difference with respect to the first wave from *A*, then general equation of the second wave is given by

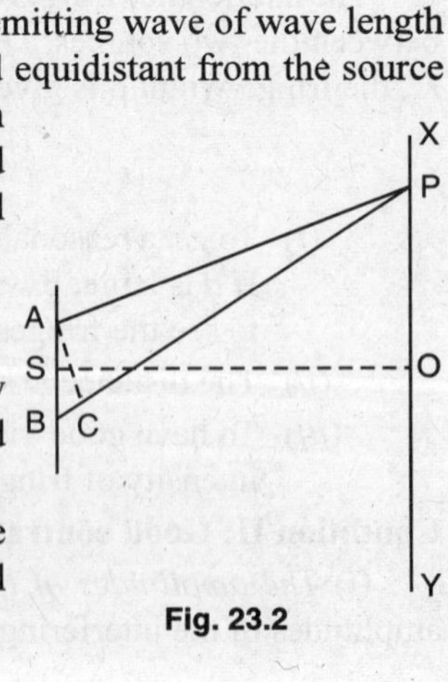

Fig. 23.2

$$y_2 = a \sin \frac{2\pi}{\lambda}(vt + x)$$

When of two waves superimpose, the resultant displacement Y is given by

$$Y = y_1 + y_2 = a \sin \frac{2\pi}{\lambda} vt + a \sin \frac{2\pi}{\lambda}(vt + x)$$

Using the relation $\sin A + \sin B = 2 \sin\ 2 \sin \frac{A+B}{2} \cos \frac{A-B}{2}$ to simplify, we get

$$Y = 2a \cos\left(\frac{2\pi}{\lambda}\frac{x}{2}\right) \sin \frac{2\pi}{\lambda}\left(vt + \frac{x}{2}\right)$$

This is the equation of a simple harmonic vibration of amplitude

$$A = 2a \cos\left(\frac{2\pi}{\lambda}.\frac{x}{2}\right) = 2a \cos\left(\frac{\pi x}{\lambda}\right) \qquad \text{... } (i)$$

For the amplitude to be *minimum*

$$\cos \frac{\pi x}{\lambda} = 0 \quad \text{or} \quad \frac{\pi x}{\lambda} = \frac{\pi}{2}, \frac{3\pi}{2}, \ldots (2n+1)\frac{\pi}{2}$$

or $$x = (2n+1)\frac{\lambda}{2} \qquad \text{... (23.6)}$$

This shows that the intensity is **minimum** *when the path difference between the two wave trains is equal to an* **odd** multiple of half a wavelength.

For the amplitude to be maximum

$$\cos \frac{\pi x}{\lambda} = 1 \text{ or } \frac{\pi x}{\lambda} = 0, \pi, 2\pi \ldots n\pi \qquad \text{... (23.7)}$$

or $$x = n\lambda$$

Thus the intensity is **maximum** *if the path-difference between the two wave trains is an even multiple of half a wavelength.*

23.7 INTERFERENCE IN TERMS OF LAW OF CONSERVATION OF ENERGY

The phenomenon of interference can also be explained with relation to law of conservation of energy.

It has been proved in article 23.6 **Eq.** (*i*) that the amplitude of the resultant wave due to two coherent sources A and B reaching a point P on the screen (Fig. 23.2) is given by

$$A = 2\, a \cos\left(\frac{\pi x}{\lambda}\right)$$

The equation for the amplitude of the resultant wave can also be written as

$$A = 2\, a \cos \delta/2$$

where δ is the phase-difference between the two waves reaching P at any instant, as the phase-difference for a path-difference λ is 2π and hence phase-difference for a path difference x is

$$\delta = \frac{2\pi}{\lambda} x = \frac{2\pi}{\lambda} \times \text{path-difference}$$

$\therefore$ $$\text{Intensity } I = A^2 = 4\, a^2 \cos^2 \delta/2$$

Hence the intensity is proportional to $\cos^2 \delta/2$

(*i*) When the phase-difference $\delta = 0, \pm 2\pi, \pm 2\,(2\pi), \ldots \pm n\,(2\pi)$

or Path-difference $x = 0, \lambda, 2\lambda, \ldots n\lambda$, the intensity

$$I = 4\, a^2$$

Hence intensity is **maximum** when the phase-difference is a whole number multiple of 2π or path-difference is a whole number multiple of wavelength.

(*ii*) When the phase-difference $\delta = \pm\pi, \pm 3\pi \ldots \pm (2n+1)\pi$

or Path-difference $x = \lambda/2, 3\lambda/2, 5\lambda/2 \ldots (2n+1)\lambda/2$

$$I = 0$$

Hence intensity is **minimum** when the phase-difference is an odd multiple of π or path- difference is an odd multiple of half a wavelength.

If the values of δ are marked along the X-axis and those of intensity I along the Y- axis through $\delta = 0$, then the graph is, as shown in Fig. 23.3.

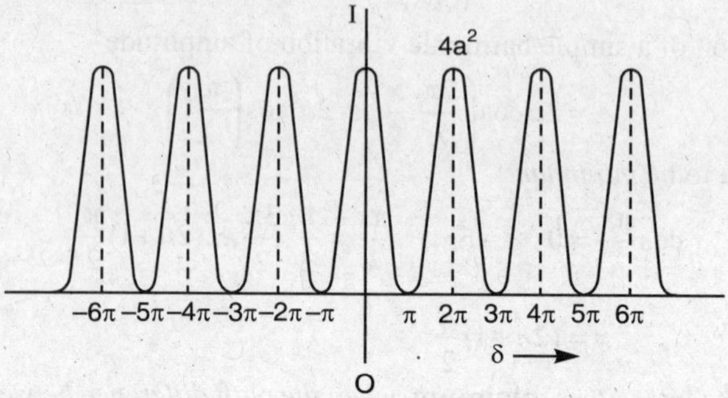

Fig. 23.3

Thus we see that intensity varies between 0 and $4a^2$ according as the phase difference δ is an odd or even multiple of π. According to the law of conservation of energy, the energy cannot be destroyed, but is transferred from points of minimum intensity to points of maximum intensity. In the absence of interference, the intensity due to the two waves should be $a^2 + a^2 = 2a^2$, but at bright points, due to interference it is $4a^2$ and at minimum intensity points it is zero. Thus the energy which apparently disappears at minima has actually been transferred to the maxima where the intensity is greater than that produced by the two beams acting separately. Hence the formation of interference fringes is in accordance with the law of conservation of energy.

23.8 YOUNG'S EXPERIMENT

Young's double slit experiment. Let A and B be two narrow slit illuminated by monochromatic light from the source S so that the light vibrations issuing from the two slits are in phase and they behave as two coherent sources. Let the distance between A and B be d and the distance of the screen XY placed perpendicular to SO be D. The point O is equidistant from A and B, hence the waves originating from A and B will reinforce each other on reaching O. Thus O will be the centre of a bright fringe (Fig. 23.4).

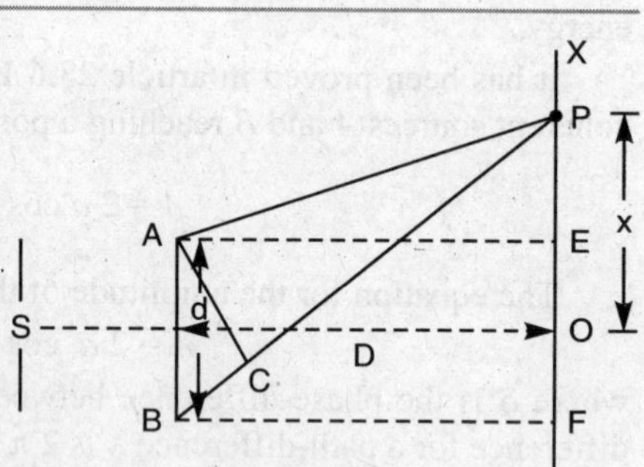

Fig. 23.4

The illumination at any other point P, distant x from O on the screen can be calculated by finding the path-difference (BP-AP). Join AP and BP and draw AC perpendicular on BP, then

$$BP^2 = BF^2 + FP^2 = D^2 + \left(x + \frac{d}{2}\right)^2$$

Also
$$AP^2 = AE^2 + EP^2 = D^2 + \left(x - \frac{d}{2}\right)^2$$

$$\therefore \quad BP^2 - AP^2 = \left[D^2 + \left(x+\frac{d}{2}\right)^2\right] - \left[D^2 + \left(x-\frac{d}{2}\right)^2\right]$$

$$= \left(x+\frac{d}{2}\right)^2 - \left(x-\frac{d}{2}\right)^2 = 2dx$$

or $$(BP + AP)(BP - AP) = 2dx$$

As P lies very closed to O, hence to a first approximation

$$BP = AP = D$$

$\therefore$ Path-difference $(BP–AP) = BC = \dfrac{2xd}{2D} = \dfrac{d}{D}x$

Condition for dark fringe. For P to lie at the centre of a dark fringe

Path-difference $$x\frac{d}{D} = (2n+1)\frac{\lambda}{2}$$

or $$x = \frac{D}{d}(2n+1)\frac{\lambda}{2}$$

Condition for bright fringe. For P to lie at the entre of a bright fringe

Path-difference $$x\frac{d}{D} = n\lambda$$

or $$x = \frac{D}{d}n\lambda$$

Thickness of bright fringe. The distance between two consecutive **bright** fringes β, is given by

$$\beta = x_n - x_{n-1} \qquad \text{... } (i)$$

where x_n is the distance of nth bright fringe and x_{n-1} the distance of $(n–1)$th bright fringe.

$$\therefore \quad \beta = \frac{D}{d}n\lambda - \frac{D}{d}(n-1)\lambda = \frac{D}{d}\lambda \qquad \text{... (23.10)}$$

Thickness of dark fringe. The distance between two consecutive dark fringes

$$\beta = \frac{D}{d}(2n+1)\frac{\lambda}{2} - \{2(n-1)+1\}\frac{\lambda}{2} = \frac{D}{d}\lambda \qquad \text{... (23.11)}$$

This shows that the distance between two consecutive dark or two consecutive bright fringes is equal. This is known as the **fringe-width.**

This proves that in this case *dark and bright bands are of equal width.*

If D and d are constants, then the fringe-width

$$\beta \propto \lambda$$

Hence fringes produced by light of shorter wavelengths will be narrow as compared to those produced by longer wavelengths.

Experiment in water medium

When Young's double slit experiment is carried out in air and then the apparatus is immersed in water the wave length of monochromatic light as well as the fringe width will decrease. The refractive index of water $\mu = 1.33$. The wave length of monochromatic light will, therefore, decrease from λ to $\lambda/\mu = \lambda/1.33$. The fringe width β will also decrease from $\beta = \frac{D}{d}\lambda$ to $\frac{D\lambda}{d\mu} = \frac{D}{d}\frac{\lambda}{1.33}$.

Similarly, if the medium in which the experiment is carried out a replaced by another optically denser medium the wave length as well as the fringe width will decrease.

Example 23.3 *Two narrow and parallel slits 0.1 cm apart are illuminated with a monochromatic light of wave length 589.3 nm. The interference pattern is observed at a distance of 25 cm from the slits. Calculate the fringe width.* **(Pbi. 2008)**

Solution Here $d = 0.1$ cm $= 10^{-3}$ m; $D = 25$ cm $= 25 \times 10^{-2}$ m

$$\lambda = 589.3 \text{ nm} = 589.3 \times 10^{-9} \text{ m}$$

Now fringe width $\beta = \frac{D}{d}\lambda = \frac{25\times10^{-2}}{10^{-3}}\times589.3\times10^{-9}$

$$= 14732.5 \times 10^{-8} \text{ m}$$

$$= 0.147 \times 10^{-3} \text{ m} = 0.147 \text{ mm}$$

Example 23.4 *Distance between two slits is 0.1 mm and the width of the fringes formed on the screen is 5 mm. If the distance between the screen and the slit is one metre calculate the wave length of light used.* **(P.U. 2001)**

Solution Here $d = 0.1$ mm $= 10^{-4}$m; $D = 1$ metre; $\beta = 5$ mm $= 5\times 10^{-3}$m

$\therefore$ Wave length λ is give by $\beta = \lambda\frac{D}{d}$

or $$\lambda = \beta\frac{d}{D} = \frac{5\times10^{-3}\times10^{-4}}{1} = 5\times10^{-7}\text{ m} = 5000\times10^{-10}\text{m}$$

$$= 5000 \text{ A°}.$$

23.9 FRESNEL'S BIPRISM

It is made up of two acute angled prisms with their bases in contact, forming a single obtuse angled prism. In actual practice the prism is ground from a single optically true plate. The acute angle α on both sides is of the order of 30' or one degree. The prism is placed with its refracting edge parallel to the line joining the source and the point O on the screen placed perpendicular to the plane of paper.

In this way, line SO is perpendicular to the base of the prism.

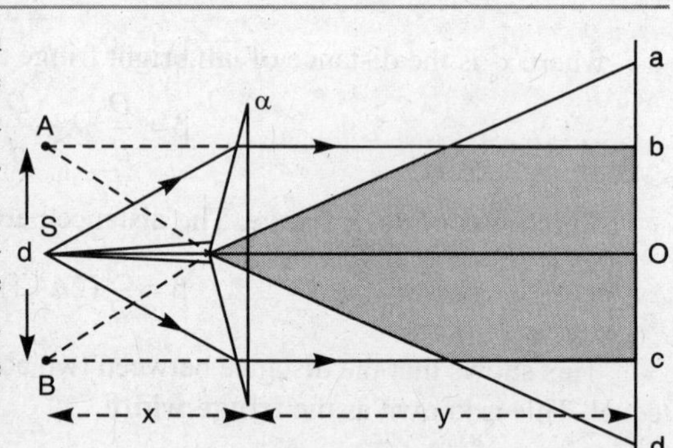

Fig. 23.5

Production of two coherent sources and interference fringes. When light from the source S falls on the lower portion of the prism, after refraction it appears to come from the virtual image B. Similarly light from the source S falling on the upper portion, after refraction appears to come from the virtual images A. Hence A and B act as two coherent sources. The distance x between the biprism and the source S is to adjusted that the virtual images A and B lie very close to each other. Interference fringes of equal width are produced between b and c on the screen placed at a large distance y from the biprism. To observe the fringes, screen is replaced by a micrometer eyepiece. If d is the distance between the two virtual sources A and B, D is the distance between the source and the screen and β is the fringe-width, then

$$\beta = \frac{D}{d}\lambda \text{ or } \lambda = \beta\frac{d}{D} = \frac{\beta d}{x+y} \qquad \text{... (23.12)}$$

If d, D and β are known, the wavelength of light can be determined.]

Proof. This equation has been proved in previous article 23.8 (see relation 23.10 and 23.11).

Experimental arrangement. *Adjustment.* A narrow adjustable slit S, biprism, and a micrometer eye-piece E are arranged in uprights on the bed of an optical bench in the same straight line and at the same height. The slit is made vertical and parallel to the edge of the biprism by rotating it in its own plane and is illuminated with monochromatic source of light the wavelength of which is to be determined. The biprism is moved parallel to the axis of the optical bench till on looking, through it along the axis of bench, two equally bright vertical sources A and B are seen. The eye-piece is moved at right angle to the length of the bench till the overlapping region is brought into field of view of the micrometer eye-piece. The slit is made narrow till the fringes appear in the focal plane F of the eye-piece. If the fringes are not clear, the biprism is rotated in its own plane till the edge of the biprism becomes exactly parallel to the slit and the fringes become very clear.

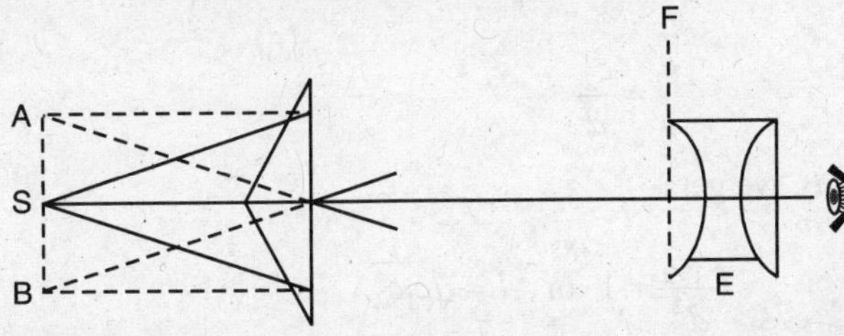

Fig. 23.6

(*i*) **Determination of fringe-width** β**.** To determine fringe-width β, the fringes are first obtained in the field of view of the micrometer eye-piece. The vertical cross-wire is made to coincide with the centre of a particular bright fringe. Considering this as zero position, the position of the eye-piece is read on the scale. The micrometer screw is moved side-ways in the direction of the arrow-head and the number of fringe that pass across the cross wire is counted, as shown in Fig. 23.7. The position of the eye-piece is again read on the scale, then

Fig. 23.7

$$\beta = \frac{\text{Total distance moved}}{\text{Number of fringes moved across the crosswire}}$$

(*ii*) **Determination of** D**.** The distance D between the slit and the focal plane of the eye-piece is found out by noting their positions on the scale of the optical bench. This gives uncorrected distance. The correction to be applied is found out by adjusting a knitting needle between the slit and the eye-piece so that its one end touches the slit and the other end is clearly in focus when seen thought the micrometer eye-piece. The correction to be applied is found in the usual way.

(*iii*) **To determine** d**.** To determine d the distance between the two virtual sources A and B, a convex lens whose focal length is less than one fourth of the distance between the slit and the eye piece *i.e.*, D is mounted between the biprism and the eye-piece without disturbing their positions. The position of the lens is adjusted, as shown by complete lines till a sharp pair of the slits is obtained in the field of view of the eye-piece. The distance between these two images is measured. Let this be denoted by c_1

It u is the distance of slit and v that of eye-piece from the lens, then

$$\text{Magnification} \qquad \frac{v}{u} = \frac{c_1}{d} \qquad \text{... (}i\text{)}$$

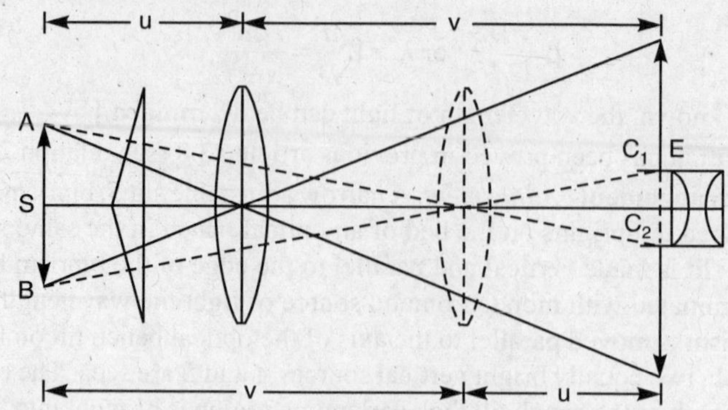

Fig. 23.8

The lens is now moved to the position shown by dotted lines till again a pair of images of the slit is seen in sharp focus. The distance between these two sharp images is again measured. Let this be c_2, then

$$\text{Magnification} \quad \frac{u}{v} = \frac{c_2}{d} \qquad \text{... } (ii)$$

Multiplying (*i*) and (*ii*), we get

$$\frac{c_1 c_2}{d^2} = 1 \text{ or } d = \sqrt{c_1 c_2} \qquad \text{... (23.13)}$$

Determination of wavelength. As β, d and D are known, the wavelength λ can be determined, using the relation $\lambda = \beta \frac{d}{D}$.

Interference Fringes with White Light

When the slit is illuminated with white light, the central fringe at O (Fig. 23.9) is achromatic as all wave-lengths will meet in phase and will produce constructive interference. The white light is a mixture of various colours with violet and red at the two ends of the spectrum in the visible region. The wavelength of violet light is about 4×10^{-7} m and that of red light about 7×10^{-7}m.

As we move on either side of O, the path difference gradually increases and will become equal to half the wavelength of the components having the smallest wavelength *i.e.,* violet.

For example if the point P is such that

$$BP - AP = \frac{\lambda_v}{2} = 2 \times 10^{-7}$$

then complete destructive interference will occur only for the violet colour and partial destructive interference will occur for other wavelengths. Consequently we will have a line devoid of violet colour and will appear reddish. The point Q which satisfies.

$$BQ - AQ = 3.5 \times 10^{-7} = \frac{\lambda_r}{2}$$

will be devoid of red colour, but will correspond almost to constructive interference for the violet colour.

This is the position of the first dark fringe for violet. Beyond this point the first minimum for various colours like blue, green, yellow and lastly for red will lie. The

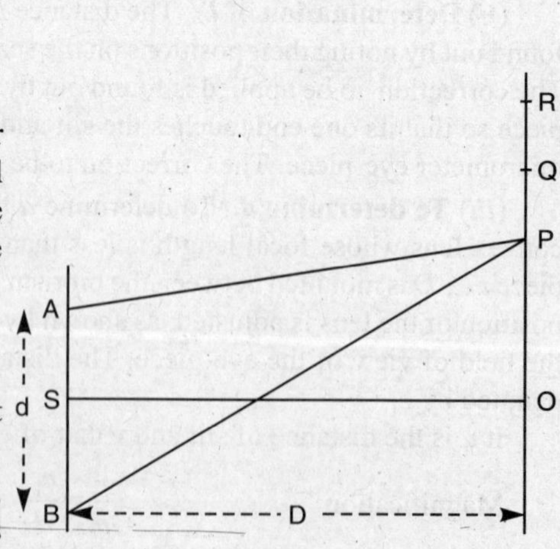

Fig. 23.5

fringe will, therefore, appear violet on its inner edge and red on its outer edge. Thus a system of coloured fringes will be seen. As the fringe width is proportional to λ, the width of red fringes is nearly double that of violet fringes. As a result of this there will be some overlapping of red and violet fringes. This overlapping increases with the order of fringes and finally results in general illumination.

Lateral Shift

If the edge of the biprism is not parallel to the slit, then the edge will not divide the incident wave front into two equal parts and the fringes observed will not be sharp but blurred. The interference pattern, in general will remain unchanged but the pattern as a whole will move sideways. This is known as **lateral shift.** To remove lateral shift the biprism is rotated in its own plane till the edge of the bi-prism becomes exactly parallel to the slit and the fringes become sharp and clear.

Location of zero order fringe

In a biprism experiment with monochromatic light all bright fringes are alike. However, with white light source zero order fringes are white. All other fringes are coloured. Thus by replacing monochromatic source by white light source we can easily place the cross wire in the position of the zero-order fringe. If now the white light is replaced with monochromatic light, the position of the cross wire in the eye-piece will give the zero-order band.

When eyepiece moved away from the source

When the eye-piece is moved away from the source in biprism experiment, the distance D between the source and the eye-piece increases. Now the fringe-width β is given by

$$\beta = \frac{D}{d}\lambda$$

As d and λ remain the same, the fringe width β increases *i.e.* the fringes become wider.

23.10 RELATION FOR d IN REFRACTING ANGLE (α) AND R.I. (μ)

The biprism has a very small refracting angle α of the order of less than 1°. The deviation of ray produced by such a prism is given by $(\mu-1)\,\alpha$, where μ is the refractive index of the material of the prism. If θ is the total deviation produced in a ray which strikes the centre of the biprism as shown in Fig. 23.10 then

$$\frac{\theta}{2} = (\mu - 1)\alpha$$

Now if x is the distance of the biprism from the source, then,

$$d = x\,\theta = 2\,x\,(\mu - 1)\,\alpha \quad \text{... (23.14)}$$

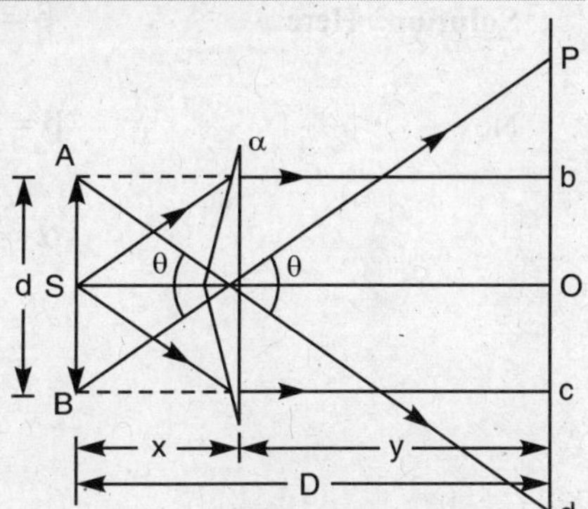

Fig. 23.10

Example 23.5 *The angle of refraction of a bi-prism is 0.5° and its refractive index 1.5. It is kept 9 cm from source of light. Calculate the distance between two coherent sources obtained.*

Solution. Here distance of the biprism from the source $x = 9$ cm

Angle of prism $\alpha = 0.5° = 0.5 \times \frac{\pi}{180}$ radian, $\mu = 1.5$

$\therefore$ Distance between the two coherent sources $d = 2x\,(\mu - 1)\,\alpha$

$$= 2 \times 9\,(1.5 - 1) \times 0.5 \times \frac{\pi}{180} = 0.0786 \text{ cm}$$

$$= 0.786 \text{ mm}$$

Example 23.6 *Calculate the separation between coherent sources formed by a bi-prism whose inclined faces make angles of 2° with the base. The slit source being 10 cm away from the bi-prism and refractive index of the material of the bi-prism being 1.5* **(K.U. 2001)**

Solution $d = 2x\ (\mu - 1)\alpha = 2 \times 10 \times 0.5 \times 2 \times \frac{\pi}{180} = 0.349$ cm

$= 3.49$ mm

Example 23.7 *The inclined faces of a biprism of refractive index 1.50 make angle of 2° with the base. A slit illuminated by a monochromatic light is placed at a distance of 10 cm from the biprism. If the distance between the two dark fringes observed at a distance of one metre from the biprism is 0.18 mm, find the wavelength of light used.*

Solution. Here $\alpha = 2° = \frac{\pi}{90}$ radian, $\mu = 1.5$

$$x = 10 \text{ cm} = 0.1 \text{ m}, y = 1.0 \text{ m}, \beta = 0.18 \times 10^{-3} \text{m}$$

$$D = x + y = 0.1 + 1.0 = 1.1 \text{ m}$$

Now $$d = 2x(\mu - 1)\alpha = \frac{2 \times 0.1 \times 0.5 \times \pi}{90} = 0.0035 \text{ m}$$

$\therefore$ $$\lambda = \frac{\beta \times d}{D} = \frac{0.18 \times 10^{-3} \times 0.0035}{1.1}$$

$$= 5727 \times 10^{-10} \text{ m} = 5727 \text{ Å}$$

Example 23.8 *In a typical biprism experiment, the fringe width is 10^{-3} m for a wavelength of 5893 Å. If $\frac{y}{x} = 20$ where y is the distance between the biprism and screen and x between the slit and the biprism, calculate the refracting angle α of the biprism whose μ = 1.5.*

Solution. Here $$\beta = 10^{-3} \text{m}, \frac{y}{x} = 20;\ \lambda = 5893 \times 10^{-10} \text{m};\ \alpha = ?; \mu = 1.5$$

Now $$\beta = \frac{\lambda \times D}{d} = \frac{\lambda(x+y)}{2x(\mu-1)\alpha} = \frac{\lambda}{2(\mu-1)\alpha}\left(1 + \frac{y}{x}\right)$$

$$\alpha = \frac{\lambda}{2(\mu-1)\beta}\left(1 + \frac{y}{x}\right)$$

$$= \frac{5893 \times 10^{-10}}{2 \times 0.5 \times 10^{-3}}(1 + 20) = 0.0123753 \text{ radian}$$

$$= \frac{0.0123753 \times 180}{\pi} = 0.709°$$

Example 23.9 *If a biprism of angle 1° and refractive index 1.5 is at 40 cm from the slit, find fringe width at 60 cm from it for sodium light of wave length 5893 Å.* **(M.D.U. 2003)**

Solution. Here $\mu = 1.5$, $\lambda = 5893$ Å $= 5893 \times 10^{-10}$ m

Distance of biprism from source $x = 40$ cm $= 0.4$ m

Distance of biprism from screen $y = 60$ cm $= 0.6$ m

$\therefore$ $$D = x + y = 40 + 60 = 100 \text{ cm} = 1 \text{ m}$$

Angle of biprism $= 1° = \frac{1 \times \pi}{180} = \frac{\pi}{180}$ radian

Distance between two coherent source $d = 2\,x\,(\mu-1)\alpha$

$\therefore$ Fringe width $$\beta = \frac{\lambda D}{d} = \frac{\lambda D}{2x(\mu-1)\alpha} = \frac{5893 \times 10^{-10} \times 1 \times 180}{2 \times 0.4(1.5-1) \times \pi}$$

$$= \frac{5893\times10^{-10}}{6.98\times10^{-3}}\text{m} = \frac{.5893}{6.98}\times10^{-3}\text{m}$$

$$= 0.0844 \text{ mm}$$

Example 23.10 *The inclined faces of a biprism (μ =1.5) make angles of 1° with the base of the prism. The slit is 10 cm from the biprism and is illuminated by light of wave length 5900 Å. Find the fringe width observed at a distance of one metre from the biprism.*

Solution. $d = 2x(\mu - 1)\alpha; D = 10\text{cm} + 1\text{m} = 1.1 \text{ m}; \alpha = \frac{\pi}{180} \text{radian}; \lambda = 5900\times10^{-10}\text{m}$

Fringe width $\beta = \frac{\lambda D}{d} = \frac{5900\times10^{-10}\times1.1\times180}{2\times0.1\times0.5\times\pi}$

$$= 0.000372 \text{ m} = 0.372 \text{ mm}$$

Example 23.11 *The distance between two virtual images of a slit formed by a biprism is 0.3 mm. If fringes of width 0.59 mm are formed on a screen placed at a distance of 30 cm, from the slit calculate the wave length of light used.* ***(Gauhati U. 2000)***

Solution. Distance between two virtual images of slit $d = 0.3 \text{ mm} = 0.3\times10^{-3}$ m

Distance between the slit and the screen $D = 30 \text{ cm} = 30 \times 10^{-2}$ m

Fringe width $\beta = 0.59 \text{ mm} = 0.59 \times 10^{-3}$ m

Now $\beta = \frac{D}{d}\lambda$ or $\lambda = \beta\frac{d}{D} = \frac{0.59\times10^{-3}\times0.3\times10^{-3}}{30\times10^{-2}} = 59\times10^{-8}\text{m}$

$$= 5900 \times 10^{-10} \text{ m}$$

$$= 5900 \text{ Å}$$

Example 23.12 *Interference fringes are produced by Fresnel's biprism in the focal plane of a reading microscope which is 1.0 m from the slit. A lens interposed between the biprism and the microscope gives two images of the slit in two positions. If the images of the slits are 4.05 mm in one position 2.90 mm in the other position and the wavelength of the sodium light is 5893 Å find the distance between the consecutive interference bands.*

Solution. Here $\lambda = 5893 \text{ Å} = 5893\times10^{-10}\text{m}, D = 1.0\text{m}$

$c_1 = 4.05 \text{ mm} = 4.05\times10^{-3}\text{m}, c_2 = 2.9 \text{ mm} = 2.9\times10^{-3}\text{m}$

$$d = \sqrt{c_1 c_2} = \sqrt{4.05\times2.9}\times10^{-3} = 3.427\times10^{-3}\text{m}$$

Now $\beta = \frac{D}{d}\lambda = \frac{1.0\times5893\times10^{-10}}{3.427\times10^{-3}} = 0.000172 \text{ m}$

$$= 0.172 \times 10^{-3} \text{ m}$$

$$= 0.172 \text{ mm}$$

Example 23.13 *What is the separation between the slit of Young's double slit experiment that gives second order maxima at a distance of 5 mm from the central maxima. The screen is at a distance of 2m from the slit and wavelength of light is 500 n.m.* ***(Pbi. U. 2003)***

Solution. Since second order maxima is formed at a distance of 5 mm, fringe width $\beta = \frac{5}{2} = 2.5 \text{ mm} = 2.5\times10^{-3}\text{m}$

Distance between the screen and the slit $D = 2$m; wavelength of light $\lambda = 500$ nm $= 500 \times 10^{-9}$ m separation between the slits $d = ?$

Now $$\beta = \frac{\lambda D}{d} \quad \therefore \; d = \frac{\lambda D}{\beta} = \frac{500 \times 10^{-9} \times 2}{2.5 \times 10^{-3}} = 0.4 \times 10^{-3} \text{ m} = 0.4 \text{ mm}$$

23.11 OPTICAL PATH RETARDATION

Lateral shift of fringes. When a things glass plate of refractive index μ is placed in the path of light from one of the two sources as shown in Fig. 23.11. The central bright fringe is shifted to some other point due to the change in the path difference. If the shift can be measured then the wave length of light used or thickness of the plate or the refractive index of the material can be found out if two out of three quantities are known. Suppose a thin glass place of thickness t and refractive index μ is placed in the path AP. Since the velocity of light in air and in the medium of the glass plate is different, the path difference changes and hence the central fringe will shift to some other position.

If c_1 is the velocity of light in air and c_2 that in the medium of the glass plate, then in the path AP, the length $(AP - t)$ is travelled in air and the length t in glass. Hence

Time taken by light to travel from A to P $$= \frac{AP - t}{c_1} + \frac{t}{c_2}$$

$$= \frac{AP - t}{c_1} + \frac{\mu t}{c_1} = \frac{AP + (\mu - t)t}{c_1} \qquad \left(\because \mu = \frac{c_1}{c_2}\right)$$

This shows that by the introduction of the glass plate, the effective optical path AP in air has been increased by $((\mu - 1)t$

$\therefore$ Path difference $= BP - \{AP + (\mu - 1)\, t\}$

$= BC - (\mu - 1)t$

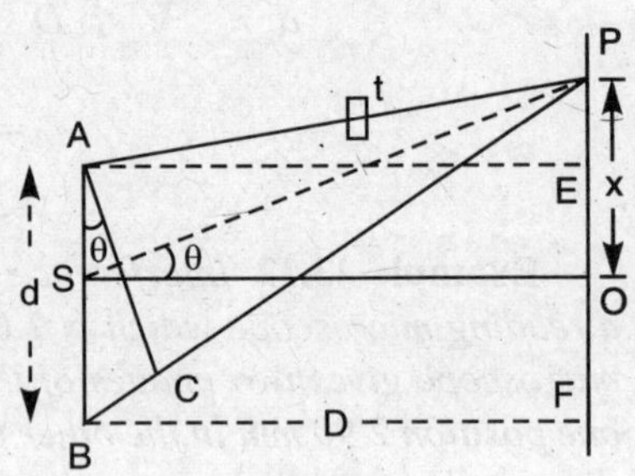

Fig. 23.11

Now in $\Delta\, ABC$, $\sin\theta = \frac{BC}{AB} = \frac{BC}{d}$

where d is the distance between the two virtual images of the slit

and in $\Delta\, OSP$ $\quad \sin\theta = \frac{OP}{OS} = \frac{x}{D}$

Hence $$\frac{BC}{d} = \frac{x}{D} \text{ or } BC = x\frac{d}{D}$$

and path difference $= x\frac{d}{D} - (\mu - 1)t$

The point P will be at the centre of nth bright fringe if path difference is equal to $n\lambda$ i.e.

$$\frac{xd}{D} - (\mu - 1)t = n\lambda$$

or $$x = \frac{D}{d}[n\lambda + (\mu - 1)t] \qquad \text{... (23.15)}$$

where x is the distance of the point P at which the nth fringe will form when the glass plate is placed in the path of one of the interfering beams.

Shift of central bright fringe. The distance x_0 where the central bright fringe is formed when the glass place is placed in the path AP, is given when $n = 0$

$$\therefore \quad x_0 = \frac{D}{d}(\mu - 1)t \qquad \text{... (23.16)}$$

Since μ is greater than one, therefore, x_0 is positive. This shows that the central bright fringe will move upward when a thin glass plate is placed in the path $A.P.$

Similarly, the first, second and nth bright fringe will be displaced upward through a distance $\frac{D}{d}(\mu-1)t$ *i.e.* there will be a uniform upward displacement of all the fringes by a distance $\frac{D}{d}(\mu-1)t$.

Measurement of thickness. In relation (23.16) if we measure x_0 the shift of the central bright fringe as explained later, D, d and μ are given, the thickness of the thin transparent plate can be calculated.

Determination of refractive index. If the thickness t is given μ can be calculated from the same relation.

Effect on fringe width. The position of the central bright fringe is given by relation (23.15) by putting $n = 0$

$$\therefore \qquad x_0 = \frac{D}{d}(\mu-1)t$$

For $n = 1$ the position x_1 of the first bright fringe can be found out, Thus

$$x_1 = \frac{D}{d}(\lambda+(\mu-1)t)$$

$\therefore$ The fringe-width

$$\beta = \frac{D}{d}(\lambda+(\mu-1)t) - \left(\frac{D}{d}(\mu-1)t\right) = \frac{D}{d}\lambda \qquad \text{... (23.16)}$$

which is the same for all fringes. Hence the introduction of the glass plate causes no change in the width of the fringes.

Determination of wavelength From relation (23.16), we have

$$\beta = \frac{D}{d}\lambda$$

or

$$\frac{D}{d} = \frac{\beta}{\lambda}$$

Substituting the value of $\frac{D}{d}$ in (23.16) we have

$$x_0 = \frac{\beta}{\lambda}(\mu-1)t \qquad \text{... (23.17)}$$

If t, x_0, β and μ are known, the wavelength λ can be determined from the relation (23.17).

The fringe-width β and the displacement x_0 of the central bright fringe are determined experimentally.

Use of white light to measure shift of central bright fringe. When monochromatic light is used, fringes observed are all alike and hence it is not possible to measure the shift x_0 of the central bright fringe.

This is usually achieved by replacing the monochromatic light with white light. With white light, except the central fringe, all other fringes are coloured, hence the position to which the central fringe is shifted can be located. The shift is then, measured accurately by using monochromatic light.

Example 23.14 *Fringes are produced by monochromatic light of wavelength 5.45×10^{-7} m. A thin plate of glass of refractive index 1.50 is then placed normally in the path of one of the interfering beams and the central bright band of the fringe system is found to move into the position previously occupied by the third bright band from the centre, Calculate the thickness of the glass plate.*

(Nagpur U. 2010; Kerala U. 2001)

Solution Here $\lambda = 5.45 \times 10^{-7}$ m; $\mu = 1.50$; t =?

$x_0 = 3\beta$ where β is the fringe-width

Now $$x_0 = \frac{\beta}{\lambda}(\mu - 1)t$$

or $$3\beta = \frac{\beta}{\lambda}(\mu - 1)t$$

$\therefore$ $$t = \frac{3\lambda}{\mu - 1} = \frac{3 \times 5.45 \times 10^{-7}}{0.5} = 3.27 \times 10^{-6} \text{ m}$$

Example 23.15 *Determine the thickness of a glass plate which when introduced in the path of either beam shifts interference pattern through one fringe width.* $\mu = 1.5$ $\lambda = 6000$ Å **(P.U. 2002)**

Solution $$x_0 = \frac{\beta}{\lambda}(\mu - 1)t = 1\beta$$

$\therefore$ $$t = \frac{\lambda}{\mu - 1} = \frac{6000 \times 10^{-10}}{(1.5 - 1)} = 1.2 \times 10^{-6} \text{ m}$$

Example 23.16 *When a thin sheet of transparent material of thickness* 7.2×10^{-6} *m is introduced in the path of one of the interfering beams, the central fringe shifts to a position occupied by the sixth bright fringe. If* $\lambda = 6 \times 10^{-7}$ *m, find the refractive index of the sheet.*

Solution. Here $\lambda = 6 \times 10^{-7}$ m , $t = 7.2 \times 10^{-6}$ m, μ = ?

Distance through which central fringe shifts $x_0 = 6\beta$

Now $$x_0 = \frac{\beta}{\lambda}(\mu - 1)t$$

or $$6\beta = \frac{\beta}{\lambda}(\mu - 1)t$$

or $$\mu = \frac{6\lambda}{t} + 1 = \frac{6 \times 6 \times 10^{-7}}{7{,}2 \times 10^{-6}} + 1 = 1.5$$

Example 23.17 *In a biprism experiment fringes were first observed with sodium light of wavelength 5890 Å and fringe width was measured to be 0.342 mm. Sodium light was then replaced with white light and central fringe was located. On introducing a thin glass sheet in half of the beam, the central fringe was shifted by 2.143 mm. Calculate the thickness of the glass sheet if the refractive index of glass is 1.542.* **(B.H.U. 2001)**

Solution Here $\lambda = 5890 \times 10^{-10}$ m, $\mu = 1.542$

$x_0 = 2.143$ mm $\qquad \beta = 0.342$ mm

If n is the number of fringes contained in x_0, then

$$n = \frac{2.143}{0.342}$$

Now $(\mu - 1)\, t = n\lambda$

$\therefore$ $$t = \frac{n\lambda}{\mu - 1} = \frac{2.143 \times 5890 \times 10^{-10}}{0.342 \times 0.542}$$

$$= 6.71 \times 10^{-6} \text{ m}$$

23.12 PHASE CHANGE ON REFLECTION

When a ray of light is reflected at the surface of a medium, which is optically denser than the medium through which the ray is travelling, a change of phase equal to π or a path-difference $\lambda/2$ is introduced.

When reflection takes place at the surface of a rarer medium, no change in phase or path-difference takes place.

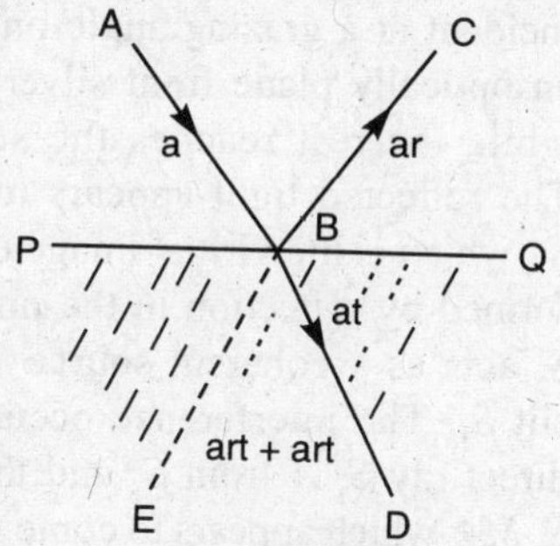

Fig. 23.12

Let PQ be the surface separating the denser medium below it from the rarer medium above it, as shown in Fig 23.12. A ray of light AB of amplitude a incident on this surface is partly reflected along BC and partly refracted into the denser medium along BD. If r is the co-efficient of reflection at the surface of a denser-medium, *i.e.*, the fraction of the incident light which is reflected, then

Amplitude of the reflected ray $BC = ar$.

If t is the co-efficient of transmission from the rarer into the denser medium, i.e. the fraction of the incident light transmitted, then

Amplitude of the refracted ray $BD = at$.

If there is no absorption of light, then $r + t = 1$

If the reflected and the refracted rays are reversed, the resultant should have the same amplitude a as that of the incident ray.

When CB is reversed it is partly reflected along BA with amplitude ar^2 and partly refracted along BE.

The amplitude of the refracted ray along $BE = art$.

Similarly when the ray DB is reversed it is partly refracted along BA and partly reflected along BE. If r' is the co-efficient of reflection at the surface of a rarer medium, then

Amplitude of the reflected ray along $BE = atr'$

If t' is the co-efficient of transmission from the denser into the rater medium, then

Amplitude of the reversed refracted ray along $BA = att'$

The two amplitudes along BA will combine together to produce the original amplitude, *only* if the total amplitude along BE is zero.

$$art + ar't = 0 \quad \text{... (i)}$$

In such a case

$$ar^2 + att' = a \quad \text{... (ii)}$$

Equation's (*i*) and (*ii*) are known as *Stokes relations*

From relation (*i*) we have

$$art = -ar't$$

or

$$r = -r'$$

The negative sign shows that when one ray has a positive displacement, the other has a negative displacement. Hence the two rays, one reflected in reaching a denser medium and the other reflected on reaching a rarer medium, differ in phase by π from each other.

This explains the presence of a central dark spot in Newton's rings and is also responsible for the reversal of the conditions of darkness and brightness produced in the reflected and transmitted systems in colours of thin films and in the fringes produced by Lloyd's single mirror.

23.13 LIOYD'S SINGLE MIRROR

It is a simple arrangement of producing interference fringes, in which one part of the incident wave-front reaches the screen directly, while the other reaches the screen after suffering reflection from an optically plane front silvered mirror MN. The arrangement of apparatus is shown in

Fig. 23.13. Light from a narrow slit S_1, illuminated by monochromatic light is partly incident at a grazing angle on the surface of an optically plane front silvered mirror *MN*, while the rest reaches the screen directly. The reflected light appears to diverge from S_2, which is the virtual image of the source S_1 formed by reflection in the mirror *MN*. Thus S_2 acts as a coherent source with respect to slit S_1. The interference occurs between the direct ray $S_1 A$ from S_1 and the reflected ray $S_1 MA$ which appears to come from the virtual source S_2.

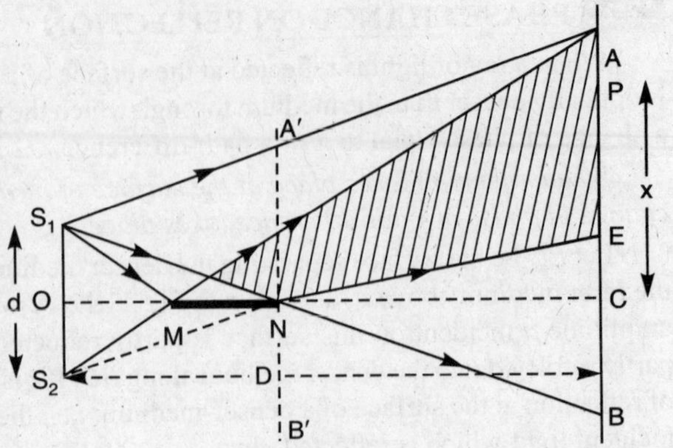

Fig. 23.13.

The light reflected from the extremities of *MN* lies in the region *AE*, thus interference between direct and reflected rays occurs only in this region. The point *C* lying on the right bisector of $S_1 S_2$ gives the positions of the central fringe of zero path-difference. It is not visible, as no reflected light reaches *C*. Not only this, even less than half the interference pattern is visible with the arrangement shown. However, this can be brought into view by displacing the screen to the position *A′NB′*, so that it just touches the end *N* which is equidistant from S_1 and S_2 and passes through the centre of the fringe of zero geometrical path-difference.

The central fringe can also be brought into view by introducing a thin mica plate in the path of the direct ray, which causes, the central fringe to move in the upward direction towards *A*.

Why central fringe is black. It is found that the central fringe is black, because a phase change of π takes place in the ray $S_1 MA$ or $S_1 NE$ on reflection at the surface of a denser medium at *M* or *N*. We can conclude from this that the light from coherent sources S_1 and S_2 differ in phase by π at all times, instead of being in equal phase, as in double slit Fresnel's biprism and double mirror experiments.

Fringe-width. If *P* is a point on the screen, then

Path-difference = $S_2P = S_1P$

If *d* is the distance between the sources S_1 and S_2 and *D* the distance between the centre of the screen and the centre of the line joining S_1 and S_2, then as proved in Young's double slit experiment (article 23.8)

$$S_2P - S_1P = x\frac{d}{D}$$

The point *P* will have **minimum intensity,** if

$$\frac{xd}{D} = n\lambda \quad \text{for } n = 0, 1, 2, 3$$

or
$$x = \frac{D}{d} n\lambda \qquad \text{... (i)}$$

The point *P* will have **maximum intensity,** if

$$\frac{xd}{D} = (2n+1)\frac{\lambda}{2} \quad \text{for } n = 0,1,2,3....$$

or
$$x = \frac{D}{d}(2n+1)\frac{\lambda}{2} \qquad \text{... (ii)}$$

Fringe-width
$$\beta = x_n - x_{n-1}$$

For minimum intensity
$$\beta = \frac{D}{d}\lambda(n-n+1) = \frac{D}{d}\lambda \qquad \text{...(23.18)}$$

For maximum intensity $\beta = \frac{D}{d}(2n+1)\frac{\lambda}{2} - \{2(n-1)+1\}\frac{\lambda}{2}$

$$= \frac{D}{d}\lambda \qquad \text{... (23.19)}$$

This shows that fringes are equally spaced

Achromatic fringes with LIoyd's mirror

In LIoyd's single mirror it is possible to produced achromatic fringes with white light. It is because lateral inversion of the image of the source occurs, as shown in Fig. 23.14. If a spectrum *RV* is formed, its reflected image *V'R'* is produced. If

$$\frac{RR'}{VV'} = \frac{\lambda_r}{\lambda_v}$$

or

$$\frac{\lambda_r}{RR'} = \frac{\lambda_v}{VV'}$$

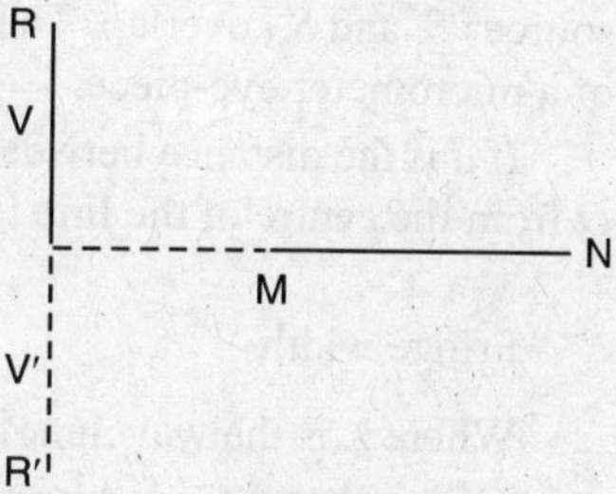

Fig. 23.14

then $\frac{\lambda}{d}$ will be constant for all wavelengths. If a grating spectrum is used as one of the sources, then the fringes produced are achromatic, because in the grating spectrum *dispersion is proportional to the wavelength* and hence $\frac{D\lambda}{d}$ *i.e.*, fringe-width will be constant for all colours. Thus if red light is used, red and black fringes of fringe-width β are produced and if blue light is used, blue and black fringes of the same fringe width are produced. If, therefore, white light is used, black and white achromatic fringes are produced.

Difference between Biprism fringes and LIoyd's single mirror fringes:

(*i*) In Fresnel's biprism a number of fringes on both sides of the central fringe are seen with monochromatic light, whereas in LIoyd's mirror less than half the field is in view.

(*ii*) With monochromatic light, the central fringe in Fesnel's biprism is bright, whereas with LIloyd's mirror it is dark. In the latter case, it is not ordinarily visible but can be located by placing a thin mica sheet in the path of the direct beam.

(*iii*) With white light, Fesnel's biprism produces a very limited number of coloured fringes with central white fringe, whereas LIoyd's mirror produces a large number of black and white fringes, the central fringe being dark.

Example 23.18 *A LIyod's mirror of length 5 cm is illuminated by light of λ = 600 n.m. from a narrow slit 0.1 cm above the plane of mirror. Calculate the fringe width on a screen 120 cm from the slit.*

(Kerala U. 2001)

Solution. The reflected image in a plane mirror is as far behind the mirror as the object is in front of it.

∴ Distance between the source and its virtual image

$$d = 2 \times 0.1 = 0.2 \text{ mm} = 2 \times 10^{-4} \text{ m}$$

Distance between the slit (source) and the screen

$$D = 120 \text{ cm} = 1.2 \text{ m}$$

$$\lambda = 600 \text{ mm} = 600 \times 10^{-9} \text{ m}$$

∴ Fringe width $\beta = \frac{\lambda D}{d} = \frac{600\times10^{-9}\times1.2}{2\times10^{-4}} = 3.6\times10^{-3}\text{m}$

$$= 3.6 \text{ mm.}$$

23.14 FRESNEL'S DOUBLE MIRROR METHOD

Fresnel used two plane mirror OM_1 and OM_2 silvered in the front, placed close together, so that their planes make a very small angle θ with each other, to produce two coherent sources. Light from a narrow slit S placed parallel to the mirror and illuminated by monochromatic light is made to fall on the mirrors. The mirrors OM_1 and OM_2 give rise to two virtual images S_1 and S_2 respectively. If a screen XY is placed, as shown in Fig. 23.15 interference fringes are obtained in the region CD where the two reflected beams reaching A and B, and C and D appear to come respectively from the virtual sources S_1 and S_2, overlap. These fringes are equally spaced and can be observed in the field of view of a micrometer eye-piece.

If d is the distance between the two virtual images of the sources S_1 and S_2 and x the distance of O from the centre of the line joining S_1 and S_2 and y the distance OO' from the screen, then

Fringe width $$\beta = \frac{\lambda D}{d} = \frac{\lambda(x+y)}{d} \quad \text{... }(i)$$

Where λ is the wavelength of monochromatic light

The angle subtended by $S_1 S_2$ at O is 2 θ.

$$\therefore \quad d = x\,2\theta = 2\,x\,\theta$$

Substituting the value of d in (*i*), we have

$$\beta = \frac{(x+y)\lambda}{2x\theta} \quad \text{...(23.20)}$$

As θ is small and y is large, the fringes are seen as separate

$$\therefore \quad \lambda = \frac{2x\theta}{x+y}\beta \quad \text{... (23.21)}$$

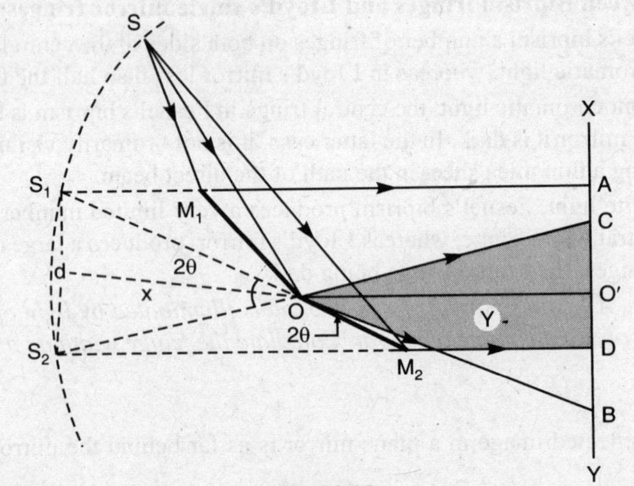

Fig. 23.15

Example 23.19 *A pair of Fresnel's double mirrors, making an angle of 2°, are held at a distance of 10 cm from a slit source emitting light of wavelength 600 Å. Find the fringe-width on a screen distant one metre from the intersection of the mirrors.*

Solution. Here $\theta = 2° = \pi/90$ radian, $\lambda = 6000$ Å $= 6 \times 10^{-7}$ m

$x = 10$ cm $= 0.1$ m, $y = 1$m

Now $$\beta = \frac{(x+y)\lambda}{2x\theta} = \frac{1.1 \times 6 \times 10^{-7} \times 90}{2 \times 0.1 \times \pi}$$

$$= 0.945 \times 10^{-4} \text{ m}$$

Example 23.20 *In interference with white light fringe width of red coloured fringes is double than that of violet coloured. Why?* **(P.U. 2000)**

Solution. The wave length of violet light is about 4000 Å and that of red light is about 8000 Å *i.e.* the wave length of red light is about twice that of violet light. The fringe width is given by $\beta = \lambda \frac{D}{d}$ *i.e.* β is proportional to λ.

Hence thickness of red fringe $\beta_r = \lambda_r \frac{D}{d}$

and thickness of violet fringe $\beta_v = \lambda_v \frac{D}{d}$

$$\therefore \quad \frac{\text{Thickness of red fringe}}{\text{Thickness of violet fringe}} = \frac{\beta_r}{\beta_v} = \frac{\lambda_r}{\lambda_v} = \frac{8000\,\text{Å}}{4000\text{Å}} = 2$$

Hence fringe width of red coloured fringes is double than that of violet coloured.

23.15 RAYLEIGH REFRACTOMETER

Rayleigh devised the arrangement to measure refractive Index of a gas. The principle is explained by Fig. 23.16.

The slits S, S_1 and S_2 make Young's double-slit device, and the two coherent beams pass through separate tubes A and B, in one of which the gas pressure may be varied. Interference fringes are observed in the focal plane of lens L_2. Pressure change in any tube makes fringes shift laterally across the cross-wire of the eyepiece E. If tube length is L, then

$$L.\,\Delta\,\mu = m\,\lambda$$

where m is the number of fringes passing. The refractive index for any pressure can thus be obtained, since $(\mu - 1)$ is proportional to the pressure. It may be noted that the tubes may have to be about a metre long (unlike the very thin sheet t in Fig. 23.11). Since $(\mu-1)$ for gas is of the order of 10^{-4}. This is also known as Rayleigh interferometer.

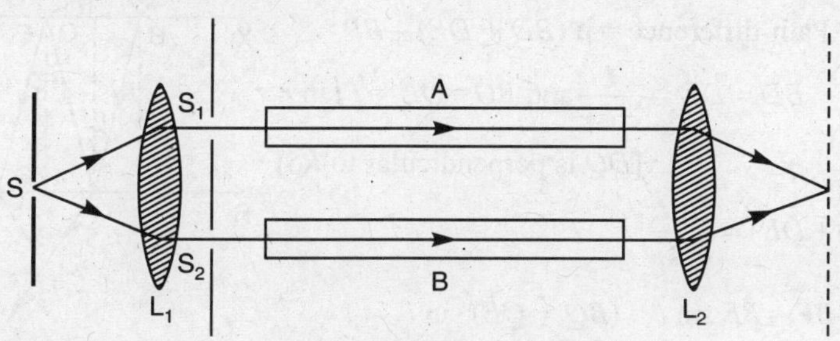

Fig. 23.16. Rayleigh Refractometer

23.16 INTERFERENCE BY DIVISION OF AMPLITUDE

Consider a thin plate *PQRS* of a transparent material. A ray of light *AB* is incident on its surface *PQ*. It is partially reflected along *BC* and partially refracted along *BD*. The refracted ray *BD* is again partially reflected along *DB′* at the surface *RS* and partially refracted along *DE*. The reflected ray *DB′* is again reflected and refracted partially, and the process continues as shown in Fig. 23.17. Finally we get a number of parallel rays *BC*, *B′C′*, *B″C″*... etc in the reflected system and *DE*, *D′E′*, *D″E″* ... etc in the transmitted system. The process of division of amplitude begins with the first reflection at *B* and at every further reflection or refraction the amplitude (and hence the intensity) of the light wave goes on decreasing *i.e.*, *the amplitude goes on suffering further and further division.*

If a lens L_1 is placed in the path of parallel reflected beam and a lens L_2 in the path of parallel transmitted beam, these rays combine to produce interference at I_1 or I_2 respectively. Interference can also be observed without using the lenses. In that case the eye serves the purpose of the lens.

Division of amplitude. *The process of dividing the amplitude of an incoming wave into two or more parts by partial reflection and refraction thereby producing two or more rays forming a parallel beam which can produce interference is known as division of amplitude.*

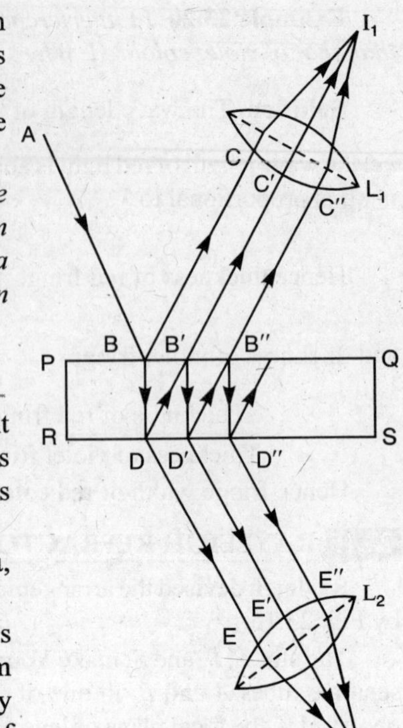

Fig. 23.17

23.17 INTERFERENCE IN THIN FILMS

The production of colours, when thin films of transparent medium like *soap film or oil spread over water surface,* is exposed to an extended source of light, can be explained in terms of interference phenomenon.

Consider a thin film of thickness t and refractive index μ, bounded by two plane surfaces XY and RS.

A ray of light AB incident on the surface XY at an angle i is partly reflected along BC and partly refracted into the medium along BD, making an angle of refraction r. At D it is again partly reflected along DE inside the medium and partly refracted out of the medium along DK parallel to AB, as the medium above and below is the same. Similar reflections and refractions occur at E, G, etc. as shown in Fig. 23.18.

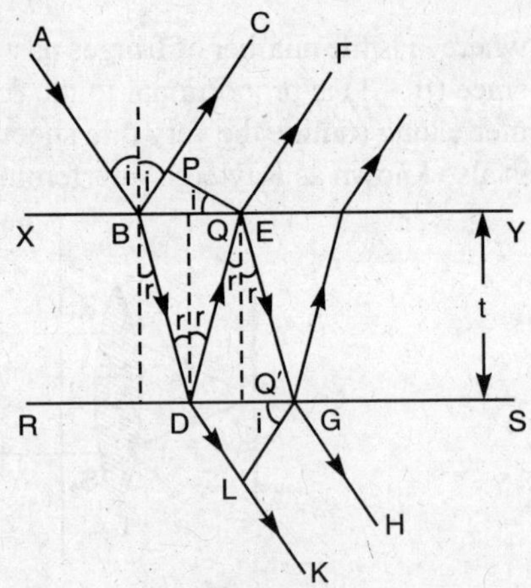

Fig. 23.18

Reflected system. To find the path-difference between the rays BC and EF in the reflected system draw EP perpendicular to BC, then

$$\text{Path-difference} = \mu\,(BD + DE) - BP$$

Now $$BD = DE = \frac{t}{\cos r} \text{ and } BQ = QE = t \tan r$$

... [DQ is perpendicular to RS]

$$\therefore \quad \mu(BD + DE) = \frac{2\mu t}{\cos r}$$

Also $$BP - BE \sin i = (BQ + QE) \sin i$$

$$= 2\,t \tan r.\ \mu \sin r = \frac{2\mu t \sin^2 r}{\cos r}$$

$$(\because \sin r = \mu \sin r)$$

$$\therefore \quad \text{Path difference} = \frac{2\mu t}{\cos r} - \frac{2\mu t \sin^2 r}{\cos r} = \frac{2\mu t(1-\sin^2 r)}{\cos r} = \frac{2\mu t \cos^2 r}{\cos r} \qquad ...(23.23)$$

$$= 2\ \mu t \cos r$$

Monochromatic light. Let us consider a monochromatic light of wave length λ. The path difference between the rays BC and EF in the reflected system = $\mu\ t \cos r$.

Since the ray BC suffers reflection at air-medium interfaces, it undergoes a phase change of π or a path increase of $\lambda/2$, while the ray $BDEF$ does not, as it is reflected at D, the surface of a denser medium. Hence the net path-difference between the two rays BC and $BDEF$

$$= 2\mu t \cos r - \lambda/2$$

(*i*) **Extremely thin film appears black.** When the film is exceedingly thin as compared to the wavelength of light, $2\mu t \cos r$ can be neglected and the net path-difference is $\lambda/2$. *The two rays will produce destructive interference and the film will appear dark.*

(*ii*) **Condition for maxima (bright band).** When the thickness is increased so that $2\mu t \cos r$ cannot be neglected, the film will appear **bright** if the path-difference

$$2\mu t \cos r - \frac{\lambda}{2} = n\lambda$$

or $$2\mu t \cos r = (2n+1)\frac{\lambda}{2} \quad \text{... (i)}$$

where $n = ...\ 0, 1, 2, 3, ...$

(*iii*) **Condition of minima (dark band).** The film will appear dark if the path-difference

$$2\mu t \cos r - \frac{\lambda}{2} = (2n+1)\frac{\lambda}{2}$$

or $$2\mu t \cos r = (n+1)\lambda \quad \text{... (ii)}$$

where $$n = 0, 1, 2, 3 ...$$

Transmitted system. The path difference between two transmitted rays DK and GH is similarly equal to

$$\mu(DE + EG) - DL$$

Now $DE = EG = \dfrac{t}{\cos r}$ and $DQ' = Q'G = t$ an r [EQ' is perpendicular to XY)

$\therefore$ $$\mu(DE + EG) = \frac{2\mu t}{\cos r}$$

Also $$DL = DG \sin i = (DQ' + Q'G) \sin i$$

$$= 2t \tan r \sin i = 2\mu t \tan r \sin r$$ [since $\sin i = \mu \sin r$)

$$= \frac{2\mu t \sin^2 r}{\cos r}$$

$\therefore$ Path difference $$= \frac{2\mu t}{\cos r} - \frac{2\mu t \sin^2 r}{\cos r} = \frac{2\mu t(1-\sin^2 r)}{\cos r} = \frac{2\mu t \cos^2 r}{\cos r}$$

$$= 2\mu t \cos r$$

In this case there is no phase change on reflection at E as it takes place at the surface of a rarer medium.

(*i*) **Extremely thin film appears bright.** When the thickness of the film is very small as compared to the wavelength, so that $2\mu t \cos r$ is negligible, the two rays will be in the same phase and thus will reinforce each other. The film will, therefore, appear bright.

(*ii*) **Condition for maxima (bright-band).** As the thickness of the film increases, the film will appear bright if the path-difference.

$$2\mu t \cos r = n\lambda \text{ where } n = 0, 1, 2, 3... \quad \text{... (iii)}$$

(*iii*) **Condition for minima (dark band)**. The film appear dark, if the path-difference

$$2\mu t \cos r = (2n+1)\frac{\lambda}{2} \quad \text{where } n = 0, 1, 2, 3... \quad \text{... (iv)}$$

Reflected and transmitted systems are complementary. It is clear from the relations (*i*), (*ii*), (*iii*) and (iv) that the conditions of interference in the reflected and transmitted systems for fringes are quite opposite to each other. For a certain path-difference if a bright fringe in produced in the reflected system, a dark fringe will be produced in the transmitted system for the same path-difference. In other words, the part of the film that appears dark in the reflected system will appear bright in the transmitted system and vice-versa.

This shows that the interference pattern in the reflected and transmitted system are complementary to each other.

Necessity of a broad source for observing colours in thin films. In the case of interference in thin films, the narrow source limits the visibility of the film.

To explain this consider a ray *SA* starting from a narrow source *S* (Fig. 23.19 (a). It after suffering reflection and internal reflection will enter the eye, whereas the ray *SB* incident at a different angle after reflection and internal reflection will not reach the eye. Hence only a limited portion of the film is visible.

Thus a point source is not suitable for observing interference of light simultaneously from whole of the film.

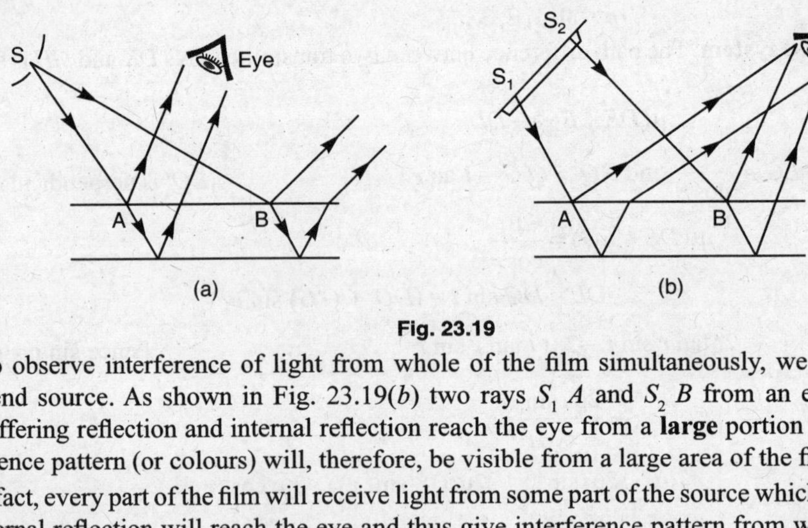

Fig. 23.19

To observe interference of light from whole of the film simultaneously, we have to use an extendend source. As shown in Fig. 23.19(*b*) two rays $S_1\,A$ and $S_2\,B$ from an extended source after suffering reflection and internal reflection reach the eye from a **large** portion of the film. The interference pattern (or colours) will, therefore, be visible from a large area of the film.

In fact, every part of the film will receive light from some part of the source which after reflection and internal reflection will reach the eye and thus give interference pattern from whole of the film simultaneously.

23.18 THIN FILMS WITH WHITE LIGHT

When interference of light takes place in a thin film with monochromatic light alternate bright and dark fringes are obtained both in the reflected as well as in the transmitted system. The path difference between the rays—one reflected at the upper surface of the film and the other refracted into the film and then reflected at the lower surface is equal to 2 $\mu\, t \cos r$ where t is the thickness of the film, μ its refractive index, and r the angle of refraction corresponding to an angle of incidence i.

In **reflected system, bright fringes** are obtained for

$$2\mu t \cos r = (2n+1)\frac{\lambda}{2}$$ and **dark** fringes for

$$2\mu\, t \cos r = (n+1)\lambda$$

In **transmitted** system **bright** fringes are obtained for

$$2\mu t \cos r = n\lambda$$ and **dark** fringes for

$$2\mu t \cos r = (2n+1)\frac{\lambda}{2} \text{ where } n = 0,1,2,3 \ldots$$

Production of colours with white light. When white light is used in place of monochromatic light, the fringes obtained are coloured. It is because the path difference 2 $\mu t \cos r$ depends upon μ, t and r.

(*i*) Even if t and r are kept constant, the path difference will change with μ or wavelength λ of light used. White light is composed of various colours from violet ($\lambda_v = 4000$Å) to red ($\lambda_r = 8000$Å)

In the reflected system the condition for a dark fringe is 2 μ t cos r = $(n + 1)$ λ. When this condition is satisfied by a particular wavelength the *reflected light will not include the wavelength* or the colour corresponding to it. Other colours (or wavelengths) will, however, be present.

Similarly, when a wave length λ satisfies the condition for a bright fringe *i.e.* $2\mu\ t\ \cos r = (2n+1)\frac{\lambda}{2}$, *this particular wave length (or colour) will be present* in the reflected light but other colours will be absent as the wavlengths corresponding to these colours do not satisfy the condition for brightness.

In the transmitted system the conditions for a dark fringe is $2\mu\ t \cos r = (2n+1)\frac{\lambda}{2}$. When this condition is satisfied *this particular wavelength* (or colour) *will be absent* but other colours will be present.

Similarly, the condition for a bright fringe is 2 μ t cos r = $n\lambda$. When this condition is satisfied *this particular wavelength* (*or colour*) *will be present* in the transmitted light but all other colours will be absent.

Hence in reflected as well as in the transmitted system coloured fringes will be obtained and the thin film will show various colours from violet to red. As $\lambda_v < \lambda_r$ the distance between two bright bands will be less for violet than for red rays.

(*ii*) If the thickness of the film varies uniformly, the same order of colour will be observed beginning with the thin end of the film which will appear black. As the path difference varies with thickness of the film, it passes through various colours for the same angle of incidence when seen in white light.

(*iii*) Using a broad source of light, if the angle of incidence changes, the angle of refraction also changes, so that with white light the film assumes various colours when viewed from different directions.

Colours by reflected and transmitted light are complementary. As explained in (*i*) the condition for a particular colour (or λ) to be absent from the reflected system is the same as for same colour (or λ) to be present in the transmitted system.

Similarly the condition for a particular colour (or λ) to be present in the reflected system is the same as for the same colour to be absent from the transmitted system.

Thus the colours in the reflected and transmitted system are complimentary.

Example 23.21 *When a thin soap film of refractive index 1.33 is seen by normally reflected light of sodium of wavelength 5893 Å, it appears black. Find the minimum thickness of the film.*

(Meerut U.2002)

Solution. As the incidence is normal, $i = 0$,

$\therefore$ $r = 0$ and $\cos r = 1$; $\mu = 1.33$; $\lambda = 5893$ Å $= 5893 \times 10^{-10}$ m

For minimum thickness $2\,\mu\,t\cos r = \lambda$

or $$2 \times 1.33 \times t = 5893 \times 10^{-10}$$

$$\therefore \quad t = \frac{5893\times10^{-10}}{2\times1.33} = 2215\times10^{-10} = .2215\times10^{-6}\,\text{m}$$
$$= 0.2215\ \mu\text{m}$$

23.19 NON-REFLECTING FILMS

Non-reflecting glass surfaces can be prepared by depositing a thin layer or film of a transparent material. The refractive index of the material is so chosen that it has an intermediate value between glass and air. The thickness of the film is so chosen that it introduces a path-difference of $\lambda/2$. For example, the refractive index of *magnesium flouride* is 1.38. This value is greater than the refractive index of air and smaller than the refractive index of glass ($\mu = 1.5$). The ray AB suffers reflection at B on the surface of a denser medium and proceeds along BC. A part of it moves along BD and after suffering reflection again at the surface of a denser medium (glass) emerges out along EF. As both reflections at B as well as at D occur at a denser medium, so due to reflection the total phase difference introduced between the rays BC and DE is $\pi + \pi = 2\pi$ which is equivalent to zero phase difference.

Minimum thickness of anti-reflection coating. The optical path-difference for normal incidence will be $2\mu t$, where μ is the refractive index of the material of film and t its thickness. There will be destructive interference if

$$2\mu t = (2n+1)\frac{\lambda}{2} \quad \text{where } n = 0, 1, 2 \ldots$$

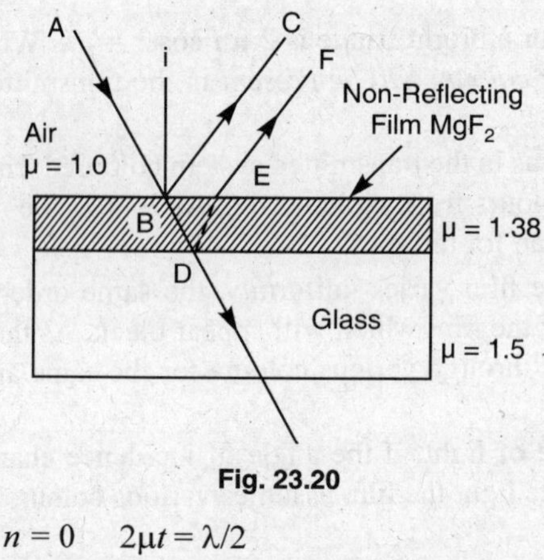

Fig. 23.20

For $\quad n = 0 \quad 2\mu t = \lambda/2$

Hence the minimum thickness of the coating required for no reflection $t = \dfrac{\lambda}{4\mu}$

It means that the optical thickness of the anti-reflection coating should be one quarter wavelength ($\lambda/4$). Such quarter wavelength coating supresses the reflection and cause the light to pass into the transmitted component.

At the centre of the visible spectrum ($\lambda = 55 \times 10^{-5}$ cm) for MgF_2 ($\mu = 1.38$), the minimum thickness of anti-reflection coating is given by

$$t = \frac{\lambda}{4\mu} = \frac{5.5\times10^{-5}}{4\times1.38} = 0.996\times10^{-5}\,\text{cm}$$

It may be noted that some reflection does take place on both the longer and shorter wavelengths and reflected light has a purple colour. By coating the surface of a lens or a prism, the overall reflection can be reduced from 4 to 5 per cent to a fraction less than one per cent.

This method is highly useful in reducing loss of light by reflection in instruments like periscope, which has a number of airglass surfaces.

Refractive index of anti-reflection coating. To find the refractive index of the material of the film to be chosen for making a non-reflecting layer let the refractive index of the transparent slab (say glass) be μ_g, that of the film μ_f and the medium in which the slab is placed (say air) μ_a.

The first reflection takes places at *B* which lies on the surface separating the film from air. Therefore, according to electro-magnetic theory the coefficient of reflection for the electric wave vector

$$r_1 = \frac{1-\mu_f/\mu_a}{1+\mu_f/\mu_a}$$

The second reflection takes place at *D* which lies on the surface separating the glass slab from the film. Therefore the coefficient of reflection for the electric wave vector

$$r_2 = \frac{1-\mu_g/\mu_f}{1+\mu_g/\mu_f}$$

The interference between the rays *BC* and *DEF* will be completely destructive *i.e.* the reflected light-will be completely absent if the amplitudes of the electric wave vector of the two interfering beams are equal *i.e.*

$$r_1 = r_2$$

or $$\frac{\mu_f}{\mu_a} = \frac{\mu_g}{\mu_f}$$

or $$\mu f^2 = \mu_a \mu_g$$

or $$\mu_f = \sqrt{\mu_a \mu_g} \qquad \text{... (23.24)}$$

For air $\mu_a = 1$ $$\therefore \mu_f = \sqrt{\mu_g} \qquad \text{... (23.25)}$$

For glass refractive index $\mu_g = 1.5$, $\mu_f = \sqrt{1.5} = 1.22$. In practice the transparent material having refractive index close to 1.22 is $M_g F_2$ with $\mu = 1.38$.

Example 23.22 *What should be the thickness of a non-reflecting layer to be deposited on glass surface corresponding to wavelength 6000 Å? Refractive index of the layer is 1.35.*

(H.P.U. 2000)

Solution. For minimum thickness of the coating of refractive index μ required for no rflection at the centre of visible spectrum corresponding to the given wavelength λ is

$$t = \frac{\lambda}{4\mu}$$

Here $$\mu = 1.35,\ \lambda = 6000\text{Å} = 6000 \times 10^{-10}\text{m}$$

$$t = \frac{6000 \times 10^{-10}}{4 \times 1.35} = 1.11 \times 10^{-7}\text{ m}$$

Example 23.23 *Find the thickness of a non-reflecting film for light of wave length 4000 Å*

(M.D.U. 2003)

Solution. The thickness of non-reflecting film $t = \frac{\lambda}{4\mu} = \frac{4000 \times 10^{-10}}{4\mu} = \frac{10^{-7}}{\mu}$ m

Example 23.24 *Consider a non-reflecting film of refractive index 1.38. Assume that its thickness is 9×10^{-6} cm. Calculate the wavelength in the visible region for which the film be non-reflecting.*

Solution For minimum thickness of the coating of refractive index μ required for no reflection at the centre of visible spectrum corresponding to the given wavelength λ is

$$t = \frac{\lambda}{4\mu}$$

Here $\lambda = ?\ \mu = 1.38$ and $t = 9\times10^{-6}$ cm $= 9 \times 10^{-8}$ m

$\therefore$ $\lambda = 4\ \mu t = 4 \times 1.38 \times 9 \times 10^{-8}$

$= 4968 \times 10^{-10}$ m $= 4968$ Å

Example 23.25 *A parallel beam of light strikes an oil film (μ=1.40) floating on surface of water (μ = 1.33). When viewed at an angle of 30° from the normal sixth dark fringe is seen. Find the thickness of the film. (Given wave length of light 589 n.m)* **(H.P.U. 2002)**

Solution. As the film is viewed at an angel of 30° from the normal, angle of reflection = 30°

$\therefore$ Angle of incidence $i = 30°$

It r is the angle of refraction from air to oil film, then

$$\frac{\sin i}{\sin r} = 1.40 \quad \text{or} \quad \sin r = \frac{\sin i}{1.40} = \frac{\sin 30°}{1.40} = \frac{0.5}{1.40} = 0.3571$$

$\therefore$ $r = \sin^{-1}\ 0.3571 = 20.92°$

and $\cos r = \cos 20.92° = 0.934$

For nth dark fringe (in the reflected system) $2\ \mu t \cos r = n\ \lambda$

Here $n = 6;\ \lambda = 589 \times 10^{-9}$ m; $\mu = 1.40$

$\therefore$ $2\ \mu t \cos r = 6 \times 589 \times 10^{-9} =$ or $t = \dfrac{6\times589\times10^{-9}}{2\times1.4\times0.934}$

$= 1351.3 \times 10^{-9} = 1.351 \times 10^{-6}$ m

Example 23.26 *A parallel beam of light of wave length 5890 Å is incident on a glass plate having refractive index μ = 1.5 such that the angle of refraction in the plate is 60°. Calculate the smallest thickness of glass plate which will appear dark by reflected light.* **(Nag. U. 2008, 2002)**

Solution. The film will appear dark by reflected light when path difference $2\ \mu$ t $\cos r = (n+1)$ λ and $n = 0, 1, 2\ldots$

For minimum thickness $n = 0$

$\therefore$ $2\ \mu t \cos r = \lambda$

Here $r = 60°$

$\therefore$ $\cos r = 0.5$; $\mu = 1.5$; $\lambda = 5890$ Å

$\therefore$ $2\times1.5 \times t \times 0.5 \times 5890 \times 10^{-10}$

or $t = \dfrac{5890\times10^{-10}}{2\times1.5\times0.5} = 3927 \times 10^{-10}$ m

$= .3927 \times 10^{-6}$m $= .3927$ μm

Example 23.27 *A soap film of refractive index 1.333 is illuminated by white light incident at an angle of 45°. The light refracted by it is examined by a spectroscope and two consecutive bright bands are focussed corresponding to wavelegth 6.1 × 10 cm and 6.0 × 10^{-5} cm. Find the thickness of the film.* **(Nagpur U. 2001)**

Solution. The light refracted into the soap film finally emerges out of it forming the transmitted system of fringes. In transmitted system the film will appear bright when $2\ \mu\ t \cos r = n\lambda$.

As the two wave lengths $\lambda_1 = 6.1 \times 10^{-5}$ cm and $\lambda_2 = 6.0 \times 10^{-5}$cm form two consecutive bright bands

$$2\mu t \cos r = n\lambda_1 = (n+1)\lambda_2$$

$\therefore$ $$n \times 6.1 \times 10^{-5} = (n+1) \times 6.0 \times 10^{-5}$$

or $$n(6.1 - 6.0) = 6 \quad \text{or} \quad n = 60$$

Here $$i = 45°, \quad \mu = 1.333,$$

But $$\frac{\sin i}{\sin r} = \mu$$

$\therefore$ $$\frac{\sin 45°}{\sin r} = 1.333 \quad \text{or} \quad \sin r = \frac{0.707}{1.333} = 0.53$$

and $$\cos r = \left(1 - \sin^2 r\right)^{\frac{1}{2}} = \sqrt{1 - (0.53)^2} = 0.848$$

Path difference $$2\mu t \cos r = n\lambda_1 = 60 \times 6.1 \times 10^{-5}$$

or $$t = \frac{60 \times 6.1 \times 10^{-5}}{2 \times 1.333 \times 0.848} = 161.9 \times 10^{-5} \text{ cm}$$

$$= 1.62 \times 10^{-4} \text{ mm}$$

23.20 INTERFERENCE FRINGES AT WEDGE-SHAPED FILM

Let *ABC* be a wedge shaped film of refractive index μ, having a very small angel θ at *A*, as shown in Fig. 23.21. If a parallel beam of monochromatic light is allowed to fall on the upper surface and the surface is viewed by reflected light, then alternate dark and bright fringes become visible.

Fringe width. Consider a point *P* at a distance x_1 from *A*, where the thickness of the film is *t*. When light is incident normally, the total path-difference between the light reflected at *R* from the upper face *AB* and that reflect at *P* from the lower face *AC* is $2\mu t - \frac{\lambda}{2}$ as a path-difference of $\frac{\lambda}{2}$ is produced in the beam reflected from the upper face *AB* at *R* where reflection takes place at the surface of a denser medium. The point *P* will appear dark and a dark band will be observed across the wedge, if

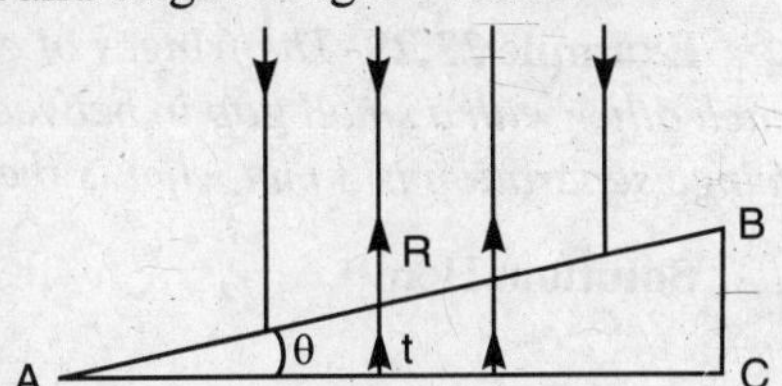

Fig. 23.21

$$2\mu t - \frac{\lambda}{2} = (2n+1)\frac{\lambda}{2}$$

or $$2\mu t = (n+1)\lambda = n\lambda$$

Here *n* is an integer, hence (*n* +1) can also be taken as an integer.

As θ is small, $t = x_1 \theta$

$\therefore$ $$2\mu x_1 \theta = n\lambda \qquad \text{... (i)}$$

Now if x_2 is the distance of the (*n* + *m*) th dark band, then

$$2\mu x_2 \theta = (n+m)\, m\lambda \qquad \text{... (ii)}$$

$\therefore$ $$2\mu\theta(x_2 - x_1) = m\lambda$$

or Fringe width $$\beta = \frac{x_2 - x_1}{m} = \frac{\lambda}{2\mu\theta} \qquad \text{... (iii)}$$

The same relation is obtained if we consider bright fringes.

Difference of the film thickness between two points. If t_1 is the thickness of the film at P and t_2 that at Q and m is the number of bands observed between these point, then

$$x_2 = \frac{t_2}{\theta} \text{ and } x_1 = \frac{t_1}{\theta}$$

$$\therefore \qquad x_2 - x_1 = \frac{t_2 - t_1}{\theta}$$

Substituting in (*iii*), we get

$$\frac{t_2 - t_1}{m\theta} = \frac{\lambda}{2\mu\theta}$$

or

$$t_2 - t_1 = \frac{m\lambda}{2\mu} \qquad \text{... (23.26)}$$

Example 23.28 *A wedge–shaped air film, having an angle of 40 seconds is illuminated by monochromatic light and fringes are observed vertically through a microscope. The distance measured between the consecutive bright fringes is 0.12 × 10⁻² m. Calculate the wavelength of light used.*

Solution. Here

$$\theta = 40'' = \frac{40 \times \pi}{180 \times 60 \times 60} = 0.194 \times 10^{-3} \text{ radian}$$

$$\beta = 0.12 \times 10^{-2} \text{ m and } \mu = 1$$

$$\therefore \quad \lambda = 2\mu\beta\theta = 2 \times 1 \times 0.12 \times 10^{-2} \times 0.194 \times 10^{-3}$$

$$= 4656 \times 10^{-10} \text{ m} = 4656 \text{ Å}$$

Example 23.29 *The fringes of equal thickness are formed when two glass plates are kept over each other with a small gap in between. If a parallel beam of light of wavelength 6000 Å is used and fringe separation is 3 mm what is the angle between the plates in seconds.*

Solution. Here

$$\beta = 3 \text{ mm} = 3 \times 10^{-3} \text{ m}, \mu = 1$$

$$\lambda = 6000 \text{ Å} = 6000 \times 10^{-10} \text{ m}, \theta = ?$$

Now

$$\beta = \frac{\lambda}{2\mu\theta}$$

or

$$\theta = \frac{\lambda}{2\mu\beta} = \frac{6000 \times 10^{-10}}{2 \times 1 \times 3 \times 10^{-3}} = 10^{-4} \text{ radian}$$

$$= \frac{10^{-4} \times 180 \times 3600}{\pi} = 20.62''$$

Example 23.30 *Using light of $\lambda = 5.9 \times 10^{-7}$ m, it is found that in a thin film of air 7.4 fringes occur between two points. Deduce the difference of film thickness between these point.*

Solution. Here number of fringes between two points where thickness are t_2 and t_1 = 7.4

$$\lambda = 5.9 \times 10^{-7} \text{ m}, \mu = 1$$

Now

$$t_2 - t_1 = \frac{\lambda \times \text{No. of fringes}}{2\mu}$$

$$= \frac{5.9 \times 10^{-7} \times 7.4}{2 \times 1} = 0.2183 \times 10^{-5} \text{ m}$$

Example 23.31 *When a wedge-shaped air film is viewed by a mono-chromatic source of light incident normally, the interference fringes 4 mm apart are observed. If the air space is filled with water (μ = 1.33), how far apart will the fringes be observed?*

Solution. If λ is the wavelength of light used, θ the angle of the wedge shaped film of refractive index μ, then

$$\text{Fringe width } \beta = \frac{\lambda}{2\mu\theta}$$

For air μ = 1 and β = 4mm = 4× 10^{-3} m

$$\therefore \quad 4\times10^{-3} = \frac{\lambda}{2\theta}$$

In the second case μ = 1.33

$$\therefore \quad \beta = \frac{\lambda}{2\theta}\times\frac{1}{\mu} = \frac{4\times10^{-3}}{1.33} = 3\times10^{-3}\text{ m} = 3\text{ mm}$$

23.21 NEWTON'S RINGS (in Reflected light)

Circular interference fringes can be produced by enclosing a very thin film of air or any other transparent medium of varying thickness between a plane glass plate and a convex lens of a large radius of curvature. Such fringes were first obtain by Newton and are known as Newton's rings.

Experimental arrangement. To produce Newton's rings monochromatic light from an extended sources *S* is rendered parallel by a convex lens *L*. It falls on a glass plate *G* inclined at an angle of 45° to the incident beam and is reflected normally on to a convex lens *N* of large radius of curvature placed on a plane glass plate *P* in a holder, as shown in Fig. 23.22.

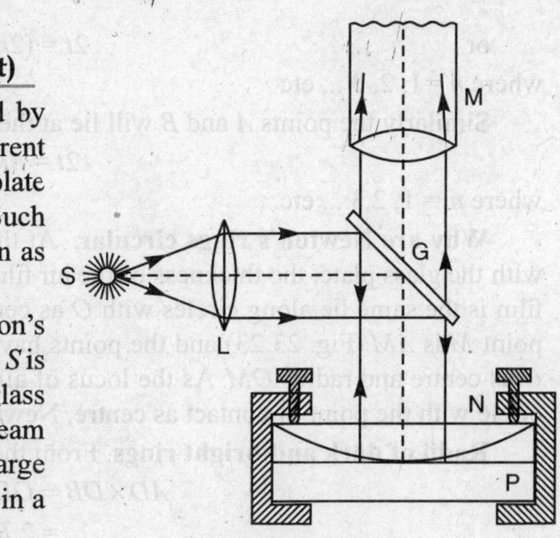

Fig. 23.22

Light rays reflected upward from the top and bottom surfaces of the air film formed between the lens and the glass plate *P*, superimpose each other with a path-difference depending upon the air thickness in between. Due to interference of these rays, dark and bright circular fringes are seen with monochromatic light. *The fxinges are circular because the air film is symmetrical about the point of contact of the lens N with the glass plate P.* These fringes can be observed by means of a low power travelling microscope *M* and their diameters can also be measured in this manner.

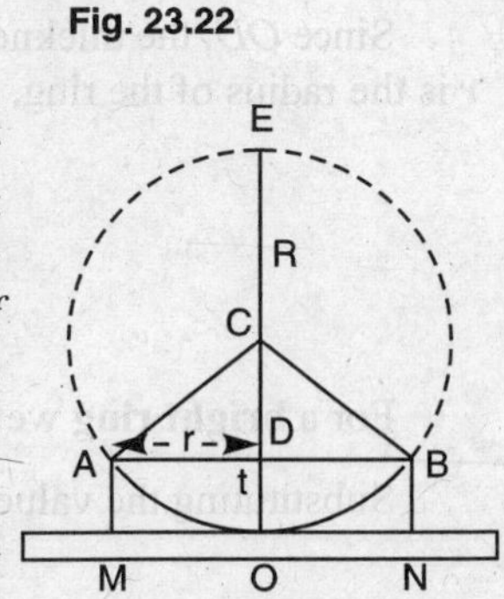

Fig. 23.23

Theory. Let *AOB* be the vertical section of the lens surface through its centre of curvature *C* having a radius of curvature equal to *R*. The lens is in contact at *O* with the plane glass plate *MON*, in such a manner that the point *B* and *A* are equidistant from *O*.

Complete the circle *AOBE* and draw the diameter *ODCE*. Draw also *BN* and *AM* perpendiculars to the plate *MN* and let

$$AM = BN = t$$

The thickness of the air film will be zero at the point of contact *O*. The points at which the thickness of the air film is the same will be along circles with *O* as centre. The thickness of the air film at the point M is AM and the points having the same thickness as AM will lie along the circle with O as centre and radius OM. The fringes obtained are, therefore, circular.

The path-difference between the two rays, one reflected from A and the other M as shown in Fig. 23.23 is $2\mu t \cos r$. Now since μ for air is unity and incidence is normal, hence the angle of refraction r is zero.

$$\therefore \quad \text{Path-difference} = 2t$$

As one of the rays MKK' (Fig. 23.24) suffers reflection at a denser medium, *i.e.*, at M or N, a further phase change of π or path-difference $\lambda/2$ takes place. Hence

$$\text{Total path-difference} = 2t + \frac{\lambda}{2}$$

The points A and B equidistant from O will, therefore, lie on a **bright ring** of diameter AB according as the path-difference.

$$2t + \frac{\lambda}{2} = n\lambda$$

or
$$2t = (2n-1)\frac{\lambda}{2} \quad \text{... (i)}$$

where n = 1, 2, 3 ... etc

Similarly the points A and B will lie at the centre of a **dark ring** when the path-difference

$$2t = n\lambda \quad \text{... (ii)}$$

where n = 1, 2,3 ... etc.

Why are Newton's rings circular. At the point O where the plano-convex lens is in contact with the glass plate, the thickness of the air film is zero. The points at which the thickness of the air film is the same lie along circles with O as centre. For example, the thickness of the air film at the point M is AM (Fig. 23.23) and the points having the same thickness as AM lie along a circle with O as centre and radius OM. As the locus of all points having the same thickness of the air film is a circle with the point of contact as centre, Newton's rings produced are circular.

Radii of dark and bright rings. From the geometry of the circle in Fig. 23.23

$$AD \times DB = OD \times DE = OD(2R - OD)$$

$$= 2Rt - OD^2 = 2Rt \quad \text{... (iii)}$$

Since OD, the thickness of air film is very small, OD^2 can be neglected, as compared to $2Rt$. If r is the radius of the ring, then

$$AD = DB = r$$

$$r^2 = 2Rt$$

$$2t = \frac{r^2}{R} \quad \text{... (iv)}$$

For a **bright ring** we have $2t = (2n-1)\frac{\lambda}{2}$ where n = 1, 2, 3 ... etc.

Substituting the value of $2t$ in Eq. (*iv*), we have

$$\frac{r^2}{R} = (2n-1)\frac{\lambda}{2} \quad \text{... (v)}$$

or Radius of n th **bright** ring $r = \sqrt{\frac{(2n-1)\lambda R}{2}}$ ... (vi)

As λ and R are constant $r \propto \sqrt{(2n-1)}$. *Hence the radius or diameter of bright rings are proportional to square root of odd integral numbers.*

For a *dark rings* we have $2t = n\lambda$ where $n = 0, 1, 2, 3, ...$ etc

Substituting the value of $2t$ in equation (*iv*), we have

$$\frac{r^2}{R} = n\lambda \qquad ...(vii)$$

or Radius of nth **dark** ring $r = \sqrt{n\lambda R}$ ...(*viii*)

As λ and R are constant $r \propto \sqrt{n}$. *Hence the radius or diameter of* **dark** *rings are proportional to square root of natural integral numbers.*

Central ring. For $n = 0$, $r = 0$ for a dark ring. Hence in the reflected system the central ring of zero order is dark.

For $n = 1$, The radius of the dark ring = $\sqrt{\lambda R}$ and

For $n = 1$, The radius of the bright ring = $\sqrt{\frac{\lambda R}{2}}$

Hence, around the central dark ring there are alternate bright and dark rings.

Fringe width decreases with order of rings.

If d is the diameter of the nth dark ring, then

$$d = 2r = 2\sqrt{nR\lambda}$$

For the central dark ring $n = 0$

$$\therefore \qquad d = 2\sqrt{nR\lambda} = 0$$

This corresponds to the centre of Newton's rings. The central ring is not counted while counting the order of dark rings.

The difference between the diameters of 4th and 1st dark rings

$$d_4 - d_1 = 2\sqrt{4\lambda R} - 2\sqrt{\lambda R} = 2\sqrt{\lambda R} \qquad ...(ix)$$

Similarly the difference between d_{25} and d_{16} is

$$d_{25} - d_{16} = 2\sqrt{25\lambda R} - 2\sqrt{16\lambda R}$$

$$= 2\sqrt{\lambda R} \qquad ...(x)$$

From (*ix*) and (*x*) it is clear that difference in diameter of 3 rings and 9 rings is the same, which shows that the width of rings decreases with the order of rings.

Wavelength (λ) in terms of radius of curvature of lens.

In the production of Newton's rings by enclosing a very thin film of air or any other transparent medium of varying thickness between a plane glass plate and convex lens of large radius of curvature, it has been assumed that the convex surface of the lens is in perfect contact with the glass plate and the thickness of air-film is zero at the contact point. It is possible that the two surfaces may have some imperfections or some dust particles may be present and thus the thickness of air-film may not be zero. Hence we are not sure of the order of the central fringe which may be bright or dark. To avoid this, the diameters of two bright fringes say nth and mth are measured with the help of a low power travelling microscope, then for A and B to lie on the nth bright ring (Fig. 23.23), we have from eq. (*v*)

$$\frac{d_n^2}{4R} = (2n-1)\frac{\lambda}{2} \qquad ...(xi)$$

and for A and B to lie on the mth bright ring

$$\frac{d_m^2}{4R} = (2m-1)\frac{\lambda}{2} \qquad ...(xii)$$

Subtracting (*xii*) from (*xi*) we get

$$\frac{d_n^2 - d_m^2}{4R} = (n-m)\lambda$$

or

$$\lambda = \frac{d_n^2 - d_m^2}{4R(n-m)} \qquad \text{... (23.27)}$$

The same result can be obtained by measuring the diameters of *n*th and *m*th dark rings.

The radius of curvature *R* of the lower surface of the convex lens can be found out with the help of a spherometer by using the relation

$$R = \frac{l^2}{6h} + \frac{h}{2} \qquad \text{... (23.28)}$$

Where the letters have their usual meaning. Thus the wavelength of monochromatic light used can be found out.

23.22 NEWTON'S RINGS IN TRANSMITTED LIGHT

Newton's rings can also be obtained by transmitted light. The interference in this case takes place between two rays *MM′* and *KLL′* as shown in Fig. 23.24.

The ray *MM′* suffers no reflection at a denser medium. The ray *KLL′* suffers first reflection at denser medium at *M* and a second reflection at *K*. At each reflection a phase change of π take place. Thus the total phase change due to both the reflections is π + π = 2π which physically means zero phase change.

The path difference between the two interfering beams is, therefore, $2\mu t \cos r$. For air and normal incidence μ = 1 and cos *r* = cos 0 = 1 and the path difference is 2*t*

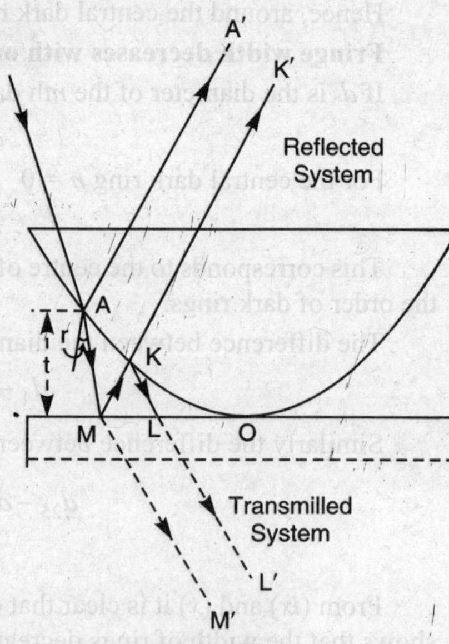

Fig. 23.24

For **bright** ring $2t = n\lambda$ and

For **dark** ring $2t = (2n+1)\frac{\lambda}{2}$ where *n* = 0, 1, 2, 3 ... etc

Now $2t = \frac{r^2}{R}$ where *r* is the radius of the ring and *R* the radius of the spherical surface of the lens. [For proof see article 23.21 (eq. *iv*)].

∴ Radius of *n*th bright ring $r = \sqrt{n\lambda R}$ and

Radius of *n*th dark ring $r = \sqrt{\frac{(2n+1)\lambda R}{2}}$

For *n* = 0, *r* = 0 for a bright ring and $r = \sqrt{\frac{\lambda R}{2}}$ for a dark ring. Therefore, the central ring is bright. Around the central ring there are alternate dark and bright rings.

Example 23.32 *In a Newtons ring experiment, the diameter of the 15th ring was found to be 0.590 cm and that of the 5th ring was 0.336 cm. If the radius of the plano-convex lens is 100 cm, calculate the wavelength of light used.* **(*Nagpur U. 2009; KU 2002*)**

Solution. Here d_{15} = 0.590 cm; d_5 = 0.336 cm; *R* = 100 cm.

$$\therefore \qquad \lambda = \frac{d_n^2 - d_m^2}{4R(n-m)} = \frac{d_{15}^2 - d_5^2}{4R(15-5)} = \frac{(0.590)^2 - (0.336)^2}{4 \times 100 \times (15-5)}$$

$$= 5.88 \times 10^{-5} \text{ cm} = 5.88 \times 10^{-7} \text{ m} = 5880 \times 10^{-10} \text{ m}$$
$$= 5880 \text{ Å}$$

Example 23.33 *Newtons rings are observed in reflected light of wavelength 5.9 × 10⁻⁷ m. The diameter of 10th dark ring is 0.5 cm. Find the radius of curvature of the lens and thickness of the air film.* ***(Kerala U. 2001, Gauhati U. 2000)***

Solution Here $\lambda = 5.9\times10^{-7}$ m, $d_{10} = 0.5\text{cm} = 5\times10^{-3}$m

Now $$d_n^2 = 4n\lambda R$$

$$\therefore \quad R = \frac{d_n^2}{4n\lambda} = \frac{5\times10^{-3}\times5\times10^{-3}}{4\times10\times5.9\times10^{-7}} = 1.059 \text{ m}$$

Thickness of air film at 10th dark ring

$$t = \frac{r_n^2}{2R} = \frac{d_n^2}{8R} = \frac{5\times10^{-3}\times5\times10^{-3}}{8\times1.059} = 2.95\times10^{-6}\text{ m}$$

Example 23.34 *In Newton's rings experiment the diameter of the 4th and 12th rings are 0.4 cm and 0.7 cm respectively. Find the diameter of the 20th dark ring.*

Solution. Here $d_4 = 0.4$ cm $= 4 \times 10^{-3}$ m, $d_{12} = 0.7$ cm $= 7 \times 10^{-3}$ m

Now $$\lambda = \frac{d_m^2 - d_n^2}{4R(m-n)} = \frac{(7\times10^{-3})^2 - (4\times10^{-3})^2}{4R\times8}$$

or $$\lambda R = \frac{(49-16)10^{-6}}{32} = 1.031\times10^{-6}\text{ m}$$

$$\therefore \quad d_{20} = \sqrt{4R\lambda n} = \sqrt{4\times1.031\times10^{-6}\times20}$$
$$= 0.908 \times 10^{-2} \text{ m}$$

Example 23.35 *In a Newton's rings experiment the radius of 10th and 20th rings are 0.2 and 0.3 cm respectively and the focal length of the plano-convex lens is 90 cm. Calculate the wavelength of light used in nanometers.* ***(H.P.U. 2000)***

Solution. Taking $\mu = 1.5$, $f = 90$ cm, we have for the plano-convex lens

$$\frac{1}{f} = (\mu - 1)\left(\frac{1}{R_1} - \frac{1}{R_2}\right)$$

or $$\frac{1}{90} = (1.5-1)\left(\frac{1}{R_1}\right) \qquad \therefore R_1 = 90 \times 0.5 = 45 \text{ cm}$$

$$d_n = d_{20} = 0.3 \text{ cm}; \; d_m = d_{10} = 0.2 \text{ cm}$$

now $$\lambda = \frac{d_n^2 - d_m^2}{4R(n-m)} = \frac{(0.3)^2 - (0.2)^2}{4\times45\times(20-10)} = \frac{0.05}{1800} = 2.78\times10^{-5}\text{ cm}$$
$$= 278 \times 10^{-9} \text{ m} = 278 \text{ nm}$$

Example 23.36 *In Newton ring arrangement a source is emitting two wavelengths $\lambda_1 = 6\times10^{-7}$m and $\lambda_2 = 5.9 \times 10^{-7}$ m. It is found that nth dark ring due to one wavelength coincides with (n+1)th dark ring due to the other. Find the diameter of the nth dark ring if radius of curvature of the lens is 0.9m.*

Solution. Diameter of the *n*th dark ring with light of wavelength λ_1

$$d_n^2 = 4\,Rn\lambda_1$$

Diameter of the $(n + 1)^{th}$ dark ring with light of wavelength λ_2

$$d_{n+1}^2 = 4R(n+1)\lambda_2$$

Since the two rings coincide

$$\therefore \quad 4\,Rn\lambda_1 = 4\,R(n+1)\lambda_2$$

or $$n\lambda_1 = (n+1)\lambda_2$$

Here $$\lambda_1 = 6\times10^{-7}\,\text{m},\ \lambda_2 = 5.9\times10^{-7}\,\text{m and } R = 0.9\text{m}$$

$\therefore$ $$n\times6\times10^{-7} = (n+1)5.9\times10^{-7}$$

or $$n = 59$$

$$d_n = \sqrt{4\times0.9\times59\times6\times10^{-7}} = 1.1289\times10^{-2}\,\text{m}$$

Example 23.37 *Light containing two wavelengths λ_1 and λ_2 falls normally on a convex lens of radius of curvature R resting on a plane glass plate. If the nth dark ring due λ_1 coincides with (n+1) th dark ring due to λ_2, prove that radius of the nth dark ring due to λ_1 is equal to* $\sqrt{\dfrac{\lambda_1\lambda_2 R}{\lambda_1-\lambda_2}}$

(P.U. 2007 Nagpur U. 2008)

Solution. Diameter of the nth dark ring with light of wavelength λ_1 is

$$d_n^2 = 4\,Rn\lambda_1 \qquad \text{... }(i)$$

Diameter of $(n+1)^{\text{th}}$ dark ring with light of wavelength λ_2 is

$$d^2_{\,n+1} = 4R(n+1)\lambda_2 \qquad \text{... }(ii)$$

Since, these two rings coincide

$$4Rn\lambda_1 = 4R(n+1)\lambda_2$$

or $$n\lambda_1 = (n+1)\lambda_2$$

$$n(\lambda_1-\lambda_2) = \lambda_2$$

or $$n = \frac{\lambda_2}{\lambda_1-\lambda_2} \qquad \text{... }(iii)$$

Substituting in Eq (*i*)

$$d_n^2 = 4R\frac{\lambda_1\lambda_2}{\lambda_1-\lambda_2} \qquad (\because r_n = d_n/2)$$

$$r_n^2 = \frac{R\lambda_1\lambda_2}{\lambda_1-\lambda_2}$$

or $$r_n = \sqrt{\frac{R\lambda_1\lambda_2}{\lambda_1-\lambda_2}}$$

23.23 REFRACTIVE INDEX BY NEWTON'S RINGS

In the experimental arrangement Fig. 23.22 if the air space between the lens and the glass plate is replaced by the liquid whose refractive index μ is to be determined, the path difference between the two rays one reflected at A and the other at M (Fig. 23.23) is

$$= 2\,\mu\,t \qquad (\because \cos r = 1)$$

Hence for the nth bright fringe, we have

$$2\mu t = (2n-1)\frac{\lambda}{2}$$

and for the nth dark fringe

$$2\mu t = n\lambda \qquad \text{... }(i)$$

Also $$2t = \frac{d_n^2}{4R}$$

Substituting the value of 2*t* in (*i*), we get

$$\frac{\mu d_n^2}{4R} = n\lambda \qquad \text{... (ii)}$$

Similarly for the mth dark ring, we get

$$\frac{\mu d^2 m}{4R} = m\lambda \qquad \text{... (iii)}$$

Subtracting (*iii*) from (*ii*), we get

$$\mu \frac{(d_n^2 - d_m^2)}{4R} = (n-m)\lambda$$

or

$$\mu = \frac{4R(n-m)\lambda}{d_n^2 - d_m^2} \qquad \text{... (23.29)}$$

Knowing the wavelength of light λ and by measuring d_n, d_m and R in the usual manner, the refractive index μ can be calculated.

Wavelength of light not known. If the wavelength of light is not known, the diameter of a particular ring first with air and then with liquid in the intervening space is measured.

Let d_a and d_l be the diameters of the *n*th dark ring in the two cases, then

$$d_a^2 = 4Rn\lambda \text{ and } d_l^2 = \frac{4\ Rn\lambda}{\mu}$$

$$\therefore \qquad \left(\frac{d_a}{d_l}\right)^2 = \mu \qquad \text{... (23.30)}$$

Example 23.38 *In the Newton's rings experiment, the diameter of a certain ring is found to be 4 mm when there is air between the convex lens and the glass plate. However, when air is replaced by a liquid, the radius of the ring shrinks from 4 mm to 3.3 mm. Find the refractive index of the liquid.*

(Nagpur U. 2002)

Solution. Radius of 10th dark ring with air = 4.0 mm = 4×10^{-3} m

Radius of 10th dark ring with liquid = 3.3 mm = 3.3×10^{-3} m

$$\therefore \qquad \mu = \left(\frac{d_a}{d_1}\right)^2 = \left(\frac{2\times4\times10^{-3}}{2\times3.3\times10^{-3}}\right)^2 = 1.47$$

Example 23.39 *Newton's rings are formed with reflected light of wavelength 5.895 × 10^{-7} m with a liquid between the plane and the curved surface. The diameter of the 5th dark ring is 0.3 cm and the radius of curvature of the curved surface is 1 m. Calculate the refractive index of the liquid.*

(P. U. 2000)

Solution. Here λ = 5.895 × 10^{-7} m, d_5 = 0.3 cm = 3×10^{-3} m

For the dark ring system, we have

$$\frac{\mu d_n^2}{4R} = n\lambda$$

or

$$\mu = \frac{4R}{d_n^2} n\lambda = \frac{4\times1\times5\times5.895\times10^{-7}}{3\times10^{-3}\times3\times10^{-3}}$$

$$= 1.31$$

Example 23.40 *Newton's rings are formed with reflected light of wavelength 5890 × 10^{-10} m using a plano-convex lens and plane glass plate, with liquid between them; the diameter of 10th dark ring is 4.2 mm and the radius of curvature of the curved surface of the lens is 100 cm. Find the refractive index of the liquid if the ring is (i) dark and (ii) bright.* **(Nagpur U. 2001)**

Solution. Here $\lambda = 5890 \times 10^{-10}$ m, $R = 100$ cm = 1 m, $n = 10$

$$d_{10} = 4.2 \text{ mm} = 4.2 \times 10^{-3} \text{ m}$$

(*i*) When the ring is dark

$$\mu = \frac{4Rn\lambda}{d_n^2} = \frac{4 \times 1 \times 10 \times 5890 \times 10^{-10}}{4.2 \times 10^{-3} \times 4.2 \times 10^{-3}} = 1.336$$

(*ii*) When the ring is bright

$$\mu = \frac{4R(2n-1)\lambda}{2 \times d_n^2} = \frac{4 \times 1 \times 19 \times 5890 \times 10^{-10}}{2 \times 4.2 \times 4.2 \times 10^{-6}} = 1.269$$

EXERCISE

1. (*a*) Explain the phenomenon of interference of light and define interference. ***(Meerut U. 2002)***

   (*b*) Sketch the diagram showing the formation of maxima and minima due to overlapping of monochromatic light from two narrow slits of double slit experiment.

2. Explain the difference between interference by division of wave front and by division of amplitude. Give examples ***(G.N.D.U. 2007, Pbi. U. 2006, P.U. 2008, 2000)***

3. Differentiate between two methods of producing interference pattern. ***(P.U. 2004)***

4. (*a*) State principle of superposition. Define interference and derive conditions of formation of I_{max} and I_{min} ***(Napur U. 2010, 2009; Gujrat U. 2004; P.U. 2002)***

   (*b*) Two coherent sources of intensity ratio α interfere. Prove that in the interference pattern

   $$\frac{I_{max} - I_{min}}{I_{max} + I_{min}} = \frac{2\sqrt{a}}{1+a}$$ ***(M.D.U. 2003)***

5. What are coherent sources? What are the conditions for two sources to be coherent? How are they realised in practice? Can two point like independent sources become coherent? Explain. ***(Madurai U.2003, Pbi.U. 2003, 2001, H.P.U. 2002, G.N.D.U 2006, 2000, K.U. 2000, M.D.U. 2007, Luck U. 2002)***

6. State and explain in brief, the conditions for (*a*) observance (*b*) good contrast of fringes and (*c*) stationary interference pattern. ***(M.D.U. 2007, 2002; Luck U. 2001)***

7. Give the condition to produce good interference fringes. ***(Gharwal U. 2004. Bhopal. U. 2004; K.U. 2000)***

8. Explain analytically the interference of light on the basis of wave theory.

9. Discuss briefly the phenomenon of interference with relation to law of conservation of energy.

10. Find an expression for fringe width in case of Young's double slit experiment. Prove that in this case of interference dark and bright bands are of equal width. ***(Nagpur U. 2011, Pbi.2008, M.D.U. 2002 P.U. 2005)***

11. A Young's double slit (Y.D.S) interference experiment is carried out with monochromatic light in air. What will be the change in wavelength and fringe width when the apparatus is immersed in water or the medium is replaced by an optically denser medium. ***(P.U. 2002, 2001, 1999)***

12. Describe Fresnel's biprism. Describe in detail how the wavelength of monochromatic source of light can be determined with its help. ***(P.U. 2003, M.D.U. 2006, 2003, Luck U.2001, K.U. 2001; Meerut U. 2001; Kerala U. 2001; Gauhati U. 2000)***

13. What is meant by "Lateral shift of fringes". How is it used to determine refractive index of unknown material? ***(Nagpur U. 2011)***

14. Explain what happens when (*i*) white light is used to illuminate the slit.
(*ii*) the edge of the prism is not parallel to the slit in Fresnel's biprism experiment or what is lateral shift? How can you remove it in a bi-prism? ***(M.D.U. 2000)***

15. (*a*) In the Fresnel's biprism arrangement show that $d = 2x (\mu-1) \alpha$ where x is the distance of the source from the base of the biprism, α is the refracting angle of the biprism and μ the refractive index of the material of biprism. ***(P.U. 2001)***

16. Explain the effect of introducing a thin plate of glass in the path of one of the interfering beams in bi-prism experiment. Calculate the displacement of fringes. Show how the method can be used for finding the refractive index, thickness of the plate and wavelength of light.
(P.U. 2006, 2005, Meerut U. 2003; K.U. 2000, Gauhati U.2000; M.D.U.2005; H.P.U.2001; Nagpur 2010)

17. (*a*) How will you locate the fringe of zero order in biprism experiment?
(*b*) Explain what will happen in biprism experiment when eye-piece is moved away from the source? ***(H.P.U. 2001)***

18. Explain the change of phase when light is reflected from a denser surface. ***(Nagpur U. 2010)***

19. Discuss the formation of fringes by LIoyd's single mirror and explain why the central fringe is black. Find an expression for fringe-width. ***(P.U. 2004, 2002, 2000; Pbi U. 2002)***

20. Describe Fresnel's double mirror method of obtaining interference fringes and derive an expression for the distance between the consecutive dark bands. ***(Kerala U. 2001)***

21. (*a*) Describe a method of obtaining achromatic fringes with Lloyd's mirror.
(*b*) In what respects do LIyod's single mirror fringes differ from those of biprism fringes.

22. Explain the term division f amplitude. Give an example of interference by division of amplitude. Draw necessary diagram. ***(G.N.D.U. 2001; Indore U. 2001)***

23. Explain the working of Rayleigh's interferometer to measure R.I. of a gas ***(Nagpur U.2009, 2008)***

24. Discuss the phenomenon of interference in thin films. Obtain the condition for maxima and minima. Show that the interference pattern in the reflected and transmitted system are complimentary. Why an extremely thin film appears black in reflected light?
(G.N.D.U. 2006, Gujrat U. 2004, P.U. 2008, 2006, 2003, Pbi U. 2007, 2005, 2003, Meerut U.2003, H.P.U. 2001, 2002, Nagpur U. 2010, 2009, 2001; M.D.U. 2001, 2000, K.U.2000)

25. Why a broad source of light in necessary for observing colours in thin films?
(M.D.U. 2006, 2002, Pbi. U. 2000, P.U.2000; Luck U. 2002)

26. Explain analytically the colour of thin films. Why are the colours in reflected and transmitted light complementary?
(P.U. 2003, 2002, 2000, Kerala U.2001; Nagpur U.2009, 2001, M.D. U. 2000)

27. Derive the conditions of interference for bright and dark fringes due to reflected light from a thin film. ***(Nagpur 2008)***

28. (*a*) What is a non-reflecting film? Derive an expression for the required thickness of anti-reflection coating. ***(G.N.D.U. 2008, 2007; Pbi U.2008)***
(*b*) The refractive index of glass is 1.5. A glass slab is to be coated with a material such that is becomes non-reflecting. What should be the refractive index of the material to be coated?
(G.N.D.U. 2001, P U.2007, 2001; H.P.U. 2000)

29. Explain how interference fringes are formed by a thin wedge shaped film; when examined by normally reflected light ? Find the expression for fringe width. How will you estimate the difference of film thickness between two points. ***(M.D.U. 2007, 2002, 2000; K.U. 2002)***

30. (*a*) Discuss the formation of Newton's rings by reflected light. Describe the experimental arrangement and give necessary theory. Why are Newton's rings circular?

(G.N.D.U. 2005; Gharwal U. 2004; Gujrat U.2004; Luck U.2001; Indore U.2001; Nagpur 2011, 2009, 2008, 2001, P.U.2001, Meerut U.2001, 2008, 2006, 2005, M.D.U. 2001, H.P.U.2001;K.U. 2000, Pbi.U. 2008)

(*b*) What will happen if the glass plate is replaced by a plane mirror in Newton's ring experiment? ***(Nagpur U. 2008)***

31. Prove that in Newton's rings experiment

(*i*) Diameter of dark rings are proportional to square root of natural integral numbers.

(*ii*) Diameter of bright rings are proportional to square root of odd integral numbers ***(Nagpur U. 2008)***

(*iii*) Central ring of zero order is bright or dark? Why?

(P.U. 2001, 1999, M.D.U. 2001, 1999; Indore U. 2001, Meerut U. 2001, 2000, K.U. 1992)

32. Compare Newton's rings in reflected and transmitted light. ***(Kerala U. 2001)***

33. Derive an expression for the wavelength of monochromatic light source used in Newton's ring experiment in terms of diameter of rings and radius of curvature of the lens used.

(Nagpur U. 2009, Meerut U. 2002, G.N.D.U. 2001, Gauhati U. 2000)

34. Explain the method of Newton's rings for determining the refractive index of rare liquids. Derive the necessary formula. ***(Meerut U.2002, M.D.U. 2001, K.U. 2000; P.U. 1991)***

35. What happens to the rings pattern when a liquid is introduced between the plano-convex lens and plate in Newton's ring experiment? ***(Nagpur U. 2010)***

36. If the diameter of *n*th dark ring in Newton's rings experiment changes from 1.2. to 1.0 cm when air is replaced by a transparent liquid, find refractive index of the liquid.

[Ans. μ = 1.44] ***(M.D.U. 2002)***

MULTIPLE CHOICE QUESTIONS

23.1 Which of the following belongs to physical optics?

(*a*) Dispersion (*b*) Reflection

(*c*) Refraction (*d*) Interference

23.2 Can interference occur in longitudinal waves

(*a*) Yes (*b*) no

(*c*) not certain

23.3 On reflection from denser to a rarer medium the path difference introduced is

(*a*) zero (*b*) $\lambda/2$

(*c*) λ (*d*) 2λ

23.4 In Youngs double slit experiment, the distance between two consecutive dark ringes β is :

(*a*) $\frac{d}{D}\lambda$ (*b*) $\frac{D}{d}\lambda$

(*c*) $\frac{d}{\lambda D}$ (*d*) $\frac{D}{\lambda d}$

23.5 In LIoyd mirror experiment, the condition for maximum intensity is

(*a*) $\beta = \frac{\lambda d}{D}$ (*b*) $\frac{\lambda D}{d}$

(*c*) $\lambda \frac{D}{d}$ (*d*) $\frac{D}{d}\lambda$

23.6 Interference in thin films for transmitted system, the path difference is

(*a*) $\frac{2\mu}{t} \cos r$ (*b*) $\frac{t}{2\mu} \cos r$

(*c*) $2\mu t \cos r$ (*d*) $2\mu^2 t^2 \cos r$

23.7 In Newton's rings experiment with the order of rings, fringe width

(*a*) increases (*b*) decreases

(*c*) remains constant

ANSWERS

23.1 (*d*)	**23.2** (*a*)	**23.3** (*b*)	**23.4** (*b*)	**23.5** (*d*)	**23.6** (*c*)	**23.7** (*b*)

HAIDINGER FRINGES

24
Chapter

24.1 HAIDINGER FRINGES

Haidinger fringes are also called as fringes in parallel films. In these films, interference fringes are produced due to path difference $2\mu t \cos r$ between the overlapping rays. For a given film the path difference may arise due to (*i*) r, the angle of refraction inside the film or (*ii*) t, the change in thickness. We can express the change in path difference by differentiating the equation $2\mu t \cos r$.

Change in path difference, $\delta(\Delta) = 2\mu t \cdot \delta(\cos r) + 2\mu \cos r(\delta t)$ ... (24.1)

When the film is of **uniform** (*i.e.*, **constant**) thickness, then $\delta t \to 0$ and hence, the change in path difference in only due to the change in r. Thus,

$$\delta (\Delta) = 2\mu t.\delta (\cos r) \qquad ... (24.2)$$

If the thickness of the film (t) is large, the path difference will change appreciably even when r changes in a small way. Fringes are produced in this case due to superposition of rays; which are equally inclined to the normal. These fringes are called *fringes of equal inclination.* The fringes of equal inclination are known as *Haidinger fringes.*

Experimental arrangement. In order to produce the Haidinger fringes, the source must be extended source, the film thickness must be appreciably large and the observing instrument is to be focused for parallel rays.

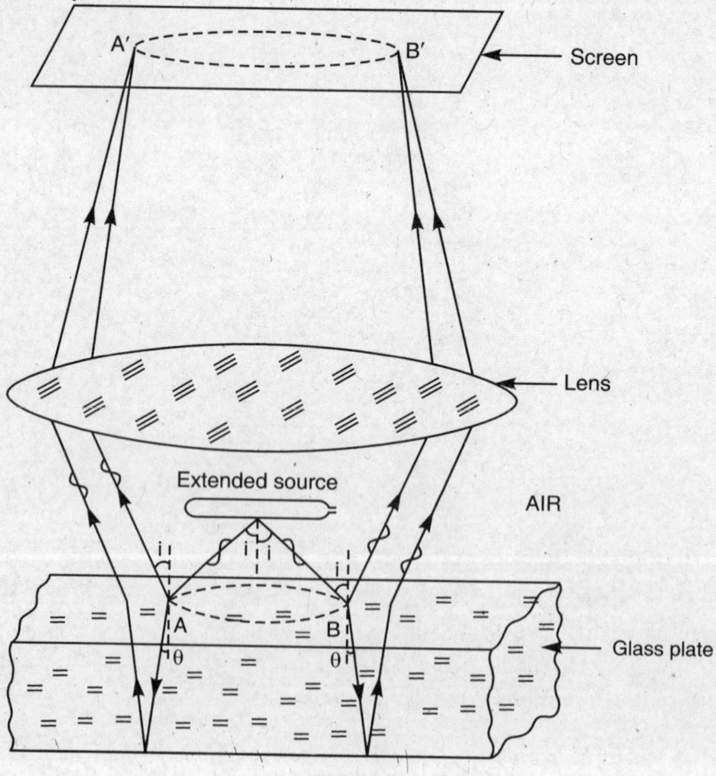

Fig. 24.1

The experimental arrangement to produce Haidinger fringes is as shown in Fig. 24.1. Let a thin plate be illuminated by an extended monochromatic light source. A lens is arranged parallel to the plate and a screen is kept in the focal plane of the lens. Light from the extended source is incident on the plate in diverse directions. The waves propagating parallel to the plane of the page and falling on the plate at an angle i at points A and B get reflected from the top and bottom surfaces of the plate. The reflected pairs of waves will meet at points A' and B' respectively on the screen due to the focussing action of the lens. Depending upon their path difference the reflected waves produce either brightness or darkness on the screen. In fact, the waves incident at the top surface of the plate at an angle i travel along the generators of a cone a shown in fig. 24.1. Each pair of parallel reflected waves interfere at diametrically opposite points. Thus, a *circular fringe is produced.* Similarly, the waves incident at a different angle will produce a collection of identical points arranged along a circle of another radius. As a result, a system of concentric (with a common centre) alternating bright and dark circular fringes are observed on the screen.

Conditions for bright and dark fringes. Each fringe is characterised by a particular value of m. Bright fringes are produced where the condition $2\mu t \cos r = m\lambda$ is satisfied; and dark fringes are produced where the condition $2\mu t \cos r = (2m + 1)\lambda/2$ is satisfied. The parallel pairs of reflected rays meet only at infinity; therefore a lens is used to focus such parallel rays. Accordingly, these fringes of equal inclination are said to be *localized at infinity.*

24.2 MICHELSON'S INTERFEROMETER

Principle. *The amplitude of the light beam from an extended source is divided into two parts of equal intensities by partial reflection and refraction. These beams are sent in two directions at right angles and are brought together after reflection from plane mirrors to produce interference fringes. This is underlying principle of Michelson's interferometer.*

Construction. The main optical parts consist of two highly polished *front silvered plane mirrors* M_1 and M_2 and two *plane parallel* glass plates G_1 and G_2 of the same thickness. The plate G_1 is *half silvered* at the back so that the incident beam is divided into a reflected and a transmitted beam of equal intensity. The plates G_1 and G_2 are held parallel to each other and are inclined at an angle of 45° to the mirrors M_1 and M_2 which are mutually perpendicular to each other. The mirror M_1 is mounted on a carriage and can be moved exactly parallel to itself with the help of a micrometer screw fitted with a graduated drum capable of reading displacement of the order of 10^{-5} cm. The mirrors M_1 and M_2 are provided with levelling screws at their backs with the help of which these can be made perfectly perpendicular to the direction of the two beams. The interference bands are observed in the field of view of the telescope T.

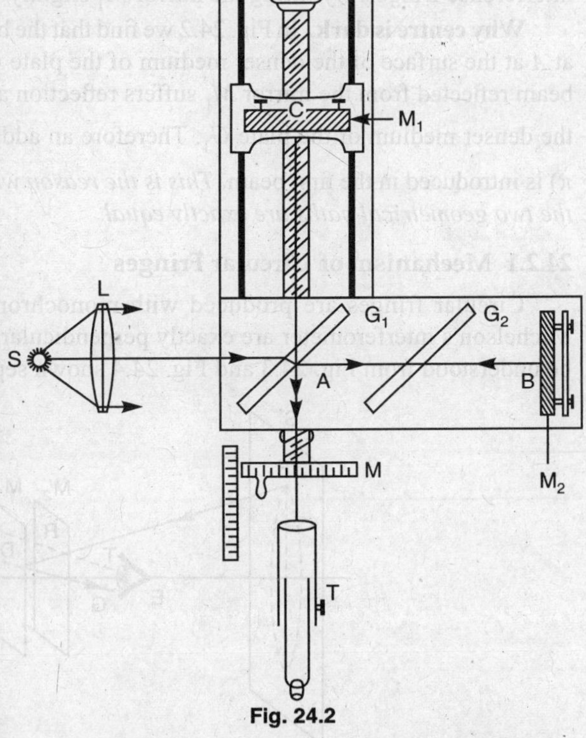

Fig. 24.2

Working. Light from the source S is rendered parallel by means of a collimating lens L and is made to fall on the glass plate G_1. It is partly reflected at the back surface of G_1 along AC and partly transmitted along AB. The ray AC is received by the plane mirror M_1 normally, so that it is reflected along the same path and emerges out along AT. The transmitted ray AB is received by the mirror M_2 normally, reflected along the same path and then moves along AT after reflection at the back surface of G_1. Thus two beams along AT are produced from a single source by division of the amplitude. These two beams produce interference under suitable conditions. The course of rays is shown in Fig. 24.2.

Compensating plate. It is clear from Fig. 24.2 that a ray starting from the source and suffering reflection at the mirror M_1 traverses the glass plate G_1 thrice, whereas the ray reflected from the mirror M_2 traverses the glass pate G_1 only one. To compensate for this an exactly similar glass plate G_2 is introduced in the path AB parallel to the glass plate G_1. This is known as *compensating plate.* It is made of the same glass from the same plate and of the same thickness as the plate G_1. The ray suffering reflection at the mirror M_2 also now traverses this compensating plate twice and thus if the two mirrors M_1 and M_2 are equidistant from the plate G_1, the two paths are exactly equal.

Adjustment. (*i*) Distances of the mirrors M_1 and M_2 are adjusted to be very nearly equal from the glass plate G_1.

(*ii*) To make the incident light from the source parallel, a tin sheet with a fine hole in it is placed in front of the source S just opposite to the bright part of it. The hole and the source of light are adjusted in line with the centre of the glass plates and mirror M_2. A lens is then placed between G_1 and the tin sheet. A plane mirror is also placed between G_1 and the lens normally. The position of the lens is adjusted till the image of the hole falls back on the tin plate very close to the hole. The light is now in the form of a parallel beam when it leaves the lens. After removing the plane mirror if now we look in the direction AT four images of the hole will be seen. The adjustment of the mirrors is made till images coincide two by two. If now the tin sheet is removed and the two paths are exactly equal, the field of view will be totally dark. A slight motion of M_1 parallel to itself will produce circular interference fringes. By tilting the mirror M_1 slightly the fringes can be made practically straight.

Why centre is dark. In Fig. 24.2 we find that the beam reflected from mirror M_2 suffers reflection at A at the surface of the denser medium of the plate G_1 into the rarer medium *i.e.*, air, whereas the beam reflected from the mirror M_1 suffers reflection at A at the surface of the rarer medium air into the denser medium of the plate G_1. Therefore an additional path difference $\frac{\lambda}{2}$ (or phase difference π) is introduced in the first beam. *This is the reason why the centre of the fringe system is dark when the two geometrical paths are exactly equal.*

24.2.1 Mechanism of Circular Fringes

Circular fringes are produced with monochromatic light when the mirrors M_1 and M_2 of Michelson's interferometer are exactly perpendicular to each other. The origin of these fringes can be understood from Fig. 24.3 and Fig. 24.4 shown separately.

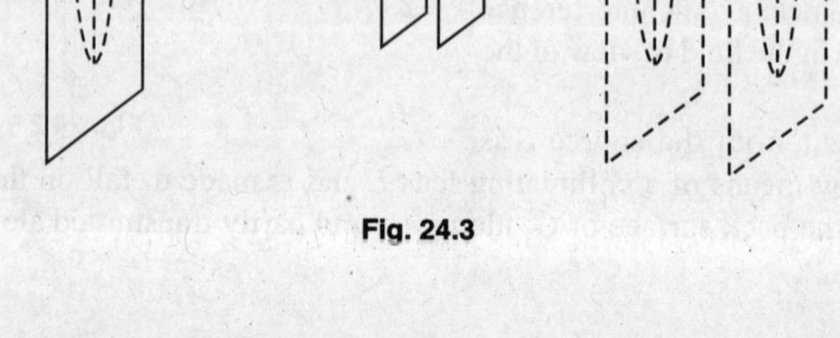

Fig. 24.3

As several reflections take place in the interferometer, we consider an extended source. When the interferometer is in normal adjustment, the image M'_2 of M_2 in glass plate G_1 is exactly parallel to M_1. The real source is also replaced by its image S' formed in the plate G_1, behind the eye E. This source S' gives rise to two virtual images S_1 in M_1 and S_2 in M'_2 respectively. If d is the distance between M_1 and M'_2, then the separation between S_1 and S_2 will be $2d$. The virtual images S_1 and S_2 behave as virtual coherent sources in the sense that the corresponding points, like P_1 and P_2, the images of point P (in S') in M_1 and M'_2 are exactly in phase in all respects at all instants. Light waves from P after reflection from M_1 and M_2 at an angle θ appear as if these have originated from P_1 and P_2. Hence the two reflected waves RT and DG are coherent and pursue parallel paths. Consequently when brought to focus by a telescope set for infinity superimpose and produce interference fringes.

The path difference between two rays coming to the eye from corresponding points P_1 and P_2 will be $2d \cos \theta$ where θ is the angle which the reflected beam makes with the normal as shown in Fig. 24.4 which is a simple way of representing Fig. 24.3. As there is an additional path difference $\frac{\lambda}{2}$ caused in the beam reflected at mirror M_2 total path difference is $2d \cos\theta + \frac{\lambda}{2}$. For minimum in the interference pattern (dark fringe)

$$2d \cos \theta + \frac{\lambda}{2} = (2n+1)\frac{\lambda}{2}$$

or
$$2d \cos \theta = (2n+1)\frac{\lambda}{2} - \frac{\lambda}{2}$$

$$= 2n\frac{\lambda}{2} = n\lambda$$

or
$$2d \cos \theta = n\lambda \qquad \text{... (i)}$$

For a maximum (bright fringe)

$$2d \cos \theta + \frac{\lambda}{2} = n\lambda$$

or
$$2d \cos \theta = n\lambda - \frac{\lambda}{2} = (2n-1)\frac{\lambda}{2} \qquad \text{... (ii)}$$

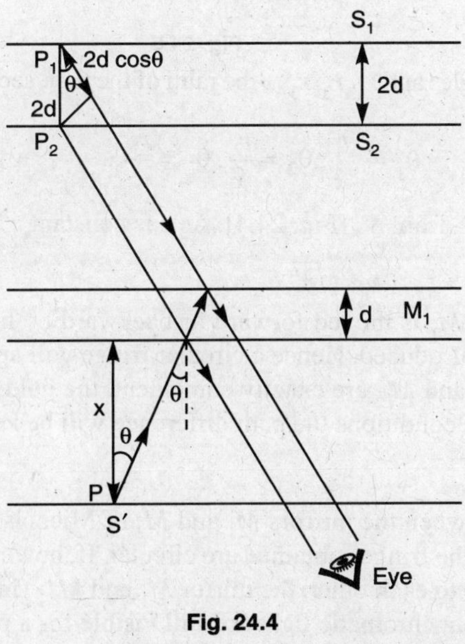

Fig. 24.4

In both the cases, for a given value of d, n and λ the angle θ will be a constant. Hence the locus of a fringe (dark or bright) is a circle.

Considering the case of a dark fringe, for normal incidence path difference = $2d$.

Let $$2d = n'\lambda \qquad \text{... } (iii)$$

where n' is an integer.

Keeping M_2 fixed if M_1 is moved either way parallel to itself widely spaced circular fringes are produced, as shown in Fig. 24.5 (*i*) and (*iii*). If the path difference is increased, the width of these fringes decreases. These fringes are visible even for a very large path difference.

24.2.2 Radii of circular fringes are proportional to the square root of natural numbers

From (*i*) and (*iii*), we have

$$2d - 2d\cos\theta = (n' - n)\,\lambda = n''\lambda \qquad (n < n')$$

or $$2d\,(1 - \cos\theta) = n''\lambda$$

where n'' is also an integer.

Now $$(1 - \cos\theta) = 1 - \left(1 - \frac{\theta^2}{2!} + \frac{\theta^4}{4!} \ldots\right)$$

Neglecting $\frac{\theta^4}{4!}$ and higher powers of θ, we have

$$\frac{\theta^2}{2} = \frac{n''\lambda}{2d} \quad \text{or} \quad \theta^2 = \frac{n''\lambda}{d}$$

$$\therefore \qquad \theta \propto \sqrt{n''} \qquad \text{... } (iv)$$

(i) (ii) (iii)

Fig. 24.5

If $\theta_1, \theta_2, \theta_3$ ... are the angles and r_1, r_2, r_3 ... the radii of the first, second, third.... rings respectively, then

$$\theta_1 = \frac{r_1}{x}, \theta_2 = \frac{r_2}{x}, \theta_3 = \frac{r_3}{x} \ldots$$

where x is the distance of M'_2 from S', [Fig. 24.4]. As x is constant $r \propto \theta$, but as $\theta \propto \sqrt{n''}$

$$\therefore \qquad r \propto \sqrt{n''} \qquad \text{... (24.3)}$$

It may be noted that as M_1 is moved forward or backward by half a wavelength, keeping M'_2 fixed a path change of λ is introduced. Hence a circular fringe will appear or disappear at the centre of the field. However, if M_1 and M'_2 are exactly coincident, the field of view will be dark as shown in Fig. 24.5 (*ii*). Under these conditions the path difference will be zero for all angles of incidence.

24.2.3 Localised fringes

It has been proved that when the mirrors M_1 and M_2 of Michelson's interferometer are exactly at right angles to each other the fringes obtained are circular. If, however, the two mirrors M_1 and M_2 are not exactly at right angle to each other, the mirror M_1 and M'_2 (image of M_2) will not be exactly parallel, the fringes with monochromatic light are still visible for a path difference not exceeding a few millimetres. In this case the air path between the two mirrors is *wedge-shaped. With such a film, the locus of points of equal thickness is a straight line parallel to the edge of the wedge.*

The fringes observed, therefore, are nearly straight. For large value of distance between the two mirrors, the fringes are not exactly straight, as the path difference changes with the angle. The fringes are in general curved and are always convex towards the thin edge of the wedge.

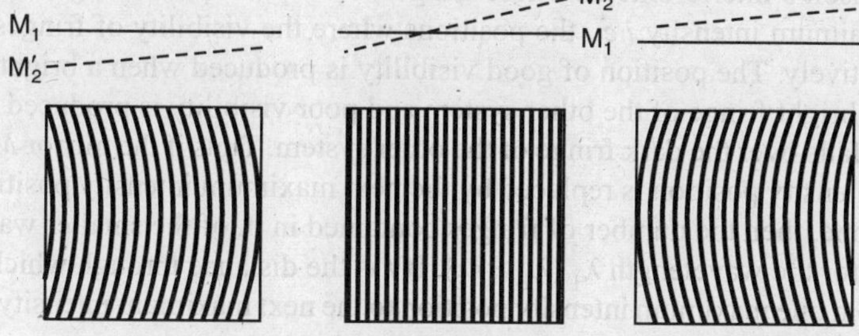

(i) (ii) (iii)

Fig. 24.6

For a certain value of d, where d is the path difference between the two rays, one reflected from the mirror M_1 and other reflected from mirror M_2 the fringes observed are as shown in Fig. 24.6 (*i*). The mirror M_1 and the image of M_2 in position M'_2, are also as shown in Fig. 24.6 (*i*). If the separation of mirrors is decreased, the fringes will move to the left, across the field of view, a new fringe crossing the centre each time d is changed by $\lambda/2$. As we approach zero path difference the fringes become straight till the point is reached when M_1 actually intersects M'_2. In this position the fringes are perfectly straight as shown in Fig. 24.6 (*ii*). Beyond this point they begin to curve in the opposite direction, as shown in Fig. 24.6 (*iii*).

24.2.4 Fringes with White Light

If a white light source is used, instead of monochromatic source of light, a few coloured fringes with a central dark fringe can be observed. In observing these fringes, the mirrors are slightly tilted as for localised fringes and position of M_1 is found where it interesects M'_2. This position is often troublesome to find with white light. The position can best be located approximately with monochromatic light when the fringes become straight. Then a very slow motion of M_1 in this region using white light will bring these fringes in view, when a central dark fringe surrounded on either side by 8 to 10 coloured fringes are observed. These fringes are useful for the determination of the zero path difference especially in the determination of the length of the standard metre.

24.3 APPLICATIONS OF MICHELSON'S INTERFEROMETER

(1) Determination of wavelength of monochromatic light. (Precision measurement of wavelength). The Michelson's interferometer is adjusted to obtain circular fringes in the field of view of the telescope. If the mirrors M_1 and M_2 (Fig. 24.2) are equidistant from the glass plate G_1, the field of view will be perfectly dark. The position of the mirror M_1 is adjusted till a particular bright fringe appears in the field of view of the telescope with its centre coinciding with the cross wire. When the mirror M_1 is moved backward or forward, each fringe in the focal plane of the telescope is displaced parallel to itself. When the mirror M_1 is moved through a distance $\lambda/2$, the path difference changes by λ and the position of a particular bright fringe is taken by the next bright fringe. If n is the number of fringes that moves across the field of view when the mirror is displaced through a distance l, then

$$n\frac{\lambda}{2} = l$$

or Wavelength $$\lambda = \frac{2l}{n} \qquad \text{... (24.4)}$$

The distance l can be measured accurately correct up to 10^{-4} mm with the help of the micrometer screw and the number of fringes n that move across the field of view is actually counted. This is the most correct method to measure the wavelength with a great precision.

(2) Width of Special Lines. If a source of light consisting of two wavelengths λ_1 and λ_2, which differ slightly, is used, then two sets of fringes corresponding to the two wavelengths λ_1 and λ_2 are produced in a Michelson's interferometer. These fringes, on superimposition, give rise to position of maximum and minimum intensity *i.e.*, the positions where the visibility of fringes is very good and very poor respectively. The position of good visibility is produced when a bright fringe of one system falls over the bright fringe of the other system and poor visibility is produced when a bright fringe of one system falls over the dark fringe of the other system. Thus if the mirror M_1 is moved so that one maximum intensity position is replaced by the next maximum intensity position in the field of view of the telescope, then the number of fringes contained in it, of the smaller wavelength λ_2 is one more than those for the wavelength λ_1 ($\lambda_1 > \lambda_2$). If l is the distance through which the mirror is moved, in passing from one maximum intensity position to the next maximum intensity position, then

$$\frac{2l}{\lambda_2} - \frac{2l}{\lambda_1} = 1 \quad \text{or} \quad 2l\left(\frac{\lambda_1 - \lambda_2}{\lambda_1\lambda_2}\right) = 1$$

or
$$\lambda_1 - \lambda_2 = \frac{\lambda_1\lambda_2}{2l} \qquad \text{... (24.5)}$$

Taking λ as the mean wavelength of the two wavelengths λ_1 and λ_2, the small difference $\delta\lambda$ is given by

$$\delta\lambda = \frac{\lambda^2}{2l} \qquad \text{... (24.6)}$$

To determine the value of l, the interferometer is adjusted to obtain circular fringes. The position of the mirror M_1 is adjusted so that the fringes obtained have a *maximum intensity.* In this position, the bright fringe of wavelength λ_1 superimposes on the bright fringe of the wavelength λ_2. When the mirror M_1 is moved, the two sets of fringes get out of step and a position is achieved where a bright fringe of one falls on the dark fringe of the other. In this position the intensity of fringes is the least. The position of *maximum indistinctness* is noted. The mirror M_1 is moved further and again the position of *maximum intensity* first and then the position of *maximum indistinctness* is noted. The difference between the readings of the micrometer screw for the two consecutive positions of *maximum intensity or maximum indistinctness* gives the value of l.

(3) Determination of thickness of a thin transparent sheet. If a sheet of thickness t of a transparent medium of refractive index μ is introduced into the path of one of the interfering beams of the interferometer, the optical path in this beam is increased as light travels more slowly in the medium than in air. The optical path in the medium now becomes μt instead of t in the corresponding thickness of air. This results in the increase of optical path by $(\mu - 1)\, t$ in passing through the medium. Since the beam traverses the medium twice, extra path difference of $2\, (\mu - 1)\, t$ between the two beams is introduced. If n is the number of fringes by which the fringe system is displaced and λ the wavelength of light used, then

$$2\,(\mu - 1)\, t = n\lambda \qquad \text{... (24.7)}$$

Since the fringes with monochromatic light are exactly alike, it is not possible to distinguish between the fringes and as there is a sudden shift of fringes on inserting the thin plate in one of the beams, it is not possible to count the number of fringes shifted. To overcome this difficulty the instrument is set for localised white light fringes as explained earlier so that the central dark fringe coincides with the cross wire of the telescope. The paths AB and AC are then exactly equal. (Fig. 24.2).

The thin film is now introduced between the glass plate G_1 and the mirror M_2 such that light passes through it normally. Position of mirror M_1 is adjusted till again a dark fringe of zero path difference coincides with the crosswire of the telescope. The distance l through which the mirror M_1 is moved is noted.

The white light is now replaced with monochromatic light of wavelength λ. The mirror M_1 is moved back slowly and number of fringes n contained in l is found out.

Thus
$$t = \frac{n\lambda}{2(\mu - 1)} \qquad \text{....(24.8)}$$

Determination of refractive index. To find the refractive index of a thin transparent sheet (say of mica) with a Michelson's interferometer the procedure in the same as for finding the thickness of a transparent material. In this case the thickness t and wavelength λ are known.

$$\therefore \quad 2(\mu - 1)\, t = n\lambda$$

or
$$(\mu - 1) = \frac{n\lambda}{2t}$$

$$\therefore \quad \mu = 1 + \frac{n\lambda}{2t} \qquad \text{... (24.9)}$$

Using Eq. 24.9, the refractive index (μ) of a thin transparent film can be determined to a greater accuracy.

(4) Measurement of a standard metre in terms of standard wavelength. Michelson determined the number of times the wavelength of monochromatic light like *Cd* red, blue or green, was contained in a standard metre. The counting of number of fringes which cross the field of view when the movable mirror is displaced through a distance of one metre is not practicable. In the first place, interference fringes cannot be obtained for a path difference greater than 24 cm *i.e.*, the mirror-displacement should not exceed more than 12 cm and in the second place, a displacement of 10 cm involves a transit of about 3×10^3 fringes, a number which was impossible to count accurately with the mechanical aids available at that time. Hence a series of **nine** *etalons i.e.*, intermediate standards of length each twice the length of the preceding one were constructed as shown in Fig. 24.7. The longest etalon was 10 cm and the shortest 0.390625 mm, which contained 600 *Cd* red light waves or 1200 fringes in the distance between the two parallel surfaces by which the etalon was defined.

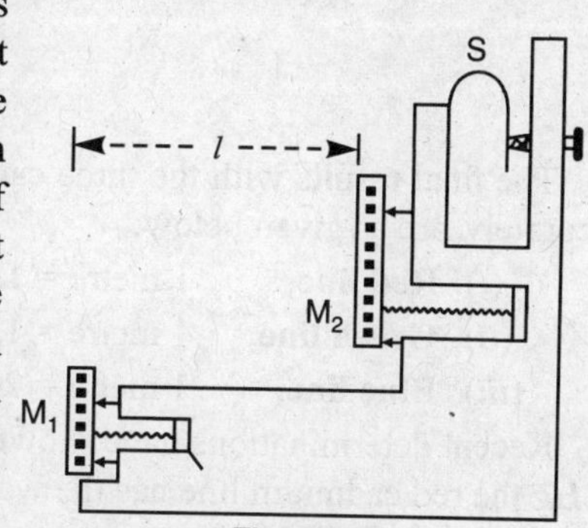

Fig. 24.7

The **etalon** consists of two plane mirrors M_1 and M_2 which are rigorously parallel to each other and are silvered on their front. The distance between the two silvered surfaces being equal to the length l of the etalon. The two mirrors can be made perfectly parallel with the help of the screw T and the springs S.

The length of the shortest etalon, in terms of the wavelength of cadmium red lines was determined. For this purpose Michelson modified the apparatus, as shown in Fig. 24.8. The two shortest etalons were mounted with a field of view covering the four mirrors M_1, M_2 and M'_1 and M'_2, as shown. With the help of white light the distances of M, M_1 and M'_1 from the eye were made equal, so that a black fringe was visible in the field of view. The mirror M was then moved to the position B and when this is exactly coplanar with M_2, again a black white light fringe will appear in its field of view. The distance AB was exactly equal to $M_1 M_2$. The white light was then replaced by red light of cadmium and the mirror M was slowly moved from position A to B, counting the number of fringes which moved across the reference line. The fraction of the fringe in excess of the integral number required to reach this position was also determined by rotating the compensating glass plate G_3 which is calibrated previously. In this way distance $M_1 M_2$ was determined in terms of the wavelength. The smallest etalon was then moved through its own length, without counting the fringes till again the white light black fringe reappeared in M_1. Finally M was moved to C, when again white light black fringe appeared in M'_2 as well as in M_2. In this way, the number of waves contained in $M'_1 M'_2$ was obtained. The second etalon was then compared with the third etalon and in this way the number of waves were found out in the etalon of 10 cm length. The 10 cm long etalon is taken as a substandard and is compared directly with the prototype standard metre.

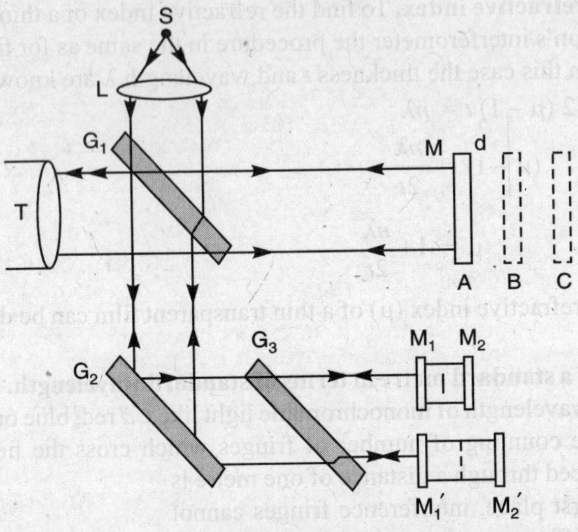

Fig. 24.8

The final results with the three cadmium lines whose wavelengths have been determined very accurately, are as given below:

(*i*) **Red line,** 1 metre = 1,553, 163.5 λ_r, where λ_r = 6438.4722Å

(*ii*) **Green line,** 1 metre = 1,966,249.7 λ_g, where λ_g = 5985.8241 Å

(*iii*) **Blue line,** 1 metre = 2,083,372.1 λ_b, where λ_b = 4799.9107 Å

Recent determinations have shown that in dry atmospheric air at 15°C and pressure of 760 mm of *Hg* the red cadmium line has the wavelength

$$\lambda_r = 6438.4696 \text{ Å}$$

A still more satisfactory line for use as a standard wavelength has been discovered. This is the *green line* of mercury given out by the single isotope Hg^{198}. This is a very sharp line better than the cadmium lines and has a wavelength of 5460.7532 Å.

(5). Determining Earth's Motion through Ether

Michelson's interferometer was originally designed to prove the existence of hypothetical medium 'Ether'. For this the velocity of light along the earth's motion and along a direction perpendicular to it was calculated and found to differ. The intention of performing this experiment was to test the existence of *absolute space*. The interferometer was placed on a rock about 1m^3 size, which was floated on mercury. One arm was set parallel to the earth's motion in space and interference one fringes were obtained. On rotating the instrument through a right angle, a small fringe-shift was expected, if there existed an absolute space. The experiment of Michelson and Morley (1881 to 1887) showed a null result. And this shook the foundation of much philosophical thought which led Einstein to give birth to his 'Theory of relativity'.

Example 24.1 *A Michelson interferometer is set for the white light straight fringes. When a mica sheet of thickness 0.005 cm is put in front of the fixed mirror, then in order to bring back the coloured fringes to their original position, the movable mirror is moved by 0.0025 cm. Calculate the refractive index of mica.*

Solution. If t is the thickness of the mica sheet and μ its refractive index, then

Increase in the path = $(\mu - 1)\, t$

Let l be the distance through which the movable mirror has to be moved to bring back the coloured fringes to their original position, then

$$(\mu - 1)\, t = l$$

Now, thickness of mica sheet $t = 0.005 \text{ cm} = 5 \times 10^{-5}$ m

Distance through which mirror is moved, $l = 0.0025 \text{ cm} = 25 \times 10^{-6}$ m

$$(\mu - 1) = \frac{25 \times 10^{-6}}{5 \times 10^{-5}} = 0.5$$

or Refractive index $\mu = 1.50$.

Example 24.2 *In Michelson's interferometer a thin plate is introduced in the path of one of the beams and it is found that 50 bands had crossed the line of observation. If the wavelength of light used is 5896 Å and $\mu = 1.4$, determine the thickness of the plate.*

Solution. Here $\mu = 1.4$, $n = 50$, $\lambda = 5896 \times 10^{-10}$ m

The thickness t is found from the relation

$$t = \frac{n\lambda}{2(\mu - 1)} = \frac{50 \times 5896 \times 10^{-10}}{2(1.4 - 1)}$$

$$= 0.3685 \times 10^{-4} \text{ m.}$$

Example 24.3 *A thin film of a material whose refractive index is 1.45 on being introduced in one of the arms of Michelson's interferometer causes a shift of 6.5 fringes. If the wave length of light used is 5890Å, calculate the thickness of film.* **(M.D.U 2002)**

Solution.
$$t = \frac{n\lambda}{2(\mu - 1)} = \frac{6.5 \times 5890 \times 10^{-10}}{2 \times 0.45} = 4.254 \times 10^{-6} \text{ m}$$

Example 24.4 *A sheet of CaF_2 of refractive index 1.434 is inserted normally in one arm of a Michelson's interferometer. At $\lambda = 5896$ Å, 40 fringes are observed to be displaced. Calculate the thickness of the sheet.* **(G.N.D.U 2000)**

Solution.
$$t = \frac{n\lambda}{2(\mu - 1)} = \frac{40 \times 5896 \times 10^{-10}}{2 \times 0.434} = 27 \times 10^{-6} \text{ m}$$

Example 24.5 *In Michelson's interferometer 100 fringes cross the field of view when the movable mirror is displaced through 0.02948 mm. Calculate the wavelength of monochromatic light used.* **(Nagpur U. 2011)**

Solution. Here $l = 0.02948 \text{ mm} = 0.02948 \times 10^{-3}$ m

$n = 100$ and $\lambda = ?$

Now
$$\lambda = \frac{2l}{n} = \frac{2 \times 0.02948 \times 10^{-3}}{100}$$

$$= 5896 \times 10^{-10} \text{ m} = 5896 \text{ Å}$$

Example 24.6 *In Michelson's interferometer 200 fringes cross the field of view when the movable mirror is displaced through .05896 mm. Calculate the wavelength of monochromatic light used.* **(P.U. 2007; Kerala U. 2001; M.D.U. 2001)**

Solution. $\lambda = \dfrac{2l}{n} = \dfrac{2\times 0.05896\times 10^{-3}}{200} = 5896\times 10^{-10}$ m

Example 24.7 *In Michelson's interferometer 500 fringes cross the field of view when the movable mirror is displaced through 0.147 mm. Calculate the wavelength of monochromatic light used.* **(M.D.U. 2003)**

Solution. $\lambda = \dfrac{2l}{n} = \dfrac{2\times 0.147\times 10^{-3}}{500} = 5880\times 10^{-10}$ m

Example 24.8 *In Michelson's interferometer 790 fringes cross the field of view when the movable mirror is displaced through 2.33 × 10^{-4} m. Calculate the wavelength of monochromatic light used.* **(M.D.U. 2006)**

Solution. $\lambda = \dfrac{2l}{n} = \dfrac{2\times 2.33\times 10^{-4}}{790} = 5898\times 10^{-10}$ m.

Example 24.9 *The wavelength of two components of D-lines of sodium are 5890 Å and 5896 Å. By how much distance one of the mirrors of Michelson's interferometer be moved so as to obtain consecutive positions of maximum distinctness.* **(P.U. 2003; H.P.U. 2001)**

Solution. Here $\lambda_1 = 5896$ Å $= 5896 \times 10^{-10}$ m

$\lambda_2 = 5890$ Å $= 5890 \times 10^{-10}$ m

If l is the distance through which the mirror is moved in passing from one maximum intensity position to the next maximum position, then

$$\lambda_1 - \lambda_2 = \frac{\lambda_1 \lambda_2}{2l}$$

$$\therefore \quad l = \frac{5896\times 10^{-10}\times 5890\times 10^{-10}}{2\times 6\times 10^{-10}}$$

$$= 0.2894 \times 10^{-3} \text{ m}$$

Example 24.10 *In an experiment with Michelson's interferometer, the distance travelled by the mirror for two successive positions of maximum distinctness was 0.2945 mm. If the mean wavelength for the two components of sodium D-line is 5893Å, calculate the difference between the two wavelengths.* **(Meerut, U. 2001)**

Solution. If l is the distance through which the mirror is moved for maximum distinctness, then total path-difference introduced is $2l$. If λ_1 and λ_2 are the two wavelengths, then maximum distinctness will be produced if maximum of one wavelength falls on the maximum of the other wavelength. Thus

$$\frac{2l}{\lambda_2} - \frac{2l}{\lambda_2} = 1 \quad \text{or} \quad \frac{\lambda_1 - \lambda_2}{\lambda_1 \lambda_2} = \frac{1}{2l}$$

Here $l = 0.2945 \times 10^{-3}$ m

Mean wavelength $\lambda = 5893$ Å $= 5893 = 10^{-10}$ m

$$\therefore \quad \lambda_1 - \lambda_2 = \frac{\lambda_1 \lambda_2}{2l} = \frac{\lambda^2}{2l} = \frac{(5893\times 10^{-10})^2}{2\times 0.2945\times 10^{-3}}$$

$$= 5.896 \times 10^{-10} \text{ m} = 5896 \text{ Å}$$

Example 24.11 *In an experiment for determining the refractive index of a gas using Michelson's interferometer a shift of 140 fringes is observed when all the gas is removed from the tube. If the wavelength of light used is 5460 Å and the length of the tube is 20 cm, calculate the refractive index of the gas.* **(Delhi Hons. 2000)**

Solution. Here $\quad n = 140, \lambda = 5460 \times 10^{-10}$ m

$l = 20$ cm $= 0.2$ m, $\mu = ?$

Now $\quad 2(\mu - 1)\, l = n\lambda$

or $$\mu - 1 = \frac{n\lambda}{2l} = \frac{140 \times 5460 \times 10^{-10}}{2 \times 0.2}$$

$$= 0.000191$$

$\therefore \quad \mu = 1 + 0.000191 = 1.000191.$

24.4 TWYMAN AND GREEN INTERFEROMETER

Twyman and Green interferometer is the technically modified version of the Michelson's interferometer which is widely used. The construction of Twyman and Green interferometer is shown in Fig. 24.9. Instead of a broad source, a plane wavefront coherent over the whole field obtained from a pin-hole S with collimating lens L_1 is used. The fringes are viewed with eye placed at the focus of lens L_2, so that one sees not S but the field of plate P'.

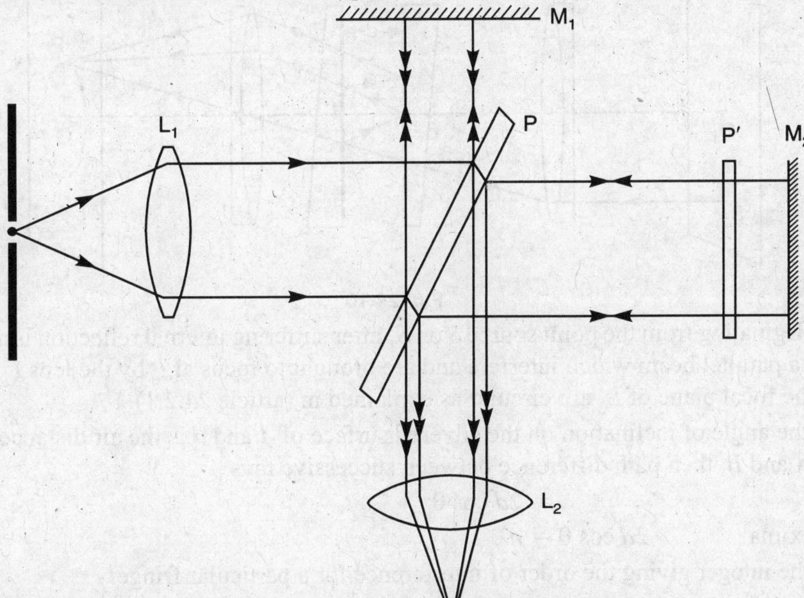

Fig. 24.9. Twyman-Green Interferometer

Working. With mirror M_2 set parallel to M_1 the whole surface of M_1 shows a uniform intensity. Now a test plate P' is introduced. If it is uniform in thickness and structure, the field of view remains of uniform intensity. But if there are irregularities in P', fringes show up around these irregularities and one can correct them (by polishing etc.) while viewing the improvements. If P' is a convex lens, then M_2 is replaced by a convex mirror of radius of curvature equal to the focal length of the lens. If mirrors are to be tested and corrected, P' is an ideal lens and M_2 is the mirror under test. In effect, the arrangement is used to make the test plates, lenses and mirrors accurate to better than $\lambda/10$ or so in precision.

24.5 MULTIPLE BEAM INTERFEROMETRY

In fig. 23.16, we showed how in multiple reflection at ordinary surfaces only the first two beams are of appreciable amplitudes. The situation changes altogether if the *surfaces of the film are partially silvered.* Then all the beams become of comparable amplitude and the intensity distribution in the interference pattern is very different. This forms the basis of 'multiple beam interferometry'.

24.6 FABRY-PEROT INTERFEROMETER

In this interferometer interference fringes of constant inclination called **Haidinger fringes are** obtained by transmitted light after multiple reflections between two glass plates separated by a distance.

Construction. The apparatus consists of two glass plates *A* and *B* accurately parallel, separated by a distance *d*. The inner surfaces of these plates are optically plane and thinly silvered so that they can reflect 70% of the incident light. **The outer faces of these plates are also parallel to each other, but inclined to their respective inner surfaces to avoid interference effect due to multiple reflection and refraction.**

Light from an extended source S_1 of monochromatic light is rendered parallel by the collimating lens L_1. Each of the parallel rays suffers multiple reflection in the air film enclosed between the plates *A* and *B* and emerge from the plate *B*, fig. 24.10.

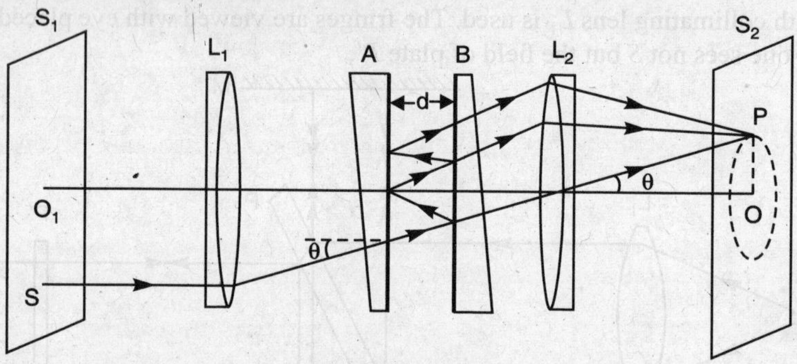

Fig. 24.10

A ray originating from the point source *S* on S_1 after suffering internal reflection issues from *B* in the form of a parallel beam which interfere and are brought to focus at *P* by the lens L_2. The fringes formed in the focal plane of L_2 are circular as explained in (article 24.2.1).

If θ is the angle of inclination on the silvered surface of *A* and *d* is the air distance between the two plates *A* and *B*, then path difference between successive rays

$$= 2d \cos \theta \qquad \text{... (i)}$$

For maxima $\qquad 2d \cos \theta = n\lambda \qquad$ ... (*ii*)

where *n* is the integer giving the order of interference for a particular fringe.

This condition is satisfied for all the points on a circle through *P* with centre at *O*. The value of path difference depends upon the angle θ, so that we can have a series of concentric circles with *O* as centre on the screen (or eye-piece).

In actual practice the thickness of the air film can also be changed by keeping one of the plates fixed and providing a micrometer screw to move the other either way. In this manner the thickness of the air film can be changed.

Intensity of transmitted beam. The two plates *A* and *B* a Fabry Perot interferometer are shown in Fig. 24.11. It is clear from this figure that the intensity of successive parallel beams goes on decreasing and depends upon the reflecting and transmitting powers of the silver-coated surfaces. Suppose transmitted amplitudes are c, cr^2, cr^4, cr^6 ... where *r* is the reflection coefficient and *c* is the product of the original amplitude and transmission coefficient at the two plates. The transmitted beams have a constant phase difference δ given by

$$\delta = \frac{2\pi}{\lambda}(2d \cos \theta)$$

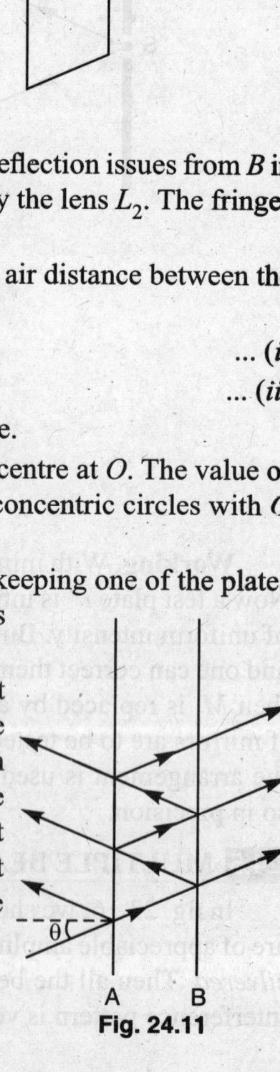

Fig. 24.11

because as explained above, the path difference between successive rays is $2d \cos \theta$ according to Eq. (*i*).

Applying the law of polygon of vectors, the intensity is proportional to the square of the resultant amplitude, *i.e.*,

$$I \propto (c + cr^2 \cos\delta + cr^4 \cos 2\delta + ...)^2 + (cr^2 \sin \delta + cr^4 \sin 2\delta + ...)^2$$

or
$$I \propto c^2 \{(1 + r^2 \cos \delta + r^4 \cos 2\delta + ...)^2 + (r^2 \sin \delta + r^4 \sin 2\delta + ...)^2\} \quad ... (iii)$$

To evaluate we, however, use exponential, instead of trigonometrical expression's *i.e.*,

$$1 + r^2 e^{i\delta} + r^4 e^{i2\delta} + r^6 e^{i3\delta} + ... = \frac{1}{1 - r^2 e^{i\delta}}$$

Multiply the numerator and the denominator of R.H.S. by $1 - r^2 e^{-i\delta}$

$$\therefore \quad 1 + r^2 e^{i\delta} + r^4 e^{i2\delta} + r^6 e^{i3\delta} + ... = \frac{1 - r^2 e^{-i\delta}}{1 - r^2 (e^{i\delta} + e^{-i\delta}) + r^4} \quad ... (iv)$$

Also we know $e^{\pm i\delta} = \cos \delta \pm i \sin \delta$

Substituting this value in (*iv*), we get

$$1 + r^2 (\cos \delta + i \sin \delta) + r^4 (\cos 2\delta + i\sin 2\delta) + ..$$

$$= \frac{1 - r^2 (\cos \delta - i \sin \delta)}{1 - r^2 (\cos \delta + i \sin \delta + \cos \delta - i \sin \delta) + r^4} \quad ... (v)$$

Equating the real and imaginary terms, we get

$$(1 + r^2 \cos \delta + r^4 \cos 2\delta + ...) = \frac{1 - r^2 \cos \delta}{1 - 2r^2 \cos \delta + r^4} \quad ... (vi)$$

and
$$(r^2 \sin \delta + r^2 \cos 2\delta + ...) = \frac{r^2 \sin \delta}{1 - 2r^2 \cos \delta + r^4} \quad ... (vii)$$

Substituting these values in (*iii*), we get

$$I \propto c^2 \frac{(1 - r^2 \cos \delta)^2 + (r^2 \sin \delta)^2}{(1 - 2r^2 \cos \delta + r^4)^2}$$

$$\propto c^2 \left[\frac{1 - 2r^2 \cos \delta + r^4}{(1 - 2r^2 \cos \delta + r^4)^2}\right]$$

$$\propto \frac{c^2}{(1 - 2r^2 \cos \delta + r^4)}$$

$$I = k \left(\frac{c^2}{(1 - r^2)^2 + 4r^2 \sin^2 \left(\frac{\delta}{2}\right)}\right) \quad ... (24.10)$$

When
$$\delta = 0, 2\pi, 4\pi..., \sin\left(\frac{\delta}{2}\right) = 0$$

The denominator in (24.10) will have a minimum value and hence intensity is maximum.

$$\therefore \quad I_{max} = k \left\{\frac{c^2}{(1 - r^2)^2}\right\} \quad ... (24.11)$$

When $\delta = \pi, 3\pi, 5\pi, \ldots$ etc. $\sin\frac{\delta}{2} = 1$

The denominator in (24.10) will be maximum and hence intensity is minimum.

$$\therefore \quad I_{min} = k\left\{\frac{c^2}{(1+r^2)^2}\right\} \qquad \ldots (24.12)$$

Hence
$$\frac{I_{\max}}{I_{\min}} = \frac{(I+r^2)^2}{(1-r^2)^2} \qquad \ldots (24.13)$$

This gives the ratio of maximum intensity to minimum intensity.

Visibility and sharpness of fringes

1. Visibility of fringes. The visibility V of fringes is given by

$$V = \frac{I_{\max} - I_{\min}}{I_{\max} + I_{\min}}$$

Substituting the values of $I_{\max}$ and $I_{\min}$ from Eq. (24.11) and (24.12)

$$V = \frac{2r^2}{1+r^4} \qquad \ldots (24.14)$$

It is clear that visibility of fringes is a function of the coefficient of reflection r only. If r decreases, the visibility decreases.

If $\quad r = 1$ Visibility $V = \frac{2+1}{1+1} = 1$

For $\quad r = 0.8$ Visibility $V = \frac{2\times 0.64}{1+0.4096} = 0.908$

For $\quad r = 0.5$ Visibility $V = \frac{2\times 0.25}{1+0.0625} = 0.47$

2. Sharpness of fringes. If for a given value of r a graph is plotted between I and δ, then we get a graph of the form shown in Fig. 24.12. From the graph it is noticed that half width is quite small and is very nearly equal to the peak width. The *half width* is defined as the width of the graph at half the peak value of I *i.e.*, at

$$I = \frac{I_{\max}}{2} \qquad \ldots (24.15)$$

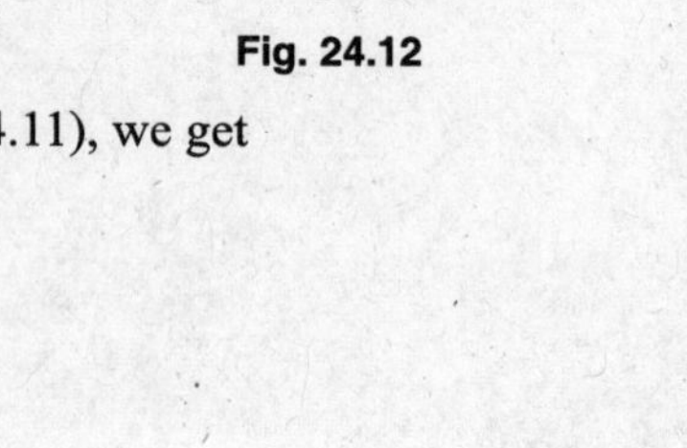

Fig. 24.12

Substituting the value of I and I_{max}, from Eq,. (24.10) and (24.11), we get

$$\frac{I}{I_{\max}} = \frac{(1-r^2)^2}{(1-r^2)^2 + 4r^2\sin^2\left(\frac{\delta}{2}\right)}$$

or
$$I = \frac{I_{\max}}{1+\frac{4r^2}{(1-r^2)^2}\left[\sin^2\frac{\delta}{2}\right]} \qquad \ldots (24.16)$$

Comparing (24.15) and (24.16), we get

$$1+\frac{4r^2}{(1-r^2)^2}\left[\sin^2\left(\frac{\delta}{2}\right)\right]=2$$

or $$\sin^2\frac{\delta}{2} = \frac{(1-r^2)^2}{4r^2}$$

or $$\sin\frac{\delta}{2} = \frac{1-r^2}{2r}$$

$\therefore$ $$\delta = 2\sin^{-1}\left(\frac{1-r^2}{2r}\right) \quad \text{... (24.17)}$$

If $r = 1$, then $\delta = 0$

i.e., the visibility will be maximum.

This shows that δ at half-width increases with decrease in the value of r *i.e.*, sharpness decreases with decrease in the value of r. Theoretically sharpness is maximum for $r = 1$.

3. Coefficient of Finesse

If R represents the **energy reflectivity,** then $R = r^2$ where r is the **amplitude reflection coefficient.** In terms of R, we have

$$I_{max} = k\left[\frac{c^2}{(1-r^2)^2}\right] = k\left[\frac{c^2}{(1-R)^2}\right] \quad (\because \text{Eq. 24.11})$$

$$I_{min} = k\left[\frac{c^2}{(1+r^2)^2}\right] = k\left[\frac{c^2}{(1+R)^2}\right] \quad (\because \text{Eq. 24.12})$$

$$I = I_{\max}\left[1+\frac{4r^2}{(1-r^2)^2}\sin^2\frac{\delta}{2}\right]^{-1}$$

$$= I_{\max}\left[1+\frac{4R}{(1-R)^2}\sin^2\frac{\delta}{2}\right]^{-1} \quad (\because \text{Eq. 24.16})$$

Visibility $V = \dfrac{2r^2}{1+r^4} = \dfrac{2R}{1+R^2}$

From Eq. 24.16, the quantity $\dfrac{4r^2}{(1-r^2)^2}$ is called the coefficient of Finesse (F)

$\therefore$ Coefficient of Finesse $F = \dfrac{4r^2}{(1-r^2)^2} = \dfrac{4R}{(1-R)^2}$

and phase difference $$\delta = 2\sin^{-1}\left[\frac{1-r^2}{2r}\right] = 2\sin^{-1}\left[\frac{1-R}{2\sqrt{R}}\right]$$

Note. The outer faces of the two plates of Febry-perot interferometer are inclined to their respective inner faces to avoid interference effects due to multiple reflections and refractions. This is why all the faces are not made parallel.

24.7 RESOLVING POWER OF FABRY-PEROT INTERFEROMETER

As shown in Fig. 24.10 a ray originating from the point source S on S_1 after suffering internal reflection issues from B in the form of a parallel beam which interfere and are brought to focus at P by the lens L_2. The fringes formed in the focal plane of L_2 are circular.

If θ is the angle of inclination on the silvered surface of A and d is the air distance between the two plates A and B.

The path difference between successive rays = $2d \cos \theta$

For maxima, $$2d \cos \theta = n\lambda \qquad ...(i)$$

For small angles of incidence, $\cos \theta = 1$ and Eq. (*i*) becomes

$$2d = n\lambda$$

The change in the order n due to change in λ is given by

$$n\, d\lambda + \lambda\, dn = 0 \qquad [\because 2d = \text{a constant}]$$

Hence Resolving Power $$\frac{\lambda}{d\lambda} = -\frac{n}{dn}$$

It can be shown that $$dn = \frac{2}{\pi} \sin^{-1}\left[\frac{0.367\,(1-R)^2}{R}\right]^{\frac{1}{2}} \qquad ...(ii)$$

Now, according to Eq. (*i*) n is proportional to d and according to Eq. (*ii*) dn depends upon R. Therefore, the resolving power of a Fabry-Perot interferometer depends upon

(*i*) d – The separation between the two plates and

(*ii*) R – The fraction of incident light energy reflected at each of the surfaces

24.8 ADVANTAGES OF FABRY-PEROT INTERFEROMETER

The Fabry-Perot interferometer has the following advantages over Michelson's interferometer.

(*i*) The fringes obtained in it are much sharper than those obtained in Michelson's interferometer.

(*ii*) The resolving power of this interferometer is very high so that it can directly observe the fine structure of spectral lines. This is not possible with Michelson's interferometer.

(*iii*) Due to reasons given above Fabry-Perot interferometer is more suited to the study of hyper fine structure of spectral lines.

24.9 FABRY-PEROT ETALON

The Fabry-Perot interferometer, designed in 1899 by C. Fabry and A. Perot, represents a significant improvement over the Michelson interferometer, the basic setup maintains the same general principle of the Michelson interferometer. The difference between the two lies in the fact that the Fabry-Perot design contains plane surfaces that are all partially reflecting so that multiple rays of light are responsible for creation of the observed interference patterns. The general theory behind interferometry still applies to the Fabry-Perot model, however, these multiple reflections reinforce the areas where constructive and destructive effects occur making the resulting fringes much more clearly defined.

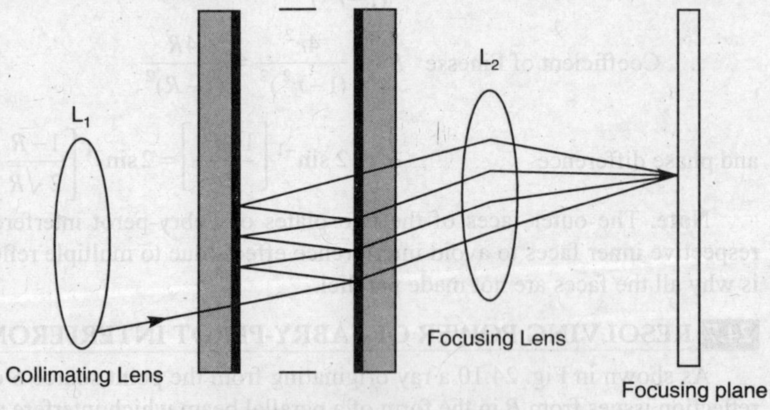

Fig. 24.13. Fabry-Perot Etalon

The Fabry-Perot interferometer consists of two flat, planar, surfaces that are almost, if not exactly, parallel (remember that circular fringes only occur when these plates are parallel!). The distance between these partially reflecting surfaces may be changed as well, just as with the Michelson interferometer. This characteristic actually represents the reason for deeming it an interferometer, for if the plates were immovable, it would be referred to as an Etalon. These differences can be seen as diagrammed below:

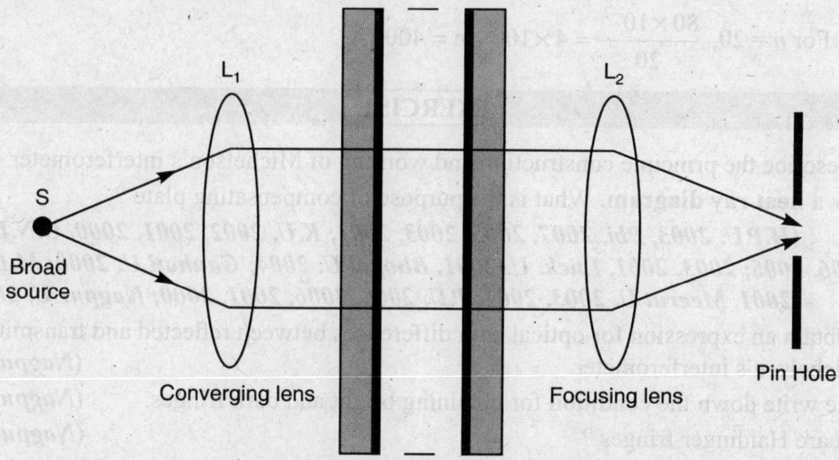

Fig. 24.14. Fabry-Perot Interferometer

Note that for the interferometer, the source is a broad one, while for the etalon it is a point source. In the interferometer picture (bottom), one of the planes will move

Example 24.12 *White light is incident normally on a Fabry-Perot interferometer with plate separation of 4×10^{-6} m. Calculate the wavelengths for which there are interference maxima transmitted beam in the range of 4000 Å to 5000Å.* ***(Nagpur U. 2010)***

Solution. As light is incident normally

$$\theta = 0 \text{ or } \cos\theta = 1$$

If d is the distance between the two plates, then for maxima the path-difference

$$2d\cos\theta = n\lambda$$

or

$$2d = n\lambda$$

$$\therefore \quad \lambda = \frac{2d}{n} = \frac{2 \times 4 \times 10^{-6}}{n} = \frac{80 \times 10^{-7}}{n} \text{ m}$$

For values of λ to be in the range of 4000 Å to 5000 Å *i.e.*, 4×10^{-7} m to 5×10^{-7} m; the values of n for

$$\lambda = 4 \times 10^{-7}, n = \frac{80 \times 10^{-7}}{4 \times 10^{-7}} = 20$$

and

$$\lambda = 5 \times 10^{-7}, n = \frac{80 \times 10^{-7}}{5 \times 10^{-7}} = 16$$

Thus n lies between 16 and 20.

(*i*) For $n = 16$, $\lambda_1 = \frac{80 \times 10^{-7}}{16} = 5 \times 10^{-7} \text{ m} = 5000 \text{ Å}$

(*ii*) For $n = 17$, $\lambda_2 = \frac{80 \times 10^{-7}}{17} = 4.7059 \times 10^{-7} \text{ m} = 4706 \text{ Å}$

(*iii*) For $n = 18$, $\lambda_3 = \frac{80 \times 10^{-7}}{18} = 4.4444 \times 10^{-7} \text{ m} = 4444 \text{ Å}$

(iv) For $n = 19$, $\frac{80\times10^{-7}}{19} = 4211\times10^{-7}\ \text{m} = 4211\ \text{Å}$

(v) For $n = 20$, $\frac{80\times10^{-7}}{20} = 4\times10^{-7}\ \text{m} = 4000\ \text{Å}$

EXERCISE

1. (a) Describe the principle construction and working of Michelson's interferometer. **Draw a neat ray diagram.** What is the purpose of compensating plate ? ***(H.P.U. 2003, Pbi. 2007, 2005, 2003, 2001; K.U. 2002, 2001, 2000, G.N.D.U. 2008, 2006, 2005; 2003, 2001, Luck. U. 2001, Bhopal U. 2004; Gauhati U. 2000; M.D.U. 2003, 2001, Meerut U. 2003, 2001; P.U. 2008, 2006, 2001, 2000; Nagpur U. 2001, 2008)***

   (b) Obtain an expression for optical path differences between reflected and transmitted beams in Michelson's interferometer. ***(Nagpur U. 2008)***

   Hence write down the condition for obtaining bright and dark fringes. ***(Nagpur U. 2011)***
2. What are Haidinger fringes ? ***(Nagpur U. 2011)***
3. Explain how circular fringes are produced in Michelson's interferometer ? Show that the radii of circular fringes obtained by the interferometer are proportional to the square root of natural numbers. ***(G.N.D.U. 2003; Pbi. U. 2005)***
4. How do you obtain localised fringes in Michelson's interferometer ? How straight fringes are obtained ? ***(Kerala, U. 2001)***
5. How white light fringes are produced in Michelson's interferometer ?
6. How will you find the wavelength of monochromatic light with Michelson's interferometer ? ***(Madurai U.M.D.U. 2003)***
7. How will you measure small difference in the wavelength of two waves with Michelson's interferometer ?
8. How will you use Michelson's interferometer for finding the difference between two D-lines of Sodium ? ***(H.P.U. 2003, 2002; Pbi. U. 2007, 2003, 2001; Meerut 2005; Luck. U. 2001; P.U. 2004, 2005; G.N.D.U. 2006)***
9. How will you use Michelson's interferometer to determine the thickness of a thin transparent sheet of refractive index μ. ***(Gauhati. U. 2000; K.U. 2001)***
10. How will you use a Michelson's interferometer to find the refractive index of a thin transparent sheet. ***(Meerut U. 2006; H.P.U. 2000; K.U. 2001)***
11. What are the applications of Michelson's interferometer ?
12. Describe Michelson's interferometer. How will you use it to calibrate a metre in terms of a standard wavelength ? ***(H.P.U. 2002; M.D.U. 2001)***
13. Describe Twyman and Green interferometer ? Mention its merits in terms of precision.
14. What do you mean by multiple beam interferometry ? Given one example of such interferometer.
15. Describe the principle, construction and production of interference fringes with Fabry-Perot interferometer. ***(G.N.D.U. 2008, 2007, 2006; Nagpur U. 2011, 2010, 2008; Pbi. U. 2002; Luk. U. 2001; P.U. 2007, 2006, 2001, 2000, Gauhati U. 2000; Utkal U. 2003)***
16. Discuss the intensity distribution of Fabry-Perot interference fringes and find the ratio of I_{max} to I_{min}. ***(Pbi. U. 2006; P.U. 2006; Luck. U. 2001)***
17. Explain the visibility and sharpness of the fringes obtained with Fabry-Perot interferometer. ***(P.U. 2000)***

18. Explain the finesse of Fabry-Perot interferometer.

19. How Fabry-Perot interferometer is used to determine the difference between two closely situated wavelengths ? ***(Pbi. U. 2006, 2002; P.U. 2000; Gujrat U. 2004, Nagpur U. 2009, 2008)***

20. Why the outer surfaces of the two plates of Fabry-Perot interferometer are not exactly parallel to the reflecting surfaces ? ***(Pbi. U. 2007)***

21. On what factors does the resolving power of Fabry-Perot interferometer depends ?

22. What are the advantages of Fabry-Perot interferometer over Michelson's interferometer ? ***(Nagpur U. 2010)***

23. Why Fabry-Perot interferometer is more suited for study of hyperfine structure of spectral lines. ***(P. U. 2001)***

MULTIPLE CHOICE QUESTIONS

24.1 In Michelson's interferometer, when M_1 and M_2 mirrors are perfectly perpendicular to the direction of two beams, the fringes obtained are:
(*a*) elliptical (*b*) straight (*c*) circular (*d*) inclined

24.2 In Michelson's interferometer, when M_1 and M_2 mirrors are made slightly inclined, the fringes obtained are :
(*a*) elliptical (*b*) straight (*c*) circular (*d*) inclined

24.3 In Fabry-Perot interferometer fringes are formed due to:
(*a*) division of wavefront (*b*) division of rays
(*c*) division of amplitude (*d*) None of these

24.4 In Michelson's interterometer, a compensating plate is used for
(*a*) to make paths equal between two rays
(*b*) to make frequency equal between two rays
(*c*) to make apparatus sturdy
(*d*) to make the apparatus lookwise good.

24.5 Michelson's interferometer was originally designed to test:
(*a*) the existence of earth itself (*b*) the existence of absolute space
(*c*) the existence of matter on earth (*d*) none of these

ANSWERS

24.1 (*c*)	**24.2** (*b*)	**24.3** (*c*)	**24.4** (*a*)	**24.5** (*b*)

FRESNEL DIFFRACTION

25

Chapter

25.1 DIFFRACTION

To all outward appearance, light seems to travel in a straight line. A very careful observation, however, reveals that light does suffer some deviation from its straight path when it passes close to the edges of opaque obstacles or narrow slits. It is found that some light bends into the region of geometrical shadow. The deviation of light is very small when the dimensions of the obstacle or aperture are large, as compared to the wavelength of light and becomes much pronounced when the dimensions of the obstacle or the aperture are comparable with the wavelength of light. Even in such a case the phenomenon is observed on careful investigation.

The bending of light round corners and spreading of light waves into the geometrical shadow of an object is called **diffraction.**

Difference between interference and diffraction

Interference	*Diffraction*
1. Interference is the result of interaction of light coming from two different wave fronts originating from the same source.	Diffraction pattern is the result of interaction of light coming from different parts of the same wave front.
2. Interference fringes *may* or *may not* be of the same width.	Diffraction fringes are *not* of the same width.
3. Points of minimum intensity are perfectly dark.	Points of minimum intensity are not perfectly dark.
4. All bright bands are of uniform intensity.	All bright bands are not of same intensity.

25.2 TYPES OF DIFFRACTION

There are two types of diffraction phenomenon, known as *Fresnel's diffraction* and *Fraunhofer diffraction.*

1. Fresnel's diffraction. In this type of diffraction, the source of light or screen or both are at finite distances from the obstacle or aperture. No lenses are used to make the rays parallel or convergent. The incident wave front is not plane but is either spherical or cylindrical. As a result, the phase of secondary wavelets is not the same at all points in the plane of the aperture or the obstacle causing diffraction. The resultant amplitude at any point of the screen is obtained by the mutual interference of secondary wavelets from different elements of unblocked portions of wave front.

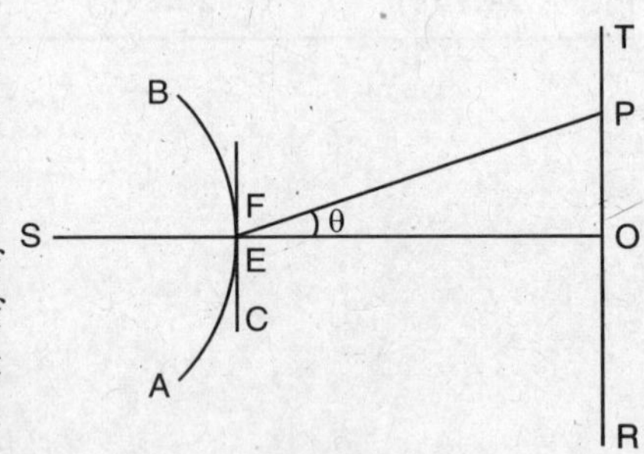

Fig. 25.1

To explain this, Fresnel made the following assumptions:

(*i*) A wave front could be divided into a large number of small elements or zones called Fresnel's zones, each of small area. The resultant effect at any point *P* on the screen will depend on the combined effect of all the secondary waves originating from the various zones.

(*ii*) The effect at any point due to a particular zone will depend on the distance of the point from the zone.

(*iii*) The effect at P will also depend upon the obliquity of the point with respect to the zone. The obliquity factor is proportional to $(1 + \cos\theta)$, where θ is the angle which PE makes with EO. Considering an elementary wave front at E, the effect is maximum at O as $\theta = 0$ and $\cos\theta = 1$. As we move from O towards P the value of θ increases and hence resultant effect decreases. Similarly in a direction EF tangential to the wave front, the resultant effect is one half of that at O as $\theta = 90°$ and $\cos\theta = 0$. In the direction ES, the resultant effect is zero as $\theta = 180°$ and $\cos\theta = -1$. This explains the non-existence of the wave in the backward direction.

Fresnel type of diffraction is produced when light suffers diffraction at a straight edge, a narrow slit, a thin wire, a small circular hole or an obstacle. An application of this type of diffraction is met with in the construction of a zone plate.

2. Fraunhofer diffraction. In this type of diffraction, the source of light and the screen are effectively at infinite distances from the aperture or obstacle, which causes diffraction. This may be achieved by using two convex lenses, one to make the light from the source parallel before it falls on the aperture and the other to focus the light after diffraction on the screen. This arrangement in fact removes the source and the screen to infinity.

The incident wave front is plane and *the secondary wavelets, which originate from the unblocked portions of the wave front, are in the same phase at every point in the plane of the aperture.* Thus diffraction is produced by the interference between parallel rays which are brought into focus with the help of a convex lens.

The most important example of Fraunhofer type of diffraction is met with in the case of a plane transmission grating or concave reflection grating.

As stated above, it is in *Fraunhofer type* of diffraction that secondary wavelets are in the same phase at all points in the plane of the aperture.

25.3 FRESNEL HALF-PERIOD ZONES

Consider a plane $ABCD$ which represents a plane wave front of monochromatic light of wavelength λ travelling from left to right. According to *Huygen's principle* every point on the wave front is regarded as the origin of secondary wavelets and at a given instant every one of these secondary wavelets passes through the point O. The resultant effect at O due to the whole wave front will be equal to the total effect of all the disturbances reaching from different portions of the wave front. This is known as **Huygen's Fresnel theory.** To find the resultant effect we divide the entire wave front into concentric zones as follows.

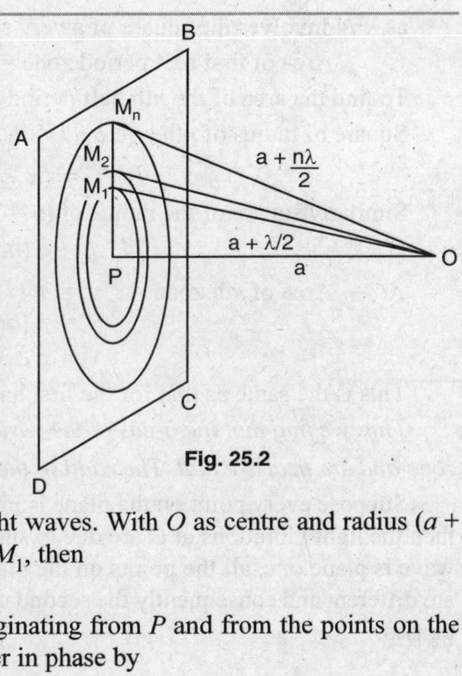

Fig. 25.2

From O drop a perpendicular on $ABCD$ at the point P. This point is the foot of the perpendicular from O on $ABCD$ and is called the *pole* of the wave with respect to O. Let OP be equal to a and λ be the wavelength of light waves. With O as centre and radius $(a + \lambda/2)$ draw a sphere, cutting the wave front in a circle at M_1, then

$$OM_1 = a + \lambda/2 \quad \text{or} \quad OM_1 - OP = \lambda/2$$

This simply means that the secondary wavelets originating from P and from the points on the circumference of the circle M_1, on reaching O, will differ in phase by

$$\frac{2\pi}{\lambda}(OM_1 - OP) = \pi$$

Now π radian is equivalent to phase difference $T/2$. Hence the area enclosed by the circle M_1 is called **Fresnel's first half-period zone.**

Similarly we can draw other spheres of radii $(a + 2\,\lambda/2)$ $(a + 3\,\lambda/2)$ $(a + 4\,\lambda/2)$... which will intersect the wavefront in circles M_2, M_3.... This construction divides the wave front into a number of Fresnel's half-period zones. It may be noted that the first half-period zone is a circle, while the second half-period zone is an annular ring between M_1 and M_2 and so on.

Why these are so called ? These are known as **half-period zones,** because the waves reaching O from P and M_1 differ in phase by π or $T/2$ or $\lambda/2$. Similarly the waves reaching O from M_1 and M_2 also differ in phase by π or $T/2$ or $\lambda/2$. Thus the phase difference between wavelets from successive half-period zones is π radian or *half-period.* This justifies their nomenclature as *HPZ's.*

25.4 RECTILINEAR PROPAGATION OF LIGHT

To explain rectilinear propagation of light we have to find the resultant effect of the whole wave front at the point O, (fig. 25.2). To do this divide the whole wave front into **half-period zones.** The problem is then reduced to find the resultant effect of a large number of disturbances originating from the various annular rings into which the whole wave front is divided. *The amplitude due to wavelets produced by each zone reaching O depends upon the following factors:*

(*i*) It is directly proportional to the area of the zone.

(*ii*) It varies inversely as the distance of zone from the point O. Greater the distance smaller is the amplitude reaching O.

(*iii*) It varies with the *obliquity factor* $(1 + \cos\theta)$. The amplitude decreases with increasing obliquity.

The areas of each of the half-period zones are approximately equal, as shown below.

The square of the radius of the *first half-period zone*

$$PM_1^2 = (a+\lambda/2)^2 - a^2 = (a^2+\lambda^2/4 + a\lambda - a^2) = a\lambda \quad \text{... } (i)$$

as $\lambda^2/4$ involves the square of a very small quantity and can be neglected as compared to $a\,\lambda$.

∴ Area of first half period zone $= \pi\, a\, \lambda$

To find the area of the nth half period zone.

Square of radius of nth circle

$$PM_n^2 = (a + n\lambda/2)^2 - a^2 = an\lambda$$

Similarly square of the radius of $(n-1)$th circle

$$PM_{n-1}^2 = a\,(n-1)\lambda$$

∴ Area of nth zone

$$= \pi\,(PM_n^2 - PM_{n-1}^2)$$
$$= \pi\,[an\lambda - a(n-1)\lambda]$$
$$= \pi\, a\, \lambda$$

This is the same as that for the first half period zone.

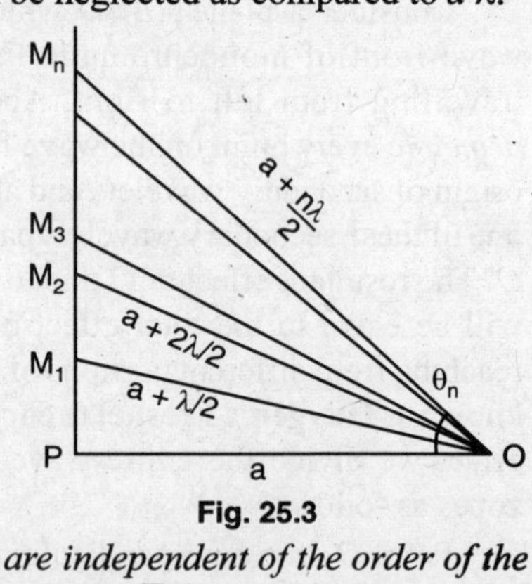

Fig. 25.3

Thus we find that the areas of the various half period zones are independent of the order of the zone and are nearly equal. The radii of these zones are proportional to $\sqrt{n}$ *where* $n = 1, 2, 3$ etc.

Suppose every point on the plane is in a state of vibration and sends out secondary wave trains, then the light vibrations at O are due to superposition of these waves (Fig. 25.3). Since the incident wave is plane one, all the points on the plane $ABCD$ are in the same phase. Their distances from O are different and consequently the secondary waves which these points send out, reach O in different phases.

The phase of the wavelets coming from P is zero. The distance of O from points in the first zone lies between a and $(a + \lambda/2)$. Hence the phase of wavelets from intermediate points between P and M_1 will vary from 0 to π. Thus

Average phase of all wavelets from first zone $= \frac{0+\pi}{2} = \frac{\pi}{2}$ radian.

Similarly the phase difference of wavelets from M_1 and M_2 will lie between π and 2π. Hence Average phase-difference of wavelets from second zone $= \frac{\pi+2\pi}{2} = \frac{3\pi}{2}$ radian. This is opposite to that due to the wavelets from the first half-period zone.

Similarly the average phase difference of wavelets from the third zone will be $5\,\pi/2$ and fourth zone will be $7\,\pi/2$ and so on.

It is clear that the resultant phase-difference between two consecutive zones is π radian, while that between alternate zones is 2π radian. Thus if the resultant amplitude of wavelets from first half-period zone is *positive,* that from the second half-period zone will be *negative,* that from third zone positive and fourth zone negative and so on (Fig. 25.4).

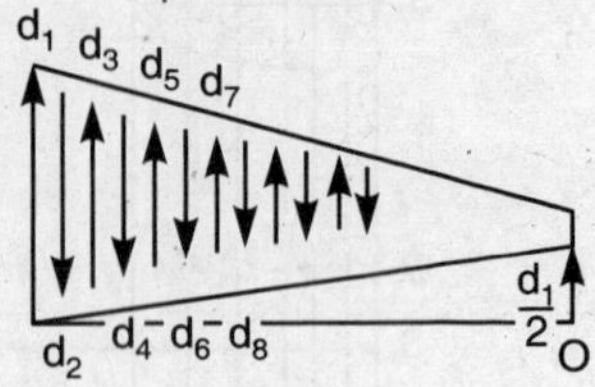

Fig. 25.4

Let d_1, d_2, d_3 etc., be the resultant amplitudes at O due to first, second, third etc., half-period zones respectively, then

Resultant amplitude $D = d_1 - d_2 + d_3 - d_4 \ldots\ldots \pm d_n$

The amplitudes of the successive zones depend upon the obliquity factor $(1 + \cos\theta_n)$ where θ_n is the angle which the direction of O from the nth zone makes with PO. It is assumed that the obliquity factor varies slowly so that it may be regarded constant over a single half-period zone. As θ increases from zero, $\cos\theta$ decreases very slowly first, but more rapidly for larger values of θ. Thus, successive amplitudes decrease a first, slowly but more rapidly for higher values of n. Hence the amplitude goes on decreasing with the order of zones. Thus each term in the above series is slightly less than the one preceding it and is greater than the one succeeding it. To a first approximation, we can write

$$d_2 = \frac{d_1+d_3}{2},\ d_4 = \frac{d_3+d_5}{2} \text{ and so on.}$$

The above series can now be rewritten in the following form

$$D = \frac{d_1}{2} + \left(\frac{d_1+d_3}{2} - d_2\right) + \left(\frac{d_3+d_5}{2} - d_4\right) + \ldots + \frac{d_n}{2}$$

$$= \frac{d_1}{2} + \frac{d_n}{2} \text{ (if } n \text{ is odd)}$$

$$D = \frac{d_1}{2} + \frac{d_{n-1}}{2} - d_n \text{ (if } n \text{ is even)}$$

If n is sufficiently large, the effect due to the nth zone becomes negligible and resultant amplitude due to the whole wave

$$D = \frac{d_1}{2}$$

Thus the amplitude due to complete wave front is half of what it would be caused by the first half-period zone.

As intensity I is proportional to the square of the amplitude

$$I = \frac{d_1^2}{4}$$

Hence the intensity at O due to the wavelets from all the zones is equal to *one fourth* of the intensity due to the waves from the *first half-period zone.*

If an obstacle is placed at P, (Fig. 25.2) the resultant disturbance at O is equal to half the disturbance due to the *first exposed zone.* As the displacement decreases rapidly with the order of

the zone, if the obstacle at P blocks a considerable number of half-period zones the effect will be negligible and practically no light will be received at O. In other words, *light travels approximately in a straight line.*

25.5 CYLINDRICAL WAVE FRONT

Consider a long and narrow slit S_1S_2 held perpendicular to the plane of the paper and illuminated by monochromatic light of wavelength λ. It gives rise to cylindrical wave front with the slit as the axis. To find the effect of the wave front at any point O in front of it, the wave front may be divided into half-period strips or elements as described above.

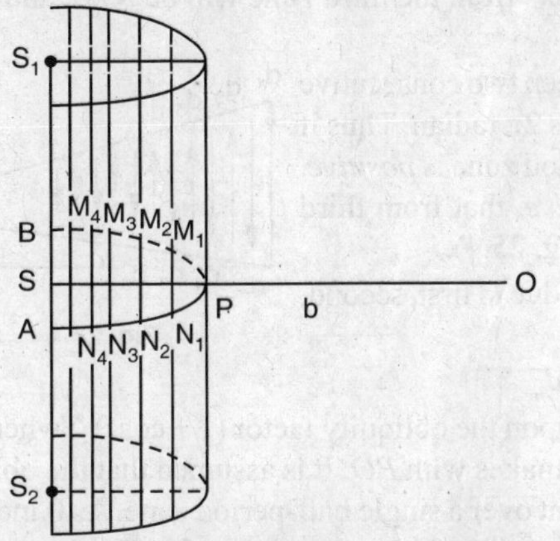

Fig. 25.5

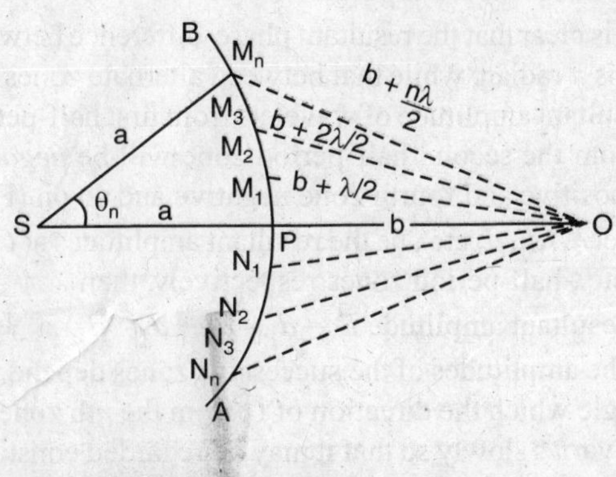

Fig. 25.6

Fresnel's construction. Draw a perpendicular OS from O on $S_1\,S_2$ cutting the wave front at P, then P is the pole of the wave front with respect to the point O. Consider an equatorial section APB through P in the plane of the paper. Let $OP = b$.

With O as centre and radius $\left(b+\dfrac{\lambda}{2}\right)\left(b+\dfrac{2\lambda}{2}\right)\left(b+\dfrac{3\lambda}{2}\right)$.... etc. draw arcs cutting the section APB of the wave front at points M_1N_1, M_2N_2, M_3N_3.... etc. respectively. Through the points M_1, N_1, M_2, N_2 ... etc. draw lines parallel to the axis of the wave front *i.e.*, the length of the slit, dividing the wavefront into similar halves, as shown in Fig. 25.7. Considering one half of the wavefront the area of the wave front between the line through P and M_1 is the first **half period strip.**

The area between the lines through M_1 and M_2 is the **second half-period** strip and so on.

The amplitude of the waves reaching O, due to a half-period strip depends upon.

(*i*) area of the strip

(*ii*) average distance from O and

(*iii*) obliquity of the strip

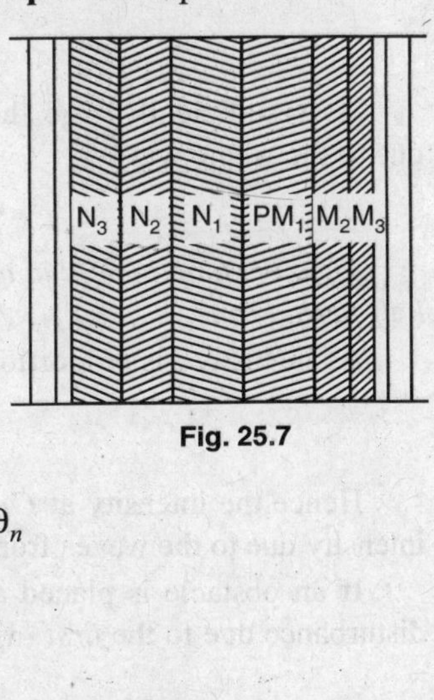

Fig. 25.7

Resultant amplitude. As the lengths of the strips are the same, the areas of successive strips are proportional to the arcs PM_1, M_1M_2, M_2M_3...etc. To calculate the width of the strip, we know from Fig. 25.6.

$$OM_n = b+\frac{n\lambda}{2},$$

In $\Delta\, OM_nS$, we have

$$OM_n^2 = SM_n^2 + OS^2 - 2OS \cdot SM_n \cos\theta_n$$

or
$$\left(b+\frac{n\lambda}{2}\right)^2 = a^2 + (a+b)^2 - 2(a+b)a\cos\theta_n$$

$$= a^2 + (a+b)^2 - 2(a+b)a\left(1 - \frac{\theta_n^2}{2}\right)$$

$$\because \quad \cos\theta_n = 1 - \frac{\theta_n^2}{2!} + \frac{\theta_n^4}{4!} - \frac{\theta_n^6}{6!} + \ldots\ldots\ldots$$

$$\therefore \quad b^2 + \frac{n^2\lambda^2}{4} + bn\lambda = a^2 + a^2 + b^2 + 2ab - 2a^2 - 2ab + a(a+b)\theta_n^2$$

Neglecting λ^2, we get

$$bn\lambda = a(a+b)\,\theta_n^2$$

or

$$\theta_n = \sqrt{\frac{bn\lambda}{a(a+b)}} = k\sqrt{n}$$

Now

$$\text{arc } PM_n = a\theta_n = ak\sqrt{n}$$

$$PM_1 = ak, PM_2 = ak\sqrt{2}, PM_3 = ak\sqrt{3} \ldots$$

Hence

$$\text{arc } PM_1 = ak$$

$$M_1M_2 = ak(\sqrt{2}-1) = 0.414\,ak$$

$$M_2M_3 = ak(\sqrt{3}-\sqrt{2}) = 0.318\,ak$$

$$M_3M_4 = ak(\sqrt{4}-\sqrt{3}) = 0.268\,ak$$

$$M_4M_5 = ak(\sqrt{5}-\sqrt{4}) = 0.236\,ak \text{ and so on.}$$

This shows that the areas of the strips decrease rapidly first and then more slowly as the order of the strips increases. Hence successive higher order strips cancel the effect of each other in pairs, being in opposite phase and having practically equal amplitudes. Thus the effect O is only due to the first few half-period elements.

If d_1, d_2, d_3 .... etc. be the amplitudes at O due to the first, second, third etc. half-period strips on one side, then the resultant amplitude at O due to either half of the wave front

$$D = d_1 - d_2 + d_3 - d_4 + d_5 \ldots.$$

The amplitude from the alternate strips have opposite signs, because the waves from two consecutive strips differ in path by $\frac{\lambda}{2}$ and hence reach O in opposite phase. The resultant amplitude due to each half of the wavefront

$$D = \frac{d_1}{2}$$

$\therefore$ Resultant amplitude due to the whole wavefront

$$= \frac{d_1}{2} \pm \frac{d_1}{2} = d_1 \qquad \ldots (25.1)$$

Plus sign is taken when n is odd and minus sign when n is even.

25.6 DIFFRACTION AT A STRAIGHT EDGE

Consider a straight edge C and an illuminated narrow slit S parallel to each other and perpendicular to the plane of the paper.

It is required to determine the illumination of the screen TR placed perpendicular to the plane of paper. Join SC and produce it to meet the screen at O. A straight line on the screen through O perpendicular to the plane of the paper defines the limits of the geometrical shadow.

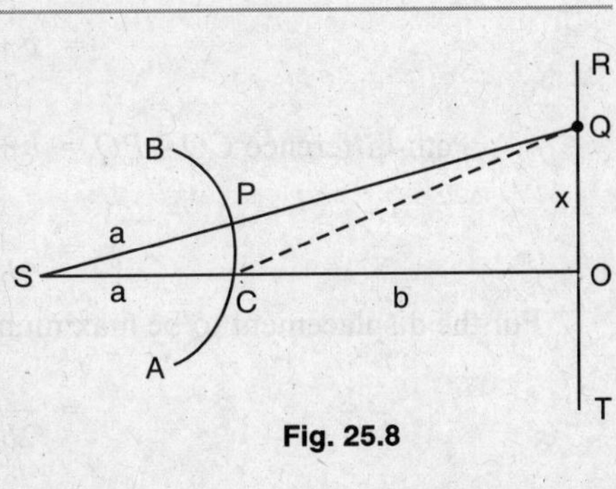

Fig. 25.8

According to the laws of geometrical optics the illumination should start abruptly at O. In practice however dark and bright bands of unequal width and of varying intensity are observed in the region above O. The intensity goes on decreasing gradually, as we move into the geometrical shadow below O.

To study the intensity at any point Q distant x above O on the screen, let ACB be the incident wavefront at C originating from the source S. Join QS, cutting the wavefront at P, then P is the pole of the wavefront with respect to the point Q. With P as the pole construct *Fresnel's half-period* elements. The effect at Q depends upon the number of half-period elements contained in PC and the effect of the upper half of the wave PB.

Intensity at *O*. The effect at O is due to the upper *half* of the wavefront only. Hence the displacement at O is *half of the displacement* which would have been there if the whole wavefront was effective. The intensity at O, therefore, is $\frac{1}{4}$ of the intensity at a point far removed from O where the whole of the wavefront is effective.

Geometrical shadow. As we move into the geometrical shadow, the pole P moves from C towards A and first, second, third etc. half-period elements are intercepted respectively. The intensity, therefore, goes on decreasing gradually.

Illuminated part. As we move from O towards R, first, second, third, etc., half-period elements are exposed. The illumination at a point Q, distant x from O on the screen, is due to *the complete half of the wave* surface PB together with the resultant of the number of half-period elements contained in CP.

The amplitude at Q will be maximum or minimum, according as *CP contains odd* or *even number of half-period elements.* The intensity at a maximum goes on decreasing and at a minimum goes on increasing and finally when we move a large distance from O, we get a uniform illumination. This is shown in Fig. 25.9.

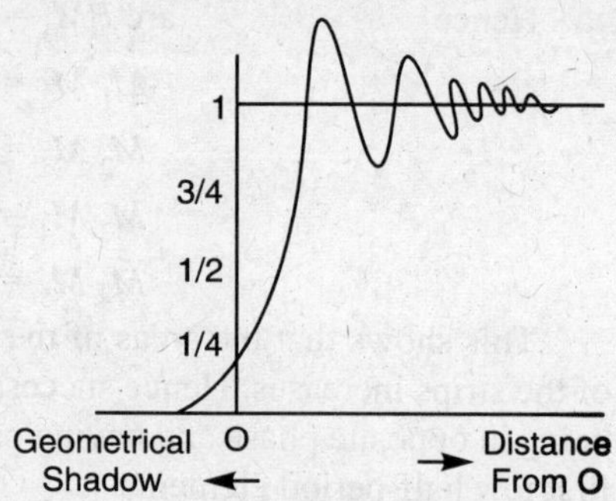

Fig. 25.9

The number of half-period elements contained in CP depends upon the path-difference $CQ - PQ$.

Suppose $SC = a$ and $CO = b$, then

$$CQ = \sqrt{b^2 + x^2}$$

$$= b\left(1 + \frac{x^2}{b^2}\right)^{\frac{1}{2}} = b + \frac{1}{2}\cdot\frac{x^2}{b}$$

Since x is very small as compared to b, therefore, other terms are negligible.

Similarly $$SQ = \sqrt{(a+b)^2 + x^2} = a + b + \frac{x^2}{2(a+b)}$$

Hence $$PQ = SQ - SP = a + b + \frac{x^2}{2(a+b)} - a$$

$$= b + \frac{x^2}{2(a+b)}$$

$\therefore$ Path-difference $$CQ - PQ = b + \frac{x^2}{2b} - b - \frac{x^2}{2(a+b)}$$

$$= \frac{ax^2}{2b(a+b)} \qquad \text{... (25.2)}$$

For the displacement to be maximum

$$= \frac{ax^2}{2b(a+b)} = (2n+1)\frac{\lambda}{2}$$

or $$x = \sqrt{\frac{b(a+b)\,(2n+1)\lambda}{a}} \quad \text{where } n = 0, 1, 2 \ldots \qquad \ldots (25.3)$$

For displacement to be minimum

$$\frac{ax^2}{2b(a+b)} = n\lambda$$

or $$x = \sqrt{\frac{b(a+b)\,2n\lambda}{a}} \quad \text{where } n = 1, 2, 3 \ldots \qquad \ldots (25.4)$$

Since $x \propto \sqrt{n}$ for a dark band and $x \propto \sqrt{2n+1}$ for a bright band the fringe width goes on decreasing as we move away from O. This will be clear from the following example.

When $n = 1 \quad x_1 = k\sqrt{1} = k$

where $$k = \sqrt{\frac{2b(a+b)\lambda}{a}}$$

For $n = 2; \quad x_2 = k\sqrt{2} = 1.414\,k$

Width of the first bright fringe $= 1.414\,k - k = 0.414\,k$

Similarly width of 2nd bright fringe $= k\sqrt{3} - k\sqrt{2}$

$= 1.732k - 1.414k = 0.318\,k$

and width of 3rd bright fringe $= k\sqrt{4} - k\sqrt{3} = 2\,k - 1.732\,k = 0.268\,k.$

This shows that *fringe width or the separation between successive maxima goes on decreasing as we move away from the region of geometrical shadow.*

Determination of wavelength. A straight edge, say a sharp razor blade is set with its edge parallel to the slit on one of the stands on the optical bench. The slit is illuminated with a monochromatic light. Diffraction bands of unequal width and decreasing intensity are observed in the field of view of the micrometer eye-piece. The position of the first maximum intensity band and the one (say nth) most distant and clearly visible band is noted, then

$$x_1 = \sqrt{\frac{b(a+b)\lambda}{a}} \qquad (\because n = 0)$$

For the nth maximum intensity band, we get

$$x_n = \sqrt{\frac{b(a+b)\,(2n+1)\lambda}{a}}$$

$$\therefore \quad x_n - x_1 = \sqrt{\frac{b(a+b)\lambda}{a}}\,[\sqrt{2n+1} - 1] \qquad \ldots (25.5)$$

The distance b from the straight edge to the eye-piece and the distance a between the slit and the straight edge are measured. Hence knowing $(x_n - x_1)$, we can find the wavelength of light.

Example 25.1 *A narrow slit illuminated by light of wavelength 5890 Å is located at a distance of 0.1 m, from a straight edge. If the measurements are made at a distance of 0.5m from the edge, calculate the distance between the first and the second dark band.* ***(Nagpur U. 2008)***

Solution. For the nth dark band

$$x_n = \sqrt{\frac{b(a+b)\,2n\lambda}{a}}$$

where a is the distance between the slit and the straight edge and b that between the straight edge and the screen.

or $$x_2 - x_1 = \sqrt{\frac{b(a+b)\,2\lambda}{a}}\,(\sqrt{2} - 1)$$

Now $a = 0.1$ m, $b = 0.5$ m and $\lambda = 5890 \times 10^{-10}$ m

$$\therefore \quad x_2 - x_1 = \sqrt{\frac{0.5(0.1+0.5)\,2\times 5890\times 10^{-10}}{0.1}}\,(1.4142-1)$$

$$= 0.7786 \times 10^{-3}\ \text{m}$$

25.7 DIFFRACTION AT A WIRE

Let S represent a narrow rectangular slit placed parallel to the wire of thickness XY and perpendicular to the plane of the paper. This will give rise to a geometrical shadow MN on the screen RT.

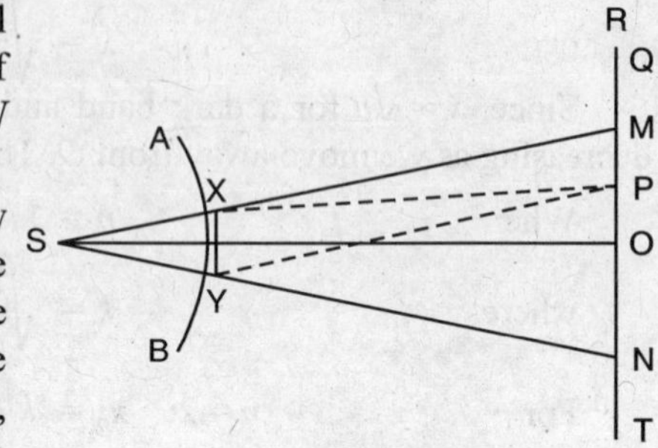

Fig. 25.10

The effect at a point Q outside the geometrical shadow is the same as that due to a straight edge at X and hence diffraction bands of unequal width and intensity will be seen above M and similarly below N. These bands are independent of the thickness of the wire, as on either side, the effect of the other half of the wave is negligible, because most important half-period elements are cut off due to the finite width of the wire.

Within the geometrical shadow, interference fringes are observed. The effect due to the portion AX of the cylindrical wavefront, at any point P within the geometrical shadow, is due entirely to a few half-period elements at X, so that it may be regarded as a small luminous source at X. Similarly the effect due to the portion BY of the wavefront is equal to a small source at Y.

The effect at any point P within the geometrical shadow MN will depend upon the path-difference $PY - PX$. The point P will be bright if $PY - PX = n\lambda$ and dark if it is equal to $(2n+1)\frac{\lambda}{2}$. These fringes are of equal width and the fringe width $\beta = \frac{D}{d}\lambda$ where D is the distance of the screen from the obstacle XY and d the thickness of the obstacle, *i.e.*, the diameter of the wire.

The centre O will be bright because at O the waves from X and Y will always meet in phase.

The variation of intensity of the interference fringes produced within the geometrical shadow is shown in Fig. 25.11 (*a*).

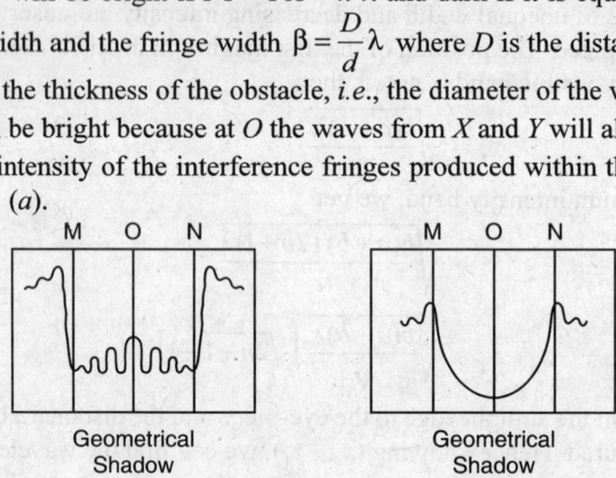

Fig. 25.11

As the diameter of the wire is increased the fringe width decreases and when the thickness of the wire is sufficiently large, the interference fringes will disappear and only the diffraction bands outside the limits of the geometrical shadow will be visible as shown in Fig. 25.11 (*b*).

To measure thickness of a wire. Focus the micrometer eye-piece on the interference fringes obtained within the geometrical shadow of the wire. The fringes are of equal width. Find the mean fringe width by measuring the shift of the cross-wire for a known number of fringes. The diameter of the wire can be found out from the expression—

$$\beta = \frac{D}{d}\lambda. \qquad \text{... (25.6)}$$

Here d is the diameter of the wire, D the distance between the obstacle and cross-wires of micrometer eye-piece and λ the wavelength of monochromatic light used.

25.8 DIFFRACTION DUE TO A NARROW SLIT

Let C and F be the edges of a rectangular aperture and AB the section of a cylindrical wavefront of which the axis coincides with S. The edges C and F of the slit will cast a geometrical shadow above M and below N on the screen held normal to SO. The space between M and N will be illuminated.

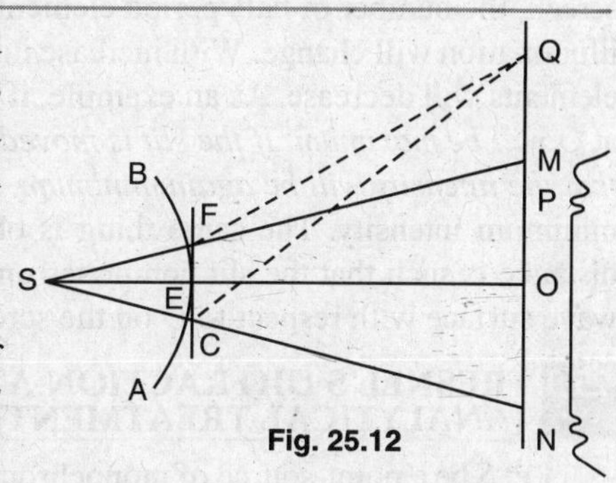

Fig. 25.12

1. Wide slit. When the slit is wide enough so that it allows *about a hundred half-period elements* from each half of the wavefront to pass through the aperture, the edges C and F will act like independent straight edges. This will result in producing diffraction bands of unequal width at M and N inside the geometrical bright portion, similar to the bands produced by a straight edge. As we move from M and N towards O, the general illumination increases and the visibility of the bands decreases, so that there is a uniform illumination about O. As we proceed beyond M and N the illumination rapidly falls to zero. This is shown on the side of Fig. 25.12.

2. Narrow slit. When the slit is narrow so that it allows only a few half-period elements, say *five* (odd) to pass through each half of the wavefront with respect to O, the illumination at O is again maximum. If the number of half-period elements contained in each half of the wavefront that passes through the slit is even, the point O will have a minimum intensity.

Keeping the screen fixed if we move to a point P such that *four* half-period elements from the upper half of the wavefront and *six* half-period elements from the lower half of the wavefront pass through the aperture, the resultant amplitude of the wavelets from the upper half of the wavefront will be $d_1 - d_2 + d_3 - d_4 = 0$ and that of the wavelets from the lower half will be $d_1 - d_2 + d_3 - d_4 + d_5 - d_6 = 0$. Hence the illumination at P will be minimum.

If the point P is moved a little further, so that *three* half-period elements from the upper half and *seven* from the lower half of the wavefront pass through the aperture, then the amplitude of the waves from three half-period elements will be $d_1 - d_2 + d_3 = d_3$ and that from the *seven* half-period element will be

$$d_1 - d_2 + d_3 - d_4 + d_5 - d_6 + d_7 = d_7 \qquad \text{... (25.7)}$$

Hence the resultant amplitude is $d_3 + d_7$ which will produce a maximum intensity. In general–

"*As P moves up, the illumination is maximum or minimum according as odd or even number half-period elements, from each half of the wavefront pass through the aperture and reach P*".

This shows that a set of alternately dark and bright bands parallel to the edges of the geometrical shadow will be seen inside MN.

For a point Q within the geometrical shadow, one-half of the wavefront is entirely cut off. The intensity at Q will be maximum or minimum according as CF contains odd or even number of half period elements with respect to Q. If the path difference $CQ - FQ = (2n + 1)\,\lambda/2$ *i.e., an odd multiple of* $\lambda/2$, there will be odd number of half-period elements in CF and hence the point Q will be bright as one half-period element will be in excess. If the path difference $CQ - FQ = 2n\,\lambda/2$ there will be even half-period elements in CF and hence Q will have minimum intensity. Thus bands within the geometrical shadow beyond M and N are also *diffraction bands* of decreasing intensity and width.

3. Aperture very narrow. When the aperture comprises *one* or *less than one half*-period element with respect to O, it is always bright and there are no bands between M and N. For a point Q in the geometrical shadow the number of elements contained in the slit may be more than one. Hence the intensity at such a point is maximum or minimum according as the number of half-period elements contained in the aperture with respect to O is odd or even.

Effect of movement of slit away from screen. When the slit is moved gradually away from the screen, the number of half-period elements in each half of the wavefront will change and hence the illumination will change. With increase in distance of screen from the slit, the number of half-period elements will decrease. As an example, if the number decreases from *four* to *three* the displacement at O *will be maximum. If the slit is moved further away the number will become two with the result that the intensity will be again minimum. Thus the point* O will undergo a change of maximum and minimum intensity. The same thing is observed when the distance is decreased gradually. If the distance is such that the slit comprises one or less than one half-period element in each half of the wave surface with respect to O on the screen, there will be a broad maximum at O.

25.9 FRESNEL'S DIFFRACTION AT CIRCULAR APERTURE (ANALYTICAL TREATMENT)

Let S be a point-source of monochromatic light and CF a narrow circular aperture as shown in Fig. 25.13. Let AB be the section of a spherical wavefront which touches the aperture at any instant. The aperture will cast a geometrical shadow above M and below N on the screen held normal to SO which is perpendicular to the aperture and passes through its centre P. The amplitude at O is the resultant of the amplitudes of a large number of secondary wavelets originating from the wavefront CPF. Divide the wavefront into Fresnel's half-period zones about P, the pole of the wavefront with respect to the point O.

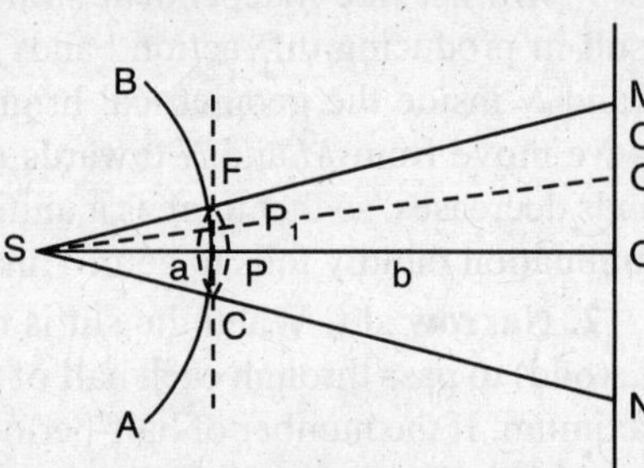

Fig. 25.13

Intensity at the axial point. Suppose the screen is placed at such a distance from the aperture that the axial point O exposes the first half-period zone only, then

The amplitude at $O = d_1$

This is twice the amplitude if the whole of the wavefront were exposed. The intensity at O is, therefore, **four** times the intensity obtained when the whole of the wavefront is exposed. The point O is extremely bright and gives the position of maximum intensity.

If the screen is moved towards the aperture, so that first and the second half-period zones are exposed, then

The amplitude at $O = d_1 - d_2$.

This has a minimum value as d_1 is approximately equal to d_2. The intensity at O is zero and this gives the position of a minimum. If the screen is moved nearer three, four ... etc. half-period zones are exposed and the point O passes alternately through positions of maximum and minimum intensity according as the number of half-period zones exposed is odd or even.

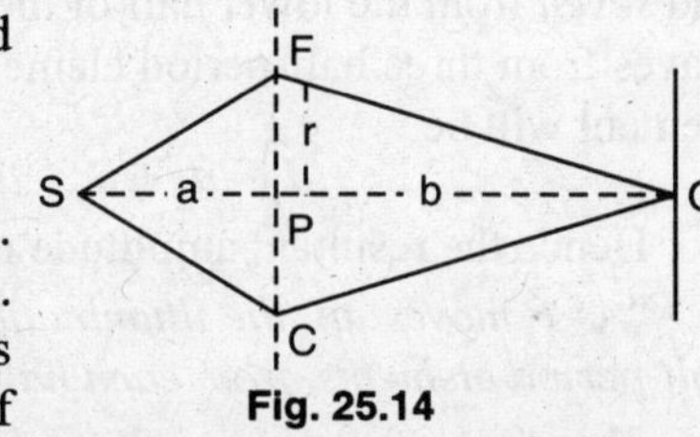

Fig. 25.14

The distance of any point of maximum and minimum intensity from the aperture can be derived as follows:

Let the distance of the aperture from the source be a and that of the screen from the aperture be b. If the radius of the aperture is r, then path-difference δ for the waves reaching O along the paths SFO and SPO (Fig. 25.14) is given by

$$\delta = (SF + FO) - (SP + PO) = (a^2 + r^2)^{1/2} + (b^2 + r^2)^{1/2} - (a + b)$$

$$= a\left\{1+\frac{r^2}{a^2}\right\}^{\frac{1}{2}} + b\left\{1+\frac{r^2}{b^2}\right\}^{\frac{1}{2}} - (a+b)$$

$$= a\left\{1+\frac{1}{2}\frac{r^2}{a^2}+...\right\} + b\left\{1+\frac{r^2}{2b^2}+...\right\} - (a+b)$$

$$= \frac{r^2}{2}\left\{\frac{1}{a}+\frac{1}{b}\right\}$$

$$\therefore \quad \frac{1}{a}+\frac{1}{b} = \frac{2\delta}{r^2}$$

If the position of the screen is such that the aperture contains n half-period zones, then

Path-difference $\quad \delta = \frac{n\lambda}{2}$

or $\quad 2\delta = n\lambda$

$$\therefore \quad \frac{1}{a}+\frac{1}{b} = \frac{2\delta}{r^2} = \frac{n\lambda}{r^2}$$

The point O on the screen will have minimum intensity if n is even and maximum intensity if n is odd.

If the incident light is parallel, the source lies at infinity and the incident wavefront is plane, then

$$a = \infty \quad \text{or} \quad 1/a = 0$$

$$\therefore \quad \frac{1}{b} = \frac{n\lambda}{r^2}$$

or
$$n = \frac{r^2}{b\cdot\lambda} = \frac{\pi r^2}{\pi b\lambda} \qquad \text{... (25.8)}$$

πr^2 is the area of the aperture and $\pi\, b\, \lambda$ is the area of half-period zone, constructed for a point distant b from the wavefront.

If n is *odd* we get a maximum at O and if n is even we get a minimum.

Intensity at the non-axial point. Suppose the screen is placed such that the intensity at O is maximum and the aperture contains five half-period zones with respect to O. As we move away from the axis from O say to O_1, the pole P shifts to P_1 where the line SO_1 meets the wavefront (Fig. 25.13). In this position let the first four zones be completely exposed along with about half of the 5th and 6th zones as shown in Fig. 25.15. The resultant displacement at O_1.

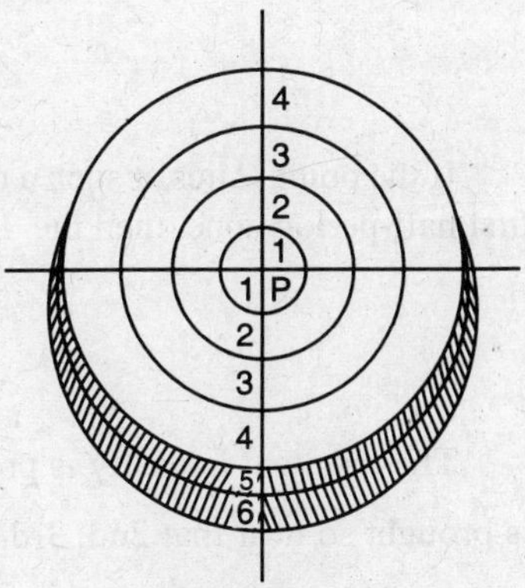

Fig. 25.15

$$= d_1 - d_2 + d_3 - d_4 + \frac{d_5}{2} - \frac{d_6}{2}$$

$$= \frac{d_1}{2} + \left(\frac{d_1+d_3}{2} - d_2\right) + \left(\frac{d_3+d_5}{2} - d_4\right) - \frac{d_6}{2}$$

$$= \frac{d_1}{2} - \frac{d_6}{2} = \text{Minimum} \qquad \text{... (25.9)}$$

Hence the point O_1 will have minimum intensity.

As we proceed further to O_2, so that first three zones are completely exposed along with nearly half of the 4th, 5th, 6th and 7th zones, the intensity will again be maximum. Thus there are a series of points along OM at which the intensity is alternately maximum and minimum. If we rotate the figure about OS as the axis, the points O_1, O_2 etc. will describe bright and dark rings of unequal width about O.

In the case of a large aperture, these rings are visible only near the limits of the geometrical shadow and intensity falls off rapidly within the geometrical shadow, being zero at a point well within the shadow.

If the aperture is so small that it contains only a fraction of the first half-period zone with respect to O, the point O will be bright and there will be no rings within the geometrical '**image**' of the aperture on the screen. In such a case, this bright point will be surrounded by alternately *dark* and *bright* rings, according as the aperture contains an *even* or *odd* number of the fractional half-period zones with respect to a point on the screen within the geometrical shadow of the aperture.

25.10 DIFFRACTION AT A CIRCULAR OBSTACLE

Let S be a point source of monochromatic light, XY an opaque circular disc and O a point on the screen lying on the line SO perpendicular to XY and passing through its centre. Consider a portion AB of a spherical wavefront originating from S and touching the obstacle XY at its centre, as shown in Fig. 25.16. To find the effect of the wavefront at O divide it into a number of half-period zones with respect to O.

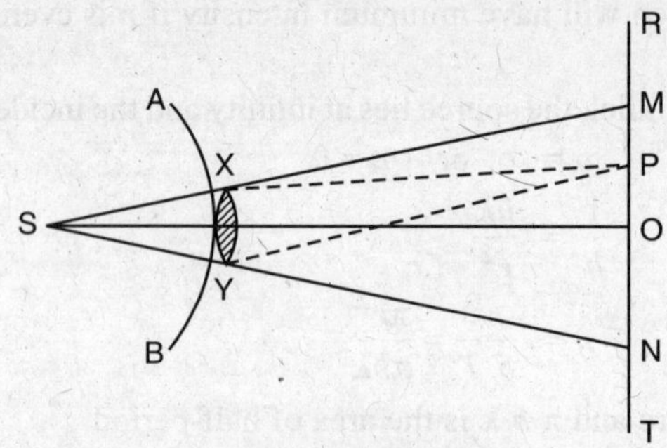

Fig. 25.16

If the point O lies at such a distance and the size of the disc is small, so that it obstructs only the first half-period zone, then the displacement at O

$$= -d_2 + d_3 - d_4 + d_5 \ldots\ldots$$

$$= -\frac{d_2}{2}$$

The resultant intensity is proportional to $\frac{d_2^2}{4}$. If the size of the disc is increased or the point O is brought so near that 2nd, 3rd... etc. half-period zones are obstructed, then the displacement at O will be respectively $\frac{d_3}{2}, -\frac{d_4}{2}$ ... etc., and intensity is proportional to $\frac{d_3^2}{4}, \frac{d_4^2}{4}$, etc. Thus we find that the point O always remains bright and its displacement is equal to half that due to the *first exposed zone.* In other words, the centre of the geometrical shadow of the disc is always bright.

At any other point P within the geometrical shadow MN of the disc, the diffracted waves meet with a phase-difference, thereby producing interference bands. The bright spot O is thus surrounded by circular interference bands within the geometrical shadow of the disc, exactly similar to the bands observed within the geometrical shadow of a narrow obstacle like a wire.

Just outside the geometrical shadow, broad circular diffraction bands of unequal width, similar to those produced by diffraction at a straight edge, are obtained.

Example 25.2 *A circular aperture of 1.2 mm diameter is illuminated by plane waves of monochromatic light; the diffracted light is received on a distant screen which is gradually moved towards the aperture. If the centre of the circular patch of light first becomes dark, when the screen is 30 cm from the aperture, find the wavelength of light used.*

Solution. Diameter of aperture = 1.2 mm = 1.2×10^{-3} m

∴ Radius $r = 0.6 \times 10^{-3}$ m, $b = 30$ cm = 0.3 m.

For the first minimum on the axial point the aperture should contain two half-period zones.

$$\therefore \qquad \frac{\pi r^2}{\pi b\lambda} = 2$$

$$\text{or} \qquad \lambda = \frac{r^2}{2b} = \frac{0.6\times10^{-3}\times0.6\times10^{-3}}{2\times0.3}$$

$$= 6\times10^{-7}\text{ m} = 6000\text{ Å}$$

Example 25.3 *Light of wavelength 6000 Å passes through a narrow circular aperture of radius 0.9×10^{-3} m. At what distance along the axis will the first maximum intensity be observed ?*

Solution. Here $\lambda = 6000 \times 10^{-10}$ m, $r = 0.9 \times 10^{-3}$ m, $b = ?$

$$b^2 + r^2 = \left(a+\frac{\lambda}{2}\right)^2$$

$$\text{or} \qquad b^2 + r^2 = b^2 + b\lambda + \frac{\lambda^2}{4}$$

Neglecting $\frac{\lambda^2}{4}$, we have

$$r^2 = b\lambda$$

$$\text{or} \qquad b = \frac{r^2}{\lambda} = \frac{0.9\times10^{-3}\times0.9\times10^{-3}}{6000\times10^{-10}} = 1.35\text{ m}$$

25.11 ZONE PLATE

It is a transparent plate on which circles whose radii, proportional to the square roots of natural numbers 1, 2, 3, ... are drawn. The alternate annular zones thus formed are blocked. Such a plate behaves like a convex lens and produces an image of a source of light on the screen placed at a suitable distance.

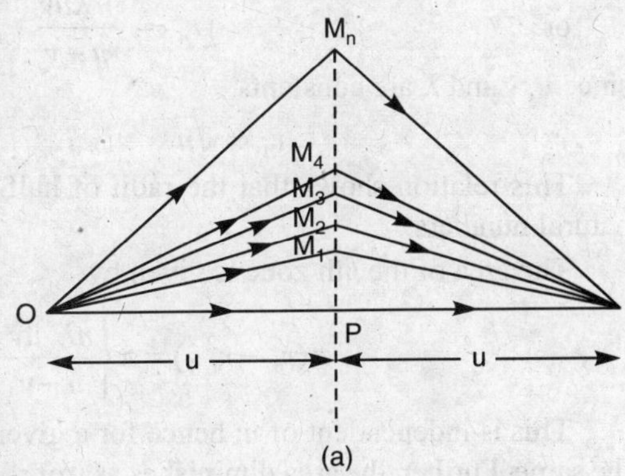

Fig. 25.17

Theory. Let O be a luminous point-object, emitting spherical waves of wavelength λ whose effect at the point I on the screen is required. Consider an imaginary plane through P of a transparent medium perpendicular to the plane of the paper and the line joining OI. Divide this plane into zones bounded by circles having centres at P and radii $PM_1 = r_1, PM_2 = r_2$ .... $PM_n = r_n$ such that

$$OM_1 + IM_1 = OP + IP + \frac{\lambda}{2}$$

$$OM_2 + IM_2 = OP + IP + \frac{2\lambda}{2}$$

....

$$OM_n + IM_n = OP + IP + \frac{n\lambda}{2}$$

The annular rings thus formed are half-period zones for the image I because length of the path of light through the corresponding points of any two consecutive zones differs by $\lambda/2$. To find the radius r_n of the nth circle, we have

$$OM_n + IM_n = OP + IP + \frac{n\lambda}{2} \qquad \text{... }(i)$$

$$\text{Let} \qquad OP = u \text{ and } IP = v$$

$$\text{Now} \qquad OM_n = (OP^2 + PM_n^2)^{\frac{1}{2}} = (u^2 + r_n^2)^{\frac{1}{2}}$$

$$= u\left(1+\frac{r_n^2}{u^2}\right)^{\frac{1}{2}} = u+\frac{r_n^2}{2u}$$

because u is very large as compared to r_n and higher powers of $\frac{r_n^2}{2u}$ can be neglected.

Similarly $\qquad IM_n = (v^2+r_n^2)^{\frac{1}{2}} = v+\frac{r_n^2}{2v}$

Substituting these values in (*i*), we have

$$u+\frac{r_n^2}{2u}+v+\frac{r_n^2}{2v} = u+v+\frac{n\lambda}{2}$$

or $$r_n^2\left(\frac{1}{u}+\frac{1}{v}\right) = n\lambda$$

or $$\frac{1}{u}+\frac{1}{v} = \frac{n\lambda}{r_n^2}$$

Applying the sign convention, we have

$$\therefore \quad \frac{1}{v}-\frac{1}{u} = \frac{n\lambda}{r_n^2} \quad \text{... (ii)}$$

or $$r_n^2 = \frac{n\lambda uv}{u-v} \quad \text{... (25.10)}$$

since u, v and λ are constants

$$\therefore \quad r_n \propto \sqrt{n} \quad \text{... (25.11)}$$

This relation shows that the radii of half-period zones are proportional to the square root of natural numbers.

The area of the nth zone is given by

$$\pi(r_n^2 - r_{n-1}^2) = \pi\left\{\frac{n\lambda.\, uv}{u-v} - \frac{(n-1)\lambda uv}{u-v}\right\} = \frac{\pi\lambda uv}{u-v} \quad \text{...(25.12)}$$

This is independent of n; hence for a given object and image, the areas of all the zones remain the same. Further, the area diminishes as v or u decreases *i.e.*, as the plate is approached by the object or the image. The numerical magnitudes of the displacements d_1, d_2, d_3... at I due to wavelets from the various zones diminish only slightly with the order of the zone. Hence the resultant displacement at I

$$D = d_1 - d_2 + d_3 - d_4 + d_5 \ldots$$

How zone plate acts like a convex lens. If now the alternate zones, say even,are blocked, then the resultant displacement at I will become

$$D = d_1 + d_3 + d_5 \ldots .$$

which is many times greater than that due to all the zones and hence the intensity at I is very much increased. This explains the focusing action of the zone plate. Under these conditions I is said to be the image of the object O.

Focal length. The relation between u *and* v. the respective distance of the object and the image, is given by relation (*ii*), *i.e.*,

$$\frac{1}{v}-\frac{1}{u}=\frac{n\lambda}{r_n^2} \quad \text{...(25.13)}$$

This is a result similar to the one found for a convex lens, *i.e.*,

$$\frac{1}{v}-\frac{1}{u}=\frac{1}{f}.$$

Thus the focal length of the zone plate $f=\frac{r_n^2}{n\lambda}$ ... (25.14)

If the radius of the first half-period zone is 0.316 mm and the wavelength $\lambda = 5\times10^{-5}$ cm, then the zone plate will behave like a lens of focal length 20 cm.

Zone plate has multiple foci. The focal length of a zone plate is given by the relation $f=\frac{r_n^2}{n\lambda}$.

Focal length of a zone plate is the distance measured along its axis at which the maximum is produced when the source of light is located at infinity *i.e.* when a plane wavefront is incident on the zone plate. In actual practice between this *brightest maximum* and the zone plate there are other points at distances *f*/3, *f*/5, *f*/7 etc. at which fainter maxima are produced. Thus unlike a converging lens, which has only one focus, a *zone plate has multiple foci.*

Let F be the focus of the zone plate where the brightest maximum is produced. If we move from F towards the zone plate we get a point F' such that three zones can be constructed in each of the half-period zones of the given zone plate *i.e.* with respect to F' the whole of the zone plate can be divided into $3n$ half-period zones.

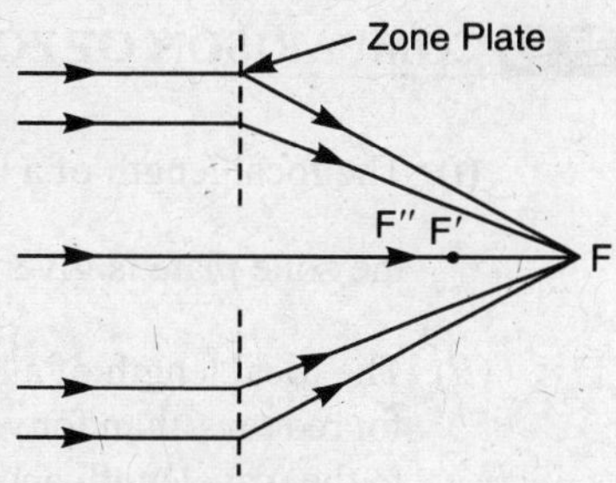

(b)

Fig. 25.18

As the first half-period zone of the original zone plate is divided into three sub-zones contributing displacements (or amplitudes) d_{11}, d_{12}, d_{13} respectively, their net contribution to displacement by the first half-period zone (of the original zone plate)

$$d_1' = d_{11} - d_{12} + d_{13} = d_{13}$$

Similarly net contribution to displacement by the second half-period zone

$$d_2' = d_{21} - d_{22} + d_{23} = d_{23}$$

The contribution due to different zones of the original zone plate will again superimpose at F' to produce a maximum. This maximum will be less bright than that at F. Since the total number of zones into which the zone plate is now divided is $3n$, therefore

$$f' = \frac{r_n^2}{3n\lambda} = \frac{f}{3}$$

Moving further towards the zone plate we can have another point F'' such that now each half-period zone of the original zone plate can be divided into 5 sub-zones. As explained above, F'' will also be a maximum but will be of a brightness even less than that of F'.

Also $$f'' = \frac{r_n^2}{5n\lambda} = \frac{f}{5}$$

Thus zone plate has multiple foci given by

$$f=\frac{r_n^2}{n\lambda},\ f'=\frac{r_n^2}{3n\lambda},\ f''=\frac{r_n^2}{5n\lambda} \text{ and so on.} \quad ...(25.15)$$

25.12 CONSTRUCTION OF ZONE PLATE

Zone plate is a system of areas corresponding to the half-period zones. To construct a zone plate, concentric circles whose radii are proportional to the square roots of the natural numbers are drawn on a sheet of white paper. The alternate zones are painted black and a very much reduced photograph of this drawing is obtained on a glass plate. The negative, when held in the light path from a distant point source, produces a large intensity at a point on its axis at a distance corresponding to the size of the zone and the wavelength of light used.

Positive and negative zone plates. If odd zones are transparent and even zones are opaque, as shown in Fig. 25.19 (*b*), it is said to be a **positive zone plate**. If the even zones are transparent and odd zones are opaque, as shown in Fig 25.19(a) it is said to be a **negative zone plate.**

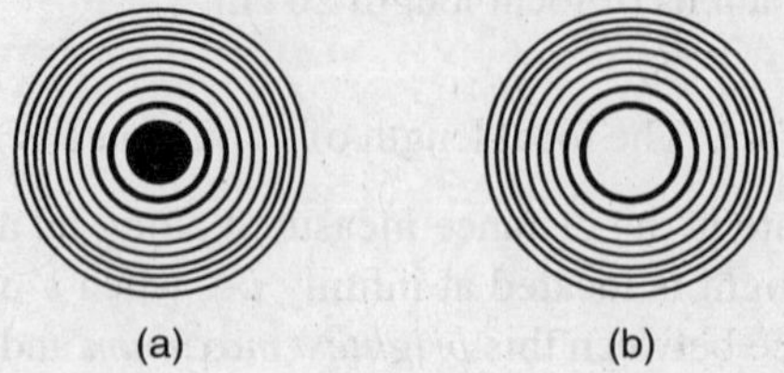

(a) (b)

Fig. 25.19

25.13 COMPARISON OF ZONE PLATE WITH A CONVERGING LENS.

(*i*) The focal length of a lens is given by the relation $\frac{1}{f} = (\mu - 1)\left(\frac{1}{R_1} - \frac{1}{R_2}\right)$ whereas that of the zone plate is given by $\frac{1}{f} = \frac{n\lambda}{r_n^2}$.

(*ii*) The focal length of a lens is directly proportional to the wavelength and hence is greater for red rays than for violet rays. The focal length of a zone plate is inversely proportional to the wavelength and hence is greater for violet rays than for red rays.

(*iii*) Light, in passing through the lens, takes the same time to go from *O* to *I* through any part of the lens, whereas in a zone plate disturbances from any transparent zone reach the point *I* one period later than the disturbances from the next inner zone.

(*iv*) **Zone plate has multiple foci.** For a fixed distance of the object, a lens produces only one image, whereas a zone plate produces a number of images. In other words, it has multiple foci. It is because the number of half-period zones contained in an area depends upon the position of the screen. If therefore, a point I_1 lies near to *P* than *I* so that the transparent rings contain *three* half-period zones, then the point I_1 will have again maximum illumination and thus will be the image of *O*. Similar will be the case when 5, 7... half-period zones are contained in a transparent ring. The position of the various foci is given by

$$f_1 = \frac{r_n^2}{n\lambda}, f_2 = \frac{r_n^2}{3n\lambda}, f_3 = \frac{r_n^2}{5n\lambda}, ...$$

The intensity of the image, however, decreases as the focal length decreases.

Example 25.4 *What is the radius of first zone of a zone plate of focal length 0.2 m for a light of wavelength 5000 Å ?* **(*Nagpur U. 2011*)**

Solution. Here $\lambda = 5000$ Å $= 5 \times 10^{-7}$ m, $f = 0.2$ m

If r_1 is the radius of the first zone, then

$$f_1 = \frac{r_n^2}{n\lambda} = \frac{r_1^2}{\lambda}$$

$$\therefore \quad r_1 = \sqrt{f\lambda} = \sqrt{0.2 \times 5 \times 10^{-7}} = 3.16 \times 10^{-4} \text{ m}$$

$$= 0.0316 \text{ cm.}$$

Example 25.5 *Determine the radius of the first half-period zone of a zone plate which behaves as a convex lens of focal length 40 cm for a wavelength of 4000 Å.* **(M.D.U. 2008 ; K.U., 2000)**

Solution. $r_1 = \sqrt{f\lambda} = \sqrt{40\times10^{-2}\times4000\times10^{-10}} = 4\times10^{-4}\,\text{m} = 0.4\text{ mm}$

Example 25.6 *A zone plate has focal length 50 cm at a wavelength 6000 Å. What will be its focal length at λ = 5000 Å?* **(Nagpur U. 2008, Indore U., 2001)**

Solution. Here $f_1 = 50$ cm; $\lambda_1 = 6000$ Å; $\lambda_2 = 5000$ Å; $f_2 = ?$

Now $$f = \frac{r_n^2}{n\lambda} \quad \therefore\ f_1 = \frac{r_n^2}{n\lambda_1} \quad \text{and } f_2 = \frac{r_n^2}{n\lambda_2}$$

$$\therefore \quad \frac{f_2}{f_1} = \frac{\lambda_1}{\lambda_2} \quad \text{or} \quad f_2 = f_1\frac{\lambda_1}{\lambda_2} = \frac{50\times6000\times10^{-8}}{5000\times10^{-8}} = 60\text{ cm.}$$

Example 25.7 *If the focal length of a zone plate is 1 m for light of wavelength 5 × 10⁻⁵ cm. what will be its focal length for wavelength 6×10⁻⁵ cm?* **(P.U. 2008)**

Solution. $f_2 = f_1\dfrac{\lambda_1}{\lambda_2} = \dfrac{1\times5\times10^{-5}}{6\times10^{-5}} = \dfrac{5}{6}\text{ m} = 83.3\text{ cm.}$

Example 25.8 *The radius of the first ring of a zone plate is 0.55 mm. If light of wavelength 5894Å falls normally on it, where should the screen be placed to get a bright spot?* **(Nagpur U. W/2008; M.D. U., 2003; K.U., 2002)**

Solution. Here $r_1 = 0.55\text{ mm} = 0.55\times10^{-3}\text{ m} = 55\times10^{-5}\text{ m}$

$$\lambda = 5894\text{ Å} = 5894\times10^{-10}\text{ m}$$

As the light falls normally, it is parallel to the axis of the zone plate and after pagging through it will meet the axis at the focus.

Now $$f = \frac{r_n^2}{n\lambda} \text{ or } \frac{r_1^2}{\lambda} = \frac{55\times10^{-5}\times55\times10^{-5}}{5894\times10^{-10}} = 0.51\text{m} = 51\text{ cm}$$

The screen should be placed at a distance of 51 cm from the zone plate.

Example 25.9 *The innermost half-period zone has a diameter of 0.5 mm. Find the focal length of the plate if λ = 550 nm* **(H.P.U., 2000)**

Solution. $d = 0.5$ mm

$$\therefore \quad r = 0.25\text{ mm};\ f = \frac{r_1^2}{\lambda} = \frac{0.25\times10^{-3}\times0.25\times10^{-3}}{550\times10^{-9}}$$

$$= 11.4\times10^{-2}\text{ m} = 11.4\text{ cm}$$

Example 25.10 *An object illuminated by 5000 Å wavelength of light is placed at 60 cm from a zone plate and its image (brightest) is obtained at 30 cm from the plate. Calculate the number of Fresnel zones in a radius of 5 cm of the plate.* **(Garhwal U., 2000)**

Solution. Here $u = 60$ cm; $v = 30$ cm; $\lambda = 5000\text{ Å} = 5000\times10^{-8}$ cm

Now $\dfrac{1}{v} - \dfrac{1}{u} = \dfrac{1}{f}$ where f is the focal length of the zone plate. According to sign convention u is negative.

$$\therefore \quad \frac{1}{30} - \left(-\frac{1}{60}\right) = \frac{1}{f} \quad \text{or} \quad \frac{1}{f} = \frac{1}{30} + \frac{1}{60} = \frac{3}{60} = \frac{1}{20}$$

$f = 20$ cm; Given $r_n = 5$ cm

Also $f = \frac{r_n^2}{n\lambda}$ or $n = n = \frac{r_n^2}{f\lambda} = \frac{5\times5}{20\times5000\times10^{-8}}$

$= 2.5 \times 10^4 = 25000.$

Example 25.11 *A zone plate is illuminated with a source of sodium light (λ = 5896 Å) placed at a distance of 1 m. If the image of the point source is obtained at a distance of 2 m on the other side, what will be the power of the equivalent lens which may replace the zone plate without disturbing the set-up? Also calculate the radius of the first zone of the plate.* **(M.D U. 2005)**

Solution. Here $u = 1$ m; $v = 2$m; $\lambda = 5896$ Å $= 5896 \times 10^{-10}$ m

Now $$\frac{1}{f} = \frac{1}{v} + \frac{1}{u} = \frac{1}{2} + 1 = \frac{3}{2}$$

∴ $$f = \frac{2}{3}\text{ m}$$

Power in Dioptre $= \frac{100}{\text{F.L. in cm}} = \frac{1}{\text{F.L. in metre}} = \frac{1}{2/3} = \frac{3}{2} = 1.5$ Dioptre

Also $f = \frac{r_n^2}{n\lambda}$ $\therefore r_n = \sqrt{n\lambda f}$

For $n = 1$, $r_1 = \sqrt{\lambda f} = \sqrt{5896\times10^{-10}\times\frac{2}{3}} = 62.7\times10^{-5}\text{m} = 0.627$ mm.

Example 25.12 *Calculate the radii of the first three clear elements of a zone plate, which is designed to bring a parallel beam of light of wavelength 6000 Å to the first focus at a distance of 2 metres.* **(P.U., 1994)**

Solution. As a parallel beam of light is incident *i.e.*, $u = \infty$ and

∴ $$v = f$$

Here $$v = f = 2.0\text{ m},\quad \lambda = 6000\text{ Å} = 6\times10^{-7}\text{m}$$

Now $$r_n = \sqrt{n\lambda f}$$

(*i*) For the first clear zone $n = 1$

∴ $$r_1 = \sqrt{1\times2\times6\times10^{-7}} = 1.095\times10^{-3}\text{ m}$$

(*ii*) For the second clear zone $n = 3$

∴ $$r_2 = \sqrt{3\times2\times6\times10^{-7}} = 1.897\times10^{-3}\text{ m}$$

(*iii*) For the third clear zone $n = 5$

∴ $$r_3 = \sqrt{5\times2\times6\times10^{-7}} = 2.449\times10^{-3}\text{ m.}$$

Example 25.13 *A plane wavefront λ = 6000 Å advancing towards a point is divided into a number of half-period zones. Amplitude contribution of these half-period zones is given by 1.00, 0.98, 0.96 ... up to zero. Compare the intensities at the point when first 31 and 36 h.p.z. only are exposed.*

Solution. $d_1 = 1.00$, $d_2 = 0.98$, $d_3 = 0.96$

∴ $$d_1 - d_2 = 1.00 - 0.98 = 0.02$$

$$d_{31} = 1.00 - 30\times0.02 = 0.40$$

$$f = \frac{r_n^2}{n\lambda} = \frac{10^{-3} \times 10^{-3}}{5000 \times 10^{-10}} = 2\text{m}$$

$$u = -2 \text{ m and } f = 2\text{m}$$

$$v = \infty$$

e the primary image lies at infinity.

second focal length $f_2 = \frac{f_1}{3} = \frac{2}{3}$ m

$$\frac{1}{v} = \frac{u+f}{uf} \quad \text{or } v = \frac{uf}{u+f}$$

$$v_2 = \frac{-2 \times \frac{2}{3}}{-2 + \frac{2}{3}} = \frac{-4 \times 3}{3 \times (-4)} = 1\text{m}$$

$$f_3 = \frac{f_1}{5} = \frac{2}{5} \text{m} = 0.4 \text{ m}, u = -2\text{m}$$

$$v_3 = \frac{-2 \times 0.4}{-2 + 0.4} = \frac{-0.8}{-1.6} = 0.5\,\text{m}$$

$$f_4 = \frac{f_1}{7} = \frac{2}{7} \text{ m} = 0.2857$$

$$u = -2\text{m}$$

$$v_4 = \frac{-2 \times 0.2857}{-2 + 0.2857}$$

$$= 0.333 \text{ m}.$$

Thus, the primary image will be formed at infinity and the other secondary images will be formed
) m, 0.5 m, 0.333 m and so on.

14 DIFFERENCE IN PROPAGATION OF SOUND AND LIGHT WAVES

The wavelength of the sound waves is very large as compared to that of light. The area of the fir
f-period zone, *i.e.*, $\pi a \lambda$ for sound waves is, therefore, very large. As an example, the wave
sodium light is nearly 6×10^{-5} cm and that of the human voice for a frequency of 332 i
nce an obstacle more than a million times big will be required to stop the effect of
ont. Thus for objects of ordinary size, sound waves bend round them and do no
a straight line.

EXERCISE

LONG QUESTIONS

1. What is diffraction? Explain clearly the difference be
(G.N.D.U. 2008, 2007
2002; P.U

2. How many types are of diffraction? Dis
diffraction. State in which type of diffraction
all points in the plane of the aperture.
(Madurai U. 2003; Pbi. U.2003, 200
Meerut U. 2005, 2002, 2000; Nag U. 20

$$d_{35} = 1.00 - 0.68 = 0.32$$

$$d_{36} = 0.32 - 0.02 = 0.30$$

Resultant amplitude of 31 zones $= \dfrac{1.00}{2} + \dfrac{0.4}{2} = 0.7$

Resultant amplitude of 36 zones $= \dfrac{1.00}{2} + \dfrac{0.32}{2} - 0.30 = 0$

$$\therefore \quad \frac{I_{31}}{I_{36}} = \frac{0.7^2}{0.36^2} = \frac{3.78}{1}$$

$$I_{31} : I_{36} :: 3.78 : 1$$

Example 25.14 *A zone plate is found to give series of images ... strongest and the second strongest images are at distances of 0.30 m ... zone plate (both on the same side remote from the source), calculat... the zone plate, principal focal length and the radius of first zone. As...*

Solution. Here $\lambda = 5 \times 10^{-7}$ m, $v_1 = 0.3$ m, $v_2 = 0.06$ m

Now
$$f_1 = \frac{r_n^2}{\lambda} \text{ and } f_2 = \frac{r_n^2}{3\lambda}$$

$$\therefore \quad f_2 = f\frac{1}{3}$$

If u is the distance of the object from the zone plate, then

$$\frac{1}{u} = \frac{1}{v_1} - \frac{1}{f_1} \text{ also } \frac{1}{u} = \frac{1}{v_2} - \frac{1}{f_2}$$

$$\therefore \quad \frac{1}{v_1} - \frac{1}{f_1} = \frac{1}{v_2} - \frac{3}{f_1}$$

or
$$\frac{3}{f_1} - \frac{1}{f_1} = \frac{1}{v_2} - \frac{1}{v_1} = \frac{1}{0.06} - \frac{1}{0.3}$$

or
$$\frac{2}{f_1} = \frac{0.3 - 0.06}{0.3 \times 0.06} = 13.33$$

or
$$f_1 = \frac{2}{13.33} = 0.15 \text{ m}$$

...w
$$r_1 = \sqrt{f_1 \times \lambda} = \sqrt{0.15 \times 5 \times 10^{-7}}$$

$$= 0.274 \times 10^{-4} \text{ m}$$

$$\frac{1}{u} = \frac{1}{v_1} - \frac{1}{f} = \frac{1}{0.3} - \frac{1}{0.15}$$

$$u = -\frac{0.3 \times 0.15}{0.15} = -0.3\text{m}$$

...5 *A zone plate is drawn and then copied on a reduced scale so that the ... 2 mm. If a source of monochromatic light of λ = 5000Å is placed 2 m ... sition of the primary image and the weaker secondary images.*

...meter = 2mm or radius $= 10^{-3}$ m

$$\lambda = 5000 \text{ Å} = 5000 \times 10^{-10} \text{ m}$$

3. (*a*) Explain the meaning of Fresnel's half-period zones. Why are they so called? What is the phase difference between wavelets from successive half-period zones ? What are the factors on which the amplitude of light waves from a half-period zone at the observation point depend?

(*b*) Show that the amplitude due to complete wavefront is half of what it would be caused by the first zone. Hence explain rectilinear propagation of light on the wave theory.

(G.N.D.U 2007, 2006, 2005, 2003, 2002, 2000; P.U. 2007, 2005, 2004, 2000, H.P.U. 2001, Kerala U. 2001; Gauhati U. 2000; Luck U. 2000; Pbi. U. 2006; Meerut U. 2007)

4. (*a*) Describe, in detail, the phenomenon of Fresnel diffraction due to a straight edge.

(Nagpur U. 2008)

(*b*) Obtain the position of dark bands on the screen due to diffraction at a straight edge.

(Nagpur U. W/2008)

5. Describe Fresnel's construction for the formation of half-period elements when the wavefront is cylindrical and find the resultant amplitude. ***(Nagpur U. 2011, 2001)***

6. Discuss the phenomenon of diffraction at a straight edge. Explain the diffraction pattern in the geometrical shadow and in the illuminated part of the screen. Show that the separation between successive maxima goes on decreasing as we move away from the region of geometrical shadow. How will you determine wavelength of light from a study of the fringes?

(M.D.U 2006, 2003, 2002, 2000; H.P.U. 2002, Luck U. 2001; Kerala U. 2001; P. U. 2004, 2000; Meerut U. 2005, 2000; Garhwal U. 2004; Nagpur U. 2011)

7. (*a*) Give an elementary treatment of diffraction of monochromatic light by a thin wire. Explain the changes produced in diffraction pattern as the diameter of the wire is increased.

(*b*) How will you use the pattern to measure the thickness of the wire?

8. Give an account of diffraction effect produced by a narrow slit. Explain what happens when

(*a*) slit width is gradually increased.

(*b*) slit is moved away from the screen. ***(Kerala U. 2001; Pbi. U., 1993)***

9. Explain diffraction at a circular obstacle.

10. What is a zone plate? Show that the radii of its half-period zones are proportional to the square root of natural numbers. Derive an expression for its focal length. Show that a zone plate has multiple foci.

(Pbi. U 2008, 2007, 2006; P.U. 2006, 2004, 2003, 2002, 2001, K.U. 2002, 2001; Bhopal U. 2004; Meerut U. 2007, 2001; Luck U. 2001; H.P.U. 2001, M.D.U. 2008, 2000, Gauhati U. 2000; G.N.D.U. 2008, 2006, 2005, Madurai U. 2003)

11. How is zone plate constructed? What are positive and negative zone plates?

(P.U. 2006, 2003, 2001; Meerut U. 2007; K.U., 2002, Pbi. U., 2008, 2007, 2006, 2002; H.P.U., 2001; M.D.U., 2008, 2000)

12. (*a*) Explain the construction of Fresnel's half period zones on a plane wavefront

(Nagpur U. 2011)

(*b*) How many zone plates are there in a portion of a plane wave front of radius 10^{-2} m? Given that wavelength is 6000Å and distance of point of observation from the wavefront is 1 m. ***(Nagapur U. 2009)***

13. Compare the performance of zone plate with that of a converging lens.

(H.P.U., 2003, 2002, 2001, P.U., 2006, 2003, M.D.U., 2002, 2000; K.U., 2002, Nagpur is 2009; Pbi. U., 2002, Indore U., 2001; Gauhati U., 2000)

14. A beam of monochromatic light on passing through a slit of width 1.6 mm wide falls on a screen held close to the slit. The screen is then gradually moved away and the middle of patch of light becomes dark when the distance of the screen from the slit is 0.5 m. Find the wavelength of light. **[Ans. 6400 Å]**

15. Light of wavelength 5000 Å passes through a narrow circular aperture of radius 0.5×10^{-3} m. Find the distance along the axis at which the first maximum intensity is observed. **[Ans.0.5 m]**

16. For a wavelength of 6000 Å and radius of first half-period zone 0.6 mm, the zone plate brings rays to a focus at its bright spot. Find the focal length of equivalent lens. **[Ans.60 cm]**

17. A very narrow vertical slit illuminated with light of wavelength 6000 Å casts a shadow of vertical wire of diameter 2 mm and 1 m away on a screen 3m away from the wire. Calculate the distance apart of bands inside the shadow and total number of bands that can be seen. **[Ans.** 0.09 cm, 9 approx.**]**

18. What is the radius of the first zone in a zone plate of focal length 15 cm for light of 6000 Å? **[Ans.** 3×10^{-4} m**]**

19. With zone plate for a point source of light of wavelength 5000 Å on the axis, the strongest and the next strongest images are formed at 30 cm and 6 cm respectively from the zone plate. Both the images are on the same side and on the other side of the source. Find (*i*) distance of source from the zone plate, (*ii*) the radius of first zone and (*iii*) principal focal length.

[Ans. (*i*) $u = -30$ cm (*ii*) $r_1 = 3 \times 10^{-3}$ cm (*iii*) 18 cm**]**

SHORT QUESTIONS WITH ANSWERS

1. *In what essential respects the case of propagation of sound waves differs from those of light waves?*

Ans. Refer article 25.14]

2. *What are the factors on which the amplitude of light waves from a half-period zone at observation point depend?*

Ans. (*i*) Resulatant of secondary wavelets from the half-period zone.

(*ii*) On obliquity factor (1+ cos θ), where θ is the angle which the line joining the centre of half period zone makes with the horizontal.

3. *Why do we get bright light spot at the centre of the image of an opaque object when the size is extremely small?*

Ans. The resultant displacement due to the whole wavefront is $\frac{d_1}{2}$. If the obstacle blocks first, second, third half-period zones, the resultant displacement will be $-\frac{d_2}{2}, \frac{d_3}{2}, -\frac{d_4}{2}$ ...etc. and the intensity will be $\frac{d_2^2}{4}, \frac{d_3^2}{4}, \frac{d_4^2}{4}$ etc. Thus the central point will always be bright if the obstacle is small.

4. *Why diffraction effects in light are difficult to observe?*

Ans. Because wavelength of light is very small as compared to the size of the obstacle.

5. *What is the relation between the wavelength of wave and size of the obstacle for diffraction to take place?*

Ans. The size of the obstacle should be comparable with the wavelength of light used.

6. *What is the cause of light streaks one sees while looking at a strong source of light with half-shut eyes?* ***(G.N.D.U. 2008; Pbi. U., 2000; H.P. U.)***

Ans. It is due to diffraction effect that we see light streaks when we look at a strong source of light with half-shut eyes.

7. *Why do radio waves diffract around buildings while visible light waves do not?* ***(Nagpur U., 2001)***

Ans. The phenomenon of diffraction is observed only when the size of the obstacle is comparable to the wavelength of the waves. The wavelength of radio waves is comparable to the size of the buildings. So radio waves can diffract around buildings. The wavelength of visible light is very much smaller than the size of the buildings. So diffraction of light is not possible and the light waves do not diffract around buildings.

8. *Why is diffraction pattern generally not observed with an extended source of light?*

Ans. An extended source can be considered to be an aggregate of a large number of point sources. Each point on the extended source gives out its own wavefront which undergoes diffraction. Various diffraction patterns so formed may overlap causing general illumination. Hence diffraction pattern is not clearly observed.

9. *Why is diffraction of sound waves more evident is our daily life than that of light waves?* **(H.P.U., 2002)**

Ans. To obtain a well-defined diffraction pattern, the size of the obstacle or aperture should be of the same order as the wavelength. The wavelength of sound is comparable to the size of most of the obstacles or apertures we come across in daily life. The diffraction of sound is, therefore, a common experience. But the wavelength of light is very small as compared to the size of the obstacles or apertures we come across in daily life. We, therefore, do not observe diffraction of light as an everyday phenomenon.

10. *Can diffraction occur for virtual images ? Explain.* **(G.N.D.U. 2007)**

Ans. No. It is not possible to observe diffraction effects with virtual images because diffraction requires the superposition of waves that must be real *i.e.* originate from real source of light.

11. *If we look at the sun through a piece of fine cloth we observe coloured spectra at the site of holes in the cloth. Why?* **(G.N.D.U. 2007)**

Ans. At the site of fine apertures or holes in the cloth, we get diffraction patterns. As the light from the sun is white light, consisting of colours from red to violet, we get coloured spectra or coloured streaks.

MULTIPLE CHOICE QUESTIONS

25.1 In Fresnel's half-period zone, the phase difference between wavelets from successive half-period zones is :

(*a*) $\frac{\pi}{4}$ radian (*b*) $\frac{\pi}{2}$ radian

(*c*) $\frac{3\pi}{4}$ radian (*d*) π radian

25.2 The radii of half-period zone are related with the order (n) of the zone by the relation

(*a*) directly proportional to n (*b*) directly proportional to $\sqrt{n}$

(*c*) inversely proportional to n (*d*) inversely proportional to $\frac{1}{\sqrt{n}}$

25.3 The area of second half-period zone S_2 and that of first half-period zone S_1 are related as:

(*a*) $S_2 = 2S_1$ (*b*) $S_2 = S_{1/2}$

(*c*) $S_2 = S_1$ (*d*) $S_2 = \sqrt{S_1}$. **(G.N.D.U., 2000)**

25.4 Which of the following is essential for observing diffraction ?

(*a*) Two coherent sources (*b*) A narrow slit

(*c*) A screen (*d*) White light **(H.P.U., 2001)**

25.5 The focal length of zone plate of is given by

(*a*) $r_n^2 / n\lambda$ (*b*) $\frac{r_n^{\lambda}}{n}$

(*c*) $\frac{r_n \lambda^2}{n}$ (*d*) $\frac{r^2 n}{n^2 \lambda^2}$

25.6 The wavelength of sound waves as compared to light waves is :

(*a*) very small (*b*) equal

(*c*) very large (*d*) infinity

ANSWERS

25.1 (*d*)	25.2 (*b*)	25.3 (*c*)	25.4 (*b*)	25.5 (*a*)	25.6 (*c*)

FRAUNHOFER DIFFRACTION

26 Chapter

26.1 ESSENTIAL CONDITION FOR FRAUNHOFER DIFFRACTION

The essential condition for having Fraunhofer diffraction is that the source and the screen are at infinite distance from the aperture. This is achieved by placing the source on the focal plane of a convex lens and placing the screen on the focal plane of another convex lens. The two lenses thus effectively move the source and the screen to infinity. The first lens makes the incident light beam parallel and the second lens effectively makes the screen receive a parallel beam of light. As a result, the incident and diffracted wavefronts are both plane wavefronts.

26.2 DIFFRACTION AT A SINGLE SLIT

Consider a narrow slit *AB* of width *a* perpendicular to the plane of paper and illuminated by a parallel beam of monochromatic light of wave length λ. The light, in passing through the slit, suffers diffraction. If a lens *L* is placed in the path of the diffracted beam, a real image of the diffraction pattern is obtained on the screen *RT* placed perpendicular to the plane of paper in the focal plane of the lens (Fig. 26.1)

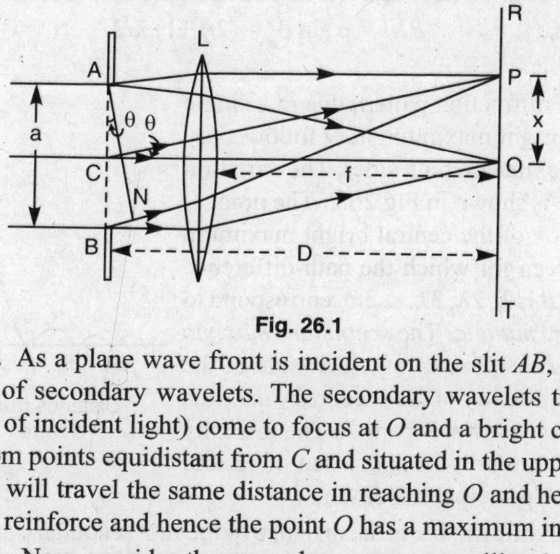

Fig. 26.1

Central maximum. As a plane wave front is incident on the slit *AB*, each point on the wave front becomes a source of secondary wavelets. The secondary wavelets travelling in a direction parallel to *CO* (direction of incident light) come to focus at *O* and a bright central image is formed. The secondary waves from points equidistant from *C* and situated in the upper and lower halves *CA* and *CB* of the wavefront will travel the same distance in reaching *O* and hence the path-difference is zero. These, therefore, reinforce and hence the point *O* has a maximum intensity.

Secondary minima. Now consider the secondary waves travelling in a direction making an angle θ with *CO*. These secondary wavelets are brought to focus by the lens at a point *P* which will have a maximum or minimum intensity, depending upon the path-difference between the secondary waves originating from the corresponding points of the wavefront. Draw *AN* perpendicular to the direction of the diffracted rays from *A*, then

Path-difference between the wavelets from the extremities of the slits *A* and *B*

$$= BN$$

$$= AB \sin\theta = a \sin\theta \qquad \text{... } (i)$$

If $BN = \lambda$, then *P* will be a point of minimum intensity. It is because the whole wave front *AB* can be considered to be made up of two halves *AC* and *BC*. If the path-difference between the secondary waves from A and B is λ, then the path-difference between the secondary waves from A and C will be $\lambda/2$. Thus, for every point in the upper half CA, there is a corresponding point in the lower half CB the path-difference between the secondary waves from which is $\lambda/2$, giving rise to destructive interference at P, which has minimum intensity. Hence for first minimum

$$BN = a \sin\theta = \lambda \qquad \text{... (ii)}$$

If the path-difference between the extreme wavelets from A and B is 2λ *i.e.*,

$$BN = a\sin\theta = 2\lambda$$

the intensity will again be minimum, because the slit can now be supposed to be divided into four equal parts and the rays from corresponding points separated by a distance $a/4$ in the two halves of each half of the slit will have a path-difference of $\lambda/2$, thus destroying each other's effect. Hence for second minimum intensity position

$$BN = a \sin\theta = 2\lambda \qquad \text{... (iii)}$$

In general, for the various secondary minima the path-difference BN between the extreme wavelets from A and B should be an even multiple of $\lambda/2$ or an integral multiple of λ. Thus, for *secondary minima.*

$$BN = a \sin\theta_n = n\lambda$$

or
$$\sin\theta_n = \frac{n\lambda}{a} \qquad \text{... (iv)}$$

where θ_n gives the direction of the nth minimum. Here n is an integer.

Secondary maxima. In addition to the central maximum at O, there are secondary maxima which lie in between the secondary minima on either side of the central maximum. These are situated in a direction in which the path-difference BN is an odd multiple of $\lambda/2$. Hence for *secondary maxima*

$$BN = a \sin\theta_n = (2n+1)\,\lambda/2 \qquad \text{... (v)}$$

where $n = 1, 2, 3$ ... etc.

Thus we see that the diffraction pattern due to a single slit consists of a central bright maximum at O followed by secondary minima and maxima on both sides. The variation of intensity with distance is shown in Fig.26.2. The point O corresponds to the position of the central bright maximum and for points on the screen for which the path-difference between the points A and B is $\lambda, 2\lambda, 3\lambda$, ... etc. correspond to the positions of *secondary minima*. The *secondary maxima* lie in between the *secondary minima* on either side of the principal maximum. The intensity of the secondary maxima falls off rapidly as we move outward from O. The intensity of the first secondary maximum is 1/22 and that of second 1/61 of the intensity of the principal maximum.

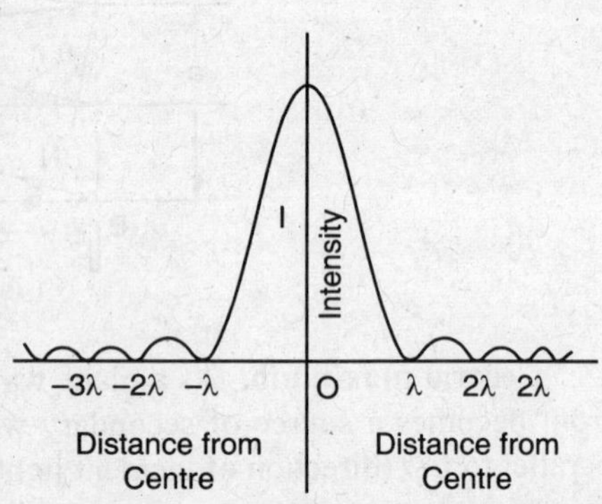

Fig. 26.2

Width of central maximum. If x is the distance of the first secondary minimum from the centre of the principal maximum, then

Width of central maximum = $2x$

If the lens L lies very close to the slit or the screen is placed at a very large distance from the lens, then $CO = D$ is very large. So that

$$\tan\theta = \sin\theta = \frac{x}{D}$$

Now for the first minimum $\sin\theta = \lambda/a$

$$\therefore \quad \frac{\lambda}{a} = \frac{x}{D}$$

or $$x = \frac{\lambda D}{a} \quad \text{... (vi)}$$

If f is the focal length of the lens, and the lens lies very close to the slit, then

$$D = f$$

$$x = \frac{\lambda f}{a} \quad \text{... (vii)}$$

or Width of central maximum $$2x = \frac{2f\lambda}{a} \quad (26.1)$$

This shows that the width of the central maximum is proportional to the wavelength of light. The width of the central maximum for red light of larger wavelength is nearly double than that for the violet light of shorter wavelength.

Effect of width of slit. With monochromatic light, the diffraction pattern consists of alternate bright and dark bands. If white light is used, central maximum is white and on either side, the diffraction bands are coloured. From the relation $\sin\theta = \lambda/a$, we find that if the width of the slit a is large, then for a given wavelength, $\sin\theta$ is small and hence θ is small. This shows that the maxima and minima lie very close to the central maximum. With a narrow slit, as a is small, θ is large. Hence diffraction maxima and minima are quite distinct on either side of O.

Example 26.1 *A single slit of width 1 mm is illuminated by light of wavelength 589 nm. Find the angular spread of the central maximum of diffraction pattern observed.* **(*Kerala U., 2001*)**

Solution The central maximum due to a single slit extends from the axis of the slit to the first minima on either side.

For the first minima $$\sin\theta_1 = \frac{\lambda}{a}$$

Here $\lambda = 589$ nm $= 589\times10^{-9}$ m; $a = 1$ mm $= 10^{-3}$ m

$$\therefore \quad \sin\theta_1 = \frac{589\times10^{-9}}{10^{-3}} = .000589$$

or $$\theta_1 = \sin^{-1}.000\,589 = 0.0337°$$

Hence angular spread of central maximum $= \theta_1 = 2\times0.0337 = .0674°$

Example 26.2 *550 nm light falls normally on a slit of width 2.2 μm. Determine the angular position of second and third minima.* **(*P.U. 2002*)**

Solution Here $\lambda = 550$ nm $= 550\times10^{-9}$ m ; $a = 2.2$ μm $= 2.2\times10^{-6}$ m

Angular position of nth minima is given by $\sin\theta_n = \dfrac{n\lambda}{a}$

$\therefore$ Angular position of second minima is given by $\sin\theta_2 = \dfrac{2\times550\times10^{-9}}{2.2\times10^{-6}}$

$\therefore \quad \sin\theta_2 = 0.5 \quad$ or $\theta_2 = 30°$

For angular position of third minima $\sin\theta_3 = \dfrac{3\times550\times10^{-9}}{2.2\times10^{-6}} = 0.75$

$$\therefore \quad \theta_3 = 58.59°$$

Example 26.3 *A convex lens of focal length 20 cm is placed after a slit of width 0.6 mm. If a plane wave of wavelength 6000 Å falls on the slit normally, calculate the separation between the second minima on either side of central maximum.* **(*P.U. 1996*)**

Solution Here $\lambda = 6000$Å $= 6\times10^{-7}$ m, $f = 20$ cm $= 0.2$ m

Width of slit $a = 0.6$ mm $= 0.6\times10^{-3}$ m

If the second minimum occurs in a direction making an angle θ_2 on either side of the central maximum, then

$$\sin\theta_2 = \frac{2\lambda}{a} = \frac{2\times6\times10^{-7}}{0.6\times10^{-3}} = 0.002$$

The diffraction pattern is formed on the screen placed in the focal plane of the lens placed close to the slit,

$$\therefore \quad \sin\theta_2 = \tan\theta_2 = \frac{x}{f} = \frac{2\lambda}{a}$$

Where x is the distance between the centre of principal maximum and the second minimum.

$$\therefore \quad x = \frac{2\lambda}{a}\times f = 0.002\times0.2 = 0.004 \text{ m}$$

Example 26.4 *In an arrangement for Fraunhofer diffraction we use a slit of width 0.2 mm and first minimum is at 5 mm on either side of central maximum. If the distance between the lens and the screen is 2 m, calculate the wavelength of light.* ***(Nagpur U. W/2008)***

Solution If x is the distance of first secondary minimum from the centre of the principal maxima and D the distance of screen from the slit, then

$$x = \lambda\frac{D}{a}$$

Here $x = 5\times10^{-3}$ m, $D = 2$ m, $a = 0.2$ mm $= 2\times10^{-4}$ m

$$\therefore \quad \lambda = \frac{ax}{D} = \frac{2\times10^{-4}\times5\times10^{-3}}{2}$$

$$= 5\times10^{-7} \text{ m}$$

Example 26.5 *A plane wave of wavelength 5893 × 10⁻⁸ cm passes through a slit 0.5 mm wide and forms a diffraction pattern on a screen placed on the focal plane of a lens of focal length 1 m. Calculate the separation of the dark band on either side of the central maxima.* ***(P. U. 1999)***

Solution Here $a = 0.5$ mm $= 5\times10^{-4}$ m, $\lambda = 5893\times10^{-8}$ cm $= 5893\times10^{-10}$ m

For the first dark band (secondary minima touching the central maxima), we have $\sin\theta_1 = \frac{\lambda}{a}$

As θ is small $\sin\theta_1 = \theta_1$

$$\therefore \quad \theta_1 = \frac{\lambda}{a} = \frac{5893\times10^{-10}}{5\times10^{-4}} = 1.179\times10^{-3} \text{ radian}$$

Hence separation betweent the dark band on either side of central maxima

$$= 2\,\theta_1 = 2\times1.179\times10^{-3} = 2.358\times10^{-3} \text{ radian}$$

and separation on the screen $x = \frac{2\lambda}{a} f = 2\theta_1 f = 2.358\times10^{-3}\times1\text{m}$

$$= 2.358 \text{ mm.}$$

26.3 THE PHASOR DIAGRAM METHOD

Let us consider a general case of two vibrations,

$y_1 = a_1 \cos 2\pi \nu t$ and $y_2 = a_2 \cos(2\pi\nu t + \delta)$

and let us examine how to draw phasor diagram to obtain the resultant vibration using this method. This method is applicable to find resultant of several vibrations acting simultaneously.

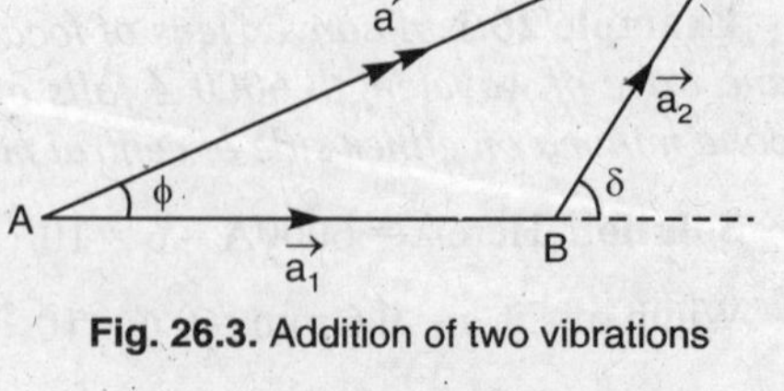

Fig. 26.3. Addition of two vibrations

From any point A, draw line AB proportional to a_1. Now draw a line BC proportional to a_2 and in a direction making an angle δ anticlock wise with AB produced as shown in Fig.26.3. Join AC. Then, the length AC represents **amplitude** a of the resultant vibration, and the $\angle CAB = \phi$ represents its phase relative to vibration y_i. Fig 26.3 is called a phasor diagram because the angles of BC and AC (relative to AB) represent phase differences, **not the directions** of vibrations.

Fig. 26.4 gives the case of adding several simple harmonic vibrations of **same frequency.** We can select any reference phase line OX, and successively draw lengths OA_1, A_1A_2, A_2A_3, A_3A_4, ... proportional to the amplitude of constituent vibrations and in directions making angles with OX equal to their phase angles ϕ_1, ϕ_2, ϕ_3 ....etc. The resultant vibration is shown **in amplitude and phase** by closing the last arm of the polygon taken in opposite sense. All positive ϕ's must be measured anticlock wise from OX.

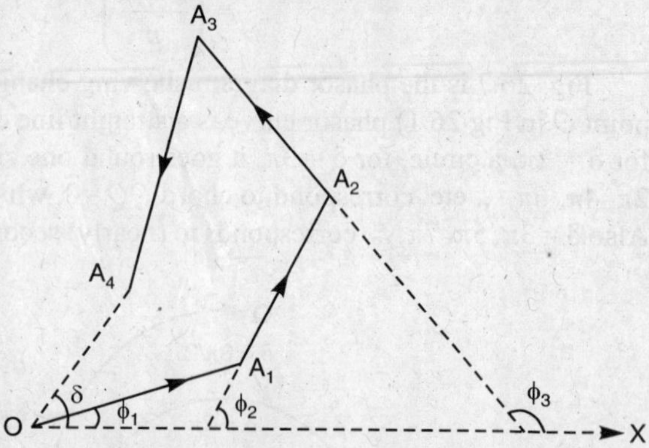

Fig. 26.4. Addition of several vibrations

Application of phasor diagram method to diffraction due to a single slit.

Let us apply the phasor diagram method discussed in Article 26.3 in case of diffraction produced by a single slit. For this, imagine that the slit AB is divided into p equal parts. Then the amplitude contributed by each element is Ao/p, where Ao is the amplitude that the entire slit sends waves in the same phase. Also the phase difference between waves for successive elements is δ/p, where δ is given by

$$\delta = \frac{2\pi}{\lambda}(a \sin\theta) \qquad \text{... (26.2)}$$

Fig. 26.5 shows the phasor diagram of p disturbances. The amplitude Ao/p of each disturbance is added vectorially along with successive phase disturbances δ/p. Fig.26.6 shows the resultant as $p \to \infty$, the phasor diagram becomes a smooth circular arc OQ. Thus, under limit $p \to \infty$, we get a continuous addition of wavelets.

Fig.26.5. Stepped addition of wavelets

Fig. 26.6. Continuous addition of wavelets

The resultant amplitude A_o at the observation point P (Fig. 26.1) is given by the chord OQ in Fig 26.6. From the geometry, we have

$$\frac{A_\theta}{A_o} = \frac{\text{chord OQ}}{\text{Arc OQ}} = \frac{2r\,\sin(\delta/2)}{(2r.\delta/2)} = \frac{\sin(\delta/2)}{(\delta/2)} \qquad \text{... (26.3)}$$

For intensities, we have

$$\frac{I_\theta}{I_o} = \left(\frac{A_\theta}{A_o}\right)^2 = \left[\frac{\sin(\delta/2)}{(\delta/2)}\right]^2 \quad \text{where } \delta = \frac{2\pi\, a \sin\theta}{\lambda} \qquad \text{... (26.4)}$$

$$= \left(\frac{\sin p}{p}\right)^2 \quad \text{where } p = \frac{\pi a \sin\theta}{\lambda} \qquad \text{... (26.5)}$$

Fig. 26.7 is the phasor diagram showing change in δ with increasing θ. For δ = 0 (observation point *O* in Fig 26.1) phasor curve is a straight line of length A_o. For δ = π, it becomes a semi-circle, for δ = 2π, a circle, for δ = 3π, it goes round one circle and a half, for δ = 4π two circles. Thus δ = 2π, 4π, 6π, ... etc. correspond to chord *OQ* =0, which means zero amplitude (hence zero intensity). Also δ = 3π, 5π, 7π, ... corresponds to (nearly) secondary maxima, the main maximum being at δ = 0

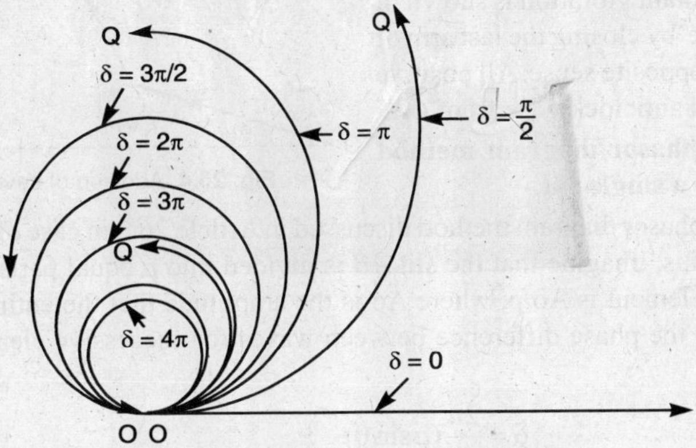

Fig. 26.7. variation in δ with increasing θ.

26.4 INTEGRAL CALCULUS METHOD

The addition of the secondary wavelets reaching *P* from different parts of slit *AB* (Fig. 26.1) can also be done by integration. Fig. 26.8 may be used for this.

Let the disturbance caused at *P* by the wavelet from unit width of the slit of *O* be

$$y_o = A \cos \omega t$$

Then the wavelet from width *dx* at *C* when it reaches *P* has amplitude *Adx* and phase $\omega t + (2\pi/\lambda)\, x \sin\theta$. Calling the disturbance *dy* we have

$$dy = Adx \cos\left\{\omega t + \frac{2\pi\, x \sin\theta}{\lambda}\right\}$$

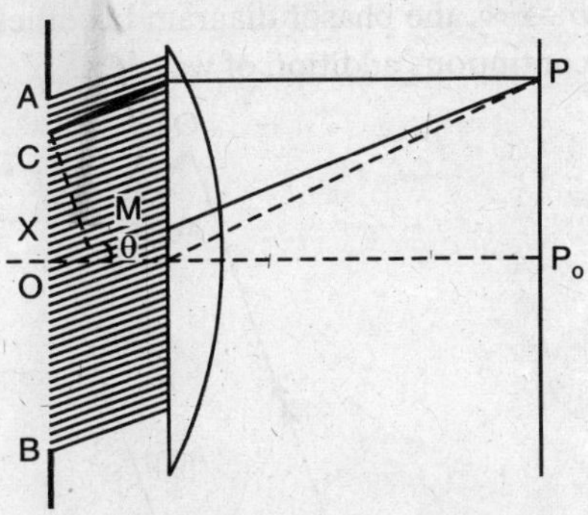

Fig 26.8. Diffraction at a slit.

For the total disturbance at the point of observation at angle θ, we get

$$y = \int_{-a/2}^{+a/2} A \cos\left\{\omega t + \frac{2\pi x \sin\theta}{\lambda}\right\}.dx$$

$$= \left[A.\frac{\sin(\pi a \sin\theta/\lambda)}{\pi \sin(\theta/\lambda)}\right] \cos\omega t \qquad \text{... (26.6)}$$

The quanity in brackets in Eq. 26.6 gives the amplitude A_θ of the resultant disturbance. For θ = 0 it becomes *Aa*, which we call A_o. Thus the result is

$$\frac{A_\theta}{A_0} = \frac{\sin[\pi a(\sin\theta/\lambda)]}{[\pi a(\sin\theta/\lambda)]}$$

$$= \frac{\sin(\delta/2)}{(\delta/2)} \quad \text{... (26.7)*}$$

where $\delta = 2\pi a \sin(\theta/\lambda)$. Eq.26.7 is the same as Eq.26.3, that was obtained using phasor diagram method.

26.5 INTENSITY DISTRIBUTION

To investigate the intensity distribution in the diffraction pattern due to a single slit, consider a plane wave to be incident on the slit *AB*, as shown in Fig. 26.1. The path-difference between the secondary waves originating from *A* and *B* and reaching *P*

$$BN = AB \sin\theta = a \sin\theta$$

where the width of slit $AB = a$

This means that the path-difference of different rays vary uniformly from 0 to $a \sin\theta$.

Suppose the aperture is divided into a large number of infinitesimally small strips, say *n*, of equal widths, parallel to its length, each strip forming an elementary source. As the incident light is parallel, the amplitude of vibration from each strip is taken to be the same. The secondary waves are considered in a direction making an angle θ with *AB*. As the path- difference increases, the phase- difference also increases. Let 2 α be the phase different between the waves from the points *A* and *B* of the slit. The resultant amplitude due to all the individual small strips can be obtained by the vector polygon method. Now the amplitude is small and the phase-difference increases by infinitesimally small value from strip to strip so that the vibration polygon coincides with the circular arc *OQ*. The tangents *OM* at *O* and *MQ* at *Q* give the directions of the first and last vector due to secondary wavelets from the first and the last vibration. These vectors are inclined at 2α which give the phase-difference between the first and the last vibration (Fig.26.9). The path-difference between the first and the last vibration coming from the points *A* and *B* of the slit respectively $= BN = a \sin\theta$ as shown in Fig. 26.1. Draw *CL* perpendicular from the centre of the circular arc on the resultant vector *OQ* (Fig.26.9)

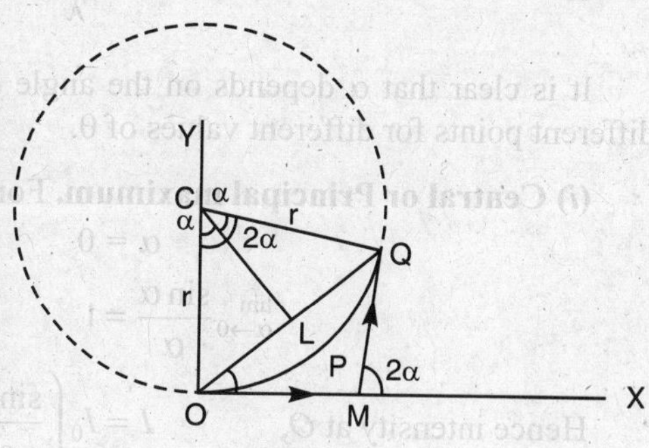

Fig. 26.9

$$\therefore \qquad \angle QMX = \angle QCO = 2\alpha$$

Also $$\sin\alpha = \frac{OL}{CO}$$

or $$OL = r \sin\alpha$$

where *r* is the radius of the circular arc

$$\therefore \qquad OQ = 2OL = 2r\sin\alpha \qquad \text{... } (i)$$

The length of the arc *OPQ* is proportional to the width of the slit.

$$\therefore \qquad \text{Length } OPQ = Ka = A_0 \text{ (say)}$$

Where *K* is a constant

Also $$2\alpha = \frac{\text{Arc } OPQ}{\text{radius}} = \frac{Ka}{r}$$

$$y = A\cos\int_{-a/2}^{+a/2} \cos\frac{2\pi x\sin\theta}{\lambda}dx - A\sin\omega t\int_{-a/2}^{+a/2} \sin\frac{2\pi x\sin\theta}{\lambda}dx$$

The second integral is zero. The gives the result given in (Eq. 26.7)

or $$2r = \frac{Ka}{\alpha} \quad \text{... (ii)}$$

Substituting the value of $2r$ in (i), we get

$$OQ = \frac{Ka}{\alpha}\sin\alpha$$

Put $$OQ = A$$

$\therefore$ Resultant amplitude $$A = (Ka)\frac{\sin\alpha}{\alpha} = A_0\frac{\sin\alpha}{\alpha} \quad \text{... (iii)}$$

As the screen is at infinity the obliquity factor (1+ cos θ) can be neglected as the amplitude of each wave is assumed to be the same.

The intensity I at a point is given by

$$I = A^2 = A_0^2\left(\frac{\sin\alpha}{\alpha}\right)^2 = I_0\left(\frac{\sin\alpha}{\alpha}\right)^2 \quad \text{... (26.8)}$$

Where $I_0 = A_0^2$ is the intensity of principal maximum at $\alpha = 0$

As the phase change 2π corresponds to a path-difference λ, the phase change corresponding to a path-difference $a \sin\theta$

$$= 2\alpha = \frac{2\pi}{\lambda}a\sin\theta \quad \text{... (iv)}$$

or $$\alpha = \frac{\pi}{\lambda}a\sin\theta$$

It is clear that α depends on the angle of diffraction θ and $I_0\dfrac{\sin^2\alpha}{\alpha^2}$ gives the intensity at different points for different values of θ.

(*i*) Central or Principal maximum. For the central point O on the screen,

$$\alpha = 0$$

$$\lim_{\alpha\to 0}\frac{\sin\alpha}{\alpha} = 1$$

Hence intensity at O, $$I = I_0\left(\frac{\sin^2\alpha}{\alpha^2}\right) = I_0$$

This is maximum as all waves reach O in phase.

Again $$\alpha = 0$$

$\therefore$ $$\frac{\pi}{\lambda}a\sin\theta = 0$$

or $$\theta = 0$$

This shows that the waves are travelling normal to the slit and O gives the position of **central maximum.**

(*ii*) Secondary minima. The directions of secondary minima are given by

$$a\sin\theta = \pm n\lambda \quad \text{... (vi)}$$

Substituting the value of $a \sin\theta$ in (v), we get

$$\alpha = \frac{\pi}{\lambda}\times n\lambda = \pm n\pi \quad \text{... (vi)}$$

If $$\alpha = \pm\pi, \sin\alpha = 0$$

$\therefore$ Intensity $$I = I_0 \frac{[\sin(\pm\pi)]^2}{(\pm\pi)^2} = 0$$

That at $a = \pm\pi$, the intensity becomes zero and then it passes through various minima on either side of central maximum, when

$$\alpha = \pm\pi, \pm 2\pi, \pm 3\pi + ... \pm n\pi$$

where n is an integer but not equal to zero.

The positions of various minima are given by

$$\frac{\pi a \sin\theta}{\lambda} = \pm n\pi$$

or $$\sin\theta = \pm n\frac{\lambda}{a} \quad ... (vii)$$

The positions of first, second, third.. etc. minima are given by $\pm\frac{\lambda}{a}, \pm\frac{2\lambda}{a}, \pm\frac{3\lambda}{a}$ ... for various values of sin θ.

(*iii*) **Secondary maxima.** The direction of nth secondary maximum is given by

$$\sin\theta_n = \pm\frac{(2n+1)\lambda}{2a} \quad ... (viii)$$

Substituting in (v), we get

$$\alpha = \pm\frac{\pi}{\lambda}.a\frac{(2n+1)\lambda}{2a} = \pm(2n+1)\frac{\pi}{2} \quad ... (ix)$$

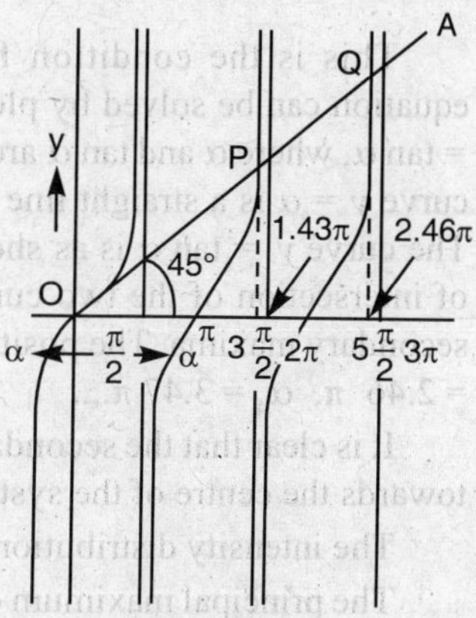

Fig. 26.10

For various values of $n = 1, 2, 3 ....$, we get

$$\alpha = \pm\frac{3\pi}{2}, \pm\frac{5\pi}{2}, \pm\frac{7\pi}{2}....$$

(*a*) For the first secondary maximum substituting in Eq. 26.8, $\alpha = 3\pi/2$, we get

$$I = I_0\left(\frac{\sin\alpha}{\alpha}\right)^2 = I_0\left(\frac{\sin 3\pi/2}{3\pi/2}\right)^2 = I_0\left(\frac{-1}{3\pi/2}\right)^2$$

$$= \frac{4}{9\pi^2}I_0 = \frac{I_0}{22}.$$

(*b*) For the second secondary maximum $\alpha = 5\pi/2$

$$\therefore \quad I = I_0\left(\frac{\sin 5\pi/2}{5\pi/2}\right)^2 = I_0\left(\frac{2}{5\pi}\right)^2 = \frac{I_0}{61}$$

Thus the intensity of secondary maxima falls off rapidly. Hence we find that secondary maxima of decreasing intensity occur on either side of the central maximum. Their position is given by equation (*viii*).

The intensity of principal maximum falls to zero at $\alpha = \pm\pi$ and then passes through a number of secondary maxima. To determine the position of secondary maxima differentiate the equation of intensity.

$$I = I_0\left(\frac{\sin^2\alpha}{\alpha^2}\right)$$

$$\therefore \quad \frac{dI}{d\alpha} = I_0 \left(\frac{\alpha^2 (2 \sin \alpha \cos \alpha) - (\sin^2 \alpha) 2\alpha}{\alpha^4} \right)$$

$$= I_0 \left(\frac{2 \sin \alpha \cos \alpha}{\alpha^2} - \frac{2 \sin^2 \alpha}{\alpha^3} \right)$$

For I to be maximum, put $\frac{dI}{d\alpha} = 0$

$$\therefore \quad \frac{2 \sin \alpha \cos \alpha}{\alpha^2} - \frac{2 \sin^2 \alpha}{\alpha^3} = 0$$

or
$$\cos \alpha - \frac{\sin \alpha}{\alpha} = 0$$

$$\therefore \quad \tan \alpha = \alpha$$

This is the condition for secondary maxima. The equation can be solved by plotting a graph for $x = \alpha$ and $y = \tan \alpha$, where α and $\tan \alpha$ are the x and y co-ordinates. The curve $y = \alpha$ is a straight line OA inclined to α-axis at 45°. The curve $y = \tan \alpha$ is as shown in Fig. 26.10. The points of intersection of the two curves, P, Q give the position of secondary maxima. The positions are $\alpha_1 = 0$, $\alpha_2 = 1.43\ \pi$, $\alpha_3 = 2.46\ \pi$, $\alpha_4 = 3.47\ \pi$....

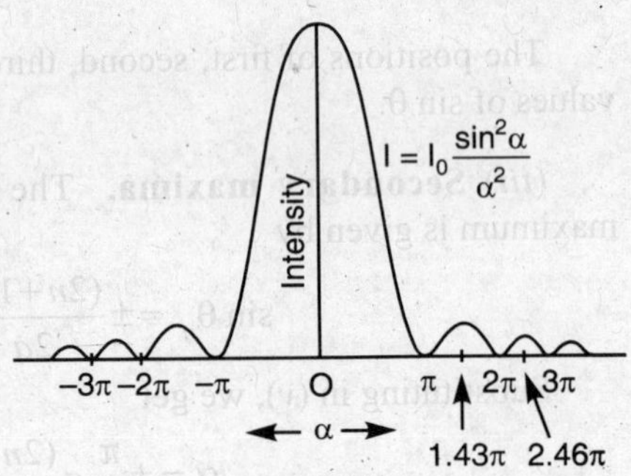

Fig. 26.11

It is clear that the secondary maxima do not fall half way between two minima but are displaced towards the centre of the system by an amount which decreases with increase of n.

The intensity distribution curve of Fraunhofer diffraction at single slit is as shown in Fig. 26.11

The principal maximum occurs at $\alpha = 0$, secondary minima occur at $\alpha = \pm \pi, \pm 2\pi$ ...., secondary maxima occur at $\alpha = 1.43\ \pi$, $2.46\ \pi$... The intensity of secondary maxima falls off rapidly.

Example 26.6 *A parallel beam of light of wavelength 5×10^{-5} cm is incident normally on a narrow slit of width 0.2 mm. The fraunhofer diffraction pattern is observed on a screen which is placed at the focal plane of a convex lens of focal length 50 cm. Calculate the distance between the first two minima and the first two maxima on the screen. Assume that the lens placed very close to the slit.*

(Nagpur U. 2010)

Solution. Here $a = 0.2$ mm $= 2 \times 10^{-4}$ m ; $\lambda = 5 \times 10^{-7}$ m ; $f = 0.5$ m

For secondary minima , we have

$$a \sin \theta = \pm n\lambda$$

As θ is small $\sin \theta = \theta$.

$\therefore$ Angular diffraction corresponding to first minima

$$\theta_1 = \frac{\lambda}{a} = \frac{5 \times 10^{-7}}{2 \times 10^{-4}} = 2.5 \times 10^{-3} \text{ radian}$$

Angular diffraction corresponding to second minima

$$\theta_2 - \frac{2\lambda}{a} = \frac{2 \times 5 \times 10^{-7}}{2 \times 10^{-4}} = 5.0 \times 10^{-3} \text{ radian}$$

$\therefore$ Separation between the 1st and 2nd minima $x = f(\theta_2 - \theta_1)$

or
$$x = (5 - 2.5)\ 10^{-3} \times 0.5 = 0.125 \times 10^{-3} \text{ m}$$

The first and second maxima will occur when

$$\alpha = 1.43\ \pi \text{ and } 2.46\ \pi$$

∴ Angular diffraction corresponding to 1st maximum

$$\theta_1' = 1.43 \times 2.5 \times 10^{-3}$$

For 2nd diffraction maximum

$$\theta_2' = 2.46 \times 2.5 \times 10^{-3} \qquad \left[\because \alpha = 1.43\ \pi = \frac{\pi}{\lambda} a\, \theta_1' \text{ or } \theta_1' = 1.43 \frac{\lambda}{a}\right]$$

∴ Separation between the first two secondary maxima $x' = f(\theta_2' - \theta_1')$

or $$x' = (2.46 - 1.43) \times 2.5 \times 10^{-3} \times 0.5$$

$$= 0.1288 \times 10^{-2} \text{ m}$$

26.6 HALF ANGULAR WIDTH OF A PRINCIPAL MAXIMA

Half angular width of principal maxima. *It is defined as the angular width of the principal maxima at which the intensity of light falls to one half of the peak value.*

If I_0 is the maximum or peak value of the intensity, then the intensity of light at any point is given by

$$I = I_0 \left(\frac{\sin \alpha}{\alpha}\right)^2 \qquad (\therefore \text{ Eq. } 26.8)$$

Where $$\alpha = \frac{\pi}{\lambda} a \sin \theta \qquad (\therefore \text{ Eq. } 26.9)$$

θ = Angle of diffraction; λ = wave length of light and

α = width of the slit

If Δθ represents the half angular width of the principal maxima, then $\theta = \frac{\Delta\theta}{2}$

For this value of θ, the intensity of light falls from the peak value $I_0 = I_0/2$

or $$\left(\frac{\sin \alpha}{\alpha}\right)^2 = \frac{1}{2}$$

This relation is satisifed for α = 1.4 radian

Now $$\alpha = \frac{\pi}{\lambda} a \sin \theta = \frac{\pi}{\lambda} a \sin\left(\frac{\Delta\theta}{2}\right) \qquad \left[\because \theta = \frac{\Delta\theta}{2}\right]$$

∴ $$\frac{I_0}{2} = I_0 \left(\frac{\sin \alpha}{2}\right)^2 \text{ or } \sin\left(\frac{\Delta\theta}{2}\right) = \frac{\alpha\lambda}{\pi\alpha} = \frac{1.4\lambda}{\pi a}$$

or $$\frac{\Delta\theta}{2} = \sin^{-1}\left(\frac{1.4\lambda}{\pi a}\right)$$

or $$\Delta\theta = 2 \sin^{-1}\left(\frac{1.4\lambda}{\pi a}\right) \qquad \text{... (26.10)}$$

Example 26.7 *A 12000 Å wide single slit is illuminated by a parallel beam of monochromatic light of wavelength 6000Å. Determine half-angular width of the central maximum in the Fraunhofer diffraction pattern.*

Solution. Here $$\lambda = 6000 \text{ Å} = 6000 \times 10^{-10} \text{ m} ; a = 12000 \text{ Å} = 12000 \times 10^{-10} \text{ m}$$

∴ $$\sin\left(\frac{\Delta\theta}{2}\right) = \frac{1.4\lambda}{\pi a} = \frac{1.4 \times 6000 \times 10^{-10}}{\pi \times 12000 \times 10^{-10}} = 0.2228$$

$$\therefore \qquad \frac{\Delta\theta}{2} = \sin^{-1} 0.2228 = 12.87°$$

$$\text{or} \qquad \Delta\theta = 25.74°$$

26.7 DIFFRACTION AT A CIRCULAR APERTURE

Consider a narrow aperture AB of diameter d perpendicular to the plane of paper illuminated by a parallel beam of monochromatic light of wavelength λ. The light, in passing through the aperture, suffers diffraction. If a lens L is placed in the path of the diffracted beam very close to aperture, a real image of the diffracted pattern is obtained on the screen RT, placed perpendicular to the plane of the paper, in the focal plane of the lens.

The secondary waves travelling in the direction CO come to focus at O and a bright image is formed as the secondary waves, equidistant from C in the upper and lower half travel the same distance before reaching O and reinforce each other.

The rays reaching a point P on the screen distant r from O, after diffraction at an angle θ from the extreme ends A and B of the aperture, have a path-difference.

$$BN = AB \sin\theta = d \sin\theta$$

If this path-difference is an integral multiple of λ, then by reasoning similar so that as in the case of **single slit pattern,** the point P will have *minimum intensity*, when

$$d \sin\theta = n\lambda \qquad \text{... (i)}$$

The point P will have maximum intensity if the path-difference is equal to an odd multiple of $\lambda/2$ *i.e.,*

$$d \sin\theta = (2n+1)\lambda/2 \qquad \text{...(ii)}$$

When $n = 1, 2, 3$ ...

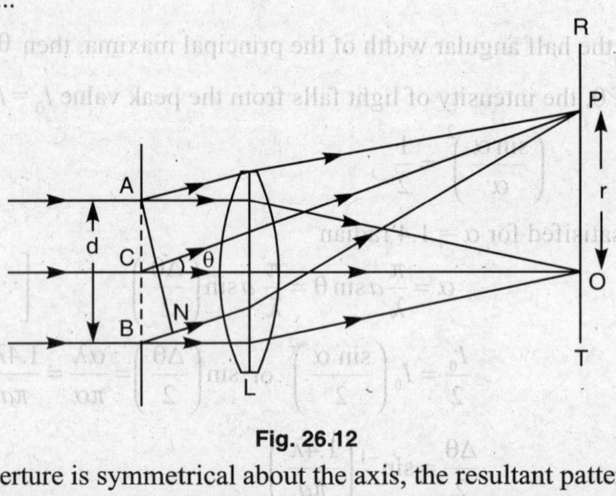

Fig. 26.12

As the circular aperture is symmetrical about the axis, the resultant pattern may be obtained by rotating Fig. 26.12 about the same axis. The point P thus will trace out a circular ring of uniform intensity, minimum or maximum, depending upon the conditions given above. Thus the diffraction pattern due to a circular aperture consists of a central bright disc called the **Airy's disc** surrounded by alternate dark and bright concentric rings called the **Airy's rings.** The intensity of dark rings is zero and that of bright rings decreases gradually outward from O.

If the collecting lens L lies very close the aperture and the screen is at a very large distance, then

$$\sin\theta = \theta = \frac{r}{f} \qquad \text{... (iii)}$$

Also, for the first secondary minimum

$$d \sin\theta = \lambda$$

or $$\sin\theta = \theta = \frac{\lambda}{d} \quad ...(iv)$$

Hence $$\frac{r}{f} = \frac{\lambda}{d}$$

or $$r = \frac{f\lambda}{d} \quad ...(26.11)$$

where r is the radius of Airy's disc.

Actually the radius of the first dark ring is slightly more than given by Eq. (26.11). According to Airy, it is given by

$$r = \frac{1.22}{d} f\lambda \quad ...(26.12)$$

The intensity distribution of the bright and dark rings is similar to the one discussed for rectangular slit.

It is clear from 26.11 that the radius of the central disc decreases with the increase of diameter of the aperture.

Importance. The diffraction of a plane wave from a narrow circular aperture is of great importance, on account of its application to resolving power of telescopes and microscopes. A lens acts like a circular aperture and produces a diffraction pattern surrounded by alternate dark and bright rings. If there are two point-objects lying close to each other, then two diffraction patterns are produced. The two objects are said to be resolved if the central maximum of one falls on the first secondary minimum of the other. *Thus diffraction imposes a limit on the resolving power of opitical instruments.*

Example 26.8 *A circular aperture of radius 10 μm is placed in front of a convex lens of focal length 1m. The aperture is illuminated by light of wavelength 6000 Å incident normally on the aperture. Calculate the radius of the central bright disc (Airy's disc.).* **(P.U. 2008)**

Solution Here, diameter of the circular aperture $d = 2r = 2\times10$ μm = 20 μm = 20×10^{-6}m

Focal length of the lens f = 1m

Wavelength of light λ = 6000 Å = 6000×10^{-10} m

∴ Radius of Airy's disc (central bright maxima) $= \frac{1.22}{d} f\lambda$

$$= \frac{1.22\times1\times6000\times10^{-10}}{20\times10^{-6}} = 36.6 \text{ mm}$$

26.8 EMISSION BY A DISC SOURCE

The result obtained by rigorous theory of Fraunhofer diffraction at a circular aperture of diameter d, the central image is a disc of angular radius θ, is given by

$$d\sin\theta = 1.22\,\lambda \quad ...(26.13)$$

This means that the intensity falls off with increasing θ, reaching a minimum value at $\theta = \theta_1$. Thereafter it rises, reaches a maximum (much less than that at θ = 0); falls again and so on. This result of Eq. (26.13) is of much wider significance than just transmission. It applies to emission as well as reception of waves. Consider a source of sound waves comprising a disc of diameter d and oscillating as a whole normal to its face. Then the maximum intensity will be along the normal, but the sound will **necessarily spread** up to a cone of semi-angle $\sin^{-1}(\lambda/d)$ nearly. If $\lambda > d$, the spread will extend up to 90°, on the other hand if $\lambda = d/20$ (Say), the spread will be about 1/20 radian about the normal. We can reverse the statement saying that in order to limit the spread of the emitted waves to a cone of semi-angle θ (considered small), the diameter over which we must have in phase (coherent) oscillations should be nearly λ/θ. The statement applies to radar antenna and receivers also.

26.9 FRAUNHOFER DIFFRACTION AT A DOUBLE SLIT

Suppose there are two parallel rectangular slits *AB* and *CD* of equal width *a* separated by an opaque distance *b* placed perpendicular to the plane of the paper. The distance between the corresponding points of the two slits will be $(a+b)$ (Fig. 26.13)

Let a plane wave front of monochromatic light of wavelength λ be incident on the double slit. Suppose each slit diffracts the beam in a direction making an angle θ with the direction of the incident beam. The Path-difference between the rays from the corresponding points *A* and *C* of the two slits.

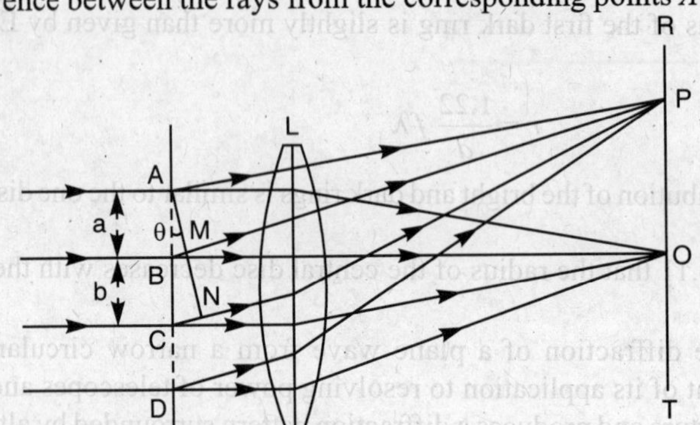

Fig. 26.13

$$CN = (a+b)\sin\theta$$

If 2 β is the phase-difference, then

$$2\beta = \frac{2\pi}{\lambda}(a+b)\sin\theta$$

or

$$\beta = \frac{\pi}{\lambda}(a+b)\sin\theta \quad \text{... } (i)$$

Now if each slit is considered to be an independent small source of monochromatic light and light from the two slits is focused on the screen *TR* by means of a convex lens *L*, then the resultant pattern will be due to interference from two sources of light placed at the corresponding points of the two slits.

If 2 α is the phase-difference between the extreme rays from the first slit *i.e.*, from *A* and *B*, then

$$2\alpha = \frac{2\pi}{\lambda}a\sin\theta \quad \textbf{[See Eq. (iv), article 26.5)}$$

or

$$\alpha = \frac{\pi}{\lambda}a\sin\theta \quad \text{... } (ii)$$

The resultant displacement y_1 due to the rays from the first slit is given by

$$y_1 = A\sin\omega t \quad \text{... } (iii)$$

where

$$A = A_0\frac{\sin\alpha}{\alpha} \quad \textbf{[(as already proved) in Eq. (iii), article 26.5]}$$

Here A_0 is the resultant amplitude of the incident rays and *A* that of the rays diffracted at an angle θ. The resultant displacement y_2 due to the rays from the second slit is given by

$$y_2 = A\sin(\omega t + 2\beta) \quad \text{... } (iv)$$

Hence the resultant displacement *Y* due to rays from both slits diffracted at an angle θ is given by

$$Y = y_1 + y_2 = A\sin\omega t + A\sin(\omega t + 2\beta)$$

$$= 2A \cos\beta \sin(\omega t + \beta) \qquad \text{... } (v)$$

The amplitude of this resultant

$$= 2A\cos\beta = 2A_0 \frac{\sin\alpha}{\alpha}\cos\beta$$

Intensity. The resultant intensity $I = Y^2$ is equal to $4\ A_0^2 \dfrac{\sin^2\alpha}{\alpha^2}\cos^2\beta$ or $I = 4\,I_0 \dfrac{\sin^2\alpha}{\alpha^2}\cos^2\beta$ where $I_0 = A_0^2$.

Thus we see that the resultant intensity in the diffraction pattern of two slits depends upon two factors:

(*i*) $A_0^2 \dfrac{\sin^2\alpha}{\alpha^2}$ which gives the diffraction pattern due to a single slit.

(*ii*) $\cos^2\beta$ which gives a system of interference fringes due to wavelets from the corresponding points of the two slits.

The set of diffraction fringes have greater dispersion while the set of interference fringes have lesser dispersion. These are superimposed on the diffraction set of fringes and are visible only in the region of central maximum.

Nature of diffraction pattern. The factor $\dfrac{A_0^2 \sin^2\alpha}{\alpha^2}$ gives diffraction pattern due to a single slit. The **central maximum** is given when

$$\alpha = 0 \text{ as } \underset{\alpha\to 0}{Lt} \frac{\sin\alpha}{\alpha} = 1$$

The position of **secondary minima** due to this factor is given when

$$\alpha = \pm n\pi \qquad \textbf{[Eq. (\textit{vi}), article 26.5]}$$

or $\quad a\sin\theta_n = \pm n\lambda \qquad$ **[Eq. (*vii*) article 26.5]**

Thus for $\quad n = 1, 2, 3$ ... we get

$$\alpha = \pm\pi, \pm 2\pi, \pm 3\pi$$

and

$$\sin\theta_1 = \pm\frac{\lambda}{a}, \sin\theta_2 = \pm\frac{2\lambda}{a}, \sin\theta_3 = \pm\frac{3\lambda}{a}$$

For **secondary maxima**

$$\alpha = \pm(2n+1)\frac{\pi}{2} \qquad \textbf{[Eq. (\textit{ix}), article 26.5]}$$

or

$$\sin\theta_n = \pm(2n+1)\frac{\lambda}{2} \qquad \textbf{[Eq. (\textit{viii}), article 26.5]}$$

Thus for $\quad n = 1, 2, 3$, we get

$$\alpha = \pm\frac{3\pi}{2}, \pm\frac{5\pi}{2}, \pm\frac{7\pi}{2}$$

and

$$\sin\theta_1 = \pm\frac{3\lambda}{2a}, \sin\theta_2 = \pm\frac{5\lambda}{2a}, \sin\theta_3 = \frac{7\lambda}{2a}$$

Interference pattern. Considering the second factor $\cos^2\beta$ for intensity variation, we find that intensity will have a **maximum** value if

$$\cos^2\beta = 1 \textit{ i.e.}, \beta = \pm n\pi, \text{ where } n = 0, 1, 2 \text{ ...}$$

$$\therefore \quad \frac{\pi}{\lambda}(a+b)\sin\theta = \pm n\pi$$

or $\quad (a+b)\sin\theta = n\lambda \qquad$ ... (*vi*)

Putting $n = 1, 2, 3...$ corresponding directions of *maxima* can be obtained.

From equation (*vi*)

$$\sin\theta_1 = \pm\frac{\lambda}{a+b},$$

$$\sin\theta_2 = \pm\frac{2\lambda}{a+b}, \sin\theta_3 = \pm\frac{3\lambda}{a+b}$$

$$\therefore \quad \sin\theta_2 - \sin\theta_1 = \pm\frac{\lambda}{a+b}$$

and $$\sin\theta_3 - \sin\theta_2 = \pm\frac{\lambda}{a+b}$$

The angular separation between the two secondary maxima is equal to $\pm\frac{\lambda}{a+b}$.

The interference **central maximum** is given when $\theta = 0$ in the direction in which the rays are incident. This is also the direction of **central maximum** of diffraction pattern.

From minima $\cos^2\beta = 0$

where $\beta = \pm(2n+1)\pi/2$

$$\therefore \quad \frac{\pi}{\lambda}(a+b)\sin\theta = \frac{(2n+1)\pi}{2}$$

or $$(a+b)\sin\theta = \pm(2n+1)\lambda/2$$

For $n = 1, 2, 3 ....$, we get

$$\sin\theta_1 = \pm\frac{3\lambda}{2(a+b)}, \sin\theta_2 = \pm\frac{5\lambda}{2(a+b)} ...$$

$$\therefore \quad \sin\theta_2 - \sin\theta_1 = \pm\frac{\lambda}{a+b}$$

The angular separation between two consecutive *minima* and two consecutive *maxima* is the same and is equal to $\frac{\lambda}{(a+b)}$.

The *minimum intensity* occurs along directions for which the path-difference between two corresponding points of the slits is an odd multiple of $\frac{\lambda}{2}$ and *maximum intensity* occurs when the path-difference between the corresponding points of the slits is an even multiple of $\frac{\lambda}{2}$.

For $\beta = 0$, the intensity of central maximum $= 4I_0$ where $\alpha = 0$. The intensity diagram corresponding to $2a = b$ is as shown in Fig. 26.14. The full line curve represents the intensity distribution due to interference of light of both the slits. The pattern consists of diffraction maxima and minima and equally spaced interference maxima and minima within each diffraction maximum.

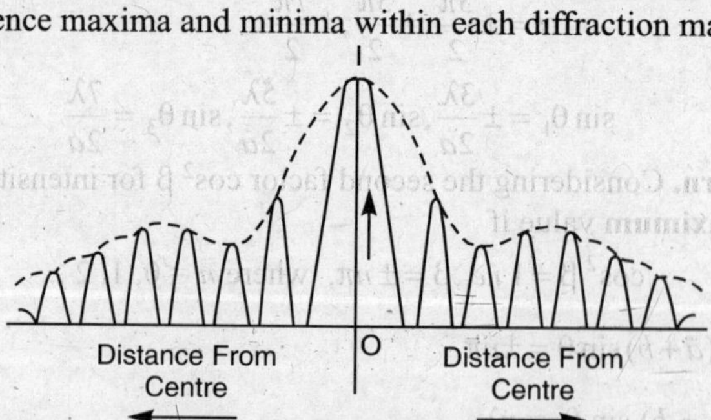

Fig. 26.14

The dotted curve represents the diffraction maxima and minima. The spacing of the interference maxima and minima is dependent on the values of a and b.

Missing orders in a double slit diffraction pattern. Let a be the width of each slit and b the distance between them.

The directions of interference maxima are given by

$$(a+b)\sin\theta = n\lambda \quad \text{... (i)}$$

The directions of diffraction minima are given by the relation

$$a\sin\theta = p\lambda \quad \text{... (ii)}$$

where n and p are integers. If the value of a and b are such that both the equations are satisfied for the same value of θ, then a certain interference maximum will overlap the diffraction minimum and hence the spectrum order will be missing.

Dividing (*i*) by (*ii*), we get

$$\frac{a+b}{a} = \frac{n}{p}$$

If $a = b$, then

$$\frac{n}{p} = 2 \text{ or } n = 2p$$

If $p = 1, 2, 3$ ... etc., then $n = 2, 4, 6$ etc.

This means that the 2,4,6 etc. orders of interference maxima will be missing in the diffraction pattern. There will be three interference maxima in the central diffraction maximum.

If $b = 2a$, then

$$\frac{a+2a}{a} = \frac{n}{p} \text{ or } n = 3p.$$

If $p = 1, 2, 3, ...,$ then $n = 3, 6, 9$ ...

This means that 3rd, 6th etc orders of interference maxima will be missing in the diffraction pattern.

On both sides of the central maximum, the number of interference maximum is 2 and hence there will be five interference maxima in the central diffraction maximum. The position of the third interference maximum corresponds to the first diffraction minimum.

If $b = 0$, the two slits join and all the interference maxima will be missing and the diffraction pattern will be same as due to a single slit of width $2a$.

DIFFERENCE BETWEEN SINGLE SLIT AND DOUBLE SLIT DIFFRACTION PATTERN

The single slit diffraction pattern consists of a central bright maximum with secondary maxima and minima on either side. The intensity of secondary maxima and minima gradually decreases.

The double slit diffraction pattern consists of a central maximum which contains equally spaced interference maxima and minima. The intensity of central diffraction maximum is four times the intensity of central maximum due to a single slit.

In a double slit diffraction pattern, the spacing of the diffraction maxima and minima depends upon the width of the slit *i.e.*, a and the spacing of the interference maxima and minima depends upon the width of slit a and the opacity b between the two slits. The intensity of the interference maxima and minima is not constant but decreases to zero on either side of the central maximum.

In a double slit diffraction pattern, the slit width is taken as a and the separation between the slits is b. If the slit width a is kept constant, the diffraction pattern remains the same. Keeping a constant, if the distance between the slit centres b is varied, the spacing between the interference maxima changes, depending upon the relative values of a and b. Certain orders of interference maxima will be missing in the resultant pattern.

The angular separation between two secondary maxima due to interference is given by

$$\sin\theta_2 - \sin\theta_1 = \frac{\lambda}{a+b}$$

If b increases, the angular separation decreases. Hence the spacing between the secondary interference maxima decreases.

(*i*) The position of secondary minima is given by

$$a\sin\theta = \pm\, n\,\lambda$$

(*ii*) As a remains constant, the diffraction pattern remains constant.

Difference between double slit diffraction pattern and interference due to two sources. The double slit diffraction pattern consists of central maximum which contains equally spaced, interference maxima and minima on either side. In addition to these, there is a central maximum with secondary maxima and minima the intensity of which gradually decreases due to each slit. The intensity of central maximum is four times the intensity of central maximum due to a single slit.

The spacing of diffraction maxima and minima depends upon the width of slit a. The spacing of interference maxima and minima depends upon $(a + b)$ where b is the distance between the two slits. The intensity of this interference maxima and minima is not constant due to overlapping of secondary maxima and minima due to diffraction and hence its intensity decreases to zero on either side of central maxima.

The interference due to two independent sources is not possible as the two sources will not emit continuous waves of the same wavelength in the same phase or with constant phase-difference having same amplitude.

If, however, two coherent sources are used, then large number of interference fringes of uniform intensity and equal spacing are produced. The points of minimum intensity are perfectly dark.

26.10 RESOLVING POWER

It is defined as the reciprocal of the smallest angle subtended at the objective (of an optical instrument) by two point objects which can just be distinguished as separate.

If $d\theta$ is the smallest angle subtended by two point objects on the objective of the instrument, then Resolving power $= \dfrac{1}{d\theta}$ and $d\theta$ gives the *limits of resolution.*

Magnifying power. *It is defined as the ratio of the angle subtended by the image at the eye, as seen through the instrument, to the angle subtended by the object at the centre of the unaided eye, This distance for a microscope is the distance of distinct vision and for a telescope it is infinity.*

Distinction. (*i*) Magnifying power is the ability of the telescope to show a magnified view of the object to the eye, whereas the resolving power is its ability to show the amount of detail or grain in the object.

(*ii*) Magnifying power increases the distance between the centres of the diffraction patterns of the two points, without giving any idea of the size of central maximum.

The resolving power not only gives us the idea about the distance between the centres of two maxima, but also gives us the idea about their width.

(*iii*) The magnifying power of a telescope is equal to F/f where F is the focal length of the objective and f that of the eye-piece. Two telescopes having objectives of the same diameter will have different magnifying powers, as the focal length can be different. The resolving power of a telescope is given by the ratio d/λ which is the same for all telescopes having the objectives of the same diameter d for the given wavelength λ.

(*iv*) Resolving power increases as the wavelength of light decreases whereas the magnifying power is independent of the wavelength of light.

26.11 RAYLEIGH CRITERION

The theory of optical instruments is based on the laws of geometrical optics and rectilinear propagation of light. These laws are only approximately true. When a beam of light passes through the objective of a telescope, the lens acts like a circular aperture and produces a diffraction pattern, instead of a point image. This diffraction pattern is a bright point surrounded by alternate dark and bright rings. If there are two point-objects lying close to each other, then two diffraction-patterns are produced which overlap.

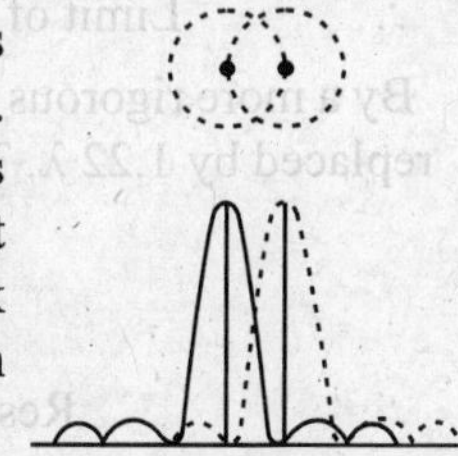

Fig. 26.15

According to *Rayleigh's criterion, the two images of such point-objects lying close to each other are said to be separated (just resolved) if the central maximum of one falls on the first secondary minimum of the other.* In other words, when the central bright image of one falls on the first dark ring of the other, as shown in Fig.26.15 the two are said to be resolved.

This criteria of resolution is equally applicable to the resolution of spectral lines of equal intensity. The two close lines in the spectrum are said to be just resolved when the central maximum of the diffraction-pattern of one falls on the first minimum of diffraction-pattern of the *other i.e., the least angular separation between the principal maxima of the diffraction-pattern of the two spectral lines in a given diffraction order should be equal to half the angular width of either principal maximum.* The resolving power in that case is $\lambda/d\lambda$ where λ is the wavelength of a line in the spectrum and $d\lambda$ is the difference in the wavelength that can just be seen as separate. Thus

$$\text{Resolving power} = \frac{\lambda}{d\lambda}. \qquad \text{... (26.14)}$$

26.12 LIMIT OF RESOLUTION AND RESOLVING POWER OF A TELESCOPE

Let us consider an image formed by a telescope having an objective of aperture $AB = d$. Consider two point-objects P and Q lying on the object plane (not shown) and having an angular separation $d\theta$. Parallel rays coming from these two point-objects will give rise to two point-images P_1 and Q_1 lying in the focal plane of the objective of the telescope as shown in Fig. 26.16

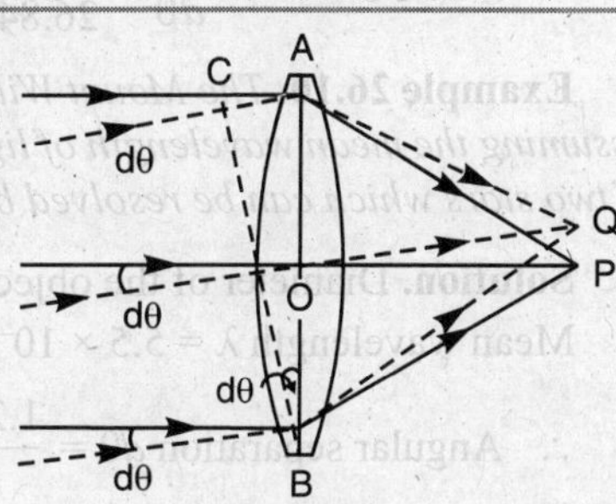

Fig. 26.16

Since light from the object P is incident normally,

$$AP_1 = BP_1$$

Thus the first maximum of the image of P will be formed at P_1.

In order that the image Q_1 may appear separate from P_1, the first maximum of image Q_1 should fall on the first secondary minimum of the image P_1. This condition is achieved if the path-difference between the extreme rays from Q and those from P is λ.

Draw BC normal to the rays coming from Q, then the path-difference is AC. In order that the first secondary minimum of the image Q_1 should fall on the first central maximum of P_1, the path-difference between the two rays should be equal to λ. It is because the lens AB can be supposed to be divided into two halves AO and OB, and the rays coming from the corresponding points like A and O from the two halves will interfere destructively, only if the path-difference is $\lambda/2$.

Now $$AC = AB \,.\, d\theta$$

where angle CBA is equal to $d\theta$ and has a very small value.

$$\therefore \qquad d\theta = \frac{AC}{AB} = \frac{\lambda}{d}$$

where AB is the aperture of the lens and is equal to d.

This gives the condition that the two-point-objects are just seen as separate. Thus $d\theta$ is a measure of **limit of resolution** of the telescope objective

$\therefore$ Limit of resolution $d\theta = \dfrac{\lambda}{d}$ ... (26.15)

By a more rigorous treatment of this condition for a circular aperture Airy showed that λ should be replaced by 1.22 λ. Thus

$$d\theta = \frac{1.22\lambda}{d}$$

$\therefore$ Resolving power $= \dfrac{1}{d\theta} = \dfrac{d}{1.22\lambda}$ ... (26.16)

It is clear from the expression that resolving power of telescope depends upon (*i*) diameter of the objective and (*ii*) the wavelength of light used. Larger the diameter and smaller the wavelength higher will be the resolving power.

Example 26.9 *The objective of an astronomical telescope is 30 cm. diameter. Assuming the mean wavelength of white light 660 nm estimate the smallest angular separation of the two stars which can be just resolved by the telescope. Hence find the resolving power.* **(P. U., 2003)**

Solution Diameter of the objective $d = 30$ cm $= 30 \times 10^{-2}$ m

Mean wavelength $\lambda = 660 \times 10^{-9}$ m

$\therefore$ Smallest angular separation of two stars for just resolution

$$d\theta = \frac{1.22\lambda}{d} = \frac{1.22\times660\times10^{-9}}{30\times10^{-2}} = 26.84\times10^{-7}\text{ radian} = 0.5535''$$

$$\text{Resolving power} = \frac{1}{d\theta} = \frac{1}{26.84\times10^{-7}} = \frac{10^7}{26.84} = 3.72\times10^5$$

Example 26.10 *The Mount Wilson observatory telescope has an objective of diameter 2.54 m. Assuming the mean wavelength of light to be 5.5×10^{-5} cm, estimate the smallest angular separation of two stars which can be resolved by it.* **(P.U., 2001)**

Solution. Diameter of the objective $d = 2.54$ m

Mean wavelength $\lambda = 5.5 \times 10^{-5}$ cm $= 5.5\times10^{-7}$ m.

$$\therefore \text{ Angular separation } d\theta = \frac{1.22\times5.5\times10^{-7}}{2.54}$$

$$= 2.642 \times 10^{-7} \text{ radian}$$

$$= 0.0544 \text{ sec.}$$

Example 26.11 *A spectrometer, having a very large circular scale reading up to one second, is fitted with a telescope of 1 inch objective. Criticise its usefulness and suggest modification required.*

Solution. The smallest angular separation $d\theta$ *that can just be resolved by a* lens of 1 inch aperture, taking the mean wavelength of white light to be that of sodium *D*-lines *i.e.*, 5893×10^{-8} cm, is given by

$$d\theta = \frac{1.22\lambda}{d} = 1.22\times5893\times\frac{10^{-8}}{2.54}\text{ radian}$$

$$= \frac{1.22\times5893\times10^{-8}\times180\times3600}{2.54\times\pi} = 5.836 \text{ sec.}$$

It is, therefore, no use having a scale reading up to one second as the angular resolution is 5.836 sec or 6 sec. Thus theoretically the minimum scale reading should not be less than 6 second. If the reading up to one second is desired, then the aperture of the telescope objective should be increased so that dθ = 1 second. Hence the least aperture should be 6 inches or 0.15 m.

Example 26.12 *Calculate the mean separation of two points on moon that can be resolved by a 500 cm telescope. The distance of the moon is 3.8 × 10⁵ km. The eye is most sensitive to wavelength 5500 A°.* **(M.D.U. 2005)**

Solution. Angular separation $d\theta = \frac{1.22\lambda}{d} = \frac{1.22\times5500\times10^{-10}}{500\times10^{-2}} = 13.42\times10^{-8}$ rad.

Distance of the moon $R = 3.8 \times 10^5$ km $= 3.8 \times 10^8$ m

∴ Mean separation of two points that can be resolved $= Rd\theta$

$= 3.8 \times 10^8 \times 13.42 \times 10^8 = 51$ metre.

Example 26.13 *Find the limit of resolution and resolving power of a telescope of 25 cm diameter when white light of 555 nm is used.* **(H.P.U. 2001)**

Solution. Diameter of the objective $d = 25$ cm $= 25\times10^{-2}$ m

Mean wavelength $\lambda = 555 \times10^{-9}$ m

∴ Limit of resolution $d\theta = \frac{1.22\lambda}{d} = \frac{1.22\times555\times10^{-9}}{25\times10^{-2}} = 27.08\times10^{-7}$ radian=0.5584″

Resolving power $= \frac{1}{d\theta} = \frac{1}{27.08\times10^{-7}} = \frac{10^7}{27.08} = 3.69\times10^5$

Example 26.14 *Calculate the aperture of the objective of a telescope which may be used to resolve stars separated by 4.88 × 10⁻⁶ radians for light of wavelength 6000 Å.* **(Nagpur U. 2011, 2009)**

Solution. Given: $d\theta = 4.88\times10^{-6}$ radians ; $\lambda = 6000\text{Å} = 6000 \times 10^{-10}$ m

$D = ?$

$$d\theta = \frac{1.22\lambda}{D} = \frac{1.22\times600\times10^{-10}}{D} = 4.88\times10^{-6}$$

$$\therefore \quad D = \frac{1.22\times6000\times10^{-10}}{4.88\times10^{-6}} = 15\times10^{-2}\text{ m}$$

26.13 EYE RING AND M.P. OF TELESCOPE

The course of rays by which the image of an object at infinity is formed by an astronomical telescope is shown in Fig.26.17. Since there are no lenses to the left of the objective *O* and light from the object directly enters the telescope through it, the aperture of the objective is the **entrance pupil** of the telescope. The eye-piece *E* forms a real image of the objective *O* at *XY*. The two pairs of rays (shown thick) starting from the top and bottom of the objective, after refraction through the eye-piece respectively meet at *X* and *Y*. Then *XY* is called **exit pupil** of the telescope or the eye -ring. In actual practice a stop with a hole of the size of *XY* is place in this position and hence the name eye-ring. The point *Z* where *XY* intersects the axis of the telescope is called *eye-point*. If the eye is placed to the right or left of this point, the entire bundle of rays will not enter the eye.

(*i*) When the magnifying power of the instrument is such that its *exit pupil* (eye-ring) coincides with the *entrance pupil* of the eye, it is called **normal magnifying power.** For this magnifying power the brightness of the image, as seen through the instruments, will be the same as observed with unaided eye and will be *maximum* for all the rays that enter the objective pass through *XY* and enter the eye. The field of view will also be larger.

(*ii*) When the magnifying power is greater than normal, the diameter of the exit *pupil* (eye-ring) is greater than the diameter of the entrance pupil of the eye. This causes a decrease in illumination of the retinal image and a decrease in the field of view.

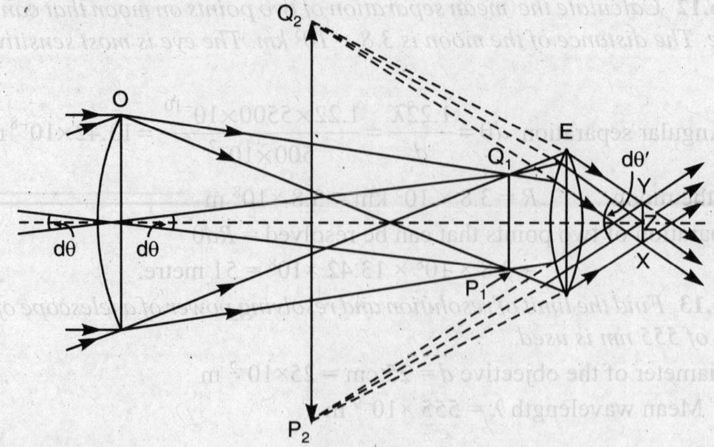

Fig. 26.17

(*iii*) When the magnrifying power is less than the normal, the diameter of the *exit pupil*(eye-ring) is smaller than the diameter of the entrance pupil of the eye. All the light rays passing through the instrument cannot enter the eye in this case. Hence the brightness of the final image and the field of view decreases.

Magnifying power. In normal adjustment, the object and the final image both lie at infinity, hence the distance between the objective and the eye-piece is $F+f$, where F is the focal length of the objective and f that of the eye-piece. The eye -piece thus produces an image of the objective of diameter D lying at a distance $F+f$ at the eye-point at a distance v from it and has a size XY equal to d.

$$\therefore \qquad \frac{1}{v} = \frac{1}{f} - \frac{1}{F+f} = \frac{F}{f(F+f)}$$

or

$$v = \frac{f(F+f)}{F}$$

Now

$$\frac{D}{d} = \frac{F+f}{v} = \frac{(F+f)F}{(F+f)f} = \frac{F}{f} = M \qquad \text{... (26.17)}$$

Hence the magnifying power

$$M = \frac{D}{d} = \frac{\text{Diameter of radius of objective}}{\text{Diameter of radius of eye-ring}} \qquad \text{... (26.18)}$$

When the diameter of the eye-ring is equal to the diameter of the pupil of the eye, the magnifying power is said to be normal, Hence

$$\text{Normal magnifying power} = \frac{D}{d_e} \qquad \text{... (26.19)}$$

where d_e is the diameter of the entrance pupil of the eye.

Let two point objects subtending an angle $d\theta$ at the objective of a telescope be just resolved by the telescope, then

$$d\theta = \frac{1.22\lambda}{D} \qquad \text{... (26.20)}$$

If these are to be further resolved by the eye, the final image must subtend an angle at least equal to $d\theta'$ which is the angle that can be just resolved by the eye and is given by

$$d\theta' = \frac{1.22\lambda}{d_e} \qquad \text{...(26.21)}$$

Thus the smallest magnifying power that the telescope must have is given by

$$\frac{d\theta'}{d\theta} = \frac{1.22\lambda}{d_e} \times \frac{D}{1.22\lambda} = \frac{D}{d_e} \qquad \text{... (26.22)}$$

= Normal magnifying power.

If the magnifying power is less than this, the angle subtended by the final image will be less than $d\theta$ and the image will cease to be resolved by the eye though they are resolved in the focal plane of the objective.

Example 26.15 *Calculate the useful magnifying power of a telescope of one inch objective, assuming the resolving power of the eye to be equal to 2 minutes and the wavelength of light 6×10^{-5}cm.*

Solution. Diameter of the objective $d = 1'' = 2.54 \text{ cm} = 2.54 \times 10^{-2} \text{ m}$

Wavelength of light $\lambda = 6 \times 10^{-5} \text{ cm} = 6 \times 10^{-7} \text{ m}$

$\therefore$ Resolving power of the telescope $\dfrac{1}{d\theta} = \dfrac{d}{1.22\lambda}$

Hence the smallest angle that can be just resolved by it

$$d\theta = \frac{1.22 \times 6 \times 10^{-7}}{2.54 \times 10^{-2}} = 2.882 \times 10^{-5} \text{ radian}$$

$$= 0.00002882 \text{ radian} = 0.09907 \text{ minutes}$$

The smallest angle that can be resolved by the eye

$$d\theta' = 2 \text{ minutes}$$

The useful magnifying power of the telescope should be such that if an object subtends an angle of 0.09907 minute at the objective, the final image must subtend an angle of 2 minutes at the eye of the observer.

$\therefore$ Useful magnifying power $\dfrac{d\theta'}{d\theta} = \dfrac{2}{.09907} = 20.19$

26.14 RESOLVING POWER OF A MICROSCOPE

The minimum distance by which two points in the object are separated from each other so that their images as produced by the microscope are just seen as separate is called its **limit of resolution.**

The reciprocal of the limit of resolution is known as the resolving power.

Consider two point objects separated by a distance x placed very close to the objective AB of the microscope and perpendicular to the axis Fig. 26.18(a). The rays starting from O after refraction through the lens meet at I in the same phase giving rise to a diffraction pattern such that central bright disc coincides with I surrounded by a system of alternately dark and bright diffraction rings. Similarly the image of O' at I' is a bright central disc surrounded by dark and bright diffraction rings.

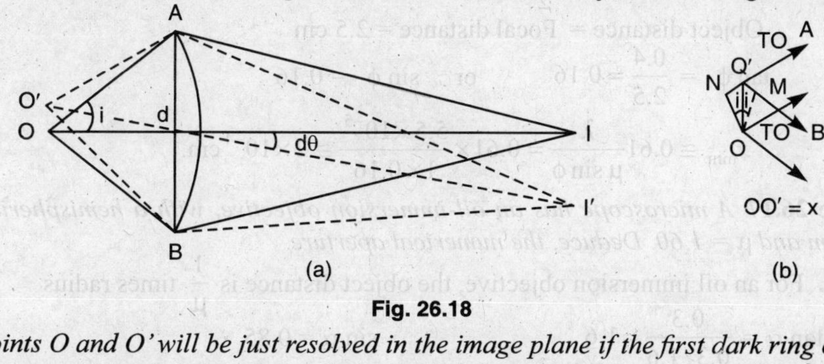

Fig. 26.18

The points O and O' will be just resolved in the image plane if the first dark ring of diffraction pattern at I' passes through I, the centre of principal bright disc of O. For this condition to be satisfied, the path-difference between the extreme rays from O' and reaching I' must be 1.22λ.

A magnified view of OO' and the rays starting from them is shown in Fig. 26.18 (*b*)

It is clear that $O'B$ is longer than OB by $O'M$

$$\therefore \quad O'M = OO' \sin i = x \sin i$$

where i is the semivertical angle of the cone of rays received by the objective AB from O.

Similarly the ray $O'A$ is shorter than OA by an amount $O'N$ where again

$$O'N = OO' \sin i = x \sin i$$

Thus the path-difference between the rays $O'BI$ and $O'AI$

$$= O'M + O'N = 2x \sin i$$

Equating this path-difference equal to 1.22 λ, we get

$$2x \sin i = 1.22\lambda$$

or
$$x = \frac{1.22\lambda}{2 \sin i}$$

In laboratories generally high resolving power microscopes used are **Oil immersion type** in which both the object and objective are immersed in *cedar wood oil* the refractive index μ of which is the same as that of glass. If the wavelength of light λ is in the liquid and λ_0 that in vacuum, then

$$\lambda = \frac{\lambda_0}{\mu}$$

$$\therefore \quad x = \frac{1.22\lambda_0}{2\mu \sin i} \quad \text{...(26.23)}$$

The quantity μ sin *i* is the *numerical aperture (N.A.)* of the microscope and is generally marked on the objective of the microscope. This quantity is of great importance. As greater is its magnitude, smaller is the value of *x i.e.*, finer is the detail that can be observed.

Thus we find that to decrease the distance *x* between two points to be resolved (*i*) λ_0 should be decreased and (*ii*) *N.A.* should be increased.

The relation 26.23 is true if the points *O* and *O′* are self luminous which is not generally the case. Abbe has proved that in such a case limit of resolution is obtained very approximately by omitting the factor 1.22. Hence.

$$\text{Limit of resolution } x = \frac{\lambda_0}{2NA} \quad \text{... (26.24)}$$

The highest value of *NA* in practice is 1.6, taking the effective wavelength of white light as 5500 Å.

Example 26.16 *A microscope objective has aperture diameter 8 mm and focal length 2.1 cm. Estimate its resolving power. We assume that effective λ = 5.5 ×10⁻⁵ cm and also that the object to be placed only a little beyond the focal distance, so this distance may be taken as nearly 2.5 cm.*

Solution. Aperture radius $= \frac{8}{2} = 4$ m.m. $= 0.4$ cm

Object distance = Focal distance = 2.5 cm

Then, $\tan\phi = \frac{0.4}{2.5} = 0.16$ or $\sin\phi = 0.16$

$$\therefore \quad x_{min} \cong 0.61 \frac{\lambda}{\mu \sin\phi} = 0.61 \times \frac{5.5 \times 10^{-5}}{1 \times 0.16} \cong 2 \times 10^{-4} \text{ cm.}$$

Example 26.17 *A microscope has an oil-immersion objective, with a hemispherical lens of diameter 6 mm and μ = 1.60. Deduce, the numerical aperture.*

Solution. For an oil immersion objective, the object distance is $\frac{1}{\mu}$ times radius

$$\therefore \quad \tan\alpha = \frac{0.3}{0.3/1.6} = 1.6 \qquad \sin\alpha = 0.85$$

$$\therefore \quad \text{N.A.} = \mu \sin\alpha = 1.6 \times 0.85 = 1.36$$

Example 26.18 *Calculate the magnifying power and also the resolving power of a laboratory microscope from the following data: N.A.= 0.12, $\lambda = 6\times10^{-7}$ m. Distance of cross-wire from the objective = 0.15 m. Focal length of objective F = Focal length of eye-piece f = 0.025 m.*

Solution. Magnifying power: The image is formed by the objective on the cross-wire.

$\therefore \quad v = 0.15$ m, $\quad F = 0.025$ m, $\quad u = ?$

Now
$$u = \frac{Fv}{F-v} = \frac{0.025\times0.15}{0.025-0.15} = -0.03\text{m}$$

Taking distance of distinct vision $D = 0.25$ m.

$$M.P. = \frac{v}{u}\left(1+\frac{D}{f}\right) = \frac{0.15}{0.03}\left(1+\frac{0.25}{0.025}\right)$$
$$= 5 \times 11 = 55$$

Resolving power. As the objects viewed by the microscope are generally not self luminous.

$$\therefore \quad \text{Limit of resolution, } x = \frac{\lambda_o}{2NA} = \frac{6\times10^{-7}}{2\times0.12} = 0.25\times10^{-5}\text{ m}$$

Example 26.19 *Deduce the missing order for a double slit Fraunhofer diffraction pattern if the slit widths are 0.16 mm and they are 0.8 mm apart.*

Solution. Here, width of slit $a = 0.16$ mm $= 0.16\times10^{-3}$ m

Space between the slits $b = 0.8$ mm $= 0.8\times10^{-3}$ m

Now
$$\frac{n}{p} = \frac{a+b}{a} = \frac{(0.16+0.8)10^{-3}}{0.16\times10^{-3}} = 6$$

Hence for values of $p = 1, 2, 3$, etc.

$$n = 6, 12, 18, \text{etc.}$$

Thus the orders 6, 12, 18 etc. of the interference maxima will be missing from the diffraction pattern.

EXERCISE

1. What is the essential condition for having Fraunhofer diffraction?
2. Discuss Fraunhofer diffraction pattern due to a single slit. Derive the condition for production of maxima and minima and state their position. Find the expression for the width of central maximum. What happens when the width of the slit is gradually increased?
   (Pbi.U. 2007, 2003, G.N.D.U. 2007; Nagpur U. 2011, 2001; Indore U. 2001, Meerut U. 2006, 2000; P.U. 2007)
3. Discuss Fraunhofer diffraction at rectangular aperture ***(P.U. 2006)***
4. What is the importance of phasor diagram method? Discuss its application in case of diffraction due to a single slit.
5. Explain the method of integral calculus to find the diffraction pattern due to a single slit.
6. Discuss analytically the distribution of intensity in the diffraction pattern due to a single slit.
   (P.U. 2001; K.U. 2000; Indore U. 2001; Meerut U. 2000)
7. Find the value of half angular width of the principal maxima for a single slit.
8. Discuss the diffraction pattern produced by a circular aperture. What is its importance?
   (Nagpur U. 2010, 2008)
9. Discuss 'Emission by a disc source'. Is it used in radar antenna?

10. Discuss the Fraunhofer diffraction pattern produced by a circular aperture. What is its importance? Or How does diffraction impose a limit on the resolving power of optical instruments? ***(M.D.U. 2007; G.N.D.U. 2005; P.U., 2007)***

11. What is the essential condition for having Fraunhofer diffraction? Discuss Fraunhofer diffraction at a double slit. Derive expression for intensity distribution and find the positions of maxima and minima. ***(Meerut U. 2003; Nagpur U. 2002, 2001; K.U. 2002; M.D.U. 2003; G.N.D.U. 2001, Gauhati U. 2000; P.U. 2006)***

12. Find the missing orders in a double slit diffraction pattern. ***(Nagpur 2011, 2008)***

13. (*a*) Distinguish between single slit and a double slit diffraction pattern.

(*b*) How does the diffraction pattern due to double slit change if the distance between the slit centres is varied, keeping slit width constant? ***(Meerut. U. 2003)***

14. Is there any difference in the diffraction pattern due to double slit and interference due to two sources? Explain.

Define resolving power and distinguish between magnifying power and resolving power ***(G.N.D.U 2006, 2005; M.D.U. 2005; Nagpur U. 2005, 2003; MK. U. 2001)***

15. State and Explain Rayleigh's criterion of resolution of two close images with neat diagram. ***(Nagpur U. 2009)***

16. What is Rayleigh's criterion for no resolution, just resolution and well resolution? ***(Nagpur 2011, 2008)***

17. Discuss the effect of diffraction on the image formed by optical instruments. Explain Rayleigh's criterion for resolution. ***(P.U. 2008, 2007, 2002, 2000; G.N.D.U. 2006, 2003; M.D.U. 2005; Pbi. U. 2005; Meerut U. 2007, 2006, 2002; Nagpur is 2002, 2001; H.P.U. 2001; Gujrat U. 2000)***

18. Using Rayleigh's criterion, state the formula for limit of resolution and hence derive an expression for the resolving power of a telescope. ***(G.N.D.U. 2006; Pbi.U. 2005; Nagpur U. 2011, 2008, 2003)***

19. Explain the function of eye-ring. Show that the magnifying power of a telescope is equal to the radius of the object-glass divided by the radius of the eye-ring. Hence or otherwise deduce that to obtain full resolution the magnifying power of telescope must not be less than the ratio of the diameter of the object-glass to the diameter of the pupil of the eye.

20. Why is meant by the resolving power of a microscope? How is it related to the numerical aperture (NA) of its objective? Obtain the equation of resolving power of a microscope. ***(Nagpur U. 2010)***

21. What is numerical aperture (N.A.) of microscope. ***(Nagpur U. 2011, 2008)***

22. Calculate the useful magnifying power of a telescope of objective 2.54 cm, assuming the resolving power of the eye to be 2 min and the wavelength of light 6000 Å. **[Ans. 20.2]**

MULTIPLE CHOICE QUESTIONS

26.1 The essential condition to produce Fraunhofer diffraction is

(*a*) Source is close to aperture and screen at infinite distance
(*b*) Source is at infinite and screen close to the aperture
(*c*) Source and screen very close to aperture.
(*d*) Source and the screen are at infinite distance from aperture

26.2 Half-angular width of the central maximum in Fraunhofer diffraction pattern does not depends on:

(*a*) wavelength (*b*) amplitude
(*c*) angle of diffraction (*d*) width of slit

26.3 The limit of resolution (dθ) is given by:

(*a*) $\dfrac{d}{\lambda}$ (*b*) $\dfrac{d}{1.22\lambda}$

(*c*) $\dfrac{\lambda}{d}$ (*d*) $\dfrac{1.22d}{\lambda}$

26.4 Resolving power of a telescope is given by:

(*a*) $\dfrac{d}{\lambda}$ (*b*) $\dfrac{d}{1.22\lambda}$

(*c*) $\dfrac{\lambda}{d}$ (*d*) $\dfrac{1.22d}{\lambda}$

26.5 The numerical aperture of a microscope is given by

(*a*) $\mu \sin i$ (*b*) $\sin i / \mu$

(*c*) $\mu / \sin i$ (*d*) $2\mu / \sin i$

26.6 Limit of resolution of a laboratory microsocope having numerical aperture

(*a*) $\dfrac{\lambda_o}{NA}$ (*b*) $\dfrac{\lambda_0^2}{NA}$

(*c*) $\dfrac{\lambda_o}{2NA}$ (*d*) $\dfrac{NA}{\lambda_0^2}$

ANSWERS

26.1 (*d*) **26.2** (*b*) **26.3** (*c*) **26.4** (*b*) **26.5** (*a*) **26.6** (*c*)

DIFFRACTION GRATINGS

27

Chapter

INTRODUCTION

Diffraction grating is an important device among laboratory instruments. Large number of equidistant parallel lines, nearly 15000 lines per inch (2.54 cm) are rulled on a glass plate and then replica is prepared. This replica works as a grating. The lines closely drawn act like multiple slits and hence results in a diffraction phenomenon. Truely speaking, the term 'diffraction' is not appropriate.3 The spectra is produced by *interference* of N wavelets, the diffraction grating only serves to send light in the required direction.

27.1 DIFFRACTION BY MULTIPLE SLITS (N-PARALLEL SLITS)

According to the results of Fraunhofer diffraction at a single slit of width a, *the wavelets proceeding from all points in the slit along a direction* θ *are equivalent to a single wave of amplitude* $a = \left(\dfrac{A_0 \sin\alpha}{\alpha}\right)$ *starting from the middle point of the slit.* [For proof sec, article 26.5 Eq. (*iii*)]

The phase-difference between the two rays from the extreme ends of the slit

$$2\alpha = \frac{2\pi}{\lambda} a \sin\theta \quad \text{or} \quad \alpha = \frac{\pi}{\lambda} a \sin\theta \qquad \text{[Eq. (}iv\text{), article 26.5]}$$

If we consider two slits, the phase-difference between two corresponding points

$$2\beta = \frac{2\pi}{\lambda}(a+b)\sin\theta \quad \text{or} \quad \beta = \frac{\pi}{\lambda}(a+b)\sin\theta \qquad \text{[Eq. (}i\text{), article 26.5]}$$

If N is the total number of slits, then the problem is reduced to find the amplitude of N vibrations in a direction θ each of amplitude $\dfrac{A_0 \sin\alpha}{\alpha}$, but the phase increases in arithmetical progression, the common phase-difference 2β being $\dfrac{2\pi}{\lambda}(a+b)\sin\theta$. By the vector addition of amplitudes, the resultant amplitude R in a direction θ is given by

$$R = a\frac{\sin(Nd/2)}{\sin d/2}$$

where
$$d = \frac{2\pi}{\lambda}(a+b)\sin\theta \text{ and } a = A_0\frac{\sin\alpha}{\alpha};$$

$$\therefore \quad \frac{Nd}{2} = \frac{N.2\pi}{2\times\lambda}(a+b)\sin\theta = N\frac{\pi}{\lambda}(a+b)\sin\theta = N\beta$$

Since
$$\beta = \frac{\pi}{\lambda}(a+b)\sin\theta$$

Hence resultant amplitude R is given by

$$R = a\frac{\sin(Nd/2)}{\sin d/2} = \frac{A_0\sin\alpha}{\alpha}\times\frac{\sin N\beta}{\sin\beta}$$

$\therefore$ Resultant intensity
$$I = R^2 = A_0^2\frac{\sin^2\alpha}{\alpha^2}\times\frac{\sin^2 N\beta}{\sin^2\beta} \qquad \text{... (}i\text{)}$$

This equation gives the distribution of intensity for finite number of beams with equal amplitudes and phases increasing in arithmetical progression.

The factor $\frac{A_0^2 \sin^2 \alpha}{\alpha^2}$ gives the intensity distribution in the diffraction pattern of a single slit as already discussed in Eq. 26.8. The factor $\frac{\sin^2 N\beta}{\sin^2 \beta}$ is supposed to represent the interference term and gives the distribution of intensity due to all the slits. Let Z denote this factor. The intensities of maxima and minima are obtained by equating its first differential co-efficient *i.e.*, $dZ/d\beta$ equal to zero. Now

$$Z = \frac{\sin^2 N\beta}{\sin^2 \beta}$$

$$\therefore \quad \frac{dZ}{d\beta} = \frac{\sin^2 \beta \times 2N \sin N\beta \cos N\beta - 2\sin^2 N\beta \sin \beta \cos \beta}{\sin^4 \beta}$$

$$= \frac{2N \sin N\beta \cos N\beta}{\sin^2 \beta} - 2\sin^2 N\beta \frac{1}{\sin^2 \beta} \frac{\cos \beta}{\sin \beta}$$

$$= \frac{2N \sin^2 N\beta \cos N\beta}{\sin^2 \beta \sin N\beta} - \frac{2\sin^2 N\beta}{\sin^2 \beta} \cdot \frac{\cos \beta}{\sin \beta}$$

$$= \frac{2\sin^2 N\beta}{\sin^2 \beta}(N \cot N\beta - \cot \beta)$$

For maxima or minima $\frac{dZ}{d\beta} = 0$

$$\therefore \quad \frac{2\sin^2 N\beta}{\sin^2 \beta}(N \cot N\beta - \cot \beta) = 0$$

Hence either $\frac{\sin N\beta}{\sin \beta} = 0$ or $N \cot N\beta - \cot \beta = 0$

Principal maxima. Considering the equation $\left(\frac{\sin N\beta}{\sin \beta}\right)$, if $\sin \beta = 0$, then $\beta = \pm n\pi$ where $n =$ 1, 2, 3 ......

Hence $\frac{\sin N\beta}{\sin \beta} = \frac{0}{0}$

This becomes indeterminate. To find the value of this limit, the numerator and the denominator are differentiated.

$$\therefore \quad \underset{\beta \to n\pi}{\text{Lt}} \frac{\sin N\beta}{\sin \beta} = \underset{\beta \to n\pi}{\text{Lt}} \frac{\frac{d}{dt}(\sin N\beta)}{\frac{d}{dt}(\sin \beta)}$$

$$= \underset{\beta \to n\pi}{Lt} \frac{N \cos N\beta}{\cos \beta} = \pm N$$

Thus the resultant amplitude in these directions is proportional to N, and resultant intensity is proportional to N^2. Substituting in (*i*), we get

Intensity of principal maxima $I = R^2 = A_0^2 \dfrac{\sin^2 \alpha}{\alpha^2} N^2$ ... *(ii)*

These maxima are intense and are called **principal maxima.** The position of these maxima corresponds to

$$\beta = \frac{\pi}{\lambda}(a+b)\sin\theta = \pm n\pi$$

or $\quad (a+b)\sin\theta = \pm n\lambda \quad$ where $n = 0, 1, 2....$

The ± sign indicates that there are two principal maxima of the same order lying on either side of the zero order maximum.

Secondary minima. When $\sin N\beta = 0$ but $\sin\beta \neq 0$, the factor $\dfrac{\sin N\beta}{\sin\beta}$ becomes zero and hence intensity is minimum. Thus for minima, we have

$$\sin N\beta = 0 \text{ or } N\beta = \pm m\pi$$

$$\therefore \quad N\frac{\pi}{\lambda}(a+b)\sin\theta = \pm m\pi$$

or $\quad N(a+b)\sin\theta = \pm m\lambda$

where m has all the integral values except $0, N, 2N ... nN$ as for these values of m, $\sin\beta$ also becomes zero and we obtain principal maxima. It is thus evident that there are $(N-1)$ minima between two consecutive principal maximum.

Secondary maxima. Consider the equation $N \cot N\beta - \cot\beta = 0.$

Its roots, other than those for which $\beta = \pi n$, give rise to another set of maxima, called the **secondary maxima.** Its intensity is much less than the intensity of the principal maxima.

To evaluate $\dfrac{\sin^2 N\beta}{\sin^2\beta}$ subject to the condition that $N\cot N\beta - \cot\beta = 0$, we have

$$\frac{\sin^2 N\beta}{\sin^2\beta} = \frac{1}{\operatorname{cosec}^2 N\beta}\frac{1}{\sin^2\beta} = \frac{1}{1+\cot^2 N\beta}\cdot\frac{1}{\sin^2\beta}$$

$$= \frac{N^2}{N^2 + N^2\cot^2 N\beta} \times \frac{1}{\sin^2\beta} = \frac{N^2}{N^2+\cot^2\beta} \times \frac{1}{\sin^2\beta} \quad (\because N\cot N\beta = \cot\beta)$$

$$= \frac{N^2}{N^2 + \dfrac{\cos^2\beta}{\sin^2\beta}} \times \frac{1}{\sin^2\beta}$$

$$= \frac{N^2}{N^2\sin^2\beta + \cos^2\beta}$$

$$= \frac{N^2}{N^2\sin^2\beta + 1 - \sin^2\beta}$$

$$= \frac{N^2}{1 + (N^2-1)\sin^2\beta}$$

$$\therefore \quad \frac{\sin^2 N\beta}{\sin^2\beta} = \frac{N^2}{1+(N^2-1)\sin^2\beta}$$

$$\therefore \quad \text{Intensity of secondary maxima} = A_0^2 \frac{\sin^2 \alpha}{\alpha} \times \frac{N^2}{1+(N^2-1)\sin^2 \beta} \quad \text{... (iii)}$$

$$\therefore \frac{\text{Intensity of secondary maxima}}{\text{Intensity of principal maxima}} = \frac{N^2}{1+(N^2-1)\sin^2 \beta} \times \frac{1}{N^2} \quad \text{[From (ii) and (iii)]}$$

$$= \frac{1}{1+(N^2-1)\sin^2 \beta}$$

It is clear that as N increases the intensity of secondary maxima relative to principal maxima decreases and becomes negligible when N is very large. This is why these secondary maxima are not visible with a grating having about 15000 or more lines per inch.

27.2 PLANE DIFFRACTION GRATING

A plane transmission diffraction grating is an arrangement which is equivalent in action to a large number of parallel and equidistant slits of the same width. It is constructed by ruling equidistant parallel lines on a transparent material, such as glass plate, by means of a fine diamond point worked with a ruling engine. The ruled lines act like opaque wires and thus do not allow the incident light to pass through them. These are called **opacities.** Light passes through the spaces in between the lines. These are called *transparencies.* The number of lines on a plane transmission grating is of the order of 15000 per inch.

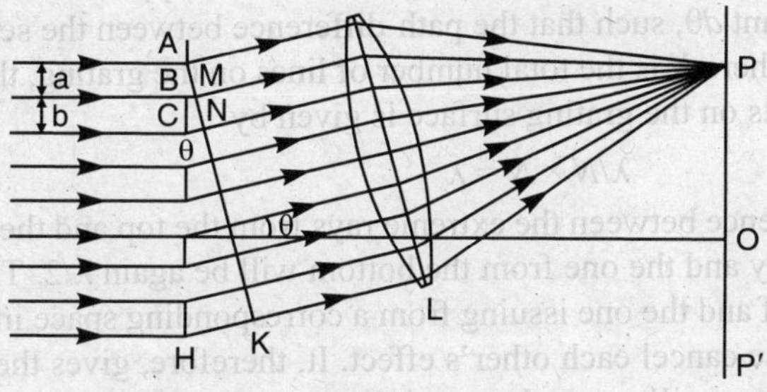

Fig. 27.1

Theory. Let $ABC....\ H$ be the section of a plane transmission grating supposed perpendicular to the plane of the paper.

Grating element. Let the width of the transparent portion AB be equal to a and opaque portion BC be equal to b. The distance $(a + b)$ is called the *grating element* or *grating constant*. The points in two consecutive spaces separated by a distance $(a + b)$ are called corresponding points.

Central maximum. Consider a parallel beam of monochromatic light (plane wave) of wavelength λ incident normally on the grating surface. Most of this light, issuing from the spaces, will go straight and is brought to focus by the lens L at O giving rise to the central maximum. As the width of slit is comparable to wavelength of light, part of the light spreads out in all directions. Let us consider the light from the corresponding points A and C, in a direction making an angle θ with the normal to the slit. (Fig. 27.1.) To find the path-difference draw AK perpendicular to the direction of diffracted rays, then $\angle HAK = \theta$. The path-difference between the rays issuing from the corresponding points A and C is CN and is given by

$$CN = AC \sin \theta = (a + b) \sin \theta$$

The point P will have maximum intensity if this path-difference is an integral multiple of the wavelength λ of light. Light from corresponding points of neighbouring slits reinforce one another and angle θ gives the direction of maximum intensity. In general

$$(a + b) \sin \theta_n = n\lambda$$

where θ_n is the direction of the nth principal maximum.

For $n = 0$, we get the central maximum at O.

Primary maxima. For $n = 1$, $(a + b) \sin \theta_1 = \lambda$

This gives the direction of first order primary (or principal) maxima on either side of O *i.e.*, at P and P'. The intensity at P is less than the intensity at O.

For $n = 2$, $(a + b) \sin \theta_2 = 2\lambda$

This gives the condition for the second order principal maximum whose intensity is much less than the intensity of the first order.

Incident light with more than one wavelength. If the incident light consists of more than one wavelength, the beam gets dispersed and angle of diffraction is different for different wavelengths. If for two nearly equal wavelengths λ and $(\lambda + d\lambda)$ the corresponding angles of diffractions are θ and $(\theta + d\theta)$, then in first order.

$$(a + b) \sin \theta = \lambda \text{ and}$$

$$(a + b) \sin (\theta + d\theta) = \lambda + d\lambda$$

Thus, for white light, the diffraction pattern on the screen will consist of central bright maximum, surrounded on both sides by spectrum corresponding to different wavelengths of the constituents of white light.

Condition for secondary maxima and minima. Now suppose the angle of diffraction is increased by a small amount $d\theta$, such that the path-difference between the secondary waves from A and C increases by λ/N, where N is the total number of lines on the grating, then the path-difference between the extreme points on the grating surface is given by

$$\lambda/N \times N = \lambda$$

Hence the path-difference between the extreme rays from the top and the central ray will be $\lambda/2$ and between the central ray and the one from the bottom will be again $\lambda/2$. Thus, light issuing from any space in the upper half and the one issuing from a corresponding space in the lower half differs in path by $\lambda/2$. Hence these cancel each other's effect. It, therefore, gives the condition for the *first secondary minimum.* There will again be a minimum when the path-difference between the corresponding points of the two consecutive spaces is $\frac{2\lambda}{N}, \frac{3\lambda}{N} \ldots \frac{(N-1)\lambda}{N}$. Thus there will be $(N - 1)$ *secondary minima* and therefore $(N - 2)$ secondary maxima between the two consecutive *principal maxima.*

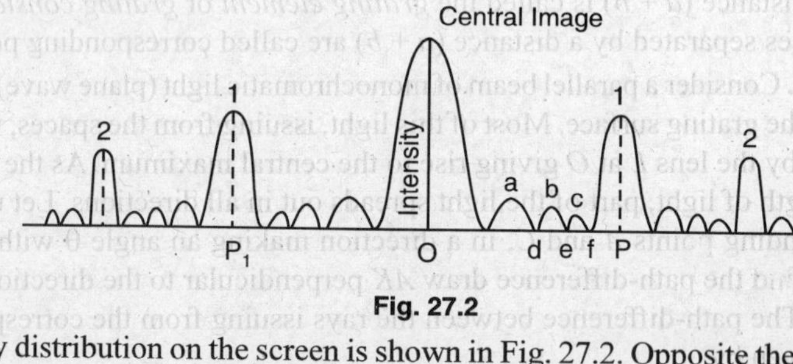

Fig. 27.2

The intensity distribution on the screen is shown in Fig. 27.2. Opposite the point O, the intensity is maximum and corresponds to the intensity of central maximum. Opposite to P and P_1 the intensity is less than the intensity of central maximum. It corresponds to the intensity of the first order principal maximum. In between the central maximum and first principal maximum, there are a number of secondary maxima (of low intensity) and secondary minima. The intensity and angular spacing of the secondary maxima and minima are so small that they cannot be observed and thus there is a uniform darkness between any two principal maxima.

Width of central maximum. The central maximum lies in between the first secondary minima on either side of it. The position of first secondary minima is given by

$$d\theta = \frac{\lambda}{N}$$

$\therefore$ width of central maximum $= 2d\theta = \dfrac{2\lambda}{N}$

Half-angular width of principal maxima. *The angular width of a principal maxima is the angle between the two first minima lying on either side of it.*

In a diffraction grating, *n*th order principal maxima is given by

$$(a + b) \sin \theta_n = n\lambda \quad \text{... (i)}$$

Let $(\theta_n + d\theta_n)$ and $(\theta_n - d\,\theta_n)$ be the angles of diffraction corresponding to the first minima on either side of the *n*th principal maxima, then

$$(a+b)\sin(\theta_n \pm d\theta_n) = n\lambda \pm \frac{\lambda}{N} \quad \text{... (ii)}$$

Dividing (*ii*) by (*i*), we have

$$\frac{\sin(\theta_n + d\theta_n)}{\sin\theta_n} = 1 \pm \frac{1}{nN}$$

or $$\frac{\sin\theta_n \cos d\theta_n + \cos\theta_n \sin d\theta_n}{\sin\theta_n} = 1 \pm \frac{1}{nN}$$

or $$\cos d\theta_n + \cot\theta_n \sin d\theta_n = 1 \pm \frac{1}{nN}$$

As $d\theta_n$ is very small, $\cos d\theta_n = 1$ and $\sin d\theta_n = d\theta_n$

$\therefore$ $$1 + \cot\theta_n \; d\theta_n = 1 \pm \frac{1}{nN}$$

or $$d\theta_n = \pm \frac{1}{nN \cot\theta_n} \quad \text{... (iii)}$$

Eq. (*iii*) gives the *angular half width of nth order principal maxima.*

As *N* increases $d\theta_n$ decreases *i.e.* principal maxima becomes sharper as *N* increases.

27.3 MAXIMUM NUMBER OF ORDERS AVAILABLE WITH A GRATING

For a plane diffraction grating, the *n*th order principal maxima is given by

$$(a + b) \sin \theta = n\lambda$$

The maximum value of θ the angle of diffraction possible is π/2 and therefore $\sin\theta = \sin \pi/2 = 1$

Thus the highest possible order n_{max} is given by

$$(a + b) \sin \pi/2 = n_{max} \lambda$$

or $$n_{max} = \frac{(a+b)}{\lambda} \qquad \left[\because \sin\frac{\pi}{2} = 1\right]$$

Example 27.1 *Show that a grating with 5000 lines per cm cannot give a spectrum in 4th or higher order for light of wavelength 5890Å.* **(G.N.D.U. 2004)**

Solution. Here number of lines on the grating = 5000 / cm

$$\therefore \quad \text{grating element } (a+b) = \frac{1}{5000} \text{ cm}$$

$$\lambda = 5890 \text{ Å} = 5890 \times 10^{-8} \text{ cm}$$

$$\therefore \quad n_{max} = \frac{a+b}{\lambda} = \frac{1}{5000 \times 5890 \times 10^{-8}} = \frac{10^4}{2945} = 3.39$$

This grating can, therefore, give spectra up to *third* and cannot give a spectrum in 4th or higher order.

Example 27.2 *Monochromative light of wavelength 5000 Å is diffracted by a grating of 2500 lines per cm. Show whether 16th order diffraction is possible.* **(Gauhati U., 2000)**

Solution. $n_{max} = \dfrac{a+b}{\lambda} = \dfrac{1}{2500 \times 5000 \times 10^{-8}} = 8$ *i.e.* 16th order diffraction is not possible. In fact the highest order diffraction possible is *eight*.

Example 27.3 *A grating with 2500 lines per cm is illuminated at normal incidence by light of wavelength 5500Å. How many orders will be visible ?* **(Nagpur U., 2001)**

Solution. $$n_{max} = \frac{a+b}{\lambda} = \frac{1}{2500 \times 5500 \times 10^{-8}} = 7.27$$

Only seven orders will be visible.

Example 27.4 *What is the highest order spectrum which may be seen with light of wavelength 5800 Å by means of a grating with 3000 lines per inch.* **(Nagpur U., 2008, Luck. U., 2001)**

Solution. $n_{max} = \dfrac{a+b}{\lambda} = \dfrac{2.54}{3000 \times 5800 \times 10^{-8}} = 14.6.$ Highest order which can be seen = 14

27.4 MISSING ORDER SPECTRA FOR A DIFFRACTION GRATING

The *n*th principal maxima in a plane diffraction grating is given by

$$(a + b) \sin \theta_n = n\lambda \qquad \text{... } (i)$$

If the value of a and θ_n are such that

$$a \sin \theta_n = \lambda \qquad \text{... } (ii)$$

then the effect of wavefronts from any particular slit will be zero. It is because in such a case the slit can be divided into two halves and hence the path-difference between the secondary waves from the corresponding points of the two halves will be $\lambda/2$ and hence the resultant intensity will be zero.

When the conditions (*i*) and (*ii*) are simultaneously obeyed, the beams from all the slits reinforce each other but the resultant intensity is zero. Hence the spectrum will be absent.

Dividing (*i*) by (*ii*), we get

$$\frac{(a+b)}{a} = n \qquad \text{... } (iii)$$

If $a \sin \theta_n = 2\lambda$, or any integral multiple of λ *i.e.*

$$a \sin \theta_n = p\lambda$$

then again the effect of wavefronts from any particular slit will be zero and that order spectrum will be missing.

$\therefore$ General condition for missing spectra is

$$a \sin \theta_n = p\lambda \qquad \text{... } (iv)$$

where $p = 1, 2, 3$ ... etc.

Dividing (*i*) by (*iv*), we have

$$\frac{a+b}{a} = \frac{n}{p} \qquad \text{... (v)}$$

The condition for any spectral order to be missing is obtained by giving various values to *n*. If $a = b$, then

$$\frac{a+a}{a} = n = 2$$

Thus if the width of opacity and transparency is equal, the second order spectrum will be absent. Similarly from relation (*v*), we have

$$\frac{a+a}{a} = \frac{n}{p} = 2$$

or $n = 2p$ where $p = 1, 2, 3$ etc.

$\therefore$ Second, fourth, sixth, eighth etc. order spectra will also be absent.

27.5 DETERMINATION OF WAVELENGTH USING PLANE TRANSMISSION GRATING

Procedure

(*i*) The eye-piece of the telescope of a spectrometer is first focused on the cross-wires so that they are clearly visible without any strain. The telescope is then directed towards a far off object and adjustments are made till there is no parallax between the image of that object and the vertical cross-wire. The slit of the collimator is now illuminated by a monochromatic light and telescope is turned to receive the light leaving the lens of the collimator. The distance of the slit is adjusted till there is no parallax between the image of slit and the vertical cross-wire. This make *the adjustment for the telescope to receive parallel rays and the collimator to give a parallel beam of light.*

(*ii*) If necessary, the telescope and the collimator are levelled so that their axes pass through the centre of the table and the image lies symmetrically about the cross-wires.

(*iii*) The grating is now set on the central table with its plane approximately parallel to the line joining two of the screws provided for levelling the table. The table is turned so that the plane of the grating is inclined at an angle of 45° to the axis of the collimator. The telescope is turned till the reflected image of the slit is seen in the field of view. If the image of the slit is not symmetrical about the horizontal cross-wire, this can be achieved by adjusting the third levelling screw. In this condition, the plane of the grating is parallel to the axis about which the telescope is rotated.

(*iv*) The table carrying the grating is turned so that the plane of the grating is approximately perpendicular to the axis of the collimator. The telescope is turned till the central bright image coincides with the vertical cross-wire. The telescope is turned till one of the first order diffracted images falls on the cross-wire. One of the screws is now adjusted to bring the middle of the diffracted image on the intersection of the cross-wires. *The rulings are now parallel to the axis of the spectrometer.*

(*v*) *To set the grating perpendicular to the axis of collimator.* The telescope is turned so that the direct image falls on the vertical cross-wire. After reading the position of the telescope, it is turned through exactly 90° so that the axis of the telescope is at right angles to the axis of the collimator. The telescope is fixed in this position and the table carrying the grating is turned till the image of the slit formed by reflection from the *plane* surface of the grating coincides with the vertical cross-wire. The grating plane is now inclined at an angle of 45° to the collimator. The table is now fixed after turning it through exactly 45°. In this position the plane of the grating is normal to the axis of the collimator and hence incident light.

(*vi*) The telescope is now turned to see the direct image on the vertical cross-wire, when the slit is illuminated with monochromatic light whose wavelength is to be determined. The telescope is then turned on to the various diffracted images on either side of the direct image and corresponding readings of the telescope are noted on the scale. If θ_1 is the mean of the two readings for the first order spectrum and θ_2 that for the second order, then

$$\lambda = (a + b) \sin \theta_1 \text{ and } 2\lambda = (a + b) \sin \theta_2$$

where $(a + b)$ is the grating constant and is found from the number of lines marked on the grating. If N is the number of lines per inch, then

$$(a + b) = \frac{2.54}{N}$$

If the source of light emits radiations of different wavelengths, then the beam is dispersed and different lines corresponding to different wavelengths are observed in the first and second order principal maxima. To find the wavelength of any line, diffraction angles for that line are noted and wavelength found as explained.

This method helps to determine the wavelength accurately as it requires only the measurement of angles of diffraction. The angles are fairly large and can be measured accurately with a good spectrometer.

Determination of grating constant

To determine the grating constant, set the spectrometer as explained above and find out the angle of diffraction θ_1 for the first order, then as

$$(a + b) \sin \theta_n = n\lambda$$

where θ_n is the angle of diffraction corresponding to the nth order spectrum for the wavelength λ and $(a + b)$ is the grating constant. For the first order spectrum, we have

$$(a+b) = \frac{\lambda}{\sin \theta_1}.$$

Example 27.5 *A parallel beam of sodium light is allowed to be incident normally on a plane grating having 4250 lines per centimetre and a second order spectral line is observed to be deviated through 30°. Calculate the wavelength of the spectral line.*

Solution. Here $\theta_2 = 30°$, $\lambda = ?$ $N = 4250$.

$$\therefore \quad (a + b) = 1/4250$$

Now $\quad (a + b) \sin \theta_2 = 2\lambda$

$$\therefore \quad \lambda = \frac{\sin 30 \times 1}{4250 \times 2} = \frac{0.5}{2 \times 4250} = 5882 \times 10^{-8} \text{ cm} = 5882 \text{Å}$$

Example 27.6 *A parallel beam of monochromatic light is allowed to be incident normally on a plane transmission grating having 5000 lines per cm and second order spectral line is found to be diffracted through 30°. Calculate the wavelength of light.* **(G.N.D.U., 2001; M.D.U., 2001)**

Solution. $\quad (a+b) = \frac{1}{5000}$; $\sin \theta = \sin 30 = 0.5$; $n = 2$

$$(a+b) \sin \theta = 2\lambda \text{ or } \lambda = \frac{(a+b) \sin \theta}{2} = \frac{1 \times 0.5}{5000 \times 2} = 5 \times 10^{-5} \text{ cm} = 5000 \times 10^{-8} \text{ cm}$$

$$= 5000 \text{ Å}.$$

Example 27.7 *In a plane transmission grating the angle of diffraction for second order maxima for wavelength 5×10^{-5} cm is 30°. Calculate the number of lines in one centimetre of the grating surface.* **(M.D.U., 2003)**

Solution. If N is the number of lines per cm, then

Grating element $(a + b) = 1/N$

Here $n = 2, \theta = 30°$ and $\lambda = 5 \times 10^{-5}$ cm

Now $(a + b) \sin\theta = n\lambda$

$$\therefore \quad \frac{1}{N} = \frac{2 \times 5 \times 10^{-5}}{0.5} \text{ or } N = \frac{0.5}{10^{-4}} = 5000 \text{ per cm}$$

Example 27.8 *A wire grating is made of 200 wires per centimetre placed at equal distance apart. The diameter of each wire is 0.025 mm. Calculate the angle of diffraction for the third order spectrum and also find the absent spectra, if any. Wavelength of light used = 6000 Å.*

Solution. Here wavelength $\lambda = 6000 \times 10^{-8}$ cm

Grating element $(a + b) = \frac{1}{200}$ cm $= 0.005$ cm and $n = 3$

Now $(a + b) \sin\theta_3 = 3\lambda$

$\therefore \quad \sin\theta_3 = 3 \times 6000 \times 10^{-8} \times 200 = 0.00360$

or $\theta_3 = 2°4'$

Width of opacity b = diameter of the wire = 0.0025 cm

$\therefore$ Width of transparency $a = 0.005 - 0.0025 = 0.0025$ cm

Order of absent spectrum $n = \frac{a+b}{a} = \frac{.005}{.0025} = 2$

This shows that the second order spectrum is missing.

Example 27.9 *Show that in a diffraction grating with grating element 1.5×10^{-6} m and light of wavelength 550 nm, third and higher order principal maxima are not visible.*

(P.U. 2003; H.P.U., 20001)

Solution. Here $\lambda = 550$ nm $= 5500 \times 10^{-10}$ m

$(a + b) = 1.5 \times 10^{-6}$ m

Let n be the maximum number of orders of the spectrum that can be observed with the grating. Since the maximum angle of diffraction is 90°, the maximum possible order n is given by

$$n = \frac{a+b}{\lambda} = \frac{1.5 \times 10^{-6}}{5500 \times 10^{-10}} = 2.72$$

Thus, only second order will be visible, as 3rd and higher orders are not possible.

Example 27.10 *If we use white light source with a diffraction grating, with 15000 lines per inch, show that the second and third order spectra overlap. Use the wavelength components 4000Å and 7000Å for studying the white light beam.*

Solution. The grating element $(a + b) = \frac{2.54}{15000} = 1.69 \times 10^{-4} = 1.69 \times 10^{-6}$ m

Wavelength of violet light $= 4000 \times 10^{-8}$ cm $= 4 \times 10^{-7}$ m

Wavelength of red light $= 7000 \times 10^{-8}$ cm $= 7 \times 10^{-7}$ m

In the second order spectrum

$$\sin\theta_v = \frac{2 \times 4 \times 10^{-7}}{1.69 \times 10^{-6}} = 0.4734$$

$$\therefore \quad \theta_v = 28.2°$$

and $$\sin \theta_r = \frac{2\times 7\times 10^{-7}}{1.69\times 10^{-6}} = 0.8284$$

$$\theta_r = 55.9°$$

In the third order spectrum

$$\sin \theta_v' = \frac{3\times 4\times 10^{-7}}{1.69\times 10^{-6}} = 0.710$$

$$\theta_v' = 45.23°$$

and $$\sin \theta_r' = \frac{3\times 7\times 10^{-7}}{1.69\times 10^{-6}} = 1.2426$$

Since θ_r is greater than θ_v', the second and third order spectra will overlap. Again $\sin \theta_r' > 1$ which is not possible. Hence third order spectrum for red light will not be observed.

Example 27.11 *A diffraction grating used at normal incidence gives a line (5400 Å) in certain order superimposed on another line (4050 Å) of the next higher order. If the angle of diffraction is 30°, how many lines/cm are there on the grating ?*

Solution. Here $\lambda_1 = 5400\text{Å} = 5400 \times 10^{-10}$ m

$\lambda_2 = 4050\text{Å} = 4050 \times 10^{-10}$ m

$\theta_1 = \theta_2 = 30°, n = ?$

The two lines in nth and $(n + 1)$th order superimpose

$\therefore$ $(a + b) \sin \theta_1 = n\lambda_1$ and $(a + b) \sin \theta_2 = (n + 1) \lambda_2$

or $$\frac{n+1}{n} = \frac{\lambda_1}{\lambda_2} = \frac{5400\times 10^{-10}}{4050\times 10^{-10}} = \frac{540}{405}$$

or $$\frac{1}{n} = \frac{540}{405} - 1 = \frac{135}{405} = \frac{1}{3}$$

$\therefore$ $n = 3$

Now $(a + b) \sin 30° = 3 \times 5400 \times 10^{-10}$

or $$a + b = \frac{3\times 5400\times 10^{-10}}{0.5} = 324\times 10^{-8}$$

Number of lines/m $$= \frac{1}{a+b} = \frac{10^8}{324} = 308642 = 3086\,/\,\text{cm}$$

27.6 NORMAL SPECTRUM

The grating spectrum is called a normal spectrum due to the following reasons—

The expression for the dispersive power of a grating is

$$\frac{d\theta}{d\lambda} = \frac{n}{(a+b)\cos\theta}$$

(*i*) The angle of diffraction θ is usually small and cos θ differs little from unity. If, therefore, a spectrum with white light is produced, the value of cos θ changes so little between the two extreme colours that it can be treated as constant. Hence the angular separation $d\theta$ *changes so little between the two extreme colours that it can be treated as constant. Thus the angular separation* $d\theta$ between any two lines for a given difference in wavelength $d\lambda$ is the same all along the spectrum, in a particular order.

(*ii*) The equation shows that for a wavelength difference $d\lambda$ the angular separation $d\theta$ is *proportional to the order n* of the spectrum. Hence, the second order spectrum is twice as wide as the first order and so on.

(*iii*) The angular separation is inversely proportional to the grating element $(a + b)$. Thus smaller the grating element, more widely spaced is the **spectrum**. If $d\theta'$ is the angular separation for another grating of grating element $(a' + b')$ for the given wavelength difference $d\lambda$ *in the same order, then*

$$\frac{d\theta'}{d\theta} = \frac{a+b}{a'+b'}.$$

Such a spectrum which satisfies all these conditions is called a **normal spectrum.**

27.7 PLANE REFLECTION GRATING

Consider an optically plane surface with its surface made reflecting by silvering or aluminizing. If a large number (N) of parallel straight lines are ruled on it, each of width 'b' and leaving equal widths 'a' clear between them, we get a *plane reflection grating.*

The plane reflection grating is used to obtain spectra just like the plane transmission grating, the theory and action being the same. Reflection grating has the particular advantage that it can be used in the ultra-violet (UV) and infra-red (IR) regions where most materials become opaque. Hence it has wide applications.

27.8 BLAZED GRATING

For good dispersion, higher order spectra are desirable. However the intensity is maximum in zero order spectra and goes on decreasing (i.e., falls) as n, the order of spectra increases. To overcome this discrepancy, blazed grating is designed. This shifts the intensity maximum towards a desired higher order spectrum. Grooves with inclined reflecting faces are made in the grating surface (fig. 27.3). The inclination λ is called the *blazing angle.*

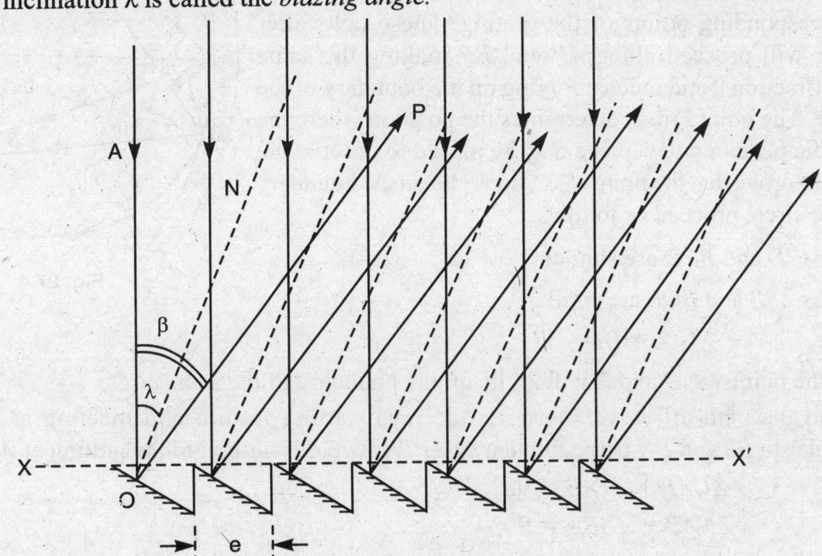

Fig. 27.3. Blazed defraction grating

Working. When light is incident along normal OA to the grating surface XX', the *central diffraction maximum* now falls not along OA, but at an angle $2y$ from OA. The spectrum seen at angle β from OA corresponds to

$$e \sin\beta = n\lambda$$

It is interesting to note that it β is very close to *ny*, then the spectrum has large intensity, despite the order of spectra (*n*) being high. If we desire to choose a grating blazed for a given order *n* for a given region λ, then β can be easily calculated and set $y = \frac{1}{2}\beta$ as the *blazing angle.*

27.9 CONCAVE GRATING

Use of plane grating needs a collimating and a focusing system. Rowland discovered that if a grating is ruled on a concave spherical reflecting surface, then under some conditions it produces focused spectra by it self.

Rowland's concave reflection grating. Rowland developed a concave reflection grating with which the use of lenses is dispensed with. It consists of a concave polished spherical mirror usually of *speculum metal,* having very fine parallel lines ruled on it. It produces diffraction of light and at the same time, focuses it without the use of an additional lens. The spectra obtained are thus free from chromatic and other aberrations. This grating can also be used to produce spectrum of the ultraviolet portion, which light is absorbed by glass.

Let *G* be the magnified view of the portion of the concave reflection grating whose rulings are perpendicular to the plane of the paper. The point *C* is the centre of curvature of the grating surface and *R* is the radius of curvature. A circle with centre *O* and radius *R*/2 passes through *C* and touches the grating at its middle point. (Fig. 27.4). This circle is called *Rowland circle* and has a diameter *R*. If a source is placed at any point on the circumference of the Rowland circle, the dispersed spectral images of the slit are obtained at points such as *P* on the circumference of the same circle. By keeping the source at *S* the spectra can be observed at other points on the circumference of the circle.

Theory. Let *a* be the width of the space and *b* that of the ruling so that $(a + b)$ is the grating element. Suppose the source of light is placed at *S* on the boundary of the circle, so that the rays *SA* and *SB* are incident at the same angle of incidence *i* at the corresponding points of the grating. These rays, after diffraction, will proceed along *AP* and *BP* making the same angle of diffraction θ and meet at *P* lying on the boundary of the same circle. The point *P* thus determines the position where the photographic plate or an eyepiece may be placed to observe the spectrum. To prove that the points *S*, *C*, and *P* lie on the boundary of the same circle proceed as follows:

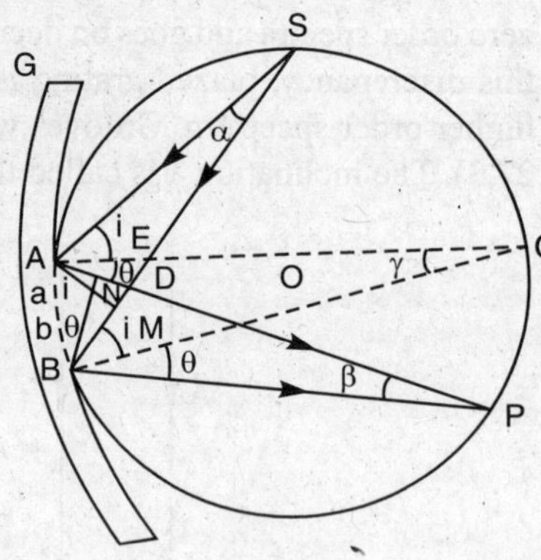

Fig. 27.4

As Δ*s* *ASD* and *BPD* are similar $\quad \therefore \quad \alpha = \beta$

Also Δ*s* *ASE* and *BCE* are similar $\quad \therefore \quad \alpha = \gamma$

Hence $\quad \alpha = \beta = \gamma$

Thus the points, *S*, *C* and *P* will all lie on the boundary of the same circle.

To find the path-difference between the rays starting from *S* and meeting at *P* draw *AM* perpendicular to *BS* and *BN* perpendicular to *AP*. Now *CA* is normal to the grating at *A*.

$$\therefore \quad \angle CAP + \angle PAB = 90°$$

$$\text{Also} \quad \angle NAB + \angle ABN = 90°$$

$$\therefore \quad \angle ABN = \theta$$

$$\text{Similarly} \quad \angle BAM = i$$

$$\begin{aligned}\text{Now path-difference} &= (SB + BP) - (SA + AP)\\ &= (SB - SA) - (PA - BP)\\ &= BM - AN = AB \sin i - AB \sin \theta\end{aligned}$$

Now $AB = (a + b)$

$\therefore$ Path-difference $= (a + b) (\sin i - \sin \theta)$ ... (*i*)

Hence the condition of the *n*th order spectrum to lie at *P* is

$$(a + b) (\sin i - \sin \theta) = n\lambda$$

If the points *S* and *P* lie on the same side of *C*, then the formula takes the form

$$(a + b) (\sin i + \sin \theta) = n\lambda \quad ... (ii)$$

The secondary waves from the corresponding points *A* and *B* reinforce at *P* and same holds good for other pairs of corresponding points and also for the whole grating surface.

Equation (*i*) and (*ii*) hold good for diffraction at every element of the grating surface. In other words, the path-difference for any pair of diffracted rays from the corresponding points of the grating surface must be constant for a given wavelength. Accordingly the rays corresponding to the particular wavelength in a given order will come to focus at a single point on the circumference of Rowland circle.

27.10 MERITS OF CONCAVE GRATING

(*i*) In a plane transmission grating, convergent lenses are required for the formation of spectra. One lens is used for making the incident light parallel and other lens is used for bringing the parallel diffracted ray to focus. The lenses used are not free from spherical and chromatic defects. These defects make the spectrum more complex. In concave reflecting grating the use of lenses is eliminated, as the rulings are made on a concave reflecting surface, instead of a plane surface. The concave mirror is a highly polished metal surface, usually *speculum* metal, which focuses the diffracted rays and thus the effect of chromatic aberration is eliminated.

(*ii*) Another advantage of a concave grating is that we can study the spectra in the ultraviolet region as ultraviolet light is absorbed by glass lenses.

(*iii*) A concave grating, when properly mounted, produces *normal spectrum,* in which the distance between lines is proportional to the wavelength difference.

(*iv*) There is a practical difficulty to rule more than 25000 lines per inch in case of a plane transmission grating, as with very small wavelengths, the spectra formed are very close to the central maximum. This difficulty is mostly solved in the concave reflection gratings.

27.11 MOUNTINGS OF CONCAVE GRATING

There are three important methods of mounting a concave grating: (*i*) Rowland mounting, (*ii*) Paschen mounting and (*iii*) Eagle mounting.

(*i*) **Rowland mounting.** This mounting, in addition to making use of the property of the *Rowland circle,* enables the observations of the spectrum to be made in a direction normal to the grating so that the dispersion is linear, giving a normal spectrum. If a source of light *S* lies on a circle the diameter of which is equal to the radius of curvature of the concave grating, the various spectra will also lie on the same circle, if the grating is placed so that this circle is a tangent to it at its middle point. Such a circle is known as *Rowland's circle.* As shown in Fig. 27.5, the grating *G* and the plate holder *P* are fixed on the opposite ends of a rigid beam of length *R*, equal to the diameter of Rowland's circle. The two ends of this beam rest on *swival trucks* C_1 and C_2 which are free to move along two tracks mutually at right angle to each other. The slit *S* is placed at the intersection of the two tracks *SA* and *SB*. The grating *G* and the plate-holder *P* are mounted in such a way that the tangent planes at their mid-points are normal to the

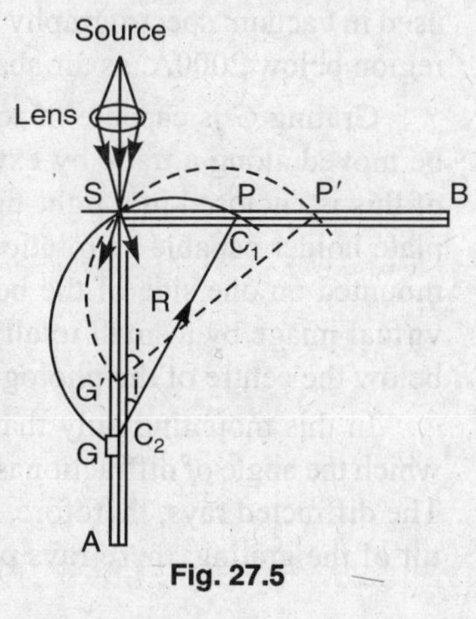

Fig. 27.5

axis of the beam R. The slit is mounted at S. Various positions of the spectra can be focused on the plate by sliding the rod R, thereby varying the angle of incidence i. The angle of diffraction is always nearly zero because the rays reaching P are almost normal to the surface of the grating. The spectrum produced in this way is normal spectrum.

Monochromatic rays of light diffracted along GP will have $\theta = 0$ and $i = \angle SGP$. Hence condition for the bright spectral line at P corresponding to the wavelength λ is given by

$$(a + b) \sin i = n\lambda \qquad (\because \sin\theta = 0)$$

or

$$\lambda = \frac{a+b}{n} \sin i = \frac{a+b}{n} \times \frac{SP}{R} \qquad (\because \sin i = SP/R)$$

$$\therefore \quad \lambda \propto SP$$

Thus the arm SP can be calibrated to give the wavelength directly. By this arrangement, we can study both ultraviolet and infrared regions of spectrum as well, without the use of lenses. As the initial direction of the beam always remains fixed, so the solar spectrum can be studied. The only drawback is that there is astigmatism with higher order spectra.

(*ii*) **Paschen mounting.** This is the most satisfactory mounting used for general work, devised by Paschen and Runge in 1902. In this arrangement, the slit S and the grating GG' are mounted in fixed positions on the Rowland circle as shown, the slit being exactly parallel to the rulings on the grating. The light is made incident on the grating at any desired angle depending upon the problem under investigation. The various order spectra will be in focus on the Rowland circle.

A semicircular steel frame coincides with region of the Rowland circle. The photographic plates are held in this semicircular frame by means of clamps, so that the plates are bent and coincide with the focal curve. As shown, C is the central image, A_1B_1 the *first order* spectrum, A_2B_2 the *second order* spectrum and A_3B_3 the *third order* spectrum. (Fig. 27.6).

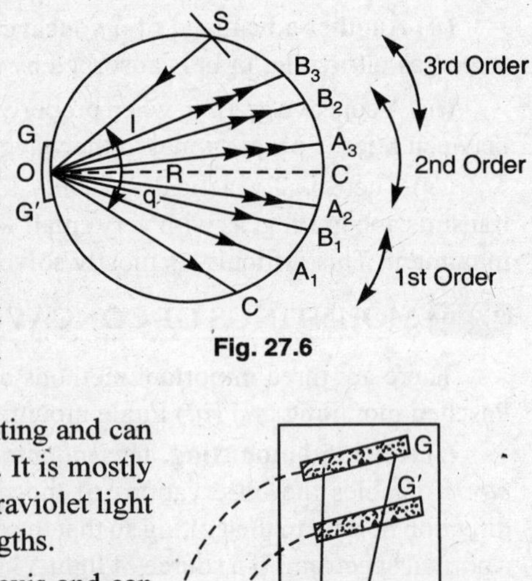

Fig. 27.6

With this mounting several orders of spectrum can be photographed simultaneously without altering the adjustment. No other mounting covers such a wide range under similar circumstances. Its disadvantages, however, are that (*i*) The spectrum is not normal so that the dispersion is non-linear, except when the rays fall back on the camera normally; (*ii*) There is high astigmatism in every order except zero order.

***(iii)* Eagle mounting.** This is a very compact mounting and can be removed from one room to another in the laboratory. It is mostly used in vacuum spectrography to study the spectra of ultraviolet light region below 2000Å, as air absorbs these short wavelengths.

Grating G is capable of rotation about the vertical axis and can be moved along a track by external mechanical devices. The whole of this is enclosed in a light-tight box at the other end of which is a plate holder capable of rotation about the vertical axis. The slit S is mounted on one side of the box in a tube at such a position that its virtual image by a small totally reflecting prism is formed at O, just below the centre of the photographic plate P. (fig. 27.7).

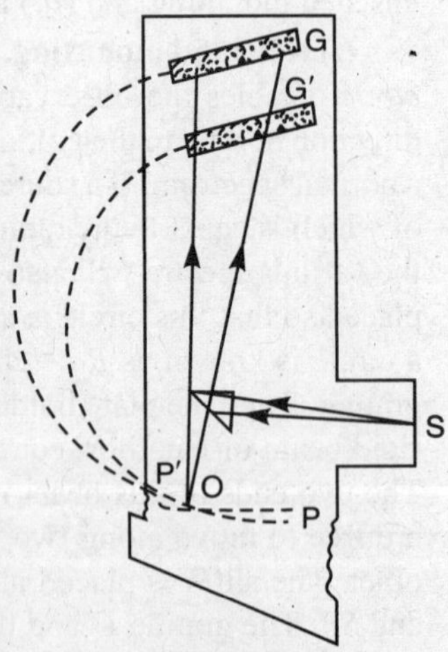

Fig. 27.7

In this mounting only that part of the spectrum is observed for which the angle of diffraction is nearly equal to the angle of incidence. The diffracted rays, therefore, nearly retrace their path and by slight tilt of the grating, these rays pass over the prism and are brought to focus on the plate. To get a sharp spectrum, the grating is moved along the track and also given a rotation about the vertical axis till the Rowland circle which moves with the grating passes through *O* where the image of the slit is formed by total internal reflection. The plate-holder whose axis of rotation also passes through *O* is rotated such that the surface of the plate also lies on the Rowland circle. This is achieved by trial focusing.

To pass from one spectral region to the other, the grating is rotated by proper amount, to bring the desired region on the plate and above adjustments are repeated to bring about focusing. Two positions of grating and plate and position of Rowland circle are shown for two different spectral regions.

Example 27.12 *What is the highest order spectrum which may be observed with monochromatic light of wavelength 5000 Å by means of a grating with 5000 lines/cm ?*

Solution. Maximum value of sin $\theta = 1$, $a+b = \dfrac{1}{5000} = 2\times10^{-4}$

$$\therefore \qquad n = \frac{a+b}{\lambda} = \frac{2\times10^{-4}}{5\times10^{-5}} = 4$$

Hence maximum order observed is 4.

27.12 ANGULAR DISPERSIVE POWER OF A GRATING

It is defined as the change in the angle of diffraction corresponding to a unit change in the wavelength.

The expression for the principal maximum in the *n*th order for a diffraction grating in normal incidence is

$$(a + b) \sin \theta = n\lambda \qquad \text{... } (i)$$

This expression shows that θ varies with wavelength. If the wavelength changes from λ to $(\lambda + d\lambda)$, then the angle of diffraction will change from θ to $(\theta + d\theta)$. The ratio $\dfrac{d\theta}{d\lambda}$ is called the angular **dispersive power.**

Expression for angular dispersive power

Differentiating $(a + b) \sin \theta = n\lambda$, we get

$$(a + b) \cos \theta \, d\theta = n.d\lambda$$

$$\therefore \qquad \text{Angular dispersive power } \frac{d\theta}{d\lambda} = \frac{n}{(a+b)\cos\theta} \qquad \text{... } (ii)$$

But from relation (*i*) $\qquad n = \dfrac{(a+b)\sin\theta}{\lambda}$

$$\therefore \qquad \frac{d\theta}{d\lambda} = \frac{(a+b)\sin\theta}{\lambda\,(a+b)\cos\theta} = \frac{\tan\theta}{\lambda} \qquad \text{... } (iii)$$

The expression given by Eq. (*ii*) shows that the angular dispersive power of a grating increases with

(*i*) the order of the spectrum and

(*ii*) the decrease of the grating element, *i.e.*, with the increase in the number of lines per cm.

27.13 RESOLVING POWER OF A PLANE TRANSMISSION GRATING

Resolving power. *It is defined as the ratio of the wavelength of a line in the spectrum to the least difference in the wavelength of the next line that can just be seen as separate.* It is thus equal to $\dfrac{\lambda}{d\lambda}$ where $d\lambda$ is the difference between the two wavelengths which can just be seen as separate.

Two lines of wavelength λ and $\lambda + d\lambda$ are said to be just resolved in a certain order of the grating spectrum if they satisfy Rayleigh's criterion *i.e.*, if the primary maximum of one falls on the first secondary minimum of the other.

Expression for resolving power of a plane transmission grating. The *n*th order principal maximum for a plane diffraction grating for normal incidence is given by

$$(a + b) \sin \theta = n\lambda$$

where $(a + b)$ is the grating element and θ the angle of diffraction corresponding to the *n*th order spectrum.

If now the angle of diffraction θ changes by a small value $d\theta$ so that the path-difference between the corresponding points of two consecutive transparencies changes by λ/N where N is the total number of lines in the grating, then the path-difference between the extreme rays is given by

$$\frac{\lambda}{N} \times N = \lambda$$

In such a case the path-difference between the extreme top ray and the central ray is $\lambda/2$ and between the extreme bottom ray and the central ray is also $\lambda/2$. Thus light issuing from any space in the upper half and the one from the corresponding space in the lower half will differ in path by $\lambda/2$. These, therefore, will cancel the effect due to each other. This gives the condition for the first secondary minimum in the *n*th order spectrum. It is given in a direction making an angle $(\theta + d\theta)$ with the normal to the grating and is given by

$$(a + b) \sin (\theta + d\,\theta) = n\lambda + \lambda / N \quad \text{... (i)}$$

If the *n*th order primary maximum for the wavelength $(\lambda + d\,\lambda)$ is produced in the same direction and falls on the first secondary minimum of wavelength λ, then

$$(a + b) \sin (\theta + d\,\theta) = n\,(\lambda + d\lambda) \quad \text{... (ii)}$$

Comparing (*i*) and (*ii*), we have

$$n(\lambda + d\lambda) = n\lambda + \lambda / N$$

or $$nd\lambda = \lambda / N$$

$\therefore$ Resolving power $$\frac{\lambda}{d\lambda} = nN$$

Thus *the resolving power of a grating is equal to the product of total number of lines in the grating and order of the spectrum.*

Now $$n = \frac{(a+b)\sin\theta}{\lambda}$$

If W is the total width of the ruled surface of the grating, then $N = \dfrac{W}{a+b}$

$\therefore$ Resolving power $$n\,N = \frac{(a+b)\sin\theta}{\lambda} \times \frac{W}{a+b} = \frac{W}{\lambda}\sin\theta$$

For maximum resolving power $\sin\theta = 1$ $\quad\therefore$ Maximum $R.P. = \dfrac{W}{\lambda}$

The relation $\dfrac{\lambda}{d\lambda} = n\,N$ shows that the resolving power increases with

(*i*) the order of the spectrum and

(*ii*) the increase in the total number of lines in the effective part of the grating.

It is further seen that the resolving power is independent of the grating element.

Alternative method. The resolving power of a grating spectrometer may also be calculated in the following manner.

The *n*th order principal maximum for a plane diffraction grating for normal incidence is given by

$$(a + b) \sin \theta = n\lambda \qquad \text{... (iii)}$$

where the constants have the usual meaning.

If the wavelength changes by a small amount $d\lambda$, the corresponding angle of diffraction changes by a small angle $d\theta$, which can be obtained by differentiating (*iii*)

$$\therefore \qquad (a + b) \cos \theta \, d\theta = n.d\lambda$$

or
$$d\theta = \frac{n \cdot d\lambda}{(a+b)\cos\theta}$$

The light, after suffering diffraction at the grating, enters the objective of the telescope in a grating spectrometer. If *d* is the diameter of telescope objective, the smallest angle $d\theta'$ that it is able to resolve for a light of wavelength λ is given by

$$d\theta' = \frac{\lambda}{d}$$

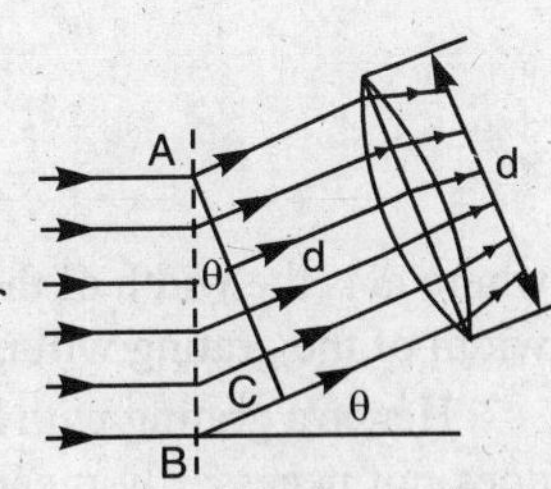

Fig. 27.8

Hence the two principal maxima in a certain order of the grating spectrum corresponding to wavelength λ and $(\lambda + d\lambda)$ having an angular separation $d\theta$ will be just resolved by the telescope, if

$$d\theta = d\theta'$$

or
$$\frac{n \cdot d\lambda}{(a+b)\cos\theta} = \frac{\lambda}{d}$$

$$\therefore \qquad \frac{\lambda}{d\lambda} = \frac{n \cdot d}{(a+b)\cos\theta} \qquad \text{... (iv)}$$

If the total width of the grating is *AB* and it contains *N* lines, then

$$AB = N(a + b)$$

Now
$$AC = d = AB \cos\theta = N(a + b)\cos\theta$$

Substituting the value of *d* in (*iv*), we have

$$\frac{\lambda}{d\lambda} = \frac{n \cdot N(a+b)\cos\theta}{(a+b)\cos\theta} = n.N$$

It is clear from this relation that in order to use full resolving power of the grating in a certain order of the spectrum, the aperture of the telescope objective should at least be equal to ($AB \cos\theta$) where θ is the angle of diffraction corresponding to that order of the spectrum.

27.14 DEPENDANCE OF DISPERSIVE AND RESOLVING POWERS OF A GRATING

The angular dispersive power of a grating is given by

$$\frac{d\theta}{d\lambda} = \frac{n}{(a+b)\cos\theta}$$

$$= \frac{nN'}{\cos\theta} \qquad \text{... (i)}$$

where *N'* is the number of lines per centimetre of the grating surface.

From (*i*) it is clear that for a given order spectrum the angular dispersive power depends upon the number of lines per centimetre. If the number of lines N' per centimetre is increased, the dispersive power is also increased.

The resolving power $$\frac{\lambda}{d\lambda} = nN \qquad \text{... (ii)}$$

where N is the total number of lines on the effective width of the grating. Substituting the value of $n = \frac{(a+b)\sin\theta}{\lambda}$ in (*ii*) we get

$$\frac{\lambda}{d\lambda} = \frac{N(a+b)\sin\theta}{\lambda}$$

$$= \frac{\omega\sin\theta}{\lambda}$$

where ω is the width of the ruled surface of grating. Thus change in the number of lines on the ruled width of the grating will not increase the resolving power.

Hence a grating with higher dispersive power *i.e.*, increase in the number of lines per centimetre does not necessarily mean that it has a higher resolving power as increase in number of lines per centimetre will decrease $(a + b)$ and ω is constant, hence this does not increase the resolving power.

Grating with greater number of lines is superior. A grating with greater number of lines is superior because

(*i*) with the increase in number of lines per cm the dispersive power of the grating increases and

(*ii*) with increase in total number of lines in the effective part of the grating resolving power increases.

Example 27.13 *Calculate the angular dispersion in degrees per angstrom for a diffraction grating having 14438 lines per inch when used in the third order at 4200 Å.*

Solution. Here $n = 3$; $\lambda = 4200$ Å $= 4200 \times 10^{-10}$ m

Number of lines per inch = 14438

$$\therefore \qquad a + b = \frac{2.54\times10^{-2}}{14438} = 0.1759\times10^{-5} \text{ m}$$

Also $$\sin\theta = \frac{n\lambda}{a+b} = \frac{3\times4200\times10^{-10}}{0.1759\times10^{-5}} = 0.7163$$

$$\therefore \qquad \cos\theta = \sqrt{1-(0.7163)^2} = 0.6978$$

Angular dispersion/Å $$d\theta = \frac{nd\lambda}{(a+b)\cos\theta} = \frac{3\times10^{-10}}{0.1759\times10^{-5}\times0.6978}$$

$[\therefore d\lambda = 1\text{Å} = 10^{-10} \text{ m}]$

$$= 0.0002444 \text{ radian}$$

$$= 0.014°.$$

Example 27.14 *Gratings A and B each having 5000 lines per cm are of width 2 cm and 4 cm respectively. Light containing wave lengths λ_1 and λ_2 falls on the two. Explain which of the two will cause better dispersion for a given order and which one will deliver a better resolution.* **(K.U., 2000)**

Solution. The dispersive power of a grating is given by $\frac{d\theta}{d\lambda} = \frac{n}{(a+b)\cos\theta}$ where n is the order of the spectrum, $(a + b)$ the grating element and θ the angle of diffraction.

As the value of n, $(a + b)$ and θ the angle of diffraction for the mean wavelength $\frac{\lambda_1 + \lambda_2}{2}$ in a given order is the same for both the gratings, the two will give the same dispersive power.

On the other hand, the resolving power is given by $\frac{\lambda}{d\lambda} = nN$ where N is the total number of lines on the grating. As a grating having a width of 4 cm has double the total number of lines as compared to the grating having a width of a 2 cm, the 4 cm wide grating will have double the resolving power.

Example 27.15 *In a second order spectrum of a plane diffraction grating a certain spectral line appears at an angle of 10°, while another line of wavelength 0.5 × 10^{-10} m greater appears at an angle 3" greater. Find the wavelength of the lines and the minimum grating width required to resolve them. Given sin 10° = 0.1736 and cos 10° = 0.9848.* **[Delhi Hons., 2001]**

Solution. Here

$$\sin 10° = 0.1736; \cos 10° = 0.9848$$

$$d\lambda = 0.5 \times 10^{-10} \text{ m} = 0.5 \text{ Å}; n = 2$$

$$d\theta = 3'' = \frac{3 \times \pi}{180 \times 3600} = 0.1455 \times 10^{-4} \text{ radian}$$

$$\therefore \quad \frac{d\theta}{d\lambda} = \frac{0.1455 \times 10^{-4}}{0.5 \times 10^{-10}} = 2.91 \times 10^5$$

Now

$$\frac{d\theta}{d\lambda} = \frac{n}{(a+b)\cos\theta}$$

$$\therefore \quad (a+b) = \frac{2}{2.91 \times 10^5 \times 0.9848} = 0.6979 \times 10^{-5} \text{ m}$$

Now if λ_1 is the wavelength of first line, then

$$(a+b)\sin\theta = n\lambda_1$$

or

$$\lambda_1 = \frac{(a+b)\sin 10°}{2} = \frac{0.6979 \times 10^{-5} \times 0.1736}{2}$$

$$= 6057.77\text{Å}$$

$$\therefore \quad \lambda_2 = \lambda_1 + 0.5 \text{ Å} = 6057.77 + 0.5$$

$$= 6058.27\text{Å}$$

If N is the total number of lines on the grating so that the two lines are just resolved, then

$$\frac{\lambda}{d\lambda} = nN$$

or

$$N = \frac{1}{n} \times \frac{\lambda}{d\lambda} = \frac{6057.77}{2 \times 0.5} = 6057.77$$

$$= 6058$$

Grating width

$$= N(a+b) = 6058 \times 0.6979 \times 10^{-5}$$

$$= 0.04228 \text{ m} = 4.228 \text{ cm}$$

Example 27.16 *A plane transmission grating having 6000 lines per cm is used to obtain spectrum of light of wavelength 5890 Å at normal incidence. Calculate the dispersive power of the grating in second order spectrum.* **(Nagpur U., 2002)**

Solution. Here $\lambda = 5890$ Å $= 5890 \times 10^{-8}$ cm; $n = 2$; $N = 6000$ lines/cm;

$$\therefore \quad (a+b) = \frac{1}{6000} \text{ cm}$$

$$\sin\theta = \frac{n\lambda}{(a+b)} = 2\times 5890\times 10^{-8}\times 6000 = 0.7068$$

$$\therefore \quad \cos\theta = \sqrt{1-\sin^2\theta} = \sqrt{1-(0.7068)^2} = 0.7074$$

$$\therefore \quad \text{Dispersive power } \frac{d\theta}{d\lambda} = \frac{n}{(a+b)\cos\theta} = \frac{nN}{\cos\theta} = \frac{2\times 6000}{0.7074} = 16963$$

Example 27.17 *Find the number of lines a grating should have to resolve the first order doublet having a wavelength difference 0.6 nm at 589.3 nm.* **(K.U., 2002; H.P.U., 2001; P.U., 2003)**

Solution. $d\lambda = 0.6 \text{ nm} = 0.6 \times 10^{-9}$ **m,** $n = 1$

$\lambda = 589.3 \text{ nm} = 589.3 \times 10^{-9}$ m

If N is the number of lines on the grating to resolve the two wavelengths, then

$$\frac{\lambda}{d\lambda} = Nn$$

or

$$N = \frac{\lambda}{n\,d\lambda} = \frac{589.3\times 10^{-9}}{1\times 0.6\times 10^{-9}} = 982$$

Example 27.18 *What should be the minimum number of lines on a grating which will just resolve Sodium light of wavelength $D_1 = 5890$ Å and $D_2 = 5896$ Å in second order ?* **(Nagpur U. 2009; K.U., 2001)**

Solution. Mean wavelength $\lambda = \dfrac{5896+5890}{2} = 5893$Å

$$d\lambda = 5896 - 5890 = 6\text{Å}$$

Here $n = 2$

$$\therefore \quad N = \frac{\lambda}{nd\lambda} = \frac{5893\times 10^{-8}}{2\times 6\times 10^{-8}} = 491$$

Example 27.19 *Calculate the minimum number of lines per cm at a diffraction grating of width 3 cm which would just resolve sodium lines of wavelength 5890Å and 5896Å in the second order spectrum.* **(P.U. 2004, Similar) (K.U., 2000)**

Solution. As calculated in the above example the minimum total number of lines required $N = 491$. Here width of grating = 3 cm.

$\therefore$ Number of lines per cm = 491/3 = 164.

Example 27.20 *In a grating the sodium doublet (5890 Å and 5896 Å) is viewed in third order at 30° to the normal and is just resolved. Find the grating spacing and the total width of the rulings.* **(P.U. 2006)**

Solution. Resolving power of the grating $\dfrac{\lambda}{d\lambda} = nN$

$$\text{Mean wavelength } \lambda = \frac{5890 + 5896}{2} = 5893 \text{ Å} = 5893 \times 10^{-10} \text{ m}$$

$\therefore$ Total number of lines on the grating $N = \dfrac{\lambda}{nd\lambda}$

$$= \frac{5893 \times 10^{-10}}{3 \times 6 \times 10^{-10}} = 328$$

Now $(a + b) \sin \theta = n\lambda$

$$\therefore \quad \text{Grating spacing } (a + b) = \frac{n\lambda}{\sin \theta} = \frac{3 \times 5893 \times 10^{-10}}{\sin 30}$$

$$= \frac{3 \times 5893 \times 10^{-10}}{0.5} = 3.5358 \times 10^{-6} \text{ m}$$

Hence total width of the rulings (on the grating) = $N(a + b)$

$$= 328 \times 3.5358 \times 10^{-6} = 1160 \times 10^{-6} \text{ m} = 1.16 \text{ mm.}$$

Example 27.21 *Light is incident normally on a grating 0.5 cm wide with 2500 lines. Find the angular separation for the principal maxima of the two sodium lines in the first order. Wavelengths of the lines are 5890 Å and 5896 Å. Are the two lines resolved ?* **(P.U. 2003)**

Solution. Mean wavelength $\lambda = 5893 \text{ Å} = 5893 \times 10^{-10}$ m

$$d\lambda = 6 \text{ Å} = 6 \times 10^{-10} \text{ m}, n = 1$$

No. of lines on 0.5 cm wide grating = 2500

$$\therefore \quad a + b = \frac{0.5}{2500} = 2 \times 10^{-4} \text{ cm} = 2 \times 10^{-6} \text{ m}$$

If θ is the mean angle of diffraction for the sodium lines in the first order, then

$$\sin \theta = \frac{\lambda}{a + b} = \frac{5893 \times 10^{-10}}{2 \times 10^{-6}} = 0.2947$$

$$\therefore \quad \cos \theta = \sqrt{1 - (0.2947)^2} = 0.9556$$

$$\text{Angular separation} \quad d\theta = \frac{nd\lambda}{(a + b) \cos \theta} = \frac{1 \times 6 \times 10^{-10}}{2 \times 10^{-6} \times 0.9556}$$

$$= 0.0003139 \text{ radian}$$

$$= 1', 4.74''$$

The two lines will be seen distinctly if they are resolved by the grating.

$$\therefore \quad \text{Required resolving power } \frac{\lambda}{d\lambda} = \frac{5893}{6} = 982$$

Actual *R.P.* $= nN = 1 \times 2500 = 2500$

As the actual resolving power is greater than the required resolving power, the two lines will be seen as separate in the first order.

Example 27.22 *A plane transmission grating has 15000 lines per inch. Find the angle of separation of 5016 A.U. and 5048 A.U. line of He in second order spectrum. Also find the minimum number of lines required to just resolve the above lines in second order.* **(Nagpur U., 2008, 2001)**

Solution. Here we shall find the angle of separation by calculating the angle of diffraction in the second order separately for the two given wavelengths.

Here Number of lines per inch = 15000

$$\therefore \quad \text{grating constant } (a+b) = \frac{2.54}{15000} \text{ cm}; \quad n = 2$$

Let θ_1 be the angle of diffraction in the second order for wavelength $\lambda = 5016$ A.U. $= 5016 \times 10^{-8}$ cm and θ_2 for $\lambda = 5048$ A.U. $= 5048 \times 10^{-8}$ cm

$$\therefore \quad (a+b)\sin\theta_1 = 2\lambda_1 \quad \text{or} \quad \sin\theta_1 = \frac{2\lambda_1}{(a+b)}$$

$$= \frac{2 \times 5016 \times 10^{-8} \times 15000}{2.54} = 0.59244$$

$$\text{or} \quad \theta_1 = 36.33°$$

$$\text{Similarly} \quad \sin\theta_2 = \frac{2\lambda_2}{(a+b)} = \frac{2 \times 5048 \times 10^{-8} \times 15000}{2.54} = 0.59622$$

$$\text{or} \quad \theta_2 = 36.60°$$

$$\therefore \quad \text{Angle of separation } d\theta = \theta_2 - \theta_1 = (36.60 - 36.33) = 0.27°$$

Let N be the minimum number of lines required to just resolve the two lines, then

$$\text{Average wavelength} \quad \lambda = \left(\frac{5016 + 5048}{2}\right) \times 10^{-8} \text{ cm} = 5032 \times 10^{-8} \text{ cm}.$$

$$d\lambda = (5048 - 5016) \times 10^{-8} \text{ cm} = 32 \times 10^{-8} \text{ cm}.$$

$$\therefore \quad \frac{\lambda}{d\lambda} = nN \quad \text{or} \quad \frac{5032 \times 10^{-8}}{32 \times 10^{-8}} = 2N$$

$$\text{or} \quad N = \frac{5032}{32 \times 2} = 78.6 \text{ or } 79 \text{ lines}.$$

Example 27.23 *Separation of Na-lines (λ = 5893 Å) in the second order spectrum of a plane transmission grating having 5000 lines cm^{-1} is 2.5′ of an arc for normal incidence. Find the difference in wavelength between the two yellow lines.*

Solution. Here $\quad N = 5000$/cm, $\quad n = 2$

Mean wavelength $\quad \lambda = 5893 \times 10^{-10}$ m

$$\text{Grating constant} \quad a + b = \frac{10^{-2}}{5000} = 2 \times 10^{-6} \text{ m}$$

$$\sin\theta_2 = \frac{2\lambda}{a+b} = \frac{2 \times 5893 \times 10^{-10}}{2 \times 10^{-6}} = 0.5893$$

$$\cos\theta = \sqrt{1 - (0.5893)^2} = 0.8079$$

$$d\theta = 2.5' = \frac{2.5 \times \pi}{60 \times 180} = 0.7273 \times 10^{-3} \text{ radian}$$

Now $$d\theta = \frac{n d\lambda}{(a+b)\cos\theta}$$

or $$d\lambda = \frac{(a+b)\cos\theta \cdot d\theta}{2} = \frac{2 \times 10^{-6} \times 0.8079 \times 0.7273 \times 10^{-3}}{2}$$

$$= 5.876 \times 10^{-10} \text{ m} = 5.876 \text{ Å}$$

Example 27.24 *Two spectral lines with average wavelength 6000 Å are resolved in second order by a grating having 500 lines per cm. The least width of the grating is 2 cm. Find the difference in wavelength of the lines.* **(*Nagpur U., 2001*)**

Solution. Here number of lines per cm = 500

Width of the grating = 2 cm

∴ Total number of lines $N = 2 \times 500 = 1000$

Order of spectrum $n = 2$

$\lambda = 6000 \text{ Å} = 6000 \times 10^{-8}$ cm $\quad d\lambda = ?$

Resolving power $$\frac{\lambda}{d\lambda} = nN \quad \text{or} \quad d\lambda = \frac{\lambda}{nN}$$

∴ $$d\lambda = \frac{6000 \times 10^{-8}}{2 \times 1000} = 3 \times 10^{-8} \text{ cm} = 3\text{Å}$$

Example 27.25 *A grating has 1100 lines on it. What is the difference between two wavelengths that can be just resolved in the first order spectrum in the region of wavelength 660 nm?* **(*P.U. 2005; Pbi. U., 2001*)**

Solution. $$d\lambda = \frac{\lambda}{nN} = \frac{660 \times 10^{-9}}{1 \times 1100} = 0.6 \times 10^{-9} \text{ m} = 6 \times 10^{-10} \text{ m} = 6\text{Å}$$

Example 27.26 *A diffraction grating has 15 cm of surface ruled with 6000 lines per cm. What is the resolving power of the grating in first order ?* **(*M.D.U. 2006*)**

Solution. Resolving power = nN. Here $n = 1$ and $N = 15 \times 6000 = 90{,}000$

∴ R.P = $nN = 90{,}000$

Example 27.27 *A plane grating of width 2.5 cm is found to just resolve two wavelengths differing by 6Å in second order. If mean wavelength is 5893Å, calculate the grating element.* **(*Nagpur U. 2011*)**

Solution. Given: $d\lambda = 6\text{Å} = 6 \times 10^{-10}$ m; $n = 2$; $\lambda = 5893\text{Å} = 5893 \times 10^{-10}$ m

Resolving power, $$\frac{\lambda}{d\lambda} = nN$$

∴ Total no. of lines, $$N = \frac{\lambda}{d\lambda} \cdot \frac{1}{n} = \frac{5893 \times 10^{-10}}{6 \times 10^{-10}} \times \frac{1}{2} = 491$$

∴ Grating element, $$(a+b) = \frac{2.5 \times 10^{-2} \text{ m}}{491} = 0.0000509 \text{ m} = 5.09 \times 10^{-5} \text{ m}$$

27.15 RESOLVING POWER OF PRISM SPECTROMETER (R.P. of Prism and Telescope)

It is defined as the ratio of the wavelength λ to the smallest difference in wavelength dλ that can be just distinguished as separate by the spectrometer.

$$\text{Resolving power} = \frac{\lambda}{d\lambda}$$

Let ABC be the section of a prism. A plane wavefront BP of wavelength λ is incident on it and meets the prism placed in the minimum deviation position. It produces a refracted wavefront CQ which in turn produces the image I at the focal plane of the telescope objective L. (Fig. 27.9) It is supposed that light passes through the whole of the prism.

If μ is the refractive index of the material of the prism for a wavelength λ, then

$$PA + AQ = \mu \times BC \qquad \text{... (i)}$$

If now another plane wavefront of wavelength $\lambda + d\lambda$ is incident along the same path and produces a refracted wavefront CQ' so that the angle between the two refracted wavefronts is $d\theta$, then the wavefront CQ' after refraction through the lens produces an image I' at the focal plane of the telescope objective.

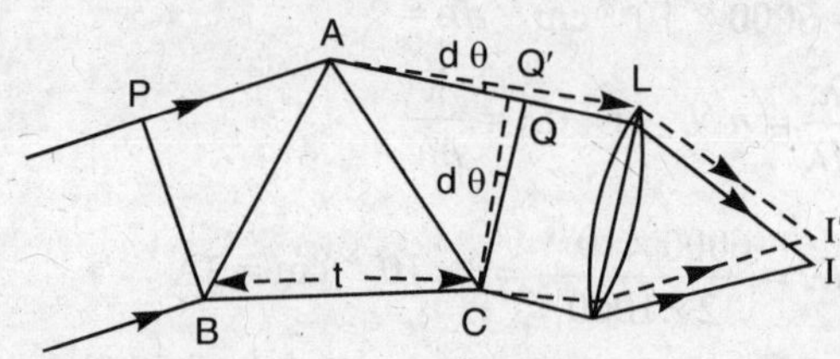

Fig. 27.9

When the wavelength increases, the refractive index decreases. Thus the refractive index for the wavelength $(\lambda + d\lambda)$ will be $(\mu - d\mu)$.

$$\therefore \quad PA + AQ' = (\mu - d\mu)\, BC \qquad \text{... (ii)}$$

Subtracting (*ii*) from (*i*), we have

$$AQ - AQ' = d\mu . BC$$

or

$$QQ' = t . d\mu$$

If d is the width of the emergent refracted wavefront and also the diameter of the telescope objective, then

$$QQ' = d . d\theta$$

$$\therefore \quad d . d\theta = t . d\mu$$

or

$$d\theta = \frac{t}{d}\, d\mu \qquad \text{... (iii)}$$

If $d\theta$ is the limit of resolution of the telescope, then

$$d\theta = \frac{\lambda}{d} \qquad \text{... (iv)}$$

Comparing (*iii*) and (*iv*), we have

$$\frac{\lambda}{d} = \frac{t}{d}\, d\mu$$

$$\text{or} \qquad \lambda = t.d\lambda$$

$$\therefore \qquad \frac{\lambda}{d\lambda} = t\frac{d\mu}{d\lambda}$$

This shows that the resolving power increases with the length of the base and depends upon the material of the prism and the wavelength of light used.

27.16 DIFFERENCE BETWEEN GRATING AND PRISM SPECTRA

Prism spectrum	*Grating spectrum*
1. There is only one spectrum.	There are a number of spectra on either side of the central maximum.
2. Deviation produced depends upon the angle of the prism, refractive index of the material of the prism and the wavelength of light used. It is more for violet light than for red light.	Deviation is directly proportional to the wavelength of light used and is inversely proportional to the grating element. It is independent of the material of the grating.
3. The dispersive power of a prism is more for a shorter wavelength than for a longer wavelength, *i.e.*, the violet end is more drawn out. Hence the violet portion is less intense than the red portion.	The dispersive power of a grating is given by $\frac{n}{(a+b)\cos\theta}$ (*i*) Since θ is usually small, therefore, the angular separation between two lines for a given spectrum is practically the same all along in a particular order of the spectrum. (*ii*) The angular separation in the second order is double of what it is in the first order. (*iii*) The angular separation increases as the grating element decreases.
4. The spectra produced by two prisms are neither similar nor regular.	The spectrum produced by grating for a given source is exactly similar to that produced by another grating, with this difference that the length is directly proportional to the number of lines per centimetre of grating.
5. The resolving power of a prism is small and is given by $t.d\mu/d\lambda$ *i.e.*, it depends upon the length of the base of prism and the wavelength of light.	The resolving power is very large and is given by Nn, where N is the total number of lines in the effective part of the grating and n the order of the spectrum. It is thus independent of the wavelength and the grating constant.

The points enumerated above are in favour of the grating spectrum and are the advantages that the latter possesses over the former. A serious disadvantage in a grating spectrum is that its intensity is low because most of the incident light is transmitted without any deviation and hence concentration at the central band is much more than at other spectra. Further, the portion of diffracted light is spread over the various order spectra. In the case of a prism there is only one spectrum and hence the entire light is concentrated at one point. Thus the intensity of prism spectrum is very good.

EXERCISE

LONG QUESTIONS

1. Using Fraunhofer's theory of single slit, explain the diffraction by multiple slits (N-parallel slits) ***(Gujarat U. 2004; G.N.D.U., 2002)***
2. What is plane diffraction grating ? Define grating element. Give the construction and discuss its theory. Derive the condition for secondary maxima and minima. Find the width of the central maxima. ***(G.N.D.U. 2008, 2007, 2005; Nagpur U. 2010, 2009, 2008; Meerut U. 2002, 2001, 2000; P.U. 2001; M.D.U. 2007, 2006, 2001; K.U. 2001; Luck U. 2001; H.P.U. 2000, Gauhati U. 2000)***
3. Derive an expression for half angular width of principal maxima in a diffraction grating. ***(Meerut U. 2003; P.U. 2001, 2000)***
4. (*a*) Derive an expression for resolving power of a grating. ***(Nagpur U. 2011, 2010, 2009)***
   (*b*) Find the maximum number of orders available with a grating. ***(Meerut U. 2003)***
5. Derive the condition for the missing order spectra for a diffraction grating. What particular order spectrum would be absent if the width of the transparencies and oapacities of a grating are equal?
6. State and explain Rayleigh's criterion for resolution of two close images with neat diagram. ***(Nagpur U. 2009, 2008)***
7. Is there any difference in the diffraction pattern due to double slit and interference due to two sources ? Explain.
8. (*a*) Describe in full detail how would you use the plane transmission grating to determine the wavelength of light. ***(Nagpur U. 2010; Cochi. U. 2002)***
   (*b*) Given a spectrometer and monochromatic source of light, how will you find the grating constant ?
9. What is a normal spectrum ?
10. Mention different types of gratings. What is plane reflection grating ? How it differs from plane transmission grating ?
11. What is blazed grating ? Explain its construction and working. What is blaze angle ?
12. (*a*) What is concave grating ? Why collimating and focusing system is not required in concave grating ?
13. Describe Rowland's concave reflecting grating, giving its theory. ***(Luck 2002; Gauhati U. 2000; Meerut U. 2006)***
14. What are merits of concave grating over a transmission grating ? ***(Meerut U. 2006)***
15. Describe three important mountings of a concave grating ?
16. Distinguish between the angular dispersive power and the resolving power of a grating. Deduce an expression for the dispensive power of a plane transmission grating and show that for normal incidence it is given by tan θ/λ. On what factors does it depend ? ***(Meerut U., 2006, 2003, 2002, 2001; Nagpur U., 2002; M.D.U., 2002; G.N.D.U., 2002; Pbi. U., 2001; Gauhati U., 2000; P.U., 2007, 2005)***
17. Derive an expression for the resolving power of a plane transmission grating. Show that the maximum value of resolving power of a grating $= \dfrac{W}{\lambda}$ where W is the total width of the ruled surface of the grating. On what factors does the resolving power of a grating depend ? ***(P.U. 2008, 2007, 2000; Gujarat. U. 2004; M.D.U., 2005, 2003, 2001; Meerut U., 2007, 2005, 2002; Nagpur U., 2002; Pbi. U., 2001, 2000; Kerala U., 2001; Gauhati U., 2000; G.N.D.U., 2008, 2007, 2005)***
18. (*a*) Discuss the statement 'A grating having higher dispersive power than another does not necessarily have a higher resolving power'.

(b) Explain why are the gratings with greater number of lines superior ? ***(Luck. U., 2001)***

19. A grating with greater number of lines is superior ? Explain.

20. Derive an expression $\frac{\lambda}{d\lambda} = t\frac{d\mu}{d\lambda}$ for the resolving power of a prism and telescope. ***(Indore U., 2001)***

21. State the difference between the grating and the prism spectra.

(P.U., 2002; Nagpur U., 2011, 2009, 2008, 2003, 2002, 2001; Kerala U., 2001)

22. In a Fraunhofer diffraction due to a narrow slit, a screen is placed 2 m away from the lens to obtain the pattern in its focal plane. Find the slit width if the first minimum lies 5 mm on either side of central maximum when plane wave, of 5000 Å are incident on the slit. **[Ans.** 0.2 mm]

23. The limits of visible spectrum are 4000 Å to 7000 Å. Find the angular separation in the first order grating spectrum having 15000 lines per inch. **[Ans.** 9° 45′]

24. How many orders will be visible if light of wavelength 5000 Å is incident normally on a grating having 7620 lines/inch? **[Ans.** 6]

25. Calculate the useful magnifying power of a telescope of objective 2.54 cm, assuming the resolving power of the eye to be 2 min and the wavelength of light 6000 Å. **[Ans.** 20.2]

26. Light is incident normally on a grating 0.5 cm wide with 2500 lines. Find the angular separation of the two sodium lines in the first order spectrum. Can they be resolved ? Wavelengths of lines are 5890 Å and 5896 Å. **[Ans.** 1′4.7″; yes]

27. The refractive index for C and F lines for a certain glass $\mu_C = 1.6545$ and $\mu_F = 1.6635$. The wavelengths of C and F lines in the solar spectrum are 6563 Å and 5270 Å respectively. Calculate the base of a 60° prism made of this glass which is just capable of resolving the two sodium lines of wave-lengths 5896 Å and 5890 Å. **[Ans.** $t = 1.41$ cm]

MULTIPLE CHOICE QUESTIONS

27.1 In a plane diffraction grating when number of line/cm is increased

(a) linear (or angular) dispersive power increases and resolving power decreases.

(b) linear (or angular) dispersive power decreases and resolving power increases.

(c) both linear (or angular) dispersive power and resolving powers increase.

(d) both linear (or angular) dispersive power and resolving powers decrease.

(e) none of the above is true.

27.2 Grating element of a plane transmission grating is:

(a) a (b) b (c) $a + b$ (d) $a - b$

27.3 Width of the central maxima obtained by a plane transmission grating is:

(a) $\frac{N}{2\lambda}$ (b) $\frac{2\lambda}{N}$ (c) $\frac{\lambda}{N}$ (d) $\frac{N}{\lambda}$

27.4 Resolving power of a plane transmission grating is:

(a) $n\lambda$ (b) nN (c) n/λ (d) n/N

27.5 Resolving power of a prism spectrometer is given by:

(a) $\frac{1}{t}\frac{d\mu}{d\lambda}$ (b) $\frac{1}{t}\frac{d\lambda}{du}$ (c) $t\frac{d\mu}{d\lambda}$ (d) $t\frac{d\lambda}{d\mu}$

27.6 Resolving power of a plane transmission grating depends upon:

(a) grating element (b) slit width

(c) total number of lines on the grating (d) order of spectra

ANSWERS

27.1 (e) **27.2** (c) **27.3** (b) **27.4** (b) **27.5** (c) **27.6** (c)

POLARIZATION OF LIGHT

28 Chapter

INTRODUCTION

We are aware that velocity of light and hence the refractive index is independent of the direction of propagation of electromagnetic wave and state of polarisation of the wave. Amorphous solids like glass, crystalline solids and certain liquids show this behaviour. These are said to be *optically isotropic.* The progress of wave trains in isotropic substances can be found out by Huygen's construction. The secondary wave trains are spherical surfaces in such a medium.

There are other crystalline substances which are *optically anisotropic i.e.*, their optical properties are different in different directions. Consider a substance which is composed of non-spherical molecules which are longer in length than in width and suppose these molecules are arranged in the substance with their long axis parallel. If now a plane polarised electromagnetic wave passes through such a substance the electrons in the substance respond more easily to oscillations in the direction parallel to the axis of the molecules than they would respond if the electric field tries to push them at right angles to that direction. If we call the direction of the axis of molecules as **optic axis,** then refractive index in the direction of optic axis will be different from that in the direction perpendicular to the direction of optic axis. Such substances are called **birefringent.** Examples of birefringent crystals are calcite and ice land spar.

28.1 PHENOMENON OF DOUBLE REFRACTION

If polarised light beam is incident on such a substance and if polarisation is parallel to the optic axis, the light will pass through with one velocity and if polarisation is perpendicular to the optic axis, the light is transmitted with a different velocity. Hence a birefringent substance will have two refractive indices. Thus a ray of light passing through such substances will suffer **double refraction** and on emergence will give rise to two images of the same object. This phenomenon is also known as double refraction and the substances are called **doubly refracting substances.**

If a beam of unpolarised light is allowed to pass through a calcite or quartz crystal, it is split up into two refracted beams instead of one. The crystal having this property is said to be a **doubly refracting** crystal and exhibits **birefringence.** This phenomenon is called **double refraction.** If a ray of light *SA* from a point source is incident on a calcite crystal making an angle of incidence *i* it is refracted along two paths *AB* and *AC* making angles of refraction r_1 and r_2 respectively. These rays emerge out along *BO* and *CE* parallel to each other as shown. Thus two images *O* and *E* can be obtained on the screen. If the crystal is rotated about the incident beam as axis one image *O* remains stationary while the other image *E* rotates round the image *O*. (Fig. 28.1)

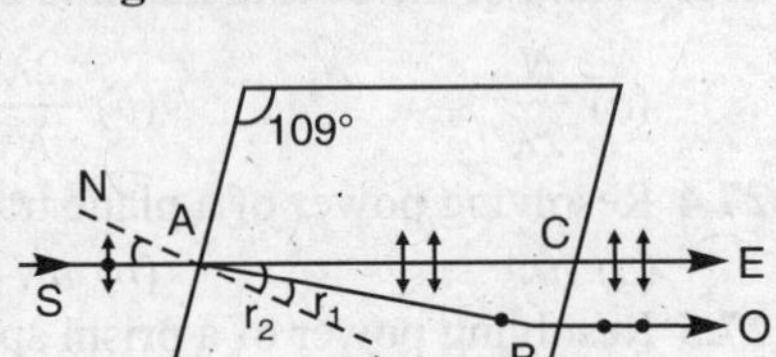

Fig. 28.1

Ordinary ray. As the ray which produces the image *O* obeys the ordinary laws of refraction it is called **ordinary ray** and the image is called the **ordinary image.**

Extraordinary ray. The other ray which produces the image *E* does not obey Snell's law. This is called the **extraordinary ray** and the image is called the **extraordinary image.**

For ordinary ray the refractive index $\mu_0 = \dfrac{\sin i}{\sin r_1}$ is constant while for the extraordinary ray $\mu_e = \dfrac{\sin i}{\sin r_2}$ is not constant but varies with the angle of incidence i.

Both are plane polarised. It may be noted that both the rays are plane polarised, but their planes of polarisation are at right angle to each other. The vibrations of *O*-ray are normal to the plane of paper while those of *E*-ray are in the plane of paper as shown in Fig. 28.2. This can be verified by viewing the image through a tourmaline crystal. On rotating the tourmaline crystal the intensities of both the images will undergo a change. If the intensity of ordinary image increases, that of extraordinary image decreases and vice-versa. The *O* and *E* images alternately are bright twice in one complete rotation of tourmaline crystal.

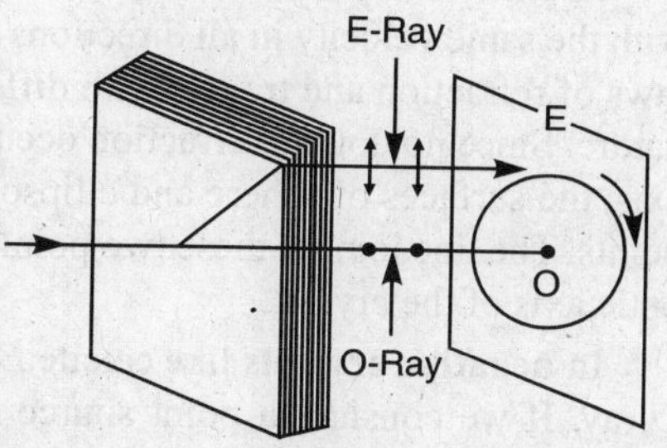

Fig. 28.2

28.2 OPTIC AXIS AND PRINCIPAL PLANE

(*i*) **Optic axis.** *For a uniaxial crystal the direction along which the velocity of the E-ray is equal to the velocity of the O-ray, is called the optic axis.*

Calcite known as iceland spar ($CaCO_3$) exists in nature in several forms all of which give a *rhombohedron* on cleavage. The six faces of the rhombohedron are parallelograms each having angles of 102° and 78°. At the two opposite corners such as *A* and *B* as shown in Fig. 28.3, the three angles of the faces meeting there are *obtuse.* These corners are known as *blunt corners.* At each of the remaining corners one of the angles is obtuse whereas the other two are acute.

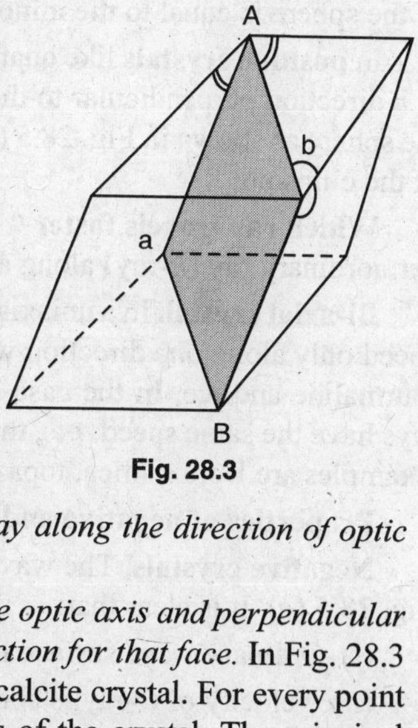

Fig. 28.3

A line equally inclined to the three edges meeting at one of the blunt corners A or B or any other line parallel to it is the direction of the **optic axis.** The properties of optic axis are:

(1) *Optic axis is a direction and not a particular line.*

(2) *The crystal is symmetrical about it.*

(3) *The velocities of the ordinary ray and the extra-ordinary ray along the direction of optic axis are the same.*

***(ii)* Principal plane or principal section.** *A plane containing the optic axis and perpendicular to the face of the crystal is called the principal plane or principal section for that face.* In Fig. 28.3 *A a B b* is the principal plane for the top and the bottom faces of the calcite crystal. For every point there are three principal planes, one for each pair of opposite faces of the crystal. The principal plane for calcite crystal is a parallelogram with angles 71° and 109°, as shown in Fig. 28.4. Since the opposite faces of a calcite crystal are always parallel, the two refracted rays emerge out parallel to each other and the incident ray.

The ordinary ray is plane polarised in the principal plane *i.e.*, it has vibrations perpendicular to the principal plane. These have been shown by *dots i.e.*, at right angles to the plane of the paper.

The extraordinary ray is also plane polarised in a plane perpendicular to the principal plane *i.e.*, the vibrations take place in the principal plane. These have been shown by *lines i.e.*, in the plane of the paper, as in Fig. 28.4.

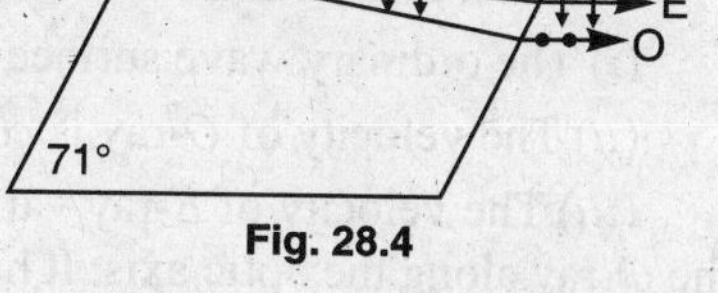

Fig. 28.4

28.3 DOUBLE REFRACTION THROUGH UNIAXIAL CRYSTAL

According to Huygens, in double refraction a single ray gives rise to two refracted rays—*O*-ray and *E*-ray.

In uniaxial crystals the wavefront of *O*-ray obeys ordinary laws of refraction and thus travels with the same velocity in all directions and hence is *spherical* in nature. The *E*-ray does not obey the laws of refraction and travels with different velocities in different directions and thus is *ellipsoid* in nature. Since no double refraction occurs in the direction of optic axis, the surfaces of sphere and ellipsoid touch each other at two points. The line joining these two points gives the direction of the optic axis of the crystal.

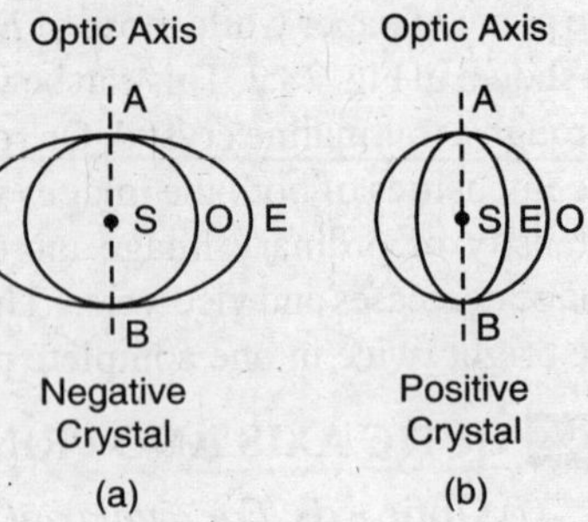

Fig. 28.5

In **negative crystals** like calcite *E*-ray travels faster than the *O*-ray. If we consider a point source *S* of monochromatic light within the crystal, then *O* and *E* wavefronts in two dimensions are as shown in Fig. 28.5. (*a*) surrounding the point *S*. The velocity of *E*-ray is maximum along a direction perpendicular to the direction of optic axis and least but equal to that of *O*-ray along the optic axis. The sphere is thus enclosed inside the ellipsoid. The diameter of the sphere is equal to the minor axis of the ellipsoid.

In **positive crystals** like quartz, the E-ray travels slower than *O*-ray and has a minimum velocity in a direction perpendicular to the direction of optic axis. The ellipsoid is therefore enclosed inside the sphere as shown in Fig. 28.5 (*b*). In this case the diameter of the sphere is equal to the major axis of the ellipsoid.

Which ray travels faster ? In a positive crystal the ordinary ray (*O*-ray) travels faster than the extraordinary ray (*E*-ray) along a direction other than the optic axis.

Bi-axial crystal. In a uniaxial crystal the two rays, ordinary and the extraordinary, have the same speed only along *one* direction which is the optic axis of the crystal. Examples are, Quartz, calcite, tourmaline and ice. In the case of bi-axial crystals there are *two* directions along whiich the two rays have the same speed. *i.e.*, there are two optic axis. This is why such crystals are called bi-axial. Examples are Borax, mica, topaz, and selenite.

Properties of negative and positive crystals

Negative crystals. The wavefronts surrounding a point source *S* in such a crystal are shown in Fig. 28.5 (*a*). It is clear that:

(*i*) Ordinary wave surface lies inside the extraordinary wave surface.

(*ii*) Velocity of *O*-ray is constant in all directions.

(*iii*) The velocity of *E*-ray varies with the direction. It is *minimum* and equal to the velocity of the ordinary ray along the optic axis. It is *maximum* in a direction perpendicular to the direction of the optic axis.

(*iv*) The refractive index for the *E*-ray is *less* than the refractive index for the *O*-ray *i.e.*, $\mu_e < \mu_0$.

Positive crystals. The wavefronts surrounding a point source *S* in such a crystal are shown in Fig. 28.5 (*b*). It is clear that:

(*i*) The ordinary wave surface lies *outside* the extraordinary wave surface.

(*ii*) The velocity of *O*-ray is *constant* in all directions.

(*iii*) The velocity of *E*-ray varies with the direction. It is *maximum* and equal to the velocity of the *O*-ray along the optic axis. It has a *minimum* value in a direction perpendicular to the optic axis.

(*iv*) The refractive index for the *E*-ray is *more* than the refractive index for the *O*-ray *i.e.*, $\mu_e > \mu_o$.

28.4 PRINCIPAL REFRACTIVE INDICES OF A CRYSTAL

It has been seen in a doubly refracting crystal that velocity of *O*-ray is the same in all directions while the velocity of *E*-ray is different in different directions, but both the rays have got the same velocity along the direction of optic axis. Thus in uniaxial crystals, there are two **principal refractive** indices: one corresponding to the velocity of *O*-ray and the other corresponding to the velocity of *E*-ray.

The velocity of *O*-ray is constant in all direction

$$\therefore \qquad \mu_o = \frac{\text{velocity of light in air (vacuum)}}{\text{velocity of O-ray}} \qquad \text{... (28.1)}$$

The **principal refractive** *index for E-ray is defined as the ratio of velocity of light in vacuum to the velocity of E-ray in a direction perpendicular to the optic axis.*

For a *negative uniaxial crystal* the velocity of *E*-ray in a direction perpendicular to the direction of optic axis is **maximum,** hence $\mu_e < \mu_o$.

For a *Positive uniaxial crystal* the velocity of E-ray in a direction perpendicular to the optic axis is *minimum,* hence $\mu_e > \mu_o$.

28.5 NICOL'S PRISM

It is an optical device made from a calcite crystal and is used in many instruments to produce and analyse the plane polarised light.

When light is passed through a doubly refracting crystal it is split up into the *O*-ray and the *E*-ray. Both these rays are plane polarised. One of these rays is cut off by total internal reflection. This prism was designed by William Nicol and is known as Nicol's prism after his name.

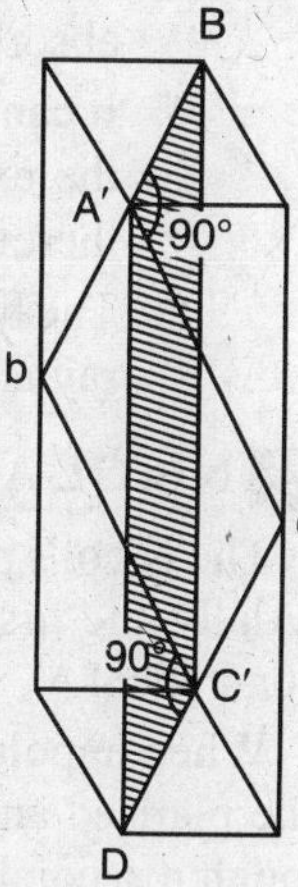

Fig. 28.6

Construction. The Nicol prism is constructed from a calcite crystal whose length is nearly three times its width. The end faces of the crystal are cut down so as to reduce the angles at *B* and *D* from 71° in the principal section to a more acute angle of 68°. The crystal is then cut *along the plane A′ b C′ d perpendicular both to the principal section A′ B C′ D and the end faces such that A′ C′ makes an angle of 90° with the end faces A′ B and C′ D, as shown in Fig. 28.6.* The two cut surfaces are ground, polished optically flat and then cemented together with canada balsam, a transparent cement, so that the crystal is just as transparent as it was previously to its having been sliced. The refractive index of canada balsam μ_b has a value which lies midway between the refractive index of calcite μ_o for the *O*-ray and μ_e for the E-ray. The values of these for the sodium light of mean wavelength $\lambda = 5893$ Å are $\mu_o = 1.658$, $\mu_b = 1.55$ and $\mu_e = 1.486$.

The sides of the prism are blackened to absorb the totally reflected rays.

Action. If a ray of light *SM* (Fig. 28.7) is incident nearly parallel to *BC′* in the plane of the paper on the face *A′B*, it suffers double refraction and gives rise to:

(*i*) the extraordinary beam *ME*, and

(*ii*) the ordinary beam *MN*.

The *E*-ray passes through along *ME* which is plane polarised and has vibrations in the plane of the paper.

The *O*-ray which is also plane polarised suffers total internal reflection at the canada balsam layer for nearly normal incidence.

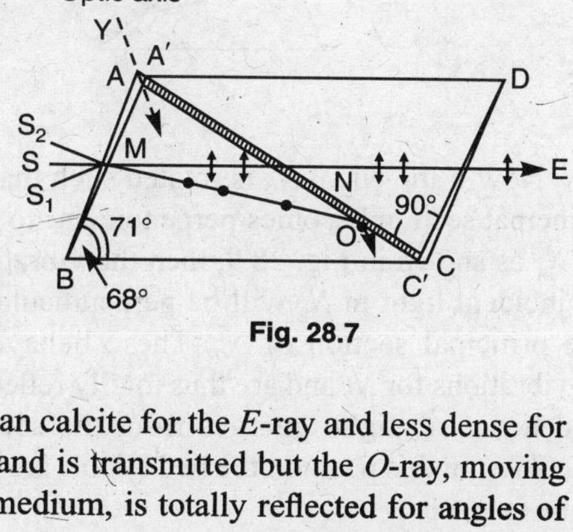

Fig. 28.7

It is because canada balsam is optically more dense than calcite for the *E*-ray and less dense for the *O*-ray. The *E*-ray is refracted through canada balsam and is transmitted but the *O*-ray, moving from denser calcite medium to the rarer Canada balsam medium, is totally reflected for angles of incidence greater than the critical angle. The value of critical angle for the ordinary ray for calcite to canada balsam

$$= \sin^{-1}\left[\frac{1.550}{1.658}\right] = 69.2°$$

If the incident ray makes an angle much smaller than the angle *BMS* with the surface *A'B* the ordinary ray will strike the balsam layer at an angle less than the critical angle and hence will be transmitted.

If the incident ray makes an angle greater than the angle *BMS* the extraordinary ray will become more and more parallel to the otic axis *A'Y* and hence its refractive index will become nearly equal to that of calcite for the ordinary ray. This will then also suffer total internal reflection like the ordinary ray. Hence no light will emerge out of the Nicol's prism. A Nicol's prism, therefore, cannot be sued for highly convergent or divergent beams. The angle between the extreme rays of the incoming beam is limited to about 28°.

Drawbacks of Nicol's prism

1. Large and flawless calcite crystal from which Nicol prism is made is not easily available.
2. The process of grinding and polishing the crystal is very costly.
3. A Nicol prism cannot be used for ultra-violet light as the layer of canada balsam in it absorbs ultra-violet light.
4. It cannot be used for highly convergent and highly divergent beams. The angle between the extreme rays of the incoming beam is limited to about 28°. The field of view is, therefore, restricted.
5. The light coming out of the Nicol prism is not uniformly plane polarised and the emergent ray is always displaced to one side.

28.6 NICOL AS POLARISER AND ANALYSER

The Nicol's prism can be used both as a polariser and an analyser. When two Nicols are mounted coaxially, the first Nicol N_1 which produces plane polarised light is called the **polariser** while the second Nicol N_2 which analyses the incoming light is called the **analyser.**

When unpolarised beam of light is incident on a Nicol prism N_1, the light emerging out of it is plane plarised and has vibrations parallel to its principal section. If now this light is made to pass through a second Nicol N_2 the principal section of which is parallel to the principal section of N_1, the light vibrations in N_2 are parallel to its principal section and hence are completely transmitted as shown in Fig. 28.8. The intensity of the emergent beam is maximum.

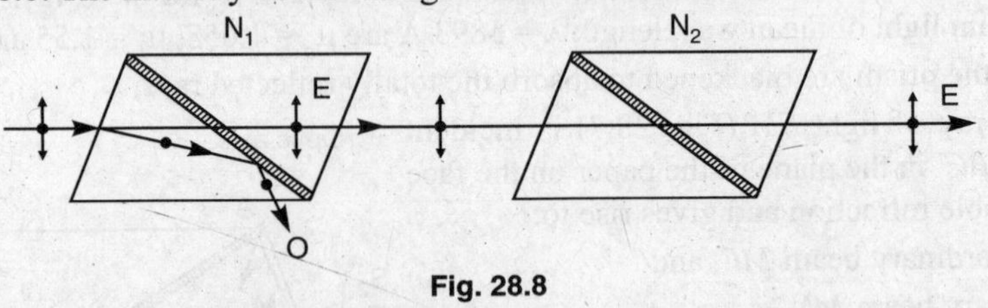

Fig. 28.8

Now if the Nicol N_2 is rotated such that its principal section becomes perpendicular to that of N_1 as shown in Fig. 28.9, then the vibrations of incident light in N_2 will be perpendicular to the principal section of N_2. These behave as *O*-vibrations for N_2 and are thus totally reflected and hence no light emerges from the second Nicol N_2. In this position the two Nicols are said to be **crossed.**

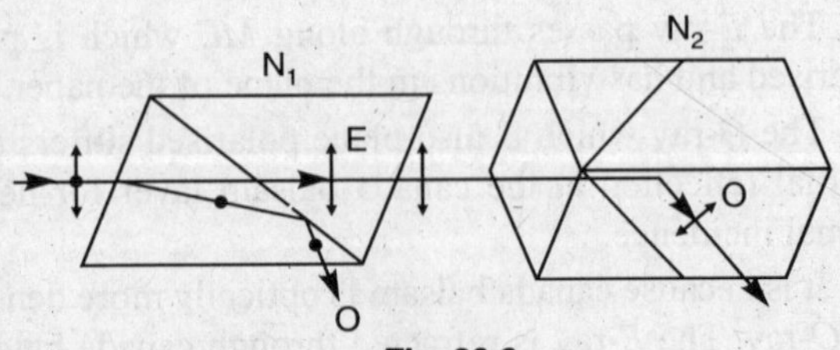

Fig. 28.9

When the Nicol N_2 (analyser) is further rotated the principal sections of the two Nicols are again in parallel position as shown in Fig. 28.10. In this position the E-ray is again transmitted through the analyser Nicol N_2.

After rotation of 270° Nicols N_1 and N_2 are again in crossed position and no light is transmitted by the combination.

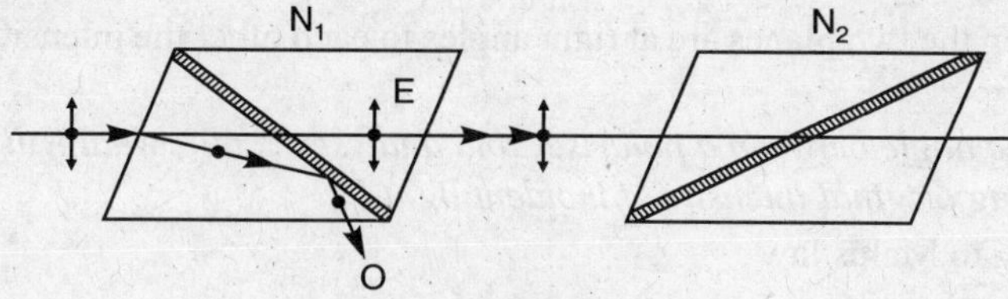

Fig. 28.10

Example 28.1 *If the refractive indices for the ordinary ray in the case of calcite and canada balsam are 1.658 and 1.550 respectively, calculate the maximum possible inclination with the canada balsam surface so that the ordinary ray is still quenched.*

Solution. Here $\mu_o = 1.658, \quad \mu_b = 1.550$

If C is the critical angle when light travels from calcite to balsam layer, then

$$\sin C = \frac{1.550}{1.658} = 0.9348$$

$$\therefore \quad C = 69.2°$$

Thus the O-ray incident on the balsam layer at an angle less than 69.2° will not be quenched. Hence maximum angle with the balsam layer for the O-ray to be quenched.

$$= 90 - 69.2 = 20.8°$$

28.7 MALUS LAW

When light is incident on a polariser, the transmitted light is plane polarised. If this light is allowed to pass through the analyser, the intensity varies with the angle between the planes of the polariser and analyser. Malus studied this problem and stated the variation of intensity in the form of a law known as Malus law.

It states that the intensity of the polarised light transmitted through the analyser varies as the square of the cosine of the angle between the plane of transmission of the analyser and the plane of polariser.

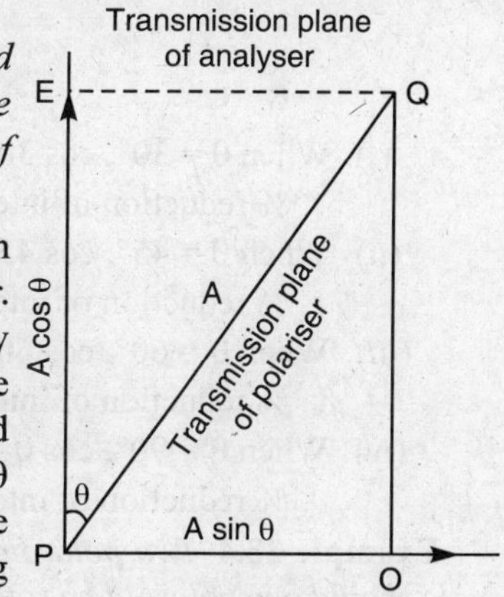

Fig. 28.11

The intensity of light transmitted by the analyser depends upon the inclination of its plane of transmission with that of the polariser. Let θ be the angle between the two transmission planes at any instant and A be the amplitude of the light vector PQ of the plane polarised light emerging from the polariser. This can be resolved into two rectangular components A cos θ along PE and A sin θ along PO perpendicular to the transmission plane of analyser. The component along PE passes through the analyser while that along PO is eliminated by the analyser (fig. 28.11). If I_1 is the intensity of the transmitted component, then

$$I_1 = A^2 \cos^2 \theta \quad \text{... (28.2)}$$

When the plane of analyser makes an angle 0 or 180° with the plane of polariser, then

$$I = A^2 = I_0$$

$$\therefore \quad I_1 = I_0 \cos^2 \theta$$

In this case the two planes are parallel.

If $\theta = \frac{\pi}{2}$ *i.e.*, the planes of polariser and analyser are at right angle to each other, then

$$I_1 = I_0\left(\cos\frac{\pi}{2}\right)^2 = 0$$

This shows that when the two planes are at right angles to each other thc intensity of transmitted light is zero.

Example 28.2 *If the angle between a polariser and analyser is 60°, what will be the intensity of light transmitted having original intensity of incident light I_0.*

Solution. According to Malus law

$$I_1 = I_0 \cos^2\theta$$

where I_0 is the original intensity, I_1 the intensity of transmitted light and θ the angle through which the analyser is rotated.

Now $\theta = 60°$ and $\cos\theta = 0.5$

$\therefore$ Intensity of light transmitted $I_1 = I_0 (\cos 60)^2$

$$= I_0 \times 0.5^2 = 0.25\, I_0$$

Example 28.3 *Two Nicol prisms are so arranged that the amount of light transmitted through them is maximum. What will be the percentage reduction in the intensity of the incident light when the analyser is rotated through (i) 30°, (ii) 45°, (iii) 60° and (iv) 90° ?*

Solution. According to Malus law

$$I_1 = I_0 \cos^2\theta$$

or

$$\frac{I_1}{I_0} = \cos^2\theta$$

where I_0 is the original intensity, I_1 the intensity of transmitted light and θ the angle through which the analyser is rotated.

$\therefore$ Percentage reduction of intensity

$$= \left(\frac{I_0 - I_1}{I_0}\right)\times 100 = \left(1 - \frac{I_1}{I_0}\right)100$$

$$= (1 - \cos^2\theta)100$$

(*i*) When $\theta = 30°$, $\cos 30 = 0.866$

$\therefore$ % reduction of intensity $= [1 - (0.866)^2]\ 100 = 25$

(*ii*) When $\theta = 45°$, $\cos 45 = 0.707$

$\therefore$ % reduction of intensity $= [1 - (0.707)^2]\ 100 = 50$

(*iii*) When $\theta = 60°$, $\cos 60 = 0.50$

$\therefore$ % reduction of intensity $= [1 - (0.5)^2]\ 100 = 75$

(*iv*) When $\theta = 90°$, $\cos\theta = 0$

$\therefore$ % reduction of intensity $= (1 - 0)\ 100 = 100$

Example 28.4 *Two polaroids are adjusted so as to obtain maximum intensity. Through what angle should one polaroid be rotated to reduce the intensity to (i) half (ii) one fourth ?*

(P.U., 2007, 2003, H.P.U., 2003, Pbi. U., 2005, 2001)

Solution. According to Malus' law

$$I_1 = I_0 \cos^2\theta$$

or

$$\frac{I_1}{I_0} = \cos^2\theta$$

Now (*i*) $\frac{I_1}{I_0} = \frac{1}{2}$

$\therefore$ $\cos^2\theta = 0.5$ or $\cos\theta = 0.7071$

Hence $\theta = 45°$

Now (*ii*) $\frac{I_1}{I_0} = 0.25$

$\therefore$ $\cos^2\theta = 0.25$ or $\cos\theta = 0.5$

Hence $\theta = 60°$

Example 28.5 *Unpolarised light is incident on two polarising sheets placed one on top of the other. What must be the angle between the characteristic directions of sheets if the intensity of transmitted light is (i) $\frac{1}{3}$ rd of maximum intensity of the transmitted beam and (ii) $\frac{1}{3}$ rd of intensity of incident beam ? Assume that sheet reduces the intensity of unpolarised light by exactly 50%.*

(P.U., 2000)

Solution. Let intensity of unpolarised light be $= I_0$

Then, intensity of polarised light transmitted by the first sheet $I = \frac{1}{2}I_0$

(*i*) In the first case; intensity of light after transmission through the second sheet $I_1 = \frac{1}{3}I$

Let θ_1 be the angle between the characteristic directions of the two sheets, then

$$I_1 = I\cos^2\theta_1$$

or $\frac{1}{3}I = I\cos^2\theta_1$ or $\cos^2\theta_1 = \frac{1}{3}$ or $\cos\theta_1 = \pm\frac{1}{\sqrt{3}}$

$\therefore$ $\theta_1 = \pm 54.73°$

(*ii*) In the second case; intensity of light after transmission through the second sheet

$$I_2 = \frac{1}{3}I_0 = \frac{2}{3}I$$

Let θ_2 be the angle between the characteristic directions of the two sheets, Then

$$I_2 = I\cos^2\theta_2$$

or $\frac{2}{3}I = I\cos^2\theta_2$ or $\cos^2\theta_2 = \frac{2}{3}$ or $\cos\theta_2 = \pm\sqrt{\frac{2}{3}}$

$\therefore$ $\theta_2 = \pm 35.26°$

28.8 VERIFICATION OF MALUS' LAW

If a beam of light is incident on a polariser, the outgoing beam is plane polarised. The lines on the polariser represent the direction of the electric vector in the transmitted light. When light is incident on it, only the components whose electric vector vibrates parallel to this direction pass through while those which vibrate at right angles to this are absorbed. If the transmitted light is allowed to fall on a photocell connected to an ammeter, the current indicated is proportional to the intensity of incident light.

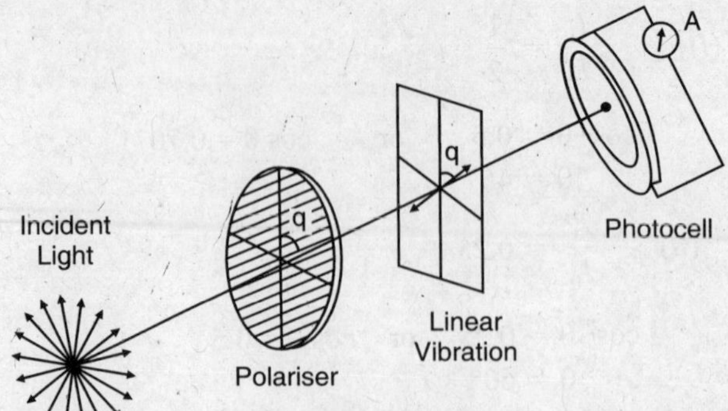

Fig. 28.12

If the incident light is unpolarised, then as the polariser is rotated about the ray as an axis, the reading of the ammeter does not change, as the polariser allows the component parallel to the lines on it only and by symmetry all the components are equal.

On the other hand if the incident light is partially polarised, the ammeter will show different readings as the polariser is rotated about the incident ray as the axis. If we note the maximum and minimum values of the current *i.e.*, I_{max} and I_{min}, then as ammeter reading corresponds to light intensity.

$$\text{Percentage polarisation} = \frac{I_{\max} - I_{\min}}{I_{\max} + I_{\min}} \times 100 \qquad \text{... (28.3)}$$

Verification of Malus' law. Now insert a second polaroid between the first polariser and the photocell, as shown in Fig. 28.13 and place it in such a manner that it allows only those components to pass through whose electric vectors vibrate vertically.

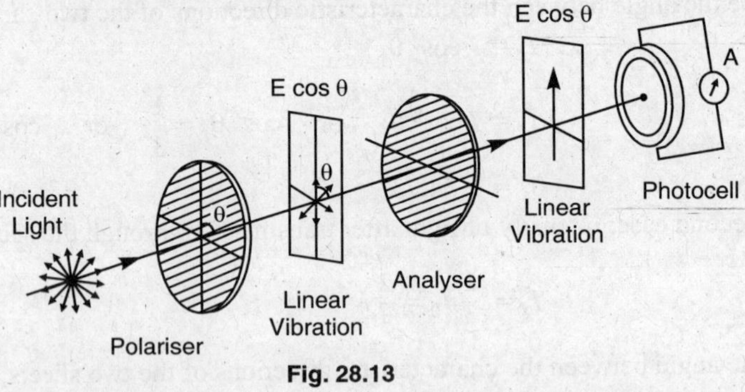

Fig. 28.13

If θ is the transmission direction of the polariser with the vertical, then vertical component E cos θ will be transmitted through the analyser. The photocell will record maximum current when θ = 0 and zero current when θ = 90°. When θ = 90°, the polariser and the analyser are said to be crossed. At intermediate angles, the current is proportional to the square of electric vector. Hence

$$I = I_0 \cos^2 \theta \qquad \text{... (28.4)}$$

where I_0 corresponds to the maximum amount of light transmitted and I corresponds to the light transmitted at an angle θ. This proves *Malus' law.*

28.9 HUYGEN'S THEORY OF DOUBLE REFRACTION

Huygen's explained the phenomenon of double refraction by extending his theory of secondary wavelets in suitable manner. According to his theory—

(*i*) Every point on the crystal surface is disturbed by the incident wavefront and becomes the origin of two wavefronts—one the *ordinary* and other the *extraordinary*—to account for the ordinary and the extraordinary rays.

(*ii*) The ordinary ray always obeys both the laws of refraction. The refractive index of crystal μ_o is given by

$$\mu_o = \frac{\sin i}{\sin r_o} = \frac{\textit{velocity in air (vacuum)}}{\textit{velocity of O-ray}} = \frac{v}{v_0} \quad \text{... (28.5)}$$

It is thus independent of the direction of its propagation within the crystal. This is only possible if v_o the velocity of *O*-wave is the same in all directions.

(*iii*) The *E*-ray does not obey ordinary laws of refraction. The ratio $\frac{\sin i}{\sin r}$ for the *E*-ray has a value which varies with the direction of its propagation relative to the optic axis. The velocity of *E*-ray is the same as that of *O*-ray along the optic axis and differs most from this value in a direction perpendicular to the optic axis. The *E*-wavefront is, therefore, not spherical but is an ellipsoid in nature,

(*iv*) the properties of uniaxial crystals are perfectly symmetrical about the optic axis. Hence *E*-ray is propagated with equal velocity in directions which are equally inclined to optic axis. Thus ellipsoid of revolution is symmetrical about the optic axis through the point of origin of the wavelets.

(*v*) As velocity of *E*-ray is equal to the velocity of *O*-ray along the optic axis, uniaxial crystals do not show double refraction along the optic axis. The two wavelets touch each other at points where the optic axis through the point of origin of the wavelets intersects them.

Thus according to Huygen's each point on a doubly refracting surface becomes the source of two wavefronts—O-wavefront and the E-wavefront. The *O*-wavefront is spherical and *E*-wavefront ellipsoidal of revolution about the optic axis. The two wave surfaces touch along the extremities of optic axis through the point of origin. Both the wave surfaces are plane polarised.

28.10 ELECTRO-MAGNETIC THEORY OF DOUBLE REFRACTION

The double refracting crystals belong to the group of non-conducting materials. They are known as *dielectric* materials. If a dielectric crystal is placed between the plates of a parallel plate capacitor and a voltage is applied, the crystal becomes polarized. This polarization is *dielectric polarization* and is different from the optical polarization. The voltage produces an electric field E and under the action of the electric field electron clouds in the atoms of the crystal deform.

As a result of this, the centres of action of negative charge and positive charge get separated by a small distance, known as *electrical dipole.* Such large number of dipoles are produced throughout the crystal. Negative charges of the dipoles lie on the side of the positively charged plate of capacitor and positive charges lie on the side of the negatively charged plate. Thus, electric charges of opposite nature are induced on the opposite faces of the crystal in response to the applied electric field as shown in Fig. 28.14. This is known as **dielectric polarization.**

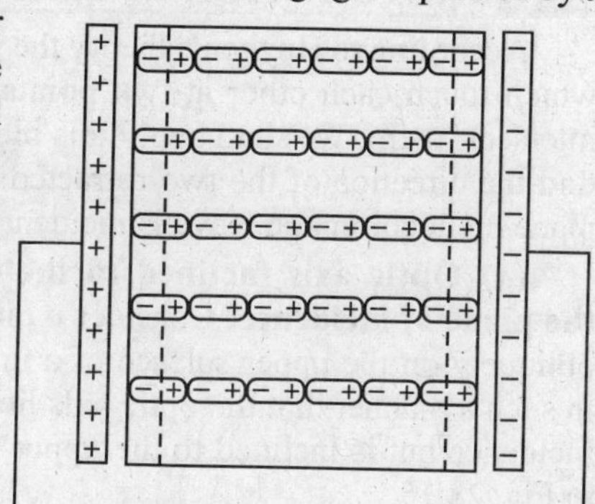

Polarization of a dielectric

Fig. 28.14

The electric field *E*, polarization *P* and dis-placement *D* in the crystal are related by

$$\vec{D} = \varepsilon_o \vec{E} + \vec{P} \quad \text{... (i)}$$

where

$$\vec{D} = \varepsilon_o \varepsilon_r \vec{E} \quad \text{... (ii)}$$

and

$$\vec{P} = \chi \varepsilon_0 \vec{E} \quad \text{... (iii)}$$

ε_r and χ are known as *relative permittivity* and *electric susceptibility* of the crystal. The relative permittivity is related to the refractive index of the material through the relation

$$\mu = \sqrt{\varepsilon_r} = \sqrt{\varepsilon / \varepsilon_o} \quad \text{... (iv)}$$

We know that refractive index is related to the velocity of light in the medium.

$$\mu = \frac{c}{v} \quad \text{... (v)}$$

It is not necessary for dielectric polarization that the crystal be placed in between the plates of a capacitor. The electric field of a light wave propagating in the crystal also can cause dielectric polarization. According to the above relations the dielectric polarization caused by the light wave is related to the velocity (direction of propagation) of light wave with in the crystal.

In case of isotropic crystals, the vectors $\vec{D}, \vec{E}$ and $\vec{P}$ are parallel and ε_r, μ and χ are scalars. In anisotropic crystals the induced polarization is in a direction different from that of the field. As a result, ε_r, μ and χ are tensor quantities. The components of **polarization** are given by

$$\begin{aligned} P_x &= \varepsilon_o (\chi_{11} E_x + \chi_{12} E_y + \chi_{13} E_z) \\ P_y &= \varepsilon_o (\chi_{21} E_x + \chi_{22} E_y + \chi_{23} E_z) \\ P_z &= \varepsilon_o (\chi_{31} E_x + \chi_{32} E_y + \chi_{33} E_z) \end{aligned} \quad \text{... (vi)}$$

The above equations can be simplified by choosing the co-ordinate axes in such a way that the off-diagonal elements vanish. Then

$$\begin{aligned} P_x &= \varepsilon_o \chi_{11} E_x \\ P_y &= \varepsilon_o \chi_{22} E_y \\ P_z &= \varepsilon_o \chi_{33} E_z \end{aligned} \quad \text{... (vii)}$$

Similarly, we can express

$$\begin{aligned} D_x &= \varepsilon_{11} E_x, \\ D_y &= \varepsilon_{22} E_y \\ D_z &= \varepsilon_{33} E_z \end{aligned} \quad \text{... (viii)}$$

These direction are called the **principal axes** of the crystal and the corresponding diagonal terms ε_{11}, ε_{22} and ε_{33} the *principal permittivities.* The variation in the permittivity and the corresponding variation in refractive index and hence in wave velocity is the origin of double refraction. μ_x, μ_y and μ_z are the three *principal refractive indices* corresponding to the three principal permittivities. **The subscripts *x, y* and *z* relate to the direction of polarization of the light** wave **and not to their direction of propagation.** The principal permitivities and refractive indices are related as

$$\mu_x = \sqrt{\varepsilon_{11} / \varepsilon_o}, \mu_y = \sqrt{\varepsilon_{22} / \varepsilon_o} \text{ and } \mu_z = \sqrt{\varepsilon_{33} / \varepsilon_o} \quad \text{... (ix)}$$

28.11 PROPOGATION OF PLANE WAVES IN UNIAXIAL CRYSTALS

According to Huygen's theory the wave surfaces in uniaxial crystals are a sphere and a spheroid which touch each other at two points where the optic axis through the points of origin of light intersects them. We shall apply this method to draw the refracted wavefronts in isotropic media, to find the direction of the two refracted rays and refracted wavefronts within calcite crystal when a plane monochromatic wave is incident upon it obliquely from air.

(*a*) Optic axis inclined to the upper face and lying in the plane of incidence. Consider a plane wavefront *AB* incident obliquely on the upper surface of a uniaxial negative crystal cut in such a manner that the optic axis lies along *AX*, in the plane of incidence but is inclined to the upper face at an angle, as shown in Fig. 28.15.

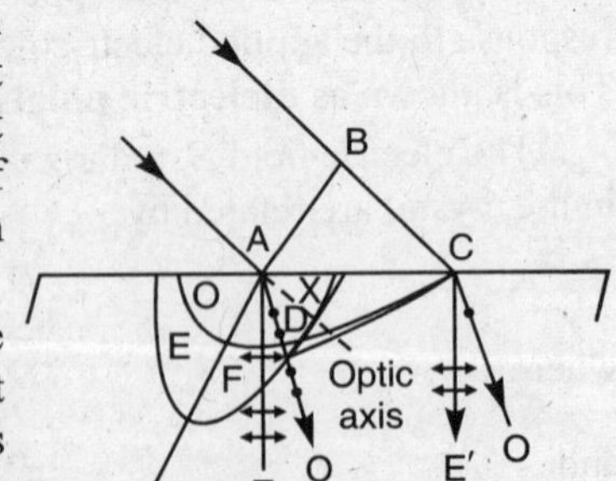

Fig. 28.15

(*i*) *To find the position of the ordinary refracted wavefront.* The point *A* of the plane wavefront *AB* where it strikes the surface first becomes the centre of the ordinary spherical secondary wavelets *O* and the extraodinary secondary wavelets *E*. By the time the disturbance from *B* reaches *C* the ordinary wave surface acquires a radius $\frac{BC}{\mu_o}$ where μ_o is the refractive index for the *O*-ray. To find the position of the *O*-wave surface draw with *A* as centre and radius $\frac{BC}{\mu_o}$, a circle cutting the optic axis at *X*. From *C* draw *CD* tangent plane touching the circle then *CD* represent the position of the ordinary refracted wavefront and *AD* the direction of the *O*-ray.

(*ii*) *To find the position of the extraordinary wavefront.* In a negative crystal the refractive index for the *E*-ray is always less than the refractive index for the *O*-ray. *The minimum value of the refractive index for the E-ray is called* **the principal refractive index.** The maximum value of μ_e is equal to the refractive index μ_o when light is incident along the optic axis. The extraordinary wavefront is thus an ellipse touching the ordinary wavefront at *X*, *AX* is thus the semi-minor axis of the ellipse. The semi-major axis is given by $\frac{BC}{\mu_e}$ where μ_e is the principal refractive index for the *E*-ray.

To find the extraordinary refracted wavefront, draw an ellipse with the given major and minor axes touching the circle at *X*. A tangent plane *CF* to the ellipse gives the position of the refracted E-wavefront and the line *AF* represents the direction of E-ray. The direction of vibration is also shown. It is thus seen that the *O*-ray and the *E*-ray travel along different directions with different velocities.

(*b*) **Optic axis parallel to the upper face and lying in the plane of incidence.** Two different cases arise, depending upon the angle of incidence.

(*i*) *Oblique incidence.* As the optic axis is parallel to the face of the crystal, the positions of the *O* and the *E* wavefronts can be drawn by making a construction similar to the one given in (*a*).

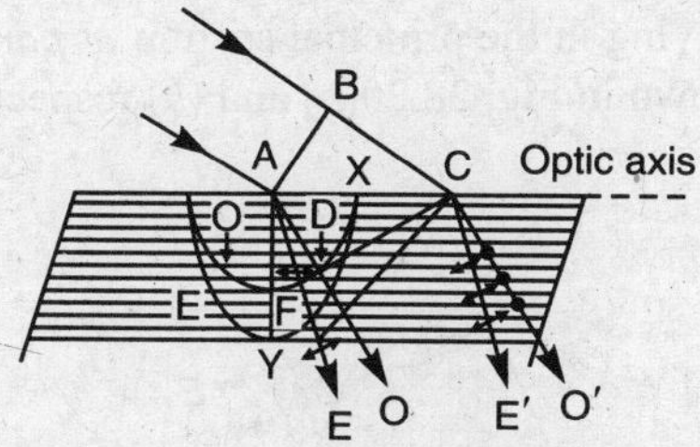

Fig. 28.16(a)

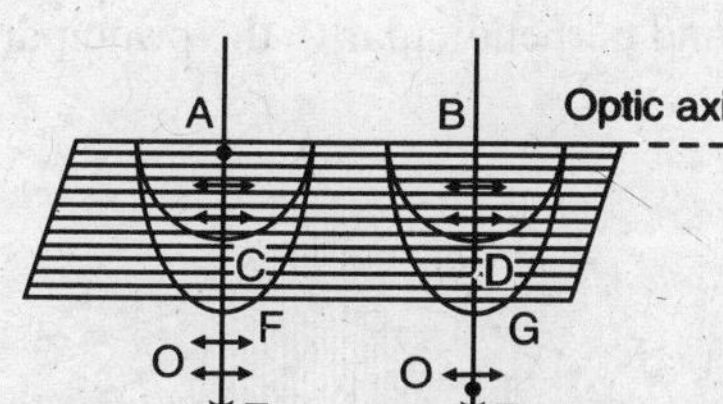

Fig. 28.16(b)

The sphere and the ellipse will touch at *X* along *AX* (Fig. 28.16(*a*)). It is also clear that the *O*-ray and the *E*-ray travel with different velocities in different directions.

(*ii*) *Normal incidence.* When the incidence is normal, the construction is as shown in Fig. 28.16(*b*). The *O* and *E* wavefronts are parallel to each other and the *O*-ray and the *E*-ray travel in the same direction but with different velocities. This case is thus of a special importance, because the path-difference created between the *O*-ray and the *E*-ray is made use of in the construction of the quarter and the half wave plates.

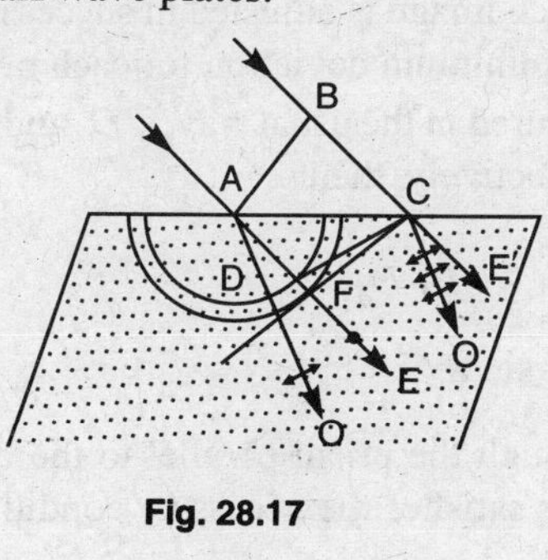

Fig. 28.17

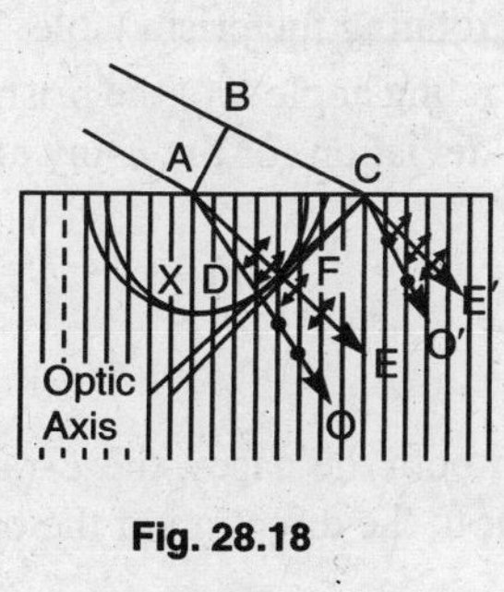

Fig. 28.18

(*c*) **Optic axis parallel to the upper face but perpendicular to the plane of incidence.** As the optic axis is perpendicular to the plane of incidence, the sections of the *O* and the *E* wavefronts in the plane of incidence are both circles. The construction is as shown in Fig. 28.17. It is seen that the *O*-ray and *E*-ray travel in different directions with different velocities, except when the incidence is normal. In that case both the rays travel in the same direction.

It is also seen that the *E*-ray has the same velocity in all directions in the plane of incidence and hence obeys both the laws of refraction.

(*d*) **Optic axis perpendicular to the upper face.** As the optic axis is perpendicular to the face of the crystal, the *O* and the *E* wavefronts can be drawn by making a similar construction as explained in (*a*). The two wavefronts touch each other at *X*. The *O*-ray and the *E*-ray travel with different velocities in different directions (Fig. 28.18).

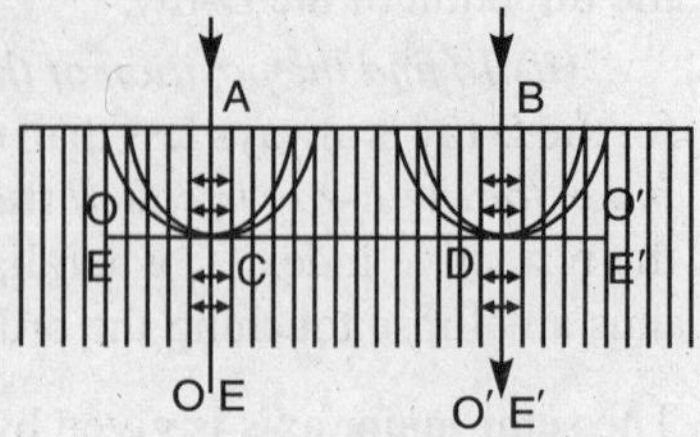

Fig. 28.19

When *the incidence is normal* there is no double refraction as the *O* and the *E*-rays travel with the same velocity in the same direction. This is shown in Fig. 28.19. The case of normal incidence with optic axis parallel to the face of the crystal as given in (*b*) (*ii*) must be carefully distinguished from this case. Although in both the cases the *O* and the *E*-rays travel in the same direction, the velocities in the former case were *different* whereas in this case these are the same.

28.12 DETERMINATION OF μ_e AND μ_O

The ordinary and extraordinary refractive indices can be found by using prisms of uniaxial material cut either with optic axis *perpendicular to the* edge and lying in the principal section or *parallel to the edge* and perpendicular to the principal section as shown in Fig. 28.20(*a*) and (*b*) respectively.

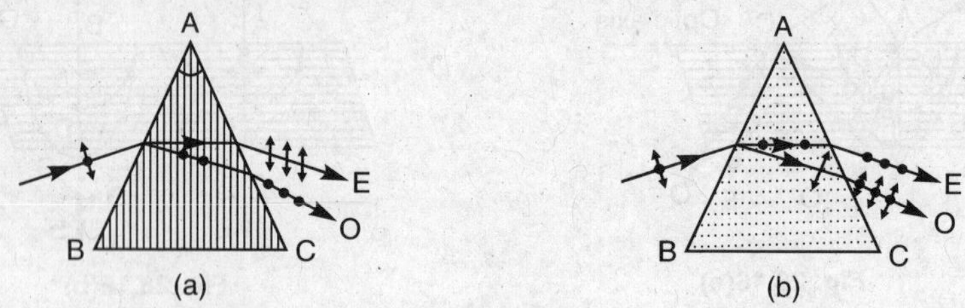

Fig. 28.20

For a given wavelength, the direction of *O* and *E* rays traversing the prism at minimum deviation must be perpendicular to the optic axis so that the fundamental condition for the determination of μ_e is satisfied. The *O*-ray and *E*-ray are refracted at minimum deviation like ordinary light with refractive index μ_o and μ_e respectively.

The prism is mounted on the prism table of the spectrometer in such a way that the two images of the slit are seen by refraction through the prism. Each image is adjusted in succession for minimum deviation by rotating the prism table. The angle of minimum deviation for each ray is measured.

The refracting angle *A* of the prism is also measured in the usual way. If D_e and D_o are the angles of minimum deviation for the *E*-ray and *O*-ray respectively, then

$$\mu_e = \frac{\sin\frac{1}{2}(A+D_e)}{\sin A/2} \text{ and } \mu_o = \frac{\sin\frac{1}{2}(A+D_o)}{\sin A/2} \qquad \text{... (28.6)}$$

Note: In both the cases, the *E*-ray travels through the prism parallel to the base and hence is perpendicular to the direction of the optic axis. This satisfies the necessary condition.

28.13 POLARIZATION BY SCATTERING

We can explain polarization of light by scattering. If a strong beam of light is made to pass through a pure liquid free from all suspended particles and observations are made in a dark room, a small amount of bluish light scattered laterally is seen. This scattering is due to the individual molecules of the liquid. The scattering increased if the liquid molecules are replaced by gas molecules.

Lord Rayleigh, from the study of scattering, found that the intensity of scattered light depends upon the wavelength and is proportional to $\frac{1}{\lambda^4}$ when the size of the suspended particles is small as compared to the wavelength of light. Thus scattering of violet light is nearly 16 times as large as the scattering of red light, because the wavelength of violet light is half that the red light.

If light is incident on a medium which contains floating particles such as dust, moisture or molecules of air, the transmitted light is reddish in colour, whereas the scattered light is bluish in colour.

The following experiment beautifully illustrates the fact that smaller wavelengths are scattered more readily than the larger wavelengths.

If a beam of white light is passed through a weak solution of sodium thiosulphate to which a few drops of sulphuric acid are added, then on viewing sideways, *i.e.*, normal to the beam of light, it will be seen that the colour of light is distinctly blue whereas the transmitted light is yellowish in tint, which gradually becomes orange and then finally red. It is because addition of sulphuric acid precipitates sulphur particles which grow in size so that light of longer wavelengths are also added to scattered light and a deficit of colours in transmitted light occurs.

If blue light in the transverse direction is examined by a Nicol prism, it is found to be partially plane polarised. Maximum polarisation is observed in a direction perpendicular to the incident light.

28.14 BLUE COLOUR OF SKY

The blue colour of sky is due to the scattering of light by small particles of the atmosphere. Lord Rayleigh has shown that the percentage of light scattered is inversely proportional to the 4th power of wavelength when the size of the scattering particles is small as compared to the wavelength of light waves. The scattering of blue light is almost 16 times as compared to the red light. Hence blue contained in the Sun light is scattered to a much greater extent and is more intense than the red.

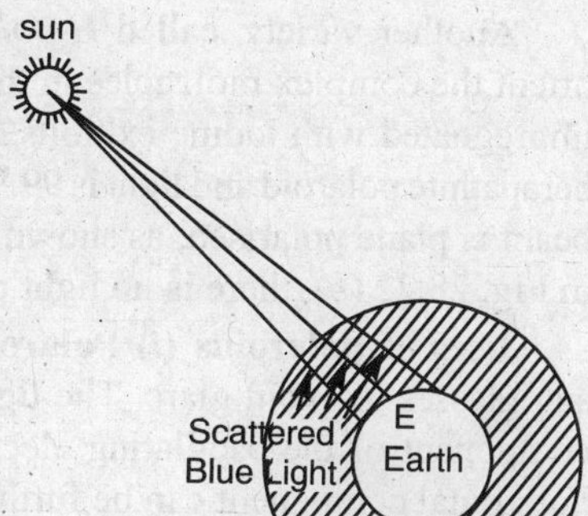

Fig. 28.21

When white light from Sun falls on the particles of the atmosphere, the blue colour is scattered and if we look over head the sky far away from the Sun appears blue. In the afternoon when the Sun is over head, the observer at E will get direct light from the Sun, the sky appears white. Blue colour is more brilliant when the atmosphere is free from dust particles and water molecules and scattering is only due to the gas molecules which are to a considerable extent responsible for the blue colour of the sky.

If the earth has no atmosphere, no light from sky will reach earth. The sky would appear perfectly black, even during day, as no colour is scattered.

Red colour at sunset and sunrise. At sunset and sunrise, the neighbouring portions and the Sun appears red because the rays coming from Sun to the observer at E have to traverse a greater distance approximately horizontally through the earth's atmosphere. A large proportion of blue is scattered to sides and lost by fine dust and smoke particles near the earth's surface. When blue colour is removed from white, the remaining colour is red. When such light is incident on the clouds near sunset or sunrise, the reflected light appears red. This red light if examined by a Nicol is only partially plane polarised.

28.15 POLAROIDS

Certain doubly refracting crystals have the property of absorbing the *O*-ray and *E*-ray unequally. The most outstanding substance is *tourmaline.* If a beam of unpolarised light is incident on a tourmaline crystal, it is broken into *O*-ray having vibrations in the plane of incidence and *E*-ray having vibrations perpendicular to the plane of incidence. On passing through the crystal, the *O*-ray is absorbed and *E*-ray whose vibrations are parallel to the optic axis passes through the crystal. The *O*-ray is completely absorbed if the crystal is cut of a proper thickness. This phenomenon is known as **dichroism.**

The phenomenon of selective absorption is made use of in the construction of polaroids. These are large sized polarising films of any cross-section which are used to produce and detect plane polarised light like Nicol prism. The use of tourmaline as polariser is limited, as the plane polarised light is coloured due to unequal absorption of *E*-rays of various wavelengths.

W.B. Herapath in 1852 discovered a synthetic crystalline material *iodosulphate of quinine* called herapathite which exhibits strong dichroism and transmits linearly polarised light of all colours and wavelengths. But these crystals are not stable and are affected by slight strain and so are quite useless as such.

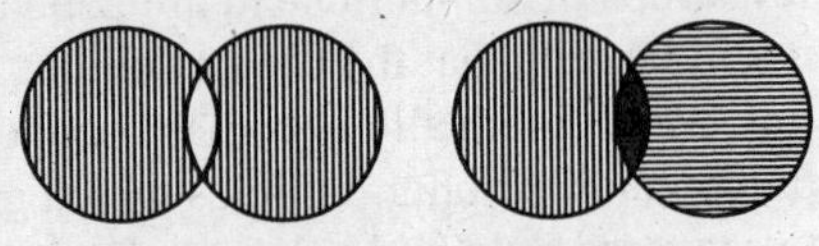

Fig. 28.22

In 1934 E.H. Lamb developed a process by which herapathite crystals are arranged side by side with their optic axes all parallel, so that they function as one single crystal of large dimensions. This was achieved by preparing a paste of crystals in nitro-cellulose and squeezing it out through a fine slit, through which only those crystals pass whose axes are parallel to the length of the slit. The sheet thus produced has millions of tiny crystals with their optic axes parallel to one another. The sheet thus prepared is mounted between two thin glass sheets, forming what is called a **polaroid.**

Another variety, called *H-polaroid,* is formed by stretching polyvinyl alcohol films so as to orient the complex molecules in the direction of stress, which makes it doubly refracting and when impregnated with iodine exhibits dichroism. This is colourless and transmits 33% more light than herapathite polaroid and light is 99.99% polarised. When the two polaroids are uncrossed, the emergent beam is plane polarised, as shown in Fig. 28.22 (*a*). When the two polaroids are crossed, as shown in Fig. 28.22 (*b*), there is no light passing through.

Uses of polaroids (*i*) Polaroids in sunglasses. Polaroids are widely used in spectacles and sun glasses to avoid glare. The light reflected from the horizontal surfaces has a higher horizontal component of the oscillating electric vector than its vertical component. By using polaroids, the horizontal component can be further cut down and thus glare reduced considerably.

(*ii*) These are fitted in motor car headlights to eliminate headlight glare.

(*iii*) These are used in windows of railways, aeroplane etc. in which there is a fixed outer and rotatable inner polaroid to control the amount of light entering it from outside.

(*iv*) They are also used to improve the colour contrast in old oil paintings.

(*vii*) They are used as polariser and analyser in laboratory.

28.16 WOLLASTON PRISM

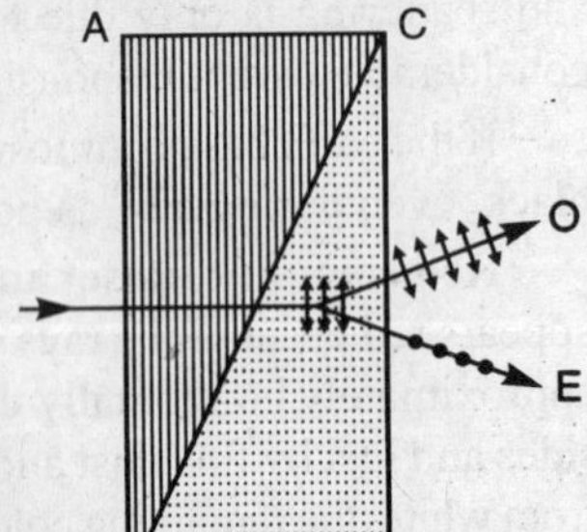

Fig. 28.23

It consists of two right-angled prisms of calcite *ABC* and *BCD* cemented together to form a rectangular block. The prism *ABC* is cut so that the optic axis of the crystal is parallel to the surface AB at which the light enters. The prism *BCD* is cut in such a manner that the optic axis is still parallel to *AB* and *CD* but turned through 90°, as shown in Fig. 28.23. A ray of light incident normally on the face *AB* travels perpendicular to the optic axis and is thus divided into the O and the E-rays vibrating parallel and perpendicular to the plane of the paper. These rays travel with different speeds in the same direction.

On crossing the section BC these rays are parallel to the principal plane of the prism BCD. The refractive index, therefore, increases for one ray and decreases by the same amount for the other ray so that both are deflected in opposite directions from the normal to AB. The divergence of the rays is further increased on refraction at the surface CD and thus they form two widely separated beams polarised in perpendicular planes.

28.17 PLANE, ELLIPTICALLY AND CIRCULARLY POLARIZED LIGHT

Plane polarised light. When light travels along a certain direction, the vibrations take place in a direction at right angles to the direction of propagation. If the vibrations of the ether particles are linear and take place parallel to a plane through the axis of the beam or the direction of propagation, light is said to be *plane polarised.*

Elliptically polarised light. When the vibrations of the ether particles are elliptical, having a constant period, and take place in a plane perpendicular to the direction of propagation (transverse plane), light is said to be *elliptically polarised.* In this case *the amplitude of the vibrations changes in magnitude as well as in direction.*

Circularly polarised light. When the vibrations of the ether particles are circular having constant period and take place in the *transverse plane,* light is said to be *circularly polarised.* In this case *the amplitude of the vibrations remains constant but the direction changes only.*

Theory. Suppose plane polarised monochromatic light is incident normally on a *calcite crystal cut with its faces parallel to the optic axis* as shown in Fig. 28.24 (*a*).

The light on entering the crystal at P normal to the surface but with vibrations making an angle θ with the direction of optic axis, will break up into the two E and O components. The E-wave will have vibrations along the direction of optic axis and O-wave in a direction perpendicular to the optic axis.

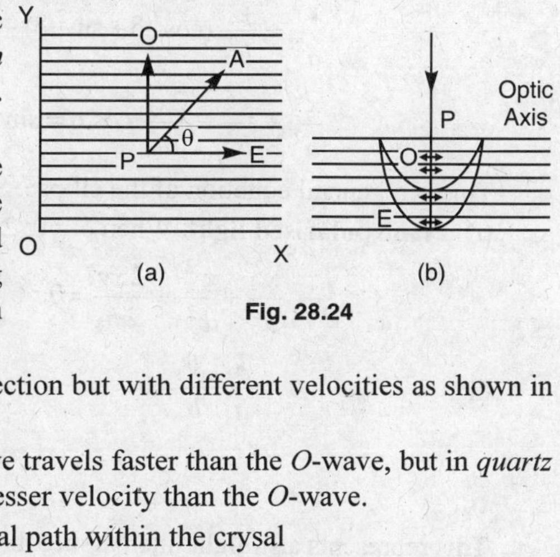

Fig. 28.24

These two waves will travel along the same direction but with different velocities as shown in Fig. 28.24(*b*).

In *calcite and other nagative crystals* the E-wave travels faster than the O-wave, but in *quartz and other positive crystals* the E-wave travels with lesser velocity than the O-wave.

If d is the thickness of the crystal, then the optical path within the crysal

$$\text{for the } O\text{-ray} = \mu_o d \text{ and}$$

$$\text{for the } E\text{-ray} = \mu_e d$$

where μ_o and μ_e is the refractive index of the crystal for the O-ray and E-ray respectively. When the two waves emerge out of the crystal, a path difference is introduced between them, given by

$$\text{Path difference} = d(\mu_o - \mu_e)$$

$$\therefore \quad \text{Phase difference } \delta = \frac{2\pi}{\lambda} d\,(\mu_o - \mu_e) \quad \text{... (28.7)}$$

Suppose A is the amplitude of the incident vibrations, then the amplitude a of the E-vibrations along PE is given by

$$\alpha = \text{A} \cos\theta$$

and amplitude b of the O-vibrations along PO is given by

$$b = \text{A} \sin\theta$$

If n is the frequency of vibrations of the ether particles, then the displacement of the O-vibrations at any time t along the Y-axis is given by

$$y = b \sin 2\pi nt \qquad \text{... (i)}$$

Similarly the displacement x of the E-vibrations at the same time t along the X-axis is given by

$$x = a \sin (2\pi nt + \delta)$$

or
$$\frac{x}{a} = \sin 2\pi nt \cos \delta + \cos 2\pi nt \sin \delta \qquad \text{... (ii)}$$

From (*i*), we have

$$\sin 2\pi nt = \frac{y}{b}$$

$\therefore$
$$\cos 2\pi nt = \sqrt{1-\frac{y^2}{b^2}}$$

Substituting these values in (*ii*), we have

$$\frac{x}{a} = \frac{y}{b}\cos\delta + \sqrt{1-\frac{y^2}{b^2}}\sin\delta$$

or
$$\left[\frac{x}{a} - \frac{y}{b}\cos\delta\right]^2 = \sin^2\delta - \frac{y^2}{b^2}\sin^2\delta$$

or
$$\frac{x^2}{a^2} + \frac{y^2}{b^2}(\cos^2\delta + \sin^2\delta) - \frac{2xy}{ab}\cos\delta = \sin^2\delta \qquad \text{... (iii)}$$

or
$$\frac{x^2}{a^2} + \frac{y^2}{b^2} - \frac{2xy}{ab}\cos\delta = \sin^2\delta \qquad \text{... (28.8)}$$

This is a general equation of the ellipse. Some special cases are discussed below :

(*i*) **Plane polarised light.** When $\delta = 0, 2\pi....$, $\cos\delta = 1$ and $\sin\delta = 0$.

$\therefore$
$$\frac{x^2}{a^2} + \frac{y^2}{b^2} - \frac{2xy}{ab} = 0$$

or
$$\frac{x}{a} - \frac{y}{b} = 0$$

$\therefore$
$$y = \frac{b}{a}x \qquad \text{... (28.9)}$$

This represents a straight line passing through the centre and having a positive slope equal to b/a. It indicates that the resultant light is plane polarised in the same plane as the original linear vibration as shown in Fig. 28.25 (*i*).

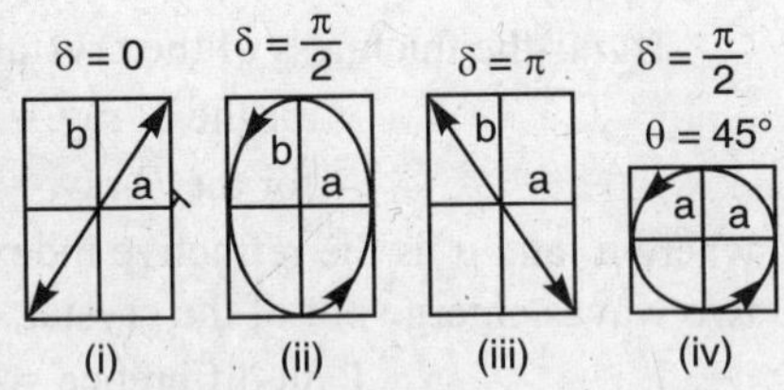

Fig. 28.25

(*ii*) when $\delta = \pi, 3\pi$ ..... $\cos\delta = -1$ and $\sin\delta = 0$.

$\therefore$
$$\frac{x}{a} + \frac{y}{b} = 0$$

or
$$y = -\frac{b}{a}x$$

This is the equation of a straight line with an equal negative slope which shows that light is plane polarised. The direction of the emergent light vibrations makes an angle equal to $2\tan^{-1}\frac{b}{a} = 2\theta$ with the incident light vibration, as shown in Fig. 28.25 (*iii*).

(*iii*) **Elliptically polarised light.** When $\delta = \frac{\pi}{2}, \frac{3\pi}{2}$ ....

$$\cos\delta = 0 \text{ and } \sin\delta = 1$$

$$\therefore \quad \frac{x^2}{a^2} + \frac{y^2}{b^2} = 1 \qquad \text{... (28.10)}$$

This is the equation of an ellipse whose major axis and minor axis coincide with the co-ordinate axes. The outgoing vibrations are elliptically polarised, as shown in Fig. 28.25 (*ii*). For the values of δ other than an integral multiple of π the ellipse is rotated with respect to X and Y-axes.

(*iv*) **Circularly polarised light.** In a special case when

$$\delta = \pi/2 \text{ or } 3\pi/2 \quad \text{and} \quad \theta = 45°$$

i.e., the incident vibrations make an angle of 45° with the optic axis, we have

$$a = A\cos 45° \quad \text{and} \quad b = A\sin 45°$$

$$\therefore \quad a = b$$

The equation (*iii*), therefore, becomes

$$x^2 + y^2 = a^2 \qquad \text{... (28.11)}$$

This is the equation of a circle. Thus the emergent light is circularly polarised, as represented in Fig. 28.25 (*iv*).

Example 28.6 *Plane polarised light is incident on a plate of quartz cut with faces parallel to optic axis. If the phase difference between the two rays is 60°, calculate the thickness of the plate. Given $\mu_E = 1.5583$; $\mu_o = 1.5442$ and $\lambda = 5000$ Å.* **(*P.U. 2005*)**

Solution. Here $\mu_o = 1.5442$, $\mu_e = 1.5583$, $\lambda = 5000 \times 10^{-10}$ *m*

Now $$\delta = \frac{2\pi}{\lambda} d(\mu_e - \mu_o) = \frac{\pi}{3} [60°]$$

$$\therefore \quad d = \frac{\lambda}{6} \frac{1}{(\mu_e - \mu_o)} = \frac{5000 \times 10^{-10}}{6 \times (1.5583 - 1.5442)} = 5.9 \times 10^{-6} \text{ m}$$

28.18 PHASE RETARDATION PLATES

It is plate cut from a doubly refracting uniaxial crystal with its optic axis parallel to the refracting faces. It can introduce a given phase difference between O and E rays which travel through it.

The phase difference $$\delta = \frac{2\pi}{\lambda} d(\mu_o - \mu_e) \qquad \text{... (28.12)}$$

where λ is the wavelength of light, d the thickness of the plate, μ_o and μ_e the refractive index of the material for O-ray and E-ray respectively.

This relation (28.12) shows that for a given material and wavelength the phase difference is proportional to the thickness of the plate.

Two main types of phase retardation plates are:

(*i*) Quarter wave plate and

(*ii*) Half wave plate

28.18.1 Quarter-Wave Plate

It is a crystal plate of a doubly refracting material of a uniaxial crystal cut with its optic axis parallel to the refracting faces. It can produce a phase difference of $\frac{\pi}{2}$ or a path difference of $\frac{\lambda}{4}$

between the *O*-ray and the E-ray when monochromatic plane polarised light of wavelength λ is incident normal to the surface. As light is incident normal to the optic axis, the *E* and *O* components travel in the same direction but with different velocities.

In calcite (a negative crystal). The velocity of *E*-ray is greater than the velocity of *O*-ray, hence $\mu_o > \mu_e$. If *d* is the thickness of the crystal plate, then the path difference between the two components on coming out.

$$= d(\mu_o - \mu_e)$$

Hence to produce a path difference of $\lambda/4$, we have

$$d(\mu_o - \mu_e) = \lambda/4$$

$$\therefore \qquad d = \frac{\lambda}{4(\mu_o - \mu_e)} \qquad \text{... (28.13)}$$

For **quartz** (a positive crystal) $\mu_e > \mu_o$

$$\therefore \quad \text{Thickness} \quad d = \frac{\lambda}{4(\mu_e - \mu_o)} \qquad \text{... (28.14)}$$

As the thickness depends upon the wavelength, it is useful only for the wavelength for which it is constructed.

Why so called? It is called a quarter wave plate because it produces a path difference of $\frac{\lambda}{4}$ – one quarter of wavelength—between *O*-ray and *E*-ray.

A quarter wave plate is used for the production and detection of circularly polarised light. In conjunction with a Nicol prism, it is used for analysing any kind of light.

Example 28.7 *Calculate the thickness of quarter wave plate for light of wavelength 5893Å, given refractive indices for ordinary ray and extraordinary ray are 1.544 and 1.553 respectively.*

(*Nagpur U. 2011, 2002; Pbi. U. 2007; Meerut. U. 2005*)

Solution. Here $\mu_o = 1.544, \ \mu_e = 1.553$

$$\lambda = 5893\text{Å} = 5893 \times 10^{-10} \text{ m}$$

As $\mu_e > \mu_o$, the quarter wave plate is made from a *uniaxial positive* crystal, the thickness *d* of which is given by

$$d = \frac{\lambda}{4(\mu_e - \mu_o)} = \frac{5893 \times 10^{-10}}{4(1.553 - 1.544)}$$

$$= \frac{5893 \times 10^{-10}}{4 \times 0.009} = 0.1637 \times 10^{-4} \text{ m}$$

Example 28.8 *Calculate the thickness of a quarter wave plate of wavelength 5890 Å.*

(i) Given $\mu_o = 1.55$ and $\mu_e = 1.50$ **(*K.U., 2000*)**

(ii) Given $\mu_o = 1.55$ and $\mu_e = 1.57$ **(*M.D.U., 2000*)**

Solution. (*i*) $d = \frac{\lambda}{4(\mu_o - \mu_e)} = \frac{5890 \times 10^{-10}}{4(1.55 - 1.50)} = 0.2945 \times 10^{-5}$ m

(*ii*) $d = \frac{\lambda}{4(\mu_e - \mu_o)} = \frac{5890 \times 10^{-10}}{4(1.57 - 1.55)} = 0.736 \times 10^{-5}$ m

Example 28.9 *If the difference between ordinary and extraordinary refractive index is 0.001, find the thickness of the quarter wave plate for* $\lambda = 5000$ *Å.* **(*H.P.U. 2000*)**

Solution. $d = \frac{\lambda}{4(\mu_o - \mu_e)} = \frac{5000 \times 10^{-10}}{4 \times 0.001} = 1.25 \times 10^{-4}$ m

Example 28.10 *Calculate the thickness of a quarter wave plate for light of wavelength 5890 Å. The refractive indices for ordinary and extraordinary components are respectively 1.658 and 1.486.* **(*Pbi. U. 2008*)**

Solution. $d = \dfrac{\lambda}{d(\mu_o - \mu_e)} = \dfrac{5890 \times 10^{-10}}{4 \times (1.658 - 1.486)} = 0.856 \times 10^{-6}$ m

Example 28.11 *Determine the wavelength of light used when a quarter wave plate of thickness 1.2×10^{-3} cm is used for detection. Given $\mu_e = 1.65$ and $\mu_o = 1.64$.*

Solution. $d = \dfrac{\lambda}{4(\mu_e - \mu_o)}$

$$\lambda = 4\,d\,(\mu_e - \mu_o) = 4 \times 1.2 \times 10^{-3}\,(1.65 - 1.64)$$
$$= 4800 \times 10^{-8} \text{ cm} = 4800 \times 10^{-10} \text{ m} = 4800 \text{ Å}$$

Example 28.12 *Linearly polarised light is changed into circularly polarised light after passing through a slice of crystal 2.5×10^{-5} m thick. If the difference in refractive index for ordinary and extraordinary ray is 0.005, find the wavelength of light used.* **(*G.N.D.U. 2003*)**

Solution. *Here* $(\mu_e - \mu_o) = 0.005$; Thickness $d = 2.5 \times 10^{-5}$ m

As linearly polarised light is changed into circularly polarised light after passing through a slice of the crystal, it is a quarter wave plate and its thickness is given by

$$d = \frac{\lambda}{4(\mu_o - \mu_e)} \quad \text{or} \quad \lambda = 4\,d\,(\mu_o - \mu_e)$$

$$\lambda = 4 \times 2.5 \times 10^{-5} \times 0.005 = 5 \times 10^{-7} \text{ m} = 5000 \times 10^{-10} \text{ m} = 5000\text{Å}$$

Example 28.13 *Plane polarised light is converted into curcularly polarised light after passing through a slice of crystal. What is the minimum thickness of the crystal if the difference in refractive index of the two rays in the crystal is 0.005 and wavelength of light is 6000 Å.* **(*G.N.D.U., 2002*)**

Solution. $d = \dfrac{\lambda}{4(\mu_e - \mu_o)} = \dfrac{6000 \times 10^{-10}}{4 \times 0.005} = 3 \times 10^{-5}$ m

28.18.2 Half-Wave Plate

This plate is also made from a doubly refracting uniaxial crystal of calcite or quartz with its refracting faces cut parallel to the optic axis. The thickness *d* of this plate is such that it introduces a phase difference of π or a path difference of $\lambda/2$ between the *O*-ray and the *E*-ray in passing through it when light is incident normally on the face of the crystal.

For a negative crystal, thickness $d = \dfrac{\lambda}{2(\mu_o - \mu_e)}$ ... (28.15)

and for a positive crystal, thickness $d = \dfrac{\lambda}{2(\mu_e - \mu_o)}$ ... (28.16)

Quarter wave plates or half wave plates are usually made of thin sheets of split mica or of quartz cut parallel to optic axis. The thickness of a half wave plate can be calculated from relations (28.15) and (28.16) respectively, using the values of μ_e and μ_o for sodium light of wavelength 5893 Å.

Example 28.14 *Find the relation between the thickness of a quarter wave plate and thickness of a half wave plate for light of wavelength λ incident on a quartz crystal.* **(*P.U. 2008*)**

Solution. Thickness of a half wave plate for light of wavelength λ incident on a quartz crystal (positive)

$$d_H = \frac{\lambda}{2(\mu_e - \mu_o)}$$

Corresponding thickness of the quarter wave plate $d_Q = \dfrac{\lambda}{4(\mu_e - \mu_o)}$

$$\therefore \quad \frac{d_H}{d_Q} = \frac{\lambda}{2(\mu_e - \mu_o)} \cdot \frac{4(\mu_e - \mu_o)}{\lambda} = 2$$

Example 28.15 *Calculate the thickness of a half wave plate for light of wavelength $\lambda = 5000$ Å, $\mu_o = 1.544$, $\mu_e = 1.533$.* **(P.U., 2003, H.P.U., 2001; Pbi. U., 2002; K.U., 2000, Cal. U., G.N.D.U. 2004)**

Solution. Here $\mu_o = 1.544$, $\mu_e = 1.533$ and $\lambda = 5000 \times 10^{-10}$ m

Thickness $d = \dfrac{\lambda}{2(\mu_o - \mu_e)} = \dfrac{5000 \times 10^{-10}}{2(1.544 - 1.533)}$

$= 0.227 \times 10^{-4}$ m.

Example 28.16 *Calculate the thickness of a doubly refracting crystal to introduce a path difference of $\lambda/2$ (or phase difference $\pi = 180°$) between ordinary and extraordinary ray when $\lambda = 5893$ Å, $\mu_o = 1.5442$ and $\mu_e = 1.5533$.* **(M.D.U., 2002; K.U., 2001; Kerala U., 2001)**

Solution. Here $\mu_o = 1.5442$, $\mu_e = 1.5533$ and $\lambda = 5893$ Å

$\therefore$ Thickness $d = \dfrac{\lambda}{2(\mu_e - \mu_o)} = \dfrac{5893 \times 10^{-10}}{2(1.5533 - 1.5442)} = 3.238 \times 10^{-5}$ m

Example 28.17 *Calculate the thickness of doubly refracting crystal required to introduce a path difference of $\dfrac{\lambda}{2}$ between ordinary and extraordinary rays when $\lambda = 6000$ Å, $\mu_o = 1.65$ and $\mu_e = 1.48$. State whether the crystal is positive or negative.* **(Nagpur. U., 2001)**

Solution. $d = \dfrac{\lambda}{2(\mu_o - \mu_e)} = \dfrac{6000 \times 10^{-10}}{2(1.65 - 1.48)} = 0.176 \times 10^{-5}$ m

The crystal is *negative* as $\mu_o > \mu_e$

Example 28.18 *The refractive index of E-ray and O-ray are 1.486 and 1.658 respectively with sodium light of wavelength 589 nm. Find the thickness of quarter and half wave plate.* **(H.P.U., 2001)**

Solution. $\mu_e = 1.486$; $\mu_o = 1.658$; $\lambda = 589 \text{ nm} \times 10^{-9}$ m

Thickness of quarter wave plate $d = \dfrac{\lambda}{4(\mu_o - \mu_e)} = \dfrac{589 \times 10^{-9}}{4(1.658 - 1.486)}$

$= 0.856 \times 10^{-6}$ m

Thickness of half wave plate $d = \dfrac{\lambda}{2(\mu_o - \mu_e)} = \dfrac{589 \times 10^{-9}}{2(1.658 - 1.486)}$

$= 1.712 \times 10^{-6}$ m

Example 28.19 *Calculate the thickness of a quarter wave plate and a half wave plate. Given that $\mu_E = 1.558$, $\mu_o = 1.544$ and $\lambda = 6500$ Å.* **(P. U. 2006)**

Solution. Thickness of quarter wave plate $d =$

$$\frac{\lambda}{4(\mu_E - \mu_o)} = \frac{6500 \times 10^{-10}}{4(1.558 - 1.544)} = 11.6 \times 10^{-6} \text{ m}$$

Thickness of half wave plate = 23.2×10^{-6} m.

Example 28.20 *A beam of linearly polarised light is changed into circularly polarised light by passing it through a slice of crystal 0.003 cm thick. Calculate the difference in refractive indices of the two rays in the crystal, assuming this to be minimum thickness that will produce the effect and that the wavelength is 6 × 10⁻⁷ m.*

Solution. As linearly polarised light is changed into circularly polarised light the crystal plate has a thickness equal to quarter wave plate given by $d = \dfrac{\lambda}{4(\mu_e - \mu_o)}$, considering it to be a positive crystal.

Here $d = 0.003 \text{ cm} = 3 \times 10^{-5}$ m, $\lambda = 6 \times 10^{-7}$ m

$$\therefore \quad \mu_e - \mu_o = \frac{\lambda}{4d} = \frac{6\times10^{-7}}{4\times3\times10^{-5}} = 0.005$$

28.19 PRODUCTION OF PLANE, ELLIPTICAL AND CIRCULARLY POLARIZED LIGHT

1. Production of plane polarised light. If a beam of monochromatic light is passed through a Nicol prism, it is split up into *E*-ray and *O*-ray. The *O*-ray is totally internally reflected at the canada balsam layer and is absorbed, while the *E*-ray passes through the Nicol prism, which is plane polarised.

2. Production of elliptically polarised light. An ellipically polarised light consists of two mutually perpendicular vibrations of unequal amplitudes with a phase difference of $\pi/2$ or a path difference of $\lambda/4$.

A parallel beam of monochromatic light is allowed to fall on a Nicol prism N_1. The outgoing plane polarised light is made to pass through the second Nicol N_2 which is gradually rotated till no light is seen through the system. The two Nicols are in a crossed position.

A quarter wave plate *P* is mounted on a tube T_1, circumference of which is graduated in degrees. The tube T_1 is capable of rotation in another tube T_2 mounted on a stand *S*. This arrangement is placed in between the two Nicols, as shown in Fig. 28.26. The plane polarised light from the Nicol N_1 falls normally on the quarter wave plate *P*. The field of view through the Nicol N_2 may now be bright. The quarter wave plate is rotated till again no light passes through the Nicol N_2. In this position *the vibrations of light incident on the quarter wave plate are parallel to the optic axis of the plate* and both the *E* and *O* components travel in the same direction but with different velocities. These are refused transmission through the Nicol prism N_2 which is in a crossed position with respect to the Nicol N_1. Keeping the quartz wave plate fixed, the tube T_1 is rotated till the zero mark of the scale coincides with the mark *X* on the *Q*-plate. The quarter wave plate *P*, keeping the tube T_1 fixed, is rotated till the mark *X* coincides with any mark other than 0, 45 and 180. The plane polarised beam consisting of *O* and *E* vibrations will have unequal amplitudes. These on coming out of the quarter wave plate will result in an elliptically polarised light. If now the Nicol N_2 is rotated the intensity varies between a maximum and a minimum but is never zero. This behaviour is similar to the case when a mixture of plane polarised and ordinary light is examined through a Nicol prism.

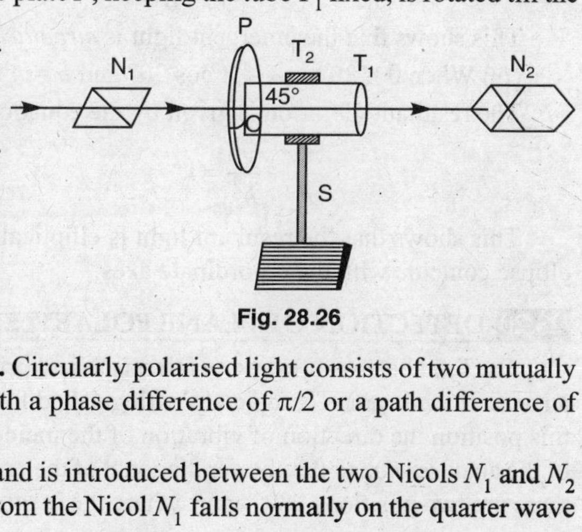

Fig. 28.26

3. Production of circularly polarised light. Circularly polarised light consists of two mutually perpendicular vibrations of equal amplitudes with a phase difference of $\pi/2$ or a path difference of $\lambda/4$. This is achieved with a quarter wave plate.

The quarter wave plate *P* mounted on the stand is introduced between the two Nicols N_1 and N_2 in a crossed position. The plane polarised light from the Nicol N_1 falls normally on the quarter wave plate and the field of view through N_2 is illuminated as the light is elliptically polarised as explained above, except when the incident vibration makes an angle of 45° with the optic axis of the quarter wave plate. To produce circularly polarised light, the quarter wave plate is rotated till again no light is seen through N_2. In this position, the incident vibration is parallel to the optic axis of the quarter wave-plate.

Keeping the quarter wave plate fixed, the tube T_1 is rotated till the zero mark of the scale coincides with the mark X on the quarter wave plate P. Keeping the tube T_1 fixed the quarter wave plate is rotated so that the mark X coincides with the division 45° on the tube T_1. In this way the vibrations in the incident plane polarised beam make an angle of 45° with the direction of optic axis of the quarter wave plate. The light is split up into E-ray and O-ray having equal amplitudes. The light is thus *circularly polarised.* If the Nicol N_2 is rotated at this state, the field of view is uniform in inten sity, the behaviour is similar to the ordinary light passing through the Nicol prism.

Example 28.21 *Plane polarised light is incident normally on a quarter wave plate. Determine the nature of emergent light in each case, when the vibration in the incident light make angles of 0°, 30°, 45° and 90° with the axis of the plate.*

Solution. When plane polarised vibration of amplitude A is incident normally on a quarter wave plate, it is split up into the E and O vibrations. If θ is the angle which the incident vibration makes with the direction of optic axis, then on emergence the resultant vibration is given by

$$\frac{x^2}{a^2} + \frac{y^2}{b^2} = 1$$

where $a = A \cos \theta$ and $b = A \sin \theta$

(*i*) When $\theta = 0$, $a = A$ and $b = 0$.

This shows that the amplitude of the vibration perpendicular to the optic axis is zero which therefore vanishes. The emergent light is *plane polarised* having vibrations in a direction parallel to the optic axis.

(*ii*) When $\theta = 90°$, $a = 0$ and $b = A$

The amplitude of the vibration along the optic axis vanishes. The outgoing light, therefore, is plane polarised having vibrations perpendicular to the optic axis.

(*iii*) When $\theta = 45°$, $a = A \cos 45°$ and $b = A \sin 45°$.

$\therefore \quad a = b$

Substituting a for both a and b, we have

$$x^2 + y^2 = a^2$$

This shows that the emergent light is *circularly polarised.*

(*iv*) When $\theta = 30°$, $a = A \cos 30°$ and $b = A \sin 30°$ *i.e.*, a is not equal to b.

The resultant vibration is given by the equation

$$\frac{x^2}{a^2} + \frac{y^2}{b^2} = 1$$

This shows that the resultant light is elliptically polarised, the major and the minor axes of the ellipse coincide with the co-ordinate axes.

28.20 DETECTION OF PLANE POLARIZED LIGHT

The light beam is allowed to fall on a Nicol prism capable of rotation about the incident beam as an axis. If the beam is completely extinguished for one setting of the Nicol prism (analyser) then in this position the direction of vibration of the incident light is *perpendicular to the principal section* of the analyser *i.e.*, to the shorter diagonal of the end face. For each complete rotation of the analyser, there are two positions 180° apart where the incoming beam is completely extinguished. There will be two maximum intensity positions 90° from each zero intensity position. In this case the vibrations of incident beam are *parallel* to the principal section of the Nicol prism and thus these will pass out. In such a case the *beam is plane polarised.*

28.21 DETECTION OF CIRCULARLY POLARIZED LIGHT

A circular vibration consists of two linear vibrations of equal amplitudes mutually perpendicular to each other. When circularly polarised light is examined through a Nicol prism the intensity of the transmitted beam is not altered as the Nicol is rotated. *The behaviour is similar to that of unpolarised beam* of light.

To distinguish between the two, allow the light to fall normally on a quarter wave plate and then examine it through the Nicol prism. If the intensity varies between maximum and zero, it is circularly polarised light. It is because the circular vibration on entering the quarter wave plate is broken up into two linear vibrations of the same amplitude mutually perpendicular to each other. The quarter wave plate further introduces a phase change of $\pi/2$ between the two. The resultant phase difference becomes 0 or π and the outgoing light is *plane polarised.* On examination through the rotating Nicol, maximum intensity is observed when the principal section of the Nicol is parallel to the vibration of the plane polarised light. Light will be cut off completely when the principal section of the Nicol is perpendicular to the direction of the resultant plane polarised vibration.

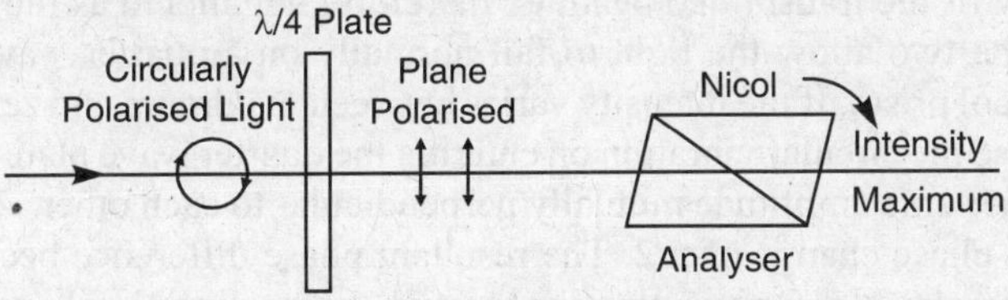

Fig. 28.27

The unpolarised light after passing through a quarter wave plate remains unpolarised and hence no change in intensity will be observed when passed through a Nicol prism which is rotated gradually.

Thus we conclude that if the beam after passing through the quarter wave plate is extinguished twice in each rotation of the Nicol prism, it is **circularly polarised.**

28.22 DETECTION OF ELLIPTICALLY POLARIZED LIGHT

When a beam of elliptically polarised light is examined through a Nicol prism rotating about the ray as an axis, its intensity varies in magnitude, but is never zero. It is because an elliptic vibration consists of a combination of two unequal rectilinear vibrations in two directions perpendicular to each other along the major and minor axis of the ellipse. When the principal plane of the Nicol is parallel to the vibration along the *major axis,* the transmitted light has a maximum intensity and when the principal plane is parallel to vibrations along the *minor axis* it has a minimum intensity.

The behaviour is similar to a mixture of unpolarised and plane polarised light when examined through a Nicol prism.

To distinguish between the two,

(*i*) Allow the light to pass through the rotating Nicol alone and adjust its position for maximum intensity.

(*ii*) Allow the light under examination to pass through the quarter wave plate normally and the optic axis of the plate parallel to the principal plane of the Nicol adjusted for maximum intensity.

The elliptic vibration is resolved into two rectangular vibrations of unequal amplitudes with a phase difference of $\pi/2$ between them. By passage through the quarter wave plate a further phase difference of $\pi/2$ is introduced between the two rectangular components along the two axes of the ellipse, so that the resultant phase difference becomes π or 0. This light on emergence becomes plane polarised. If the light is now examined through the Nicol prism, the intensity is reduced to **zero twice** in one rotation of the Nicol prism when the principal section of the Nicol is perpendicular to the plane of vibration of the emergent light. The incident light is thus elliptically polarised.

28.23 DISTINCTION BETWEEN CIRCULARLY POLARIZED AND UNPOLARISED LIGHT

When circularly or unpolarised light is passed through a Nicol prism which is gradually rotated about the ray as an axis, no variation of intensity in both the cases is observed in the nature of emergent light. It is because in the case of unpolarised light one vibration or the other is always parallel to the principal section of the Nicol and hence the intensity is uniform.

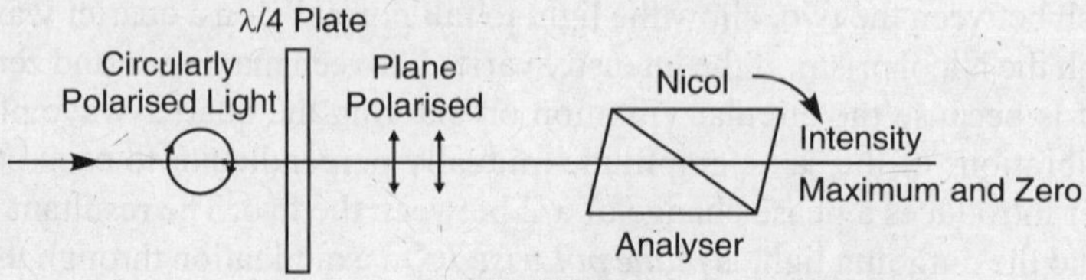

Fig. 28.28

The circular vibration consists of two linear vibrations of equal amplitude mutually perpendicular to each other. One of the components passes through the Nicol prism and the other is refused tranmission. The intensity of the transmitted beam is, therefore, not altered as the Nicol is rotated. To distinguish between the two allow the light to fall normally on a quarter wave plate and then examine it through the Nicol prism. If the intensity varies between maximum and zero, it is circularly polarised light. It is because the circular vibration on entering the quarter wave plate is broken up into two linear vibrations of the same amplitude mutually perpendicular to each other. The quarter wave plate further introduces a phase change of $\pi/2$. The resultant phase difference becomes 0 or π and the outgoing light is *plane polarised.* On examination through the rotating Nicol, maximum intensity is observed when the principal section of the Nicol is parallel to the vibration of the plane polarised light. Light will be cut off completely when the principal section of the Nicol is perpendicular to the direction of the resultant plane polarised vibration.

The unpolarised light after passing through a quarter wave plate remains unpolarised and hence no change in intensity will be observed when passed through a Nicol prism which is rotated gradually.

28.24 DISTINCTION BETWEEN ELLIPTICALLY POLARIZED AND MIXTURE OF UNPOLARIZED AND PLANE POLARIZED LIGHT

When elliptically polarised light is passed through a Nicol prism and the Nicol is rotated about the ray as an axis, its intensity varies in magnitude but is never zero. In this respect it resembles a mixture of unpolarised and plane polarised light. An elliptic vibration consists of a combination of two unequal rectilinear vibrations in two directions perpendicular to each other along the major and the minor axes. When the principal plane of the Nicol is parallel to vibrations along the *major axis* the transmitted light has a maximum intensity.

In the case of a *mixture of unpolarised and plane polarised* light the *maximum* intensity is seen when the principal plane of the Nicol is *parallel* to the direction of vibrations of the plane polarised light and *minimum* intensity is obtained when the principal plane of the Nicol is *perpendicular* to the direction of vibrations of the plane polarised light. The intensity is never zero as the unpolarised light always contributes a constant intensity.

Since the behaviour of both types of light when examined through a Nicol is similar, the distinction between them is made as follows :–

(*i*) The light is examined through a rotating Nicol alone and its position is adjusted for maximum intensity.

(*ii*) A quarter wave plate is placed between the source of light and the Nicol prism, such that the light passes through the plate normally and the optic axis of the plate is parallel to the principal section of the Nicol adjusted for maximum intensity.

(*iii*) The Nicol is then rotated about the direction of light as axis.

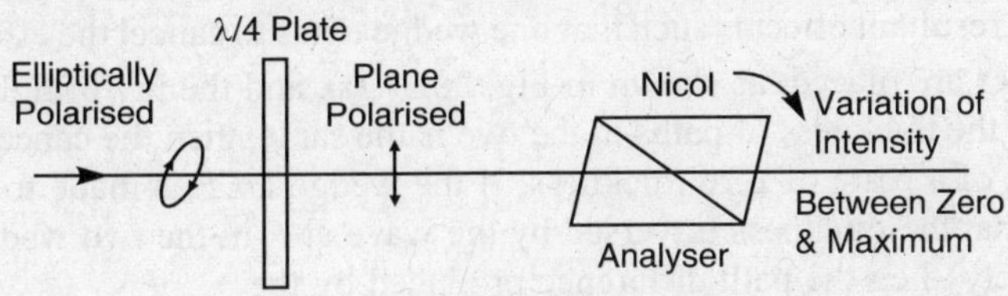

Fig. 28.29

If the finally emergent light shows variation in intensity with zero minimum, the given light is elliptically polarised. It is because in the case of elliptically polarised light the intensity is maximum when the principal section of the Nicol is parallel to the major axis of the elliptic vibration. When the quarter wave plate is inserted with optic axis parallel to the principal section of the analyser adjusted for maximum intensity, the optic axis of the quarter wave plate is also parallel to the major axis of the elliptic vibration. Under these conditions, the quarter wave plate introduces a further phase difference between the two mutually perpendicular vibrations so that the outgoing light becomes plane polarised. The rotations of the analyser will show maximum intensity when the principal section is parallel to the vibrations in the plane polarised light and zero intensity when the principal section is perpendicular to these vibrations.

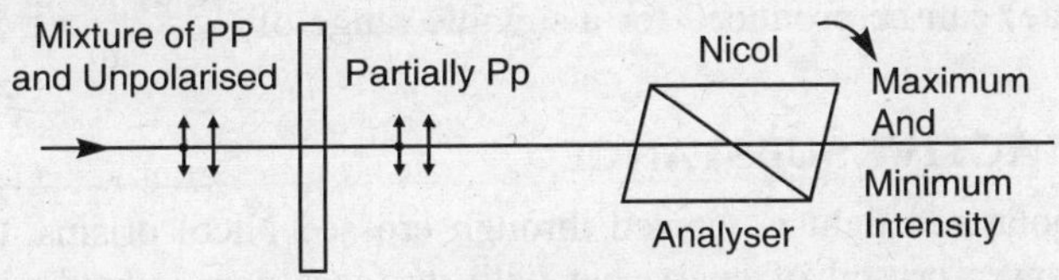

Fig. 28.30

In the case of mixture of plane polarised light and unpolarised light, the intensity through the rotating Nicol is maximum when the principal section of the Nicol is parallel to the vibrations of the plane polarised light. When the quarter wave plate is inserted with its axis parallel to the principal section of the analyser for maximum intensity, the polarised light passes through it as plane polarised with direction of vibration unchanged and the unpolarised part emerges out as unpolarised. The rotation of the analyser Nicol therefore produces vibration of intensity between maximum and minimum but is not zero. The maximum intensity is observed in the same position of the Nicol as before introducing the quarter wave plate.

28.25 BABINET'S COMPENSATOR

In the analysis of polarised light it is desirable to have a device which can produce a phase difference of any arbitrary value between two plane polarised waves in mutually perpendicular planes. A single crystal, such as a quarter wave plate, can only produce one particular phase difference, depending upon its thickness and the wavelength of light used. Babinet devised an arrangement called *Babinet's compensator* by means of which a measurable and variable path difference can be introduced. It consists of two wedge-shaped plates *ABC* and *DEF* placed with their hypotenuse adjacent so as to form a small rectangular block. One of the wedges is fixed, while the other can be moved in its own plane by means of a micrometer screw. The wedges are cut from crystalline quartz in such a manner that the optic axis in one is parallel and in the other perpendicular to the two refracting faces.

If plane polarised light is incident on the face *AB* of the compensator with the plane of vibration making an angle θ with the optic axis, it will be broken up into ordinary and extraordinary components. The *E*-component parallel to the optic axis in the crystal travels slower than the *O*-component until it reaches the point *P* on the face *ED* of the second crystal. At this point the *E*-vibration becomes the *O*-vibration as now it is perpendicular to the optic axis and the *O*-vibration becomes the *E*-vibration in the second crystal. In other words, the two vibrations exchange velocities in passing from one wedge to the other. Thus the resultant effect is such that one wedge tends to cancel the effect due to the other.

If the two wedges are placed, as shown in Fig. 28.31 (*i*), and the ray of light is incident at the centre as at P so that the thickness of paths in the two is the same, then the cancellation is complete and the effect is that of a plate of zero thickness. If the wedge DEF is made to slide, as shown in Fig. 28.31. (*ii*), so that the thickness traversed by the wave at P in the two wedges ABC and DEF is t_1 and t_2 respectively, then the path difference produced by the first wedge will be $t_1 (\mu_e - \mu_o)$ and that produced by the other will be $t_2 (\mu_e - \mu_o)$. Hence

Resultant path difference $= (t_1 - t_2)(\mu_e - \mu_o)$.

Thus by adjusting the positions of the two wedges relative to each other, any desired path difference can be produced.

Superiority over quarter wave plate. A quarter wave plate can introduce a path difference of $\lambda/4$ between the ordinary and extraordinary rays for a fixed wavelength only, but a Babinet's compensator can produce any desired path difference between ordinary and extraordinary rays by adjusting the position of the two wedges. As such a fixed path difference say $\lambda/4$ (as produced by a quarter wave plate) can be produced for a suitable range of wavelength.

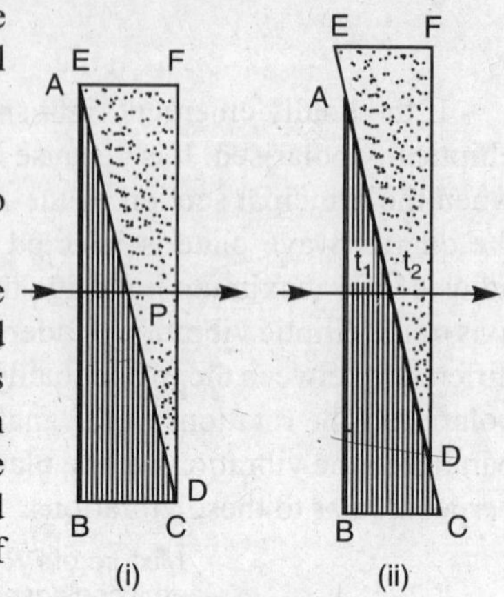

Fig. 28.31

28.26 OPTICALLY ACTIVE SUBSTANCE

If a beam of unpolarised light is viewed through crossed Nicol prisms, the field of view is completely dark. If now a crystal of quartz cut with its *faces perpendicular to the optic axis* is interposed *between the polarising and the analysing Nicols,* such that light is incident normally on quartz, it is found that light is no longer completely cut off by the analyser. The analyser has to be rotated through a certain angle in order to stop the light again. This angle of rotation depends upon the thickness of quartz crystal.

The experiment shows that light, after passing through the first Nicol prism, becomes plane polarised and remains plane polarised even after passing through the quartz plate, but the direction of vibration and, therefore, the plane of polarisation is rotated through a certain angle. This angle is equal to the angle through which the second Nicol has to be rotated to completely stop the light.

This property of rotating the plane of vibration of polarised light by certain crystals and other substances is called optical activity.

Substances which rotate the plane of polarisation are said to be **optically active.**

There are two types of substances :

(*i*) *Dextro-rotatory or right-handed.* These produce the rotation of the plane of polarisation towards the right, *i.e.*, in the clockwise direction on looking towards the source.

(*ii*) *Laevo-rotatory or left-handed.* These produce the rotation of the plane of polarisation towards the left *i.e.*, in the anti-clockwise direction on looking towards the source.

In addition to certain crystals, substances that exhibit optical activity include liquids such as turpentine oil and solutions of optically active substances in an *inactive solvent,* such as sugar solution.

28.27 SPECIFIC ROTATION

The angle through which the plane of polarisation is rotated depends upon:

(*i*) thickness of the substance,

(*ii*) concentration of the solution or density of the material,

(*iii*) wavelength of light, and

(*iv*) the temperature.

Specific rotation for a given wavelength of light at a given temperature is defined as the rotation produced by one decimetre length of the solution containing 1 gm of optically active material per c.c. of solution.

If θ is the rotation produced by l decimetres length of a solution and C the concentration in grams per c.c., then the specific rotation S at a given temperature t, corresponding to a wavelength λ, is given by

$$[S]_\lambda^t = \frac{\theta}{l \times C} = \frac{\text{Rotation in degrees}}{\text{Length in decimetre} \times \text{concentration in gm/c.c.}}$$

Example 28.22 *Calculate the specific rotation if the plane of polarisation is turned through 26.4° on traversing 20 cm length of 20% sugar solution.* **(Meerut U., 2000)**

Solution. Here l = 20 cm = 2 decimetre; θ = 26.4°.

Concentration C = 20% = 0.2 gm/c.c.

$$\therefore \quad \text{Specific rotation} \quad [S]_\lambda^t = \frac{\theta}{l \times C} = \frac{26.4}{2 \times 0.2} = 66°$$

Example 28.23 *Find the specific rotation of sugar from the following data :*

Length of solution tube = *22 cm*

Volume of solution = *80 × 10^{-6} m^3*

Amount of sugar in solution = *6 gm*

Angle of rotation = *9° – 54′* **(M.D.U., 2002)**

Solution. Here l = 22 cm = 2.2 decimetre; θ = 9° – 54′ = 9.9°

Volume of solution = 80 × 10^{-6} m^3 = 80 c.c.

Amount of sugar in solution = 6 gm

$$\text{Concentration} \quad C = \frac{6}{80} = 0.075 \text{ gm/c.c.}$$

$$\therefore \quad \text{Specific rotation} \quad [S]_\lambda^t = \frac{\theta}{l \times C} = \frac{9.9}{2.2 \times 0.075} = 60°$$

Example 28.24. *Calculate the specific rotation of sugar if plane of polarisation is rotated by 10.5°. The length of the tube containing the solution is 20 cm and 8.0 gm of sugar is dissolved in 100 c.c. of water.* **(Luck. U., 2001)**

Solution. $C = \frac{8}{100} = .08; \quad l = 20 \text{ cm} = 2 \text{ decimetre}$

$$[S]_\lambda^t = \frac{10.5}{2 \times .08} = 65.6°$$

Example 25. *A sugar solution in a tube 20 cm long produces an optical rotation of 13° with light of wavelength 6000 Å. Find the strength of the solution. Given specific rotation of sugar is 65°.* **(M.D.U., 2003; Madurai U. 2003)**

Solution. Here l = 20 cm = 2 decimetres; θ = 13°

Specific rotation $[S]_\lambda^t$ = 65°; concentration C = ?

$$\text{Now} \quad [S]_\lambda^t = \frac{\theta}{l \times C} \quad \text{or} \quad 65 = \frac{13}{2 \times C} \quad \text{or} \quad 130\,C = 13$$

$$\therefore \quad C = 0.1 \text{ gm/c.c.}$$

Example 28.26 *80 gms of impure sugar dissolved in a litre of water gives an optical rotation of 9.9° when placed in a tube of length 0.2 m. If specific rotation of sugar is 66°, find percentage purity of sugar.* **(M.D.U. 2007)**

Solution. Concentration of impure solution $= \frac{80}{1000}$ g cm^{-3} = 0.08 gm cm^{-3}

Now specific rotation $[S]_\lambda^t = \frac{\theta}{l \times C}$ where C is the concentration of pure sugar and $l = 2$ decimetre

$$\therefore \quad 66 = \frac{9.9}{2C} \quad \text{or} \quad C = \frac{9.9}{66 \times 2} = 0.075 \text{ gm cm}^{-3}$$

$$\therefore \text{ Percentage purity of sugar} = \frac{0.075}{0.08} = 93.75\%$$

Example 28.27 *A mass of 80 gms of sugar dissolved in one litre of water contained in a tube of length 20 cm causes an optical rotation of 12° with λ = 4000 Å at a temperature of 27°C. What mass of sugar should be dissolved in 2 litres of water so that the solution placed in a tube of length 30 cm would cause an optical rotation of 24° with light of same wavelength at the same temperature.* **(K.U., 2000)**

Solution. As 80 gm of sugar is dissolved in one litre of water, concentration $C_1 = \frac{80}{1000} = .08$ gm/c.c.
$l_1 = 20$ cm = 2 decimetre; $\theta_1 = 12°$

$$\therefore \quad [S]_\lambda^t = \frac{\theta}{l \times C} = \frac{12}{2 \times .08}$$

If m gm of sugar be dissolved in 2 litre of water, then concentration $C_2 = \frac{m}{2000}$ gm/ cc; $l_2 = 30$ cm = 3 decimetre; $\theta_2 = 24°$

$$\therefore \quad [S]_\lambda^t = \frac{\theta}{l \times C} = \frac{24 \times 2000}{3 \times m} \qquad \text{... (ii)}$$

Comparing (*i*) and (*ii*), we get

$$\frac{24 \times 2000}{3 \times m} = \frac{12}{2 \times .08}$$

or
$$36\, m = 2 \times .08 \times 24 \times 2000 \quad \text{or} \quad m = \frac{2 \times .08 \times 24 \times 2000}{36}$$
$$= 213.3 \text{ gm.}$$

Example 28.28 *A solution of camphor in alcohol in a tube 20 cm long is found to rotate the plane of vibration of light by 27°. What is the mass of camphor in unit volume of solution? Specific rotation of camphor is 54°.* **(K.U., 2002)**

Solution. Here $\theta = 27°$; $[S]_\lambda^t = S = 54°$; $l = 20$ cm = 2 decimetre

If C is the concentration per c.c., then

$$C = \frac{\theta}{S \times l} = \frac{27}{54 \times 2} = 0.25 \text{ gm/ cm}^3$$
$$= 0.25 \text{ kg m}^{-3}$$

Example 28.29 *A solution of camphor in alcohol contained in a tube of length 0.2 m produce a rotation of 33°. The concentration of the solution is 0.305 gm/c.c. Find the specific rotation of camphor.* **(M.D.U. 2006)**

Solution. $[S]_\lambda^t = \frac{\theta}{l \times C} = \frac{33}{2 \times 0.305} = 54°$

28.28 ORIGIN OF OPTICAL ACTIVITY

From experimental evidence, it is established that the action of rotating the frame of polarisation is a body effect and not the surface effect. This is quite clear because the angle rotation is proportional to path length l. Thus, it occurs inside the medium continuously. Moreover, the angle of rotation θ is proportional to concentration and $\theta = \Sigma\theta_i$ for a mixture shows that it is a *molecular phenomenon.*

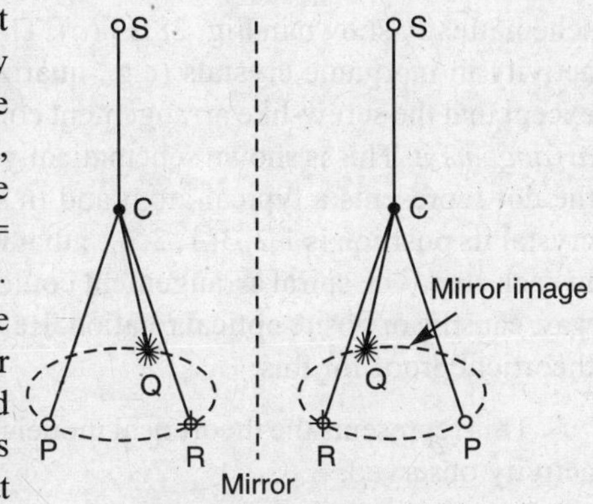

Fig. 28.32. Mirror-image symmetry

Every molecule of the active substance plays its role *largely* independent of the presence or absence of other molecules. Substances which show optical activity in liquid state or in solutions must therefore have in their molecules some rotational asymmetry. It is further interesting that most *optically active substances have both dextro and laevo-forms, *e.g.*, *d*-lactic acid, *l*-lactic acid. The two have all physical and chemical properties identical, except that the optical rotations are of opposite sign and that they crystallize into hemihedral forms in opposite directions, like mirror images. It follows that the phenomenon of optical rotation *in organic liquids and solutions* is directly associated with structural asymmetry in the molecules. Fig. 28.32 shows a carbon atom C (tetravalent) surrounded by 4 *different* atoms P, Q, R, S in tetrahedral arrangement. On the left diagram, P, Q, R are arranged clockwise (as seen along line SC), while on the right diagram they are arranged anticlockwise. Seen along PC we again find Q, R, S arranged in opposite order in the two cases, and so on. One structure is mirror image of the other. No rotation in space can make one structure coincide with the other. If this mirror-image asymmetry is the cause of optical rotation, it is apparent that a *tetravalent* atom, *e.g.*, carbon, sulphur, etc., attached to four *different* atoms or groups is an essential requirement for such asymmetry and hence for ptical activity in liquids and solutions. This is experimentally verified and hence holds good.

Examples: Tartaric acid is a simple illustrative example. Structural formulae of dextro (*d*-), laevo (*l*-) and meso tartaric acid are given below. Each of the two C atoms has four dissimilar groups (H, OH, $COOH$ and the other C) attached to it, and there are two such C atoms in each molecule. In *d*-tartaric acid, both C atoms have H, C, OH, $COOH$ in one cyclic order; in *l*-tartaric acid again both C atoms have H, C, OH, $COOH$ in one cyclic order which is opposite to that of the former; in meso-tartaric acid the orders of H, C, OH, $COOH$ in the two groups are opposite to one another. We find that in the first two the optical rotations are equal but opposite, while the third molecule (meso tartaric acid) is not optically active. Thus, we may attribute some optical activity with *each group* within the molecule. In case (*a*) the activities of the two groups add up, so also in case (*b*), but in the third case (*c*) the activities of the two groups, being opposite, cancel out.

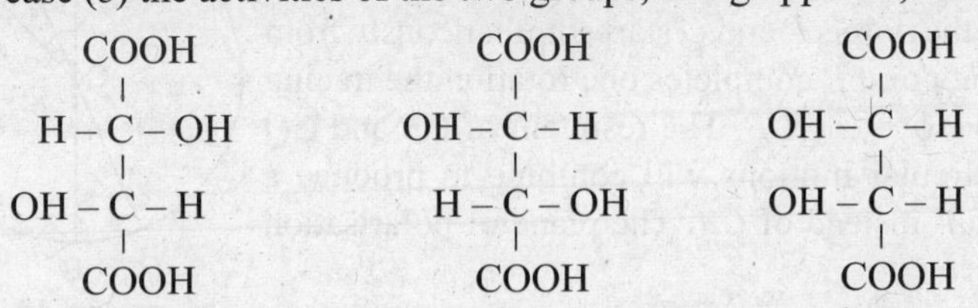

(a) (b) (c)

Structural formula of (a) dextro–, (b) laevo–, and (c) meso–tartaric acid.

* Cane-sugar is always dextro-rotatory, whatever the light source may be.

Another situation for optical activity arises when the molecules have spiral-like arrangement in themselves, as schematically shown in Fig. 28.33 (*a*). The explanation of optical activity in inorganic crystals (e.g., quartz), is on the same lines, except that the screw-like arrangement comes from the *crystalline arrangement.* This is shown schematically in Fig. 28.33 (*b*), where the dot represents a typical atom and in successive layers in the crystal its position is 1,2, 3, 1, 2, 3..., that is, turned by 2π/3 radian in each step. The spiral arrangement could as well be the opposite way, causing opposite optical rotation. Reuch gave the satisfactory theortical proof for this.

This represents the theoretical model to understand the optical activity observed.

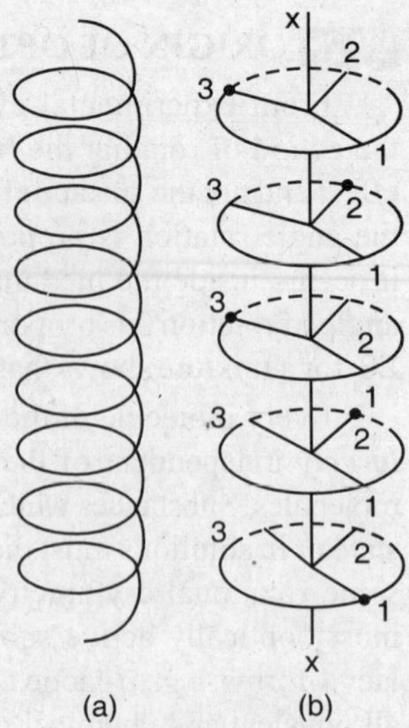

Fig. 28.33. Spiral-like arrangements

28.29 FRESNEL'S EXPLANATION OF ROTATION OF PLANE OF POLARIZATION

It is based on the principle that *any single harmonic motion along a straight line is the resultant of two circular motions of the same frequency and half the amplitude, propagated in opposite directions.*

(*i*) A plane polarised light entering a crystal along the optic axis *CA* is split up into two circularly polarised vibrations rotating in opposite directions with the same frequency. The tracing point *P* rotates in the anti-clockwise direction and its rotation is called *left-handed* (Fig. 28.34). The tracing point *Q* rotates in the clockwise direction and its rotation is called *right handed.*

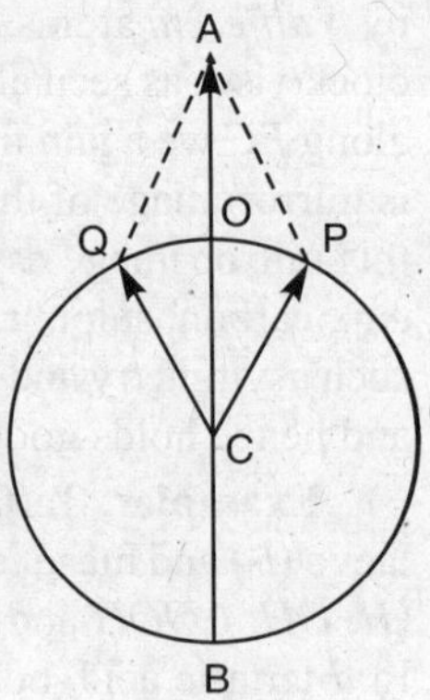

Fig. 28.34

(*ii*) In a crystal like calcite, which is not optically active, the two circular motions have the same velocity. Hence if the tracing points *P* and *Q* start simultaneously from *O* and travel in their respective directions, their resultant, when in any position, will always be along *CA*. Thus on coming out of such a crystal the two circular motions will again combine to produce a linear simple harmonic vibration along *CA*. This shows that there will be no rotation in the plane of polarisation.

(*iii*) In a crystal like quartz which is optically active, the two circular vibrations move forward with slightly different velocities in their respective directions. In right-handed quartz the *right-handed* or the *clockwise* motion will travel faster than the *left-handed* or the *anti-clockwise* motion and *vice versa.* In a right-handed crystal if the points *P* and *Q* start simultaneously from *O*, then during the time the tracing point *P* completes one rotation the tracing point *Q* will travel a further angle $\phi = \angle QCP$. The resultant of *CP* and *CQ* will act along *CA'*. Hence the circular motions will combine to produce a simple harmonic motion along *CA'* instead of *CA*. The plane of polarisation will turn through an angle.

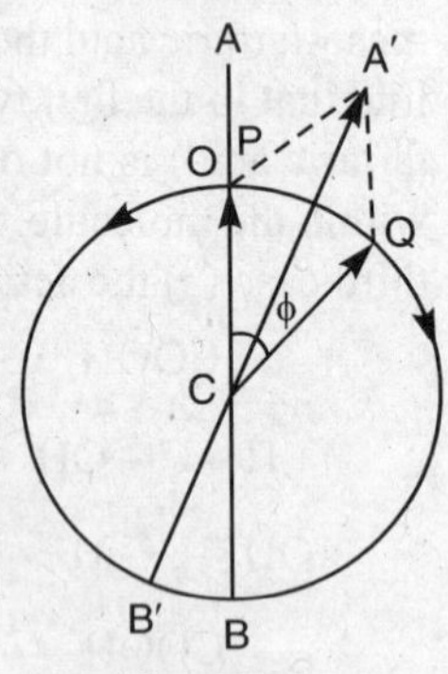

Fig. 28.35

$$ACA' = \frac{\phi}{2}$$

The angle through which the plane of polarisation is rotated is proportional to the thickness of the crystal *d*. If μ_L and μ_R denote the refractive indices for the left-handed and the right-handed circular components respectively, then the angle ϕ is given by

$$\phi = \frac{2\pi}{\lambda}(\mu_L - \mu_R)d$$

or
$$\frac{\phi}{2} = \frac{\pi}{\lambda}(\mu_L - \mu_R)d$$

Experimental demonstration. Real existence of two circularly polarised rays within a crystal was demonstrated by Fresnel as given below.

Fresnel built a compound rectangular block of alternate prisms of *right* and *left-handed* quartz, the optic axis in all of them being in the same direction perpendicular to the end faces AC and EF as shown in Fig. 28.36. A plane polarised ray normally incident on the face AC is broken up into two circularly polarised components which travel through the prism with unequal velocities but in the same direction. The right-handed circular component travels faster in the *right handed prism* and slower in the *left-handed prism and vice versa* for the left-handed circular component. When these components pass from the right-handed prism to the left-handed prism the material of the L-prism acts as a denser medium for the R-component and as a rarer medium for the L-component. In passing through the L-prism the R-component bends towards the base and the L-component away from he base. On passing from the left-handed prism to the right-handed prism the L-component bends towards the base of R-prism and the R-component bends away from it. The separation between the two components goes on increasing as they pass through various prisms. Hence the emergent rays are completely separated.

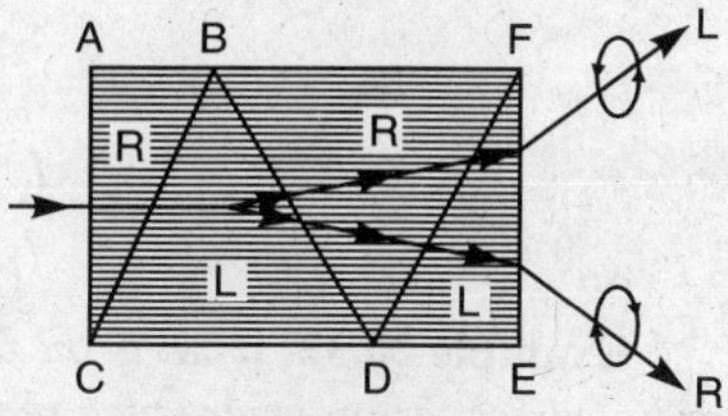

Fig. 28.36

On analysis by means of a quarter wave plate the rays are found to be cicularly polarised in opposite sense which confirms Fresnel's hypothesis.

Example 28.30 *For a given wavelength one millimetre of quartz cut perpendicular to the optic axis rotates the plane of polarisation by 20°. Find for what thickness will no light of this wavelength be transmitted when the quartz piece is interposed between a pair of parallel Nicols. Why is it necessary that the quartz plate be cut perpendicular to the optic axis ?* ***(Guahati U., 2000)***

Solution. The angle of rotation is proportional to the thickness of the plate interposed.

Here thickness of the plate = 1 mm = 10^{-3} m

Angle of the rotation = 20°

Now plane polarised light passes through the second Nicol freely if it is polarised perpendicular to its principal plane. This light will be stopped completely if it is polarised in the principal plane. Hence if the two Nicols are parallel, light will pass through the second Nicol freely. To stop the passage of light through the second Nicol, it will have to be rotated through 90° or the thickness of the quartz plate should be such that it produces a rotation of 90°.

$$\text{Thickness of quartz} = \frac{90}{20} = 4.5 \text{ mm.} = 4.5 \times 10^{-3} \text{ m}$$

The quartz plate with its face is cut perpendicular to the optic axis because the two components into which the incident light is broken up, when passing along the optic axis will travel in the same direction and with the same velocity. Hence on emergence the light will remain plane polarised.

Example 28.31 *Plane polarised light is incident on a quartz plate with faces cut parallel to the optic axis. If the vibration plane of the incident beam makes an angle of 30° with the optic axis, compare the intensities of ordinary and extraordinary vibrations.*

Solution. Suppose the incident vibration makes an angle θ with the direction of optic axis and has an amplitude A. This vibration may be regarded as the resultant of two components vibrations along OX and OY. The amplitude of the E-vibration along OX will be $A \cos \theta$ and that of the O-vibration along OY is $A \sin \theta$. As the intensity is proportional to the square of the amplitude.

$$\therefore \quad \frac{I_o}{I_e} = \frac{A^2 \sin^2 \theta}{A^2 \cos^2 \theta} = \tan^2 \theta$$

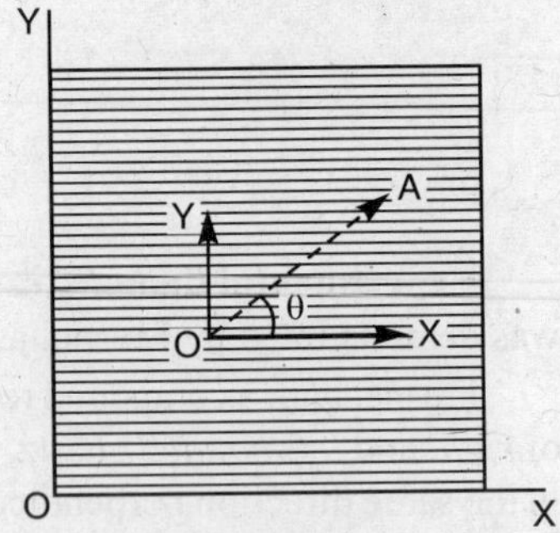

Fig. 28.37

where I_o is the intensity of O-vibration and I_e that of E-vibration.

When $\theta = 30°$, $\tan 30° = \sqrt{\frac{1}{3}}$

$$\therefore \quad \frac{I_o}{I_e} = \frac{1}{3}$$

or $\quad I_o : I_e :: 1 : 3$

Example 28.32 *What is the least thickness of quartz plate for which the extraordinary and the ordinary waves on emergence recombine to form plane polarised light ? Given* $\lambda = 6000$ Å, $\mu_e = 1.5532$ *and* $\mu_o = 1.5442$. *What is the plane of polarisation of emergent beam in this case ?*

Solution. When plane polarised light is incident normally on a quartz plate with its faces parallel to the optic axis, it is split up into the O-vibration and the E-vibration. These vibrations travel in the same direction but with different velocities. In traversing the thickness of the crystal plate, a path difference is introduced between the two vibrations. In order that the vibrations on emergence may combine to form plane polarised light, the thickness of the crystal plate should introduce a phase difference equal to π. This can be achieved by a half wave plate which introduces a path difference of λ/2. The thickness d of a half wave plate is given by

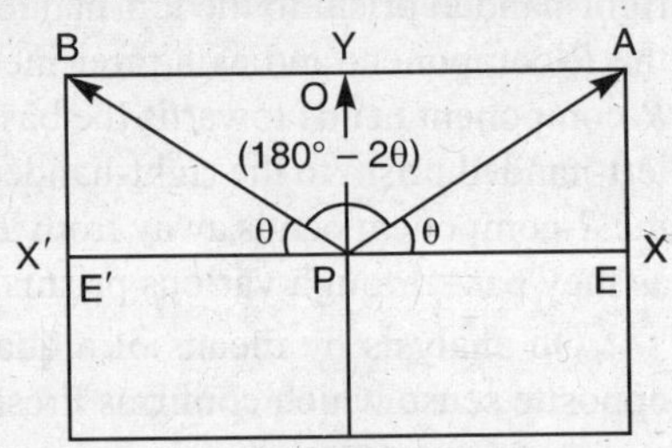

Fig. 28.38

$$\therefore \quad d = \frac{\lambda}{2(\mu_e - \mu_o)}$$

Now $\quad \lambda = 6000 \text{ Å} = 6000 \times 10^{-10}$ m, $\mu_e = 1.5532$ and $\mu_o = 1.5442$

$$\therefore \quad d = \frac{6000 \times 10^{-10}}{2(1.5532 - 1.5442)} = 0.333 \times 10^{-4} \text{ m.}$$

Plane of polarisation of emergent beam. Let the incident plane polarised vibration make an angle θ with the direction of X-axis. It is composed of E-vibration acting along PX and O-vibration acting along PY. On emergence from a crystal plate which introduces a phase difference $\delta = \pi$, the E-vibration acts along PX' while the O-vibration is still acting along PY. The resultant vibration now acts along PB. The plane of vibration is, therefore, rotated through an angle (180 – 2θ) and is along PB. The plane of polarisation is perpendicular to this plane of vibration.

28.30 LAURENT'S HALF-SHADE POLARIMETER

It is an instrument used for finding the optical rotation of certain solutions. When used for finding the optical rotation of sugar, it is called a *saccharimeter.* If the specific rotation of sugar is known, the concentration of the sugar solution can be determined.

Construction. The optical parts of a polarimeter are shown in Fig. 28.39. Monochromatic light from a source, usually a sodium lamp, after passing through a narrow slit S is rendered into a parallel beam by the lens L. This light is rendered plane polarised by the Nicol P and after passing through the half-shade device H, a glass tube T containing the solution is made to fall on the analysing Nicol A. The light is viewed through a Galilean telescope E. The analysing Nicol A can be rotated about the axis of the tube and its rotation can be measured with the help of a vernier scale on the graduated circular scale C divided in degrees.

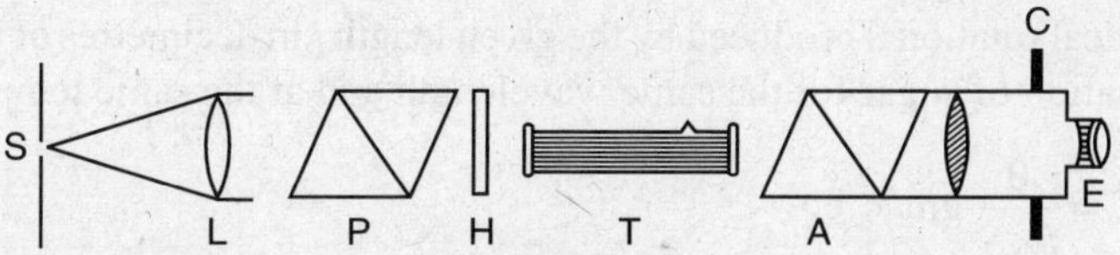

Fig. 28.39

The position of the analyser is adjusted so that the field of view is completely dark. The tube *T* is filled with the required solution and is placed in position. The field now becomes illuminated. Darkness can again be achieved by rotating the analysing Nicol *A* through a certain angle which gives the optical rotation for the solution. It is found that when *A* is rotated the total darkness of the field of view is attained rather gradually and hence it is difficult to find the exact position correctly for which complete darkness is achieved. Laurent devised an ingenious method to achieve this. The arrangement is known as *Laurent's half-shade device.*

Laurent's half-shade device. It consists of a semi-circular plate *ADB* of glass cemented to a semi-circular plate *ACB* of quartz. The quartz plate is cut with its optic axis parallel to the line of separation *AOB*. The thickness of the quartz plate is such that it introduces a phase difference of π between the *O* and *E*-vibrations. In other words, *it is half wave plate.* The thickness of the glass plate is such that it absorbs the same amount of light as is done by the quartz half wave plate.

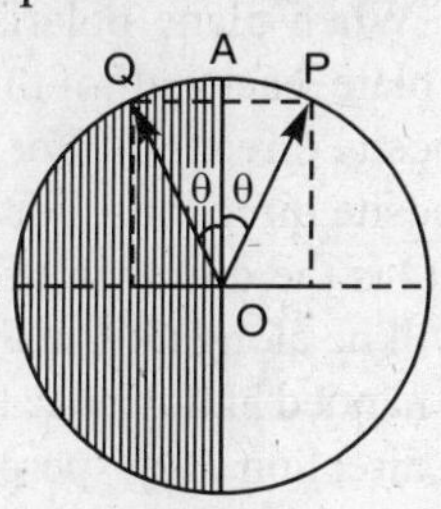

Fig. 28.40

Suppose light after passing through the polariser *P* is incident normally on the half-shade plate and has vibrations along *OP*. On passing through the glass half the vibrations will remain along OP, but on passing through the quartz half these will be split up into *E* and *O* components. The vibrations of the *O*-component are along *OD* and those of *E*-component along *OA* (Fig. 28.40).

On passing through the quartz plate a phase difference of π is introduced between these two vibrations. The *O*-vibrations will advance in phase by π and will occur along *OC* instead of *OD* on emergence. The resultant vibration on emerging from the quartz plate will, therefore, be along *OQ*, such that

$$\angle POA = \angle QOA$$

If the analysing Nicol is fixed with its principal plane parallel to *OP*, the plane polarised light through glass half will pass and hence it will appear brighter than the quartz half from which light is partially obstructed.

If the principal plane of the Nicol is parallel to *OQ*, the quartz half will appear brighter than the glass half due to the same reason.

When the principal plane of the analysing Nicol is parallel to *AOB*, the two halves will appear **equally bright.** It is because the vibrations emerging out of the two halves are equally inclined to its principal plane and hence two components will have equal intensity.

When the principal plane of the analyser is at right angle to *AOB*, again the components of *OP* and *OQ* are equal. The two halves are again equally illuminated, but as the intensity of the components passing through is small as compared to that in the previous case, the two halves are said to be **equally dark.**

The eye can easily detect a slight change when the two halves are equally dark. The readings are, therefore, taken for this position.

To find the strength of sugar solution. Fill the polarimeter tube with water and find the reading on the circular scale corresponding to equally dark position of the half shade device.

Now fill the tube completely with the given sugar solution and again find the reading on the circular scale for equally dark positions of the half shade device. The difference between the scale readings gives the optical rotation θ produced by the given length l in decimetres of the sugar solution. If S is the specific rotation of sugar for the same wavelength and at the same temperature, then

$$\text{Concentration } C = \frac{\theta}{l \times S} \text{ gm/ c.c.}$$

If the strength of sugar solution is known in gm/cc, then specific rotation $S' = \dfrac{\theta}{lC}$

28.31 BIQUARTZ POLARIMETER

A biquartz is a simple and sensitive device which can be used in place of a half shade arrangement in a polarimeter. It consists of two semi-circular quartz plates ACB and ADB each 3.75 mm thick, one of left handed L and the other of right handed R quartz cut with their axes perpendicular to their refracting faces. These are cemented together along AB to form a circular plate, as shown in Fig. 28.41 (*a*).

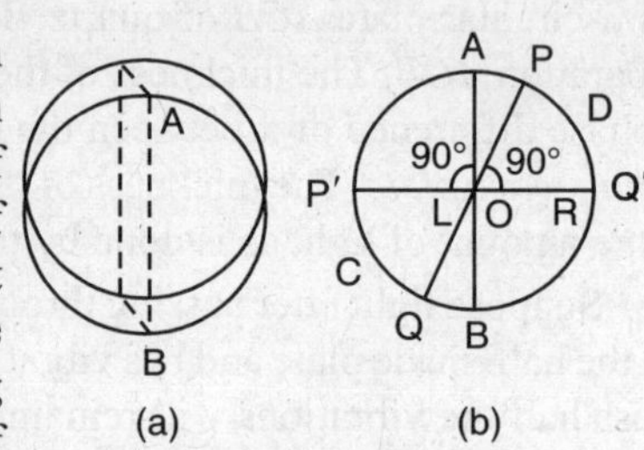

Fig. 28.41

When plane polarised white light is incident normally on the plate, each half of this plate will rotate each colour equally in opposite directions. For yellow light this rotation is about 90°. If opposite directions. For yellow light this rotation is about 90°. If POQ is the direction of incident vibration, then for yellow light it will be along OQ' in the right handed half and along OP' in the left handed half. Hence the vibration of yellow light will be along the direction $P'\,Q'$ perpendicular to the direction POQ. Thus if the principal plane of the analysing Nicol be parallel to POQ yellow light will be quenched in both halves of the field, while other colours will be present in the same proportion, thereby giving the same tint of greyish violet colour to both the halves. This is known as the tint of passage. If the analyser is rotated to one side from this position, one half of the field of view appears blue while the other half appears red. If the analyser is rotated in the opposite direction, the first half which was blue earlier appears red and the second half which was red earlier appears blue. As the transition from red to blue is very rapid, the zero position can be obtained very accurately.

28.32 MERITS AND DEMERITS OF HALF-SHADE AND BIQUARTZ POLARIMETERS

(*i*) **Laurent's half shade device** is suitable only for one wavelength for which the path difference between the E-waves and O-waves is λ/2. Usually it is constructed for yellow light of sodium.

(*ii*) When adjusted for equally dark halves, it gives fairly accurate observation as slight rotation of the analysing Nicol changes the intensity of the two haves abruptly.

Biquartz arrangement. (*i*) Zero position can be obtained with a fair degree of accuracy as slight rotation of A-Nicol changes the colour of two halves.

(*ii*) The arrangement though designed for yellow light can be used with white light when the measurement of rotation is small.

(*iii*) It furnishes a highly sensitive device for measuring optical rotation.

Disadvantages of Biquartz. (*i*) If colourless optically active substances are examined, it is not possible to achieve accurate setting for which the tint in both the halves of field is the same. It is due to rotatory dispersion produced by the optically active substance itself. This error is considerably increased when the optically active substance is used.

(*ii*) It is not possible for the colour-blind persons to use this polarimeter.

EXERCISE

1. What is birefringence ? How does it explain double refraction ?
2. What do you understand by double refraction ? What are ordinary and extraordinary rays and how will you show that these are plane polarised.
   (G.N.D.U. 2005; P.U., 2004, 2003, 2001, 2000; Meerut U., 2003, Pbi. U., 2007, 2006, 2002; K.U., 2002, 2000; M.D.U., 2000; Madurai U. 2003; Gujarat. U. 2004, 2000; Nagpur U. 2011)
3. Define optic axis of a crystal and (*ii*) Principal plane (section) of a crystal.
   (Nagpur. U. 2003; M.D.U., 2002; H.P.U.)
4. Explain double refraction through uniaxial crystal. Which of the two rays—ordinary or extraordinary—travels faster along a direction other than the optic axis in positive crystals ? What is a bi-axial crystal ?
   (Pbi. U. 2007; K.U., 2003; H.P.U., 2002; Meerut U., 2005, 2002; M.D.U., 2002; Nag. U., 2002; G.N.D.U., 2001)
5. What is the difference between negative and positive crystals ?
   (Nag. U., 2011, 2008, 2002; M.D.U., 2002; Meerut U., 2001; G.N.D.U., 2001. K.U., 2002)
6. Explain what do you understand by principal refractive indices of a crystal.
   (P.U., 2007, 1994)
7. Describe the construction and action of Nicol prism. Mention drawbacks of Nicol prism.
   (M.D.U., 2006, 2005, 2003, 2000, 1999; Indore U., 2001; Nagpur U. 2011, 2003, 2002, 2001; K.U., 2002, 2000, 1995; Gujarat U. 2004; Meerut U., 2006, 1999; G.N.D.U., 1995; P.U., 2006, 1994; H.P.U., 1994)
8. Explain how a Nicol's prism is used to produce and analyse plane polarised light. What is meant by crossed Nicols or crossed polaroids ? ***(M.D.U. 2006, 2005; Meerut. U. 2006; Nagpur U., 2002, 2001; K.U., 2002; Pbi. U., 2008, 2000; Coch. U., 1997)***
9. State and explain Malus law. ***(M.D.U. 2007, 2006, 2005, 2003; Pbi. U. 2003; G.N.D.U., 2006, 2003, 200; P.U., 2002, 2000; Meerut U., 2002; H.P.U.)***
10. State the law which gives the intensity of polarised light transmitted by a polariser in a particular orientation. ***(P.U., 2007, 2004)***
11. How will you find percentage polarisation and prove Malus' law experimentally ?
12. (*a*) What is double refraction ?
    (*b*) Give Huygen's theory of double refraction., in a uniaxial crystal.
    (Nagpur U. 2011, Pbi. U. 2006; Meerut. U., 2001; M.D.U., 2001)
13. Give electro-magnetic theory of double refraction.
14. Draw diagrams and discuss double refraction through uniaxial crystals due to plane wave when.
    (*a*) Optic axis inclined to the upper face but lying in the plane of incidence.
    (*b*) Optic axis parallel to the upper face but lying in the plane of incidence.
    (*c*) Optic axis parallel to the upper face but perpendicular to the plane of incidence.
    (*d*) Optic axis perpendicular to the upper face.
    (Nagpur U., 2001; Kerala U., 2001; Gauhati U., 2000)
15. Explain the method of finding the refractive indices for the ordinary and the extraordinary rays. (Or Explain the determination of μ_o and μ_e) ***(G.N.D.U. 2005)***
16. Explain the polarisation of light by scattering. ***(Pbi. U. 2008, 2005; Kerala U., 2001; P.U.)***
17. Explain (*a*) blue colour of the sky (*b*) red colour at sunset and sunrise.

18. What are polaroids? Give their uses. How are polaroids used for making goggles or spectacles ? ***(Pbi. U., 2005, 2003; G.N.D.U., 2007, 2005, 2002; Meerut. U., 2002)***

19. Write a short note on Wollaston prism.

20. What is meant by plane, elliptical and circularly polarised lights? Explain with necessary theory, how to obtain plane, elliptically and circularly polarised light.

(G.N.D.U. 2007, 2005; P.U., 2008, 2005, 2003; Meerut U., 2002; Luck. U., 2001; Nagpur U. 2011, 2010; 2008, H.P.U. 2002; K.U., 2000)

21. What are phase retardation plates. Name two main types of phase retardation plates. Give their uses. ***(Nagpur U. 2009; 2008; P.U. 2007; G.N.D.U. 2006)***

22. What is a quarter wave plate ? Give its theory and construction. Derive an expression for its thickness. Mention its uses. Why is it so called ? ***(G.N.D.U. 2006, 2005; M.D.U., 2002, 2001; Nagpur U., 2002, 2001; Kerala U., 2001; Meerut U., 2000)***

23. Explain the working of a half wave plate. Calculate its thickness.

(Nagpur U., 2001; K.U., 2001; Meerut U., 2006, 2001; H.P.U. 2002; P.U., 2008; G.N.D.U., 2000; M.D.U., 2007)

24. How are plane polarised, elliptically polarised and circularly polarised light produced experimentally ? ***(Nagpur U., 2009, 2002, 2001; K.U., 2002; Luck. U., 2001, Kerala U., 2001; M.D.U., 2005, 2001; Madurai. U. 2003; P.U., 2001, 2000; H.P.U., 2001, 1996; G.N.D.U., 2007, 2001; Pbi. U., 1999; Cach. U., 1997)***

25. How would you analyse plane polarised, circularly polarised and elliptically polarised light ?

(Pbi. U., 2006, 2005, 2002; Luck. U., 2001, Nagpur U. 2008, 2001; P.U., 2001; G.N.D.U., 2005)

26. Explain how the state of polarisation of a given beam of light can be determined with the help of a Nicol prism and quarter wave plate.

(Meerut. U. 2005; M.D.U. 2005; H.P.U., 2003, 2002)

27. How will you distinguish between elliptically polarised light and a mixture of plane polarised and unpolarised light ? (or partially polarised light) ***(M.D.U. 2007; Gauhati U., 2000)***

28. How is circularly polarised light distinguished from unpolarised light ?

(Meerut U., 2003, 2001; Kerala U, 2001; Gauhati U. 2000)

29. Explain the working of Babinet's compensator. Why this device is superior to quarter wave plate ? ***(Osm. U. 2004; Luck. U., 2001; Indore U. 2001; Garhwal U., 2000; P.U., 2001)***

30. Explain clearly (*i*) an optically active substance and (*ii*) specific rotation.

(Madurai U. 2003; Luck. U., 2001; K.U., 2001; Meerut U., 2006, 2001; M.D.U., 2006, 2000, Nagpur U. 2009)

31. Give Fresnel's explanation of rotation of plane of polarisation by an optically active substance.

(Meerut. U. 2007; Osm U. 2004; Bhopal U. 2004; Nagpur U. 2008)

32. (*a*) Describe the construction and principle of a half-shade polarimeter.

(*b*) How will you use it to find (1) specific rotation and (*ii*) the strength of sugar solution ?

(Meerut U., 2003, 2001; M.D.U., 2006, 2003, 2001; K.U., 2001; Luck. U., 2001; Garhwal U. 2004)

33. Give the construction and working of the biquartz arrangement in a polarimeter.

(Meerut. U. 2006, 2005; K.U., 2000; M.D.U., 2007, 2000)

34. Give relative merits of Half shade and Biquartz polarimeters. ***(M.D.U. 2007; K.U.)***

35. Explain rotatory polarization. What are dextro-and leavo-rotatory substances ?

(Nagpur U. 2008)

36. The polarising angle for a medium is 60°. What will be the critical angle for the medium ? **[Ans.** 25° 16′]

37. A beam has an intensity of 10 watt/m^2 and is linearly polarised in the vertical direction. Find the intensity of transmitted beam by a polaroid when the plane of transmission makes an angle of 60° with the vertical. **[Ans.** 2.5 watt/m^2]

38. Plane polarised light is incident on a quartz piece cut with faces parallel to the axis. Find the least thickness for which the *O*-ray and the *E*-ray will combine to form plane polarised light. Given $\mu_o = 1.5442$ and $\mu_e = 1.5533$ and $\lambda = 5000$ Å. **[Ans.** 2.748×10^{-5} m]

39. Calculate the thickness of the quartz plate for *C*-line of wavelength 6563 Å. Given $\mu_e = 1.55085$ and $\mu_o = 1.54181$. **[Ans.** 3.629×10^{-5} m]

40. A 10 cm long tube containing 20% sugar solution produces an optical rotation of 13.2°. Find the specific rotation. **[Ans.** 66°]

41. On putting a polarimeter tube 25 cm long containing sugar solution of unknown strength, the plane of polarisation is rotated through 10°. Find the strength of sugar solution. Specific rotation of sugar is 60° / decimetre / unit concentration. **[Ans.** 6.67%]

42. A 22 cm along tube containing 88 cc of sugar solution produces an optical rotation of 9.9° when placed in polarimeter. If the amount of sugar in the solution is 6 gm, find the specific rotation. **[Ans.** 66°]

SHORT QUESTIONS WITH ANSWERS

1. *Give one example each of positive and negative unioxial crystal.*

Ans. Positive-Quartz; Negative-Calcite

2. *Light is generally characterised by electric vector although it also possesses magnetic vector. Explain.* ***(G.N.D.U. 2007; P.U., 2000; H.P.U., 2002)***

Ans. Light is a form of electromagnetic wave motion. The electromagnetic waves have electric and magnetic field vectors oscillating in phase but perpendicular to each other. The interaction of electromagnetic waves with matter can be explained either with the help of electric vector of with magnetic vector. But the peak value of electric vector is much higher than that of the magnetic vector. Moreover, our eye is more sensitive to electric vector than to magnetic vector. Hence polarisation and other optical phenomenon are preferably described in terms of electric vector. This is why light is characterised by electric vector although it also possesses magnetic vector.

3. *When a beam of light is polarised, does its intensity vary ?* ***(G.N.D.U. 2007; Pbi. U., 2000)***

Ans. Yes. The polariser allows light waves oscillating only in one particular direction to pass through and cuts off those oscillating in all other planes. Hence the intensity of the polarised beam decreases.

4. *Unpolarised light is symmetrical about the direction of propagation. Comment.*

Ans. Unpolarised light contains oscillations in all directions in the plane perpendicular to the direction of propagation. This is what we mean by saying that unpolarised light is symmetrical about the direction of propagation. ***(Pbi. U., 2000; H.P.U., 2001)***

5. *Can our eye distinguish between polarised light and unpolarised light ?* ***(Pbi. U., 2001)***

Ans. No. Our eye cannot distinguish between polarised light and unpolarised light.

6. *What is meant by crossed polaroids ?* ***(G.N.D.U. 2008; Pbi. U., 2000)***

Ans. Polaroids are said to be in crossed position when the plane of polarisation of the two polaroids are at right angles to each other. In crossed position no light passes through the two combined polaroids.

7. *What are the five ways in which plane polarised light can be produced ?* **(G.N.D.U. 2000)**

Ans. Plane polarised light can be produced by

(*i*) Reflection—when light is incident on a reflecting surface at the polarising angle.

(*ii*) Wire-grid polariser

(*iii*) Double refraction

(*iv*) Scattering

(*v*) Dichroism—polarisation by selective absorption.

8. *When a polaroid is rotated the intensity of light varies but never becomes zero. It shows that the incident light is*

(a) Completely plane polarised (b) unpolarised (c) partially plane polarised (d) none of these. **(H.P.U., 2002)**

Ans. (*c*) Partially plane polarised.

9. *What do you conclude about the nature of light ?*

Ans. Light is a transvere wave motion.

10. *What is the greatest importance of polarization ?* **(Meerut U. 2003, 2002)**

Ans. It confirms the wave theory and transverse nature of light.

11. *Explain why light can be polarized, but sound can not be polarized ?* **(G.N.D.U. 2008; P.U. 2000)**

Ans. It is because light waves are transverse in nature while sound waves are longitudinal.

12. *What is the nature of emergent polarised light when circularly polarised light is passed through (i) a quarter wave plate and (ii) a half wave plate.*

Ans. (*i*) When circularly polarised light is passed through a *quarter wave plate* such that the vibrations make an angle of 45° with the direction of optic axis, the outgoing light will be plane polarised as the phase difference between the *E*-component and *O*-component becomes $\frac{\pi}{2}+\frac{\pi}{2}=\pi$.

(*ii*) When left-handed circularly polarised light is passed through a *half wave plate* such that the vibrations make an angle of 45° with the direction of optic axis, the outgoing light is right-handed circularly polarised light as $\delta=\frac{\pi}{2}+\pi=\frac{3\pi}{2}$.

13. *How will you detect plane polarised light from partially polarised light ?*

Ans. When plane polarised light is incident normally on a quarter wave plate, the vibration making an angle 45° with the direction of optic axis, it is converted into circularly polarised light. If the outgoing light is examined through a Nicol prism, it will show no change in intensity as the Nicol is rotated.

When partially polarised light is passed through a Nicol prism it becomes plane polarised. If it is now examined through the analysing Nicol prism, the outgoing light will be extinguished twice in one rotation of the analyser.

14. *Explain how will you detect beams of circularly and elliptically polarized lights.*

Ans. **Detection of circularly polarised light.** To confirm whether light emerging out of the quarter wave plate is circularly polarised, it is first passed through another quarter wave plate with optic axis in any position. The light on emergence from the second quarter wave plate will become plane polarised. If it is now examined through a Nicol prism, it will show variation of intensity with zero minimum.

Detection of elliptically polarised light. To confirm if the light coming out of the quarter wave plate is elliptically polarised, first examine it with a rotating Nicol and adjust the position of the Nicol for maximum intensity. Place another quarter wave plate between the first quarter wave plate and the analysing Nicol such that the optic axis of the second quarter wave plate is parallel to the principal section of the Nicol adjusted for maximum intensity. If the light was elliptically polarised it will now become plane polarised. If the Nicol prism is now rotated, variation of intensity from maximum to zero will be observed.

MULTIPLE CHOICE QUESTIONS

28.1 If θ is the polarizing angle, the refractive index μ of the material is given by: ***(H.P.U. 2003)***

(*a*) sin θ (*b*) tan θ (*c*) cos θ (*d*) $\cos^{-1}(\theta)$

28.2 Quartz is a:

(*a*) Negative bi-axial crystal (*b*) Negative uni-axial crystal
(*c*) Positive bi-axial crystal (*d*) Positive uniaxial crystal

28.3 Calcite is a :

(*a*) Negative bi-axial crystal (*b*) Negative uni-axial crystal
(*c*) positive bi-axial crystal (*d*) positive uniaxial crystal

28.4 The axis along which *E*-ray and *O*-ray has equal velocities is called

(*a*) optic axis (*b*) principal axis
(*c*) polarizing axis (*d*) non-polarizing axis

28.5 According to Malu's law, the intensity of transmitted light at an angle θ is proportional to:

(*a*) sin θ (*b*) $\sin^2 \theta$ (*c*) cos θ (*d*) $\cos^2 \theta$

28.6 According to Rayleigh, the intensity of scattered light is proportional to:

(*a*) λ^4 (*b*) λ^3 (*c*) λ^2 (*d*) $1/\lambda^4$

28.7 A quarter-wave plate produces a phase difference between *E*-ray and *O*-ray of:

(*a*) π (*b*) $\frac{\pi}{2}$ (*c*) $\frac{\pi}{4}$ (*d*) 2π

28.8 A half-wave plate produces a phase difference between *E*-ray and *O*-ray of:

(*a*) π (*b*) $\frac{\pi}{2}$ (*c*) $\frac{\pi}{4}$ (*d*) 2π

28.9 In a positive crystal *E*-ray has minimum velocity in a direction making an angle to the direction of optic axis.

(*a*) 0° (*b*) 45° (*c*) 90° (*d*) 135°

28.10 In a positive crystal *O*-ray travels in all directions:

(*a*) faster (*b*) slower (*c*) same velocity (*d*) zero

ANSWERS

28.1 (*b*)	**28.2** (*d*)	**28.3** (*b*)	**28.4** (*a*)	**28.5** (*d*)
28.6 (*d*)	**28.7** (*b*)	**28.8** (*a*)	**28.9** (*c*)	**28.10** (*c*)

SECTION VIII

Lasers

29. LASER SYSTEM AND NON-LINEAR OPTICS

30. TYPES OF LASERS AND THEIR APPLICATIONS

Laser light is highly directional, or collimated. This means that all the photons travel in the same direction. There is one more important quality of laser light. It is coherent, or very organized and all the waves in a laser beam have the same wave pattern. This means that the light waves are lined up with each other and moving along in step, almost like a regiment of soldiers marching in a parade. By contrast, ordinary light is incoherent. Its waves become mixed up as they move along, like the crowd of people watching the parade. These properties make laser light more intense than ordinary light and allow it to travel long distances. Ordinary light waves scatter in all directions from their source. As the beam of ordinary light travels, the more it spreads out and gets dimmer. It tends interfere with one another. On the other hand, collimated laser light can travel a great distance without losing very much of its energy and brightness. Ordinary light could never have made it to the moon mirror and back, whereas laser light did quite easily. So, the laser light is special. It is concentrated, directional, and organized. Thus, a laser beam has the most striking features like directionality, high degree of coherence and extraordinary monochromaticity. This is well illustrated in the following figure.

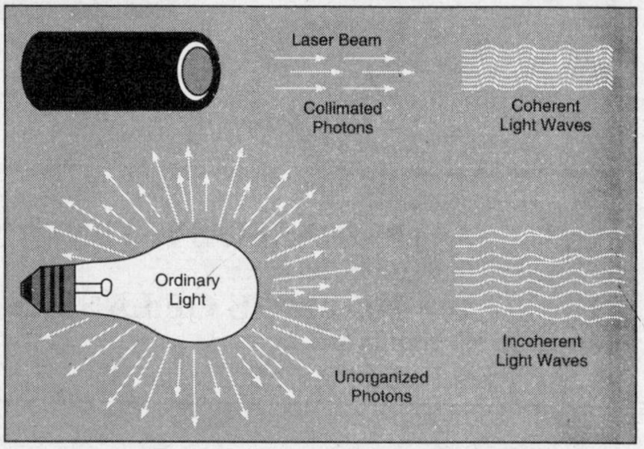

These three qualities combined together have made the laser an extremely powerful and useful tool. Laser can be used in various fields like fundamental research in **Science,** nuclear fusion, nuclear fission, in isotope separation, wide variety of interferometric techniques, Raman effect mechanical industry like surface processing, cutting, drilling, welding, photolithography used in manipulations of metals, plastics, wood, ceramics, cloth, and other materials. Laser based light detection and ranging (LIDAR) technique has applications in geology, surveying, metrology, seismology, remote sensing. In atmospheric physics, lasers have applications in atmospheric remote sensing, pollution monitoring, hologram preparation and viewing. In astronomy, lasers have been used to create artificial laser guide stars. Lasers are used in entertainment electronics, laser printing, CDs, DVDs, optical memory cards, data processing and optical communications. In consumer and industrial imaging equipment utilises barcode readers. In **Medicine,** for cosmetic surgery (removing tattoos, scars, stretch marks, sunspots, wrinkles, birthmarks, and hairs), Eye surgery and refractive surgery and ophthalmology, soft tissue surgery, general surgery, gynaecological, urology, laparoscopic, endoscopy, "No-Touch" removal of tumours, especially of the brain and spinal cord. In dentistry for caries removal, tooth whitening, and oral surgery. Lasers are used in dermatology to treat various skin conditions. In **Defence,** lasers include applications such as target designation and ranging, velocity measurement, defensive countermeasures, communications and directed energy weapons. It disrupts the trajectory of shoulder-fired missiles. They are effectively employed to study microscopic objects, including molecules, DNA testing and even to fight crime.

LASER SYSTEM AND NON-LINEAR OPTICS

29 Chapter

INTRODUCTION

LASER is one of the most important inventions of the 20th century* (early 1960). The acronym of LASER is **L**ight **A**mplification by **S**timulated **E**mission of **R**adiation. It is a light source but differs vastly from the common light sources and is not useful for illumination purposes like incandescent lamps, CFL lamps etc. By nature they are electromagnetic waves like radio and microwaves. Because of the remarkable characteristics of directionality, coherence, monochromaticity and high intensity exhibited by laser beam, Laser occupies a unique place in modern technology. Today, scientists, lab technicians, engineers and industrial technicians regularly utilise lasers to perform a wide range of important tasks. This includes, metal processing, printing, entertainment, communications, medical diagnosis, surgery, opthalmology, astronomy, geography, defence, DNA testing and even to fight crimes.

29.1 ENERGY LEVELS

Max Planck, in 1900 proposed that light consists of discrete bundles or chunks of energy. The energy of each bundle is $h\nu$. Such bundles of energy are called *quanta*. In 1905, Einstein refined the quantum hypothesis of Planck and provided theoretical justification for the features of photo-electric effect. He gave the name "photon' to the quantum of light energy. Each photon carries an amount of energy proportional to the frequency (ν) of light wave and is given by $E = h\nu$, where h is planck's constant. It is obvious that the higher the frequency of a photon, the more energy it possesses. Thus, *UV* light photons are more energetic than visible light photons. The light energy, $\rho(\nu)$ emitted by a source or propagating in space, can not take arbitrary values but must be integral multiples of the photon energy $h\nu$. Thus, *UV* light photons are more energetic than visible light photons.

$$\rho(\nu) = nh(\nu) \qquad \text{... (29.1)}$$

where $n = 1, 2, 3, ...$ This is known as "quantization of energy"

The photon energy is usually expressed in terms of electron volt (eV). The photon energy (in eV) corresponding to the light of wavelength λ is given by

$$E = \frac{12,400}{\lambda} \quad \text{(in eV)} \qquad \text{... (29.2)}$$

where λ is in Angstrom unit.

According to Bohr's postulates of permitted orbits, the electrons in an atom cannot have arbitrary amounts of energy but they take only discrete energies. The energy of electron at a very large distance from the nucleus is purely kinetic energy and it is assigned a positive value. As the distance between the nucleus and the electron decreases, the kinetic energy dereases. When the electron is just about to come under the influence of nucleus, its energy becomes zero. When the electron becomes bound to the nucleus and orbits round it, electron energy is potential in nature and treated as a negative

---

* In 1917 Albert Einstein put forward the principle of stimulated emission, in 1958 Townes, Schawlow, Basov, Prokhorov enunciated the Laser principle; in 1960 Ruby laser was demonstrated by Theodore Maiman; in 1961 first gas laser (He-Ne laser) was developed by A. Javan, W. Bennett and D. Harriott of Bell Laboratories.

quantity. Among the infinite number of mechanically possible orbits, only a certain set of orbits are actually permissible for electrons due to quantum restrictions. These permitted orbits are characterized by allowed energy values. We therefore say that electron in an atom can occupy only certain energy levels. The energy levels are schematically represented by horizontal lines drawn to an energy scale. For example, electron orbits and the corresponding energy levels of hydrogen atom are shown in Fig 29.1. The atom is in its lowest possible energy level when the electron is in the innermost orbit, closest to the nucleus. Then, the atom is said to be in the *ground state.* An electron in this ground state is stable and moves in this orbit continuously without emitting energy. If electron in ground state absorbs energy in some way, it will go to a higher energy level and the atom is said to be *excited.* The passing of an electron from one energy level to another level within the atom occurs in a jump which is called a *quantum transition.* The electron transitions may be induced by a variety of ways. Interaction with light photons is one of the means of supplying energy to orbital electrons which causes upward electron transitions and sends the atom into its excited state.

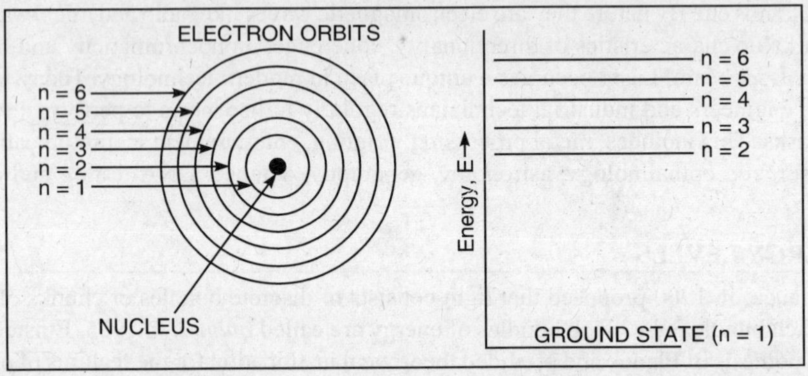

Fig. 29.1. Electron orbits and the corresponding energy levels of the hydrogen atom.

29.2 POPULATION

The atoms of each chemical element have their own characteristic system of energy levels. The energy difference between the successive energy levels of an atom is of the order of 1 eV to 5 eV. The energy levels are common to all the atoms in a system which is composed of identical atoms. We can therefore say that a certain number of atoms occupy a certain energy state. The number of atoms per unit volume that occupy a given energy state is called the *population* of that energy state. The population N of an energy level E depends on the temperature T. Thus,

$$\boxed{N = e^{-E/kT}} \quad \text{Boltzmann's equation} \qquad \text{... (29.3)}$$

where k is known as the Boltzmann constant.

In a material, atoms are distributed differently in different energy states. The atoms normally tend to be at their lowest possible energy level which need not be the ground state. At temperatures above 0 K, the atoms always have some thermal energy and therefore, they are distributed among the available energy levels according to their energy.

29.3 THERMAL EQUILIBRIUM

At thermal equilibrium, the number of atoms (population) at each energy level decreases with the increase of energy level as shown in Fig. 29.2.

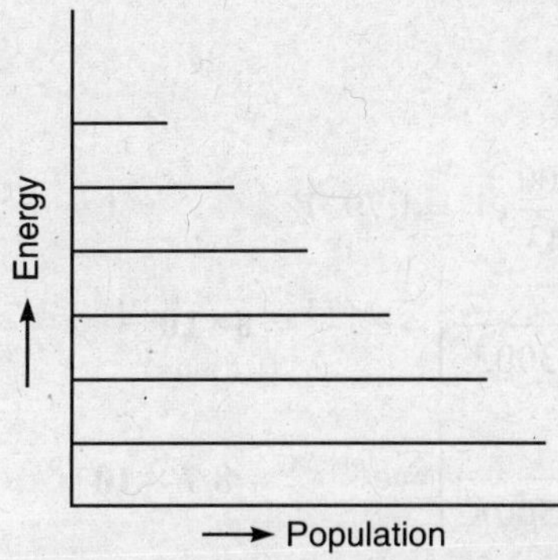

Fig. 29.2. Relative populations of energy levels above the ground state at thermal equilibrium

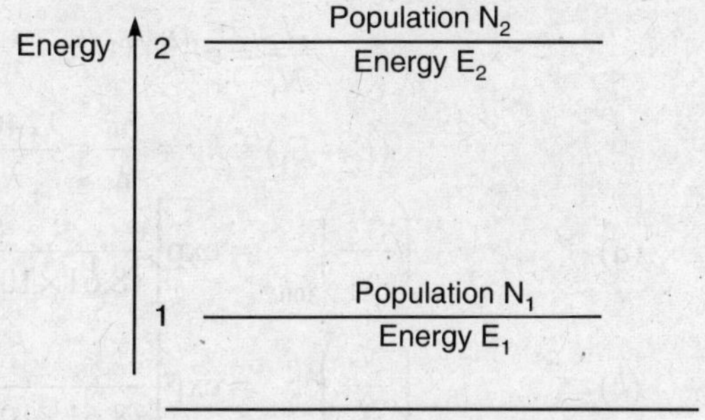

Fig. 29.3. Two energy level system

If we consider two energy levels E_1 and E_2 as shown in Fig 29.3, their populations can be computed using Boltzmann equation (29.3). Thus,

$$N_1 = e^{-E_1/kt} \quad \text{and} \quad N_2 = e^{-E_2/kt} \qquad \text{... (29.4)}$$

The ratio of populations in these two states, N_2/N_1 is called *relative population* and is given by

$$\frac{N_2}{N_1} = \frac{e^{-E_1/kt}}{e^{-E_2/kt}}$$

or

$$\frac{N_2}{N_1} = e^{-(E_2-E_1)/kT} \qquad \text{... (29.5)}$$

$$\therefore \quad N_2 = N_1\, e^{-\Delta E/kT} \qquad \text{... (29.6)}$$

where $\Delta E = (E_2 - E_1)$. The plot of eq. (29.6) is known as **Boltzmann distribution.** Eq. 29.5 suggests that relative population N_2/N_1 depends upon two factors: (*i*) the energy difference $(E_2 - E_1)$ and (*ii*) the temperature T.

For simplicity, let us take the example of hydrogen gas which is a mon atomic gas and find out the atomic population at room temperature at the fixed excited level. Here $E_1 = -13.6$ eV; $E_2 = -3.39$ eV and $T = 27°C = 300$ K. Thus,

$$\Delta E = (E_2 - E_1) = +[(-3.39) - (-13.6)] \text{ eV} = +10.21 \text{ eV}$$

$$kT = (8.6 \times 10^{-5} \text{ eV/K})(300 \text{ K}) = 0.025 \text{ eV}$$

$$\therefore \quad \frac{N_2}{N_1} = e^{\frac{-10.21}{0.025}} = e^{-408.4} \cong 0$$

This means that at room temperature all atoms are in ground state. If temperature is raised, say to 6000 K,

$$\frac{N_2}{N_1} = e^{-\frac{10.21}{0.516}} = e^{-19.79} = 2.5\times10^{-9}$$

Thus, we find that in a material at thermal equilibrium, more atoms are in the lower energy state than in the higher energy state.

Example 29.1 *Find the relative population of two states in a Ruby laser that produces a light beam of wavelength 6943 A° at 300 K and 500 K*

Solution. The population ratio is given by

$$\frac{N_2}{N_1} = e^{-(E_2-E_1)/kT}$$

$$\therefore \quad (E_2 - E_1) = h\nu = \frac{hc}{\lambda} = \frac{12400}{\lambda} eV = \frac{12400}{6943} eV = 1.79\, eV$$

(*a*) $$\left(\frac{N_2}{N_1}\right)_{300k} = \exp\left[\frac{-1.79\, eV}{(8.61\times10^{-5} eV/K)\times 300K}\right] = e^{-69.3} = \mathbf{8\times10^{-31}}$$

(*b*) $$\left(\frac{N_2}{N_1}\right)_{500k} = \exp\left[\frac{-1.79\, eV}{(8.61\times10^{-5} eV/K)\times 500K}\right] = e^{-41.58} = \mathbf{8.7\times10^{-19}}$$

29.4 ABSORPTION AND EMISSION OF LIGHT

In an atom, an electron in the ground state is stable and moves continuously in that orbit without radiating energy. When the electron receives an amount of energy equal to the difference of energy of the ground state and one of the excited states (*i.e.* outer orbits), it absorbs energy and jumps to the excited state. There are a variety of ways in which the energy may be supplied to the electron. One way is to illuminate the material with light of appropriate frequency $\nu = (E_2 - E_1)/h$. The photons of energy $h\nu = (E_2 - E_1)$ induce electron transitions from the energy level E_1 to the level E_2, as shown in Fig. 29.4

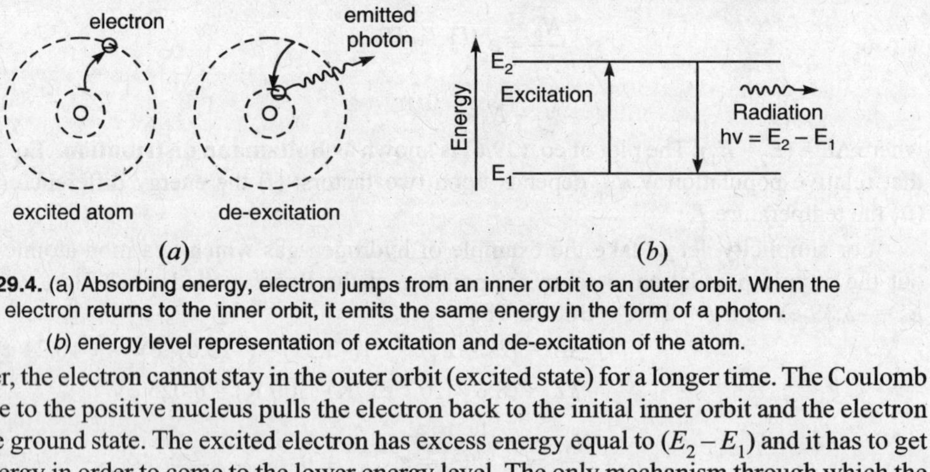

Fig. 29.4. (a) Absorbing energy, electron jumps from an inner orbit to an outer orbit. When the electron returns to the inner orbit, it emits the same energy in the form of a photon. (*b*) energy level representation of excitation and de-excitation of the atom.

However, the electron cannot stay in the outer orbit (excited state) for a longer time. The Coulomb attraction due to the positive nucleus pulls the electron back to the initial inner orbit and the electron returns to the ground state. The excited electron has excess energy equal to $(E_2 - E_1)$ and it has to get rid of this energy in order to come to the lower energy level. The only mechanism through which the electron can lose its excess energy is through the emission of a photon. Therefore, the excited electron emits a photon of energy $h\nu = (E_2 - E_1)$ and returns to the ground state. This is the visualization of emission of light according to Bohr's quantum theory.

When we see light from any source, we actually "see" electrons jumping from excited states to lower states. This type of emission of light which occurs on its own is known as *spontaneous emission* and is responsible for the light coming from candles, electric bulbs, fire, stars, sun etc. conventional sources of light.

29.5 COHERENCE: TEMPORAL AND SPATIAL

Coherence. *The term 'coherence' refers to the degree of co-relation between the phases at different points and different times in a beam of light or radiation.*

Coherence is of two types:

(*i*) Temporal or longitudianl coherence and (*ii*) spatial or transverse coherence

Temporal (or longitudinal) coherence. A one-dimensional plane wave propagating in + X direction can be represented as

$$\psi(x, t) = Ae^{i(kx - \omega t)}$$

where ψ is the displacement at any time and A is the amplitude of the wave which is in general a complex quantity. If we write

$$A = |A| e^{i\phi}$$

Then

$$\psi(x, t) = |A| e^{i(kx - \omega t + \phi)}$$

$$= |A| [\cos(kx - \omega t + \phi) + i \sin(kx - \omega t + \phi)]$$

If E is the actual displacement which is given by the real part of ψ, then

$$E = |A| \cos(kx - \omega t + \phi)$$

According to this equation the displacement remains sinusoidal for all times ($-\infty < t < +\infty$) as shown in Fig. 29.5(*a*). In actual practice no source is either a point or strictly monochromatic so that no light is a perfect sine wave extending over infinite space or time. It is because a light pulse from a source is emitted when an orbital electron in an excited atom jumps from a higher energy level to a lower energy level (or ground state) thereby releasing the energy difference between the two levels in the form of a light wave of a short duration. The duration of the pulse is of the order of 10^{-10} sec. This means the field remains sinusoidal for about 10^{-10} sec.

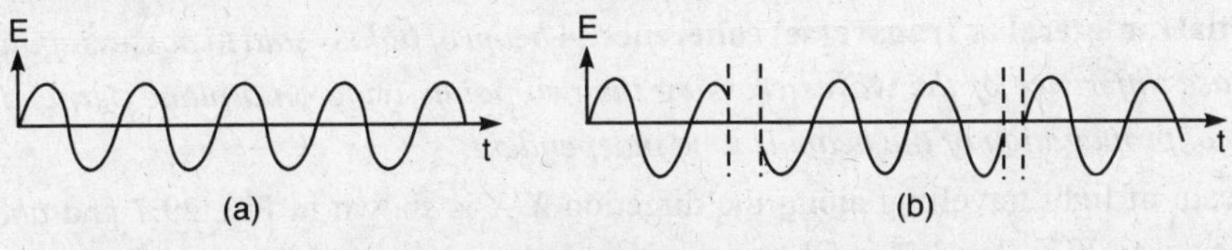

Fig. 29.5

This time is known as the average lifetime of the wave train.

The emission of light from different atoms or even from the same atom at different times is a random and arbitrary process. So the phase of the light pulses given out by the source varies randomly and independently of each other as shown in Fig. 29.5 (*b*).

Such a beam of light does not possess temporal coherence.

Temporal coherence. *A beam of light is said to possess temporal coherence if the phase difference of the wave crossing the two points lying along the direction of propagation of the beam is time independent.*

Temporal coherence is also known as *longitudinal coherence.*

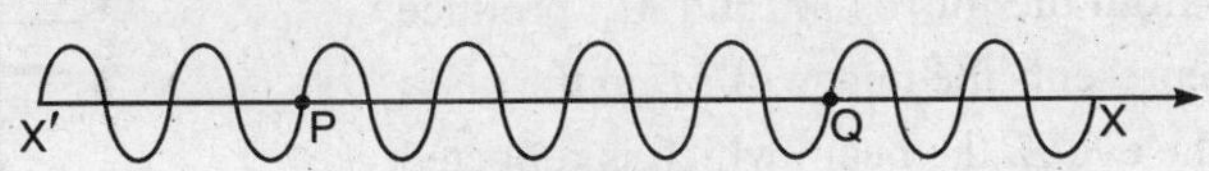

Fig. 29.6

A beam of light travelling along the direction $X'X$ is shown in Fig. 29.6. P and Q are two points lying on the line $X'X$. Suppose the phases of the wave crossing the points P and Q at any instant t_1 are ϕ_1 and ϕ_2 and the same when measured at the time t_2 are ϕ_1' and ϕ_2', then the beam is said to be coherent if $\phi_2 - \phi_1 = \phi_2' - \phi_1'$.

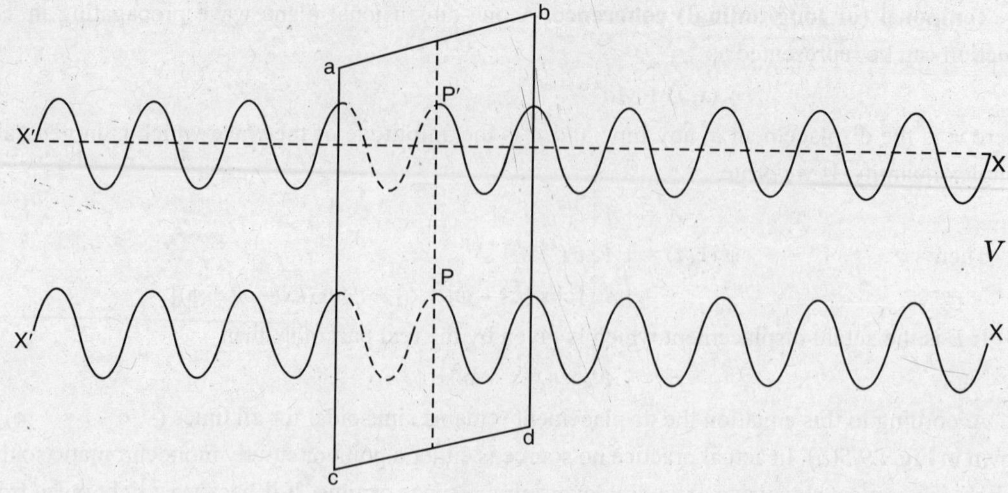

Fig. 29.7

Coherence time. *The time interval over which this phase difference remains constant is known as coherence time,* τ_c.

For a single pulse of light emitted by a source, coherence time is equal to average lifetime.

Spatial (or lateral or transverse) coherence. *A beam of light is said to possess spatial coherence if the phase difference of the waves crossing the two points lying on a plane perpendicular to the direction of propagation of the beam is time independent.*

A beam of light travelling along the direction $X'X$ is shown in Fig. 29.7 and *abcd* is a plane perpendicular to $X'X$. P and P' are the two points on the plane within the beam of light. Suppose the waves crossing P and P' at any time t_1 have the same phase ϕ_1 and when measured at a time t_2 again have the same phase ϕ_2 then the beam is said to possess spatial coherence.

Coherent length. *The distance travelled by the light pulse during the coherent time is known as coherent length or spatial interval.*

If τ_c is the coherence time of the beam, then coherent length $L = c\tau_c$ ... (29.7)

where c is the velocity of light.

Determination of coherent length. To understand the concept of coherence time and coherence length, consider the example of Michelson's interferometer, in which light beam is divided into a reflected and transmitted beams of equal intensity by the half silvered glass plate G, which after suffering reflections from the mirros M_1 and M_2, produce interference. If M_2' represents the image of M_2 in the glass plate G, as seen by the eye E, the beam which is reflected from the mirror M_2 travels an additional distance $2d$. Thus the beam reflected from M_1 interferes with the beam reflected from M_2 which originates $\frac{2d}{c}$ seconds earlier. If the distance d is such that $\frac{2d}{c} \le \tau_c$, then a definite phase-relationship exists between the two beams and so

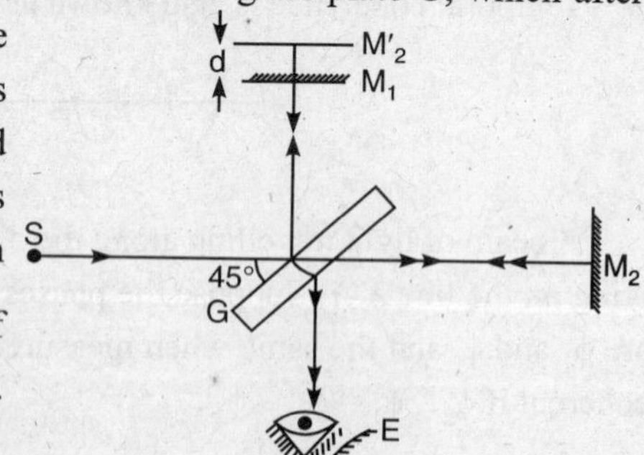

Fig. 29.8

interference fringes are observed. If $\frac{2d}{c} \geq \tau_c$, then no phase-relationship will exist between the two beams and hence interference pattern is not observed.

There is no definite distance at which the interference pattern disappears as *d* increases. However, contrast of fringes becomes poorer and poorer, eventually fringe system disappears. For neon λ = 6382 Å, the disappearance occurs when the path difference is only a few cm. The coherence time for neon is 10^{-10} sec.

$$\therefore \qquad L = 10^{-10} \times 3 \times 10^{10} = 3 \text{ cm}$$

For cadmium red λ = 6438 Å, the coherence length is of the order of 30 cm.

$$\therefore \qquad \tau_c = \frac{L}{c} = \frac{30}{3} \times 10^{-10} = 10^{-9} \text{ sec.}$$

For commercially available lasers, the coherence length can exceed a few kilometres.

29.6 LINE WIDTH AND FREQUENCY SPREAD

The line width of a laser, typically a single-frequency laser, is defined as the width (the full width at half-maximum, FWHM) of its optical spectrum.

In Michelson's interferometer, it has seen that the interference pattern produced by two close wavelenths λ_1 and λ_2 will disappear if

$$\frac{2d}{\lambda_2} - \frac{2d}{\lambda_1} = \frac{1}{2}$$

where $2d$ is the path difference between the two beams producing interference. Thus

$$2d = \frac{\lambda_1 \lambda_2}{2(\lambda_1 - \lambda_2)} = \frac{\lambda^2}{2(\lambda_1 - \lambda_2)} \qquad \text{... (29.8)}$$

where λ is the mean wavelength of λ_1 and λ_2.

If instead of two discrete wavelengths, we consider a beam having wavelengths lying between λ and $(\lambda + \delta\lambda)$, then the fringes will not be visible if $2d \leq \frac{\lambda^2}{\delta\lambda}$.

This is because if $2d = \frac{\lambda^2}{2\left(\frac{\delta\lambda}{2}\right)} = \frac{\lambda^2}{\delta\lambda}$ the interference pattern produced by the wavelengths λ and $\lambda \pm \frac{\delta\lambda}{2}$ will disappear.

$\delta\lambda$ is called the *natural line width.* It means that the light wave train which is supposed to be of wavelength λ actually has an effective wavelength between $\left(\lambda - \frac{\delta\lambda}{2}\right)$ and $\left(\lambda + \frac{\delta\lambda}{2}\right)$. As such, when the wave trains superimpose to produce an interference pattern, there will be a certain amount of uncertainty in the existence of maxima and minima at the observation point.

The line width is caused due to the uncertainty in the frequency of the wave train emitted when the orbital electron jumps from a higher orbit to a lower orbit. This uncertainty in frequency is due to Doppler effect because of the motion of the electron — the source of light. This is why line width is also known as *Doppler width.*

The temporal coherence of the beam is thus directly associated with the width of the spectral line.

Relation between frequency spread and coherence time. Since the fringes are not observed if the path difference exceeds the coherence length *L*, we can assume that if the beam contains all the wavelengths lying between λ and $(\lambda + \delta\lambda)$, then

Coherent length $L = \frac{\lambda^2}{\delta\lambda}$

or $\delta\lambda = \frac{\lambda^2}{L}$ ... (*i*)

Further, since $v = \frac{c}{\lambda}$, the *frequency spread* δv of the line would be

$$\delta v = \frac{c}{\lambda^2}\delta\lambda \text{ (neglecting the negative sign)}$$

$$= \frac{c}{\lambda^2} \times \frac{\lambda^2}{L} = \frac{c}{L}$$

$\therefore$ Frequency spread or frequency width $\delta v = \frac{c}{L}$ ... (*ii*)

Also the coherent length $L = c\tau_c$ where τ_c is the coherence time

$$\tau_c = \frac{L}{c} \quad \text{... (iii)}$$

From relation (*ii*) and (*iii*), we have $\delta v = \frac{1}{\tau_c}$ ... (29.9)

Thus the frequency spread of a line is of the order of the reciprocal of coherence time. It is clear that if frequency spread is known, the coherence time can be calculated.

Example 29.2 *Find the coherent length for white light, the wavelength of white light ranges from 400 n.m. to 700 nm.* **(*H.P.U. 2003*)**

Solution. Here $\delta\lambda = 700 \text{ n.m} - 400 \text{ n.m.} = 300 \text{ nm}.$

Average wavelength $\lambda = \frac{\lambda_1 + \lambda_2}{2} = \frac{700 + 400}{2} = 550 \text{ nm}.$

$\therefore$ Coherent length $L = \frac{\lambda^2}{\delta\lambda} = \frac{(550 \times 10^{-9})}{300 \times 10^{-9}} = 10^{-6} \text{ nm}.$

Example 29.3 *If light of wavelength 660 n.m. has wave trains 13.2 × 10^{-6} long, calculate coherent time.* **(*G.N.D. 2002*)**

Solution. Coherent length L = Length of one wave train = 13.2 × 10^{-6} m, Coherence time τ_c = ?

Now $L = c\tau_c$

$\therefore$ $\tau_c = \frac{L}{c} = \frac{13.2 \times 10^{-6}}{3 \times 10^8} = 4.4 \times 10^{-14}\ S = 44 \times 10^{-15}\ S$

Example 29.4 *The Doppler width for orange line of krypton is 550 × 10^{-15} m. If the wavelength of light is 605.8 n.m., Calculate the coherent length.* **(*G.N.D.U. 2007, 2003*)**

Solution. Doppler width = line width $\delta\lambda$ = 550 × 10^{-15} m

λ = 605.8 n.m = 605.8 × 10^{-9} m

Coherent length L = ?

Now $L = \frac{\lambda^2}{\delta\lambda} = \frac{605.8 \times 605.8 \times 10^{-18}}{550 \times 10^{-15}} = 667.3 \times 10^{-3} \text{ m}$

$= 66.73$ cm

Example 29.5 *Compute the coherence length of yellow light with 5893 Å in 10^{-12} sec. pulse duration. Find also the band width.* **(P.U. 2007)**

Solution. Coherence time = pulse duration $\tau_c = 10^{-12}$ sec.

∴ Coherence length $L = c\tau_c = 3 \times 10^8 \times 10^{-12} = 3 \times 10^{-4}$ m = 0.3 m.m

$$\text{Band width} \qquad \delta\lambda = \frac{\lambda^2}{L} = \frac{5893 \times 5893 \times 10^{-10} \times 10^{-10}}{3 \times 10^{-4}} = 11.6 \times 10^{-10} \text{ m}$$

$$= 11.6 \text{ Å}.$$

Example 29.6 *Find the coherence length of a laser beam for which band width δv = 3000 Hz.* **(M.D.U. 2008, 2003, 2000)**

Solution. Here band width δv = 3000 Hz

$$\therefore \text{ Coherence time} \qquad \tau_c = \frac{1}{\delta v} = \frac{1}{3000} \text{ s}$$

and coherence length $L = c\tau_c = 3 \times 10^8 \times \dfrac{1}{3000} = 10^5 \text{ m} = 100 \text{ km}.$

Example 29.7 *Find the coherence length of a laser source of monochromatic light with frequency width 10,000 Hz.* **(Pbi. U. 2007)**

Solution. Frequency width = band width δv = 10,000 Hz

$$\therefore \qquad \text{Coherence time } \tau = \frac{1}{\delta v} = \frac{1}{10{,}000} \text{ sec.}$$

and coherence length $L = c\tau_c = 3 \times 10^8 \times \dfrac{1}{10{,}000} = 3 \times 10^4 \text{ m} = 30 \text{ km}.$

Example 29.8 *For a red cadmium line of wavelength 6438 Å and the coherence length 38 cm deduce the order of magnitude of (a) coherence time and (b) spectral width of the line.* **(Nag. U. 2002)**

Solution. Coherence length L = 38 cm = 0.38 m; λ = 6438 Å

Coherence time τ_c ?; spectral width $\delta\lambda$ = ?

Now $\qquad L = c\tau_c$

$$\therefore \qquad \tau_c = \frac{L}{c} = \frac{0.38}{3 \times 10^8} = 1.27 \times 10^{-7} \text{ S}$$

$$\text{Again coherence length} \qquad L = \frac{\lambda^2}{\delta\lambda}$$

$$\therefore \qquad \delta\lambda = \frac{\lambda^2}{L}$$

$$\text{or} \qquad \text{spectral width} \qquad \delta\lambda = \frac{(6438)^2 \times 10^{-20}}{38 \times 10^{-2}} = 1.09 \times 10^{-12} \text{ m}$$

$$= .01 \text{ Å}$$

29.7 PURITY OF SPECTRAL LINES

Spectral purity is an important consideration in a wide variety of experiments. It is a quantification of the monochromaticity of a given light sample. In other words, it represents the degree to which a signal is monochromatic. This is a particularly important parameter in areas like laser operation and time measurement. Spectral purity is given by

$$\text{Spectral purity} = \frac{\delta\nu}{\nu} \quad \text{... (20.10)}$$

This is also known as *spectral purity factor* or *frequency stability*. Spectral purity is easier to achieve in devices that generate visible and ultraviolet light, since higher frequency light results in greater spectral purity.

Example 29.9 *One of the most ideal line of Krypton (orange) has a wavelength 6058 Å and coherence-length L = 20 cm. Calculate the line-width, the coherence time and frequency stability* $\left(\frac{\delta\nu}{\nu}\right)$.

Solution. Here $\lambda = 6058\ \text{Å} = 6.058 \times 10^{-7}$ m, $L = 20$ cm $= 0.2$ m

$c = 3 \times 10^8$ m/sec.

Line width $$\delta\lambda = \frac{\lambda^2}{L} = \frac{(6.058\times10^{-7})^2}{0.20}$$

$$= 0.01835 \times 10^{-10}\ \text{m} = 0.01835\ \text{Å}$$

Frequency-spead $$\delta\nu = \frac{c}{L} = \frac{3\times10^8}{0.20} = 1.5\times10^9\ \text{Hz}$$

Coherence time $$\tau_c = \frac{1}{\delta\nu} = \frac{1}{1.5\times10^9} = 0.67\times10^{-9}\ \text{sec.}$$

Frequency $$\nu = \frac{c}{\lambda} = \frac{3\times10^8}{6.058\times10^{-7}} = 4.952\times10^{14}\ \text{Hz}$$

$\therefore$ Frequency stability $$\frac{\delta\nu}{\nu} = \frac{1.5\times10^9}{4.952\times10^{14}} = 0.3\times10^{-6}$$

Example 29.10 *For an ordinary source coherence time is nearly 10^{-10} sec. Calculate the degree of non-chromaticity for $\lambda = 5400Å$* **(*Pbi. U. 2007, 2006*)**

Solution. Degree of non-chromaticity = Purity factor = frequency stability = $\frac{\delta\nu}{\nu}$

Again $$\lambda\nu = \frac{1}{\tau_c};\ \nu = \frac{c}{\lambda}$$

$$\therefore \quad \frac{\delta\nu}{\nu} = \frac{1}{\tau_c}\cdot\frac{\lambda}{c} = \frac{5400\times10^{-10}}{10^{-10}\times3\times10^8} = 18\times10^{-6}$$

***Example 29.11** *The coherence length of sodium D_2 line is 2.5 cm. Deduce the (i) coherence time (ii) spectral width of line and purity factor. Given $\lambda = 5890$ Å* **(*Pbi. U. 2007*)**

Solution. Coherence time $$\tau_c = \frac{L}{c} = \frac{2.5\times10^{-2}}{3\times10^8} = 0.83\times10^{-10}\text{S.}$$

---

* In optics, pertaining to a single wavelength of electromagnetic radiation or to a single color. In practice, optical radiation is never perfectly monochromatic; *i.e.*, it never consists of only one wavelength. It always has a finite spectral width, albeit narrow.

$$\therefore \quad \text{Spectral width} = \text{line width } \delta\lambda = \frac{\lambda^2}{L} = \frac{5890 \times 5890 \times 10^{-20}}{2.5 \times 10^{-2}} = 0.1387 \times 10^{-10} \text{ m.}$$

$$= 0.1387 \text{Å}$$

$$\text{Spectral purity factor} = \text{frequency stability} = \frac{\delta\nu}{\nu} = \frac{c}{L} \times \frac{\lambda}{c} = \frac{\lambda}{L} = \frac{5890 \times 10^{-10}}{2.5 \times 10^{-2}}$$

$$= 2.356 \times 10^{-5}$$

29.8 MINIMIZATION OF LASER LINE WIDTH

The line width of a laser is strongly (but not trivially) related to the temporal coherence, characterized by the coherence time and coherence length. A finite line width arises from *phase noise* if the phase undergoes unbounded drift, as in the case of free running oscillators.

The line width of a laser depends also on the type of laser. It may be minimized by optimizing the laser design and suppressing external noise influences as far as possible. The first step should be to determine whether quantum noise (can be calculated using Schawlow–Townes equation) or classical noise is dominating. The influence of quantum noise (essentially spontaneous emission noise) is small for a laser with high intracavity power, low resonator losses and a long resonator round-trip time. Classical noise may be introduced via mechanical fluctuations, which can often be kept weaker for a compact short laser resonator. It is to be noted that resonator length fluctuations of a certain magnitude have a stronger effect in a shorter resonator. Proper mechanical construction can minimize the coupling of the laser resonator to external vibrations and also minimize effects of thermal drift. There can also be thermal fluctuations in the gain-medium introduced e.g. by a fluctuating pump power.

Advantages of Narrow Line Width

Lasers with very narrow line width (high degree of monochromaticity) are required for various applications, e.g. as light sources for various kinds of fiber-optic sensors, for spectroscopy (e.g., LIDAR), in coherent optical fiber communication and for test and measurement purposes.

Disadvantages of Narrow Line Width

A narrow line width from a laser source is not always desirable.

(*i*) A large coherence length implies that interference effects can easily spoil the beam profile. In laser projection displays, speckle effects* can disturb the image quality.

(*ii*) For transmission of light in passive or active optical fibres, a narrow line width can cause problems due to Brillouins scattering.

29.9 SPATIAL COHERENCE DUE TO POINT AND EXTENDED SOURCE

Spatial coherence due to a point source. In temporal coherence the effect at a point is due to two different optical paths. In spatial coherence we consider the phase-relationship due to a source at different points in space.

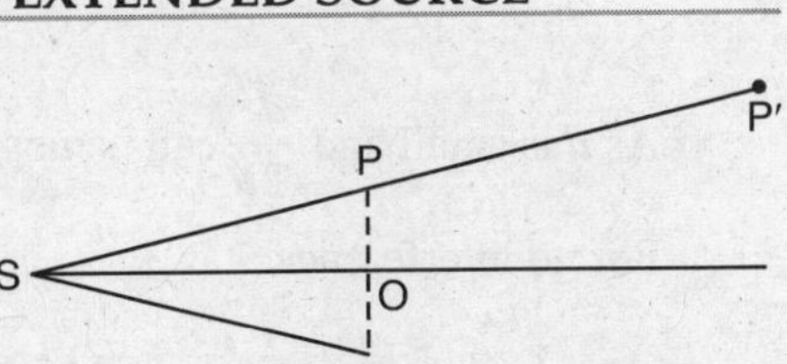

Fig. 29.9

Consider a point source S giving out waves in all directions in space. Let P and Q be two equidistant points from the source S and P' be a point along a line joining SP

* If a surface, such as of a wall or of a metal, is illuminated by a laser beam it appears to be granular. The granules are randomly distributed over the surface and appear to be shifting. Such grainy patterns are called laser speckles.

as shown in Fig. 29.9. The phase-relationship between the point P and P' depends upon the distance PP' and the temporal coherence of the beam. If $\frac{PP'}{c} \gg \tau_c$, no definite phase-relationship will exist and if $\frac{PP'}{c} \ll \tau_c$, phase-relationship will exist.

As $SP = SQ$, the points P and Q will always have the same phase.

Extended source. If the source S is an extended source, a definite phase-relationship may or may not exist between the points P and Q (Fig. 29.9). To study the spatial extent of the source, let us consider Young's double slit experiment with two independent sources S and S' separated by a distance l. Let the two slits S_1 and S_2 separated by a distance d lie at a distance a. Now each source will produce its own interference pattern. Let us assume that $SS_1 = SS_2$ and the point O' is such that $S_1O' = S_2O'$ (Fig. 29.10).

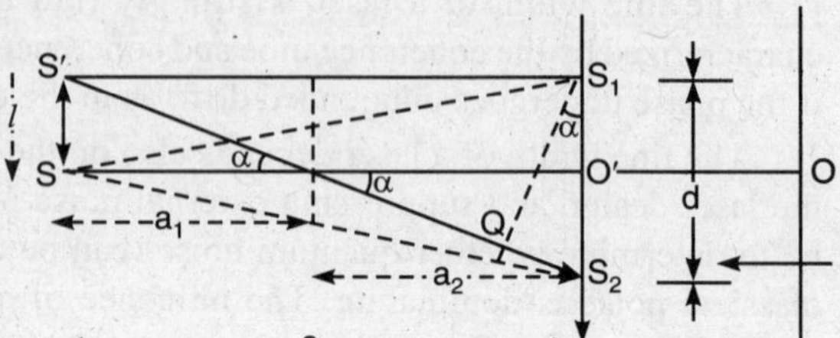

Fig. 29.10

Thus the source S will produce maximum around O' the mid-point of S_1S_2. The intensity at O' due to the source S' will depend on the optical path $S'S_2 - S'S_1$. If

$$S'S_2 - S'S_1 = \frac{\lambda}{2},$$

then minimum of interference pattern produced by S' will fall on the maximum of interference pattern produced by S. Hence the interference pattern due to S and S' will be out of step.

Now $$S'S_2 - S'S_1 = S_2Q = d\alpha$$

Also $$\alpha = \frac{d/2}{a_2} = \frac{l}{a_1}$$

$\therefore$ $$a = a_1 + a_2 = \frac{l}{\alpha} + \frac{d}{2\alpha} = \frac{1}{\alpha}\left(l + \frac{d}{2}\right)$$

or $$\alpha = \frac{l + d/2}{a}$$

$\therefore$ $$S'S_2 - S'S_1 = d\alpha = \frac{d}{a}\left(l + \frac{d}{2}\right)$$

$$= \frac{ld}{a} + \frac{d^2}{2a} = \frac{ld}{a}$$

As d is small and $\frac{d^2}{2a}$ can be neglected.

For no interference

$$\frac{ld}{a} = \frac{\lambda}{2}$$

or $$l = \frac{\lambda a}{2d} \qquad \text{... (29.11)}$$

Thus, for an extended source if the spatial extension exceeds $\frac{\lambda a}{2d}$ there will be no interference pattern on the screen.

If θ is the angle subtended by the extended source SS' at O', then

$$\theta = \frac{l}{a} = \frac{\lambda}{2d}$$

$$\therefore \quad d = \frac{\lambda}{2\theta}. \qquad \text{... (29.12)}$$

This gives the lateral distance which the beam may assume to be spatially coherent. The quantity λ/θ is called the **lateral coherent width** and is denoted by l_ω. For a circular source if the lateral coherent width.

$$l_\omega = \frac{1.22\,\lambda}{\theta} \qquad \text{... (29.13)}$$

then there is no phase-relationship between the beams and hence no interference pattern is observed

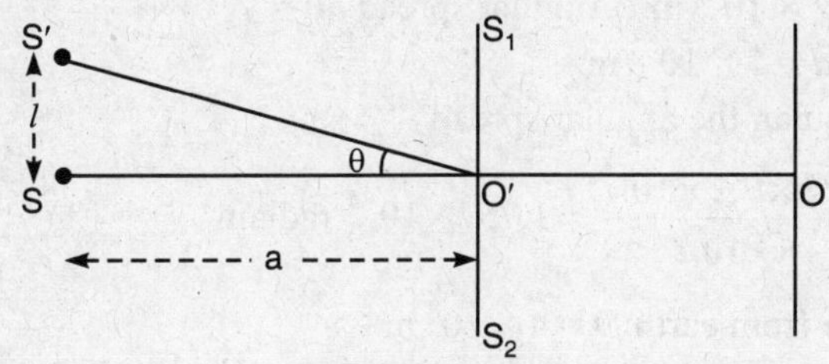

Fig. 29.11

It may be mentioned here that there is no definite distance at which interference pattern disappears, as the distance increases. However, contrast of the fringes gradually becomes poorer and eventually the fringe system disappears. For neon $\lambda = 6328$ Å the disapperance occurs when the path difference is about a few cm giving $\tau_c = 10^{-10}$ sec. For commercially available lasers, the coherence length can exceed a few kilometers.

Example 29.12 *In Young's double slit experiment the pin-hole is of 1 mm diameter. What should be the distance between S_1 and S_2 (Fig. 29.10) to get distinct interference fringes if $SS_1 = SS_2 = 100$ cm and $\lambda = 5 \times 10^{-7}$ m?*

Solution. Referring to Fig. 29.11, the angle θ is given by

$$\theta = \frac{l}{a}$$

where l is the diameter of pin-hole and a is the distance between the hole and the centre of $S_1\,S_2$

Here $\qquad l = 1 \text{ mm} = 10^{-3} \text{ m}, a = 1\text{m}$

$$\theta = \frac{10^{-3}}{1} = 10^{-3} \text{ radian}$$

Now coherent width for a circular aperture

$$l_\omega = \frac{1.22\lambda}{\theta} = \frac{1.22 \times 5 \times 10^{-7}}{10^{-3}}$$

$$= 6.1 \times 10^{-4} \text{ m}$$

Thus the distance S_1S_2 should be small, as compared to 6.1×10^{-4} m. The distance between the slits should have to be less than the lateral coherence width in order to obtain distinct interference fringes.

Example 29.13 *The coherence length for the red cadmium line of wavelength 6.438 × 10^{-7} m is 30 cm. Calculate (i) the number of oscillations corresponding to the coherence length and (ii) the coherence time.*

Solution. Here $\lambda = 6.438 \times 10^{-7}$ m and $L = 30$ cm $= 0.3$ m

(*i*) If n is the number of oscillations in length L, then

$$n = \frac{L}{\lambda} = \frac{0.30}{6.438 \times 10^{-7}} = 4.66 \times 10^5$$

(*ii*) Coherence time

$$\tau_c = \frac{L}{c} = \frac{0.30}{3 \times 10^8} = 10^{-9} \text{ sec.}$$

Example 29.14 *A laser beam has a wavelength of 7.2 × 10^{-7} m and aperture 5 × 10^{-3} m. The laser beam is sent to moon the distance of which from earth is 4 × 10^8 m. Calculate (i) the angular spread and (ii) aerial spread when the beam reaches the moon.* ***(Pbi. U. 2005)***

Solution. Here $\lambda = 7.2 \times 10^{-7}$ m, 'Angular spread $d\theta = ?$'

Diameter of aperture $d = 5 \times 10^{-3}$ m.

As the aperture is circular, the angular spread

$$(i) \quad d\theta = \frac{1.22\lambda}{d} = \frac{1.22 \times 7.2 \times 10^{-7}}{5 \times 10^{-3}} = 1.757 \times 10^{-4} \text{ radiam}$$

(*ii*) Distance of moon from earth $D = 4 \times 10^8$ m

$$\therefore \text{ Aerial spread } = (D.d\theta)^2$$

$$= (4 \times 10^8 \times 1.757 \times 10^{-4})^2$$

$$= 49.39 \times 10^8 \text{ m}^2$$

Example 29.15 *A laser beam has a power of 100 mW. It has an aperture of 5 × 10^{-3} m and emits a light of wavelength 6943Å. The beam is focused with a lens of focal length 0.1 m. Calculate the area and the intensity of the image.*

Solution. Diameter of aperture $d = 5 \times 10^{-3}$ m

Wavelength $\lambda = 6943 \times 10^{-10}$ m, $f = 0.1$ m

Power of laser beam = 100 mW = 10^{-1} watt

As the aperture is circular

$$\therefore \quad \text{Angular spread } d\theta = \frac{1.22\lambda}{d} = \frac{1.22 \times 6943 \times 10^{-10}}{5 \times 10^{-3}} = 1.694 \times 10^{-4} \text{ radian}$$

$$(i) \quad \text{Aerial spread} = (d\theta.f)^2$$

$$= (1.694) \times 10^{-4} \times 0.1)^2 = 2.89 \times 10^{-10} \text{ m}^2$$

$$(i) \quad \text{Intensity} = \frac{\text{Power}}{\text{Area}} = \frac{10^{-1}}{2.89 \times 10^{-10}} = 3.46 \times 10^8 \text{ watt/ m}^2$$

Example 29.16 *A beam has a power of 0.2 watt and has an aperture of 1 mm. It emits light of wavelength 6000Å. If it is focussed by a lens of F.L. 80 cm, calculate the area and intensity of the image.* ***(P.U. 2006, 2005)***

Solution.

$$d\theta = \frac{1.22\lambda}{d} = \frac{1.22 \times 6000 \times 10^{-10}}{10^{-3}} = 7.32 \times 10^{-4} \text{ radian}$$

$$\text{Area (Aerial spread)} = (d\theta if)^2 = (7.32 \times 10^{-4} \times 0.8)^2 = 34.3 \times 10^{-8}\ m^2$$

$$\text{Intensity} = \frac{\text{Power}}{\text{Area}} = \frac{0.2}{34.3 \times 10^{-8}} = 5.83 \times 10^5\ W/m^2$$

29.10 CRITICAL POTENTIAL, IONIZATION POTENTIAL AND EXCITATION POTENTIAL

Critical potential. *The least energy in electron volts which is necessary to excite a free neutral atom from its ground state to a higher energy state is called critical potential of the atom.*

A critical potential is called an *ionisation potential* when the absorption of energy by the excited atom *just* removes an electron from it *i.e.*, the electron is raised from the level $n = 1$ to $n = \infty$.

A critical potential is called an *excitation potential* when the absorption of energy by the unexcited atom *just* raises it to a higher energy state (*but does not ionise it*). On the basis of Bohr's theory the various energy states in a hydrogen atom are given by

$$E_n = -\frac{2\pi^2 Z^2 me^4}{n^2 h^2} = -\frac{13.6}{n^2} eV \text{ (for Hydrogen } Z = 1).$$

Thus the energy of the electron in a hydrogen atom for the First, Second, Third...etc., orbits are $-13.6\ eV\ (n = 1)$, $-3.4\ eV\ (n = 2)$, $-1.51\ eV\ (n = 3)$...., $0\ eV\ (n = \infty)$.

Hence the energy required to raise the hydrogen atom from the ground state to the first excited state is equal to $(13.6 - 3.4) = 10.2\ eV$, that required to raise it to the second excited state is $(13.6 - 1.51) = 12.09$ eV and the energy required to ionise it is $(13.6 - 0) = 13.6\ eV$. In other words, discrete amounts of energy are required to excite the hydrogen atom to different energy states; which are thus quantised. The first and second excitation potentials are 10.2 and 12.09 volts respectively and the ionisation potential is 13.6 volts.

29.11 MECHANISMS FOR BROADENING OF SPECTRAL LINES

Emission of spectral lines. Whenever an atom is excited its outermost electron goes from the ground state E_1 to some higher energy state E_2. The electron stays in the excited state only for a very short duration. Very soon it jumps back to the ground state (or lower energy state) and emits its energy in the form of a spectral line, the frequency v of which is given by

$$E_2 - E_1 = \Delta E = hv$$

However, in reality, the atoms do not emit radiation at *precise* frequencies. Each emission is more or less broadened due to various causes. These causes can be divided into two categories:

(*i*) **Homogeneous and** (*ii*) **Inhomogeneous**

(*i*) **Homogeneous broadening.** In homogeneous broadening, the broadening mechanism affects each individual atom uniformly. As a result, broadening is around a central frequency v_0. There are two important causes for homogeneous broadening: (*i*) *Natural broadening and* (*ii*) *Collision broadening.*

Homogeneous broadening is *Lorentzian* in shape as shown in the line shape graph [Fig. 29.12(1).

(*ii*) **Inhomogeneous broadening.** In inhomogeneous broadening different atoms have slightly different frequencies for the same transition. In this case the causes of broadening are random and the central frequency of response

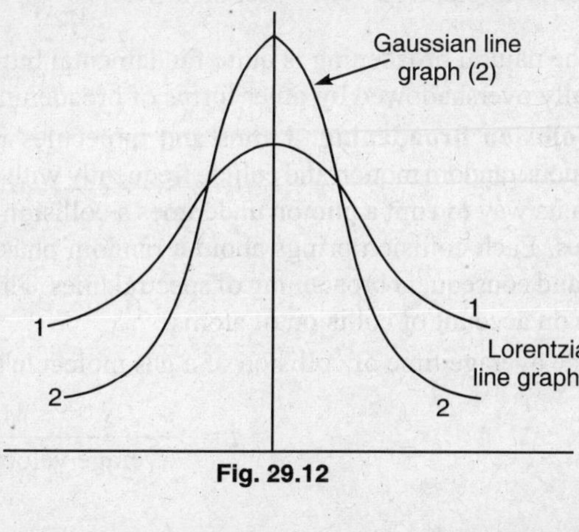

Fig. 29.12

of individual atoms shifts by different amounts, thereby leading to inhomogeneous overall broadening. For this type of broadening there are two different reasons. (*i*) *Doppler broadening and* (*ii*) **broadening due to crystal defects.**

In inhomogeneous broadening the line shape graph is *Gaussian* in shape as shown in [Fig. 29.12(2)].

Natural broadening. Natural broadening can be explained on the basis of quantum theory and Heisenberg's uncertainty principle. Even in an isolated system of stationary atoms the atomic energy levels are not indefinitely sharp.

An excited atom has a small finite lifetime Δt and the uncertainty for this interval is of the same value. The energy of the excited state is, therefore, also uncertain. If ΔE is the uncertainty in the energy of the state, then

$$\Delta E.\Delta t = \frac{\hbar}{2}$$

or

$$\Delta E = \frac{\hbar}{2\Delta t} = \frac{h}{4\pi\Delta t}$$

The ground state of the atom is sharply defined, because left to itself the atom will remain in that state for an infinite time. Thus $\Delta t = \infty$ and $\Delta E = 0$. A transition between the excited state and ground state will, therefore, have an energy uncertainty of ΔE. As $E = h\nu$, the uncertainty in the emitted photon $\Delta\nu$ is given by

$$\Delta E = h\Delta\nu$$

$$\therefore \quad \Delta\nu = \frac{\Delta E}{h} = \frac{h}{4\pi\Delta t}\cdot\frac{1}{h} = \frac{1}{4\pi\Delta t} \qquad \text{... }(i)$$

In terms of ω, we have

$$\Delta\omega = 2\pi\,\Delta\nu = \frac{1}{2\Delta t}$$

Note. If we use the uncertainty principle $\Delta E.\Delta t = \hbar = \frac{h}{2\pi}$, then

$$\Delta E = h\Delta\nu = \frac{h}{2\pi\Delta t} \quad \text{or} \quad \Delta t = \frac{h}{2\pi h\Delta\nu} = \frac{1}{2\pi\Delta\nu} = \frac{1}{\Delta\omega}$$

$$\therefore \quad \Delta\omega = \frac{1}{\Delta t} \qquad \text{... (29.14)}$$

For example, if $\Delta t = 10^{-8}$ sec, then $\Delta\nu = \frac{1}{4\pi\Delta t} \approx 10^{8}$ Hertz

The natural broadening is quite fundamental but is relatively very small in magnitude and is generally overshadowed by other forms of broadening.

Collision broadening. Atoms and molecules in liquid and gaseous phase are in a state of continuous random motion and collide frequently with each other. If an atom in the process of transition *i.e.*, on its way to emit a photon undergoes a collision with another atom, the phase of radiated wave changes. Each collision brings about a random phase change which results in shortening of wave trains and consequent broadening of spectral lines. This broadening is known as *collision broadening* as it is on account of collision of atoms.

The average time of collision of a gas molecule (or atom) is given by

$$t = \frac{\text{Mean free path}}{\text{Average velocity of molecule (or atom)}}$$

Now mean free path $$\lambda = \frac{kT}{\sqrt{2}p\pi d^2} = \frac{kT}{4\sqrt{2}p\pi a^2} \quad \text{... (29.15)}$$

where k = Boltzmann's gas constant; T = Temperature of the gas in Kelvin; p = pressure of the gas and d = diameter of the gas molecule = $2a$ where a is the radius of the molecule.

The average velocity of the molecule (or atom), $\bar{v} = \sqrt{\frac{8kT}{\pi m}}$

$$\therefore \quad t = \frac{kT}{4\sqrt{2}\, p\pi a^2} \cdot \frac{1}{\sqrt{\frac{8kT}{\pi m}}} = \frac{kT}{4\sqrt{2}p\pi a^2} \cdot \frac{\sqrt{\pi m}}{8kT} = \frac{1}{16pa^2} \cdot \sqrt{\frac{mkT}{\pi}} \quad \text{... (29.16)}$$

Hence frequency spread $$\Delta\omega = \frac{1}{t} = \frac{16\, pa^2}{\sqrt{mkT/\pi}} \quad \text{... (29.17)}$$

In terms of ν, we have, $\Delta\omega = 2\pi\,\Delta\nu$

$$\therefore \quad \Delta\nu = \frac{\Delta\omega}{2\pi} = \frac{16p\pi a^2}{2\pi\sqrt{mkT/\pi}} = \frac{8pa^2}{\sqrt{\pi mkT}} \quad \text{... (29.18)}$$

Doppler broadening. Doppler effect is the apparent change in the frequency of radiation due to relative motion between the source and the observer. The frequency as measured by the observer *increases* if the source and the observer are *approaching* each other and *decreases* if they *move away* from each other.

In a gas system the atoms move with all possible velocities in all possible directions due to random chaotic nature of thermal motion. As the velocities of atoms differ in magnitude as well as direction, the spread of frequencies due to Doppler effect is a kind of inhomogeneous broadening.

If ν_0 is the frequency of radiation emitted by the atom and ν is the observed frequency, then

$$\nu - \nu_0 = \frac{v}{c}\nu_0 \quad \text{... (29.19)}$$

where v is the relative velocity of the atom w.r.t. the observer and c the velocity of radiation. In terms of ω the above results is

$$\omega - \omega_0 = \frac{v}{c}\omega_0 \quad \text{... (29.20)}$$

$$\therefore \quad \text{Doppler broadening } \Delta\omega = \omega - \omega_0$$

$$= \frac{v}{c}\omega_0 \quad \text{... (29.21)}$$

It can be further proved that

$$\Delta\omega = 2\omega_0\sqrt{\frac{2kT \ln 2}{mc^2}} \quad \text{... (29.22)}$$

where $\Delta\omega$ is *full width at half maximum* (FWHM) and is known as *line width* as shown in Fig. 29.13.

ω_0 = Angular frequency corresponding to maximum.

k = Boltzmann's constant

T = Temperature in Kelvin

m = mass of the atom

c = velocity of light.

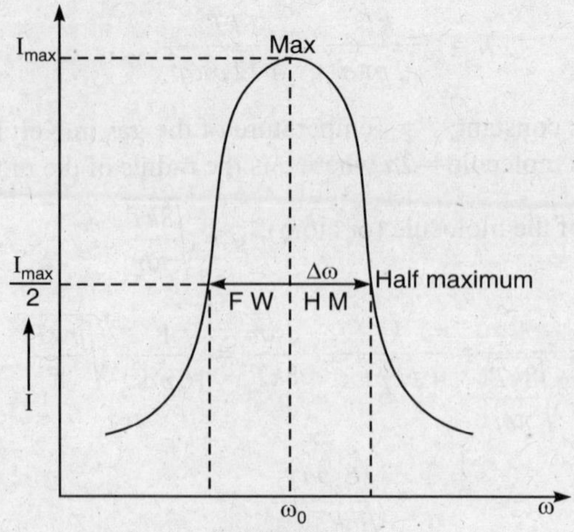

Fig. 29.13

Example 29.17 *Calculate the line width for 6328Å wavelength of He-Ne laser operating at 300 K, given that mass of neon atom = 20 × 1.66 × 10^{-27} kg.*

Solution. Line width $\Delta\omega = 2\omega_0\sqrt{\frac{2kT\ln 2}{mc^2}}$

or $$\Delta\nu = \frac{\Delta\omega}{2\pi} = \frac{\omega_0}{\pi}\sqrt{\frac{2kT\ln 2}{mc^2}}$$

$$= \frac{\omega_0}{\pi c}\left[\frac{2kT\ln 2}{m}\right]^{1/2}$$

$$= \frac{2}{\lambda_0}\left[\frac{2kT}{m}\ln 2\right]^{1/2} \quad \left[\because \frac{\omega_0}{\pi c} = \frac{2\pi\nu_0}{\pi c} = \frac{2}{\lambda_0} \text{ as } \frac{c}{\nu_0} = \lambda_0\right]$$

$$\therefore \quad \Delta\nu = \frac{2}{6328\times 10^{-10}}\left[\frac{2\times 1.38\times 10^{-23}\times 300}{20\times 1.66\times 10^{-27}}\times 0.693\right]^{1/2}$$

$$= 1.3\times 10^9\ \text{s}^{-1} = 1300\times 10^6\ \text{Hz} = 1300\ \text{MHz}$$

29.12 LINE PROFILE AND LINE WIDTH (GAIN BANDWIDTH)

Line profile. The most important property of laser light is its *monochromaticity i.e.,* single wavelength. But no laser can produce absolutely monochromatic light. The nature of laser radiations is analysed by a spectro-microphotometer and a graph is plotted between intensity I measured in W m^{-2} and corresponding wavelength λ. The graph is shown in Fig. 29.14. The line shape plot of this graph is known as *line profile.*

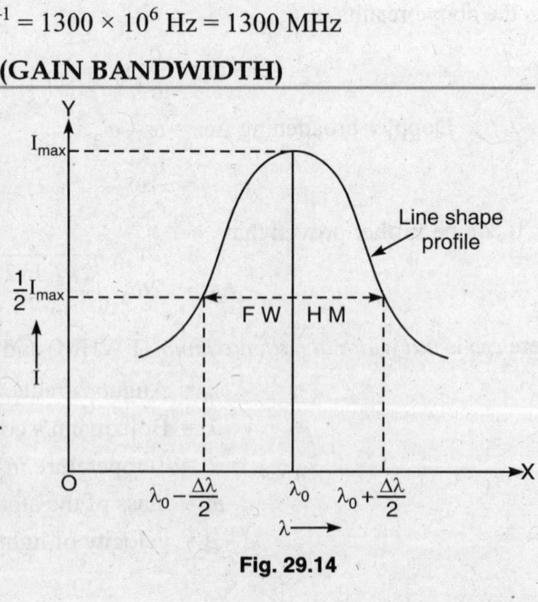

Fig. 29.14

Line width. The emitted light has a wavelength spread Δλ about a central wavelength λ_0. Most of the light is emitted at wavelength λ_0 having intensity I_{max} but some light is also emitted at wavelengths above and below λ_0, having a wavelength spread of $\lambda_0 \pm \frac{\Delta\lambda}{2}$ where $\lambda_0 \pm \frac{\Delta\lambda}{2}$ is the wavelength corresponding to which the intensity falls from I_{max} to $\frac{1}{2} I_{\max}$ on either side. Thus $\Delta\lambda$ gives the wavelength separation between the two points on the line shape plot for which I_{max} reduces to $\frac{1}{2} I_{\max}$ and is known as *Full Width at Half Maximum* or FWHM or **line width, or gain bandwidth.**

Line width may also be expressed in terms of ω where $\omega = 2\pi\nu = \frac{2\pi c}{\lambda}$. ... (29.23)

Example 29.18 *The half-width of the gain profile of a He-Ne laser material is about 2 × 10^{-3} nm. What should be the maximum length of the cavity in order to have a single longitudinal mode oscillation ?*

Solution. The separation between successive lines is given by

$$\Delta\lambda = \frac{\lambda^2}{2\mu L}$$

In order to obtain a single mode of oscillation.

$$\Delta\lambda = \text{Half-width of the gain profile}$$

$$\therefore \quad L = \frac{\lambda^2}{2\mu\Delta\lambda} = \frac{(6328\times10^{-10})^2 m^2}{2\times1\times2\times10^{-12} m} (\mu = 1)$$

$$= \frac{4\times10^{-13}}{4\times10^{-12}} m = \mathbf{10\ cm}$$

Example 29.19 *The half-width of the gain profile of a He-Na laser material is about 2 × 10^{-3} nm. If the length of the cavity is 30 cm, how many longitudinal modes can be excited ? The emission wavelength of He-Ne laser is 6328Å.*

Solution The distance of separation between successive modes is given by

$$\Delta\lambda = \frac{\lambda^2}{2\mu L} = \frac{(6328\times10^{-10} m)^2}{2\times1\times30\times10^{-2} m} = \frac{4\times10^{-13}}{6\times10^{-1} m}$$

$$= 0.66 \times 10^{-3} \text{ nm}.$$

μ is taken as 1.

$$\text{Number of modes} = \frac{\text{Half} - \text{width of the gain profile}}{\Delta\lambda}$$

$$= \frac{2\times10^{-3}\ nm}{0.66\times10^{-3}\ nm}$$

$$= \mathbf{3.}$$

29.13 DISTINCTION BETWEEN GAUSSIAN AND LORENTZIAN LINE GRAPHS

A visual comparison between Gaussian and Lorentzian line shape graphs is shown in Fig. 29.15. The broadened line is Gaussian in shape if the effects which cause inhomogeneous broadening are random in origin. On the other hand, the line graph is Lorentzian in shape if the broadening is due to homogeneous causes. Each graph shows its *peak sharpness* as well as its line broadening. Gaussian line graph is sharper and narrower.

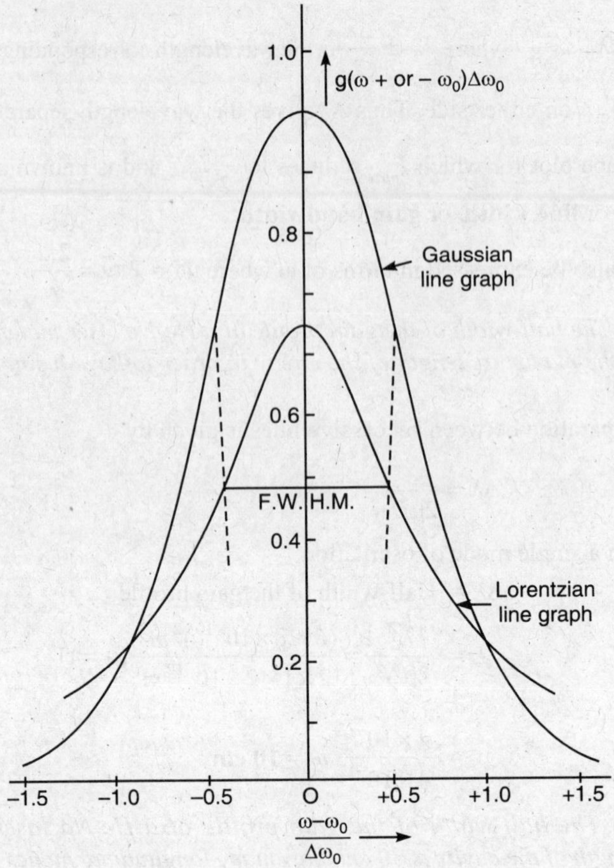

Fig. 29.15

FWHM represents the full width at half maximum value.

29.14 CONVENTIONAL LIGHT SOURCE IS INCOHERENT

Light is emitted by a source when an electron in the atom constituting the source jumps from a *higher* excited state of energy E_2 to a lower (or ground) state of energy E_1. The difference in energy $E_2 - E_1$ is radiated as light of frequency υ given by $h\upsilon = E_2 - E_1$. The light is emitted in the form of pulses of short duration and even two consecutive pulses do not bear a constant or definite phase relation with each other. The pulses are also emitted in all possible directions. As a result various pulses given out by the conventional source travel in different directions, have no constant phase relation and also have large frequency spread due to various causes, like natural broadening, collision and Doppler broadening. Hence the light is incoherent.

29.15 THE THREE PROCESSES

Let us consider a medium consisting of identical atoms capable of being excited from the energy level 1 to the energy level 2 by absorption of photons. Let the levels be denoted by E_1 and E_2 and their populations be N_1 and N_2 respectively. Let the atoms be in thermal equilibrium. In the equilibrium condition, the number of atomic transitions upward must be equal to the number of downward transitions. Thus no net photons are generated or lost. However, when the atoms are subjected to an external light of frequency υ, the following three processes occur in the medium.

(*a*) Absorption

An atom residing in the lower energy level E_1 may absorb the incident photon and jump to the excited state E_2 as depicted in Fig. 29.16. This transition is known as *induced* or *stimulated absorption* or simply as *absorption.* Corresponding to each absorption transition, one photon disappears from the incident light field and one atom adds to the population at the excited energy level E_2.

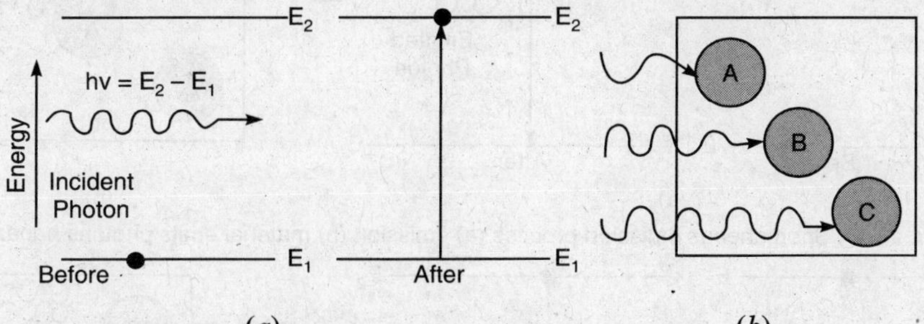

Fig. 29.16. Absorption process (*a*) Induced absorption (*b*) Material absorbs photons

This process may be represented as

$$A + h\nu \rightarrow A^*$$

where A denoted an atom in the lower state and A^* an excited atom.

This process of interaction of the atom with the radiation field is called ***stimulated*** *or* ***induced absorption.***

The *transition rate for induced absorption* by the atoms is given by

$$\left|\frac{dN_{12}}{dt}\right|_I = B_{12}\,\rho_{21}\,N_1$$

where $\left|\frac{dN_{12}}{dt}\right|_I$ = Number of atoms which absorb the incident radiation and move from energy state E_1 to E_2 per unit time. It has units of s^{-1}.

N_1 = Number of atoms availabel for excitation in the lower energy state ($n = 1$ or E_1).

ρ_{21} = The energy density of incident electromagnetic radiation per unit volume per unit frequency with energy equal to the energy difference between the two states ($E_2 - E_1$). Therefore, it has units of $J/m^3\ s^{-1}$.

B_{12} = constant of proportionality, known as *Einstein's co-efficient for induced absorption,* which depends upon the nature of two states. It has units of $J^{-1}\ m^3\ s^{-2}$.

(*b*) Spontaneous emission. When an atom in an excited state E_2 goes to the ground state E_1 by spontaneously emitting a photon of frequency $\nu = \frac{E_2 - E_1}{h}$, the process is known as *spontaneous emission.* In such a case

(*i*) The emitted photon can move in any random direction.

(*ii*) The photons emitted from various atoms in the assembly have no phase-relationship between themselves

Thus, the radiation given out by such a source is *incoherent.*

The probability or the transition rate for spontaneous emission is given by

$$\left|\frac{dN_{21}}{dt}\right|_s = A_{21}\,N_2 \qquad \text{... (29.24)}$$

where N_2 = number of atoms available with electrons in the upper excited energy state and

A_{21} = constant of proportionality, known as *Einstein's co-efficient for spontaneous emission.* It has units of s^{-1}.

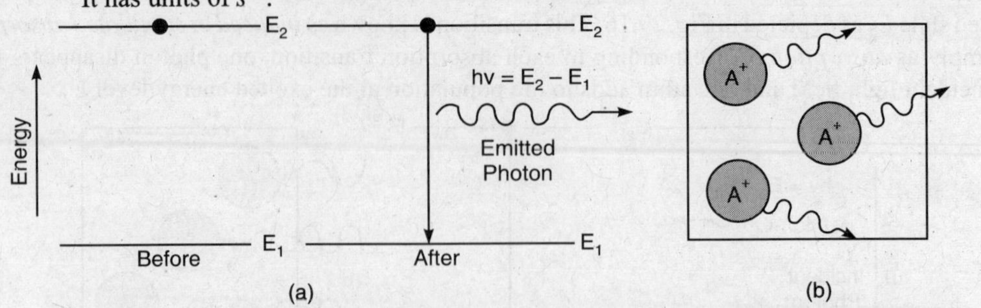

Fig. 29.17. Spontaneous emission process (a) emission (b) material emits photons hapazardly

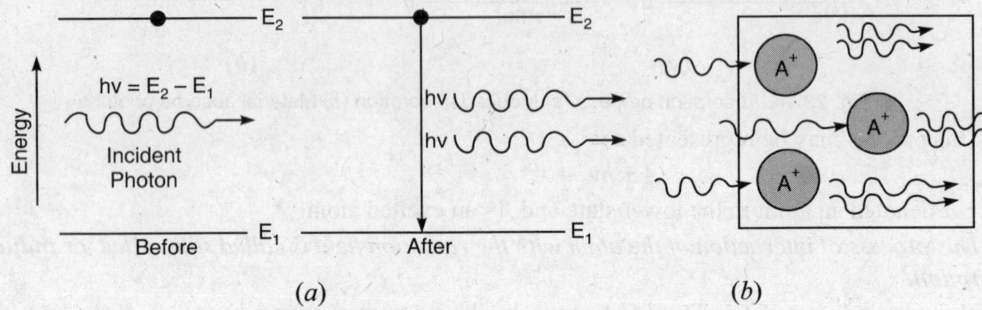

Fig. 29.18. Stimulated emission process (*a*) emission
(*b*) Material emits photons in a coordinated manner

(c) Stimulated or induced emission. When an atom is an excited state E_2 interacts with an incident photon of right or *resonant frequency* v and is thereby *induced* to move to the ground state E_1 by emitting the difference of energy $E_2 - E_1$ as a photon of the same frequency v, the process is known as *stimulated or induced emission.* In such a case

(*i*) The emitted photon travels in the same direction in which the incident photon is moving.

(*ii*) The two photons are *in phase* with each other.

Since the emitted photon has the *same frequency* and is in *phase* with the incident photon, the two further interact with other atoms and thus, the process builds up and a sort of *chain emission* starts, in which all the photons are of the *same frequency, in phase* and *travel in the same direction.* The number of *identical* photons goes on increasing with each emission and a single photon, thus, produces a very larger number of identical photons, so that the radiation beam is now highly amplified. *Hence the beam of electromagnetic radiation coming from such a source is highly intense, monochromatic, coherent and uni-directional.*

The *probability or the transition rate for stimulated emission* is given by

$$\left|\frac{dN_{21}}{dt}\right|_I = B_{21}\,\rho_{21}N_2 \qquad \text{... (29.25)}$$

where N_2 and ρ_{21} are the same as given above and B_{21} = a constant of proportionality, known as *Einstein's co-efficient for stimulated emission,* which depends upon the nature of two states. It has a same units as B_{12} *i.e.* $J^{-1}\ m^3\ s^{-2}$

Relation between Einstein's co-efficients. In the state of equilibrium the number of absorptions should be equal to the total number of emissions–spontaneous plus stimulated (or induced)- per unit volume per unit time *i.e.*,

$$\left|\frac{dN_{12}}{dt}\right|_I = \left|\frac{dN_{21}}{dt}\right|_S + \left|\frac{dN_{21}}{dt}\right|_I$$

or $$B_{12}\,\rho_{21}\,N_1 = A_{21}\,N_2 + B_{21}\,\rho_{21}\,N_2$$

or $$\rho_{21}\,(B_{12}\,N_1 - B_{21}\,N_2) = A_{21}\,N_2$$

or $$\rho_{21} = \frac{A_{21}\,N_2}{B_{12}\,N_1 - B_{21}\,N_2} = \frac{A_{21}}{\frac{N_1}{N_2}B_{12} - B_{21}} \qquad \text{... }(i)$$

According to Boltzmann's law, at thermal equilibrium at temperature T

$$N_2 = N_1\,e^{-\frac{E_2-E_1}{kT}}$$

or $$\frac{N_2}{N_1} = e^{\frac{(E_2-E_1)}{kT}} = e^{-h\nu/kT} \qquad \text{where} \qquad h\nu = E_2 - E_1$$

or $$\frac{N_1}{N_2} = e^{+h\nu/kT} \qquad \text{... }(ii)$$

Substituting the value of $\frac{N_1}{N_2}$ from Eq. (*ii*) in Eq. (*i*), we get

$$\rho_{21} = \frac{A_{21}}{B_{12}\,e^{h\nu/kT} - B_{21}} = \frac{A_{21}}{B_{12}}\,\frac{1}{e^{h\nu/kT} - \frac{B_{21}}{B_{12}}} \qquad \text{... }(iii)$$

The energy ρ_{21} per unit volume per unit frequency of incident electromagnetic radiation is given by Planck's law of black body radiation as

$$\rho_{21} = \frac{8\pi h\nu^3}{c^3}\,\frac{1}{e^{h\nu/kT} - 1} \qquad \text{... }(iv)$$

Comparing equation (*iii*) and (*iv*), we have

$$\frac{B_{21}}{B_{12}} = 1 \qquad \text{or} \qquad B_{21} = B_{12} \qquad \text{... }(v)$$

and $$\frac{A_{21}}{B_{12}} = \frac{8\pi h\nu^3}{c^3}$$

or $$A_{21} = B_{12}\,\frac{8\pi h\nu^3}{c^3} = B_{21}\,\frac{8\pi h\nu^3}{c^3} \qquad \text{... }(vi)$$

To sum up

(*i*) $\boldsymbol{B_{12} = B_{21}}$. It means that stimulated or induced emission is a process inverse of stimulated absorption. But their rates are different because stimulated absorption is proportional to N_1 and stimulated emission is proportional to N_2.

(*ii*) $\boldsymbol{B_{21} = A_{21}\,\frac{c^3}{8\pi h\nu^3}}$. It means Einstein's co-efficient for stimulated emission B_{21} is inversely proportional to the cube of the frequency of resonant radiation. Thus, higher this frequency smaller is the value of B_{21} *i.e.* it is relatively difficult to obtain stimulated emission of high frequencies (or smaller wavelengths).

(*iii*) The relation between Einstein's co-efficients has been obtained for the state of thermodynamic equilibrium. However, the results are valid even in the absence of thermodynamic equilibrium. It is because the emission process is a characteristics of the atom independent of the surrounding conditions.

Ratio of spontaneous to stimulated emission. Using Eg. (*iii*), we write

$$\rho_{21} = \frac{A_{21}}{B_{12}} \frac{1}{e^{h\nu/kT} - \frac{B_{21}}{B_{12}}} \quad \text{... } (vii)$$

and as in Eq. (*v*)

$$B_{21} = B_{12}$$

Substituting $B_{21} = B_{12}$ in above Eq. (*vii*) we get

$$\rho_{21} = \frac{A_{21}}{B_{12}} \frac{1}{e^{h\nu/kT} - 1}$$

or

$$\frac{A_{21}}{B_{21}\,\rho_{21}} = e^{h\nu/kT} - 1$$

$$= e^{\frac{\hbar\omega}{kT}} - 1 \qquad \left[\because \hbar\omega = \frac{h}{2\pi}\cdot 2\pi\nu = h\nu\right]$$

$$= \text{Exp}\,(\hbar\,\omega/kT) - 1.$$

Substituting $A_{21} = A$; $B_{21} = B$ and ρ_{21} the energy density per unit volume per unit frequency = $\rho(\omega)$, we have

$$\frac{A}{B\,\rho(\omega)} = \exp\,(\hbar\omega/\,kT) - 1. \quad \text{... (29.26)}$$

Example 29.20 *Find the ratio of spontaneous emission to stimulated emission for a cavity of temperature 50 K and wavelength 10^{-5} m.* **(P.U. 2005)**

Solution $\dfrac{\text{Rate of spontaneous emission}}{\text{Rate of stimulated emission}} = \dfrac{A_{21}}{B_{21}} = \exp\,(\hbar\omega/\,kT) - 1$

$$= \exp\left(\frac{h\nu}{kT}\right) - 1 = \exp\left(\frac{hc}{\lambda kT}\right) - 1$$

$$= \text{Exp.}\left[\frac{6.625\times10^{-34}\times3\times10^{8}}{10^{-5}\times1.38\times10^{-23}\times50}\right] - 1 = \text{Exp.}\,28.8 - 1$$

$$= \text{Exp. } 28.8 \qquad [\because \text{Exp } 28.8 \gg 1]$$

Metastable state. Normally an atom in the excited state comes back to the original or ground state within a very short time of about 10^{-8} second by the emission of a photon.

The average time for which an atom remains in an excited state is known as its mean life. The mean life of an atom is characteristic of the energy state. Some energy states have comparatively longer mean life.

Energy states having mean life of more than 10^{-3} sec. are known as metastable states.

29.16 LIGHT AMPLIFICATION

The outstanding feature of this process is the multiplication of photons. For one photon hitting an excited atom, there are two photons emerging. The two photons are in phase and travel along the same direction. These two photons stimulate two excited atoms in their path and produce a total four photons which are in phase and travel along the same direction. These four photons can in turn stimulate four excited atoms and generate eight photons and so on. The number of photons builds up in an avalanche like manner, as illustrated in Fig. 29.19.

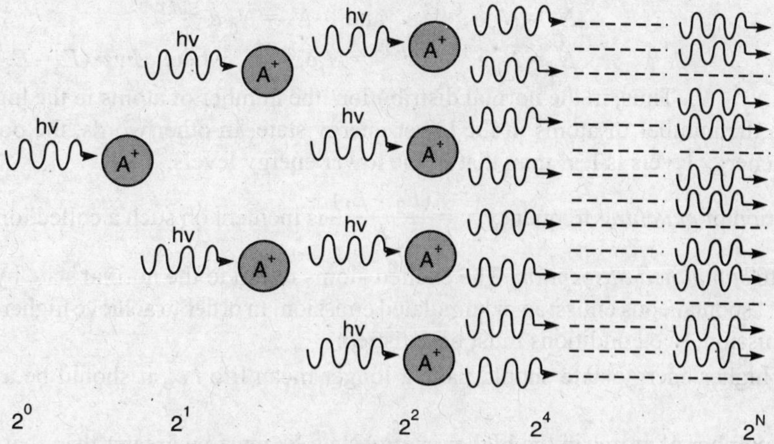

Fig. 29.19. Multiplication of Stimulated Photons into an Avalanche

All the light waves generated in the medium are due to one initial wave and all of them are in phase. Therefore, the waves are *coherent* and interfere constructively as shown in Fig. 29.20. As the resultant wave propagates through the collection of excited atoms, its amplitude continuously grows on as more and more stimulated waves join it. The result is that light gets amplified.

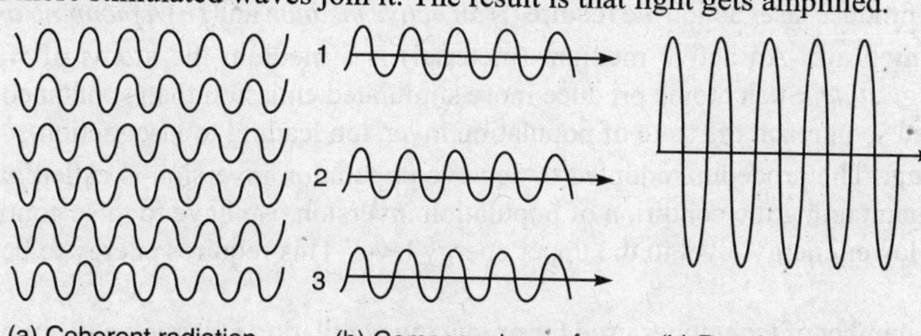

Fig. 29.20. Coherent radiation (*a*) coherent waves (*b*) component wave form (*c*) Resultant wave

High Intensity: Because of constructive interference of the waves, the net intensity of the resultant light will be proportional to the square of the number of atoms emitting light. Thus,

$$I_{\text{Total}} = N^2 I \qquad \text{... (29.27)}$$

Hence, the light produced due to stimulated emissions is of very high intensity than the light generated through spontaneous emissions.

The above three processes occur together and under steady state condition, there would be a balance achieved between the absorption and emission processes.

29.17 POPULATION INVERSION

The transition probability for induced emission depends upon

(*i*) The number of atoms in the excited state and

(*ii*) The energy density of the incident radiation $\rho(\nu)$.

To achieve a higher probability of stimulated emission we make use of the phenomenon of population inversion.

When we have a collection of large number of atoms, say N_0, in thermal equilibrium, the distribution of atoms in different energy states is given by Maxwell-Boltzmann statistics. If T K is the temperature of collection, N_1 the number of atoms in the energy state E_1 and N_2 that in the energy state E_2, then

$$N_1 = N_0\, e^{-E_1/kT} \quad \text{and} \quad N_2 = N_0\, e^{-E_2/kT}$$

or $$N_2 = N_1\, e^{-(E_2-E_1)kT} = N_1 e^{-h\nu/kT} \quad \text{where} \quad h\nu = (E_2 - E_1) \quad ...(29.28)$$

As $E_2 > E_1$, $N_2 < N_1$. Thus, in the normal distribution, the number of atoms in the higher energy state is less than the number of atoms in the lower energy state. In other words, the population of atoms in higher energy levels is less than that in the lower energy levels.

When radiation of *matching* frequency $\nu = \dfrac{(E_2 - E_1)}{h}$ is incident on such a collection, the atoms are excited due to *stimulated absorption*. The excited atoms return to the normal state by any of the two processes *i.e.*, spontaneous emission or stimulated emission. In order to achieve higher probability of stimulated emission, two conditions must be satisfied:

(*i*) The higher energy state should have a longer mean life *i.e.*, it should be a metastable state.

(*ii*) The number of atoms in the higher energy state E_2 must be greater than that in E_1.

The establishment of situation in which the number of atoms in the higher energy level is greater than that in the lower energy level is called **population inversion.**

It should be noted that population inversion is a *non-equilibrium condition.*

Requirements for population inversion and laser action. For the purpose of population inversion and to produce laser action we require (*i*) *an active medium* and (*ii*) *a pumping arrangement.*

(*i*) **Active medium.** An active medium (meterial) is a medium the atoms of which have a *metastable energy state.* Such atoms produce more stimulated emission than spontaneous emission and when excited soon reach the state of population inversion leading to laser action.

(*ii*) **Pumping.** The procedure adopted to achieve population inversion is called *pumping.* For achieving and maintaining the condition of population inversion, we have to raise continuously the atoms from the lower energy level to the upper energy level. This requires energy to be supplied to the system.

There are a number of techniques used for producing population inversion e.g. optical pumping, electrical discharge, direct conversion etc. However, optical pumping is very important and convenient method.

Optical pumping. The procedure adopted to achieve population inversion is called *pumping.* In *optical pumping,* the active material is illuminated with light of suitable frequency $\nu = \dfrac{E_2 - E_1}{h}$.

As a result, an atom in the lower energy state E_1 absorbs the incident photon of energy $h\nu$ and is raised to the highest energy level E_2.

As the excited atoms lose their energy by spontaneous emission and drop to the lower energy level in the a very short time, the process fails to produce necessary population inversion.

In practice, therefore, population inversion is brought about by (*i*) *a three level scheme* or (*ii*) *a four level scheme.*

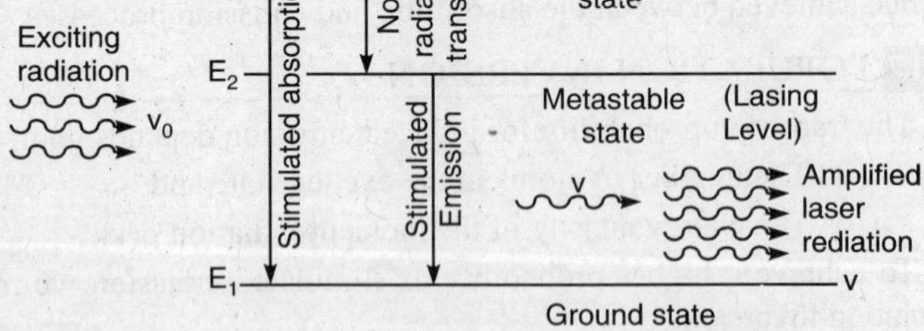

Fig. 29.21. Three level scheme

(*i*) **Three level scheme.** Consider an atom with energy levels E_1, E_2 and E_3 with increasing values of energy. E_1 is the ground state, E_3 is a short lived state and E_2 is an intermediate meta-stable state. Further, the transition from E_3 to E_1 is forbidden but transition from E_3 to E_2 is allowed. When these atoms are irradiated with an exciting radiation of the *right* frequency $\nu_0 = \frac{E_3 - E_1}{h}$, the atoms are excited to the E_3 state by the process of *stimulated absorption.* Some excited atoms quickly drop to the intermediate level E_2 by spontaneous emission or by a *non-radiative* proccess, thereby converting their excess energy into vibrational kinetic energy of the atoms forming the substance (Fig. 29.21). As E_2 is a metastable state, the stoms remain in this excited state for comparatively longer time of 10^{-3} sec. as compared to 10^{-8} sec. for the short lived state E_3 and for this time, the population of the state, E_2 is more than that of state E_1 thus resulting in population inversion of the collection of atoms.

If an atom in the state E_2 decays by spontaneous emission or stimulated emission, it emits a radiation of frequency $\nu = \frac{E_2 - E_1}{h}$. The photon may produce stimulated emission from another atom, thereby giving two chorent photons, moving in the same direction. These two photons produce two more photons and so on, producing an amplified beam of photons.

(*ii*) **Four level scheme.** In *four level scheme* there is an additional *very shortlived* energy state E_1' between the ground state (level) E_1 and *metastable* state E_2. When exciting radiation of right frequency $\nu_0 = \frac{E_3 - E_1}{h}$ is incident on the lasing medium the atoms in the ground state E_1 are excited to the level E_3 by the process of *stimulated absorption.* The atoms stay at the E_3 level for a very short time of about 10^{-8} sec. and quickly drop down to the metastable state E_2. Spontaneous transition from the metastable state E_2 to the level E_1' is forbidden and, therefore, cannot take place. As a result, the atoms accumulate at the level E_2 and the population at this level rapidly increases. Again the level E_1' is well above the ground level E_1 so that $(E_1' - E_1) >> kT$. Therefore, at normal temperature atoms cannot jump to the level E_1' from the ground level E_1 because of thermal energy. As a result, the level E_1' is almost virtually empty. The population inversion is thus maintained between the levels E_2 and E_1'.

Further, when the atom in the state E_2 decays by spontaneous emission or stimulated emission to the state E_1', it emits a photon of frequency $\nu' = \frac{E_2 - E_1'}{h}$. This photon further produces stimulated emission from another atom and thus starts the laser action.

The level E_1' being a very short lived state, the atoms immediately relax further to the ground staıe, ready for being pumped to the level E_3.

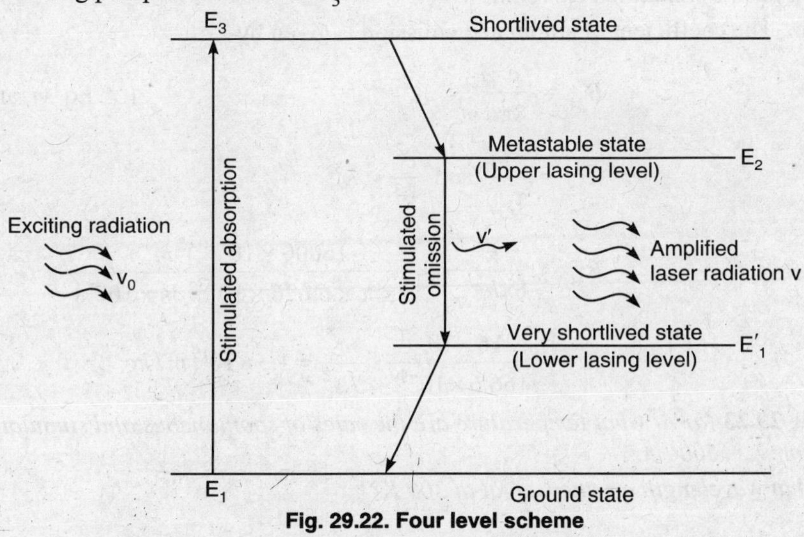

Fig. 29.22. Four level scheme

Comparison of four level pumping scheme with three level pumping scheme. 1. In three level pumping scheme the atoms finally return to the ground state directly from the metastable state. Therefore, in order to achieve population inversion pumping has to go on until more than half of the ground state atoms reach the metastable state E_2 which is the actual lasing level. As number of atoms in the ground state is very large a *very high pumping power* is required for the purpose.

But in case of four level scheme the terminal level is almost virtually empty so that the condition of population inversion is readily established even if a small number of atoms arrive at the upper lasing (or metastable) level. Consequently only a small pumping power is required to establish population inversion in four level scheme.

2. In three level pumping scheme, as soon as stimulated emission starts, the *population inversion condition* returns to *normal population condition* and lasing stops as soon as the excited atoms drop to ground level. Again lasing begins *only when* the condition of population inversion is re-established. The laser light given out is, therefore, in the form of *pulses of short duration.* But, in the case of four level scheme, the population inversion continues without interruption and laser light is continuously obtained. The laser, therefore, operates in *continuous wave (c w) mode.*

PN junction semiconductor laser is a two level laser, Ruby (Al_2O_3 crystal) laser is a three level laser and NdYAG is a four level laser.

Example 29.21 *Find the ratio of population inversion of the two states in a He – Ne laser that produces light of wavelength 6328Å at 27°C.*

Solution. $$N_2 = N_1 \cdot e^{-(E_2-E_1)/kT}$$

or $$\frac{N_2}{N_1} = e^{-(E_2-E_1)/kT}$$

But $$(E_2 - E_1) = \frac{12400}{6328}\, eV = 1.96\, eV$$

$\therefore$ $$\frac{N_2}{N_1} = \exp\left[\frac{-1.96\, eV}{(8.61\times10^{-5}\, eV/K\times300\, K)}\right]$$

$$= 1.1 \times 10^{-33}$$

Example 29.22 *The wavelength of emission is 6000 Å and the lifetime τ_{sp} is 10^{-6}s. Determine the coefficient for the stimulated emission.*

Solution. The coefficient of stimulated emission is given by

$$B_{21} = \frac{c^3 A_{21}}{8\pi h\, \nu^3}$$ (∵ Eq. *vi*, article 29.15)

But $$A_{21} = \frac{1}{\tau_{sp}} \text{ and } \frac{c^3}{\nu^3} = \lambda^3$$

$\therefore$ $$B_{21} = \frac{\lambda^3}{8\pi h\,\tau_{sp}} = \frac{(6000\times10^{-10})^3\, m^3}{8\times\pi\times6.626\times10^{-34}\text{ Js}\times10^{-6}\, s}$$

$$= \frac{216\times10^{-21}}{166.6\times10^{-40}}\cdot\frac{m^3}{J.s^2} = 1.3\times10^{19}\, m/kg$$

Example 29.23 *(a) At what temperature are the rates of spontaneous and stimulated emission equal ? Assume λ = 5000 Å.*

(b) At what wavelength are they equal at 300 K?

Solution. If the rates of spontaneous and stimulated emission are equal, then

$$R = \left[e^{h\nu/kT} - 1\right]^{-1} = 1$$

$$\lambda = 5000\,\text{Å}, \nu = \frac{c}{\lambda} = \frac{3 \times 10^8 \text{ m/s}}{5000 \times 10^{-10} \text{ m}}$$

$$= 6 \times 10^{14} \text{ Hz}$$

$$\therefore \quad \frac{h\nu}{kT} = \frac{(6.626 \times 10^{-34} \text{ J.s})\,(6 \times 10^{14} \text{ s}^{-1})}{(1.38 \times 10^{-23} \text{ J/K})T}$$

$$= \frac{28.81 \times 10^3}{T} \text{K}$$

$$(a) \quad \therefore \exp\left[\frac{28.81 \times 10^3}{T} \text{K}\right] = 2$$

$$\text{or} \quad \frac{28.81 \times 10^3}{T} \text{K} = ln2 = 0.693$$

$$\therefore \quad T = \frac{28.81 \times 10^3}{0.693} \text{K} = \mathbf{41{,}573\ K}$$

$$(b) \quad \therefore \quad \frac{h\nu}{kT} = \frac{(6.626 \times 10^{-34} \text{ j.s.})\,\nu}{(1.38 \times 10^{-23} \text{ J/K})(300 \text{ K})}$$

$$= (1.6 \times 10^{-13} \text{ s})\nu$$

$$\therefore \quad \exp\,[(1.6 \times 10^{-13} s)\nu] = 2$$

$$\text{or} \quad (1.6 \times 10^{-13} s)\nu] = 0.693$$

$$\therefore \quad \nu = \frac{0.693}{1.6 \times 10^{-13}} \text{Hz} = 4.3 \times 10^{12} \text{ Hz}$$

$$\therefore \quad \lambda = \frac{c}{\nu} = \frac{3 \times 10^8 \text{ m/s}}{4.3 \times 10^{12} s^{-1}} = \mathbf{69.8\ \mu m}$$

29.18 IMPORTANCE OF METASTABLE STATE

An atom in the ground state can be excited to a higher state by supplying energy to it. Normally, an atom in the excited state comes back to the original or ground state in a very short time of about $10^{-8} - 10^{-9}$ second by emission of a photon through spontaneous emission.

The average time for which an atom remains in an excited state is known as its mean life.

The mean life of an atom is characteristic of energy state. Some energy states have comparatively a longer mean life.

Energy states having mean life of more than 10^{-3} sec are known as metastable states.

Importance. In general, the lifetime of an excited state is small. Therefore, the atoms do not stay in the excited state long enough to be stimulated. These undergo spontaneous transitions and rapidly return to the lower energy states. Thus large number of atoms do not accumulate there and population inversion is not possible.

For population inversion to take place excited state should have a long life, i.e. it should be a metastable state. Anelectron excited to the metastable state cannot return to the ground state emitting a photon because of restrictions imposed by the conservation of angular momentum. As a result, a large number of excited atoms accumulate at the metastable state so that its population far exceeds that at the ground state, thus causing population inversion. *It is not possible to create population inversion without the presence of a metastable state.* We conclude that

No population inversion and hence laser action is possible if metastable states do not exist.

29.19 IMPORTANCE OF STIMULATED EMISSION

Just as metastable state is essential for population inversion and hence for laser action, stimulated emission is necessarily required for light amplification. In the absence of stimulated emission, light amplification is not possible. In other words, we cannot obtain light amplification in the absence of stimulated emission. It is because only in the case of stimulated emission we obtain photons of the same phase, energy, frequency (or wavelength), state of polarisation and travelling in a direction parallel to each other. These photons constitute the coherent radiation and superimpose to produce light amplifiaction and hence radiation of very high intensity.

29.20 EFFICIENCY OF A LASER

Efficiency of a laser is defined as the ratio of output energy E_{output} to the input energy E_{input} *i.e. energy used for pumping.*

$$\eta = \frac{E_{output}}{E_{input}} \qquad \text{... (29.30)}$$

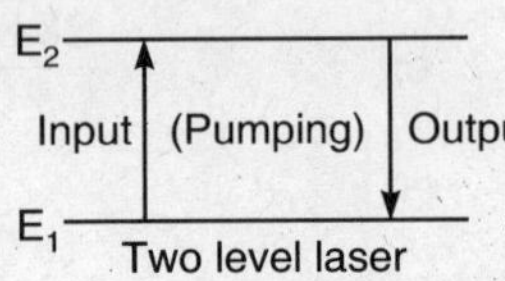

Fig. 29.23

For a two level laser output energy is equal to input energy used for pumping. Its efficiency is, therefore, 100%.

In the case of three level and four level laser, energy output is less than the input pumped energy and therefore, the efficiency is less than 100%. The efficiency of a four level laser is less than that of a three level laser. Although the efficiency of a four level laser is less than that of a three level laser, yet four level laser is preferred because in this case pumping is easier and thus population inversion can be more easily achieved.

29.21 LASER: THREE COMPONENT DEVICE

The essential requirements for a laser are (*i*) Pump, (*ii*) Laser material and (*iii*) Resonator. These are known as components of a laser. Thus, basically, laser is a three-component device.

(*i*) **Pump.** In order to realize and maintain the state of population inversion, it is necessary that atoms must be continuously promoted from the lower level to the excited level. Energy is to be supplied somehow to the laser medium to raise atoms from the lower level to the excited level and for maintaining population at the excited level at a value greater than that of the lower energy. The process by which atoms are raised from the lower level to the upper level is called *pumping*.

The usual method of exciting atoms to higher energy levels is to heat the material. From Eq. 29.28 it is seen that as long as $E_2 > E_1$, $N_2 < N_1$. Heating the material only increases the average energy of atoms but does not make N_2 greater than N_1. Therefore, population inversion cannot be achieved by heating the material.

The pump is an external source which supplies energy to bring about excitation of the atoms of the active medium from the ground state to a higher energy state either directly or through an atom to atom collision. The atoms in the higher energy state jump to a lower metasable energy state which has comparatively a very large lifetime as compared to the normal lifetime of the atom in other energy states, thus bringing about *population inversion.*

The pump can be optical, electrical or thermal. In Ruby laser optical pumping is used, whereas in *He—Ne* laser pumping is brought about by an electrical discharge. In semi-conductor laser, a direct conversion of electrical energy to light energy takes place.

(*ii*) **Laser material.** It is a material in which laser action is made to take place. It may be solid, liquid or gas.

The necessary condition is that the material should have an intermediate metastable state between the ground state and a higher energy state so that the atoms when excited to the higher energy state jump from there to the intermediate metastable state. As the number of atoms in the metasable state increases, population inversion takes place between the metastable state and a lower state or the ground state.

Laser action has been observed in almost half of the known atoms and laser wavelength may extend from ultraviolet region to infrared region.

Many lasers are named after the material used in them e.g. Ruby laser, *He–Ne* laser, CO_2 laser and *Nd-YAG* laser etc.

(*iii*) **The resonator (optical Resonant cavity).** The resonator consists of a pair of plane, or spherical mirrors with their reflecting sides facing each other. We shall consider a pair of *optically plane, parallel mirrors constituting the optical resonant cavity,* known as *Fabry Perot resonator.* The reflection coefficient of one of the mirrors is kept very nearly equal to one and that of the other a little less than one so that it allows a part of the internally reflected beam to escape out as a laser beam. The resonator is, in fact, a feedback device. It directs the photons back and forth through the laser medium and in this process the number of photons gets multiplied due to stimulated emission.

Laser action. Various steps involved in laser action are:

(*i*) **Stimulated absorption.** As shown in Fig. 29.21 the atoms are excited from the ground state E_1 to the higher energy state E_3 by supplying energy from an external source. For example, in optical pumping the laser medium is irradiated by radiation of frequency $\nu = \dfrac{E_3 - E_1}{h}$ so that the atoms are excited by *stimulated absorption.*

(*ii*) **Population inversion.** The energy state E_3 is a short lived state. Therefore, the atoms from energy state E_3 drop to the metastable energy state E_2 by spontaneous emission. This may happen almost immediately with in 10^{-8} sec. of excitation. As E_2 is a metastable state, the excited atoms stay in this state comparatively for a longer time (10^{-3} sec). As a result, the number of atoms in the energy level E_2 soon becomes much larger than the number in the energy level E_1. Thus population inversion occurs between energy levels E_2 and E_1.

(*iii*) **Stimulated emission.** A photon of energy $h\nu = (E_2 - E_1)$ may be emitted by spontaneous emission which is always going on.

This photon will pass through the laser medium and may cause stimulated absorption from energy level E_1 and E_2 or cause stimulated emission from energy level E_2 to E_1. As the rate of stimulated emission is much greater than that of stimulated absorption (due to population inversion), the emission process dominates. However, two important cases arise—

Case I. If the photon so emitted happens to be travelling in a direction inclined to the optic axis of the system as shown in Fig. 29.24 (*a*) it gets lost through the sides. Any other photon got emitted by it through stimulation is also lost along with it (Fig. 29.24b).

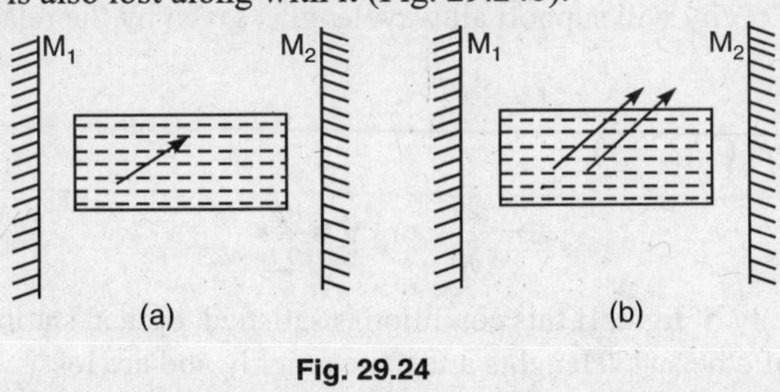

Fig. 29.24

Case II. If the photon is travelling parallel to the optic axis of the system as shown in Fig 29.25 (*a*) it acts as a *seed photon*. The other photon got emitted by it due to stimulation also travels along with it – parallel to it, in phase with it and in the same state of polarisation – as shown in Fig. 29.25

(*b*). After reflection from the mirror M_2 the photons again pass through the laser medium and further multiplication of coherent photons takes place as shown in Fig. 29.25 (*c*) and (*d*). Thus due to repeated passage of coherent photons through the laser medium because of multiple reflections at the mirrors M_1 and M_2 large scale amplification of the coherent photons takes place.

One of the mirrors M_2 is made partially reflecting. So every time the beam of coherent photons strikes the mirror M_2 a laser beam is transmitted out. In the absence of resonator cavity there would be not generation of light and no laser effect.

Nature of mirrors. Most of the resonators use parallel mirrors as described above . But practical difficulty does arise in aligning the parallel mirrors of the resonator. This difficulty can be overcome by using spherical mirrors in **confocal configuration,** with radius of curvature of each mirror being equal to the length of the optical cavity (see article 29. 23)

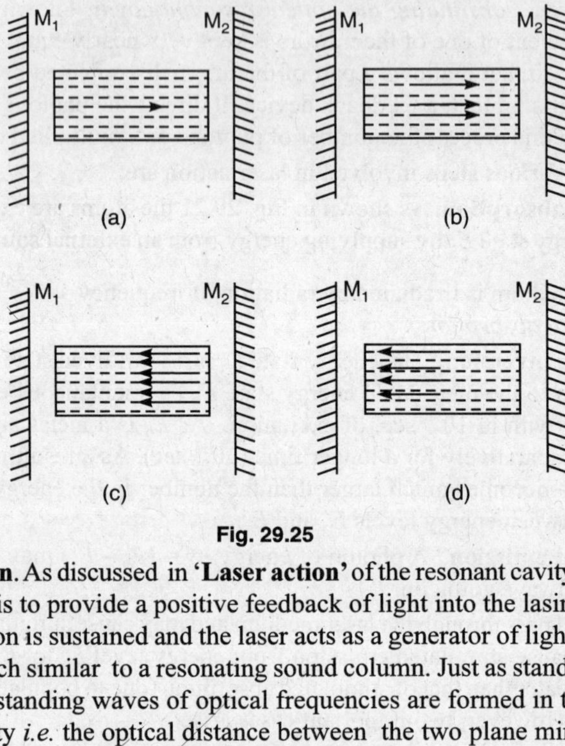

Fig. 29.25

Frequency condition. As discussed in **'Laser action'** of the resonant cavity, the primary function of the optical resonator is to provide a positive feedback of light into the lasing medium so that the stimulated emission action is sustained and the laser acts as a generator of light. For this purpose the optical cavity is very much similar to a resonating sound column. Just as standing waves are formed in a resonating column, standing waves of optical frequencies are formed in the optical cavity. If L is the length of the cavity *i.e.* the optical distance between the two plane mirrors, then the longest wavelength that will produce a standing wave pattern is $\lambda = 2L$ or $L = \frac{\lambda}{2}$.

In general, the cavity will support all wavelengths given by the relation

$$L = m\,\frac{\lambda}{2}$$

where m is an integer (1, 2, 3...)

$$\therefore \qquad \lambda = \frac{2L}{m} \quad \text{or} \quad \nu = \frac{mc}{2L} \qquad \text{... (29.31)}$$

where c is the velocity of light. If this condition is satisfied, optical stationary waves wil be set up in the cavity. Waves of other wavelenghts attenuate quickly and are lost.

This is known as **frequency condition.**

29.22 THRESHOLD CONDITION FOR LASER ACTION

As light travels in an optical resonant cavity in the forward and backward direction-it undergoes amplification and also suffers various losses. The losses take place due to (*i*) scattering and diffraction of light within the active medium and (*ii*) transmission of laser light through the output mirror.

For oscillations to build up it is necessary that the amplification between two consecutive reflections of light from rear end mirror balances all the losses.

Suppose the resonator mirrors M_1 and M_2 (Fig. 29.25) have a reflectively r_1 and r_2 respectively and the distance between the two mirror is L. If the intensity of light starting from the mirror M_1 to begin with is I_0 then as the light reaches the mirror M_2 its intensity will be

$$I_L = I_0\, e^{(\gamma - \alpha) L}$$

where γ = amplification factor of the light beam per unit length of the resonator cavity.

α = loss per unit length due to scattering, diffraction and absorption taking place as light travels through the medium in the cavity.

After reflection at mirror M_2 the intensity of light beam will become $r_2 I_0\, e^{(\gamma - \alpha)L}$

Again after returning from M_2 and reflection from M_1 i.e. after one complete round trip the intensity of light will become

$$I_{2L} = r_1\, r_2\, I_0\, e^{(\gamma - \alpha)2L}$$

$\therefore$ Amplification after one round trip $\dfrac{I_{2L}}{I_0} = r_1 r_2 e^{(\gamma - a)2L}$

For balancing all the losses, the amplification $\dfrac{I_{2L}}{I_0} \geq 1$

or $\quad r_1\, r_2\, e^{(\gamma - \alpha)2L} \geq 1$ or $e^{(\gamma - \alpha)2L} \geq \dfrac{1}{r_1 r_2}$

Taking natural logs on both sides

$$2L(\gamma - \alpha) \geq -\ln r_1\, r_2$$

or $$\gamma - \alpha \geq -\frac{1}{2L} \ln r_1\, r_2$$

or $$\gamma \geq \alpha - \frac{1}{2L} \ln r_1 r_2$$

or $$\gamma \geq \alpha + \frac{1}{2L} \ln \frac{1}{r_1 r_2} \qquad (29.30)$$

Eq. 29.30 is known as **lasing condition.** It shows that for lasing action to be maintained the initial gain should exceed the sum of all the losses in the cavity.

The threshold value of γ i.e., $\gamma_{th} = \alpha + \dfrac{1}{2L} \ln \dfrac{1}{r_1\, r_2}$ ...(29.31)

Eq. 29.31 gives the threshold condition for laser action to take place. γ, the amplification factor of the laser, depends upon how hard the laser medium is pumped. As the pump power is slowly increased the value of γ_{th} is reached and the kaser starts oscillating.

Example 29.24 *The length of a laser tube is 150 mm and the gain factor of the laser material is 0.0005/cm. If one of the cavity mirrors reflects 100% light that is incident on it. What is the required reflectance of the other cavity mirror?*

Solution $$\gamma \geq +\frac{1}{2L}\ln\frac{1}{r_1 r_2}$$

$$\therefore \quad r_2 = \frac{1}{r_1 e^{2L\gamma}} = \frac{1}{1\times e^{2\times15\times0.0005}}$$

$$= \mathbf{0.985}$$

It means the second mirror should have a reflectance of 98.5%.

29.23 CAVITY CONFIGURATIONS

The mirrors used in an optical cavity may not necessarily be flat. Several configurations are used in practice. Some of them are illustrated in fig. 29.26.

(*a*) Plane Parallel Cavity

The cavity consisting of two flat mirrors (Fig 29.26 (*a*)) is conceptually simple, but is unstable. Aligning the two mirrors strictly parallel poses a serious problem. If the mirrors are not precisely parallel, the light beam bouncing between them will walk away after a few reflections, and lead to a loss of light and consequent instability. This configuration is however very efficient. The light beam makes maximum use of the laser medium as there is no focussing of light within the cavity. This configuration is rarely used now-a-days because of alignment problem and low stability.

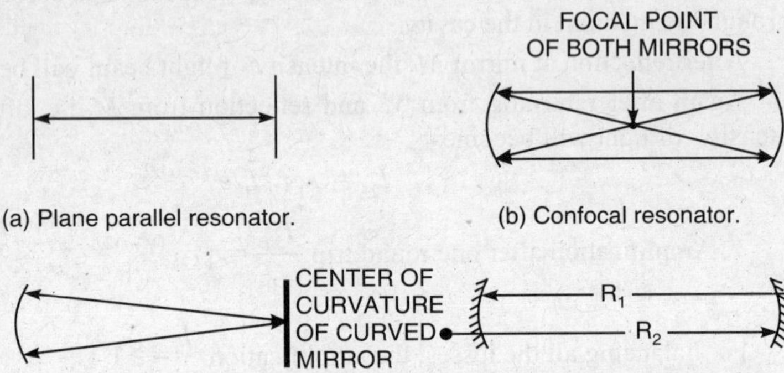

(a) Plane parallel resonator. (b) Confocal resonator.

(c) Hemispherical resonator. (d) Long radius resonator.

Fig. 29.26. Some of the laser resonator configurations

(*b*) Confocal Cavity

A confocal cavity (Fig. 29.26 (*b*)) has two concave mirrors of the same radius of curvature R. They are separated by a distance L equal to R, i.e. $L = R$. The focal length of the mirrors is $L/2$ and the focal points coincide in the centre. Hence the configuration is called "confocal". A confocal cavity is much easier to align. However, the filling is poor and only a small fraction of the medium is utilized.

(*c*) Hemispherical Cavity

A hemispherical cavity (Fig. 29.26 (*c*)) is a hybrid between the plane parallel and the confocal type. It has a concave mirror at one end and a plane mirror at the other. The plane mirror is placed at the centre of curvature of the spherical mirror, *i.e.* $L = R$. Even if the light is not emitted precisely along the axis of the system, the curved mirror can often reflect it back into the cavity. However, in this case also the filling is poor and the output is low.

(*d*) Long Radius Cavity

A long radius cavity (Fig. 29.26 (*d*)) has two concave mirrors. Their radii of curvature are significantly longer than the distance between them. That is $R_1 = R_2 > L$. This is a good compromise between the plane parallel and the confocal configurations. Slight misalignment of the cavity mirrors would not cause severe problems because their curvature would automatically focus light back toward the other mirror. This configurations is a stable resonant cavity because the light rays reflected from one mirror to the other will keep bouncing back and forth indefinitely. It is this type of cavity which is used in commercial lasers now-a-days.

29.24 MODES

A wave of frequency ν that travells along the axis of the cavity forms a series of **standing waves** within the cavity. They are discrete resonant conditions determined by the physical dimensions of the cavity. The laser modes governed by the axial dimension of the resonant cavity are called the **longitudinal modes** or **axial modes.** The laser modes determined by the cross-sectional dimension of the optical cavity are called the *transverse modes*.

1. **Longitudinal (or Axial) modes.** According to frequency condition only those waves of light produce constructive interference in an optical resonance cavity, which satisfy the condition

$$L = m\frac{\lambda}{2} \quad \text{... } (i)$$

where L is the length of the cavity, λ the wavelength of light and m an integer (1, 2, 3 ...)

It is clear from this equation that only those waves which can form *standing wave pattern* as shown in Fig. (29.27) can exist inside the cavity in the steady state. Waves of other wavelengths undergo destructive interference and die out quickly. As the length of the cavity is very large as compared to the wavelength of light, the optical resonator can simultaneously maintain a very large number of standing waves of multiple wavelength. The frequency of various modes is given by the relation.

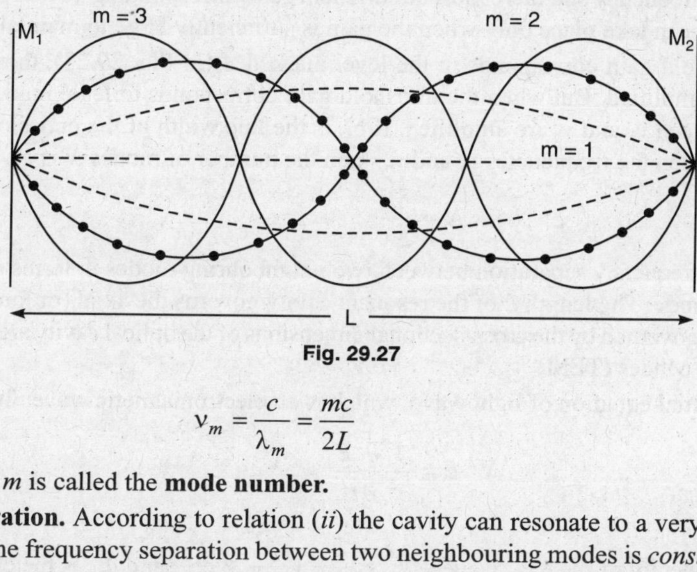

Fig. 29.27

$$\nu_m = \frac{c}{\lambda_m} = \frac{mc}{2L} \quad \text{... } (ii)$$

The number m is called the **mode number.**

Mode separation. According to relation (*ii*) the cavity can resonate to a very large number of frequencies but the frequency separation between two neighbouring modes is *constant* and given by

$$\Delta\nu = \nu_{m+1} - \nu_m = \frac{c}{2L}(m+1-m) = \frac{c}{2L} \quad \text{... } (iii)$$

In terms of wavelength,

$$\nu = \frac{c}{\lambda} \text{ and } \Delta\nu = -\frac{c}{\lambda^2}\Delta\lambda$$

$$\therefore \quad \Delta\lambda = \frac{\lambda^2}{c}\Delta\nu = \frac{\lambda^2}{2L} \quad (29.32)$$

Number of modes. The laser does not operate on all the frequencies given by the relation $\nu_m = \frac{mc}{2L}$. This relation gives the frequency of the modes belonging to *passive cavity i.e.* a cavity in which active medium is not present.

When the active medium is present between the two mirrors the cavity is known as *active cavity* and then the stationary waves formed in the cavity are due to the light waves emitted by the stimulated atoms of the medium. In an ideal case, the emission of light should take place at a *single* frequency because all identical atoms radiate at the same frequency. But because of a number of *line broadening factors* there is a *spread of* frequencies about a central frequency of the emission line. The shape of a typical emission line with ν_0 as *central frequency* is shown in Fig. 29.28. This curve is known as **gain curve** or **gain profile** and it indicates the range of frequencies over which stimulated emission can provide sufficient gain.

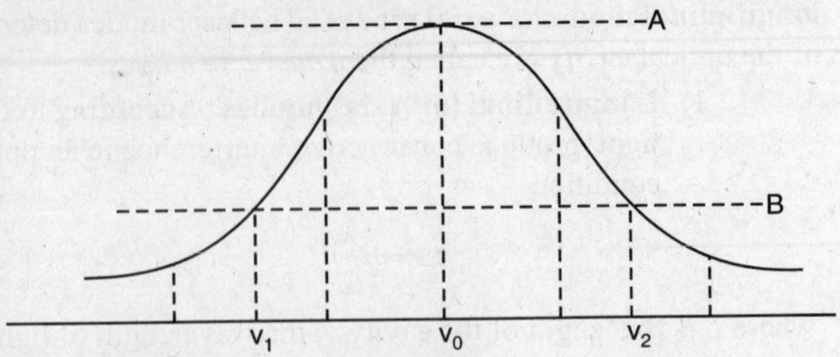

Fig. 29.28

Two factors determine the laser operating frequencies.

(*i*) Resonant frequency of the cavity and

(*ii*) Laser emission line width

For the output due to a given frequency to be present in the laser beam the cavity must be resonant to that particular frequency and there must be sufficient gain corresponding to that frequency because laser oscillations can take place only when the gain is sufficiently large to maintain resonance.

If the threshold gain corresponds to the level marked *A* (in Fig. 29.28), then only the central frequency ν_0 is amplified. But when the threshold gain corresponds to level marked *B*, then all the frequencies between ν_1 and ν_2 are amplified. If $\delta\lambda$ is the line width of the emission line in terms of wavelength between the frequencies ν_1 and ν_2, then the number of modes *N* is given by

$$N = \frac{\delta\lambda}{\Delta\lambda}$$

where $\Delta\lambda$ is the frequency separation between two neighbouring modes in terms of wavelength.

2. **Transverse mode.** The length *L* of the resonant cavity governs the axial (or longitudinal) modes. The laser modes governed by the cross-sectional dimensions of the optical cavity are called Transverse Electromagnetic Modes (TEM)

The differential equation of light wave, which is an electromagnetic wave, in the cavity is

$$\nabla^2 \vec{E} = \frac{1}{c^2}\frac{\partial^2 E}{\partial t^2} \qquad ...(i)$$

where *c* is the velocity of the e.m. waves, $\nabla^2 = \frac{\partial^2}{\partial x^2} + \frac{\partial^2}{\partial y^2} + \frac{\partial^2}{\partial z^2}$ and $\vec{E}$ is the electric field vector at the instant *t*.

The stationary waves formed in the cavity are given by the solution of Eq. (*i*)

i.e.,
$$\vec{E}\left(\vec{r}, t\right) = \vec{E}_0\, e^{i\left(\vec{k}\cdot\vec{r} - \omega t\right)} \qquad ...(ii)$$

where $\vec{E}_0$ is the amplitude of the electric field vector, $\vec{k}$ the wave vector $\left(\vec{k} = \frac{2\pi}{\lambda}\right.$, where λ is the wavelength of *E.M.* wave$\left.\right)$ ω the angular velocity $\left(\omega = 2\pi\nu = 2\pi\frac{c}{\lambda} = c\vec{k}\right)$

In terms of components along the three co-ordinate axes the complete solution with real spatial parts, obtained from Eq. (*ii*) is as under:

$$E_x = E_{ox} \cos k_x x \sin k_y y \sin k_z z$$
$$E_y = E_{oy} \sin k_x x \cos k_y y \sin k_z z$$
$$E_z = E_{oz} \sin k_x x \sin k_y y \cos k_z z$$

where $\vec{k} = \hat{i}\, k_x + \hat{j}\, k_y + \hat{k}\, k_z$

or $k^2 = k_x{}^2 + k_y{}^2 + k_z{}^2$

$$\vec{E}_0 = \hat{i}\, E_{ox} + \hat{j}\, E_{oy} + \hat{k}\, E_{oz}$$

and $\vec{E} = \hat{i}\, E_x + \hat{j}\, E_y + \hat{k}\, E_z$

Suppose the resonator cavity is in the form of a cube of length L and the *e.m.* wave is travelling in it along the Z-axis, then the tangential components of the electric field must vanish at the walls. In such a case

(*i*) $E_x = 0$ on plane $y = 0$; $y = L$ and $z = 0$; $z = L$

$\therefore$ $\sin k_y L = 0$; $\sin k_z L = 0$

or $k_y = \dfrac{n\pi}{L}$; $k_z = \dfrac{q\pi}{L}$

where n and q are integers (0, 1, 2.... etc.)

(*ii*) Similarly $E_y = 0$ on plane $x = 0$; $x = L$ and $z = 0$; $z = L$

$\therefore$ $k_x = \dfrac{m\pi}{L}$; $k_z = \dfrac{q\pi}{L}$

where $m = 0, 1, 2$ ... an integer.

The boundary conditions make the wave vector $\vec{k}$ take only discrete values specified by *m*, *n* and *q*. Each set of integers (*m, n, q*) defines a particular mode of vibration. The modes are known as **Transverse Electromagnetic Modes** and are designated TEM_{mnq}. As the wave is travelling along the Z-axis, *m* and *n* are *transverse indices* and *q* is *longitudinal index*. We, therefore, represent the transverse modes as TEM_{mn}. The transverse modes characterise the intensity distribution across the cross-section of the laser beam. If the laser beam is focussed on a screen, several bright patches are seen on the screen. The patches are separated by the intervals called *nodal lines*. The integers *m* and *n* represent the intensity minima in two perpendicular directions of the laser beam.

The lowest order transverse mode is TEM_{00}. It is the simplest mode and has a smooth cross-section profile with a peak in the middle [Fig. 29.29 (a)]. TEM_{01} beam has a single minima dividing the beam into two bright spots [Fig. 29.29 (*b*)]. The TEM_{11} beam has two perpendicular minima dividing the beam into four quadrants (Fig. 29.29 *c*).

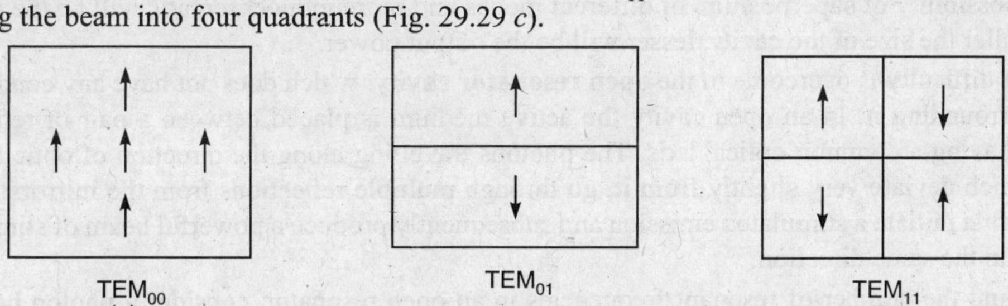

Fig. 29.29

Operating the laser in multimode form provides considerably more power than in a single mode operation.

29.25 OPEN CAVITY RESONATORS PREFERRED

Difficulty with closed cavity resonator. In a closed cavity resonator of cubic shape of each side L, the vibrational modes in therms of wave vector $\vec{k}$ along the three co-ordinate axes are given by

$$k_x = \frac{m\pi}{L}; k_y = \frac{n\pi}{L} \text{ and } k_z = \frac{q\pi}{L}$$

where $m, n,$ and q are integers (1, 2, 3 ...). If we consider points separated by $\frac{\pi}{L}$ in the *wave vector space,* then each point has a vacant space equal in volume to $\left(\frac{\pi}{L}\right)^3$

Now, consider a spherical shell in *wave vector space* of radius k and thickness dk, then volume of the shell = $4\pi k^2 dk$

Hence the volume between consecutive shells in the positive octant = $\frac{1}{8}.4\pi k^2 dk = \frac{\pi}{2} k^2 dk$

∴ Number of vibrational modes of standing waves for wave vector between, k and $k+dk$

$$= \frac{\frac{\pi}{2} k^2 dk}{(\pi / L)^3} = \frac{L^3 k^2 dk}{2\pi^2}$$

Since there are two directions of polarisation of wave,

$$\text{Number of modes} = 2.\frac{L^3 k^2 dk}{2\pi^2} = \frac{L^3 k^2 dk}{\pi^2}$$

∴ For a cubic cavity of each side L, (volume L^3) normal modes of vibration per unit volume of cavity = $\frac{k^2 dk}{\pi^2}$

$$\text{Now} \quad k = \frac{2\pi}{\lambda} = \frac{2\pi\nu}{c} \quad \therefore \quad dk = \frac{2\pi}{c} d\nu$$

Hence number of modes *per unit volume* having frequency between ν and $(\nu + d\nu)$

$$= \frac{1}{\pi^2}.\frac{4\pi^2 \nu^2}{c^2}.\frac{2\pi}{c} d\nu = \frac{8\pi}{c^3} \nu^2 d\nu$$

Thus the number of modes in the optical cavity can be restricted by restricting the volume of the cavity. Smaller the volume of the cavity, lesser will be the number of modes in it and hence lesser will be possibility of superposition of different modes and more monochromatic will be the output. But, smaller the size of the cavity, lesser will be the output power.

This difficulty is overcome in the **open resonator cavity** which does not have any conducting walls surrounding it. In an open cavity the active medium is placed between a pair of reflecting mirrors having a common optical axis. The photons travelling along the direction of optic axis or those which deviate very slightly from it, go through multiple reflections from the mirrors. These photons first initiate a stimulated emission and subsequently produce a powerful beam of stimulated photons in the same direction.

To find the number of resonant frequencies in an open resonator, consider a photon beam of wave vector $\vec{k}$ which makes a very small angle with the optic axis, then

$$k = \sqrt{k_x^2 + k_y^2 + k_z^2} = \frac{\pi}{L}\sqrt{m^2 + n^2 + q^2}$$

$$\left[\because k_x = \frac{m\pi}{L}; k_y = \frac{n\pi}{L}; \text{and } k_z = \frac{q\pi}{L}\right]$$

As the beam travels close to the resonator axis (taken along Z-axis) k_x and k_y and hence m and n have a very small value as compared to q and may be ignored.

$$\therefore \qquad k = \frac{\pi q}{L} \text{ or } L = \frac{q\pi}{k} = q\pi\frac{\lambda}{2\pi} = q\frac{\lambda}{2}$$

The difference between two consecutive resonant frequencies in the cavity $= \nu_{q+1} - \nu_q = \dfrac{c}{2L}$

[Refer Eq. (*iii*), article 29.24].

If $\Delta\nu$ is the line width of radiation, number of resonant frequencies within it are

$$N = \frac{\Delta\nu}{\nu_{q+1} - \nu_q} = \frac{L}{2c}\Delta\nu \qquad \text{... (29.32a)}$$

i.e. the open resonator transforms a gain line into a set of narrow lines. From the above discussion we find that:

A resonator is characterised by various modes of vibration with different frequencies. In a closed resonator the number of modes which can get amplified and which can oscillate in a resonator of practical dimensions becomes so large that the output becomes far from monochromatic. In the open resonator, on the other hand, the number of modes which can oscillate are only a few and it is possible to reduce it even to a single mode.

Furthermore, the open sides of a resonator can be used for optical pumping as in a ruby laser. This is why open cavity resonators are preferred over closed cavity resonators.

29.26 QUALITY FACTOR OF A CAVITY

Quality factor or Q-factor of a resonance cavity is defined as

$$Q = -\omega \times \frac{\text{Energy stored in the mode}}{\text{Energy lost per second in the mode}}$$

The loss in energy of a beam of light with time, as it travels through the cavity, is given by

$$U = U_0 e^{-t/\tau}$$

where U_0 is the energy in the beginning, U the energy after a time t, and τ the time during which energy reduces by $1/e$ of its original value.

$$\therefore \qquad Q = -\omega\frac{U}{dU/dt} = -\omega\frac{U_0 e^{-t/\tau}}{\frac{d}{dt}U_0 e^{-t/\tau}} = \frac{-\omega U_0 e^{-t/\tau}}{U_0 e^{-t/\tau}\frac{d}{dt}\left(-\frac{t}{\tau}\right)} = \omega\tau$$

Now $\qquad \tau = \dfrac{1}{\Delta\omega}$ where $\Delta\omega$ is the line width

$$\therefore \qquad Q = \frac{\omega}{\Delta\omega} \qquad \text{... (29.33)}$$

Thus smaller the line width $\Delta\omega$, higher is the quality factor. In case of lasers, Q is very high and is of the order of 10^8. This is so because in lasers the gain medium is supplying energy to the oscillating modes. Theoretically, energy dissipation or loss can be zero and Q infinite. However, there are always losses which limit Q to a high value. Q-factor is a measure of the mirror losses.

If L is the length of the optical cavity, σ is the cross-sectional area of the laser beam, and v is the volume density of energy, then the energy stored in the resonator is given by

$$W = \sigma L v$$

The volume density of energy v consists of two equal densities of energy fluxes moving in opposite directions. The energy lost during one cycle is given by

$$\Delta W = \left(\frac{v}{2}\right)\sigma L(1 - e^{-2f})$$

When $f << 1$, we can approximate $e^{-2f} \approx 1 - 2f$

$$\therefore \qquad \Delta W = v\sigma Lf$$

If T is the period of laser radiation and $2L/v$ is the duration of one cycle, then

$$\Delta W = \frac{\sigma L v}{\left(\frac{1}{2}\right) v v \sigma f T} = \frac{2L}{f\lambda} \qquad \text{... (i)}$$

But $$\frac{2L}{\lambda} = m \qquad \text{... (ii)}$$

$$\therefore \qquad Q = \frac{m}{f} \qquad \text{... (iii)}$$

We can express the condition necessary for lasing, using $\alpha_0 L = f$, in Eq. (*iii*), we then obtain

$$Q = \frac{m}{\alpha_0 L}.$$

Using Eq. (*ii*) we get $$\frac{\alpha_0 \lambda}{2} = \frac{1}{Q} \qquad \text{... (iv)}$$

Thus, the threshold condition depends on the reciprocal of the quality factor. The higher the Q-factor, the lower the lasing threshold. Q is also related to the laser line width in a laser cavity and is given by

$$Q = \frac{\nu_0}{\Delta\nu} \qquad \text{... (29.34)}$$

29.27 Q-SWITCHING FOR GIANT PULSES

The power of a laser beam can be drastically increased if it can be arranged that a very large number of atoms of the active medium participate in the process of stimulated emission. This is possible if a very high population inversion density is established in the active medium. In the normal course, when the laser starts oscillating, the high value of the population inversion density drops to the threshold value of the steady state condition. To have a high power laser beam it is necessary that process of oscillation is delayed until the highest possible number of atoms accumulate at the upper lasing level. One of the methods to achieve this is to make the resonant mirrors slightly non-parallel. If one of the mirrors is misaligned it will not be able to reflect the incident photons into the active medium and therefore, stimulated emission will not be able to take place. Actually, it means that Q-factor of the resonator will be reduced to a very low value, thereby greatly increasing the optical losses. As a result, the pumping arrangement will be able to build up the population inversion to a very high value in the active medium.

If now the alignment of the two mirrors is suddenly restored, these will reflect photons into the medium. The feedback of the photons, now triggers a chain of stimulated emissions and a photon avalanche is rapidly built up. As a result the Q-factor of the cavity is abruptly increased and suddenly the laser oscillations set in. All the energy stored in the cavity is released in a single giant pulse with peak power very much greater than the laser can produce normally. This pulse lasts for a very short duration of time, thus quickly depopulating the upper energy level and hence the laser action stops. This method of controlling the laser output power is known as **Q-switching.**

In short, *Q*-switching in a laser system means *closing* the optical cavity or resonant box by a suitable method until a *very strong population inversion*, much higher than the threshold, is built up. When this is accomplished the optical cavity is suddenly opened so that a large fraction of the stored energy is released into a *single short pulse. This intermittent high power* laser pulse thus gives way to the smooth uninterrupted laser output.

Arrangement for Q-switching. To switch on the resonant cavity from a passive to an active state, a voltage controlled electro-optic shutter is used. It consists of a crystal which becomes doubly refracting when an electric field is applied across it. Such an arrangement is known as "*electro-optic cell*" and is shown in Fig. 29.30.

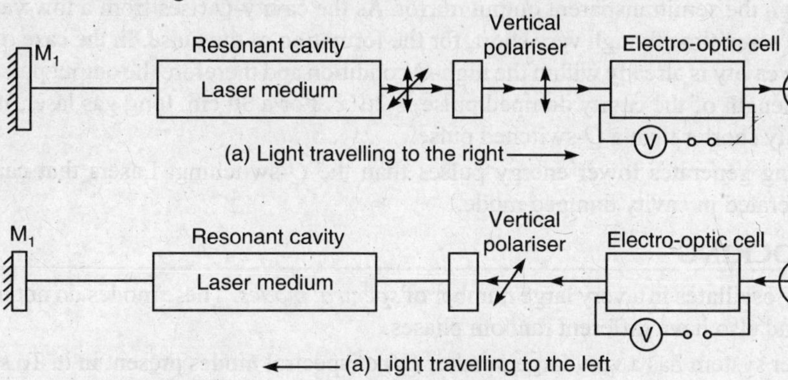

Fig. 29.30

During the pumping of the laser, a voltage is applied to the crystal (of the electro-optic cell) by a light flash from a flash lamp. The magnitude of the voltage is adjusted to such a value that it transforms the electro-optic crystal into a quarter wave plate. As shown in Fig. 29.30 (a) the laser light coming out of the cavity gets *linearly polarised* on passing through the polariser. The linearly polarised light is then incident on the crystal of the electro-optic cell and gets split into two mutually perpendicular components. As the two components travel through the electro-optic cell, a relative phase retardation of $\frac{\pi}{2} (= 90°)$ is produced between them due to the action of the crystal as a quarter wave plate. After coming out of the electro-optic crystal the two components combine to produce circularly polarised light. The light beam gets reflected from the mirror M_2 and travelling in the opposite directions returns to the cavity. On reflection from M_2 the sense of rotation of circularly polarised light reverses and on again passing through the crystal the two components of circularly polarised light go through a further retardation of 90° with respect to each other and on coming out of the crystal at the other end the two components again combine to produce linearly polarised light. The direction of polarisation of this light is inclined at an angle of 90° w.r.t. its original direction of polarisation and transmission axis of the polariser. The polariser does not allow this light to pass through. As a result the light does not come back to the laser cavity and hence the cavity gets switched off. In this way the *Q*-factor of the cavity is reduced to a very low value.

However, when the voltage applied to the electro-optic crystal is switched off, double refraction cannot take place in the crystal and the state of polarisation of light passing through the crystal remains unaffected. Light, therefore, travels freely through the crystal in both the directions and returns to the laser cavity where it is further amplified. The cavity, therefore, gets *switched on* and regains its high *Q*-value. The *Q*-switching process is synchronised with pumping mechanism in such a way that the voltage applied to the electro-optic crystal drops to zero at the time when the population inversion in the laser medium attains peak value.

29.28 CAVITY DUMPING

A cavity dumped laser employs two full reflective mirrors, one on each end of the cavity. When pumping of the lasing medium is done, power inside the cavity builds up due to stimulated emissions. The continuously operating pump maintains the population inversion above the threshold. Oscillations commence but the radiation is confined within the cavity, because the mirrors prevent the light from escaping. Thus, the cavity operates at high-Q. The cavity contains a beam deflector or other component that briefly deflects the beam outside of the laser. Thus, the energy within the cavity is dumped off the cavity in a single pulse.

In Q-switching mode, the energy within the cavity is low and Q is low. Oscillations do not start in the initial state. When the cavity Q is switched to the high value, oscillations start and a short intense pulse passes through the semitransparent output mirror. As the cavity-Q rises from a low value to a high value, it takes some time though very short, for the formation of the pulse. In the case of cavity dumping mode, the cavity is already within the high-Q condition and therefore the output pulse forms very quickly. The length of the cavity dumped pulse is $2\mu Lc$. For a 30 cm. long gas laser, it would be 2 ns, significantly shorter than a Q-switched pulse.

Cavity dumping generates lower energy pulses than the Q-switching. Lasers that cannot be Q-switched are operated in cavity dumped mode.

29.29 MODE LOCKING

A laser usually oscillates in a very large number of *spectral modes*. These modes do not oscillate at the same time and also have different random phases.

Suppose a laser system has a very large number (N) of spectral modes present in it. To simplify matters, let each mode have the same amplitude A. Then, the electric field at a time t inside the resonator cavity is given by

$$E(t) = A \sum_{n=1}^{n=N} \exp[i(\omega_n t + \delta_n)]$$

where ω_n is the angular frequency ($= 2\pi\, v_n$) of the nth mode (v_n being frequency) and δ_n the relative phase. As A is the amplitude of the mode, its intensity is A^2. For these uncoupled (or unlocked N modes, the total intensity is Fgiven by the sum of the intensities of individual modes *i.e.*

Total intensity $\quad I = N\,(A^2)$

When these modes are coupled *i.e. locked*, their individual phases are frozen and all these have one and the same common value of phase δ.

$\therefore \quad \delta_n = \delta$

The laser so operated becomes **mode locked laser.** For such a **mode locked** (ML) system having N modes, all in the same phase, the total amplitude is given by the sum of the amplitudes of N modes.

$\therefore$ Total amplitude of N mode locked modes $= NA$

Hence the total intensity of N mode locked modes

$$I_{ML} = (NA)^2 = N.\,N(A^2) = NI \qquad \text{... (29.35)}$$

Thus the intensity of the mode locked laser system is N times the intensity (or output) of the same laser system with modes uncoupled. As N has very large value I_{ML} is very much higher than I.

Mode locking is of two types:

(*i*) *Active mode locking* and (*ii*) *Passive mode locking*

(*i*) **Active mode locking.** In active mode locking a polished quartz crystal is placed inside the laser resonance cavity. It is made to oscillate in acoustic resonance with the help of a piezo-electric crystal which when subjected to pressure develops positive and negative charges on its opposite faces. These acoustic waves induce periodic variations of refractive index in the quartz crystal which acts like a diffraction grating and establishes a constant fixed phase relationship among all the modes, thereby producing *mode-locking*.

(*ii*) **Passive mode locking.** In passive mode locking a *bleachable dye cell* using suitable concentration of dye solution is placed inside the laser cavity. The dye absorbs the low power components and transmits the high power components. For this purpose absorbers with *very small relaxation time* are used so that the absorber *recovers* in a time which is short compared with duration of the pulse.

29.30 CHARACTERISTICS OF LASER LIGHT

Laser is an abbreviated form of '*Light Amplification by Stimulated Emission of Radiations*'. Laser light differs from ordinary light in following respects:

(*i*) Laser beam is highly monochromatic and sharp.

(*ii*) Laser light is highly intense and travels as a highly concentrated parallel beam along a particular direction.

(*iii*) Laser beam is highly coherent with all the waves exactly in phase with each other and in the same state of polarisation. The beam has both spatial and temporal coherence.

On the other hand.

(*i*) Ordinary light or light from a filament lamp is not monochromatic. Light spectrum from a source may extend over a wide range of wavelength.

(*ii*) The light from an ordinary source (filament lamp) travels and spreads in all directions. Ordinary light is diffused.

(*iii*) Ordinary light is incoherent *i.e.* there is a wide phase difference between the light observed at a point at different times and at a time at different points in space.

Important properties of laser light are:

(*i*) **Directionality.** Laser emits radiation in the form of a highly directional collimated beam with only a very small angle of divergence. It means that the energy carried by the laser beam can be collected easily and focussed on to a small area. The directionality of a laser beam is expressed in terms of beam divergence.

Divergence

Light from a laser diverges very little. Upto a certain distance, the beam shows little spreading and remains essentially a bundle of parallel light rays. The distance from the laser over which the light rays remain parallel is known as *Rayleigh range.* The laser beam diverges beyond the Rayleigh range as shown in Fig. 29.31. There are two parameters which cause beam divergence. They are (*i*) the size of the beam waist and (*ii*) diffraction. The *divergence angle* is measured from the centre of the beam to the edge of the beam, where the edge is defined as the location in the beam where the intensity decreases to $1/e$ of that at the centre.

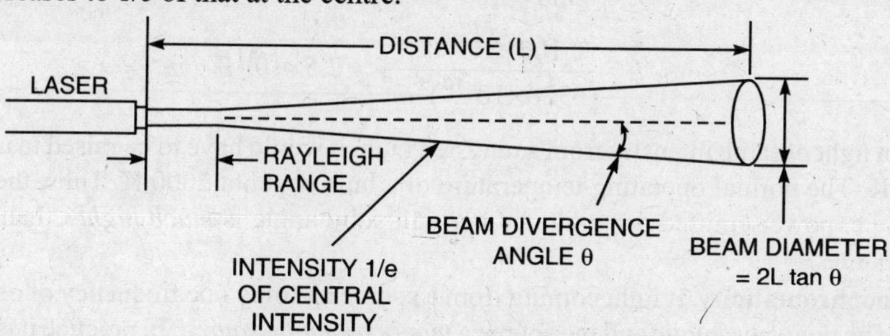

Fig. 29.31. Divergence of a laser beam

Twice the angle of divergence is known as the *full angle beam divergence*. This angle tells us how much the beam will spread as it travels through space. The full angle divergence is given by

$$2\theta = \frac{4\lambda}{\pi d_0} \qquad \text{... (29.36)}$$

where $d_0 = 2W_0$ is the diameter of the beam waist (Fig.29.32). It is seen from Eq. 29.36 that the divergence is inversely proportional to d_0. Thus, divergence is large for a beam of small waist.

The beam divergence due to diffraction is determined from Rayleigh's criterion.

$$d\theta = 1.22\frac{\lambda}{D} \qquad \text{... (29.37)}$$

where D is the diameter of the laser's aperture.

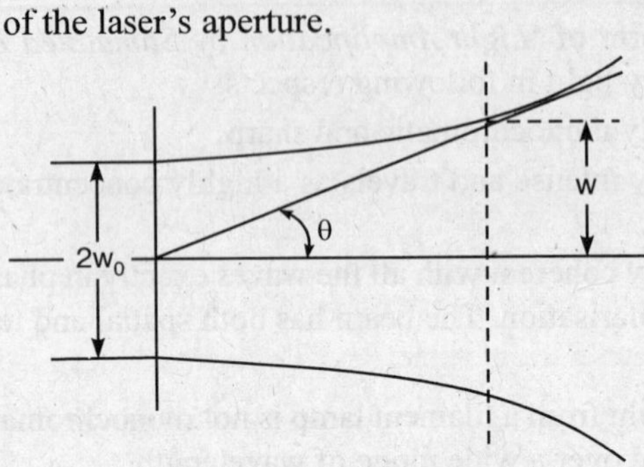

Fig. 29.32. Relationship between beam waist and divergence of a laser beam

In case of gas lasers, the diffraction divergence is about twice as large as the beam-waist divergence. A typical value of divergence for a He-Ne laser is 10^{-3} rad. It means that the diameter of the laser beam increases by about 1 mm for every metre it travels. Beam divergence of large lasers may be as small as a micro-dgree. A laser beam of 5 cm. diameter (divergence = 10^{-6} degree) when focussed from Earth will spread to a diameter of only about 10 m on reaching the surface of the moon. This extreme collimation of the beam makes lasers a very useful tool for surveying.

***(ii)* Intensity**

The power output of a laser may vary from a few milliwatts to a few kilowatts. But this energy is concentrated in a beam of very small cross-section. The intensity of a laser beam is approximately given by

$$I = \left(\frac{10}{\lambda}\right)^2 P \quad W/m^2 \qquad \text{... (29.38)}$$

where P is the power radiated by the laser. In case of 1 mW He-Ne laser,

$$\lambda = 6328 \times 10^{-10} \text{ m and}$$

$$I = \frac{100 \times 10^{-3} W}{(6328 \times 10^{-10})^2 m^2} = 2.5 \times 10^{11} W/m^2$$

To obtain light of same intensity from a tungsten bulb, it would have to be raised to a temperature of 4.6×10^6 K. The normal operating temperature of a bulb is about 2000 K. Thus, the intensity of a laser beam i.e. power emitted per unit area per unit solid angle is *much higher* than that of light from other sources.

***(iii)* Monochromaticity.** If light coming from a source has only one frequency of oscillation, the light is said to be *monochromatic* and the source a *monochromatic source.* In practice it is not possible to produce light with only one frequency. Light coming out from any source consists of a band of frequencies closely spaced around the central frequency, v_o. The band of frequency Δv is called the *line width* or *bandwidth*.

A laser beam is highly monochromatic. But the spectral content of the laser raditation extends over a range equivalent to the fluorescent line width of the laser medium. Although the line width of an individual cavity mode is extremely small, there may be many modes present in the laser output.

The light from conventional sources has large line widths of the order of 10^{10} Hz or more. On the other hand, light from lasers is more monochromatic, having linewidth of the order of 100 Hz.

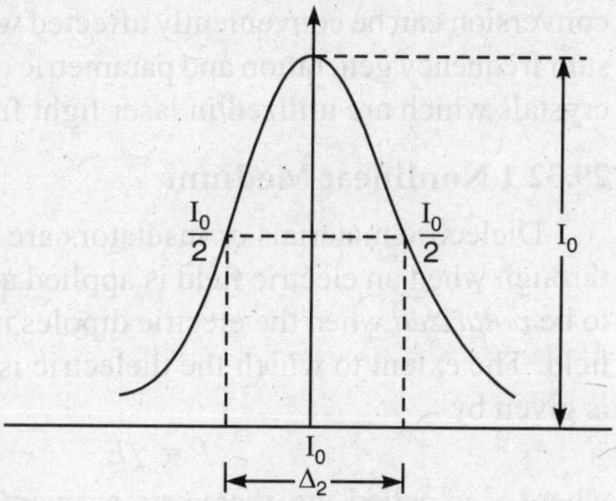

Fig. 29.33

From the theory of interference, the linewidth of mirrored cavity is given by

$$\Delta\gamma = \frac{c}{2\pi L}\left(\frac{1-R}{\sqrt{R}}\right) \qquad \text{... (29.39)}$$

where L is the length of the cavity and R is the reflectance of the out put mirror.

(*iv*) **Coherence.** A laser beam is highly coherent. It is because one important property of stimulated emission is that the wave or phton produced by stimulated emission is exactly in phase with the stimulating wave or photon. As a result the spatial and temporal variation of the electric field vector of the two waves are exactly similar. For a perfect laser the electric field varies with time in an identical manner for every point on the cross-section of the beam. (For details refer article 29.5)

29.31 POLARIZATION

Light waves are electromagnetic waves consisting of electric and magnetic fields vibrating perpendicular to each other and to the direction of propagation. The alignment of the electric and magnetic fields in a light wave is known as polarization. There are several types of polarization, out of which the simplest is linear (or plane) polarization. In a linearly polarized light beam, the orientation of the electric field remains in one plane while its magnitude changes with time. Any other type of polarized light can be viewed as a result of superpositions of two linearly polarized waves that are having electric fields perpendicular to each other. For example, unpolarized light can be divided into two components with linear polarization, one with a vertical field and one with a horizontal field.

The conventional light sources produce unpolarized light. External devices such as polarizers and crystal plates are used to convert the unpolarized light into light of desired polarization. In case of lasers, most of them also emit unpolarized light. Laser output can be made linearly polarized light by adding a suitable device. Normally, Brewster window is used to obtain linearly polarized light from a laser. It is a surface arranged at an angle, called Brewster angle, with the resonator axis. Light incident on a surface at the Brewster angle will be reflected or transmitted depending on its polarization. Light polarized parallel to the plane of incidence is transmitted through the Brewster window while light polarized *perpendicular* to the plane of incidence is partly transmitted and partly reflected. The component that undergoes reflection will be lost to the resonator and will not contribute to the oscillation of the laser. The result is that the output beam obtained is plane polarized having polarization parallel to the plane of incidence.

29.32 FREQUENCY CONVERSIONS USING NON-LINEAR CRYSTALS

Although a large number of active materials are available and lasers can be built using them, it is not possible to produce light covering all wavelengths in interest. It becomes therefore, necessary to transform the frequency of light generated by lasers into light of desired frequency. The frequency conversion can be conveniently affected with the help nonlinear optical media. Harmonic generation, sum frequency generation and parametric oscillation are three important effects exhibited by nonlinear crystals which are utilized in laser light frequency transformations.

29.32.1 Nonlinear Medium

Dielectric materials or insulators are nonconductors. They do not allow electric current to pass through when an electric field is applied across them. Instead, they get polarized. A dielectric is said to be *polarized* when the electric dipoles in it orient in the direction of the externally applied electric field. The extent to which the dielectric is polarized is denoted by the polarization vector, P, which is given by

$$P = \chi E \quad \text{... (29.40)}$$

where χ is called the *dielectric susceptibility*. The stronger the electric field the greater will be polarization. Materials, in which this kind of linear relationship holds, are known as *linear dielectrics*. Dielectrics are polarized, for example, when they are placed between the charged plates of a capacitor. In a similar way, they are polarized when electromagnetic wave propagates through them. Light waves are electromagnetic waves and the electric field of the waves polarizes the dielectric. Hence the optical parameters are closely related to the dielectric polarization. The refractive index of the medium is given by

$$\mu = \sqrt{\varepsilon} = \sqrt{1 + 4\pi\chi} \quad \text{... (29.41)}$$

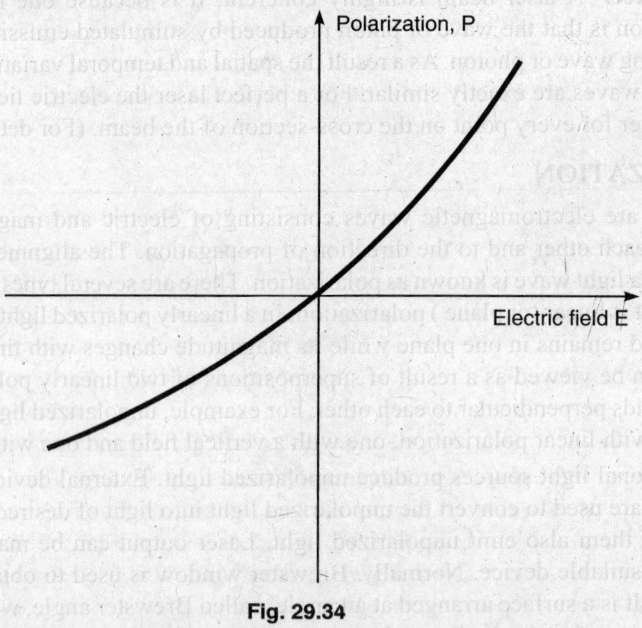

Fig. 29.34

As long as the intensity of the light propagating in a dielectric medium is small, the parameters χ and μ are constant quantities and are independent of the intensity of light. Ordinary light sources generate light of field strengths of the order of 10^5 V/m which are very small compared to atomic fields, and therefore cannot affect the optical parameters of the medium. Lasers produces light of field strengths of the order of 10^7 to 10^{11} V/m, which are of the order of atomic field strengths. Therefore, the intense light of lasers is in a position to influence the optical parameters of a medium. When the electric field E in the light wave is very large, the parameters χ, ε and μ become functions of E. Thus we can express

$$\chi(E) = \chi_0 + \chi_1 E + \chi_2 E^2 + ... \quad \text{... (29.42)}$$

and
$$P = \chi(E)\,E$$
$$\therefore \quad P = \chi_0 E + \chi_1 E^2 + \chi_2 E^3 + ... \quad ...(29.43)$$
where χ_0, χ_1, χ_2 ... are parameters of the medium characterizing its polarizability. The second and higher order terms represent the *nonlinear* part of the polarization. Materials, in which polarization exhibits nonlinear dependence on the strength of the electric field, are called *monlinear medium.* Fig. 29.34 shows the variation of electric polarization with the electric field strength in a nonlinear medium.

29.32.2 Second Harmonic Generation

Strictly speaking, any medium becomes "nonlinear" provided the electric field of the incident radiation is very high. In case of optically isotropic materials, the coefficients of even powers of E in Eq. (29.43) are zero. But in case of anisotropic materials, coefficients of both even and odd powers of E exist. However, the third order term is substantially smaller than the second order term. Hence, Eq. (29.43) can be rewritten in case of anisotropic crystal as
$$P = \chi_0 E + \chi_1 E^2 \quad ...(29.44)$$
Fig. 29.35 shows the response of a nonlinear medium, in which a symmetrical electric field produces an asymmetrical polarization. A Fourier analysis of the polarization shows that it consists of components having frequencies ν, and 2ν as well as *d.c.* components. Consider a plane monochromatic light wave of frequency ν travelling at the velocity υ_1 in a nonlinear medium. The electric field of the wave may be of the form.
$$E = E_0 \sin[2\pi\nu(t - x/\upsilon_1)] + \chi_1 E_0^2 \sin^2[2\pi\nu(t - x/\upsilon_1)]$$
$$= E_0 \sin[2\pi\nu(t - x/\upsilon_1)] + \frac{1}{2}\chi_1 E_0^2 - \frac{1}{2}\chi_1 E_0^2 \cos[4\pi\nu(t - x/\upsilon_1)] \quad ...(29.45)$$

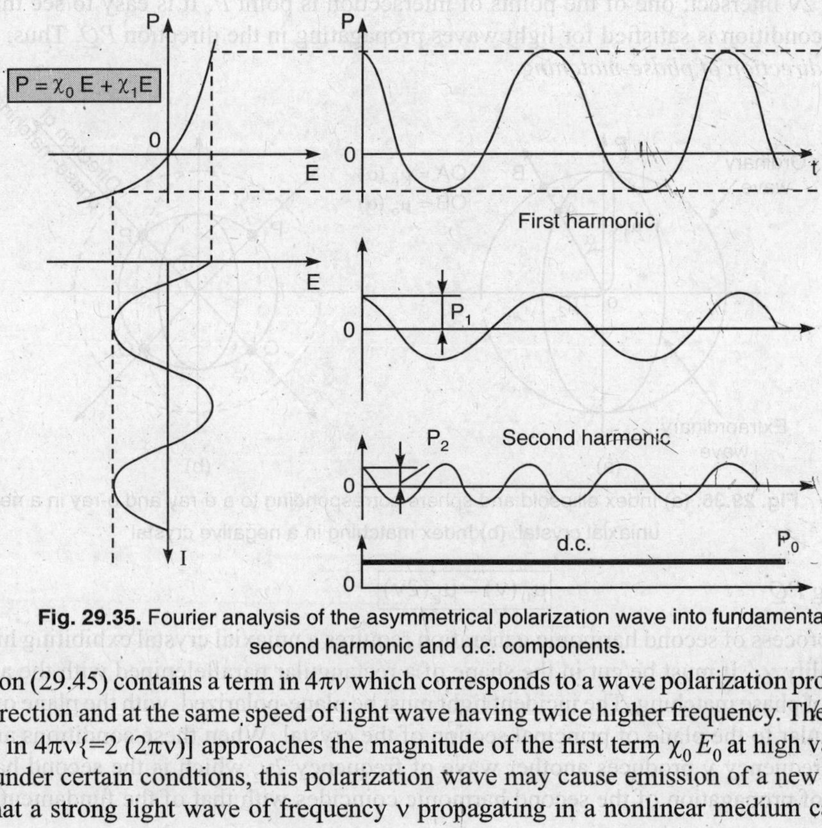

Fig. 29.35. Fourier analysis of the asymmetrical polarization wave into fundamental, second harmonic and d.c. components.

Equation (29.45) contains a term in 4πν which corresponds to a wave polarization propagating in the same direction and at the same speed of light wave having twice higher frequency. The magnitude of the term in 4πν{=2 (2πν)] approaches the magnitude of the first term $\chi_0 E_0$ at high values of E_0. Secondly, under certain condtions, this polarization wave may cause emission of a new light wave. It means that a strong light wave of frequency ν propagating in a nonlinear medium can generate a new light wave at frequency 2ν. It is a second harmonic and hence the phenomenon is known as *second harmonic generation.*

While the polarization wave propagates at the velocity υ_1, the second harmonic propagates at the velocity υ_2 in the medium. The velocities are given by

$$v_1 = \frac{c}{\mu(\nu)}$$

and
$$v_2 = \frac{c}{\mu(2\nu)}$$

The velocity υ_2 of the generated wave differs from that the incident wave, υ_1, because the refractive index is a function of frequency. Efficient transfer of energy from the polarization wave to the second harmonic requires that the two waves must be matched in velocity, *i.e.* they should have identical refractive indices. Thus, the condition is that

$$\mu(\nu) = \mu(2\nu) \qquad \text{... (29.46)}$$

The above condition is known as the *phase-matching condition.*

This condition is not satisfied in general cases. It is satisfied only in anisotropic crystals known as *uniaxial crystals*. When a light ray is incident on uniaxial crystals, it splits into two rays namely, an *ordinary ray* (*o*-ray) and an *extra ordinary* ray (*e*-ray). The refractive index of *e*-ray depends on the direction of propagation. The refractive index surface is a sphere in case of *o*-ray and an ellipsoid in case of *e*-ray. The refractive index surfaces in case of negative uniaxial crystal is shown in Fig 29.36 (*a*). Fig. 29.36 (*b*) shows the refractive index surfaces for frequency ν and for frequency 2ν. It is seen that the refractive index surface of the *o*-ray with frequency ν and that of the *e*-ray with frequency 2ν intersect; one of the points of intersection is point *P*. It is easy to see that the phase-matching condition is satisfied for light waves propagating in the direction *PQ*. Thus, the direction *PQ* is the *direction of phase-matching.*

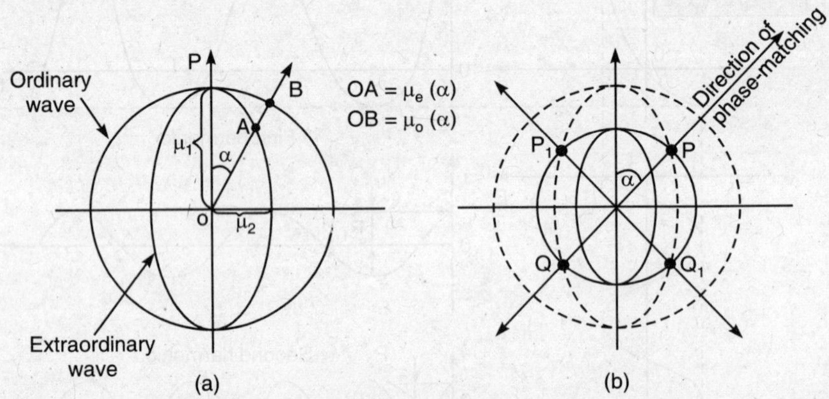

Fig. 29.36. (a) Index ellipsoid and sphere corresponding to a e-ray and o-ray in a negative uniaxial crystal. (b) Index matching in a negative crystal

Along *PQ*
$$\boxed{\mu_0(\nu) = \mu_e(2\nu)} \qquad \text{... (29.47)}$$

The process of second harmonic generation requires a uniaxial crystal exhibiting high nonlinear susceptibility χ_2. It must be cut in the shape of a rectangular parallelepiped with the axis along the direction of phase matching. The incident light must be plane-polarized, with the plane of polarization perpendicular to the plane of principal section of the crystal. When these conditions are satisfied, a wave of frequency ν produces another wave of frequency 2ν, which is the second harmonic. The direction of propagation of the second harmonic coincides with that of the fundamental wave. The plane of polarization of second harmonic will be perpendicular to that of the fundamental wave, as shown in Fig. 29.37.

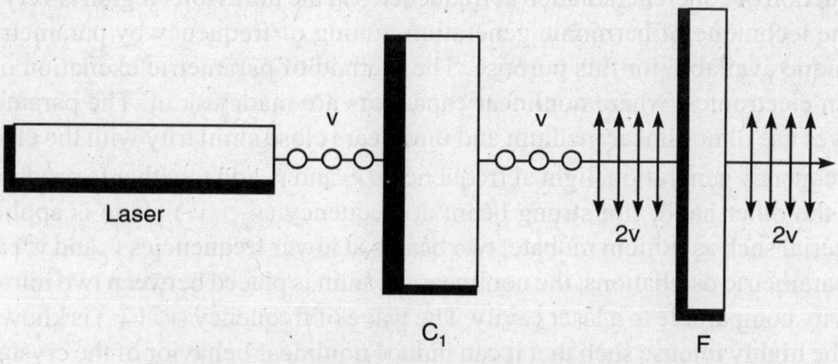

Fig. 29.37. Arrangement for second harmonic generation

There are crystals like KDP which offer a conversion efficiency of 20 to 50%. Second harmonic generation helps us in extending the range of laser wavelengths into the blue and ultraviolet part of the spectrum, which are not rich in naturally occurring laser lines.

Second harmonic generation was first demonstrated by Franken and his coworkers in 1961. They focussed a 3 kW ruby laser pule ($\lambda = 6943$Å) onto a quartz crystal and obtained a very low intensity ouptut at a wavelength of 3471.5 Å.

29.32.3 Sum Frequency Conversion

Besides second harmonic generation, nonlinear crystals make it also possible to generate beams of sum and difference frequencies when two light waves of different frequencies interact with them. Thus, interaction of waves with frequencies v_1 and v_2 generates light waves with frequencies $v_1 + v_2$ and $v_1 - v_2$.

Let us consider two coherent waves of frequencies v_1 and v_2 propagating simultaneously in a nonlinear medium. Let the waves be represented by $E_1 \sin 2\pi v_1 t$ and $E_2 \sin 2\pi v_2 t$, respectively. The superposition of the waves leads to

$$E = E_1 \sin 2\pi v_1 t + E_2 \sin 2\pi v_2 t$$

Using the equation (29.47) into Eq. (29.44), we obtain the following expression for the nonlinear term in the polarization of the medium.

$$\begin{aligned}P_{nl} &= \chi_1 E^2\\ &= \chi_1 [E_1 \sin 2\pi v_1 t + E_2 \sin 2\pi v_2 t]^2\\ &= \chi_1 E_1^{\ 2} \sin^2 2\pi v_1 t + \chi_1 E_2^{\ 2} \sin^2 2\pi v_2 t + 2\,\chi_1 E_1 E_2 \sin (2\pi v_1 t) \sin (2\pi v_2 t)\\ &= \chi_1 E_1^{\ 2} (1 - \cos 4\pi v_1 t + \chi_1 E_2^{\ 2} (1 - \cos 4\pi v_2 t)\\ &\qquad + \chi_1 E_1 E_2 \{\cos 2\pi\, (v_1 - v_2) t - \cos 2\pi v\, (v_1 + v_2) t\}\end{aligned}$$

or $$\begin{aligned}P_{nl} &= \chi_1 (E_1^{\ 2} + E_2^{\ 2}) - \chi_1 E_1^{\ 2} \cos 2\pi\, (2v_1) t - \chi_1 E_2^{\ 2} \cos 2\pi\, (2v_2) t\\ &\qquad + \chi_1\, E_1\, E_2 \{\cos 2\pi\, (v_1 - v_2) t - \cos 2\pi\, (v_1 + v_2) t\}\end{aligned} \quad ...(29.48)$$

The eq. (29.48) shows that the interaction of light waves of frequencies v_1 and v_2 propagating in a nonlinear medium generates new light waves of sum $(v_1 + v_2)$, difference $(v_1 - v_2)$ and double $2v_1$ and $2v_2$ frequencies.

In sum frequency generation, v_1 is taken as a weak infrared wave that sums with a strong pump wave v_2 to generate a high frequency wave at $v_3 = (v_1 + v_2)$. Thus, the light of wavelength of 10.6 μm produced by CO_2 laser is made to interact with a laser beam of 1.06 μm wavelength produced by Nd : YAG laser to yield a light beam of wavelength of 0.964 μm. Such upconversion is done because it is difficult to detect the radiation at 10 μm while the radiation at 0.964 μm can be easily detected.

29.32.4 Parametric oscillation

The production of coherent radiation at frequencies in the ultraviolet region is very much essential. Apart from the technique of harmonic generation, tuning of frequency by parametric oscillation is another technique available for this purpose. The method of parametric excitation of oscillations is widely used in electronics, where nonlinear capacitors are made use of. The parametric generation of light involves use of nonlinear medium and thus bears close similarity with the electronic process.

In sum frequency generation, light at frequencies v_1 and v_2 add together to produce the frequency $(v_1 + v_2)$. On the other hand, if a strong beam at frequency $(v_1 + v_2)$ alone is applied to a suitable nonlinear material such as lithium niobate, two beams at lower frequencies v_1 and v_2 can be generated. For exciting parametric oscillations, the nonlinear medium is placed between two mirrors, which form a resonant cavity comparable to a laser cavity. The wave of frequency $(v_1 + v_2)$ is known as *pump wave* and it should be highly intense such that it can induce nonlinear behavior of the crystal. Secondly, the process requires that the phase-matching condition is satisfied. The wave of higher frequency, say v_1, is called the *signal wave* and the lower frequency wave, say at v_2, is called the *idler wave.* Both the idler and signal waves are weak and are always present in the crystal in the form of *noise* which arises due to spontaneous photons. If the pump frequency $(v_1 + v_2)$ is fixed, then the two frequencies v_1 and v_2 are free to spread over a wide range of values. This effect is known as parametric.

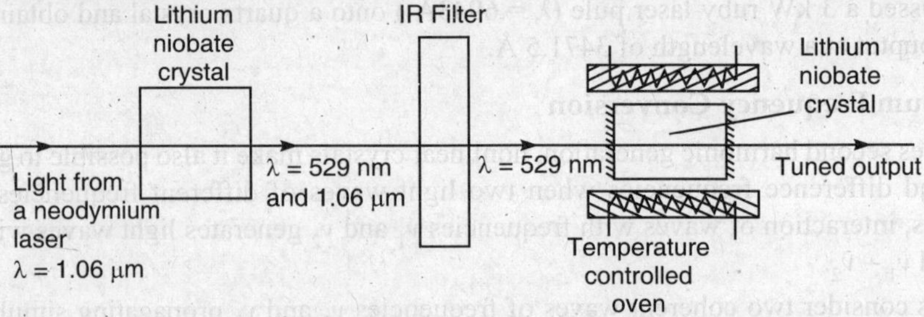

Fig. 29.38. A typical arrangement for observing parametric oscillation in a nonlinear medium

amplification, By varying the orientation of the crystal, the values of v_1 and v_2 can be varied. If the mirrors of the resonant cavity are reflective at v_1 and v_2 and not at the $(v_1 + v_2)$ frequency, resonance of these waves in the cavity enhances the interaction in the nonlinear crystal and builds up the waves. Thus, the system acts as a parametric oscillator. Tuning can be achieved by varying the phase-matching conditions through mechanical or temperature control of the cavity.

Parametric oscillation was first observed by Giordmaine and Miller in 1965. The schematic of set up used by them is shown in Fig. 29.38. The output was tuned by changing the temperature of lithium niobate crystal. A temperature range of about 11°C produced wavelengths in the range 9680 Å to 11540 Å. The conversion efficiency was as low as 1%. Currently, conversion efficiency is improved to 40% to 50%.

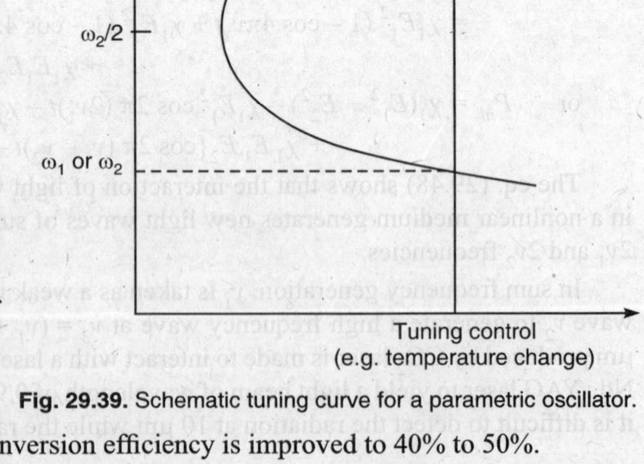

Fig. 29.39. Schematic tuning curve for a parametric oscillator.

29.33 COMPARISON OF LASER AND MASER

Laser. A laser is a device which produces a highly concentrated, monochromtic, coherent and unidirectional beam of light. The term LASER stands for light Amplification by stimulated Emission of Radiation. The typical operating frequency of laser is 10^{15} Hz in visible region.

Principle of a laser. Light emited from conventional light sources is incoherent as radiations emitted from different atoms are not in phase with each other. To understand let us consider that atoms are continuously pumped to an excited state. A short time later, of the order of 10^{-8} second, spontaneous emission occurs and the excited atom becomes normal after emitting a photon of the same frequency which was originally absorbed but in a random direction and random phase. Thus light emitted is *spatially incoherent*.

In the emission of laser beam, physical phenomenon involved is that of stimulated emission. If a photon is absorbed by an atom, the energy of the photon is converted into the internal energy of the atom, due to which it is raised to an excited quantum state. It radiates this energy spontaneously emitting a photon and reverting to the ground state or to some in-between state called the metastable state. During the period in which the atom is still excited, if it is struck by an outside phton having the same energy as that would be emitted spontaneously, it is stimulated to emit a photon. In this stimulated emission, the photon forces the excited atom to emit another photon of the same frequency in the same direction and in the same phase. The two photons go together as coherent radiation. These can cause further stimulated emission and this process leads to coherent radiation. However for amplification to take place in the system, there should be a large number of atoms in the higher energy state than in the lower energy state. This is known as population inversion.

Further multiplication of laser beam takes place due to repeated reflections in a resonance cavity formed by two parallel mirrors. One of the mirrors has a reflection co-efficient slightly less than one and a small fraction of the laser beam emerges out of this mirror every time reflection takes place on it.

Maser. A maser is the modified from of laser, which uses microwaves instead of light. Thus the term MASER stands for microwave Amplification by stimulated Emission of Radiation. The typical operating frequency for a maser is 10^{10} Hz.

Like laser, maser is also based on the principle of simulated emission proposed by Albert Einstein in 1917. When atoms have been put into an excited energy state, they can amplify radiation at the proper frequency. By putting such an amplifying medium in a resonant cavity, feedback is created that can produce coherent radiation.

Some common types of masers are:

(*i*) Atomic beam masers: Ammonia maser, free electron maser; Hydrogen maser.

(*ii*) Gas masers, Rubidium maser

(*iii*) Solid state masers, Ruby maser,

The dual noble gas maser is an example of a maser medium which is non-polar.

Uses (*i*) Masers serve as high precision frequency references. These atomic frequency standards are one form of *atomic clock*.

(*ii*) They are also used as *electronic amplifiers in radio telescopes*.

(*iii*) Masers are being developed as *directed-energy weapons*.

EXERCISE

1. The light emitted from a conventional source is said to be incoherent. Why ? ***(G.N.D.U. 2008)***
2. What is coherence ? Discuss in detail the terms (*i*) temporal coherence and (*ii*) spatial coherence. Define coherence time and coherence length and give the relation between them.
(P.U. 2008, 2007, 2006, 2005, 2004, 2003, 2001, 2001; Pbi. U. 2006, 2005, 2000; M.D.U. 2003; Nagpur U. 2002; G.N.D.U., 2007, 2006, 2005, 2004, 2002, 2001, 2000; H.P.U. 2000; Meerut U. 2006; Garhwal. U. 2004; Nagpur U. 2010)
3. Explain what you understand by line width and frequency spread. What is the relation between frequency width and coherence time of a monochromatic light source ?
(Pbi. U. 2007; G.N.D.U. 2006; M.D.U. 2003; 2000; Pune U. 2010)
4. Explain the term "lateral coherence length". Sunlight has mean wavelength of 5400 AU. Considering sun as a source of light in a double slit experiment with $\theta = 32$, find the lateral coherence length. ***(Pune U. 2010)***
5. What is the difference between coherence and Monochromatism of laser. ***(Pune U. 2010)***
6. Explain the terms (*i*) Coherence length and (*ii*) Coherence time. ***(Nagpur U. 2010)***
7. What are the characteristics of laser light. ***(Pune U. 2010)***
8. Why do we say light is amplified in a laser ? ***(Pune U. 2010)***
What is a Resonator in a Laser ? ***(Pune U. 2010)***
9. What is brightness or intensity ? Explain directionality.
10. Describe how you will determine coherent length using Michelson's interferometer.
11. A pinhole of diameter 1 mm is placed from two pinholes at a distance of 1m. Find the lateral coherence width of the pinholes which is illuminated with light of wavelength 5000 Å.
[Ans. 0.06 cm]
12. A laser beam has a power of 50 mW. It has an aperture of 5×10^{-3} m and it emits light of wavelength 7200 Å. The beam is focused with a lens of focal length 0.1 m. Calculate the area and the intensity of the image. ***(P.U. 2004)*** **[Ans.** 2.074×10^{-10} m^2, 2.411×10^3 watt/m^2**]**
13. Explain spatial coherence due to a point source and extended source.
14. What do you mean by thermal equilibrium?
15. Explain the mechanism for broadening of spectral lnes.
16. Explain natural broadening and Doppler broadening of a spectral line.
(Nagpur U 2011/s 2010)
17. Explain the concept of a 'purity of spectral line'. ***(Nagpur U. 2010, 2011/s)***
18. Explain the term 'pulse width' of a laser. ***(Pune U 2010)***
19. Distinguish between critical potential, ionisation potential and excitation potential.
20. What are the advantages and disadvantages of a narrow line width ?
21. Graphically distinguish between Gaussian line graph and Lorentizian line graph. ***(P.U. 2005)***
22. Define stimulated absorption and metastable state. Explain the terms 'spontaneous emission' and 'stimulated emission'. Write down the relation for transition probabilities of stimulated absorption as well as spontaneous and stimulated emission. What are Einstein co-efficients? Derive relation between them and explain.
(P.U. 2008, 2007; Meerut U. 2003, 2000; M.D.U. 2008, 2007, 2004, 2002, 2000; G.N.D.U., 2008, 2007, 2006, 2004, 2001; Kerala U. 2001, K.U. 2001; Pbi. U. 2008, 2007, 2006, 2005; Pune U. 2010, Nagpur U. 2010)
23. What is metastable state ? What is its importance for working of a laser ? ***(P.U. 2008, 2007)***

24. Explain Einstein's co-efficients and derive an expression showing that the ratio of spontaneous to stimulated emission is given by *(K.U. 2002, Pune U. 2010)*

$$\frac{A}{B\rho(\omega)} = \exp(\hbar\omega / kT) - 1$$

25. What is the need of population inversion in laser ? *(Pune U. 2010)*

26. What is meant by the action of 'Pumping'. ? Give methods of pumping briefly. *(Pune U. 2010)*

27. What is continuous-wave (CW) operation in terms of lasers ?

28. Explain in brief Q-factor of a cavity. *(G.N.D.U. 2007, 2006)*

29. Write a short note on mode locking. *(G.N.D.U. 2008; 2007; P.U. 2006)*

30. What do you mean by Q-switching ? Describe different types of Q-switches. *(G.N.D.U. 2007, Pbi. U. 2007)*

31. (*a*) What are the characteristics that distinguish laser beam from ordinary light ?
OR How does laser light differ from the light of a filament lamp ?
(*b*) Discuss in detail the concept of directionality, monochromacity, intensity and coherence or laser light. *(K.U. 2001; G.N.D.U. 2000; P.U. 2007, 2006, 2005, 2004, 2003, Pbi. U 2006, 2000; M.D.U. 2001, 2000)*

32. What do you mean by a Laser and a Maser ? Describe briefly the principle of a laser. Name some important types of lasers. *(P.U. 2007; Meerut U. 2001; Pbi. U. 2001, G.N.D.U. 2006, 2001; Guwahati U. 2000)*

33. Explain the advantage of frequency conversion. How it is achived using non-linear crystals.

34. Explain with graph the variation of electric polarization with electric field in a non-linear medium.

35. Explain the experimental arrangement for obtaining second harmonic generation.

36. What is parametric oscillation. Draw the suitable experimental arrangement for observing parametric oscillations in a non-linear medium.

37. Open cavity resonators are preferred over closed cavity resonators. Why ? *(P.U. 2006)*

38. What do you mean by thermal equilibrium ?

39. (*a*) Discuss the essential requirements and various steps involved for producing laser action. Laser is basically a three-component device. Explain.
(*b*) Explain the principle and working of a laser.
(P.U. 2007, 2006; Meerut U. 2003, 2001; Pbi. U. 2000; G.N.D.U. 2001)

40. (*a*) Explain the elementary theory of optical cavity. Explain the nature of mirrors in case of resonance cavity. *(P.U. 2007, 2006; G.N.D.U. 2007, 2006; Pbi. U. 2000)*

41. Describe in detail longitudinal and transverse modes. *(G.N.D.U. 2007)*

42. Open cavity resonators are preferred over closed cavity resonators. Why ? *(P.U. 2006)*

43. Explain the following:
(*i*) Active mode locking, (*ii*) Passive mode lacking.

TYPES OF LASERS AND THEIR APPLICATIONS

30

Chapter

INTRODUCTION

There are several ways to classify the different types of lasers. The classification can be done according to what material or element is used as the *active medium*. It can be done according to the operation of laser in a pulse mode or in a continuous wave (CW) mode. The classification can also be done depending upon the parameters such as gain of the laser medium; power delivered by the laser, efficiency or applications. We will prefer here to classify the lasers on the basis of material used as active medium. Accordingly, lasers are broadly divided into four categories–solid lasers, gas lasers, liquid lasers and semiconductor lasers. We will discuss here some of the important and practical lasers.

30.1 TYPES OF LASERS

Various types of lasers in use are:

(*i*) **Solid lasers.** The most commonly used solid laser is a *Ruby laser*. The other solid laser used is $Y_2\,Al_5\,O_{12}$ doped with Nd^{+3} ions in place of Y (Yttrium). Its most useful operating wavelength is the infra red light $\lambda = 1.064$ μm.

(*ii*) **Gas lasers.** Helium- Neon gas lasers consisting of about 10 parts of *Ne* and 1 part of (*He*) is the best known gas laser. It operates on three wavelengths, 6328 Å in the red and 1.15 and 3.3 μm in the infra-red region.

The CO_2 gas laser is also an important example of this type from the point of view of technological applications. In it, the energy levels are provided by the quantisation of the energy of vibration and rotation of the constituent gas molecules. CO_2 lasers have efficiency up to 30% and a power output up to 100 W. Another example is nitrogen gas laser.

(*iii*) Other important types include semi-conductor lasers, chemical laser and dye lasers.

In a semi-conductor laser, the conduction band population of electrons is increased by injecting free electrons into the negative side of an ordinary junction diode. The operating wavelength for *GaAs* laser is usually in the range 8400 –8500 Å.

In a **chemical laser,** population inversion results from a chemical reaction.

In a **dye-laser**, laser action is obtained by pumping certain dye solutions with micro-second long flashes from high intensity lamps. The active material is an organic fluorescent dye dissolved in a common solvent. Optical excitation and emission in organic dyes takes place betweeen the rotational vibrational levels of the *singlet ground state* S_0 and the *singlet excited state* S_1.

30.2 RUBY LASER

It is one of the first lasers to be constructed. The first laser was fabricated by Marman in 1960. It makes use of the *three level scheme* of population inversion with a ruby as a solid state laser element. For this purpose, the ruby crystals are grown in special furnaces and then shaped into cylindrical rod 2 to 30 cms. in length and 0.5 to 2 cms. in diameter. Both the end faces of the crystal are made optically *flat* and *silvered*. One face *A* is *thickly coated* to become *fully reflecting*. while the other face *B* is *partially coated*, so that an intense beam can emerge out of it. The curved surface is also coated to make it reflecting. The ruby cylinder, prepared in this manner, is called *resonant cell* or *cavity*. The ruby rod is surrounded by a cylindrical glass tube, through which liquid Nitrogen is circulated to keep the rod cool. A *Xenon* flash tube is wound round the glass cylinder, so that the ruby rod lies along its axis. It is connected to a suitable source of power supply and made to flash for a few milli-seconds.

STRUCTURE

The schematic of a ruby laser is shown in Fig 30.2. Ruby rod is taken in the form of a cylindrical rod of 4 cm. in length and 0.5 cm. in diameter. Its ends are grounded and polished such that the end faces are exactly parallel and also perpendicular to the axis of the rod. One face is silvered to achieve 100% reflection while the other is silvered to give 10% transmission. The silvered faces form the Fabry -perot resonator. The rod is surrounded by a helical photographic flash lamp filled with xenon. The lamp produces flashes of white light whenever activated by the power supply. The system is cooled with the help of a coolant circulating around the ruby rod.

In practical lasers, flash lamps of helical design are no longer used. The most common types are linear lamps. The laser rod is generally placed in an elliptical cavity with a linear flash lamp mounted at the focus of internally reflecting cavity, as shown in Fig.30.1, to ensure an efficient transfer of pumping energy to the ruby crystal. The mirrors are arranged outside.

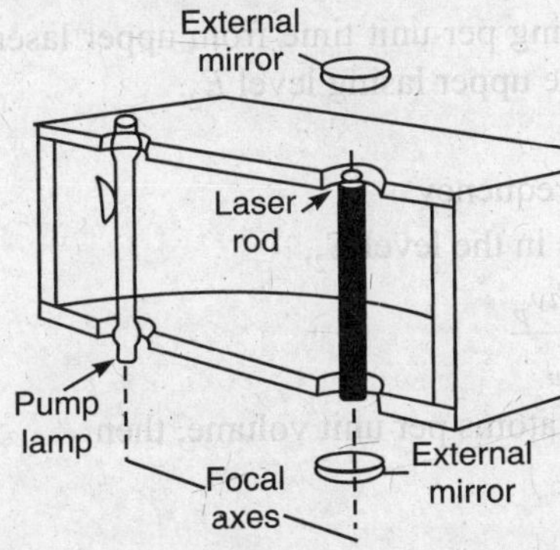

Fig. 30.1. Elliptic cylinder housing for concentrating light onto the laser rod

Working. Ruby consists of a crystal of Al_2O_3 (Aluminium oxide) doped with 0.03 to 0.05% of Cr_2O_3 (Chromium oxide), so that Cr^{+3} ions replace some of the *Al* atoms. The crystal field splits up the energy levels of Cr^{+3} ions in such a way that it has a shortlived energy state E_3= 2.26 eV and an intermediate metastable energy state E_2 =1.79 eV above the ground state E_1.

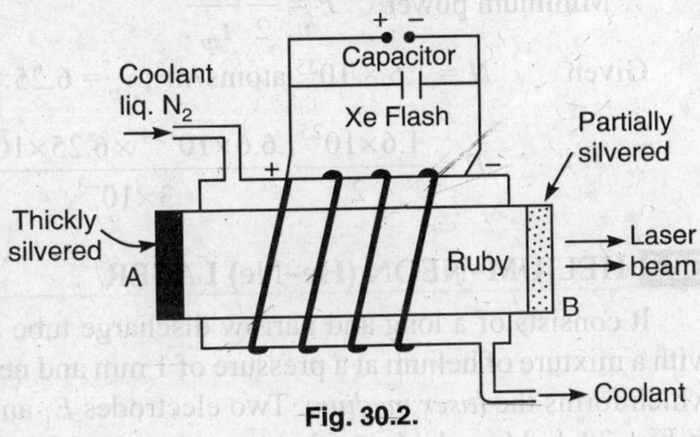

Fig. 30.2.

[The energy state E_3 for Cr^{+3} consists of two wide energy bands and the state E_2 consists of a closely spaced double level, but most of the contribution to laser action comes from the lower energy level 1.79 eV.]

The *Xe* flash tube acts as an optical pump and only a small fraction of the energy emitted by it in the form of blue green radiation of $\lambda = 5500Å$ corresponding to energy level $E_3 - E_1$ is absorbed by the ruby and used to excite the Cr^{+3} ions. The rest of the energy is used up in heating the apparatus, which therefore has to be kept cold by circulating liquid nitrogen through the glass tube surrounding the ruby rod. The Cr^{+3} ions excited to the shortlived 2.26 eV level jump to the 1.79 eV metastable level (lifetime 4×10^{-3} sec.) in the first step by losing 0.47 eV of energy. As a result the level E_2 becomes more populated than the level E_1.

Such a system with population inversion is very unstable. Once the stimulated emission is started, photons of energy 1.79 eV corresponding to red light of $\lambda = 6943$Å are continuously produced. The length of the ruby rod is made an exact multiple of half wavelength so that the light waves trapped in it form an optical standing wave. Thus, the photons undergo multiple reflections from the silvered ends of the crystal till the beam becomes sufficiently intense to emerge out of the partially silvered end. Thus, a highly intense, monochromatic, coherent and unidirectional beam is obtained.

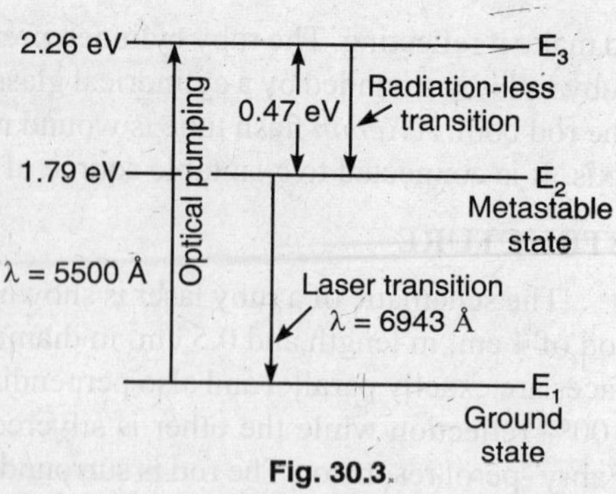

Fig. 30.3.

Example 30.1 *Calculate threshold pumping power per unit volume in case of Ruby laser. Given $N=1.6\times 10^{25}$ atoms/m³, frequency $v_p = 6.25\times 10^{14}$ and $t_{sp} = 3\times 10^{-3}$ sec.* ***(Pbi. U. 2008)***

Solution. In a three level laser scheme, as in the case of a Ruby laser, let N_2 be the number of atoms in the upper laser level E_2 (metastable state) and t_{sp} the spontaneous relaxation time at this level, then Number of atoms decaying per unit time from upper laser level = $\frac{N_2}{t_{sp}}$. For each atom lifted from the ground state E_1 to the upper lasing level E_2.

Energy required = hv_p

where v_p is the average pump frequency.

Therefore to maintain N_2 atoms in the level E_2,

The minimum power $P = \dfrac{N_2 hv_p}{t_{sp}}$

Now, if N is the toal number of atoms per unit volume, then

$$N_2 = \frac{N}{2}$$

$\therefore$ Minimum power $P = \dfrac{N}{2}\dfrac{hv_p}{t_{sp}}$

Given $N = 1.6\times 10^{25}$ atoms/m³; $v_p = 6.25\times 10^{14}$; $t_{sp} = 3\times 10^{-3}$ s

$$\therefore \quad P = \frac{1.6\times10^{25}}{2}.\frac{6.6\times10^{-34}\times6.25\times10^{14}}{3\times10^{-3}} = 1100\times10^{6}\,W/m^3$$

30.3 HELIUM–NEON (He–Ne) LASER

It consists of a long and narrow discharge tube about 80 cm long and 1 cm in diameter, filled with a mixture of helium at a pressure of 1 mm and neon at a pressure of 0.1 mm of mercury column, which forms the *laser medium.* Two electrodes E_1 and E_2 are fitted to the discharge tube, as shown in Fig. 30.4. M_1 and M_2 are the two mirrors which form a resonant cavity. The mirror M_1 is fully reflecting, whereas the mirror M_2 is partially reflecting, so as to allow the laser beam to pass out of it.

Working. As soon as the electric discharge passes through the mixture of *He* and *Ne* gases, electrons in the tube are accelerated. These collide with helium atoms and excite them to higher energy levels S_2 and S_3, as shown in Fig. 30.5. These levels of helium are metastable and the excited helium atoms remain in these levels for a 'long' time before being de-excited. On the other hand, some of the excited energy states, say E_4 and E_6 of neon correspond very approximately to the energy state S_2 and S_3 of helium.

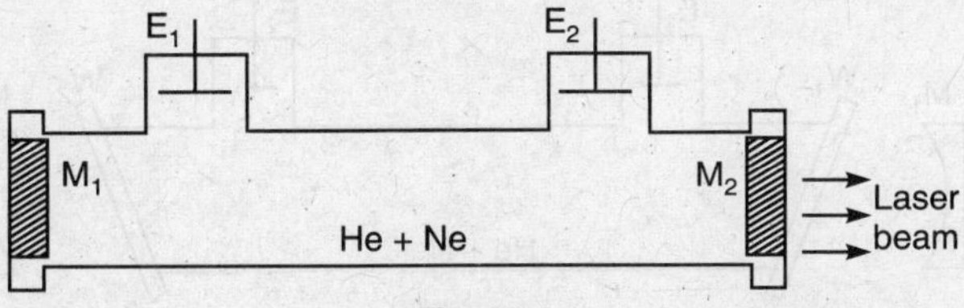

Fig. 30.4

Because of this, when helium atoms in the energy states S_2 and S_3 collide with neon atoms in the ground state E_1, *Ne* atoms absorb energy and are excited to the energy states E_4 and E_6, whereas He atoms lose energy and are de-excited to the ground state S_1. This process continuously transfers more and more neon atoms from the ground state to the excited energy states E_4 and E_6

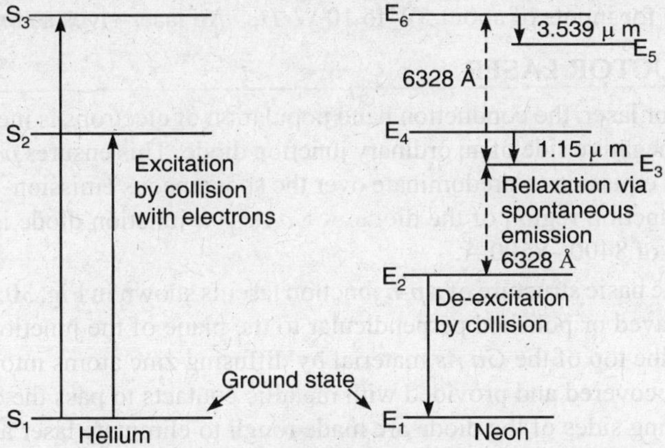

Fig. 30.5

There are two more energy states E_3 and E_5 slightly below the states E_4 and E_6 and another energy state E_2 in *Ne*. Since the energy state E_4 and E_6 are *highly populated*, there is a *population inversion* between the states E_4 and E_6 and the lower energy states E_3 and E_5. As shown in Fig 30.5 transition between E_6 and E_5, E_4 and E_3 and E_6 and E_2 respectively lead to emission of wavelength 3.539 μm, 1.15 μm and 6328 Å. The first two lie in the infra red region and the last corresponds to the red light from the *He-Ne* laser in the visible region, which is generally used as a laser beam for experimental and other purposes. Selection of a suitable frequency is made by using mirrors which reflect only a small range of frequency band.

In an improved form of the *He-Ne* laser, the resonator mirrors M_1 and M_2 are used *externally* to the laser cavity, because the mirrors sealed inside the discharge tube are eroded by the gas discharge and have to be replaced. These are spherical in shape and the radius of curvature of each mirror is equal to the cavity length.

Further, to minimise the loss due to reflection at the ends of the discharge tube, Brewster windows W_1 and W_2 are used, as shown. These are fitted at an angle known as Brewster angle θ_B given by

$$\tan \theta_B = \mu$$

where μ is the refractive index of the material of the window

To understand the action of Brewster's windows W_1 and W_2, we consider an unpolarised wave incident on a plate as a resultant of two superimposed plane polarised waves, one of which is polarised in a plane passing through the normal to the window and the tube axis and the other at right angle to this plane. The wave polarised normal to the plane of oscillation is completely reflected by the window plate and, therefore, does not play any part. The other wave, polarised in the plane of oscillation, is transmitted in the same direction and is completely reflected by the mirrors M_1 and M_2 again and **again, thus producing a strong** *plane polarised* laser beam.

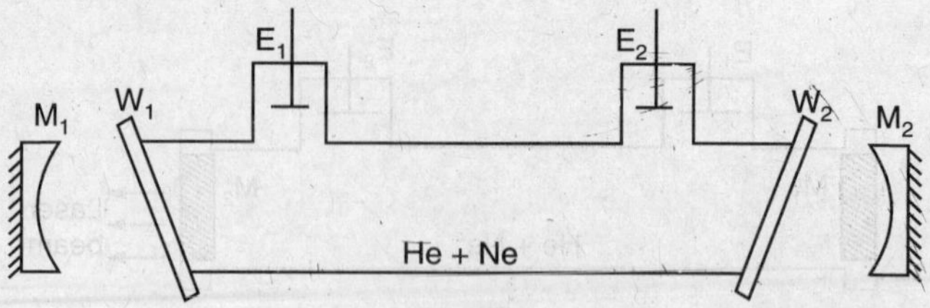

Fig. 30.6

Operational wavelength. The emission wavelengths of the *He–Ne* laser are 1.15 μm, 3.539 μm and 6328 Å. The first two lie in the infra-red region and the last corresponds to the red light in the visible region. The 6328 Å wavelength is , therefore, the operational wavelength.

Output power. The power output of *He –Ne* laser lies between one 1 m W and 50 mW of continuous wave (c.w) for inputs of about 5W to 10 W. *He Ne* laser gives a *continuous operation.*

30.4 SEMI-CONDUCTOR LASER

In a semi-conductor laser, the conduction band population of electrons is increased by injecting free electrons into the negative side of an ordinary junction diode. This ensures *population inversion* required for stimulated emission to predominate over the spontaneous emission. The recombination takes place in *active* junction region of the diode. A *Ga As p-n* junction diode is generally used. It has an operating range of 8400 –8500 Å.

Construction. The basic structure of a *p-n* junction laser is shown in Fig. 30.7. A pair of parallel planes *A* and *B* are cleaved or polished perpendicular to the plane of the junction. A heavily doped *p*-region is formed on the top of the *Ga As* material by diffusing zinc atoms into it. The top and the bottom faces are metal covered and provided with metallic contacts to pass the current through the diode. The two remaining sides of the diode are made rough to eliminate laser action in a direction other than the main. The structure is called *Fabry-Perot cavity.*

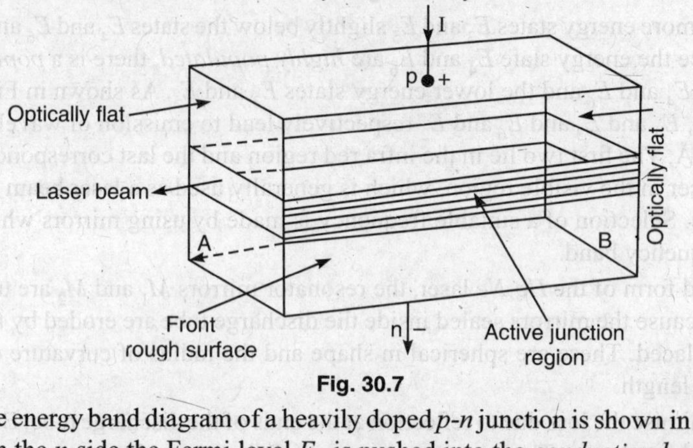

Fig. 30.7

Working. The energy band diagram of a heavily doped *p-n* junction is shown in Fig.30.8 (a). Due to heavy doping on the *n*-side the Fermi level E_F is pushed into the *conduction band*. The electrons occupy the portion of the conduction band below the *Fermi* level between E_F and E_C. Similarly on the heavily doped *p*-side, *Fermi level* E_F lies within the valence band. The holes occupy the portion of the valence band above the Fermi level between E_F and E_v. At the condition of thermal equilibrium, *Fermi level is uniform* across the junction as shown.

When the junction is forward biased electrons from the *n* region and holes from the *p*-region are injected into the junction region. This is how *pumping action* takes place in a *p-n* junction semiconductor laser. For small values of forward bias the electron holes recombine and the recombination brings about spontaneous, non-coherent *emission* of photons in all directions. At this stage the junction acts a *light emitting* diode (LED). As the forward bias is increased the intensity of light increases but when the forward bias reaches a *threshold value* the carrier concentration in the junction region rises to a very high value. The junction region, therefore, attains a very high concentration of electrons in the conduction band and at the same time a very high concentration of holes in the valence band. As a result the upper energy level in the narrow junction region has a high electron population whereas the lower energy levels in this region are vacant, thereby giving rise to the condition of *population inversion* in the region. This narrow region in which population inversion takes place is known as "*Inversion region*' or '*Active region*'

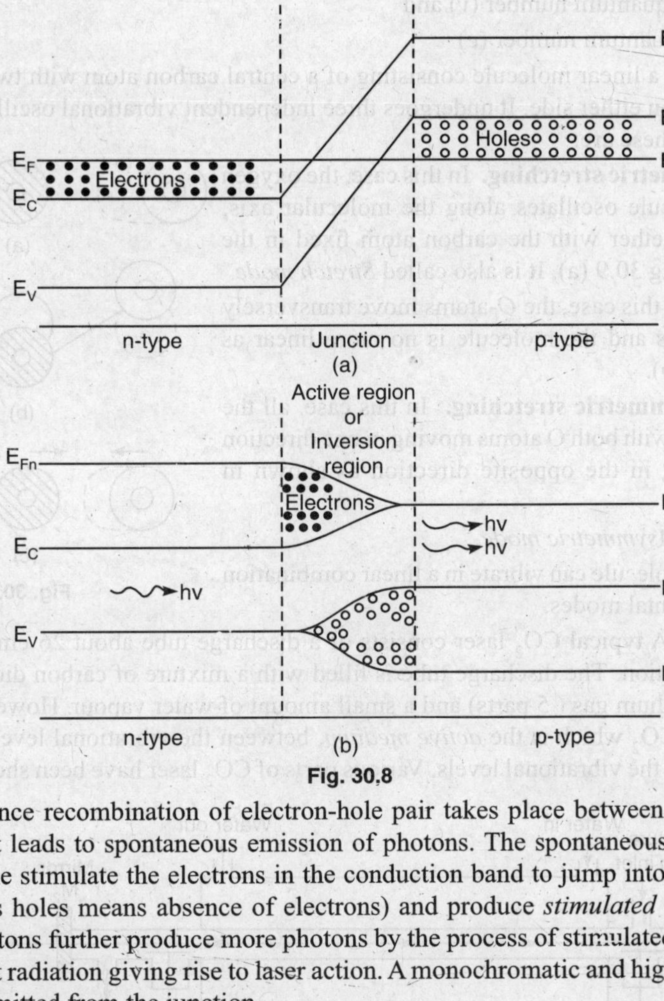

Fig. 30.8

When a chance recombination of electron-hole pair takes place between the conduction and valence bands, it leads to spontaneous emission of photons. The spontaneous photons moving in the junction plane stimulate the electrons in the conduction band to jump into the vacant states of *valence band* (as holes means absence of electrons) and produce *stimulated* or *induced photons*. The induced photons further produce more photons by the process of stimulated emisssion and thus produce coherent radiation giving rise to laser action. A monochromatic and highly directional beam of light is then emitted from the junction.

Semi-conductor lasers are very compact, efficient and can be easily fabricated. But their spectral purity is low and emission pattern broad. Their monochromatic property, coherence and directional character are, therefore, not as good as that of other type of lasers.

30.5 CARBON-DIOXIDE (CO_2) LASER

Carbon dioxide (CO_2) laser is a molecular gas laser. It is a *four level* laser and works at 10.6 μm wavelength in the infra-red region.

Principle. Just as in atoms, electrons in the molecule can also be excited to higher energy levels. The molecules have *electronic energy levels* like the atoms and in addition *vibrational* and *rotational* energy levels (or states). The rotational energy levels arise from the rotation of the molecule about an axis passing through the centre of mass of the molecule and perpendicular to the line joining the atoms. The vibrational energy levels arise from the vibration of the atoms about their equilibrium position in the molecule. Each vibrational state consists of a number of rotational states and each electronic energy level consists of a number of equally spaced vibrational states. The separation between the rotational and vibrational states are much smaller on the energy scale as compared to that between electronic energy states. The energy state of the molecule is represented by three quantum numbers:

(*i*) Rotational quantum number (*J*),

(*ii*) Vibrational quantum number (*v*) and

(*iii*) Electronic quantum number (*e*)

CO_2 molecule is a linear molecule consisting of a central carbon atom with two oxygen atoms attached to it— one on either side. It undergoes three independent vibrational oscillations known as vibrational modes. These are:

(*i*) **Linear symmetric stretching.** In this case, the oxygen atoms in CO_2 molecule oscillates along the molecular axis, getting in or out together with the carbon atom fixed in the centre as shown in Fig 30.9 (a). It is also called *Stretch mode.*

(*ii*) **Bending.** In this case, the *O*-atoms move transversely to the molecular axis and the molecule is no more linear as shown in Fig. 30.9 (*b*).

(*iii*) **Linear asymmetric stretching.** In this case, all the three atoms oscillate with both O atoms moving in one direction and *C* atom moving in the opposite direction as shown in Fig. 30.9 (*c*).

It is also called *Asymmetric mode.*

At a time CO_2 molecule can vibrate in a linear combination of the three fundamental modes.

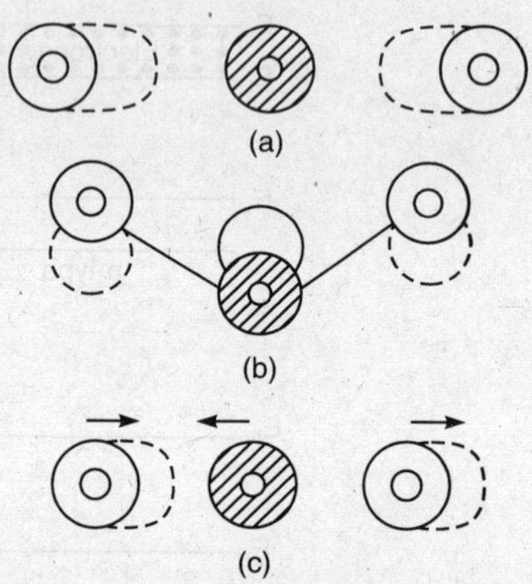

Fig. 30.9

Construction. A typical CO_2 laser consists of a discharge tube about 26 cm long and about 1.5 mm^2 in cross-section. The discharge tube is filled with a mixture of carbon dioxide (one part), nitrogen (4 parts), helium gas (5 parts) and a small amount of water vapour. However, laser action only takes place in CO_2 which is the *active medium*, between the vibrational levels and rotational levels accompanying the vibrational levels. Various parts of CO_2 laser have been shown and labelled in Fig. 30.10.

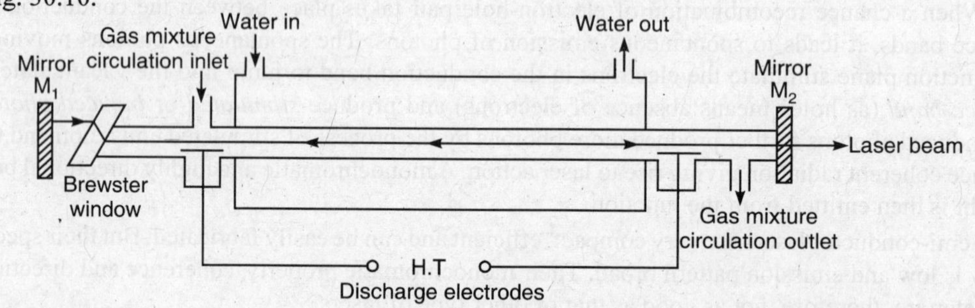

Fig. 30.10

Working . The lowest vibrational levels of *ground* electron energy state of CO_2 molecule and N_2 molecule are shown in Fig. 30.11. E_1 (000) is the ground state, E_5 (001) is the upper lasing level, whereas E_4 (100) and E_3 (020) levels act as lower lasing levels.

The excited state of N_2 molecule is metastable and it is identical in energy with E_5 (001) vibrational level of CO_2 molecule. As the electric discharge passes through the mixture of gases in the discharge tube the N_2 molecules get excited to the metastable state. The excited N_2 molecules cannot spontaneously lose their energy and as a result the number of N_2 molecules at this level goes on increasing. These N_2 molecules undergo inelastic collisions with CO_2 molecules and themselves return to the ground state taking the CO_2 molecules to the excited state E_5 (001).

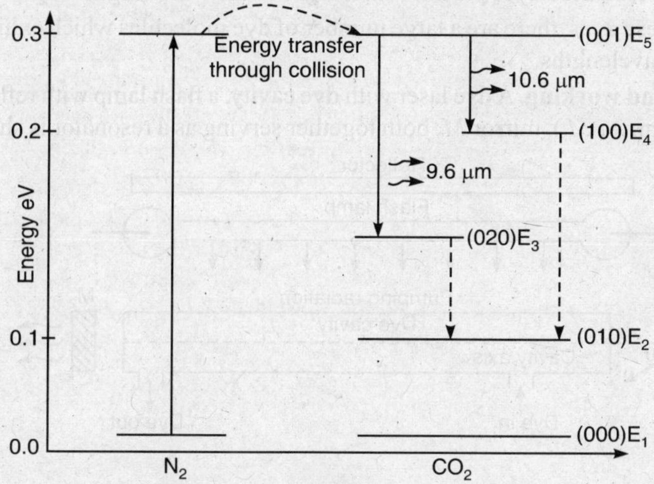

Fig. 30.11

As the population of CO_2 molecules builds up at E_5 level, *population inversion* is achieved between E_5 level and the levels at E_4 and E_3. A substantial population builds up at the E_5(001) level due to *radiation trapping* and E_4(100) and E_3 (020)states remain almost vacant through decay to E_2 (010) level. The laser transition between E5 and E4 level produces radiation of wavelength 10.6 μm and transition between E_5 and E_3 produces radiation of wavelength 9.6 μm, both in the infra-red region. E_3 and E_4 levels are also metastable and the CO_2 molecules at these two levels fall to the lower E_2 (010) level through inelastic collision with normal unexcited CO_2 molecules. This process leads to accumulation of population at E_2 level. Further the E_2 level being close to the ground level E_1 also gets populated due to thermal excitation. As a result the de-excitation of CO_2 molecules at the lower lasing levels E_4 and E_3 cannot take place which inhibits the laser action. But the presence of Helium along with CO_2 helps to reduce the population density at E_2 level by de-exciting CO_2 molecules by inelastic collisions. The presence of Helium also helps in cooling the gaseous mixture by heat conduction. Helium has high thermal conductivity and thus keeps temperature of CO_2 low so that the population rise of lower laser levels through thermal excitation is checked. In short, while N_2 helps to build the population of upper levels in CO_2, Helium helps to keep the population of lower levels low. Both these factors help in population inversion and thus the CO_2 laser operates in CW mode. It can generate power of several kilowatts at a sufficiently high efficiency of about 40%. It is, therefore, widely used in communication, defence and industry.

30.6 DYE LASER (TUNABLE LASER)

A dye laser is a type of *liquid medium* laser. Although dye lasers can be used in solid, liquid or gas phase, liquid solutions of dye are convenient as their concentration can be controlled. An organic dye dissolved in a suitable liquid solvent like ethyl alcohol, benzene, acetone or even water serves as the active medium. The laser beam output from a dye laser has a strong absorption and fluorescent band in the wavelength range lying between 3500 Å in the visible region to 13000 Å in the far infra-red region. Because of suitable line width of the luminiscence line in a dye it is possible to tune continuously the wavelength of the laser beam over a range of about 1000 Å.

Out of a large number of dyes used in a dye laser, following are most common:

***(i)* Coumarin dyes.** These dyes emit light in the *blue green region*. The absorption maxima of these dyes in methanol occurs at 3510 Å, 3640 Å, 3730 Å and 3900 Å respectively.

(*ii*) **Xanthene dyes.** These dyes cover the yellow red region with wavelength range 5000 Å–7000 Å. The xanthene dyes are soluble in water.

(*iii*) **Oxazine dyes.** These dyes produce laser light in the wavelength range 5500–7000 Å.

In addition to these dyes, there are a large number of dye molecules which exhibit efficient lasing action at different wavelengths.

Construction and working. A dye laser with dye cavity, a flash lamp with reflector, a diffraction grating (in place of mirror M_1), mirror M_2 both together serving as a resonator is shown in Fig. 30.12.

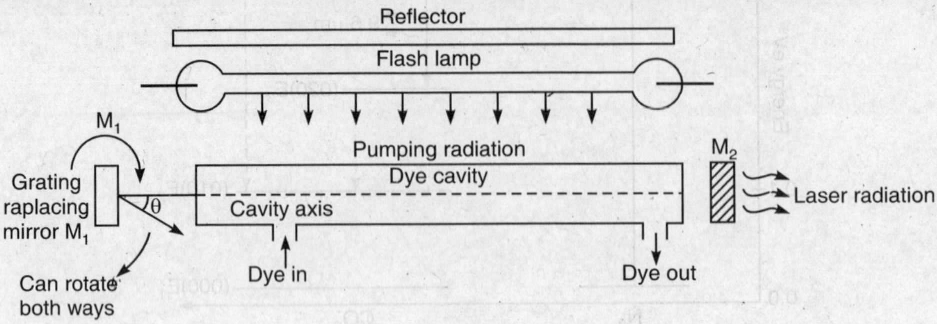

Fig. 30.12

The flash lamp is a Xenon lamp. It is commercially available. The reflector serves to concentrate the light emitted by the extended source on to the dye solution in the cavity. For minimum use of the pumplight the length of the dye cavity is kept equal to that of the flash lamp.

Dye lasers have *fast* relaxation time and hence can be subjected to *rapid* and *intense pumping.*

During the working of the laser large amount of thermal energy is transferred to the dye solution which creates a problem. The dye solution is, therefore, allowed to flow continuously. It is a necessity because even a small thermal gradient seriously affects the resonator action.

In order that the radiation of a particular wavelength is only reflected back along the cavity axis, the angle θ between the normal to the plane of the grating (which has replaced the mirror M_1) and the cavity axis satisfies the Bragg's diffraction condition.

$$2d\sin\theta = n\lambda \ (n = 1, 2, 3 \ldots \text{etc})$$

where d stands for the grating spacing. It is, therefore, clear that for a given value of θ there is only a single wavelength λ. We can have a number of values of λ, one by one, by slowly rotating the grating. This process is called *wavelength tuning.* As the grating allows only one value of λ at a time the arrangement is known known as *spectral selection element.* We can also use a *dispersive prism* in place of the grating.

Uses. Dye lasers have been used in the study of '*atomic and molecular spectroscopy*'. Dye lasers have broad spectral band width and therefore can be used for production of ultra short pulses of half width smaller than that with any other type of laser. Dye laser system is ideal for research applications where *continuously tunable monochromatic* light is required. This system is also suitable for study of hyperfine structure, atomic beam deflection, isotope separation and other similar applications.

Advantages. (*i*) Dye lasers have fast relaxation time and hence can be subjected to rapid and intense pumping.

(*ii*) By adjusting the angle between the normal to the plane of the grating and cavity axis, radiation of only a particular wavelength is reflected and thus we get a highly monochromatic laser beam. This is why it is known as a *tunable laser*.

(*iii*) Dye lasers have broad spectral band width and can produce ultra short pulses of very small half width.

(*iv*) Dye lasers can use the dye as a laser medium in solid, liquid or vapour phase.

30.7 ND: YAG LASER

The neodymium ion is a rare earth ion. It produces emission at about 1 μm; when doped into a solid-state proper host crystal. Most commonly used host materials are Yttrium aluminium garnet (YAG) crystal and glass. When doped in YAG, Nd^{3+} ions take the place of yttrium ions. Doping concentrations are typically of the order of 0.725% by weight which corresponds to about 1.4×10^{26} atoms per cubic meter.

Principle and Working

The basic principle of operation of the Nd: YAG laser is the same as that of a Ruby laser. It also has a *four level system*. The energy level diagram of Nd^{3+} is shown in Fig. 030.13.

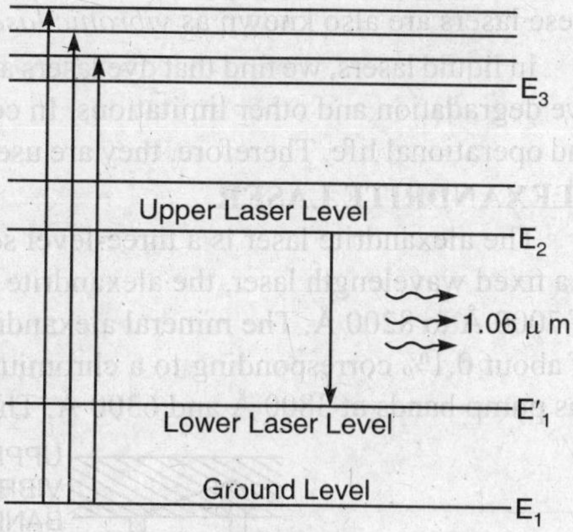

Fig. 30.13

The pumping of Nd^{3+} ions to the upper states is achieved by using intense flash of white light from a Xenon flash tube or a Krypton arc lamp giving light of wavelength 5000 – 8000 Å. This light excites the Nd^{3+} ions from the ground state E_1 to various higher energy states marked E_3. These ions immediately relax down to the level E_2 which is a *metastable level* with a life time of 0.25m sec. Laser action takes place between the *upper laser level* E_2 and *lower laser level* E^1_1 at a wavelength of about 1.06 μm in the infra-red region.

The active medium for the laser is the rare earth metal ion (Neodymium) Nd^{3+} doped with Yattrium Aluminium Garnet ($Y_3 Al_5 O_{12}$). The *YAG* itself does not take part in the laser process. It only provides a host crystal lattice so that when Nd^{3+} ion is placed in the *YAG* lattice, it is subjected to the electrostatic field of the surrounding ions—called the *crystal field.*

This crystal field modifies the transition probabilities between various energy levels of Nd^{3+} ions so that some transitions which normally are forbidden become allowed due to crystal field. The pumping flash lasts only for a short time of 1ms. The laser output, therefore, is in the form of a pulse which starts about 0.5 ms after the pumping flash is produced. This represents the time for the population to build up. The stimulation process builds up even more rapidly. It, therefore, depopulates the upper level much faster than the pumping can replace the excited atoms. Thus the laser action momentarily stops until population inversion builds up again and the process repeats itself.

Nd: YAG laser is used in many machine operations like drilling, welding etc, in medical surgery and many other fields.

30.8 ND: GLASS LASER

Glass is an excellent host material for neodymium. The chief attraction of glass is the well developed technology for making large size laser glass with good optical quality. While Nd: YAG laser can be operated in a cw mode, Nd: glass laser is operated only in a pulsed mode, because of the low thermal conductivity of glass. Nd: glass lasers give very high output of energy per unit volume of the material. The Nd: glass laser produces light at almost the same frequency as Nd:YAG laser, i.e., at around 1.06 μm.

Nd: glass laser materials are primarily used as amplifiers for very large pulsed lasers. The NOVA laser developed for nuclear fusion by Lawrence Livermore National Laboratories, USA employed a large number of Nd: glass amplifiers to produce very large powers of the order of 10^{13} W. Neodymium lasers are widely used in material processing and resistor trimming. It is used in medical application in association with optical fiber delivery systems to deliver energy to the appropriate location in the body. They are used in nuclear fusion and in military applications such as range finding.

30.9 TUNABLE SOLID-STATE LASERS

There is an important group of solid state lasers which produce output over a range of wavelengths. These lasers are tunable over a range and are therefore very popular. The tunability arises because of the existence of a cluster of vibrationally excited terminal levels near the ground state. Therefore, these lasers are also known as *vibronic lasers*.

In liquid lasers, we find that dye lasers are tunable (see 30.10). However, the dye lasers suffer from dye degradation and other limitations. In contrast solid state tunable lasers have unlimited shelf life and operational life. Therefore, they are useful in applications like remote sensing and in space craft.

ALEXANDRITE LASER

The alexandrite laser is a three-level solid state laser similar to a ruby laser. While a ruby laser is a fixed wavelength laser, the alexandrite laser is a tunable laser and lases in the wavelength range of 7000 Å to 8200 Å. The mineral alexandrite ($BeAl_2O_4$) is doped with chromium in concentrations of about 0.1% corresponding to a chromium ion density of 3.5×10^{25} per cubic meter. Alexandrite has pump bands at 3800 Å and 6300 Å. Therefore, pumping can be done using a flash lamp.

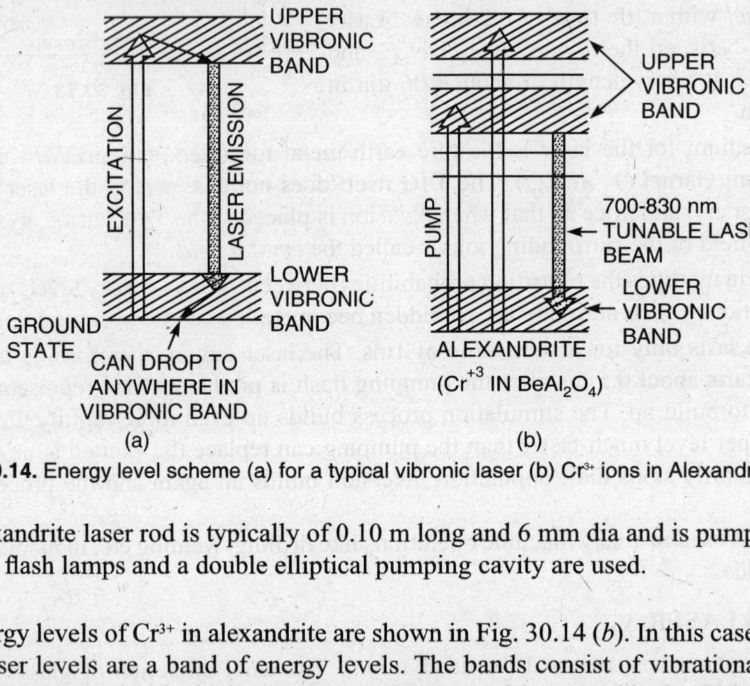

Fig. 30.14. Energy level scheme (a) for a typical vibronic laser (b) Cr^{3+} ions in Alexandrite crystal.

Structure:

The alexandrite laser rod is typically of 0.10 m long and 6 mm dia and is pumped using a flash lamp. Linear flash lamps and a double elliptical pumping cavity are used.

Working:

The energy levels of Cr^{3+} in alexandrite are shown in Fig. 30.14 (*b*). In this case, both the upper and lower laser levels are a band of energy levels. The bands consist of vibrational sublevels of a single electronic energy level which arise due to vibrations of the crystal lattice. When Cr^{3+} ions drop from the upper laser level to the lower band of energy levels, they undergo a compound transition where both electronic and vibrational energy change. This compound transition is called a vibrational electronic transition or in short a *vibronic transition*. The interesting feature of vibronic transition is that it can occur over a range of energies; because the excited ion can drop from the upper level to any level within the lower vibronic band. This means that a vibronic laser can be made to operate over a range of wavelengths within its gain bandwidth. Therefore, *vibronic laser is tunable* to any desired wavelength within its emission spectrum. This characteristic of vibronic laser makes it possible to produce wavelengths that are not obtained from other solid-state lasers. An alexandrite laser emits light in the range of 7000Å to 8300 Å. Alexandrite rod can store more energy than an equal sized Nd: YAG rod. Alexandrite laser can operate in a pulsed or CW mode.

The alexandrite laser is widely used in cancer therapy. Other applications are in population detection and kidney stone removal.

30.10 TUNABLE DYE LASERS

Dye lasers belong to the family of liquid lasers. The active material is a dye dissolved in a host medium of a liquid solvent, such as ethylene glycol. The solution is very similar to solid-state lasers where Cr^{3+}, Nd^{3+} or Ti^{3+} ions are used in a solid host. The advantage of liquid host is that the concentration of the active ions can be easily varied. Typical dye concentration of 10^{-4} to 10^{-3} molar solution is used which corresponds to 10^{24} and 10^{25} dye molecules per cubic meter. There are over 200 laser dyes, the most important one being rhodamine 6G. When used sequentially, they can produce tunable laser output over wavelengths ranging from 0.32 to 1.2 μm. Dye lasers operate both in pulsed and CW modes. Pulsed dye lasers are pumped by a flash lamp or other lasers and can produce upto 400 J output in a 10μs-pulse. CW dye lasers are pumped by other CW lasers such as argon ion laser and produce power of about 2 Watt. The gain of the dye lasing medium is very high. Consequently, a small volume of dye solution is sufficient to sustain the lasing action.

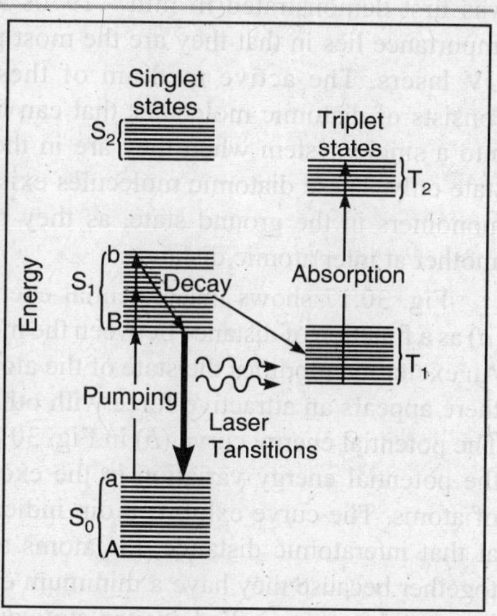

Fig. 30.15. Schematic representation of the energy levels of an organic dye molecule.

The energy band diagram of a typical dye laser is shown in Fig. 30.15. The organic dye molecule is known to have two sets of excited states: singlet states denoted by S_0, S_1 and S_2 and triplet states denoted by T_1 and T_2. The transitions from singlet to triplet states are forbidden. The optical pumping excites the dye molecules from the lowest vibronic level of the ground state S_0 to one of the upper vibronic level of the excited states S_1. From there, they undergo a non-radiative transition to the lower vibronic level of S_1 state which acts as the upper laser level. The role of lower laser level is played

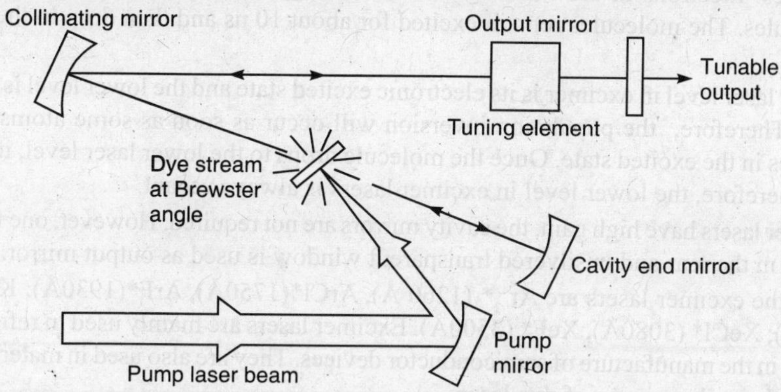

Fig. 30.16. Schematic diagram of a laminar-flow dye laser

by one of the upper vibronic levels of S_0. The levels at S_0 are so closely spaced that they form a continuum. It may be seen from fig. 30.15 that the laser transition can be to various levels within a range defined by the vibrationally excited sublevels of the ground state. Thus, the laser operates over a wide range of wavelengths. Thus, dye lasers differ from other lasers that emit light at fixed wavelengths. The schematic of a laminar-flow dye laser is shown in fig. 30.16.

30.11 EXCIMER LASERS

Excimer lasers constitute an interesting and important class of molecular lasers. Excimer laser was first demonstrated in mid - 1970s and their importance lies in that they are the most powerful UV lasers. The active medium of these lasers consists of diatomic molecules that can be bound into a single system when they are in the excited state only. These diatomic molecules exist only as monomers in the ground state, as they repel one another at inter atomic distance.

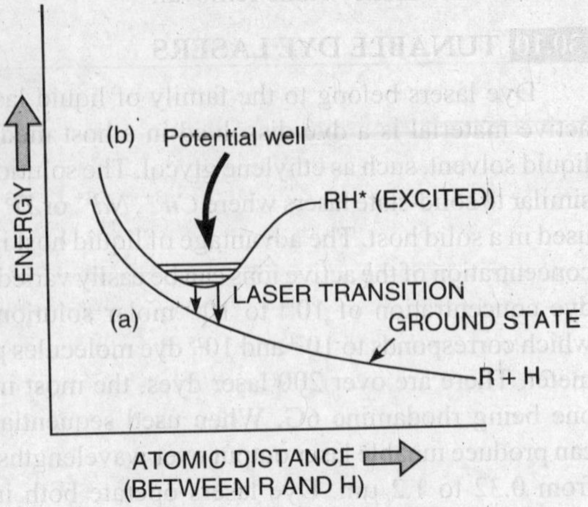

Fig. 30.17. Internal energy of a rare gas halide molecule in excited and ground

Fig. 30.17 shows that potential energy curve (*a*) as a function of distance between the monomers. An excitation modifies the state of the atom so that there appears an attractive force with other atoms. The potential energy curve (*b*) in Fig. 30.17 shows the potential energy variation in the excited state of atoms. The curve exhibits a dip indicating that at that interatomic distance, the atoms are bound together because they have a minimum energy at that separation. Such molecules which exist only in excited state are called *excited-state dimers* or *excimers* in short. For example, the atoms of inert gases can be bound into molecules by imparting energy to them. The excimer state is a metastable state. When the atoms are bound together in the excited state, they can occupy several vibrational levels (horizontal lines) in the potential well of curve (*b*) Fig. 30.17.

The active medium can be an excited rare gas dimer such as Ar_2*, Kr_2*, Xe_2*, ..., a rare gas oxide such as ArO*, KrO*, XeO*, ..., or a rare gas atom in combination with a halide atom such as ArF*, KrF*, XeCl*, ... (Here * means excited).

Excimer lasers are excited by passing a short, intense electric discharge through a mixture of desired gases. Electrons in the discharge transfer energy to the laser gas causing formation of excited molecules. The molecules remain excited for about 10 ns and then drop to the ground state and dissociate.

The upper laser level in excimer is its electronic excited state and the lower level is the electronic ground state. Therefore, the population inversion will occur as soon as some atoms are bound to form molecules in the excited state. Once the molecule drops to the lower laser level, it separates out into atoms. Therefore, the lower level in excimer lasers is always vacant.

As excimer lasers have high gain, the cavity mirrors are not required. However, one fully reflective mirror is used in the rear and unsilvered transparent window is used as output mirror.

Some of the excimer lasers are Ar_2* (1260 Å), ArCl*(1750Å), ArF*(1930Å); KrCl*(2220Å); KrF* (2490 Å); XeCl* (3080Å), XeF* (3500Å). Excimer lasers are mainly used in refractive surgery on the eye and in the manufacture of semiconductor devices. They are also used in material processing, photolithography and pumping of dye lasers.

30.12 SOME IMPORTANT CONCLUSIONS

1. Gas lasers, in general, are found to emit light which is more unidirectional and monochromatic as compared to solid-state lasers. This is because of the absence of the effects crystalline imperfections (crystal defects); thermal distortion and scattering which are present in solid state lasers. These are, **therefore also known as continuous wave (CW) output lasers.**

2. In lasers, we get an intense beam of monochromatic and coherent light. You may sometimes, think that energy conservation is violated in a laser. But the answer is that the energy conservation law is not at all violated in laser. In a laser, we only multiply the number of coherent photons through stimulated emission. The process of stimulated emission takes place only after obtaining *'population inversion'* in the active medium. Population inversion is achieved through suitable *'pumping'* process. The energy spent in pumping is recovered through laser emission.

30.13 APPLICATION OF LASERS

A laser beam has the following most striking features:

1. *Narrow band width*, which implies extraordinary monochromaticity or high temporal coherence,
2. *Narrow angular spread*, which implies high directionality or high spatial coherence,
3. *High intensity.*

These special features have made Lasers to find wide applications in different branches of science, engineering, technology, industry, surgery, defence, crime detection and so many.

30.13.1 Laser in Scientific Research

In basic scientific research, laser has been used in exploring the molecular structure, nature of chemical reactions, Raman spectroscopy, precision measurement of length and accurate measurement of velocity of light. It is used in nuclear fusion, in nuclear fission, in isotope separation.

30.13.2 Laser in Mechanical Industry

Drilling. Laser beam can project intense energy into a small area approximately 0.001 mm in diameter and produce at the focal point temperature of the order of 10,000 to 16,000 K. Therefore, laser has been used in mechanical industry as a "Tool Box". Laser drilling is a process used to drill alloys and composites that usually cannot be drilled by conventional methods. In many cases, laser drilling is the only solution for many drilling challenges. By laser drilling very high precision and diameter tolerances, less than 1 μm (micron) can be achieved. Some of the benefits of utilizing laser hole drilling technique include: Low dimensional variability, Low distortion, Minimal mechanical force on work-piece, Highly repeatable, Faster than any conventional method, 5- or 6-axis laser control, Creating entrance angles not possible with contact machining, Low tool cost, no tool wear, Drills angled holes and difficult access holes, Economic setup time.

Laser drilling can be used to produce highly accurate micro-holes in nozzles, orifices, photovoltaic cells, fuel injection components, vertical probe cards, metered dose inhaler products, pinholes and slits for scientific instrumentation, inkjet printer nozzles, detectors, sensors, high resolution circuitry, fuel cells, fiber optic interconnects and medical devices etc. Micro-drilling micro-holes using laser in metals, ceramics, diamond, silicon and other semiconductors, polymers, glass, sapphire and in almost any material is very common today.

Cutting. Laser cutting is technology that uses a laser to cut materials, and is typically used for industrial manufacturing applications. Laser cutting works by directing the output of a high-power laser, by computer, at the material to be cut. The material then either melts, burns, vaporizes away, or is blown away by a jet of gas, leaving an edge with a high-quality surface finish. Industrial laser cutters are used to cut flat-sheet material as well as structural and piping materials.

There are three main types of lasers used in laser cutting. The CO_2 laser is suited for cutting, boring and engraving. CO_2 lasers are used for industrial cutting of many materials including mild steel, aluminium, stainless steel, titanium, paper, wax, plastics, wood, and fabrics. YAG lasers are primarily used for cutting and scribing metals and ceramics. CO_2 lasers are used for selective removal of insulation from electrical wires and cables. While the insulation is stripped off, the electrical wire inside is not affected because of its high reflectivity. The neodymium (Nd) and neodymium yttrium-aluminum-garnet (Nd-YAG) lasers are identical in style and differ only in application. Nd is used for boring and where high energy but low repetitions are required. The Nd-YAG laser is used where very high power is needed and for boring and engraving.

There are many different methods in cutting using lasers, with different types used to cut different material. Some of the methods are vaporization, melt and blow, melt blow and burn, thermal stress cracking, scribing, cold cutting and burning stabilized laser cutting.

Welding. Like electron beam welding (EBW), laser beam welding has high power density (of the order of 1 megawatt/cm^2(MW)) resulting in small heat-affected zones and high heating and cooling rates. The spot size of the laser can vary between 0.2 mm and 13 mm, though only smaller sizes are used for welding. The depth of penetration is proportional to the amount of power supplied, but is also dependent on the location of the focal point: penetration is maximized when the focal point is slightly below the surface of the work-piece.

A continuous wave (CW) or pulsed laser beam may be used depending upon the application. Milliseconds long pulses are used to weld thin materials such as razor blades while continuous laser systems are employed for deep welds.

Laser beam welding (LBW) is a versatile process, capable of welding carbon steels, HSLA steels, stainless steel, aluminium, and titanium. Due to high cooling rates, cracking is a concern when welding high-carbon steels. The weld quality is high, similar to that of electron beam welding. The speed of welding is proportional to the amount of power supplied but also depends on the type and thickness of the work-pieces. The high power capability of gas lasers make them especially suitable for high volume applications. LBW is particularly dominant in the automotive industry. Here the welding of the curved contours under the body of automobiles is carried out. The automobile is made to move while computer-controlled beam deflectors control the laser beam which perform the welding.

Surface Treatment. Heat treating is a process which consists of heating metals and certain other materials for sometimes to harden them. Heat treating converts the surface layer to a crystalline state that is harder and more resistant to wear. CO_2 lasers of about 1 kW output power operating in CW mode are used for heat treating.

30.13.3 Laser in Electronics Industry

In electronics industry, lasers are mainly employed in wafer scribing, soldering and trimming.

Scribing. Scribing of silicon and ceramic substances is very important in the electronics industry. Lasers are extensively used for breaking silicon chips having hundreds of ICs on them. Scribing involves drawing fine lines in brittle ceramic and semiconductor wafers. When silicon or ceramic wafer is bent, it breaks along the weakened line scribed by the laser. Computer controlled scribing with laser is a *non-contact process* and hence does not damage the surface. Low pressure single mode CO_2 lasers are generally used for scribing operations.

Soldering. Lasers are used in welding of leads of materials such as platinum, silver, palladium which are usually difficult to solder. Nd: YAG laser is used for this process of fluxless soldering. The main advantage of laser soldering is that we can solder the areas that are visible but difficult to reach.

Trimming. Resistors of precise value are not possible by using the existing methods of manufacturing the thick and thin film resistors. Using laser technique, the resistance dependent area is trimmed by removing some surface material and the value of resistance is continuously monitored. 50% cost saving is possible by this method as compared to other conventional methods. Pulsed Nd: YAG laser is used for this process, as most metals have high absorption at 1.06 μm. Pulsed operation is preferred since it minimizes the thermal damage to the substrate.

30.13.4 Laser in Nuclear Energy

Isotope Separation. Isotopes are chemically almost identical. The differences in the nuclear mass shift the electronic energy slightly and therefore each isotope absorbs light at different characteristic wavelength. The absorption bands are fairly narrow and lie close to each other. If the mixture of isotopes is irradiated by a source of narrow bandwidth, it is possible to excite one isotope without disturbing the other. The desired energy may be obtained by tuning a dye laser to the precise wavelength with a very narrow line-width. The selectively excited U-235 atoms can be ionized by applying another short wavelength light to the mixture. The ionized U-235 atoms can be separated from the neutral U-238 atoms using electrostatic fields.

Nuclear Fusion. In nuclear fusion, enormous amount of energy is released through the fusion of nucleus of lighter elements to form nuclei of higher elements. It's a alternative source of energy as compared to nuclear fission due to mainly the following reasons (a) it offers a low cost and pollution free energy, (b) it produces less radioactive nuclear waste materials etc.

Nuclear Fission. Laser-induced fission of ^{238}U was observed at the VULCAN laser facility at Rutherford Appleton Laboratory (RAL) in 1999. Uranium nucleus has a rugby ball shape-range nuclear force balances the repulsive Coulomb force due to the 92 protons to give the nucleus its prolate deformation. Because uranium is such a large nucleus, it behaves like a drop of liquid. Supply of energy from the outside (from, e.g., an incident neutron or γ-ray) causes the nucleus to vibrate and rotate. The vibrations produce increased average distance between the nucleons resulting in a reduction of the binding effects of the short range nuclear force, while the infinite ranged Coulomb force is little affected. At some stage in the vibrations, the attractive forces are no longer able to hold the nucleus together and it flies apart with two main fragments, a few evaporated neutrons and gamma rays. The fragments are more tightly bound than the original uranium nucleus and the excess energy is released as kinetic energy of motion.

Three 80 J pulses from the VULCAN laser irradiated a 1.75 mm Ta target backed with a 2 mm thick ^{238}U sample. The ^{238}U was shrink wrapped in plastic to contain any gaseous activity and enclosed in an aluminium container. The laser pulses produced a flux of high energy gamma rays which induced fission in uranium nuclei as described above. Unambiguous characteristic gamma rays from the fission fragments ^{134}I, ^{138}Cs and ^{92}Sr were observed. From the measured intensities, it is concluded that about a million fission events are generated by a 10^{19} Wcm^{-2} laser shot in a 2 mm thick ^{238}U target.

30.13.5 Laser in Medicine and Surgery

Laser are playing nowadays a very important role both in medicine and surgery. In medicine, cosmetic surgery (removing tattoos, scars, stretch marks, sunspots, wrinkles, birthmarks, and hairs), eye surgery and refractive surgery and opthalmology, soft tissue surgery, general surgery, gynecological, urology, laparoscopic, endoscopy. Laser surgery is surgery using a laser to cut tissue instead of a scalpel. Examples include the use of a laser scalpel in otherwise conventional surgery, and soft tissue laser surgery, in which the laser beam vaporizes soft tissue with high water content. Laser resurfacing is a technique in which molecular bonds of a material are dissolved by a laser. Lasers have become in association with optical fibers a powerful tool in diagnostics. Interiors of body which otherwise be difficult or impossible to see, now be seen very easily with the help of optical fibers. Fiber-optic sensors are very useful both before and during surgery. The focused laser beam proved to be a new and unique 'scalpel' in the hands of surgeons. Laser surgery is a highly sterile process since contact does not occur between surgical tool and tissues being cut. Hence, laser surgery is much safer than normal surgery as it makes no physical contact so no infections are spread. Further advantage is that the laser does not cut but also "welds" blood vessels being cut. Operations are done very fast and the patient does not feel pain and can be discharged very soon from the hospital and can resume his/her job in a normal way. Green laser surgery is used for the treatment/reduction of enlarged prostates. We discuss here a few interesting applications of lasers in the medical field.

Eye Laser Surgery: The first big success of lasers in medicine was in the treatment of eye. Nowadays, laser eye surgery has become very common commonly. Techniques used include LASIK, which is used to correct near- and far- sightedness in vision, and photorefractive keratectomy, a procedure which permanently reshapes the cornea using an excimer laser to remove a small amount of tissue.

Photocoagulation: Diabetic retinopathy is retinopathy (damage to the retina) caused by complications of diabetes mellitus, which can eventually lead to blindness, Laser photocoagulations of retinopathies is a common treatment. In people suffering from diabetes, for some unknown reasons, abnormal blood vessels (streamers) spread across the surface of the retina. If these blood streamers are not removed, the patient can become blind. Laser photocoagulation is used to destroy areas of new blood vessels and areas of hypoxic retina which is believed to further increase blood vessel proliferation. It is an ocular manifestation of systemic disease which affects up to 80% of all patients who have had diabetes for 10 years or more. Despite these intimidating statistics, research indicates that at least 90% of these new cases could be reduced if there was proper and vigilant treatment and monitoring of the eyes.

Radial Keratomy: Another interesting eye surgery is the radial keratomy or refractive surgery, which is done for correcting the refractive defects of eye such as nearsightedness and astigmatism. The cornea is the outer transparent layer of the eye and is responsible for most of the eye's refractive power. Lasers are used to change the shape of the cornea by making small incisions on it. ArF excimer laser (λ = 193 nm) with a pulse energy of 250 mJ and pulse length of 14 ns is typically used to alter shape of cornea.

Skin Conditioning using Erb-YAG Laser: Cosmetic Surgery (removing tattoos, scars, stretch marks, sunspots, wrinkles, birthmarks and hairs) has become popular now a days. The Erb-YAG Laser is the first clinically proven laser for skin resurfacing and management of wrinkles and scars. It's a gentle treatment that ensures minimal discomfort and rapid recovery. It works by removing skin layer by layer. Once the damaged skin is removed it is replaced by the new skin which grows quickly. Conditions treatable with this laser are acne, chicken pox and surgical scars, birth marks, wrinkles around the mouth and eyes. It can also be used for skin grafting & hair transplantation.

Laser in Dentistry: Dentistry has entered an exciting era of high technology. The laser technique has become popular in most branches of dentistry because of its ability to vaporize tissue, coagulate and seal small blood vessels. In soft tissue procedures (gignivectomy, frenectomy, curettage, crown lengthening, lesion removal) CO_2 laser, Nd: YAG laser, Ar (Argon) laser, Er: YAG (Erbium doped Yattrium Aluminium Garnet) laser and diode laser are used. CO_2 laser was a valuable aid in direct pulp capping in human patients. While in hard tissue procedures (cutting of enamel) Er. YAG laser is preferable. In composite curing: Argon laser is used.

Laser Angioplasty: An important area of application of lasers is the laser angioplasty. Angioplasty (Greek word) is the technique of mechanically widening a narrowed or obstructed blood vessel, typically as a result of atherosclerosis (the buildup of fatty deposits called plaque in human arteries). The principal method of dealing with coronary artery blockage is bypass surgery, in which clogged blood vessels are replaced by high cost procedure with some risk. A non-surgical alternative is available to some patients is balloon angioplasty. An empty and collapsed balloon on a guide wire, known as a balloon catheter, is passed into the narrowed locations and then inflated to a fixed size using water pressures some 75 to 500 times normal blood pressure (16 to 20 atmospheres). The balloon crushes the fatty deposits, opening up the blood vessel for improved flow, and the balloon is then deflated and withdrawn.

They are effectively employed to study microscopic objects, including molecules, DNA testing and even to fight crime.

Advantages and Disadvantages of laser therapy

Lasers are more precise than standard surgical tools (scalpels), so they do less damage to normal tissues. As a result, patients usually have less pain, bleeding, swelling, and scarring. With laser therapy, operations are usually shorter. In fact, laser therapy can often be done on an outpatient basis. It takes less time for patients to heal after laser surgery, and they are less likely to get infections. Patients should consult with their health care provider about whether laser therapy is appropriate for them.

Laser therapy also has several limitations. Surgeons must have specialized training before they can do laser therapy, and strict safety precautions must be followed. Also, laser therapy is expensive and requires bulky equipment. In addition, the effects of laser therapy may not last long, so doctors may have to repeat the treatment for a patient to get the full benefit.

30.13.6 Laser in LIDAR (Light Detection And Ranging)

The acronym LADAR (Laser Detection and Ranging) is often used in military contexts. Large distances can be measured using pulse echo techniques. It is an optical remote sensing technology that can measure the distance to, or other properties of a target by illuminating the target with light, often using pulses from a laser.

It consists of a Q-switched ruby laser yielding giant pulses, a telescope, a photodetector and an accurate timer. The sharply collimated laser pulse is reflected by the target and by measuring the round trip transit time, the distance can be determined.

LIDAR uses ultraviolet, visible, or near infrared light to image objects and can be used with a wide range of targets, including non-metallic objects, rocks, rain, chemical compounds, aerosols, clouds and even single molecules. A narrow laser beam can be used to map physical features with very high resolution. LIDAR has been used extensively for atmospheric research and meteorology. Downward-looking LIDAR instrument fitted to aircraft and satellites are used for surveying and mapping. A recent example being the NASA Experimental Advanced Research Lidar. Wavelengths in a range from about 10 micrometers to the UV ($\lambda = 250$ nm) are used to suit the target.

LIDAR technology has application in geomatics (geodesy and geoinformatics), archaeology, geography, geology, geomorphology, seismology, forestry, remote sensing and atmospheric physics.

30.13.7 Laser in Defence

Military uses of lasers include applications such as target designation and ranging, defensive countermeasures, communications and directed energy weapons. This involves mainly ranging, guiding weapons to the intended target and laser beam itself acting as a weapon. Directed energy weapons are also in use, such as Boeing's Airborne Laser which was constructed inside a Boeing 747. It disrupts the trajectory of shoulder-fired missiles.

On March 18, 2009 Northrop Grumman announced that its engineers in Redondo Beach had successfully built and tested an electric laser capable of producing a 100-kilowatt ray of light, powerful enough to destroy cruise missiles, artillery, rockets and mortar rounds. An electric laser is theoretically capable, according to Brian Strickland, manager for the United States Army's Joint High Power Solid State Laser program, of being mounted in an aircraft, ship, or vehicle because it requires much less space for its supporting equipment than a chemical laser. On 19 July 2010 an anti-aircraft laser was unveiled at the Farnborough Airshow. It was described as the Laser Close-In Weapon System. On April 6, 2011, the U.S. Navy successfully tested a laser gun, manufactured by Northrop Grumman, that was mounted on the former USS Paul Foster, which is currently used as the navy's test ship.

30.13.8 Laser in Holography

Principle: The working of holography is based on the principle of interference. Hence the light waves of high degree coherence is very much essential for its realization.

A laser beam is splitted using a splitter S into two beams A and B. Beam A recognizes the object whose holograph is to be recorded and a part of light scattered by the object impinges on a photographic plate P. The reflected beam B, also called as reference beam, falls on the same photographic plate (Fig. 30.18).

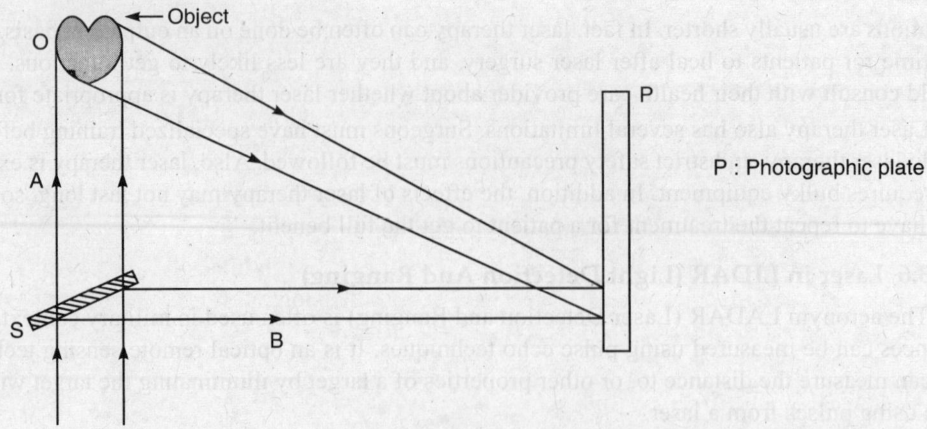

Fig. 30.18

The superposition of these two beams produces an interference pattern that is recorded on the plate. The plate is then developed. The developed plate is known as Holograph. The pattern is very fine and the spacing between the fringes are of the order of 0.001 mm.

30.13.9 Laser in Supermarket Scanners (The Bar Code Reader)

This is a direct application of holography.

A barcode also known as the Universal Product Code (UPC) is a series of lines at various widths that is a representation of information, which is read by an optic scanner called a barcode reader or scanned for an image by special software. Initially, barcodes contained the data within the widths and spacing of the parallel lines, today that's no longer true Today, the information is concealed in patterns of dots, concentric circles, and hidden in images. The information can be price, destination, or other vital information concerning the product.

30.13.10 Laser in Compact Disc (CD)

Compact Discs (CDs) are similar to photograph records in their appearance. Music is recorded in analog form on phonograph records, while it is recorded in digital form on the compact discs. As in the case of phonograph records, the master is first prepared. An optical flat glass disc is coated with a photoresist material layer of about $\lambda/4$ thickness. A He-Cd laser whose intensity is modulated by the digitized music is directed onto the photoresist layer keeping the glass disc rotating. The encoding pattern follows a continuous, spiral path covering the entire disc surface and extending from the innermost track to the outermost track. The data is stored on the disc with a laser (while writing), and can be accessed when the data path is illuminated (while reading) with a laser diode in an optical disc drive which spins the disc at speeds of about 200 rpm up to 4000 rpm or more depending on the drive type, disc format, and the distance of the read head from the center of the disc (inner tracks are read at a faster disc speed). For both, writing and reading purpose, optical lens is used.

The pits or bumps distort the reflected laser light, hence most optical discs (except the black discs of the original Play Station video game console) characteristically have an iridescent appearance created by the grooves of the reflective layer. The reverse side of an optical disc usually has a printed label, generally made of paper but sometimes printed or stamped onto the disc itself. This side of the disc contains the actual data and is typically coated with a transparent material, usually lacquer. Unlike the 3½-inch floppy disk, most optical discs do not have an integrated protective casing and are therefore susceptible to data transfer problems due to scratches, fingerprints, and other environmental problems.

After recording, the disc is developed which causes hardening of the unexposed area of the photoresist layer. Then the photoresist layer is etched to remove the exposed area. This process leaves a pattern of pits in the photoresist. The surface is then coated with silver and electroplated with nickel. Master is thus ready. It is used to transfer the pit pattern to a plastic disc by injecting moulding. The plastic disc is then typically coated with a transparent material, usually lacquer.

30.13.11 Optical Data Storage

Optical discs which are used for playback of music are also useful for high volume digital data storage. We have seen that a beam of laser focused to a high intensity spot causes a localized change in a photosensitive medium to record a 'bit' (either logic 0 logic 1) which is an essential part of digitized information data. High capacity, short access time, high reliability, and absence of mechanical motion make optical storage an attractive alternative to the conventional magnetic disc memories. The optical discs known as CD ROMs are random access, read only memory devices used in computer based applications.

For computer data backup and physical data transfer, optical discs such as CDs and DVDs are gradually being replaced with faster, smaller, and more reliable solid-state devices, especially the USB flash drive. This trend is expected to continue as USB flash drives continue to increase in capacity and drop in price. Similarly, personal portable CD players have been supplanted by portable solid-state digital audio player (MP3 players), and MP3 music purchased or shared over the Internet has significantly reduced the number of audio CDs sold annually.

30.13.12 Laser in Optical Communication

Lasers occupy a very important role in communications. The frequency used for satellite communication should be selected from bands that are most favourable in terms of power efficiencies, minimal propagation or distortion, and reduced noise and interference effects. Terrestrial systems tend to favor these same bands. So, concern for interference effect between the satellite and terrestrial systems must be made. Satellite use from space must be regulated and shared on a worldwide basis. For this reason, frequencies to be used by the satellite are established by a world body known as the International Telecommunications Union (ITU) with broadcast regulations controlled by a subgroup known as World Administrative Radio Conference (WARC). The basic objective is to allocate particular frequency bands for different types of satellite services, and also to provide international regulations in the areas of maximum radiation's level from space, co-ordination with terrestrial systems and the use of specific satellite locations in a given orbit. Within these allotments and regulations an individual country can make its own specific frequency selections based on intended uses and desired satellite services. The electromagnetic frequency spectrum allocated internationally accepted for various bands in satellite communication.

Solid state lasers have offered higher power levels and the ability to operate in high peak power modes for the acquisition. When laser diodes are used to optically pump the lasing media graceful degradation and higher overall reliability is achieved. A variety of materials have been proposed for laser transmitters: Neodyminium doped Yttrium Aluminium garnet (Nd: YAG) is the most widely used. Operating at wavelength 1064 nm, these lasers require an external modulator leading to a slight increase in the complexity and reliability.

Some advantages of laser communications over RF are smaller antenna size, lower weight, lower power and minimal integration impact on the satellite. Laser communication is capable of much higher data rates than RF. The laser beam width can be made as narrow as the diffraction limit of the optic allows. This is given by beam width = 1.22 times the wavelength of light divided by the radius of the output beam aperture. The antennae gain is proportional to the reciprocal of the beam width squared. To achieve the potential diffraction limited beam width a single mode high beam quality laser source is required; together with very high quality optical components throughout the transmitting sub system.

EXERCISE

1. Discuss the types of lasers on the basis of medium. Discuss their relative merits and demerits.
2. Briefly explain the classification of lasers. ***(Pure cs. 2010)***
3. Discuss with suitable diagram the principle of construction and working of a Ruby laser. ***(Garhwal U. 2004; P.U. 2007, 2006, 2003; G.N.D.U. 2007, 2006, 2005, 2004, 2002; Kerala U. 2001; Indore U. 2001; Meerut U. 2001, 2000; Pbi. U. 2005, 2000; M.D.U. 2007, 2000, Pune U. 2010)***
4. Name the active material in Ruby laser? ***(Pune U. 2010)***
5. Explain the concept of Gas laser. Explain He-Ne lasers.
6. Explain the principle, construction and diagram of He-Ne laser. What is the operational wavelength and approximate output power? ***(Pune U. 2010; P.U. 2008, 2006; G.N.D.U. 2007, 2004; Pbi. U. 2008, 2007, 2005, 2005; M.D.U. 2008, 2006, 2005; Meerut U. 2002; K.U. 2001, 2000; Garhwal U. 2004; Bangalore U. 2000)***
7. Discuss the construction and working of a CO_2 laser. ***(P.U. 2008, 2006; G.N.D.U. 2007, 2006, Pbi. U. 2006; Pune CS. 2010)***
8. Explain the main features and conditions for laser action in a semiconductor. ***(G.N.D.U. 2008, 2006, 2005, P.U. 2008, 2007, 2005; Pbi U. 2006; K.U. 2002; M.D.U. 2002)***
9. Describe briefly a semiconductor laser. ***(K.U. 2002; M.D.U. 2002)***
10. Discuss dye laser and its relative advantages. Why is it known as tunable laser! ***(G.N.D.U. 2007, 2006; P.U. 2006)***
11. Write short note on a Nd : YAG laser.
12. Write the applications of semiconductor lasers.
13. What are the various applications of a LASER? ***(G.N.D.U. 2002, Pune U. 2010, Gharwal U. 2004, Nagpur U. 2011/s, 2010)***
14. Explain laser used as a tool in
    (*i*) Welding; (*ii*) cutting and (*iii*) drilling.
15. Write short note on Laser-Diagnostic.
16. Explain the use of laser in angioplasty and dentistry in medical science.
17. How lasers can be used in material processing? ***(Pune U. 2010)***
18. What is LIDAR ? Write its applications. ***(Pune U. 2010)***
19. Explain the use of Laser in 'data-storage'.
20. Discuss the applications of lasers in defence.
21. What is holography ? Give the principle of holography technique. How it differs from photography? Write some applications of holography. ***(Pune cs. 2010)***
22. Explain the importance of laser in optical communication.

Notes

Notes